Thirteenth Edition

WILLIAMS' NUTRITION

FOR HEALTH, FITNESS & SPORT

Eric S. Rawson
Messiah University

J. David Branch
Old Dominion University

Tammy J. Stephenson
University of Kentucky

WILLIAMS' NUTRITION FOR HEALTH, FITNESS AND SPORT, THIRTEENTH EDITION

Published by McGraw Hill LLC, 1325 Avenue of the Americas, New York, NY 10019. Copyright ©2024 by McGraw Hill LLC. All rights reserved. Printed in the United States of America. Previous editions ©2020, 2017, and 2013. No part of this publication may be reproduced or distributed in any form or by any means, or stored in a database or retrieval system, without the prior written consent of McGraw Hill LLC, including, but not limited to, in any network or other electronic storage or transmission, or broadcast for distance learning.

Some ancillaries, including electronic and print components, may not be available to customers outside the United States.

This book is printed on acid-free paper.

1 2 3 4 5 6 7 8 9 LWI 28 27 26 25 24 23

ISBN 978-1-260-70236-1 (bound edition)
MHID 1-260-70236-7 (bound edition)
ISBN 978-1-266-12990-2 (loose-leaf edition)
MHID 1-266-12990-1 (loose-leaf edition)

Portfolio Manager: *Lauren Vondra*
Product Developers: *Krystal Faust, Monica Toledo*
Marketing Manager: *Tami Hodge*
Content Project Managers: *Jeni McAtee, Rachael Hillebrand*
Manufacturing Project Manager: *Laura Fuller*
Content Licensing Specialist: *Melissa Homer*
Cover Image: *Elena Veselova/Shutterstock, MichaelSvoboda/Getty Images, dolomite-summits/Shutterstock, Erik Isakson/Getty Images*
Compositor: *MPS Limited*

All credits appearing on page or at the end of the book are considered to be an extension of the copyright page.

Library of Congress Cataloging-in-Publication Data

Names: Rawson, Eric S., author. | Branch, J. David, 1956- author. | Stephenson, Tammy J., author.
Title: Williams' nutrition for health, fitness & sport / Eric S. Rawson,
  Messiah College, J. David Branch, Old Dominion University,
  Tammy J. Stephenson, University of Kentucky.
Other titles: Williams' nutrition for health, fitness and sport
Description: Thirteenth edition. | New York, NY : McGraw Hill LLC, [2024] |
  Includes bibliographical references and index.
Identifiers: LCCN 2022030990 | ISBN 9781260702361 (hardcover ; acid-free paper) |
  ISBN 9781266129902 (spiral bound ; acid-free paper) |
  ISBN 9781266131899 (ebook)
Subjects: LCSH: Nutrition. | Physical fitness. | Sports–Physiological aspects.
Classification: LCC QP141 .W514 2024 | DDC 612.3–dc23/eng/20220729
LC record available at https://lccn.loc.gov/2022030990

The Internet addresses listed in the text were accurate at the time of publication. The inclusion of a website does not indicate an endorsement by the authors or McGraw Hill LLC, and McGraw Hill LLC does not guarantee the accuracy of the information presented at these sites.

mheducation.com/highered

Dedication

> **In memory of Melvin H. Williams**
>
> We dedicate this thirteenth edition to the founding author, Melvin H. Williams. We are honored to carry on the legacy of a friend and respected colleague in the nutrition field.

To Debbie, Christopher, Matthew, and Erica
—*Eric S. Rawson*

To Carol, David, Molly, Catherine Bruce, Anne Randolph, Tip, Brooks, Gracie, Banks, and the rest of my family
—*J. David Branch*

To Brian, Bailey, Kylie Mae, and Ansley
—*Tammy J. Stephenson*

and

To our teachers, colleagues, and students
Eric, David, and Tammy

About the Authors

Courtesy of Eric Foster, Bloomsburg University

Eric S. Rawson, PhD, FACSM, CSCS, is Chair and Professor of Health, Nutrition, and Exercise Science at Messiah University in Mechanicsburg, Pennsylvania. Dr. Rawson received his PhD from the University of Massachusetts, Amherst, where he studied under the direction of Dr. Priscilla Clarkson. Over the past two decades, Dr. Rawson's research has focused on the interactions between nutrition and skeletal muscle. In particular, Dr. Rawson has extensively studied the effects of the dietary supplement creatine on muscle and brain function. Dr. Rawson has been an active member in the American College of Sports Medicine (ACSM) since 1996, has served on the ACSM Board of Trustees, on the ACSM Annual Meeting Program Committee, as Chair of the ACSM National Chapter Nutrition Interest Group, and is a past president of the Mid-Atlantic ACSM regional chapter. Dr. Rawson has delivered more than 180 professional presentations, is coeditor of the text *Nutrition for Elite Athletes,* coauthor of the 10th, 11th, and 12th editions of *Nutrition for Health, Fitness & Sport,* and has authored/coauthored numerous articles and book chapters. His research has been funded by the National Institutes of Health and various foundations.

Courtesy of Chuck Thomas, University Photographer, Old Dominion University

J. David Branch, earned a BA degree from Furman University, and MS and PhD degrees in Exercise Science from the University of South Carolina. Since 1994, he has been at Old Dominion University in the Department of Human Movement Sciences, where he has taught exercise physiology, exercise testing, research methods, and other courses in the undergraduate and graduate exercise science programs. Prior to that, he was a lecturer at Furman University and worked for many years in a facility specializing in health and fitness testing of South Carolina law enforcement personnel. He is a Fellow in the American College of Sports Medicine. Dr. Branch enjoys reading, running, the beach, spending time with wife Carol, dog Gracie, grand dog Banks, and the accomplishments of his adult children, David and Anne Randolph, and his grandchildren.

Sarah Caton Walters

Tammy J. Stephenson, PhD, received her BS degree in Food Science and Human Nutrition, and PhD in Nutritional Sciences from the University of Kentucky. She has taught a variety of food, nutrition, and wellness courses, including sports nutrition and introductory nutrition, in the Department of Dietetics and Human Nutrition at the University of Kentucky for the past two decades. Currently, Dr. Stephenson serves as Department Chair, while continuing to teach both in person and online classes. Prior to assuming the Chair position, she served as Director of Undergraduate Studies for the Dietetics and Human Nutrition degree programs, as Director of the Undergraduate Certificate in Food Systems and Hunger Studies, and as Co-Director of the Undergraduate Certificate in Nutrition for Human Performance. She has been recognized with multiple teaching and advising awards at the university, state, and national levels, including the University of Kentucky Alumni Association's Great Teacher Award and the Provost Office's Outstanding Teacher Award. Dr. Stephenson is an active member of the Academy of Nutrition and Dietetics, having served as Chair of the Nutrition Educators of Health Professionals practice group. She is also a member of the American College of Sports Medicine and the Sports and Human Performance Nutrition practice group of the Academy. Dr. Stephenson is coauthor of *Human Nutrition: Science for Healthy Living,* now in its third edition. Outside of the classroom, she enjoys yoga, cycling, hiking, reading, gardening, watching her daughters play sports, traveling, and spending time with her family.

Brief Contents

CHAPTER ONE	Introduction to Nutrition for Health, Fitness, and Sports Performance 1
CHAPTER TWO	Healthful Nutrition for Fitness and Sport 36
CHAPTER THREE	Human Energy 76
CHAPTER FOUR	Carbohydrates: The Main Energy Food 111
CHAPTER FIVE	Fat: An Important Energy Source during Exercise 161
CHAPTER SIX	Protein: The Tissue Builder 208
CHAPTER SEVEN	Vitamins: Fat-Soluble, Water-Soluble, and Vitamin-Like Compounds 246
CHAPTER EIGHT	Minerals: The Inorganic Regulators 278
CHAPTER NINE	Water, Electrolytes, and Temperature Regulation 313
CHAPTER TEN	Body Weight and Composition for Health and Sport 373
CHAPTER ELEVEN	Weight Maintenance and Loss through Proper Nutrition and Exercise 430
CHAPTER TWELVE	Gaining Lean Body Mass through Proper Nutrition and Exercise 457
CHAPTER THIRTEEN	Nutritional Supplements and Ergogenic Aids 490

Contents

Preface xv

CHAPTER ONE

Mauro Grigollo/E+/Getty Images

Introduction to Nutrition for Health, Fitness, and Sports Performance 1

Introduction 2

Fitness and Exercise 4
- How are health-related fitness and sports-related fitness different? 4
- What are the basic principles of exercise training? 4

Exercise and Health Promotion 6
- How does exercise enhance health? 6

Physical Activity Guidelines 8
- Key principles for developing physical activity programs 8
- Are Americans meeting physical activity guidelines? 10
- Am I exercising enough? 10
- Can too much exercise be harmful to my health? 12

Nutrition and Fitness 13
- What is nutrition? 13
- What is the role of nutrition in health promotion? 13
- Do most Americans eat a well-balanced diet? 14
- What are some general guidelines for healthy eating? 15
- Are there additional health benefits when both exercise and dietary habits are improved? 16

Sports-Related Fitness: Exercise and Nutrition 17
- What is sports-related fitness? 17
- What is sports nutrition? 18
- Is sports nutrition a profession? 18
- Are athletes today receiving adequate nutrition? 19
- Why are some athletes malnourished? 20
- How does nutrition affect athletic performance? 20
- What should athletes eat to help optimize sport performance? 21

Ergogenic Aids and Sports Performance: Beyond Training 21
- What is an ergogenic aid? 21
- Why are nutritional ergogenics so popular? 22
- Are nutritional ergogenics effective? 22
- Are nutritional ergogenics safe? 23
- Are nutritional ergogenics legal? 23
- Where can I find more detailed information on sports supplements? 24

Nutrition and Health Misinformation in Sports 24
- What is nutrition and health misinformation? 24
- Why is nutrition misinformation so prevalent in athletics? 25
- How do I recognize nutrition misinformation in health and sports? 26
- Where can I get sound nutritional information to combat nutrition misinformation in health and sports? 26

Research and Evidence-Based Recommendations 28
- What types of research provide valid information? 28
- Why do we often hear contradictory advice about the effects of nutrition on health or physical performance? 30
- What is the basis for the dietary recommendations presented in this book? 30
- How does all this relate to me? 31

Application Exercise 32

Review Questions—Multiple Choice 32

Critical Thinking Questions 33

References 33

CHAPTER TWO

Ken Welsh/age fotostock

Healthful Nutrition for Fitness and Sport 36

Introduction 37

Essential Nutrients and Recommended Nutrient Intakes 38
- What are essential nutrients? 38
- What are nonessential nutrients? 39
- How are recommended dietary intakes determined? 39

The Balanced Diet and Nutrient Density 42
- What is a balanced diet? 42
- What foods should I eat to obtain the nutrients I need? 42
- What is the MyPlate food guide? 42
- How has the COVID-19 pandemic impacted dietary practices? 44
- What is the key-nutrient concept for obtaining a balanced diet? 45
- What is the concept of nutrient density? 46
- Will using the MyPlate food guide guarantee me optimal nutrition? 46

Healthful Dietary Guidelines 48
- What is the basis underlying the development of healthful dietary guidelines? 48

What are the recommended dietary guidelines for reducing the risk of chronic disease? 48

Plant-Based Diets 53
What types of foods are included in plant-based diets? 53
What are some of the nutritional concerns with plant-based diets? 53
Is following a plant-based diet beneficial for health? 54
What are recommendations for following a plant-based diet? 56
Will a following a plant-based diet affect physical performance potential? 56

Consumer Nutrition—Food Labels and Health Claims 57
What nutrition information do food labels provide? 58
How can I use this information to select a healthier diet? 58
What health claims are allowed on food products? 60
What are functional foods? 61

Consumer Nutrition—Dietary Supplements and Health 62
What are dietary supplements? 62
Will dietary supplements improve my health? 63
Can dietary supplements harm my health? 64

Healthful Nutrition: Recommendations for Better Physical Performance 65
What should I eat during training? 65
When and what should I eat just prior to competition? 66
What should I eat during competition? 67
What should I eat after competition? 68
Should athletes use commercial sports foods? 68
How can I eat more nutritiously while traveling for competition? 69
How do gender and age influence nutritional recommendations for enhanced physical performance? 70

Application Exercise 71

Review Questions—Multiple Choice 72

Critical Thinking Questions 72

References 73

Corbis/Glow Images

CHAPTER THREE

Human Energy 76

Introduction 77

Measures of Energy 77
What is energy? 77
What terms are used to quantify work and power during exercise? 78
How do we measure physical activity and energy expenditure? 78
What is the most commonly used measure of energy? 81

Human Energy Systems 83
How is energy stored in the body? 83
What are the human energy systems? 84
What nutrients are necessary for operation of the human energy systems? 87

Human Energy Metabolism during Rest 88
What is metabolism? 88
What factors account for the amount of energy expended during rest? 88
What effect does eating a meal have on the metabolic rate? 88
How can I estimate my daily resting energy expenditure (REE)? 89
What genetic factors affect my REE? 89
How do dieting and body composition affect my REE? 90
What environmental factors may also influence the REE? 90
What energy sources are used during rest? 90

Human Energy Metabolism during Exercise 91
How do my muscles influence the amount of energy I can produce during exercise? 91
What effect does muscular exercise have on the metabolic rate? 92
How is energy expenditure of the three human energy systems measured during exercise? 92
How can I convert the various means of expressing exercise energy expenditure into something more useful to me, such as kcal per minute? 93
How can I tell what my metabolic rate is during exercise? 95
How can I determine the energy cost of my exercise routine? 95
What are the best types of activities to increase energy expenditure? 96
Does exercise affect my resting energy expenditure (REE)? 97
Does exercise affect the thermic effect of food (TEF)? 98
How much energy should I consume daily? 98

Human Energy Systems and Fatigue during Exercise 101
What energy systems are used during exercise? 101
What energy sources are used during exercise? 102
What is the "fat burning zone" during exercise? 103
What is fatigue? 104
What causes acute fatigue in athletes? 105
How can I delay the onset of fatigue? 106
How is nutrition related to fatigue processes? 106

Application Exercise 108

Review Questions—Multiple Choice 108

Critical Thinking Questions 109

References 109

CHAPTER FOUR

Carbohydrates: The Main Energy Food 111

Introduction 112

Dietary Carbohydrates 112
- What are the different types of dietary carbohydrates? 112
- What are some common foods high in carbohydrate content? 114
- How much carbohydrate do we need in the diet? 115

Metabolism and Function 116
- How are dietary carbohydrates digested and absorbed and what are some implications for sports performance? 116
- What happens to the carbohydrate after it is absorbed into the body? 117
- What is the metabolic fate of blood glucose? 118
- How much total energy do we store as carbohydrate? 121
- Can the human body make carbohydrates from protein and fat? 122
- What are the major functions of carbohydrate in human nutrition? 122

Carbohydrates for Exercise 124
- In what types of activities does the body rely heavily on carbohydrate as an energy source? 124
- Why is carbohydrate an important energy source for exercise? 124
- What effect does endurance training have on carbohydrate metabolism? 125
- How is hypoglycemia related to the development of fatigue? 125
- How is lactic acid production related to fatigue? 127
- How is low muscle glycogen related to the development of fatigue? 127
- How are low endogenous carbohydrate levels related to the central fatigue hypothesis? 129
- Will eating carbohydrate immediately before or during an event improve physical performance? 130
- When, how much, and in what form should carbohydrates be consumed before or during exercise? 133
- What is the importance of carbohydrate replenishment after prolonged exercise? 137
- Will a high-carbohydrate diet enhance my daily exercise training? 139

Carbohydrate Loading 140
- What is carbohydrate, or glycogen, loading? 140
- What type of athlete would benefit from carbohydrate loading? 140
- How do you carbohydrate load? 141
- Will carbohydrate loading increase muscle glycogen concentration? 142
- How do I know if my muscles have increased their glycogen stores? 143
- Will carbohydrate loading improve exercise performance? 143
- Are there any possible detrimental effects relative to carbohydrate loading? 144

Carbohydrates: Ergogenic Aspects 145
- Do the metabolic by-products of carbohydrate exert an ergogenic effect? 145

Dietary Carbohydrates: Health Implications 147
- How do refined sugars and starches affect my health? 147
- Are artificial sweeteners safe? 148
- Why are complex carbohydrates thought to be beneficial to my health? 149
- Why should I eat foods rich in fiber? 150
- Do some carbohydrate foods cause food intolerance? 152

Application Exercise 153

Review Questions—Multiple Choice 154

Critical Thinking Questions 155

References 155

CHAPTER FIVE

Fat: An Important Energy Source during Exercise 161

Introduction 162

Dietary Fats 162
- What are the different types of dietary fats? 162
- What are triglycerides? 162
- What are some common foods high in fat content? 163
- How do I calculate the percentage of fat kcal in a food? 164
- What are fat substitutes? 165
- What is cholesterol? 166
- What foods contain cholesterol? 166
- What are phospholipids? 167
- What foods contain phospholipids? 167
- How much fat and cholesterol do we need in the diet? 167

Metabolism and Function 169
- How does dietary fat get into the body? 169
- What happens to the lipid once it gets in the body? 169
- What are the different types of lipoproteins? 170
- Can the body make fat from protein and carbohydrate? 172
- What are the major functions of the body lipids? 172
- How much total energy is stored in the body as fat? 173

Fats and Exercise 174
  Are fats used as an energy source during exercise? 174
  Does sex influence the use of fats as an energy source during exercise? 175
  What effect does exercise training have on fat metabolism during exercise? 176

Fats: Ergogenic Aspects 177
  High-fat diets 177
  High-fat diets and weight loss 179
  Does exercising on an empty stomach or fasting improve performance? 180
  Can the use of medium-chain triglycerides improve endurance performance or body composition? 181
  Is the glycerol portion of triglycerides an effective ergogenic aid? 182
  Omega-3 fatty acid and fish oil supplements 182
  Can carnitine improve performance or weight loss? 183
  Can hydroxycitrate (HCA) enhance endurance performance? 184
  Can conjugated linoleic acid (CLA) enhance exercise performance or weight loss? 184
  Can ketone supplements improve endurance performance? 185
  What's the bottom line regarding the ergogenic effects of fat burning diets or strategies? 185

Dietary Fats and Cholesterol: Health Implications 186
  How does cardiovascular disease develop? 186
  How do the different forms of serum lipids affect the development of atherosclerosis? 187
  Can I reduce my serum lipid levels and possibly reverse atherosclerosis? 190
  What should I eat to modify my serum lipid profile favorably? 190
  What is a heart healthy diet? 190
  Can exercise training also elicit favorable changes in the serum lipid profile? 198

Application Exercise 200

Review Questions—Multiple Choice 200

Critical Thinking Questions 201

References 201

Don Hammond/Design Pics

CHAPTER SIX

Protein: The Tissue Builder 208

Introduction 209

Dietary Protein 209
  What is protein? 209
  Is there a difference between animal and plant protein? 210
  What are some common foods that are good sources of protein? 211
  How much dietary protein do I need? 211
  How much of the essential amino acids do I need? 213
  What are some dietary guidelines to ensure adequate protein intake? 213

Metabolism and Function 214
  What happens to protein in the human body? 214
  Can protein be formed from carbohydrates and fats? 215
  What are the major functions of protein in human nutrition? 215

Proteins and Exercise 216
  Are proteins used for energy during exercise? 217
  Does exercise increase protein losses in other ways? 218
  What happens to protein metabolism during recovery after exercise? 218
  What effect does exercise training have upon protein metabolism? 219
  Does exercise increase the need for dietary protein? 219
  What are some general recommendations relative to dietary protein intake for athletes? 221
  Are protein supplements necessary? 225

Protein-Related Supplements 227
  Strong Evidence of Efficacy 228
  Less or Controversial Evidence of Efficacy 232
  Supplements with Little to No Evidence of Efficacy, or, Data That Do Not Support an Ergogenic Effect 233
  Amino Acid Supplements 234

Dietary Protein: Health Implications 236
  Does a deficiency of dietary protein pose any health risks? 237
  Does excessive protein intake pose any health risks? 237
  Does the consumption of individual amino acids pose any health risks? 237

Application Exercise 238

Review Questions—Multiple Choice 238

Critical Thinking Questions 239

References 239

BananaStock/Getty Images

CHAPTER SEVEN

Vitamins: Fat-Soluble, Water-Soluble, and Vitamin-Like Compounds 246

Introduction 247

Basic Facts 248
  What are vitamins and how do they work? 248
  What vitamins are essential to human nutrition? 248
  In general, how do deficiencies or excesses of vitamins influence health or physical performance? 249
  How are vitamin needs determined? 249

Fat-Soluble Vitamins 252
- Vitamin A (retinol) 252
- Vitamin D (cholecalciferol) 253
- Vitamin E (alpha-tocopherol) 256
- Vitamin K (menadione) 258

Water-Soluble Vitamins 259
- Thiamin (vitamin B1) 259
- Riboflavin (vitamin B2) 260
- Niacin 260
- Vitamin B6 (pyridoxine) 261
- Vitamin B12 (cobalamin) 262
- Folate (folic acid) 264
- Pantothenic acid 265
- Biotin 266
- Vitamin C (ascorbic acid) 266
- Vitamin-like compounds: Choline 268

Vitamin Supplements: Ergogenic Aspects 269
- Should physically active individuals take vitamin supplements? 269
- Can the antioxidant vitamins prevent fatigue or muscle damage during training? 269
- How effective are the multivitamin supplements marketed for athletes? 271

Application Exercise 273

Review Questions—Multiple Choice 273

Critical Thinking Questions 274

References 274

Erik Isakson/Blend Images LLC

CHAPTER EIGHT

Minerals: The Inorganic Regulators 278

Introduction 279

Basic Facts 279
- What are minerals, and what is their importance to humans? 279
- What minerals are essential to human nutrition? 280
- In general, how do deficiencies or excesses of minerals influence health or physical performance? 280

Major Minerals 281
- Calcium (Ca) 281
- Phosphorus (P) 289
- Magnesium (Mg) 290

Trace Minerals 292
- Iron (Fe) 292
- Copper (Cu) 299
- Zinc (Zn) 300
- Chromium (Cr) 302
- Selenium (Se) 304

Mineral Supplements: Exercise and Health 305
- Does exercise increase my need for minerals? 305
- Can I obtain the minerals I need through my diet? 306
- Are mineral megadoses or some nonessential minerals harmful? 306
- Should physically active individuals take mineral supplements? 306

Application Exercise 308

Review Questions—Multiple Choice 308

Critical Thinking Questions 309

References 309

Greg Epperson/Shutterstock

CHAPTER NINE

Water, Electrolytes, and Temperature Regulation 313

Introduction 314

Water 315
- How much water do you need per day? 315
- What else is in the water we drink? 315
- Where is water stored in the body? 317
- How is body water regulated? 318
- How do I know if I am adequately hydrated? 320
- What are the major functions of water in the body? 320
- Can drinking more water or fluids confer any health benefits? 320

Electrolytes 321
- What is an electrolyte? 321
- Sodium (Na) 321
- Chloride (Cl) 323
- Potassium (K) 324

Regulation of Body Temperature 325
- What is the normal body temperature? 325
- What are the major factors that influence body temperature? 326
- How does the body regulate its own temperature? 326
- What environmental conditions may predispose an athletic individual to hyperthermia? 327
- How does exercise affect body temperature? 328
- How is body heat dissipated during exercise? 329

Exercise Performance in the Heat: Effect of Environmental Temperature and Fluid and Electrolyte Losses 330
- How does environmental heat affect physical performance? 330
- How do dehydration and hypohydration affect physical performance? 332
- How fast may an individual become dehydrated while exercising? 333
- How can I determine my sweat rate? 334
- What is the composition of sweat? 335
- Is excessive sweating likely to create an electrolyte deficiency? 335

Exercise in the Heat: Fluid, Carbohydrate, and Electrolyte Replacement 335
- Which is most important to replace during exercise in the heat—water, carbohydrate, or electrolytes? 336
- What are some sound guidelines for maintaining water (fluid) balance during exercise? 336
- What factors influence gastric emptying and intestinal absorption? 338
- How should carbohydrate be replaced during exercise in the heat? 340
- How should electrolytes be replaced during or following exercise? 341
- What is hyponatremia and what causes it during exercise? 342
- Are salt tablets or potassium supplements necessary? 343
- What are some prudent guidelines relative to fluid replacement while exercising under warm or hot environmental conditions? 344

Ergogenic Aspects 348
- Does oxygen water enhance exercise performance? 349
- Do precooling techniques help reduce body temperature and enhance performance during exercise in the heat? 349
- Does sodium loading enhance endurance performance? 349
- Does glycerol supplementation enhance endurance performance during exercise under warm environmental conditions? 350

Health Aspects: Heat Illness 351
- Should I exercise in the heat? 351
- What are the potential health hazards of excessive heat stress imposed on the body? 351
- What are the symptoms and treatment of heat injuries? 354
- Do some individuals have problems tolerating exercise in the heat? 355
- How can I reduce the hazards associated with exercise in a hot environment? 357
- How can I become acclimatized to exercise in the heat? 357

Health Aspects: High Blood Pressure 359
- What is high blood pressure, or hypertension? 359
- How is high blood pressure treated? 359
- What dietary modifications may help reduce or prevent hypertension? 360
- Can exercise help prevent or treat hypertension? 363

Application Exercise 365

Review Questions—Multiple Choice 365

Critical Thinking Questions 366

References 366

CHAPTER TEN

Body Weight and Composition for Health and Sport 373

Introduction 374

Body Weight and Composition 375
- What is the ideal body weight? 375
- What are the values and limitations of the BMI? 375
- What is the composition of the body? 376
- What techniques are available to measure body composition and how accurate are they? 378
- What problems may be associated with rigid adherence to body fat percentages in sport? 383
- How much should I weigh or how much body fat should I have? 384

Regulation of Body Weight and Composition 386
- How does the human body normally control its own weight? 386
- How is fat deposited in the body? 392
- What are the contributing factors to obesity? 392
- Can the set point change? 397
- Why is prevention of childhood obesity so important? 398

Weight Gain, Obesity, and Health 399
- What health problems are associated with overweight and obesity? 400
- How does the location of fat in the body affect health? 402
- Does having obesity increase health risks in youth? 403
- Does losing excess body fat reduce health risks and improve health status? 404
- Does being physically fit negate the adverse health effects associated with being overweight? 404

Excessive Weight Loss and Health 406
- What health problems are associated with improper weight-loss programs and practices? 406

Body Composition and Physical Performance 410
- What effect does excess body weight have on physical performance? 410
- Does excessive weight loss impair physical performance? 412

Disordered Eating and Eating Disorders 413
- What is anorexia nervosa? 413
- What is bulimia nervosa? 414
- What is binge eating disorder? 414
- What are examples of "other specified" or "unspecified" feeding or eating disorders? 415
- What eating disorders are most commonly associated with sports? 415
- How can eating disorders be prevented, detected, and treated in athletes? 417

Application Exercise 419

Review Questions—Multiple Choice 419

Critical Thinking Questions 420

References 421

Application Exercise 454

Review Questions—Multiple Choice 454

Critical Thinking Questions 455

References 455

Mara Zemgaliete/Magone/123RF

CHAPTER ELEVEN

Weight Maintenance and Loss through Proper Nutrition and Exercise 430

Introduction 431

Health Consequences of Obesity 432
 What social and psychological consequences are associated with obesity? 432
 What is weight bias? 433
 What are evidence-based strategies to support weight loss? 434

Nutrition and Weight Loss 434
 How is kcal intake related to weight loss? 435
 What are general dietary recommendations to support weight loss? 436
 How does consumption of ultra-processed foods impact body weight? 436
 What are practical recommendations for following a diet to support weight loss? 437

Mindful Eating Practices 438
 What is mindful eating? 438
 How can mindfulness practices support both mental and physical health? 440

Physical Activity and Weight Loss 441
 How is physical activity related to obesity? 441
 What are strategies to increase physical activity? 442
 How does someone get started with a physical activity program? 445

Weight-Loss Programs 446
 What factors should be considered when selecting a weight-loss program? 446
 What types of weight-loss programs are available? 446
 Are there specific weight-loss programs for children and adolescents? 449
 Intermittent fasting, high-protein diets, high-fat (keto) diets . . . so many options! Is there one that is the "best" for weight loss? 449

Weight Loss and Athletic Performance 451
 What are the recommendations for exercise before and after bariatric surgery? 451
 How does weight loss impact sport-specific performance in athletes? 452
 Does weight cycling increase risk for obesity later in life? 453

Liquidlibrary/Getty Images

CHAPTER TWELVE

Gaining Lean Body Mass through Proper Nutrition and Exercise 457

Introduction 458

Basic Considerations 459
 Why are some individuals underweight? 459
 What steps should I take if I want to gain weight? 459

Nutritional Considerations 460
 How many kcal are needed to form 1 pound of muscle? 460
 How can I determine the amount of kcal I need daily to gain 1 pound per week? 461
 Is protein supplementation necessary during a weight-gaining program? 461
 Are dietary supplements necessary during a weight-gaining program? 464
 What is an example of a balanced diet that will help me gain weight? 465
 Would such a high-kcal diet be ill advised for some individuals? 467

Exercise Considerations 467
 What are the primary purposes of resistance training? 468
 What are the basic principles of resistance training? 469
 What is an example of a resistance-training program that may help me to gain body weight as lean muscle mass? 471
 Are there any safety concerns associated with resistance training? 473
 How does the body gain weight with a resistance-training program? 478
 Is any one type of resistance-training program or equipment more effective than others for gaining body weight? 480
 If exercise burns kcal, won't I lose weight on a resistance-training program? 481
 Are there any contraindications to resistance training? 482
 Are there any health benefits associated with resistance training? 483
 Can I combine aerobic and resistance-training exercises into one program? 484

Application Exercise 485

Review Questions—Multiple Choice 485

Critical Thinking Questions 486

References 486

CHAPTER THIRTEEN

Nutritional Supplements and Ergogenic Aids 490

Introduction 491

Alcohol: Ergogenic Effects and Health Implications 492
    What is the alcohol and nutrient content of typical alcoholic beverages? 492
    What is the metabolic rate of alcohol clearance in the body? 492
    Is alcohol an effective ergogenic aid? 493
    What effect can drinking alcohol have upon my health? 495

Caffeine: Ergogenic Effects and Health Implications 502
    What is caffeine, and in what food products is it found? 502
    What effects does caffeine have on the body that may benefit exercise performance? 503
    Does caffeine enhance exercise performance? 504
    Does drinking coffee, tea, or other caffeinated beverages provide any health benefits or pose any significant health risks? 508

Ephedra (Ephedrine): Ergogenic Effects and Health Implications 513
    What is ephedra (ephedrine)? 513
    Does ephedrine enhance exercise performance or promote weight loss? 513
    Do dietary supplements containing ephedra pose any health risks? 513

Sodium Bicarbonate: Ergogenic Effects, Safety, and Legality 515
    What is sodium bicarbonate? 515
    Does sodium bicarbonate, or soda loading, enhance physical performance? 515
    Is sodium bicarbonate supplementation safe and legal? 518

Anabolic Hormones and Dietary Supplements: Ergogenic Effects and Health Implications 518
    Is human growth hormone (hGH) an effective, safe, and legal ergogenic aid? 518
    Are testosterone and anabolic-androgenic steroids (AAS) effective, safe, and legal ergogenic aids? 519
    Are anabolic prohormone dietary supplements effective, safe, and legal ergogenic aids? 521

Ginseng and Selected Herbals: Health and Ergogenic Effects 523
    Does ginseng or ciwujia enhance exercise or sports performance? 524
    What herbals are effective ergogenic aids? 525

Sports Supplements: Efficacy, Safety, and Permissibility 526
    What sports supplements are considered to be effective, safe, and permissible? 526

Application Exercise 527

Review Questions—Multiple Choice 528

Critical Thinking Questions 529

References 529

APPENDIX A Energy Pathways of Carbohydrate, Fat, and Protein 540

APPENDIX B Determination of Healthy Body Weight 544

APPENDIX C Units of Measurement: English System—Metric System Equivalents 549

APPENDIX D Approximate Energy Expenditure (Kcal/Min) by Body Weight Based on the Metabolic Equivalents (METs) for Physical Activity Intensity 551

Glossary 555

Index 567

Preface

According to the World Health Organization, better health is the key to human happiness and well-being. Many factors influence one's health status, including some shared by various government and health agencies, such as safe living environments and access to proper health care. However, in general, one's personal health over the course of a lifetime is dependent more upon personal lifestyle choices, two of the most important being proper exercise and healthy eating.

In the twenty-first century, our love affair with fitness and sports continues to grow. Worldwide, although rates of physical inactivity are still prevalent in developed nations, there are millions of children and adults who are active in physical activities such as bicycling, running, swimming, walking, and weight training. Improvements in health and fitness are major reasons more and more people initiate an exercise program, but many may also become more interested in sports competition, such as age-group road racing; running and walking race competitions have become increasingly popular, and every weekend numerous road races can be found within a short drive. Research has shown that adults who become physically active also may become more interested in other aspects of their lifestyles—particularly nutrition—that may affect their health in a positive way. Indeed, according to all major health organizations, proper exercise and a healthful diet are two of the most important lifestyle behaviors to help prevent chronic disease.

Nutrition is the study of foods and their effects upon health, development, and performance. Over the years, nutrition research has made a significant contribution to our knowledge of essential nutrient needs. During the first part of the twentieth century, most nutrition research focused on identification of essential nutrients and amounts needed to prevent nutrient-deficiency diseases, such as scurvy from inadequate vitamin C. As nutrition science evolved, medical researchers focused on the effects of foods and their specific constituents as a means to help prevent the major chronic diseases, such as heart disease and cancer, that are epidemic in developed countries. *Nutriceutical* is a relatively new term used to characterize the drug, or medical, effects of a particular nutrient. Recent research findings continue to indicate that our diet is one of the most important determinants of our health status. Although individual nutrients are still being evaluated for possible health benefits, research is also focusing on dietary patterns, or the totality of the diet, and resultant health benefits. However, we should note that research relative to the effects of diet, including specific nutrients, on health is complex and dietary recommendations may change with new research findings. For example, as noted later in the text, the guidelines regarding dietary intake of cholesterol have been modified after being in effect for more than 50 years.

Other than the health benefits of exercise and fitness, many physically active individuals are also finding the joy of athletic competition, participating in local sports events such as golf tournaments, tennis matches, triathlons, and road races. Individuals who compete athletically are always looking for a means to improve performance, be it a new piece of equipment or an improved training method. In this regard, proper nutrition may be a very important factor in improving sports performance. Various sports governing agencies indicate today's athletes need accurate sports nutrition information to maximize sports performance. Although the effect of diet on sports and exercise performance was studied only sporadically prior to 1970, subsequently numerous sports scientists and sports nutritionists have studied the performance-enhancing effects of nutrition, such as diet composition and dietary supplements. Results of these studies have provided nutritional guidance to enhance performance in specific athletic endeavors. In the United States, many universities and professional sports teams, such as those in Major League Baseball, the National Hockey League, and the National Football League, employ registered dietitian nutritionists as well as culinary chefs to provide dietary guidance to their athletes.

With the completion of the Human Genome Project, gene therapies are being developed for the medical treatment of various health problems. Moreover, some contend that genetic manipulations may be used to enhance sports performance. For example, gene doping to increase insulin-like growth factor, which can stimulate muscle growth, may be applied to sport.

Our personal genetic code plays an important role in determining our health status and our sports abilities, and futurists speculate that one day each of us will carry our own genetic chip that will enable us to tailor food selection and exercise programs to optimize our health and sports performance. Such may be the case, but for the time being we must depend on available scientific evidence to provide us with prudent guidelines.

Each year thousands of published studies and reviews analyze the effects of nutrition on health or exercise and sports performance. The major purpose of this text is to evaluate these scientific data and present prudent recommendations for individuals who want to modify their diet for optimal health or exercise/sports performance.

Textbook Overview

This book uses a question-answer approach, which is convenient when you may have occasional short periods to study, such as between classes. In addition, the questions are arranged in a logical sequence, the answer to one question often leading into the question that follows. Where appropriate, cross-referencing within the text is used to expand the discussion. No deep scientific background is needed for the chemical aspects of nutrition and energy expenditure, as these have been simplified. Instructors who use this book as a course text may add details of biochemistry as they feel necessary.

Chapter 1 introduces you to the general effects of exercise and nutrition on health-related and sports-related fitness, including the importance of well-controlled scientific research. Chapter 2 provides a broad overview of sound guidelines relative to nutrition for optimal health and exercise performance. Chapter 3 focuses on energy and energy pathways in the body, the key to all exercise and sports activities.

Chapters 4 through 9 explore the six basic nutrients—carbohydrate, fat, protein, vitamins, minerals, and water—with emphasis on the health and performance implications for the physically active individual. Chapters 10 through 12 review concepts of body composition and weight control, with suggestions on how to gain or lose body weight through diet and exercise, as well as the implications of such changes for health and athletic performance. Content on disordered eating and eating disorders is covered in Chapter 10, including a new section on relative energy deficiency in sport (RED-S). Chapter 13 summarizes dietary supplements and ergogenic aids and potential impacts on health and exercise performance. The textbook has been updated with the most recent information available on theses supplements, including preworkout supplements, creatine, and caffeine. Four appendices complement the text, providing detailed metabolic pathways for carbohydrate, fat, and protein, methods to determine healthy body weight, units of measurement: English System–Metric System equivalents, and approximate energy expenditure by body weight.

New to the Thirteenth Edition

The first edition of this textbook, titled *Nutrition for Fitness and Sport,* was published in 1983. As one would expect, much has changed in the fields of nutrition and exercise science over the past 40 years. This edition of the textbook has been updated with the most current research available from evidence-based sources regarding the effects of nutritional choices on health, fitness, and sports performance. New features and updated assessments, including critical thinking questions, make the textbook user-friendly and help students learn and apply content. The *Training Table* feature is embedded throughout the chapters and provides practical and relevant examples and content on a variety of topics related to physical activity and nutrition. As instructors ourselves, we hope that both faculty and students find the textbook engaging, informative, relevant, and interesting.

As you read through the thirteenth edition of the textbook, the following updates have been made.

Chapter 1—Introduction to Nutrition for Health, Fitness, and Sports Performance

- Updated statistics on the leading causes of death in the United States with an expanded discussion of those related to diet and/or physical activity
- New and revised *Training Tables* on current and interesting topics such as *Healthy People 2030* objectives, examples of physical activity options at different intensities, and more
- Physical activity guidelines section updated with the 2018 *Physical Activity Guidelines for Americans* recommendations and specific examples
- New content on the physical activity habits of Americans with updated figure 1.5 map of the United States showing the percentage of the population who are physically inactive in each state
- The most current information available on fitness trackers and heart rate monitors, including a new figure 1.6 showing different options
- Specific recommendations from the *2020-2025 Dietary Guidelines for Americans,* including an expanded discussion of those guidelines
- New information on nutrition and health misinformation and recommendations for evaluating nutrition and health content shared online, through social media, and more
- An introduction to ergogenic aids and general advice about their use, with specific details embedded throughout subsequent chapters
- Updated guidelines on evaluating and understanding different types of research studies and making evidence-based recommendations
- Application Exercise based on a case-study scenario
- Innovative Critical Thinking Questions that challenge students to go beyond memorizing content, and to truly apply the material
- New and revised references

Chapter 2—Healthful Nutrition for Fitness and Sport

- Updates throughout the chapter to reflect the recommendations of the *2020-2025 Dietary Guidelines for Americans* and associated MyPlate resources
- Training Tables on topics including food sources of empty calories, healthy eating on a budget, and limiting sodium intake
- Revised section with new information on how dietary recommendations are set, showing the relationship between RDAs, AIs, ULs, and others
- Significantly revised section on plant-based diets, including nutrient intake, health effects, and more
- Expanded discussion of dairy alternatives and recommendations for meeting MyPlate food group needs for those who do not consume dairy products
- Updated figures and photos showing food labels and associated Nutrition Facts panels
- Specific dietary advice based on the most currently available literature and recommendations from evidence-based sources
- Updated content related to classification and monitoring of dietary supplements with practical advice on how supplements can be a healthy addition to a well-balanced diet

- Introduction of key concepts of sports nutrition with practical recommendations and guidance, including specific examples of precompetition meals
- Application Exercise based on a case-study scenario
- Updated Critical Thinking Questions
- New and revised references

Chapter 3—Human Energy
- Improved terminology for energy expenditure assessment
- New data on ultraprocessed foods, energy intake, and absorption
- New data on standing desks and energy expenditure
- Introduction of Relative Energy Deficiency in Sport (RED-S)
- Updated images
- New and revised references

Chapter 4—Carbohydrates: The Main Energy Food
- Updated data on the health effects of low glycemic-index diets and on glycemic index values
- Updates on energy metabolism and adenosine triphosphate production from carbohydrate
- Updates on the effects of altitude on carbohydrate metabolism and energy production
- Updates on non-nutritive sweeteners and dietary fiber and the microbiome
- Updated images
- New and revised references

Chapter 5—Fat: An Important Energy Source during Exercise
- Updated food label for finding hidden fats in foods
- Updated websites related to dietary fats and health
- Updates on fat-related dietary supplements
- Updated data on ergogenic and health-related effects of omega-3 fatty acid supplements
- Updated data on ergogenic effects of ketone supplements
- Updated cardiovascular risk calculator
- New image of recommended changes using the DASH eating plan to help lower blood pressure and LDL
- New table on guidelines to manage blood cholesterol
- Updated section on dietary recommendations to improve serum lipids and reduce the risk of atherosclerotic cardiovascular disease
- Updated images
- New and revised references

Chapter 6—Protein: The Tissue Builder
- Updated section on muscle protein synthesis and exercise
- New section on plant-based vs. animal proteins
- Updates on presleep protein ingestion
- New section on protein-related supplements, including data on creatine monohydrate, beta-alanine, dietary nitrate/beetroot, taurine, gelatin/collagen, beta-hydroxy beta-methylbutyrate, inosine, colostrum, and glucosamine/chondroitin
- Updated section on amino acid supplements including, branched-chain amino acids, glycine, glutamine, tyrosine, arginine/citrulline, arginine/lysine/ornithine, and tryptophan
- Updated section on protein ingestion and health risks
- Updated images
- New and revised references

Chapter 7—Vitamins: Fat-Soluble, Water-Soluble, and Vitamin-Like Compounds
- Relevant content on the vitamins with updates based on the most current position paper from the Academy of Nutrition and Dietetics and the American College of Sports Medicine
- Expanded overview of vitamins with updated table 7.1 showing a summary of each vitamin
- Updated content with the latest research on the effects of specific vitamins on health and physical activity performance
- Photos provide a visual representation of good food sources for each vitamin
- Figure 7.5 showing the role of folate and vitamin B12 in red blood cell formation
- Revised *Training Tables,* one listing the classification of fat-soluble and water-soluble vitamins and vitamin-like substances, and another providing practical advice about how to read a Supplement Facts label and make prudent vitamin supplement choices
- Specific information on the health aspects of vitamin supplements now integrated within the discussion of each vitamin
- Updated Multiple Choice and Critical Thinking Questions
- New and revised references

Chapter 8—Minerals: The Inorganic Regulators
- Relevant content on the minerals with updates based on the most current position paper from the Academy of Nutrition and Dietetics and the American College of Sports Medicine and the *2020-2025 Dietary Guidelines for Americans*
- Expanded overview of minerals with additional content on the difference between major, trace, and possibly essential minerals, including tables 8.2 and 8.4 summarizing each of the major and trace minerals
- Updated content with the latest research on the effects of specific minerals on health and physical activity performance
- New photos to break up the text and provide a visual of good food sources for each mineral
- Four Training Tables on topics including factors that increase or decrease calcium absorption, how to reduce one's risk for osteoporosis and improve bone health, common signs and symptoms of iron-deficiency anemia, and a summary of two possibly essential minerals
- Table 8.5 differentiating factors that influence iron bioavailability and an expanded section on iron-deficiency anemia
- Application Exercise is provided for students to evaluate minerals with potential ergogenic benefits and to develop informational handouts on one of those minerals
- Updated Multiple Choice and Critical Thinking Questions
- New and revised references

Chapter 9—Water, Electrolytes, and Temperature Regulation
- Extensive updating of chapter content with 55 deleted references and 64 new references
- Extensive revision of Table 9.1 Sodium Content of Common Foods
- Addition of age-related recommended intakes of sodium, chloride, and potassium based on *2020-2025 Dietary Guidelines for Americans*
- Extensive revision of Table 9.4 Potassium content in some common foods in the major Food Groups.

- Updated Figure 9.5 Sources of heat gain and heat loss
- New Training Table – Calculation of Sweat Rate
- Updated information in Table 9.5 Fluid-replacement and high-carbohydrate* beverage comparison chart per 8-oz serving
- Updated Table 9.6 Fluid consumption (milliliters) at a given percent carbohydrate concentration to obtain desired grams of carbohydrate
- New Training Table – Calculation of hydration volume
- Addition of plant-based dairy alternatives in Table 9.12 The DASH eating plan (servings per day based on a 2,000-kcal diet)
- Critical Thinking Question
- New and revised references

Chapter 10—Body Weight and Composition for Health and Sport

- Chapter organization has been updated to enhance flow and readability of the chapter. This includes a re-organization of the section on "Disordered Eating and Eating Disorders" and including that at the end of the chapter, rather than embedded in the middle. Throughout the chapter, significant modifications have been made based on the most current science in the areas of body composition, obesity, and disordered eating and eating disorders.
- Given the rapidly evolving nature of the content covered in this chapter, we have made extensive updates to the references, including the most current science and evidence-based recommendations.
- Multiple tables and figures have been updated for clarity and to support student learning. This includes:
  ○ New Table 10.2 Body mass index percentile categories for children and adolescents ages 2-20
  ○ New Figure 10.2 Hydrostatic Weighing
  ○ Revised Table 10.3 Methods used to determine body composition using 2- and 3-component models
  ○ New Figure 10.3 Air-displacement plethysmography
  ○ New Figure 10.5 Bioelectrical impedance
  ○ New Figure 10.6 DEXA
  ○ Updated Table 10.10 Weight-loss drugs approved by the US Food and Drug Administration
  ○ Elimination of Training Table/Replacement with Table 10.11 Symptoms of anorexia nervosa
  ○ Elimination of Training Table/Replacement with Table 10.12 DSM-V criteria for bulimia nervosa
  ○ Elimination of Training Table/Replacement with Tables 10.13 Behaviors associate with binge eating disorders
  ○ Elimination of Training Table/Replacement with Tables 10.14 Other selected disordered eating and body image disorders
  ○ New Table 10.17 Body fat percentage ranges by sport and gender
- Extensive updates to the section of the chapter focusing on eating disorders. New content added contrasting disordered eating and eating disorders. Most current scientific evidence provided related to eating disorders as well as new content on risk factors for eating disorders and an overview of treatment options.
- Significant revisions to the section on the female athlete triad and new content added on the male athlete triad.
- The chapter has been updated to best support diversity, equity, inclusion, and accessibility (DEIA) as it relates to the field of body weight and body composition. This includes changes to language and terminology (e.g. individual with obesity replaces obese individual) as well as how content is presented.

Chapter 11—Weight Maintenance and Loss through Proper Nutrition and Exercise

- Chapter has been significantly revised and re-organized with new section headings and new topics introduced
- New figures and tables throughout the chapter displaying the most current data on overweight and obesity and evidence-based recommendations to support weight loss and maintenance
- Figure 11.2 shows the relationship between overweight and obesity and risk for several chronic health conditions
- Weight bias is introduced with new content on how health-care providers can minimize weight bias in their practice
- New content on ultra-processed foods and how intake impacts weight loss efforts
- Section dedicated to mindful eating practices, including Table 11.4 summarizing general characteristics of mindful eating
- Updates based on the most current evidence on the impact of physical activity on weight loss and maintaining that weight loss long-term
- Key recommendations of the American College of Sports Medicine, Academy of Nutrition and Dietetics and the *2018 Physical Activity Guidelines for Americans*
- Overview of factors to be considered when selecting a weight-loss program and summary of weight-loss program types
- Section dedicated to specific recommendations for weight loss in children and adolescents with overweight or obesity
- Table 11.11 summarizes the three main types of intermittent fasting and the chapter explores specific dietary patterns that may support weight loss, including high-protein diets, keto diets, and more
- New recommendations for exercise before and after bariatric surgery are provided
- Table 11.12 summarizes common symptoms of low energy availability in athletes
- All new end-of-chapter materials - application exercise, multiple-choice questions, and critical thinking questions.
- The reference list has been updated to include the most current research available

Chapter 12—Gaining Lean Body Mass through Proper Nutrition and Exercise

- Extensive updating of chapter content with 12 deleted references and 22 new references
- Case study of a diet and exercise plan for a 20 year old male college basketball guard to increase lean body mass for the next season.
- Updated Table 12.1 Estimated kilocalorie intake needed for a 170-pound, 76" tall, sedentary, 20-year-old low-active (Physical

Activity Coefficient=1.1) male basketball player to gain 1 pound per week.
- Updated Table 12.2. A sample high-kcal 24-hr meal plan diet with sufficient protein intake combined with progressive resistance training to support muscle synthesis.
- New Table 12.3 Volumes for eight resistance exercises, total volumes for separate upper-body and back/lower body sessions, and weekly volume.
- Table 12.4 Weekly record for resistance-training program of nine exercises
- New art and photographs for Figures 12.7, 12.8, 12.9, 12.10, 12.11, 12./12, 12.13, and 12.14
- Summary points of 2020 American Academy of Pediatrics (AAP) position statement regarding resistance training for children and adolescents.

Chapter 13—Nutritional Supplements and Ergogenic Aids
- Extensive updating of chapter content with 102 deleted references and 117 new references
- Edited Figure 13.2 Simplified pathways for alcohol metabolism
- Updated text summarizing use of alcohol as an ergogenic aid
- Added "driving while impaired" and "aggressive behavior" subheadings under psychological effects of alcohol
- New heading entitled "Alcohol Consumption in College Students and Athletes"
- New text added on moderate alcohol consumption and kidney cancer
- Extensive revision of Table 13.6 Caffeine content in selected products
- Updated text summarizing use of caffeine as an ergogenic aid
- Updated Table 13.8 Caffeine content in selected energy drinks (n = 7) and shots (n = 5) with descriptive information on caffeine content for 220 energy drinks and 36 energy shots
- Added text on pre-workout supplements
- New Table 13.10. Summary of position stands from the American College of Sports Medicine (ACSM), the National Strength and Conditioning Association (NSCA), and the National Athletic Trainers' Association (NATA) regarding use and abuse of Anabolic-Androgenic Steroids (AAS) and/or human Growth Hormone (hGH).
- Deleted discussion of nitrates

Enhanced Pedagogy

Each chapter contains several features to help enhance the learning process. **Learning Outcomes** are presented at the beginning of each chapter, highlighting the key points and serving as a studying guide for students and an assessment tool for faculty. **Key Terms** also are listed at the beginning of each chapter and definitions are included both in the chapter and in the glossary. A new Training Table feature has also been added to this edition of the textbook. The **Training Tables** emphasize practical and current concepts relevant to each chapter. **Key Concepts** provide a summary of essential information presented throughout each chapter. Bulleted lists are utilized to help students focus on the key information. **Check for Yourself** includes individual activities, such as checking food labels at the supermarket or evaluating one's own body fat percentage. The **Application Exercise** at the end of each chapter may require more extensive involvement, such as a case study or a survey of an athletic team. **Multiple Choice Questions** and **Critical Thinking Questions** are also included at the end of each chapter for students to self-assess their knowledge of the chapter content. The Critical Thinking Questions require students to apply the knowledge they've learned in each chapter.

The reference lists have been completely updated for this edition with the inclusion of hundreds of new references that provide the scientific basis for the new concepts or additional support for those concepts previously developed. These references provide greater in-depth reading materials for the interested student. Although the content of this book is based on appropriate scientific studies, a reference-citation style is not used, that is, each statement is not referenced by a bibliographic source. However, names of authors may be used to highlight a reference source where deemed appropriate.

This book is designed primarily to serve as a college text in professional preparation programs in dietetics and human nutrition, health and physical education, exercise science, athletic training, sports medicine, and sports nutrition. It is also directed to the physically active individual interested in the nutritional aspects of physical and athletic performance.

Those who desire to initiate a physical training program may also find the nutritional information useful, as well as the guidelines for initiating a training program. This book may serve as a handy reference for coaches, trainers, and athletes. With the tremendous expansion of youth sports programs, parents may find the information valuable relative to the nutritional requirements of their active children.

In summary, the major purpose of this book is to help provide a sound knowledge base relative to the role that nutrition, complemented by exercise, may play in the enhancement of both health and sports performance. We hope the information provided in this text will help inspire the reader to make health-promoting choices related to diet and physical activity.

Acknowledgments

This book would not be possible without the many medical/health scientists and exercise/sports scientists throughout the world who, through their numerous studies and research, have provided the scientific data that underlie its development. We are fortunate to have developed a friendship with many of you, and we extend our sincere appreciation to all of you. We would like to thank the nutrition educators who reviewed this text.

Eric S. Rawson
J. David Branch
Tammy J. Stephenson

Instructors
The Power of Connections

A complete course platform

Connect enables you to build deeper connections with your students through cohesive digital content and tools, creating engaging learning experiences. We are committed to providing you with the right resources and tools to support all your students along their personal learning journeys.

65% Less Time Grading

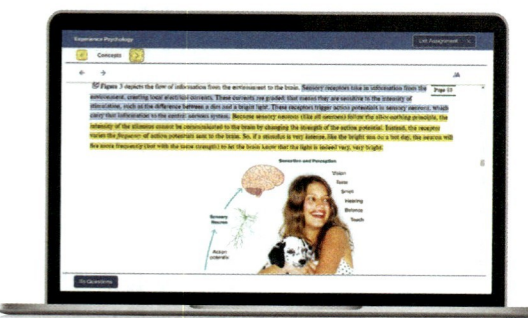

Laptop: Getty Images; Woman/dog: George Doyle/Getty Images

Every learner is unique

In Connect, instructors can assign an adaptive reading experience with SmartBook® 2.0. Rooted in advanced learning science principles, SmartBook 2.0 delivers each student a personalized experience, focusing students on their learning gaps, ensuring that the time they spend studying is time well-spent.
mheducation.com/highered/connect/smartbook

Affordable solutions, added value

Make technology work for you with LMS integration for single sign-on access, mobile access to the digital textbook, and reports to quickly show you how each of your students is doing. And with our Inclusive Access program, you can provide all these tools at the lowest available market price to your students. Ask your McGraw Hill representative for more information.

Solutions for your challenges

A product isn't a solution. Real solutions are affordable, reliable, and come with training and ongoing support when you need it and how you want it. Visit **supportateverystep.com** for videos and resources both you and your students can use throughout the term.

Students
Get Learning that Fits You

Effective tools for efficient studying

Connect is designed to help you be more productive with simple, flexible, intuitive tools that maximize your study time and meet your individual learning needs. Get learning that works for you with Connect.

Study anytime, anywhere

Download the free ReadAnywhere® app and access your online eBook, SmartBook® 2.0, or Adaptive Learning Assignments when it's convenient, even if you're offline. And since the app automatically syncs with your Connect account, all of your work is available every time you open it. Find out more at **mheducation.com/readanywhere**

"I really liked this app—it made it easy to study when you don't have your text-book in front of you."

- Jordan Cunningham, Eastern Washington University

iPhone: Getty Images

Everything you need in one place

Your Connect course has everything you need—whether reading your digital eBook or completing assignments for class—Connect makes it easy to get your work done.

Learning for everyone

McGraw Hill works directly with Accessibility Services Departments and faculty to meet the learning needs of all students. Please contact your Accessibility Services Office and ask them to email accessibility@mheducation.com, or visit **mheducation.com/about/accessibility** for more information.

Digital Tools for Your Success

Prep for Nutrition. A challenge nutrition instructors often face is students' lack of basic math, chemistry, and/or biology skills when they begin the course. To help you level-set your classroom, we've created Prep for Nutrition. This question bank highlights a series of questions to give students a refresher on the skills needed to enter and be successful in their nutrition course! By having these foundational skills, you will feel more confident your students can begin class, ready to understand more complex concepts and topics. Prep for Nutrition is course-wide for ALL nutrition titles and can be found in the Question Bank dropdown within Connect®.

NutritionCalc Plus is a powerful dietary analysis tool featuring over 106,000 foods from the ESHA Research nutrient database, which is comprised of data from the latest USDA Standard Reference database, manufacturer's data, restaurant data, and data from literature sources. NutritionCalc Plus allows users to track food and activities and then analyze their choices with a robust selection of intuitive reports. An updated mobile-friendly interface has been developed according to WCAG guidelines for further accessibility.

Auto-graded, case study-based assignments in Connect correspond with NutritionCalc Plus reports for students to apply their knowledge and gain further insight into dietary analysis.

Assess My Diet: Auto-graded personal dietary analysis in Connect. One of the challenges many instructors face when teaching the nutrition course is having the time to grade individual dietary analysis projects. To help overcome this challenge, we've created auto-graded assignments in Connect that complement the NutritionCalc Plus tool. Students are directed to answer questions about their dietary patterns based on generated reports from NutritionCalc Plus. These assignments were created and reviewed by instructors just like you, who use them in their own teaching. Designed to be relevant, current, and interesting, you will find them easy to implement and use in your classroom.

Proctorio

Remote Proctoring & Browser-Locking Capabilities

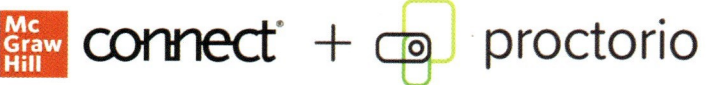

Remote proctoring and browser-locking capabilities, hosted by Proctorio within Connect, provide control of the assessment environment by enabling security options and verifying the identity of the student.

Seamlessly integrated within Connect, these services allow instructors to control the assessment experience by verifying identification, restricting browser activity, and monitoring student actions.

Instant and detailed reporting gives instructors an at-a-glance view of potential academic integrity concerns, thereby avoiding personal bias and supporting evidence-based claims.

Read or study when it's convenient for you with McGraw Hill's free ReadAnywhere® app. Available for iOS or Android smartphones or tablets, ReadAnywhere gives users access to McGraw Hill tools, including the eBook and SmartBook® 2.0 or Adaptive Learning Assignments in Connect. Take notes, highlight, and complete assignments offline—all of your work will sync when you open the app with Wi-Fi access. Log in with your McGraw Hill Connect username and password to start learning—anytime, anywhere!

OLC-Aligned Courses

Implementing High-Quality Instruction and Assessment through Preconfigured Courseware

In consultation with the Online Learning Consortium (OLC) and our certified Faculty Consultants, McGraw Hill has created preconfigured courseware using OLC's quality scorecard to align with best practices in online course delivery. This turnkey courseware contains a combination of formative assessments, summative assessments, homework, and application activities and can easily be customized to meet an individual instructor's needs and desired course outcomes. For more information, visit www.mheducation.com/highered/olc.

Tegrity: Lectures 24/7

Tegrity in Connect is a tool that makes class time available 24/7 by automatically capturing every lecture. With a simple one-click start-and-stop process, you capture all computer screens and corresponding audio in a format that is easy to search, frame by frame. Students can replay any part of any class with easy-to-use, browser-based viewing on a PC, Mac, or mobile device.

Educators know that the more students can see, hear, and experience class resources, the better they learn. In fact, studies prove it. Tegrity's unique search feature helps students efficiently find what they need, when they need it, across an entire semester of class recordings. Help turn your students' study time into learning moments immediately supported by your lecture. With Tegrity, you also increase intent listening and class participation by easing students' concerns about note-taking. Using Tegrity in Connect will make it more likely you will see students' faces, not the tops of their heads.

Test Builder in Connect

Available within Connect, Test Builder is a cloud-based tool that enables instructors to format tests that can be printed, administered within a Learning Management System, or exported as a Word document. Test Builder offers a modern, streamlined interface for easy content configuration that matches course needs without requiring a download.

Test Builder allows you to:

- access all test bank content from a particular title.
- easily pinpoint the most relevant content through robust filtering options.
- manipulate the order of questions or scramble questions and/or answers.
- pin questions to a specific location within a test.
- determine your preferred treatment of algorithmic questions.
- choose the layout and spacing.
- add instructions and configure default settings.

Test Builder provides a secure interface for better protection of content and allows for just-in-time updates to flow directly into assessments.

Writing Assignment

Available within Connect and Connect Master, the Writing Assignment tool delivers a learning experience to help students improve their written communication skills and conceptual understanding. As an instructor, you can assign, monitor, grade, and provide feedback on writing more efficiently and effectively.

Create

Your Book, Your Way

McGraw Hill's Content Collections Powered by Create® is a self-service website that enables instructors to create custom course materials—print and eBooks—by drawing upon McGraw Hill's comprehensive, cross-disciplinary content. Choose what you want from our high-quality textbooks, articles, and cases. Combine it with your own content quickly and easily, and tap into other rights-secured, third-party content such as readings, cases, and articles. Content can be arranged in a way that makes the most sense for your course, and you can include the course name and information as well. Choose the best format for your course: color print, black-and-white print, or eBook. The eBook can be included in your Connect course and is available on the free ReadAnywhere® app for smartphone or tablet access as well. When you are finished customizing, you will receive a free digital copy to review in just minutes! Visit McGraw Hill Create®—www.mcgrawhillcreate.com—today and begin building!

Reflecting the Diverse World Around Us

McGraw Hill believes in unlocking the potential of every learner at every stage of life. To accomplish that, we are dedicated to creating products that reflect, and are accessible to, all the diverse, global customers we serve. Within McGraw Hill, we foster a culture of belonging, and we work with partners who share our commitment to equity, inclusion, and diversity in all forms. In McGraw Hill Higher Education, this includes, but is not limited to, the following:

- Refreshing and implementing inclusive content guidelines around topics including generalizations and stereotypes, gender, abilities/disabilities, race/ethnicity, sexual orientation, diversity of names, and age.
- Enhancing best practices in assessment creation to eliminate cultural, cognitive, and affective bias.
- Maintaining and continually updating a robust photo library of diverse images that reflect our student populations.
- Including more diverse voices in the development and review of our content.
- Strengthening art guidelines to improve accessibility by ensuring meaningful text and images are distinguishable and perceivable by users with limited color vision and moderately low vision.

Introduction to Nutrition for Health, Fitness, and Sports Performance

CHAPTER ONE

Mauro Grigollo/E+/Getty Images

LEARNING OUTCOMES

After studying this chapter, you should be able to:

1. List the leading causes of death in the United States and identify those that may be related to lifestyle factors, including diet and/or physical activity.
2. Explain the importance of genetics, diet, and physical activity in the determination of optimal health and successful sport performance.
3. Describe the components of health-related fitness and identify the potential health benefits associated with each.
4. Compare and contrast sports-related fitness and health-related fitness.
5. Summarize the seven key principles of exercise training.
6. Explain the importance of diet choices and proper nutrition in promoting optimal health and wellness.
7. Summarize the role of dietary supplements as ergogenic aids to promote sports performance.
8. Explain nutrition and health misinformation and provide strategies that can be utilized to determine whether claims regarding a dietary recommendation are valid.
9. Explain what types of research have been used to evaluate the relationship between nutrition and health or sport performance, and evaluate the pros and cons of each type.

KEY TERMS

antipromoters
cytokines
doping
epidemiological research
epigenetics
epigenome
ergogenic aids
exercise
experimental research
health-related fitness
high-intensity interval training (HIIT)
malnutrition
meta-analysis
nutrient
nutrition
nutrition and health misinformation
physical activity
physical fitness
promoters
Prudent Healthy Diet
risk factor
sarcopenia
Sedentary Death Syndrome (SeDS)
sports nutrition
sports-related fitness
sports supplements
structured physical activity
unstructured physical activity

Introduction

There are two major focal points of this book. One is the role that nutrition, complemented by physical activity and exercise, may play in the enhancement of one's health status. The other is the role that nutrition may play in the promotion of fitness and sports performance. Many individuals today are physically active, and athletic competition spans all ages. Healthful nutrition is important throughout the life span of the physically active individual because suboptimal health status may impair training and competitive performance. In general, as we shall see, the diet that is optimal for health is also optimal for exercise and sports performance.

Nutrition, fitness, and health. Health care in most developed countries has improved tremendously over the past century. With modern health care, once deadly diseases are no longer a major source of concern. Rather, the treatment and prevention of chronic diseases, such as diabetes and obesity, are now the emphasis of much research and health recommendations.

Table 1.1 lists the ten leading causes of death in the United States in 2020 and the approximate percentage of deaths associated with each. For both males and females, heart disease is the leading cause of death, accounting for death in over one in five Americans. In 2020, COVID-19 was the third leading cause of death in the United States, accounting for over 10% of deaths. Of the leading causes of death, risk for heart disease, cancer, COVID-19, stroke, Alzheimer's disease, diabetes, and kidney disease have been linked to a person's diet, physical activity habits, and/or body composition. According to the U.S. Department of Health and Human Services (HHS), unhealthy eating and physical inactivity are primary contributors to death in the United States.

In addition to lifestyle choices, family history also impacts risk for chronic disease. According to Simopoulos, all diseases have a genetic predisposition. The Human Genome Project, which deciphered the DNA code of our 80,000 to 100,000 genes, has identified various genes associated with many chronic diseases, such as breast and prostate cancer. Genetically, females whose mothers had breast cancer are at an increased risk for breast cancer, while males whose fathers had prostate cancer are at an increased risk for prostate cancer.

TABLE 1.1 Leading causes of death in the United States (2020)

	Approximate percentage of deaths
Heart disease*	20.6
Cancer*	17.8
COVID-19*	10.4
Unintentional injuries (accidents)	5.9
Stroke*	4.7
Chronic lower respiratory diseases	4.5
Alzheimer's disease*	4.0
Diabetes mellitus*	3.0
Influenza and pneumonia	1.6
Kidney disease*	1.6
All other causes	25.9

*Cause of death for which diet, physical activity, and/or body composition may impact risk.

Source: U.S. Department of Health and Human Services, National Center for Health Statistics. *Mortality in the United States, 2020.* www.cdc.gov/nchs/data/databriefs/db427.pdf. Accessed September 24, 2022.

Completion of the Human Genome Project is believed to be one of the most significant medical advances of all time. Although multiple genes are involved in the etiology of most chronic diseases and research regarding the application of the findings of the Human Genome Project to improve health is still in its initial stages, the future looks bright. For individuals with genetic profiles predisposing them to a specific chronic disease, such as cancer, genetic therapy eventually may provide an effective treatment or cure.

Our genes harbor many secrets to a long and healthy life, but genes alone are unlikely to explain all the secrets of longevity. The role of a healthful diet and exercise are intertwined with your genetic profile. What you eat and how you exercise may influence your genes. **Epigenetics** is a relatively new field of research involving the role of the **epigenome,** a multitude of specialized chemical compounds that influence the human genome by activating or deactivating DNA and subsequent genetic and cellular activity. Various factors in our environment, such as substances in the foods we eat, may interact with the epigenome and thus modify cell functions—either in a positive or negative manner. Exercise, as noted later, also stimulates release of substances from muscle cells that may affect the epigenome. Genomics represents the study of genetic material in body cells, and the terms *nutrigenomics* and *exercisenomics* have been coined to identify the study of the genetic aspects of nutrition and exercise, respectively, as related to health benefits. *Sportomics* involves study of the metabolic response of the athlete in an actual sport environment, not in a laboratory.

Preventing chronic disease. Many forms of chronic disease are preventable through proper nutrition and physical activity and recognizing risk factors for a particular health condition. A **risk factor** is a lifestyle behavior that has been associated with a particular disease, such as cigarette smoking being linked to lung cancer. As described previously, diet and physical activity choices are also key risk factors for chronic disease. For example, a sedentary lifestyle and having obesity are risk factors for heart disease and some forms of diabetes.

To help improve the health of Americans, the United States Office of Disease Prevention and Health Promotion (ODPHP) publishes health-related reports and goals every ten years. Many of the goals outlined in *Healthy People 2030* address issues specific to physical activity and diet choices. The *Training Table* in this section provides examples of such objectives.

Nutrition, fitness, and sport. *Sport* is most commonly defined as a competitive athletic activity requiring skill or physical prowess, for example, baseball, basketball, soccer, football, track, wrestling, tennis, and golf.

To be successful at high levels of competition, athletes must possess the appropriate biomechanical, physiological, psychological, and genetic characteristics associated with success in a given sport. International-class athletes have such genetic traits. In recent reviews, Tucker and others highlighted the genetic basis for elite running performance while Eynon and others discussed the role of genes for elite power and sprint performance. Moreover, Wolfarth and others have assembled a human gene map for performance and health-related fitness.

For optimal performance, athletes must also develop their genetic characteristics maximally through proper biomechanical, physiological, and psychological coaching and training. Whatever the future holds for genetic enhancement of athletic performance, specialized exercise training will still be the key to maximizing genetic potential for a given sport activity. Training programs at the elite level have become more intense and individualized, sometimes based on genetic predispositions. Modern scientific training results in significant performance gains, and world records continue to improve. David Epstein, in his book *The Sports Gene,* provides a fascinating account of the role both genes and the training environment play relative to elite sport performance.

Proper nutrition is also an important component in the total training program of the athlete. Certain nutrient deficiencies can seriously impair performance, whereas supplementation of other nutrients may help delay fatigue and improve performance. Over the past 50 years, research has provided us with many answers about the role of nutrition in athletic performance, yet there is still much to be learned as research in sports nutrition continues to expand.

The purpose of this chapter is to provide a broad overview of the role that exercise and nutrition may play relative to health, fitness, and sport, and to provide evidence-based recommendations. More detailed information regarding specific relationships of nutritional practices to health and sports performance is provided in subsequent chapters.

Training Table

Examples of some of the **Healthy People 2030** objectives related to physical activity:

- Reduce the proportion of adults who do no physical activity in their free time.
- Increase the proportion of adults who do enough aerobic physical activity for substantial health benefits.
- Increase the proportion of adults who walk or bike to get places.
- Increase the proportion of adolescents who do enough aerobic and muscle-strengthening activities.
- Increase the proportion of children and adolescents who play sports.

Visit www.healthypeople.gov to see how progress is being made toward these goals.

www.health.gov/healthypeople Check for the full report of *Healthy People 2030.*

www.who.int/dietphysicalactivity/en/ The World Health Organization report provides global recommendations related to diet and physical activity for health.

www.ncbi.nlm.nih.gov/genome/guide/human/ Access the human genome map and the National Institutes of Health Epigenetics Roadmap.

Key Concepts

▶ Many chronic diseases in major developed countries (heart disease, cancer, stroke, and diabetes) may be prevented by appropriate lifestyle behaviors, particularly maintaining a healthy body composition, proper exercise and a healthy diet.
▶ The two primary determinants of health status are genetics and lifestyle.
▶ Several of the key health promotion objectives in *Healthy People 2030* are increased levels of physical activity, a healthier diet, and reduced rates of overweight and obesity.
▶ Sports success is dependent on biomechanical, physiological, and psychological characteristics specific to a given sport, but proper training, including nutrition, is essential to maximizing one's genetic potential.

Check for Yourself

▶ Discuss with your parents, grandparents, and other relatives any health problems they have experienced, such as high blood pressure or diabetes, to determine whether you may be predisposed to such health problems in the future. Use the "My Family Health Portrait" tool at http://kahuna.clayton.edu/jqu/FHH/html/index.html to create a family health history.

Fitness and Exercise

Physical fitness may be defined, in general terms, as a set of abilities individuals possess to perform specific types of physical activity. The development of physical fitness is an important concern of many professional health organizations, including the Society of Health and Physical Educators (SHAPE), which has classified fitness components into two different categories. In general, these two categories may be referred to as health-related fitness and sports-related fitness. Both types of fitness may be influenced by nutrition and exercise.

How are health-related fitness and sports-related fitness different?

As summarized in the introduction to this chapter, lifestyle behaviors, including appropriate physical activity and a high-quality diet, may influence one's health status and wellness. Proper physical activity may improve one's health status by helping to prevent excessive weight gain, but it may also enhance other facets of health-related fitness as well. **Health-related fitness** includes not only a healthy body weight and body composition, but also cardiovascular-respiratory fitness, adequate muscular strength and muscular endurance, and sufficient flexibility (**figure 1.1**). As one ages, other measures used as markers of health-related fitness include blood pressure, bone strength, postural control and balance, and various indicators of lipid and carbohydrate metabolism.

In contrast to health-related fitness, **sports-related fitness** is the fitness an athlete develops specific to their sport. Dependent on the sport, this may include strength, power, speed, endurance, and/or neuromuscular motor skills. Through proper physical and mental training, athletes may maximize their genetic potential, thus preparing both their body and mind for intense competition. Compared to health-related fitness, training for sports performance is often more intense, prolonged, and frequent than training for health.

What are the basic principles of exercise training?

Several health professional organizations, such as the American College of Sports Medicine (ACSM) and American Heart Association (AHA), have indicated that various forms of physical activity may be used to enhance health. In general, **physical activity** involves any bodily movement caused by muscular contraction that results in the expenditure of energy. For the purpose of studying its effects on health, some epidemiologists classify physical activity as either unstructured or structured.

Unstructured physical activity, also known as leisure-time activity, includes many of the usual activities of daily living, such as leisurely walking and cycling, climbing stairs, dancing, gardening and yard work, various domestic and occupational activities, and games and other childhood pursuits. These unstructured activities are not normally planned to be exercise. However, as will be noted throughout the textbook, these types of activities may play an important role in body weight control.

Nicholas Monu/Vetta/Getty Images

Structured physical activity, as the name implies, is a planned program of physical activities usually designed to improve fitness. For the purpose of this book, we shall refer to structured physical activity as **exercise,** particularly some form of planned moderate or vigorous exercise, such as brisk, not leisurely, walking.

Exercise training programs may be designed to provide health-related and/or sports-related fitness benefits. However, no matter what the purpose, several general principles are used in developing an appropriate exercise training program.

Principle of Overload Overload is the basic principle of exercise training, and it represents the altering of the intensity, duration, and frequency of exercise. For example, a running program for cardiovascular-respiratory fitness could involve training at an intensity of 70 percent of maximal heart rate, a duration of 30 minutes, and a frequency of 5 times per week. The adaptations the body makes are based primarily on the specific exercise overload.

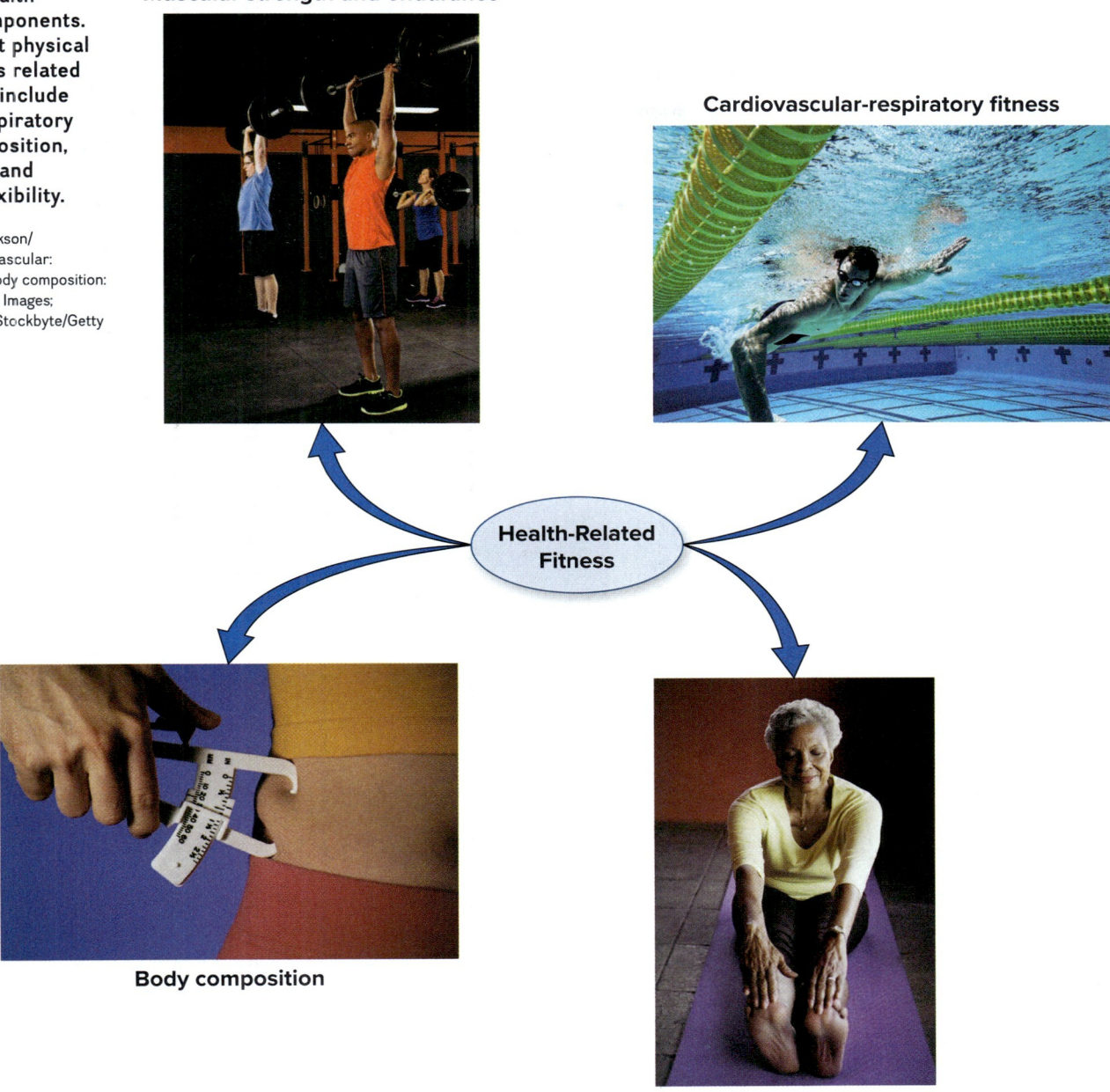

FIGURE 1.1 Health-related fitness components. The most important physical fitness components related to personal health include cardiovascular-respiratory fitness, body composition, muscular strength and endurance, and flexibility.

Muscular strength: Erik Isakson/Blend Images LLC; Cardiovascular: epicstockmedia/123RF; Body composition: ComStock/Stockbyte/Getty Images; Flexibility: Jupiterimages/Stockbyte/Getty Images

The terms *moderate* exercise and *vigorous* exercise are often used to quantify exercise intensity and are discussed later in this chapter.

Principle of Progression Progression is an extension of the overload principle. As your body adapts to the original overload, the overload must be increased if further beneficial adaptations are desired. For example, you may start lifting a weight of 20 pounds, increase the weight to 25 pounds as you get stronger, and so forth. The overloads are progressively increased until the final health-related or sports-related goal is achieved or exercise limits are reached.

Principle of Specificity Specificity of training represents the specific adaptations the body will make in response to the type of exercise and overload. For example, running and weight lifting impose different demands on muscle energy systems, so the body adapts accordingly. Both types of exercise may provide substantial, yet different, health benefits. Exercise training programs may be designed specifically for certain health or sports-performance benefits.

Principle of Recuperation Recuperation is an important principle of exercise training. Also known as the principle of recovery, it represents the time in which the body rests after exercise. This principle may apply within a specific exercise period, such as including rest periods when doing multiple sets during a weight-lifting workout. It may also apply to rest periods between bouts of exercise, such as a day of recovery between two long cardiovascular workouts.

Principle of Individuality Individuality reflects the effect exercise training will have on each individual, as determined by genetic characteristics. The health benefits one receives from a specific exercise training program may vary tremendously among individuals. For example, although most individuals with high blood pressure may experience a reduction during a cardiovascular-respiratory fitness training program, some may not.

Principle of Reversibility Reversibility is also referred to as the principle of disuse, or the concept of *use it or lose it*. Without the use of exercise, the body will begin to lose the adaptations it has made over the course of the exercise program. Individuals who suffer a lapse in their exercise program, such as a week or so, may lose only a small amount of health-related fitness gains. However, a total relapse to a previous sedentary lifestyle can reverse all health-related fitness gains.

Principle of Overuse Overuse represents an excessive amount of exercise that may induce some adverse, rather than beneficial, health effects. Overuse may be a problem during the beginning stages of an exercise program if one becomes overenthusiastic and exceeds her capacity, such as developing shin splints by running too much or too far. As described in chapter 3, overuse may also occur in elite athletes who become overtrained.

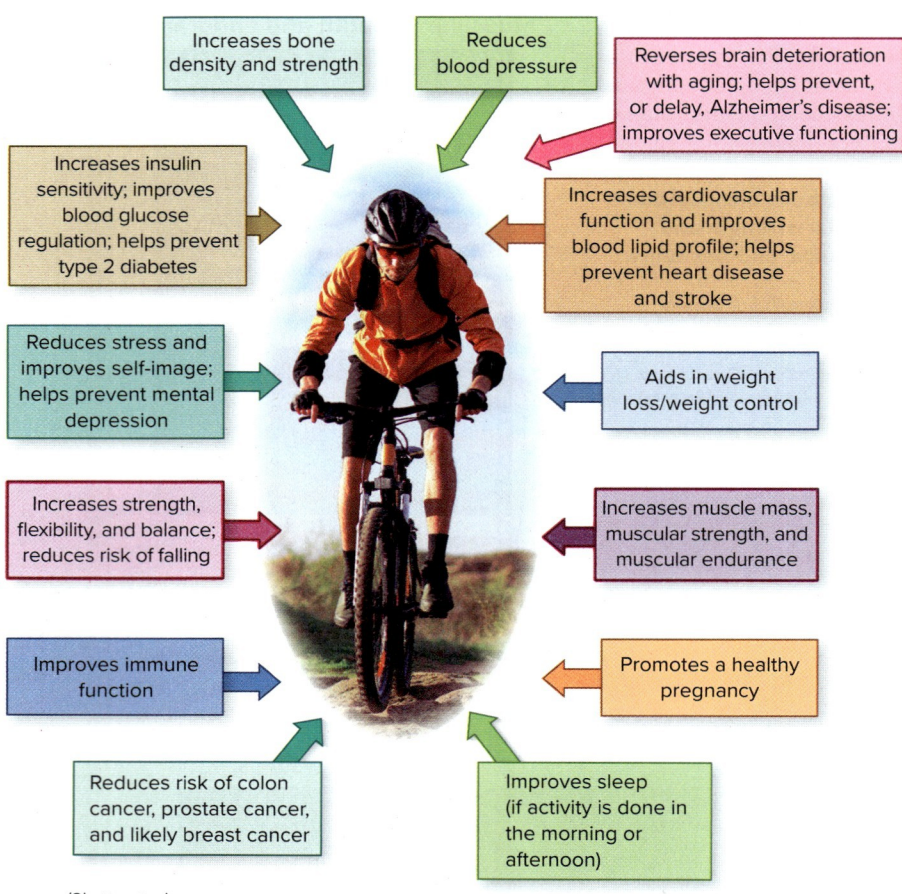

maxpro/Shutterstock

FIGURE 1.2 Exercise is medicine. Researchers have identified more than three dozen specific health benefits associated with engaging in regular physical activity. This figure summarizes some of those key health benefits.

Exercise and Health Promotion

The beneficial effect of exercise on health has been known for centuries. For example, Plato noted that "lack of activity destroys the good condition of every human being while movement and methodical physical exercise save and preserve it." Plato's observation is even more relevant in contemporary society. Frank Booth, a prominent exercise scientist at the University of Missouri, has coined the term **Sedentary Death Syndrome,** or **SeDS,** and he and his colleagues have noted that physical inactivity is a primary cause of most chronic diseases. Slentz and others discussed the cost of physical inactivity over time. The *short-term* cost of physical inactivity is metabolic deterioration and weight gain; the *intermediate-term* cost is an increased risk for disease, such as type 2 diabetes, whereas the *long-term* cost is increased risk for premature mortality.

To help promote the health benefits of physical activity, the ACSM and the American Medical Association (AMA) launched a program, entitled *Exercise Is Medicine®*, designed to encourage physicians and other health-care professionals to include exercise as part of the treatment for every patient. Clinical, epidemiological, and basic research evidence clearly supports the inclusion of regular physical activity as a tool for the prevention of chronic disease and the enhancement of overall health. **Figure 1.2** summarizes some of the specific health benefits that have been associated with regular physical activity.

In essence, physically active individuals enjoy a higher quality of life, a *joie de vivre,* because they are less likely to suffer the disabling symptoms often associated with chronic diseases, such as loss of ambulation experienced by some stroke victims. As noted in the next section, physical activity may also increase the quantity of life. James Fries, who studied healthy aging at the Stanford University School of Medicine's Center on Longevity, stated, "If you had to pick one thing to make people healthier as they age, it would be aerobic exercise."

www.exerciseismedicine.org Website provides practical recommendations and guidance, including on physical activity during the COVID-19 pandemic.

How does exercise enhance health?

The specific mechanisms whereby exercise may help to prevent the development of various chronic diseases are not completely understood. However, such benefits are likely related to changes in gene expression that modify cell structure and function following physical activity. As noted previously, research by Booth and Neufer found that physical inactivity causes genes to misexpress proteins,

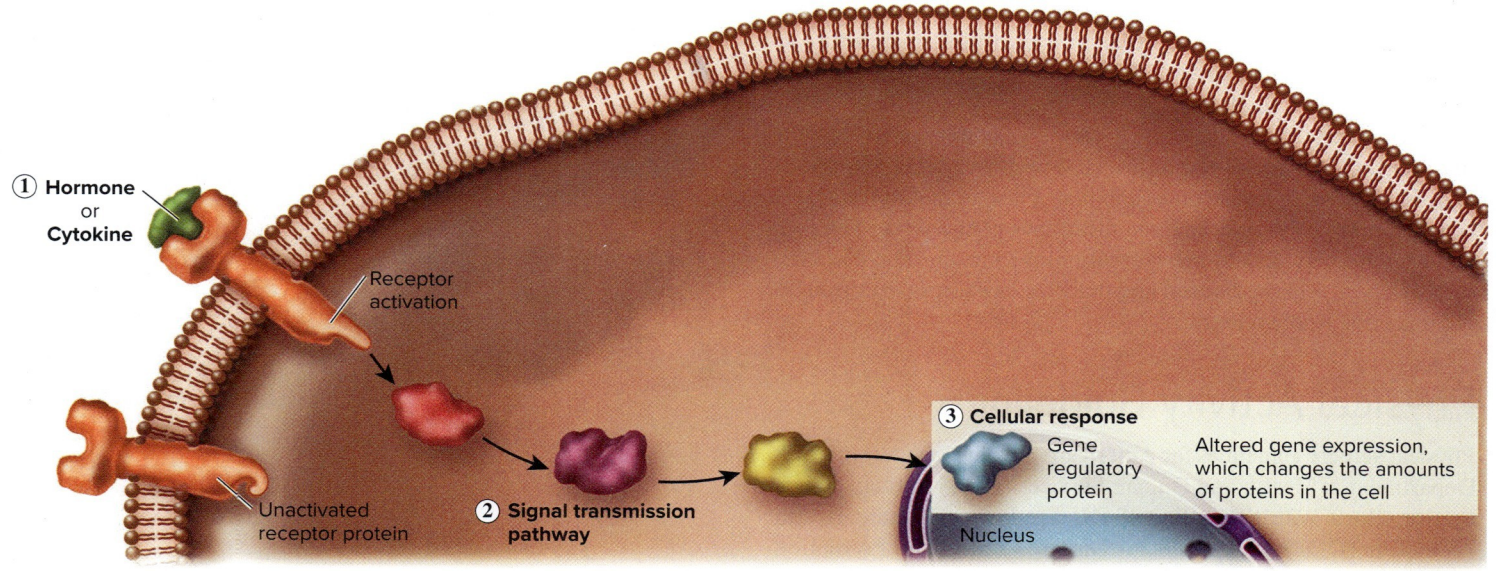

FIGURE 1.3 Exercise may induce adaptations that have favorable health effects in various body tissues. One suggested mechanism is the effect that various hormones or cytokines, which are produced during exercise, may have on gene regulation in body cells. (1) The hormone or cytokine binds to a cell receptor that activates a signal within the cell, (2) the signal is transmitted along a specific pathway, (3) the signal may alter gene expression and induce changes within the cell. Cell signals may also affect enzymes or other cell structures that may induce beneficial health effects.

producing the metabolic dysfunctions that result in overt clinical disease if continued long enough. In contrast, exercise may cause the expression of genes with favorable health effects.

Most body cells can produce and secrete small proteins known as **cytokines,** which are similar to hormones and can affect tissues throughout the body. Cytokines enter various body tissues, influencing gene expression that may induce adaptations either favorable or unfavorable to health (**figure 1.3**). Two types of cytokines are of interest to us. Muscle cells produce various cytokines called *myokines* (referred to as *exerkines* when produced during exercise), whereas fat (adipose) cells produce cytokines called *adipokines.* Muscle cells also produce *heat shock proteins (HSPs),* which may have beneficial health effects. **Table 1.2** lists important cytokines produced in muscle and fat cells.

Overall, Brandt and Pederson theorize that exercise-induced cytokine effects on genes reduce many of the traditional risk factors associated with development of chronic diseases; Geiger and others note similar effects for HSPs. According to McAtee, one of the common causes of various chronic diseases is an inflammatory environment created by the presence of excess fat, particularly within blood vessels. Local inflammation is thought to promote the development of several types of chronic disease, including heart disease, cancer, diabetes, and dementia. Work by Nimmo and others suggests that exercise produces an anti-inflammatory cytokine that may help cool inflammation and reduce such health risks. They note that the most marked improvements in the inflammatory profile are conferred with exercise performed at higher intensities, with combined aerobic and resistance exercise training potentially providing the greatest benefit. Cytokines and heat shock proteins may also prevent chronic diseases by increasing the number of glucose receptors in muscle cells, improving insulin sensitivity, and helping to regulate blood glucose and prevent type 2 diabetes.

There are also other health-promoting mechanisms of exercise. One of the most significant contributors to health problems with aging is **sarcopenia,** or loss of muscle tissue. Regular exercise is the only strategy that has been found to consistently prevent frailty and improve sarcopenia and physical function in older adults. Additional mechanisms associated with exercising lowering risk for chronic disease include:

- Loss of excess body fat may reduce production of cytokines that may impair health.
- Loss of excess body fat may reduce estrogen levels, reducing risk of breast cancer.
- Reduction of abdominal obesity may decrease blood pressure and serum lipid levels.
- Increased mechanical stress on bone with high-impact exercise may stimulate increases in bone density.
- Production of some cytokines, such as BDNF, may enhance neurogenesis and brain function.

TABLE 1.2	Major cytokines produced in muscle and fat cells
Muscle cells	**Fat cells**
Interleukin-6 (IL-6)	Tumor Necrosis Factor-alpha (TNF-α)
Brain-Derived Neurotropic Factor (BDNF)	Adiponectin

Some healthful adaptations may occur with even just a single bout of exercise. Nimmo and others reported that single bouts of exercise have a potent anti-inflammatory influence, while others have noted that a single exercise session can acutely improve the blood lipid profile, reduce blood pressure, and improve insulin sensitivity, all beneficial responses. However, such adaptations will regress unless exercise becomes habitual. Thus, to maximize health benefits, exercise should be done most days of the week. The role that exercise may play in the prevention of some chronic diseases, such as obesity, heart disease, and diabetes, are discussed throughout this book where relevant.

Physical Activity Guidelines

Physical activity guidelines for Americans are developed by the U.S. Department of Health and Human Services (HHS) through collaborative efforts with the Office of Disease Prevention and Health Promotion (ODPHP), Centers for Disease Control and Prevention (CDC), National Institutes of Health (NIH), and the President's Council on Fitness, Sports, and Nutrition (PCFSN). The *Training Table* in this section summarizes the key physical activity guidelines for healthy adults from the *2018 Physical Activity Guidelines for Americans*. You can learn more about the process of developing the physical activity guidelines and follow the latest on these recommendations at https://health.gov/paguidelines/.

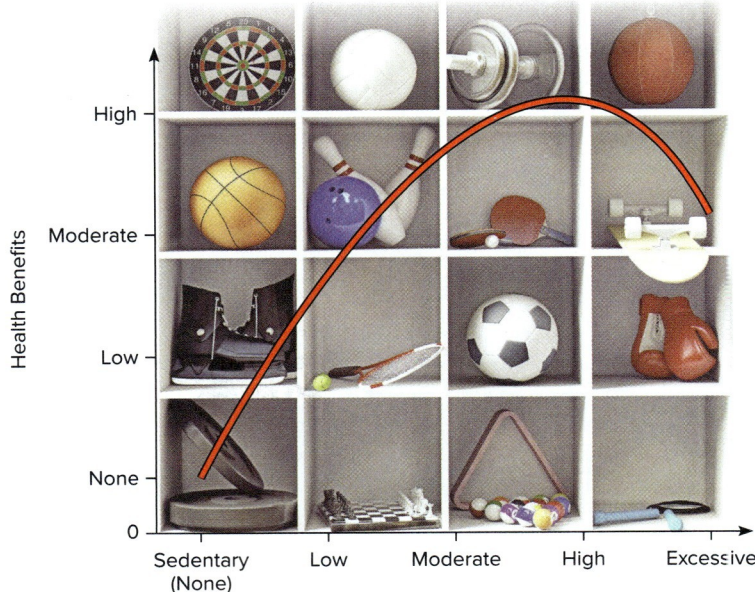

FIGURE 1.4 Significant health benefits may occur at low to moderate levels of physical activity with diminishing returns as the amount of exercise becomes excessive. Dependent on the individual, exercising too much or, in some cases, at an excessive intensity, may actually be detrimental to health.

Training Table

According to the *2018 Physical Activity Guidelines for Americans*, healthy adults should avoid inactivity. For optimal health benefits, healthy adults should:

- Participate in at least 150 minutes (2 hours and 30 minutes) a week of moderate-intensity physical activity or 75 minutes (1 hour and 15 minutes) a week of vigorous-intensity physical activity. Aerobic activity should be spread out throughout the week.
- Participate in muscle-strengthening activities, such as lifting weights or using resistance bands, that are moderate- or high-intensity and involve all major muscle groups at least two days a week.
- Include flexibility exercises, such as stretching, yoga, or pilates, as part of the exercise program. Time spent doing such activities should not be counted toward meeting the aerobic- or muscle-strengthening guidelines, but such flexibility exercises may reduce risk for injury and support optimal health and aging.

Visit https://health.gov/paguidelines/ for a complete list of the physical activity guidelines, including those specific for active children and adolescents as well as for older adults.

Key principles for developing physical activity programs

To reap the health benefits of exercise, most health professionals recommend a comprehensive program of physical activity, including aerobic exercise and resistance training. Flexibility and balance exercises become increasingly important for older adults to prevent falls and maintain mobility as one ages. In general, there is a curvilinear relationship between the amount of physical activity (dose) and related health benefits (response). As shown in **figure 1.4,** a sedentary lifestyle is thought to offer no health benefits. However, health benefits increase rapidly with low to moderate levels of weekly activity. When a person goes beyond moderate levels of weekly physical activity, the increase in health benefits will rise gradually and then plateau. Excessive exercise may actually begin to have adverse effects on some health conditions, including unhealthy weight loss. For this reason, engaging in enough, but not too much, exercise appears to be optimal for promoting health.

The following guidelines should be considered when developing physical activity plans to promote health and wellness:

- *Individualization.* Exercise programs should be individualized based on physical fitness level and health status. Claude Bouchard, an expert in genetics, exercise, and health, noted that due to genes, physical activity may benefit some, but not others. For example, although most individuals who are sedentary will respond favorably to an aerobic exercise training program, such as an improved insulin sensitivity, others will not respond and have no change in insulin sensitivity. Currently, there is no gene profile for responders and nonresponders to exercise training, but that may change in the future so that specific exercise programs may be designed for individuals.

- *Leisure-time activity.* A key component of a fitness plan is simply to reduce the amount of daily sedentary activity. One important modification to your daily lifestyle is to sit less and move more. The *Training Table* in this section provides recommendations for building light physical activity into your daily schedule. Accumulating more daily unstructured physical activity may be very helpful in maintaining a healthy body weight and body composition. Additionally, leisurely walking may be adequate physical activity for older adults with compromised health status or very low fitness levels.

Training Table

Time spent on sedentary activities, such as sitting at the computer or driving, should be limited as much as is reasonably possible. As a college student, you may spend many hours sitting in class or working on online assignments. When possible, try to get up and stretch and move around for short breaks. Additional recommendations for light activity include:

- If possible where you live, bike or walk to campus or work, rather than driving.
- Use cleaning your dorm room/suite, apartment, or house as an opportunity to get some exercise.
- Stand instead of sitting when you can.
- Take the stairs instead of the elevator.
- When driving to the supermarket or mall, park at the edge of the parking lot so you can get in more steps walking.
- Walk your dog instead of letting him or her out into the backyard (your dog needs exercise too).
- Play ultimate Frisbee, disc golf, or another similar activity with your friends.

- *Aerobic exercise.* For important health benefits, both adults and older adults should engage in moderate-intensity aerobic (endurance) exercise, such as brisk walking, for a minimum of 150 minutes every week, or about 30 minutes for 5 days. Alternatively, both may engage in vigorous-intensity exercise, such as jogging or running, for 75 minutes every week. **High-intensity interval training (HIIT)** is used to describe protocols in which the training stimulus is "near maximal" or the target intensity is between 80 and 100 percent of maximal heart rate. Comparatively, *sprint interval training (SIT)* describes protocols that involve supramaximal efforts, in which target intensities correspond to workloads greater than what is required to elicit 100 percent of maximal oxygen uptake (VO_2 max). These supramaximal exercise tasks may be accomplished in much less time as compared to moderate-intensity exercise. Additionally, adults may engage in an equivalent mix of moderate- and vigorous-intensity exercise over the course of the week.

Children and adolescents should participate in 60 minutes of moderate-to-vigorous physical activity daily. Short bursts of vigorous activity in games are included. Exergames, interactive video games that promote physical activity, may hold promise to promote aerobic physical activity in youth.

Wavebreak Media Ltd/123RF

Health benefits may be achieved whether the daily minute allotment for exercise is done continuously, or as three 10-minute *exercise snacks* done throughout the day, such as three brisk walks. Aerobic exercise programs, including the determination of moderate- and vigorous-intensity exercise and discussion of HIIT, are detailed in chapters 3 and 11. In brief, exercise intensity is based on the MET, a term associated with the metabolic rate that will be explained in detail in chapter 3. Your resting metabolic rate, such as when you are sitting quietly, is 1 MET. Moderate-intensity exercise is about 3–6 METs, and vigorous-intensity exercise is greater than 6 METs. You may access the MET values for a wide variety of physical activities at the following website.

https://sites.google.com/site/compendiumofphysicalactivities/. Click on Activity Categories, such as bicycling, and the METs value will be provided for a wide variety of bicycling activities.

Table 1.3 provides examples of moderate- and vigorous-intensity exercise. You might also use the "talk test" when exercising to determine your level of exercise intensity. For the present, the following characteristics of the *talk test* while exercising may be may sufficient to determine exercise intensity.

Light: You can carry on a normal conversation.
Moderate: You can talk, but not sing but a few notes before taking a breath.
Vigorous: You cannot say more than a few words.

- *Muscle-strengthening exercise.* Resistance exercise also conveys significant health benefits. Both adults and older adults should engage in muscle-strengthening activities on 2 or more days a week that work all major muscle groups (legs, hips, back, abdomen, chest, shoulders, and arms). Children and adolescents should do the same at least 3 days a week. The recommendation includes about 8–10 exercises that stress these major muscle groups. Individuals should perform about 8–12 repetitions of each exercise at least twice a week on nonconsecutive days. Older adults may lift lighter weights or use less resistance, but do more repetitions. Resistance exercises may include use of weights or other resistance modes or weight-bearing activities such as stair climbing, push-ups, pull-ups,

TABLE 1.3 Some examples of moderate-intensity and vigorous-intensity exercise

Moderate-intensity exercise	Vigorous-intensity exercise
Leisurely bicycling, 5–8 mph	Bicycling, 12 mph and faster
Walking, leisurely, 3–4 mph	Walking, 4.5 mph and faster
Dancing, slow ballroom	Dancing, aerobic, with 6- to 8-inch step
Jogging, slow on a mini-tramp	Jogging/running, 4 mph and faster
Swimming, slow leisurely	Swimming, fast crawl, 50 yards/minute
Tennis, doubles	Tennis, singles
Golf, walking, carrying clubs	Basketball, competitive game
Pilates, general	Exergaming, vigorous effort

and various other calisthenics that stress major muscle groups. Resistance exercise programs will be discussed in chapter 12.

- *Flexibility and balance exercises.* Adults should perform activities that help maintain or increase flexibility on at least 2 days each week for at least 10 minutes. Flexibility exercises are designed to maintain the range of joint motion for daily activities and physical activity. In addition, activities that support a strong core and improve balance help reduce risk of injury and improve posture, thus supporting the spinal cord and reducing risk for back pain.
- *A little extra may be beneficial.* The *2018 Physical Activity Guidelines for Americans* notes that more exercise time, particularly increasing the weekly amount of moderate-intensity aerobic activity to 300 minutes or vigorous-intensity aerobic activity to 150 minutes, or an equivalent combination of the two, equals more health benefits. The *Guidelines* also note that going beyond this 300 or 150 minutes a week may provide even more health benefits and increase life expectancy.

Michael DeYoung/Blend Images LLC

For those who have the time and energy, exceeding the recommended amounts of physical activity may provide additional health benefits. In particular, exercise may lead to greater improvements in body composition, lower stress levels, and improved sleep quality. However, as shown in **figure 1.4,** more is not always better and care should be taken not to exercise excessively.

www.health.gov/paguidelines   Provides details on the Physical Activity Guidelines for children, adults, and older adults.
www.shapeamerica.org   The Society for Health and Physical Educators provides physical activity guidelines for children.
www.cdc.gov/physicalactivity/everyone/guidelines/adults.html Provides details on complete exercise programs for adults.
www.who.int/dietphysicalactivity/pa/en/   The World Health Organization provides recommendations on diet and physical activity to promote health.
www.fitness.gov/be-active/   The President's Council on Sports, Fitness & Nutrition provides ideas to help you become more physically active.
www.cdc.gov/physicalactivity/basics/measuring/   This video provides information on exercise intensity.

Are Americans meeting physical activity guidelines?

According to the CDC, only 1 in 5 adults meets the recommendations of the *2018 Physical Activity Guidelines*. Compared to those living in the West, Northeast, and Midwest regions of the country, those living in the South are less likely to be physically active. As well, non-Hispanic white adults are more likely to meet the recommendations than non-Hispanic black adults and Hispanic adults. **Figure 1.5** provides a map showing physical inactivity at the state-level across the United States. Use the interactive map at www.cdc.gov/physicalactivity/data/databases.htm to look up the physical activity characteristics for your state or county.

Am I exercising enough?

Several approaches may be used to answer this question. One approach is to track all of your physical activity for a week, such as how many minutes you walk; engage in some type of aerobic physical activity such as swimming, cycling, or jogging; or perform resistance exercise such as lifting weights or using resistance bands. Tallying your totals for the week and comparing them to the previously mentioned recommendations for aerobic and resistance exercise will give you a good idea as to whether you are meeting current recommendations.

Today, there are a plethora of exercise gadgets that can be used to monitor and record your daily levels of physical activity. Such gadgets started with the basic pedometer, but now include numerous gadgets you can wear on a finger or wrist that will effortlessly synchronize with your smartphone and provide you data on heart rate, blood pressure, energy (kcal) expended, and other health-related variables. **Figure 1.6** provides examples of fitness trackers useful for monitoring and tracking physical activity. The cost varies, and the fitness tracker business continues to expand with new products on the market.

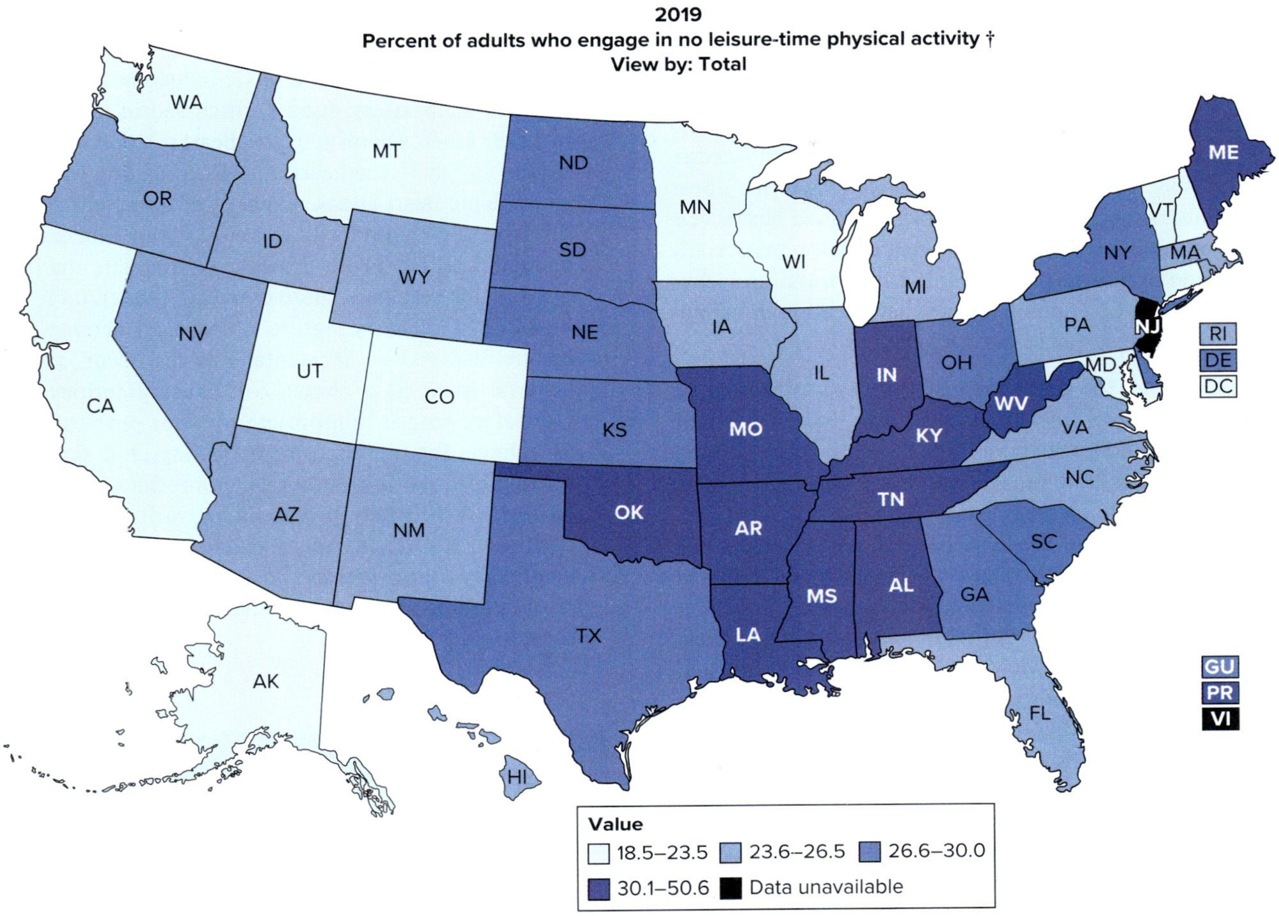

FIGURE 1.5 This map of the United States (2019) shows the percentage of the population who are physically inactive. The darker the color indicates a greater proportion of those living in that area who do not engage in regular physical activity.

Source: CDC

Pedometer: andreypopov/123RF; Fitness band: Sasils/Shutterstock; Smart watch: Andrey_Popov/Shutterstock; Heart rate monitor: suedhang/Cultura/Getty Images

FIGURE 1.6 There are numerous gadgets available to track physical activity and/or heart rate during exercise.

Can too much exercise be harmful to my health?

In general, the health benefits far outweigh the risks of exercise. Although individuals training for sport may need to undergo prolonged, intense exercise training, such is not the case for those seeking health benefits of exercise. Given our current state of knowledge, adhering to the guidelines presented above, preferably at the upper time and day limits, should be safe and provide optimal health benefits associated with physical activity. However, exercise, particularly when excessive and in individuals with preexisting health problems, may increase health risks. Training for and participating in various sports may also predispose one to various health problems.

- *Orthopedic problems.* Too much exercise may lead to orthopedic problems, such as stress fractures in the lower leg in those who run, particularly in those with poor biomechanics. Injuries to tendons and bones are common in some sports. Proper rest is often recommended for such injuries.
- *Impaired immune functions.* While moderate physical activity may enhance immune function, prolonged, high-intensity exercise temporarily impairs immune competence, which may be associated with an increased incidence of upper respiratory tract infections. Moreover, according to a review by Nijs and others, individuals with chronic fatigue syndrome may have an altered immune response to exercise and other reports link it to excessive exercise.
- *Exercise-induced asthma.* Some endurance athletes, such as runners and cross-country skiers, particularly when exercising in cold weather, may be more prone to exercise-induced asthma. Excessive lung ventilation may dry the airways with subsequent release of inflammatory mediators that cause contraction of the airways, making breathing more difficult. In severe cases, exercise-induced asthma may be fatal.
- *Exercise addiction.* Exercise is known to release various brain chemicals, including endorphins, which may elicit euphoric feelings such as the *runners high.* However, experts note that exercise addiction may also have an obsessive-compulsive dimension and may be linked to other psychiatric disorders, such as substance abuse and eating disorders.
- *Osteoporosis.* When coupled with inadequate dietary energy intake, exercise that leads to excessive weight loss may contribute to the menstrual irregularities in female athletes that may exacerbate loss of bone mass, or osteoporosis. Known as the female athlete triad, this topic is discussed in chapters 8 and 10.
- *Heat illness and kidney failure.* Exercising in the heat may cause heat stroke or other heat illnesses with serious consequences, such as kidney failure and death, as discussed in chapter 9.
- *Brain damage.* As noted previously, exercise exerts multiple beneficial effects on the brain, such as improved psychological health and reduced risk of mental decline with aging. However, participation in some sports may be associated with mild traumatic brain injury (mTBI) and, rarely, catastrophic traumatic injury and death. Repetitive mTBIs, such as concussions, can lead to neurodegeneration, or chronic traumatic encephalopathy (CTE). CTE has been reported most frequently in American football players and boxers but is also associated with other sports such as ice hockey, soccer, rugby, and baseball.
- *Heart attacks and sudden death.* Although sudden death among young athletes is very rare, it is still two to three times more frequent than in the age-matched control population and attracts significant media attention.

Sudden death in older athletic individuals may be associated with coronary artery disease, discussed in detail in chapter 5. In brief, atherosclerosis in the heart's blood vessels may limit oxygen supply to the heart muscle, triggering what is known as an ischemic heart attack. Experts recommend that heart attack survivors use caution with exercise, noting moderate levels may be beneficial but higher levels may attenuate the benefits. For heart attack survivors, more exercise is better, up to a point.

- *Accidents.* Given the nature of physical activity, particularly competitive sports, accidental injuries occur, and some may be fatal, such as a concussion causing serious head injury. Use safety gear as appropriate for your physical activity, such as helmets for bicycling, rollerblading, and skiing, as well as other protective sportswear as appropriate for any given activity. Adhere to safety protocols for various activities, such as cycling in traffic. In recent years, reports indicate increasing emergency room visits by those who walk and talk on their cell phones and experience an accident, either by falls or being hit by motor vehicles.

Key Concepts

- Health-related fitness includes a healthy body weight, cardiovascular-respiratory fitness, adequate muscular strength and muscular endurance, and sufficient flexibility.
- Overload is the key principle underlying the adaptations to exercise that may provide a wide array of health benefits. The intensity, duration, and frequency of exercise represent the means to impose an overload on body systems that enable healthful adaptations.
- Physical inactivity may be dangerous to your health. Some contend "Sitting is the new smoking." Exercise, as a form of physical activity, is becoming increasingly important as a means to achieve health benefits, by preventing the development of many chronic diseases.
- Physical activity need not be strenuous to achieve health benefits, but additional benefits may be gained through more vigorous and greater amounts of physical activity.
- In general, more exercise is better, up to a point. Excessive exercise may cause some minor and major health problems in some individuals. You should be aware of personal health issues or other factors that may be related to exercise-associated health risks.

Check for Yourself

- Using a fitness tracker, track your physical activity for one week. Tally the number of minutes spent engaged in aerobic activities and muscle-strengthening activities. How do these levels compare to the 2018 *Physical Activity Guidelines for Americans?* Based on these findings, are there any changes you should make to your fitness?

Nutrition and Fitness

What is nutrition?

Nutrition is the science of food, the sum total of the processes involved in the intake and utilization of food substances by living organisms, including ingestion, digestion, absorption, transport, and metabolism of nutrients found in food. This definition stresses the biochemical or physiological functions of the food we eat, particularly in relation to health and disease. Additionally, the Academy of Nutrition and Dietetics (AND) notes that nutrition may be interpreted in a broader sense and be affected by a variety of psychological, sociological, and economic factors.

The primary purpose of the food we eat is to provide us with a variety of nutrients. A **nutrient** is a specific substance found in food that performs one or more physiological or biochemical functions in the body. There are six major classes of essential nutrients found in foods: carbohydrates, fats, proteins, vitamins, minerals, and water. However, as noted in chapter 2, food contains substances other than essential nutrients that may affect body functions.

As illustrated in **figure 1.7,** the essential nutrients perform three basic functions. First, they provide energy for human metabolism (see chapter 3). Carbohydrates and fats are the prime sources of energy. Protein may also provide energy, but this is not its major function. Vitamins, minerals, and water are not energy sources. Second, all nutrients are used to promote growth and development by building and repairing body tissue. Protein is the major building material for muscles, other soft tissues, and enzymes, while certain minerals such as calcium and phosphorus make up the skeletal framework. Third, all nutrients are used to help regulate and maintain the diverse physiological processes of human metabolism.

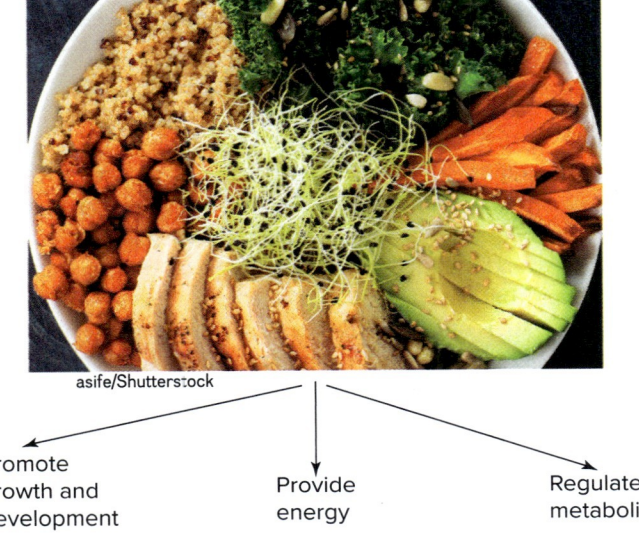
asife/Shutterstock

Promote growth and development | Provide energy | Regulate metabolism

FIGURE 1.7 Foods provide a mix of nutrients. This Buddha bowl with kale salad, quinoa, roasted chickpeas, grilled chicken breast, avocado, baked sweet potatoes, leek sprouted seeds, pine nuts, and sesame seeds is an example of a meal containing multiple essential nutrients.

In order for our bodies to function effectively, we need more than 40 specific nutrients, and we need these nutrients in various amounts as recommended by nutrition scientists. Dietary Reference Intakes (DRI) represent the current recommendations in the United States and include the Recommended Dietary Allowances (RDA), Adequate Intakes (AI), and Tolerable Upper Intake Levels (UL). These recommendations are explained in detail in chapter 2. Nutrient deficiencies or excesses may cause various health problems, some very serious.

What is the role of nutrition in health promotion?

As noted previously, your health is dependent upon the interaction of your genes and your environment, and the food you eat is part of your personal environment.

> *Let food be your medicine and medicine be your food.*

This statement by Hippocrates, made over two thousand years ago, is becoming increasingly meaningful as the preventative and therapeutic health values of food relative to the development of chronic diseases are being unraveled. Nutrients and other substances in foods may influence gene expression, some having positive and others negative effects on our health. For example, adequate amounts of certain vitamins and minerals may help prevent damage to DNA, the functional component of your genes, while excessive alcohol may lead to DNA damage.

Most chronic diseases have a genetic basis; if one of your parents has had coronary artery disease or cancer, you have an increased probability of contracting that disease. Such diseases may go through three stages: initiation, promotion, and progression. Your genetic predisposition may lead to the initiation stage of the disease, but factors in your environment that influence your epigenome may promote its development and eventual progression. In this regard, some nutrients are believed to be **promoters** that lead to progression of the disease, while other nutrients are believed to be **antipromoters** that deter the initiation process from progressing to a serious health problem.

What you eat plays an important role in the development or progression of a variety of chronic diseases. For example, the CDC indicates that consuming nutritious foods lowers people's risk for many chronic diseases, including heart disease, stroke, some types of cancer, diabetes, and osteoporosis (**figure 1.8**). The National Cancer Institute (NCI) estimates that one-third of all cancers are linked in some way to diet, ranking just behind tobacco smoking as one of the major causes of cancer. Research suggests that high adherence to a healthy diet, such as the Mediterranean diet, is associated with a significant reduction in the risk of overall cancer mortality, particularly colorectal, prostate, and other cancers of the digestive tract.

As noted previously, *exercise is medicine.* In a like manner, *food is medicine* may also be an appropriate phrase, not only attributable to the quote from Hippocrates but also based on modern medicine as well. The types and amount of nutrients and phytochemicals found in our foods, the source of our food, and the method of food

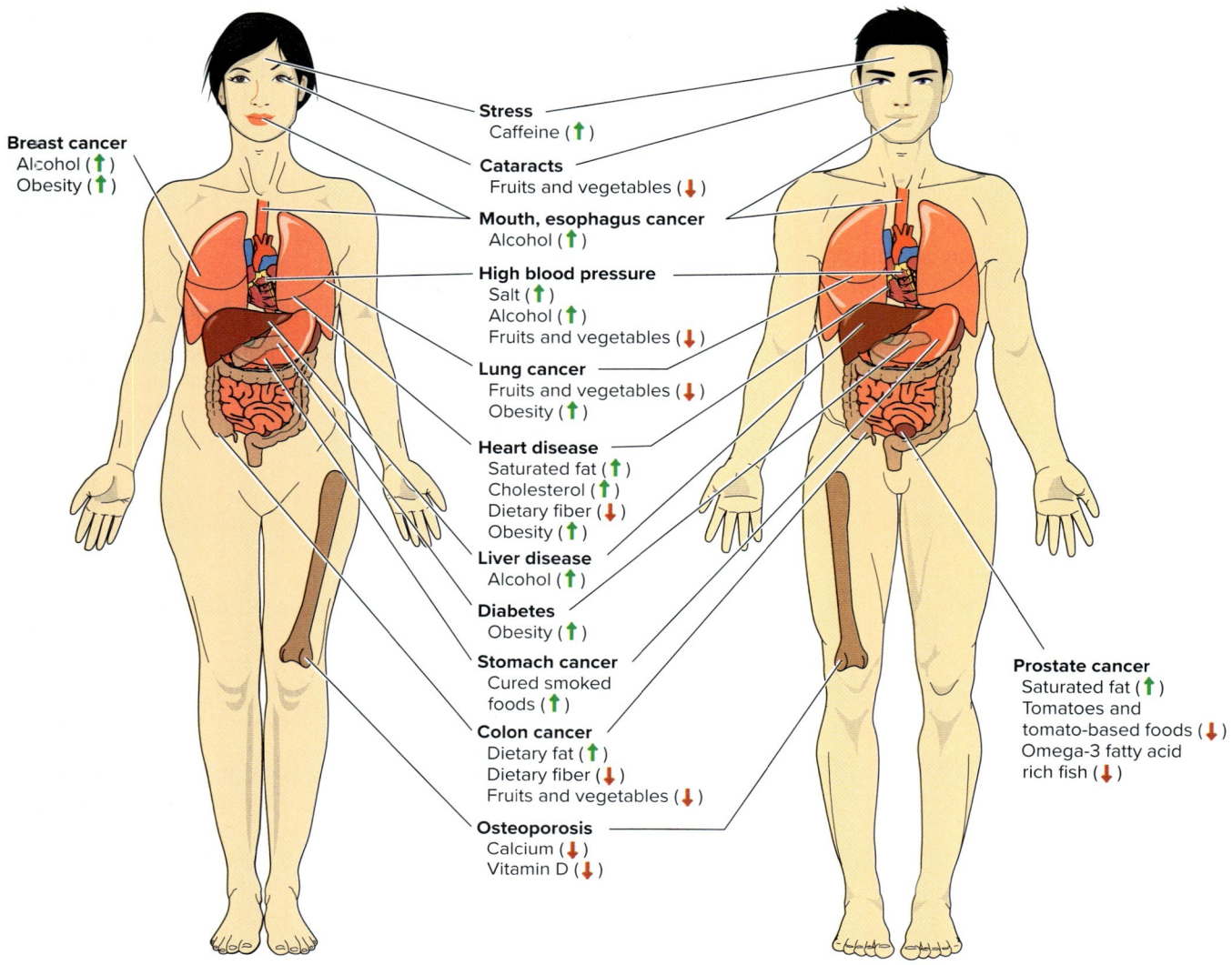

FIGURE 1.8 Some possible health problems associated with poor dietary habits. An upward arrow (↑) indicates excessive intake, while a downward arrow (↓) indicates low intake or deficiency.

preparation are all factors that may influence the epigenome and subsequent gene expression or other metabolic functions that may affect our health status. The *Training Table* in this section summarizes some of the key ways by which nutrients may impact health and risk for chronic disease.

The beneficial, or harmful, effects of specific nutrients and various dietary practices on mechanisms underlying the development of chronic diseases will be discussed as appropriate in later sections of this book.

Do most Americans eat a well-balanced diet?

Surveys indicate that most people are aware of the role of nutrition in health and want to eat better to support health, but they do not translate their desires into appropriate action. Poor eating habits span all age groups. According to the *2020–2025 Dietary Guidelines for Americans* report, on average, Americans of all ages consume too few vegetables, fruits, high-fiber whole grains, low-fat milk products, and seafood and they eat too much added sugars, solid fats, refined grains, and sodium. **Table 1.4** summarizes the five overarching guidelines from the *2020–2025 Dietary Guidelines for Americans*.

One of the goals of *Healthy People 2030* is to "improve health by promoting healthy eating and making nutritious foods available." Fruits, vegetables, and whole grains are examples of foods that, when consumed in recommended amounts, may improve health and reduce a person's risk for chronic disease. In contrast, consuming too many foods high in saturated fat and added sugars may increase a person's risk for obesity and other health problems. To support these healthy eating recommendations, public health interventions are focused on ensuring all people, including those with limited resources, have access to affordable healthy foods and beverages.

TABLE 1.4 Overarching guidelines from the *2020–2025 Dietary Guidelines for Americans*

- Follow a healthy dietary pattern at every life stage.
- Customize and enjoy nutrient-dense foods and beverages to reflect personal preferences, cultural traditions, and budgetary considerations.
- Focus on meeting food group needs with nutrient-dense foods and beverages, and stay within calorie limits.
- Limit foods and beverages higher in added sugars, saturated fat, and sodium, and limit alcoholic beverages.

Training Table

The following are some of the proposed effects of various nutrients and appropriate energy intake that may help promote good health:

- Inactivate carcinogens or kill bacteria that damage DNA.
- Help repair DNA.
- Increase insulin sensitivity.
- Relax blood vessels and improve blood flow.
- Reduce blood pressure.
- Optimize serum lipid levels.
- Reduce inflammation.
- Inhibit blood clotting.
- Enhance immune system functions.
- Prevent damaging oxidative processes.
- Dilute harmful chemicals in the intestines.
- Promote more frequent bowel movements.

Some advances are being made in the battle against unhealthy eating, obesity, and poor health in the United States. For example, some food manufacturers have reduced the amounts of solid fats, added sugars, and salt in their products. Some fast-food restaurants are offering healthier alternatives, such as oatmeal with fruit for breakfast, and menu labeling laws now require posting of certain nutrition information. The National School Lunch Program has promoted a program to incorporate more fresh fruit and vegetables into daily school lunches. Although these are worthwhile endeavors, many more are needed to support optimal health and wellness.

What are some general guidelines for healthy eating?

Because the prevention of chronic diseases is of critical importance, thousands of studies have been and are being conducted to discover the intricacies of how various nutrients may affect our health. Particular interest is focused on nutrient function within cells at the molecular level, the interactions between various nutrients, and the identification of other protective factors in certain foods. All of the answers are not in, but sufficient evidence is available to provide us with some useful, prudent guidelines for healthful eating practices.

In response to the need for healthier diets, a variety of public and private health organizations analyzed the research relating diet to health and developed some basic guidelines for the general public. The details underlying these recommendations may be found in several governmental reports, including the scientific report accompanying the *2020–2025 Dietary Guidelines for Americans* and *Healthy People 2030*. These reports serve as the basis for dietary recommendations provided in the United States Department of Agriculture (USDA) "MyPlate" recommendations. According to MyPlate recommendations, approximately half of a person's "plate" should include fruits and vegetables. Grains and proteins should make up the other half of the plate, and low-fat or fat-free dairy options (or nondairy alternatives) should be included with most meals. **Figure 1.9** shows the MyPlate graphic representation of the guidelines. The www.myplate.gov website provides details about each of the food groups and includes resources specific to certain age groups and recommendations for meal preparation and eating on a budget.

Although we do have considerable research to support dietary recommendations to promote health, the research is incomplete. Moreover, inconsistencies in research findings, such as the health effects of saturated fat, discussed later in this chapter, may affect recommendations. Thus, the following recommendations may be considered to be prudent, and throughout this book we will refer to these recommendations as a **Prudent Healthy Diet.** These recommendations are in accordance with the *total diet* approach of the AND and the various governmental and professional health organizations noted previously. Each specific dietary recommendation may convey some health benefit, so the more of these dietary guidelines you adopt, the greater should be your overall health benefits. The *Training Table* summarizes the key Prudent Healthy Diet recommendations.

An expanded discussion of these guidelines along with practical recommendations to help you implement them is presented

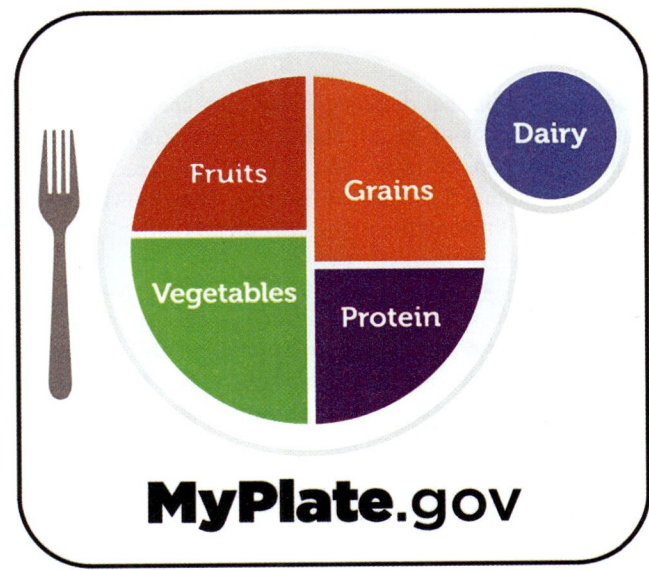

FIGURE 1.9 MyPlate visual showing the relative breakdown of food groups as part of a healthy diet.
Source: U.S. Department of Agriculture (USDA)

Training Table

The following are key recommendations of the Prudent Healthy Diet:

1. Balance the food you eat with physical activity to maintain or achieve a healthy body weight. Consume only moderate food portions. Be physically active every day.
2. Eat a nutritionally adequate diet consisting of a wide variety of nutrient-rich foods. Eat more whole foods in their natural form. Eat fewer highly processed foods.
3. Choose a plant-rich diet with plenty of fruits and vegetables, whole-grain products, and legumes, which are rich in complex carbohydrates, phytonutrients, and fiber.
4. Choose a diet moderate in total fat, but low in saturated (solid), *trans* fat and cholesterol.
5. Choose beverages and foods to moderate or reduce your intake of added sugars and highly refined carbohydrates.
6. Choose and prepare foods with less salt and sodium.
7. Maintain protein intake at a moderate yet adequate level, obtaining much of your daily protein from plant sources, complemented with smaller amounts of fish, skinless poultry, and lean meats.
8. Choose a diet adequate in calcium and iron.
9. Practice food safety, including proper food storage, preservation, and preparation.
10. Consider the possible benefits and risks of food additives and dietary supplements.
11. If you drink alcoholic beverages, do so in moderation.
12. Enjoy your food. Eat what you like, but balance it within your overall healthful diet.

in chapter 2. Additional details on how each specific recommendation may affect your health status, including specific considerations for women, children, and older adults, are presented in appropriate chapters throughout this book. The following websites present detailed information on healthy dietary guidelines:

www.dietaryguidelines.gov The *2020–2025 Dietary Guidelines for Americans* focus on the total diet and how to integrate all of the recommendations into practical terms, encouraging personal choice but result in an eating pattern that is nutrient-dense and kcal-balanced.

www.myplate.gov MyPlate offers personalized eating plans and interactive tools to help you plan your food choices.

https://food-guide.canada.ca/en/ Canada's Food Guide provides excellent information on healthy eating.

www.eatright.org The Academy of Nutrition and Dietetics website provides evidence-based recommendations related to diet and health.

Are there additional health benefits when both exercise and dietary habits are improved?

A poor diet and sedentary lifestyle are individual major risk factors for the development of chronic diseases. Collectively, however, they may pose additional risks, particularly increasing risk for pre-diabetes, a condition preceding type 2 diabetes, and for the two leading causes of death in the United States—heart disease and cancer. Recent research also indicates certain that dietary factors may complement exercise for enhanced brain function. Thus, combining a recommended exercise program with a healthy diet may have additive effects on one's health.

Pre-diabetes Several factors, such as excess body weight, impaired fasting blood glucose, and glucose intolerance, may be associated with pre-diabetes and predispose one to type 2 diabetes. Prevention interventions that include diet and both aerobic and resistance exercise training have been found to be modestly effective in reducing risk factors associated with pre-diabetes in adults, which help in the prevention of type 2 diabetes.

Onoky/SuperStock

Heart Disease In a recent publication, Angell and colleagues discuss the AHA 2030 Impact Goal to "equitably increase healthy life expectancy beyond current projections, with global and local collaborators." The AHA provides recommendations for individuals to understand their risk for heart disease and strategies to lower risk for coronary artery disease. As shown in **table 1.5,** the key lifestyle behaviors that may be effective in favorably modifying heart disease risk factors are proper nutrition, exercise, and stress management. Moreover, several of the risk factors for heart disease are diseases themselves, such as diabetes, obesity, and high blood pressure, all of which may benefit from the combination of proper nutrition and exercise.

Cancer In an extensive worldwide report on the means to prevent cancer, the American Institute of Cancer Research highlighted the three most important means to prevent a wide variety of cancers, and all are related to exercise and nutrition:

- Choose mostly plant foods, limit red meat, and avoid processed meat.
- Be physically active every day in any way for 30 minutes or more.
- Aim to be a healthy weight throughout life as much as possible.

www.heart.org The American Heart Association provides information on a variety of hearthealth related topics and recommendations for healthy habits to reduce risk for heart disease.

www.cancer.org The American Cancer Society provides answers, guidance, and support for individuals diagnosed with cancer as well as their family and support network.

Brain Health Exercise and nutrition are both powerful means to positively influence the brain and may influence brain health through several mechanisms that create new neurons (neurogenesis). Specifically, exercise collaborates with other aspects of

TABLE 1.5 Modifiable risk factors associated with coronary artery disease

Risk factor	Classification	Positive health lifestyle modification
High blood pressure	Major	Proper nutrition, exercise, stress management, quality sleep, maintain or achieve healthy body weight and body composition
High blood lipids	Major	Proper nutrition, exercise
Cigarette smoking	Major	Stop smoking
Sedentary lifestyle	Major	Exercise, stress management, quality sleep
ECG abnormalities	Major	Proper nutrition, exercise
Obesity	Major	Proper nutrition, exercise, stress management, quality sleep
Diabetes	Major	Proper nutrition, maintain or achieve healthy body weight and body composition, stress management, quality sleep
Stressful lifestyle	Contributory	Stress management
Dietary intake	Contributory	Proper nutrition

lifestyle to influence cognition. In particular, select dietary factors share brain-enhancement mechanisms similar to exercise, and in some cases can complement the action of exercise. Experts suggest that exercise and diet appear to be effective strategies to counteract neurological and cognitive disorders.

Key Concepts

- The primary purpose of the food we eat is to provide us with nutrients essential for the numerous physiological and biochemical functions that support life.
- Dietary guidelines developed by major professional health organizations are comparable, and collectively help prevent major chronic diseases such as heart disease, cancer, diabetes, high blood pressure, and obesity.
- The *2020–2025 Dietary Guidelines for Americans* and the *Healthy People 2030* report note that poor nutrition is a major health problem in the United States.
- Basic guidelines for a Prudent Healthy Diet include maintenance of a proper body weight and consumption of a wide variety of natural foods rich in nutrients associated with health benefits. The more healthful dietary guidelines that you adopt, the greater will be your overall health benefits.
- Although engaging in regular exercise and sound nutrition habits may confer health benefits separately, health benefits may be maximized when both healthy exercise and nutrition lifestyles are adopted.

Check for Yourself

- Using a diet analysis program, such as NutritionCalc Plus, create a profile with your personal information and look at your specific dietary recommendations. Then, track your diet for 24 hours, comparing your intake with recommended intakes. Are there key nutrients you are lacking? Are there key nutrients you are consuming in excess?

Sports-Related Fitness: Exercise and Nutrition

As with health, genetic endowment plays an important underlying role in the development of success in sport. In his book *The Sports Gene*, David Epstein notes that nature and nurture are both essential ingredients for superior performance in a given sport. Nature is in the genes, the hardware, whereas nurture is in the environment, the software. Nurture involves not only exposure to the sport at a specific time but expert training as well. Ahmetov and Rogozkin suggest that optimal responses to training are also dependent on possession of appropriate genes. Genes explain why some individuals benefit while others do not from the same sport training program. Elite athletes are not only born with the right genes for a given sport but must also have the right genes to benefit from proper training. Moreover, Joyner and Coyle note that complex motivational and sociological factors also play important roles in who does or does not become a sport champion. For example, one is more likely to be successful in ice hockey if born in Canada rather than Brazil, but the Brazilian child may be more successful in soccer.

What is sports-related fitness?

One of the key factors determining success in sport is the ability to maximize your genetic potential with appropriate physical and mental training to prepare both mind and body for intense competition. As described earlier in this chapter, athletes develop sports-related fitness by training specifically for their sport. For example, strength and power are key to success in sports such as football, bobsled, and shot put. In comparison, endurance training is essential for success in long-distance cycling, running, or swimming. Proper training to develop neuromuscular skills is important for gymnastics, archery, and cross-country skiers participating in the biathlon event. For many sports, such as soccer and basketball, athletes must train appropriately to develop multiple fitness components.

The principles of exercise training introduced earlier, such as overload and specificity, are as applicable to sports-related fitness as they are to health-related fitness. However, training for sports performance is more intense, prolonged, and frequent than training for health, and training is specific to the energy demands and skills associated with each sport. We will discuss energy expenditure for sports performance in chapter 3. The *Training Table* in this section provides examples of different sports-related fitness components and examples of each.

Training Table

The following are examples of general categories of sports, with an example included for each:

- Explosive "power" sports
  - Olympic weight lifting
- Very high-intensity sports
  - 100-meter run
- High-intensity, short-duration sports
  - 5,000-meter run (3.1 miles)
- Intermittent high-intensity sports
  - Soccer
- Endurance sports
  - Marathon running (26.2 miles, 42.2 kilometers)
- Low-endurance, precision skill sports
  - Golf
- Weight-control and body-image sports
  - Bodybuilding

Training of elite athletes at the United States Olympic and Paralympic Training Center focuses on three attributes:

- Physical power
- Mental strength
- Mechanical edge

Coaches and scientists work with athletes to maximize physical power production for their specific sport, to optimize mental strength in accordance with the psychological demands of the sport, and to provide the best mechanical edge by improving specific fitness and sport skills, sportswear, and sports equipment. Sports science and technology provide elite competitors with the tiny margins needed to win in world-class competition.

Athletes at all levels of competition, whether an elite international competitor, a college wrestler, a high school volleyball player, a seniors age-group distance runner, or a youth league soccer player, can best improve their sports-related fitness and performance by intense training appropriate for their age, physical and mental development, and sport. To paraphrase Theodore Roosevelt, "Do the best with what you got." While proper training and hard work are essential, sports and exercise scientists have investigated a number of means to improve athletic performance beyond that attributable to training, and one of the most extensively investigated areas has been the effect of nutrition.

What is sports nutrition?

At high levels of athletic competition, athletes generally receive excellent coaching to enhance their biomechanical skills (mechanical edge), sharpen their psychological focus (mental strength), and maximize the physiological functions (physical power) essential for optimal performance. Clyde Williams, a renowned sports scientist from England, notes that, in addition to specialized training, from earliest times certain foods were regarded as essential preparation for sports competition, including the Olympics in ancient Greece.

As we shall see, there are various dietary factors that may influence biomechanical, psychological, and physiological considerations in sport. For example, losing excess body fat will enhance biomechanical efficiency; consuming carbohydrates during exercise may maintain normal blood sugar levels for the brain and prevent psychological fatigue; and providing adequate dietary iron may ensure optimal oxygen delivery to the muscles. All these sports nutrition factors may favorably affect athletic performance.

Sports nutrition involves the application of nutritional principles to enhance athletic performance. Louise Burke, an internationally renowned sports nutritionist from Australia, defined sports nutrition as the application of eating strategies with several major objectives:

- To promote good health
- To promote adaptations to training
- To recover quickly after each training session
- To perform optimally during competition

Sports nutritionists may meet these objectives in various ways, such as developing meal plans for training, recovery, and competition; coordinating training tables; providing appropriate information about healthy diets; teaching cooking skills; discussing the efficacy, safety, and permissibility of sports supplements; counseling individual athletes with special diets, such as those with Celiac disease; and monitoring athletes for intentional or unintentional weight loss and eating disorders.

Although investigators have studied the interactions between nutrition and various forms of sport or exercise for more than a hundred years, it is only within the past several decades that extensive research has been undertaken regarding specific recommendations for athletes.

Adie Bush/Getty Images

Is sports nutrition a profession?

Sports nutrition is recognized as an important factor for optimal athletic performance. Sports nutrition is sometimes referred to as *performance nutrition* when coupled with exercise designed for health- and sport-related fitness. Registered dietitian nutritionists (RDs/RDNs) with an expertise in sports nutrition play an integral role in many athletic programs, including those at the collegiate and professional levels.

Professional Associations Several professional associations, such as the Sports and Human Performance Nutrition (SHPN) practice group of the AND, Professionals in Nutrition for Exercise and Sport (PINES), the Collegiate & Professional Sports Dietitians Association (CPSDA), and the International Society of Sports Nutrition (ISSN), are involved in the application of nutrition to sport, health, and wellness.

Certification Programs Several professional and sports-governing organizations have developed a recognized course of study or certification program to promote the development of professionals who can provide athletes with sound information about nutrition. For example, AND has established a program for certification as a Specialist in Sports Dietetics (CSSD), while the International Olympic Committee offers a diploma in sports nutrition.

Research Productivity Numerous exercise-science/nutrition research laboratories at major universities are dedicated to sports nutrition research. Almost every scientific journal in sport/exercise science, and even in general nutrition, appears to contain at least one study or review in each issue that is related to sports nutrition. Several journals, such as the *International Journal of Sport Nutrition and Exercise Metabolism,* focus almost exclusively on sports nutrition.

International Meetings International meetings have focused on sports nutrition, some meetings highlighting nutritional principles for a specific sport, such as soccer or track and field, while others may focus on a specific sport supplement, such as creatine.

Consensus Statements and Position Stands Several international sports-governing organizations have developed consensus statements on nutrition for their specific sport. For example, the International Swimming Federation (Fédération Internationale de Natation, FINA) publishes a practical guide on nutrition for aquatic sports, which is designed to provide sound nutrition information for aquatic athletes worldwide. A more generalized position stand entitled "Nutrition and Athletic Performance" was issued jointly by the AND, the ACSM, and the Dietitians of Canada.

Career Opportunities Sports dietitians are employed by professional sport teams and athletic departments of colleges and universities to design optimal nutritional programs for their athletes. In addition, professionals with an expertise in performance nutrition work with individuals of all ages to support their exercise- and sport-related goals.

---

www.acsm.org You may access the position paper entitled "Nutrition and Athletic Performance" by clicking on Education & Resources.

www.shpndpg.org/professional-development/cssd The SHPN site provides information on what is necessary to become a Board Certified Specialist in Sports Dietetics (CSSD).

www.sportsoracle.com Check this PINES site to see what is needed to become a member and the requirements for the IOC Diploma in Sports Nutrition.

www.sportsrd.org The CPSDA site provides information on membership.

Are athletes today receiving adequate nutrition?

Numerous survey studies regarding dietary intake of athletes have been conducted over the course of the past two decades and, in general, present mixed results. Based on recommended dietary practices for athletes, the following is a brief summary of the findings.

- Many athletes do not consume adequate amounts of energy, particularly healthy carbohydrates.
- Many athletes consume more dietary fat than recommended, particularly saturated fat.
- Intake of micronutrients, such as vitamins and minerals, varies. Some athletes exceed current recommended intakes, while others have inadequate intakes.
- Athletes involved in weight-control sports who may restrict energy intake may be at high risk for micronutrient deficiencies. Iron and calcium deficiencies may be common in female athletes.
- Many athletes, including youth athletes, take dietary supplements designed to enhance performance.

This brief review indicates that some athletic groups are not receiving the recommended allowances for a variety of essential nutrients or may not be meeting certain recommended standards. It should be noted, however, that these surveys have analyzed the diets of the athletes only in reference to a standard, such as the RDA, and many studies have not analyzed the actual nutrient or biochemical status (such as by a blood test) of the athlete or the effects that the dietary deficiency exerted on exercise performance capacity or sport performance. The RDA for vitamins and minerals incorporates a safety factor, so an individual with a dietary intake of essential nutrients below the RDA may not necessarily suffer a true nutrient deficiency. However, athletes who do develop a nutrient deficiency may experience a deterioration in athletic performance and poor health as a result. Examples discussed in later chapters include impaired aerobic endurance capacity associated with iron deficiency and premature decreases in bone density with calcium deficiency.

Why are some athletes malnourished?

Recent studies have indicated a variety of factors that may contribute to poor dietary habits in many athletes, including the following:

- Athletes may not possess sufficient knowledge to make appropriate food choices.
- Athletes have misconceptions about the roles of specific nutrients in sport performance; if they choose foods based on these misconceptions, then sports performance may suffer.
- Athletes may not be getting sound sports nutrition information. Although some college varsity athletes receive nutrition information from reliable sources, such as sports dietitians, considerable nutrition information was obtained from less reliable sources such as websites and coaches with an inadequate education in sports nutrition.
- Finances and time may limit preparation of healthier meals, particularly with college athletes. Healthy meal preparation may take a back seat to time needed for sport practices, classes, and study time.

However, the future looks bright. SHPN and the CPSDA have partnered with the NCAA Sports Science Institute to publish nutrition information monthly on its website. Along with increased emphasis on sports nutrition education for collegiate strength and conditioning coaches, such endeavors may help improve nutrition among collegiate athletes. Various education programs also are being developed for professional and youth sports by various groups, such as, respectively, the National Football League (NFL) and the PCFSN. Such programs should help. For example, Valliant and others reported a nutrition education program was useful in improving dietary intake and nutrition knowledge of female athletes.

How does nutrition affect athletic performance?

The nutrients in the foods we eat can affect exercise and sports performance in accordance with the three major functions of nutrients. First, nutrients may provide energy for the different energy-producing systems discussed in chapter 3. Second, nutrients also help regulate metabolic processes important to energy production and temperature regulation during exercise. Third, nutrients support the growth and development of specific body tissues and organs as they adapt to exercise training; **figure 1.10** highlights some of the roles diet and nutrients play during exercise. A well-planned sport-specific diet will help optimize sports performance, while a poor diet plan may lead to fatigue and impaired performance.

Malnutrition represents unbalanced nutrition and may exist as either undernutrition or overnutrition, that is, an individual does not receive an adequate intake (*undernutrition*) or consumes excessive amounts of single or multiple nutrients (*overnutrition*). Either condition can hamper athletic performance. An inadequate intake of certain nutrients may impair athletic performance due to an insufficient energy supply, an inability to regulate exercise

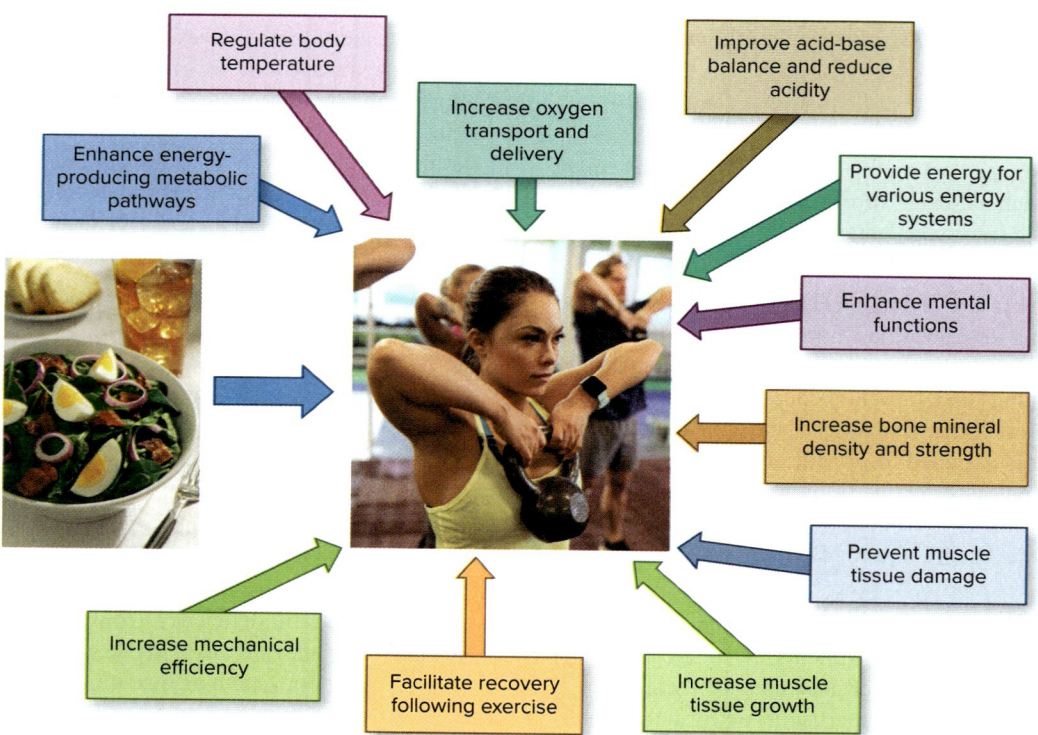

Salad: TheCrimsonMonkey/Getty Images; Woman: Syda Productions/Shutterstock

FIGURE 1.10 Nutrients in the foods we eat and dietary strategies may influence exercise or sport performance in a variety of ways. This figure summarizes some of the key effects of nutrient intake (diet) on physical activity performance.

metabolism at an optimal level, or a decreased synthesis of key body tissues or enzymes. In contrast, excessive intake of some nutrients may also impair athletic performance, and even the health of the athlete, by disrupting normal physiological processes or leading to undesirable changes in body composition.

What should athletes eat to help optimize sport performance?

Sports nutrition experts agree that the type, amount, composition, and timing of food intake can dramatically affect exercise performance, recovery from exercise, body weight and composition, and health. The importance of nutrition to your athletic performance may depend on a variety of factors, including your gender, your age, your body weight status, your eating and lifestyle patterns, the environment, the type of training you do, and the type of sport or event in which you participate. As an example of the last point, the carbohydrate needs of a golfer or baseball player may vary little from those of the nonathlete, whereas those of a marathon runner or ultraendurance triathlete may be altered significantly during training and competition.

The opinions offered by researchers in the area of exercise and nutrition relative to optimal nutrition for the athlete run the gamut. At one end, certain investigators note that the daily food requirement of athletes is quite similar to the nutritionally balanced diet for everyone else, and therefore no special recommendations are needed. At the other extreme, some, such as sports supplement companies, state that it is almost impossible to obtain all the nutrients the athlete requires from the normal daily intake of food, and for that reason nutrient supplementation is absolutely necessary. Other reviewers advocate a compromise between these two extremes, recognizing the importance of a nutritionally balanced diet but also stressing the importance of increased consumption of specific nutrients or dietary supplements for athletes in certain situations.

The review of the scientific literature presented in this book supports the later point of view. In general, athletes who consume enough kcal to meet their energy needs and who meet the requirements for essential nutrients should be obtaining adequate nutrition. Dietary guidelines for better health, as discussed previously and expanded upon in chapter 2, are the same for optimal physical performance. The key to sound nutrition for the athletic individual is to eat a wide variety of healthful foods.

Although a healthy diet is the foundation of a dietary plan for athletes, modifications may be important for training and competition in various sports. For example, adequate carbohydrate is important as an energy source for aerobic endurance athletes, adequate protein may help optimize muscle development in strength/power athletes, and adequate iron may help ensure adequate oxygen delivery in female athletes. Some basic guidelines regarding eating for training and for competition are presented in chapter 2, whereas details regarding the use of specific nutrients, such as carbohydrate and protein, are presented in the chapter highlighting that nutrient.

Some athletes believe that there are *super* foods or diets that provide a competitive advantage in sports. Numerous *sports supplements* are marketed to athletes with this premise in mind and have been the subject of considerable research by sports nutrition scientists. The following section discusses the general role of such supplements in the enhancement of sports performance, while more details on specific sports supplements are presented in the chapter highlighting that nutrient.

> ### Key Concepts
>
> ▸ Success in sports is primarily dependent on genetic endowment and proper training, but nutrition can also be an important contributing factor.
> ▸ The major objectives of sports nutrition are to promote good health and adaptations to training, to recover quickly after each training session, and to perform optimally during competition.
> ▸ Although athletes desire to eat a diet that may enhance sports performance, their knowledge of nutrition is often inadequate, and some are not meeting the dietary recommendations of sports dietitians.
> ▸ In general, the diet that is optimal for health is optimal for sports performance. However, athletes involved in certain sports may benefit from specific dietary modifications.

Ergogenic Aids and Sports Performance: Beyond Training

Since early times, athletes have attempted to use a wide variety of techniques or substances to enhance sports performance beyond the effects that could be obtained through training. In sport and exercise science terminology, such techniques or substances are referred to as **ergogenic aids.**

What is an ergogenic aid?

As mentioned previously, the two key factors important to athletic success are genetic endowment and state of training. At certain levels of competition, the contestants generally have similar genetic athletic abilities and have been exposed to similar training methods, and thus they are fairly evenly matched. Given the emphasis placed on winning, many athletes training for competition are always searching for the ultimate method or ingredient to provide that extra winning edge. Indeed, some suggest that two of the key factors leading to better athletic records in recent years are improved diet and ergogenic aids.

The word *ergogenic* is derived from the Greek words *ergo* (meaning work) and *gen* (meaning production of) and is usually defined as *to increase potential for work output*. In sports, various ergogenic aids, or ergogenics, have been used for their theoretical ability to improve sports performance by enhancing physical power, mental strength, or mechanical edge. There are several different classifications of ergogenic aids, grouped according to the general nature of their application to sport. Mechanical and psychological aids are often referred to as *performance-enhancing techniques,* whereas physiological, pharmacological, and nutritional aids involve taking some substance into the body and are known as *performance-enhancing substances.* The *Training Table* in this section lists the major categories of ergogenic aids with an example of one theoretical ergogenic aid for each.

Training Table

Ergogenic aids are now a multibillion-dollar industry. Athletes, from the weekend runner to the elite professional athlete, are often looking for that "extra" competitive edge. The five main categories of ergogenic aids, with an example of each, are provided below.

Mechanical: Mechanical, or biomechanical, aids are designed to increase energy efficiency, to provide a mechanical edge. Lightweight racing shoes may be used by a runner in place of heavier ones so that less energy is needed to move the legs and the economy of running increases.

Psychological: Psychological aids are designed to enhance psychological processes during sport performance, to increase mental strength. Hypnosis, through posthypnotic suggestion, may help remove psychological barriers that limit physiological performance capacity.

Physiological: Physiological aids are designed to augment natural physiological processes to increase physical power. Blood doping, or the infusion of blood into an athlete, may increase oxygen transport capacity and thus increase aerobic endurance. However, its use is illegal.

Pharmacological: Pharmacological aids are drugs designed to influence physiological or psychological processes to increase physical power, mental strength, or mechanical edge. Anabolic steroids, whose use is prohibited in sports, are still used by some athletes to help increase muscle mass, strength, and power.

Nutritional: Nutritional aids are nutrients designed to influence physiological or psychological processes to increase physical power, mental strength, or mechanical edge. Protein supplements may be used by strength-trained athletes in attempts to increase muscle mass because protein is the major dietary constituent of muscle.

making such ergogenic aids the most commonly utilized. Sports nutrition supplements can be placed in the following three categories:

- Sports supplements, including powders, pills, and ready-to-drink products
- Nutrition bars and gels
- Sports and energy drinks and shots

Marekuliasz/iStock/Getty Images

Sports supplements are popular world-wide and are used by all types of athletes: male and female, young and old, professional and amateur. Reports indicate that 90 percent or more of elite, international-class athletes consume dietary supplements. Other surveys document significant use among high school and collegiate athletes, military personnel in elite groups, and fitness club members.

Sports supplements are popular for several reasons. Athletes have believed that certain foods may possess magical qualities, so it is no wonder that a wide array of nutrients or special preparations have been used since time immemorial in attempts to run faster, jump higher, or throw farther. Shrewd advertising and marketing strategies promote this belief, enticing many athletes and physically active individuals to try sports supplements. Many of these products may be endorsed by professional athletes, giving the product an aura of respectability. Specific supplements also may be recommended by coaches and fellow athletes. Additionally, as drug testing in sports gets increasingly sophisticated, leading to greater detection of pharmacological ergogenics, many athletes may resort to sports supplements, believing them to be natural, safe, and legal. However, as noted later, this may not be the case.

Are nutritional ergogenics effective?

There are a number of theoretical nutritional ergogenic aids in each of the six major classifications of nutrients, and athletes have been known to take supplements of almost every nutrient in attempts to improve performance. Here are a few examples:

Carbohydrate. Special metabolites of carbohydrate have been developed to facilitate absorption, storage, and utilization of carbohydrate during exercise.
Fat. Certain fatty acids have been used in attempts to provide an alternative fuel to carbohydrate.
Protein. Special amino acids derived from protein have been developed and advertised to be more potent than anabolic steroids in stimulating muscle growth and strength development.
Vitamins. Special vitamin mixtures and even "nonvitamin vitamins," such as vitamin B15, have been ascribed ergogenic qualities ranging from increases in strength to improved vision for sport.
Minerals. Special mineral supplements, such as chromium, vanadium, and boron, have been advertised to promote muscle anabolism.

Why are nutritional ergogenics so popular?

Dietary supplements marketed to physically active individuals are commonly known as sports nutrition supplements, or simply **sports supplements.** Companies market their products as "Supplements for the Competitive Athlete," and overall sales exceed $25 billion,

Water. Special oxygenated waters have been developed specifically for aerobic endurance athletes, theoretically designed to increase oxygen delivery.

In addition to essential nutrients derived from foods, there are literally hundreds of nonessential substances or compounds that are classified as dietary supplements and targeted to athletes as potent ergogenics. These include creatine, L-carnitine, coenzyme Q10, inosine, octacosonal, and ginseng. Moreover, many products contain multiple ingredients, each purported to enhance sports performance. For example, several of the "energy" drinks on the market include carbohydrates, amino acids, vitamins, minerals, metabolites, herbs, and caffeine.

Supplementation with essential nutrients above and beyond the RDA is not necessary for the vast majority of well-nourished athletes. In general, consumption of specific nutrients above the RDA has not been shown to exert any ergogenic effect on human physical or athletic performance. As well, many sports supplements sold on the market are labeled with various performance-enhancement claims without any scientific evidence. However, there are some exceptions. As noted in chapters 4 through 10, there may be some justification for nutrient supplementation or dietary modification in certain athletes under specific conditions, particularly in cases where nutrient deficiencies may occur. Some specific dietary supplements and food drugs may also possess ergogenic potential under certain circumstances.

The effectiveness of almost all of the popular nutritional ergogenics, including the essential nutrients, the nonessential nutrients, caffeine, the steroid precursor androstenedione, and other agents, will be covered in this book.

Are nutritional ergogenics safe?

The majority of over-the-counter dietary supplements, particularly those containing essential nutrients, appear to be safe for the general population when taken in recommended dosages. However, some dietary supplements, including sports supplements, may contain ingredients that pose serious health risks in several ways.

The FDA has noted that some sports supplements contain chemicals that have been linked to numerous serious illnesses and even death, particularly when taken in excess. For example, although perceived as safe, many different herbal and dietary supplements have been reported to cause liver injury, but the exact component that is responsible for injury is difficult to discern. In the United States, products used for bodybuilding and weight loss are the most commonly implicated.

Supplements that are mislabeled and contain unlisted substances pose a serious health threat. Some companies are unscrupulous and may add chemicals, such as stimulants or steroids, to help make the product more effective, but which also may have adverse health effects. Another potential problem is that some, particularly younger, athletes may have the mentality that "if one is good, then ten is better" and thus may overdose, increasing the potential health risk of a potentially harmful ingredient.

Fortunately, in the United States, the government is working to require that all ingredients be listed on dietary supplement labels, and hopefully appropriate warnings of any potential health risks will be provided as new laws take effect. Currently, some companies are voluntarily adding warnings in their advertisements and product labels.

Are nutritional ergogenics legal?

The use of pharmaceutical agents to enhance performance in sport has been prohibited by the governing bodies of most organized sports. The use of drugs in sports is known as **doping,** and the World Anti-Doping Agency (WADA) has an extensive list of drugs and doping techniques that have been prohibited.

At present, all essential nutrients are not classified as drugs and are considered to be legal for use in conjunction with athletic competition. Most other food substances and constituents sold as dietary supplements are also legal. However, some dietary supplements are prohibited, such as androstenedione, because they are classified as anabolic steroids, which are prohibited drugs. Nevertheless, such supplements may still be obtained via online sales. Other dietary supplements may contain substances that are prohibited; for example, Chinese ephedra and some forms of ginseng may contain ephedrine, a stimulant prohibited in competition. Various athletic governing associations have addressed the issue of sports supplements. For example, the NFL, partnering with the NSF Certified for Sports program, has developed strict requirements for the manufacturing of dietary supplements approved for use by its players. The National Collegiate Athletic Association (NCAA) places supplements for student athletes into three categories:

- Permissible—may be provided by the university. Supplements include vitamins, minerals, sports drinks, energy bars, and similar products.
- Impermissible—may not be provided by the university but may be purchased by the student athlete. These supplements are mainly high-protein products, such as those rich in whey protein.
- Banned—mainly drugs, such as those banned by WADA, including stimulants and anabolic agents. Some prescription drugs may also be banned unless under guidance of a physician.

Evidence suggests that contamination of sports supplements that may cause an athlete to fail a doping test is widespread. Some studies of sports supplements targeted for muscle building and marketed on the internet have reported that up to 25 percent were contaminated with prohibited substances and note that many athletes, including Olympic champions, who have claimed they have not taken drugs, but only dietary supplements, have tested positive for doping.

It is hoped that, with pending legislation, all ingredients will be listed in correct amounts on dietary supplement labels. In the meantime, athletes should consult with appropriate authorities before using any sports nutrition supplements marketed as performance enhancers.

Some organizations, such as NSF International, have created programs such as NSF Certified for Sport®, designed to minimize the risk of contaminated sports supplements. Another such group is Informed Sport. You can learn more about each of these programs by visiting www.nsfsport.com and https://sport.wetestyoutrust.com/.

Where can I find more detailed information on sports supplements?

Elnur/Shutterstock

Although details on various sports supplements are presented in later chapters, space limitations prevent detailed accounts of each and every supplement available on the market. The following resources may provide detailed information regarding efficacy and safety of numerous supplements, including sports supplements.

This USDA website provides numerous links to ergogenic aids and dietary supplements marketed to athletes, and you can search for specific supplements. You may access this information at www.nutrition.gov/dietary-supplements/dietary-supplements-athletes.

The Australian Institute of Sport (AIS) provides a comprehensive coverage of sports supplements in four categories based on scientific evidence:

A—Strong scientific evidence for use in specific situations in sport using evidence-based protocols.
B—Emerging scientific support, deserving of further research. Considered for use by athletes under a research protocol or case-managed monitoring situation.
C—Scientific evidence not supportive of benefit amongst athletes or no research undertaken to guide an informed opinion.
D—Banned or at high risk of contamination with substances that could lead to a positive doping test

www.ais.gov.au/nutrition/supplements The AIS website provides specific details on the benefits and risks of using supplements and sports foods.

Key Concepts

- The most prevalent ergogenic aids used to increase sports performance are those classified as nutritional, for theoretical nutritional aids may be found in all six classes of nutrients.
- Although many sports supplements are safe and legal, most are not effective ergogenic aids, and some are unsafe or illegal. Before using a sports supplement, athletes should try to determine if it is effective, if it is safe, and if it is legal.

Check for Yourself

- Search online for dietary supplements that may help enhance your sports performance, such as increasing muscle mass or losing body fat. Review how the products are being marketed, including statements about their effectiveness. Then, research the supplements on the evidence-based websites noted in this chapter and compare the findings.

Nutrition and Health Misinformation in Sports

Increasing numbers of dietary supplements are being marketed to the general population as products to improve health and well-being and exercise performance. Unfortunately, many of the products that advertise extravagant claims of enhanced health or performance are promoted by unscrupulous entrepreneurs, have no legitimate basis, and may be regarded as misinformation. In addition, with over 80% of adults in the United States using social media, more and more health and nutrition information is being shared through social media platforms. As of 2022, use of Instagram, Snapchat, and TikTok were particularly common among adolescents and young adults.

What is nutrition and health misinformation?

Nutrition and health misinformation results in the spread of claims that are not supported by well-controlled research studies. In some instances, this information can be spread intentionally, for example, a salesperson trying to sell a particular supplement with the single goal of making a greater profit. Or, this misinformation can be spread more unintentionally. This could occur when an individual shares their personal story of using a supplement and, while the supplement was beneficial to the individual, research studies have not been conducted and/or have not reported similar results.

Knowledge relative to nutrition, health, and sports performance has increased significantly in recent years. Hundreds of thousands of studies have been conducted, revealing facts to help unravel some of the mysteries of human nutrition. The AND indicates that consumers are taking greater responsibility for self-care and are eager to receive food and nutrition information. However, that creates opportunities for nutrition misinformation and health fraud to flourish. This is particularly the case as social media use has increased and nutrition information is so readily available to consumers.

Promotion of products marketed to enhance health and/or sports performance is big business. Reports suggest that Americans spend approximately $1.5 billion on dietary supplements and vitamins annually. Experts note that the amount of misinformation about nutrition is overwhelming, and it is circulated widely, particularly by those who may profit from it. For example, social media influencers who are being compensated to promote a particular product or paid advertising on social media platforms.

As noted previously, there are some well-researched health benefits associated with the foods we eat. Federal legislation in the United States allows for the placement of FDA-approved health claims on food packaging. One such health claim is that the consumption of macadamia nuts may reduce risk of heart disease. Such may not be the case, however, with dietary supplements, which are not regulated in the same way as packaged food products. Therefore, claims posted on dietary supplement labels have not been through the rigorous approval process required of food products.

Before the passage of the 1994 Dietary Supplement Health and Education Act (DSHEA), many extravagant health claims were made by some unscrupulous companies in the food supplement industry. As an example, the deceptive label of one secret formula noted that it would help you lose excess body fat while sleeping, a false claim. Although the DSHEA was designed to eradicate such fraudulent health claims, dietary supplements today appear to have more leeway than packaged foods to imply health benefits. Technically, labels on dietary supplements are not permitted to display scientifically unsupported claims. However, companies are allowed to make general health claims like "boosts the immune system" if, for example, the product contains a nutrient, such as zinc, that has been deemed important in some way to immune functions in the body. Although companies may not claim that the product prevents a particular health condition, the consumer may erroneously make such an assumption.

Many companies now use a disclaimer for general health claims on their labels, noting "These statements have not been evaluated by the Food and Drug Administration" and "This product is not intended to diagnose, treat, cure, or prevent any disease." Companies may also circumvent government regulations by using *freedom of the press*. They may provide information in the form of a reprint of an article, a brochure with highlighted research, or other materials that are distributed in connection with the sale of the product.

Although these advertising strategies may contain fraudulent information, the federal agencies that monitor such practices are understaffed and cannot litigate every case of misleading or dishonest advertising. Thus, unsuspecting consumers may be lured into buying an expensive health-food supplement that has no scientific support of its effectiveness.

Why is nutrition misinformation so prevalent in athletics?

As with nutrition and health misinformation in general, hope and fear are the motivating factors underlying the use of nutritional supplements by athletes. They hope that a special nutrient concoction will provide them with a slight competitive edge, and they fear losing if they do not do everything possible to win.

Various factors within the athletic environment help nurture these hopes and fears, but the most significant factor contributing to misinformation is direct advertising, as caricatured by the fabricated advertisement in **figure 1.11**. If you scan through various websites and social media accounts targeting bodybuilders or endurance athletes, you will see dozens of advertisements suggesting enhancement of strength, endurance, and sports performance. Such advertisements often use endorsements by star athletes. However, in most cases, there is little or no research supporting the purported ergogenic effects of the advertised supplement.

Additionally, many sports magazines will run articles on the ergogenic benefits of a particular nutrient and in close proximity to the article (either digital or print copy) place an advertisement for a product that contains that nutrient. Freedom of

FIGURE 1.11 Simulated nutritional supplement advertisement aimed at athletes.

speech guaranteed by the First Amendment permits the author of the article to make sensational and deceptive claims about the nutrient. However, freedom of speech does not extend to advertising, so fraudulent or deceptive claims may be grounds for prosecution by the FDA or the Federal Trade Commission (FTC). Thus, by cleverly positioning the article and the advertisement, the promoter can make the desired claims about the value of the product and yet avoid doing so illegally. Classic examples of this technique may be found with protein and amino acid supplement advertising on online platforms for bodybuilders. Moreover, many advertisements now appear in a format designed to look like a scientific review, though in actuality they are deceptive advertisements for sports supplements. Check the top of the page of such articles and you will find *Advertisement* in small print.

Most of these advertised products are economic frauds. The prices are exorbitant in comparison to the same amount of nutrients that may be obtained in ordinary foods. Besides being an economic fraud, these products are an intellectual fraud, for there is very little scientific evidence to support their claims. Simple basic facts about the physiological functions of the nutrients in these products are distorted, magnified, and advertised in such a way as to make one believe that they will increase athletic performance. Unfortunately, in the area of nutrition and sport, it is very easy to distort the truth and appeal to the psychological emotions of the athlete. In many cases, supplements are manufactured by a third party and sold to many different companies, which market them under their personal brand and slick advertising.

https://medlineplus.gov/webeval/intro1.html The National Library of Medicine provides a tutorial for consumers on how to evaluate health information found on the internet.

How do I recognize nutrition misinformation in health and sports?

It is often difficult to differentiate between nutrition misinformation and reputable nutritional information. The *Training Table* in this section provides practical advice for consumers when evaluating advertisements for nutritional supplements.

Training Table

How can you tell if the claims made on advertisements for a nutritional supplement are true or are nutrition misinformation? Read through this list and, if the answer is "yes," to any of the questions, you should be skeptical of such supplements and investigate their value before investing any money.

- Does the product promise quick improvements in health or physical performance?
- Does it contain a secret or magical ingredient or formula?
- Is it advertised mainly by use of anecdotes, case histories, or testimonials?
- Are currently popular personalities or star athletes featured in its advertisements?
- Does it take a simple truth about a nutrient and exaggerate that truth in terms of health or physical performance?
- Does it question the integrity of the scientific or medical establishment?
- Is it advertised on a health or sports-magazine website whose publishers also sell nutritional aids?
- Does the person who recommends it also sell the product?
- Does it use the results of a single study or dated and poorly controlled research to support its claims?
- Is it expensive, especially when compared to the cost of equivalent nutrients that may be obtained from ordinary foods?
- Is it a recent discovery not available from any other source?
- Is its claim too good to be true? Does it promise the impossible?

Should you have questions about a specific nutritional supplement, contact a trained and certified professional, such as a registered dietitian nutrition (RD/RDN). Visit www.eatright.org/find-a-nutrition-expert to search for such professionals.

Where can I get sound nutritional information to combat nutrition misinformation in health and sports?

The best means to evaluate claims of enhanced health or sports performance made by dietary supplements or other nutritional practices is to possess a good background in nutrition and a familiarity with related high-quality research. Unfortunately, many individuals, including most athletes, coaches, and health professionals, have not been exposed to such an educational program, so they must either take formal course work in nutrition or sports nutrition, develop a reading program in nutrition for health and sport, or consult with an expert in the field.

This book has been designed to serve as a text for a college course in nutrition for health-related and sports-related fitness, but it may also be read independently. It is an attempt to analyze and interpret the available scientific literature as to how nutrition may affect health and sports performance and to provide some simple guidelines for physically active individuals to help improve their health or athletic performance. It should provide the essential science-based (evidence-based) information you need to plan an effective nutritional program, either for yourself, other physically active individuals, or athletes, and to evaluate the usefulness of many nutritional supplements or practices designed to improve health or sports performance. Here are some key resources.

Books Numerous reputable books that detail the relationship of nutrition to health and sports performance are available. However, some books, such as diet books based on an author's personal experiences, may not contain reputable information. A good guide is to check the author's credentials.

Government, Health Professional, Consumer, and Commercial Organizations and Related Websites Accurate information relating nutrition to health is published by governmental agencies such as the FDA and USDA; health professional groups such as the AND, ACSM, Dietitians of Canada, and American Medical Association; consumer groups such as Consumers Union and Center for Science in the Public Interest; and some commercial groups such as the National Dairy Council and the PepsiCo's Gatorade Sports Science Institute. As noted previously, the AIS provides detailed, accurate information on a wide variety of sports supplements. Excellent materials relative to nutrition may be obtained free or at small cost from some of these organizations.

www.hsph.harvard.edu/nutritionsource Website for nutrition information from the Harvard School of Public Health.

www.gssiweb.com Website for the Gatorade Sports Science Institute, providing detailed reviews on various topics in sports nutrition.

www.healthfinder.gov U.S. Department of Health and Human Services website for information on various health topics, including nutrition.

http://medlineplus.gov National Library of Medicine, a comprehensive health-information retrieval website.

Scientific Journals Many scientific journals publish reputable findings about nutrition, exercise, and health. These technical journals may not be readily available in public libraries but may be found in university and medical libraries. Examples of such publications include *Medicine & Science in Sports & Exercise, Journal of the Academy of Nutrition and Dietetics, American Journal of Clinical Nutrition, Sports Medicine,* and *International Journal of Sport Nutrition and Exercise Metabolism.*

> *www.pubmed.gov* National Library of Medicine website provides abstracts of original research studies and excellent reviews and meta-analyses published in scientific medical journals. Free full-text articles are provided for some journals. Many colleges and universities also subscribe to a wide variety of health-related professional journals. Check with the library at your college or university for access to journal articles.
>
> *www.eatright.org* Search the AND website for the names of local dietitians, as well as other sources of sound nutrition information.

Popular Magazines Articles in popular health and sports magazines may or may not be accurate. The credentials of the author, if listed, should be a good guide to an article's authenticity. A Ph.D. listed after the author's name may not guarantee accuracy of the content of the article. Be wary of publications emanating from organizations or publishers that also sell nutritional supplements.

Consultants Nutritional consultants are another source of information. Such consultants should have a solid background in nutrition, particularly sports nutrition, if they are to advise athletes. The consultant should be a registered dietitian nutritionist (RD or RDN) or possess appropriate professional certification, such as the Certified Nutrition Specialist (CNS). He or she should be a member of a reputable organization of nutrition experts, such as the AND, which can be contacted at its website address to provide you with the name of a local dietitian. Other recognized nutritional organizations include the American Society for Nutrition, the American College of Nutrition, and the Dietitians of Canada.

As noted previously, the AND Commission on Dietetic Registration has developed a certification program for RDN's who work in sports to achieve the status of Board Certified Specialist in Sports Dietetics (CSSD). A qualified sports dietitian will be able to assess your nutritional status, including variables such as body composition, dietary analysis, and eating and lifestyle patterns, and relate these nutritional factors to the physiological and related nutritional demands of your sport or exercise program, providing you with a plan to help you reach your performance goals.

Be wary of individuals who do not possess professional degrees or appropriate certification, such as "experts" in nutrition or fitness. Many states do not have regulations restricting the use of various terms, such as *nutritionist* or *fitness professional*. Although these individuals may have some practical experience with helping people change their diets and initiate exercise programs, they normally do not have the depth of knowledge required in some cases. For proper nutritional advice, be certain to ask for proof of certification from recognized nutrition professional groups as cited previously. For fitness professionals, check for certification by such groups as the ACSM, the American Council on Exercise (ACE), or the National Strength and Conditioning Association (NSCA).

Cautions on Using the Internet The U.S. Department of Health and Human Services has recommended caution in using the internet to find health information. Along with others, here are some of its major points:

- No one regulates information on the internet. Thus, anyone can set up a home page and claim anything.
- Some official-sounding websites, such as Wikipedia, permit anyone to enter or modify the information presented.
- Search engines host paid advertisements which usually have priority listing and may contain biased information.
- Compare the information you find on the internet with other resources, such as medical journals and textbooks.
- Check the author's or organization's credentials. Unfortunately, there are many so-called nutritionists and other health professionals making false claims on the internet.
- Be wary of websites advertising and selling products that claim to improve your health.
- Be cautious when using information found on discussion boards or during chat sessions with others.
- Don't believe everything you read.

Several websites listed previously provide reputable information. Although some commercial (.com) and organization (.org) websites provide trustworthy information and may be cited in this text, others may not be as reputable, as they may be sponsored by unethical supplement companies. In general, education (.edu) and government (.gov) websites provide trustworthy information. The websites cited in this text are deemed to be reliable.

Key Concepts

- Nutrition and health misinformation is widespread as related to the purported benefits of specific dietary supplements. This is particularly the case with dietary supplements marketed to physically active individuals.
- There are a number of guidelines to help identify nutrition misinformation and false claims regarding dietary supplements, but one of the critical points to consider is if the claim simply appears to be too good to be true.
- The best means to counteract nutrition misinformation is to possess a good background in nutrition. Reputable sources of information are available to help provide contemporary viewpoints on the efficacy, safety, and legality of various dietary supplements for health or sport.

Research and Evidence-Based Recommendations

As discussed throughout this chapter, nutrition and exercise may influence health and sports performance. But how do we know what effect a nutrient, food, or dietary supplement we consume or exercise program we undertake will have on *our* health or performance? To find answers to specific questions, we should rely on the findings derived from scientific research, which is the heart of *evidence-based* medicine. As sophisticated sciences, nutrition and exercise science have a relatively short history. Not too long ago, nutrition scientists were concerned primarily with identifying the major constituents of the foods we eat and their general functions in the human body, while those investigating exercise concentrated more on its application to enhance sports performance. Over time, however, numerous scientists have turned their attention to the possible health benefits of certain foods and various forms of exercise, and, in the case of sports scientists, the possible applications to athletic performance. These scientists are not only attempting to determine the general effects of diet and exercise on health and performance, but also investigating the effects of specific nutrients at the molecular and genetic levels to determine possible mechanisms of action to improve health or performance in sport.

Because this book provides *evidence-based* recommendations relative to sports and health, it is important to review briefly the nature and limitations of nutritional and exercise research with humans. For the purpose of this discussion, our emphasis will be on nutritional research, although the same research considerations apply to exercise as well.

What types of research provide valid information?

Several research techniques have been used to explore the effects of nutrition on health or athletic performance. The two major general categories have been epidemiological research and experimental research.

Epidemiological research, also known as *observational research,* involves studying large populations to find relationships between two or more variables, such as dietary fat and heart disease. However, the treatment of interest, such as dietary fat, is not assigned to the subjects. Their normal diet and its relationship to the development of heart disease is the main variable of interest. There are various forms of epidemiological research. One general form uses retrospective techniques. In this case, individuals who have a certain disease are identified and compared with a group of their peers, called a *cohort,* who do not have the disease. Researchers then trace the history of both groups through interviewing techniques to identify dietary practices that may have increased the risk for developing the disease. Another general form of epidemiological research uses prospective techniques. In this case, individuals who are free of a specific disease are identified and then followed for years, during which time their diets are scrutinized. As some individuals develop the disease and others do not, the investigators then attempt to determine what dietary behaviors may increase the risk for the disease.

Epidemiological research helps scientists identify important relationships between nutritional practices and health. For example, years ago several epidemiological studies reported that individuals who consumed a diet high in fat were more likely to develop heart disease. One should note that such epidemiological research does not prove a cause-and-effect relationship. Although these studies did note a deleterious association between a diet high in fat and heart disease, they did not actually prove that fat consumption (possible cause) leads to heart disease (possible effect), but only that some form of relationship between the two existed. However, in some cases, the relationship between a lifestyle behavior and a disease is so strong that causality is inferred. In this regard, epidemiologists often calculate and report relative risks (RR) or odds ratios (OR), which are probability estimates of getting some disease by practicing some unhealthful behavior. An RR of 1.0 is normal probability, so if a study reports an RR of 2.5 for developing heart disease in individuals who consumed a diet rich in saturated fatty acids, such diets may increase one's risk 2.5 times normal. Conversely, if a study reports an RR of 0.5 for developing heart disease by consuming a purely vegetarian diet, such diets may cut heart disease risk in half. Epidemiological research is useful in identifying relationships between variables and generating hypotheses and is often a precursor to experimental research, but it does not prove a cause-and-effect relationship.

Experimental research is essential to establishing a cause-and-effect relationship (**figure 1.12**). In human nutrition research, experimental studies are often referred to as *randomized clinical trials (RCTs)* or *intervention studies,* usually involving a treatment group and a control, or placebo, group. RCTs may involve studying a smaller group of subjects under tightly controlled conditions for a short time frame or larger population groups living freely over a long time frame. In RCTs, an independent variable (cause) is manipulated so that changes in a dependent variable (effect) can be studied. If we continue with the example of fat and heart disease, a large (and expensive) clinical intervention study could be designed to see whether a low-saturated fat diet could help prevent heart disease. Two groups of subjects would be matched on several risk factors associated with the development of heart disease, and over a certain time, say ten years, one group would receive a low-saturated fat diet (treatment, or cause) while the other would continue to consume their normal high-saturated fat diet (control or placebo). At the end of the experiment, the differences in the incidence of heart disease (effect) between the two groups would be evaluated to determine whether or not the low-fat diet was an effective preventive strategy. Bouchard presents an excellent, detailed overview of the quality of different research-based sources of evidence, noting that RCTs with large populations represent one of the richest sources of data. If the results of an RCT showed that consumption of a low-saturated fat diet had no effect upon the incidence rate of heart disease, should you continue to

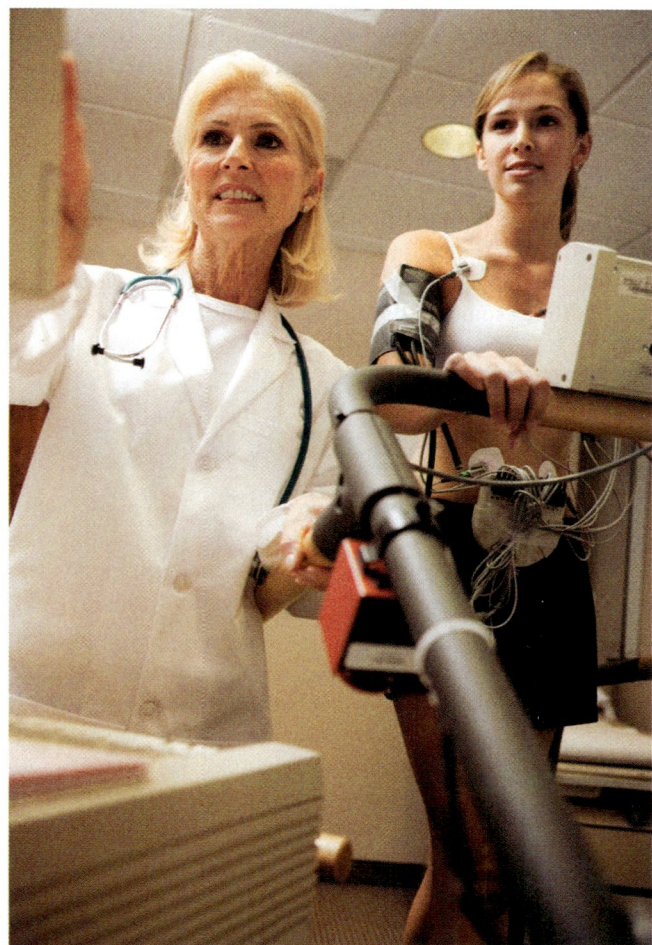

MBI/Alamy Stock Photo

FIGURE 1.12 Well-controlled experimental research serves as the basis underlying recommendations for the use of nutritional strategies to enhance health status or sports performance.

consume a high-saturated fat diet? The answer to this question, as we shall see later, is "not necessarily."

Most of the research designed to explore the effect of nutrition on sports performance is experimental in nature, and of a much shorter duration than studies investigating the relationship of nutrition and health. Additionally, most sports nutrition studies are conducted in a laboratory with tight control of extraneous variables. Very few studies have actually investigated the effect of nutritional strategies on actual competitive sports performance. Nevertheless, although most of our information about the beneficial effects of various nutritional strategies on sports performance is derived from laboratory-based research, many of these studies use laboratory protocols designed to mimic the physiological demands of a specific sport. In later chapters, as we discuss the effects of various nutritional strategies or dietary supplements on sports performance, we will often refer to studies that have problems with their experimental methodology, but we will also note studies that

were well controlled. The *Training Table* in this section provides examples of some major questions you should ask when evaluating the experimental methodology of a study. We use creatine supplementation as an example.

Training Table

How can you tell if a research study was well-designed? For this example, we will use research investigating creatine supplementation as a means to increase muscular strength and power in athletes. The following are some major questions you should ask about the experimental methodology:

- **Is there a legitimate reason for creatine supplementation in athletes?** Yes! Theoretically, creatine may add to the stores of creatine phosphate in the muscle and serve as a source of energy.
- **Were appropriate subjects used?** Yes! As creatine phosphate may theoretically benefit power performance, trained strength exercises would be ideal subjects.
- **Was creatine the only ingredient in the supplement?** Yes! This allows us to limit the investigation to creatine, and not other ingredients that may be present in a supplement.
- **Are the performance tests valid?** Validated tests should be used to collect data on the dependent variable, in this case valid strength and power tests.
- **Was a placebo control used?** A placebo similar in appearance and taste to the creatine supplement should be used as the control.
- **Were the subjects randomly assigned to treatments?** Subjects should be randomly assigned to separate groups, either the treatment (creatine) or control (placebo) group.
- **Was the study double-blind?** Neither the investigators nor the subjects should know which groups received the treatment or the placebo until the conclusion of the study.
- **Did the study have a crossover design with sufficient washout?** Subjects in a crossover study receive both the treatment and placebo, but at different time points with a period of time, the washout period, between each.
- **Were extraneous factors controlled?** Investigators should try to control other factors that may influence power, such as physical training, diet, and activity prior to testing.
- **Was the data properly analyzed?** Appropriate statistical techniques should be used to reduce the risk of statistical error. Using a reasonable number of subjects also helps to minimize statistical error.

Why do we often hear contradictory advice about the effects of nutrition on health or physical performance?

It is very difficult to conduct nutritional research about health and athletic performance with human subjects. For example, many diseases, such as cancer and heart disease, are caused by the interaction of multiple risk factors and may take many years to develop. It is not an easy task to control all of these risk factors in freely living human beings so that one independent variable, like dietary fat, can be isolated to study its effect on the development of heart disease over 10 or 20 years. In a similar manner, numerous physiological, psychological, and biomechanical factors also influence athletic performance on any given day. Why can't athletes match their personal records day after day? Because their physiology and psychology vary from day to day and even within the day.

Although well-designed studies in peer-reviewed scientific journals serve as the basis for making an informed decision as to whether or not to use a particular nutritional strategy or dietary supplement to enhance health or sports performance, it is important to realize that the results from a single study with humans do not prove anything. For example, Ioannidis noted that even the most highly cited RCTs, particularly small ones with a limited number of subjects, may be challenged and refuted over time. While most investigators attempt to control extraneous factors that may interfere with the interpretation of the results of their study, there may be some unknown factor that leads to an erroneous conclusion. For example, investigators studying the effect of creatine supplementation need to control dietary intake prior to testing. If not, consumption by some subjects of beverages containing caffeine, an effective ergogenic aid, could confound the results. Consequently, for this and other reasons, the results of single studies, whether epidemiological or experimental, should be considered with caution.

The Center for Science in the Public Interest published an article entitled "Behind the Headlines," noting that headlines often neglect to consider important limitations to the study. In this regard, Wellman and others indicated that, unfortunately, all too often the media make bold headlines based on the findings of an individual study, and often these headlines inadvertently exaggerate the findings of the study and their importance to health or physical performance. For example, a newspaper headline might blare "Coffee drinking causes heart disease" after a study is published indicating that coffee drinking could increase blood cholesterol levels slightly. The study did *not* show that coffee drinking caused heart disease, but only that it may have adversely affected one of its risk factors. A year or so later one may read headlines that report "Coffee drinking does not cause heart disease" because a more recent individual study did not find an association between coffee use and serum cholesterol levels. Is it no wonder consumers are often confused about nutrition and its effects on health or sports performance? Overall, most experts agree that nutrition scientists should be more involved in helping the media accurately convey diet and health messages.

https://cspinet.org/resource/behind-headlines The Center for Science in the Public Interest provides resources to help consumers make informed decisions when reading health-related news stories.

For the purpose of improving public understanding, the National Cancer Institute provided some guidelines that journalists and others in the communications business can use for reporting health-related nutrition research. The *Training Table* in this section outlines some of the key recommendations.

Training Table

Journalists and health professionals are often responsible for relaying health-related nutrition to the general public and patients. To improve public understanding of the research, the following key points should be considered:

- **The quality and credibility of the study.** Was it well-designed and published in a high-quality journal?
- **Peer-reviewed study or presentation at a meeting.** Was it presented at a meeting, which normally does not require a review by other scientists?
- **Comparison of findings to other studies.** Was the study compared to other studies reporting contrasting findings?
- **Putting findings into context, such as a risk–benefit assessment.** Are the health risks meaningful? An increased health risk from one in a million to three in a million, if reported as a threefold increase, may appear to be more meaningful than it really is.
- **Funding sources.** Was it funded by a company that could benefit financially from the results?

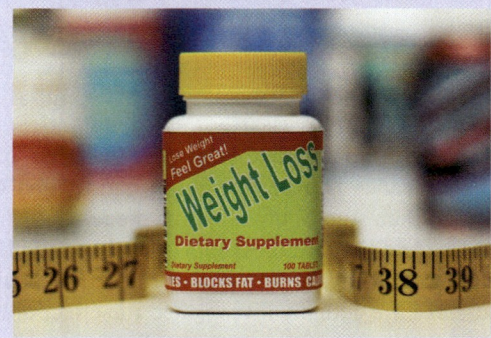

Don Farrall/Photodisc/Getty Images

What is the basis for the dietary recommendations presented in this book?

Scientists consider each single study as only one piece of the puzzle in attempting to find the truth. To evaluate the effects of nutritional strategies or dietary supplements on health or sports performance, individual studies should be repeated by other scientists and, if possible, a consensus developed. Reviews and meta-analyses provide a stronger foundation than the results of an individual study.

In reviews, an investigator analyzes most or all of the research on a particular topic and usually offers a summarization and conclusion. However, the conclusion may be influenced by the studies reviewed or by the reviewer's orientation. There have been instances in which different reviewers evaluated the same studies and came up with diametrically opposed conclusions.

Meta-analysis, a review process that involves a statistical analysis of previously published studies, may actually provide a quantification and the strongest evidence available relative to the effect of nutritional strategies or dietary supplements on health or sports performance. According to Binns and others, the meta-analysis is the gold standard for evidence-based clinical practice guidelines.

The value of reviews and meta-analyses is based on the quantity and quality of studies reviewed. If the number of studies is limited and they are not well controlled, or if improper procedures are used in analyzing and comparing the findings of each study, the conclusions may be inaccurate. For example, Hart and Dey noted that three meta-analyses of the use of Echinacea for the prevention of colds had somewhat different conclusions, as selection criteria for studies used in the analysis varied. Nevertheless, well-designed reviews and, in particular, meta-analyses provide us with valuable data to make prudent decisions. Position statements and position stands are developed using an *evidence-based* approach, which includes an evaluation of the quality of the studies reviewed. Such groups normally use only RCTs to support their position on specific topics. A number of such position statements are cited throughout this text where relevant.

Comparable to the science of other human behaviors, the science of human exercise and nutrition is not, as many may believe, exact. Although in many cases we still do not have absolute proof that a particular nutritional practice will produce the desired effect, we do have sufficient information to make a recommendation that is prudent, meaning that it is likely to do some good and cause no harm. Thus, the recommendations offered in this text should be considered evidence-based; they are based upon a careful analysis and evaluation of the available scientific literature, primarily comprehensive reviews and meta-analyses of the pertinent research by various scientists or public and private health or sports organizations.

> *https://nesr.usda.gov/* The USDA Nutrition Evidence Systematic Review (NESR) collaborates with leading scientists using state-of-the-art methodology to review, evaluate, and synthesize research to answer important diet-related questions.

How does all this relate to me?

Remember that we all possess biological individuality and thus might react differently to a particular nutritional or exercise intervention. For example, relative to health, many of us have little or no reaction to an increase in dietary salt, but some individuals are very sensitive to salt intake and will experience a significant rise in blood pressure with increased dietary salt. Relative to athletic performance, Mann and others note there are high responders and low responders to the same standardized exercise training program, some individuals improving markedly but others less so. Such individual reactions have been noted in some research studies and are discussed where relevant in the following chapters. With advances in genetic technology, diets and exercise training may one day be individualized to conform to our genetically determined favorable responses to particular dietary strategies. However, to our knowledge, individualized diets and exercise training for health or sports performance based on one's genetic profile have not yet been developed. For example, Sales and others note that the science of nutrigenomics seeks to explain the interactions between genes and nutrients in order to customize diets according to each individual's genotype, which may help prevent some chronic diseases. Moreover, in a major review of the genomics of elite sporting performance, Wang and others noted that progress has been made, such as identifying single genes with sprint or endurance performance, but they note that only after a lengthy and costly process will the true potential of genetic testing in sport be determined.

Westend61/Getty Images

Thus, recommendations offered in this text should not be regarded as medical advice. Individuals should consult an appropriate health professional for advice on taking any dietary supplement for health purposes. Additionally, although information presented in this book may help athletes make informed decisions regarding the use of nutritional strategies as a means to improve sports performance, athletes should confer with an appropriate health professional before using sports supplements or nutritional ergogenics.

Key Concepts

▶ Epidemiological research helps to identify relationships between nutritional practice and health or sports performance and may be helpful in developing hypotheses for experimental research. However, experimental studies, such as randomized controlled trials, are needed to establish a cause-effect relationship. Such experimental studies should adhere to appropriate research design protocols.

▶ Nutritional recommendations for enhancement of health or athletic performance are based on reputable evidence-based research.

Check for Yourself

▶ Use PubMed to search for and read a scientific article that involves the use of a dietary supplement to improve some facet of sports performance. To get a list of studies, you may go to www.pubmed.gov, and type in the name of the supplement and the term "exercise" in the search column, or simply scan some online sports medicine and nutrition journals. Compare the methodology to the recommended criteria presented in this section. Develop a short synopsis of the research article that you could post on a reputable sports-nutrition blog.

APPLICATION EXERCISE

Jada is a 20-year-old college sophomore who is taking a full load of classes, volunteers at a local school about 5 hours per week, and works at the campus gym about 10 hours per week. Working at the gym has motivated Jada to train for her first 5K, which she will do in 8 weeks.

Jada works out a couple of times per week, most commonly doing fitness classes such as cardio Zumba and Pilates. She has started to walk and jog outdoors to prepare for the 5K.

In terms of her diet, Jada eats most of her meals on campus as she is a Residential Advisor in one of the dorms. She has a small refrigerator and kitchenette in her dorm suite where she can prepare meals, but she tends to eat most frequently at the dining hall.

Your Turn:

1. Use a diet analysis program, such as NutritionCalc Plus, to create a meal plan for Jada. You will need to create a profile for her. She is 20 years old, 5'8" tall, 140 lbs, engages in 30–60 minutes of moderate physical activity daily, and she would like to stay at her current weight.
2. Develop a one-day meal plan for Jada to follow. Organize the plan by meal and include specifics on the type of food/beverage and serving size.
3. What general advice do you have for Jada in terms of food and beverage items to store in her kitchenette area? List three snack options that she can take with her to eat between classes.
4. Look at reputable online sources to find training plans for a new runner who is training for a 5K. Develop that training plan into a table for Jada.

Samuel Borges Photography/Shutterstock

Review Questions—Multiple Choice

1. What is the leading cause of death in the United States (2020)?
   a. heart disease
   b. diabetes
   c. cancer
   d. accidents
   e. pneumonia
2. According to the *2018 Physical Activity Guidelines for Americans*, healthy adults should engage in at least _____ minutes of moderate-intensity physical activity each week.
   a. 60
   b. 75
   c. 100
   d. 150
   e. 210
3. Which of the following is NOT a health benefit associated with engaging in regular physical activity?
   a. increased bone density and strength
   b. reduced risk for type 1 diabetes
   c. enhanced immune function
   d. reduced stress
   e. prevention of brain deterioration that occurs with aging
4. A person trying to follow the Prudent Healthy Diet recommendations might do all of the following except _____.
   a. eat only low-acid foods
   b. eat foods with less salt
   c. eat a wide variety of foods
   d. eat a diet low in saturated fat
   e. choose a diet rich in plant foods
5. Poor nutrition may contribute to the development of numerous chronic diseases. For example, obesity, high blood pressure, diabetes, and heart disease are most associated with which of the following nutritional problems?
   a. diets rich in vitamins and minerals
   b. diets rich in dietary fiber
   c. diets rich in fat and kcal
   d. diets rich in complex carbohydrates
   e. diets rich in plant proteins
6. Which of the following is considered an intermittent, high-intensity sport?
   a. marathon running
   b. Olympic weight lifting
   c. golf
   d. sprint speed skating
   e. soccer
7. Based on the *2018 Physical Activity Guidelines for Americans*, which of the following statements is false?
   a. Moderate-intensity aerobic exercise should be done for a minimum of 150 minutes each week.
   b. Vigorous-intensity exercise may be done for a minimum of 75 minutes each week.
   c. Each daily exercise bout of aerobic exercise may be done continuously or in smaller segments, such as three 10-minute bouts.
   d. In general, more is better, as exceeding the minimum recommended amounts of exercise may provide additional health benefits.
   e. Resistance exercise, including exercises for the major muscle groups in the body, is recommended at least 5, and preferably 7, days per week.
8. Which of the following statements regarding ergogenic aids is false?
   a. They are designed to enhance sports performance.
   b. Use of any aid that enhances sports performance is illegal and is grounds for disqualification.
   c. Although most nutritional ergogenics are safe, some dietary supplements pose significant health risks.
   d. Endorsement of a nutritional ergogenic by a professional athlete does not necessarily mean that it is effective as advertised.
   e. Some nutritional supplements marketed as ergogenics may contain prohibited drugs.

9. In an experimental study to evaluate the effect of caffeine supplementation on endurance, which of the following would not be considered acceptable for the research methodology to be followed in the conduct of the study?
    a. Use well-trained power sport athletes.
    b. Use a double-blind protocol.
    c. Use a placebo control group.
    d. Use a sport-related performance task.
    e. Use participants of a similar age and training level.
10. A meta-analysis is _____.
    a. an ergogenic aid for mathematicians
    b. a technique to evaluate the presence of drug metabolites in athletes
    c. a statistical evaluation of a collection of studies in order to derive a conclusion
    d. an evaluation of the daily metabolic rate
    e. an analytical technique to evaluate biomechanics in athletes

Answers to multiple choice questions: 1. a; 2. d; 3. b; 4. a; 5. c; 6. e; 7. e; 8. b; 9. a; 10. c

Critical Thinking Questions

1. List at least eight potential health benefits of a person engaging in a regular, comprehensive exercise program. Then, describe at least two possible mechanisms by which exercise may enhance health status.
2. List two *Healthy People 2030* objectives related to physical activity and summarize the progress toward achieving those goals (www.healthypeople.gov).
3. Describe five general dietary strategies that an athlete may follow to enhance sports performance.
4. Search online for a dietary supplement advertised for building lean body mass (muscle). Evaluate the information provided on the website for that supplement. Would you recommend this supplement to a friend? Why or why not?
5. Use an online search tool (e.g., PubMed) to find an original research study related to sports nutrition. Describe the type of research study, including the number and characteristics of participants and study design. Based on that information, evaluate the credibility of the study. Was it well designed?

References

Academy of Nutrition and Dietetics. 2016. Position of the Academy of Nutrition and Dietetics, Dietitians of Canada, and the American College of Sports Medicine: Nutrition and athletic performance. *Journal of the Academy of Nutrition and Dietetics* 116(3):501-28.

Academy of Nutrition and Dietetics. 2014. Position of the Academy of Nutrition and Dietetics: Nutritional genomics. *Journal of the Academy of Nutrition and Dietetics* 114(2):299-312.

Ahmetov, I., and Rogozkin, V. 2009. Genes, athlete status and training—An overview. *Medicine and Sport Science* 54:43-71.

Angell, S., et al. 2020. The American Heart Association 2030 Impact Goal: A presidential advisory from the American Heart Association. *Circulation* 141(9):e120-38.

Beck, K., et al. 2015. Role of nutrition in performance enhancement and postexercise recovery. *Journal of Sports Medicine* 6:259-67.

Ben-Zaken, S., et al. 2022. Genetic characteristics of competitive swimmers: A review. *Biology of Sport* 39(1):157-70.

Binns, C., et al. 2008. Tea or coffee? A case study on evidence for dietary advice. *Public Health Nutrition* 11:1132-41.

Birkenhead, K., and Slater, G. 2015. A review of factors influencing athletes' food choices. *Sports Medicine* 45:1511-22.

Booth, F., and Lees, S. 2007. Fundamental questions about genes, inactivity, and chronic diseases. *Physiological Genomics* 28:146-57.

Booth, F. W., and Neufer, P. D. 2006. Exercise genomics and proteomics. In *ACSM's Advanced Exercise Physiology*, ed. C. M. Tipton. Philadelphia: Lippincott Williams & Wilkins.

Booth, F., et al. 2017. Role of inactivity in chronic diseases: Evolutionary insight and pathophysiological mechanisms. *Physiological Reviews* 97(4):1351-1402.

Bouchard, C., et al. 2000. Genomic scan for maximal oxygen uptake and its response to training in the HERITAGE family study. *Journal of Applied Physiology* 88:551-59.

Braakhuis, A., et al. 2021. Consensus report of the Academy of Nutrition and Dietetics: Incorporating genetic testing into nutrition care. *Journal of the Academy of Nutrition and Dietetics* 121(3):545-52.

Brandt, C., and Pedersen, B. 2010. The role of exercise-induced myokines in muscle homeostasis and the defense against chronic diseases. *Journal of Biomedicine & Biotechnology* 520258.

Cho, J., et al. 2016. Cross-national comparisons of college students' attitudes toward diet/fitness apps on smartphones. *Journal of the American College of Health* 65(7):437-49.

Christensen, N., et al. 2021. Diet quality and mental health status among division 1 female collegiate athletes during the COVID-19 pandemic. *International Journal of Environmental Research and Public Health* 18(24):13377.

Clark, L., et al. 2017. Cytokine response to exercise and activity in patients with chronic fatigue syndrome: Case control study. *Clinical and Experimental Immunology* 190(3):360-71.

Colberg, S., et al. 2010. Exercise and type 2 diabetes: American College of Sports Medicine and the American Diabetes Association: Joint Position Statement. Exercise and type 2 diabetes. *Medicine & Science in Sports & Exercise* 42:2282-303.

Coppetti, T., et al. 2017. Accuracy of smartphone apps for heart rate measurement. European *Journal of Preventive Cardiology* 24(12):1287-93.

deBoer, Y., and Sherker, A. 2017. Herbal and dietary supplement-induced liver injury. *Clinics in Liver Disease* 21(1):135-49.

Denham, B. 2017. Athlete information sources about dietary supplements: A review of extant research. *International Journal of Sports Nutrition and Exercise Metabolism* 27(4):325-34.

Donnelly, J., et al. 2009. American College of Sports Medicine Position Stand. Appropriate physical activity intervention strategies for weight loss and prevention of weight regain for adults. *Medicine & Science in Sports & Exercise* 41:459-71.

Durnin, J.V. 1967. The influence of nutrition. *Canadian Medical Association Journal* 96:715-20.

Ehlert, T., et al. 2013. Epigenetics in sports. *Sports Medicine* 43:93-110.

Erdman, K. 2006. Influence of performance level on dietary supplementation in elite Canadian athletes. *Medicine & Science in Sports & Exercise* 38:348-56.

Eynon, N., et al. 2013. Genes for elite power and sprint performance: ACTN3 leads the way. *Sports Medicine* 43:803-17.

Federation Internationale De Natation (FINA). 2022. Nutrition for aquatic athletes. https://resources.fina.org/fina/document/2021/02/04/5c14b311-7eba-4d2b-9114-acf13d300683/nutrition_for_aquatic_athletes_booklet_v5_final.pdf. Accessed September 24, 2022.

Franklin, B., et al. 2020. Exercise-related acute cardiovascular events and potential deleterious adaptations following long-term exercise training: Placing the risks into perspective—an update: A scientific statement from the American Heart Association. *Circulation* 141(13):e705-36.

Freeland-Graves, J., et al. 2013. Position of the Academy of Nutrition and Dietetics: Total diet approach to healthy eating. *Journal of the Academy of Nutrition and Dietetics*. 113:307-17.

Geiger, P., et al. 2011. Heat shock proteins are important mediators of skeletal muscle insulin sensitivity. *Exercise and Sport Sciences Reviews* 39:34-42.

Goedecke, J., and Micklesfield, L. 2014. The effect of exercise on obesity, body fat distribution and risk for type 2 diabetes. *Medicine and Sport Science* 60:82-93.

Grundy, Q., et al. 2017. Tracing the potential flow of consumer data: A network of prominent health and fitness apps. *Journal of Medical Internet Research* 19(6):e233.

Haines, J., et al. 2019. Nurturing children's healthy eating: Position statement. *Appetite* 137:124-33.

Hart, A., and Dey, P. 2009. Echinacea for prevention of the common cold: An illustrative overview of how information from different systematic reviews is summarised on the internet. *Preventive Medicine* 49:78-82.

Harvey, J., et al. 2013. Prevalence of sedentary behavior in older adults: A systematic review. *International Journal of Environmental Research and Public Health*. 10:6645-61.

Heaton, L., et al. 2017. Selected in-season nutritional strategies to enhance recovery for team sport athletes: A practical overview. *Sports Medicine* 47(11):2201-18.

Higgins, J., et al. 2013. Sudden cardiac death in young athletes: Preparticipation screening for underlying cardiovascular abnormalities and approaches to prevention. *The Physician and Sports Medicine* 41(1):81-93.

Hongu, N., et al. 2014. Mobile technologies for promoting health and physical activity *ACSM's Health & Fitness Journal* 18 (4):8-15.

Ioannidis, J. 2005. Contradicted and initially stronger effects in highly cited clinical research. *Journal of the American Medical Association* 294:218-28.

Iwai, K., et al. 2022. Usefulness of aerobic exercise for home blood pressure control in patients with diabetes: Randomized crossover trial. *Journal of Clinical Medicine* 11(3):650.

Joyner, M., and Coyle, E. 2008. Endurance exercise performance: The physiology of champions. *Journal of Physiology* 586:35-44.

Kantor, E., et al. 2016. Trends in dietary supplement use among US adults from 1999-2012. *Journal of the American Medical Association* 316(14):464-74.

Kitamura, K., et al. 2022. Leisure-time and non-leisure-time physical activities are dose-dependently associated with a reduced risk of dementia in community-dwelling people aged 40-74 years: The Murakami cohort study. *Journal of the American Medical Directors Association* 15:S1525-8610(22)00079-2.

Knapik, J., et al. 2021. Prevalence, factors associated with use, and adverse effects of sport-related nutritional supplements (sport drinks, sport bars, sport gels): The US military dietary supplement use study. *Journal of the International Society of Sports Nutrition* 18(1):59.

Kraschnewski, J., and Schmitz, K. 2017. Exercise in the prevention and treatment of breast cancer: What clinicians need to tell their patients. *Current Sports Medicine Reports* 16(4):263-7.

Krebs-Smith, S., et al. 2010. Americans do not meet federal dietary recommendations. *Journal of Nutrition* 140:1832-8.

Kreider, R., et al. 2017. International Society of Sports Nutrition position stand: Safety and efficacy of creatine supplementation in exercise, sport, and medicine. *Journal of the International Society of Sports Nutrition* 14:18.

Landry, B., and Driscoll, S. 2012. Physical activity in children and adolescents. *PM&R The Journal of Injury, Function, and Rehabilitation* 4(11):826-32.

LeLorier, J., et al. 1997. Discrepancies between meta-analyses and subsequent large, randomized, controlled trials. *New England Journal of Medicine* 337:559-61.

Loughrey, D., et al. 2017. The impact of the Mediterranean Diet on the cognitive functioning of healthy older adults: A systematic review and meta-analysis. *Advances in Nutrition* 8(4):571-86.

Mann, T., et al. 2014. High responders and low responders: Factors associated with individual variation in response to standardized training. *Sports Medicine* 44:1113-24.

Marra, M., and Boyar, A. 2009. Position of the Academy of Nutrition and Dietetics: Nutrient supplementation. *Journal of the American Dietetic Association* 109(12):2073-85.

Mathews, N. 2018. Prohibited contaminants in dietary supplements. *Sports Health* 10(1):19-30.

McAtee, C. 2013. Fitness, nutrition and the molecular basis of chronic disease. *Biotechnology & Genetic Engineering Reviews* 29:1-23.

McKee, A., et al. 2014. The neuropathology of sport. *Acta Neuropathologica* 127:29-51.

Meadows, M. 2005. Genomics and personalized medicine. *FDA Consumer* 39(6):12-17.

Meeusen, R. 2013. Exercise, nutrition & the brain. *Sports Science Exchange* 26 (112):1-6.

Minihane A., et al. 2015. Low-grade inflammation, diet composition and health: Current research evidence and its translation. *British Journal of Nutrition* 114(7):999-1012.

Monteiro R., et al. 2018. Effect of exercise on inflammatory profile of older persons: Systematic review and meta-analyses. *Journal of Physical Activity and Health* 15(1):64-71.

Montuori, P., et al. 2021. Bodybuilding, dietary supplements and hormones use: Behaviour and determinant analysis in young bodybuilders. *BMC Sports Science, Medicine and Rehabilitation* 13(1):147.

Naseeb, M., and Volpe, S. 2017. Protein and exercise in the prevention of sarcopenia and aging. *Nutrition Research* 40:1-20.

National Cancer Institute. 1998. Commentary: Improving Public Understanding: Guidelines for communicating emerging science on nutrition, food safety, and health. *Journal of National Cancer Institute* 90 (3):194-99.

Newman, J., et al. 2017. Primary prevention of cardiovascular disease in diabetes mellitus. *Journal of the American College of Cardiology* 70(7):883-93.

Nijs, J., et al. 2014. Altered immune response to exercise in patients with chronic fatigue syndrome/myalgic encephalomyelitis: A systematic literature review. *Exercise Immunology Review* 20:94-116.

Nimmo, M., et al. 2013. The effect of physical activity on mediators of inflammation. *Diabetes, Obesity and Metabolism* 15 (Supplement 3):51-60.

Parnell, J., et al. 2016. Dietary intakes and supplement use of pre-adolescent and adolescent Canadian athletes. *Nutrients* 8(9):e526.

Raman, G., et al. 2013. Tai chi improves sleep quality in healthy adults and patients with chronic conditions: A systematic review and meta-analysis. *Journal of Sleep Disorders and Therapy* 2(6):141-55.

Rankinen, T., et al. 2010. Advances in exercise, fitness, and performance genomics. *Medicine & Science in Sports & Exercise* 42:835-46.

Sales, N., et al. 2014. Nutrigenomics: Definitions and advances of this new science. *Journal of Nutrition and Metabolism*. doi: 10.1155/2014/202759. Epub 2014 Mar 25.

Sarzynski, M., et al. 2016. Advances in exercise, fitness, and performance genomics in 2015. *Medicine and Science in Sports and Exercise* 48(10):1906-16.

Simopoulos, A. 2010. Nutrigenetics/nutrigenomics. *Annual Review Public Health* 21:53-68.

Slawson, D., et al. 2013. Position of the Academy of Nutrition and Dietetics: The role of nutrition in health promotion and chronic disease prevention. *Journal of the Academy of Nutrition and Dietetics* 113(7):972-79.

Slentz, C., et al. 2007. Modest exercise prevents the progressive disease associated with physical inactivity. *Exercise and Sport Sciences Reviews* 35:18-23.

Smith, J., et al. 2014. The health benefits of muscular fitness for children and adolescents: A systematic review and meta-analysis. *Sports Medicine* May 1. [Epub ahead of print]

Starr, R. 2015. Too little, too late: Ineffective regulation of dietary supplements in the United States. *American Journal of Public Health* 105(3):478-85.

Stewart, L., et al. 2007. The influence of exercise training on inflammatory cytokines and C-reactive protein. *Medicine & Science in Sports & Exercise* 39:1714-19.

Sun, Y., et al. 2017. The effectiveness and cost of lifestyle interventions including nutrition education for diabetes prevention: A systematic review and meta-analysis. *Journal of the Academy of Nutrition and Dietetics* 117(3):404-21.

Tam, R., et al. 2022. Recent developments in the assessment of nutrition knowledge in athletes. *Current Nutrition Reports* February 16, 2022. Online ahead of print.

Thornton, J., et al. 2016. Physical activity prescription: A critical opportunity to address a modifiable risk factor for the prevention and management of chronic disease: A position statement by the Canadian Academy of Sport and Exercise Medicine. *Clinical Journal of Sports Medicine* 26(4):259-65.

Tucker, R., et al. 2013. The genetic basis for elite running performance. *British Journal of Sports Medicine* 47:545-49.

Valliant, M., et al. 2012. Nutrition education by a registered dietitian improves dietary intake and nutrition knowledge of a NCAA female volleyball team. *Nutrients* 4:506-16.

Vandercappellen, E., et al. 2022. Sedentary behaviour and physical activity are associated with biomarkers of endothelial dysfunction and low-grade inflammation - relevance for (pre)diabetes: The Maastricht Study. *Diabetologia* February 4, 2022. Online ahead of print.

Virani, S., et al. 2021. Heart disease and stroke statistics - 2021 update: A report from the American Heart Association. *Circulation* 143(8):e254-743.

Wansink, B. 2006. Position of the American Dietetic Association: Food and nutrition misinformation. *Journal of the American Dietetic Association* 106:601-7.

Wang, G., et al. 2013. Genomics of elite sporting performance: What little we know and necessary advances. *Advances in genetics* 84:123-49.

Wellman, N., et al. 1999. Do we facilitate the scientific process and the development of dietary guidance when findings from single studies are publicized? An American Society for Nutritional Sciences controversy session report. *American Journal of Clinical Nutrition* 70:802-5.

Westberg, K., et al. 2022. Promoting healthy eating in the community sport setting: A scoping review. *Health Promotion International* 37(1):daab030.

Williams, M., and Branch, J. D. 2000. Ergogenic aids for improved performance. In *Exercise and Sport Science*, eds. W. E. Garrett and D. T. Kirkendall. Philadelphia: Lippincott Williams & Wilkins.

Wolfarth, B., et al. 2014. Advances in exercise, fitness, and performance genomics in 2013. *Medicine & Science in Sports & Exercise* 46:851-59.

Yang, Y., and Koenigstorfer, J. 2021. Determinants of fitness app usage and moderating impacts of education-, motivation-, and gamification-related app features on physical activity intentions: Cross-sectional survey study. *Journal of Medical Internet Research* 23(7):e26063.

Healthful Nutrition for Fitness and Sport

CHAPTER TWO

Ken Welsh/age fotostock

KEY TERMS

Acceptable Macronutrient Distribution Range (AMDR)
Adequate Intake (AI)
chronic training effect
complementary proteins
Daily Value (DV)
dietary guidance systems
Dietary Reference Intakes (DRIs)
dietary supplement
empty kcal
essential nutrients
Estimated Average Requirement (EAR)
Estimated Energy Requirement (EER)
functional foods
health claims
key-nutrient concept
lactovegetarian
liquid meals
macronutrient
micronutrient
MyPlate
nonessential nutrient
nutraceuticals
nutrient content claims
nutrient density
Nutrition Facts panel
nutritional labeling
ovolactovegetarian
ovovegetarian
pescovegetarian
phytonutrients
plant-based diet
Recommended Dietary Allowance (RDA)
semivegetarians
sports bars
Tolerable Upper Intake Level (UL)
vegan diet
vegetarian diet

LEARNING OUTCOMES

After studying this chapter, you should be able to:

1. List the six major classes of nutrients that are essential for human nutrition and identify specific nutrients within each class.
2. Explain the development of the DRIs and explain the meaning of their various components, including the RDA, AI, AMDR, UL, EAR, and EER.
3. Discuss the concept of a balanced diet and the basic components and recommendations of the MyPlate food guide.
4. Explain the key recommendations of the *2020–2025 Dietary Guidelines for Americans*, including recommendations for saturated fat, added sugars, and sodium intake.
5. Describe key principles of healthy eating and provide examples for how food might be selected or prepared in order to adhere to these guidelines.
6. Describe the potential health benefits of following a well-planned plant-based diet.
7. List the nutrients that must be included on a nutrition facts panel and explain how reading food labels may help one consume a healthier diet.
8. Identify the various types of dietary supplements and discuss, in general, the potential benefits and risks associated with taking dietary supplements.
9. Provide practical recommendations to athletes in relation to optimal nutrition for training and competition to maximize sports performance.

Introduction

Ken Welsh/age fotostock

What you eat can have a significant effect on your health. Hippocrates, the Greek physician known as the father of medicine, recognized the value of nutrition and the power of food to enhance health when he declared that you should, "Let food be your medicine, and medicine be your food." As noted in chapter 1, the foods we eat contain various nutrients to sustain life by providing energy, promoting growth and development, and regulating metabolic processes. Basically, healthful nutrition is designed to optimize these life-sustaining properties of nutrients and other substances found in food.

Although the Paleolithic (Paleo) diet has been popularized in recent years, the hunter/gatherer diet was actually not all meat and marrow; fossil research suggests humans may have been eating grains and tubers for at least 100,000 years, suggesting Neanderthals were omnivores, not total carnivores. A natural diet of animal and plant foods provided the nutrients necessary to sustain the lives of our hunter/gatherer ancestors. Through trial and error in selecting and consuming various foods, our ancient ancestors were able to obtain the numerous specific nutrients essential to life.

As science evolved, human food consumption patterns gradually changed. For example, food scientists identified specific nutrients, foods rich in such nutrients, and the amounts of each necessary to life. Food hazards, such as bacteria, were also uncovered and methods to combat them were developed. Foods could be packaged, refrigerated, and shipped thousands of miles. Overall, modern developments in the food industry have improved food quality and safety. For example, provision of a wide variety of foods has helped to eradicate most nutrient-deficiency diseases in industrialized nations. However, at the same time, in the twenty-first century, a wide variety of foods rich in solid fats and added sugars and low in dietary fiber are readily available. Regular consumption of such types of foods and beverages has contributed to rising rates of obesity and chronic disease in the United States.

Three keys to a healthful diet are *balance, variety,* and *moderation*. In general, a healthful diet is one that provides a balanced proportion of foods from different food groups, a variety of foods from within the different food groups, and moderation in the consumption of any food. Such a diet should provide us with the nutrients we need to sustain life and, as noted later, food scientists have determined the amounts of each we need.

In past years, nutrition research focused on the determination of how specific nutrients in our diet, such as saturated fat, affect our health, primarily as related to the development of chronic diseases. Although this research continues to be important, in recent years an increased focus has involved the health effects of the overall diet and whole foods within such diets, rather than individual nutrients. For example, the United States Department of Agriculture (USDA), in its *2020–2025 Dietary Guidelines for Americans,* provides guidance for making healthy and well-balanced dietary choices. In its recent position statement, the Academy of Nutrition and Dietetics (AND) advocated a *total diet* approach to healthy eating.

In general, a common core of these dietary recommendations underlies the Prudent Healthy Diet, a dozen guidelines used in this text to promote healthful eating. Although the basic guidelines underlying the Prudent Healthy Diet are rather simple, selecting the appropriate foods in modern society may be somewhat confusing. Fortunately, nutrition labels should provide the knowledgeable consumer with sufficient information to make intelligent choices and select high-quality foods. Food safety is also another consumer concern, and appropriate food selection and preparation practices may help minimize most of the health risks associated with certain foods.

The Prudent Healthy Diet also serves as the basic diet for those interested in optimal physical performance, although it may be modified somewhat for specific types of athletic endeavors, as shall be noted as appropriate throughout the book.

Several smartphone applications (apps) to help you eat a healthier diet are presented in this chapter and throughout the book. Many such apps are available, and new versions continue to be developed.

SDI Productions/iStock/Getty Images

37

Essential Nutrients and Recommended Nutrient Intakes

"You are what you eat" is a popular phrase that contains some truth, particularly in its implications for both health and athletic performance. The foods you eat contain a wide variety of nutrients, both essential and nonessential, as well as other substances that may affect your body functions. These nutrients are synthesized by plants from water, carbon dioxide, and various elements in the soil, and they also become concentrated in animals that consume plant foods. Various nutrients may also be added to foods in the manufacturing process. Careful selection of wholesome, natural foods will provide you with the proper amounts of nutrients to optimize energy sources, to build and repair tissues, and to regulate body processes. However, as we shall see in later chapters, poor food selection with an unbalanced intake of some nutrients may contribute to the development of significant health problems and impair sports performance.

What are essential nutrients?

As noted in chapter 1, six classes of nutrients are considered necessary in human nutrition: carbohydrates, fats (also called lipids), proteins, vitamins, minerals, and water. Within most of these general classes are a number of specific nutrients necessary for life. For example, more than a dozen vitamins are needed for optimal physiological functioning.

In relation to nutrition, the term **essential nutrients** describes nutrients that the body needs but cannot produce at all or cannot produce in adequate quantities. Thus, in general, essential nutrients must be obtained from the food we eat or through appropriate dietary supplements. Essential nutrients also are known as *indispensable nutrients*.

Table 2.1 lists the specific nutrients currently known to be essential to humans. Curing a nutrient-deficiency disease by providing that specific nutrient as part of the diet has been the key factor underlying the determination of nutrient essentiality. However, the concept of nutrient essentiality has evolved to include substances that may help prevent the development of chronic diseases. For example, most recently, the essentiality of choline was included with the B vitamin group.

Some foods, such as whole wheat bread, may contain all six general classes of nutrients, whereas others, such as table sugar, contain only one nutrient class. However, whole wheat bread cannot be considered a "complete food" because it does not contain a proper balance of all essential nutrients.

The human body requires substantial amounts of some nutrients, particularly those that may provide energy and support growth and development of the body tissues, namely carbohydrate, fat, and protein. These nutrients are referred to as **macronutrients** because the body needs them in relatively large amounts. Most nutrients that help to regulate metabolic processes, particularly vitamins and minerals, are needed in much smaller amounts (usually measured in milligrams or micrograms) and are referred to as **micronutrients**. You will notice that water is classified in neither the macronutrient nor the micronutrient category because, while water is needed in large amounts, the nutrient does not provide energy.

Essential nutrients are necessary for human life. An inadequate intake may lead to disturbed body metabolism, certain disease

TABLE 2.1 Nutrients essential or probably essential to humans

Carbohydrates

Glucose

Lipids (fats)

Linoleic acid
Alpha-linolenic acid

Proteins (essential amino acids)

Histidine	Phenylalanine
Isoleucine	Threonine
Leucine	Tryptophan
Lysine	Valine
Methionine	

Vitamins

Water soluble	Fat soluble
B1 (thiamin)	A (retinol)
B2 (riboflavin)	D (calciferol)
Niacin	E (tocopherol)
B6 (pyridoxine)	K
Pantothenic acid	
Folic acid (folate)	
B12 (cyanocobalamin)	
Biotin	
C (ascorbic acid)	
Choline*	

Minerals

Major	Trace	Possibly essential
Calcium	Chromium	Arsenic
Chloride	Fluoride	Boron
Magnesium	Copper	Lithium
Phosphorus	Iodine	Nickel
Potassium	Iron	Silicon
Sodium	Manganese	Vanadium
Sulfur	Molybdenum	
	Selenium	
	Zinc	

Water

*Technically not classified as a vitamin (see chapter 7).

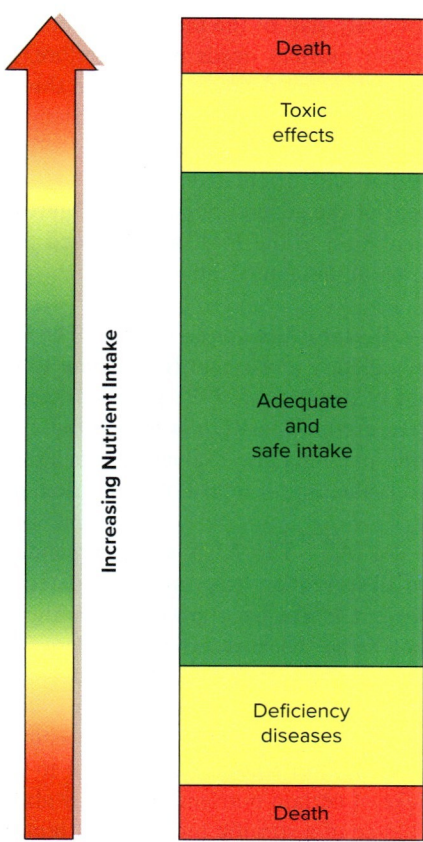

FIGURE 2.1 One model of the possible relationship between nutrient intake and health status. An inadequate intake may lead to nutrient-deficiency diseases, while excessive amounts may cause various toxic reactions. Both deficiencies and excesses can be detrimental to health and, in some cases, may lead to death.

states, or death. Conversely, an excess of certain nutrients may also disrupt normal metabolism and may even be lethal (**figure 2.1**).

What are nonessential nutrients?

Those nutrients found in food but that also may be formed in the body are known as **nonessential nutrients**, or dispensable nutrients. A good example of a nonessential nutrient is creatine. Although we may obtain creatine from food, the body can also manufacture creatine from several amino acids when necessary. Thus, we need not necessarily consume creatine but must consume adequate amounts of the amino acids from which creatine is made. As we shall see later, creatine combined with phosphate is a very important nutrient for energy production during very high intensity exercise.

Other than nonessential nutrients, foods contain nonessential substances that may be involved in various metabolic processes in the body. These substances, sometimes referred to as *non-nutrients,* include those found naturally in foods and those added either intentionally or inadvertently during the various phases of food production and preparation. Classes of these substances include phytochemicals, extracts, herbals, food additives, drugs, and even antinutrients (substances that may adversely affect nutrient status). Many of these nonessential nutrients and other substances are marketed as a means to enhance health or sports performance. Nielsen discusses the issue of whether some nonessential nutrients,

Chayapon Bootboonneam/123RF

particularly various phytochemicals found in plants, may eventually be classified as essential nutrients if they are found to provide significant health benefits. Carotenoids, isoflavonoids, and quercetin are examples of phytochemicals that may have positive impacts on health. Fruits, vegetables, green and black tea, and soybeans are examples of foods that contain these phytochemicals. The image in this section shows workers harvesting green tea leaves in a tea plantation.

How are recommended dietary intakes determined?

In the United States, the Food and Nutrition Board, Institute of Medicine, of the National Academy of Sciences has established recommendations for the amounts of each nutrient that should be consumed. The first set of recommendations, known as the *Recommended Dietary Allowances (RDAs),* was published in 1941 and revised periodically over the years. In general, scientists with considerable knowledge of a specific nutrient met to evaluate the totality of scientific data concerning the need for that nutrient in the diet. Based on their analysis, specific dietary intake recommendations were made.

The early RDAs were established at levels appropriate to prevent deficiency diseases. For example, the RDA for vitamin C was set to prevent scurvy. However, more recently, scientists have discovered that higher amounts of some nutrients may confer specific health benefits. Therefore, newer recommendations take into account the importance of vitamin C in optimizing health of the immune system and preventing cancer, not just preventing scurvy. However, can a person consume too much? Definitely! And, for this reason, scientists have also established recommendations for upper limits of intake. These upper limits are particularly important given research to suggest that over half of adults in the United States take a multivitamin-multimineral.

Expert scientists still meet to evaluate the available scientific data that serve as the basis for dietary recommendations, but the basis for such recommendations has been expanded beyond the objective of simply preventing deficiency diseases. The current philosophy relative to the development of recommendations for dietary intake focuses on a continuum of nutrient intake. Several points along this continuum for a specific nutrient may be (1) the amount that prevents a nutrient-deficiency disease, (2) the amount that may reduce the risk of a specific health problem or chronic diseases, and (3) the amount that may increase health risks.

Based on this concept, the Food and Nutrition Board of the Institute of Medicine developed new standards for nutrient intake, the **Dietary Reference Intakes (DRIs)**, for Americans. As shown

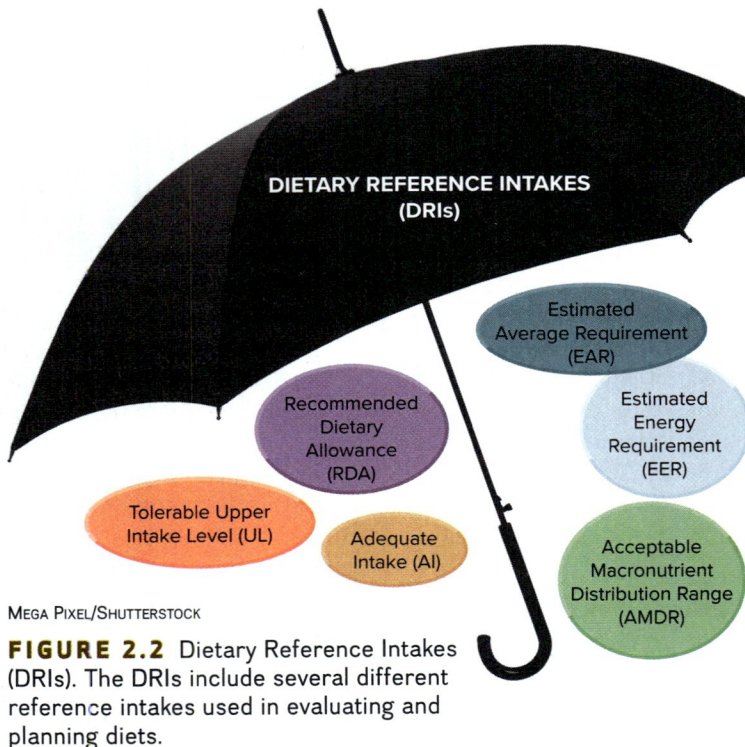

FIGURE 2.2 Dietary Reference Intakes (DRIs). The DRIs include several different reference intakes used in evaluating and planning diets.

in **figure 2.2**, *DRI* is an umbrella term, consisting of various reference intakes, including the **Estimated Average Requirement (EAR), Estimated Energy Requirement (EER), Acceptable Macronutrient Distribution Range (AMDR), Recommended Dietary Allowance (RDA), Adequate Intake (AI),** and **Tolerable Upper Intake Level (UL)**. The following sections summarize each of these DRI standards.

Estimated Average Requirement (EAR) The EAR represents a nutrient intake value that is estimated to meet the requirement of half the healthy individuals in a life-stage/sex group (e.g., 20-year-old males, or 35-year-old females who are not pregnant or lactating). Conversely, half of the individuals consuming the EAR will not meet their nutrient needs and, therefore, EARs are not used as nutrient recommendations. Rather, the EAR is used to establish the RDA for a particular nutrient. Depending on the data available, the RDA is some multiple of the EAR, mathematically calculated to provide adequate amounts to 97-98 percent of the general population. **Figure 2.3** provides a graphic representation of the EAR as compared to the RDA (or AI) recommendation for a nutrient.

Recommended Dietary Allowance (RDA) The RDA represents the average daily dietary intake that is sufficient to meet the nutrient requirement of nearly all (97-98 percent) healthy individuals in a life-stage/sex group. The RDA is to be used as a goal for the individual in diet planning and evaluation. For example, you may use the RDA to evaluate your intake of a specific nutrient.

Adequate Intake (AI) The AI is a recommended daily intake level based on observed or experimentally determined approximations of nutrient intake by a group of healthy people. When an RDA cannot be set because extensive scientific data are not available, an AI may be established instead based on the recommendations of scientific experts. You might be surprised to learn that sodium and vitamin K are nutrients that, as of 2022, have an AI, instead of RDA. Like RDAs, AIs are used to plan and evaluate dietary intake of certain nutrients. You may also use the AI to evaluate your intake of a specific nutrient, but remember that it is not as well established as the RDA.

Tolerable Upper Intake Level (UL) The UL is the highest level of daily nutrient intake that is likely to pose no risks of adverse health effects to most individuals in the general population. The UL is given to assist in advising individuals what levels of intake may result in adverse effects if habitually exceeded. The UL is not intended to be a recommended dietary intake; you should consider it as a maximum for your daily intake of a specific nutrient on a long-term basis. The UL for nutrients is cited throughout this book when data are available.

Table 2.2 provides an example of the RDA/AI and UL recommendations of some key vitamins and minerals for a 20-year-old female (not pregnant or lactating). You can look up your individual recommendations using the DRI tables available at www.nal.usda.gov/fnic/dietary-reference-intakes.

Estimated Energy Requirement (EER) The EER is an estimate of the amount of energy needed to sustain requirements for daily physical activity. The EER is covered in detail in chapter 3.

Acceptable Macronutrient Distribution Range (AMDR) The AMDR is defined as a range of intakes for a particular energy source that is associated with reduced risk of chronic disease while

> www.nal.usda.gov/fnic/dietary-reference-intakes You may access the DRI, including RDA, AI, AMDR, and UL for all age groups. Click on Dietary Reference Intakes: Recommended intakes for individuals.

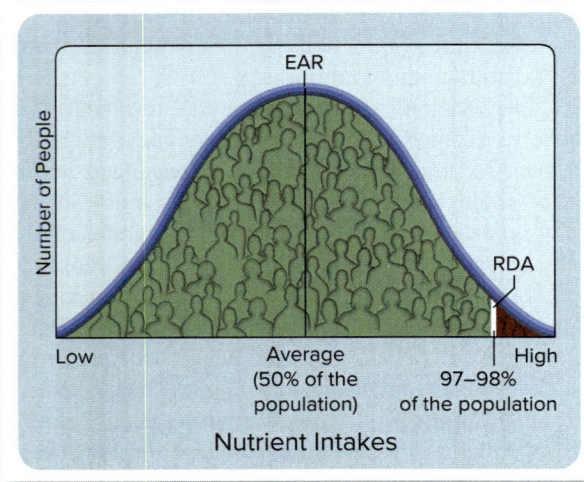

FIGURE 2.3 Establishing RDAs.

TABLE 2.2 Recommended intake of select vitamins and minerals for a 20-year-old female (not pregnant or lactating)

Micronutrient	Recommended Intake (RDA or AI)	Tolerable Upper Intake Level (UL)
Vitamin A	700 mcg	3,000 mcg
Vitamin D	15 mcg	100 mcg
Vitamin E	15 mg	1,000 mg
Vitamin K	90 mcg*	Not determined
Vitamin B6	1.3 mg	100 mg
Vitamin B12	2.4 mcg	Not determined
Folate	400 mcg	1,000 mcg
Niacin	14 mg	35 mg
Vitamin C	75 mg	2,000 mg
Calcium	1,000 mg	2,500 mg
Potassium	4,700 mg*	Not determined
Sodium	1,500 mg*	2,300 mg

*AI established for this nutrient.
Source: U.S. Department of Agriculture: *DRI Tables and Application Reports*. www.nal.usda.gov/fnic/dri-tables-and-application-reports. Accessed September 24, 2022.

providing adequate intakes of essential nutrients. The AMDR is expressed as a percentage of total energy intake and has both an upper level and a lower level. Individuals consuming below or above this range are at more risk for inappropriate intake of essential nutrients and development of chronic diseases. AMDRs have been set for carbohydrate, fat, and protein. **Table 2.3** shows AMDR recommendations for adults, and the *Training Table* in this section shows how the recommendations can be used in diet planning.

TABLE 2.3 Acceptable Macronutrient Distribution Ranges (AMDRs) for adults

	Percent of total kcal
Carbohydrate	45–65
Fat	20–35
Protein	10–35

Source: Adapted from Otten, J. J., et al. 2006. *Dietary Reference Intakes: The Essential Guide to Nutrient Requirements*. Institute of Medicine of the National Academies. Washington, DC: National Academies Press.

Training Table

The AMDRs can be used to plan and evaluate a person's diet. To calculate how many grams of carbohydrate, fat, and protein an adult should consume, you first need to know the number of kcal recommended. Then, using the AMDR ranges, determine the percentage of kcal to apply toward carbohydrate, fat, and protein. Keep in mind that the percentages must add up to 100%.

For this example, we will assume that a healthy adult needs **2,400 daily kcal** and wants to consume 55% of kcal from carbohydrate, 30% of kcal from fat, and 15% of kcal from protein.

Carbohydrate:
2,400 kcal × 0.55 = 1,320 kcal
1,320 kcal ÷ 4 kcal per gram = 330 grams of carbohydrate

Fat:
2,400 kcal × 0.30 = 720 kcal
720 kcal ÷ 9 kcal per gram = 80 grams of fat

Protein:
2,400 kcal × 0.15 = 360 kcal
360 kcal ÷ 4 kcal per gram = 90 grams of protein

For this example, we could recommend that an adult who consumes 2,400 kcal should consume 330 grams of carbohydrate, 80 grams of fat, and 90 grams of protein.

If this person was exercising regularly, we might consider increasing the protein slightly, to perhaps 20% of kcal from protein. By increasing protein by 5%, we would have to reduce the percent of kcal from carbohydrate or fat by 5%.

Key Concepts

▶ Balance, variety, and moderation are three key aspects of a healthy diet.
▶ The principal purposes of the nutrients we eat are to provide energy, build and repair body tissues, and regulate metabolic processes in the body.
▶ More than 40 specific nutrients are essential to life processes. They may be obtained in the diet through consumption of the six major nutrient classes: carbohydrates, fats, proteins, vitamins, minerals, and water.
▶ The Dietary Reference Intakes (DRIs) provide us with a set of standards for our nutritional needs and have been developed with the goal of promoting optimal health.

Check for Yourself

▶ Use the interactive DRI toolbox at www.nal.usda.gov/fnic/interactiveDRI/ to determine your personal DRIs for the macronutrients, vitamins, and minerals.

The Balanced Diet and Nutrient Density

One of the major concepts advanced by nutritionists over the years to teach proper nutrition is that of the balanced diet, one containing a variety of foods that provide us with the wide range of nutrients essential to life. Early food guides, in the 1940s, had little focus on healthful nutrition, mainly because the role of nutrition in chronic disease prevention was not a major research focus. However, as such research developed in the 1950s, food guides began to evolve and promoted healthier eating practices. In accordance with this concept the AND, in its recent position statement on the total diet approach to healthy eating, indicated the diet should focus on variety, moderation, and proportionality in the context of a healthy lifestyle.

What is a balanced diet?

As noted previously, the human body needs more than 40 different nutrients to function properly. The concept of the balanced diet is that by eating a wide variety of foods in moderation you will obtain all the nutrients you need to support growth and development of all tissues, regulate metabolic processes, and provide adequate energy for proper weight control. You should obtain the RDA or AI for all essential nutrients and adequate food energy to achieve a healthy body weight.

Although everyone's diet requires the essential nutrients and adequate energy, the proportions differ at different stages of the life cycle. The infant has needs differing from those of his grandfather, and the pregnant or lactating female has needs differing from those of her adolescent daughter. There also are differences between the needs of males and females, particularly in regard to the iron content of the diet. Moreover, individual variations in lifestyle may impose different nutrient requirements. A long-distance runner in training for a marathon has some distinct nutritional needs compared to a sedentary peer. The individual trying to lose weight while building lean body mass needs to balance kcal losses with nutrient adequacy. And, a person with diabetes needs to control carbohydrate intake while maintaining a balanced diet. Thus, a number of different conditions may influence nutrient needs and the concept of a balanced diet.

The food supply in the United States is extremely varied, and most individuals who consume a wide variety of foods do receive an adequate supply of nutrients. However, there appears to be some concern that many Americans are not receiving optimal nutrition because they consume excessive amounts of highly processed foods. In general, Americans consume more than recommended amounts of sugar-sweetened beverages, high-fat dairy foods (e.g., cheese), red meat, and refined grains. At the same time, many Americans do not eat recommended amounts of whole grains, low-fat dairy, fish, and/or fruits and vegetables.

An unbalanced diet is due not to the unavailability of proper foods but rather to our choice of foods. To improve our nutritional habits we need to learn to select our foods more wisely.

What foods should I eat to obtain the nutrients I need?

Although the RDA, AI, and AMDR recommendations provide us with information relative to the nutrients we need, they don't guide us in appropriate food selection. Thus, over the years a number of educational approaches have been used to convey the concept of a balanced diet to help individuals select foods that will provide sufficient amounts of all essential nutrients. **Dietary guidance systems**, also called food guides, provide practical ways for consumers to select certain foods and food components to promote health and meet dietary recommendations. Such food guides help consumers to plan and evaluate their diet using a relatively simplistic approach.

In the United States, the USDA develops and distributes dietary guidance systems. The first USDA food guide was published in 1943 when the USDA introduced the "Basic 7." The Basic 7 included food groups for green and yellow vegetables; oranges, tomatoes, and grapefruit; milk and milk products; meat, poultry, fish, or eggs; bread, flour, and cereals; and butter and fortified margarine. At the time, the guide was developed to provide nutrition standards under World War II food rationing. In 1956, the "Basic 4" replaced the Basic 7, simplifying foods into four simple groups: vegetables and fruits; milk; meat; and cereals and breads. The Basic 4 was utilized for nearly 20 years, until the development of the Food Guide Pyramid in 1992. The Food Guide Pyramid was controversial and was quickly replaced in 2005 with MyPyramid. Some consumers and health professionals found MyPyramid difficult to use, and that led to the establishment of the current food guide in the United States, MyPlate.

What is the MyPlate food guide?

MyPlate is a comprehensive program including many features to help promote health through proper nutrition and physical activity. The MyPlate graphic (see figure 1.9) depicts the distribution of fruits, vegetables, grains, protein foods, and dairy as part of a typical meal. In addition to this simplistic representation of healthy eating, **figure 2.4**

FIGURE 2.4 The MyPlate recommendations provide guidance on how to "make every bite count" as part of a healthy eating routine.
U.S. Department of Agriculture (USDA)

provides additional recommendations for each food group. Visit www.myplate.gov for additional resources, including guidance specific for certain stages of the life cycle, such as children, adolescents, and older adults. There are even resources specific to college students!

The MyPlate guidelines focus on five different food groups: fruits, vegetables, protein foods, grains, and dairy (or fortified dairy alternatives). Consumers are provided with basic advice when selecting food options from each food group. For example, when selecting foods from the "grain" food group, make at least half of those grains whole grains. As well, limit intake of fruit juices and focus on whole fruits.

In addition, the MyPlate website provides nutritionally adequate daily food plans, each plan supplying between 1,000 and 3,200 daily kcal. The plans provide recommendations for how much food to consume from each of the food groups. For example, 3 cups of vegetables or 6 ounces of protein foods. **Table 2.4** provides a sample MyPlan for a person consuming a 2,000 kcal diet.

Using the MyPlan recommendations, consumers can plan nutritionally adequate diets meeting each of the food group needs. Once those are met, people can "make up" the remaining kcal by selecting foods that are sources of empty kcal. **Empty kcal** are those kcal provided by added sugars, solid fats, and, for those who consume alcohol, alcohol-containing beverages. The most common sources of empty kcal foods in the typical American diet are sugar-sweetened beverages such as sodas, energy drinks, sports drinks, and fruit juices; "dessert-type" foods such as cakes, cookies, and pastries; cheese and pizza; ice cream; and meats high in saturated fat such as bacon, ribs,

TABLE 2.4 Sample MyPlan recommendations, 2,000-kcal diet

	Key tip	Recommended amount to be consumed from food group	Sample serving equivalent
Grains 	Eat at least half of all grains as whole grains.	6 ounces	**1 ounce of grains:** 1 slice of bread ½ cup cooked pasta, rice, or cereal 1 tortilla 1 pancake 1 cup ready-to-eat cereal
Vegetables 	Select a variety of vegetables from the five subgroups: dark green, red and orange, beans and peas, starchy, and other.	2½ cups	**1 cup of vegetables:** 1 cup raw or cooked vegetable 1 cup 100% vegetable juice 2 cups leafy salad greens
Fruits 	Select fresh, frozen, canned, and dried fruit more than juice.	2 cups	**1 cup of fruit:** 1 cup raw or cooked fruit 1 cup 100% fruit juice ½ cup dried fruit
Dairy 	Choose fat-free or low-fat milk or yogurt more than cheese.	3 cups	**1 cup of dairy:** 1 cup milk (or fortified milk alternative) 1 cup yogurt 1½ ounces natural cheese 2 ounces processed cheese
Protein foods 	Eat a variety of protein-rich foods, selecting lean options.	5½ ounces	**1 ounce of protein foods:** 1 ounce lean meat, poultry, or seafood 1 egg 1 Tbsp peanut butter ½ ounces nuts or seeds ¼ cup cooked beans or peas

Pancake Stack: kellyschulz/123RF; Fresh salad: Kudryashova Alla/Shutterstock; Watermelon: Corbis; Dairy products: Oleksandra Naumenko/Shutterstock; Chicken on grill: Lew Robertson/Brand X Pictures/Getty Images

Source: U.S. Department of Agriculture. MyPlate. www.myplate.gov Accessed: September 24, 2022.

hot dogs, and sausage. Foods that are rich sources of empty kcal most often do not need to be entirely avoided, but should be consumed in moderation and only after MyPlate food group needs are met. The *Training Table* in this section provides more information on sources of empty kcal and how, for some people, these foods make up a significant proportion of the diet.

Training Table

Research suggests that one-third of Americans eat fast-food on any given day with adults ages 20–39-year-olds consuming fast-food most frequently. Fast-food consumption is highest at lunchtime, and most Americans consume fast-food one to three times per week. These numbers suggest that fast-food consumption is high in the United States, often leading to overconsumption of foods rich in empty kcal.

Food sources of empty kcal are typically those high in solid fats and/or added sugars. These types of foods should be consumed in moderation; if not, they may contribute to unwanted weight gain and increase risk for chronic disease.

Examples of foods with significant empty kcal:

- 16 oz Vanilla Frapppucino
  400 kcal, 10 g saturated fat, 57 g sugars
- Fast-Food Deluxe Burger
  590 kcal, 11 g saturated fat, 8 g sugars

Use an app or online website, such as https://fastfoodnutrition.org/fast-food-meal-calculator, to look up the nutrition information for some of your favorite fast-food items. Are they sources of empty kcal?

Brent Hofacker/Shutterstock

achieve improved health and wellness. For example, the *Training Table* in this section provides recommendations from the MyPlate resources for eating healthy on a budget.

 www.myplate.gov Provides evidence-based recommendations on diet and exercise.

Training Table

You CAN eat healthy on a budget. The MyPlate website has resources to help consumers to plan, purchase, and prepare healthy food and beverage items. Tips to eat healthy on a budget include:

- When possible, choose seasonal fruits and vegetables. Instead of buying pre-cut produce, purchase the whole fruit or vegetable because the pre-packaged items tend to be more expensive.
- Look for in-store specials on fresh chicken, fish, and lean cuts of meat. Buy in larger quantities, repackage into smaller portions, and freeze for future use.
- Purchase whole-grain cereals, rice, and similar types of grain foods in bulk at warehouse stores.
- Buy bags of frozen fruits and vegetables when budget-friendly fresh produce is not available.
- Legumes, such as black beans, pinto beans, and garbanzo beans tend to be low-cost options.
- Try to meal-prep for the week. Double or triple recipes and freeze meal-sized portions for later.

For more tips to eat healthy on a budget, visit www.myplate.gov/eat-healthy/healthy-eating-budget.

In addition to recommendations related to diet, the MyPlate guidelines also emphasize the importance of engaging in physical activity. MyPlate resources provide information about different types of exercises and recommendations for including physical activity as part of a healthy lifestyle.

Overall, the MyPlate resources can help support efforts to make changes to one's diet and level of physical activity, thereby enhancing personal health and reducing risk for chronic disease. While there are many barriers to healthy eating and exercise, such as time, money, and resources, you can take steps to

How has the COVID-19 pandemic impacted dietary practices?

The COVID-19 pandemic has had a significant impact on the lives of those in the United States as well as across the globe. Since the onset of the pandemic in early 2020, there have been disruptions to the food system and changes to dietary habits. As well, as noted by Boaz and colleagues, the pandemic has been associated with increased anxiety, poor quality dietary intake, and unhealthy weight gain. Rates of food insecurity have increased during the pandemic, with many households having less money available for food than prior to the pandemic.

Access to food has also been impacted, particularly early on in the pandemic when many restaurants were forced to shut down or take to-go orders only and supermarkets limited in-person shopping. This resulted in more consumers preparing meals at home, some of

Gorodenkoff/Shutterstock

which were healthy, but many relying on nonperishable foods due to limited access to fresh foods, including fruits and vegetables. The pandemic has changed the hospitality industry, with more restaurants now offering curb-side service and meal delivery services on the rise. In January 2022, approximately one-half of consumers in the United States reported having ordered from a meal delivery service.

Resources available on the MyPlate website provide recommendations for food planning during a pandemic — www.myplate.gov/eat-healthy/healthy-eating-budget/covid-19.

What is the key-nutrient concept for obtaining a balanced diet?

As already noted, humans require many diverse nutrients, including 20 amino acids, 13 vitamins, and more than 15 minerals. To plan our daily diet to include all of these nutrients would be mind-boggling, so simplified approaches to diet planning have been developed.

The nutritional composition of foods varies tremendously. The USDA website at http://fdc.nal.usda.gov/nhb/ provides a detailed analysis of foods, including nearly 120 nutrients and food components. If you examine a food-composition table, you will quickly see that no two foods are exactly alike in nutrient composition. However, certain foods are similar enough in nutrient content to be grouped accordingly. As described in this chapter, this fact is the basis for approaching nutrition education with MyPlate.

Eight nutrients are particularly central to human nutrition: protein, thiamin, riboflavin, niacin, vitamins A and C, iron, and calcium. When found naturally in plant and animal sources, these nutrients are usually accompanied by other essential nutrients. The central theme of the **key-nutrient concept** is simply that if these eight key nutrients are adequate in your diet, you will probably receive an ample supply of *all* nutrients essential to humans. It is important to note that for the key-nutrient concept to work, you must obtain the nutrients from a wide variety of minimally processed whole foods. For example, highly processed foods to which some vitamins have been added will not contain all of the trace elements, such as chromium, that were removed during processing.

Table 2.5 presents the eight key nutrients and some significant plant and animal sources. You can see that the food groups can be a useful guide to securing these eight key nutrients. Keep in mind, however, that there is some variation in the proportion of the nutrients, not only between the food groups but also within

TABLE 2.5 Eight key nutrients and significant food sources from plants and animals

Nutrient	RDA or AI		Plant source	Animal source	Food group
	M	F			
Protein	58 g	46 g	Dried beans and peas, nuts	Meat, poultry, fish, cheese, milk	Protein foods, dairy
Vitamin A	900 mcg	700 mcg	Dark-green leafy vegetables, orange-yellow vegetables, margarine	Butter, fortified milk, liver	Vegetables, fruits, protein foods, dairy
Vitamin C	90 mg	75 mg	Citrus fruits, broccoli, potatoes, strawberries, tomatoes, cabbage, dark-green leafy vegetables	Liver	Fruits, vegetables
Thiamin	1.2 mg	1.1 mg	Breads, cereals, pasta, nuts	Pork, ham	Grains, protein foods
Riboflavin	1.3 mg	1.1 mg	Breads, cereals, pasta	Milk, cheese, liver	Grains, dairy
Niacin	16 mg	14 mg	Breads, cereals, pasta, nuts	Meat, fish, poultry	Grains, protein foods
Iron	8 mg	18 mg	Dried peas and beans, spinach, breads, cereals	Meat, liver	Protein foods, vegetables, grains
Calcium	1,000 mg	1,000 mg	Turnip greens, okra, broccoli, spinach, kale	Milk, cheese, sardines, salmon	Dairy, protein foods, vegetables

Recommended Dietary Allowance (RDA) or Adequate Intake (AI) for males (M) and females (F) age 19–50.

Source: U.S. Department of Agriculture: *DRI Tables and Application Reports*. www.nal.usda.gov/fnic/dri-tables-and-application-reports. Accessed September 24, 2022.

each food group. For example, the grain food group does contain some protein, but it is not as good a source as the protein foods or dairy groups. Within the fruit group, oranges are an excellent source of vitamin C, but peaches are not, although peaches are high in vitamin A. If you select a wide range of foods within each group, the nutrient intake should be balanced over time.

What is the concept of nutrient density?

As mentioned previously, the nutrient content of foods varies considerably, and the differences between food groups are more distinct than the differences between foods in the same group. **Nutrient density** is an important concept relative to the proportions of essential nutrients such as protein, vitamins, and minerals that are found in specific foods. In essence, a food with high nutrient density possesses a significant amount of a specific nutrient or nutrients per serving compared to its caloric content.

Let's look at an extreme example between two different food groups. Consider the nutrient differences between 6 ounces of baked yellowfin tuna and 6 strips of fried bacon, each containing about 220 kcal. The tuna fish would provide a young adult female with 100 percent of her requirement for two key nutrients (protein and niacin) along with substantial amounts of several other vitamins, and minerals, but very little fat. The bacon would contain less than 25 percent of the protein requirement and about 10 percent of the niacin requirement, with greater amounts of total and saturated fat. Hence, the tuna fish has greater nutrient density and considerably greater nutritional value.

Let's also look at a comparison of two types of seafood. Consider the following nutritional data for 3 ounces of fresh yellowfin tuna and 3 ounces of raw Eastern oysters:

	kcal	Protein	Iron
3 oz. yellowfin tuna	93	22 g	0.65 mg
3 oz. Eastern oysters	43	5 g	3.87 mg

Per serving, both of these seafood choices are relatively low in kcal and good sources of protein and iron. Although the tuna contains more than twice the kcal as oysters per serving, it also contains more than four times the amount of protein. And, although the tuna is a relatively good source of iron, the oysters contain nearly 6 times as much iron per serving and more than 12 times as much based on kcal. Another example is presented in **figure 2.5**, which compares the nutrient density of cow's milk and sugar-sweetened cola. These examples illustrate the need to consume a wide variety of foods among food groups and within each food group to satisfy your nutrient needs.

Will using the MyPlate food guide guarantee me optimal nutrition?

If you use the key-nutrient and nutrient density concepts, MyPlate may be an effective means to obtain optimal nutrition and help sustain a healthful body weight. However, although the MyPlate food guide represents a significant improvement over previous food

FIGURE 2.5 Comparison of the nutrient density of a sugar-sweetened cola soft drink with that of fat-free cow's milk. Both contribute fluid to the diet. However, choosing a glass of fat-free cow's milk makes a significantly greater contribution to nutrient intake in comparison with a sugar-sweetened soft drink. An easy way to determine nutrient density is to see how many of the nutrient bars in the graph are longer than the energy (kcal) bar. The soft drink has no long nutrient bars. Fat-free cow's milk has long nutrient bars for protein, vitamin A, thiamin, riboflavin, and calcium. Including many nutrient-dense foods in your diet aids in meeting nutrient needs.

guides to help ensure proper nutrition, flaws may exist if foods are not selected carefully. For example, individuals who predominantly choose the foods high in solid fats and added sugars from among the food lists may be more susceptible to the development of chronic health problems.

Developed by scientists and clinicians at Harvard University, the Healthy Eating Plate is one of many alternatives to MyPlate for use in diet planning. Compared to MyPlate, the Healthy Eating Plate (figure 2.6) puts more emphasis on whole grains, healthy proteins, healthy oils, and drinking water right on the "plate" itself. Other examples of meal planning tools include the Mediterranean Food Guide and the Dietary Approaches to Stop Hypertension (DASH) diet. Overall, most healthy eating plans attempt to reduce one's consumption of and/or modify the type of fat, to increase the consumption of whole-grain products, and to increase the consumption of plant products, particularly beans and other legumes, fruits, and vegetables.

www.health.harvard.edu/plate/healthy-eating-plate The Harvard Medical School modified MyPlate to present more specific recommendations for healthy eating.

Key Concepts

- If most healthy individuals in a given population consume wholesome, natural foods in amounts adequate to meet their RDAs, there will be very little likelihood of nutritional inadequacy or impairment of health.
- The MyPlate food guide should be viewed as an educational approach to help individuals obtain proper nutrition. Foods are grouped as grains, protein foods, fruits, vegetables, dairy, and oils, and recommendations for energy intake are provided. Individualized plans are available at www.myplate.gov.
- According to the key-nutrient concept, there are eight key nutrients (protein, vitamin A, thiamin, riboflavin, niacin, vitamin C, calcium, and iron) that, if adequate in the diet and obtained from wholesome foods, should provide an ample supply of all nutrients essential to human nutrition.
- Some foods contain a greater proportion of these key essential nutrients than other foods and thus have a greater nutrient density or nutritional value.

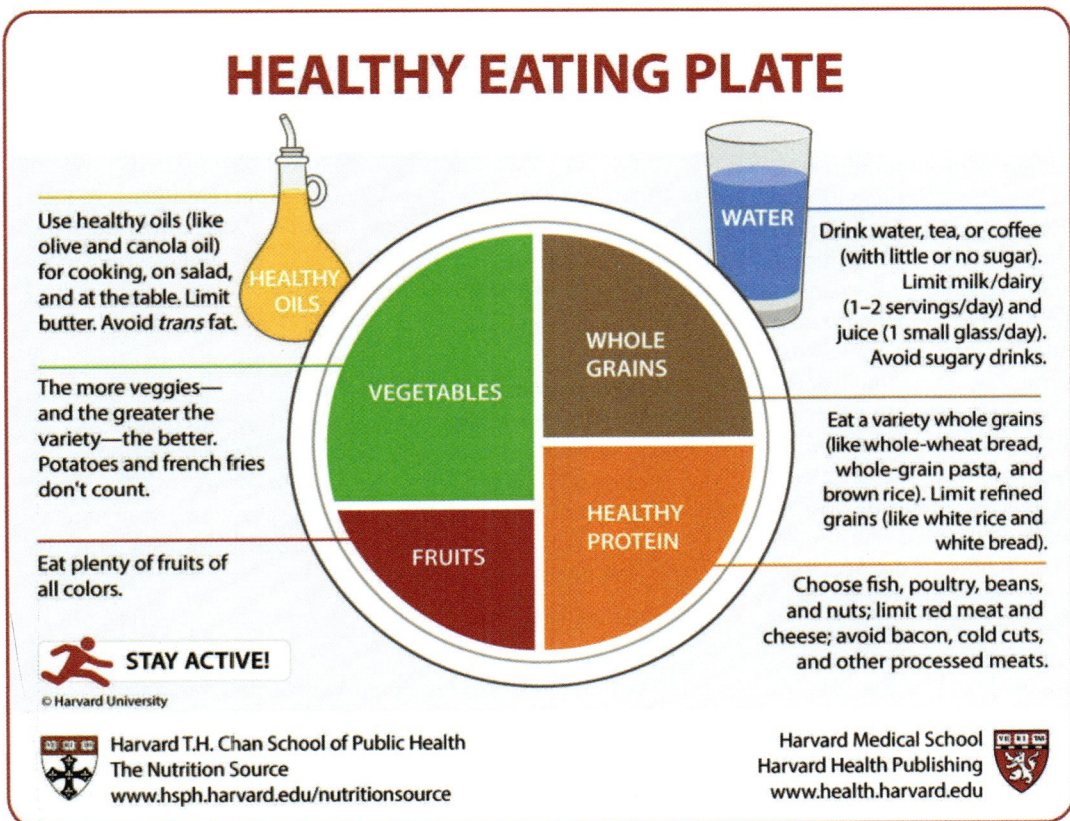

FIGURE 2.6 Healthy Eating Plate.

Source: www.health.harvard.edu/plate/healthy-eating-plate.

> **Check for Yourself**
>
> ▶ Use a diet analysis program, such as NutritionCalc Plus, to create a profile for yourself or a friend/family member. Then, using the recommendations provided, develop a one-day ideal meal plan that meets the requirements for each of the food groups and total kcal. Were you able to meet your food group needs and include any foods or beverages that are considered sources of empty kcal? If so, what foods did you add?

Healthful Dietary Guidelines

In past centuries, most morbidity and mortality in industrialized nations were caused by nutrient-deficiency diseases and infectious diseases. Advances in nutritional and medical science have almost eliminated most of the adverse health consequences associated with these diseases. Today, morbidity and mortality are often associated with chronic diseases (e.g., coronary artery disease, stroke, cancer, diabetes, osteoporosis, obesity), and most dietary guidelines for healthful nutrition are targeted to prevent these chronic diseases.

Nutrition scientists use both epidemiological and experimental research in attempts to determine what types of diet, specific foods, and specific nutrients or food constituents may either cause or prevent the development of chronic diseases.

What is the basis underlying the development of healthful dietary guidelines?

In general, healthful dietary guidelines are based on appropriate research. Over the years, epidemiologists have attempted to determine the relationship between diet and the development of chronic diseases. In early research, the focus was simply on the overall diet and its relationship to disease, such as comparing the typical American diet to the Mediterranean (Greece, Italy, Spain) or Japanese diet. If a significant relationship was found between the diets of two nations, for example, more heart disease among Americans compared to those consuming the Mediterranean diet, scientists then attempted to determine what specific foods, particularly which macronutrients (carbohydrate, fat, and protein) in those foods, may have been related to either an increased or a decreased risk for heart disease. In more recent years, scientists have been investigating the roles of specific nutrients or food constituents and their potential to prevent or deter chronic diseases.

Based on the evaluation of current research findings, nutritional scientists believe that the development of most chronic diseases may be associated with either deficiencies or excesses of various nutrients or food constituents in the diet. Many Americans eat more food than they need, yet eat less of the types of foods their bodies need.

To help prevent chronic diseases, numerous governmental and professional health organizations have developed general dietary guidelines for good health. Some of these guidelines have been criticized, possibly because they were not based on the best science. Most of the early research relating diet to disease was epidemiological in nature, which may have led to some erroneous dietary recommendations. However, experimental studies involving randomized controlled trials (RCTs) have predominated in recent years with rigorous procedures involving evidence-based approaches currently being used to develop dietary guidelines to promote health. Nevertheless, as noted in chapter 1, some dietary recommendations based on research, even using meta-analysis of numerous studies, may be the subject of debate.

Although there is no absolute proof that dietary changes will enhance the health status of every member of the population, scientists involved in the development of healthful dietary guidelines believe that they are *prudent* recommendations for most individuals and are based on the available scientific evidence.

What are the recommended dietary guidelines for reducing the risk of chronic disease?

In chapter 1 we introduced the concept of the Prudent Healthy Diet, a dozen guidelines to healthier eating. These prudent dietary recommendations represent a synthesis of various recent reports from governmental, educational, and professional health organizations. These recommendations also align with the *2020–2025 Dietary Guidelines for Americans.* Key aspects of the dietary guidelines are provided in the *Training Table* in this section. The Prudent Healthy Diet guidelines are not considered to be static and may be modified somewhat as we gain more knowledge through research. In particular, healthful dietary recommendations may be individualized in the future as scientists and healthcare providers learn more about the role of individual differences in genetic make-up on how the body metabolizes different nutrients.

> **Training Table**
>
> The *2020–2025 Dietary Guidelines for Americans* provide healthy eating recommendations with an emphasis on the fact that dietary components are interconnected, and we eat foods rather than individual nutrients. According to the guidelines, a healthy eating pattern includes:
>
> - A variety of vegetables from all of the subgroups: dark green, red and orange, legumes, starchy, and others
> - Fruits, emphasizing whole fruits
> - Grains, with at least 50% which are whole grains
> - Low-fat or fat-free dairy foods such as milk, yogurt, cheese, and/or fortified dairy alternatives
> - A variety of protein foods, including seafood, lean meats and poultry, eggs, legumes, nuts, seeds, and soy products
> - Healthy oils
>
> At the same time, saturated fats, *trans* fats, added sugars, and sodium should be limited.
>
> - Less than 10% of daily kcal from added sugars
> - Less than 10% of daily kcal from saturated fats
> - Less than 2,300 mg of daily sodium
>
> If alcohol is consumed, it should be consumed in moderation. Moderation is defined as up to one drink per day for adult females and up to two drinks per day for adult males.

Taken together, these recommendations may be helpful in preventing most chronic diseases, including cardiovascular diseases and cancer. The rationale as to how these dozen healthful dietary recommendations may promote good health is presented in later chapters where appropriate. These guidelines do, however, come with several caveats.

Remember, diet is only one factor that may influence the development of chronic diseases. As noted by the American Heart Association (AHA), other positive lifestyle behaviors, such as exercise, avoiding tobacco use, and managing stress, are also important.

The Mediterranean diet has been the focus of increased research, and results suggest significant health benefits to following this diet. Evidence suggests that following a Mediterranean diet could increase secretion of anti-inflammatory cytokines and antioxidant activity, which could counter diseases associated with chronic inflammation, including metabolic syndrome, atherosclerosis, cancer, diabetes, obesity, pulmonary diseases, and cognition disorders. The Mediterranean diet is a holistic dietary approach that emphasizes plant-based foods, olive oil, fresh fish, and red wine (in moderation, optional), while enjoying meals with family and friends. Some contend there is no such thing as an official Mediterranean diet, but it has certain characteristics. The Prudent Healthy Diet is based on these characteristics as well as other guidelines for healthful eating.

The Prudent Healthy Diet

1. *Balance the food you eat with physical activity to maintain or achieve a healthy body weight.* Consume only moderate food portions. Be physically active every day. Preventing obesity helps to reduce the risk of numerous chronic diseases, such as heart disease and cancer. Chapter 11 provides specific guidelines for maintaining or achieving a healthy body weight and body composition. An appropriate exercise program and adherence to the concept of nutrient density, which includes a number of the following recommendations, serves as the basis for a sound weight-control program.

2. *Eat a nutritionally adequate diet consisting of a wide variety of nutrient-rich foods.* Build a healthy base. Eating a wide variety of natural foods from within (and among) the MyPlate food groups will assure you of obtaining a balanced and adequate intake of all essential nutrients. Focus on foods that are nutrient dense, particularly those that are rich in the key nutrients.

    However, keep the concept of variety in perspective. Research has shown that the more food choices we have, the more likely we are to eat more. Our supermarket society, with more than 50,000 food products in the typical store, provides us with so many choices that it is very easy to gain unhealthy weight. Thus, although consuming a wide variety of foods is a good strategy to get the nutrients we need, we must keep in mind the first point, to maintain or achieve a healthy body weight.

Shutterstock/Yuliia Kononenko

Bartosz Luczak/Getty Images

FIGURE 2.7 The Prudent Healthy Diet emphasizes plant-based foods, including whole grains and vegetables. Vegetables, such as broccoli in this grain bowl, provide complex carbohydrates, phytonutrients, and fiber.

3. *Choose a plant-rich diet with plenty of fruits and vegetables, whole-grain products, and legumes, foods that are rich in complex carbohydrates, phytonutrients, and fiber* (**figure 2.7**). In general, healthy carbohydrates are the ones that nature makes and are found naturally in whole fruits, whole vegetables, and whole grains. About 45–65 percent or more of your daily kcal should come from carbohydrates, about 35–55 percent from complex carbohydrates, and the other 10 percent or less from simple, naturally occurring carbohydrates. To accomplish this, you need to eat more whole-grain breads, cereal, rice, and pasta; more legumes such as beans and peas; more fruits and vegetables; and fewer refined grain products, such as white bread, rice, and pasta.

    Numerous epidemiological studies have shown that diets rich in plant foods confer significant health benefits, which are attributable mainly to the complex mixture of phytonutrients and dietary fibers. The health benefits of phytonutrients and dietary fiber are discussed, respectively, in the next section, on plant-based diets, and in chapter 4.

    Health professionals recommend that 50 percent or more of our total grain intake should be whole grains, but currently Americans consume less than 15 percent. A study of more than 13,000 adults by O'Neil and colleagues showed that, on average, Americans consumed less than one serving of whole grains per day and that those who consumed the most whole grains had better diet quality and nutrient intakes. When buying cereal, bread, rice, and other grain products, ensure that they contain whole grains, preferably stating 100 percent whole grain on the food label (**figure 2.8**). Whole-grain products are rich in dietary fiber and phytonutrients as well as several key nutrients. Select whole wheat bread in which you can see the grain; whole-grain cereals; and whole-grain pasta. Check food labels that list grams of fiber per serving.

4. *Choose a diet moderate in total fat, but low in saturated (solid) fat, trans fat, and cholesterol.* For several decades, experts pushed

Mark Dierker/McGraw Hill

FIGURE 2.8 Look for "100 percent whole grain" on the label as that indicates the food is a source of fiber. The label on this loaf of bread says "100 percent whole wheat." Wheat is a type of grain, so this bread is a nutrient-dense choice.

low-fat diets, with little emphasis on the type of fats being consumed. More recent research suggests that we should focus on eating a diet moderate in fat, with an emphasis on "healthy" fats such as olive oil, canola oil, and the fats found naturally in avocados and nuts. The *Training Table* in this section provides additional advice for dietary fat intake. In a position statement authored by Vannice and Rasmussen, the AND, within the context of the rapidly evolving science relative to dietary fats and human health, made some recommendations. In general, the AND recommended the following, which it considered a prudent approach:

- 20–35 percent of kcal from fat
- Increased consumption of omega-3 polyunsaturated fatty acids
- Limited intake of saturated fats
- Limited intake of *trans* fat

The *Training Table* in this section reminds us that not all fats are "bad," and provides practical recommendations for limiting overall fat intake and emphasizing "healthy" sources of fat.

5. *Choose beverages and foods to moderate or reduce your intake of added sugars and highly refined carbohydrates.* In general, unhealthy carbohydrates include added sugars and highly refined white flour or rice found in such foods as bread, cereal, pasta, crackers, and pastry. Sugar is somewhat addictive because it may stimulate the same pleasure centers of the brain that respond to drugs, such as cocaine, and thus may be consumed in excess. In some aspects, refined carbohydrates may have effects comparable to sugar in human physiology. Research suggests that the typical American diet provides over 22 teaspoons daily, including sucrose, high-fructose corn syrup, and other sweeteners. On average, Americans age 2 years and older consume 22 percent of their total kcal from sugars.

 Training Table

The Prudent Healthy Diet recommendation "Choose a diet moderate in total fat, but low in saturated (solid) fat, trans fat, and cholesterol," emphasizes the importance of both the amount and type of fat consumed. This *Training Table* provides advice for limiting overall fat intake and eating "healthy" fats:

- Choose plant oils or other healthy oils. Fish, nuts, seeds, and vegetable oils, particularly olive and canola oil, are good options.
- Eat less meat with a high fat content. For example, limit intake of hot dogs, luncheon meats, red meat, sausage, and bacon.
- Eat more fish, particularly those rich in healthy omega-3 fatty acids. Salmon, tuna, and mackerel are good options. White fish, such as flounder, is low in fat and kcal. Children and pregnant women should limit intake of fish that may be polluted with mercury (chapter 5).
- Consume foods that contain cholesterol, such as eggs, in moderation.
- Eat fewer dairy products that are high in fat. For example, select low-fat milk instead of whole milk or low-fat cheese instead of regular cheese.
- Avoid foods that contain *trans* fats. The human body does not need *trans* fats and too much has been associated with an increased risk of chronic disease.
- Limit consumption of fast foods because many options are high in saturated fats. If you must eat at a fast-food restaurant, choose leaner options, such as baked fish, grilled skinless chicken, and salads with vegetable-based dressing (no bacon and cheese on the salad).
- Read food labels to evaluate not only the amount, but types of fats. Look at the ingredient list for "hidden" sources of fat.
- Limit frying of foods. When preparing meat, poultry, or fish, bake, broil, or grill the protein food.

The major sources of added sugars in the diet are soda, energy and sports drinks, grain-based desserts, sugar-sweetened fruit drinks, dairy-based desserts, and candy. According to the *2020–2025 Dietary Guidelines for Americans,* Americans should limit their added sugar intake to less than 10 percent of their total daily kcal. **Figure 2.9** provides an example of the amount of added sugars in a chocolate candy bar.

6. *Choose and prepare foods with less salt and sodium.* Restrict sodium intake to less than 1,500 milligrams daily. For the average healthy adult, this amount of sodium is sufficient for normal physiological functioning. Sodium is found naturally in a wide variety of foods, so it is not difficult to get an adequate

PKruger/Shutterstock

FIGURE 2.9 Consumers are advised to limit added sugars to less than 10 percent of their total daily kcal. A 1.6 ounce milk chocolate bar provides around 235 kcal, approximately 70 of those kcal from added sugars. As such, around 30 percent of the kcal in the candy bar are from added sugars.

supply. The *Training Table* in this section provides key suggestions to help you reduce the sodium content in your diet.

7. *Maintain protein intake at a moderate yet adequate level, obtaining much of your daily protein from plant sources, complemented with smaller amounts of fish, skinless poultry, and lean meats.* The recommended dietary intake is 0.8 gram of protein per kilogram body weight, which averages out to about 50–60 grams per day for the average adult male and somewhat less for the average adult female, or about 10 percent of daily kcal, which is about 100 kcal on a 2,000-kcal diet. The current recommendation of the National Academy of Sciences is to obtain 10–35 percent of daily kcal from protein. Since the average daily American intake of protein is about 100 grams, we appear to be staying within the guidelines. A healthy breakfast is a good way to get a jumpstart on daily protein intake. A

mackoflower/123RF

Training Table

©PM Images/Getty Images

According to recommendations of the American Heart Association, Academy of Nutrition and Dietetics, and associated professional organizations, sodium is a nutrient of public health concern given the large amounts consumed by many as part of the typical American diet. The following are suggestions from the Prudent Healthy Diet recommendation to "Choose and prepare foods with less salt and sodium."

- Get rid of your salt shaker and put less salt on your food, both in your cooking and on the table.
- Reduce consumption of obviously high-salt foods such as pretzels, potato chips, pickles, and other snack foods.
- Check food labels carefully for the sodium content.
- Eat more fresh fruits and vegetables, which are naturally very low in sodium.
- If you are using canned vegetables, rinse the vegetables before preparing to reduce the sodium content.
- Use fresh herbs such as cilantro, basil, and oregano in your cooking instead of salt.

The DASH Diet will be discussed in chapter 9.

bowl of high-protein cereal in a bowl of milk with a quarter cup of almonds on top, along with an egg, will provide about 30 grams of protein.

Much of the protein Americans eat is found in the meat from various animals. Meat is an excellent source of complete protein and, compared to plant foods, is a better source of dietary iron and other minerals such as zinc and copper. Three to six ounces of lean meat, fish, or poultry, together with two glasses of cow's milk, will provide the average individual with the daily RDA for protein, totaling about 40–60 grams of high-quality protein. Combining this animal protein intake with plant foods high in protein, such as whole-grain products, seeds, beans and peas, and vegetables, will substantially increase your protein intake and more than meet your needs. The photo in this section shows an authentic Korean cuisine bowl with an egg, grains, and vegetables.

For health promotion, the general recommendation is to consume red meat and processed meat in moderation. Like many foods and beverages, consuming too much of any one food or beverage can have a negative impact on health. Rather, consuming a variety of different sources of foods and nutrients, including protein, is recommended.

8. *Choose a diet adequate in calcium and iron.* Adequate calcium and iron are particularly important for women and children. Fat-free or low-fat milk and other low-fat dairy products are excellent sources of calcium. For example, one glass of cow's milk provides nearly one-third the RDA for calcium. Milk substitutes, such as soy, almond, coconut, and rice milk, may have similar or lower kcal and comparable amounts of calcium but may contain much lower amounts of protein. Certain vegetables, such as broccoli, are also good sources of calcium. Iron is found in good supply in many protein foods and grains. Small amounts of lean or very-lean meats should be selected and whole-grain or enriched products should be chosen over those made with bleached, unenriched white flour. Some foods rich in calcium and iron are listed in **table 2.6**.

Although adequate amounts of calcium and iron, as well as other minerals, are important for optimal health, excessive intakes may cause health problems, as noted in chapter 8.

9. *Practice food safety, including proper food storage, food preservation, and preparation.* While food safety is a key responsibility of food producers, government agencies, and food distributors, we, the consumers, also play a key role. Using recommended protocols during food preservation and preparation may help prevent food poisoning and decrease the formation of potential risk factors for chronic disease.

To prevent food poisoning, the Check Your Steps Program (www.foodsafety.gov) recommends a four-step protocol.

- **Clean.** Wash hands, surfaces, cutting boards, cooking utensils, dishes, and produce.
- **Separate.** When shopping for, storing, and preparing foods, prevent cross-contamination of harmful bacteria from juices of some foods, such as meat and chicken, to other ready-to-eat foods, such as fruit and vegetables.
- **Cook.** Use a food thermometer for cooking foods to the safe minimum temperatures to kill harmful bacteria.
- **Chill.** Refrigerate foods quickly to a temperature of 40°F or below to slow the growth of bacteria that could cause food poisoning.

10. *Consider the possible benefits and risks of food additives and dietary supplements.* Food additives are any substances that become incorporated into a food during production, packaging, transport, or storage. Some additives are intentional, such as those for color, while others are unintentional, such as compounds from a food's container or wrapper. Additives are used by food manufacturers to enhance their products for a variety of reasons, including texture, color, shelf life, taste, and/or nutritional quality. The use of any additive in food must be tested for safety and approved by the Food and Drug Administration (FDA). Eating fresh, natural foods is one of the best approaches to avoiding additives. However, nutrients such as vitamins and minerals are often used as food additives to increase the nutritional quality of a food product, which may benefit some individuals.

11. *If you drink alcoholic beverages, do so in moderation.* The current available scientific evidence suggests that light to moderate daily alcohol consumption will likely not cause any health problems to the healthy, nonpregnant adult. In fact, moderate consumption of certain alcoholic beverages, including red wine, may even have positive health effects in some people. A drink is defined as one 12-ounce bottle of beer, one 4-ounce glass of wine, or 1.5 ounces of 80-proof distilled spirits. Current guidelines for moderate alcohol intake recommend no more than two drinks per day for males and one for females. Excessive alcohol consumption is a serious health problem. An expanded discussion is presented in chapter 13.

12. *Enjoy your food. Eat what you like, but balance it within your overall healthful diet.* It is important to note that, within reason, you can eat whatever you want with the Prudent Healthy Diet. There are no unhealthy foods, only unhealthy diets. The dietary advice regarding moderate intake of certain foods, such as high-fat meats and ice cream, does not mean that they have to be eliminated from the diet, only that their intake should be limited and balanced with other nutrient-dense foods in the total diet. Portion control is an important concept. Balance, variety, and moderation in the overall diet are most important, not any single food.

Enjoy eating; it is one of life's pleasures. To make healthier foods such as chicken, fish, and vegetables taste better, season them with spices. Spices not only enhance the flavor of some foods but also contain various phytonutrients that may have beneficial health effects.

TABLE 2.6	Foods rich in calcium and iron
Mineral	**Food source**
Calcium	Dairy products: cow's milk, cheese, yogurt; fortified dairy alternatives; egg yolk; dried peas and beans; dark-green leafy vegetables such as beet greens, spinach, and broccoli; cauliflower
Iron	Organ meats such as liver; meat, fish, and poultry; shellfish, especially clams and oysters; dried beans and peas; whole-grain products such as breads and cereals; dark-green leafy vegetables such as spinach and broccoli; dried fruits such as figs, raisins, apricots, and dates

Hoxton/Sam Edwards/Getty Images

> **Key Concept**
>
> ▶ The Prudent Healthy Diet provides 12 recommendations for healthy eating.
>   1. Balance the food you eat with physical activity to maintain or achieve a healthy weight.
>   2. Consume a wide variety of natural, whole, nutrient-rich foods.
>   3. Choose a plant-rich diet with plenty of whole-grain products, legumes, fruits, and vegetables.
>   4. Choose a diet moderate in total fat and low in saturated fats *trans* fats, and cholesterol.
>   5. Choose a diet moderate in sugars.
>   6. Choose a diet with less salt and sodium.
>   7. Eat protein at a moderate yet adequate level.
>   8. Choose a diet adequate in calcium and iron.
>   9. Practice food safety.
>   10. Consider the benefits and risks of food additives and dietary supplements.
>   11. Drink alcoholic beverages in moderation, if at all.
>   12. Enjoy your food!
>
> **Check for Yourself**
>
> ▶ Think about your dietary habits and then determine how many of the 12 healthful dietary guidelines you follow.

Plant-Based Diets

As the name implies, **plant-based diets** are those that rely heavily, or even exclusively, on plant foods. A **vegetarian diet** is one that focuses on plant foods. While some vegetarian diets completely avoid animal foods, others include eggs, cow's milk, and/or other animal products. According to recent reports, about 3 percent of American adults follow a vegetarian diet and nearly half of vegetarians follow an exclusively plant-based diet. For young adults (18–34 years), rates of vegetarianism are higher with 6 percent of young adults following a vegetarian diet.

What types of foods are included in plant-based diets?

There are a variety of different types of plant-based diets. A person who follows a **vegan** diet eats no animal products at all. Most nutrients are obtained from fruits, vegetables, breads, cereals, legumes, nuts, and seeds. **Ovovegetarians** include eggs in their diet, while **lactovegetarians** include foods from the dairy group such as cheese and milk. An **ovolactovegetarian** eats both eggs and milk products. These later classifications are not strict vegetarians, because eggs and milk products are derived from animals.

Mizina/iStock/Getty Images

Others may call themselves *flexitarians*, because they occasionally may eat meat, fish, or poultry. Flexitarians are also known as **semivegetarians.** Those who eat fish, but not poultry, are known as **pescovegetarians.** In practice, then, vegetarians range on a continuum from those who eat nothing but plant foods to those who eat a typical American diet.

What are some of the nutritional concerns with plant-based diets?

In its position statement, the AND notes that appropriately planned vegetarian diets are healthful, nutritionally adequate, and provide health benefits in the prevention and treatment of certain diseases. However, if foods are not selected carefully, a person following a plant-based diet may suffer nutritional deficiencies involving kcal, vitamins, minerals, and protein.

Vegetarian diets, particularly vegan diets, should be well-planned, especially for children. The AND states that vegetarian diets can be appropriate for all stages of the life cycle. Yet, although exclusively plant-based diets can be adequate for children, the lack of food variety and too much reliance on vegetarian convenience foods, which may not possess the nutritional quality of unprocessed plant foods, can be a concern.

Energy Because plant products are generally low in kcal, a vegetarian may be on a diet with insufficient kcal for proper body weight maintenance. This may be particularly true for children, who need energy for growth and development, and for the active individual who may be expending more than 1,000 kcal per day through exercise. The solution is to eat greater quantities of the foods that constitute the diet, and to include some of the higher-kcal foods like nuts, beans, corn, green peas, potatoes, sweet potatoes, avocados, orange juice, raisins, dates, figs, whole wheat bread, and other whole grain products. These foods may be included in both meals and as snacks.

On the other hand, the low kcal content of vegetarian diets may be a desirable attribute for some, as it may be useful in weight-reduction programs or helpful in maintenance of proper body weight. However, Martins and others have voiced concern that some individuals may adopt a vegetarian dietary style in an attempt to mask their dieting behavior from others. Adopting a plant-based diet may be a very useful strategy for weight control, but may be detrimental if associated with eating disorders, a topic discussed in chapter 10.

Vitamins An individual following a vegan diet may incur a vitamin B12 deficiency because this vitamin is not found naturally in plant foods. In a review, Pawlak and others indicated that vegetarians, especially vegans, have a relatively high rate of vitamin B12 deficiency. Vitamin B12 is found in many animal products such as meat, eggs, fish, and dairy products, so the addition of these foods to the diet will help prevent a deficiency state. An ovolactovegetarian should have no problem getting the required amounts. A vegan will need a source of B12, such as fortified soy milk, fortified breakfast cereal, or a B12 supplement. If not exposed to regular sunlight, vegans will also need dietary supplements of vitamin D, which is not found in significant amounts in plant foods.

Minerals Mineral deficiencies of iron, calcium, and zinc may occur. During the digestion process, some plant foods form compounds known as phytates and oxalates that can bind these minerals so that they cannot be absorbed into the body. Avoidance of unleavened bread helps reduce this effect, as does thorough cooking of legumes such as beans. In general, research has revealed that a balanced intake of grains, legumes, and vegetables will not significantly impair mineral absorption. Foods rich in iron, calcium, and zinc should also be included in plant-based diets.

Iron-rich plant foods include nuts, beans, split peas, dates, prune juice, raisins, green leafy vegetables, and many iron-enriched grain products. Special attention should be given to dietary practices that promote absorption of iron and zinc from plant foods. For example, consuming foods rich in vitamin C will increase the absorption of dietary iron from plant sources. Semivegetarians may obtain high-quality iron in fish and poultry.

Calcium-rich plant foods include many green vegetables such as broccoli, cabbage, mustard greens, and spinach. Dairy products added to the diet supply significant amounts of calcium, as do calcium-fortified plant-based milks, such as soy milk. According to the AND, the calcium intake of some vegans is below recommended levels, while calcium intake in ovolactovegetarians is similar to that of nonvegetarians. Increased bone fracture risk in vegans may be a consequence of low calcium intake.

Protein A major concern of many following a plant-based diet is to obtain adequate amounts of quality protein, particularly in the case of young children. Generally speaking, obtaining sufficient protein on a vegetarian diet is not difficult. Consuming enough kcal to maintain an optimal body weight will provide adequate amounts of protein.

As will be noted in chapter 6, proteins are classified as either complete or incomplete. A protein is complete if it contains all of the essential amino acids that the human body cannot manufacture. Animal products generally contain complete proteins, whereas many plant proteins are incomplete. Certain vegetable products may also provide good sources of protein. Grain products such as wheat, rice, and corn, as well as beans, peas, and nuts, have a substantial protein content. However, most vegetable products lack one or more essential amino acids in sufficient quantity. They are incomplete proteins and, eaten individually, are generally not adequate for maintaining proper human nutrition. But, if certain plant foods are eaten together, they may supply all the essential amino acids necessary for human nutrition and may be as good as animal protein (**figure 2.10**).

An individual following a vegan diet must receive nutrients from breads and cereals, nuts and seeds, legumes, fruits, and vegetables. To receive a balanced distribution of the essential amino acids, the vegan must eat plant foods that possess **complementary proteins**. In essence, a plant food that is low in a particular amino acid is eaten with a food that is high in that same amino acid. For example, grains and cereals, which are low in lysine, are complemented by legumes, which have adequate amounts of lysine. The low level of methionine in the legumes is offset by its high concentration in the grain products. These types of food combinations are practiced throughout the world. Through the proper selection of foods that contain complementary proteins, a person following a plant-based diet can get an adequate intake of the essential amino acids. Because all amino acids must be present for tissue formation, a deficiency of one or two essential amino acids will limit the proper development of protein structures in the body.

nadianb/Shutterstock

FIGURE 2.10 It is important for a person following a plant-based diet to eat protein foods that complement each other (e.g., nuts and bread, rice and beans) so that all the essential amino acids are obtained in the diet.

Is following a plant-based diet beneficial for health?

Numerous epidemiological studies have suggested that following a well-planned plant-based diet may reduce a person's risk for chronic disease, including obesity, hypertension, diabetes, heart disease, stroke, and some types of cancer. As noted previously in the Prudent Healthy Diet, research supports that increased fruit and vegetable consumption is associated with reduced risk of major diseases, particularly cardiovascular diseases, as well as increased longevity. Specific aspects of plant-based diets that may reduce risk for chronic disease include the following.

Nutrient Density Well-planned plant-based diets often contain substantial amounts of nutrient-dense foods, particularly fruits and vegetables. For example, one 54-kcal cup of cooked broccoli contains 4 grams of protein, 5 grams of dietary fiber, 170 percent of the DV for vitamin C, 50% of the DV for vitamin A, 42% of the DV for folate, and 30% of the DV for potassium.

Low Saturated Fat The total fat and saturated fat content in a plant-based diet is usually low because the small amounts of fats found in plant foods are generally monounsaturated or polyunsaturated. This may account for the finding that those following a plant-based diet generally have lower blood triglycerides and cholesterol than meat eaters, and these lower levels may be important to the prevention of coronary artery disease.

High Fiber Plant foods possess a high content of fiber, which may help reduce levels of serum cholesterol and help in the prevention of heart disease. Diets rich in fiber may also prevent certain disorders in the gastrointestinal tract. Moreover, increased fiber intake may help maintain normal blood glucose levels and a healthy body weight, two factors involved in the prevention of diabetes. More details on the health benefits of fiber are presented in chapter 4.

Low Energy If the proper foods are selected, a plant-based diet supplies more than an adequate amount of nutrients and is rather low in kcal content. Plant foods can be high in nutrient density, providing bulk in the diet without the added kcal of saturated fat. Hence, a plant-based diet can be an effective dietary regimen for losing excess body weight. However, vegetarians who consume excess amounts of high-kcal foods, such as cheese, whole milk, and processed meat alternatives containing added sugars and saturated fat, may be at risk for excess weight gain.

High Vitamin and Phytonutrient Content Plant foods are rich in antioxidant vitamins, particularly vitamin C and beta-carotene, a precursor to vitamin A. Polyunsaturated plant oils provide substantial amounts of vitamin E. Selenium, an antioxidant mineral, is found in other plant foods.

Other than nutrients, plants also contain numerous **phytonutrients** (plant nutrients), such as phenols, plant sterols, and terpenes, which are not considered essential nutrients but may still influence various metabolic processes in the body. Collectively, these antioxidant nutrients and phytonutrients are referred to as **nutraceuticals**, parts of food that may provide a medical or health benefit. As suggested in a position statement by the AND, foods containing such phytonutrients may be classified as functional foods, a topic covered in the next section of this chapter. **Table 2.7** provides a list of some antioxidant nutrients and phytochemicals and their common plant sources.

Although the exact mechanisms whereby antioxidant nutrients and phytonutrients may help prevent chronic diseases, such as cancer or heart disease, have not been identified, several hypotheses are being studied. Potential health benefits of antioxidants and phytonutrients may be related to one or more of their possible roles in human metabolism:

- Affect enzyme activity
- Detoxify carcinogenic compounds
- Block cell receptors for natural hormones
- Prevent formation of excess oxygen-free radicals
- Alter cell membrane structure and integrity
- Suppress DNA and protein synthesis

Some of these actions may favorably affect health, as in the following two examples. Antioxidants, such as carotenoids, may block the oxidation of certain forms of serum cholesterol, reducing their potential to cause atherosclerosis and possible heart disease. Also, phytochemicals known as phytoestrogens may compete with natural forms of estrogen in the body for estrogen receptors in various tissues, blocking estrogen's natural proliferative activity and possibly suppressing cancer development.

TABLE 2.7 Some antioxidant nutrients and phytonutrients with common food sources

Antioxidant nutrients	Common plant sources
Vitamin C	Citrus fruits
Vitamin E	Potatoes
	Strawberries
	Dark-green leafy vegetables
	Margarine
	Vegetable oils
	Wheat germ
	Whole grains

Phytonutrients	Common plant sources
Allium sulfides	Garlic
	Onions
Anthocyanins	Blueberries
Capsaicin	Hot peppers
Carotenoids	Carrots
Beta-carotene	Dark-green leafy vegetables
Lycopene	Sweet potatoes
Lutein	Tomatoes
Flavonoids	Citrus fruits
Quercetin	Apples
Catechin	Tea
Indoles	Cruciferous vegetables
	Broccoli
	Brussels sprouts
	Cabbage
	Cauliflower
	Kale
Isoflavones	Soybeans
Phytoestrogens	Peanuts
Genistein	Soy milk
Isothiocyanates	Cruciferous vegetables
Sulforaphane	Brocoli
	Brussels sprouts
	Cabbage
	Cauliflower
	Kale
Phenolic acids	Carrots
	Citrus fruits
	Tomatoes
	Whole grains
Polyphenols	Grapes
Resveratrol	Red wines
	Grapes
Saponins	Beans
	Legumes
Terpenes	Cherries
Limonene	Citrus fruits

Most nutrition scientists indicate that many of these nutrients and phytonutrients share the same food sources, so the protective health effect associated with a plant-rich diet may not be attributed to a single food constituent, but may be due to the collective effect

A. Astes/Alamy Stock Photo

TABLE 2.8 Food colors of various fruits and vegetables with examples of phytonutrients in that "color" category

Red (lycopene)	Yellow/orange (carotenoids)	Green (sulforaphane)	Blue/purple (anthocyanins)	White/brown (allicin)
Cherries	Apricots	Artichokes	Blackberries	Bananas
Cranberries	Cantaloupe	Avocados	Blueberries	Cauliflower
Raspberries	Corn	Collards	Dried plums	Garlic
Red cabbage	Carrots	Cucumbers	Eggplant	Jicama
Red grapes	Lemons	Green grapes	Plums	Onions
Strawberries	Mangos	Kiwi fruit	Purple cabbage	Pears
Tomatoes	Oranges	Lettuce	Purple grapes	Potatoes
Watermelon	Pineapple	Spinach	Raisins	Turnips

of multiple nutraceuticals. Thus, health professionals currently recommend that consuming natural plant foods, rather than supplements, is the best way to obtain these purported nutraceuticals.

To help ensure that you obtain a wide variety of healthful phytochemicals in your diet, eating foods of many different colors is one strategy recommended by health professionals. **Table 2.8** lists the key food colors and some examples of foods found within each color group.

In summary, following a plant-based diet may offer benefits to health and help to prevent risk for chronic disease. A plant-based diet need not be 100% plants to offer health benefits but should be one that emphasizes fruits, vegetables, whole grains, nuts, seeds, lean sources of protein, and healthy fats.

www.nal.usda.gov/fnic/phytonutrients The USDA provides facts sheets and nutrition information on a wide variety of phytonutrients. Use one of the USDA phytonutrient databases to determine the amounts of phytonutrients in some of your favorite plant foods.

What are recommendations for following a plant-based diet?

People follow plant-based diets for a number of reasons, including improved health, religion, love for all animals, protecting the environment, and taste preferences. Choosing to adopt a plant-based diet is up to the individual and may represent a significant change in dietary habits. When deciding to follow an exclusively plant-based diet, it is important to first learn more about this dietary pattern and recommendations for meeting nutrient needs. Some examples of ways to gradually phase into a plant-based diet include:

- Start slowly and gradually start to include more plant-based foods in your diet. Choose at least one meal a day that is exclusively plants and, when eating meat, consume in moderation.
- Select leaner white meats such as baked or grilled turkey or chicken for meats that are higher in saturated fat. You may become a pescovegetarian, eating fish as your main animal food.
- You may wish to become an ovolactovegetarian, eating eggs and dairy products. These excellent sources of complete protein can be blended with many vegetable products or eaten separately.
- Learn how to purchase and prepare snacks and meals that contain complementary proteins and supply adequate amounts of essential nutrients. There are many print and online cookbooks available with plant-based recipes.

The *Training Table* in this section provides simple suggestions that may help you incorporate more fruits, vegetables, and whole grains in your diet.

Will a following a plant-based diet affect physical performance potential?

As noted previously, a plant-based diet is considered to be more healthful than the typical American diet that contains more saturated fat, added sugars, and sodium than is recommended. But will such a diet have any significant impact upon physical performance? Shaw and colleagues recently published a review article with recommendations for athletes following a plant-based diet. Based on this review and others, the following observations can be made:

- Well-planned, appropriately supplemented plant-based diets appear to effectively support athletic performance. Including fortified foods, such as plant-based dairy alternatives and whole-grain cereals, may help provide adequate amounts of some vitamins and minerals that may be low in plant-based diets.
- Plant and animal protein sources appear to provide equivalent support to athletic training and performance provided protein intakes are adequate to meet needs for total nitrogen and the essential amino acids.
- Vegan athletes, particularly females, are at an increased risk for non-anemic iron deficiency, which may limit endurance performance.
- Plant-based diets are often lower in creatine than diets supplying animal-based foods, and thus vegan athletes may have lower muscle creatine concentrations than meat-eating athletes. Lower creatine levels may impair very high intensity exercise. Details are presented in chapter 6.
- Plant-based diets may be high in healthful carbohydrates and be effective for weight control, which could be to the advantage of some athletes, particularly endurance runners. However, if used improperly as a strategy for weight control, plant-based diets

Training Table

Are you ready to start including more fruits, vegetables, and whole grains in your diet? If so, try some of these suggestions to get started:

- Buy only whole-grain bread with "whole wheat/grain" listed as the first ingredient on the food label.
- Buy large bags of frozen fruits, such as blueberries and raspberries, at club warehouses. Then, use the berries to make smoothies.
- Keep a variety of raw fruits on hand to have as snacks. For example, bananas, apples, grapes, and oranges.
- Buy whole vegetables and then "prep" them by cutting up and storing in the refrigerator. Celery and carrot sticks make an excellent snack—try dipping in hummus.
- Use frozen vegetables for quick stir-fry meals.
- Does your student union or dining hall have a salad bar? If so, load up on salad for a meal—just watch the dressing and choose options lower in solid fats.
- Use a microwave to cook sweet potatoes and baked potatoes. The microwave can also work well for steaming vegetables.
- Eat fruits for dessert, such as baked apples or fresh strawberries.

Ingram Publishing/SuperStock

If you want to shift toward a plant-based diet, you should research the diet using evidence-based sources beforehand and then initiate the process gradually. During the process, you should listen to your body—a common phrase among many athletes today. If you are active, how do you feel during your workouts? Do you have more or less stamina? Are you gaining or losing weight? Is your physical performance getting better or worse? The answers to these questions, together with other body reactions, may offer you some feedback as to whether the dietary change is beneficial.

Remember, there is nothing magical about a plant-based diet that will increase your physical performance capacity. It can be a healthful way to obtain the nutrients your physically active body needs, but so too is a well-balanced diet containing animal products.

www.vndpg.org/resources/vegetarian-dietitian-resources The Vegetarian Nutrition practice group of the Academy of Nutrition and Dietetics provides useful information on all aspects of plant-based diets.

www.vrg.org The Vegetarian Resource Group provides recipes, meal plans, and nutrition information related to vegetarian nutrition.

Key Concepts

- Individuals following a plant-based diet must be careful in selecting foods in order to obtain a balanced mixture of amino acids and adequate amounts of B12, calcium, iron, and zinc.
- Following a well-planned plant-based diet may reduce risk for chronic disease and support optimal health. Plant-based diets support sports performance but should be planned carefully to ensure nutrient needs are being met.

Check for Yourself

- Follow an exclusively plant-based (vegan) diet for a day, recording your food and beverage intake. How did you feel on the vegan diet? Is this a type of diet that may fit in your lifestyle?

could lead to eating disorders and impaired performance and health, a topic discussed in chapter 10.

As shall be noted later in this text, the amino acid leucine appears to be critical to promote muscle protein synthesis, and increased muscle mass, when consumed during resistance training programs. Whey protein, from cow's milk, contains substantial amounts of leucine, while soy protein contains slightly lower amounts. In one study, Tang and others found that, although both whey and soy protein supplementation increased muscle protein synthesis, whey protein was about 30 percent more effective, possibly due to its slightly greater leucine content. Such a finding may be of interest to individuals attempting to maximize muscle mass for sport competition.

Consumer Nutrition—Food Labels and Health Claims

Guidelines for a healthful diet will not be effective unless people change their behavior to buy and eat healthier foods. A model often used to explain the development of a set of behaviors involves a sequence of (1) acquisition of knowledge, (2) formation of an attitude or set of values, and (3) development of a particular behavior. In this sequence, knowledge is the first step that may enhance the development of proper health behaviors. Knowing how to interpret food labels may guide you in developing a nutritious, safe, and healthful diet.

What nutrition information do food labels provide?

Food manufacturers view labels as a device for persuading you to buy their product instead of a competitor's product. Just walk down the cereal aisle next time you visit the supermarket and notice the bewildering number of choices. As manufactured food products multiplied over the years, and as competition for your food dollar intensified, food companies began to manipulate their labels to enhance sales. Unfortunately, many of these practices were deceptive, and the consumer had a difficult time determining the nutritional quality of many processed foods. Thus, Congress passed a law designed to establish a set of standards to help Americans base their food choices on sound nutritional information.

This set of standards resulted in **nutritional labeling**, whereby major nutrients found in a food product must be listed on the label. It is not the total solution to the problem of poor food selection existing among many Americans, but combined with an educational program to increase nutritional awareness, it may effectively improve the nutritional health of our nation.

Initial food labeling legislation was passed in 1973, but it contained numerous flaws. Because of pressure from a variety of consumer interest groups, a major overhaul of the nutritional labeling program was signed into law as the Nutrition Labeling and Education Act in 1990, and it was in full effect in 1994. Under this law, nutrition labeling is mandatory for almost all foods regulated by the FDA. However, there are some exceptions, including small business restaurants and airplane food. Additionally, providing nutrition information is currently voluntary for many raw foods such as fresh fruits, vegetables, meat, and fish, but may become mandatory in the future.

The food label illustrated in **figure 2.11** includes a **Nutrition Facts panel** that is designed to provide information on the nutrients that are of major concern for consumers. The Nutrition Facts panel must include the serving size, the number of servings provided in an entire container, and specific nutrition information. As well, the listed serving sizes are based on standardized serving sizes utilized for all similar products. For example, a serving size of ice cream must be listed as ⅔ cup on the panel.

The Nutrition Facts panel must contain information on the food item's kcal, total fat, saturated fat, *trans* fat, cholesterol, sodium, total carbohydrate, fiber, total sugars, added sugars, protein, vitamin D, potassium, calcium, and iron content per serving. The kcal per serving must be listed in a large, bold font to be highly visible. Percent daily values must also be listed on the panel.

How can I use this information to select a healthier diet?

The **Daily Value (DV)**, which is based on dietary standards discussed earlier in this chapter, represents how much of a specific nutrient you should obtain in your daily diet. DVs have been established for macronutrients and micronutrients that may affect our health. In essence, a food label indicates how much of a given nutrient is present in that product, and for some nutrients, what percentage of the DV is provided by one serving.

The DVs cover the macronutrients that are sources of energy, consisting of carbohydrate (including fiber), fat, and protein. The DVs for the energy-producing nutrients are based on the number of kcal consumed daily. On the food

FIGURE 2.11 The Nutrition Facts panel on a food label. This nutrition information is required on virtually all processed food products. The Percent Daily Value listed on the label is the percentage of the generally accepted amount of a nutrient needed daily that is present in one serving of the product.

label, the percent of the DV that a single serving of a food contains is based on a 2,000-kcal diet, which has been selected because it is believed to have the greatest public health benefit for the nation. However, the DV may be higher or lower depending on your kcal needs. Values for some of the macronutrients are also provided for a 2,500-Calorie diet on the food label.

The DVs for select macronutrients, vitamins, and minerals, based on previously established RDAs, are presented in **table 2.9**. Daily values listed on most food products are for "adults and children 4 or more years of age." Daily values for "infants, children less than 4 years of age, and pregnant and lactating females," are also available and may be included on foods most commonly eaten by those populations (e.g., toddler foods).

Do you read food labels when making food selections? Some important points to consider in reading a food label are as follows:

1. The DV for a nutrient represents the percentage contribution one serving of the food makes to the daily diet for that nutrient based on current recommendations for healthful diets. A lower DV is desirable for total fat, saturated fat, cholesterol, and sodium; a DV of 5 percent or less is a good indicator. There is no DV for *trans* fat; only the number of grams is listed, and intake should be as low as possible, preferably 0 grams. A higher DV is desirable for dietary fiber, iron, calcium, vitamins A and C, and other vitamins and minerals that may be listed, with 10 percent or more representing a good source.
2. Related to carbohydrates, sugars include both natural and added sugars. The new food label requires percent daily values to be listed for both dietary fibers and added sugars. There is no daily value for total sugars.
3. Be aware of serving size tricks. A serving size for a cola drink may be listed as 8 ounces (100 kcal), so a 20-ounce bottle of soda is 2.5 servings (250 kcal). However, most people drink it all at one time, thinking it is only one serving and may consume more than twice the kcal as expected.
4. Check the ingredient list. Although in small type, it may provide very useful information. The list of ingredients is in order by weight. As noted, *trans* fat may be listed as 0 on the label, but the product contains some *trans* fat if hydrogenated or partially hydrogenated oils are in the ingredient list. The ingredient list for a variety of peanut butter is provided in **figure 2.12**. What is the ingredient in the highest concentration in the peanut butter? Does this variety of peanut butter contain *trans* fat?

In the past, many terms used on food labels, such as "lean" and "light," had no definite meaning. However, the FDA now permits **nutrient content claims** that reflect the levels of nutrients found in processed foods. For example, the FDA has established requirements for a food product to be labeled as low-fat, high-fiber, or sugar-free. **Table 2.10** provides examples of FDA-approved nutrient content claims. For more information on such nutrient content claims, go to www.fda.gov, then search for nutrient content claims and search for "nutrient content claims."

TABLE 2.9	DVs for select macronutrients, vitamins, and minerals
	DV (for adults and children 4 years or more of age)
Total fat	65 g
Saturated fat	20 g
Cholesterol	300 mg
Total carbohydrate	300 g
Dietary fiber	25 g
Protein	50 g
Sodium	2,400 mg
Potassium	3,500 mg
Calcium	1,000 mg
Iron	18 mg
Vitamin A	5,000 IU
Vitamin C	60 mg
Vitamin D	400 IU
Folate	400 mcg

Source: U.S. Food and Drug Administration: *Changes to the Nutrition Facts Panel*. Updated March 2022. www.fda.gov/food/food-labeling-nutrition/changes-nutrition-facts-label. Accessed September 24, 2022.

Elite Images/McGraw Hill

FIGURE 2.12 Ingredients are listed in order by weight, with the ingredient in the highest concentration listed first. Roasted peanuts are the first ingredient in this variety of peanut butter. Note that the peanut butter contains hydrogenated vegetable oils, which indicates the product contains *trans* fat.

TABLE 2.10 Examples of FDA-approved nutrient content claims on food labels

Nutrient content claim	Legal definition
Kcal free	Food provides fewer than 5 kcal per serving.
Reduced or fewer kcal	Food contains at least 25% fewer kcal than the reference food.
Sugar free	Product provides less than 0.5 g sugar per serving.
Reduced sugar	Food contains at least 25% less sugar per serving than the reference food.
Fat free	Produce provides less than 0.5 g fat per serving.
Low fat	Food contains 3 g or less fat per serving.
Reduced or less fat	Food contains at least 25% less fat per serving than the reference food.
High fiber	Food contains 5 g or more fiber per serving. Foods listing this high-fiber claim must also meet the definition of low fat.
Good source of fiber	Food supplies 2.5 to 4.9 g fiber per serving.
Extra lean meat or poultry	Food provides less than 5 g total fat, 2 g saturated fat, and 95 mg cholesterol per serving.
Lean meat or poultry	Food contains less than 10 g total fat, 4.5 g saturated fat, and 95 mg cholesterol per serving.
Low sodium	Food provides 140 mg or less sodium per serving.

Source: U.S. Food and Drug Administration: *Label Claims for Food and Dietary Supplements.* Updated March 2022. www.fda.gov/food/food-labeling-nutrition/label-claims-food-dietary-supplements. Accessed September 24, 2022.

The FDA continually monitors these claims and, at times, companies may be asked to remove claims from their labeling. For example, in 2014, the FDA issued a final ruling prohibiting certain nutrient content claims for foods that contain omega-3 fatty acids. According to the report summary, the nutrient content claims made by three separate seafood companies did not meet the requirements for such claims and the companies had until January 1, 2016 to change their food labels.

www.fda.gov Type Food Label in the search box to access numerous links to use the food label for healthy eating.

What health claims are allowed on food products?

Food manufacturers want your business. Given the public's growing awareness of the relationship between nutrition and health, many food labels now list various health claims or use terminology to entice you to buy their product. For example, one snack product has a label claiming it is made with 5 grams of whole grain, but it is mostly added sugars and all-purpose flour. And consumers do view a food product as healthier if it carries a health claim or uses *health* terminology.

The FDA permits food manufacturers to make specific **health claims** on food labels only if the food meets certain minimum standards. These health claims are permitted because the FDA believes there may be significant scientific agreement supporting a relationship between consumption of a specific nutrient and possible prevention of a certain chronic disease. However, there are several requirements, such as not stating the degree of risk reduction, using only terms such as "may" or "might" in reference to reducing health risks, and indicating that other foods may provide similar benefits. **Figure 2.13** provides an example.

The FDA approves the use of health claims on food labels based on the underlying research, usually only approving the claim if it is supported by significant scientific evidence. **Table 2.11** provides examples of FDA-approved health claims. The most recently approved health claim for macadamia nuts lowering risk for coronary artery disease was given the green

Mark Dierker/McGraw Hill

FIGURE 2.13 An example of a Nutrition Facts food label with an approved health claim: *"3 grams of soluble fiber from oatmeal daily in a diet low in saturated fat and cholesterol may reduce the risk of heart disease."*

TABLE 2.11 Qualified health claims. FDA-approved model health claims for foods that have significant scientific evidence supporting the claim

Calcium and vitamin D and osteoporosis: Regular exercise and a healthy diet with enough calcium helps teen and young adult white and Asian women maintain good bone health and may reduce their high risk of osteoporosis later in life.

Dietary fat and cancer: Development of cancer depends on many factors. A diet low in total fat may reduce the risk of some cancers.

Dietary saturated fat and cholesterol and risk of coronary artery disease: Development of heart disease depends upon many factors, but its risk may be reduced by diets low in saturated fat and cholesterol and healthy lifestyles.

Dietary noncariogenic carbohydrate sweeteners and dental caries: Frequent eating of foods high in sugars and starches as between-meal snacks can promote tooth decay. The sugar alcohol used to sweeten this food may reduce the risk of dental caries.

Fiber-containing grain products, fruits, and vegetables and cancer: Low fat diets rich in fiber-containing grain products, fruits, and vegetables may reduce the risk of some types of cancer, a disease associated with many factors.

Folate and neural-tube defects (spina bifida): Healthful diets with adequate folate may reduce a woman's risk of having a child with a brain or spinal cord birth defect.

Fruits and vegetables and cancer: Low-fat diets rich in fruits and vegetables (foods that are low in fat and may contain dietary fiber, vitamin A, and vitamin C) may reduce the risk of some types of cancer, a disease associated with many factors.

Fruits, vegetables, and grain products that contain fiber, particularly soluble fiber, and risk of coronary artery disease: Diets low in saturated fat and cholesterol and rich in fruits, vegetables, and grain products that contain some types of dietary fiber, particularly soluble fiber, may reduce the risk of heart disease, a disease associated with many factors.

Sodium and hypertension: Diets low in sodium may reduce the risk of high blood pressure, a disease associated with many factors.

Soluble fiber from certain foods and risk of coronary artery disease: Soluble fiber from foods such as oats and barley, as part of a diet low in saturated fat and cholesterol, may reduce the risk of heart disease. A serving of this food supplies 0.75 gram of the soluble fiber from oats necessary per day to have this effect.

Soy protein and risk of coronary artery disease: Diets low in saturated fat and cholesterol that include 25 grams of soy protein a day may reduce the risk of heart disease. One serving of this food provides 8 grams of soy protein.

Plant sterol/stanol esters and risk of coronary artery disease: Foods containing at least 0.65 g per serving of *plant sterol* esters eaten twice a day with meals for a daily total intake of at least 1.3 g (or 1.7 g per serving of *plant stanol* esters for a total daily intake of at least 3.4 g) as part of a diet low in saturated fat and cholesterol may reduce the risk of heart disease. A serving of this food supplies 0.75 gram of vegetable oil sterol (stanol) esters.

Macadamia nuts and risk of coronary artery disease: Supportive but not conclusive research shows that eating 1.5 ounces per day of macadamia nuts, as part of a diet low in saturated fat and cholesterol and not resulting in increased intake of saturated fat or kcal may reduce the risk of coronary artery disease.

Source: U.S. Food and Drug Administration: *Qualified Health Claims*. Updated March 2022. www.fda.gov/food/food-labeling-nutrition/qualified-health-claims. Accessed September 24, 2022.

light by the FDA in July 2017. The decision of the FDA to approve this claim for macadamia nuts was notable because it deviates from the agency's requirement that foods making claims related to coronary artery disease must be low in saturated fat. According to the FDA, approving such a claim is in line with the *2015–2020 Dietary Guidelines for Americans*—not all fats are created equal, and some fats are indeed healthy when consumed in moderation.

Appropriate food labeling may help us select healthier foods, and the forthcoming changes hopefully will do so. A simple, yet effective, approach is needed. Some countries, such as Sweden and Great Britain, have a national system that uses traffic light symbols (Red = High; Yellow = Medium; Green = Low) to instantly highlight the contents of less healthy ingredients, such as fat, saturated fat, sugars, and sodium. This traffic light system, or similar food scoring systems, may be a useful approach to help us select healthier foods.

What are functional foods?

In 1994, the FDA permitted dietary supplement manufacturers to make structure and function claims on their products. Basically, a structure and function claim simply means that the food product may affect body physiology in some way, usually in some way beneficial to health or performance. These claims may not be as authoritative as the FDA-approved claims cited above (such as reducing the risk of heart disease or cancer), but these claims may use such terminology as "helps to maintain healthy cholesterol levels" or "supports your immune system," which the consumer may interpret as preventing heart disease or cancer. Technically, these health claims must be correct, but they need not have as much supportive scientific evidence, nor do they have to have approval from the FDA.

Based partly on such health claims, dietary supplement sales have skyrocketed in the past 30 years, and the food industry jumped on the bandwagon. In recent years, numerous food manufacturers, have marketed products that have been referred to as functional foods.

In its recent position statement, the AND indicated that all food is essentially functional, such as providing energy, promoting growth and development, and regulating body metabolism. However, the AND indicates increasing scientific evidence suggests that some food components, not considered nutrients in the traditional sense, may provide positive health benefits. Foods containing these food components, including natural, fortified, and enriched products, are called **functional foods**. Functional foods include many fruits, vegetables, legumes, whole grains, fish such as salmon, and some Greek yogurts.

The AND further notes that functional food research holds many promises for improving the quality of life for consumers. Indeed, Claus indicates that by using modern genomic technology, individual metabolic profiles may be used to develop personalized functional foods for enhanced health. However, as outlined by the AND in their position paper, there remain some complexities in defining functional foods and their application to promote health. One major complexity is the fact there is no legal or governmental definition of functional food. Thus, the food industry may use health claims to market their products.

Fortification with nutrients or nutraceuticals is a current technique to make functional foods. In a sense, functional foods have been around for nearly a century, as salt was fortified with iodine, and milk was fortified with vitamins A and D, to help prevent nutrient deficiencies. More recently, calcium-fortified juices and multivitamin/mineral-fortified cereals are breakfast mainstays designed, in part, to help us obtain adequate amounts of specific nutrients. Some of these products may be worthwhile, for they may be in accord with the principles underlying FDA approval of food health claims. For example, calcium-fortified orange juice may be an excellent source of calcium for someone who does not drink milk. Cereals fortified with psyllium may be an excellent source of soluble dietary fiber. On the other hand, a sugar-sweetened beverage with added vitamins is a different story, as it is simply a vitamin pill with added sugar.

Some functional foods are designed to satisfy the criteria for qualified health claims on food labels. Many other products marketed as functional foods are simply dietary supplements in disguise, and use structure and function claims to suggest health benefits. Such products include soups with St. John's wort, snack foods with kava kava, cereals with ginkgo biloba, and energy drinks with caffeine.

In the next section we discuss dietary supplements and health claims. However, in the meantime, remember that fruits, vegetables, whole grains, and other plant foods are the optimal functional foods. Their health benefits have been well established.

Key Concepts

- Information provided through nutritional labeling on most food products may serve as a useful guide in finding foods that have a high nutrient density and are healthy choices.
- The Percent Daily Value (DV) on a food label represents the percentage of a nutrient, such as saturated fat or carbohydrate, provided in a single serving based on standard recommendations for healthy adults and children age 4 years and older.
- Terms used on food labels, such as *fat free*, must meet specific standards. In this case, use of *fat free* indicates that a serving of the food contains less than 0.5 gram of fat.
- Health claims may be placed on food labels only if they are supported by adequate scientific data and have been approved by the Food and Drug Administration (FDA).
- Some functional foods may provide some health benefits, adhering to qualified health claims. Other products marketed as functional foods may use structure and function claims, which do not have the scientific support of qualified health claims.

Check for Yourself

- Go to a supermarket and compare food labels for various products. For example, compare the fiber content of different types of cereal and the amount of added sugars in different fruit juices.

Consumer Nutrition—Dietary Supplements and Health

According to nutrition experts, foods, particularly fruits and vegetables, contain numerous nutrients or other food substances, such as vitamins, minerals, and phytonutrients, that may have pharmaceutical properties when taken in appropriate dosages. The potential health benefits of specific nutrients and phytonutrients will be covered in the following chapters as appropriate. The purpose of this section is to provide a broad overview of possible health effects of such supplements when marketed as dietary supplements.

What are dietary supplements?

The dietary supplement industry is a multibillion-dollar business. According to the National Center for Health Statistics, 58 percent of American adults report using a dietary supplement in the past month—mostly vitamins and minerals but also

Isadora Getty Buyou/Image Source, all rights Reserved

62 CHAPTER 2 Healthful Nutrition for Fitness and Sport

other substances marketed for potential health benefits. Americans have been using such supplements since the 1940s, when they were regulated as either food or drugs. However, in 1994, the United States passed the Dietary Supplements Health and Education Act (DSHEA), which defined a **dietary supplement** as a food product, excluding tobacco that contains at least one vitamin, mineral, herb or other plant product, amino acid, or a dietary substance that supplements the diet by increasing total intake.

It is important to note that the DSHEA stipulates that a dietary supplement cannot be represented as a conventional food or as the sole item of a meal or diet.

As noted by this definition, dietary supplements may contain essential nutrients such as essential vitamins, minerals, and amino acids, but also other nonessential substances such as ginseng, ginkgo, yohimbe, ma huang, and other herbal products. The technical definition of a supplement is something added, particularly to correct a deficiency. Theoretically, then, dietary supplements should be used to correct a deficiency of a specific nutrient, such as vitamin C. However, numerous dietary supplements contain substances other than essential nutrients and are marketed not to correct a deficiency but rather to increase the total dietary intake of some food or plant substance that allegedly may enhance one's health status. Like foods, dietary supplements must carry labels, or *supplement facts;* an example is presented in **figure 2.14**.

FIGURE 2.14 Dietary supplements must have a label that includes an ingredient list and facts about the content of the supplement. Dosing recommendations and safety information will also be included.

Will dietary supplements improve my health?

Dietary supplements are usually advertised to the general public as a means to improve some facet of their health and are usually under governmental regulation. In the United States, dietary supplements are regulated under food law by the FDA and thus are eligible for FDA-authorized health claims, as discussed previously. Dietary supplement health claims in Canada are governed by Health Canada's Natural Health Products Directorate. In some countries, such as Germany, a medical prescription is needed to obtain some dietary supplements containing strong herbal products. Dickinson and MacKay note that the evidence from numerous surveys shows that supplement users are making a greater effort to seek health and wellness. Can dietary supplements improve your health? Possibly, but there are several caveats.

The AND states that most people don't need supplements and recommends that eating a wide variety of nutrient-rich foods is the best way for most people to obtain the nutrients they need to be healthy and reduce their risk of chronic disease. However, the AND notes there are some people who may need supplements to help meet their nutritional needs, indicating your health-care provider might recommend a vitamin/mineral supplement when

- You are on a restrictive diet.
- You are an adult over the age of 50.
- You follow a vegan diet.
- You are pregnant or a female of child-bearing age.
- You have a medical condition that limits your food choices (e.g., certain digestive disorders) or absorption of nutrients is impaired (e.g. due to medications or surgery).

Consumers should be aware of exceeding recommended upper limits of some vitamins and minerals. Some prudent recommendations for vitamin and mineral supplementation will be presented in chapters 7 and 8.

Of the other classes of nonvitamin, nonmineral dietary supplements (e.g., herbs or other plant products), numerous products are marketed for their purported health benefits. Although we have no specific requirement for these substances, as they are not essential for normal physiological function, some may affect physiological functions in the body associated with health benefits. Using a broad interpretation of the FDA health claim regulations for dietary supplements, some supplement companies advertise their supplements as "miracle products" that can produce "magical results" in a short period of time.

Under current federal law, any dietary supplement can be marketed without advance testing. The only restriction is that the label cannot claim that the product will treat, prevent, or cure a disease. However, as noted previously, the label may make vague claims, referred to as structure/function claims, such as "enhances energy" or "supports testosterone production." Unfortunately, for most of these dietary supplements there is little research to support their claims. Most advertisements are based on theory alone, testimonials or anecdotal information, or the exaggeration or misinterpretation of research findings relative to the health effects of specific nutrients or other food constituents. Many labels carry a notice stating *This statement has not been evaluated by the FDA,* a disclaimer regarding the health claim. Moreover, although advertisers may not make unsubstantiated health claims, the 1994 DSHEA stipulates that the burden of proving the claims false rests with the government. Currently, under the DSHEA, the FDA must show in court that an unreasonable risk is posed by consumption of a dietary supplement.

Various governmental and other agencies have developed guidelines for the supplement industry to help the consumer have confidence in the composition, labeling, and safety of dietary supplement products. The FDA has established current good manufacturing practice requirements, noting that manufacturers are required to evaluate the identity, purity, quality, strength, and composition of dietary supplements. The Federal Trade Commission (FTC) has indicated that marketers of dietary supplements must have above-board scientific evidence to support any health claims. Additionally, U.S. Pharmacopeia, a respected nonprofit medical agency, launched a certification program for dietary supplements. If the dietary supplement contains what the label indicates, then it may carry the USP seal of approval. However, the USP seal does not mean the product is effective or even that it is safe to use, just that it contains what the label promises.

Ultimately, dietary supplements may exert some beneficial healthful effects in certain cases; however, as for most of us the substances found in most dietary supplements are readily available in familiar and attractive packages called fruits, vegetables, legumes, fish, and other healthy foods. Although the Prudent Healthy Diet is an optimal means to obtain the nutrients we need, dietary supplements may be recommended under certain circumstances. When appropriate, such recommendations will be provided at specific points in this text.

Can dietary supplements harm my health?

Although dietary supplements may be beneficial in certain cases and to some individuals, their use may be harmful in some ways. Key issues to consider with dietary supplement use include:

1. Nutrition is only one factor that influences health, well-being, and resistance to disease. Individuals who rely on dietary supplements to guarantee their health may disregard other very important lifestyle behaviors, such as appropriate exercise, stress management, and a well-balanced diet.
2. Dietary supplements may provide a false sense of security to some individuals who may use them as substitutes for a healthful diet, believing they are eating healthfully and not attempting to eat right.
3. Taking supplements of single nutrients in large doses may have detrimental effects on nutritional status and health. Although large doses of some vitamins or minerals may be taken to prevent some conditions, excesses may lead to other health problems.
4. Individuals who use dietary supplements as an alternative form of medicine may avoid seeking effective medical treatment.
5. Dietary supplements vary tremendously in quality, including best-selling store brands marketed by national retailers. Numerous independent analyses of specific dietary supplements, such as those by ConsumerLab.com, reveal that some may contain less than that listed, sometimes even none of the main ingredient. Some products contain substances not listed on the label. This may pose a health risk.
6. Research studies have shown that the use of some dietary supplements may impair health, and may even be fatal.

https://dsld.od.nih.gov/ The National Institutes of Health database of dietary supplements contains thousands of supplements. Use the search box to obtain product information.

http://ods.od.nih.gov/factsheets/list-all/ The Office of Dietary Supplements provides detailed information, including safety, for most common supplements.

www.fda.gov/Food/DietarySupplements/ Use the search box to find reports on the safety and effectiveness of specific dietary supplements

www.ConsumerLab.com Check the content analysis of various brands of popular dietary supplements. Fee charged for some reports.

Although most dietary supplements are safe, some may induce adverse health effects. Where appropriate, the effectiveness and safety of various dietary supplements will be discussed in later chapters of this book. The *Training Table* in this section provides some general safeguards recommended by consumer health organizations to protect your health when using dietary supplements.

Training Table

For some people, dietary supplements can be a healthy addition to a well-balanced diet. However, consumers should be aware that dietary supplements do not come without risk, and it is important to discuss supplement use with your health-care provider. If you are considering using a supplement, keep the following advice in mind:

- Before trying a dietary supplement to "treat" a health problem, try changing your diet or lifestyle first.
- Check with your health-care provider before taking any dietary supplement, particularly herbal supplements. This is especially important for pregnant and nursing women, children, and individuals taking prescribed drugs whose effects may be impaired by herbal interactions.
- Buy standardized products from reputable sources. Most dietary supplements in the United States should be standardized according to federal regulations. Supplement Facts labels should provide information comparable to the Nutrition Facts food label.
- Avoid multi-ingredient supplements which may make it difficult to determine the cause of any side effects.

- Be alert to both the positive and negative side effects of the supplement. Try to keep an objective record of the effects.
- Stop taking the supplement immediately if you experience any health-related problems. Contact your health-care provider and local health authorities to report the problem. This may help establish a database for the safety of dietary supplements.

Key Concepts

▶ As defined by the Dietary Supplement Health and Education Act (DSHEA), a dietary supplement is a food product, added to the total diet, that may contain a vitamin, mineral, herb or other plant product, amino acid, or a dietary substance that supplements the diet by increasing total intake.

▶ Although some people may need dietary supplements for various reasons, particularly vitamins and minerals, the use of supplements should not be routine practice for most individuals. Obtain nutrients through natural foods.

▶ Some dietary supplements may be dangerous. Check with your health-care provider.

Check for Yourself

▶ Visit a store that sells dietary supplements or look online to compare the content of two similar dietary supplements (e.g., two brands of multivitamins/multiminerals). How do the two products compare? Which one would you recommend and why?

Healthful Nutrition: Recommendations for Better Physical Performance

Sports nutrition for the physically active person may be viewed from two aspects: nutrition for training and nutrition for competition. Of the three basic purposes of food—to provide energy, to regulate metabolic processes, and to support growth and development—the first two are of prime importance during athletic competition, while all three must be considered during the training period in preparation for competition.

Much of the information relayed to athletes from online sources and influencers suggests that those who engage in regular physical activity have different nutritional requirements above those of nonathletes. In general, however, the diet that is optimal for health is also optimal for physical or sports performance. The Prudent Healthy Diet will provide adequate food energy and nutrients to meet the need of almost all athletes in training and competition.

Nevertheless, modifications to the Prudent Healthy Diet may help enhance performance for certain athletic endeavors, and subsequent chapters will focus on specific recommendations relative to the use of various nutrients and dietary supplements to enhance physical performance. The purpose of this section is to provide some general recommendations regarding use of the Prudent Healthy Diet by the athlete for training and competition.

However, it is very important for athletes to individualize their dietary practices. The nutrient needs and dietary practices of athletes may vary significantly, such as daily carbohydrate intake of a golfer as compared to that of a marathon runner. All athletes should keep track of what, how much, and when they eat and drink during training and competition and experiment with dietary strategies to find those that are optimal. Several prominent sports organizations have developed nutrition guides for athletes. The United States Olympic Committee published a comprehensive guide to sports nutrition entitled the *Athlete's Plate,* providing details on a wide variety of topics such as caution when traveling internationally. The United States Anti-Doping Agency also published a pamphlet, *TrueSport® Nutrition Guide,* with a focus on a healthy diet designed to provide optimal dietary intake for sport and for life.

www.teamusa.org/nutrition The US Olympic and Paralympic Committees (USOPC) provide general sports nutrition information. Sports dietitians with the USOPC also work with individual teams and athletes to support their performance nutrition needs.

www.usada.org/athletes/substances/nutrition/ The United States Anti-Doping Agency provides information on carbohydrate, protein, vitamins, and other nutrients relative to health and sport performance.

What should I eat during training?

Both sport scientists and registered dietitan nutritionists (RD/RDNs) stress the importance of proper nutrition during training. Ron Maughan, an expert in exercise metabolism and sports nutrition, notes that the main role of nutrition for the athlete may be to support consistent intensive training. As noted in chapter 1, optimal training is the most important factor contributing to improved sport performance.

Tammy Stephenson

Because energy expenditure increases during a training period, the caloric intake needed to maintain body weight may increase considerably—an additional 500–1,000 kcal or more per day in certain activities. By selecting these additional kcal wisely from a wide variety of foods, you should obtain an adequate amount of all nutrients essential for the formation of new body tissues and proper functioning of the energy systems that work harder during exercise. A balanced intake of carbohydrate, fat, protein, vitamins, minerals, and water is all that is necessary. For endurance athletes, dietary carbohydrates should receive even greater emphasis.

However, there may be some circumstances during sport training that make particular attention to the diet important. For example, during the early phases of training, the body will begin to make adjustments in the energy systems so that they become more efficient. This is the so-called **chronic training effect**, and many of the body's adjustments incorporate specific nutrients. For example, one of the chronic effects of long-distance running is an increased hemoglobin content in the blood and increased myoglobin and cytochromes in the muscle cells; all three compounds require iron in order to be formed. Hence, the daily diet would have to contain adequate amounts of iron not only to meet normal needs but also to make effective body adjustments due to the chronic effects of training.

Nutrient timing, the intake of carbohydrate and protein just before or after an intense training session, has been advocated to promote recovery and anabolism for several decades. Consuming an appropriate carbohydrate/protein combination shortly before and/or after strenuous exercise may be a recommended procedure for some athletes. More details on the carbohydrate and protein needs of specific athletes are presented in chapters 4 and 6.

Breakfast may be especially important during training. A balanced breakfast provides a significant amount of kcal and other nutrients in the daily diet of the physically active person. A breakfast of low-fat milk, a poached egg, whole-grain toast, fortified high-fiber cereal, and orange juice will help provide a substantial part of the RDA for protein, calcium, iron, fiber, vitamin C, and other nutrients and is relatively high in complex carbohydrates. A balanced breakfast high in fiber with an average amount of protein also will help prevent the onset of mid-morning hunger. The fiber and protein may help maintain a feeling of satiety throughout the morning, whereas a breakfast of refined carbohydrates, such as doughnuts, may trigger an insulin response and produce hypoglycemia (low blood sugar) in the middle of the morning. The resultant hunger is typically satisfied by eating other refined carbohydrates, which will satisfy the hunger urge only until about lunchtime. A balanced breakfast having a high nutrient density is therefore preferable to a breakfast based on refined carbohydrate products.

Moreover, for young athletes, experts have noted that breakfast consumption is an important factor in their nutritional well-being and enhancing their academic performance. Although individual preferences should be taken into account, a balanced breakfast could provide a good source of some major nutrients to the individual who is involved in a physical conditioning program. For those on a tight time schedule, a bowl of ready-to-eat, fortified high-fiber, protein-rich cereal with low-fat milk and fruit may be an ideal choice. Nancy Clark, a nationally acclaimed sports nutritionist, notes that this breakfast is not only quick, easy, and convenient but also rich in carbohydrate, fiber, iron, calcium, and vitamins. Athletes should focus on a nutrient-dense breakfast rich in complex carbohydrates, lean protein, and healthy fats.

Proper nutrition should enhance the physiological responses to training, and thus enhance competitive sports performance. The nutrient needs of athletes in training will be highlighted throughout the remainder of this text where relevant.

When and what should I eat just prior to competition?

In competition an athlete will utilize specific body energy sources and systems, depending upon the intensity and duration of the exercise. The three human energy systems will be discussed in detail in chapter 3. Briefly, however, high-energy compounds stored in the muscle are utilized during very short, high-intensity exercise; carbohydrate stored in the muscle as glycogen may be used without oxygen for intense exercise lasting about 1 to 3 minutes; and the oxidation of glycogen and fats becomes increasingly important in endurance activities lasting longer than 5 minutes. The release of energy in each of these three systems may require certain vitamins and minerals for optimal efficiency.

If an individual is well nourished, athletic competition normally will not impose any special demands for any of the six major classes of nutrients. Body energy stores of carbohydrate and fat are adequate to satisfy the energy demands of most activities lasting less than 1 hour. Protein is not generally considered a significant energy source during exercise. The vitamin and mineral content of the body will be sufficient to help regulate the increased levels of metabolic activity, and body-water supply will be adequate under normal environmental conditions.

However, content and timing of the precompetition intake may be critical. A number of special meals have been utilized throughout the years because of their alleged benefits to physical performance, and special products have been marketed as pre-event nutritional supplements. Although research has not substantiated the value of any one particular precompetition meal, some general guidelines have been developed from practical experience over the years.

There are several major goals of the precompetition meal that may be achieved through proper timing and composition. In general, the precompetition meal should do the following:

1. *Allow the stomach to be relatively empty at the start of competition.* In general, a solid meal should be eaten about 3 to 4 hours prior to competition. This should allow ample time for digestion to occur so that the stomach is relatively empty yet hunger sensations are minimized. However, pre-event emotional tension or anxiety may delay digestive time, as will a meal with a high-fat or high-protein content. Hence, the composition of the meal is critical. It should be high in carbohydrate, low in fat, and low to moderate in protein, providing for easy digestibility.

2. *Help to prevent or minimize gastrointestinal distress.* The composition of the precompetition meal should not contribute to any gastrointestinal distress, such as flatulence, increased acidity in the stomach, heartburn, or increased bulk that may stimulate the need for a bowel movement during competition. In general, foods to be avoided include gas formers such as beans,

spicy foods that may elicit heartburn, and bulk foods such as bran products. Foods and beverages high in added sugars may delay gastric emptying or create a reverse osmotic effect, possibly increasing the fluid content of the stomach, which may lead to a feeling of distress, cramps, or nausea. High-sugar loads, particularly fructose, may also lead to other forms of gastrointestinal distress, such as diarrhea. Individuals with known food intolerances, such as lactose intolerance, should use due caution. Through experience, you should learn what foods disagree with you during performance and, of course, avoid these prior to competition.

3. *Help avoid sensations of hunger, light-headedness, or fatigue.* A small amount of protein in a carbohydrate meal will help delay the onset of hunger. Large amounts of concentrated sugars can cause a reactive drop in blood sugar in susceptible individuals, which may cause light-headedness and fatigue.

4. *Provide adequate energy supplies, primarily carbohydrate, in the blood and muscles.* A wide variety of foods may be selected for the precompetition meal. The meal should consist of foods that are high in complex carbohydrates with moderate to low amounts of protein. Examples of such foods are presented in later chapters.

5. *Provide an adequate amount of body water.* Adequate fluid intake should be assured prior to an event, particularly if the event will be of long duration or conducted under hot environmental conditions. Diuretics such as alcohol and caffeine, which increase the excretion of body water, should be avoided. As well, large amounts of protein may increase the water output of the kidneys and thus should be avoided. Fluids may be taken up to 15 to 30 minutes prior to competition to help ensure adequate hydration.

Table 2.12 provides examples of two precompetition meals, each containing about 500–600 kcal with substantial amounts of carbohydrate.

Keep in mind that meals should not be skipped on competition days. They should adhere to the basic principles set forth earlier in this chapter. The ACSM and AND have published recommendations on pre-event meals scheduled at different times of the day. The *Training Table* in this section summarizes key recommendations for those meals.

Pre-event nutritional strategies will vary somewhat for athletes involved in prolonged exercise tasks, such as running a marathon. As noted by prominent Australian exercise scientist Mark Hargreaves, body stores of both carbohydrate and fluids should be optimized. To achieve this goal, athletes may engage in practices such as carbohydrate loading and water hyperhydration, which will be detailed in chapters 4 and 9.

Experts note that no one food or group of foods works for everybody; individuals may need to experiment to find which foods and amount of food that work best. Food choices may vary based on the type of exercise, as well as the intensity and duration of the exercise. However, it is important to experiment with new foods during training rather than around competition.

What should I eat during competition?

There is no need to consume anything during most types of athletic competition with the possible exception of carbohydrate and water. Carbohydrate may provide additional supplies of the preferred energy source during prolonged, high-intensity intermittent and endurance exercise, while water intake may be critical for regulation of body temperature when exercising in warm environments. In ultra-distance competition, a hypotonic salt solution also may be recommended. Appropriate details are presented in chapters 4 and 9.

Training Table

Well-planned precompetition meals can optimize athletic performance and promote recovery following competition. The ACSM and AND provide the following recommendations:

Morning events: The night before, eat a high-carbohydrate meal. Early morning, eat a light breakfast or snack: cereal and nonfat milk; fresh fruit or 100% fruit juice, toast, bagel, or English muffin; pancakes or waffles; nonfat or low-fat fruit yogurt; or a liquid pre-event meal.

Afternoon events: Eat a high-carbohydrate meal both the night before and for breakfast. Follow with a light lunch: salad with low-fat dressing; turkey sandwich with a small portion of turkey; fresh fruit; low-fat crackers; high-carbohydrate nutritional bars, pretzels, or rice cakes.

Evening events: Eat a high-carbohydrate breakfast and lunch, followed by a light meal or snack: pasta with marinara sauce; rice with vegetables; light-cheese pizza with vegetable toppings; noodle or rice soup with crackers; or baked potato with vegetables.

TABLE 2.12 Two examples of precompetition meals containing 500–600 kcal

Meal A	Meal B
Glass of orange juice	One cup low-fat yogurt
One bowl of oatmeal	One banana
Two pieces of toast with jelly	One toasted bagel
Sliced peaches with low-fat milk	One ounce of turkey breast
	One-half cup of raisins

Terry Vine/Blend Images LLC

What should I eat after competition?

In general, a balanced diet is all that is necessary to meet your nutrient needs and restore your nutritional status to normal following competition or daily, hard physical training. Carbohydrate and fat are the main nutrients used during exercise. The increased caloric intake that is needed to replace your energy expenditure also will help provide you with the additional small amounts of protein, vitamins, minerals, and electrolytes that may be necessary for effective recovery. Thirst will normally help replace water losses on a day-to-day basis; you can check this by recording your body weight each morning to see if it is back to normal.

As noted previously, nutrient timing may be an important consideration. Simple sugars eaten immediately after a hard workout may help restore muscle glycogen fairly rapidly. Consuming a small amount of high-quality protein may also be prudent. Specific guidelines are presented in chapters 4 and 6.

For those who must compete several times daily and eat between competitions, such as in tennis tournaments or swim meets, the principles relative to pregame meals may be relevant, with a focus on carbohydrate-rich foods or fluids and moderate protein intake.

Should athletes use commercial sports foods?

The sports nutrition industry is booming. Numerous products are marketed to athletes, including meal replacement powders, sports drinks, sports bars, sports gels, sports candy, and sports supplements. It is important to note that although many of these products may be convenient and appropriate for a pregame, post-training, or postcompetition meal, they do not contain all the healthful nutrients found in natural foods and thus should not be used on a long-term basis to replace the Prudent Healthy Diet.

Liquid Sports Meals **Liquid meals**, many of them designed specifically for athletes, usually contain high-quality sources of carbohydrate and protein, a low-to-moderate fat content, vitamins and minerals, and various other supplements. The food label will provide the amounts of each. They are very convenient for precompetition meals as well as for recovery nutrition after training or competition.

Liquid meals available include Nutrament®, Ensure®, Slim-Fast®, Boost®, Gatorade Nutrition Shake®, and PowerBar ProteinPlus®. Some liquid meals come premixed, while others come as powders. You can make your own liquid sports meal, or smoothie, from high-quality carbohydrate/protein powders, such as whey powder, nonfat dry milk powder, and/or other healthful sources of carbohydrate and protein, such as yogurt and fruits. The *Training Table* in this section provides a sample smoothie recipe. Search online for other recipes, including those tailored for athletes.

A liquid meal may be assimilated more readily than a solid meal, and thus may be useful as a precompetition meal because it may be taken closer to competition—say, 2–3 hours before. Research has shown that there are no differences between a liquid and a solid meal relative to subsequent hunger, nausea, diarrhea, or physical performance.

Training Table

A homemade smoothie can provide high-quality carbohydrate and protein to serve as a liquid sports meal. Try the following recipe for a delicious and nutrient-dense smoothie:

- 6 oz Greek yogurt
- 1 medium banana
- ½ cup fat-free or low-fat milk
- 1 cup fresh berries (e.g., blueberries, strawberries, raspberries, blackberries)
- Ice cubes/crushed ice

Combine and blend to desired consistency. If desired, for additional protein, add 1 small scoop of whey protein.

Ken Karp/McGraw Hill

Sports Bars Sports bars have become increasingly popular in recent years, and several dozen products are targeted to physically active individuals. **Sports bars** vary in composition. Some are high carbohydrate, some are high protein, and some have nearly equal mixtures of carbohydrate, protein, and fat. Many are vitamin and mineral fortified, and some are designed to serve as a meal replacement. Others contain drugs, such as caffeine. As with liquid meals, the food label on the sports bar will describe its contents. When compared to comparable energy sources from ordinary food, sports bars do not possess any magical qualities to enhance physical performance, but they possess some advantages similar to liquid meals, such as convenience. Because the major ingredient in many sports bars is carbohydrate, an expanded discussion is presented in chapter 4.

Sports Drinks Sports drinks are generally referred to as carbohydrate and electrolyte replacement fluids and may be consumed by athletes before, during, and after training and competition. Examples include Gatorade® and Powerade®. They are designed to provide carbohydrate, water, and electrolytes, and their role in sport is discussed in chapters 4 and 9.

Sports Gels and Candy Sports gels and candy normally provide carbohydrate but may contain other substances such as vitamins, minerals, and caffeine. Their primary purpose is to provide a source of easily digested carbohydrates for energy during exercise.

Sports Supplements As noted in chapter 1, numerous sports supplements are marketed to athletes, including various forms of carbohydrates, fats, and protein; many vitamins and minerals; several food drugs; and selected herbal or botanical products. Based on the available scientific data, the use of most sports supplements does not appear to be necessary for the well-nourished athlete during training. However, nutrient supplementation may be warranted in some cases. For example, in activities where excess body weight may handicap performance, a loss of some body fat may be helpful. During such weight loss, vitamin-mineral supplements may be recommended to prevent a nutrient deficiency. Athletes may use sports supplements in attempts to enhance performance, a practice that is quite common. Use of several sports supplements has been supported by research because they may enhance physical performance, may not pose any health risks, and may be legal. Research evaluating the effectiveness of purported sport ergogenics is presented throughout the book.

How can I eat more nutritiously while traveling for competition?

Athletes who must travel to compete are often faced with the problem of obtaining proper pre-event and postevent nutrition. After reading this chapter, you should be aware of how to select foods that are high in carbohydrate, low in fat, and moderate in protein.

For athletes traveling for competition, one option is to pack your own food and fluids in a traveling bag or cooler. Foods from each of the MyPlate food groups can be easily packed or kept on ice, such as low-fat milk; precooked low-fat meats; whole grain bagels and cereal; fruits and vegetables; sports drinks; and high-carbohydrate snacks, including whole wheat crackers, pretzels, or snack mix. Small containers of condiments can also be easily transported in the cooler, along with proper eating utensils. Taking your own food means you can eat your pre-event or postevent meal as planned, and you may save money as well. Such an approach may be very effective for short, one-day trips and may be used to complement other meals on longer journeys. Some easily packed snack foods are presented in **table 2.13**.

While traveling, you have a variety of eating places from which to select your food, including full-service restaurants, restaurants with all-you-can-eat buffets, steakhouses and fishhouses, fast-food restaurants, pizza parlors, sub shops, supermarkets, convenience stores, and even vending machines. With a solid background on the nutritional principles presented in this chapter, you should be able to select healthful, high-carbohydrate and low-fat foods at any of these establishments but, of course, the variety of food choices will vary depending on the place you choose. Keep in mind that you can always ask to see if they will create a meal for you. For example, order a salad and ask to have extra vegetables and a lean source of protein, such as fish, chicken, or beans, added with dressing on the side.

You can eat fast food and stay within the recommended nutrition guidelines for a healthy diet, but obtaining a healthful diet requires careful selection of foods. Fast-food restaurants provide materials detailing the nutrient content of each of their products. In some cases the materials may be obtained in the restaurant, and all have websites detailing nutrient analysis of their products. Many smartphone apps also have nutrition information for fast-food restaurants, making data readily available when traveling. The *Training Table* in this section provides examples of food choices athletes may consider when traveling and relying on fast-food or family-style restaurants for their meals.

TABLE 2.13 Easily packed snacks for traveling or on-the-go lunches

Grains	Protein foods	Vegetables
Bagels	Small can of baked beans	Sliced carrots
Pita bread	Cooked chicken or turkey, small 2-ounce commercial packages, packed in airtight plastic bags	Broccoli stalks
Muffins		Cauliflower pieces
Fig Newtons	Small can of tuna fish, salmon, or sardines	Celery sticks
Plain popcorn	Peanut butter	Edamame
Whole wheat crackers and pretzels	Reduced-fat cheese slices	Cherry tomatoes
Graham crackers	Hummus	
Dry cereals	Turkey breast	

Fruits	Dairy (or Dairy Alternatives)
Small container of fruit in own juice	Small containers of low-fat milk; chocolate milk; aseptic packaging if available
Small containers of 100% fruit juice, aseptic packages	Dried skim milk powder, to be reconstituted
Oranges	Yogurt
Apples	String cheese
Other raw fruits	
Dried fruits	

Training Table

While eating healthy can be more challenging when relying on fast-food or family-style restaurants, you can still make nutrient-dense choices that support optimal athletic performance. Consider the following when eating out:

Breakfast selections:
Whole-grain bagel and jelly
English muffins with Canadian bacon
Whole wheat pancakes with syrup
French toast with fresh fruit
Bran muffins
Hot whole-grain cereal, oatmeal
Read-to-eat, high-fiber cereal
Fat-free or low-fat milk
100% fruit juice

Lunch or dinner selections:
Sandwich with lean meat and on whole-grain bread (hold the mayonnaise)
Grilled chicken breast on whole-grain bun
Baked or broiled fish sandwich
Lean roast beef sandwich
Single, plain hamburger
Baked potato with broccoli
Rice dishes
Lo mein noodles, not chow mein (fried noodles)
Soups, rice and noodle
Chicken or seafood tostado, made with corn tortillas
Garden salad with low-fat dressing and a lean source of protein, such as baked/grilled chicken or beans
Veggie bowl containing brown rice and sautéed vegetables

Athletes who travel internationally for their sport should use special caution in making food and beverage choices. Specific suggestions include the following:

- Drink only bottled water and other beverages.
- Avoid use of tap water (even for brushing teeth).
- Eat only food that is fully cooked.
- Do not eat raw foods.
- Do not eat food from street vendors.

How do gender and age influence nutritional recommendations for enhanced physical performance?

The diet that is optimal for health is the optimal diet for physical performance. This is the key principle of sports nutrition and it applies to physically active males and females of all ages. However, as shall be noted at certain points in this text, specific nutrient needs may vary by gender and age, and various forms of exercise training may influence nutrient requirements as well.

Gender Seiler and others note that exercise performance is, in general, about 10 percent greater in males than females, mainly because males have greater levels of muscle mass and strength, anaerobic power and capacity, and maximal aerobic capacity, as well as lower levels of body fat. Nevertheless, most physiological adaptations to exercise are similar for males and females. Moreover, Maughan and Shirreffs, discussing nutrition and hydration needs of soccer players, noted that the differences between males and females are smaller than differences between individuals, so principles developed for male players also apply to women. However, some physiological differences between males and females could influence nutritional requirements.

Adolescent and adult premenopausal females may need more dietary iron than males. Female athletes, especially those participating in aerobic endurance sports such as distance running, must include iron-rich foods in their diet or risk incurring iron-deficiency anemia and impaired running ability.

Some female athletes may not consume adequate energy and may develop disordered eating practices as they attempt to lose body mass for competition purposes. Disordered eating is more prevalent in female athletes and may contribute to the development of premature osteoporosis due to calcium imbalance. A discussion of the topic, relative to energy deficiency in sport, which is also referred to as the female athlete triad, is presented in chapters 8 and 10.

> http://reference.medscape.com/article/108994-overview#aw2aab6b3 This site provides some detailed information on the nutrition needs of the female athlete.

Age Youth sports competition is worldwide, ranging from community-based games to Olympic competition, and proper nutrition is important for these young athletes. Petrie and others have noted that child and adolescent athletes typically consume more food to meet their energy expenditure and thus are more likely to obtain an adequate supply of nutrients. However, Oded Bar-Or noted that while nutritional considerations are similar for all athletes irrespective of age, children have several physiological characteristics that may require specific nutritional considerations. For example, their relative protein needs may be greater to support growth, and

Tammy Stephenson

their relative calcium needs may be greater for optimal bone development. Young athletes may experience greater thermal stress during exercise. Roberts indicates it may be unwise to allow children to exercise hard in high heat and humidity conditions, but also notes that there are few data to support this concern. Those who participate in weight-control sports involving excessive exercise and inadequate energy intake may be at risk for nutrient deficiencies and impaired growth and development. The American Academy of Pediatrics (AAP) has developed a policy stand on promotion of healthy weight-control practices in young athletes.

Sport participation is also very popular at the other end of the age spectrum, and older athletes may also have special nutrient needs. In general, resting metabolism declines in older age, so caloric need may decrease. Older people often eat less, so they need to make wiser food choices, that is, foods with high nutrient density. Campbell and Geik noted that nutrition is a tool that the older athlete should use to enhance exercise performance and health. In particular, they noted that older athletes may need to focus on obtaining sufficient micronutrients, such as the B vitamins and vitamin D. Supplements may be recommended to obtain adequate vitamin B12 and calcium if not obtained from the diet, such as from fortified foods. Female athletes over age 50, because of decreased estrogen levels associated with menopause, need to focus on obtaining adequate calcium. However, they may need less dietary iron compared to their younger counterparts. Older individuals also need to ensure adequate fluid intake because of increased susceptibility to dehydration.

The special nutrient requirements of females, the young, and older adults, as they relate to physical activity, will be incorporated into the text where relevant. In general, most of the nutritional principles underlying exercise and sports performance that are presented in this text apply to most physically active individuals.

https://kidshealth.org/en/teens/eatnrun.html Recommendations for healthy eating for teen athletes are provided.

www.healthychildren.org/english/healthy-living/sports/pages/default.aspx The American Academy of Pediatrics shares guidelines and advice for children and adolescents participating in sports.

www.sportsdietitians.com.au/factsheets/across-the-lifespan/nutrition-for-masters-athletes/ The Sports Dietitians of Australia provides nutrition recommendations for athletes as they age.

Key Concepts

▸ The precompetition meal should be easily digestible, high in carbohydrates, moderate in protein, and low in fat, and it should be consumed about 3–4 hours prior to competition. Athletes should determine what types of foods are compatible with their sport.

▸ Liquid meals and sports bars may be convenient as an occasional meal replacement, including use as a precompetition meal, but should be used only occasionally and not serve as a substitute for healthful whole foods.

▸ A healthful diet is the key to nutrition for male and female athletes at all age levels. However, female, young, and older athletes may have some specific nutritional needs in certain circumstances.

Check for Yourself

▸ Interview a coach or some student-athletes at your school about their meal strategies prior to competition. How do their strategies compare with general recommendations?

APPLICATION EXERCISE

Karl is a 21-year-old student-athlete who competes with his university's dive team. He has been diving since childhood and has always been competitive but recently was asked to compete in an international meet that will require him to travel out of the country for three weeks.

Overall, Karl has a fairly decent diet, but he would like to be in top shape for the international competition next month. When training with his college team, Karl mostly relies on food served at the dining hall, but he does have an apartment where he can store food and prepare meals.

Your Turn:

1. Provide Karl with two precompetition meal suggestions.
2. List at least five snacks Karl can carry with him to eat before training.
3. What suggestions do you have for Karl as he prepares for his international travels and competition? What should he consider when making food and beverage choices abroad?

sirtravelalot/Shutterstock

Review Questions—Multiple Choice

1. Which of the following is NOT considered a food group as part of the MyPlate guidelines?
   a. grains
   b. meats
   c. fruits
   d. dairy
   e. vegetables
2. Which of the following is not an acceptable definition for food labels with the listing "free"?
   a. fat free—less than 0.5 gram of total fat per serving
   b. cholesterol free—less than 2 milligrams per serving
   c. sugar free—less than 0.5 gram per serving
   d. kcal free—less than 40 kcal per serving
   e. sodium free—less than 5 milligrams per serving
3. Which of the following is an example of a serving that counts toward 1 ounce of the MyPlate grains food group?
   a. 1 slice bread
   b. 1 cup pasta
   c. 1 cup rice
   d. 2 cups ready-to-eat cereal
   e. 2 pancakes
4. Which of the following substances is considered a dietary supplement?
   a. herbal substance
   b. vitamin
   c. mineral
   d. botanical substance
   e. all of the above
5. According to the *2020-2025 Dietary Guidelines for Americans,* added sugars should account for less than _____ of a person's total daily kcal intake.
   a. 5 percent
   b. 10 percent
   c. 15 percent
   d. 20 percent
   e. 25 percent
6. According to the AMDR for protein, a person's total dietary protein intake should be between _____ of their total daily kcal intake.
   a. 5-20 percent
   b. 10-35 percent
   c. 20-40 percent
   d. 30-45 percent
   e. 45-65 percent
7. Which key nutrient is not usually found in substantial amounts in the meat group?
   a. vitamin C
   b. iron
   c. protein
   d. niacin
   e. thiamin
8. A food label lists the amount of complex carbohydrates as 5 grams, the amount of simple sugars as 10 grams, the amount of protein as 5 grams, and the amount of fat as 10 grams. Which of the following is true?
   a. Simple sugars make up the majority of the kcal.
   b. Carbohydrate makes up the majority of the kcal.
   c. The amount of kcal from protein and carbohydrate is equal.
   d. The majority of the kcal are derived from fat.
   e. None of the above statements is true.
9. A well-planned plant-based diet may be more healthful than the current typical American diet for all of the following reasons *except* which?
   a. higher in iron
   b. higher in fiber
   c. lower in saturated fats
   d. a higher polyunsaturated to saturated fat ratio
   e. lower in cholesterol
10. Which of the following is a recommendation for the precompetition meal for an endurance athlete who will be competing in warm environmental conditions?
    a. water
    b. sports drinks
    c. high carbohydrate content
    d. moderate protein content
    e. all of the above

Answers to multiple choice questions: 1. b; 2. d; 3. a; 4. e; 5. b; 6. b; 7. a; 8. d; 9. a; 10. e

Critical Thinking Questions

1. Contrast RDAs and AIs and provide examples of at least two vitamins or minerals that are assigned a RDA and two vitamins or minerals that are assigned an AI.
2. Discuss the five food groups depicted in the MyPlate design in respect to the concepts of variety and moderation.
3. Describe the importance of a precompetition meal for an athlete and provide general recommendations for the composition of a precompetition meal.
4. Contrast the dietary intakes of vegans, ovovegetarians, and ovolactovegetarians. What are three specific health benefits that have been associated with following a well-planned plant-based diet?
5. You have a friend who exercises several days a week and who recently started taking multiple dietary supplements. What general recommendations would you make to your friend about the use of dietary supplements?

References

American Academy of Pediatrics. 2005. Promotion of healthy weight-control practices in young athletes. *Pediatrics* 116:1557-64.

Academy of Nutrition and Dietetics. 2013. Position of the Academy of Nutrition and Dietetics: Functional foods. *Journal of the Academy of Nutrition and Dietetics* 113(8):1096-1103.

Academy of Nutrition and Dietetics. 2016. Position of the Academy of Nutrition and Dietetics: Vegetarian diets. *Journal of the Academy of Nutrition and Dietetics* 116(12):1970-80.

Academy of Nutrition and Dietetics. 2009. Position of the Academy of Nutrition and Dietetics: Nutrient supplementation. *Journal of the Academy of Nutrition and Dietetics* 109(12):2073-85.

Academy of Nutrition and Dietetics. 2016. Position of the Academy of Nutrition and Dietetics, Dietitians of Canada, and the American College of Sports Medicine: Nutrition and athletic performance. *Journal of the Academy of Nutrition and Dietetics* 116(3):501-28.

Andrews, A., et al. 2016. Sports nutrition knowledge among mid-major Division I university student-athletes. *Journal of Nutrition and Metabolism* 2016:31724260.

Aragon, A., and Schoenfeld, B. 2013. Nutrient timing revisited: Is there a post-exercise anabolic window: Post-exercise nutrient timing. *Journal of the International Society of Sport Sciences* 10:5.

Aridi Y., et al. 2017. The association between the Mediterranean dietary pattern and cognitive health: A systematic review. *Nutrients* 9(7):E674.

Bar-Or, O. 2001. Nutritional considerations for the child athlete. *Canadian Journal of Applied Physiology* 26:S186-91.

Boaz, M., et al. 2021. Dietary changes and anxiety during the coronavirus pandemic: Differences between the sexes. *Nutrients* 13(12):4193.

Bawa, A., and Anilakumar, K. 2013. Genetically modified foods: Safety, risks and public concerns—A review. *Journal of Food Science and Technology* 50:1035-46.

Campbell, W., and Geik, R. 2004. Nutritional considerations for the older athlete. *Nutrition* 20:603-8.

Castro, G., and Castro, J. 2014. Alcohol drinking and mammary cancer: Pathogenesis and potential dietary preventive alternatives. *World Journal of Clinical Oncology* 5:713-29.

Change, S., and Koegel, K. 2017. Back to basics: All about MyPlate food groups. Journal of the *Academy of Nutrition and Dietetics* 117(9):1351-53.

Clark, N. 1987. Breakfast of champions. *Physician and Sportsmedicine* 15(January):209-12.

Claus, S. P. 2014. Development of personalized functional foods needs metabolic profiling. *Current Opinion in Clinical Nutrition and Metabolic Care* 17:567-73.

Craddock, J., et al. 2016. Vegetarian and omnivorous nutrition—Comparing physical performance. *International Journal of Sports Nutrition and Exercise Metabolism* 26(3):212-20.

Dangour, A., et al. 2010. Nutrition-related health effects of organic foods: A systematic review. *American Journal of Clinical Nutrition* 92:203-10.

Dickinson, A., and MacKay, D. 2014. Health habits and other characteristics of dietary supplement users: A review. *Nutrition Journal* 13:14.

Dinu, M., et al. 2018. Mediterranean diet and multiple health outcomes: An umbrella review of meta-analyses of observational studies and randomized trials. *European Journal of Clinical Nutrition* 72(1):30-43.

Dwyer, J., et al. 2014. Fortification: New findings and implications. *Nutrition Reviews* 72:127-41.

Eckel, R. 2014. Association Task Force on Practice 2013 AHA/ACC guideline on lifestyle management to reduce cardiovascular risk. *Journal of the American College of Cardiology* 63(25 Pt B):2960-84.

Ewy, M., et al. 2022. Plant-based diet: Is it as good as an animal-based diet when it comes to protein? *Current Nutrition Reports* 11:337-346.

Fletcher, R., and Fairfield, K. 2002. Vitamins for chronic disease prevention in adults: Clinical applications. *Journal of the American Medical Association* 287:3127-29.

Freeland-Graves, J., and Nitzke, S. 2013. Position of the Academy of Nutrition and Dietetics: Total diet approach to healthy eating. *Journal of the Academy of Nutrition and Dietetics* 113:307-17.

Fuhrman, J., and Ferreri, D. 2010. Fueling the vegetarian (vegan) athlete. *Current Sports Medicine Reports* 9:233-41.

Fulton, S., et al. 2014. The effect of increasing fruit and vegetable consumption on overall diet: A systematic review and meta-analysis. *Critical Reviews in Food Science and Nutrition* 56(5):802-16.

Garcia-Roves, P., et al. 2014. Nutrient intake and food habits of soccer players: Analyzing the correlates of eating practice. *Nutrients* 6(7):2697-717.

Garcia-Hermoso, A., et al. 2021. Effects of physical education interventions on cognition and academic performance outcomes in children and adolescents: A systematic review and meta-analysis. *British Journal of Sports Medicine* 55(21):1224-32.

Hargreaves, M. 2001. Pre-exercise nutritional strategies: Effects on metabolism and performance. *Canadian Journal of Applied Physiology* 26:S64-70.

Hawley, J., et al. 2006. Promoting training adaptations through nutritional interventions. *Journal of Sports Sciences* 24:709-21.

Howe, S., et al. 2014. Exercise-trained men and women: Role of exercise and diet on appetite and energy intake. *Nutrients* 6(11):4935-60.

Hull, M., et al. 2016. Gender differences and access to a sports dietitian influences dietary habits of collegiate athletes. *Journal of the International Society of Sports Nutrition* 13:38.

Hull, M., et al. 2017. Availability of a sports dietitian may lead to improved performance and recovery of NCAA Division I baseball athletes. *Journal of the International Society of Sports Nutrition* 14:29.

International Food Information Council Foundation. 1999. Myths and facts about food biotechnology. *Food Insight* (September/October):2-3.

Katz, D., et al. 2015. Effects of egg ingestion on endothelial function in adults with coronary heart disease: A randomized, controlled, crossover trial. *American Heart Journal* 169:162-69.

Kent, K., et al. 2022. Food insecure households faced greater challenges putting food on the table during the COVID-19 pandemic in Australia. *Appetite* 169:105815.

Kostadinova, A., et al. 2013. Immunotherapy—Risk/benefit in food allergy. *Pediatric Allergy and Immunology* 24:633-44.

Lieberman, H., et al. 2015. Patterns of dietary supplement use among college students. *Clinical Nutrition* 34:976-85.

Liebman, B. 2013. Six reasons to eat less red meat. *Nutrition Action Health Letter* 40(5):3-7.

Lynch, H., et al. 2016. Cardiorespiratory fitness and peak torque differences between vegetarian and omnivore endurance athletes: A cross-sectional study. *Nutrients* 8(11):E726.

Martins, Y., et al. 1999. Restrained eating among vegetarians: Does a vegetarian eating style mask concerns about weight? *Appetite* 32:145-54.

Mathews N. 2018. Prohibited contaminants in dietary supplements. *Sports Health* 10(1):19-30.

Maughan, R. 2002. The athlete's diet: Nutritional goals and dietary strategies. *Proceedings of the Nutrition Society* 6:87-96.

Maughan, R., and Shirreffs, S. 2007. Nutrition and hydration concerns of the female football player. *British Journal of Sports Medicine* 41 (Supplement 1):60-63.

Misra, R., et al. 2018. Red meat consumption (heme iron intake) and risk for diabetes and comorbidities? *Current Diabetes Reports* 18(11):100.

Mishra, S., et al. 2021. Dietary supplement use among adults: United States, 2017-2018. *NCHS Data Brief, no 399.* Hyattsville, MD: National Center for Health Statistics.

Nestel, P., et al. 2021. Dietary management of cardiovascular risk including type 2 diabetes. *Current Opinions in Endocrinology, Diabetes, and Obesity* 28(2):134-41.

Nicklas, T., et al. 1998. Nutrient contribution of breakfast, secular trends, and the role of ready-to-eat cereals: A review of data from the Bogalusa Heart Study. *American Journal of Clinical Nutrition* 67:757S-763S.

Nielsen, F. H. 2014. Should bioactive trace elements not recognized as essential, but with beneficial health effects, have intake recommendations. *Journal of Trace Elements in Medicine and Biology* 28:406-8.

Noll, M., et al. 2017. Determinants of eating patterns and nutrient intake among adolescent athletes: A systematic review. *Nutrition Journal* 16(1):46.

O'Neil, C., et al. 2010. Whole-grain consumption is associated with diet quality and nutrient intake in adults: The National Health and Nutrition Examination Survey, 1999-2004. *Journal of the American Dietetic Association* 110:1461-68.

Ormiston, C., et al. 2022. Heart-healthy diets and the cardiometabolic jackpot. *Medical Clinics of North America* 106(2):235-47.

Ortega, R. 2006. Importance of functional foods in the Mediterranean diet. *Public Health Nutrition* 9:1136-40.

Palupi, E., et al. 2012. Comparison of nutritional quality between conventional and organic dairy products: A meta-analysis. *Journal of the Science of Food and Agriculture* 92:2774-81.

Parekh, N., et al. 2021. Food insecurity among households with children during the COVID-19 pandemic: Results from a study among social media users across the United States. *Nutrition Journal* 20(1):73.

Pawlak, R., et al. 2014. The prevalence of cobalamin deficiency among vegetarians assessed by serum vitamin B12: A review of literature. *European Journal of Clinical Nutrition* 68:541-48.

Petrie, H., et al. 2004. Nutritional concerns for the child and adolescent competitor. *Nutrition* 20:620-31.

Quatromoni, P. 2017. A tale of two runners: A case report of athletes' experiences with eating disorders in college. *Journal of the Academy of Nutrition and Dietetics* 117(1):21-31.

Roberts, W. O. 2007. Can children and adolescents run marathons? *Sports Medicine* 37:299-301.

Rock, C. 2007. Multivitamin-multimineral supplements: Who uses them? *American Journal of Clinical Nutrition* 85:277S-79S.

Rogeri, P., et al. 2021. Strategies to prevent sarcopenia in the aging process: Role of protein intake and exercise. *Nutrients* 14(1):52.

Rong, Y., et al. 2013. Egg consumption and risk of coronary heart disease and stroke: Dose-response meta-analysis of prospective cohort studies. *BMJ* 346:e8539.

Ronis, J., et al. 2018. Adverse effects of nutraceuticals and dietary supplements. *Annual Review of Pharmacology and Toxicology* 58:583.

Rosenbloom, C., et al. 2006. Special populations: The female player and the youth player. *Journal of Sports Sciences* 24:783-93.

Rothman, R., et al. 2006. Patient understanding of food labels: The role of literacy and numeracy. *American Journal of Preventive Medicine* 31:391-98.

Rudrapal, M., et al. 2022. Dietary polyphenols and their role in oxidative stress-induced human diseases: Insights into protective effects, antioxidant potentials and mechanism(s) of action. *Frontiers in Pharmacology* 13:806470.

Seiler, S., et al. 2007. The fall and rise of the gender difference in elite anaerobic performance, 1952-2006. *Medicine and Science in Sports & Exercise* 39:534-40.

Shaw, K., et al. 2022. Benefits of a plant-based diet and consideration for the athlete. *European Journal of Applied Physiology* 122(5):1163-1178.

Singh, R., et al. 2022. Why and how the Indo-Mediterranean Diet may be superior to other diets: The role of antioxidants in the diet. *Nutrients* 14(4):898.

Swaminathan, A., and Jicha, G. 2014. Nutrition and prevention of Alzheimer's dementia. *Frontiers in Aging Neuroscience* 6:282; 10.3389/fnagi.2014.00282.

Tagtow, A., and Raghavan, R. 2017. Assessing the reach of MyPlate using National Health and Nutrition Examination Survey Data. Journal of the *Academy of Nutrition and Dietetics* 117(2):181-83.

Tang, J., et al. 2009. Ingestion of whey hydrolysate, casein, or soy protein isolate: Effects on mixed muscle protein synthesis at rest and following resistance exercise in young men. *Journal of Applied Physiology* 107:987-92.

Torgerson, P., et al. 2014. The global burden of foodborne parasitic diseases: An update. *Trends in Parasitology* 30:20-6.

van den Brandt, P., and Schulpen, M. 2017. Mediterranean diet adherence and risk of postmenopausal breast cancer: Results of a cohort study and meta-analysis. *International Journal of Cancer* 140(10):2220-31.

Vannice, G., and Rasmussen, H. 2014. Position of the Academy of Nutrition and Dietetics: Dietary fatty acids for healthy adults. *Journal of the Academy of Nutrition and Dietetics* 114:136-53.

Venderley, A., and Campbell, W. 2006. Vegetarian diets: Nutritional considerations for athletes. *Sports Medicine* 36:293-305.

Volek, J., et al. 2006. Nutritional aspects of women strength athletes. *British Journal of Sports Medicine* 40:742-8.

Wang, X., et al. 2014. Fruit and vegetable consumption and mortality from all causes, cardiovascular disease, and cancer: Systematic review and dose-response meta-analysis of prospective cohort studies. *BMJ* 349:g4490.

Weaver, C., et al. 2014. Processed foods: Contributions to nutrition. *American Journal of Clinical Nutrition* 99:1525-42.

Wildmer R., et al. 2015. The Mediterranean diet, its components, and cardiovascular disease. *American Journal of Medicine* 128(3):229-38.

Williams, C. 2006. Nutrition to promote recovery from exercise. *Sports Science Exchange* 19(1):1-6.

Williams, M. 2008. Nutrition for the school aged child athlete. In *The Young Athlete*, eds. H. Hebestreit and O. Bar-Or. Oxford: Blackwell.

Zanini, S., et al. 2021. A review of lifestyle and environmental risk factors for pancreatic cancer. *European Journal of Cancer* 145:53-70.

Human Energy

CHAPTER THREE

Corbis/Glow Images

KEY TERMS

- adenosine triphosphate (ATP)
- aerobic glycolysis
- aerobic lipolysis
- anabolism
- anaerobic glycolysis
- ATP-PCr system
- basal energy expenditure (BEE)
- basal metabolic rate (BMR)
- Calorie
- calorimetry
- catabolism
- chronic fatigue syndrome (CFS)
- crossover concept
- dietary-induced thermogenesis (DIT)
- electron transfer system
- energy
- ergometer
- Estimated Energy Requirement (EER)
- exercise metabolic rate (EMR)
- fatigue
- glycolysis
- joule
- kilojoule
- Krebs cycle
- lactic acid system
- maximal oxygen uptake
- metabolic aftereffects of exercise
- metabolism
- METs
- mitochondria
- nonexercise activity thermogenesis (NEAT)
- onset of blood lactic acid (OBLA)
- oxygen system
- phosphocreatine (PCr)
- physical activity energy expenditure (PAEE)
- Physical Activity Level (PAL)
- power
- relative energy deficiency in sport (RED-S)
- resting energy expenditure (REE)
- resting metabolic rate (RMR)
- steady-state threshold
- thermic effect of food (TEF)
- total daily energy expenditure (TEE)
- VO$_2$ max
- work

LEARNING OUTCOMES

After studying this chapter, you should be able to:

1. Understand the interrelationships among the various forms of chemical, thermal, and mechanical energy, and be able to perform mathematical conversions from one form of energy to another.
2. Identify the three major human energy systems, their major energy sources as stored in the body, and various nutrients needed to sustain them.
3. List the components of total daily energy expenditure (TEE) and how each contributes to the total amount of caloric energy expended over a 24-hour period.
4. Describe the various factors that may influence resting energy expenditure (REE).
5. List and explain the various means whereby energy expenditure during exercise, or physical activity energy expenditure (PAEE), may be measured, and be able to calculate conversions among the various methods.
6. Describe the three different muscle fiber types and the major characteristics of each in relation to energy production during exercise.
7. Explain the relationship between exercise intensity and energy expenditure and how the MET values in the Compendium of Physical Activities may be used to design an exercise program.
8. Discuss the concept of the Physical Activity Level (PAL) and how it relates to estimated energy expenditure (EER). Calculate your EER based on an estimate of your PAL and the Physical Activity Coefficient (PA).
9. Describe the role of the three energy systems during exercise.
10. Explain the various causes of fatigue during exercise and discuss nutritional interventions that may help delay the onset of fatigue.

Introduction

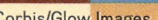
Corbis/Glow Images

As noted in chapter 1, the body uses the food we eat to provide energy, to build and repair tissues, and to regulate metabolism. Of these three functions, the human body ranks energy production first and will use food for this purpose at the expense of the other two functions in time of need. Energy is the essence of life.

Through technological processes, humans have harnessed a variety of energy sources, such as wind, waterfalls, the sun, wood, and oil, to operate the machines invented to make life easier. However, humans cannot use any of these energy sources for their own metabolism but must rely on food sources found in nature. The food we eat must be converted into energy forms that the body can use. Thus, the human body is equipped with a number of metabolic systems to produce and regulate energy for its diverse needs, such as synthesis of tissues, movement of substances between tissues, and muscular contraction.

The underlying basis for the control of movement in all sports is human energy, and successful performance depends upon the ability of the athlete to produce the right amount of energy and to control its application to the specific demands of the sport. Sports differ in their energy demands. In some events, such as the 100-meter dash, success is dependent primarily on the ability to produce energy very rapidly. In others, such as the 26.2-mile marathon, energy need not be produced so rapidly, but must be sustained at an optimal rate for a much longer period. In still other sports, such as soccer, the athlete need not only produce energy at varying rates but must carefully control the application of that energy. Thus, each sport imposes specific energy demands upon the athlete.

A discussion of the role of nutrition as a means to help provide and control human energy is important from several standpoints. For example, inadequate supplies of nutrients needed as a source of fuel, such as muscle glycogen or blood glucose, may cause fatigue. Fatigue also may be caused by the inability of the energy systems to function optimally because of a deficiency of other nutrients, such as selected vitamins and minerals. In addition, the human body is capable of storing energy reserves in a variety of body forms, including body fat and muscle tissue. Excess body weight in the form of fat or decreased body weight due to losses of muscle tissue may adversely affect some types of athletic performance.

One purpose of this chapter is to review briefly the major human energy systems and how they are used in the body under conditions of exercise and rest. Following this, chapters 4 through 9 discuss the role of each of the major classes of nutrients as they relate to energy production in the humans. Chapter 13 details the effects of various food drugs and supplements on human energy systems. Another goal of this chapter is to discuss the means by which humans store and expend energy. Chapters 10 through 12 focus on weight-control methods and expand on some of the concepts presented in this chapter.

Measures of Energy

What is energy?

For our purposes, **energy** represents the capacity to do work. **Work** is one form of energy, often called *mechanical* or *kinetic energy*. When we throw a ball or run a mile, we have done work; we have produced mechanical energy.

Energy exists in a variety of other forms in nature, such as the light energy of the sun, nuclear energy in uranium, electrical energy in lightning storms, heat energy in fires, and chemical energy in oil. The six forms of energy—mechanical, chemical, heat, electrical, light, and nuclear—are interchangeable according to various laws of thermodynamics. We take advantage of these laws every day. One such example is the use of the chemical energy in gasoline to produce mechanical energy—the movement of our cars.

In the human body, four of these types of energy are important. Our bodies possess stores of *chemical energy* that can be used to produce *electrical energy* for creation of electrical nerve impulses, to produce *heat energy* to help keep our body temperature at 37°C (98.6°F) even on cold days, and to produce *mechanical energy* through muscle shortening so that we may move about.

For earthlings, the sun is the ultimate source of energy. Solar energy is harnessed by plants, through photosynthesis, to produce either plant carbohydrates, fats, or proteins, all forms of stored chemical energy. When humans consume plant and animal foods, the carbohydrates, fats, and proteins undergo a series of metabolic changes and are utilized to develop body structure, to regulate body processes, or to provide a storage form of chemical energy (**figure 3.1**).

The optimal intake and output of energy is important to all individuals, but especially for the physically active person. To perform to capacity, body energy stores must be used in the most efficient manner possible.

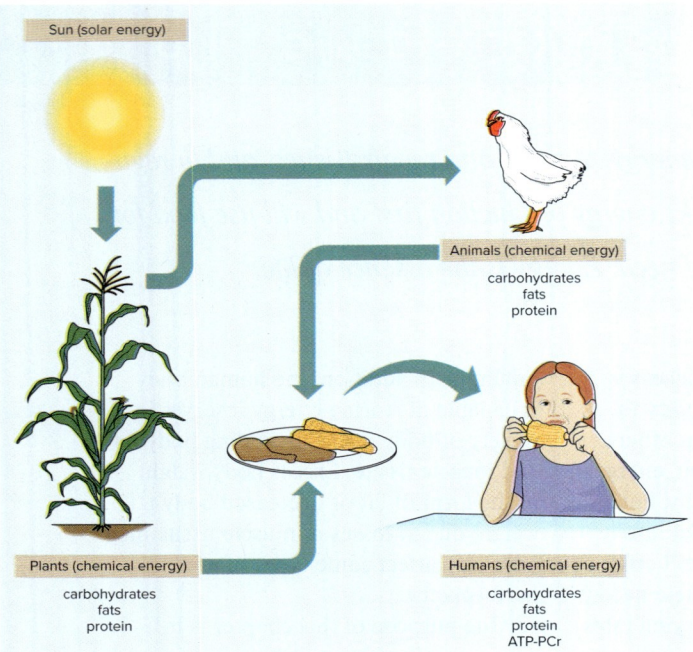

FIGURE 3.1 Through photosynthesis, plants utilize solar energy and convert it into chemical energy in the form of carbohydrates, fats, or proteins. Animals eat plants and convert the chemical energy into their own stores of chemical energy—primarily fat and protein. Humans ingest food from both plant and animal sources and convert the chemical energy for their own stores and use.

What terms are used to quantify work and power during exercise?

Energy has been defined as the ability to do work. According to the physicist's definition, work is simply the product of force times vertical distance, or in formula format, Work = Force × Distance. When we speak of how fast work is done, the term **power** is used. Power is work divided by time, or Power = Work/Time.

Two major measurement systems have been used in the past to express energy in terms of either work or power. The metric system has been in use by most of the world, while England, its colonies, and the United States have used the English system. In an attempt to provide some uniformity in measurement systems around the world, the International Unit System (*Systeme International d'Unites*, or SI) has been developed. Most of the world has adopted the SI, which is comparable to the metric system. Although legislation has been passed by Congress to convert the United States into the SI, and terms such as *gram, kilogram, milliliter, liter,* and *kilometer* are becoming more prevalent, it appears that it will take some time before this system becomes part of our everyday language.

The SI is used in most scientific journals today, but the other two systems appear in older journals. Terms that are used in each system are presented in **table 3.1**. For our purposes in this text, we shall use several English terms that are still in common usage in the United States, but if you read scientific literature, you should be able to convert values among the various systems if necessary. For example, work may be expressed as either foot-pounds, kilogram-meters (kgm), joules, or watts. If you weigh

TABLE 3.1	Terms in the English, metric, and international systems		
Unit	English system	Metric system	International system
Mass	slug	kilogram (kg)	kilogram (kg)
Distance	foot (ft)	meter (m)	meter (m)
Time	second (s)	second (s)	second (s)
Force	pound (lb)	newton (N)	newton (N)
Work	foot-pound (ft-lb)	kilogram-meter (kgm)	joule (J)
Power	horsepower (hp)	watt (W)	watt (W)

150 pounds and climb a 20-foot flight of stairs in 1 minute, you have done 3,000 foot-pounds of work, which maybe converted to the metric system and International system. One kgm is equal to 7.23 foot-pounds, so you would do about 415 kgm (3,000 foot pounds ÷ 7.23 foot pounds = 415 kgm). One **joule** is equal to about 0.102 kgm, so you have done about 4,062 joules of work (415 kgm ÷ 0.102 kgm = 4,062 joules). One watt is equal to one joule per second, so you have generated about 68 watts of power (4,062 joules ÷ 60 = 67.7 watts). Some basic interrelationships among the measurement systems are noted in **table 3.2**.

How do we measure physical activity and energy expenditure?

For research purposes, exercise and nutrition scientists are interested in measuring work output and energy expenditure under two conditions. One condition involves specific techniques in controlled laboratory research, whereas the other condition involves normal daily activities, including actual sports performance. Other than researchers, many physically active individuals, for various reasons such as body weight control, want to know how much exercise they have done and how much energy they have expended.

Over the years, numerous techniques have been used to quantify physical activity and energy expenditure. Given the worldwide increase in obesity and related diseases, such as diabetes, quantifying physical activity is very important, not only for medical scientists but also for the typical individual who exercises for its manifold health benefits. Independent research and reviews, such as those by Bort-Roig, *Consumer Reports,* and Psota and Chen, as well as Hills, Hongu, Lee, Liu, Montoye, and Schoeller and their colleagues, have documented various methods used and provided an analysis of their accuracy and usefulness, and they serve as the basis for the following discussion.

Hills and others indicate that it is important to appreciate that physical activity and energy expenditure are different constructs. Physical activity involves bodily movement, which increases energy expenditure above resting levels. In many cases, objective measures of physical activity may be used to predict energy expenditure. All of the following techniques have been used to collect data on physical activity and energy expenditure specifically for research purposes,

TABLE 3.2 Some interrelationships between work measurement systems

Weight	Distance	Work	Power
1 kilogram = 2.2 pounds	1 meter = 3.28 feet	1 kgm = 7.23 foot-pounds	1 watt = 1 joule per second
1 kilogram = 1,000 grams	1 meter = 1.09 yards	1 kgm = 9.8 joules	1 watt = 6.12 kgm per minute
			1 watt = 0.0013 horsepower
454 grams = 1 pound	1 foot = 0.30 meter	1 foot-pound = 0.138 kgm	1 horsepower = 550 foot-pounds per second
1 pound = 16 ounces	1,000 meters = 1 kilometer	1 foot-pound = 1.35 joules	1 horsepower = 33,000 foot-pounds per minute
1 ounce = 28.4 grams	1 kilometer = 0.6215 mile	1 newton = 0.102 kg	1 horsepower = 745.8 watts
3.5 ounces = 100 grams	1 mile = 1.61 kilometers	1 joule = 1 newton meter	
	1 inch = 2.54 centimeters	1 kilojoule = 1,000 joules	
	1 centimeter = 0.39 inch	1 megajoule = 1,000,000 joules	
		1 joule = 0.102 kgm	
		1 joule = 0.736 foot-pound	
		1 kilojoule = 102 kgm	

but many may be useful for individuals interested in monitoring their personal exercise habits. A detailed discussion of such measurement techniques is beyond the scope of this text but, based on the research cited previously, here are some highlights of the various methods.

Physical Activity Questionnaires The use of questionnaires was one of the first methods to assess physical activity and is still widely used. Although some researchers note that physical activity questionnaires may have potential as a source of rich descriptive data, most contend the data obtained are subjective, not objective, in nature and thus the validity of the data obtained is questionable. Further, physical activity questionnaires are susceptible to recall bias, meaning that some people may not remember their physical activity accurately. This could be on purpose, or accidental, but Matthews and others have noted that quantifying physical activity energy expenditure with questionnaires can have substantial errors.

Heart Rate Monitoring Heart rate monitoring is an easy, effective means to monitor exercise intensity in aerobic exercise. In and by itself heart rate does not measure energy expenditure, but if calibrated with oxygen consumption in an exercise protocol, it may be used to assess energy expenditure in a given individual. The use of heart rate in planning an exercise program is presented in chapter 11.

Ergometers and Exercise Equipment When conducting research, exercise scientists need accurate measures of work output and energy expenditure. An **ergometer**, such as a cycle or arm ergometer, is designed to provide accurate measurement of work, including measures of power and total work output over specific periods of time. For example, bicycle ergometers provide work output in joules and watts, measures of work and power, respectively. However, ergometers do not directly measure energy expenditure, such as kcal expended.

Exercise machines, particularly machines such as treadmills, stationary cycling devices, elliptical trainers, and step machines are examples of ergometers. Many commercial machines are configured to provide an estimate of exercise intensity, such as heart rate monitoring, as well as energy expenditure, such as kcal expended. Some of these machines use computerized standard formulas to predict energy expenditure based on a *typical* individual, while others incorporate data input from the user, such as gender, age, body weight, fitness level, and heart rate. In general, the more personal data the machine uses, the more accurate will be the prediction of energy expenditure.

Calorimetry Our bodies produce heat at all times, even at rest. When we exercise, heat production increases rapidly, and in proportion to energy expenditure. **Calorimetry** is the measurement of heat production by the body. **Figure 3.2** illustrates a bomb calorimeter, which may be used to measure the energy content of a given

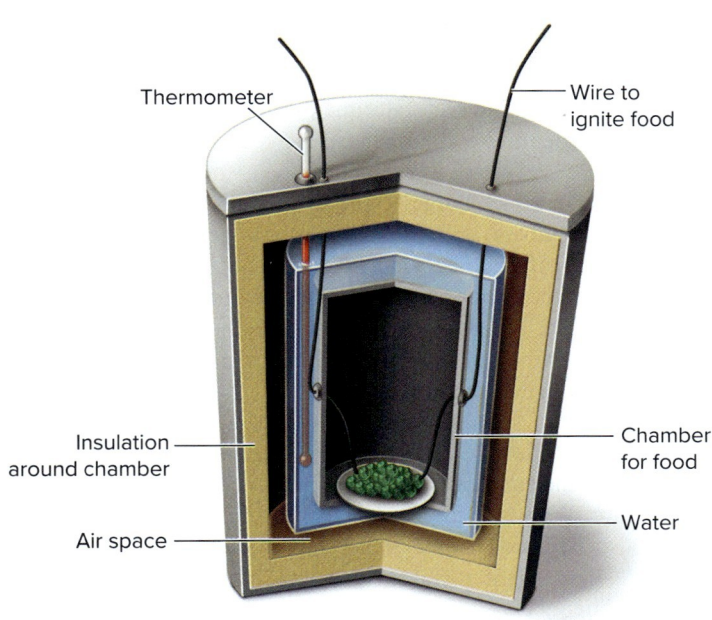

FIGURE 3.2 A bomb calorimeter. The food in the calorimeter is combusted via electrical ignition. The heat (kcal) given off by the food raises the temperature of the water, thereby providing data about the kcal content of specific foodstuffs.

FIGURE 3.3 Indirect calorimetry may be used to measure metabolism by determining the amount of oxygen consumed and the carbon dioxide produced. The test may also be used to measure VO_2 max and other measures of cardiovascular and respiratory function.

substance. For example, a gram of fat contains a certain amount of chemical energy. When placed in the calorimeter and oxidized completely, the heat it gives off can be recorded. We then know the heat energy of one gram of fat and can equate it to various units of energy. Direct and indirect calorimetry are two methods to measure energy expenditure in humans.

Direct calorimetry Large, expensive whole-room calorimeters (metabolic chambers) are available that can accommodate human beings and a water-cooled gradient layer can measure their heat production directly under normal home activities and some conditions of exercise. The limited size of such units may limit the type of physical activity, but they are equipped comparable to a hotel room for daily living activities.

Indirect calorimetry This method determines the amount of oxygen consumed and carbon dioxide produced to calculate energy expenditure. Whole-room calorimeters also use this method, but it may also be used in other laboratory conditions (**figure 3.3**). Moreover, lightweight portable oxygen analyzers are also available to record energy expenditure in freely moving individuals, including those involved in some sport activities. In essence, oxygen is needed to metabolize carbohydrate, fat, and protein for energy production in the body, and if we can accurately measure the oxygen consumption (and carbon dioxide production) of an individual, we can get an accurate assessment of energy expenditure.

Doubly Labeled Water (DLW) Technique Scientists indicate that the DLW technique is the "gold standard" technique for measuring total daily energy expenditure in individuals in a free-living environment. It is safe, noninvasive, and convenient for users over the course of a week or more. In brief, the subject ingests water with stable isotopes of hydrogen and oxygen ($^2H_2{}^{18}O$), which emit no radiation. Analysis of urine and blood samples provides data on ^2H and ^{18}O excretion. The labeled oxygen is eliminated from the body as water and carbon dioxide, whereas the hydrogen is eliminated only as water, which may provide an estimate of carbon dioxide fluctuation that may be converted to energy expenditure. Despite the advantages of this technique, it does not provide information on the type, intensity, or duration of physical activity accounting for some of the total energy expenditure.

Physical Activity Level (PAL) As a measure of physical activity and energy expenditure, the Institute of Medicine has developed four Physical Activity Level (PAL) categories based on the equivalent of walking various distances daily. The PAL can be used in exercise prescription for body weight control and as a factor to improve the assessment of body composition. This procedure is discussed in more detail later in this chapter and in chapter 11 for its role in body weight control.

Metabolic Equivalents (METs) The metabolic equivalent, or MET, of physical activity is based on the amount of oxygen consumed and thus may be converted into energy expenditure. Details on the MET are presented later in this chapter, and as discussed in chapter 11, may be a useful procedure when calculating energy expenditure during exercise for purposes of body weight control.

Motion Sensors Motion sensors are electronic devices worn on various parts of the body and are designed to monitor body motion during physical activity. The pedometer and accelerometer are two commonly used devices. These devices generate electrical signals, the pedometer through the pendular motion of walking and the accelerometer when piezoelectric transmitters are stressed by acceleration forces.

The pedometer has been in use for many years. It is easy to use, is inexpensive, and measures step counts when walking or jogging throughout the day. Thus, it may be very useful in determining the PAL mentioned previously, which is based on daily walking distance. However, pedometers cannot measure exercise intensity. Various models are available, but some may not be reliable. The Yamax Digi-Walker® SW-200 has been highly recommended for its consistent accuracy in laboratory and field tests. By providing feedback to the user, pedometers may also be useful as a means to encourage individuals to do more exercise. Although the basic pedometer may serve its intended purpose, some consider it old-fashioned because it lacks the advanced features of more modern devices.

The accelerometer is relatively small, is usually worn at the waist, and can record data for days. It measures

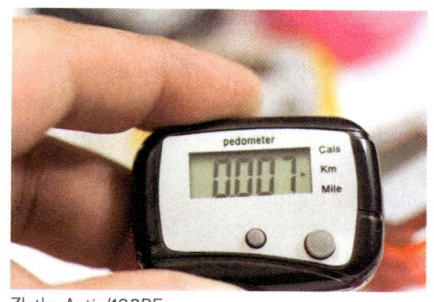

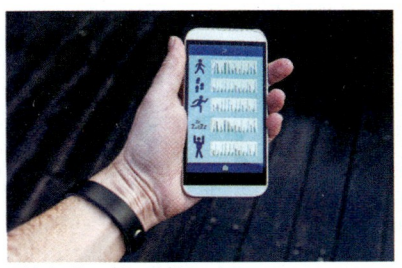
mikkelwilliam/iStock/Getty Images

changes in acceleration in one to three planes, can be used to assess physical activity and energy expenditure, and can estimate exercise intensity. The tri-axial accelerometer can evaluate motion in all three planes (vertical, anterior-posterior, and medial-lateral) and is considered to provide a more detailed analysis of motion than pedometers or uni-axial accelerometers. Moreover, wearing multiple accelerometers, on the wrist, thigh, and ankle, significantly improves the prediction of energy expenditure as compared to a single accelerometer on the hip. However, scientists note accelerometers have some shortcomings, including difficulty estimating movement during activities such as weight training, swimming, and cycling. Some accelerometers have an altimeter, which can distinguish walking or running on level ground with uphill activity. Various companies manufacture accelerometers for personal use, and many are included with mobile phones, which, when coupled with the appropriate apps, makes the assessment of energy expenditure easier for consumers and **nonresearchers**. The ActiGraph is one accelerometer that is reported to have high validity and reliability.

Combination Devices Combining the use of two or more techniques to quantify physical activity and energy expenditure may help improve accuracy as compared to a single measure. For example, Hills and others indicated that a potentially powerful approach to quantifying energy expenditure is the simultaneous use of accelerometry and heart rate monitoring; matching the accelerometer data with the heart rate data could verify that the increase in heart rate was due to physical activity. Other devices combine an accelerometer with a heat sensor. This is an active area of research, and each new device must be validated against more established techniques, such as DLW and indirect calorimetry. Not every additional measurement improves the ability to estimate energy expenditure.

Smartphone and Smartwatch Applications (Apps) for Physical Activity As mentioned in previous chapters, applications (apps) for exercise and nutrition are becoming increasingly popular. In a review, Hongu and others noted more than 100,000 mobile phone health apps currently exist on the market, and the list is increasing. Numerous wearable, wireless gadgets are available, including fashion designer clothing, jewelry, and smart watches, incorporating various technologies such as global positioning system (GPS) navigation, accelerometers, heart rate monitors, and cameras that can interact with your smartphone and provide a wealth of data on your physical activity. Bort-Roig noted that although smartphone use is a relatively new field of study in physical activity research, the few studies that evaluated the validity of physical activity assessment found average-to-excellent levels of accuracy for different behaviors.

Numerous smartphone apps are available to assess one's level of physical activity. You can use Google to access information on *Fitness Apps* or *Free Fitness Apps* to evaluate their features and their reliability. *Consumer Reports* provides periodic analyses of fitness applications, and a recent review indicated the Fitbit, which calculates kcal expended and progress toward your daily health goals, may be worth your money. When possible, select devices that can be tailored for individual characteristics, such as gender, age, and body weight.

All of these measures, from the questionnaire to the smartphone, may be useful as a means to measure and promote physical activity. Some, such as direct calorimetry and the DLW technique, are relatively expensive and used primarily for research purposes, but many, such as smartphone apps, are becoming increasingly available and effective with reasonable cost for use by the general population.

What is the most commonly used measure of energy?

Although there are a number of different ways to express energy, the most common term used in the past and still most prevalent and understood in the United States by most people is **Calorie**.

A calorie is a measure of heat. One gram calorie represents the amount of heat needed to raise the temperature of 1 gram of water 1 degree Celsius. A kilocalorie is equal to 1,000 small calories. It is the amount of heat needed to raise 1 kg of water (1 L) 1 degree Celsius. In human nutrition, because the gram calorie is so small, the kilocalorie is the main expression of energy. It is usually abbreviated as kcal, kc, or C, or capitalized as Calorie. Throughout this book, we will refer to the kilocalorie, or kcal.

According to the principles underlying the first law of thermodynamics, energy may be equated from one form to another. Thus, the kcal, which represents thermal or heat energy, may be equated to other forms of energy. Relative to our discussion concerning physical work such as exercise and its interrelationships with nutrition, it is important to equate the kcal with mechanical work and the chemical energy stored in the body. As will be explained later, most stored chemical energy must undergo some form of oxidation in order to release its energy content as work.

The following represents some equivalent energy values for the kcal in terms of mechanical work and oxygen utilization. Some examples illustrating several of the interrelationships will be used in later chapters.

1 kcal = 3,086 foot-pounds
1 kcal = 427 kgm
1 kcal = 4.2 kilojoules (kJ), or 4,200 joules
1 kcal = 200 ml oxygen (approximately)

Although the kcal is the most commonly used expression in the United States for energy, work, and heat, the **kilojoule** is the proper term in the SI and is used by the rest of the world. It is important for you to be able to convert from kcal into kilojoules, and vice versa. To convert kcal into kilojoules, multiply the number of kcal by 4.2 (4.186 to be exact); to convert kilojoules into kcal, divide the number of kilojoules by 4.2. Simply multiplying or dividing by 4 for each respective conversion will provide a ballpark estimate. In some cases, megajoules (MJ), a million joules, are used to express energy. One MJ equals about 240 kcal, or 4.2 MJ is the equivalent of about 1,000 kcal.

Through the use of a calorimeter, the energy contents of the basic nutrients have been determined. Energy may be derived from the three major foodstuffs—carbohydrate, fat, and protein—plus alcohol. The caloric value of each of these three nutrients may vary somewhat, depending on the particular structure of the different forms. For example, carbohydrate may exist in several forms—as glucose, sucrose, or starch—and the caloric value of each will differ slightly. In general, 1 gram of each of the three nutrients and alcohol, measured in a calorimeter, yields the following kcal:

$$1 \text{ gram carbohydrate} = 4.30 \text{ kcal}$$
$$1 \text{ gram fat} = 9.45 \text{ kcal}$$
$$1 \text{ gram protein} = 5.65 \text{ kcal}$$
$$1 \text{ gram alcohol} = 7.00 \text{ kcal}$$

Unfortunately, or fortunately if one is trying to lose weight, humans do not extract all of this energy from the food they eat. The human body is not as efficient as the calorimeter. For one, the body cannot completely absorb all the food eaten. Only about 97 percent of ingested carbohydrate, 95 percent of fat, and 92 percent of protein are absorbed. In addition, a good percentage of the protein is not completely oxidized in the body, with some of the nitrogen waste products being excreted in the urine. Further, food processing alters the absorption of energy. As an example, described by Traoret and others, more energy is absorbed from peanut flour than peanut butter, as it is more highly processed. Similarly, more energy is absorbed from peanut butter than peanuts, as it is more highly processed. New research from obesity expert Kevin Hall and others demonstrates that ultra processed foods encourage increased energy intake and gains in body fat, adding more evidence to the concept that counting kcal and energy balance is a complex process. Although the following values are not exactly precise, they are approximate enough to be used effectively in determining the caloric values of the foods we eat. Thus, the following caloric values are used throughout this text as a practical guide:

For our purposes, the kcal in food represent a form of potential energy to be used by our bodies to produce heat and work (**figure 3.4**). However, the fact that fat has about twice the amount of energy per gram as carbohydrate (**figure 3.5**) does not mean that it is a better energy source for the active individual, as we shall see in later chapters when we talk of the efficient utilization of body fuels.

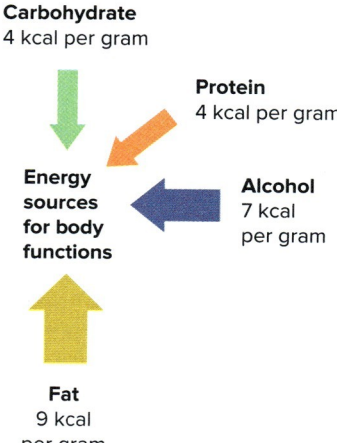

FIGURE 3.4 Eight ounces of orange juice will provide enough chemical energy to enable an average man to produce enough mechanical energy to run about 1 mile.

Pixtal/SuperStock

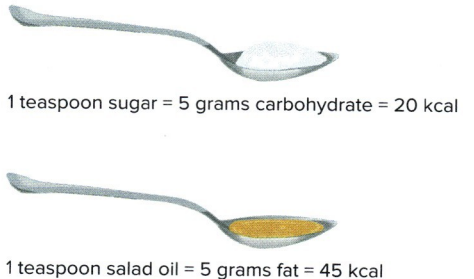

FIGURE 3.5 The kcal as a measure of energy.

Key Concepts

- Energy represents the capacity to do work, and food is the source of energy for humans.
- Energy expenditure in humans may be estimated in a variety of ways, including both direct and indirect calorimetry, doubly labeled water, accelerometers, and pedometers. Smartphones may contain devices to measure physical activity and energy expenditure. Each of these techniques has advantages and disadvantages.
- The kilocalorie, or kcal, is a measure of chemical energy stored in foods; this chemical energy can be transformed into heat and mechanical work energy in the body. A related measure is the kilojoule. One kcal is equal to 4.2 kilojoules.
- Carbohydrates and fats are the primary energy nutrients, but protein may also be an energy source during rest and exercise. In the human body 1 gram of carbohydrate = 4 kcal, 1 gram of fat = 9 kcal, and 1 gram of protein = 4 kcal. Alcohol is also a source of energy; 1 gram = 7 kcal.

> **Check for Yourself**
>
> ▸ Measure the height of a step on a flight of stairs or bleachers and convert it into feet (9 inches = 0.75 foot). Stepping in place, count the total number of steps, both up and down, you do in 1 minute. Multiply your count by the step height to determine the number of feet you have climbed. Next, multiply this value by your body weight in pounds to determine the number of foot-pounds of work you have done. Then, convert this number of foot-pounds into the equivalent amount of kilogram-meters (kgm), kilojoules (kJ), and kilocalories (kcal).

Human Energy Systems

How is energy stored in the body?

The ultimate source of all energy on earth is the sun. Solar energy is harnessed by plants, which take carbon, hydrogen, oxygen, and nitrogen from their environment and manufacture either carbohydrate, fat, or protein. These plant foods contain not only the energy stored in carbohydrate, fat, and protein but also various vitamins and minerals the body needs to process them during energy production. When we consume these foods, our digestive processes break them down into simple compounds that are absorbed into the body and transported to various cells. One of the basic purposes of body cells is to transform the chemical energy of these simple compounds into forms that may be available for immediate use or other forms that may be available for future use.

Energy in the body is available for immediate use in the form of **adenosine triphosphate (ATP)**. It is a complex molecule constructed with high-energy bonds, which, when split by enzyme action, can release energy rapidly for a number of body processes, including muscle contraction. ATP is classified as a high-energy compound and is stored in the tissues in small amounts. It is important to note that ATP is the immediate source of energy for all body functions, and the other energy stores are used to replenish ATP at varying rates. Myburgh notes that muscle contraction is totally dependent on ATP, so the body has developed an intricate system to help replenish ATP as rapidly as needed.

Another related high-energy phosphate compound, **phosphocreatine (PCr)**, is also found in the tissues in small amounts. Although it cannot be used as an immediate source of energy, it can rapidly replenish ATP.

ATP also may be formed from either carbohydrate, fat, or protein after those nutrients have undergone some complex biochemical changes in the body. **Figure 3.6** represents a basic schematic of how ATP is formed from each of these three nutrients.

Because ATP and PCr are found in very small amounts in the body and can be used up in a matter of seconds, it is important to have adequate energy stores as a backup system. Your body stores of carbohydrate, fat, and protein can provide you with ample amounts of ATP, enough to last for many weeks even on a starvation diet. The digestion and metabolism of carbohydrate, fat, and protein are discussed in their respective chapters, so it is unnecessary to present that full discussion here. However, you may wish to preview **figure 3.12** to visualize the metabolic interrelationships

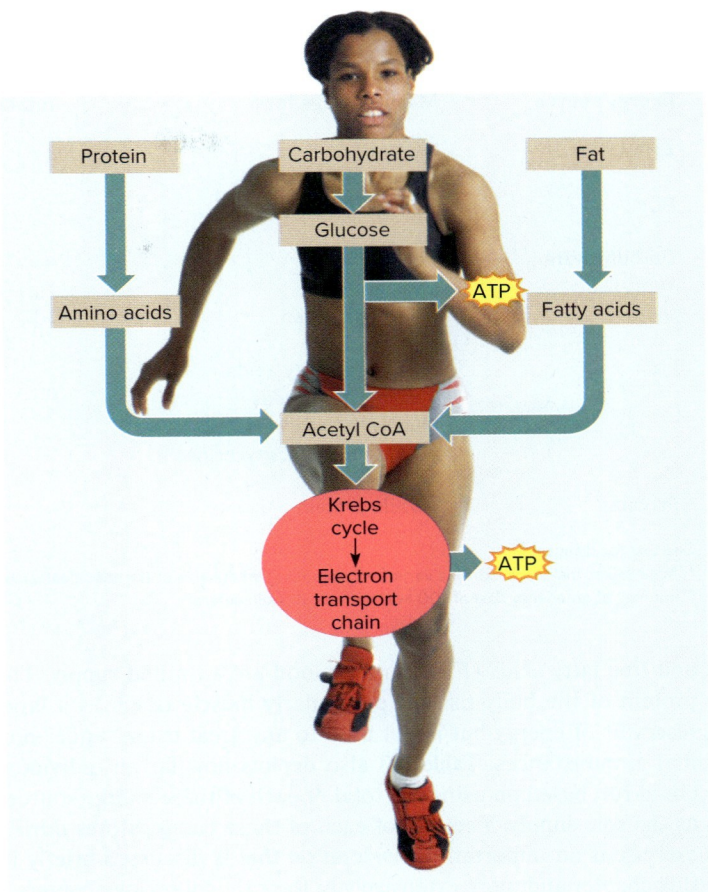

Rubberball/Getty Images

FIGURE 3.6 Simplified schematic of ATP formation from carbohydrate, fat, and protein. All three nutrients may be used to form ATP, but carbohydrate and fat are the major sources via the aerobic metabolism of the Krebs cycle. Carbohydrate may be used to produce small amounts of ATP under anaerobic conditions, thus providing humans with the ability to produce energy rapidly without oxygen for relatively short periods. For more details, see appendix A.

among the three nutrients in the body. For those who desire more detailed schematics of energy pathways, appendix A provides some of the major metabolic pathways for carbohydrate, fat, and protein.

It is important to note that parts of each energy nutrient may be converted into the other two nutrients in the body under certain circumstances. For example, protein may be converted into carbohydrate during prolonged exercise, whereas excess dietary carbohydrate may be converted into fat in the body during rest.

Table 3.3 summarizes how much energy is stored in the human body as ATP, PCr, and various forms of carbohydrate, fat, and protein. The total amount of energy, represented by kcal, is approximate and may vary considerably between individuals depending on body size and composition, diet, and physical fitness level. Carbohydrate is stored in limited amounts as blood (serum) glucose, liver glycogen, and muscle glycogen. The largest amount of energy is stored in the body as fats. Fats are stored as triglycerides in both muscle tissue and adipose (fat) tissue; triglycerides

TABLE 3.3 Major energy stores in the human body with approximate total caloric value*

Energy source	Major storage form	Total body kilocalories	Total body kilojoules	Distance covered**
ATP	Tissues	1	4.2	17.5 yards
PCr	Tissues	4	16.8	70 yards
Carbohydrate	Serum glucose Liver glycogen Muscle glycogen	20 400 1,500	88 1,680 6,300	350 yards 4 miles 15 miles
Fat	Serum-free fatty acids Serum triglycerides Muscle triglycerides Adipose tissue triglycerides	7 75 2,500 80,000	29.2 315 10,500 336,000	123 yards 0.75 mile 25 miles 800 miles
Protein	Muscle protein	30,000	126,000	300 miles

See text for discussion.
*These values may have extreme variations depending on the size of the individual, amount of body fat, physical fitness level, and diet.
**Running at an energy cost of 100 kcal per mile (1.6 kilometers).

and free fatty acids (FFA) in the blood are a limited supply. The protein of the body tissues, particularly muscle tissue, is a large reservoir of energy but is not used to any great extent under normal circumstances. Table 3.3 also depicts how far an individual could run based on using the total of each of these energy sources as the sole supply. The role of each of these energy stores during exercise is an important consideration that is discussed briefly in this chapter and more extensively in their the following chapters.

What are the human energy systems?

Why does the human body store chemical energy in a variety of forms? If we look at human energy needs from an historical perspective, the answer becomes obvious. Sometimes humans needed to produce energy at a rapid rate, such as when sprinting to safety to avoid dangerous animals. Thus, a fast rate of energy production was an important human energy feature that helped ensure survival. At other times, our ancient ancestors may have been deprived of adequate food for long periods, and thus needed a storage capacity for chemical energy that would sustain life throughout these times of deprivation. Hence, the ability to store large amounts of energy was also important for survival. These two factors—rate of energy production and energy capacity—appear to be determining factors in the development of human energy systems.

One need only watch weekend programming for several weeks to realize the diversity of sports popular throughout the world. Each of these sports imposes certain requirements on humans who want to be successful competitors. For some sports, such as weight lifting, the main requirement is brute strength, while for others such as tennis, quick reactions and hand/eye coordination are important. A major consideration in most sports is the rate of energy production, which can range from the explosive power needed by a shot-putter to the tremendous endurance capacity of an ultramarathoner. The physical performance demands of different sports require specific sources of energy.

As noted previously, the body stores energy in a variety of ways—in ATP, PCr, muscle glycogen, and so on. In order for this energy to be used to produce muscular contractions and movement, it must undergo certain biochemical reactions in the muscle. These biochemical reactions serve as a basis for classifying human energy expenditure by several energy, or power, systems.

In his 1979 text *Sports Physiology,* one of the first to discuss the application of human energy systems to sport, Edward L. Fox named three human energy systems—the ATP-PCr system, the lactic acid system, and the oxygen system. As noted in the following sections, other terminology may be used to describe the metabolic relationships to these three energy systems, but the original classification is still useful when discussing the application of human energy to sports performance.

ATP-PCr Energy System The **ATP-PCr system** is also known as the *phosphagen system* because both adenosine triphosphate and phosphocreatine contain phosphates. ATP is the immediate source of energy for almost all body processes, including muscle contraction. This high-energy compound, stored in the muscles, rapidly releases energy when an electrical impulse arrives in the muscle. **Figure 3.7** shows a graphical representation of ATP breakdown. No matter what you do, scratch your nose or lift 100 pounds, ATP breakdown makes the movement possible. ATP must be present for the muscles to contract. The body has a limited supply of ATP and must replace it rapidly if muscular work is to continue. The main purpose of every other energy system, including PCr, is to help regenerate ATP to enable muscle contraction to continue at the optimal desired rate.

PCr, which is also a high-energy compound found in the muscle, can help form ATP rapidly as ATP is used. Energy released when PCr splits is used to form ATP from ADP and P. PCr is also in short supply (but more than ATP) and has to be replenished if used. PCr breakdown to help resynthesize ATP is illustrated in **figure 3.8**.

The ATP-PCr system is critical to energy production. Because these phosphagens are in short supply, any all-out exercise for 5 to 10 seconds could deplete the supply, particularly PCr, in a given

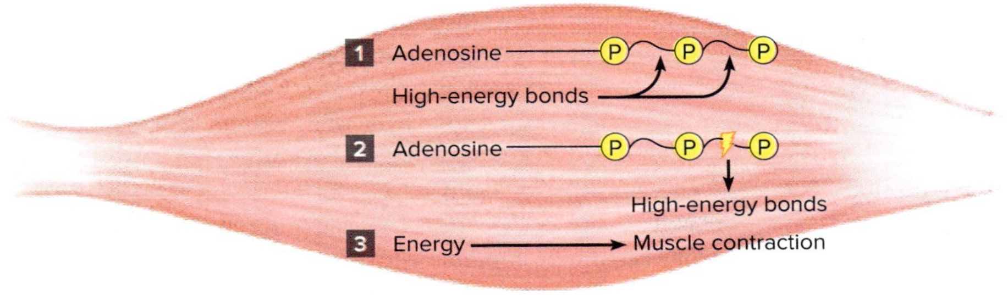

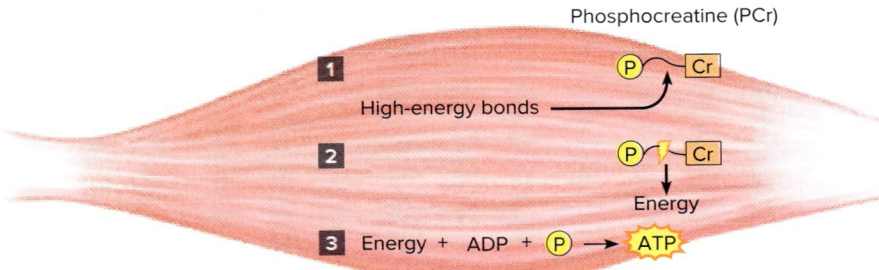

FIGURE 3.7 ATP, adenosine triphosphate. *(a)* ATP is stored in the muscle in limited amounts. *(b)* Splitting of a high-energy bond releases adenosine diphosphate (ADP), inorganic phosphate (P), and energy, which *(c)* can be used for many body processes, including muscular contraction. The ATP stores may be used maximally for fast, all-out bursts of power that last about 1 second. ATP must be replenished from other sources for muscle contraction to continue.

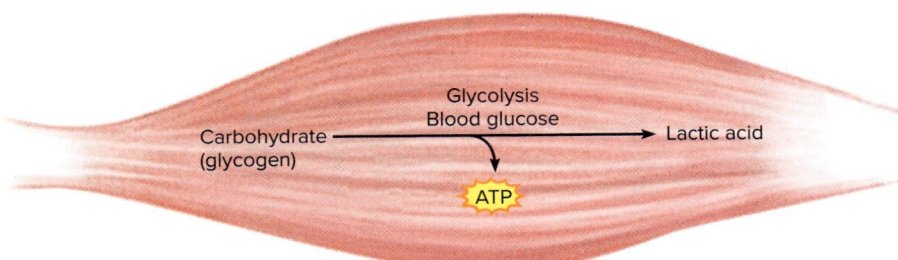

FIGURE 3.8 Phosphocreatine (PCr). *(a)* PCr is stored in the muscle in limited amounts. *(b)* Splitting of the high-energy bond releases energy, which *(c)* can be used to rapidly synthesize ATP from ADP and P. ATP and PCr are called phosphagens and together represent the ATP-PCr energy system. This system is utilized primarily for quick, maximal exercises lasting about 1 to 6 seconds or more, such as sprinting.

FIGURE 3.9 The lactic acid energy system. Muscle glycogen and glucose can break down without the utilization of oxygen. This process is called anaerobic glycolysis. (See appendix A for more details.) ATP is produced rapidly, but lactic acid is the end product. Lactic acid may be a major cause of fatigue in the muscle. The lactic acid energy system is utilized primarily during exercise bouts of very high intensity, those conducted at maximal rates for about 30–120 seconds.

muscle. Hence, the phosphagens must be replaced, and this is the function of the other energy sources. Increasing the level of PCr in the muscle would provide a greater fuel reserve and possibly enhance the performance of brief, high-intensity exercise. Creatine supplementation, discussed in later chapters, is one way athletes are increasing muscle PCr in an effort to enhance performance. In summary, the value of the ATP-PCr system is its ability to provide energy rapidly, for example, in sport events such as competitive weight lifting or sprinting 100 meters. *Anaerobic power* is a term often associated with the ATP-PCr energy system.

Lactic Acid Energy System The **lactic acid system** cannot be used directly as a source of energy for muscular contraction, but it can help replace ATP rapidly when necessary. If you are exercising at a high intensity level and need to replenish ATP rapidly, the next best source of energy besides PCr is glucose. Glucose may enter the muscle from the bloodstream or may be derived from the breakdown of glycogen stored in the muscle. The glucose molecule undergoes a series of reactions to eventually form ATP, a process called **glycolysis**. One of the major factors controlling the metabolic fate of glucose is the capacity of the **mitochondria**, which are cell organelles that *need oxygen* to process glucose to ATP. If the muscle cell mitochondria can process the available glucose, then adequate oxygen is assumed to be available. This is known as **aerobic glycolysis**. Conversely, if the rate of glycolysis surpasses the capacity of mitochondrial oxidation to meet the energy demands of the exercise task or to maintain a high level of aerobic glycolysis, then insufficient ATP is formed and lactic acid is a by-product of the process necessary to increase ATP production. This process has been referred to as anaerobic because of inadequate aerobic processing by the mitochondria, and the term **anaerobic glycolysis** has been used as a scientific term for the lactic acid energy system.

The lactic acid system is diagrammed in **figure 3.9**. It is used in sports events in which energy production is near maximal for 30–120 seconds, such as a 200- or 800-meter run. *Anaerobic capacity* is a term often associated with the lactic acid energy system.

The lactic acid system has the advantage of producing ATP rapidly. Its capacity is limited in comparison to aerobic glycolysis, for only about 5 percent of the total ATP production from muscle glycogen can be released. Moreover, the lactic acid produced as a by-product may be associated with the onset of fatigue, as discussed later in this chapter. In brief, a prevailing hypothesis suggests that anaerobic glycolysis releases hydrogen ions, increasing the acidity within the muscle cell and disturbing the normal cell environment. The processes of energy release and muscle contraction in the muscle cell are controlled by enzymes whose functions may be impaired by the increased acidity in the cell. However, it is important to note that the lactate present after loss of the hydrogen ion still has considerable energy content, which may be used by other tissues for energy or converted back into glucose in the liver.

Oxygen Energy System The third system is the **oxygen system**. It is also known as the oxidative or aerobic system. *Aerobics* is a term used by Dr. Kenneth Cooper in 1968 to describe a system of exercising that created an exercise revolution in this country. In essence, aerobic exercises are designed to stress the oxygen system and provide benefits for the heart and lungs. **Figure 3.10** represents the major physiological processes involved in the oxygen system. The oxygen system, like the lactic acid system, cannot be used directly as a source of energy for muscle contraction, but it does produce ATP in rather large quantities from other energy sources in the body. Muscle glycogen, liver glycogen, blood glucose, muscle triglycerides, blood FFA and triglycerides, adipose cell triglycerides, and body protein all may be ultimate sources of energy for ATP production and subsequent muscle contraction. To do this, glycogen, fats, and protein must be present within the muscle cell or must enter the muscle cell as glucose, FFA, or amino acids. Through a complex series of reactions, metabolic by-products of carbohydrate, fat, or protein combine with oxygen to produce energy, carbon dioxide, and water. These reactions occur in the energy powerhouse of the cell, the mitochondrion. The whole series of events of oxidative energy production primarily involves aerobic processing of carbohydrates and fats (and small amounts of protein) through the **Krebs cycle** and the **electron transfer system**. The oxygen system is depicted in **figure 3.11**. The Krebs cycle and the electron transfer system represent a highly structured array of enzymes designed to remove hydrogen, carbon dioxide, and electrons from substrates such as glucose. At different steps in this process, energy is released and ATP is formed, with most of the ATP produced during the electron transfer process. The hydrogen and electrons eventually combine with oxygen to form water.

Although the rate of ATP production is lower, the major advantage of the oxygen system over the other two energy systems is the production of large amounts of energy in the form of ATP. However, oxygen from the air we breathe must be delivered to the muscle cells deep in the body and enter the mitochondria to be used. This process may be adequate to handle mild and moderate levels of exercise but may not be able to meet the demand of very strenuous exercise. The oxygen system is used primarily in sports emphasizing endurance, such as distance runs ranging from 5 kilometers (3.1 miles) to the 26.2-mile marathon and beyond.

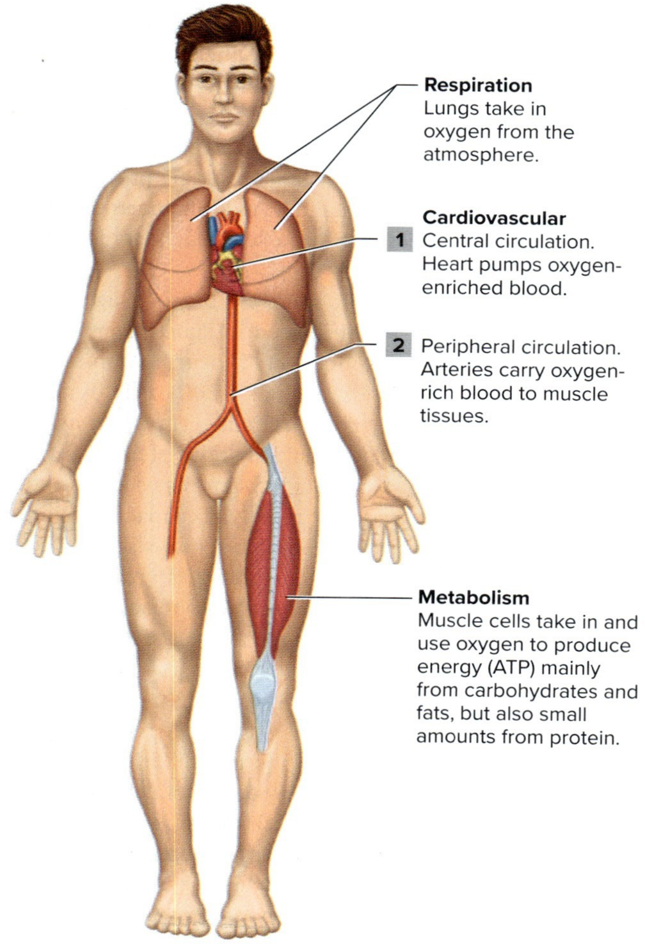

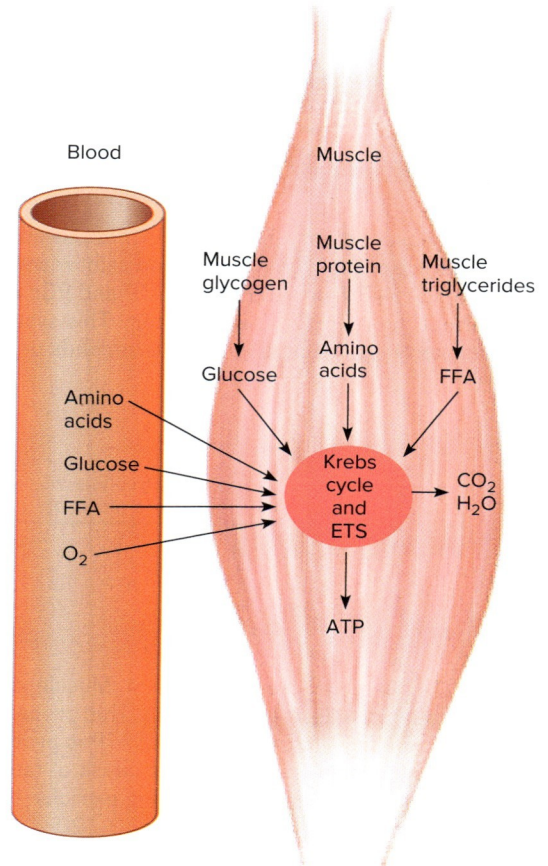

FIGURE 3.10 Physiological processes involved in oxygen uptake.

FIGURE 3.11 The oxygen energy system. The muscle stores of glycogen and triglycerides, along with blood supplies of glucose and free fatty acids (FFA), as well as small amounts of muscle protein and amino acids, undergo complex biochemical changes for entrance into the Krebs cycle and the associated electron transfer system (ETS). In this process, in which oxygen is the final acceptor to the electron, large amounts of ATP may be produced. The oxygen energy system is utilized primarily during endurance-type exercises, those lasting longer than 4 or 5 minutes.

Hawley and Hopkins subdivided the oxygen energy system into two systems. The scientific terms for these two subdivisions are aerobic glycolysis, which uses carbohydrates (muscle glycogen and blood glucose) for energy production, and **aerobic lipolysis**, which uses fats (muscle triglycerides, blood FFA). As discussed in the next two chapters, carbohydrate is the more efficient fuel during high-intensity exercise, whereas fat becomes the predominant fuel used at lower levels of exercise intensity. Thus, aerobic glycolysis provides most of the energy in high-intensity aerobic running events such as 5 kilometers (3.1 miles), 10 kilometers (6.2 miles), and even races up to 2 hours, while aerobic lipolysis may contribute significant amounts of energy in more prolonged aerobic events, such as ultramarathons of 50-100 kilometers (31-62 miles). Aerobic glycolysis and aerobic lipolysis may respectively be referred to as *aerobic power* and *aerobic capacity*. Details relative to the role of these energy systems during exercise are presented in chapters 4 and 5. **Figure 3.12** presents a simplified schematic reviewing the three human energy systems.

Summary   The energy systems for exercise are discussed later in this chapter, but in brief, human energy systems for exercise may be classified as anaerobic or aerobic, and each may be subdivided into energy systems for power and capacity as follows:

- Anaerobic power—ATP-PCr energy system
- Anaerobic capacity—lactic acid energy system (anaerobic glycolysis)
- Aerobic power—aerobic glycolysis
- Aerobic capacity—aerobic lipolysis

As noted, protein may be used as an energy source during exercise and has been referred to as aerobic *proteolysis*, but its contributions are considered minor and it is not classified here as a separate energy system. Its contributions to energy production during exercise are covered in chapter 6.

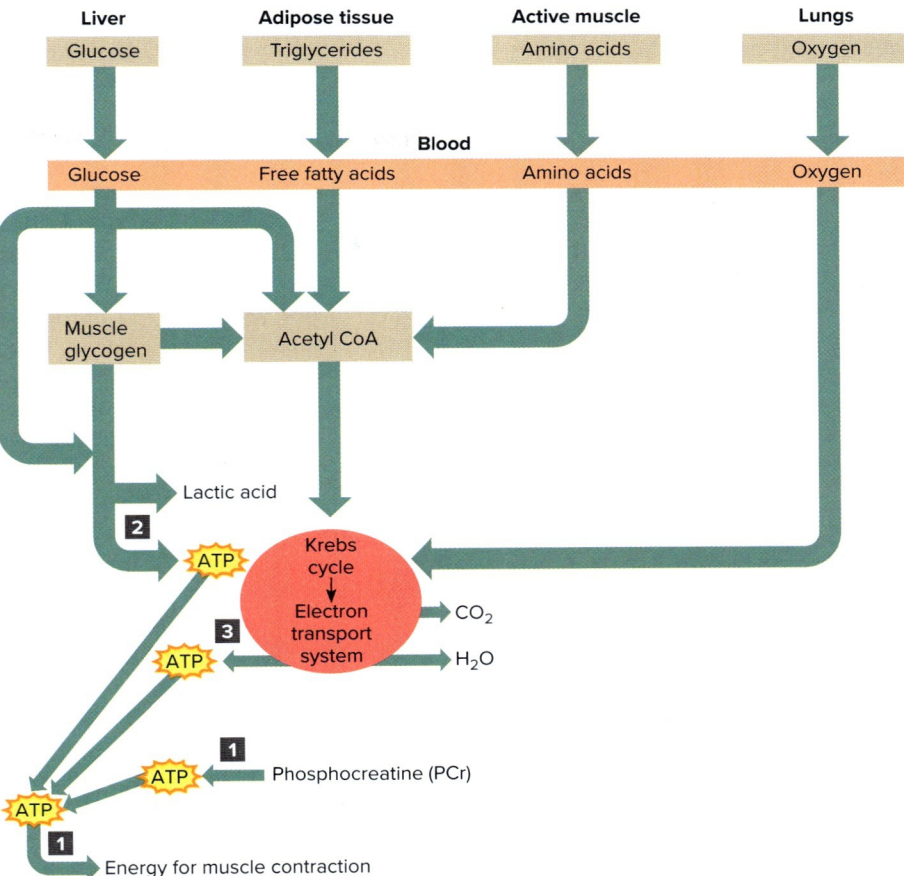

FIGURE 3.12 Simplified flow diagram of the three energy systems. The major nutrients and oxygen are transported to the cells for energy production. In the muscles, ATP is the immediate source of energy for muscle contraction. *(a)* The ATP-PCr system is represented by muscle stores of ATP and phosphocreatine (PCr); PCr can replenish ATP rapidly. *(b)* Glucose or muscle glycogen can produce ATP rapidly via the lactic acid system. *(c)* The oxygen system can produce large amounts of ATP via the aerobic processes in the Krebs cycle, primarily oxidation of carbohydrate and fatty acids. Numerous other energy pathways exist, and some are described in chapter 4 (carbohydrates), chapter 5 (fats), and chapter 6 (protein).

What nutrients are necessary for operation of the human energy systems?

Although the energy for the formation of ATP is derived from the energy stores in carbohydrate, fat, and sometimes protein, this energy transformation and utilization would not occur without the participation of the other major nutrients—water, vitamins, and minerals. These three classes of nutrients function very closely with protein in the structure and function of numerous enzymes, many of which are active in the muscle-cell energy processes.

Water is used to help break up and transform some energy compounds by a process known as *hydrolysis*. Several vitamins are needed for energy to be released from the cell sources. For example, niacin serves an important function in glycolysis, thiamin is needed to convert glycolytic end products to acetyl CoA for entrance into the Krebs cycle, and riboflavin is essential to forming ATP through the Krebs cycle and electron transfer system. A number of other B vitamins are also involved in facets of energy transformation within the cell.

Minerals, too, are essential for cellular energy processes. Iron is one of the more critical compounds. Aside from helping hemoglobin deliver oxygen to the muscle cell, it is also a component of myoglobin and the cytochrome part of the electron transfer system. It is needed for proper utilization of oxygen within the cell itself. Other minerals such as zinc, magnesium, potassium, sodium, and calcium are involved in a variety of ways, either as parts of active enzymes, in energy storage, or in the muscle-contraction process.

Proper utilization of body energy sources requires attention not only to the major energy nutrients but also to the regulatory nutrients—water, vitamins, and minerals. In addition, other nutrients and non-nutrients (such as creatine and caffeine) found in food may affect energy metabolism.

> ### Key Concepts
>
> ▶ The potential energy sources in the body include ATP and PCr; serum glucose; glycogen in the liver and muscle; serum-free fatty acids (FFA); triglycerides in the muscle and in adipose tissue; and muscle protein.
> ▶ Three human energy systems have been classified on the basis of their ability to release energy at different rates of speed; they are the ATP-PCr, lactic acid, and oxygen energy systems. The ATP-PCr system is for anaerobic power; lactic acid system for anaerobic capacity; aerobic glycolysis system for aerobic power; and aerobic lipolysis system for aerobic capacity.

Human Energy Metabolism during Rest

What is metabolism?

Human **metabolism** represents the sum total of all physical and chemical changes that take place within the body. The transformation of food to energy, the formation of new compounds such as hormones and enzymes, the growth of bone and muscle tissue, the destruction of body tissues, and a host of other physiological processes are parts of the metabolic process.

Metabolism involves two fundamental processes, anabolism and catabolism. **Anabolism** is a building-up process, or constructive metabolism. Complex body components are synthesized from the basic nutrients. For the active individual, this may mean an increased muscle mass through weight training or an increased amount of cellular enzymes to better use oxygen following endurance-type training. Energy is needed for anabolism to occur. **Catabolism** is the tearing-down process. This involves the disintegration of body compounds into their simpler components. The breakdown of muscle glycogen to glucose and eventually CO_2, H_2O, and energy is an example of a catabolic process. The energy released from some catabolic processes is used to support the energy needs of anabolism.

Metabolism is life. It represents human energy. The metabolic rate reflects how rapidly the body is using its energy stores, and this rate can vary tremendously depending on a number of factors. For all practical purposes, the **total daily energy expenditure (TEE)** may be accounted for by three factors:

- Energy for basal metabolism
- Energy for processing food intake
- Energy for physical activity

Basal energy expenditure accounts for the largest component of TEE, whereas physical activity is the most variable. We shall examine basal energy expenditure and the effect of eating in this section, while the role of physical activity, or exercise, will be covered in the section "Human Energy Metabolism during Exercise."

What factors account for the amount of energy expended during rest?

PBNJ Productions/Blend Images LLC

The body is constantly using energy to build up and tear down substances within the cells. Certain automatic body functions, such as contraction of the heart, breathing, secretion of hormones, and the constant activity of the nervous system, also are consuming energy.

Basal metabolism, or the **basal metabolic rate (BMR)**, represents the energy requirements of the many different cellular and tissue processes that are necessary to continuing physiological activities in a resting, postabsorptive state throughout most of the day. Other than sleeping, it is the lowest rate of energy expenditure. The determination of the BMR is a clinical procedure conducted in a laboratory or hospital setting. The individual fasts for 12 hours. Then, with the subject in a reclining position, the individual's oxygen consumption and carbon dioxide production are measured. Through proper calculations, the BMR is determined. **Basal energy expenditure (BEE)** represents the BMR extrapolated over a 24-hour period.

The **resting metabolic rate (RMR)** is slightly higher than the BMR. It represents the BMR plus small amounts of additional energy expenditure associated with eating and previous muscular activity. According to the National Academy of Sciences, the BMR and RMR differ by less than 10 percent. Consequently, although there are some fine differences in the two terms, they are often used interchangeably. Additionally, the term **resting energy expenditure (REE)** is used to account for the energy processes at rest when extrapolated over 24 hours. In general, we shall use REE to also represent RMR.

Although some of the energy released during oxidative processes at rest supports physiological functions, such as pumping activity of the heart muscle, the majority of energy is released as heat, a thermal effect that keeps our body temperature at about 98.6°F (37°C). Eating a meal and exercising are two other factors that induce a thermal effect.

White and Kearney note that BMR, and hence RMR, shows substantial variation between individuals, which may be a major factor in body weight control, a topic covered in chapter 10.

What effect does eating a meal have on the metabolic rate?

The significant elevation of the metabolic rate that occurs after ingestion of a meal was previously known as the *specific dynamic action of food* but is now often referred to as **dietary-induced thermogenesis (DIT)** or **thermic effect of food (TEF)**. This elevation is usually highest about 1 hour after a meal and lasts for about 4 hours, and it is due to the energy necessary to absorb, transport, store, and metabolize the food consumed. The greater the caloric content of the meal, the greater this TEF effect. Also, the type of food ingested may affect the magnitude of the TEF. The TEF for protein approximates 20-30 percent, carbohydrate approximates 5-10 percent, and the effect of fat is minimal (0-5 percent). Crovetti and others noted that a very high protein meal (68 percent of kcal) elicited a greater TEF for 7 hours post eating than did corresponding diets high in carbohydrate and fat. Even though the increased TEF amounts to only about 6 kcal more per hour, small daily changes in energy balance can lead to weight gain (e.g., 6 kcal per hour for 7 hours × 365 days per year = 15,339 additional kcal per year). As described in chapter 6, there may be some advantage from incorporating more protein into the diet for weight-loss purposes.

The TEF is expressed as a percent of the energy content of the ingested meal. The normal increase in the BMR due to TEF from a mixed meal of carbohydrate, fat, and protein is about 5-10 percent. A TEF of 10 percent will account for 50 kcal of a 500-kcal meal. The remaining 450 kcal are available for energy use by other body processes. The TEF effect accounts for approximately 5-10 percent of the total daily energy expenditure.

The role of TEF in obesity is somewhat controversial. Overfeeding may increase TEF, whereas underfeeding will decrease it. In a recent review, Westerterp indicated that alternating overfeeding and underfeeding may result in a positive energy balance, which may be one of the explanations for the increasing incidence of obesity in our current society.

How can I estimate my daily resting energy expenditure (REE)?

There are several ways to estimate your REE, but whichever method is used, the value obtained is an estimate and will have some error associated with it. To get a truly accurate value you would need a clinical evaluation, such as a standard BMR test. Accurate determination of REE is important for clinicians dealing with obesity patients, for such testing is needed to rule out hypometabolism. However, a number of formula estimates may give you an approximation of your daily REE.

Table 3.4 provides a simple method for calculating the REE of males and females of varying ages. Examples are provided in the table along with calculation of a 10 percent variability. Keep in mind that this is only an estimate of the daily REE, and additional energy would be expended during the day through the TEF effect and the effect of physical activity, as noted later.

A very simple, rough estimate of your REE is 1 kcal per kilogram body weight per hour. Using this procedure, the estimated value for the male in table 3.4 is 1,680 kcal per day (1 × 70 kg × 24 hours) and for the female is 1,320 kcal (1 × 55 kg × 24 hours), values that are not substantially different from those calculated by the table procedure.

www.globalrph.com/harris-benedict-equation.htm Various website calculators are available to estimate your REE. Enter your age, height, weight, and sedentary as the Physical Activity Level to obtain your REE.

TABLE 3.4 Estimation of the daily resting energy expenditure (REE)

Age (years)	Equation
Males	
3-9	(22.7 × body weight*) + 495
10-17	(17.5 × body weight) + 651
18-29	(15.3 × body weight) + 679
30-60	(11.6 × body weight) + 879
>60	(13.5 × body weight) + 487

Example

154-lb male, age 20
154 lbs/2.2 = 70 kg
(15.3 × 70) + 679 = 1,750

Females	
3-9	(22.5 × body weight*) + 499
10-17	(12.2 × body weight) + 746
18-29	(14.7 × body weight) + 496
30-60	(8.7 × body weight) + 829
>60	(10.5 × body weight) + 596

Example

121-lb female, age 20
121 lbs/2.2 = 55 kg
(14.7 × 55) + 496 = 1,304

To get a range of values, simply add or subtract a normal 10 percent variation to the RMR estimate.

Male example: 10 percent of 1,750 = 175 kcal
Normal range = 1,575-1,925 kcal/day
Female example: 10 percent of 1,304 = 130 kcal
Normal range = 1,174-1,434 kcal/day

*Body weight is expressed in kilograms (kg).

What genetic factors affect my REE?

Your REE is directly related to the amount of metabolically active tissue you possess. At rest, tissues such as the heart, liver, kidneys, and other internal organs are more metabolically active than muscle tissue, but muscle tissue is more metabolically active than fat. Changes in the proportion of these tissues in your body will therefore cause changes in your REE.

Wu and others identified almost 2,400 genes significantly correlated with REE. Many factors influencing the REE, such as age, sex, natural hormonal activity, body size and surface area, and to a degree, body composition, are genetically determined. The effect of some of these factors on the REE is generally well known. Because infants have a large proportion of metabolically active tissue and are growing rapidly, their REE is extremely high. The REE declines through childhood, adolescence, and adulthood as full growth and maturation are achieved. Individuals with naturally greater muscle mass in comparison to body fat have a higher REE; the REE of women is about 10–15 percent lower than that of men, mainly because women have a higher proportion of fat to muscle tissue. Genetically lean individuals have a higher REE than do stocky individuals because their body surface area ratio is larger in proportion to their weight (body volume) and they lose more body heat through radiation.

How do dieting and body composition affect my REE?

Body composition may be changed so as to alter REE. Losing body weight, including both body fat and muscle tissue, generally lowers the total daily REE. The REE may be decreased significantly in obese individuals who go on a very low-kcal diet of less than 800 kcal per day. The decrease in the REE, which is greater than would be due to weight loss alone, may be caused by lowered levels of thyroid hormones. In one study, the REE of obese subjects dropped 9.4 percent on a diet containing only 472 kcal per day. The possibility of decreased REE in some athletes who follow very low kcal diets and maintain low body weight through exercise, such as distance runners and wrestlers, has been the subject of debate but has not been shown conclusively.

In contrast, maintaining normal body weight while reducing body fat and increasing muscle mass may raise the REE slightly because muscle tissue has a somewhat higher metabolic level than fat tissue or because the ratio of body surface area to body weight is increased. The decline in the REE that occurs with aging may be attributed partially to physical inactivity with a consequent loss of the more metabolically active muscle tissue and an accumulation of body fat. Methods to lose body fat and increase muscle mass are covered in chapters 11 and 12.

The type of body fat may also influence REE. White fat, or white adipose tissue, is metabolically different from brown fat, or brown adipose tissue. Park and others note that white fat is a major source of health problems associated with obesity, primarily via its role in promoting inflammation and metabolic dysfunction in the body. On the other hand, Lee and others note that brown fat plays a key role in energy homeostasis and, via its function to burn kcal to generate heat, may help protect against diet-induced obesity.

What environmental factors may also influence the REE?

Several lifestyle and environmental factors, including some foods we eat or drink, may influence our metabolism. For example, although caffeine is not a food, it is a common ingredient in some of the foods we may eat or drink. Caffeine is a stimulant and may elicit a significant rise in the REE. One study reported that the caffeine in two to three cups of regular coffee increased the REE 10–12 percent. Hot, spicy foods, such as hot peppers containing capsaicin, can also exert a modest stimulant effect on the metabolism.

Smoking cigarettes also raises the REE. Apparently the nicotine in tobacco stimulates the metabolism similarly to caffeine. This may be one of the reasons some individuals gain weight when they stop smoking. A long time ago, cigarettes were advertised as a means to lose weight. Although some may still smoke cigarettes for weight-control purposes, such practices are strongly discouraged, given the many associated adverse health effects.

Climatic conditions, especially temperature changes, may also raise the REE. Exposure to the cold may stimulate the secretion of several hormones and muscular shivering, which may stimulate heat production up to 400 percent to help us stay warm. Exposure to warm or hot environments will increase energy expenditure through greater cardiovascular demands and the sweating response. Altitude exposure will also increase REE due to increased ventilation.

Many of these factors influencing the REE are important in themselves but may also be important considerations relative to weight-control programs and body temperature regulation. Thus, they are discussed further in later chapters.

As we shall see in the next section, the most important factor that can increase the metabolic rate is exercise.

What energy sources are used during rest?

The vast majority of the energy consumed during a resting situation is used to drive the automatic physiological processes in the body. Because the muscles expend little energy during rest, there is no need to produce ATP rapidly. Hence, the oxygen system is able to provide the necessary ATP for resting physiological processes.

The oxygen system can use carbohydrates, fats, and protein as energy sources. But, as noted in chapter 6, protein is not used as a major energy source under normal dietary conditions. Carbohydrates and fats, when combined with oxygen in the cells, are the major energy substrates during rest. Several factors may influence which of the two nutrients is predominantly used. In general, though, on a mixed diet of carbohydrate, protein, and fat, about 40 percent of the REE is derived from carbohydrate and about 60 percent comes from fat. However, eating a diet rich in carbohydrate or fat will increase the percent of the REE derived, respectively, from carbohydrate and fat. Also, when carbohydrate levels are low, such as after an overnight fast, the percentage of the REE derived from fat increases.

Key Concepts

▶ Human metabolism represents the sum total of all physiological processes in the body, and the metabolic rate reflects the speed at which the body utilizes energy.

▶ The basal metabolic rate (BMR) represents the energy requirements necessary to maintain physiological processes in a resting, postabsorptive state, while the resting metabolic rate (RMR) is a little higher due to the effects of prior eating and physical activity. The terms *BEE* and *REE* represent basal energy expenditure and resting energy expenditure, respectively, totaled over a 24-hour period.

▶ Eating a meal increases the metabolic rate as the digestive system absorbs, metabolizes, and stores the energy nutrients, a process termed the thermic effect of food (TEF).

A meal will increase the RMR by about 5–10 percent, and some meals will elevate the RMR more than others.
- Various methods may be used to estimate daily resting energy expenditure (REE), but one simple means is to use the value of 1 Calorie per kilogram body weight per hour. For a 60-kilogram (132-pound) individual, this would represent 1,440 kcal over the course of 24 hours (60 × 1.0 × 24).
- A number of different factors may affect the REE, including body composition, drugs, climatic conditions, and prior exercise.
- Fats stored in the body serve as the main source of energy during rest.

Check for Yourself

- Using the formula in **table 3.4**, estimate your daily resting energy expenditure (REE) in kcal. Keep this record for later comparisons.

Human Energy Metabolism during Exercise

Exercise is a stressor to the body, and almost all body systems respond. If the exercise is continued daily, the body systems begin to adapt to the stress of exercise. As noted previously and as we shall see in later chapters, these adaptations may have significant health benefits. The two body systems most involved in exercise are the nervous system and the skeletal muscular system. The nervous system is needed to activate muscle contraction, but it is in the muscle cell itself that the energetics of exercise occur. Most other body systems are simply designed to serve the needs of the muscle cell during exercise.

How do my muscles influence the amount of energy I can produce during exercise?

Muscles constitute a significant percentage of our body weight, approximating 45 percent in the typical adult male and 35 percent in the typical adult female. However, in any given individual, these percentages may vary tremendously depending on various factors, such as type and intensity of physical activity. We shall discuss the potential health and sports performance benefits associated with modifying muscle mass in later chapters, but our focus here is on energy production for exercise.

The skeletal muscle cell, or muscle fiber, is a rather simple machine in design but extremely complex in function. It is a tube-like structure containing filaments that can slide by one another to shorten the total muscle. The shortening of the muscle moves bones, and hence work is accomplished, be it simply the raising of a barbell as in weight training or moving the whole body as in running. Like most other machines, the muscle cell has the capability of producing work at different rates, ranging from very low levels of energy expenditure during sleep to nearly a 90-fold increase during maximal, short-term anaerobic exercise.

The human body possesses several different types of skeletal muscle fibers, and their primary differences are in the ability to produce energy. Various types of proteins are found in muscle cells, and the production of energy is dependent on the specific type of proteins present. In general, three different types of skeletal muscle fiber types have been differentiated based on their rate of energy production, and **table 3.5** presents various characteristics associated with each. For comprehensive details on muscle fiber types, see the review by Schiaffino and Reggiani.

Type I muscle fiber is also known as the *slow-twitch red fiber*, and, as this name implies, is used for slow muscle contractions, such as during rest and light aerobic physical activity. It is often referred to as the *slow-oxidative (SO) fiber*. The characteristics associated with it, such as high mitochondria and myoglobin content, support its high oxidative capacity and resistance to fatigue. Use of

TABLE 3.5 Characteristics associated with the three types of skeletal muscle fibers

Type	I	IIa	IIb (IIx)
Twitch speed	Slow	Faster	Fastest
Color	Red	Red	White
Size (diameter)	Small	Medium	Large
Fatigability	Slow	Moderate	Fast
Force production	Low	High	Highest
Oxidative processes	Highest	Moderate	Lowest
Mitochondria	Highest	Moderate	Low
Myoglobin	Highest	Moderate	Low
Blood flow	Highest	Moderate	Lowest
Triglyceride use	Highest	Moderate	Lowest
Glycogen use	Lowest	Moderate	Highest
Phosphocreatine levels	Lowest	Higher	Higher
Energy for sports	Aerobic capacity; aerobic power	Aerobic power; anaerobic capacity	Anaerobic power; anaerobic capacity

the type I fiber is important during events associated with aerobic capacity and aerobic power.

The type IIa muscle fiber, also known as the *fast-twitch red fiber*, also possesses good aerobic capacity, but not as high as the type I fiber. However, it may also produce energy anaerobically via the lactic acid energy system. Hence, it is often referred to as the fast-oxidative glycolytic (FOG) fiber. It also has high ATP-PCr capacity. Use of the type IIa fiber is important during events associated with aerobic power and anaerobic capacity, but it fatigues sooner than the type I muscle fiber.

The type IIb (IIx) muscle fiber, also known as the *fast-twitch white fiber*, possesses poor aerobic capacity and is used primarily for anaerobic energy production. It is often referred to as the fast glycolytic (FG) fiber. Like the type IIa fiber, it also has high ATP-PCr capacity. Use of the type IIb muscle fiber is important during events associated with anaerobic power and anaerobic capacity, but it fatigues very rapidly.

Most muscles contain all three types of muscle fibers, and all fibers are used during exercise tasks of varying intensity. However, the use of one fiber type will usually predominate, dependent on the intensity of the exercise task and the associated human energy system. Physical training can improve the efficacy of each muscle fiber type, and the benefits that accrue to each depend on the type and extent of exercise training. Moreover, the distribution of muscle fiber types will vary among different individuals due to genetic predisposition, and such differences may influence the level of success in certain sport endeavors. Wilson and others indicate that type I muscle fibers are found in abundance in elite endurance athletes, while type IIa and IIb fibers are proportionally higher in elite strength and power athletes.

What effect does muscular exercise have on the metabolic rate?

As noted in the previous section, the REE is measured with the subject at rest in a reclining position. Any physical activity will raise metabolic activity above the REE and thus increase energy expenditure. Accounting for changes in physical activity over the day may provide a reasonable, although imprecise, estimate of the total daily energy expenditure. Very light activities such as sitting, standing, playing cards, cooking, and typing all increase energy output above the REE, but we normally do not think of them as exercise, as noted later in this chapter. For purposes of this discussion, the **exercise metabolic rate (EMR)** represents the increase in metabolism brought about by moderate or strenuous physical activity such as brisk walking, climbing stairs, cycling, dancing, running, and other such planned exercise activities. The EMR is known more appropriately as **physical activity energy expenditure (PAEE)**.

The most important factor affecting the metabolic rate is the intensity or speed of the exercise. To move faster, your muscles must contract more rapidly, consuming proportionately more energy. Use of type I muscle fibers predominates during low-intensity exercise, and type II fibers are increasingly recruited with more intense exercise. The following represents approximate energy expenditure in kcal per minute for increasing levels of exercise intensity for an average-sized adult male. However, for most of us, it would be impossible to sustain the higher levels of energy expenditure for long, less than a minute or so, and the highest level could be sustained for only a second or so.

Level of intensity	Caloric expenditure per minute
Resting metabolic rate	1.0
Sitting and writing	2.0
Walking at 2 mph	3.3
Walking at 3 mph	4.2
Running at 5 mph	9.4
Running at 10 mph	18.8
Running at 15 mph	29.3
Running at 20 mph	38.7
Maximal power weightlift	>90.0

Although the intensity of the exercise is the most important factor affecting the magnitude of the metabolic rate, there are some other important considerations. In some activities, the increase in energy expenditure is not directly proportional to speed, for the efficiency of movement will affect caloric expenditure. Very fast walking becomes more inefficient, so the individual burns more kcal per mile walking briskly compared to more leisurely walking. A beginning swimmer wastes a lot of energy, whereas one who is more accomplished may swim with less effort, saving kcal when swimming a given distance. Swimming and cycling at very high speeds exponentially increase water or air resistance, so caloric expenditure also increases exponentially. Moreover, the individual with a greater body weight will burn more kcal for any given amount of work in which the body has to be moved, as in walking, jogging, or running. It simply costs more total energy to move a heavier load.

How is energy expenditure of the three human energy systems measured during exercise?

As discussed earlier in this chapter, physical activity and energy expenditure can be measured in a variety of ways, such as with the use of ergometers and accelerometers. In this section, we discuss measurement techniques to quantify energy production from the three human energy systems.

Ward-Smith noted that due to accurate measurements of oxygen uptake and carbon dioxide output, the energy contributions from aerobic metabolism are readily quantifiable, whereas the energy contribution from anaerobic metabolism is far more difficult to determine.

ATP-PCr Energy System
Energy production from the ATP-PCr energy system has been measured by several procedures. One procedure involves a muscle biopsy with subsequent analysis for ATP and PCr levels to determine use following exercise, but the small muscle biopsy may not represent ATP-PCr use in other muscles. ATP and PCr levels may also be determined by computerized

imaging procedures, a noninvasive procedure, but the exercise task must be confined to specific movements due to the nature of the imaging equipment. Thus, Lange and Bury indicate that it is difficult to obtain precise physiological or biochemical data during common explosive-type exercise tests, such as short sprints.

Lactic Acid Energy System Laboratory techniques are also available to measure the role of the lactic acid system in exercise, primarily by measuring the concentration of lactic acid in the blood or in muscle tissues. One measure of exercise intensity is the so-called anaerobic threshold, or that point where the metabolism is believed to shift to a greater use of the lactic acid system. This point is often termed the **onset of blood lactic acid (OBLA)**, or *lactate threshold*. The anaerobic threshold may also be referred to as the **steady-state threshold**, indicating that endurance exercise may continue for prolonged periods if you exercise below this threshold value. Other procedures, such as the maximal accumulated oxygen deficit (MAOD), are used in attempts to quantify anaerobic energy expenditure, but Noordhof and others note that unlike aerobic capacity, anaerobic capacity cannot be easily quantified.

Oxygen Energy System Laboratory tests also are necessary to measure the contribution of the oxygen system during exercise, and this is the most commonly used technique for measuring exercise intensity (**figure 3.3**). The most commonly used measurement is the **maximal oxygen uptake**, which represents the highest amount of oxygen that an individual may consume under exercise situations. In essence, the technique consists of monitoring the oxygen uptake of the individual while the exercise intensity is increased in stages. When oxygen uptake does not increase with an increase in workload, the maximal oxygen uptake has been reached. Maximal oxygen uptake is usually expressed as VO_2 **max**, which may be stated as liters per minute or milliliters per kilogram body weight per minute. An example is provided in **figure 3.13**. A commonly used technique to indicate exercise intensity is to report it as a certain percentage of an individual's VO_2 max, such as 50 or 75 percent. If blood samples are taken periodically to measure serum levels of lactic acid, the percent of VO_2 max at which the steady-state threshold occurs may be determined. Additionally, measurement of oxygen during recovery from exercise may be used to calculate the MAOD, as noted in the previous section, an indirect marker for anaerobic contributions to energy expenditure during exercise. Proper training may increase both VO_2 max and the steady-state threshold, as illustrated in **figure 3.14**.

How can I convert the various means of expressing exercise energy expenditure into something more useful to me, such as kcal per minute?

A number of research studies have been conducted to determine the energy expenditure of a wide variety of sports and other physical activities.

The energy costs have been reported in a variety of ways, including kcal, kilojoules (kJ), oxygen uptake, and **METs**. The MET

FIGURE 3.13 Maximal oxygen uptake (VO_2 max). The best way to express VO_2 max is in milliliters of oxygen per kilogram (kg) of body weight per minute (ml O_2/kg/min). As noted in the figure, the leaner individual has a lower VO_2 max in liters but a higher VO_2 max when expressed relative to weight. In this case, the leaner individual has a higher degree of aerobic fitness, at least as measured by VO_2 max per unit body weight.

VO_2 max: liters/minute	3.6 L (3,600 ml)	4.0 L (4,000 ml)
Kg body weight	60	80
VO_2 max: ml O_2/kg/minute	60	50

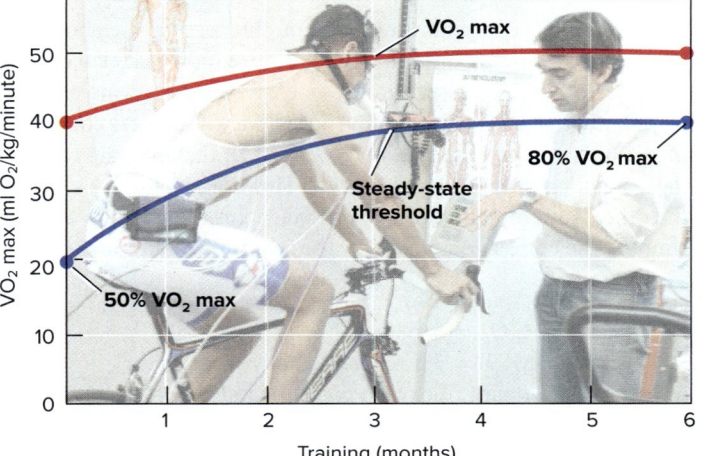

Universal Images Group/Getty Images

FIGURE 3.14 The effect of training upon VO_2 max and the steady-state threshold. Training increases both your VO_2 max and your steady-state threshold, which is the ability to work at a greater percentage of your VO_2 max without producing excessive lactic acid—a causative factor in fatigue. For example, before training the VO_2 max may be 40 ml while the steady-state threshold is only 20 ml (50 percent of VO_2 max). After training, VO_2 max may rise to 50 ml, but the steady-state threshold may rise to 40 ml (80 percent of the VO_2 max).

	Rest	Slow walk (2 mph)	Fast walk (5 mph)	Run (8 mph)
Liters of oxygen/minute	.25	.5–.75	1.5–1.75	2.5–3.0
kilocalories/minute	1.25	2.5–3.75	7.5–8.75	12.5–15.0
kilojoules/minute	5	10–15	30–35	50–60
METs	1	2–3	6–7	10–12

FIGURE 3.15 Energy equivalents in oxygen consumption, kilocalories, kilojoules, and METs. This figure depicts four means of expressing energy expenditure during four levels of activity. These approximate values are for an average male of 154 pounds (70 kg). If you weigh more or less, the values will increase or decrease accordingly.

is a unit that represents multiples of the resting metabolic rate (**figure 3.15**). These concepts are, of course, all interrelated, so an exercise can be expressed in any one of the four terms and converted into the others. For our purposes, we will express energy cost in kcal per minute based on body weight, as that appears to be the most practical method for this book. However, just in case you see the other values in another book or magazine, here is how you may simplify the conversion. We know the following approximate values:

$$1 \text{ kcal} = 4 \text{ kJ}$$
$$1 \text{ L } O_2 = 5 \text{ kcal}$$
$$1 \text{ MET} = 3.5 \text{ ml } O_2/\text{kg body weight/min}$$
(amount of oxygen consumed during rest)

These values are needed for the following calculations:
Example: Exercise cost = 20 kJ/minute
To get kcal cost, divide kJ by the equivalent value for kcal.

$$20 \text{ kJ/min}/4 = 5 \text{ kcal/min}$$

Example: Exercise cost = 3 L of O_2/min
To get kcal cost, multiply liters of O_2 × kcal per liter.

$$\text{Caloric cost} = 3 \times 5 = 15 \text{ kcal/min}$$

Example: Exercise cost = 25 ml O_2/kg body weight/min
You need body weight in kg, which is weight in pounds divided by 2.2. For this example 154 lbs = 70 kg. Determine total O_2 cost/min by multiplying body weight times O_2 cost/kg/min.

$$70 \times 25 = 1{,}750 \text{ ml } O_2$$

Convert ml into L: 1,750 ml = 1.75 L
Multiply liters O_2 × kcal per liter
Caloric cost = 1.75 × 5 = 8.75 kcal/min

Example: Exercise cost = 12 METs
You need body weight in kg—for this example, 70 kg.
Multiply total METs times O_2 equivalent of 1 MET.

$$12 \times 3.5 \text{ ml } O_2/\text{kg/min} = 42.0 \text{ ml } O_2/\text{kg/min}$$

Multiply body weight times this result
70 × 42 ml O_2/kg/min = 2,940 ml O_2/min
Convert ml into L: 2,940 ml O_2/min = 2.94 L O_2/min
Multiply liters O_2 × kcal per liter
Caloric cost = 2.94 × 5 = 14.70 kcal/min

We shall use this METs procedure later, using a simplified formula to calculate caloric expenditure:

$$\text{METs} \times 3.5 \text{ ml } O_2/\text{kg/min} \times \text{kg body weight} \div 200$$
$$12 \times 3.5 \times 70 \text{ kg} \div 200 = 14.70 \text{ kcal/min}$$

How can I tell what my metabolic rate is during exercise?

The human body is basically a muscle machine designed for movement. Almost all of the other body systems serve the muscular system. The nervous system causes the muscles to contract. The digestive system supplies nutrients. The cardiovascular system delivers these nutrients along with oxygen in cooperation with the respiratory system. The endocrine system secretes hormones that affect muscle nutrition. The excretory system removes waste products. When humans exercise, almost all body systems increase their activity to accommodate the increased energy demands of the muscle cell. In most types of sustained exercises, however, the major demand of the muscle cells is for oxygen.

Erik Isakson/Blend Images LLC

As noted previously, the major technique for evaluating metabolic rate is to measure the oxygen consumption of an individual during exercise. Athletes may benefit from such physiological testing. Measurements of VO_2 max, maximal heart rate, and the anaerobic threshold may help in planning an optimal training program, and subsequent testing may illustrate training effects. Such testing is becoming increasingly available at various universities and comprehensive fitness/wellness centers, but very useful data, such as heart rate, may be obtained with use of the various gadgets and apps discussed previously.

Given the relationships among exercise intensity, oxygen consumption, and heart rate, the average individual may be able to get a relative approximation of the metabolic rate during exercise. A more or less linear relationship exists between exercise intensity and oxygen uptake. As the intensity level of work increases, so does the amount of oxygen consumed. The two systems primarily responsible for delivering the oxygen to the muscles are the cardiovascular and respiratory systems. There is also a fairly linear relationship between their responses and oxygen consumption. In general, maximal heart rate (HRmax) and VO_2 max coincide at the same exercise intensity level. A simplified schematic is presented in **figure 3.16**.

Because the heart rate (HR) generally is linearly related to oxygen consumption (the main expression of metabolic rate), and because it is easy to measure this physiological response during exercise either manually at the wrist or neck pulse or with a gadget that monitors heart rate, it may prove to be a practical guide to your metabolic rate. The higher your heart rate, the greater your metabolic rate. However, a number of factors may influence your specific heart rate response to exercise, such as the type of exercise (running vs. swimming), your level of physical fitness, your gender, your age, your skill efficiency, your percentage of body fat, and a number of environmental conditions. Thus, it is difficult to predict your exact metabolic rate from your exercise HR. As we shall see in chapter 11, however, the HR data during exercise may be used as a basis for establishing a personal fitness program for health and weight control.

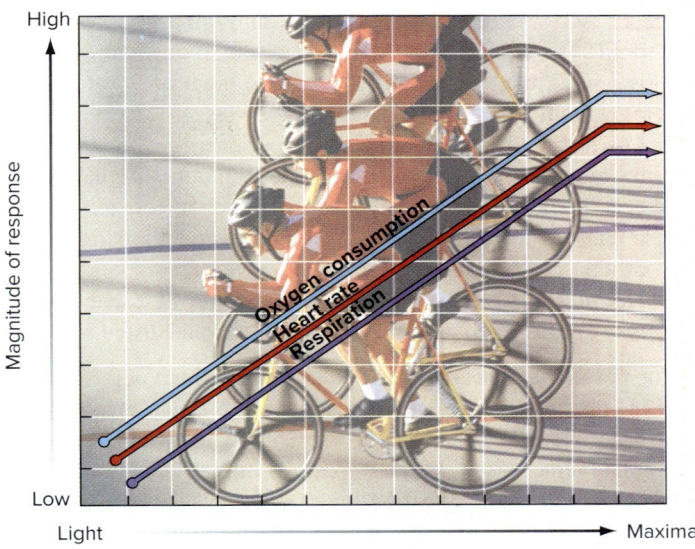
John Kelly/Media Bakery

FIGURE 3.16 Relationships among oxygen consumption, heart rate, and respiration responses to increasing exercise rates. In general, as the intensity of exercise continues, there is a rise in oxygen consumption, which is accompanied by proportional increases in heart rate and respiration. VO_2 max and HRmax usually occur at the same exercise intensity.

How can I determine the energy cost of my exercise routine?

As noted previously in this chapter, there are a variety of ways to determine and express the energy cost of exercise. Unless we are conducting research, most of us are interested in the caloric cost of exercise, primarily for purposes of body weight control. Various devices and smartphone applications may provide us with a good estimate, but simply knowing the metabolic cost in kcal per minute for our usual physical activities may also be helpful. Bushman notes that the MET levels provided in the Compendium of Physical Activities can be a useful means to quantify the caloric expenditure of exercise, as 1 MET = 1 kcal/kg/hr.

> https://sites.google.com/site/compendiumofphysicalactivities/
> Click on Activity Categories and then select the appropriate activity, such as Walking. Scan the list to find the most appropriate type of walking you do to obtain the MET level.

Although the Compendium was not developed to determine the exact energy expenditure of physical activity for individuals, it may be a useful guide with the following considerations:

1. The MET value includes the REE. Thus, the total cost of the exercise includes not only the energy expended during the exercise itself but also the resting energy expenditure, or 1 MET, during the same time frame. Suppose you ran for 1 hour and expended a total of 800 kcal, but your REE during that hour was 1 MET, or 75 kcal. The net cost of the exercise was 725 kcal.

2. The MET values in the table are only for the time you are doing the activity. For example, if your total time exercising is 1 hour, but you take three 5-minute breaks for water, count only 45 minutes for the actual exercise task.
3. The MET values are not precise. Your energy expenditure may be more or less than the estimated amount. Actual caloric cost might vary somewhat because of such factors as your skill level, your training status, the environmental temperature, and others.
4. Not all body weights or MET levels could be listed, but you may approximate by going to the closest value listed.

As one example, we can find the energy expenditure of a 154-pound (70-kg) man who walked 4.0 miles in 1 hour, a very brisk pace on a level, firm surface. Consulting the Compendium online, we find his MET value is 5.0. Next, we can estimate the rate of energy expended (1 MET = 1 kcal/kg/hour) to be 5 kcal/kg/hour or total kcal expended to be 350 kcal/hour. Selecting appropriate activities with individually appropriate MET levels is an important consideration in body weight control, as discussed in chapter 11.

What are the best types of activities to increase energy expenditure?

Purestock/SuperStock

Activities that use the large muscle groups of the body and are performed continuously usually will expend the greatest amount of kcal. Intensity and duration are the two key determinants of total energy expenditure. Activities in which you may be able to exercise continuously at a fairly high intensity for a prolonged period will maximize your total caloric loss. Although this may encompass a wide variety of physical activities, popular modes include walking, running, swimming, bicycling, and aerobic dance. Walking and running are most popular because they are so practical to do.

However, high-intensity interval training (HIIT) has become increasingly popular, particularly when time is an issue. A few general comments about some common modes of exercising would appear to be in order.

Walking Walking at a slow pace is more economical than walking at a faster pace or running. Kuo and others note that walking is a pendular motion, with the stance leg behaving as an inverted pendulum and the swing leg as a regular pendulum. Thus, the pendular motion of slow walking saves energy and reduces the metabolic cost. A good rule of thumb is that you expend about 1 kcal per kilogram/body weight per mile walking at a speed of 2-3 miles per hour on a level, smooth surface.

However, walking faster may increase energy expenditure exponentially. The MET level for walking or jogging 5 mph is the same, 8.3 METs. At high walking speeds (above 5 mph), you may expend more energy than if you jogged at the same speed. Fast, vigorous walking, is an effective way to expend kcal. However, as with other exercise activities, it takes practice to become a fast walker.

Various terms, including the following, have been used to describe walking based on speed.

Strolling–about 2 mph (30 minutes/mile)
Leisurely walking–about 3 mph (20 minutes/mile)
Aerobic or brisk walking–about 4 mph (15 minutes/mile)
Power walking–about 5 mph (12 minutes/mile)
Race walking, beginner–about 6 mph (10 minutes/mile)
Race walking, elite–about 10 mph (6 minutes/mile)

Walking intensity can be increased in other ways. Climbing stairs, at home, at work, in an athletic stadium, or on step machines, is one means to make walking more vigorous. Carrying loads, such as backpacks or hand weights, is another. The Compendium of Physical Activities lists more than 60 modes of walking, with a range of 2-12 METs. Walking leisurely less than 2 miles per hour would be 2 METs, whereas climbing a hill while carrying a heavy load would approximate 12 METs.

For health purposes, walking may be as good as running, if you have the time. In a study comparing exercise and health benefits, Williams and Thompson noted that the more runners ran and the walkers walked, the better off they were in health benefits. If the amount of energy expended was the same between the two groups, then the health benefits were comparable. But the walkers need to spend about twice the amount of time as the runners to get the same benefits.

Running As a general rule, the caloric cost of running a given distance does not depend on the speed. It will take you a longer time to cover the distance at a slower running speed, but the total caloric cost will be similar to that expended at a faster speed. The MET levels for running at a pace of 4 mph, 8 mph, and 12 mph are, respectively, 6.0, 11.8, and 19, resulting in comparable calculated energy expenditures within a range of 108-116 kcal per mile for a 70-kilogram runner. The Compendium of Physical Activities lists more than 25 levels for running and jogging, ranging from 4.5 to 23 METs.

Swimming Because of water resistance, swimming takes more energy to cover a given distance than does either walking or running. Although the amount of energy expended depends somewhat on the type of swimming stroke used and the ability of the swimmer, swimming a given distance takes about four times as much energy as running. For example, swimming a quarter-mile is the energy equivalent of running a mile. Aquatic exercise, such as water aerobics and water running (doing aerobics or running in waist-deep, chest-deep, or deep water), may be effective exercise regimens that help prevent injuries due to impact. The MET values for swimming different strokes may be found in the Compendium of Physical Activities, listed under Water Activities.

Cycling Bicycling takes less energy to cover a given distance in comparison to running on a level surface. The energy cost of bicycling depends on a number of factors such as body weight, the type of bicycle, hills, and body position on the bike (assuming a streamlined position to reduce air resistance). Owing to rapidly increasing air resistance at higher speeds such as 20 mph, the energy cost of bicycling increases at a much faster rate at such speeds. A detailed method for calculating energy expenditure during bicycling is presented in the article by Hagberg and Pena. In general, cycling 1 mile is approximately the energy equivalent of running one-third the distance. The MET values for bicycling at different speeds and under different conditions are listed in the Compendium of Physical Activities.

Group Exercise Various types of group exercise classes have been popular for more than 30 years. These classes can include high- and low-impact aerobic dance, step aerobics, Zumba, Pilates, spin classes, and cardio-kickboxing. All of these classes vary in intensity based on participant effort but have been shown to burn up to about 10 kcal per minute. Though the energy expenditure of group exercise can be comparable to individual exercise tasks such as running or cycling, the greatest benefit of group exercise might be improved exercise adherence. Burke and colleagues examined 44 studies and showed that when exercisers were given the opportunity to interact with others, as when exercising in a group, adherence was better and the exercise program was more effective. The MET equivalents for various types of group exercise are listed in the Compendium of Physical Activities under the category Conditioning Exercise.

Home Aerobic Exercise Equipment Home exercise equipment may also provide a strenuous aerobic workout. Recent research suggests that for any given level of perceived effort, treadmill running burned the most kcal. Exercising on elliptical trainers, cross-country ski machines, rowing ergometers, and stair-climbing apparatus also expended significant amounts of kcal, more so than bicycling apparatus. Many modern pieces of exercise equipment are electronically equipped with small computers to calculate approximate energy cost as kcal per minute and total caloric cost of the exercise. However, as noted previously, research shows that exercise adherence is better when there is contact with fellow exercisers, as in group exercise settings. Comparable to group exercise, the MET equivalents for various types of home exercises are listed in the Compendium of Physical Activities under the category Conditioning Exercise.

Resistance, or Weight, Training Resistance training, or weight training, may be an effective way to expend energy, but it is not as effective as aerobic types of exercise. For example, Bloomer compared energy expenditure during resistance training (free-weight squatting at 70 percent maximal) to aerobic training (cycling at 70 percent VO_2 max) for 30 minutes. Although the heart rates were the same for both types of exercise, the cycling protocol expended 441 kcal while the squatting protocol expended only 269 kcal, a 64 percent difference. Although this is a significant difference, Bloomer noted that the resistance exercise, if performed 4–5 days a week, would meet the recommendations for energy expenditure as suggested by the ACSM. The MET equivalents for resistance-type exercises are listed in the Compendium of Physical Activities under the category Conditioning Exercise.

Sports Activity One of the most enjoyable ways to increase energy expenditure is sports participation. As noted above, sports such as running, race walking, swimming, and bicycling provide opportunities to expend considerable amounts of energy, as do other sports such as soccer, basketball, handball, martial arts, singles tennis, and others. The MET equivalents of participation in a wide variety of sports are presented in the Compendium of Physical Activities.

Passive and Occupational Energy Expenditure Advances in technology have changed the way people accomplish their jobs. Specifically, people are spending more time sitting than ever, which has been implicated in decreased daily energy expenditure and as a contributing factor to the obesity epidemic. One way people are increasing energy expenditure at work is to sit on a Physioball instead of a desk chair, or to abandon sitting entirely and simply do their job while standing at their desk or even walking slowly on a treadmill. In their review of 32 studies, Torbeyns and others concluded that active workstations could increase Physical Activity Levels. One study reported an increased energy expenditure approximating 100 kcal per hour when using a "walk and work" treadmill. Now that sitting itself has been identified as an independent risk factor for mortality, researchers are focusing on how to increase passive energy expenditure to combat the decrease in occupational energy expenditure and to supplement daily energy expenditure to help maintain a healthy body weight. Although these types of interventions do increase energy expenditure, in some cases the differences are very small and unlikely to assist in weight loss. For example, Betts and colleagues reported that standing desks increase energy expenditure, but only by about 20 kcal/day. Gonzalez, an expert in energy balance research and a co-author on the study, equated the kcal in one medium latte to 20 hours of working at a standing desk.

Table 3.6 provides a classification of some common physical activities based on rate of energy expenditure. The implications of these types of exercises for weight-control programs are discussed in later chapters.

Does exercise affect my resting energy expenditure (REE)?

Hero/Corbis/Glow Images

Exercise not only raises the metabolic rate during exercise but also, depending on the intensity and duration of the activity, will keep the REE elevated during the recovery period. The increase in body temperature and in the amounts of circulating hormones such as adrenaline (epinephrine) will continue to influence some cellular activity, and some other metabolic processes, such as circulation and respiration, will remain elevated for a limited time. This effect, which has been labeled the **metabolic aftereffects of exercise**, is calculated by monitoring the oxygen consumption for several hours during the recovery period after the exercise task. The amount of oxygen in excess of the pre-exercise REE, often called excess postexercise oxygen consumption (EPOC), reflects

TABLE 3.6 Classification of selective physical activities based on rate of energy expenditure*

Light, mild aerobic exercise (<5 kcal/min)

Archery
Badminton, social
Baseball
Bicycling (5 mph)
Bowling
Croquet
Dancing, slow ballroom
Golf (using power cart)
Horseback riding (walk)
Swimming (20–25 yards/min)
Tai Chi
Walking (2–3 mph)

Moderate aerobic exercise (5–10 kcal/min)

Badminton, competitive
Basketball, game
Bicycling (10 mph)
Dancing, aerobic
Racquetball
Rope skipping (60 rpm)
Running (5 mph)
Step aerobics (10 inches)
Skateboarding, moderate speed
Tennis, recreational singles
Volleyball, competitive
Walking (3-4 mph)

Moderately heavy to heavy aerobic exercise (>10 kcal/min)

Bicycling, mountain
Bicycling (15–20 mph)
Calisthenics, vigorous
Handball, competitive
In-line skating (10–15 mph)
Racquetball, competitive
Rope skipping (120–140 rpm)
Running (6–9 mph)
Swimming (50–70 yards/min)
Volleyball, competitive
Walking (5-6 mph)
Water jogging, vigorous

*kcal per minute based on a body weight of 70 kg, or 154 pounds. Those weighing more or less will expend more or fewer kcal, respectively, but the intensity level of the exercise will be the same. The actual amount of kcal expended may also depend on a number of other factors, depending on the activity. For example, bicycling into or with the wind will increase or decrease, respectively, the energy cost. See the Compendium of Physical Activities online, for more details.

the additional caloric cost of the exercise above and beyond that expended during the exercise task itself.

Research suggests that if the exercise task is sufficiently intense, the postexercise metabolic rate may remain elevated to burn additional kcal. Knab and others reported male subjects who cycled for 45 minutes at a high-intensity level approximating 75 percent of VO_2 max experienced an elevated EPOC for nearly 14 hours, totaling about 190 kcal. High-intensity interval training (HIIT), discussed in chapter 1, has also been promoted as a means to elevate EPOC. Skelly and others recently evaluated oxygen consumption over a 24-hour period after subjects performed either HIIT or continuous moderate-intensity training. The 20-minute HIIT exercise session consisted of ten 1-minute interval exercise bouts at 90 percent of maximal heart rate, with each interval interspersed with 1 minute of active recovery. The continuous moderate-intensity bout involved cycling at 70 percent of maximal heart rate for 45 minutes. Although the total oxygen cost during HIIT exercise was lower than during the continuous exercise, the total oxygen consumption over 24 hours was similar. For individuals who are aerobically fit but have limited time to exercise, for body weight-control purposes the HIIT protocol may be as effective as more prolonged moderate exercise. Resistance training may also increase the EPOC, but the increase is relatively small. For example, Haddock and Wilkin found that although a bout of resistance training increased the resting metabolic rate for 120 minutes afterwards, subjects expended only about 23 more kcal above the normal resting level for that time frame.

Although the metabolic aftereffects of exercise may be relatively modest, they may add up over time. Moreover, exercise may help mitigate the decrease in the REE often seen in individuals on very low-kcal diets. This point is explored further in chapter 11.

Does exercise affect the thermic effect of food (TEF)?

Many studies have been conducted to investigate the effect of exercise on the thermic effect of food. Unfortunately, no clear answer has been found. Some studies have reported an increase in TEF when subjects exercise either before or after the meal, whereas others revealed little or no effects. Some research even suggests that exercise training decreases the TEF. Warwick reported that prior low-intensity exercise had no effect on the TEF of a meal containing about 560 kcal. Binns and others found that exercising after consuming a high-protein meal increased TEF more so than exercising after fasting, but there were no differences in TEF when compared to a low-protein meal. As noted in chapter 11, the TEF associated with high-protein diets may play a role in body weight control.

Other studies have investigated differences between exercise-trained and untrained individuals relative to TEF, and although some preliminary research noted a decreased TEF in endurance-trained athletes, Tremblay and others also noted that it is still unclear if training causes any significant alterations in TEF. In any case, the increases or decreases noted in the TEF due to either exercise or exercise training were minor, averaging about 5–9 kcal for several hours.

How much energy should I consume daily?

The National Academy of Sciences, through the Institute of Medicine, has released its DRI for energy in conjunction with DRI for carbohydrate, fat, and protein, as noted in chapter 2. Because of possible problems in developing obesity, no RDA or UL were developed for energy. Instead, the Institute of Medicine uses the term **Estimated Energy Requirement (EER)**, which it defines as the dietary intake that is predicted to maintain energy balance in a healthy adult of a defined age, gender, weight, height, and level of physical activity consistent with good health. In essence, the EER estimates your REE based on age, sex, weight, and height, and then modifies this value depending on your daily level of physical activity, which we refer to in this book as physical activity energy expenditure (PAEE).

Your total daily energy expenditure (TEE) is the sum of your BEE, your TEF, and your PAEE. **Figure 3.17** provides some approximate values for the typical active individual, indicating that BEE accounts for 60-75 percent of the total daily energy expenditure, TEF represents 5-10 percent, and PAEE explains 15-30 percent. These values are approximate and may vary tremendously, particularly PAEE, which may range from near 0 percent in the totally sedentary individual to 50 percent or more in ultraendurance athletes.

To illustrate the effect that physical activity, or PAEE, may have on your TEE, the Institute of Medicine developed four **Physical Activity Level (PAL)** categories, which are presented in **table 3.7**. The PAL describes the ratio of the TEE divided by the BEE over a 24-hour period. The PAL (calculated as 2,000/1,400) in **figure 3.17** is 1.43, or low active. The higher the ratio, the greater the amount of daily physical activity.

Sedentary Category The energy expenditure in individuals in the Sedentary category represents their REE, including the TEF, plus various physical activities associated with independent living, such as walking from the house or work to the car, typing, and other forms of very light activity. Levine has coined the term **nonexercise activity thermogenesis (NEAT)** for these very light activities, which represent all the energy we expend daily that is not sleeping, eating, or sports-related exercise. NEAT includes such activities as playing the piano, dancing, housework, washing the car, and similar daily physical activities.

For the other three categories, the Institute of Medicine bases the PAL on the amount of daily physical activity that is the equivalent of walking at a rate of 3-4 miles per hour.

Low Active Category An adult male who weighs 154 pounds (70 kg) and who, in addition to the normal daily activities of independent living, expended the physical activity equivalent of walking 2.2 miles per day would be in the Low Active category, with a PAL of 1.5.

Active Category To be in the Active category with a PAL of 1.75, he would need to expend the physical activity energy equivalent of walking 7.0 miles per day.

Very Active Category To be in this category, with a PAL of 2.2, he would need to expend the energy equivalent of 17 miles per day.

Keep in mind that you do not need to walk this many miles per day, but simply do a multitude of physical activities, such as climbing stairs, golfing, swimming, and jogging, that add up to this energy equivalent. Table 3.6 provides examples of physical activities ranging from light to heavy that may be used to total the required energy equivalents of walking.

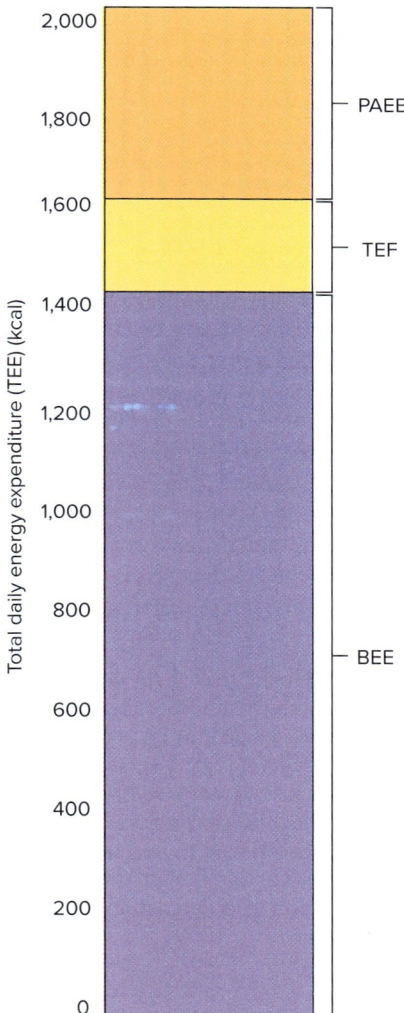

FIGURE 3.17 Total daily energy expenditure (TEE). Three major factors account for the total daily energy expenditure. Basal energy expenditure (BEE) accounts for 60–75 percent, the thermic effect of food (TEF) accounts for 5–10 percent, and 15–30 percent is accounted for by physical activity energy expenditure (PAEE). However, all of these percentages are variable in different individuals, with exercise being the most modifiable component. In the figure, the BEE is 70 percent, the TEF is 10 percent, and the PAEE is 20 percent.

TABLE 3.7	The Physical Activity Level Categories	
Category	Physical Activity Level (PAL)	Physical Activity Coefficient (PA) males/females
Sedentary	≥ 1.0 – < 1.4	1.00/1.00
Low active	≥ 1.4 – < 1.6	1.11/1.12
Active	≥ 1.6 – < 1.9	1.25/1.27
Very active	≥ 1.9 – < 2.5	1.48/1.45

Note that walking at a speed of about 3-4 mph is considered to be moderate aerobic exercise, so equivalent amounts of the types of exercise in this category may serve as substitutes for actual walking. Light and heavy exercise activities may also be done in place of walking. Also note that the energy equivalent of moderate exercise is about 5-10 kcal per minute. Perusal of the Compendium of Physical Activities will provide you with a wide variety of activities that may cost about 5-10 kcal per minute and serve as substitutes for walking.

Based on a number of doubly labeled water studies, the Institute of Medicine developed equations, utilizing the Physical Activity Coefficient (PA) described below, to determine the Estimated Energy Requirement (EER).

Males, 19 years and older:

EER = 662 − 9.53 × age + [PA × (15.91 × weight + 539.6 × height)]

Females, 19 years and older:

EER = 354 − 6.91 × age + [PA × (9.361 × weight + 726 × height)]

Age: In years.

Weight: In kilograms (kg). To convert weight in pounds into kilograms, multiply by 0.454.

Height: In meters (m). To convert height in inches into meters, multiply by 0.0254.

PA: PA is the Physical Activity Coefficient, which is based on the PAL. Based on mathematical consideration to equate energy expenditure between the various PAL categories, the PA coefficient for the Sedentary category was set at 1.0 and the PA for the other categories adjusted accordingly. The PAs for the four PAL categories are presented for adult males and females in **table 3.7**.

Although there may be variances in this estimate of your EER, the estimate may provide you with a ballpark figure of your daily energy needs. Let's look at an example, as depicted in **figure 3.18**, of the difference that physical activity may have on the daily energy needs of a sedentary and very active adult female. Both are 20 years old, weigh 132 pounds (60 kg), and are 55 inches (1.4 m) tall.

Sedentary:

EER = 354 − 6.91 × 20 + [1.0 × (9.361 × 60 + 726 × 1.4)]
EER = 354 − 138.2 + [1.0 × (561.66 + 1,016.4)]
EER = 215.8 + [1,578.06] = 1,794 kcal

Very active:

EER = 354 − 6.91 × 20 + [1.45 × (9.361 × 60 + 726 × 1.4)]
EER = 354 − 138.2 + [1.45 × (561.66 + 1,016.4)]
EER = 2,15.8 + [2,288.19] = 2,504 kcal

The total caloric difference between the sedentary and very active women approximates 700 kcal per day, which may be important in several ways for the very active female. First, as noted in chapter 1, increased physical activity is an important aspect of a healthy lifestyle to prevent a variety of chronic diseases. Second, this additional 700 kcal of energy expenditure daily could have a significant impact on her body weight over time, approximating a loss of more than a pound per week if not compensated for by increased food intake. Third, if she is at an optimal body weight, she may consume an additional 700 kcal per day without gaining weight.

In the meantime, you may wish to calculate your EER not only with the method described above but also with other procedures. The Institute of Medicine (IOM) has provided a link featuring five of the most used equations to predict your REE and TEE, including the Harris-Benedict equation, noted as being the most widely used equation for calculating BMR and TEE, as well as the latest IOM equation for similar purposes. One recommendation is to calculate your total daily energy expenditure with all five equations and compare the findings. Note that all procedures are estimates and there will be some differences among the various

FIGURE 3.18 Estimated Energy Requirement (EER) for two 20-year-old females. Both weigh 132 pounds (60 kg) and are 55 inches (1.4 m) tall. The sedentary female has a Physical Activity Coefficient (PA) of 1.0, whereas the very active female has a PA of 1.45. Compared to her sedentary counterpart, the very active female needs about 700 additional kcal to sustain her physically active lifestyle.

Sedentary lifestyle	Very active lifestyle
PA = 1.0	PA = 1.45
EER = 1,794 kcal	EER = 2,504 kcal

Happy woman: Antonio Guillem/Shutterstock; Girl jogging: Ferli/123RF

estimates, but the data should provide some useful information regarding your daily energy expenditure. You may also compare the findings with your personal wearable gadget or smartphone app that estimates your daily energy expenditure.

> www.globalrph.com/estimated_energy_requirement.htm
> Use this website to calculate your EER with five different methods.

If you are interested in increasing your PAL, then your best bet is to incorporate more light, moderate, and moderately heavy to heavy physical activities into your daily lifestyle. Some additional guidelines for estimating your daily TDEE and EER, particularly in the design of a proper weight-control program, are presented in chapter 11.

Key Concepts

- The three major muscle fiber classifications are type I, type IIa, and type IIb. Type I, known as a slow-oxidative fiber, produces ATP aerobically. Type IIa, also known as a fast-oxidative glycolytic fiber, produces ATP both aerobically and anaerobically. Type IIb, also known as a fast glycolytic fiber, produces ATP anaerobically.
- Physical activity energy expenditure (PAEE), or exercise metabolic rate (EMR), provides us with the most practical means to increase energy expenditure.
- The metabolic rate during exercise is directly proportional to the intensity of the exercise, and the exercise heart rate may serve as a general indicator of the metabolic rate.
- Activities that use the large muscle groups of the body, such as running, swimming, bicycling, and aerobic dance, facilitate energy expenditure. Resistance training of sufficient intensity and duration may also help expend enough energy to satisfy exercise recommendations for caloric expenditure.
- The total daily energy expenditure (TEE) is accounted for by BEE (60–75 percent), TEF (5–10 percent), and PAEE (15–30 percent), although these percentages may vary considerably among individuals.
- The Estimated Energy Requirement (EER) is defined as the dietary intake that is predicted to maintain energy balance in a healthy adult of a defined age, gender, height, weight, and level of physical activity consistent with good health. Changing from a sedentary Physical Activity Level (PAL) to a very active PAL is a very effective means to increase TEE and EER.

Check for Yourself

- Record the types and amounts (in minutes) of your daily physical activity. The application exercise at the end of this chapter may be useful. Consult the Compendium of Physical Activities online to determine your total amount of daily energy expenditure through physical activity and exercise. Compare your findings with the other website estimates of your TEE.

Human Energy Systems and Fatigue during Exercise

In sport, energy expenditure can vary tremendously. For example, Asker Jeukendrup and his associates noted that in one sport, World Class Cycling, events may range in duration from 10 seconds to 3 weeks, involving race distances between 200 meters and 4,000 kilometers. Exercise intensity in a 200-meter event would be extremely high, and much lower during the prolonged event.

What energy systems are used during exercise?

The most important factor determining which energy system will be used is the intensity of the exercise, which is the rate, speed, or tempo at which you pursue a given activity. In general, the faster you do something, the higher your rate of energy expenditure and the more rapidly you must produce ATP for muscular contraction. Very rapid muscular movements are characterized by high rates of power production. If you were asked to run 100 meters as fast as you could, you would exert maximal speed for a short time. On the other hand, if you were asked to run 5 miles, you certainly would not run at the same speed as you would for the 100 meters. In the 100-meter run your energy expenditure would be very rapid, characterized by a high-power production. The 5-mile run would be characterized by low-power production, or endurance.

As noted previously in this chapter, the requirement of energy for exercise is related to a power-endurance continuum. On the power end, we have extremely high rates of energy expenditure that a sprinter might use; on the endurance end, we see lower rates that might be characteristic of a marathon runner. The closer we are to the power end of the continuum, the more rapidly we must produce ATP. As we move toward the endurance end, our rate of ATP production does not have to be as great, but we need the capacity to produce ATP for a longer time.

It should be noted from the outset that all three energy systems—ATP-PCr, lactic acid, and oxygen—are used in one way or another during most athletic activities. (Hargreaves and Spriet provide an excellent overview.) However, one system may predominate, depending primarily on the intensity level of the activity. In this regard, the three human energy systems may be ranked according to several characteristics, which are displayed in **table 3.8**. You may recall that use of the ATP-PCr energy system is referred to as *anaerobic power,* and use of the lactic acid system is referred to as *anaerobic capacity,* whereas the terms *aerobic power* and *aerobic capacity* are used when the oxygen system uses, respectively, carbohydrate and fat as the main energy source.

Both the ATP-PCr and the lactic acid systems are able to produce ATP rapidly and are used in events characterized by high intensity levels that occur for short periods mainly because their capacity for total ATP production is limited. Because both of these systems may function without oxygen, they are called anaerobic. Relative to running performance, the ATP-PCr system predominates in short, powerful bursts of muscular activity such as the short dashes like the 100-meter dash, whereas the lactic acid system begins to predominate during the longer sprints and middle

TABLE 3.8 Major characteristics of the human energy systems*

	ATP-PCr (Anaerobic power)	Lactic acid (Anaerobic capacity)	Oxygen (Aerobic power)	Oxygen (Aerobic capacity)
Main energy source	ATP; phosphocreatine	Carbohydrate	Carbohydrate	Fat
Intensity level	Highest	High	Lower	Lowest
Rate of ATP production	Highest	High	Lower	Lowest
Power production	Highest	High	Lower	Lowest
Capacity for total ATP production	Lowest	Low	High	Highest
Endurance capacity	Lowest	Low	High	Highest
Oxygen needed	No	No	Yes	Yes
Anaerobic/aerobic	Anaerobic	Anaerobic	Aerobic	Aerobic
Characteristic track event	100-meter dash	200–800 meters	5,000-meter (5-km) run	Ultradistance
Time factor	1–10 seconds	30–120 seconds	5 minutes or more	Hours

*Keep in mind that during most exercises, all three energy systems will be operating to one degree or another. However, one system may predominate, depending primarily on the intensity of the activity. See text for further explanation.

distances such as 200, 400, and 800 meters. In any athletic event where maximal power production lasts about 1–10 seconds, the ATP-PCr system is the major energy source. The lactic acid system begins to predominate in events lasting 30–120 seconds, but studies have noted significant elevations in muscle lactic acid in maximal exercise even as brief as 10 seconds.

The oxygen system possesses a lower rate of ATP production than the other two systems, but its capacity for total ATP production is much greater. Although the intensity level of exercise while using the oxygen system is by necessity lower, this does not necessarily mean that an individual cannot perform at a relatively high speed for a long time. The oxygen system can be improved through a physical conditioning program so that ATP production may be able to meet the demands of relatively high-intensity exercise, as discussed previously and highlighted in **figure 3.14**. Endurance-type activities, such as those that last 5 minutes or more, are dependent primarily upon the oxygen system, but the oxygen system makes a very significant contribution even in events as short as 30–90 seconds, as documented by Spencer and Gastin.

In summary, we may simplify this discussion by categorizing the energy sources as either anaerobic or aerobic. Anaerobic sources include both the ATP-PCr and lactic acid systems, whereas the oxygen system is aerobic. **Table 3.9** illustrates the approximate percentage contribution of anaerobic and aerobic energy sources, depending on the level of maximal intensity that can be sustained for a given time period. Thus, for a 100-meter dash covered in 10 seconds, 85 percent of the energy is derived from anaerobic sources. For an elite marathoner (26.2 miles) with times of approximately 125–130 minutes in international-level competition, the aerobic energy processes contribute 99 percent. Although Ward-Smith, using a mathematical approach to predict aerobic and anaerobic contributions during running, noted that these percentage values may be modified slightly for elite athletes, the concept is correct. For example, using track athletes as subjects, Spencer and Gastin found that the relative contribution of the aerobic energy system was 29 percent in the 200-meter run and increased progressively to 84 percent in the 1,500-meter run, noting that the contribution of the aerobic energy system during track running events is greater than traditionally thought. These values are somewhat higher than the aerobic percentage values presented in **table 3.9** but support the concept. The key point is that the longer you exercise, the less your intensity has to be, and the more you rely on your oxygen system for energy production.

What energy sources are used during exercise?

The ATP-PCr system can use only adenosine triphosphate and phosphocreatine, but as noted previously, these energy sources are in short supply and must be replaced by the other two energy systems.

The lactic acid system uses only carbohydrate, primarily the muscle glycogen stores. At high-intensity exercise levels that may be sustained for 1–2 minutes or less, such as exercising well above your VO_2 max, carbohydrate will supply more than 95 percent of the energy. However, the accumulation of lactic acid may be associated with the early onset of fatigue.

TABLE 3.9 Percentage contribution of anaerobic and aerobic energy sources during different time periods of maximal work*

Time	10 sec	1 min	2 min	4 min	10 min	30 min	60 min	120 min
Anaerobic	85	70	50	30	15	5	2	1
Aerobic	15	30	50	70	85	95	98	99

*Percentages are approximate and may vary between sedentary individuals and elite athletes.

In contrast, the oxygen system can use a variety of energy sources, including protein, although carbohydrate and fat are the primary ones. The carbohydrate is found as muscle glycogen, liver glycogen, and blood glucose. The fats are stored primarily as triglycerides in the muscle and adipose cells, but small amounts are also present in the blood. As we shall see in this section and in chapters 4, 5, and 6, a number of different factors can influence which energy source is used by the oxygen system during exercise, but exercise intensity and duration are the two most important factors.

Under normal conditions, exercise intensity is the key factor determining whether carbohydrate or fat is used. Hoppeler and Weibel noted that as one does mild-to-moderate exercise, say up to 50 percent of one's VO_2 max, blood glucose and fat may provide much of the needed energy. However, the transfer of glucose and fat from the vascular system to the muscles becomes limited at about 50 percent of VO_2 max. Thus, as you start to exceed 50 percent of your VO_2 max, you begin to rely more on your intramuscular stores of glycogen and triglycerides. As you continue to increase your speed or intensity, you begin to rely more and more on carbohydrate as an energy source. Apparently, the biochemical processes for fat metabolism are too slow to meet the increased need for faster production of ATP, and carbohydrate utilization increases. The major source of this carbohydrate is muscle glycogen.

The transition from use of fat to carbohydrate as the primary fuel source during increasing intensity of exercise has been referred to as the **crossover concept**, and although the technicalities of specific fuel contributions are the subject of debate, exercise scientists agree that at some specific point in the increase of exercise intensity an individual will begin to derive more energy from carbohydrate than fat (**figure 3.19**). At high levels of energy expenditure, 70-80 percent of VO_2 max, carbohydrates may contribute more than 80 percent of the energy sources. Houston notes that elite marathoners burn about 19-20 kcal per minute and need about 4-5 grams of carbohydrate per minute. This speaks to the need for adequate muscle glycogen stores when this level of exercise is to be sustained for long periods, say in events lasting more than 60-90 minutes.

In events of long duration, when body stores of carbohydrate are nearly depleted, the primary energy source is fat. In the later stages of ultramarathoning events, fat may become the only fuel available, which may necessitate a slower pace because fat is a less efficient fuel. That is why Cermak and van Loon indicated improving carbohydrate availability during prolonged exercise through carbohydrate ingestion has dominated the field of sports nutrition research. More detail on carbohydrate use during exercise is presented in chapter 4. Moreover, protein may become an important energy source in these circumstances; its role is detailed in chapter 6.

What is the "fat burning zone" during exercise?

As noted in chapter 11, the goal in weight-loss programs is to lose fat, not muscle. Various fitness-related internet sites and the consoles of many pieces of cardiovascular exercise equipment instruct

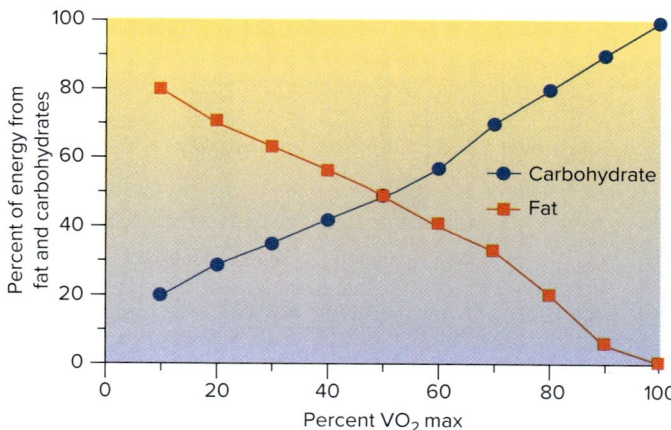

FIGURE 3.19 The crossover concept of carbohydrate and fat utilization during exercise. During low-intensity exercise (20 percent VO_2 max), the majority of energy production is derived from fat, but as exercise intensity increases, the use of carbohydrate for energy production increases and may be 80 percent or higher at very intense exercise. Highly trained endurance athletes will use less carbohydrate and more fat during submaximal exercise as compared to untrained individuals.

exercisers on how to train in the "fat burning zone" for weight loss, which as noted is most often low-intensity exercise, maybe only 40-50 percent of maximal heart rate. However, as shall be discussed in chapter 11, the best recommendation for most exercisers who want to achieve a healthy body weight and improve their fitness is to exercise at the highest intensity appropriate for their age, health, motivation, and current fitness level. **Figure 3.20** illustrates this concept. In the top panel, about 80 percent of total kcal expenditure comes from fat, and only about 20 percent from carbohydrate, so based on these percentages one is exercising in the "fat burning zone" while walking at 2 mph. In the bottom panel the reverse is true; about 80 percent of kcal expenditure comes from carbohydrate and 20 percent comes from fat while walking at 9 mph.

However, one of the key concepts underlying exercise as a means for weight loss is to exercise as intensely as possible for a given time frame. Total caloric expenditure during exercise is the key to promote weight loss. At the bottom in **figure 3.20**, a typical average adult male walking for 30 minutes at a pace of 2 mph will burn about a total of 100 kcal, about 80 from fat and 20 from carbohydrate. Running at 9 mph for the same time frame will burn about 450 kcal, about 90 from fat and 360 from carbohydrate, more than four times the caloric expenditure in the same time frame and similar amounts of fat. Nevertheless, exercising in the "fat burning zone" may be very appropriate for some individuals, such as those beginning an exercise program, the elderly, and others. Additional details regarding exercise and weight loss are presented in chapter 11.

Other than exercise intensity and duration, a number of different factors are known to influence the availability and use

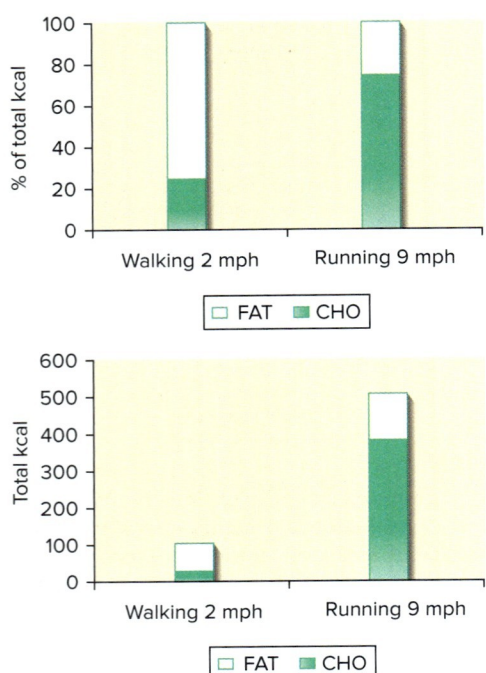

FIGURE 3.20 Top: Energy expenditure from fat and carbohydrate (CHO) during 30 minutes of low-intensity and high-intensity exercise expressed as a percentage of total kcal. Bottom: Energy expenditure from fat and carbohydrate (CHO) during 30 minutes of low-intensity and high-intensity exercise expressed in absolute total kcal. Although a greater percentage of fat is oxidized during low-intensity exercise, many more kcal are expended during high-intensity exercise.

of human energy sources during exercise. Gender, hormones, state of training, composition of the diet, time of eating prior to competition, nutritional status, nutrient intake during exercise, environmental temperature, and drugs are some of the more important considerations. For example, warm environmental temperatures may increase the use of carbohydrates, whereas fasting may facilitate the use of fats. These considerations will be incorporated into the following chapters where appropriate. For the interested reader, Hargreaves provides an excellent detailed review.

What is fatigue?

Fatigue is a very complex phenomenon. It may be chronic, or it may be acute. Both types may affect the athlete.

Chronic Fatigue Chronic fatigue syndrome (CFS), or myalgic encephalomyelitis, is a medical condition characterized by numerous symptoms, the most prevalent being prolonged, incapacitating fatigue lasting at least six months. Moss-Morris and others note that the etiology of CFS is complex and unlikely to be understood through a single mechanism. Multiple factors may be involved, such as viral illnesses, sleep disturbances, immune system dysfunction, excessive mental stress, and prolonged overwork, factors which may be observed in athletes engaged in excessive physical training.

Chronic fatigue in the athlete may develop over time, usually in endurance athletes involved in prolonged, intense training that may involve conditions known as overreaching and overtraining. *Overreaching* is a condition of physical and mental stress that may impair physical performance, but it may be a planned phase of training in elite athletes followed by short-term recovery with return to previous or improved levels of performance. *Overtraining* is a term often used to characterize a syndrome in athletes involving prolonged periods of fatigue. However, Halson and Jeukendrup note that although some scientific and anecdotal evidence support its existence, there appear to be no clear markers for overtraining and more research is needed to establish its existence with certainty. Some contend that the term *overtraining* is misleading, and may actually be related to *underrecovery*, particularly involving inadequate nutrition.

Roy Shephard, the renowned Canadian sport scientist, indicated that overtraining and/or a negative energy balance may be related to the development of CFS in athletes. In such cases, training would be adversely affected and performance certainly would suffer. Given the debilitating effects on exercise and sports performance, scientists are attempting to identify the causes, prevention, and treatment of both overtraining and chronic fatigue syndrome. Recently, Stellingwerff and others noted significant overlap between overtraining and **Relative Energy Deficiency in Sport (RED-S)**, including shared pathways, symptoms, and diagnostic complexities. Importantly, both can be influenced by low carbohydrate and energy availability. Although chronic fatigue syndrome is a serious medical condition, its prevalence in the general population, including athletes, appears to be very low. In a recent meta-analysis, Johnston and others reported the incidence of CFS is low, less than 1 percent of the population with clinical assessment, but over 3 percent with self-diagnosis. CFS also affects children and adolescents.

> www.cdc.gov/cfs/general/index.html For more details on CFS, use the following Centers for Disease Control and Prevention website.

Acute Fatigue Acute mental or physical fatigue is experienced by most athletes at one time or another during maximal efforts. For purposes of the present discussion, **fatigue** will be defined as the inability to continue exercising at a desired level of intensity. Relative to this definition, fatigue may be due to a failure of the rate of energy production in the human body to meet the demands of the exercise task. In simple terms, ATP production rates are unable to match ATP utilization rates.

As acute fatigue can adversely affect sports performance, it has been the subject of considerable research. In general, sports scientists classify the site of acute fatigue in the body as either central or peripheral (**figure 3.21**). *Central fatigue* involves the

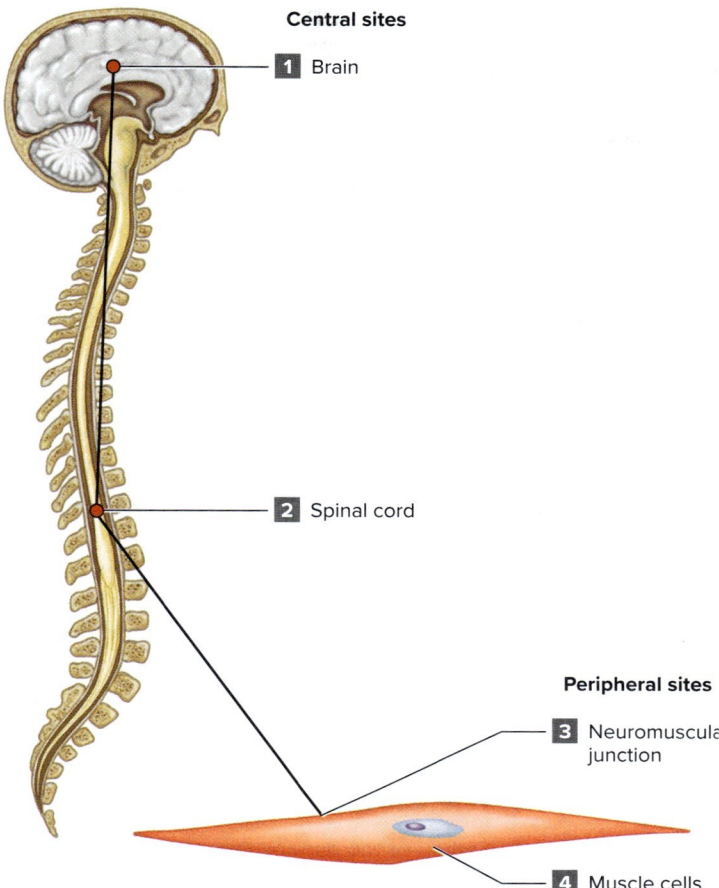

FIGURE 3.21 Fatigue sites. The causes of fatigue are complex and may involve central sites such as the brain and spinal cord or peripheral sites in the muscles. Hypoglycemia, or low blood sugar, could adversely affect the functioning of the brain, while the acidity associated with the production of lactic acid could interfere with optimal energy production in the muscle cells.

brain or spinal cord of the central nervous system (CNS), while *peripheral fatigue* is associated primarily with the muscles and, under some conditions, other body organs such as the heart or lungs.

What causes acute fatigue in athletes?

Robert Fitts, an expert on muscular fatigue during prolonged exercise, noted that the etiology of fatigue is not fully understood despite more than a century of research. He indicated that the mechanisms of muscle fatigue are complex and depend on the type of exercise, one's state of fitness, and the fiber type composition of the muscle.

Although the cause of acute exercise-induced fatigue has not been determined, numerous hypotheses exist, involving both peripheral and central fatigue.

Peripheral Fatigue Fatigue may develop in the muscle for various reasons, including depletion of energy sources or accumulation of fatigue-causing metabolites. For example, Hargreaves indicates that fatigue occurs when the compounds needed to produce ATP are depleted. Depletion of phosphocreatine could decrease the rapid replenishment of ATP in sprint-type events, such as the 200-meter track event. Depletion of muscle glycogen could impair marathon running performance. Relative to the accumulation of metabolites, Ament and Verkerke indicate that some, such as hydrogen ions and inorganic phosphates, may accumulate within the muscle cell during intense exercise, disrupting biochemical equilibrium and causing fatigue. For example, Fitts indicates that hydrogen ion accumulation can decrease sensitivity of the myofibrils to calcium. Some potential factors involve the following:

Substrate depletion	Metabolic by-product accumulation
ATP	Hydrogen ions
Phosphocreatine	ADP and inorganic phosphate
Muscle glycogen	Ammonia
Blood glucose	Reactive oxygen species

Central Fatigue In a review, Sidhu and others reported substantial evidence supporting the finding that fatigue during exercise is accompanied by changes within the central nervous system that may reduce muscle force production. For example, Noakes suggests that central fatigue may be associated with the depletion of critical neurochemicals or the accumulation of "toxic" concentrations of neurotransmitters at neural synapses, which could reduce neural impulses to the muscles. In this regard, Edgerton and Roy noted that elevated levels of serotonin, a neurotransmitter in the brain, are associated with fatigue, and exercise may increase levels of serotonin in the brain. In contrast, the neurotransmitter norepinephrine is a stimulant and may help prevent fatigue; its reduction may help induce fatigue. Relative to central fatigue, Noakes highlights the fact that the single variable that is always maximal at exhaustion during all forms of exercise is the rating of perceived exertion, a mental perception.

Peripheral and Central Fatigue Roelands and others have indicated the current focus on exercise fatigue research involves a complex interplay between peripheral and central limitations of performance. For example, both peripheral and central factors may be involved in setting the pace for a runner in a distance event. Changes within the muscle, such as increased acidity, provide feedback to the central nervous system. At the same time, anticipation by the central nervous system of forthcoming consequences, such as increased heat stress, provide feedforward input to the muscles. Both factors may be involved in setting of a race pace to help in the prevention of premature fatigue.

Dempsey and others note that most scientists agree that the *decision* to reduce power output during exercise clearly involves

TABLE 3.10 Some possible causes of fatigue during exercise

Increased formation of depressant neurotransmitters

Increased serotonin levels

Decreased levels of energy substrates

Decrease in phosphocreatine levels
Depletion of muscle glycogen
Decrease in blood sugar levels
Decrease in plasma branched-chain amino acids

Increased levels of reactive oxygen and nitrogen species

Impaired calcium recycling
Decreased muscle contraction force

Disturbed acid-base balance

Increase in hydrogen ions due to excess lactic acid production

Decreased oxygen transport

Decreased blood volume due to dehydration

Increased core body temperature resulting in hyperthermia

Decreased cooling effect due to dehydration

Disturbed electrolyte balance

Increased or decreased concentration due to sweat losses and inadequate water replacement

---

the higher areas of the central nervous system, but the real mystery is to identify those sources of input that trigger these decisions.

Some possible causes of fatigue are listed in **table 3.10**. Keep in mind that fatigue is complex and several of these factors may be involved simultaneously in the etiology of fatigue in certain sport-type events. For example, Tiidus indicates that fatigue in the 400-meter dash may involve depletion of PCr and muscle glycogen, but may also be associated with increases in hydrogen ion concentration. For the interested reader, Tornero-Auguilera and colleagues have authorized a detailed review on peripheral and central fatigue.

How can I delay the onset of fatigue?

The most important factor in the prevention of premature fatigue is proper training, including physiological, psychological, and biomechanical training.

Physiologically, athletes must train specifically on the energy system or systems that are inherent to their event. Under the guidance of sport scientists and coaches, appropriate physiological training for each specific energy system may increase its energy stores, enzymatic activity, and metabolic efficiency, thus enhancing energy production. Physiological training enhances physical power.

Psychologically, athletes must train the mind to tolerate the stresses associated with their specific event. Sport psychologists may help provide the athlete with various mental strategies, such as inducing either a state of relaxation or arousal, whichever may be appropriate for their sport. Physiological training may also confer some psychological advantages, such as tolerating higher levels of pain associated with intense exercise. Psychological training enhances mental strength.

Biomechanically, athletes must maximize the mechanical skills associated with their sport. For any sport, sport biomechanists can analyze the athlete's skill level and recommend modifications in movement patterns or equipment to improve energy production or efficiency. In many cases, modification of the amount of body fat and muscle mass may provide the athlete with a biomechanical advantage. Biomechanical training helps provide a mechanical edge.

Proper physiological, psychological, and biomechanical training represents the best means to help deter premature fatigue. However, what you eat may affect physiological, psychological, and biomechanical aspects of sports performance. Thus, nutrition is an important consideration in delaying the onset of fatigue during sport training and competition.

How is nutrition related to fatigue processes?

As noted in our discussion of the power-endurance continuum, we can exercise at different intensities, but the duration of our exercise is inversely related to the intensity. We can exercise at a very high intensity for a short time or at a lower intensity for a long time. The importance of nutrition to fatigue is determined by this intensity–duration interrelationship.

In very mild aerobic activities, such as distance walking or low-speed running in a trained ultramarathoner, the body can sustain energy production by using fat as the primary fuel when carbohydrate levels diminish. Because the body has large stores of fat, energy supply is not a problem. However, low blood sugar levels, dehydration, and excessive loss of minerals may lead to the development of both mental and physical fatigue in very prolonged activities.

In moderate-to-heavy aerobic exercise, the body needs to use more carbohydrate as an energy source and thus will run out of muscle glycogen faster. As we shall see later, carbohydrate is a more efficient fuel than fat, so the athlete will have to reduce the pace of the activity when liver and muscle carbohydrate stores are depleted, such as during endurance-type activities lasting more than 90 minutes. Thus, energy supply may be critical. Low blood sugar, changes in blood constituents such as certain amino acids, and dehydration also may be important factors contributing to the development of mental or physical fatigue in this type of endeavor.

In very high-intensity exercise lasting only 1 or 2 minutes, the probable cause of fatigue is the disruption of cellular metabolism caused by the accumulation of hydrogen ions resulting from excess lactic acid production. There is some evidence to suggest that beta-alanine and sodium bicarbonate (discussed in chapters 6 and 13, respectively), which promote intracellular buffering, may help reduce the disruptive effect of lactic acid to some extent. Furthermore, a very low supply of muscle glycogen in fast-twitch muscle fibers may impair this type of performance.

In extremely intense exercise lasting only 5–10 seconds, a depletion of phosphocreatine (PCr) may be related to the inability to maintain a high force production. Supplementation with creatine monohydrate, discussed in chapter 6, has been shown to increase muscle PCr levels and many studies have reported improved performance in high-intensity exercise tasks.

In summary, a deficiency of a number of different nutrients may be a contributing factor in the development of fatigue. A poor diet can hasten the onset of fatigue. Proper nutrition is essential to assure the athlete that an adequate supply of nutrients is available in the diet, not only to provide the necessary energy, such as through carbohydrate and fat, but also to ensure optimal metabolism of the energy substrate via protein, vitamins, minerals, and water.

The role of specific nutrients or dietary supplements relative to fatigue processes will be discussed in later chapters of the book where appropriate. **Table 3.11** provides some examples of how some nutrients or dietary supplements are thought to delay fatigue. Beneficial effects of some of these supplements are well documented, but for others, research does not support a beneficial effect on exercise-related fatigue.

Key Concepts

- The ATP-PCr and lactic acid energy systems are used primarily during fast, anaerobic, power-type events, while the oxygen system is used primarily during aerobic, endurance-type events.
- Fats serve as the primary source of fuel during mild levels of aerobic exercise intensity, but carbohydrates begin to be the preferred fuel as exercise intensity increases.
- Fatigue may be classified as central (neural) and/or peripheral (muscular) fatigue. Fatigue may also be caused by a variety of factors, including the depletion of energy substrate or the accumulation of fatigue-causing metabolites.
- A sound training program and proper nutrition are important factors in the prevention of fatigue during exercise.

Check for Yourself

- Check the world records in running for 100 meters, 400 meters, 1,500 meters, and the marathon (42,200 meters). Calculate the average speed for each distance. Can you relate your findings to the human energy systems and their relationship to fatigue?

TABLE 3.11 Examples of some nutritional ergogenic aids and, theoretically, how they may influence physiological, psychological, or biomechanical processes to delay fatigue

Provide energy substrate

Carbohydrate: Energy substrate for aerobic glycolysis
Creatine: Substrate for formation of phosphocreatine (PCr)

Enhance energy-generating metabolic pathways

B vitamins: Coenzymes in aerobic and anaerobic glycolysis
Carnitine: Enzyme substrate to facilitate fat metabolism

Increase cardiovascular-respiratory function

Iron: Substrate for hemoglobin formation and oxygen transport
Nitrates: Promote utilization of oxygen for energy production

Increase size or number of energy-generating cells

Arginine and ornithine: Amino acids that stimulate production of human growth hormone, an anabolic hormone
Chromium: Mineral to potentiate activity of insulin, an anabolic hormone

Attenuate fatigue-related metabolic by-products

Beta-alanine: Amino acid that acts as an intracellular buffer and attenuates acidosis
Sodium bicarbonate: Buffer to reduce effects of lactic acid

Prevent catabolism of energy-generating cells

Antioxidants: Vitamins to prevent unwanted oxidation of cell membranes
HMB: By-product of amino acid metabolism to prevent protein degradation

Ameliorate psychological function

BCAA: Amino acids that favorably modify neurotransmitter production
Caffeine: Reduces the sensation of psychological effort during exercise

Improve mechanical efficiency

Ma huang: Stimulant to increase metabolism for fat loss
Hydroxycitrate (HCA): Supplement to increase fat oxidation for fat loss

Note: These examples as to how nutritional aids may delay fatigue are based on theoretical considerations. As shall be shown in respective chapters, supplementation with many of these nutritional ergogenic aids has not been shown to enhance exercise or sports performance.

APPLICATION EXERCISE

	Distance logged Sunday	Distance logged Monday	Distance logged Tuesday	Distance logged Wednesday	Distance logged Thursday	Distance logged Friday	Distance logged Saturday
12:00–2:00 A.M.							
2:00–4:00 A.M.							
4:00–6:00 A.M.							
6:00–8:00 A.M.							
8:00–10:00 A.M.							
10:00–12:00 A.M.							
12:00–2:00 P.M.							
2:00–4:00 P.M.							
4:00–6:00 P.M.							
6:00–8:00 P.M.							
8:00–10:00 P.M.							
10:00–12:00 P.M.							

Use a pedometer, accelerometer, smartwatch, smartphone, or other types of fitness tracker and keep a record of your daily movement (recording the amount every 2 hours). This will provide you with an estimate of your daily physical activity involving movement and will be useful in determining your Estimated Energy Requirement (EER) and maintaining an optimal body weight as discussed in chapter 11.

Review Questions—Multiple Choice

1. Which energy system would predominate in an all-out, high-intensity, 400-meter dash in track?
   a. ATP-PCr
   b. lactic acid
   c. oxygen–carbohydrate
   d. oxygen–fat
   e. oxygen–protein

2. If a 50-kilogram body-weight athlete was exercising at an oxygen consumption level of 2.45 liters (2,450 ml) per minute, approximately how many METs would she be attaining?
   a. 8
   b. 10
   c. 11
   d. 12
   e. 14

3. Which of the following classifications of physical activity is rated as light, mild aerobic exercise—because it is likely to burn less than 7 kcal per minute?
   a. competitive racquetball
   b. running at a speed of 7 miles per hour
   c. walking at a speed of 2.0 miles per hour
   d. competitive singles tennis
   e. bicycling at a speed of 15 miles per hour

4. Which of the following statements relative to the basal metabolic rate or resting metabolic rate is false?
   a. The BMR is high in infancy but declines throughout adolescence and adulthood.
   b. The BMR is higher in women than in men due to the generally higher levels of body fat in women.
   c. The resting metabolic rate is the equivalent of 1 MET.
   d. The resting metabolic rate is higher than the BMR.
   e. Dietary-induced thermogenesis raises the resting metabolic rate.

5. Which of the following is not likely to be a cause of fatigue?
   a. depletion of PCr in fast-twitch fibers in a 200-meter dash
   b. depletion of muscle glycogen in fast-twitch fibers in a 400-meter dash
   c. depletion of adipose cell fatty acids in a marathon
   d. depletion of muscle glycogen in a marathon
   e. accumulation of hydrogen ions in a 400-meter dash

6. Of the following statements concerning the interrelationships between various forms of energy, which one is false?
   a. A kilojoule is greater than a kilocalorie.
   b. A kilogram-meter is equal to 7.23 foot-pounds.
   c. A gram of fat has more kcal than a gram of carbohydrate.
   d. A gram of fat has more kcal than a gram of protein.
   e. A liter of oxygen can release more than 1 kilocalorie when metabolizing carbohydrate.

7. Approximately how many kcal will a 200-pound individual use while jogging a mile?
   a. 70
   b. 145
   c. 200

108   **CHAPTER 3**   Human Energy

d. 255
e. 440

8. Which of the following statements relative to exercise and metabolic rate is false?
   a. The intensity of the exercise is the most important factor to increase the metabolic rate.
   b. Increased efficiency for swimming a set distance will decrease the energy cost.
   c. A heavier person will burn more kcal running a mile than a lighter person.
   d. Oxygen consumption and heart rate are two ways to monitor the metabolic rate.
   e. Walking a mile slowly and jogging a mile cost the same amount of kcal.

9. Which energy system has the greatest capacity for energy production (i.e., endurance?)
   a. ATP-PCr
   b. lactic acid
   c. anaerobic glycolysis
   d. oxygen
   e. phosphagens

10. Which of the following is *not* needed to calculate the Estimated Energy Requirement (EER)?
    a. body fat percentage
    b. age
    c. height
    d. weight
    e. Physical Activity Level (PAL)

Answers to multiple choice questions:
1. b; 2. e; 3. c; 4. b; 5. c; 6. a; 7. b; 8. e; 9. d; 10. a

Critical Thinking Questions

1. If an individual performed 5,000 foot-pounds of work in 1 minute, how many kilojoules of work were accomplished?
2. Name the sources of energy stored in the human body and discuss their role in the three human energy systems.
3. Differentiate among BMR, RMR, BEE, REE, TEF, TEE, EER, and TDEE as defined in this text.
4. Explain the role of the three energy systems during exercise and provide an example using track running events. Which muscle fiber types are the major source of energy production during these track events?
5. List the major causes of fatigue during exercise and indicate how various nutritional interventions may help prevent premature fatigue.

References

Books
Fox, E. L. 1979. *Sports Physiology.* Dubuque, IA: Wm. C. Brown.

French, D., and Torres, L. 2021. *NSCA's Essentials of Sport Science.* Champaign, IL: Human Kinetics.

McArdle, W., et al. 2015. *Exercise Physiology.* Baltimore: Wolters Kluwer.

National Academy of Sciences. 2005. *Dietary Reference Intakes for Energy, Carbohydrates, Fiber, Fat, Protein, and Amino Acids (Macronutrients).* Washington, DC: National Academies Press.

Salway, J. 2017. *Metabolism at a Glance.* 4th edition. Hoboken, NJ: Wiley-Blackwell.

Tiidus, P. et al. 2012. *Biochemistry Primer for Exercise Science.* Champaign, IL: Human Kinetics.

Reviews and Specific Studies
Ainsworth, B., et al. 2011. 2011 Compendium of Physical Activities: A second update of codes and MET values. *Medicine and Science in Sports and Exercise* 43:1575-81.

Ament, W., and Verkerke, G. J. 2009. Exercise and fatigue. *Sports Medicine* 39:389-422.

Binns, A., et al. 2015. Thermic effect of food, exercise, and total energy expenditure in active females. *Journal of Science and Medicine in Sport* 18:204-8.

Bloomer, R. 2005. Energy cost of moderate-duration resistance and aerobic exercise. *Journal of Strength and Conditioning Research* 19:878-82.

Bort-Roig, J. 2014. Measuring and influencing physical activity with smartphone technology: A systematic review. *Sports Medicine* 44:671-86.

Burke, S., et al. 2006. Group versus individual approach? A meta-analysis of the effectiveness of interventions to promote physical activity. *Sport & Exercise Psychology Review* 2:19-35.

Bushman, B. 2012. How can I use METS to quantify the amount of aerobic exercise? *ACSM's Health & Fitness Journal* 16 (2):5-7.

Cermak, N., and van Loon, L. 2013. The use of carbohydrates during exercise as an ergogenic aid. *Sports Medicine* 43:1139-55.

Consumer Reports. 2014. Get-healthy gadgets that really work. *Consumer Reports on Health* 26 (9):8.

Coyle, E. 2007. Physiological regulation of marathon performance. *Sports Medicine* 37:306-11.

Crovetti, R., et al. 1998. The influence of thermic effect of food on satiety. *European Journal of Clinical Nutrition* 52:482-88.

Dempsey, J., et al. 2008. Respiratory system determinants of peripheral fatigue and endurance performance. *Medicine & Science in Sports & Exercise* 40:457-61.

Derman, W. et al. 2021. Risk factors associated with acute respiratory illnesses in athletes: A systematic review by a subgroup of the IOC consensus on acute respiratory illness in the athlete. *British Journal of Sports Medicine* 1-13.

Edgerton, V., and Roy, R. 2006. The nervous system and movement. In *ACSM's Advanced Exercise Physiology,* ed. C. M. Tipton. Philadelphia: Lippincott Williams & Wilkins.

Fitts, R. 2006. The muscular system: Fatigue processes. In *ACSM's Advanced Exercise Physiology,* ed. C. M. Tipton. Philadelphia: Lippincott Williams & Wilkins.

Fitts, R. 2008. The cross-bridge cycle and skeletal muscle fatigue. *Journal of Applied Physiology* 104:551-58.

Haddock, B., and Wilkin, L. 2006. Resistance training volume and post-exercise energy expenditure. *International Journal of Sports Medicine* 27: 143-48.

Hagberg, J., and Pena, N. 1989. Bicycling's exclusive calorie counter. *Bicycling* 30:100-103.

Hall, K. et al. 2022. The energy balance model of obesity: Beyond calories in, calories out. *American Journal of Clinical Nutrition* 2022 115(5):1243-1254.

Halson, S., and Jeukendrup, A. 2004. Does overtraining exist? An analysis of overreaching and overtraining research. *Sports Medicine* 34:967-81.

Hargreaves, M. 2005. Metabolic factors in fatigue. *Sports Science Exchange* 18 (3):1-6.

Hargreaves, M., and Spriet, L. 2020. Skeletal muscle energy metabolism during exercise. *Nature Metabolism* 2:817-28.

Hawley, J., and Hopkins, W. 1995. Aerobic glycolytic and aerobic lipolytic power systems. *Sports Medicine* 19:240-50.

Hills, A., et al. 2014. Assessment of physical activity and energy expenditure: An overview of objective measures. *Frontiers in Nutrition.* 1:5.

Holloszy, J., et al. 1998. The regulation of carbohydrate and fat metabolism during and after exercise. *Frontiers in Bioscience* 3:D1011-27.

Hongu, N., et al. 2014. Mobile technologies for promoting health and physical activity *ACSM's Health & Fitness Journal* 18 (4):8-15.

Hoppeler, H., and Weibel, E. 2000. Structural and functional limits for oxygen supply to muscle. *Acta Physiologica Scandinavica* 168:445-56.

Jeukendrup, A., et al. 2000. The bioenergetics of World Class Cycling. *Journal of Science & Medicine in Sport* 3:414-33.

Johnston, S., et al. 2013.The prevalence of chronic fatigue syndrome/myalgic encephalomyelitis: A meta-analysis. *Clinical Epidemiology* 5:105-10.

Knab, A., et al. 2011. A 45-minute vigorous exercise bout increases metabolic rate for 14 hours. *Medicine and Science in Sports and Exercise* 43:1643-48.

Kuo, A., et al. 2005. Energetic consequences of walking like an inverted pendulum: Step-to-step transitions. *Exercise & Sports Sciences Reviews* 33:88-97.

Lange, B., and Bury, T. 2001. Physiologic evaluation of explosive force in sports. *Revue Medicale de Liege* 56:233-38.

Lee, J.-M., et al. 2014. Validity and utility of consumer-based physical activity monitors. *ACSM's Health & Fitness Journal* 18 (4):16-21.

Lee, P., et al. 2013. Brown adipose tissue in adult humans: A metabolic renaissance. *Endocrine Reviews* 34:413-38.

Levine, J. 2004. Non-exercise activity thermogenesis. *Nutrition Reviews* 62:S82-S97.

Levine, J. 2005. Measurement of energy expenditure. *Public Health Nutrition* 8:1123-32.

Liu, S., et al. 2012. Computational methods for estimating energy expenditure in human physical activities. *Medicine & Science in Sports & Exercise* 44:2138-46.

Matthews, C., et al. 2012. Improving self-reports of active and sedentary behaviors in large epidemiologic studies. *Exercise and Sport Sciences Reviews* 40:118-26.

Montoye, A., et al. 2014. Use of a wireless network of accelerometers for improved measurement of human energy expenditure. *Electronics* 3:205-20.

Moss-Morris, R., et al. 2013. Chronic fatigue syndrome. *Handbook of Clinical Neurology* 10:303-14.

Myburgh, K. 2004. Protecting muscle ATP: Positive roles for peripheral defense mechanisms. *Medicine & Science in Sports & Exercise* 37:16-19.

Noakes, T. 2011. Time to move beyond a brainless exercise physiology: The evidence for complex regulation of human exercise performance. *Applied Physiology Nutrition and Metabolism* 36:23-35.

Noakes, T. 2007. The central governor model of exercise regulation applies to the marathon. *Sports Medicine* 37:374-77.

Noordhof, D., et al. 2013. Determining anaerobic capacity in sporting activities. *International Journal of Sports Physiology and Performance* 8:475-82.

Park, Y., et al. 2014. Adipose tissue inflammation and metabolic dysfunction: Role of exercise. *Missouri Medicine* 111:65-72.

Psota, T., and Chen, K. 2013. Measuring energy expenditure in clinical populations: Rewards and challenges. *European Journal of Clinical Nutrition* 67:436-42.

Roelands, B., et al. 2013. Neurophysiological determinants of theoretical concepts and mechanisms involved in pacing. *Sports Medicine* 43:301-11.

Schiaffino, S., and Reggiani, C. 2011. Fiber types in mammalian skeletal muscles. *Physiological Reviews* 91:1447-531.

Schoeller, D., et al. 2013. Self-report-based estimates of energy intake offer an inadequate basis for scientific conclusions. *American Journal of Clinical Nutrition* 97:1413-15.

Shephard, R. 2001. Chronic fatigue syndrome: An update. *Sports Medicine* 31:167-94.

Sidhu, S., et al. 2013. Corticospinal responses to sustained locomotor exercises: Moving beyond single-joint studies of central fatigue. *Sports Medicine* 43:437-49.

Skelly, L., et al. 2014. High-intensity interval exercise induces 24-h energy expenditure similar to traditional endurance exercise despite reduced time commitment. *Applied Physiology, Nutrition, and Metabolism* 39:845-48.

Smith, E. et al. 2020. A systematic review and meta-analysis comparing heterogeneity in body mass responses between low-carbohydrate and low-fat diets. *Obesity* 28:1833-42.

Spencer, M., and Gastin, P. 2001. Energy system contribution during 200- to 1500-m running in highly trained athletes. *Medicine & Science in Sports & Exercise* 33:157-62.

Stellingwerff, T. 2021. Overtraining syndrome (OTS) and relative energy deficiency in sport (RED-S): Shared pathways, symptoms and complexities. *Sports Medicine* 51:2251-80.

Torbeyns, T., et al. 2014. Active workstations to fight sedentary behaviour. *Sports Medicine* 44:1261-73.

Tornero-Aguilera, J. 2022. Central and peripheral fatigue in physical exercise explained: A narrative review. *International Journal of Environmental Research and Public Health* 19:3909.

Traoret, C. et al. 2008. Peanut digestion and energy balance. *International Journal of Obesity* 32:322-8.

Tremblay, A., et al. 1985. The effects of exercise-training on energy balance and adipose tissue morphology and metabolism. *Sports Medicine* 2:223-33.

Tremblay, A., et al. 1990. Long-term exercise training with constant energy intake. 2: Effect of glucose metabolism and resting energy expenditure. *International Journal of Obesity* 14:75-81.

Ward-Smith, A. J. 1999. Aerobic and anaerobic energy conversion during high-intensity exercise. *Medicine & Science in Sports Exercise* 31:1855-60.

Warwick, P. M. 2007. Thermic and glycemic responses to bread and pasta meals with and without prior low-intensity exercise. *International Journal of Sport Nutrition and Exercise Metabolism* 17:1-13.

Westerterp, K. 2013. Metabolic adaptations to over–and underfeeding–Still a matter of debate? *European Journal of Clinical Nutrition* 67:443-5.

White, C., and Kearney, M. 2013. Determinants of inter-specific variation in basal metabolic rate. *Journal of Comparative Physiology* 183:1-26.

Williams, P., and Thompson, P. 2013. Walking versus running for hypertension, cholesterol, and diabetes mellitus risk reduction. *Arteriosclerosis, Thrombosis and Vascular Biology* 33:1085-91.

Wilson, J., et al. 2012. The effects of endurance, strength, and power training on muscle fiber type shifting. *Journal of Strength and Conditioning Research* 26:1724-29.

Wu, X., et al. 2010. Genes and biochemical pathways in human skeletal muscle affecting resting energy expenditure and fuel partitioning. *Journal of Applied Physiology* 110:746-55.

Carbohydrates: The Main Energy Food

Nice One Productions/Corbis/SuperStock

CHAPTER FOUR

KEY TERMS

active transport
added sugars
carbohydrate loading
carbohydrates
complex carbohydrates
Cori cycle
cortisol
dietary fiber
disaccharide
epinephrine
facilitated diffusion
fructose
functional fiber
galactose
glucagon
gluconeogenesis
glucose
glucose-alanine cycle
glucose polymers
gluten intolerance
glycemic index (GI)
glycemic load (GL)
hyperglycemia
hypoglycemia
insulin
lactose intolerance
millimole
monosaccharides
polysaccharide
reactive hypoglycemia
simple carbohydrates
total fiber

LEARNING OUTCOMES

After studying this chapter, you should be able to:

1. List the different types of dietary carbohydrates and identify foods typically rich in the different types.
2. Calculate the approximate number of kcal from carbohydrate that should be included in your daily diet.
3. Describe how dietary carbohydrate is absorbed, how it is distributed in the body, and what its major functions are in human metabolism.
4. Explain the role of carbohydrate in human energy systems during exercise.
5. Describe the various mechanisms whereby inadequate amounts of dietary carbohydrate may contribute to fatigue during exercise.
6. Understand the mechanisms whereby various dietary strategies involving carbohydrate intake (amount, type, and timing) before, during, and after exercise may help to optimize training for, and competition in, sport.
7. Identify athletes for whom carbohydrate loading may be appropriate and describe the full carbohydrate loading protocol, highlighting dietary intake and exercise training considerations.
8. Evaluate the efficacy of metabolic by-products of carbohydrate metabolism as ergogenic aids.
9. Identify carbohydrate-containing foods that are considered to be more healthful and explain why.
10. Describe the effects of chronic endurance exercise training on the subsequent use of carbohydrate as an energy source during exercise, including underlying mechanisms and potential health benefits.

Introduction

Nice One Productions/Corbis/SuperStock

One of the most important nutrients in your diet, from the standpoint of both health and athletic performance, is carbohydrate.

Over the years the reputation of carbohydrate, particularly as a component of a weight-control diet, has seesawed between friend and foe. Most recently, some have dubbed carbohydrate as foe, alleging that high-carbohydrate diets are contributing to the epidemic of obesity and diabetes within industrialized nations. Recently, interest in, and research on, low-carbohydrate diets as a means to improve health and even endurance exercise performance have seen a resurgence. Low-carbohydrate diets are most often low-carbohydrate high-fat diets, and so they will be discussed in depth in chapter 5. In general, researchers and dietitians consider consumption of carbohydrate-rich foods to be one of the most important components of a healthful diet, but choosing the right carbohydrates is key. The possible health benefits of a diet high in complex carbohydrates and fiber and low in added sugars were introduced in chapter 2 and are explained further in this chapter.

As noted in chapter 3, the major role of carbohydrate in human nutrition is to provide energy, and scientists have long known that carbohydrate is one of the prime sources of energy during exercise. Of all the nutrients we consume, carbohydrate has received the most research attention in regard to a potential influence upon athletic performance, particularly in exercise tasks characterized by endurance, such as long-distance running, cycling, and triathloning. Such research is important to athletes who are concerned about optimal carbohydrate nutrition during training and competition. Indeed, continued research over the past quarter century has enabled sports nutritionists to provide more specific and useful responses to athletes' questions. For example, compared to the first edition of this book, published in 1983, readers of this edition will note several significant differences concerning dietary carbohydrate recommendations to athletes.

In this chapter, we explore the nature of carbohydrates, their metabolic fates and interactions in the human body, their possible influence upon health status, and their potential application to physical performance, including the following: the value of carbohydrate intake before, during, and after exercise; the efficacy of different types of carbohydrates; the role of carbohydrate loading; and carbohydrate foods or compounds with alleged ergogenic properties. Although the role of sports drinks containing carbohydrate, such as Gatorade® and PowerAde®, is introduced in this chapter, additional detailed coverage of these beverages and their effect upon performance is presented in chapter 9: Water, Electrolytes, and Temperature Regulation.

Dietary Carbohydrates

What are the different types of dietary carbohydrates?

Carbohydrates represent one of the least expensive forms of kcal and hence are one of the major food supplies for the vast majority of the world's peoples. They are one of the three basic energy nutrients formed when the energy from the sun is harnessed in plants through the process of photosynthesis. Although the energy content of the various forms of carbohydrate varies slightly, each gram of carbohydrate contains approximately 4 kcal.

Carbohydrates are organic compounds that contain carbon, hydrogen, and oxygen in various combinations. A wide variety of forms exist in nature and in the human body, and novel manufactured carbohydrates, including sports drinks, have been developed for the food industry. In general terms, the major categories of importance to our discussion are simple carbohydrates, complex carbohydrates, and dietary fiber.

Simple carbohydrates, which are usually known as sugars, can be subdivided into two categories: disaccharides and monosaccharides. *Saccharide* means "sugar" or "sweet." The three major **monosaccharides** (single sugars) are **glucose, fructose**, and **galactose**. Glucose and fructose occur widely in nature, primarily in fruits, as free monosaccharides. Glucose is often called *dextrose* or *grape sugar,* while fructose is known as *levulose* or *fruit sugar.* Galactose is found in milk as part of lactose. **Figure 4.1** presents two configurations illustrating the structure of monosaccharides.

The combination of two monosaccharides yields a **disaccharide**. The disaccharides (double sugars) include maltose (malt sugar), lactose (milk sugar), and sucrose (cane sugar or table sugar). Upon digestion, these disaccharides yield the monosaccharides as follows.

Monosaccharides and disaccharides, such as glucose and sucrose, may be isolated from foods in purified forms known as

112 **CHAPTER 4** Carbohydrates: The Main Energy Food

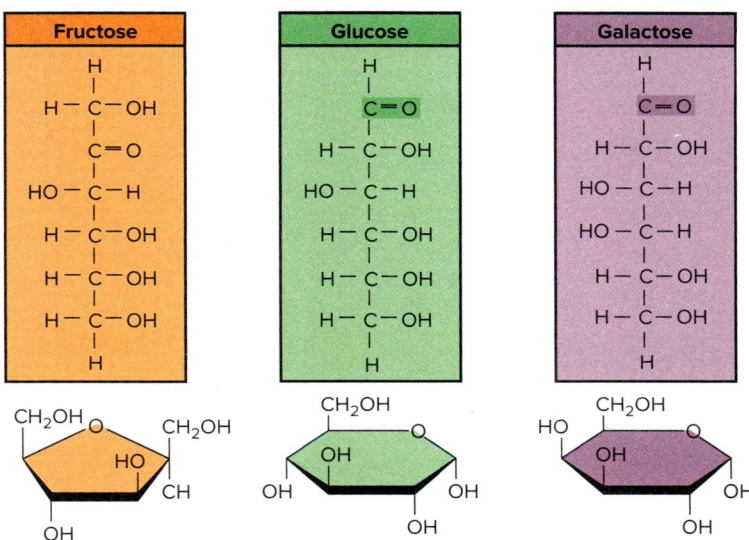

FIGURE 4.1 Chemical structure of the three monosaccharides is depicted in both the linear and ring configurations. One corner of the Haworth projection ring structure contains the oxygen of the aldehyde (glucose, galactose) or ketone (fructose) functional group, for a total of six carbons for each monosaccharide.

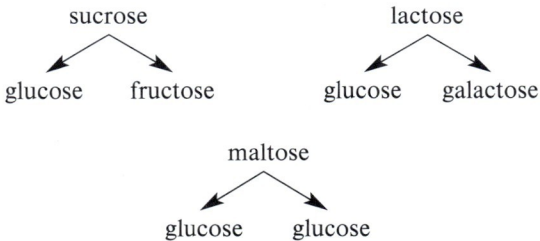

refined sugars. Trisaccharides and higher saccharides also exist and may be found in commonly used sweeteners such as corn syrup. For example, high-fructose corn syrup, a common food additive, is a manufactured carbohydrate derived from the conversion of glucose in corn starch to fructose. Other food additives that are primarily sugar include honey, brown sugar, maple syrup, molasses, and fruit juice concentrate.

Complex carbohydrates, commonly known as *starches,* are generally formed when three or more glucose molecules combine. This combination is known as a **polysaccharide** when more than ten glucose molecules are combined and may contain thousands of linked glucose molecules. Starches, which exist in a variety of forms such as amylose, amylopectin, and resistant starch, are the storage form of carbohydrates. The vast majority of carbohydrates that exist in the plant world are in polysaccharide form. Of prime interest to us are the plant starches, through which we obtain a good proportion of our daily kcal along with a wide variety of nutrients, and the animal starch, glycogen, about which we shall hear more later in relation to energy for exercise. Additionally, **glucose polymers** are polysaccharides prepared commercially by controlled hydrolysis of starch. Maltodextrins are common glucose polymers used in sports drinks, which are discussed later.

Unfortunately, because of disagreements over the classification of various forms of carbohydrate, the term "complex carbohydrate" does not appear on the Nutrition Facts panel of a food label. You may obtain a rough estimate of the complex carbohydrate content by subtracting the grams of sugar from the grams of total carbohydrate. In some cases, the term "other carbohydrates" is used, which could include complex carbohydrates and other forms of carbohydrate as well.

Fiber is a complex carbohydrate. In its report on Dietary Reference Intakes, the National Academy of Sciences settled on three terms to define fiber. **Dietary fiber** consists of nondigestible carbohydrates and lignin that are intrinsic and intact in plants; this would include resistant starch. **Functional fiber** consists of isolated, nondigestible carbohydrates that have beneficial physiological effects in humans. **Total fiber** is the sum of dietary fiber and functional fiber. These nondigestible substances, which means that they are not digested and absorbed in the human small intestine, are usually a mixture of polysaccharides found in the plant wall or intracellular structures.

The National Academy of Sciences carefully defines both dietary fiber and functional fiber, particularly as they may affect health. In essence, dietary fiber is consumed as part of intact foods (even if mechanically altered) containing other macronutrients, such as digestible carbohydrate and protein. For example, cereal brans derived from whole grains contain carbohydrate and protein, along with nondigestible fiber. Some examples of dietary fiber are cellulose, hemicellulose, lignin, pectin, and gums. When studying the health effects of dietary fiber, it is difficult to determine if health benefits are attributed to dietary fiber or to other potential healthful substances, such as certain phytochemicals, found in the food. The definition of dietary fiber includes the phytochemicals that come with it.

Functional fibers, in contrast, may be isolated or extracted from foods by chemical or other means, and may be manufactured synthetically. For example, various gums may be extracted from seeds and used as food ingredients for various purposes. Some examples of functional fiber are pectin, gums, and resistant starch. The specific fiber has to demonstrate a physiological effect in the body to be classified as a functional fiber, and this effect may be associated with a specific health benefit.

Total fiber is the sum of dietary fiber and functional fiber. A specific fiber may be classified as both dietary fiber and functional fiber. For example, cellulose can be classified as dietary fiber as a natural ingredient of an intact food, or it may be considered to be a functional fiber if extracted from a natural source and added to another food.

Previously, the various components of dietary and functional fiber have been classified on the basis of their solubility in water. Some fibers have been referred to as water soluble because they have been found to dissolve or swell in water and may be metabolized by bacteria in the large intestine. Others that do not possess these characteristics have been referred to as water-insoluble fibers. Common water-soluble fibers include gums, beta-glucans, and pectins. Common water-insoluble fibers include cellulose, hemicellulose, and lignin. Each of these types of fibers may confer specific health benefits, which will be discussed later in this chapter.

Sugar substitutes are designed to provide the sweetness of sugars, but with no or fewer kcal. A commonly used sugar substitute is sorbitol, a sugar alcohol. Health aspects of sugar substitutes are discussed later.

A summary of the different types of carbohydrates is presented in **table 4.1**, and the effects of different forms of carbohydrate on physical performance or health are presented later in this chapter.

What are some common foods high in carbohydrate content?

Table 4.2 shows the carbohydrate contents of different foods based on typical serving sizes. Foods such as vegetables are typically nutrient dense, low in kcal per serving, and also contribute to total daily carbohydrate intake. Other foods, such as milk, yogurt, beans, and split peas, contribute to daily carbohydrate intake, but are also high in protein. Fruits and starches can be healthful choices, and typically are a more concentrated source of carbohydrate, while sports food products and candy can be very dense sources of carbohydrate but, at times, do not provide high levels of other nutrients. Each of these foods in the correct proportion can be part of a healthy diet.

Added sugars is the term applied to refined sugars that are added to foods during production. Reading food labels will

TABLE 4.1 Types of dietary carbohydrates***

Monosaccharides	Disaccharides	Polysaccharides	Other carbohydrates
Glucose	Sucrose	Plant starch	Sorbitol (sugar alcohol)
Fructose	Maltose	Amylose	Ribose (a five-carbon sugar)
Galactose	Lactose	Amylopectin	
		Resistant starch	
		Animal starch	
		Glycogen	

Dietary fiber	Functional fiber	Dietary/functional fiber**	
Hemicellulose	Polydextrin	Beta-glucans	
Resistant starch*	Psyllium	Cellulose	
	Resistant starch*	Gums	
		Pectins	

*Certain forms.
**Dietary fiber if found intact in food; functional fiber if extracted and added to foods.
***See text for food sources of the types of dietary carbohydrates.

TABLE 4.2 Foods high in carbohydrate content (grams per serving)

Grains (15 grams)	Fruits (15 grams)	Vegetables (5 grams)	Dairy/dairy alternative (12 grams)	Protein foods (15 grams)	Sports drinks/gels/bars/chews (14–52 grams)	Miscellaneous (10–40 grams)
Whole grain	Apples	Asparagus	Rice milk	Kidney beans	Drinks:	Candy
Brown rice	Apricots	Broccoli	Cow's milk	Navy beans	Gatorade®	Cookies
Corn tortillas	Bananas	Carrots	Soy milk	Split peas	PowerAde®	Fruit ades
Granola	Blueberries	Mushrooms	Yogurt, fruit	Lentils	Gels:	Soft drinks
Oatmeal	Cantaloupe	Radishes			ReLode®	
Ready-to-eat cereal	Cherries	Rutabaga			GU Energy Gel®	
Rye crackers	Dried fruits	Squash, summer			Power Gel®	
Whole wheat bread	Fruit juices	Tomatoes			Bars:	
Enriched grains	Oranges	Zucchini			Power Bar®	
Bagels	Peaches				Clif Bar®	
English muffins	Pineapple				Balance Bar®	
Pasta	Plums				Chews:	
Ready-to-eat cereal	Raspberries				GU Chomps®	
White bread					Clif Shot Blocks®	
White rice					Power Bar Blasts®	
Starchy vegetables					Honey Stinger® Energy Chews	
Corn						
Green peas						
Potatoes						

For more information about high carbohydrate foods, visit www.myplate.gov/.

provide you with the total amount of carbohydrate in grams, grams of sugar, grams of added sugar, and grams of fiber (**figure 2.11** in chapter 2). The label will also provide you with the percentage of your recommended Daily Value for total carbohydrate and fiber.

As carbohydrates are the major fuel for most exercise tasks, several products have been marketed to athletes, particularly sports drinks, sports gels, and sports bars. Sports drinks, such as Gatorade® and PowerAde®, usually contain about 6-8 percent carbohydrates, or about 14-18 grams of carbohydrate per 8 fluid ounces. The carbohydrate source in each varies but usually contains a mixture of one of the following: glucose, fructose, sucrose, and glucose polymers. More details on sports drinks are presented in chapter 9. Sports gels, such as ReLode®, GU, Power Gel®, and Clif-Shot®, contain forms of carbohydrates similar to those used in sports drinks, but in a more solid, gel form. They come in small squeeze containers (usually about 1 ounce) and contain roughly 20-30 grams of carbohydrate. Energy chews are a recent alternative for endurance athletes. They come in packets of three to ten chews containing 2.3-8 grams of carbohydrate per piece. Sports bars are also high in carbohydrate, usually with small amounts of protein and fats. The amount and type of carbohydrate per bar vary, the amount generally ranging between 20-45 grams. So-called energy drinks may contain substantial amounts of carbohydrate, often 30-50 or more grams in 8 ounces. Again, Nutrition Facts panels on food labels on sports and "energy" drinks, sports gels, and sports bars provide information on the nutrient content, including amount of total carbohydrate, sugar, added sugars, and fiber, as well as other ingredients such as caffeine, amino acids, vitamins, and minerals. By careful label reading and price comparisons, you may be able to get a better buy on some products. For example, sports bars can be rather expensive, so purchasing or making your own lower-cost granola bars that contain similar nutrient value may provide some financial savings.

Foods high in dietary fiber include most vegetables and fruits, whole grains, and dried beans and peas, which contain more protein. Whole wheat products are good sources of insoluble fiber, while oats, beans, dried peas, fruits, and vegetables are excellent sources of soluble fiber. Because of the purported health benefits of fiber, cereal manufacturers have released new products containing 13-14 grams of fiber per serving. Psyllium, rich in both soluble and insoluble fiber, is now added to several breakfast cereals. **Table 4.3** presents the average fiber content in some common foods. Food labels also document fiber content per serving.

How much carbohydrate do we need in the diet?

As we shall see, the human body can convert part of dietary protein and fat to carbohydrate. Thus, the National Academy of Sciences indicated that the lower limit of dietary carbohydrate compatible with life apparently is zero, provided that adequate amounts of protein and fat are consumed. Several populations around the world have subsisted on a diet with minimal amounts of carbohydrate without any apparent adverse effects. However, given that a very low-carbohydrate diet may be associated with various micronutrient deficiencies and health problems in Western societies, Dietary Reference Intakes (DRI) have been developed for carbohydrate.

The RDA for carbohydrates is set at 130 grams per day for adults and children, which is based on the average minimum amount of glucose utilized by the brain. Fehm and others noted that although the brain constitutes only 2 percent of body mass, its metabolism accounts for 50 percent of total body glucose utilization. The Academy developed an Acceptable Macronutrient Distribution Range (AMDR), recommending that 45-65 percent of the daily energy intake be derived from carbohydrate. The AMDR for individuals has been set for carbohydrate based on scientific evidence suggesting such an intake may play a role in the prevention of increased risk of chronic diseases and may also ensure sufficient intakes of essential nutrients.

Most of us consume carbohydrate within this range and meet the RDA, as men consume an average 200-330 grams per day, while women average 180-230 grams per day. The Daily Value (DV) for carbohydrate on the food label is based on a recommendation of 60 percent of the caloric intake. The DV for a 2,000-kcal diet is 300 grams of carbohydrate, which represents 60 percent of the daily caloric intake ($4 \times 300 = 1,200$ carbohydrate kcal; $1,200/2,000 = 60$ percent).

The *2020-2025 Dietary Guidelines for Americans* also recommended that no more than 10 percent of total kcal come from added sugars. This standard was set based on evidence that individuals who exceed this level may not be obtaining adequate amounts of essential micronutrients that are not present in foods and beverages that contain added sugars. Current estimates of sugar intake approximate 20 percent of daily energy intake.

The Academy recommends an Adequate Intake (AI) for total fiber as follows:

- Men
  - 38 grams up to age 50
  - 30 grams over age 50
- Women
  - 25 grams up to age 50
  - 21 grams over age 50

The DV for a 2,000-kcal diet is 25 grams of fiber, or 12.5 grams per 1,000 kcal. Thus, the DV is in accord with the AI for women but somewhat lower than the AI for men. The current average amount of total fiber intake is approximately 15 grams per day, so most individuals are not meeting this recommendation. An increased intake of complex carbohydrates would help meet this recommendation.

TABLE 4.3	Fiber content in some common foods**
Legumes (beans)	7–9 grams per ½ cup, cooked
Vegetables	3–5 grams per ½ cup, cooked
Fruits	1–3 grams per piece
Breads and cereals*	1–3 grams per serving
Nuts and seeds	2–5 grams per ounce

*Fiber content may vary considerably in bran-type cereals, ranging up to 13–14 grams per serving. Select whole-grain products for higher fiber content.

**Check food labels for grams of fiber per serving.

Sports nutritionists also recommend a high-carbohydrate diet for individuals engaged in athletic training programs. The general recommendation for most athletes parallels the recommended dietary goals noted above. For an athlete consuming 3,000 kcal per day, 55–60 percent from carbohydrate would be 1,650–1,800 kcal or about 400–450 grams. Ron Maughan and Louise Burke, experts in sports nutrition, have described how carbohydrate intake varies considerably based on specific needs related to physical activity levels, sport, and individual goals. Thus, the amount of carbohydrate needed by an individual exercising to lose weight will be quite different from the amount needed by an individual in training to run a marathon. These considerations will be discussed later in this chapter and in chapter 11.

Many recent dietary surveys conducted with athletes, including endurance athletes, often reveal a carbohydrate intake significantly lower than these recommendations. Inadequate intake of dietary carbohydrate can have a negative impact on physical performance.

Key Concepts

▶ Dietary carbohydrates include monosaccharides and disaccharides (simple sugars) and polysaccharides (complex carbohydrates). Most foods in the starch, fruit, and vegetable exchanges contain a high percentage of carbohydrate, primarily complex carbohydrates.

▶ The RDA for carbohydrate is 130 grams daily, while the AMDR is 45–65 percent of energy intake. The AI for total fiber is 38 and 25 grams per day for young men and women, respectively, and somewhat lower for older men and women. However, the Daily Value (DV) for carbohydrate is 300 grams and for dietary fiber 25 grams, based on a 2,000-kcal diet.

Check for Yourself

▶ Peruse food labels of some of your favorite foods and check carbohydrate. Pay particular attention to percent of Daily Value for total carbohydrates, added sugars, and fiber.

Metabolism and Function

The food we eat must be processed before the nutrients it contains may be used in the body for their various purposes. This process includes digestion, absorption, and excretion of nutrients, which is the responsibility of the gastrointestinal (GI) system, or alimentary canal, which extends from the mouth to the anus. **Figure 4.2** presents the main organs involved in the GI system, some of their basic digestive functions, and the sites of nutrient absorption, which will be discussed in the following chapters.

Digestion is the process by which food is broken down mechanically and chemically in the digestive tract and converted into absorbable forms. Specific enzymes break down foods into smaller substances that may be absorbed. Specific enzymes of interest regarding carbohydrates, fats, and protein will be discussed where appropriate. Absorption of nutrients, as noted in **figure 4.2**, may occur in the stomach and large intestine, but the vast majority of nutrients are absorbed through the millions of villi lining the small intestine. Some substances are absorbed by diffusion, which may be passive or facilitated. In passive diffusion the substance simply diffuses across the cell membrane. Osmosis is the passive diffusion of water. In **facilitated diffusion**, a receptor in the cell membrane is needed to transport the substance from the intestine into the villi. No energy is required for diffusion, but other substances need energy supplied by the villi cells in order to be absorbed, a process known as **active transport**. **Figure 4.3** presents a cross section of a villus and highlights some of the key nutrients and the associated process of absorption.

How are dietary carbohydrates digested and absorbed and what are some implications for sports performance?

Carbohydrates usually are ingested in the forms of polysaccharides (starches), disaccharides (sucrose, maltose, and lactose), and monosaccharides (glucose and fructose). In addition, special carbohydrate compounds, such as glucose polymers, have been developed for athletes. To be useful in the body, these carbohydrates must be digested, absorbed, and transported to appropriate cells for metabolism.

The digestion and absorption of dietary carbohydrates are highlighted in **figure 4.4**. The enzyme that digests complex carbohydrates is amylase, which is secreted by the salivary glands and pancreas. Saliva amylase initiates digestion of the polysaccharides to disaccharides, but most digestion is done in the small intestine by pancreatic amylase. Enzymes then digest the disaccharides to monosaccharides, which are absorbed by specific receptors in the villi.

The composition of the dietary carbohydrate may influence delivery into the body. For example, as noted later in this chapter, sports drinks may be designed to take advantage of the different monosaccharide receptors in the villi. Moreover, sports drinks containing carbohydrate and sodium may enhance water absorption via a co-transport mechanism, a topic that is discussed in chapter 9.

Optimal functioning of the GI tract following carbohydrate intake has been studied extensively because improper functioning may impair athletic performance. For example, although there is very little digestion of carbohydrate in the stomach, the rapidity with which carbohydrate leaves the stomach, and its impact upon the absorption of water, may be important considerations for athletes involved in prolonged exercise under warm environmental conditions.

Certain dietary practices may predispose individuals to gastrointestinal distress, which will compromise exercise performance. High concentrations of simple sugars, particularly fructose, may exert a reverse osmotic effect in the intestines, drawing water from the circulatory system into the intestinal lumen. The resulting symptoms are referred to as the dumping syndrome and include weakness, sweating, and diarrhea. Lactose may present a problem to some athletes, and its possible effects are discussed later in the chapter.

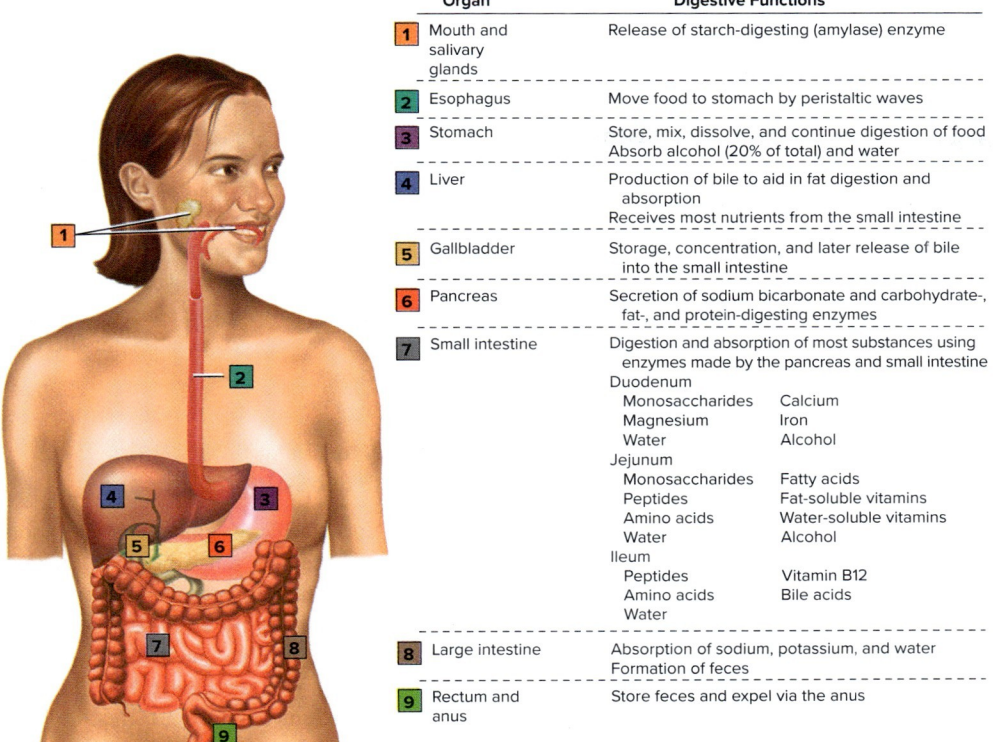

FIGURE 4.2 The alimentary canal. The alimentary canal includes the mouth, the esophagus, the stomach, the small intestine (duodenum, jejunum, ileum), the large intestine (colon), the rectum, and the anal canal. Various glands and organs, including the salivary glands, the gallbladder, and the pancreas, secrete enzymes and other constituents into the digestive tract. Most blood draining from the intestines goes to the liver for processing. Note the general sites in the digestive tract where principal nutrients and other substances are absorbed (although there is some overlap in the exact site of absorption). Most absorption of nutrients occurs throughout the small intestine.

What happens to the carbohydrate after it is absorbed into the body?

Of the three monosaccharides, glucose is of most importance to human physiology. Most dietary carbohydrates are broken down to glucose for absorption into the blood, while the majority of the absorbed fructose and galactose are converted to glucose by the liver. Glucose is the blood sugar.

The **glycemic index (GI)** represents a ranking system relative to the effect that consumption of 50 grams of a particular carbohydrate food has upon the blood glucose response over the course of 2 hours. The normal baseline measure is 50 grams of glucose, and the resultant blood glucose response is scored as 100. In general, the following values are used to rank the glycemic index of foods:

- 70 or more—high GI foods
- 69–55—medium GI foods
- 55 or less—low GI foods

Many factors other than a food's carbohydrate content, such as the physical form (coarse or fine) and serving mode (raw or cooked), may influence the glycemic index of any given food. As an example, a comprehensive study of the glycemic index of pasta conducted by Di Pede and others shows that the glycemic index of pasta, listed as 40 for spaghetti in Table 4.4, actually ranges from 20 to 93. Overall, pasta is considered a medium or low glycemic index food, but there is high variability between types. Moreover, the glycemic index for any given food may vary considerably between individuals, so although some general values of the glycemic index may be given, the effects of different foods should be tested individually in those who are concerned about their blood sugar levels. In general, foods containing high amounts of refined sugars have a high glycemic index because they lead to a rapid rise in the blood sugar, but some starchy foods also have a high glycemic index. In contrast, foods high in fiber, such as beans, generally have a low glycemic index. Interestingly, fructose has a low glycemic index, which is one of the reasons its use as the primary carbohydrate source in sports drinks had been advocated for endurance athletes. We shall discuss the role of fructose later in this chapter. **Table 4.4** classifies some common foods according to their glycemic index.

The **glycemic load (GL)** also represents a ranking system relative to the effect that eating a carbohydrate food has on the blood glucose level, but GL also includes the portion size. While the glycemic index is based on 50 grams of a particular food, a typical serving size for that food may be 6–8 ounces (180–240 grams). The GL is calculated by the following formula:

$$GL = \frac{(\text{glycemic index}) \times (\text{grams of nonfiber carbohydrate in 1 serving})}{100}$$

The following values are used to rank the glycemic load of foods:

- 20 or more—high GL foods
- 19–11—medium GL foods
- 10 or less—low GL foods

The GL for some foods is presented in **table 4.4**.

The usefulness of glycemic index and glycemic load is the subject of debate. In an umbrella review of 18 meta-analyses, Jayedi and colleagues reported a positive association between dietary glycemic index and risk of type 2 diabetes, coronary heart disease, and colorectal, breast, and bladder cancers, and between dietary glycemic load and the risk of coronary heart disease, type 2 diabetes, and stroke. On the other hand, the usefulness of glycemic index/load to

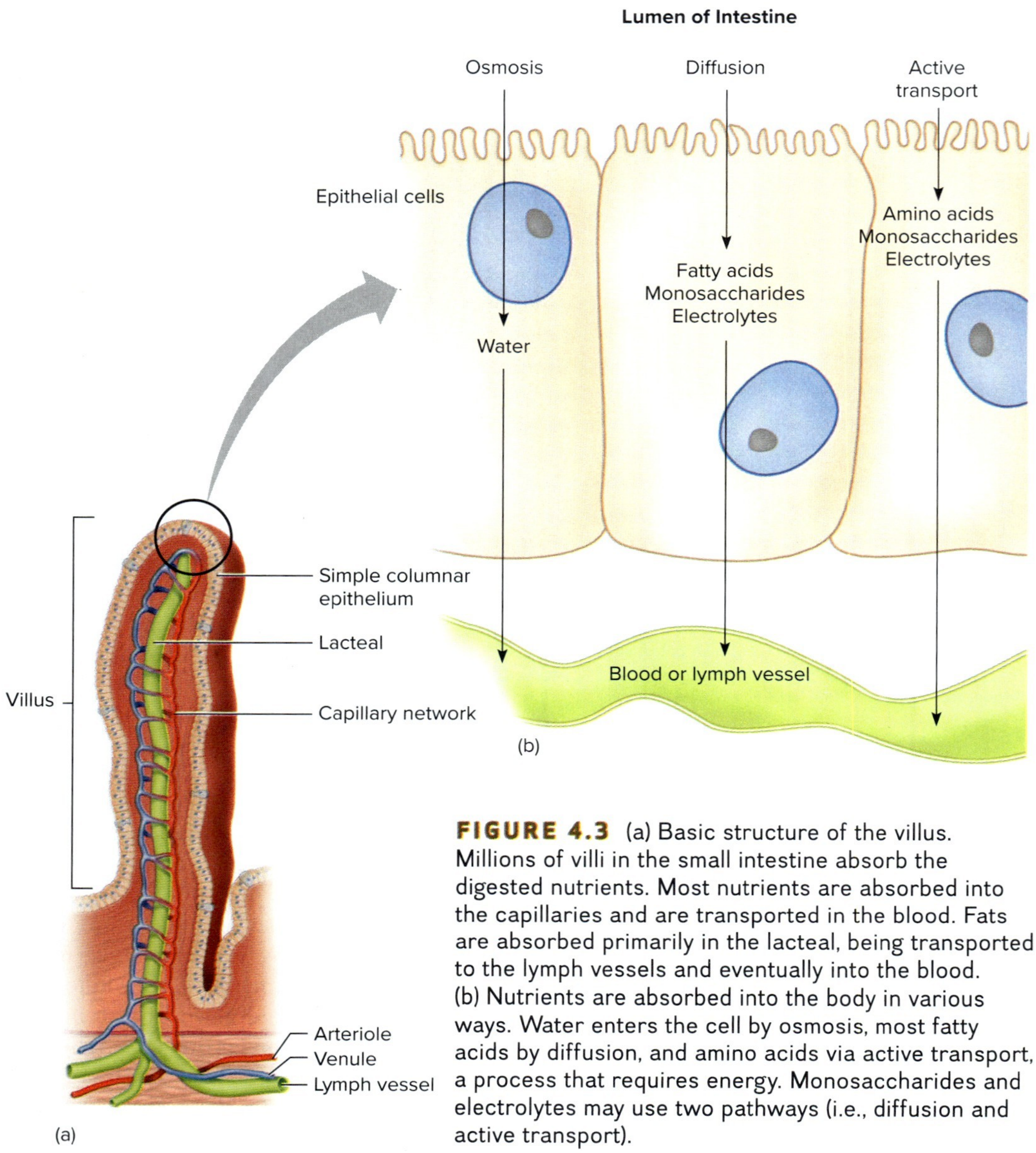

FIGURE 4.3 (a) Basic structure of the villus. Millions of villi in the small intestine absorb the digested nutrients. Most nutrients are absorbed into the capillaries and are transported in the blood. Fats are absorbed primarily in the lacteal, being transported to the lymph vessels and eventually into the blood. (b) Nutrients are absorbed into the body in various ways. Water enters the cell by osmosis, most fatty acids by diffusion, and amino acids via active transport, a process that requires energy. Monosaccharides and electrolytes may use two pathways (i.e., diffusion and active transport).

by themselves, but usually with other foods containing fat and protein such as a hamburger on a bun. The addition of fat and protein will usually reduce the glycemic index and glycemic load. The glycemic index and glycemic load can be used to help select more healthful carbohydrates. In general, the lower the glycemic index or glycemic load, the more healthful the source of carbohydrate, as discussed later in this chapter.

What is the metabolic fate of blood glucose?

Normal blood glucose levels (normoglycemia) range between 80 and 100 milligrams per deciliter of blood (80–100 mg/ml, or 80–100 milligram percent). The maintenance of a normal blood glucose level is very important for proper metabolism. Thus, the human body possesses a variety of mechanisms, primarily hormones, to help keep blood glucose levels under precise control. The rise in blood glucose, also known as *serum glucose,* stimulates the pancreas to secrete insulin into the blood. **Insulin** is a hormone that facilitates the uptake and utilization of glucose (facilitated diffusion) by various tissues in the body, most notably the muscles and adipose (fat) tissue. Cell membranes contain receptors to transport glucose into the cell. The primary receptors in muscle and fat cell membranes are known as GLUT-4 receptors, which are directly activated by insulin (**figure 4.5**). Exercise also activates these receptors to transport blood glucose into the muscle cell, independently of the effect of insulin. Other hormones, discussed later in this chapter, are also involved in regulating blood glucose. With normal amounts of carbohydrate intake in a mixed meal, blood glucose levels remain normal.

However, foods with a high glycemic index may lead rapidly to high blood glucose levels, possibly **hyperglycemia** (>140 mg percent), which will cause an enhanced secretion of insulin from the pancreas. High serum levels of insulin will then lead to a rapid, and possibly excessive, transport of blood glucose into the tissues. This may lead in turn to **hypoglycemia** (<40–50 mg percent), or low blood glucose level. This insulin response and **reactive hypoglycemia** following carbohydrate intake may be an important consideration for some athletes and is discussed later.

improve health is debated. For instance, in their umbrella review of meta-analyses, Churuangsuk and colleagues reported that low glycemic index diets had no effect on weight loss in adults with type 2 diabetes. O'Reilly notes that research on glycemic index and exercise performance, discussed later in the chapter, is at an early stage but may have some benefit. As noted, one person's glycemic response to a given food may be very different from someone else's response. Moreover, individuals do not normally consume carbohydrate foods

> The 2021 glycemic index and glycemic load values updated for 2021 can be found in the supplementary tables from Atkinson, F. 2021. International tables of glycemic index and glycemic load values 2021: A systematic review. *American Journal of Clinical Nutrition* 114(5): 1625–1632. https://doi.org/10.1093/ajcn/nqab233

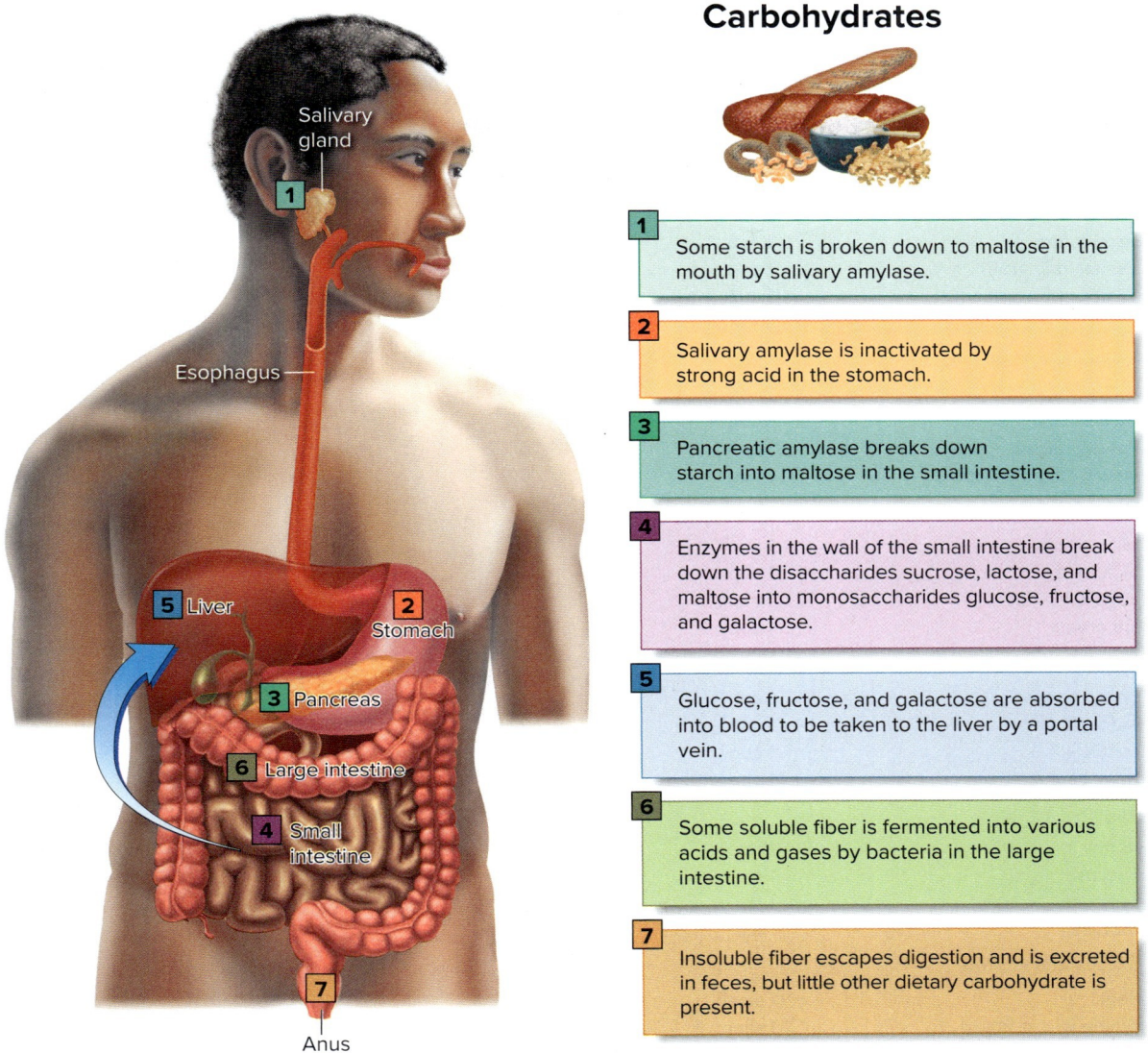

FIGURE 4.4 Carbohydrate digestion and absorption. Enzymes made by the mouth, pancreas, and small intestine participate in the process of digestion. Most carbohydrate digestion and absorption take place in the small intestine.

Foods with a low glycemic index, particularly soluble fiber forms, lead to a slower insulin response and a more stable blood glucose level. Consuming a diet based on the glycemic index has been studied as a possible means to enhance health, as well as sports performance, as noted later in this chapter.

The fate of blood glucose is dependent upon a multitude of factors, and exercise is one of the most important. The following points represent the major fates of blood glucose. **Figure 4.6** schematically represents these fates.

1. Blood glucose may be used for energy, particularly by the brain and other parts of the nervous system that rely primarily on glucose for their metabolism. Hypoglycemia can impair the normal function of the brain. Although hypoglycemia as a clinical condition is quite rare in the general population, transitory hypoglycemia may occur in very prolonged endurance exercise.

2. Blood glucose may be converted to either liver or muscle glycogen. It is important to note that liver glycogen may later be reconverted to blood glucose. However, this does not occur to any appreciable extent with muscle glycogen. In essence, glucose is locked in the muscle once it enters, owing to the lack of a specific enzyme needed to change its form so that it can cross the cell membrane back into the bloodstream. Most of the muscle glycogen is converted to this locked form of glucose during the production of energy. Researchers discovered two forms of muscle glycogen, proglycogen and macroglycogen. Hargreaves indicates that the functional significance of these glycogen forms remains to be fully elucidated. Houston indicates that proglycogen seems to be preferentially used during muscle activity, while the macroglycogen may be more of a reserve supply of carbohydrate for prolonged exercise. In this text we shall use the term *glycogen* to represent both forms.

TABLE 4.4 Glycemic index (GI) and glycemic load (GL) of common foods

Reference food glucose = 100
Low GI foods—below 55
Intermediate GI foods—between 55 and 69
High GI foods—more than 70

Low GL foods—below 10
Intermediate GL foods—between 11 and 19
High GL foods—more than 20

	Serving size (grams)	Glycemic index (GI)*	Carbohydrate (grams)	Glycemic load (GL)
Pastas/grains				
Brown rice	1 cup	55	46	25
White, long grain	1 cup	56	45	25
White, short grain	1 cup	72	53	38
Spaghetti	1 cup	41	40	16
Vegetables				
Carrots, boiled	1 cup	49	16	8
Sweet corn	1 cup	55	39	21
Potato, baked	1 cup	85	57	48
New (red) potato, boiled	1 cup	62	29	18
Dairy foods				
Milk, whole	1 cup	27	11	3
Milk, skim	1 cup	32	12	4
Yogurt, low-fat	1 cup	33	17	6
Ice cream	1 cup	61	31	19
Legumes				
Baked beans	1 cup	48	54	26
Kidney beans	1 cup	27	38	10
Lentils	1 cup	30	40	12
Navy beans	1 cup	38	54	21
Sugars				
Honey	1 tsp	73	6	4
Sucrose	1 tsp	65	5	3
Fructose	1 tsp	23	5	1
Lactose	1 tsp	46	5	2
Breads and muffins				
Bagel	1 small	72	30	22
Whole wheat bread	1 slice	69	13	9
White bread	1 slice	70	10	7
Croissant	1 small	67	26	17
Fruits				
Apple	1 medium	38	22	8
Banana	1 medium	55	29	16
Grapefruit	1 medium	25	32	8
Orange	1 medium	44	15	7
Beverages				
Apple juice	1 cup	40	29	12
Orange juice	1 cup	46	26	13
Gatorade®	1 cup	78	15	12
Coca-Cola®	1 cup	63	26	16
Rice milk	1 cup	93	23	23
Oat milk	1 cup	69	17	25
Snack foods				
Potato chips	1 oz.	54	15	8
Vanilla wafers	5 cookies	77	15	12
Chocolate	1 oz.	49	18	9
Jelly beans	1 oz.	80	26	21

*Based on a comparison to glucose.
Source: Atkinson, F. 2021. International tables of glycemic index and glycemic load values 2021: A systematic review. *American Journal of Clinical Nutrition* 114(5): 1625–1632. https://doi.org/10.1093/ajcn/nqab233

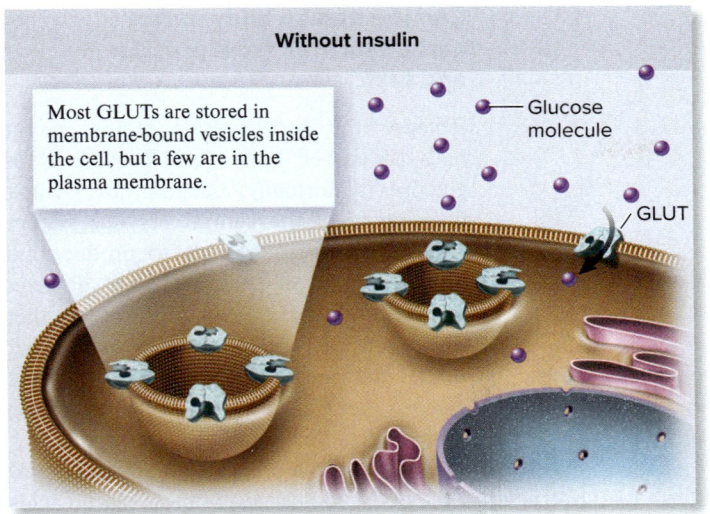
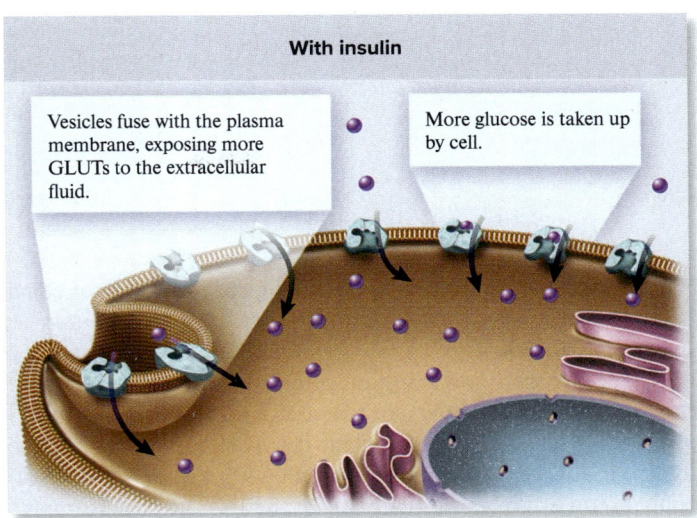

FIGURE 4.5 Insulin promotes the transport of glucose across plasma membranes. This is done by the recruitment of intracellular vesicles containing glucose transporter (GLUT) proteins to the plasma membrane, where they facilitate glucose diffusion into the cell.

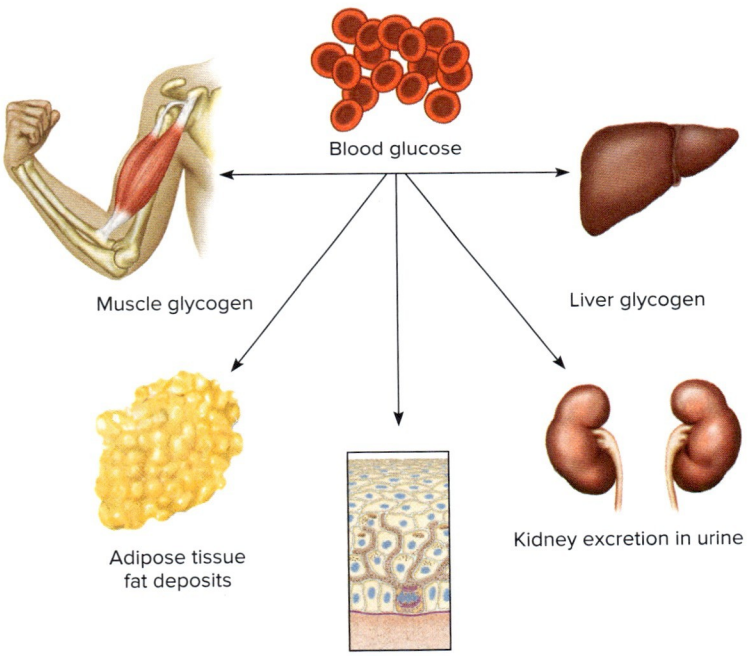

FIGURE 4.6 Fates of blood glucose. After assimilation into the blood, glucose may be stored in the liver or muscles as glycogen or be utilized as a source of energy by these and other tissues, particularly the nervous system. Excess glucose may be partially excreted by the kidneys, but major excesses are converted to fat and stored in the adipose tissues.

3. Blood glucose may be converted to and stored as fat in the adipose tissue. This situation occurs when the dietary carbohydrate, in combination with caloric intake of other nutrients, exceeds the energy demands of the body and the storage capacity of the liver and muscles for glycogen.

4. Some blood glucose also may be excreted in the urine if an excessive amount occurs in the blood because of rapid ingestion of simple sugars.

How much total energy do we store as carbohydrate?

A common method to express the concentration of carbohydrate stored in the body is in millimoles (mmol). A **millimole** is 1/1,000 of a mole, which is the term representing gram molecular weight. In essence, a mole represents the weight in grams of a particular substance such as glucose. The chemical formula for glucose is $C_6H_{12}O_6$, so it contains 6 parts of carbon and oxygen and 12 parts of hydrogen. The atomic weight of carbon is 12, hydrogen is 1, and oxygen is 16. If you multiply the number of parts by the respective atomic weights of each of the elements for glucose [$(6 \times 12) + (12 \times 1) + (6 \times 16)$], you get a total of 180. Thus, 1 mole of glucose is 180 grams, or about 6 ounces. One millimole is 1/1,000 of 180 grams, or 180 milligrams (**figure 4.7**).

As an illustration, the normal glucose concentration is about 5 mmol per liter of blood, or 90 mg/100 ml (90 mg/dL). To calculate, 5 mmol × 180 mg = 900 mg/liter, which is the same as 90 mg/100 ml. The normal individual has about 5 liters of blood. Thus, this individual would have a total of 25 mmol of glucose in the blood, or a total of 4,500 milligrams (25 × 180), or 4.5 grams.

These calculations have been presented here because this is the means whereby concentrations of glucose, glycogen, and other nutrients are expressed in contemporary scientific literature. A knowledge of these mathematical relationships should help you interpret research more effectively.

For our purposes, the body has three major energy sources of carbohydrate: blood glucose, liver glycogen, and muscle glycogen. Some glucose (about 10–15 grams) is also found in the lymph and intercellular fluids. Initial stores of blood glucose are rather limited, totaling only about 5 grams, or the equivalent of 20 kcal. However, blood glucose stores may be replenished from

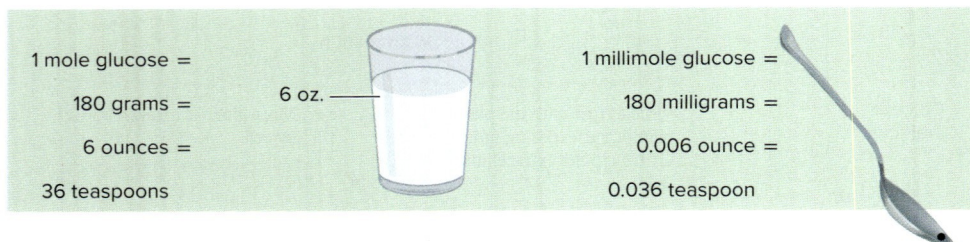

FIGURE 4.7 The millimole concept. The concentration of some nutrients in the body is often expressed in millimoles. See text for explanation.

either liver glycogen or absorption of glucose from the intestine. The liver has the greatest concentration of glycogen in the body. However, because its size is limited, the liver normally contains only about 75-100 g of glycogen, or 300-400 kcal. One hour of aerobic exercise uses more than half of the liver glycogen supply. It is also important to note that the liver glycogen content may be decreased by starvation or increased by a carbohydrate-rich diet. Fifteen hours or more of starvation will deplete the liver glycogen, whereas certain dietary patterns may nearly double the glycogen content of the liver, a condition that may be useful in certain tasks of physical performance.

The greatest amount of carbohydrate stored in the body is in the form of muscle glycogen. This is because the muscles compose such a large proportion of the body mass as contrasted to the liver. One would expect large differences in total muscle glycogen content between different individuals because of differences in body size. However, for an average-sized, untrained man with about 30 kg of his body weight consisting of muscle tissue, one could expect a total muscle glycogen content of approximately 360 g, or 1,440 kcal. This would represent a concentration of about 66 mmol, or 12 grams, per kg of muscle tissue. As with liver glycogen, the muscle glycogen stores also may be decreased or increased, with considerable effects on physical performance. For example, a trained endurance athlete may have twice the amount of stored muscle glycogen that an untrained, sedentary individual has.

If we calculate the body storage of carbohydrate as blood glucose, liver glycogen, and muscle glycogen, the total is only about 1,800-1,900 kcal, not an appreciable amount. One full day of starvation could reduce it considerably. Some normal ranges of carbohydrate stores are presented in **table 4.5**, although these normal ranges may be increased or decreased considerably by diet or exercise.

TABLE 4.5	Approximate carbohydrate stores in the body of a normal, sedentary adult	
Source	Amount in grams	Equivalent amount in kcal
Blood glucose	5	20
Liver glycogen	75-100	300-400
Muscle glycogen	300-400	1,200-1,600

Can the human body make carbohydrates from protein and fat?

Because the carbohydrate stores in the body are rather limited, and because blood glucose is normally essential for optimal functioning of the central nervous system, it is important to be able to produce glucose internally if the stores are depleted by starvation or a zero-carbohydrate diet. This process in the body is called **gluconeogenesis**, meaning the new formation of glucose. A number of different substrates from each of the three energy nutrients may be used and are depicted graphically in **figure 4.8**.

Protein may be a significant source of blood glucose. It breaks down to amino acids in the body, and certain of these amino acids, notably alanine, may be converted to glucose in the liver. This is referred to as the **glucose-alanine cycle**, which is explained further in chapter 6. A number of other amino acids also are gluconeogenic. Glucose is essential for the brain and several other tissues. If about 130 grams of carbohydrate are not consumed daily, then the body will produce the glucose it needs, primarily from protein in the body.

Fats in the body break down into fatty acids and glycerol. Although there is no mechanism in human cells to convert the fatty acids to glucose, glycerol may be converted to glucose through the process of gluconeogenesis in the liver.

Through gluconeogenesis, 1 gram of protein will yield about 0.56 gram of glucose. Triglycerides are about 10 percent glycerol, and each gram of glycerol may be converted to a gram of glucose.

In addition, certain by-products of carbohydrate metabolism, notably pyruvate and lactate, may be converted back to glucose in the liver. Some of the lactic acid produced in the muscle during intense exercise may be released into the blood and carried to the liver for reconversion to glucose. The glucose may then return to the muscles to be used as an energy source or stored as glycogen. This is the **Cori cycle**. **Figure 4.9** illustrates some of the basic interrelationships among carbohydrate, fat, and protein in human nutrition.

What are the major functions of carbohydrate in human nutrition?

The major function of carbohydrate in human metabolism is to supply energy. Some body cells, such as the nerve cells in the brain and retina and the red blood cells, are normally totally dependent upon glucose for energy and require a constant source. Through a series of biochemical reactions in the body cells, glucose is oxidized, eventually producing water, carbon dioxide, and energy. Zierler notes that although carbohydrate is an excellent source of energy, it is not the major fuel when the body is at rest; fats are. Carbohydrate is the main fuel for certain tissues during rest, such as the brain, central nervous system, and red blood cells but provides only about 15-20 percent of muscle energy needs during rest. Thus, the body conserves its limited carbohydrate stores by using fats as the primary energy source during resting conditions. As noted in chapter 3, carbohydrate can be used to

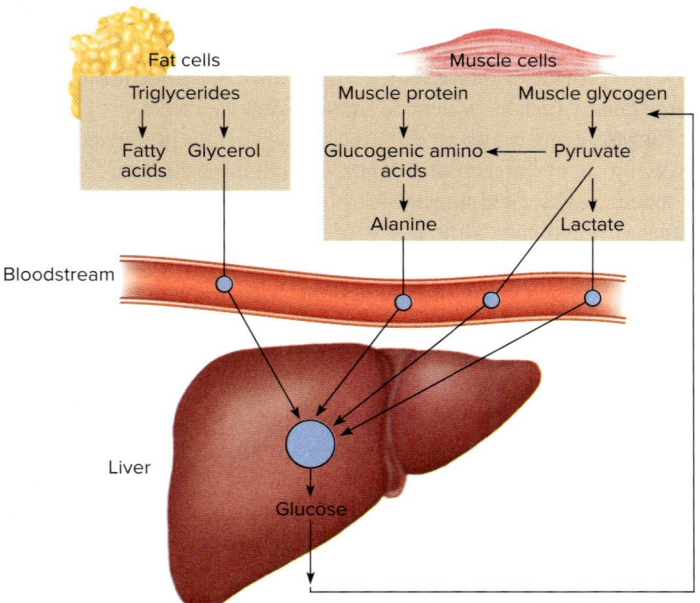

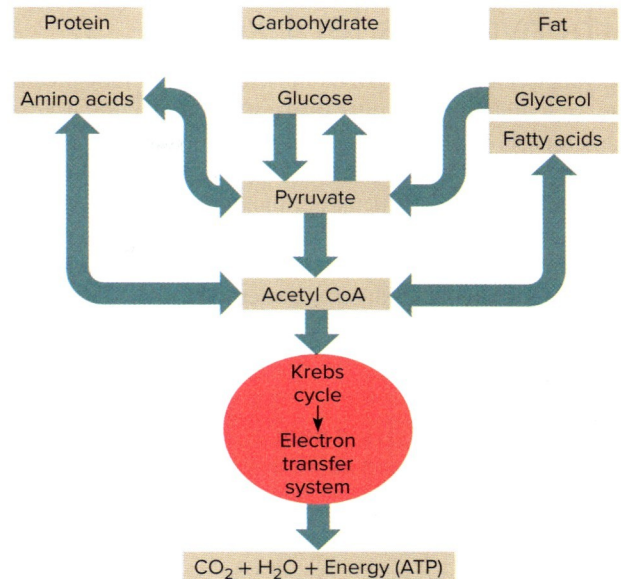

FIGURE 4.8 Gluconeogenesis. The liver is the major site for gluconeogenesis in the body. The breakdown products of fats, protein, and carbohydrate from other parts of the body may be transported to the liver by the blood for eventual reconversion into glucose. Glycerol, glucogenic amino acids, lactate, and pyruvate may be important sources for the new formation of glucose. The Cori cycle involves the transport of lactate to the liver and conversion to glucose, which may then be transported back to the muscle.

FIGURE 4.9 Interrelationships among carbohydrate, fat, and protein metabolism in humans. All three nutrients may be utilized for energy, although the major energy sources are carbohydrate and fat. Excess carbohydrate may be converted to fat; the carbohydrate structure also may be used to form protein, but nitrogen must be added. Fat cannot be used to generate carbohydrate to any large extent because acetyl CoA cannot be converted to pyruvate. The glycerol component of fat may form very small amounts of carbohydrate. Fats may serve as a basis for the formation of protein, but again, nitrogen must be added. Excess protein cannot be stored in the body but can be converted to either carbohydrate or fat.

produce energy either aerobically or anaerobically. Recall that in the lactic acid system, ATP is produced rapidly via anaerobic glycolysis, but for this system to continue functioning, the end product of glycolysis, pyruvic acid or pyruvate, must be converted into lactic acid. In the oxygen system, aerobic glycolysis predominates and pyruvic acid is converted into acetyl CoA, which enters into the Krebs cycle and electron transfer system for complete oxidation and the production of relatively large amounts of ATP. For the same amount of carbohydrate, anaerobic metabolism yields 3 ATP when starting with glycogen and only 2 ATP when starting with glucose, whereas aerobic metabolism yields about 32 ATP. Historically, textbooks often refer to the total ATP yield from the complete oxidation of one glucose to be 36–38 ATP, but due to leaky membranes and the energy cost of moving pyruvate and ADP into the mitochondrial matrix, about 29–32 ATP is a more accurate estimate. See appendix A for more details.

Carbohydrates have some functions in the body other than energy production. Monosaccharides can be used to form other, smaller carbohydrate molecules such as trioses and pentoses. These substances may combine with other nutrients and form body chemicals essential to life, such as glycolipids or glycoproteins. Glycoproteins are very important components of cell membranes, serving as receptors to help regulate cell function. Ribose is a key pentose (5-carbon sugar) that is a part of a number of indispensable compounds in the body. One of those compounds is RNA, or ribonucleic acid, which plays an important part in anabolic processes in the cells.

Key Concepts

▶ Most ingested carbohydrates are initially converted into blood glucose and used for energy or stored as liver and muscle glycogen, but excess carbohydrates may be converted into fat.

▶ The glycemic index (GI) is a measure of the rate of digestion and absorption of 50 grams of a carbohydrate food and the resultant effect on the blood glucose level. The glycemic load (GL) is similar but is based on the typical serving size for carbohydrate foods. Different numerical rankings are used for the GI and GL.

▶ The three sources of carbohydrate in the body of an average adult male are blood glucose (5 grams; 20 kcal), liver glycogen (75–100 grams; 300–400 kcal), and muscle glycogen (350–400 grams; 1,400–1,600 kcal).

▶ The liver is the main organ that can produce glucose from certain by-products of protein and fat, a process known as gluconeogenesis.

▶ The major function of carbohydrates in human metabolism is to supply energy; blood glucose is essential for optimal functioning of the nervous system, whereas muscle glycogen is essential for both anaerobic capacity and aerobic power exercise.

Carbohydrates for Exercise

Both hypoglycemia and depleted muscle glycogen may precipitate fatigue, so maintaining optimal levels of blood glucose, liver glycogen, and muscle glycogen is essential in various athletic endeavors, particularly prolonged exercise tasks. Some sports authorities indicate that carbohydrate is the master fuel for athletes.

In this section we discuss the role of carbohydrate as an energy source during exercise, the effect of training to enhance carbohydrate use for energy, and various methods to provide adequate carbohydrate nutrition to the athlete before, during, and after competition and carbohydrate intake during training.

In what types of activities does the body rely heavily on carbohydrate as an energy source?

Carbohydrate supplies approximately 40 percent of the body's total energy needs during rest, with about 15-20 percent used by the muscles. During very light exercise fat is an important energy source, but van Loon and others found that muscle glycogen and plasma glucose oxidation rates increased with every increment in exercise intensity. When exercise becomes more intense, such as when a person is working at 65-85 percent of capacity, carbohydrate becomes the preferred energy source; this is the crossover concept discussed in chapter 3. At maximal or supramaximal exercise levels, carbohydrate is used almost exclusively. Thus, carbohydrate may be the prime energy source for high-intensity anaerobic events lasting for less than 1 minute and high-intensity aerobic events lasting more than an hour or two.

As Holloszy and others summarized, carbohydrate use, then, is associated with the intensity level of the exercise. The more intense the exercise, the greater the percentage contribution of carbohydrate. Of course, the more intense the exercise, the sooner exhaustion occurs. A fairly well-conditioned person may be able to exercise for many hours at 40-50 percent of VO_2 max, for 1-2 hours or so at 70-80 percent of VO_2 max, but only for minutes at maximal or supramaximal levels of VO_2 max. As noted by Hargreaves, the fatigue that occurs in very high-intensity exercise of short duration, such as a 400-meter run, may be associated with the accumulation of hydrogen ions, a by-product of lactic acid production. In contrast, the fatigue associated with more prolonged exercise may be connected with depleted supplies of liver and muscle glycogen, both of which may be affected by dietary practices and exercise intensity and duration.

Carbohydrate intake is most important for prolonged endurance events lasting more than 90-120 minutes. Data from such endurance tasks as the Tour de France, the bicycle Race Across America, and the Ironman Triathlon, illustrate the importance of dietary carbohydrate in sustaining high energy output for prolonged periods. Most of the athletes in these events consume high-caloric diets rich in carbohydrates both before and during competition. A classic example is the ultradistance runner from Greece, Yannis Kouros, who won the Sydney to Melbourne race in Australia, a distance of approximately 600 miles, in 5 days and 5 hours, or about 114 miles of running per day. He consumed up to 13,400 kcal per day, with up to 98 percent being derived from carbohydrates. Cyclists in major multiday races consume more than 800 grams of carbohydrate daily. Data obtained from exercise tasks of lesser magnitude, such as marathons (26.2 miles; 42.2 kilometers) and ultramarathons (50 kilometers and longer), also provide evidence for the importance of carbohydrate as the prime energy fuel, as evidenced in the review by Peters.

Carbohydrate is also an essential energy fuel for prolonged sports involving many intermittent bouts of high-intensity exercise, such as soccer, rugby, field hockey, ice hockey, and tennis. For example, Bangsbo noted that although soccer players are engaged in low-intensity exercise for about 70 percent of the time, there may be 150-250 bouts of brief, intense actions during a game. Athletes in these sports repeatedly use muscle glycogen stored in their fast-twitch muscle fibers, which may lead to a selective depletion in these fibers. Bangsbo identified muscle glycogen as the most important substrate for energy production in sports such as soccer.

Environmental conditions may also increase carbohydrate use during exercise. Carbohydrate oxidation, particularly muscle glycogen, is increased during exercise in the heat. Sawka and Young indicated that the increased glycogen utilization is probably mediated by elevated epinephrine and muscle hyperthermia. They also note that lactate uptake and oxidation by the liver are impaired during exercise heat stress. However, heat acclimatization helps reduce muscle glycogen use and lactate accumulation. Pasiakos and others note that exercise performance is decreased when low-landers are acutely exposed to altitude, which may be partially explained by increased endogenous carbohydrate oxidation. However, carbohydrate ingestion may not provide an ergogenic effect in un-acclimated individuals, as exogenous carbohydrate oxidation appears to be impaired. This could be because acute hypoxia alters other aspects of carbohydrate metabolism including increasing peripheral insulin resistance, which decreases exogenous carbohydrate oxidation; increasing glycogenolysis; and suppressing fat oxidation. While it is clear that several aspects of carbohydrate metabolism are altered when exercising at altitude, Stellingwerff and others note that much of what is known was learned from studies conducted at mountaineering altitudes (>3000 meters) and not at altitudes more typical of elite athlete training (about 1600-2400 meters). Thus, more research is needed in this area.

lynx/iconotec.com/ Glowimages

Why is carbohydrate an important energy source for exercise?

Carbohydrate is the most important energy food for exercise. Besides being the only food that can be used for anaerobic energy production in the lactic acid system, it is also the most efficient fuel for the oxygen system. If we look at the caloric value of carbohydrate (1 gram = 4 kcal) and fat (1 gram = 9 kcal), we might think that fat is a better source of energy. Indeed, this is so if we only look at kcal per gram. However, more oxygen is needed to metabolize the fat, and if we look at how many kcal we get from 1 liter of oxygen, we will find that carbohydrate yields about 5.05 kcal and fat gives only 4.69. Thus, carbohydrate appears to be a more efficient fuel than fat, by about 7 percent. Houston notes an even greater benefit if we look at ATP production. You get more ATP from glucose than you do from a fatty acid. For each unit of oxygen consumed, glucose produces 2.7 ATP, whereas palmitic fatty acid produces 2.3 ATP, a 17 percent difference. The metabolic pathways for carbohydrate are also more efficient than those for fat. In essence, during aerobic glycolysis, carbohydrate is able to produce ATP for muscle contraction up to three times more rapidly than fat and even faster during anaerobic glycolysis.

The primary carbohydrate source of energy for physical performance is muscle glycogen, specifically the glycogen in the muscles that are active. Elite marathon runners may use about 4–5 grams of carbohydrate per minute. As the muscle glycogen is being used during exercise, blood glucose enters the muscles and the energy pathways. In turn, the liver will release some of its glucose to help maintain or elevate blood glucose levels and prevent hypoglycemia. Coyle noted that during moderate exercise, muscle glycogen and liver glycogen contribute equally to carbohydrate oxidation. At higher intensities, muscle glycogen use increases.

Thus, all body stores of carbohydrate—blood glucose, liver glycogen, and muscle glycogen—are important for energy production during various forms of exercise. Proper physical training is essential to optimize carbohydrate utilization during exercise, as is proper carbohydrate nutrition.

What effect does endurance training have on carbohydrate metabolism?

Because carbohydrate is a primary fuel for exercise, as you initiate an endurance exercise program, a major proportion of your energy will be derived from your muscle glycogen stores. A single bout of exercise can activate the genes that produce GLUT-4 receptors, which can exert an insulin-type effect by facilitating the transport of blood glucose into the muscle both during and immediately following the exercise bout.

As you continue your endurance exercise program, such as running or bicycling, the physical activity serves as a metabolic stressor to the body, and various tissues begin to adapt to better accommodate the exercise stress. Houston reported that as few as five days of training can exert favorable effects, such as decreasing the production of lactic acid for a standardized exercise task. These adaptations have implications for physical performance and the fuels used. **Figure 4.10** schematically represents some of these changes at the cellular level. The following have been noted to occur in both males and females after several months of endurance training:

1. You will increase your VO_2 max.
2. Of equal or greater importance, you will be able to work at a greater percentage of your VO_2 max without fatigue.
3. Endurance-trained muscles may use less glucose at low-intensity exercise but have the capacity to use more during intense, maximal exercise.
4. Endurance-trained muscle has an increased maximal capacity to utilize carbohydrates. As muscle cell mitochondria density increases, the enzymes that metabolize carbohydrate in the muscle cells will increase, especially oxidative enzymes associated with the Krebs cycle.
5. As we shall see in chapter 5, training also enhances the use of fat during exercise. By doing so, there is less reliance on carbohydrate oxidation during submaximal exercise.
6. More glycogen is stored in the muscle. Synthesis of muscle glycogen may be twofold faster in trained versus untrained individuals.

What do all these changes mean? As an example, you may be able to run a 10-kilometer (6.2-mile) road race at a 7-minute-per-mile pace instead of 8 minutes. You can cruise in high gear for longer periods because you have increased your ability to produce energy from carbohydrates. Also, by reducing your reliance on carbohydrates at lower running speeds, you may compete in more prolonged races, such as marathons, without becoming hypoglycemic.

How is hypoglycemia related to the development of fatigue?

As noted previously, blood glucose is in very short supply, so as it is being used during exercise it must be replenished from liver glycogen stores. A depletion of liver glycogen may lead to hypoglycemia during high-intensity aerobic exercise because gluconeogenesis normally cannot keep pace with glucose utilization by the muscles.

Hypoglycemia is known to impair the functioning of the central nervous system and is often accompanied by acute feelings of dizziness, muscular weakness, and fatigue. The normal blood glucose level usually ranges from 80 to 100 mg of glucose per 100 ml of blood (4.4–5.5 mmol per liter). As this level gets progressively lower, hypoglycemic symptoms may develop. The point usually used to identify hypoglycemia during research studies with exercise is 45 mg per 100 ml, or 2.5 mmol per liter, although some investigators have used higher levels.

Because hypoglycemia may disrupt functioning of the central nervous system (brain and spinal cord), the body attempts to maintain an optimal blood glucose level. Zierler noted that exercise increases muscle glucose uptake, in part due to increased cell membrane GLUT-4 receptors. Exercise may also increase the sensitivity to insulin, so more glucose is transported into the muscle for the same level of insulin. Thus, insulin levels normally drop during exercise so as to help maintain normal serum glucose. Other hormones—epinephrine (adrenaline), glucagon, and cortisol—also help maintain, and even increase, blood glucose levels during exercise.

Epinephrine is secreted from the adrenal gland during exercise, particularly intense exercise, and stimulates the liver to release glucose; it also accelerates the use of glycogen in the muscle. However, during the early phases of exercise, liver glucose output may exceed muscle glucose uptake, which may result in hyperglycemia. Glucagon and cortisol levels generally increase during the stress of exercise, cortisol particularly during prolonged exercise. **Glucagon** is released from the pancreas and generally increases the rate of gluconeogenesis in the liver. Kjaer noted that liver gluconeogenesis may contribute substantially during prolonged aerobic exercise when liver glycogen levels decline and gluconeogenic substrate becomes more abundant. **Cortisol** is secreted from the adrenal gland and facilitates the breakdown and release of amino acids from muscle tissue to provide some substrate to the liver for gluconeogenesis. Blood glucose normally increases during the initial stages of exercise and is normally well maintained by these hormonal mechanisms. A summary of hormonal actions in the regulation of blood glucose is presented in **table 4.6**.

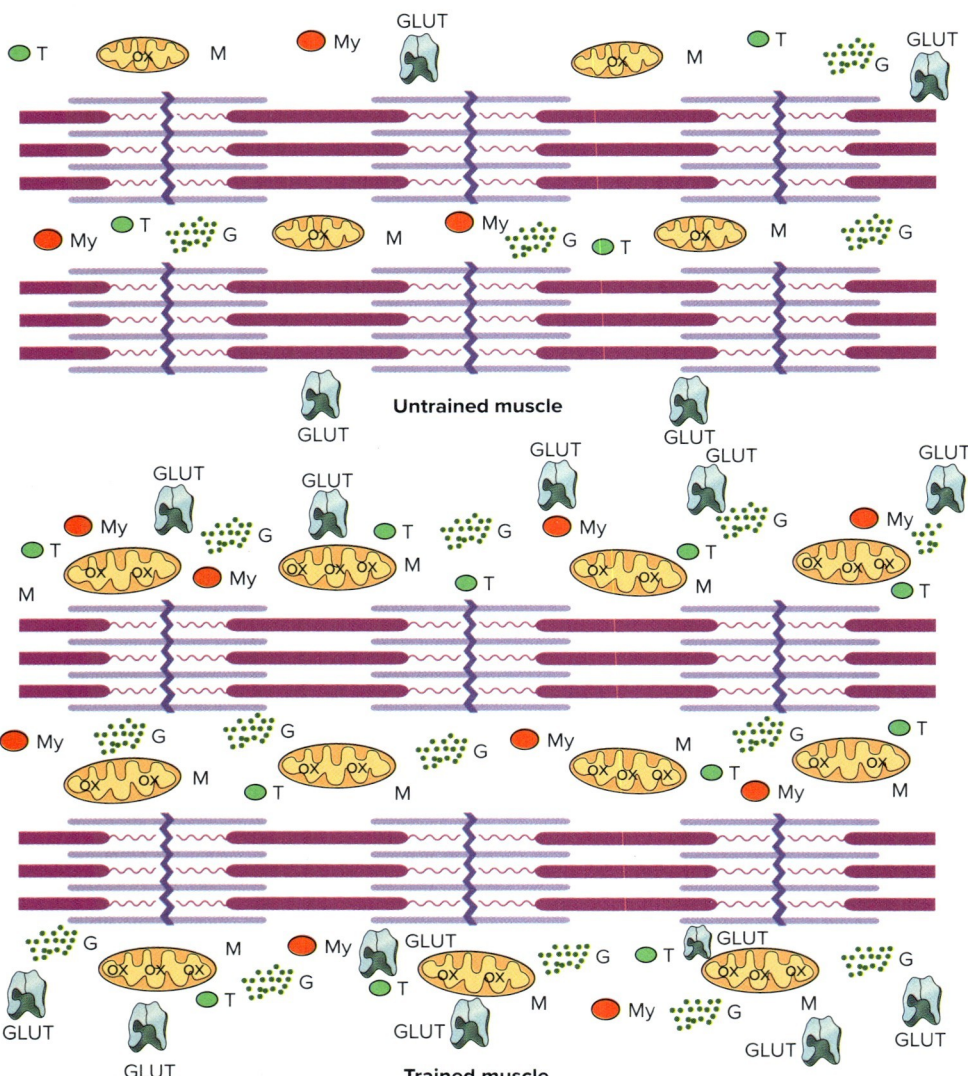

FIGURE 4.10 Some of the effects of aerobic or endurance training upon skeletal muscle. Increases in glycogen (G) and triglyceride (T) provide a greater energy store, increased levels of GLUT-4 receptors (GLUT) provide the potential to increase glucose delivery to the muscle as needed, while the increase in mitochondria size and number (M), myoglobin content (My), oxidative enzymes (ox), and slow-twitch muscle fiber size facilitates the use of oxygen for production of energy.

TABLE 4.6	Major hormones involved in regulation of blood glucose levels		
Hormone	**Gland**	**Stimulus**	**Action**
Insulin	Pancreas	Increase in blood glucose	Helps transport glucose into cells; decreases blood glucose levels
Glucagon	Pancreas	Decrease in blood glucose; exercise stress	Promotes gluconeogenesis in liver; helps increase blood glucose levels
Epinephrine	Adrenal	Exercise stress; decrease in blood glucose	Promotes glycogen breakdown and glucose release from the liver; helps increase blood glucose levels
Cortisol	Adrenal	Exercise stress; decrease in blood glucose	Promotes breakdown of protein and resultant gluconeogenesis; helps increase blood glucose levels

Hypoglycemia may be a concern of athletes in several situations. One possibility is a reactive hypoglycemia following the consumption of a high-carbohydrate meal 30–60 minutes or more prior to an athletic event. If hypoglycemia develops just prior to or during the early stages of the event, the effect could impair performance. At one time, hypoglycemia as a consequence of pre-exercise carbohydrate ingestion was believed to be a large concern. However, as described by Jeukendrup and Killer, and as discussed in depth later in the chapter, there appears to be no good reason to avoid pre-exercise carbohydrate consumption.

Hypoglycemia may also develop during prolonged exercise tasks, but it may be dependent upon the intensity level of the exercise. In low-intensity exercise, such as 30–50 percent of VO_2 max, the primary fuel is fat, and hence the use of carbohydrate is minimized. Moreover, at this low intensity level, gluconeogenesis can help maintain blood glucose above hypoglycemic levels.

During the early part of prolonged moderate- to high-intensity aerobic exercise, muscle glycogen is the major source of energy derived from carbohydrate, although some blood glucose is utilized. However, Weltan and others found that as muscle glycogen levels get low in the later stages of an endurance task, the muscle increases its fat oxidation and blood glucose, derived from liver glycogen, now accounts for most of the muscular energy from carbohydrate. Gluconeogenesis is increased, but Hargreaves indicates that it cannot completely compensate for decreased liver glycogen availability. Thus, the blood glucose levels fall toward hypoglycemia.

Whether hypoglycemia impairs physical performance may depend upon the individual. In their study, Utter and others reported that a lower rating of perceived exertion (RPE) during prolonged running was associated with higher blood glucose, suggesting that maintenance of optimal blood glucose levels could help prevent psychological distress. Some earlier research reported that exercise-induced hypoglycemia led to the expected symptoms, including dizziness and partial blackout. However, more contemporary research has revealed that a number of subjects may become hypoglycemic during the later stages of a prolonged exercise task to exhaustion at 60–75 percent of their VO_2 max and yet are able to continue exercising while hypoglycemic, even at levels as low as 25 mg per 100 ml. It appears that the role hypoglycemia plays in the etiology of fatigue in prolonged exercise has not been totally elucidated, although there appears to be individual susceptibility.

Nevertheless, it is well documented that with increased exercise duration, there is a progressive decrease in muscle glycogen and a progressive increase in blood glucose uptake by the muscle. However, liver supplies of glycogen are limited, and thus blood glucose levels eventually fall. Prevention of hypoglycemia is one of the major objectives of carbohydrate consumption during prolonged exercise, because all individuals will eventually suffer ill-effects when they reach their minimum blood glucose threshold.

How is lactic acid production related to fatigue?

In review, as noted in chapter 3, lactic acid is an end product of anaerobic glycolysis. Anaerobic glycolysis represents the lactic acid energy system, found primarily in the type IIb white muscle fibers, but also the type IIa red fibers. Anaerobic glycolysis is increased at the onset of high-intensity exercise, such as 200- to 400-meter track events, as a means to rapidly replenish ATP for muscle contraction or for the rapid resynthesis of PCr. The process of aerobic glycolysis can also generate ATP, but it is too slow during high-intensity exercise. Thus, lactic acid, or lactate, is produced when the production of pyruvate from glycolysis exceeds the oxidative capacity of the mitochondria.

Blood lactic acid (lactate) levels increase during high-intensity exercise and for years have been thought to be the cause of fatigue. However, Coyle notes that the lactate molecule *per se* does not cause fatigue, but rather that its accumulation in blood reflects a disturbance of muscle cell homeostasis. One factor that may disturb muscle cell homeostasis is an increased concentration of hydrogen ions, reflecting an increased acidity. Debold notes that even after 100 years of research on muscle fatigue, the molecular basis is still poorly understood. Using in vitro studies, Debold has demonstrated that high levels of hydrogen ions and phosphate inhibit the molecular motions of muscle proteins such as myosin, troponin, and tropomyosin, which leads to decreased force production and potentially fatigue. Fitts noted that it is now generally thought that the component of fatigue correlated with lactate results from the effects of an increased free hydrogen ion rather than lactate or the undissociated lactic acid. The hydrogen ion and associated increased acidity could elicit fatigue by inhibiting a number of physiological processes in the muscle cell associated with contraction. As noted in chapter 13, sodium bicarbonate, a buffer of acidity, may be an effective means to enhance performance in high-intensity exercise tasks.

The lactate produced during exercise is not a waste product. Lactate is a carbohydrate; one molecule of lactate contains about half the energy of one molecule of glucose. Years ago, George Brooks, from the University of California at Berkeley, proposed the lactate shuttle, whereby the lactate produced during exercise in white muscle fibers would be shuttled to other tissues, such as the heart and red oxidative muscle fibers, where it could be oxidized for energy. Hashimoto and Brooks also hypothesize the lactate produced in the cell cytoplasm may be oxidized by mitochondria in the same cell indicating that lactate is a link between glycolytic and aerobic pathways. Research supports the lactate shuttle hypothesis and, as discussed later in this chapter, some sports scientists have actually tested lactate salt supplementation as a means to provide energy to enhance exercise performance.

How is low muscle glycogen related to the development of fatigue?

Low muscle glycogen may impair both aerobic and anaerobic exercise performance.

Low Muscle Glycogen and Aerobic Exercise Muscle glycogen is the major energy source for prolonged, moderately high- to high-intensity aerobic exercise. Elite marathon runners can maintain a fast race pace primarily through oxidation of carbohydrate, primarily muscle glycogen, and to a lesser extent fatty acids stored in the muscle. Coyle indicates that exercise at 70–85 percent of VO_2 max cannot be maintained without sufficient carbohydrate oxidation, and thus the severe lowering of muscle glycogen, often coupled with hypoglycemia, results in the need to reduce exercise intensity to 40–60 percent of VO_2 max.

A number of studies have shown that physical exhaustion was correlated with very low muscle glycogen levels, but others have shown some glycogen remaining even though subjects were exhausted. Several mechanisms have been postulated to explain the development of fatigue even with some muscle glycogen remaining.

- Location of glycogen. Costill has indicated that performance would be adversely affected only when muscle glycogen levels went below 40 mmol/kg of muscle tissue. It may be that complete depletion of muscle glycogen is not necessary for performance to suffer, for glycolysis may be impaired with lower glycogen levels or the glycogen in the muscle fiber may be located where it is not readily available for glycolysis. Ørtenblad and colleagues have described that glycogen is not homogeneously distributed in muscle fibers; it is localized in separate pools. Additionally, glycogen granules have their own glycolytic enzymes and regulating proteins and it may be the depletion of these complexes localized in the myofibrils that is related to fatigue.
- Rate of energy production. Shulman and Rothman propose a model in which energy is supplied in milliseconds via glycogenolysis, and indicate that one possible mechanism for muscle fatigue is that at low but nonzero glycogen concentrations, there is not enough glycogen to supply millisecond energy needs. In a related vein, Fitts notes that low levels of muscle glycogen may interfere with maintenance of optimal levels of Krebs cycle intermediates, which can reduce the rate of aerobic ATP production, and further notes that the metabolism of blood-borne substrates (blood glucose and free fatty acids [FFA]) is simply too slow to maintain heavy exercise intensities.
- Muscle fiber type. The fatigue that develops may be related to the depletion of muscle glycogen from specific muscle fiber types. In prolonged exercise at 60–75 percent of VO_2 max, type I fibers (red, oxidative slow twitch) and type IIa fibers (red, oxidative-glycolytic fast twitch) are recruited during the early stages of the task, but as muscle glycogen is depleted, the athlete must recruit type IIb fibers (white, glycolytic fast twitch) to maintain the same pace. However, it takes more mental effort to recruit the type IIb fibers, which will be more stressful to the athlete. Type IIb fibers also are more likely to produce lactic acid, increasing the acidity, which may increase the perceived stress of the exercise. In a study, Krustrup and others reported that glycogen depletion of the slow-twitch muscle fibers, necessitating recruitment of fast-twitch muscle fibers and increased energy demands, is a factor that may predispose to fatigue.
- Use of fat for energy. As muscle glycogen becomes depleted in the slow-twitch muscle fibers, the muscle cell will rely more on fat as the primary energy source. Because fat is a less-efficient fuel than carbohydrate, the pace will slow down.
- Role of the brain. Signals sent from peripheral tissues to the brain may regulate energy metabolism. A low glycogen level in exercising muscle may be such a signal and may invoke neural responses causing fatigue. Matsui and colleagues recently showed decreased liver and muscle glycogen after 30 and 60 minutes of running, and a decrease in glycogen in five areas of the brain after 120 minutes of running. The interactions between decreased muscle and brain glycogen and fatigue need to be further studied.

Low Muscle Glycogen and Anaerobic Exercise Fatigue in very high-intensity, anaerobic-type exercise generally is attributed to the detrimental effects of the acidity in the muscle cell associated with lactic acid production. Research has now shown that maximal high-intensity exercise, lasting only about 60 seconds, is not impaired by a very low muscle glycogen concentration, approximately 30 mmol/kg muscle. However, it is possible that performance in such very high-intensity, short-term exercise tasks may be impaired with extremely low muscle glycogen in the fast-twitch muscle fibers. Moreover, with somewhat longer anaerobic tasks, approximating 3 minutes, one laboratory study reported a reduced performance in the time to exhaustion test after 4 days of a low-carbohydrate, high-fat diet when compared to a normal, mixed diet and a high-carbohydrate diet. Although muscle glycogen levels were not measured, a logical assumption is that they were lower on the low-carbohydrate diet.

In addition, field research has suggested that slower overall sprint speed, such as in the later parts of prolonged athletic contests like soccer and ice hockey, may be due to muscle glycogen depletion. Muscle biopsies of these athletes revealed very low glycogen levels, which were attributed not only to the strenuous exercise in the contest but also to the fact that these athletes were consuming diets low in carbohydrates. In support of these field studies, Balsom and others reported that low muscle glycogen levels impaired laboratory exercise performance in repeated bouts of very high-intensity intermittent exercise, 6-second cycle ergometer performance followed by 30 seconds of rest. Krustrup and others reported that almost 50 percent of muscle fibers were completely or almost empty of glycogen following a soccer game, and suggested that slower sprint performance in the later part of a game may be explained by low glycogen levels in individual muscle fibers. Also, low muscle glycogen stores may lead to a decrease in exercise intensity during training.

In summary, low levels of glycogen in the white, fast-twitch IIb muscle fibers may limit performance in intermittent, anaerobic-type exercise tasks. Both hypoglycemia and low glycogen in the red muscle fiber types, most likely a combination of the two, may be contributing factors to fatigue in prolonged endurance exercise.

Train low, compete high Although the importance of dietary carbohydrate to support exercise performance is clear, alternating between periods of high and low carbohydrate availability may offer advantages as well. John Hawley, an expert in promoting training adaptation by manipulating carbohydrate availability, a concept that has been referred to as "train low," has recently reviewed the research in this area. The concept is based on the theory that making carbohydrate less available at critical times will improve adaptations to training better than if carbohydrate was available at all times.

One way to reduce carbohydrate availability is to train twice-a-day with the second session conducted under "low glycogen availability." Essentially, an athlete would train intensely, not replace carbohydrate postexercise, and then conduct a second bout of exercise while muscle glycogen was decreased. Research supports that this method will result in positive training adaptations as measured by an increase in several enzymes involved in mitochondrial biogenesis. However, although this has been proven effective in improving training adaptations (e.g., increased muscle enzyme activity), it can result in decreased power output during subsequent training sessions, an outcome that is undesirable for athletes. Another approach is to extend the duration of low carbohydrate availability during the overnight period (train high and sleep low). Essentially, an athlete would train intensely in the evening (high glycogen availability), go to bed fasted and complete a submaximal morning exercise session before refeeding (low glycogen availability). This method is being actively studied. A final method is to reduce exogenous carbohydrate availability before or during exercise. In this scenario, exercise occurs fasted or without carbohydrate available during exercise. Studies that have investigated this concept have shown similar improvements in enzymatic activities, suggesting that training adaptations are not impaired by decreased exogenous carbohydrate availability. However, the effects on performance outcomes are unclear.

According to Hawley, there are can be discrepancies between training adaptation (assessed with cellular variables) and performance (assessed with whole body exercise) outcomes. There are several reasons to support this perspective. As Hawley notes, there may not be a direct relationship between athletic performance and training-induced cellular adaptations. Through training, well-trained athletes may have already maximized the muscle adaptations that support athletic performance, and remaining adaptations may not play an important role. Small gains in performance (<1 percent) are difficult if not impossible to measure in a laboratory. Finally some "train low" approaches might impede performance during training, causing unfavorable muscular adaptations. For instance, reduced training intensity, due to low glycogen availability, may alter muscle fiber recruitment and substrate utilization. Thus, athletes and sports dietitians must consider both physiological and performance outcomes when evaluating a dietary intervention.

Hawley reported that during short-term training programs (3 to 10 weeks) in which half of the workouts occurred when muscle glycogen or exogenous carbohydrate availability was low, training adaptations were improved or similar to training with normal muscle glycogen. Despite the supportive research on muscular adaptations, Hawley states that there is no clear evidence that "training low" improves athletic performance, and that it is probably unlikely that an athlete would choose a diet plan that compromised training intensity. At this time, many studies are under way to further investigate the effects of manipulating carbohydrate availability on exercise performance.

How are low endogenous carbohydrate levels related to the central fatigue hypothesis?

As noted previously, hypoglycemia and low muscle glycogen levels may impair exercise performance, and one mechanism for each involves adverse effects on brain function. Collectively, they may contribute to central fatigue in a different way.

In the later stages of prolonged exercise bouts, low muscle glycogen, in combination with decreased blood glucose levels, will stimulate gluconeogenesis from muscle protein. In particular, branched-chain amino acids (BCAAs) in the muscle will be catabolized to provide energy. Because BCAA release from the liver may be decreased, or uptake by the muscle may increase, blood levels of BCAA decline. The *central fatigue hypothesis* during prolonged exercise suggests that this decline in blood BCAA may contribute to fatigue. In general, fatigue is hypothesized to occur when BCAA levels drop and the concentration of another amino acid—tryptophan—increases in its free form, or free tryptophan (fTRP). BCAAs compete with fTRP for similar receptors that facilitate their entry into the brain, so high BCAA levels prevent brain uptake of fTRP. With an increased fTRP:BCAA ratio, entry of fTRP into the brain cells will be facilitated. Increased brain levels of tryptophan may stimulate the formation of serotonin, a neurotransmitter in the brain that may be related to fatigue sensations (**figure 4.11**). Preventing the increase in the fTRP:BCAA ratio is theorized to prevent the premature development of fatigue, and the use of BCAA supplements in this regard will be covered in

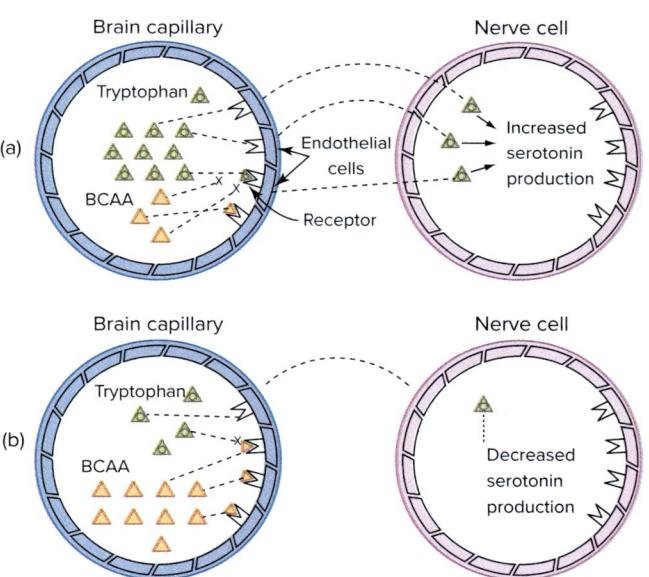

FIGURE 4.11 BCAA and the central fatigue hypothesis. (a) An increased level of serum free-tryptophan (fTRP ▲) resulting in a high fTRP:BCAA ratio will bind to receptors in the capillary for transport to nerve cells and increase serotonin formation, which may be associated with central fatigue. (b) An increased level of BCAAs (▲) will compete with the fTRP for receptors, blocking them from entering the nerve cell and decreasing production of serotonin, helping to prevent central fatigue.

chapter 6. Meeusen has noted that postponing central fatigue with nutritional interventions, including amino acids, water, carbohydrate, and caffeine, has been studied. Although carbohydrate (discussed later in this chapter) and caffeine show promise in reducing central fatigue, other nutrients, including BCAAs, do not appear as promising. However, carbohydrate intake during exercise may also be helpful, as discussed later in this section.

Will eating carbohydrate immediately before or during an event improve physical performance?

Tobias Titz/fStop/Getty Images

Because hypoglycemia or muscle glycogen depletion may be a cause of fatigue during endurance exercise, supplementation with glucose or other forms of carbohydrate before or during exercise may be theorized to delay the onset of fatigue and improve performance. Thousands of studies have been conducted on this topic ever since carbohydrates were identified as the most efficient energy source for exercise, and researchers' interest in this topic remains unabated today. In recent years, the research designs have usually been highly sophisticated as investigators have attempted to provide specific answers relative to the type, amount, and timing of carbohydrate ingestion before and during performance. Although some problems remain in providing quantitative data, the use of stable isotopes of ingested carbohydrates (referred to as *exogenous* carbohydrates in comparison to *endogenous* stores in the body), as detailed by Wolfe and George, has enhanced our understanding of their metabolic fate when ingested prior to or during exercise.

However, the reviewer attempting to synthesize the available research is confronted with a difficult task, as the experimental designs varied considerably. The amount and type of carbohydrate ingested, the use of liquid or solid forms, the method of administration (oral ingestion or venous infusion), the time prior to or during the exercise that it was taken, the diet of the subject several days prior to the study, the amount of glycogen in the muscle and liver, the intensity and duration of the exercise, the type of exercise task (running, swimming, cycling, etc.), the fitness level of the subjects, the environmental temperature, and the method used to evaluate blood glucose and muscle glycogen utilization are some of the important differences between studies.

Although the results from all of these studies were not similar, some general consistencies have evolved. The role of carbohydrate supplementation on exercise performance has been the subject of numerous reviews, and a number of contemporary reviews may be found in the reference list at the end of this chapter. Based on these reviews and an overall review of specific studies, the following generalizations appear to be logical. More specific information relative to practical recommendations is provided following this discussion.

Use of the Ingested Carbohydrate

Using labeled carbohydrate sources and analyzing the expired carbon dioxide for radioactivity, investigators have shown that some of the ingested, or exogenous, carbohydrate may be used as an energy source within 5-10 minutes, indicating that it may empty rapidly from the stomach, be absorbed from the small intestine into the blood, and enter into metabolic pathways. Peak use of exogenous carbohydrate appears to occur 75-90 minutes after ingestion. A number of studies have shown that the ingested carbohydrate may contribute a significant percentage of the carbohydrate energy source during exercise, ranging from 20 to 40 percent in some studies, but as much as 60-70 percent during the later stages of exercise, when endogenous liver and muscle glycogen stores become depleted.

Possible Fatigue-Delaying Mechanisms

The precise mechanism whereby glucose ingestion helps delay the onset of fatigue during moderate- to high-intensity exercise (i.e., >65 percent VO_2 max) has not been totally elucidated, but several theories have been studied.

Maintenance of blood glucose levels The available data suggest that the ability of carbohydrate intake to delay fatigue may be related to the maintenance of higher blood glucose levels, possibly by sparing liver glycogen until late in the exercise and the prevention of hypoglycemia in susceptible individuals; blood glucose would be available to enter the muscle and provide a source of energy for aerobic glycolysis and may provide glucose to the brain to prevent premature central fatigue. As noted, exogenous glucose is used increasingly as the exercise task becomes prolonged.

Reduction of psychological effort Some research has shown that glucose ingestion could make an endurance task psychologically easier and suggested that the physiological effects of the glucose, either in the brain or in the muscles, reduced the stressful effects of exercise. Several studies by Utter and others found that carbohydrate ingestion reduced the ratings of perceived exertion during prolonged running and cycling. For example, they reported that marathoners ingesting carbohydrate compared to placebo beverages were able to run at a higher intensity during a competitive marathon, and yet the ratings of perceived exertion (RPE) were similar in both groups of runners, suggesting that the carbohydrate may have permitted them to run at a faster rate with similar psychological effort. However, it has not been determined whether these ergogenic effects may be attributed to the effect of glucose as a source of energy in the muscle or to its effect on the central nervous system, either as a direct energy source for brain metabolism or through its effect on BCAA levels. In an effort to improve endurance performance, Jeukendrup pioneered a method of rinsing, but not ingesting, carbohydrate in the mouth. It has been shown that the presence of carbohydrate in the mouth activates several areas of the brain. In a recent review, he highlighted 12 studies that examined the effects of carbohydrate mouth rinse on endurance exercise performance (30-70 minutes) and reported that 9 studies showed significant improvements. These improvements were of a similar magnitude as would be expected with carbohydrate ingestion. The carbohydrate mouth rinse methods might be particularly useful for athletes who experience gastrointestinal upset with carbohydrate ingestion or in those attempting to reduce energy intake to manage their body mass.

In several studies, carbohydrate supplementation during exercise has been reported to prevent the decrease in serum BCAA during the later stages of prolonged exercise, possibly by mitigating secretion of cortisol. According to the central fatigue hypothesis, preventing an increase in the fTRP:BCAA ratio would deter the onset of mental fatigue.

Sparing of muscle glycogen Although sparing of muscle glycogen could be another benefit of carbohydrate ingestion before or during moderate- to high-intensity exercise, research findings are equivocal.

In a unique experiment from the University of Texas, subjects received venous glucose infusions to maintain a hyperglycemic state during 2 hours of exercise at 73 percent of VO_2 max, but the net rate of muscle glycogen utilization was not affected compared to control conditions. Arkinstall and Chryssanthopoulos and their colleagues also noted no muscle glycogen sparing effect with carbohydrate supplementation during the exercise task.

In contrast, Yaspelkis and others, also from the University of Texas, noted that during low-intensity exercise (i.e., <50 percent VO_2 max) or during low- to moderate-intensity exercise tasks, carbohydrate supplementation during exercise could spare use of muscle glycogen in slow-twitch muscle fibers and enhance performance. They noted that during low-intensity exercise the serum levels of both glucose and insulin were elevated, which could promote muscle use of serum glucose and sparing of muscle glycogen. Bosch and others, as well as Tsintzas and others, reported that carbohydrate intake during prolonged exercise at about 70 percent VO_2 max did spare muscle glycogen use, and the Tsintzas group reported that the sparing effect occurred in the type I, slow-twitch muscle fibers, but not in the type II, fast-twitch fibers. Although others concluded that consuming carbohydrate during exercise did not spare muscle glycogen in cyclists, Williams did find glycogen sparing early on in runners who consumed carbohydrate drinks throughout prolonged exercise. Hargreaves noted also that the breakdown of muscle glycogen may be slowed because the supply of blood glucose is improved when carbohydrate is consumed.

Limitations to Prevent Fatigue Although glucose ingestion may help delay fatigue during moderately high-intensity exercise, it cannot totally prevent the onset of fatigue. It appears that a maximum of about 1.5–1.7 grams of the ingested carbohydrate may be available each minute, which is much lower than the required energy needs at 65–85 percent of VO_2 max. Jeukendrup and Jentjens indicate that the intestines and liver may be limiting factors, the intestines being unable to absorb the ingested carbohydrate at a faster rate, while the liver may limit the amount of glucose released into the blood. When blood glucose, BCAA, and/or muscle glycogen levels are eventually reduced to a critical level, fatigue occurs.

Initial Endogenous Stores If the individual has normal liver and muscle glycogen stores, glucose feedings are unnecessary for continuous, moderately high-intensity exercise bouts lasting 60–90 minutes or less but may be beneficial for high-intensity exercise tasks of similar duration, as noted in the following text. Because the body can store carbohydrate in the muscles and liver, the usefulness of glucose or other carbohydrate intake before or during exercise depends on the adequacy of those supplies already in the muscle and liver to meet energy needs. For competition, the muscle and liver glycogen stores should be adequate to meet carbohydrate energy needs. The critical point is to consume substantial amounts of carbohydrates a day or two prior to the event and to decrease the duration and intensity of training to assure ample endogenous glycogen supplies.

The available research has shown that the consumption of glucose, fructose, sucrose, maltodextrin (a glucose polymer), or other carbohydrate combinations immediately prior to events of short or moderate duration has a negligible effect upon performance. Adding a gallon of gas to a full tank will not make a car go faster during a short ride. The same is true of sugar to a muscle already filled with glycogen. If, however, muscle glycogen levels are low and the exercise task is somewhat prolonged, then ingestion of carbohydrate just prior to the exercise bout may improve performance. It is important to note, however, that to enhance performance, the exogenous carbohydrate source must be able to delay the onset of fatigue that might otherwise occur as a result of premature depletion of endogenous carbohydrate sources, a viewpoint also proposed by Tsintzas and Williams.

Exercise Intensity and Duration The potential beneficial effects of carbohydrate supplementation depend on the interaction of exercise intensity and duration, which, of course, are interrelated. The shorter the duration, the greater the exercise intensity can be. Stellingwerff and Cox have

Robert Daly/Caiaimage/Getty Images

written the most comprehensive review of the effects of carbohydrate ingestion on exercise performance that takes into account the influence of exercise intensity and duration. They reported that of 61 total studies, 50 (82 percent) showed a performance-enhancing effect of carbohydrate ingestion. Of 10 studies that assessed the effects of carbohydrate ingestion on exercise tasks lasting less than 1 hour, 6 showed a performance enhancement. Of 13 studies of the effects of carbohydrate mouth rinse on exercise performance lasting less than 1 hour, 10 showed a beneficial effect. During exercise tasks lasting from 60 to 120 minutes, 15 of 18 studies showed a performance-enhancing effect of carbohydrate ingestion. Finally, in studies with exercise tests lasting more than 120 minutes in duration, 16 of 17 studies showed a performance-enhancing effect of carbohydrate supplementation. Overall, there was a positive relationship between total exercise time and percent increase in performance, indicating that carbohydrate ingestion was more effective in longer tasks. The authors noted that the primary mechanisms through which carbohydrate improves performance probably differ based on exercise intensity and duration. For instance, during brief, intense exercise carbohydrate acts on the receptors in the mouth and works through the central nervous system to enhance performance; thus, the type of carbohydrate may not be important. In longer exercise tasks, rapidly absorbed carbohydrates, possibly greater than 90 grams/hour, are needed to replenish depleting muscle glycogen stores. The authors stressed that recommendations should be tailored to individual tolerance.

The following sections provide more details about some of the individual studies that have evaluated the effects of carbohydrate ingestion on exercise performance. The time frames are representative of those that have been well studied, and science-based recommendations for carbohydrate intake pre-, during, and postexercise are presented in **table 4.7**.

TABLE 4.7 Some recommended guidelines for carbohydrate intake as a means to help optimize exercise/sports performance

Carbohydrate intake before and during exercise may help optimize performance, whereas carbohydrate intake after exercise may facilitate recovery for subsequent training or competition. In general, the lower the intensity and the shorter the duration of exercise, the less the need for additional carbohydrate. Recommendations are based on scientific studies. Physically active individuals should consume various types and amounts of carbohydrate before and during exercise and sport training to determine personal optimal dietary strategies. See the text for additional information and guidelines.

Exercise intensity/sport	Duration	Before exercise*	During exercise**	After exercise***
Very high-intensity aerobic (5K run)	<30 minutes	None needed	None needed	None needed
High-intensity aerobic (10K run)	30–90 minutes	20–25 grams	30–60 grams carbohydrate. Mouth rinse 1.5 grams/ml for 5 to 10 sec every 8 to 10 min of exercise	60–80 grams/hour for 3–4 hours
Intermittent high-intensity (team sports, such as soccer)	60–90 minutes	20–25 grams	30–60 grams carbohydrate. Mouth rinse 1.5 grams/ml for 5 to 10 sec every 8 to 10 min of exercise	60–80 grams/hour for 3–4 hours
Moderate- to high-intensity aerobic (half-marathon; marathon)	>90 minutes	20–50 grams carbohydrate loading	60–80 grams/hour	60 grams/hour for 3–4 hours
Moderate-intensity aerobic (ironman-distance triathlon; 140 miles; 226 kilometers)	>6 hour	20–50 grams carbohydrate loading	60–80 grams/hour	60–80 grams/hour for 3–4 hours
High-intensity resistance training (lifting weights)	1–2 hours	None needed	None needed	60–80 grams/hour for 3–4 hours

*Within 10–15 minutes of the start of exercise. All individuals should consume a pregame meal sometime between 1–4 hours prior to exercise in order to ensure normal muscle glycogen and blood glucose levels. Carbohydrate intake 3–4 hours before should average about 3–4 grams per kilogram body weight, but only 1–2 grams per kilogram body weight if consumed within 1–2 hours of exercise.
**Fluids such as sports drinks are recommended. Drinking about 8 ounces (240 milliliters) of a typical sports drink every 15 minutes will provide about 60 grams of carbohydrate per hour.
***For most athletes, consuming a diet with substantial amounts of healthful carbohydrates over the course of 24 hours will replace muscle glycogen levels to normal. For those who want a speedy recovery of muscle glycogen for subsequent intense training or competition the same or following day, using this rapid muscle glycogen replacement protocol may be recommended. Athletes may also benefit by consuming some protein with the carbohydrate, about 1 gram of protein for every 4 grams of carbohydrate.

Very high-intensity exercise for less than 30 minutes Research suggests that carbohydrate supplementation will not enhance performance in high-intensity exercise bouts less than 30 minutes in length. For example, Palmer and others noted that consuming carbohydrate 10 minutes before competing in a 20-kilometer cycle time trial did not enhance performance in well-trained cyclists. Nevertheless, if carbohydrate supplements could ameliorate a muscle or liver glycogen deficiency, performance could improve. For example, Walberg-Rankin reported that carbohydrate supplementation improved high-intensity anaerobic exercise performance in wrestlers following a drastic weight-reduction program with very limited carbohydrate intake. Presumably, the carbohydrate supplement, consumed in a 5-hour period before testing, increased muscle glycogen levels, particularly in fast-twitch fibers, enhancing carbohydrate utilization and subsequent performance.

Very high-intensity resistance exercise training While the performance-enhancing benefits of carbohydrate ingestion on endurance exercise performance are well accepted, less is known about the effects of carbohydrate on resistance exercise performance. As strength and conditioning training programs are an integral part of success in most sports, properly fueling resistance training workouts is essential. A bout of resistance exercise can reduce muscle glycogen by up to 40 percent, indicating the importance of glycogen as a fuel for resistance training and raising the possibility that carbohydrate ingestion or mouth rinsing may improve resistance exercise performance. In the past, carbohydrate consumption before or during resistance exercise has not been recommended. Although this was based on a small number of studies, some did indicate a benefit of pre-resistance exercise carbohydrate intake. For instance, Haff and colleagues have shown that carbohydrate supplementation prior to (1.0 gram/kg body mass) and during (0.5 gram/kg body mass) resistance exercise enhanced resistance exercise performance. However, the increase in resistance training and sport nutrition has prompted a number of new investigations, recently summarized by Henselmans and colleagues. In their review, it was noted that 39 of 49 studies show that acute or chronic carbohydrate manipulation does not improve resistance exercise outcomes, including strength and power. The studies were diverse in nature; included volunteers who were athletes, recreationally trained, and untrained; and investigated the manipulation of carbohydrate intake under various conditions, including acute supplementation (19 studies), carbohydrate supplementation following glycogen depletion (6 studies), short-term (2–7 days) carbohydrate supplementation (7 studies), and longer-term supplementation, combined with a resistance training program (17 studies). While several studies found improved performance, for example, 6 of 19 acute studies, most did not. None of the short-term or longer-term studies showed improved performance. Overall, this indicates that under typical conditions (e.g., fed state, traditional volume exercise session), additional carbohydrate does not improve resistance exercise outcomes. Additionally, it demonstrates how a body of evidence needs to develop over time before it can be systematically reviewed and well-informed conclusions can be made.

Several groups have examined the effects of carbohydrate mouth rinse on strength and power production, but at this point, the results are equivocal. For example, Jensen and colleagues were able to show an improvement in maximal strength in athletes who were fatigued by prior exercise; Phillips and others showed improved peak power output during a single 30-second cycle sprint, and Bastos-Silva reported increased muscular endurance and total lifting volume in the bench press. However, both Dunkin and Phillips, and Painelli and colleagues reported no improvements in maximal strength or muscular endurance. At this point in time, there are no specific recommendations for athletes to consume or rinse with carbohydrate before or during resistance exercise (**table 4.7**).

High-intensity exercise for 30 to 90 minutes Within this time frame, the potential benefit of carbohydrate consumption may depend on the duration of the exercise, intensity of exercise, and training level of the athlete. For example, two studies found that neither consumption of a 6 percent carbohydrate solution nor infusing glucose at the rate of 1 gram per minute improved performance in a 1-hour maximal cycling protocol. Additionally, Burke and others reported no effect of a commercial gel supplying about 1.1 grams of carbohydrate per kilogram body weight on half-marathon performance compared to the placebo. However, in a review, Karelis and colleagues examined studies with exercise times ranging between 30 and 60 minutes. They noted a beneficial effect of carbohydrate ingestion with exercise lasting at least 40-50 minutes. Moreover, supplementation may benefit well-trained athletes who may be able to exercise at high intensity for about an hour. For example, Jeukendrup and el-Sayed, with their associates, reported that cyclists exercising for about an hour at high intensity significantly improved their performance following ingestion of a carbohydrate supplement, as compared to a placebo. Also, Ball and others reported that carbohydrate intake during a simulated time trial improved performance in a sprint at the end of 50 minutes of high-intensity cycling. In such cases, it is possible that the ingested carbohydrates may help provide glucose to the fast-twitch muscle fibers or prevent premature depletion in the slow-twitch fibers.

As indicated earlier, the presence of carbohydrate in the oral cavity has been shown to improve the performance by stimulating the central nervous system. Exercises lasting about an hour have been the most well studied, and it appears that the type or sweetness of carbohydrate is not relevant. Currently, the most practical advice points toward swishing 1.5 grams of carbohydrate per milliliter of water for 5-10 seconds during every 8-10 minutes of exercise. At this time, a great deal more work needs to be done to learn how carbohydrate mouth rinsing improves exercise performance and what conditions (e.g., fasting, fed, inclusion of other nutrients) may enhance or reduce these effects. Importantly, as noted by Stellingwerff and Cox, several practical concerns need to be considered. For instance, can an athlete find time to swish with carbohydrate for 10 seconds every 8 to 10 minutes, and how will this affect breathing? The authors suggested a "sports confectionary" (concentrated carbohydrate, like a jelly bean) in the cheek cavity, but depending on the sport, this might be against the rules or present a choking hazard.

Intermittent, high-intensity exercise for 60 to 90 minutes Research has shown that individuals engaged in endurance-type contests with intermittent bouts of sprinting, such as soccer, ice hockey, or tennis, may benefit from carbohydrate supplements taken before and during the game. In a controlled laboratory protocol representative of a 60-minute intermittent, high-intensity competitive sport such as soccer or field hockey, Welsh and colleagues, from Mark Davis's laboratory, reported that carbohydrate intake before and during the exercise task resulted in significant improvements in various tests of physical and mental functions performed throughout the experimental trial. Toward the end of the 60-minute period, the carbohydrate trial resulted in faster 20-meter sprint time, longer time to fatigue in a shuttle run, enhanced whole body motor skills, and decreased self-reported perceptions of fatigue. The results suggested a beneficial role of carbohydrate-electrolyte ingestion on physical and mental functions during intermittent exercise similar to that of many competitive team sports. Similar findings were reported by Winnick and others. Other field research studies, although not universally supportive, have shown similar benefits of carbohydrate intake under game conditions. Several reviewers, such as Kirkendall and Kovacs, indicate that carbohydrate intake before and during prolonged, intermittent, high-intensity exercise sports may enhance performance.

High- to moderate-intensity exercise greater than 90 minutes Research generally supports a beneficial effect of carbohydrate intake on exercise performance tasks greater than 90 minutes (if the exercise intensity is high enough), particularly so when the task is more prolonged, such as 2 hours or more. For example, Kimber and others reported a significant inverse correlation between the amount of carbohydrate consumed and the finishing time of male triathletes in an Ironman triathlon, suggesting that increasing carbohydrate consumption during such a prolonged event may enhance performance. Jeukendrup indicates that the performance benefits of carbohydrate ingestion are likely achieved by maintaining or raising plasma glucose concentrations to help sustain high rates of carbohydrate oxidation.

When, how much, and in what form should carbohydrates be consumed before or during exercise?

The most common athletic events or physical performance activities that may benefit from carbohydrate feedings are those associated with long duration (90-120 minutes or more) at moderate- to high-intensity levels. Marathon running, cross-country skiing, and endurance cycling are common sports of this kind. Other sports that require intermittent bouts of intensive activity over a prolonged period, such as soccer, may also benefit. However, the individual participating in these activities, particularly under warm or hot environmental conditions, also needs to replenish fluid losses incurred through sweating. In such cases, fluid replenishment is more critical than carbohydrate. The topic of fluid replacement during exercise is covered in more detail in chapter 9, but because carbohydrate is one of the contents in the majority of the sports drinks developed as fluid replacements for athletes, its role is discussed briefly here.

Many studies have been conducted to determine the best carbohydrate feeding regimen to prevent fatigue during prolonged exercise. A number of different variables have been studied, such as the timing of the feeding and the type, amount, and concentration of carbohydrate.

Again, based on current reviews by the primary investigators regarding carbohydrate supplementation for exercise performance and a careful analysis of individual studies, the following points represent the general conclusions and recommendations for individuals who may be exercising at 60–80 percent of their VO_2 max or greater for 1–2 hours or longer. These points may also be applicable to athletes engaged in intermittent, high-intensity exercise sports that last an hour or more. But remember, individuals may have varied reactions to carbohydrate intake, so athletes should experiment in training before using these recommendations in actual competition. Refer to **table 4.7** for a concise summary.

Pre-exercise: When and How Much?

Four hours or less before exercise Carbohydrate intake 60–240 minutes prior to prolonged exercise tasks (longer than 90 minutes) may enhance performance. Research has demonstrated improved performance when adequate carbohydrate was consumed either 1, 3, or 4 hours prior to a prolonged exercise task involving simulated racing conditions during the later stage. Other research revealed no significant differences in 30-kilometer run performance when equal amounts of carbohydrate were supplemented either 4 hours before or during the run, suggesting the ingested carbohydrate was available for energy production using either strategy.

The amount of carbohydrate ingested 4 hours prior to performance should be based upon body weight. Several studies have used 4–5 grams/kg (1.8–2.3 grams/pound) with good results. For an athlete who weighs 60 kg (132 pounds), the recommended amount would be 240–300 grams. The carbohydrates could be consumed in any of several forms, including fluids such as juices or glucose polymer solutions, or solid carbohydrates such as fruits or starches. The fiber content should be minimized to prevent possible intestinal problems during exercise. Keep in mind that 300 grams of carbohydrate is about 1,200 kcal a somewhat substantial meal. **Table 4.8** presents a quick estimate of carbohydrates in several common foods and sports products.

Less than 1 hour before exercise Jeukendrup and Killer reviewed the myths surrounding pre-exercise carbohydrate feedings. They found that the ingestion of carbohydrates in the hour before exercise either improved performance or had no impact on performance. Based on these findings, there is little evidence to abstain from carbohydrate ingestion in the hour prior to exercise, for those who do not experience symptoms of hypoglycemia. Prudence suggests that individuals who may be prone to reactive hypoglycemia should avoid carbohydrate intake, particularly high-glycemic-index foods, 15–60 minutes prior to performance. Simple sugars ingested within this time frame may actually impair physical performance in such individuals because of the adverse effects of reactive hypoglycemia, such as muscular weakness. Moreover, this same insulin response may speed up muscle glycogen utilization. This may be a disadvantage to the marathoner, whose glycogen levels may be depleted too early in the race. Several earlier studies showed that run time to exhaustion was shorter by about 20–25 percent after athletes consumed 2–3 ounces of glucose within an hour before the endurance test.

However, not all individuals experience reactive hypoglycemia. Kuipers and others noted that about one-third of well-trained subjects experienced hypoglycemia following the ingestion of 50 grams of glucose after a 4-hour fast. However, the hypoglycemia was transient, as blood glucose levels returned to normal after 20 minutes of exercise at 60 percent VO_2 max. No performance data were measured. In a study by Seifert and others, subjects were given various carbohydrate solutions to raise their insulin levels; when their insulin levels peaked, they undertook an exercise task at 60 percent of VO_2 max for 50 minutes. No hypoglycemia developed, nor were there any adverse sensory or psychological responses.

If carbohydrate is consumed approximately 1 hour prior to performance, about 1–2 grams/kg (60–120 grams for a 60-kg athlete) may be recommended, for these levels have been shown to enhance performance in several studies. One study using only 12 grams 1 hour prior to performance showed no beneficial effect. Both glucose polymers and foods with a low glycemic index have been used successfully.

Immediately before exercise As noted previously, consuming carbohydrate immediately before exercise of short duration, and even exercise tasks of less than 90 minutes or so, normally will not enhance performance. For example, Marjerrison and others found that the ingestion of a carbohydrate solution 30 minutes before undertaking four 30-second anaerobic Wingate tests had no effect on power output; Smith and others reported no significant effect on swim time performance of ingesting a 10 percent glucose solution 5 minutes before a 4-kilometer swim of approximately 70 minutes. However,

---

TABLE 4.8 Grams of carbohydrate in selected foods and sports products

1 Fruits group = 15 grams carbohydrate
   1 apple
   1 orange
   ½ banana
   4 ounces orange juice
1 Grains group = 15 grams carbohydrate
   1 slice bread
   ½ cup cereal
   ¼ large bagel
   ½ cup cooked pasta
   1 small baked potato
Sports drinks: 7–8 ounces = 15 grams carbohydrate
   Gatorade®
   PowerAde®
   SportAde®
Sports bars = 20–50 grams carbohydrate
   1 PR Bar®
   1 Power Bar®
Sports gels = 20–30 grams carbohydrate
   1 Power Gel® packet
   1 ReLode® packet
Energy drinks: 8 ounces = 25–50 grams carbohydrate
   Gatorade® Energy Drink
   SoBe Energy®

carbohydrate intake immediately prior to (within 5-10 minutes) prolonged endurance exercise tasks of 2 hours or more may help delay the development of fatigue and improve performance if the athlete is exercising at a level greater than 50 percent VO_2 max, such as 60-75 percent. The majority of the studies, including controlled laboratory investigations and field research involving different types of endurance athletes, support this point of view. At this level of exercise intensity, the insulin response to glucose ingestion is suppressed; in addition, the secretion of epinephrine is increased. These two hormonal responses interact to help maintain or elevate the blood glucose level and prevent the hypoglycemic response that typically may occur in reactive individuals if more time elapses between the ingestion of the carbohydrate and the initiation of exercise.

If carbohydrates are consumed immediately before exercise, that is, within 10 minutes of the start, about 50-60 grams of a glucose polymer in a 40-50 percent solution has been used effectively in some studies. Dry glucose polymers are available commercially. One tablespoon is about 15 grams. To make a 50 percent solution containing 50 grams of the polymer, put about 3 level tablespoons of the polymer into 100 milliliters (about 3-4 ounces) of water. To make a 7.5 percent solution containing 15 grams, put 1 tablespoon of the polymer into 200 milliliters (about 7 ounces) of water. Several commercial "energy" drinks contain 25-50 grams of carbohydrate per 8 fluid ounces, which are about 10-20 percent solutions.

During exercise Carbohydrate ingested during prolonged exercise can help maintain blood glucose levels and reduce the psychological perception of effort, as measured by the ratings of perceived exertion, during the later stages of an endurance task. As the exercise task continues and the muscle glycogen level falls, the amount of energy derived from the ingested carbohydrates increases. Most research supports the benefits of consuming carbohydrates early in and throughout the exercise task, but even a single carbohydrate feeding late in a prolonged exercise bout may help replenish blood glucose levels, increase carbohydrate oxidation, and delay fatigue.

All major investigators, including Asker Jeukendrup, Trent Stellingwerff, Louise Burke, and Naomi Cermak, conclude that carbohydrate intake during prolonged exercise enhances performance. Studies have been undertaken in both laboratory and field settings, using different types of exercise modalities, and on both men and women.

During Exercise: When and How Much? During exercise, feedings every 15-20 minutes appear to be a reasonable schedule, but possibly more frequently when attempting to maximize carbohydrate intake or to obtain fluids when exercising under warm or hot environmental conditions. Although you may consume considerable quantities of carbohydrate during exercise, your ability to use this exogenous source for energy is limited. The reason is not known, but as noted previously, may be related to insufficient intestinal absorption or impaired delivery from the liver.

Sports drinks averaging 6-10 percent carbohydrate have been found to enhance prolonged endurance performance. A typical serving of a sports drink (8 ounces) would contain about 14-24 grams of carbohydrate, so depending on the concentration, an athlete who wanted to maximize utilizable carbohydrate intake would need to drink about 32-56 ounces to obtain about 100 grams of carbohydrate per hour. Drinking 8 ounces every 15 minutes would provide 32 ounces (1 quart) over the course of an hour, but consumption would have to be more frequent to obtain 56 ounces. Consuming 56 ounces of fluid over the course of an hour might be difficult and may pose a potential health risk for some individuals. Although the fluids could provide the desired amount of carbohydrate, excessive fluid consumption could lead to overhydration and a serious medical condition known as hyponatremia, as shall be discussed in chapter 9.

Several other protocols have been effective. One involved consumption of a high concentration (about 1 gram carbohydrate/kg body weight) immediately before or during the first 20 minutes of the exercise, and then use of lower concentrations such as found in commercial sports drinks at regular intervals. Other investigators have noted that taking a single, more concentrated dose of carbohydrate, such as 100-200 grams total, in the later stages of prolonged exercise may be beneficial. Additionally, because of the nature of their sport, soccer players and other such athletes may need to consume a high concentration before the game and during halftime, or breaks in the game as they occur.

It should be noted that consumption of carbohydrate solutions above 10 percent during exercise may cause gastrointestinal distress, as may other high concentrations of simple carbohydrates. However, some athletes may learn to tolerate larger concentrations, such as 15-20 percent. Ultradistance athletes, who exercise at a lower intensity, may tolerate even higher concentrations ranging from 20 to 50 percent.

Asker Jeukendrup, Naomi Cermak, and Luc van Loon recommend the following guidelines for carbohydrate intakes during exercise to maximize performance and minimize gastrointestinal distress. For athletes engaged in very short, high-intensity exercise lasting less than 0.5 hour: no carbohydrate is necessary during exercise. For short, high-intensity exercise or intermittent team sports lasting 0.5-1.25 hours: very little carbohydrate is required, possibly just a mouth rinse. For intermittent team sports of moderate duration from 1 to 1.5 hours: up to 60 grams per hour is recommended. For intermittent team sports of long duration beyond 2 hours: up to 90 grams per hour of multiple transportable carbohydrates (mixed sugars) are recommended. For endurance/continuous exercise lasting 1-3 hours, up to 60 grams per hour of carbohydrate are recommended. Finally, for prolonged endurance exercise beyond 2.5 hours in duration: up to 90 grams per hour of multiple transportable carbohydrates are recommended. Additionally, it is recommended that if an athlete is going to exercise for more than 2 hours, but is starting exercise with suboptimal carbohydrate levels, they consume up to 90 grams of multiple transportable carbohydrates per hour.

Athletes may learn to tolerate higher amounts of carbohydrate (i.e., train the gut), so these recommendations, although based in research, allow for some adjustment. Athletes should experiment with different doses of carbohydrate during training or practice, but not during competition.

Optimal Supplementation Protocol Several studies have indicated that although the intake of carbohydrate either before or during exercise may separately enhance performance, the best effect was observed when carbohydrate was consumed both before and during exercise. For example, Chryssanthopoulos and others had subjects run to exhaustion at 70 percent VO_2 max and found that although a high-carbohydrate meal 3 hours prior to performance improved endurance time, the combination of the meal and a carbohydrate-electrolyte solution during exercise further improved endurance running capacity.

Type of Carbohydrate A number of different types of carbohydrates have been studied, including glucose, fructose, galactose, sucrose, maltose, glucose polymers such as maltodextrins, both individually and in various combinations, as well as soluble starch (a very long polymer), high-glycemic-index foods such as potatoes, and low-glycemic-index foods such as legumes. In general, there appears to be no difference between these different types of carbohydrates as a means to enhance endurance performance when used appropriately. However, there may be some important considerations relative to the use of various carbohydrate combinations, fructose, solid carbohydrates, and low-glycemic-index foods.

Carbohydrate combinations Jeukendrup noted studies showing that a single carbohydrate ingested during exercise will be oxidized at rates up to about 1 gram/minute, even when large amounts of carbohydrate are ingested. However, combinations of carbohydrate (coined multiple transportable carbohydrates), particularly glucose and fructose, that use different intestinal transporters for absorption have been shown to result in higher oxidation rates (**figure 4.12**). This seems to be the way to increase exogenous carbohydrate oxidation rates, up to 1.7 grams/minute. Combinations of carbohydrates are recommended to increase glucose oxidation at a rate of more than 60 grams per hour. Currell and Jeukendrup reported that ingestion of a drink containing multiple carbohydrates (glucose and fructose), as compared to either a glucose drink or water, improved performance by 8 and 19 percent, respectively, in time-trial performance following a 2-hour bout of cycling. Power output was greater throughout the approximate 1-hour time trial with the glucose/fructose drink. Single sources of carbohydrate, such as glucose, appear to be adequate when less is needed. Thus, check food labels for ingredients to ensure that the chosen sports drink contains such combinations.

Fructose Most people can tolerate small amounts of fructose. It is found naturally in fruits and is an ingredient in some sports drinks. However, consuming larger amounts may pose problems. Because fructose is absorbed slowly from the intestinal tract, it can create a significant osmotic effect in the intestines, leading to diarrhea and gastrointestinal distress in some individuals. Research has indicated that a 6 percent solution of fructose, when compared to similar solutions of glucose and sucrose, caused significant gastrointestinal distress and an impairment in exercise performance. The athlete should be cautious in using fructose as the sole source of carbohydrate before or during exercise. Sports drinks containing fructose include it in small concentrations.

Fructose is generally not the preferred carbohydrate for postexercise muscle glycogen resynthesis. However, relative to muscle glycogen repletion, little is known about liver glycogen repletion. One classic human study from Nilsson and Hultman showed that, when compared to glucose, a fructose infusion caused greater liver glycogen repletion rates and no increase in hepatic glucose output. More recently, Décombaz and others showed that, following glycogen-depleting exercise, fructose or galactose plus maltodextrin ingestion caused twice the liver glycogen repletion rate as glucose plus maltodextrin ingestion. So although there are few data available on liver glycogen repletion, it is possible that postexercise carbohydrate consumption that includes some fructose and galactose may enhance liver glycogen accumulation. Additionally, high-fructose corn syrup, although rich in fructose, is treated as an added sugar. The health risks of added sugars are covered later in this chapter.

Carbohydrate form In a review, Cermak and van Loon concluded that there is no influence of the form in which carbohydrates are ingested (i.e., liquid, solid, semisolid [slurry]) on exercise performance. A small number of studies have been conducted that support this viewpoint. For instance, Pfeiffer and colleagues found no differences in oxidation rates between liquid, solid, and semisolid carbohydrates. Campbell and others studied the effect of different forms of carbohydrate (liquid, gel, and jellybeans) on endurance exercise performance and found no differences in blood glucose maintenance during exercise or exercise performance. Carbohydrates that have been modified to improve absorption, oxidation, or gastrointestinal tolerance, such as hydrogels, are now available and actively being studied. These modified carbohydrates are available in drinks, bars, and gels. In some cases, these modified carbohydrates show promise as they may influence desirable mechanisms such as speeding carbohydrate delivery during exercise or causing a larger insulinemic response post-exercise. However, as Baur and Saunders indicated in their extensive review, the value of these products remains questionable in terms of performance improvement in real-life competition.

Importantly, an athlete's preference of carbohydrate form could have a greater effect on performance than carbohydrate form, so the important factors to consider in carbohydrate choice have little to do with form and a great deal to do with preference, dose, gastric emptying, and fluid and electrolyte needs.

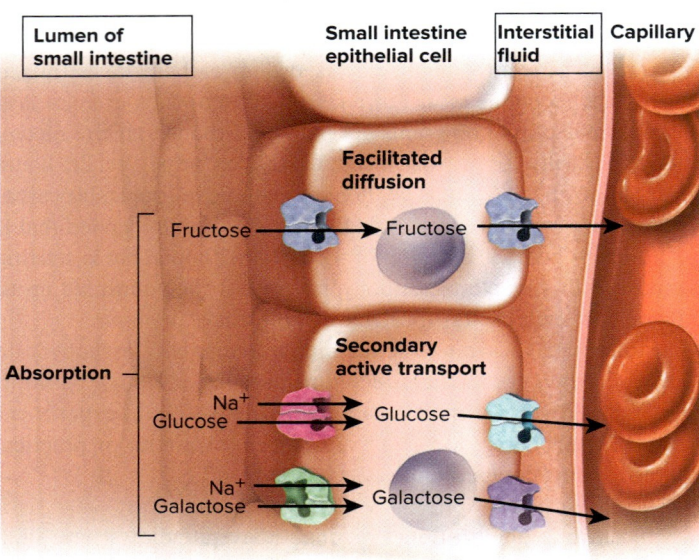

FIGURE 4.12 The villi in the intestines contain different receptors for the monosaccharides, which may increase carbohydrate absorption when multiple sources of carbohydrate are ingested.

Low-glycemic-index foods Although research on glycemic index (GI) and performance has been conducted for many years, the data are far from conclusive. Several reviews from experts in the topic are available. Donaldson and colleagues concluded in a review that there is still a lack of agreement on the benefits of consuming low versus high GI carbohydrates on exercise performance. In their review, they noted that only 5 of 13 relevant investigations of pre-exercise carbohydrate consumption measured performance and that of the 5 that did, performance only improved in the low GI condition in 2 studies. In a similar analysis, O'Reilly and others evaluated 13 studies on pre-exercise high versus low GI carbohydrate intake. Only 2 demonstrated a beneficial effect of consuming a low GI carbohydrate prior to exercise; however, this effect is attenuated if carbohydrate is consumed during exercise, which is common in most sports.

Ormsbee and others reported that consuming a low GI carbohydrate before exercise does cause a reduced blood glucose/insulin response (7 studies). Metabolically, this may cause an increase in fat oxidation or help to maintain normal blood glucose levels during exercise (7 studies); however, this was not shown in 2 other studies. So a measurable metabolic benefit is commonly found, although researchers have found it difficult to translate this into a clear performance benefit. For instance, Ormsbee notes that time to exhaustion and time trial performance has been enhanced by ingestion of low GI carbohydrate (5 studies), but this has not been shown in other studies (6 studies). There are considerable methodological differences between these studies, which makes interpretation of a small body of literature (<20 studies) difficult. Of note is that there are no reports of performance decrements due to pre-exercise consumption of low GI carbohydrates, so athletes may experiment in practice to determine if there is a benefit. Additional research is needed to evaluate the effect of pre-exercise low GI carbohydrate consumption on endurance exercise performance.

Carbohydrate with protein Studies have been conducted to determine the impact on performance of consuming protein with carbohydrate during exercise. McCleave and others found time to exhaustion to be 15.2 percent longer for female competitive cyclists and triathletes when a 3 percent carbohydrate/1.2 percent protein supplement was consumed every 20 minutes, compared to a 6 percent carbohydrate supplement. Using similar supplementation, Ferguson-Stegall found that total time to exhaustion was greater for cyclists and triathletes for intensities at or below the ventilatory threshold. Subjects cycling to exhaustion above the ventilatory threshold demonstrated no difference in performance with the supplements. In a review, van Loon noted that of 13 studies that investigated the effects of protein ingestion during exercise, only three showed a significant improvement in endurance performance. Still, some of the improvements were quite large (up to 30 percent) so at least in some individuals, during exercise protein ingestion may be ergogenic. However, the mechanism for improved performance is unknown. More importantly, protein ingestion during exercise does increase protein synthesis, before the exercise session is finished. So small amounts of protein consumed before or during exercise might aid in recovery.

Individuality Probably the most important recommendation is for the athlete to experiment with different types and amounts

Training Table

Dr. Asker Jeukendrup, a leading sport nutrition researcher, has developed a chart with evidence-based recommendations for carbohydrate intake during exercise. To use the chart, he states the following:

- First, establish the duration of your event; this will largely determine the amount of carbohydrate you need.
- Practice your strategy before using it in a race; this will give your gut a chance to adapt.
- Mix and match carbohydrate sources (i.e., drinks, gels, chews, etc.) based on your personal preferences.

Visit Dr. Jeukendrup at www.mysportscience.com to learn more about optimizing performance using evidence based science.

Visit https://www.mysportscience.com/post/2015/05/27/recommendations-for-carb-intake-during-exercise to access the carbohydrate recommendation chart and links to many other resources relating to nutrition, health, and performance.

of carbohydrate during training before using them in competition. Just as it is important for you to know your optimal race pace for an endurance event, so, too, must you know how well you can tolerate different amounts and concentrations and types of carbohydrates. Williams noted that the type of event may influence the amount of carbohydrate ingested, as runners may be more prone to gastrointestinal distress than cyclists. "Runner's trots" is a form of diarrhea that may be associated with excess consumption of highly concentrated sugar solutions, such as "energy" drinks.

Just as you train your muscles to learn their capacity, you may also be able to train your digestive system to know its limits. During training, experiment with various types and concentrations of carbohydrate, both before and during exercise. Ron Maughan, an internationally respected authority in sports nutrition, indicated that the optimal strategy relative to carbohydrate utilization is to use your own subjective experience, which you can gain during training.

What is the importance of carbohydrate replenishment after prolonged exercise?

There are several possible applications of this question. One is the athlete who may be involved in a prolonged exercise bout, have a rest period of 1–4 hours, and then must exercise again, such as athletes who train two or three times daily. Benefits may accrue to anaerobic endurance-, aerobic endurance-, and resistance-trained individuals. A second application is the athlete who trains intensely every day and must have an adequate recovery in the 1-day rest interval. A third application, covered in the next section, is the technique of carbohydrate loading.

After prolonged exercise, increased levels of GLUT-4 receptors in the muscle cell membrane help move available blood glucose into the muscle for resynthesis to muscle glycogen. Several studies have shown that ingesting carbohydrate during the rest interval between two prolonged exercise bouts improves performance in the second bout. This finding is comparable to the beneficial effects of carbohydrate intake during prolonged exercise bouts. The carbohydrate can help restore blood glucose levels but may also be used to resynthesize muscle glycogen. In cases such as this, where the rate of muscle glycogen resynthesis is important, high-glycemic-index foods, such as potatoes, bread, glucose, or glucose polymers, would be the preferred source of carbohydrate, for they apparently lead to a faster restoration of muscle glycogen than does a meal rich in low-glycemic-index foods. For repeat prolonged exercise tasks with about a 4-hour interval, a general recommendation is to consume 1 gram of carbohydrate per kilogram body weight immediately after the first event and again 2 hours prior to the second event. Additional carbohydrate may also be consumed immediately before and during the second event.

Carbohydrate with protein Carbohydrate has been combined with other nutrients, particularly protein, in attempts to enhance muscle glycogen resynthesis. Protein and some amino acids, such as arginine, may stimulate the release of insulin, which, if added to the effects of carbohydrate-mediated insulin release, could increase the rate at which glucose is transported into the muscle cell. In a review, Beelen and colleagues concluded that the inclusion of protein or amino acids with carbohydrates does not further enhance postexercise muscle glycogen synthesis when an adequate amount of carbohydrate (1.2 g/kg/hr) is ingested at frequent intervals (every 15–30 minutes). They added that this combination may accelerate postexercise muscle glycogen synthesis rates when less carbohydrate is provided (<1.0 g/kg/hr). Kerksick and others recently published the International Society of Sports Nutrition position stand on nutrient timing. It states that adding protein to carbohydrate (ratio of 1:3–4) may increase endurance performance and maximally promotes glycogen synthesis during acute and subsequent endurance exercise. Several studies, such as those by Baty and Romano-Ely and their colleagues, have shown that carbohydrate/protein supplementation may also reduce the incidence of muscle soreness following exercise, including lower levels of serum enzymes used as markers of muscle tissue damage.

Several studies have compared the effects of carbohydrate supplementation alone to carbohydrate/protein supplementation on performance following recovery from previous exercise. Betts and others had subjects complete a 90-minute run at 70 percent of VO_2 max, followed by a 4-hour recovery period, during which they consumed either a carbohydrate or a carbohydrate/protein mixture. The subjects then ran to exhaustion at 85 percent of VO_2 max, but there was no difference between treatments. Romano-Ely and others investigated the effect of a carbohydrate-protein-antioxidant drink, as compared to an isocaloric carbohydrate drink, on cycling time to exhaustion at 70 percent VO_2 max and, 24 hours later, at 80 percent VO_2 max. The drinks were consumed every 15 minutes during exercise and immediately afterward. There were no significant differences between the treatments on performance time during either of the cycling tests. Conversely, Berardi and others had subjects complete a 60-minute time trial followed by a 6-hour recovery period, during which the subjects consumed either carbohydrate or carbohydrate/protein. The subjects then repeated the 60-minute time-trial ride. Ingestion of carbohydrate/protein increased fat oxidation, increased recovery, and improved performance relative to isoenergetic carbohydrate ingestion. In a meta-analysis, Stearns and colleagues demonstrated an average 9 percent improvement in subsequent endurance performance with co ingestion of protein and carbohydrate compared to carbohydrate alone. They also found an ergogenic effect when supplements were matched for carbohydrate content.

The use of chocolate milk has been reviewed by Amiri and others as a more economical alternative to sports drinks during recovery from exercise. Benefits to this beverage include a ratio of carbohydrates to protein in the range recommended by the International Society of Sports Nutrition and its naturally high concentrations of electrolytes; it also affords a greater feeling of fullness compared to water or carbohydrate beverages. Karp and colleagues had highly trained cyclists perform an interval workout followed by 4 hours of recovery and then an endurance trial to exhaustion at 70 percent VO_2 max. During recovery, subjects consumed the same volume of chocolate milk, fluid replacement drink, or carbohydrate replacement drink. The carbohydrate content was equivalent for the chocolate milk and the carbohydrate replacement drink. Total time to exhaustion and total work were significantly greater for chocolate milk and fluid replacement subjects compared to that of the carbohydrate replacement drink group. These results are supported by research done by Pritchett, Thomas, and Gilson, who also found at least as effective muscle recovery responses with chocolate milk and carbohydrate recovery drinks.

If rapid resynthesis of muscle glycogen is not important, it is good to note that studies have shown that consumption of adequate amounts of carbohydrate over a 24-hour period will restore muscle glycogen levels to normal. For athletes who train intensely on a daily basis with either resistive or aerobic exercise that leads to muscle glycogen depletion, sports nutritionists normally recommend that approximately 8–10 grams of carbohydrate per kilogram body weight should be consumed daily to restore muscle glycogen levels to normal. For an individual who weighs 70 kilograms, this approximates 560–700 grams of carbohydrate, or 2,240–2,800 carbohydrate kcal. This amount of carbohydrate would represent about 65–80 percent of the daily caloric intake of an athlete consuming 3,500 kcal. Over the 24-hour period, the rate of muscle glycogen recovery is approximately 5–7 percent per hour. Sports drinks may be a convenient means to consume carbohydrate immediately after exercise. The remaining carbohydrate should be derived from other natural sources in the diet, including both simple carbohydrates in fruits and complex carbohydrates in grains, potatoes, and other foods with adequate dietary fiber and other nutrients. The inclusion of high-glycemic-index foods in the daily diet will help speed resynthesis of muscle glycogen over the 24-hour period and may be very compatible with the Prudent Healthy Diet. Regular meals consumed during the 24-hour recovery period should include healthful low-glycemic-index foods with adequate amounts of protein.

Following prolonged, high-intensity competitive exercise performance, such as running a marathon, the resulting muscle damage will limit muscle glycogen replenishment for several days. Rest is important during this time, and muscle glycogen levels may return to normal following seven or more days of high-carbohydrate meals.

Image Source/SuperStock

Will a high-carbohydrate diet enhance my daily exercise training?

Most scientists and sports nutritionists who study carbohydrate metabolism in athletes recommend a high-carbohydrate diet for most athletes, particularly endurance athletes, because success in athletic competition is contingent upon optimal training, and for the endurance athlete, optimal training may be contingent upon adequate nutrition, primarily the ingestion of sufficient carbohydrate every day. Louise Burke, a prominent sports nutritionist, and her associates recommended that athletes in general training consume daily approximately 5–7 grams of carbohydrate per kilogram body weight, but endurance athletes should consume about 7–10 grams per kilogram, or greater. These recommendations are comparable to those of Melinda Manore, another sports nutrition expert. Burke and others noted that most male athletes may be meeting these needs, but many female endurance athletes, particularly those attempting to lose weight for competition, may not.

Training Table

Carbohydrate confusion!

- When planning a high carbohydrate diet relative to kcal intake (percent of daily kcal) or relative to body weight (g of carbs/kg of body weight), remember, these two approaches may not produce the same recommendations.
- For example, if a 70 kg man who consumes 1,500 kcal/day aims to consume 65% of daily kcal from carbohydrate (a high carbohydrate diet), that equals 975 kcal of carbohydrate/day (244 g/day). If the same person followed the body weight adjusted recommendation for an athlete (about 10 g carbohydrate/kg bodyweight/day and also a high carbohydrate diet), that equals 700 kcal of carbohydrate/day (about 175 g)!
- The solution might be to calculate carbohydrate both ways, consider the objective (e.g., performance versus weight loss), and relate carbohydrate intake to daily energy expenditure.
- For instance, if weight loss were the goal, carbohydrate intake could be based on 45–65% of estimated daily energy intake, considering energy expenditure, and planned rate of weight loss.
- If performance were the goal, carbohydrate intake could be based on body weight, and energy expenditure, which could range from light to very high depending on the commitments of the sport (exercise intensity, duration, hours/day, days/week, etc.).
- See chapter 3 for different ways to estimate daily energy expenditure.

There are some limited data supporting the concept of enhanced training following a high-carbohydrate diet. A number of field and laboratory studies with athletes have attempted to mimic actual sport conditions. For example, one group of soccer players improved performance on an intermittent exercise task designed to mimic physical activity in a game, while another group improved performance in a standardized intermittent running task and a run to exhaustion. In other studies, runners were able to endure longer on a treadmill run to exhaustion; swimmers were better able to maintain 400-meter swim velocity; and triathletes experienced a significant improvement in treadmill endurance following 30 minutes of swimming, cycling, and running. Based on the available data, Edward Coyle, an international authority in carbohydrate metabolism during exercise, indicated that physical performance seems better maintained with a high-versus moderate-carbohydrate diet. In general, the normal carbohydrate intake of athletes in training studies was increased from approximately 40–45 percent to 55–70 percent of the daily kcal for varying periods, but usually a week or more. This level of carbohydrate approximates the upper levels of the AMDR for carbohydrate.

Not all athletes, including endurance athletes, need high-carbohydrate diets all the time. In a meta-analysis, Erlenbusch and others indicated that subjects following a high-carbohydrate diet could exercise longer until exhaustion, but this finding applied more to untrained individuals than trained individuals. As you may recall, aerobic exercise training improves the ability of the muscles to use fat as an energy source, so they may be somewhat less dependent on carbohydrate for a given training protocol. Thus, a moderate-carbohydrate diet may be adequate for trained athletes. For example, obtaining 45 percent of daily energy needs from carbohydrate might be considered moderate, as it is at the lower end of the AMDR. On such a diet, an endurance athlete who consumes 3,000 kcal per day during training would derive 1,350 kcal from carbohydrate (0.45 × 3,000), which is about 340 grams of carbohydrate. For a 60-kilogram athlete, this is about 5.6 grams of carbohydrate per kilogram body weight. Although this is slightly less than that recommended for endurance athletes, research has shown that such amounts may be sufficient to maintain training on a daily basis.

However, training may appear more stressful psychologically. In several studies, the psychological status of athletes, as measured by the vigor and fatigue components of the Profile of Mood States (POMS) questionnaire and their rating of perceived exertion (RPE) during exercise, was improved when they switched from moderate- to higher-carbohydrate diets. In his review, Coyle concluded that mood state seems better maintained with a high- rather than moderate-carbohydrate diet. Utter and others reported reduced ratings of perceived exertion in subjects involved in prolonged (2.6 hours), intermittent cycling following ingestion of carbohydrate before and during the exercise bout.

Coyle also indicates that a high-carbohydrate diet may help reduce symptoms of overreaching and, possibly, overtraining. Gleeson and others note that heavy, prolonged exertion is associated with numerous hormonal and biochemical changes in the body, many of which may have detrimental effects on immune function. A well-balanced diet helps promote optimal immune function, and as reviewed by Rawson and colleagues, carbohydrate intake appears to limit the degree of exercise-induced immunosuppression. An impaired immune response is one possible factor associated with the overtraining syndrome. Carbohydrate intake following exercise also promotes protein synthesis via the insulin effect, which may enhance muscle and overall recovery.

Some sports dietitians indicate that many athletes do not eat high-carbohydrate diets because it may be impractical for them to do so. Selecting foods high in carbohydrate content, highlighted earlier in this chapter, provides a sound guide to increase the carbohydrate content of the diet, as do some of the recommendations in the following section regarding carbohydrate loading. Chapter 11 will provide additional information specific to daily caloric intake for planning a diet.

In summary, as Coyle notes, athletes do not train hard every day, so they do not require a high intake of carbohydrate every day of training. Nevertheless, a diet rich in healthful carbohydrates not only may have several major health benefits but may also help guarantee optimal energy sources for daily exercise training. Moreover, as Louise Burke points out, experts in the field of energy metabolism indicate that there is no evidence that diets which are restricted in carbohydrate enhance training. Carbohydrate is the major fuel for most athletes in training.

> ### Key Concepts
>
> - Carbohydrate is the most important energy source for moderately high- to high-intensity exercise.
> - Regular training increases the ability of the muscles to store and use carbohydrate for energy production.
> - Low levels of blood glucose or muscle glycogen may be contributing factors in the premature onset of fatigue in prolonged exercise.
> - Low levels of muscle glycogen may contribute to impaired performance in prolonged, moderate- to high-intensity endurance exercise and in sports involving intermittent, high-intensity exercise for 60–90 minutes.
> - Consuming carbohydrate before and during prolonged, intermittent, high-intensity or continuous exercise may help delay the onset of fatigue, but unless carbohydrate intake corrects a muscle glycogen deficiency, such practices will not improve performance in most athletic events of shorter duration.
> - Combinations of carbohydrates, such as glucose and fructose, consumed during exercise appear to optimize the amount of exogenous carbohydrate that can be oxidized.
> - Athletes should experiment with different carbohydrate supplementation strategies during training to help determine the amount, type, and timing of intake that may be suitable for them in competition.
> - Glucose, sucrose, glucose polymers, and solid carbohydrates appear to be equally effective as a means of enhancing performance, but fructose may be more likely to cause gastrointestinal distress if used alone.
> - Carbohydrates with a high glycemic index may facilitate muscle glycogen replenishment when consumed immediately after exercise and every 2 hours thereafter.
> - To maintain the quality of training, athletes who train at moderate to high intensity on a daily basis should eat a healthful diet rich in complex carbohydrates, complemented with some high-glycemic-index foods, to replenish muscle glycogen.

> ### Check for Yourself
>
> ▸ Given the recommendation to consume about 1.0–1.5 grams of carbohydrate per kilogram body weight per hour for 4–5 hours after exercise in order to rapidly replenish muscle glycogen, calculate how much carbohydrate you would need per hour and list specific foods, and amounts, that you might need to consume each hour.

Carbohydrate Loading

What is carbohydrate, or glycogen, loading?

Because carbohydrate becomes increasingly important as a fuel for muscular exercise as the intensity of the exercise increases, and because the amount of carbohydrate stored in the body is limited, muscle and liver glycogen depletion could be factors that limit performance capacity in distance events characterized by high levels of energy expenditure for prolonged periods. **Carbohydrate loading**, also called *glycogen loading* and *glycogen supercompensation,* is a dietary technique designed to promote a significant increase in the glycogen content in both the liver and the muscles in an attempt to delay the onset of fatigue. It is generally used for 3–7 days in preparation for major athletic competitions.

What type of athlete would benefit from carbohydrate loading?

In general, carbohydrate loading is primarily suited for individuals who will sustain high levels of continuous energy expenditure for prolonged periods, such as long-distance runners, swimmers, bicyclists, triathletes, cross-country skiers, and similar athletes. In addition, athletes who are involved in prolonged stop-and-go activities, such as soccer, lacrosse, and tournament-play sports like tennis and handball, may benefit. For example, Rico-Sanz and others concluded that exhaustion during soccer-specific performance is related to the capacity to use muscle glycogen, underlying the importance of glycogen loading. In essence, carbohydrate loading may be effective for athletes engaged in events that use muscle glycogen as the major energy source and that may lead to a depletion of glycogen in the muscle fibers. Athletes who compete in sports involving high-intensity, short-duration energy expenditure will not benefit from carbohydrate loading. For example, Hatfield and others reported no effects of carbohydrate loading on performance in resistance training involving multiple sets of maximal jump squats. However, bodybuilders carbohydrate load prior to competition, as it is believed that increased muscle glycogen and intramuscular water retention will create a more muscular looking appearance.

Recall from chapter 3 that humans have several different types of skeletal muscle fibers. In general, the slow-twitch red and fast-twitch red fibers are used mainly during long, continuous activities and are aerobic in nature, whereas the fast-twitch white fibers are used for short, fast activities and are anaerobic in nature. Consider the differences between a distance runner and a soccer player. The former may run at a steady pace for hours, whereas the later will constantly be changing speeds, with many bouts of full speed interspersed with recovery periods of slower running. Research has shown that glycogen depletion patterns of the two different muscle fiber types are related to the type of exercise. Long, continuous exercise depletes glycogen principally in the slow-twitch red and fast-twitch red fibers, whereas fast, intermittent bouts of exercise with periods of rest—actually a form of interval training—primarily deplete glycogen in the fast-twitch white fibers. However, it should be noted that glycogen depletion may occur in all types of fibers in either prolonged, continuous or intermittent exercise and may be quite appreciable, depending upon intensity and duration of the exercise bouts. If carbohydrate loading works for the specific muscle fiber involved, then both types of athletes may benefit. Both should have greater glycogen stores in the later stages of their respective athletic contests.

How do you carbohydrate load?

As you might suspect, the key to carbohydrate loading is to switch from the normal, balanced diet to one very high in carbohydrate content. The original, classic carbohydrate loading technique, emanating from earlier Scandinavian research, involved a glycogen depletion stage induced by prolonged exercise and a restricted diet. For example, a runner might go for an 18- to 20-mile run to use as much stored glycogen as possible, and then ingest very little carbohydrate in the following 2- to 3-day period. Exercise is continued during this 2- to 3-day period to keep glycogen stores low. Following the depletion stage, the loading stage began. During this phase, carbohydrate may contribute 70 or more percent of the caloric intake. The intensity and duration of exercise during this phase were reduced considerably. The usual case was to rest fully for 2 to 3 days. Thus, the classic carbohydrate loading pattern involved three stages: depletion, carbohydrate deprivation (high-fat/protein diet), and carbohydrate loading. However, this original method may be particularly difficult to tolerate, especially if one tries to exercise at high levels during the depletion phase. The lack of carbohydrate in the diet combined with the exercise bouts may elicit symptoms of hypoglycemia (weakness, lethargy, irritability). Moreover, prolonged exhaustive exercise may lead to muscle trauma, which may actually impair the storage of extra glycogen. This classic, original method is presented in **table 4.9**.

Although some early research supported this technique, more recent data suggest that this strict routine may be unnecessary, particularly the total program of depletion. For example, in trained runners, research has shown that simply changing to a very high-carbohydrate diet, combined with 1 or 2 days of rest or reduced activity levels (tapering), will effectively increase muscle and liver glycogen. Well-controlled research has revealed that exhaustive running is not necessary to achieve muscle glycogen supercompensation. It appears to be important to continue endurance training, or other high-intensity training specific to the sport, during the 7-14 days prior to competition. Such training will maintain adequate levels of GLUT-4 receptors to transfer blood glucose into the muscle cell and of glycogen synthase, the enzyme in the muscle that synthesizes glycogen from glucose. Evidence also suggests that if the total carbohydrate content is consumed over the entire week, in contrast to concentrating it in 2-3 days, there will be little difference in the muscle glycogen content between the two techniques.

Although there may be a number of variations in the carbohydrate loading protocol, a generally recommended format is also presented in **table 4.9**. The interested athlete may want to experiment with both techniques and make adjustments through experience.

Sports scientists have generally recommended that carbohydrate intake during carbohydrate loading be about 8-10 grams per kilogram body weight, and Louise Burke, from Australian Catholic University, recommended that marathon runners consume about 10-12 grams per kilogram body weight over the 36-48 hours prior to the race. These recommendations could total about 400-800 grams per day, depending on the size of the

TABLE 4.9	Different methods for carbohydrate loading		
A recommended method		**Original, classic method**	
1st day:	tapering exercise	1st day:	depletion exercise
2nd day:	mixed diet, moderate carbohydrate; tapering exercise	2nd day:	high-protein/fat diet; low carbohydrate; tapering exercise
3rd day:	mixed diet, moderate carbohydrate; tapering exercise	3rd day:	high-protein/fat diet; low carbohydrate; tapering exercise
4th day:	mixed diet, moderate carbohydrate; tapering exercise	4th day:	high-protein/fat diet; low carbohydrate; tapering exercise
5th day:	high-carbohydrate diet; tapering exercise	5th day:	high-carbohydrate diet; tapering exercise
6th day:	high-carbohydrate diet; tapering exercise or rest	6th day:	high-carbohydrate diet; tapering exercise or rest
7th day:	high-carbohydrate diet; tapering exercise or rest	7th day:	high-carbohydrate diet; tapering exercise or rest
8th day:	competition	8th day:	competition

High-carbohydrate diet: 400–800 g per day depending on body weight; about 70–80 percent of dietary kcal should be carbohydrate.

individual, which is not too different from the generally recommended dietary content of carbohydrate for the endurance athlete in regular training; Burke recommends that marathoners consume 7–12 grams of carbohydrate per kilogram body mass during training. It is important to note that the athlete should not change his or her diet drastically prior to competition. Consuming a high-carbohydrate diet during training will condition the body to metabolize carbohydrate properly during this loading phase. **Table 4.10** represents a general dietary plan for carbohydrate loading. The total caloric value and grams of carbohydrate should be adjusted to individual needs. They are dependent upon the size of the individual and daily energy expenditure in exercise. It is important not to consume excess kcal, for they may be converted into body fat if in excess of the maximal storage capacity of the muscle and liver for glycogen.

Some guidelines for replenishment of glycogen were presented earlier. Because glycogen loading for long-distance events occurs over 2–3 days, it would be wise to stress complex carbohydrates in the diet because of their higher nutrient content. However, simple carbohydrates may also be used effectively to increase muscle glycogen stores, as can high-carbohydrate sports drinks such as Gatorade energy drink. Moreover, the diet should also include the daily requirements for protein and fat.

If, for some reason, the athlete cannot carbohydrate load over the 3- to 7-day period, a rapid protocol may be effective. Fairchild and others found that one day of a high-carbohydrate intake, approximately 10 grams of high-glycemic-index carbohydrate per kilogram body mass, nearly doubled the muscle glycogen concentration, from 109 to 198 mmol/kg wet weight muscle. The carbohydrate feeding was preceded by a short bout of near maximal-intensity exercise for 3 minutes. They reported that these muscle glycogen levels were comparable to those achieved over a 2- to 6-day regimen.

Most prolonged endurance events begin in the morning. The last large meal should be about 15 hours prior to race time, possibly topped off with a simple carbohydrate snack before retiring for the night. Some athletes drink a glucose polymer for the last major meal to avoid the presence of intestinal residue the morning of competition. A carbohydrate breakfast such as orange juice, toast, jelly, or other carbohydrates along with some protein may be eaten 3 to 4 hours prior to competition. This overall dietary regimen should help maximize muscle and liver glycogen stores. The athlete should then follow the guidelines presented previously relative to carbohydrate intake before and during performance.

Will carbohydrate loading increase muscle glycogen concentration?

Most, but not all, studies show that an appropriate carbohydrate loading protocol, compared to normal or low dietary carbohydrate intake, will substantially increase muscle glycogen

TABLE 4.10 Daily food plan for carbohydrate loading

Dietary sources of fats, proteins, and carbohydrates	Amount and kcal	Grams of carbohydrate, protein, and fat
Meat, fish, poultry, eggs, cheese, select low-fat items	6–8 oz. kcal: 330–440	0 grams carbohydrate* 42–56 grams protein 18–24 grams fat
Breads, cereals, and grain products	10–20 servings kcal: 800–1,600	150–300 grams carbohydrate 24–60 grams protein
Vegetables, high kcal (such as corn)	4 servings kcal: 280	60 grams carbohydrate 8 grams protein
Fruits	4 servings kcal: 240	60 grams carbohydrate
Fats and oils	2–4 teaspoons kcal: 90–180	10–20 grams fat
Milk, skim	2 servings kcal: 180	24 grams carbohydrate 16 grams protein
Desserts, such as pie	2 servings kcal: 700	102 grams carbohydrate 6 grams protein 30 grams fat
Beverages, naturally sweetened	8–24 ounces kcal: 80–240	20–60 grams carbohydrate
Water	8 or more servings kcal: 0	
Total kcal	2,700–3,860	
Total grams and approximate percent of dietary kcal		
Carbohydrate	416–606	65%
Protein	96–146	15%
Fat	58–74	20%

Consult **table 4.2** for specific high-carbohydrate foods in each of the food sources.

*Beans are high in protein but also low in fat and high in carbohydrates. Substitution of beans for meat will increase the total grams of carbohydrate and the percentage of dietary kcal from carbohydrate.

Including high-carbohydrate drinks, such as glucose polymers, can add significant amounts of carbohydrate to the diet and may substitute for other foods, such as desserts.

Source: Forgac, M. 1979. Carbohydrate loading: A review. *Journal of the American Dietetic Association* 75:42–5.

levels. Although some previous research found that muscle glycogen levels in the early phases of loading did not increase as much in females as in males, more recent, better-controlled research by James, Paul, and Tarnopolsky, with their associates, revealed that carbohydrate loading increased muscle glycogen concentration in both men and women, provided that total energy intake was adequate. No gender differences were noted. McLay and others noted that women generally have lower resting muscle glycogen levels during the midfollicular phase of the menstrual cycle, as compared to the midluteal phase, but the lower glycogen storage in the midfollicular phase could be overcome by carbohydrate loading. In general, carbohydrate intakes of 8 grams or more per kilogram body weight will provide optimal muscle glycogen concentrations for both males and females.

Glycogen content in the muscle has been reported to increase about two to three times beyond normal and liver glycogen content nearly doubled following a carbohydrate loading regimen, and this increase may last at least 3 days in a rested athlete. However, it may be important to taper and rest about 2 days prior to the event. Fogelholm and others reported no increase in muscle glycogen following the classic loading protocol if athletes continued to train 45–60 minutes per day, even though the training was easy. This finding merits confirmation, as other studies have shown muscle glycogen supercompensation when individuals tapered, although most studies use at least 1 day of rest before the competitive exercise test.

Carbohydrate loading has been shown to increase muscle glycogen stores after exhaustive exercise, but apparently the process does not work repeatedly within a short time frame. McInerney and others reported that muscle glycogen supercompensation did not occur when subjects attempted to increase muscle glycogen levels repeatedly during a 5-day period while performing exhaustive exercise every other day.

In general, the full carbohydrate loading procedure should be used sparingly, mainly in preparation for a peak event. However, athletes should experiment with the procedure, or at least experiment with various forms of carbohydrate to be used, sometime during training before using it in competition.

How do I know if my muscles have increased their glycogen stores?

The most accurate way would be to have a muscle biopsy taken (a needle is inserted into the muscle and a small portion is extracted and analyzed), but this is not very practical. A practical method that has been recommended is to monitor changes in body weight. Kreitzman and others noted that glycogen is stored in the liver, muscles, and fat cells in hydrated form (three to four parts water), and weight gain may occur with carbohydrate loading. Thus, keeping an accurate record of your body weight, which should be recorded every morning as you arise and after you urinate, may help you determine the answer to this question. Approximately 3 grams of water are bound to each gram of stored glycogen. If your body stores an additional 300–400 grams of glycogen, along with 900–1,200 grams of water, your body weight will increase about 1,200–1,600 grams, or 2.5–3.5 pounds, above your normal training weight during the loading phase. The weight gain would be greater with additional glycogen storage. This is indicative that the carbohydrate loading has been effective, because rapid weight gains from one day to another are usually due to changes in body water content.

Will carbohydrate loading improve exercise performance?

Although athletes in most sports may benefit from an increased carbohydrate content in the diet, the full procedure of carbohydrate loading is not necessary for the vast majority of athletes.

In general, carbohydrate loading has not been found to enhance performance in single, high-intensity exercise tasks ranging up to 60 minutes or so. For example, Vandenberghe and others found no effect of muscle glycogen levels (manipulated by carbohydrate loading) on muscle glycolytic rate during very high-intensity exercise (125 percent VO_2 max) or on all-out performance at this exercise intensity, a time approximating 3 minutes. Various other studies have reported that carbohydrate loading does not increase the speed of runners in events ranging from 10 kilometers to the half-marathon or in cycling time trials up to 60 minutes. However, Pizza and others, using an exercise task consisting of a 15-minute submaximal run followed by a run to exhaustion at 100 percent VO_2 max, reported an increase in performance associated with carbohydrate loading. The run to exhaustion approximated 5 minutes. In general, this finding is an exception to the rule.

Carbohydrate loading may benefit athletes involved in prolonged, intermittent, high-intensity exercise tasks. Akermark and others, using elite Swedish ice hockey players on two competitive teams as subjects, reported that the team that carbohydrate loaded between two games had higher muscle glycogen levels, which were associated with improvement in distance skated, number of shifts skated, and skating speed in the second game.

Carbohydrate loading has been studied most extensively as a means to improve performance in more prolonged aerobic endurance exercise tasks. In general, the results are supportive of an ergogenic effect. Laboratory studies have shown that exercise time to exhaustion is closely associated with the amount of muscle glycogen available or the amount of carbohydrate in the diet. When endurance performance is compared after subjects have been on either a high-fat/high-protein diet, a mixed, balanced diet, or a high-carbohydrate diet for 4–7 days, performance on the high-fat/high-protein diet is worse than on the other two. However, research findings comparing a mixed, balanced diet with a high-carbohydrate diet have been equivocal, with some results favoring the high-carbohydrate diet and others revealing no difference between the two.

A number of studies have shown that carbohydrate loading, as compared to a normal carbohydrate intake, does not enhance endurance performance. Interestingly, in many of these studies the performance tests may not have been long enough for the individual to derive the full benefit from carbohydrate loading, as the duration was less than 2 hours. However, one of the best-designed placebo-controlled studies also found no beneficial effect of carbohydrate loading. Using a cycling exercise protocol designed to be similar to a competitive 100-km road race (about 2.5 hours), Burke and others found that a 3-day carbohydrate loading regimen (9 grams carbohydrate/kg), as compared to a moderate-carbohydrate diet (6 grams carbohydrate/kg), did not enhance performance even though muscle glycogen content increased significantly. They also provided carbohydrate (1 gram per kilogram body mass) during the exercise test and suggested that the availability of blood glucose during exercise may offset any detrimental effects on performance of lower pre-exercise muscle and liver glycogen concentrations. Additionally, the authors noted that carbohydrate loading may be effective in prolonged endurance events in which the exercise intensity is relatively more constant. In this study, the exercise task included repeated high-intensity sprints, which are common in cycling races but not in other events, such as marathon running. Additionally, the cyclists consumed more than 60 grams of carbohydrate per hour, which though recommended for runners as well, is often more difficult to do in running than cycling. Finally, the investigators also indicated that although the time to finish the 100-km ride (about 1.6 minutes faster with carbohydrate loading) was not statistically significant, such an effect, if real, could make a difference in the finishing order of top cyclists.

In contrast, a number of studies suggest that carbohydrate loading may be an effective technique to enhance endurance exercise performance. However, it should be noted that most of these studies have not used a true placebo, as was done in the aforementioned study by Burke and others.

Hargreaves and others note that the increased muscle glycogen associated with carbohydrate loading may be used more readily in exercise tasks approximating 65–70 percent VO_2 max, which might be a reasonable pace for an average runner competing in a marathon. If muscle glycogen were used more rapidly during the early stages of a marathon, theoretically carbohydrate loading would provide no advantage during the later stages of the race. However, Bosch and others note that although carbohydrate loading may reduce the relative contribution of blood glucose to overall carbohydrate oxidation, the improved performance may be attributed to the initially greater amount of muscle glycogen as a means to spare the premature use of blood glucose and liver glycogen.

A supportive study was conducted by Clyde Williams in England. Runners performed a 30-kilometer (18.6-mile) run on a treadmill and then were divided into two groups. One used a carbohydrate loading technique for a week, while the other group maintained their normal carbohydrate intake. Although there were no significant differences between the groups for overall performance time in the 30 kilometers, the carbohydrate loading group ran the last 5 kilometers significantly faster compared to their initial trial.

In another cycling study, Walker and others studied the effect of a carbohydrate loading and exercise tapering regimen in well-trained women on performance of a cycling test to exhaustion at 80 percent VO_2 max. The high-carbohydrate diet (approximately 78 percent carbohydrate), as compared to the moderate diet (approximately 48 percent carbohydrate), induced a 13 percent greater muscle glycogen content and an 8 percent improvement in cycling time.

Moreover, several field studies with runners and cross-country skiers have shown improved performances with carbohydrate loading. In general, carbohydrate loading, in contrast to a mixed diet, did not enable these athletes to go faster during the early stages of their events, but the high glycogen levels enabled them to perform longer at a given speed. The end result was an overall faster time. Failure to carbohydrate load has also been identified as one of the factors contributing to collapse of runners in an ultramarathon. Indeed, in a review, Peters noted that current evidence continues to support high carbohydrate intakes for ultraendurance athletes to increase muscle glycogen stores before the event.

Based on studies published prior to that of Burke and others, several major reviews support the performance-enhancing effectiveness of carbohydrate loading. Clyde Williams reported that an International Consensus Conference on sports nutrition concluded that the most significant influence on performance was the amount of carbohydrate stores in the athlete's body prior to heavy endurance exercise, which is the purpose of carbohydrate loading. Hawley and others concluded that carbohydrate loading would postpone fatigue in endurance events lasting more than 90 minutes and may improve performance in events where a set distance is covered as fast as possible, such as cycling and running, by about 2–3 percent according to some scientists. In their review, Williams and Lamb generally support these viewpoints relative to male athletes. However, they note that although carbohydrate loading can increase muscle glycogen stores in women, it appears to offer no benefit to their endurance performance; Tarnopolsky indicates that women oxidize more lipid and less carbohydrate compared to men during endurance exercise, a finding that may underlie this viewpoint.

Although carbohydrate loading may be an effective technique to enhance performance in prolonged aerobic endurance events, research suggests the most effective protocol is to carbohydrate load and use carbohydrate supplements during the event. Kang and others noted that this method can exert an additional ergogenic effect by preventing a decline in blood glucose levels and maintaining carbohydrate metabolism during the later stages of prolonged aerobic exercise. Given the findings of Burke and others, this method appears to be the most appropriate, as it will help provide increases in muscle glycogen before exercise as well as replenishment of blood glucose during exercise, two factors that may be associated with enhancement of endurance performance.

Are there any possible detrimental effects relative to carbohydrate loading?

From a performance standpoint, the extra body weight associated with the increased water content may be a disadvantage. In activities where moving the body weight is important, extra energy will be required to lift the extra 2–3 pounds of body water. However, in most performance events for which carbohydrate loading is advocated, the benefits from the energy aspects of the increased glycogen should more than offset the additional water weight. Moreover, if the individual is performing in a hot environment, the extra water, even small amounts, could be available as a source of sweat and may be helpful in controlling body temperature during exercise in the heat. Although one study suggested that the water stored with glycogen did not confer any advantage in regulation of body temperature while exercising in heat, the duration of the exercise, only 45 minutes, would not be sufficient to benefit from the increased water levels. Another study conducted in South Africa revealed no beneficial or detrimental effects of carbohydrate loading on body temperature during 2.5 hours of exercise in a moderate environment (70°F; 21°C). However, performance in longer exercise tasks with greater levels of water losses might be enhanced. Additional research is needed to study the potential effects of increased muscle glycogen levels on body water availability and temperature regulation during prolonged exercise under warm environmental conditions.

From a health standpoint, there may be some hazards to individuals with certain conditions. Although diabetics have been known to carbohydrate load, they should consult their physicians prior to using the technique. Individuals with high blood lipid or cholesterol levels might avoid the high-fat/high-protein diet phase of the depletion stage if, for some reason, they prefer the original, classic method of carbohydrate loading. Blood serum lipids and cholesterol have been reported to rise significantly during this phase. In addition, these individuals should eat mostly low-glycemic-index carbohydrates during the loading phase, because an increased intake of high-glycemic-index carbohydrates may raise blood lipid levels. Furthermore, hypoglycemia may occur during the high-fat/high-protein phase.

Electrocardiographic (ECG) abnormalities have been reported in individuals following the classic carbohydrate loading technique. Although no cause-and-effect relationship was determined, it was speculated that hypoglycemia or glucose intolerance may have been involved. However, other research on marathoners and joggers revealed no ECG changes following the classic method of carbohydrate loading. It has also been theorized that carbohydrate loading could lead to destruction of muscle fibers due to excessive glycogen storage, but there are no data to support this contention. Muscle has a maximal capacity to store glycogen, approximately 4 grams/100 grams of muscle, and beyond that level excess carbohydrate would be converted to fat and stored in adipose tissues.

Other potential problems with the high-carbohydrate phase are diarrhea, nausea, and cramping, particularly when the diet is changed drastically or large amounts of simple carbohydrates are consumed. Individuals who wish to carbohydrate load should experiment with such diets during their training and not just before competition.

In general, however, the recommended carbohydrate loading technique presented in **table 4.9**, which at the most is only a 7-day dietary regimen, poses no significant health hazards to the normally healthy individual.

Key Concepts

▸ Carbohydrate loading is not a technique for all types of athletes, but it may benefit athletes involved in long-distance competition such as marathoning.
▸ Various carbohydrate loading techniques may effectively increase muscle glycogen stores, but tapering exercise or rest and a high-carbohydrate diet are the essential points.
▸ During prolonged endurance exercise, the efficacy of carbohydrate loading is enhanced by exogenous carbohydrate consumption.

Check for Yourself

▸ If you know any marathon runners or prolonged endurance athletes, interview them about their dietary strategies prior to a major competition. Do they practice carbohydrate loading?

Carbohydrates: Ergogenic Aspects

Throughout this chapter you have learned that carbohydrate intake, in a variety of ways, may be used to enhance physical performance. Truly, carbohydrates represent one of our most important ergogenic nutrients. In this brief section, we discuss the potential ergogenic effects of several carbohydrate metabolic by-products. Many of the studies described in this section were based on a sound theoretical basis but, ultimately, the particular metabolite examined did not prove ergogenic.

Steve Hamblin/Alamy Stock Photo

Do the metabolic by-products of carbohydrate exert an ergogenic effect?

Recall that the primary mechanism in the transformation of muscle glycogen into energy is glycolysis. The end product during aerobic metabolism is normally pyruvate. However, glycolysis leading to the formation of pyruvate involves production of a number of metabolic by-products in a chain of about a dozen sequential steps, each step being controlled by an enzyme (see appendix A). One theory of fatigue is that if one of these steps is blocked by inactivation of an enzyme, glycolysis may not continue at an optimal rate, since a necessary metabolic by-product may be in short supply. This blocked step could represent a weak link in the chain, possibly reducing the formation of pyruvate and subsequent ATP production. Several metabolic by-products of carbohydrate metabolism, including pyruvate, dihydroxyacetone phosphate (DHAP), lactate, and RNA, have been studied as potential ergogenic aids.

Pyruvate and DHAP Pyruvate is the penultimate product (prior to lactate) in the glycolytic pathway and, as a supplement, is theorized to provide energy more efficiently, subsequently improving performance or accelerating weight loss. DHAP is a combination of dihydroxyacetone and inorganic phosphate and is another 3-carbon metabolic by-product of glycolysis upstream of pyruvate. Pyruvate is poorly absorbed, so any potential effects might require high doses of the supplement. This increases the likelihood of gastrointestinal distress or diarrhea, which could be very problematic for an athlete. It is unclear if pyruvate even reaches and enters skeletal muscle, making direct effects on metabolism unlikely. A small number of investigations on pyruvate supplements and weight loss are available and, with the exception of obese women taking high-dose pyruvate during severe caloric restriction, the data do not support a beneficial effect. In a recent review, Onakpoya and colleagues concluded that the evidence to show that pyruvate supplementation can help reduce body weight is not convincing. Thus, pyruvate cannot be recommended as a weight loss supplement.

Similarly, the investigations into the effects of pyruvate and DHAP on exercise performance have not been supportive of an ergogenic effect. Stanko and colleagues showed improved time to exhaustion following 7 days of 75 g of DHAP/25 g pyruvate supplementation in untrained males, but mechanistic studies to explain this finding are unavailable. In a later study, Morrison and others found no effect on endurance performance in well-trained volunteers. Further, in two studies investigating combined pyruvate and creatine (discussed in chapter 6) supplements, creatine plus pyruvate was no more effective than creatine alone in one study, and the combined supplement was ineffective in the other study. Lawrence Spriet, an expert in exercise metabolism, has noted that there has not been any recent research in this area, and concludes there is no scientific basis for use of pyruvate/DHAP supplements as ergogenic aids.

Lactate Salts As noted previously, lactic acid is a metabolic by-product of anaerobic glycolysis. We also indicated that although lactic acid is often associated with fatigue, most sport scientists theorize it is the hydrogen ion release that increases the acidity and impairs performance, not the lactate itself. Lactate is actually a small metabolite of glucose; its formula is $C_3H_5O_3$, about half of that of glucose, $C_6H_{12}O_6$. Thus, lactate still possesses considerable energy and, as Van Hall noted, may be converted back to pyruvate to enter the energy pathway in the skeletal muscles. Lactate may also be converted back to glucose by the liver. However, ingested lactate can cause gastrointestinal disturbances, so it must be consumed in very low amounts compared with carbohydrate. This makes lactate supplements usefulness as a fuel source somewhat limited.

The results of studies by Fahey, Brouns, Azevedo, and others demonstrate that lactate supplements are ineffective at improving endurance cycling performance. Lactate supplements are also theorized to work as buffering agents, and, according to Morris, oral ingestion of lactate can increase blood pH or bicarbonate. Morris states that a small number of studies show improved performance of high-intensity exercise to exhaustion with lactate ingestion, but it is not known if this translates into better sports or athletic performance or if the improvement is comparable to ergogenic supplements such as beta-alanine or sodium bicarbonate. More research on the potential ergogenic buffering effects of lactate supplements is needed, but at this time, it seems unlikely that lactate supplements would be a better choice for athletes than carbohydrate, beta-alanine, or sodium bicarbonate, all of which are supported by many more research studies.

Ribose Ribose is a 5-carbon monosaccharide found throughout body cells as part of various compounds, such as RNA (ribonucleic acid) in the cell nucleus. Ribose also comprises the sugar portion of adenosine, the nucleotide found in ATP (adenosine triphosphate). ATP, as you recall, is the immediate source of energy for muscle contraction, both in the heart and in skeletal muscles.

Although found in nature, very little ribose is consumed in a natural diet. Instead, a specific metabolic pathway (pentose phosphate pathway) produces ribose from glucose to meet our body needs. Ribose supplements (made from corn sugar) have been marketed to physically active individuals as a means to promote faster recovery in heart and skeletal muscles, presumably by facilitating the formation of adenosine, one of the major components of ATP.

Research indicates that strenuous exercise may necessitate rapid recovery of adenosine within muscle cells, which might benefit from adequate ribose. Some benefits of high-dose ribose supplements have been shown in patients with congestive and ischemic heart failure, and stable coronary artery disease. However, data on exercise performance of young, healthy adults does not support any ergogenic effect. Collectively, Kerksick, Peveler, Berardi and Ziegenfuss, Hellsten, Op 'T Eijnde, Dunne, and Kreider, and their colleagues have evaluated the effects of ribose supplements on various measures of exercise performance and found no ergogenic effect. In their review, Dhanoa and Housner note that although ribose manufacturers claim it provides an ergogenic benefit, scientific research does not support this claim. Current data do not support an ergogenic effect of ribose supplementation.

Multiple Carbohydrate By-Products Sports supplements manufacturers often combine multiple nutrients in a single supplement on the theory that each may exhibit an ergogenic effect, but the effect will be amplified with multiple components. Limited research is available with multiple by-products of carbohydrate metabolism, but Brown and others investigated the ergogenic potential of a multinutrient supplement composed primarily of intermediates of the Krebs cycle, which may be derived from carbohydrate, fat, or protein. Three weeks of supplementation did not improve cycling time to exhaustion at approximately 70–75 percent VO_2 max, nor did it improve the rate of recovery.

Key Concepts

▶ Metabolic by-products of carbohydrate metabolism have been tested as ergogenics. Most have been found to be ineffective, although research with physically trained individuals is limited with several purported ergogenic products.

Check for Yourself

▶ Go to a health food store or search on-line for carbohydrate related sports supplements, including metabolites, that may help you train or compete more effectively. Evaluate the supplement fact labels for content and performance claims. What is your judgment?

Dietary Carbohydrates: Health Implications

The diet of the typical American and Canadian still appears to be unbalanced. In general, we consume too many kcal for the level of physical activity we do, and we eat less-than-recommended amounts of nutrient-dense sources of carbohydrates and fats. Such a diet may pose several health problems. As we shall see in chapter 5, excessive consumption of total and saturated fat appears to be of major concern relative to the development of several chronic diseases. In this section, we discuss the health aspects of dietary carbohydrates. In general, the health effects associated with various sugars and starches are not in the substances themselves, but rather in the nutrients that accompany them in the foods we eat. For example, sugar in orange juice is little different from sugar in a soda, but the orange juice contains substantial amounts of vitamin C, potassium, and other nutrients, whereas the soda has none unless fortified. Whole grains contain more fiber and more of some micronutrients than refined grains.

Nutritional objectives in *Healthy People 2030* and in the *2020–2025 Dietary Guidelines for Americans* recommend that we consume more grains, making whole grains half of all grains consumed. We should also reduce the consumption of refined carbohydrates and added sugars, often referred to as *bad carbs*. Although no foods, or carbohydrates, are inherently good or bad, following these two general guidelines may produce some significant health benefits. Additionally, an appropriate exercise program may have a healthful influence on carbohydrate metabolism.

How do refined sugars and starches affect my health?

As noted previously, sugars may be found naturally in foods, or they may be manufactured from starches, such as high-fructose corn syrup, and added to foods. Refined starches are predominant in many foods, such as white bread, pasta, and rice. Consumption of refined sugars and starches in excess may be associated with various health risks, attributed mainly to their high glycemic index.

Dental Caries One of the most common health problems that has been associated with dietary sugar is tooth decay, or dental caries. However, the National Institutes of Health, in its consensus statement on management of dental caries throughout life, noted that effective preventive practice involves a number of factors, including proper oral hygiene (brushing, flossing, use of fluoride) and dietary modifications (use of sugarless products). Tooth decay is not necessarily a matter of how much sugar one eats, but in what form and how often. Dental erosion is increasing and is associated with dietary acids, a major source of which is soft drinks. Sticky, chewy, sugary foods eaten often between meals increase the risk of developing dental caries. Starchy foods that adhere to teeth, such as bread, are also cariogenic. Such foods may increase the presence of dental plaque, which may lead to periodontal infection. Seymour and others cite epidemiological research supporting a relationship between periodontal infection and various systemic diseases, such as heart disease, stroke, and diabetes. The infection may lead to systemic inflammation, which may induce adverse effects, such as atherosclerosis. These authors indicate that the control of oral disease is essential in the prevention and management of these systemic conditions.

Of particular interest to athletes, von Fraunhofer and Rogers reported far greater enamel dissolution in flavored and energy (sports) drinks than previously noted for water. They noted that sipping sports drinks over long periods of time may erode tooth enamel; therefore, drink quickly. In contrast, Mathew and others reported no relationship between consumption of sports drinks and dental erosion in university athletes. Nevertheless, scientists have developed a prototype sports drink, containing substantial amounts of calcium and maltodextrins, which is alleged to cause less dental enamel erosion than the typical commercial sports drink.

Chronic Diseases Over the years, dietary intake of refined sugar has been alleged to contribute to a wide variety of health problems, including obesity, diabetes, heart disease, and cancer, as well as various psychological afflictions such as hyperactivity in children, premenstrual syndrome (PMS), and seasonal affective disorder (SAD). Such allegations have been based mainly on theoretical considerations, but with support from some recent epidemiological studies. A habitual diet rich in high-glycemic-index foods theoretically may lead to insulin resistance and high serum triglyceride levels, risk factors for diabetes and heart disease, respectively. This may be especially so in individuals who are obese, and will be discussed in detail in chapter 10. Bantle also indicated that fructose, which is a low-glycemic-index sugar, may increase serum triglycerides and may be a contributing factor to obesity. Individuals should avoid high-fructose corn syrup, but eating fruits with naturally occurring fructose is not a cause for concern. Added sugars can increase caloric intake and predispose an individual to obesity.

As a part of the National Health and Nutrition Examination Survey (NHANES) of healthy adults, fructose intake was calculated and blood pressure was directly measured. Jalal and colleagues determined a median fructose intake of 74 g/d (equivalent to 2.5 sugary soft drinks per day). They also found this fructose intake to be associated with a 26, 30, and 77 percent higher risk for blood pressure values of $\geq 135/85$, $\geq 140/90$, and $>160/100$ mmHg, respectively. Fung and others conducted a 24-year follow-up with the Nurses' Health Study cohort and identified a significant positive association between sugar-sweetened beverage intake and coronary heart disease risk.

High sugar intake has been associated with development of cancer. Two large epidemiological studies by Larsson and Stattin and their associates found that increased consumption of sugar and high-sugar foods, particularly sugar-sweetened sodas, increases the risk of pancreatic cancer. The increased sugar intake may cause the pancreas to produce more insulin, which may cause hyperinsulinemia and increased insulin-like growth factor, factors that may stimulate cell division in the pancreas and lead to cancer.

Additionally, as discussed previously, a high-carbohydrate diet can affect the fTRP:BCAA ratio and formation of the neurotransmitter serotonin. Serotonin may influence mood and behavior associated with PMS and SAD or other psychological states.

The National Academy of Sciences, in its DRI recommendations for carbohydrate, noted that, given the currently available scientific evidence relative to the effect of dietary sugar on dental caries, psychological behavior, cancer, risk of obesity, and risk of hyperlipidemia, there is insufficient evidence to set a UL for total or added sugar in the diet. Nevertheless, the Academy noted that the theory linking a high glycemic index to certain health problems, such as diabetes and CHD, appears to be valid and supported by some studies, but the evidence at this time appears to be insufficient to substantiate the theory. Furthermore, the Academy noted that individuals who consume excess amounts of added sugars may not obtain sufficient amounts of various micronutrients, and that this may lead to adverse health effects. Johnson and others, in the American Heart Association scientific statement on dietary sugars intake and cardiovascular health, recommend an upper limit of half the discretionary kcal allowance from added sugars. For most American women, this is no more than 100 kcal per day, and for most American men it is no more than 150 kcal per day from added sugars.

Given these considerations, and the fact that many health organizations recommend a reduced intake of refined sugars to about 10 percent or less of the daily caloric intake, it appears to be prudent to moderate your consumption of refined sugars and starches.

Suggestions to decrease intake of refined starches and sugars were presented in chapter 2.

Jill Braaten/McGraw Hill

Are artificial sweeteners safe?

Artificial sweeteners are products designed to provide sweetness but little or no kcal. Theoretically, these sweeteners could be used to reduce intake of refined sugars, but as Liebman noted, the consumption of both artificial sweeteners and refined sugars has increased over the past ten years. A number of artificial sweeteners have been produced and approved, and they have been incorporated into foods, dietary supplements, sports nutrition products, energy drinks, and diet products. **Table 4.11** provides a list of artificial sweeteners currently approved for use in the United States. Acesulfame-K is a naturally occurring potassium salt; Advantame is derived from aspartame and vanillin; Aspartame is made from the two amino acids aspartic acid and phenylalanine; Neotame is derived from the same amino acids; Saccharin is a non caloric derivative of coal tar; *Siraitia grosvenorii* is an extract of Swingle fruit (also known as Luo Han Guo or monk fruit); Stevia is a derivative of the plant *Stevia rebaudiana;* and Sucralose is produced by altering the sugar molecule with chlorine.

Some artificial sweeteners have kcal (for instance, Aspartame has 4 kcal per gram); however, the amount typically consumed to provide the desired sweetness usually reduces the kcal to near zero per serving. Artificial sweeteners are ubiquitous and are often combined with natural sugars to provide sweetness with fewer kcal. For example, Coca-Cola® markets Coca-Cola Life®, a reduced-sugar soda that is partially sweetened with Stevia. Many reduced-kcal yogurts are flavored with Stevia; and Arnold Whole Wheat Sandwich Thins® also contain Stevia.

Each of these sweeteners has undergone the necessary safety testing to be approved by the U.S. government; however, some still question the safety. As new artificial sweeteners are developed and incorporated into foods, sports nutrition, low-kcal, and low-carbohydrate products, and because the use and approval of artificial sweeteners differ between countries, it is wise for consumers to consult with a regulatory body for the most up-to-date information. The interested reader should consult the website of the United States Food and Drug

TABLE 4.11 Artificial sweeteners approved for use in the United States

Name	Brand name	Times sweeter than sugar	Acceptable daily intake in packets per day
Acesulfame-K	Sunett® Sweet One®	200	165
Advantame	*	20,000	4,000
Aspartame	Equal® NutraSweet®	200	165
Neotame	Newtame®	7,000–13,000	200
Saccharin	Sweet and Low® Sweet Twin® Sweet'N Low® Necta Sweet®	200–700	250
Siraitia grosvenorii: Swingle (Luo Han Guo) fruit extracts (SGFE)	Nectresse® Monk Fruit in the Raw® PureLo®	100–250	Not determined
Stevia: Certain high purity steviol glycosides purified from *Stevia rebaudiana* leaves	Truvia®	200–400	29
Sucralose	Splenda®	600	165

*No brand name yet.
Source: https://www.fda.gov/Food/IngredientsPackagingLabeling/FoodAdditivesIngredients/ucm397725.htm.

Administration (www.fda.gov) and search under the food additives section. Here one can easily find currently approved artificial sweeteners, a summary of available safety data, acceptable daily intake (ADI) levels, and whom to contact to report adverse effects.

Sugar alcohols, which are mentioned early in the text, are another class of sweeteners. Common examples include sorbitol, xylitol, mannitol, and maltitol, but there are several others. The sweetness of sugar alcohols varies from one-quarter to one-half as sweet as sugar up to about the same sweetness as sugar. Sugar alcohols do contain energy, about 1.5 to 2.5 kcal per gram, but cause a smaller increase in blood glucose, which is why they are often found in diabetic candies, gums, and sugarfree chocolate.

Sugar alcohols are only partially digested and metabolized and can lead to flatulence, diarrhea, and other gastrointestinal symptoms or discomfort. While these issues are typically associated with high intakes of sugar alcohols, some athletes may inadvertently consume large amounts of sugar alcohols because they are unaware that they are used in a wide variety of sports nutrition and diet products. For instance, on a given day, an athlete may consume chewable vitamin C with breakfast and have a low-carbohydrate protein bar as a snack, a low-sugar energy drink before a workout, and a protein shake postworkout. Chewable vitamins, low-carbohydrate sports bars, low-sugar energy drinks, and protein powders often contain several artificial sweeteners and sugar alcohols to make them more palatable. Athletes are encouraged to read labels closely to identify sugar alcohols and to avoid intake prior to or during competition. One helpful tip is to search labels for the term "net carbohydrates." For example, the EAS® AdvantEDGE® Carb Control™ Bar contains 26 grams of carbohydrate with 2 to 4 grams of "net carbs." This product contains 26 grams of carbohydrate as 5 grams of fiber, 2 grams of sugar, and 19 grams of sugar alcohols. Although this product is marketed as a snack or meal replacement, an athlete who ingests 10 to 20 grams of sugar alcohols per day may experience gastrointestinal distress. In terms of sweeteners, the product contains maltitol syrup, fructo-oligosaccharides, maltitol, xylitol, and sucralose.

The Academy of Nutrition and Dietetics has published a position stand on non-nutritive sweeteners. Perhaps the most hotly debated question about artificial sweeteners at this point in time is do they increase appetite or body weight? In the position stand, Fitch and Keim did not find evidence to suggest that non-nutritive/artificial sweeteners cause unwanted weight gain or increased appetite, but many questions remain to be answered. For instance, Harrington and others reported that changes in the microbiome in response to non-nutritive sweeteners but that the mechanisms and effects on metabolism are unclear. Wilk and colleagues stated that replacing sugar with non-nutritive sweeteners may support weight control but the effects on appetite are not clear and further research is required on the long-term effects related to body weight management.

Why are complex carbohydrates thought to be beneficial to my health?

To increase consumption of total carbohydrate in the diet while reducing the consumption of refined sugars, one must increase the consumption of complex carbohydrates. Some diet plans developed for health, such as the Pritikin program, recommend that 80 percent of the dietary kcal be supplied by carbohydrates, mostly complex and unrefined. More recently, the OmniHeart diet focused on healthy carbohydrates, fats, and proteins. The OmniHeart diet plan includes the following tips for increasing consumption of healthier carbohydrates:

- Eat 1-2 servings of fruit at every meal and have an extra fruit at breakfast.
- Have 2-3 servings of vegetables at lunch and dinner.
- Create a fruit and nut trail mix for snacks: ¼ cup dried fruit with 1 oz. unsalted nuts.
- Use whole grains rather than refined grains as often as possible.
- Select legumes for a carbohydrate and protein source several times a week.

A diet rich in complex carbohydrates may reduce the percentage contribution from fats if excessive, which may confer significant health benefits, as noted in chapter 5. Complex carbohydrates are found primarily in starchy vegetables, whole grains, and legumes, but small amounts are also found in fruits.

Whole-grain products are one of the best sources of healthy carbohydrates. As defined by the FDA, whole grains contain all three ingredients of a cereal grain, namely the outer bran and the inner germ and endosperm, and in the same proportion as found in nature (**figure 4.13**). Seal noted that an increasing body of

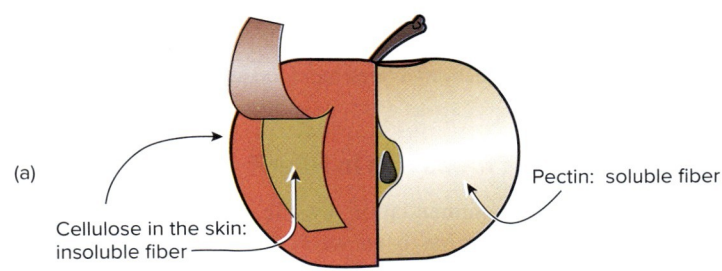

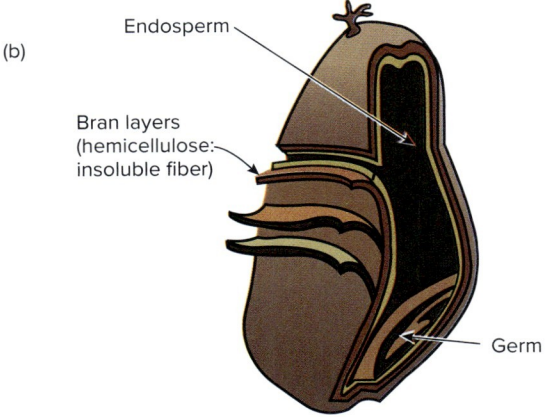

FIGURE 4.13 Various forms of fiber. (a) The skin of an apple consists of the insoluble fiber cellulose, which provides structure for the fruit. The soluble fiber pectin "glues" the fruit cells together. (b) The outside layer of a wheat kernel is made of layers of bran—insoluble fiber—making this grain a good source of fiber. Fruits, vegetables, whole grains, and legumes such as beans are rich in fiber.

evidence from both epidemiological and prospective studies supports an inverse relationship between consumption of whole-grain foods and risk of coronary heart disease. In his review, Temple noted that refined carbohydrates and, in particular, sugar sweetened beverages do increase the risk of CHD. He also notes that an extra single or double serving of whole grains and/or cereal fiber are protective and decrease risk by 10–20 percent.

When shopping, look for products labeled 100 percent Whole Wheat or 100 percent Whole Grains. Products labeled 100 percent wheat, multigrain, or stone ground may be made primarily from refined grains. The first ingredient listed should be whole oats, whole rye, whole wheat, or other whole grains such as brown rice, bulgur, or oatmeal. Many vitamins, minerals, phytonutrients, antioxidants, and fiber may be lost in processing of whole grain to refined grain. Some, but not all, may be replaced during processing, so you may not get the synergistic health effects of the multiple nutrients. In your efforts to increase fiber intake, look at the 10 Tips to Help You Eat Whole Grains on the www.ChooseMyPlate.gov site.

Current thinking supports the concept that the beneficial health effects of a diet rich in complex carbohydrates, which may be considered to be a low-glycemic-index diet, may be linked to several important attributes of such a diet, including the collective presence of various phytochemicals, vitamins, minerals, and dietary fiber. The health benefits of phytochemicals were discussed in chapter 2, while the benefits of vitamins and minerals will be covered in chapters 7 and 8. Most research relative to the healthful benefits of complex carbohydrates has focused on dietary fiber.

Why should I eat foods rich in fiber?

The current AI recommendation by the National Academy of Sciences for total fiber has been set at 14 grams per 1,000 kcal, or 38 grams for men and 25 grams for women up to age 50 and slightly lower amounts thereafter. Recall that total fiber consists of both dietary fiber and functional fiber, and various specific forms of fiber are found in each category; some specific forms of fiber may be classified as both dietary and functional fiber. Although the Academy indicates that specific forms of fiber have properties that result in different physiological effects that may impact health, it did not feel that the evidence was sufficient to establish separate recommendations for each type of fiber. Nevertheless, some feel that the use of the water-solubility classifications system, as discussed previously, might be useful conceptually to illustrate the potential health benefits of certain forms of fiber. As noted earlier, dietary fiber is found naturally in various plant foods.

Exactly how dietary fiber may be protective is not known, but several mechanisms have been proposed that may help in the prevention of certain forms of cancer, coronary heart disease, obesity, diabetes, hypertension, and various disorders of the gastrointestinal tract. Here are some of the theories relative to the potential health benefits of total fiber:

1. Water-insoluble fibers are considered to be those with the greatest effect on fecal bulk. Adding bulk to the contents of the large intestine stimulates peristalsis and speeds up the transit time of food through the intestines. The increased bulk has been shown to dilute any possible cancer-causing (carcinogens) that might attack cell walls, while faster transit diminishes the time carcinogens may have to act. Increased bulk—and peristalsis—also decreases the incidence rate of diverticulitis, an inflammatory disorder in the large intestine that may cause rupture, leading to serious complications.

2. Fiber-rich foods are low-glycemic-index foods, and they may increase insulin sensitivity and prevention of weight gain. Fiber slows down gastric (stomach) emptying and thereby slows glucose absorption in the small intestine. The high viscosity of soluble fiber also decreases intestinal absorption. These effects may lead to better control of blood sugar and may lengthen the sensation of fullness, or satiety, which may be important to individuals on weight-loss diets. Fiber-rich diets are frequently lower in fat and added sugars, and thus contain fewer kcal. So high-fiber diets may be useful in the prevention or treatment of obesity and obesity-related chronic diseases such as diabetes and hypertension.

3. Fiber, particularly gummy forms of water-soluble fiber such as beta-glucans in oats, may bind with various substances in the gastrointestinal tract. Soluble fiber may bind with carcinogens so that they are excreted by the bowel. Soluble fibers may also bind with and lead to the excretion of bile salts, which contain cholesterol; normally bile salts are reabsorbed into the body, but excretion of bile salts, along with their cholesterol content, may help reduce serum cholesterol levels. This effect may decrease the risk of coronary heart disease. (Lower serum cholesterol levels decrease the risk of atherosclerosis, a major cause of heart disease.)

4. Some water-soluble fibers may be fermented in the large intestine to form short-chain fatty acids (SCFAs). Zeng and others noted that several of these SCFAs are theorized to help prevent gastrointestinal disorders, cancer, and cardiovascular disease. Some act in the colon; others are absorbed into the blood, are delivered to the liver, and may help decrease synthesis of cholesterol. Recent data, reviewed by Guan and colleagues, support that water-soluble fibers are easily metabolized, produce beneficial metabolites, and improve the gut microbiota, which is associated with reduced risk of chronic diseases.

Although it may be illustrative to view the health benefits of fiber based on its water solubility, Joanne Slavin, a scholar on the health effects of dietary fiber, notes that it is difficult to generalize as to the physiological effects of fiber based on this classification system. For example, she notes that rice bran, which is devoid of soluble fiber, has been shown to reduce serum cholesterol, while recent research has also supported the effect of insoluble fiber to reduce the risk of heart disease. Thus, health benefits may be attributable to total fiber.

Numerous studies, including major epidemiological studies and clinical trials, have investigated the effect of total fiber on reducing the risk or incidence of chronic diseases. Based on these studies, the National Academy of Sciences established the AI for total fiber because it may reduce the risk of coronary heart disease, and more recent research supports this concept.

The Academy cited studies showing a beneficial effect of total fiber and a low-glycemic-index diet on other health problems, particularly in individuals with diabetes or hyperlipidemia, but the evidence was not as convincing as that for prevention of heart disease.

Marilyn Barbone/123RF

Although there is not complete consistency in the research, a large number of studies demonstrate beneficial effects of a diet that is high in fiber. In terms of the effects of dietary fiber on mortality, in the most important meta-analysis to date, Yang and colleagues analyzed 17 prospective studies that included 982,411 individuals. When comparing persons with dietary fiber intakes in the top third with persons whose intakes were in the bottom third, there was a 16 percent decrease in risk of mortality. Further, for every 10 grams/day increase in dietary fiber intake, there was a 10 percent reduction in mortality risk. There are many other meta-analytical reviews that demonstrate the benefits of dietary fiber. For instance, as reported in a meta-analysis by Kim and Je, cereal and vegetable, but not fruit, fiber intake was associated with reduced risk of mortality. In their meta-analysis, Threapleton and colleagues reported that dietary fiber intake from cereal and vegetable sources was associated with lower risk of cardiovascular and coronary heart disease, while fruit fiber intake was associated with decreased risk of cardiovascular disease. In a meta-analysis of randomized controlled trials of patients with type 2 diabetes, Silva and others reported that high-fiber diets (from food or supplement sources) reduced glycated hemoglobin and fasting blood glucose. The authors acknowledged that increased fiber intake improved glycemic control and should be considered in the treatment of type 2 diabetes. In their systematic review and meta-analysis, Aune and co-authors reported that at least two servings of whole grains per day should be consumed to reduce the risk of type 2 diabetes. In this analysis, a decreased risk was associated with whole-grain intake, while an increased risk was found for refined grains such as white rice. Aune and colleagues also reported that a high intake of dietary fiber was associated with decreased risk of colon cancer but that future research should include information on fiber subtypes. As Clark and Slavin have indicated, dietary fiber intake is associated with a lower body weight in many epidemiological studies. However, the majority of studies that investigate the effects of fiber intake on satiety or food intake do not support an effect.

Even though there may be debate over the usefulness of the glycemic index as a means to design a healthful diet for chronic diseases other than heart disease, consuming a fiber-rich diet is certainly prudent dietary behavior, given that coronary heart disease is the number one cause of death in Western societies. And although the National Academy of Sciences indicates that the relationship of fiber intake to other health problems is the subject of ongoing investigation and currently unresolved, there appear to be no adverse health effects associated with a fiber-rich or low-glycemic-index diet. Indeed, a fiber-rich diet may be useful for other possible health benefits.

Prevention of colon cancer has been one of the main theories underlying the promotion of a high-fiber diet. Rock has indicated that the relationship between dietary fiber intake and colorectal cancer has been inconsistent. However, she noted that no significant relationship between fiber intake (or major food sources of fiber) and risk for colorectal cancer was observed in a recently reported, large, pooled analysis of several studies. Nevertheless, she notes limitations in epidemiological studies and indicates that the effect of increased dietary fiber intake on risk for colorectal cancer has not been adequately addressed in studies conducted to date. More research, including longer-term trials and higher levels of fiber intake, is needed to increase knowledge in this area.

As the different types of fiber appear to convey health benefits in different ways, a balanced intake of total fiber appears to be the best approach. The Academy notes that although the AI are based on total fiber, the greatest health benefits may come from the ingestion of cereal fibers and various viscous fibers, including gums and pectins, which are found in fruits and vegetables. Obtaining 25–38 grams of fiber daily is not difficult, but you have to eat more whole grains, fruits, vegetables, and legumes. Total fiber is listed on food labels. In particular, check food labels on breads and cereals, staples of the daily diet. According to the *2020-2025 Dietary Guidelines for Americans,* although Americans eat adequate amounts of total grains, most of these come from refined sources rather than whole grains. Additionally, on average Americans eat less than 1 ounce-equivalent of whole grains per day, and less than 5 percent consume the minimum amount of whole grains. Refined grains contain little or no fiber, whereas whole grains are usually rich in fiber. Some brands of bread contain 3 grams of fiber per slice, and only 50 kcal. Breakfast cereals may also be very high in fiber, some containing 10 or more grams per serving.

Training Table

Ten easy ways to eat more fiber from *Consumer Reports on Health*:

1. Look for "good sources" of fiber (~3 grams per serving).
2. Choose whole grains (100 percent whole wheat or 100 percent whole grain).
3. Start the day right (oatmeal or a high-fiber cereal).
4. Choose fiber-filled snacks (raw carrots or celery, popcorn, or fiber-rich crackers).
5. Drink it up (blend chunks of fruit or vegetables with yogurt, juice, or soymilk).
6. Look for fiber-fortified pastas.
7. Don't forget legumes (add beans or lentils).
8. Keep the skin on (wash but don't peel skins of fruits and vegetables).
9. Bake it in (add crushed bran cereal or flax or sesame seeds to baked items).
10. Consider a supplement (if foods are not enough).

A good way to see if you are eating enough fiber is to observe the buoyancy of your stool in the toilet. It should float, or at least appear flaky and break apart. If it sinks or does not break apart, you are not eating enough fiber.

There appear to be few or no health disadvantages to a high-fiber diet. As we shall see in chapter 8, there has been some concern that high-fiber diets could lead to increased losses of certain minerals, such as iron and zinc, but research has shown that such concerns are generally unwarranted if one follows the recommendations just given.

Schneeman, discussing development of a scientific consensus on the importance of dietary fiber, notes that fiber serves as a marker for diets rich in plant foods, which provide additional benefits for maintaining health. Liebman notes that it is important to recognize that the health benefits attributed to dietary fiber may be associated with the form in which the fiber is consumed—as part of a whole, natural food containing other potential health-promoting nutrients such as vitamins and phytochemicals, rather than by consumption of a purified supplement form. This is in accord with the position stand of the Academy of Nutrition and Dietetics, written by Dahl and Stewart. Get more plant foods in your diet!

Do some carbohydrate foods cause food intolerance?

©Image Source

About one in nine Americans may develop gastrointestinal distress when they consume dairy products containing substantial amounts of lactose, particularly milk. African-Americans are more likely to suffer lactose intolerance. Such individuals lack the enzyme lactase and hence cannot metabolize lactose in the digestive tract. The most common symptoms of **lactose intolerance** are gas, bloating, abdominal pain, and diarrhea, although headache and fatigue may also occur.

Individuals may be diagnosed as being lactose intolerant through a lactose tolerance test administered by a physician. A self-detection technique may also be an effective approach. If you experience problems such as gas and diarrhea after consuming milk, abstain from all dairy products for 2 weeks and then evaluate the results. If the symptoms resolve, and then reoccur when you resume dairy food consumption, you may need to reduce the amount of lactose in your diet. Unfortunately, usually this means a reduced intake of dairy foods, which are considered to be the main dietary source of not only lactose but calcium as well. Di Stefano and others indicated that lactose intolerance may prevent the achievement of adequate peak bone mass in young adults and may, therefore, predispose them to severe osteoporosis.

Calcium will be discussed in detail in chapter 8, but here are several strategies lactose-intolerant individuals may use to obtain adequate calcium intake. In an extensive review of 33 studies, including nearly 3,000 children, MacGillivray and others concluded that lactose-free foods can help reduce the duration of diarrhea when compared to lactose-containing foods. Thus, changing to lactose-free milk and milk products may be a viable alternative that allows calcium intake to remain adequate. In a meta-analysis of well-designed studies, Savaiano and others indicated that symptoms of lactose intolerance may be minimal with small amounts of dairy foods, such as 1 cup or less. Consuming small amounts of milk over the course of the day may provide significant amounts of calcium. Dairy products that have been fermented, such as yogurt, may be tolerated and provide a good calcium source. Cheese may also be a good source of calcium, although it is high in fat. Dark green, leafy vegetables, tofu, sardines, salmon, and calcium-fortified fruit juices are all nondairy sources of calcium. Additionally, calcium supplements may be useful either in tablet form or as added to various foods, such as soy milk, rice milk, or orange juice.

Wheat products may produce gastrointestinal symptoms comparable to lactose intolerance. The problem is not the wheat itself but rather a protein called gluten. Gluten is found in wheat, rye, and barley, which are the main constituents in most of the grain-based products we eat, such as cereals, breads, and pasta. **Gluten intolerance** represents a sensitivity to gluten; the immune system recognizes gluten as a foreign substance, but does not induce an allergic response. Symptoms may vary from none to severe. Simple gluten intolerance could be problematic for endurance athletes. Alternate sources of carbohydrate, such as corn, potatoes, rice, soybeans, and similar foods, will be needed to replace grain-based foods. Gluten-free products are currently available in the marketplace.

Celiac disease, which, as described by Guandalini and Assiri, is the most common genetically based food intolerance in the world, with a prevalence of about 70 million people. In severe cases, the gluten damages the lining of the small intestine, which can lead to impaired nutrient absorption and a variety of nutrient-deficiency diseases, including weight loss, anemia, and osteoporosis. Celiac disease necessitates medical treatment and a lifelong gluten-free diet.

Lis and others reported that 41 percent of non celiac athletes follow a gluten-free diet 50–100 percent of the time, even though a gluten-free diet may only benefit about 5–10 percent of the general population. Most of these athletes self-diagnosed their gluten sensitivity. In a follow-up study from the same group of researchers, competitive cyclists who had no positive clinical screening for celiac disease and no history of irritable bowel syndrome were placed on a 7-day gluten-free or gluten-containing diet in a crossover design study. Exercise performance, well-being, GI symptoms, inflammatory markers, and intestinal fatty acid binding protein were not different between diets, indicating no benefits of a gluten-free diet in athletes without celiac disease.

Data are emerging on low FODMAP (Fermentable Oligosaccharides, Disaccharides, Monosaccharides and Polyols) diets, which may reduce GI distress in athletes. Low FODMAP diets were designed in Australia to treat irritable bowel syndrome, and are diets that reduce intake of a group of poorly absorbed short-chain carbohydrates. The identified carbohydrates increase the osmotic load in the intestine, which draws water into the intestine, potentially leading to diarrhea. When FODMAPS reach the colon, they are fermented and gas is produced. Lis and colleagues reported on an athlete who was placed on a low FODMAP diet for 6 days and had remarkable decreases in GI symptoms. More research, including clinical trials, must be conducted in this area before this style of diet can be widely prescribed.

https://www.mysportscience.com/post/what-are-fodmaps
Visit this website for a complete discussion of a low FODMAP diet, links to research, and the Monash University Low FODMAP Diet app.

Key Concepts

▶ Added sugars should be limited in the diet. The maximal recommended amount is 25 percent of daily energy intake, but some health professionals recommend lower amounts of about 10 percent. Intake of refined starches should also be limited.
▶ An increase in the amount of total fiber to about 25–38 grams per day may be helpful as a protective measure against the development of heart disease, and possibly other chronic diseases. Consuming more whole grains, more fresh fruits, more nonstarchy vegetables, and more legumes, which are low GI foods, will help ensure adequate fiber intake.

Check for Yourself

▶ Check the food labels of various breads for fiber content. Do some brands have significantly more than others? What impact could switching breads have on meeting the recommended daily fiber intake of 25 grams for females and 38 grams for males?

APPLICATION EXERCISE

If you are not currently eating enough fiber, try this experiment. Keep a record of your appetite and your bowel movements for a week or so, and then switch to a high-fiber diet, consuming fruits and vegetables, whole wheat and whole-grain breads, and other high-fiber foods as documented by food labels. Record approximate grams of total fiber consumed daily. Also record an increase (↑), a decrease (↓), or no change (NC) for appetite, and record the number of daily bowel movements. Compare your appetite and bowel movements to the previous week. Did the high-fiber diet influence either?

Week 1 (normal diet)

	Sunday	Monday	Tuesday	Wednesday	Thursday	Friday	Saturday
Breakfast							
Lunch							
Dinner							
Snacks							
Appetite							
Bowel movement(s)							

Week 2 (high-fiber diet)

	Sunday	Monday	Tuesday	Wednesday	Thursday	Friday	Saturday
Breakfast							
Lunch							
Dinner							
Snacks							
Appetite							
Bowel movement(s)							

Review Questions—Multiple Choice

1. Which of the following statements relative to carbohydrate loading is false?
   a. It is beneficial primarily for athletes involved in prolonged endurance events, such as the typical marathon (26.2 miles).
   b. It involves the intake of about 500–600 grams of carbohydrate each day for several days prior to competition.
   c. Research generally supports its effectiveness as a means of improving performance by helping to delay the onset of fatigue in the later stages of prolonged exercise tasks.
   d. The major advantage of carbohydrate loading is the increased storage of glycogen in the adipose cells for use during exercise.
   e. The increase in carbohydrate stores in the body may be detected by the increased body weight attributed to the water-binding effect of stored glycogen.

2. If you were to recommend to a runner a fluid replacement protocol, including carbohydrate content, for use during a marathon, which of the following would you *not* recommend?
   a. Use a 50–60 percent solution of galactose.
   b. Provide approximately 10–15 grams of carbohydrate per feeding.
   c. Provide feedings about every 15–20 minutes.
   d. Limit the amount of fructose in the solution.
   e. Choose a combination of carbohydrates.

3. Which of the following statements relative to the intake of carbohydrates and physical performance is false?
   a. If an individual has normal glycogen levels in the muscle and liver, carbohydrate feedings are usually not necessary if the exercise task is only about 60–90 minutes.
   b. The intake of concentrated sugar solutions may actually impair performance if they lead to osmosis of fluids into the stomach and precipitate the feeling of gastric distress.
   c. Carbohydrate intake may help delay the onset of fatigue in prolonged exercise by either preventing the early onset of hypoglycemia or delaying the depletion of muscle glycogen levels.
   d. Carbohydrate intake prior to and during endurance exercise tasks lasting more than 2 hours may be helpful as a means of enhancing performance.
   e. If consumed during exercise, it takes approximately 60–90 minutes for the carbohydrate to find its way into the muscle and be used as an energy source.

4. Following a meal high in simple carbohydrate, which of the following is most likely to occur in the next 1–2 hours?
   a. suppression of insulin with a resultant hyperglycemia
   b. hyperglycemia, which stimulates insulin secretion followed by possible hypoglycemia
   c. hypoglycemia, which stimulates insulin secretion and a return to normal blood glucose levels
   d. hyperglycemia with a suppression of insulin and movement of blood glucose into the liver and muscle tissues
   e. no change in blood glucose level

5. Which of the following is *not* one of the potential health benefits of dietary fiber?
   a. It may increase the bulk in the large intestine and dilute possible carcinogens.
   b. It may increase the bulk in the large intestine and help speed up intestinal transit.
   c. It may bind with carcinogens and help to excrete them.
   d. It may help excrete bile salts and reduce serum cholesterol levels.
   e. It may bind with certain minerals such as zinc and help to excrete them.

6. Which of the following food groups is least likely to be high in dietary fiber?
   a. vegetables
   b. grains
   c. dairy
   d. fruits
   e. protein

7. What two tissues in the body store the most carbohydrate?
   a. adipose and kidney
   b. kidney and liver
   c. liver and muscles
   d. muscles and kidney
   e. adipose and muscles

8. Common table sugar is _____.
   a. glucose.
   b. dextrose.
   c. fructose.
   d. sucrose.
   e. maltose.

9. The total amount of carbohydrate, as a percentage of the daily kcal, that represents the Acceptable Macronutrient Distribution Range (AMDR) for Americans and Canadians is _____.
   a. 12–15.
   b. 20–30.
   c. 30–45.
   d. 45–65.
   e. 85–90.

10. The glycemic index represents _____.
    a. the degree to which an athlete suffers from hypoglycemia.
    b. the amount of glucose released into the blood in response to exercise.
    c. the effect a particular food has on the rate and amount of increase in the blood glucose level.
    d. the amount of stored glycogen in the muscle and liver.
    e. the total amount of insulin released in response to food intake.

Answers to multiple choice questions: 1. d; 2. a; 3. e; 4. b; 5. e; 6. c; 7. c; 8. d; 9. d; 10. c

Critical Thinking Questions

1. You have eaten a high-carbohydrate meal for lunch. Explain the digestion and metabolic fate of this carbohydrate over the next 5 hours, including an hour of running at the end of this time frame.

2. Explain three possible mechanisms of fatigue due to inadequate carbohydrate intake prior to and during the running of a 26.2-mile marathon.

3. Identify athletes who might benefit from carbohydrate loading, and present details of the dietary and exercise training protocol.

4. Discuss the possible health benefits associated with a diet rich in complex carbohydrates and low to moderate in refined carbohydrates.

References

Books

American Institute of Cancer Research. 2007. *Food, Nutrition, Physical Activity and the Prevention of Cancer: A Global Perspective.* Washington, DC: American Institute of Cancer Research.

Fitts, R. 2006. The muscular system: Fatigue processes. In: *ACSM's Advanced Exercise Physiology,* ed. C. M. Tipton. Philadelphia: Lippincott Williams & Wilkins.

Houston, M. 2006. *Biochemistry Primer for Exercise Science.* Champaign, IL: Human Kinetics.

Maughan, R., and Murray, R. 2001. *Sports Drinks.* Boca Raton, FL: CRC Press.

National Academy of Sciences. 2005. *Dietary Reference Intakes for Energy, Carbohydrates, Fiber, Fat, Fatty Acids, Cholesterol, Protein and Amino Acids.* Washington, DC: National Academies Press.

U.S. Department of Health and Human Services Public Health Service. 2010. *Healthy People 2020.* Washington, DC: U.S. Government Printing Office.

Reviews and Specific Studies

Achten, J., et al. 2004. Higher dietary carbohydrate content during intensified running training results in better maintenance of performance and mood state. *Journal of Applied Physiology* 96:1331–40.

Akermark, C., et al. 1996. Diet and muscle glycogen concentration in relation to physical performance in Swedish elite ice hockey players. *International Journal of Sport Nutrition* 6:272–84.

Amiri, M., et al. 2019. Chocolate milk for recovery from exercise: A systematic review and meta-analysis of controlled clinical trials. *European Journal of Clinical Nutrition* 73(6):835–849.

Appel, L., et al. 2005. Effects of protein, monounsaturated fat, and carbohydrate intake on blood pressure and serum lipids: Results of the OmniHeart randomized trial. *Journal of the American Medical Association* 294:2455–64.

Arkinstall, M., et al. 2001. Effect of carbohydrate ingestion on metabolism during running and cycling. *Journal of Applied Physiology* 91:2125–34.

Atkinson, F., et al. 2021. International tables of glycemic index and glycemic load values 2021: A systematic review. *American Journal of Clinical Nutrition* 114:1625–32.

Aune, D., et al. 2013. Whole grain and refined grain consumption and the risk of type 2 diabetes: A systematic review and dose-response meta-analysis of cohort studies. *European Journal of Epidemiology* 28:845–58.

Azevedo, J., et al. 2007. Lactate, fructose and glucose oxidation profiles in sports drinks and the effect on exercise performance. *PLoS ONE* 26; 2 (9):e927.

Ball, T., et al. 1995. Periodic carbohydrate replacement during 50 min of high intensity cycling improves subsequent sprint performance. *International Journal of Sport Nutrition* 5:151–58.

Balsom, P., et al. 1999. High-intensity exercise and muscle glycogen availability in humans. *Acta Physiologica Scandinavica* 165:337–45.

Bangsbo, J., et al. 2006. Physical and metabolic demands of training and match-play in the elite football player. *Journal of Sports Sciences* 24:665–74.

Bantle, J. 2006. Is fructose the optimal glycemic index sweetener? *Nestle Nutrition Workshop Series: Clinical & Performance Programme* 11:83–91.

Bastos-Silva, V., et al. 2017. Effect of carbohydrate mouth rinse on training load volume in resistance exercises. *Journal of Strength & Conditioning Research*

Baty, J. 2007. The effect of a carbohydrate and protein supplement on resistance exercise performance, hormonal response, and muscle damage. *Journal of Strength & Conditioning Research* 21 (2): 321–29.

Baur, D., and Saunders, M., 2021. Carbohydrate supplementation: A critical review of recent innovations. *European Journal of Applied Physiology* 121:23–66.

Beelen, M., et al. 2010. Nutritional strategies to promote postexercise recovery. *International Journal of Sport Nutrition & Exercise Metabolism* 20:515–32.

Berardi, J., and Ziegenfuss, T. 2003. Effects of ribose supplementation on repeated sprint performance in men. *Journal of Strength Conditioning Research* 17:47–52.

Berardi, J., et al. 2008. Recovery from a cycling time trial is enhanced with carbohydrate-protein supplementation vs. isoenergetic carbohydrate supplementation. *Journal of the International Society of Sports Nutrition* 5:24.

Betts, J., et al. 2005. Recovery of endurance running capacity: Effect of carbohydrate-protein mixture. *International Journal of Sport Nutrition* 15:590–609.

Blair, S., et al. 1980. Blood lipid and ECG response to carbohydrate loading. *Physician & Sports Medicine* 8:69–75.

Bosch, A., et al. 1994. Influence of carbohydrate ingestion on fuel substrate turnover and oxidation during prolonged exercise. *Journal of Applied Physiology* 76:2364–72.

Brooks, G. 2007. Lactate: Link between glycolytic and oxidative metabolism. *Sports Medicine* 37:341–43.

Brouns, F., et al. 1995. Chronic oral lactate supplementation does not affect lactate disappearance from blood after exercise.

International Journal of Sport Nutrition 5:117–24.

Brown, A., et al. 2004. Tricarboxylic-acid-cycle intermediates and cycle endurance capacity. *International Journal of Sport Nutrition & Exercise Metabolism* 14:720–29.

Brown, L., et al. 1999. Cholesterol-lowering effects of dietary fiber: A meta-analysis. *American Journal of Clinical Nutrition* 69:30–42.

Bryner, R., et al. 1998. Effect of lactate consumption on exercise performance. *Journal of Sports Medicine & Physical Fitness* 38:116–23.

Burke, L., et al. 2000. Carbohydrate loading failed to improve 100-km cycling performance in a placebo-controlled trial. *Journal of Applied Physiology* 88:1284–90.

Burke, L., et al. 2001. Guidelines for daily carbohydrate intake: Do athletes achieve them? *Sports Medicine* 31:267–99.

Burke, L. 2001. Nutritional practices of male and female endurance cyclists. *Sports Medicine* 31:521–32.

Burke, L., et al. 2004. Carbohydrates and fat for training and recovery. *Journal of Sports Sciences* 22:15–30.

Burke, L., et al. 2005. Effect of carbohydrate intake on half-marathon performance of well-trained runners. *International Journal of Sport Nutrition & Exercise Metabolism* 5:573–89.

Burke, L., et al. 2006. Energy and carbohydrate for training and recovery. *Journal of Sports Sciences* 24:675–85.

Burke, L. 2007. Nutrition strategies for the marathon. *Sports Medicine* 37:344–47.

Burke, L. M., 2014. Carbohydrate needs of athletes in training. In *Sports Nutrition. The Encyclopedia of Sports Medicine*, ed. R. J. Maughan. Oxford, UK: Wiley Blackwell.

Campbell, C., et al. 2008. Carbohydrate-supplement form and exercise performance. *International Journal of Sport Nutrition & Exercise Metabolism* 18:179–90.

Carter, J., et al. 2004. The effect of carbohydrate mouth rinse on 1-h cycle time trial performance. *Medicine & Science in Sports & Exercise* 36:2107–11.

Carter, J., et al. 2004. The effect of glucose infusion on glucose kinetics during a 1-h time trial. *Medicine & Science in Sports & Exercise* 36:1543–50.

Cermak, N., and van Loon, L. 2013. The use of carbohydrates during exercise as an ergogenic aid. *Sports Medicine* 43:1139–55.

Chryssanthopoulos, C., et al. 2002. Influence of a carbohydrate-electrolyte solution ingested during running on muscle glycogen utilization in fed humans. *International Journal of Sports Medicine* 23:279–84.

Chryssanthopoulos, C., et al. 2002. The effect of a high carbohydrate meal on endurance running capacity. *International Journal of Sport Nutrition & Exercise Metabolism* 12:157–71.

Churuangsuk, C., et al. 2022. Diets for weight management in adults with type 2 diabetes: An umbrella review of published meta-analyses and systematic review of trials of diets for diabetes remission. *Diabetologia* 65:14–36.

Clark, M. J., and Slavin, J. L. 2013. The effect of fiber on satiety and food intake: A systematic review. *Journal of the American College of Nutrition* 32:200–11.

Consumers Union. 2005. Sweeteners can sour your health. *Consumer Reports on Health* 17 (1):8–9.

Consumers Union. 2009. 10 easy ways to eat more fiber. *Consumer Reports on Health* 21 (12):7.

Costill, D. 1988. Carbohydrates for exercise: Dietary demands for optimal performance. *International Journal of Sports Medicine* 9:1–18.

Coyle, E. 1995. Substrate utilization during exercise in active people. *American Journal of Clinical Nutrition* 61 (Supplement): 968S–79S.

Coyle, E. 2000. Physical activity as a metabolic stressor. *American Journal of Clinical Nutrition* 72:512S–20S.

Coyle, E. 2004. Highs and lows of carbohydrate diets. *Sports Science Exchange* 17 (2):1–6.

Coyle, E. 2007. Physiological regulation of marathon performance. *Sports Medicine* 37:306–11.

Currell, K., and Jeukendrup, A. 2008. Superior endurance performance with ingestion of multiple transportable carbohydrates. *Medicine & Science in Sports & Exercise* 40:275–81.

Dahl, W., and Stewart, M., 2015. Position of the Academy of Nutrition and Dietetics: Health implications of dietary fiber. *Journal of the Academy of Nutrition and Dietetics* 115:1861–70.

Davis, J. 1996. Carbohydrates, branched-chain amino acids and endurance: The central fatigue hypothesis. *Sports Science Exchange* 9(2):1–6.

Debold, E. 2012. Recent insights into the molecular basis of muscular fatigue. *Medicine and Science in Sports and Exercise* 44:1440–52.

Décombaz, J., et al. 2011. Fructose and galactose enhance postexercise human liver glycogen synthesis. *Medicine and Science in Sports and Exercise* 43:1964–71.

Desbrow, B., et al. 2004. Carbohydrate-electrolyte feedings and 1-h time trial cycling performance. *International Journal of Sport Nutrition & Exercise Metabolism* 14:541–49.

Dhanoa, T., and Housner, J. 2007. Ribose: More than a simple sugar? *Current Sports Medicine Reports* 6:254–57.

Di Pede, G., et al. 2021. Glycemic index values of past products: An overview. *Foods* 10:2541.

Di Stefano, M., et al. 2002. Lactose malabsorption and intolerance and peak bone mass. *Gastroenterology* 122:1793–99.

Donaldson, C., et al. 2010. Glycemic index and exercise endurance. *International Journal of Sport Nutrition & Exercise Metabolism* 20:154–65.

Dunkin, J., and Phillips, S. 2017. The effect of carbohydrate mouth rinse on upper-body muscular strength and endurance. *Journal of Strength and Conditioning Research* 31:1948–53.

Dunne, L., et al. 2006. Ribose versus dextrose supplementation, association with rowing performance: A double-blind study. *Clinical Journal of Sport Medicine* 16:68–71.

Edgerton, V., and Roy, R. 2006. The nervous system and movement. In *ACSM's Advanced Exercise Physiology*, ed. C. M. Tipton. Philadelphia: Lippincott Williams & Wilkins.

el-Sayed, M., et al. 1997. Carbohydrate ingestion improves performance during a 1-h simulated cycling trial. *Journal of Sports Sciences* 15:223–30.

Erlenbusch, M., et al. 2005. Effect of high-fat or high-carbohydrate diets on endurance exercise: A meta-analysis. *International Journal of Sport Nutrition and Exercise Metabolism* 15:1–14.

Fahey, T., et al. 1991. The effects of ingesting polylactate or glucose polymer drinks during prolonged exercise. *International Journal of Sport Nutrition* 1:249–56.

Fairchild, T., et al. 2002. Rapid carbohydrate loading after a short bout of near maximal-intensity exercise. *Medicine & Science in Sports & Exercise* 34:980–86.

Febbraio, M. 2001. Alterations in energy metabolism during exercise and heat stress. *Sports Medicine* 31:47–59.

Fehm, H., et al. 2006. The selfish brain: Competition for energy resources. *Progress in Brain Research* 153:129–40.

Ferguson-Stegall, L., et al. 2010. The effect of a low carbohydrate beverage with added protein on cycling endurance performance in trained athletes. *Journal of Strength & Conditioning Research* 24 (10): 2577–86.

Fitch, C., et al. 2012. Use of nutritive and non-nutritive sweeteners. *Journal of the Academy of Nutrition and Dietetics* 112:739–58.

Fogelholm, M., et al. 1991. Carbohydrate loading in practice: High muscle glycogen concentration is not certain. *British Journal of Sports Medicine* 25:41-44.

Food and Drug Administration. 2006. Artificial sweeteners: No calories . . . Sweet! *FDA Consumer* 40 (4):27-28.

Forgac, M. 1979. Carbohydrate loading: A review. *Journal of the American Dietetic Association* 75:42-5.

Fung, T., et al. 2009. Sweetened beverage consumption and risk of coronary heart disease in women. *American Journal of Clinical Nutrition* 89:1037-42.

Gallus, S., et al. 2007. Artificial sweeteners and cancer risk in a network of case-control studies. *Annals of Oncology* 18:40-44.

Gilson, S., et al. 2010. Effects of chocolate milk consumption on markers of muscle recovery following soccer training: A randomized cross-over study. *Journal of the International Society of Sports Nutrition* 7:19.

Gleeson, M. 2006. Can nutrition limit exercise-induced immunodepression? *Nutrition Reviews* 64:119-31.

Gleeson, M., et al. 2004. Exercise, nutrition and immune function. *Journal of Sports Sciences* 22:115-25.

Grotz, V., and Munro, I. 2009. An overview of the safety of sucralose. *Regulatory Toxicology & Pharmacology* 55:1-5.

Guan, Z., et al. 2021. Soluble dietary fiber, one of the most important nutrients for the gut microbiota. *Molecules* 26:6802.

Guandalini, S., and Assiri, A. 2014. Celiac disease: A review. *Journal of the American Medical Association Pediatrics* 168:272-78.

Haff, G., et al. 1999. The effect of carbohydrate supplementation on multiple sessions and bouts of resistance exercise. *Journal of Strength and Conditioning Research* 13:111-17.

Haff, G., et al. 2001. The effects of supplemental carbohydrate ingestion on intermittent isokinetic leg exercise. *Journal of Sports Medicine and Physical Fitness* 41:216-22.

Haff, G., et al. 2003. Carbohydrate supplementation and resistance training. *Journal of Strength and Conditioning Research* 17:187-96.

Hargreaves, M., et al. 1995. Influence of muscle glycogen on glycogenolysis and glucose uptake during exercise in humans. *Journal of Applied Physiology* 78:288-92.

Harrington, V., et al. 2022. Interactions of non-nutritive artificial sweeteners with the microbiome in metabolic syndrome. *Immunometabolism* 4:e220012.

Hashimoto, T., and Brooks, G. 2008. Mitochondrial lactate oxidation complex and an adaptive role for lactate production. *Medicine & Science in Sports & Exercise* 40:486-94.

Hatfield, D., et al. 2006. The effects of carbohydrate loading on repetitive jump squat power performance. *Journal of Strength Conditioning & Research* 20:167-71.

Hawley, J. A. 2014. Manipulating carbohydrate availability to promote training adaptation. *Gatorade Sports Science Exchange* 27:1-7.

Hawley, J., et al. 1997. Carbohydrate-loading and exercise performance: An update. *Sports Medicine* 24:73-81.

Hellsten, Y., et al. 2004. Effect of ribose supplementation on resynthesis of adenine nucleotides after intense intermittent training in humans. *American Journal of Physiology. Regulatory, Integrative, & Comparative Physiology* 286:R182-88.

Henselmans, M., et al. 2022. The effect of carbohydrate intake on strength and resistance training performance: A systematic review. *Nutrients* 14:856.

Holloszy, J. 2005. Exercise-induced increase in muscle insulin sensitivity. *Journal of Applied Physiology* 99:338-43.

Holloszy, J., et al. 1998. The regulation of carbohydrate and fat metabolism during and after exercise. *Frontiers in Science* 3:D1011-27.

Jalal, D., et al. Increased fructose associates with elevated blood pressure. *Journal of the American Society of Nephrology* 21:1543-49.

James, A., et al. 2001. Muscle glycogen supercompensation: Absence of a gender-related difference. *European Journal of Applied Physiology* 85:533-58.

Jayedi, A. et al., 2020. Dietary glycemic index, glycemic load, and chronic disease: An umbrella review of meta-analyses of prospective cohort studies. *Critical Reviews in Food Science in Nutrition* 62:2460-69.

Jensen, M., et al. 2015. Carbohydrate mouth rinse counters fatigue related strength reduction. *International Journal of Sport Nutrition and Exercise Metabolism* 25:252-61.

Jentjens, R., et al. 2006. Exogenous carbohydrate oxidation rates are elevated after combined ingestion of glucose and fructose during exercise in the heat. *Journal of Applied Physiology* 100:807-16.

Jeukendrup, A. 2004. Carbohydrate intake during exercise and performance. *Nutrition* 20:669-77.

Jeukendrup, A. 2007. Carbohydrate supplementation during exercise: Does it help? How much is too much? *Sports Science Exchange* 20 (3):1-6.

Jeukendrup, A. 2014. Carbohydrate ingestion during exercise. In *Sports Nutrition. The Encyclopedia of Sports Medicine,* ed. R. J. Maughan. Oxford, UK: Wiley Blackwell.

Jeukendrup, A., and Jentjens, R. 2000. Oxidation of carbohydrate feedings during prolonged exercise: Current thoughts, guidelines and directions for future research. *Sports Medicine* 29:407-24.

Jeukendrup, A., and Killer, S. 2010. The myths surrounding pre-exercise carbohydrate feeding. *Annals of Nutrition & Metabolism* 57 (Supplement 2):18-25.

Jeukendrup, A., et al. 1997. Carbohydrate-electrolyte feedings improve 1-h time trial cycling performance. *International Journal of Sports Medicine* 18:125-29.

Jeukendrup, A., et al. 2006. Exogenous carbohydrate oxidation during ultraendurance exercise. *Journal of Applied Physiology* 100:1134-41.

Jeukendrup, A., et al. 2013. Multiple Transportable Carbohydrates and Their Benefits. *Gatorade Sports Science Exchange.* 26:1-8.

Johnson, R., et al. 2009. Dietary sugars intake and cardiovascular health: A scientific statement from the American Heart Association. *Circulation* 120:1011-20.

Kalman, D., et al. 1999. The effects of pyruvate supplementation on body composition in overweight individuals. *Nutrition* 15:337-40.

Kang, J., et al. 1995. Effect of carbohydrate ingestion subsequent to carbohydrate supercompensation on endurance performance. *International Journal of Sport Nutrition* 5:329-43.

Karelis, A., et al. 2010. Carbohydrate administration and exercise performance: What are the mechanisms involved? *Sports Medicine* 40 (9):747-63.

Karp, J., et al. 2006. Chocolate milk as a post-exercise recovery aid. *International Journal of Sport Nutrition & Exercise Metabolism* 16:78-91.

Kavouras, S., et al. 2004. The influence of low versus high carbohydrate diet on a 45-min strenuous cycling exercise. *International Journal of Sport Nutrition & Exercise Metabolism* 14:62-72.

Keim, N., et al. 2006. Carbohydrates. In *Modern Nutrition in Health and Disease,* eds. M. Shils, et al. Philadelphia: Lippincott Williams & Wilkins.

Kerksick, C., et al. 2005. Effects of ribose supplementation prior to and during intense exercise on anaerobic capacity and metabolic markers. *International Journal of Sport Nutrition & Exercise Metabolism* 15:653-64.

Kerksick, C. M., et al. 2017. International society of sports nutrition position stand: Nutrient timing. *Journal of the International Society of Sports Nutrition* 14:33.

Kim, Y., and Je, Y. 2014. Dietary fiber intake and total mortality: A meta-analysis of prospective cohort studies. *American Journal of Epidemiology* 180:565-73.

Kimber, N., et al. 2002. Energy balance during an Ironman triathlon in male and female triathletes. *International Journal of Sport Nutrition & Exercise Metabolism* 12:47-62.

Kirkendall, D. 2004. Creatine, carbs, and fluids: How important in soccer nutrition? *Sports Science Exchange* 17 (3):1-6.

Kjaer, M. 1998. Hepatic glucose production during exercise. *Advances in Experimental Medicine & Biology* 441:117-27.

Koh-Banerjee, P., et al. 2005. Effects of calcium pyruvate supplementation during training on body composition, exercise capacity, and metabolic responses to exercise. *Nutrition* 21:312-19.

Kovacs, M. 2006. Carbohydrate intake and tennis: Are there benefits? *British Journal of Sports Medicine* 40 (5):e13.

Kreider, R. 2003. Effects of oral D-ribose supplementation on anaerobic capacity and selected metabolic markers in healthy males. *International Journal of Sports Nutrition and Exercise Metabolism* 13:76-86.

Kreider, R., et al. 2010. International Society of Sports Nutrition exercise and sport nutrition review: Research and recommendations. *Journal of the International Society of Sports Nutrition* 7:7.

Kreitzman, S., et al. 1992. Glycogen storage: Illusions of easy weight loss, excessive weight regain, and distortions in estimates of body composition. *American Journal of Clinical Nutrition* 56:292S-93S.

Krustrup, P., et al. 2004. Slow-twitch fiber glycogen depletion elevates moderate-exercise fast-twitch fiber activity and O_2 uptake. *Medicine & Science in Sports & Exercise* 36:973-82.

Krustrup, P., et al. 2006. Muscle and blood metabolites during a soccer game: Implications for sprint performance. *Medicine & Science in Sports & Exercise* 38:1165-74.

Kuipers, H., et al. 1999. Pre-exercise ingestion of carbohydrate and transient hypoglycemia during exercise. *International Journal of Sports Medicine* 20:227-31.

Larsson, S., et al. 2006. Consumption of sugar and sugar-sweetened foods and the risk of pancreatic cancer in a prospective study. *American Journal of Clinical Nutrition* I84:1171-76.

Liebman, B. 1998. Sugar: The sweetening of the American diet. *Nutrition Action Health Letter* 25 (9):1-8.

Liebman, B. 2008. Fiber free-for-all: Not all fibers are equal. *Nutrition Action Health Letter* 35 (6):1-7.

Lim, U., et al. 2006. Consumption of aspartame-containing beverages and incidence of hematopoietic and brain malignancies. *Cancer Epidemiology, Biomarkers & Prevention* 15:1654-59.

Lis, D., et al. 2015. Exploring the popularity, experiences, and beliefs surrounding gluten-free diets in nonceliac athletes. *International Journal of Sport Nutrition and Exercise Metabolism* 25: 37-45.

Lis, D., et al. 2015. No effect of a short-term gluten-free diet on performance in nonceliac athletes. *Medicine and Science in Sports and Exercise* 47: 2563-70.

Lis, D., et al. 2016. Case study: Utilizing a low FODMAP diet to combat exercise-induced gastrointestinal symptoms. *International Journal of Sport Nutrition and Exercise Metabolism* 26:481-87.

Lupton, J., and Trumbo, P. 2006. Dietary fiber. In *Modern Nutrition in Health and Disease,* eds. M. Shils, et al. Philadelphia: Lippincott Williams & Wilkins.

Ma, Y., et al. 2005. Association between dietary carbohydrates and body weight. *American Journal of Epidemiology* 161:359-67.

MacGillivray, S., et al. 2013. Lactose avoidance for young children with acute diarrhoea. *Cochrane Database of Systematic Reviews.* Oct 31;10:CD005433.

Magnuson, B., et al. 2007. Aspartame: A safety evaluation based on current use levels, regulations, and toxicological and epidemiological studies. *Critical Reviews in Toxicology* 37:629-727.

Marjerrison, A., et al. 2007. Preexercise carbohydrate consumption and repeated anaerobic performance in pre- and early-pubertal boys. *International Journal of Sport Nutrition & Exercise Metabolism* 17:140-51.

Marlett, J., et al. 2002. Position of the American Dietetic Association: Health implications of dietary fiber. *Journal of the American Dietetic Association* 102:993-1010.

Mathew, T., et al. 2002. Relationship between sports drinks and dental erosion in 304 university athletes. *Caries Research* 36:281-87.

Matsui, T., et al. 2011. Brain glycogen decreases during prolonged exercise. *Journal of Physiology* 589:3383-93.

Maughan, R., and Burke, L. 2015. Carbohydrates. In *Nutritional Supplements in Sport, Exercise, and Health. An A-Z Guide,* ed. L. M. Castell, S. J. Stear, and L. M. Burke. New York: Routledge.

Maughan, R., et al. 1997. Diet composition and the performance of high-intensity exercise. *Journal of Sports Sciences* 15:265-75.

McCleave, E., et al. 2011. A low carbohydrate-protein supplement improves endurance performance in female athletes. *Journal of Strength & Conditioning Research* 25 (4):879-88.

McInerney, P., et al. 2005. Failure to repeatedly supercompensate muscle glycogen stores in highly trained men. *Medicine & Science in Sports & Exercise* 37:404-11.

McLay, R., et al. 2007. Carbohydrate loading and female endurance athletes: Effect of menstrual-cycle phase. *International Journal of Sport Nutrition & Exercise Metabolism* 17:189-205.

Meeusen, R. 2014. Exercise, nutrition and the brain. *Sports Medicine* 44:S47-S56.

Morris, D. 2012. Effects of oral lactate consumption on metabolism and exercise performance. *Current Sports Medicine Reports* 11:185-88.

Morrison, M., et al. 2000. Pyruvate ingestion for 7 days does not improve aerobic performance in well-trained individuals. *Journal of Applied Physiology* 89:549-56.

Murray, R., et al. 1989. The effects of glucose, fructose, and sucrose ingestion during exercise. *Medicine & Science in Sports & Exercise* 21:275-82.

National Institutes of Health. 2001. Diagnosis and management of dental caries throughout life. *NIH Consensus Statement* 18(1):1-23.

Nieman, D. 2007. Marathon training and immune function. *Sports Medicine* 37:412-15.

Nieman, D., and Bishop, N. 2006. Nutritional strategies to counter stress to the immune system in athletes, with special reference to football. *Journal of Sports Sciences* 24:763-72.

Nilsson L., and Hultman, E. 1974. Liver and muscle glycogen in man after glucose and fructose infusion. *Scandinavian Journal of Laboratory and Clinical Investigation* 33:5-10.

O'Reilly, J., et al. 2010. Glycaemic index, glycaemic load and exercise performance. *Sports Medicine* 40:27-39.

Onakpoya, I. 2014. Pyruvate supplementation for weight loss: A systematic review and meta analysis of randomized clinical trials. *Critical Reviews in Food Science and Nutrition* 54:17-23.

Op 'T Eijnde, B., et al. 2001. No effects of oral ribose supplementation on repeated maximal exercise and de novo ATP resynthesis. *Journal of Applied Physiology* 91:2275-81.

Ormsbee, M., et al. 2014. Pre-exercise nutrition: The role of macronutrients, modified starches and supplements on metabolism and endurance performance. *Nutrients* 6:1782-808.

Ørtenblad, N., et al. 2013. Muscle glycogen stores and fatigue. *Journal of Physiology* 15:4405-13.

Ostojic, S., and Ahmetovic, Z. 2009. The effect of 4 weeks treatment with a 2-gram daily dose of pyruvate on body composition in healthy trained men. *International Journal for Vitamin & Nutrition Research* 79 (3):173-79.

Painelli, V., et al. 2011. The effect of carbohydrate mouth rinse on maximal strength and strength endurance. *European Journal of Applied Physiology* 111:2381-86.

Palmer, G., et al. 1998. Carbohydrate ingestion immediately before exercise does not improve 20 km time trial performance in well-trained cyclists. *International Journal of Sports Medicine* 19:415-18.

Pasiakos, S. M., et al. 2021. Challenging traditional carbohydrate intake recommendations for optimizing performance at high altitude. *Current Opinions in Clinical Nutrition and Metabolic Care* 24:483-89.

Paul, D., et al. 2001. Carbohydrate during the follicular phase of the menstrual cycle: Effects on muscle glycogen and exercise performance. *International Journal of Sport Nutrition & Exercise Metabolism* 11:430-31.

Peters, H., et al. 1995. Exercise performance as a function of semi-solid and liquid carbohydrate feedings during prolonged exercise. *International Journal of Sports Medicine* 16:105-13.

Peveler, W., et al. 2006. Effects of ribose as an ergogenic aid. *Journal of Strength & Conditioning Research* 20:519-22.

Pfeiffer, B., et al. 2010. CHO oxidation from a CHO gel compared with a drink during exercise. *Medicine & Science in Sports & Exercise* 42 (11):2038-45.

Phillips, S., et al. 2014. The influence of serial carbohydrate mouth rinsing on power output during a cycling sprint. *Sports Science Medicine* 13:252-8.

Pi-Sunyer, F. 2002. Glycemic index and disease. *American Journal of Clinical Nutrition* 76:290S-98S.

Pizza, F., et al. 1995. A carbohydrate loading regimen improves high intensity, short duration exercise performance. *International Journal of Sport Nutrition* 5:110-16.

Pliml, W., et al. 1992. Effects of ribose on exercise-induced ischaemia in stable coronary artery disease. *Lancet* 340:507-10.

Pritchett, K., et al. 2009. Acute effects of chocolate milk and a commercial recovery beverage on postexercise recovery indices and endurance cycling performance. *Applied Physiology, Nutrition, & Metabolism* 34:1017-22.

Rauch, H., et al. 2005. A signalling role for muscle glycogen in the regulation of pace during prolonged exercise. *British Journal of Sports Medicine* 39:34-38.

Rawson E., et al. 2018. Dietary supplements for health, adaptation, and recovery in athletes. *International Journal of Sport Nutrition & Exercise Metabolism* 28:188-199.

Richter, E., et al. 2001. Regulation of muscle glucose transport during exercise. *International Journal of Sport Nutrition & Exercise Metabolism* 11:S71-77.

Rico-Sanz, J., et al. 1999. Muscle glycogen degradation during simulation of a fatiguing soccer match in elite soccer players examined noninvasively by 13C-MRS. *Medicine & Science in Sports & Exercise* 31:1587-93.

Rock, C. 2007. Primary dietary prevention: Is the fiber story over? *Recent Results in Cancer Research* 173:171-77.

Romano-Ely, B., et al. 2006. Effect of an isocaloric carbohydrate-protein-antioxidant drink on cycling performance. *Medicine & Science in Sports & Exercise* 38:1608-16.

Roy, B. 2008. Milk: The new sports drink? A review. *Journal of the International Society of Sports Nutrition* 5:15.

Saris, W. 1989. Study of food intake and energy expenditure during extreme sustained exercise: The Tour de France. *International Journal of Sports Medicine* 10:S26-S31.

Savaiano, D., et al. 2006. Lactose intolerance symptoms assessed by meta-analysis: A grain of truth that leads to exaggeration. *Journal of Nutrition* 136:1107-13.

Sawka, M., and Young, A. 2006. Physiological systems and their responses to conditions of heat and cold. In *ACSM's Advanced Exercise Physiology,* ed. C. M. Tipton. Philadelphia: Lippincott Williams & Wilkins.

Schneeman, B. 1999. Building scientific consensus: The importance of dietary fiber. *American Journal of Clinical Nutrition* 69:30-42.

Seal, C. 2006. Whole grains and CVD risk. *Proceedings of the Nutrition Society* 65:24-34.

Seifert, J., et al. 1994. Glycemic and insulinemic response to preexercise carbohydrate feedings. *International Journal of Sport Nutrition* 4:46-53.

Seymour, G., et al. 2007. Relationship between periodontal infections and systemic disease. *Clinical Microbiology & Infection* 13 (Supplement 4):3-10.

Short, S. 1993. Surveys of dietary intake and nutrition knowledge of athletes and their coaches. In *Nutrition in Exercise and Sport,* ed. I. Wolinsky and J. Hickson. Boca Raton, FL: CRC Press.

Shulman, R., and Rothman, D. 2001. The "glycogen shunt" in exercising muscle: A role of glycogen in muscle energetics and fatigue. *Proceedings of the National Academy of Sciences* 98:457-61.

Silva, F. M., et al. 2013. Fiber intake and glycemic control in patients with type 2 diabetes mellitus: A systematic review with meta-analysis of randomized controlled trials. *Nutrition Reviews* 71:790-801.

Smith, G., et al. 2002. The effect of pre-exercise glucose ingestion on performance during prolonged swimming. *International Journal of Sport Nutrition & Exercise Metabolism* 12:136-44.

Spriet, L. 2007. Regulation of substrate use during the marathon. *Sports Medicine* 37:332-36.

Spriet, L. 2015. Dihydroxyacetone phosphate and pyruvate. In *Nutritional Supplements in Sport, Exercise, and Health: An A-Z Guide.* Eds: L. M. Castell, S. J. Stear, and L. M. Burke. New York: Routledge.

Stanko, R., et al. 1990. Enhanced leg exercise endurance with a high-carbohydrate diet and dihydroxyacetone and pyruvate. *Journal of Applied Physiology* 69:1651-56.

Stanko, R., et al. 1990. Enhancement of arm exercise endurance capacity with dihydroxyacetone and pyruvate. *Journal of Applied Physiology* 68:119-24.

Stattin, P., et al. 2007. Prospective study of hyperglycemia and cancer risk. *Diabetes Care* 30:561-67.

Stearns, R., et al. 2010. Effects of ingesting protein in combination with carbohydrate during exercise on endurance performance: A systematic review with meta-analysis. *Journal of Strength & Conditioning Research* 24 (8):2192-202.

Stellingwerff, T., and Cox, G. 2014. Systematic review: Carbohydrate supplementation on exercise performance or capacity of varying durations. *Applied Physiology Nutrition and Metabolism* 39:998-1011.

Stellingwerff, T., et al. 2019. Nutrition and altitude: Strategies to enhance adaptation, improve performance and maintain health: A narrative review. *Sports Medicine* 49:S169-S184.

Tarnopolsky, M. 2000. Gender differences in substrate metabolism during endurance exercise. *Canadian Journal of Applied Physiology* 25:312-27.

Tarnopolsky, M. 2008. Sex differences in exercise metabolism and the role of 17-beta estradiol. *Medicine & Science in Sports & Exercise* 40:648-54.

Tarnopolsky, M., et al. 2001. Gender differences in carbohydrate loading are related to energy intake. *Journal of Applied Physiology* 91:225-30.

Temple, N. 2018. Fat, sugar, whole grains, and heart disease: 50 years of confusion. *Nutrients* 10:1-9.

Thomas, K., et al. 2009. Improved endurance capacity following chocolate milk consumption compared with 2 commercially available sport drinks. *Applied*

Threapleton, D., et al. 2013. Dietary fibre intake and risk of cardiovascular disease: Systematic review and meta-analysis. *British Medical Journal* 347:f6879.

Tsintzas, K., and Williams, C. 1998. Human muscle glycogen metabolism during exercise: Effect of carbohydrate supplementation. *Sports Medicine* 25:7-23.

Tsintzas, K., et al. 2001. Phosphocreatine degradation in type I and type II muscle fibers during submaximal exercise in man: Effect of carbohydrate ingestion. *Journal of Physiology* 537:305-11.

Tufts University. 2007. Five new reasons to get whole grains. *Health & Nutrition Letter* 25 (6):1-2.

Utter, A., et al. 2007. Carbohydrate attenuates perceived exertion during intermittent exercise and recovery. *Medicine and Science in Sports and Exercise* 39:880-5.

Utter, A., et al. 2005. Carbohydrate supplementation and perceived exertion during resistance exercise. *Journal of Strength and Conditioning Research* 19:939-43.

Utter, A., et al. 2004. Carbohydrate supplementation and perceived exertion during prolonged running. *Medicine and Science in Sports and Exercise* 36:1036-41.

Utter, A., et al. 2002. Effect of carbohydrate ingestion on ratings of perceived exertion during a marathon. *Medicine and Science in Sports and Exercise* 34:1779-84.

Van Hall, G. 2000. Lactate as fuel for mitochondrial regeneration. *Acta Physiologica Scandinavica* 168:643-56.

Physiology, Nutrition, & Metabolism 34:78-82.

van Loon, L. C. 2013. Is there a need for protein ingestion during exercise? *Gatorade Sports Science Exchange* 109:1-6.

van Loon, L., et al. 2001. The effects of increasing exercise intensity on muscle fuel utilisation in humans. *Journal of Physiology* 536:295-304.

Vandenberghe, K., et al. 1995. No effect of glycogen level on glycogen metabolism during high intensity exercise. *Medicine & Science in Sports & Exercise* 27:1278-83.

Venables, M., et al. 2005. Erosive effect of a new sports drink on dental enamel during exercise. *Medicine & Science in Sports & Exercise* 37:39-44.

von Fraunhofer, J., and Rogers, M. 2005. Effects of sports drinks and other beverages on dental enamel. *General Dentistry* 53:28-31.

Walberg-Rankin, J. 2000. Dietary carbohydrate and performance of brief, intense exercise. *Sports Science Exchange* 13 (4):1-4.

Walker, J., et al. 2000. Dietary carbohydrate, muscle glycogen content, and endurance performance in well-trained women. *Journal of Applied Physiology* 88:2151-58.

Wallis, G., et al. 2007. Dose-response effects of ingested carbohydrate on exercise metabolism in women. *Medicine & Science in Sports & Exercise* 39:131-38.

Welsh, R., et al. 2002. Carbohydrates and physical/mental performance during intermittent exercise to fatigue. *Medicine & Science in Sports & Exercise* 34:723-31.

Weltan, S., et al. 1998. Influence of muscle glycogen content on metabolic regulation. *American Journal of Physiology* 274:E72-E82.

Weltan, S., et al. 1998. Preexercise muscle glycogen content affects metabolism during exercise despite maintenance of hyperglycemia. *American Journal of Physiology* 274:E83-E88.

Williams, C. 2006. Nutrition to promote recovery from exercise. *Medicine & Science in Sports & Exercise* 19:1-6.

Williams, C., and Lamb, D. 2008. Do high carbohydrate diets improve exercise performance? *Sports Science Exchange* 21(1):1-6.

Wilk, K., et al. 2022. The effect of artificial sweeteners use on sweet taste perception and weight loss efficacy: A review. *Nutrients* 14:1261.

Winnick, J., et al. 2005. Carbohydrate feedings during team sport exercise preserve physical and CNS function. *Medicine & Science in Sports & Exercise* 37:306-15.

Wolfe, R., and George, S. 1993. Stable isotopic tracers as metabolic probes in exercise. *Exercise & Sports Science Reviews* 21:1-31.

Wong, J., et al. 2006. Colonic health: Fermentation and short chain fatty acids. *Journal of Clinical Gastroenterology* 40:235-43.

Yang, Y., et al. 2015. Association between dietary fiber and lower risk of all-cause mortality: A meta-analysis of cohort studies. *American Journal of Epidemiology* 181:83-91.

Yaspelkis, B., et al. 1993. Carbohydrate supplementation spares muscle glycogen during variable-intensity exercise. *Journal of Applied Physiology* 75:1477-85.

Zeng, H. 2014. Mechanisms linking dietary fiber, gut microbiota and colon cancer prevention. *World Journal of Gastrointestinal Oncology* 6:41-51.

Zierler, K. 1999. Whole body glucose metabolism. *American Journal of Physiology* 276:E409-26.

Fat: An Important Energy Source during Exercise

Floresco Productions/Punchstock/Corbis

LEARNING OUTCOMES

After studying this chapter, you should be able to:

1. List the different types of dietary fatty acids and identify general types of foods in which they are found.
2. Calculate the approximate amount, in grams or milligrams, of total fat, and saturated fat that should be included in your daily diet.
3. Describe how dietary fat is absorbed, how it is distributed in the body, and what its major functions are in human metabolism.
4. Explain the role of fat in human energy systems during exercise and how endurance exercise training affects exercise fat metabolism.
5. Explain the theory underlying the role of increased fat oxidation to enhance prolonged aerobic endurance performance.
6. List the various dietary fat strategies and dietary supplements that have been investigated as a means of enhancing exercise performance, and highlight the major findings.
7. Describe the proposed process underlying the development of atherosclerosis and cardiovascular disease, including the role that dietary fat and cholesterol may play in its etiology.
8. List and describe at least eight of the ten dietary strategies that are proposed to help treat or prevent the development of atherosclerosis and cardiovascular disease.
9. Explain the role of exercise as a means of helping prevent the development of atherosclerosis and cardiovascular disease.

KEY TERMS

adipokines
alpha-linolenic acid
angina
apolipoprotein
arteriosclerosis
atherosclerosis
beta-oxidation
carnitine
cholesterol
chylomicron
cis fatty acids
conjugated linoleic acid (CLA)
coronary artery disease (CAD)
coronary heart disease (CHD)
coronary occlusion

CHAPTER FIVE

coronary thrombosis
eicosanoids
ester
fat loading
fat substitutes
fatty acids
glycerol
High-density lipoprotein (HDL)
hidden fat
ischemia
ketoacidosis
ketogenic
ketones
ketosis
Low-density lipoprotein (LDL)
lecithin
linoleic acid
lipids
lipoprotein
lipoprotein (a)
medium-chain triglycerides (MCTs)
monounsaturated fatty acid
myocardial infarction
omega-3 fatty acids
omega-6 fatty acids
partially hydrogenated fats
phospholipids
plaque
polyunsaturated fatty acid
saturated fatty acid
trans fatty acids
triglycerides
VLDL

Introduction

From a health standpoint, dietary fat is the nutrient of greatest concern to the American Heart Association because excessive consumption of certain types of fat has been associated with the development of coronary heart disease. Excessive consumption of dietary fat may also contribute to the development of obesity, a risk factor for several other chronic diseases, such as diabetes. Thus, one of the major recommendations of the 2020-2025 Dietray Guidelines for Americans for a healthier diet is to reduce the amount of total and saturated dietary fat to a reasonable level. Part of the rationale for this recommendation and some general guidelines for implementing it are presented in this chapter, but additional information may also be found in chapters chapters 1 and 2.

We need some fat in our diet. Despite its potential health hazards when consumed in excess, dietary fat contains several essential nutrients that serve a variety of important functions in human nutrition. The National Academy of Sciences has set an Acceptable Macronutrient Distribution Range (AMDR) for total fat of 20-35 percent of daily energy intake. Because some types of fat may confer various health benefits, the Academy also set Adequate Intakes (AI) for several fatty acids. Moreover, in its position statement regarding nutrition and athletic performance, the Academy of Nutrition and Dietetics, Dietitians of Canada, and the American College of Sports Medicine noted that, overall, diets should provide moderate amounts of energy from fat (20-35 percent of energy) and that there is no health or performance benefit to consuming a diet containing less than 20 percent energy from fat. For the endurance athlete, one of the most important functions of fat is to provide energy during exercise, and researchers have explored a variety of techniques in attempts to improve endurance performance by increasing the ability of the muscle to use fat as a fuel.

To clarify the role of dietary fat in health and its possible relevance to sports, this chapter presents information on the basic nature of dietary fats and associated lipids, the metabolic fate and physiological functions of fats and cholesterol in the body, the role of fat as an energy source during exercise, the use of various dietary practices or ergogenic aids in attempts to improve fat metabolism and endurance performance, and possible health problems associated with excessive dietary fat.

Dietary Fats

What are the different types of dietary fats?

What we commonly call fat in our diet actually consists of several substances classified as lipids. **Lipids** represent a class of organic substances that are insoluble in water but soluble in certain solvents such as alcohol or ether. The three major dietary lipids of importance to humans are triglycerides, cholesterol, and phospholipids. All three have major functions in the body.

What are triglycerides?

The **triglycerides**, also known as the true fats or the neutral fats, are the principal form in which fats are eaten and stored in the human body. Triglycerides are composed of two different compounds—fatty acids and glycerol. When an acid (fatty acid) and an alcohol (glycerol) combine, an **ester** is formed, the process being known as *esterification*. (Three fatty acids are attached to each glycerol molecule.) **Figure 5.1** is a diagram of a triglyceride. Another term used for triglyceride is *triacylglycerol*. Some triglycerides may be modified commercially to contain only two fatty acids, known as diglycerides, or diacylglycerols.

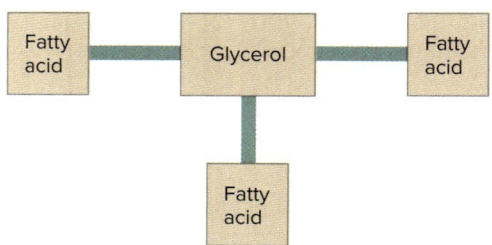

FIGURE 5.1 Structure of a triglyceride. Three fatty acids combine with glycerol to form a triglyceride.

Fatty acids, one of the components of fat, are chains of carbon, oxygen, and hydrogen atoms that vary in length and in the degree of saturation of carbon with hydrogen. Short-chain fatty acids (SCFAs) contain fewer than 6 carbons, medium-chain fatty acids (MCFAs) have 6-12 carbons, and long-chain fatty acids (LCFAs) have 14 or more carbons.

Fatty acids may be saturated or unsaturated. A **saturated fatty acid** contains single bonds between adjacent carbon atoms and the maximum number of hydrogen atoms that could be bound to the carbon atoms; saturated fats such as butter are solid at room temperature. Carbon molecules in unsaturated fatty acids may incorporate more hydrogen because they have some unfilled bonds, or double bonds. These unsaturated fatty acids may be classified as **monounsaturated**, having a single double bond and capable of incorporating two hydrogen ions, and **polyunsaturated**, having two or more double bonds and capable of incorporating four or more hydrogen ions; monounsaturated and polyunsaturated fats such as oils are liquid at room temperature.

Polyunsaturated fatty acids are further identified according to the location of the first carbon double bond from the last, or omega, carbon. **Omega-3** and **omega-6 fatty acids** are the two major types, and the numeric represents the location of the first double bond. Other terminology to identify these two fatty acids are ω-3 and n-3 and ω-6 and n-6. Adequate amounts of omega-3 and omega-6 fatty acids may confer health benefits, as noted later in the chapter. At room temperature, saturated fats are usually solid, while unsaturated fats are usually liquid. **Partially hydrogenated fats** or oils have been treated by a process that adds hydrogen to some of the unfilled bonds, thereby hardening the fat or oil. In essence, the fat becomes more saturated. During the hydrogenation process, the normal position of hydrogen ions at the double bond is on the "same side" (*cis*). This is known as a *cis* **fatty acid**, but if partially hydrogenated so that hydrogen ions are on opposite sides of the double bond, it results in a *trans* **fatty acid**. Figure 5.2 represents the structural difference among a saturated, a monounsaturated, a polyunsaturated (*cis* and *trans*), and an omega-3 polyunsaturated fatty acid. The health implications of these different types of fats are discussed later in this chapter. In general, though, excess intake of saturated and *trans* fatty acids is associated with increased health risks, whereas adequate intake of monounsaturated, polyunsaturated, and omega-3 fatty acids may be associated with neutral or some beneficial health effects.

Glycerol is an alcohol, a clear, colorless, syrupy liquid. It is obtained in the diet as part of triglycerides, but it also may be produced in the body as a by-product of carbohydrate metabolism. On the other hand, glycerol can be converted back to carbohydrate in the process of gluconeogenesis in the liver.

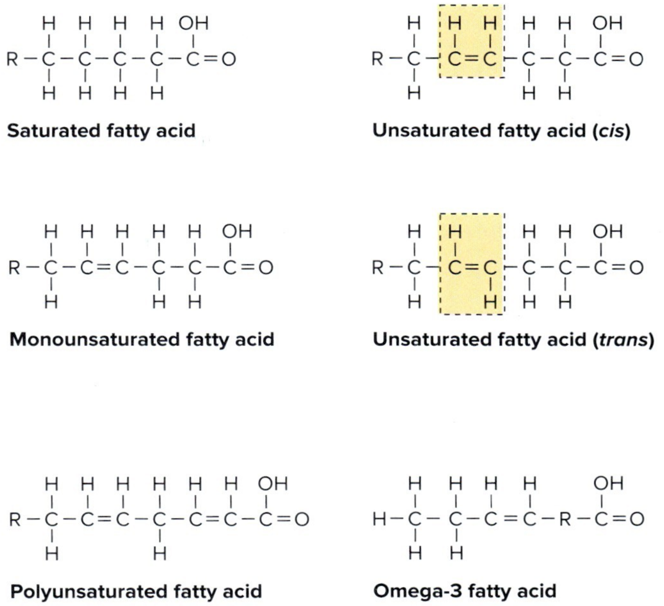

What are some common foods high in fat content?

The fat content in foods can vary from 100 percent, as found in most cooking oils, to minor trace amounts, less than 5–10 percent, as found in most fruits and vegetables. Some foods obviously have a high-fat content: butter, oils, shortening, mayonnaise, margarine, and the visible fat on meat. However, in other foods, the fat content may be high but not as obvious. This is known as **hidden fat**. Whole milk, cheese, nuts, desserts, crackers, potato chips, and a wide variety of commercially prepared foods may contain considerable amounts of hidden fat. For example, a 5-ounce baked potato contains 145 kcal with about 3 percent fat, while a 5-ounce serving of potato chips contains 795 kcal, more than 60 percent of them from fat.

In general, animal foods tend to contain more total fat and saturated fat than plant foods. Hamburger meat contributes the most

FIGURE 5.2 Structural differences between saturated, monounsaturated, and polyunsaturated fatty acids (including the *cis* and *trans* forms) and the omega-3 fatty acid. Note there is a single double bond between carbon atoms in the monounsaturated fatty acid and two or more in the polyunsaturated fatty acid. In the omega-3 fatty acid, the double bond is located three carbons from the last, or omega, carbon. The R represents the radical, or the presence of many more C—H bonds. In the *trans* configuration, one of the hydrogen ions is moved to the opposite side.

saturated fat to the typical American diet. However, careful selection and preparation of foods in these groups will considerably reduce fat content. The percentage of fat in meat and dairy products may vary considerably; beef, pork, and cheese products usually contain considerable amounts of fat, up to 70 percent or more fat kcal. The meat and dairy industries are responding to dietary modifications by many Americans and are making low-fat red meats and low-fat cheeses available to consumers. For example, 3 ounces of beef eye of round or pork tenderloin contain about 140 kcal, 4 grams of total fat, and 1.5 grams of saturated fat; both cuts of meat contain fewer than 30 percent of their kcal as fat. Lean cuts of poultry and fish have much lower levels of fat. Trimming the fat from meats or removing the skin from poultry drastically reduces the fat content. Some fish, such as flounder and tuna, are remarkably low in fat, whereas others, such as salmon and mackerel, are higher in total fat content but contain greater amounts of omega-3 fatty acids. In the dairy group, whole cow's milk contains about 8 grams of fat per cup; skim milk contains about 0.5–1.0 gram, which is much less than whole milk. Dairy alternatives, such as almond, soy, or rice milk and related products, often contain less fat than cow's milk. However, the nutrient profile is different such that these alternatives often provide less calcium and vitamin D.

Small amounts of *trans* fatty acids are found naturally in beef, butter, and cow's milk, but deep-fried foods and commercially prepared products, particularly stick margarine and snack foods such as chips, cakes, and cookies, have contained substantial amounts of artificial trans fats. Today, artificial trans fats have been banned from food, although complete removal from the food supply takes many years. The health risks associated with *trans* fatty acids are discussed later in this chapter.

Most plant foods, such as vegetables, fruits, beans, and natural whole-grain products, generally are low in fat content, and the fat they do contain is mostly unsaturated. On the other hand, some plant foods, such as nuts, seeds, and avocados, are very high in fat, but again primarily unsaturated fats. However, coconuts and palm kernels are extremely high in both total and saturated fats.

All fats contain a mixture of saturated, monounsaturated, and polyunsaturated fatty acids. Later in this chapter, we discuss some health implications relating to the types of fats we eat; **figure 5.3** presents an approximate percentage of the amount of saturated, monounsaturated, and polyunsaturated fatty acids found in some common oils and fats. Several high-content sources for each of the various types of fatty acids are noted.

How do I calculate the percentage of fat kcal in a food?

It is important to realize that a product advertised as 95 percent fat free (or only 5 percent fat) may contain a considerably higher percentage of its kcal as fat: The advertised percentage refers to the *weight of the product, not its caloric content*. The product may contain a considerable amount of water, which contains no kcal. Thus, luncheon meat advertised as 95 percent fat free may actually contain more than 40 percent of its kcal from fat depending on the water weight. Foods with a high water content contain even higher percentages of fat kcal. A striking example is whole milk, which is only 3.5 percent fat by weight; however, one glass of milk contains about 150 kcal and 8 grams of fat, which accounts for 48 percent of the caloric content ($8 \times 9 = 72$; $72/150 = 48$). Even low-fat milk (2 percent fat) contains about 37 percent fat kcal.

If you want to calculate the percentage of fat kcal in most foods you eat, you can get the information you need from the food label, which will include the total kcal and the kcal from fat; additional mandatory information is the total fat, saturated fat, and *trans* fatty acid content. The label optionally may list the kcal from saturated fat and the amounts of polyunsaturated and monounsaturated fat. **Figure 5.4** presents a food label to help illustrate the hidden fat in a hot dog (wiener). If you do not have a food label, then you will need to know the kcal and grams

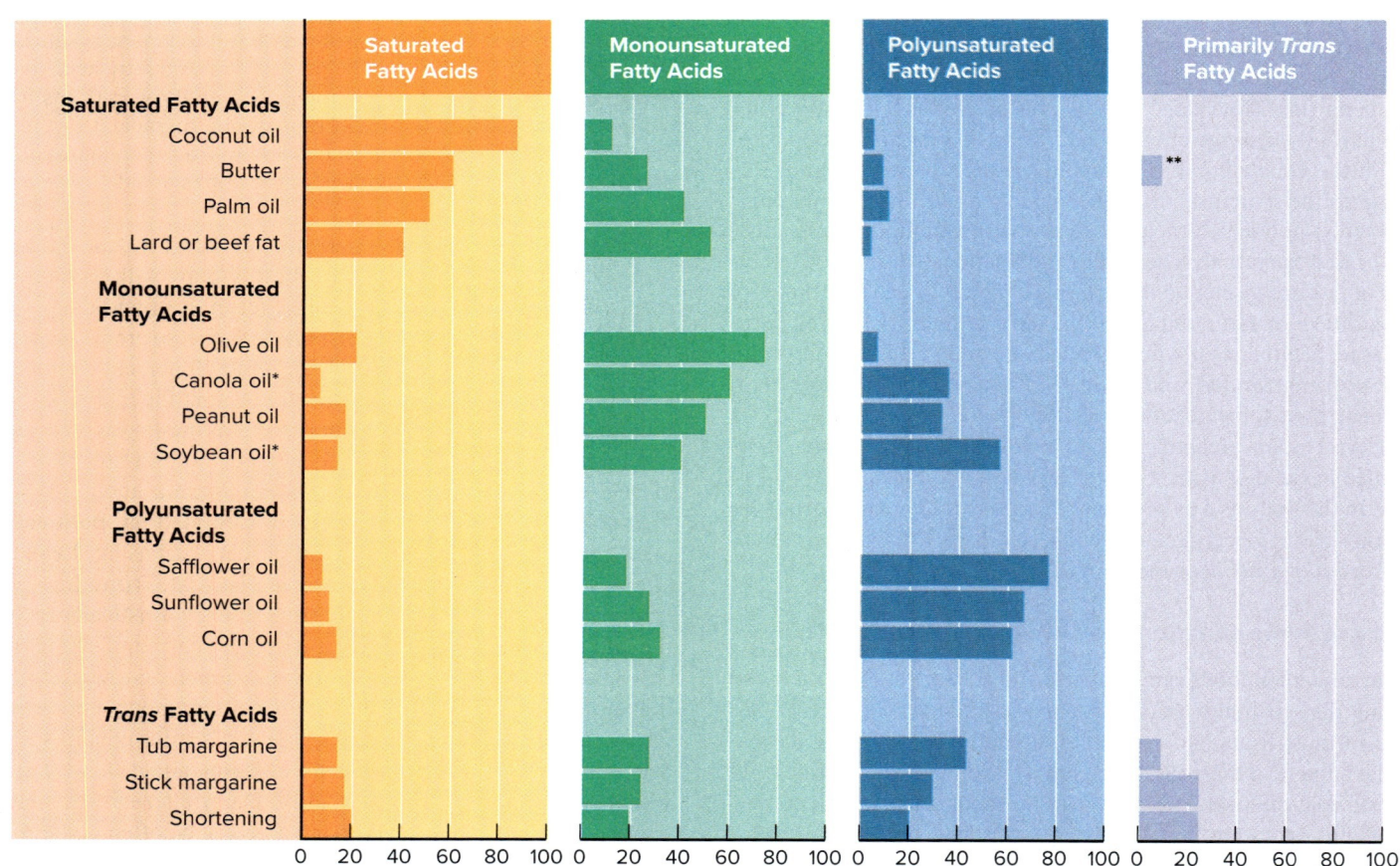

FIGURE 5.3 Saturated, monounsaturated, polyunsaturated, and *trans* fatty acid composition of common fats and oils (expressed as percent of all fatty acids in the product).

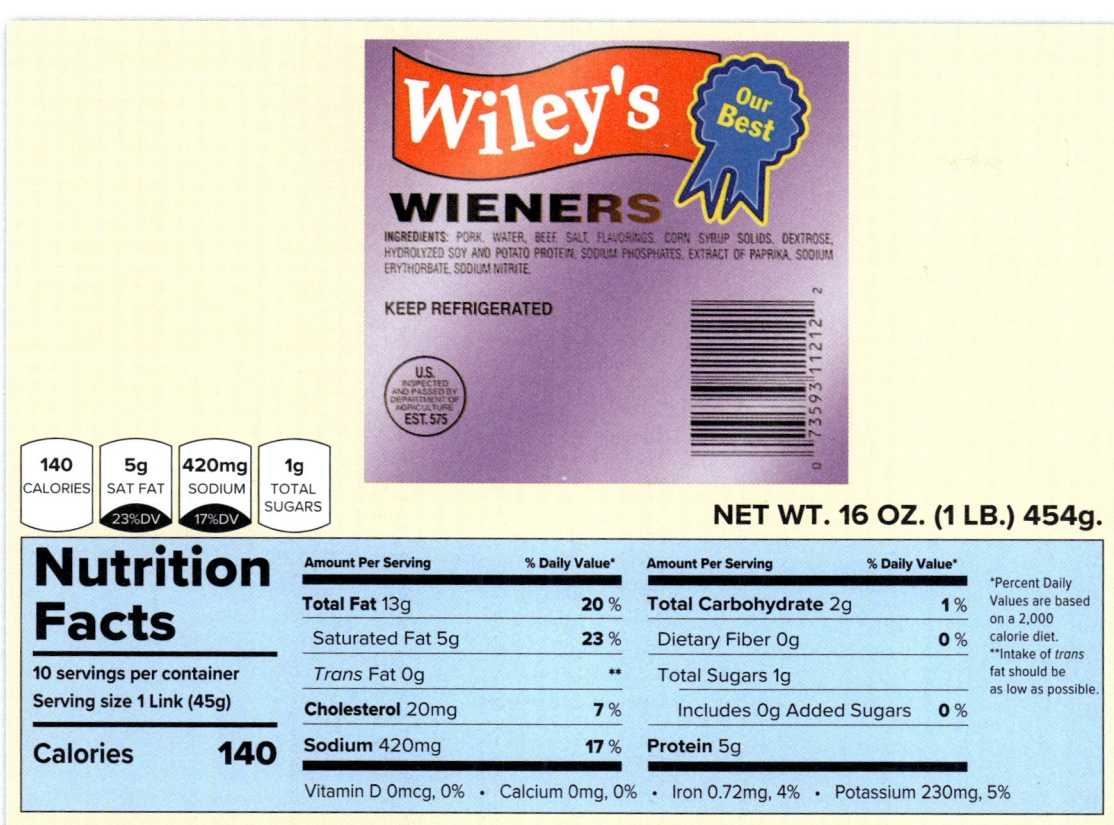

FIGURE 5.4 Reading labels helps locate hidden fat. Who would think that wieners (hot dogs) can contain about 85 percent of energy content as fat? Looking at the hot dog itself does not suggest that almost all its energy content comes from fat, but the label shows otherwise. Do the math: 120/140 kcal = 0.85, or 85 percent.

of fat for the food. Both of these values can be obtained from a food composition table found in most basic nutrition texts or online. Table 5.1 presents the methods for calculating the percent of fat kcal and percent of saturated fat kcal from either a food label or food composition table. Table 5.2 represents the percentage of food energy that is derived from fat in some common foods; the percentages are indicated for both total fat and saturated fat.

What are fat substitutes?

Fat substitutes, or fat replacers, are designed to provide the taste and texture of fats, but without the kcal (9 kcal per gram), saturated fat, or cholesterol. They are found in many normally high-fat products, such as ice cream, that are marketed as fat free. Fat substitutes may be manufactured from carbohydrate, protein, or fats. Although a number of fat substitutes are under development, the following are commonly used.

Some carbohydrates, such as starches and gums, provide thickness and structure and are useful as fat substitutes. Guar gum, gum arabic, and cellulose gel are examples. Oatrim®, made from oats, is being used to replace fat in milk. Depending on the form used, the caloric content may range from 0 to 4 kcal per gram. Simplesse® is manufactured from milk or egg protein by a microparticulation process so that it has the taste and texture of fat. The caloric value of Simplesse is only 1.3 kcal per gram. The use of Simplesse has been approved by the FDA. Salatrim®, which is an acronym for *short- and long-chain fatty*

TABLE 5.1 Calculation of the percentage of kcal in foods that are derived from fat

Method A. Data from food label

Amount per serving

kcal = 90
kcal from fat = 30
To calculate percentage of food kcal that consist of fat, simply divide the kcal from fat by the kcal per serving and multiply by 100 to express as a percent.
   30/90 = 0.33   0.33 × 100 = 33 percent fat kcal

Method B. Data from food composition table

Amount per serving

kcal = 90
Total fat, grams = 8
Saturated fat, grams = 3
To calculate percentage of food kcal that consist of total fat or saturated fat, use the caloric value for fat of 1 gram = 9 kcal.
Total fat = 8 grams
Total fat kcal = 8 grams × 9 kcal/gram = 72 kcal
Use the same procedure as in Method A.
   72/90 = 0.80   0.80 × 100 = 80 percent fat kcal
Saturated fat = 3 grams
Saturated fat kcal = 3 grams × 9 kcal/gram = 27 kcal
   27/90 = 0.30   0.30 × 100 = 30 percent saturated fat kcal

TABLE 5.2 Percentage of total fat kcal and saturated fat kcal in some common foods*

Food	% kcal total fat	% kcal saturated fat	Food	% kcal total fat	% kcal saturated fat
Protein foods			*Vegetables*		
Bacon	80	30	Broccoli	12	1.5
Beef, lean and fat (untrimmed)	70	32	Carrots	4	<1
Beef, lean only (trimmed)	35	15	Potatoes	1	<1
Hamburger, regular	62	29	*Fruits*		
Chicken, breast (with skin)	35	11	Apples	5	<1
Chicken, breast (without skin)	19	5	Bananas	5	2
Luncheon meat (bologna)	82	35	Oranges	4	<1
Salmon	37	7	*Grains*		
Flounder, tuna	8	2	Bread		
Egg, white and yolk	67	22	White	12	2
Egg, white	0	0	Whole wheat	12	2
Beans, dry, navy	4	<1	Crackers	30	12
Beans, navy, canned with pork	28	12	Doughnuts	43	7
Peanuts	77	17	Macaroni	5	<1
Peanut butter	76	19	Macaroni and cheese	46	20
Dairy			Oatmeal	13	2
Cow's milk, whole	45	28	Pancakes, wheat	30	7
Cow's milk, skim	5	2.5	Spaghetti	5	<1
Cheese, cheddar	74	47	*Fats and oils*		
Cheese, mozzarella, part skim	56	35	Butter	99	62
Ice cream	49	28	Lard	99	40
Greek yogurt, full fat	36	27	Margarine	99	21
Coconut milk	90	90	Oil, corn	100	13
Vegetables			Oil, coconut	100	87
Asparagus	8	2	Salad dressings		
Beans, green	7	1.5	French	95	15
			French, special dietary low fat	14	<1

<1 = less than 1 percent
*Percentages may vary. See food labels for specific information when available.

acid triglyceride molecule, is a modified fat containing only 5 kcal per gram. Olestra® is an ester of sucrose with long-chain fatty acids, a structure that cannot be hydrolyzed by digestive enzymes or absorbed by the gastrointestinal tract and therefore supplies no kcal to the body.

What is cholesterol?

Cholesterol is one of the lipids known as *sterols*. It is not a fat, but it is a fat-like, pearly substance found in animal tissues. Cholesterol is not an essential nutrient for humans because it is manufactured naturally in the liver from fatty acids and from the breakdown products of carbohydrate and protein, glucose and amino acids, respectively.

What foods contain cholesterol?

Cholesterol is found only in animal products and is not found in fruits, vegetables, nuts, grains, or other nonanimal foods. **Table 5.3** presents some foods from the protein foods and dairy groups with the cholesterol content in milligrams. Several foods from the grains group are also included, indicating that the preparation of some

TABLE 5.3	Cholesterol content, in milligrams, for some common foods	
	Amount	Cholesterol
Protein foods		
Beef, pork, ham	1 oz.	25
Poultry	1 oz.	23
Fish	1 oz.	21
Shrimp	1 oz.	45
Lobster	1 oz.	25
Eggs	1	220
Liver	1 oz.	120
Dairy		
Cow's milk, whole	1 cup	27
Cow's milk, 2%	1 cup	15
Cow's milk, skim	1 cup	7
Butter	1 tsp	12
Margarine	1 tsp	0
Cream cheese	1 tbsp	18
Ice milk	1 cup	10
Ice cream	1 cup	85
Grains		
Bread	1 slice	0
Biscuit	1	17
Pancake	1	40
Sweet roll	1	25
French toast	1 slice	130
Doughnut	1	28
Cereal, cooked	1 cup	0

Fruits, vegetables, and nuts have no cholesterol.

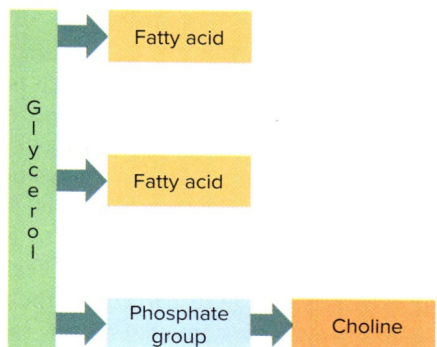

FIGURE 5.5 Simplified diagram of the phospholipid lecithin.

bread/cereal products may add cholesterol by including some animal product containing cholesterol, mainly eggs.

What are phospholipids?

Chemically, **phospholipids** are somewhat comparable to triglycerides. They have a glycerol base, one or two attached fatty acids, and an additional structure that contains a phosphate group. One of the most common phospholipids is **lecithin**, whose structure is depicted as a simple diagram in **figure 5.5**. Phospholipids are not essential nutrients, as the body can make them from triglycerides.

What foods contain phospholipids?

Egg yolks provide substantial amounts of lecithin, and other good sources include liver, wheat germ, and peanuts. However, lecithin may be degraded in the digestive tract to smaller constituents. Your body can make all of the phospholipids it needs. Because dietary phospholipids are not associated with any health risks, there is little concern with dietary intake.

How much fat and cholesterol do we need in the diet?

Dietary fat is most necessary for three reasons: to meet energy needs, to provide essential fatty acids, and to provide essential fat-soluble vitamins. As a concentrated source of energy, adequate intake of dietary fat is very important during the growth and development years. Dietary fats also provide several essential fatty acids, and the fat-soluble vitamins A, D, E, and K, without which various health problems would develop.

These factors, along with possible implications for health, were taken into consideration in the development of Dietary Reference Intakes for dietary fats, which may be found at: https://ods.od.nih.gov/HealthInformation/Dietary_Reference_Intakes.aspx. For the purpose of developing its DRI, the National Academy of Sciences classified fat into the following categories:

Total fat
Saturated fatty acids (SFAs)
Cis monounsaturated fatty acids (MUFAs)
Cis polyunsaturated fatty acids (PUFAs)
 n-6 fatty acids (omega-6)
 n-3 fatty acids (omega-3)
Trans fatty acids

The abbreviations for the various fatty acids, as well as the omega classification, will be used interchangeably with the respective fat or fatty acids.

Total Fat The Academy developed an AMDR of 20–35 percent of daily energy intake from total fat, which is an estimate based on adverse effects that may occur from consuming either a low-fat or high-fat diet. Individuals should obtain sufficient but not excessive amounts of dietary fat within this range of energy intake. No RDA, AI, or UL have been developed for total fat because of insufficient evidence, but diets higher than 35 percent dietary fat are not recommended because this could cause saturated fat intake to be increased beyond a healthful level.

Saturated Fatty Acids and *Trans* Fatty Acids The Academy did not develop an RDA, AI, or UL for saturated fat or *trans* fat. However, increased intake of both of these fats is associated with increased risk of coronary heart disease. Although the Academy notes that because most diets contain fats and because most fats contain a mixture of fatty acids, it is not possible to consume a diet devoid of saturated and *trans* fats. Nevertheless, the prevailing undertone in the DRI report is to minimize the dietary intake of these two types of fat. Other health organizations suggest a maximum of 7-10 percent of the daily energy intake be derived from the combination of saturated and *trans* fats.

Cis Monounsaturated Fatty Acids The Academy did not develop an RDA, AI, or UL for monounsaturated fats, indicating that they are not essential fatty acids because they may be synthesized by the body. About 20-40 percent of the fat we consume is monounsaturated, and primarily olive oil. Although no DRI has been set for monounsaturated fat, the Academy notes that they may have some benefit in the prevention of chronic disease. Olive oil is a staple in the Mediterranean diet and its alleged health benefits will be discussed later in this chapter.

Cis Polyunsaturated Fatty Acids The Academy developed an AI for polyunsaturated fatty acids because there may be some health benefits associated with such dietary intakes. AI were set for both omega-6 and omega-3 fatty acids.

Omega-6 fatty acids **Linoleic acid**, an essential omega-6 polyunsaturated fatty acid, must be supplied in the diet because the body cannot produce it from other fatty acids. The AI for adult males age 19-50 is 17 grams of linoleic acid daily, 12 g/day for females. For males and females age 51 and over, the AI are 14 and 11 grams daily, respectively. Somewhat smaller AI have been developed for children and adolescents age 9-18. Linoleic acid is found in vegetable and nut oils—such as corn, sunflower, peanut, and soy oils—that constitute food products such as margarine, salad dressings, and cooking oils. **Conjugated linoleic acid (CLA)**, an isomer of linoleic acid, has been suggested to be ergogenic and possess health benefits, which will be discussed in later sections of this chapter.

Omega-3 fatty acids **Alpha-linolenic acid**, an omega-3 polyunsaturated fatty acid, is also considered to be an essential fatty acid. The AI for adult males age 19 and older is 1.6 grams of alpha-linolenic acid daily, 1.1 g/day for females. Somewhat smaller AI have been developed for children and adolescents age 9-18. Alpha-linoleic acid is found in green leafy vegetables, canola oil, flaxseed oil, soy products, some nuts, and fish. The potential health benefits of omega-3 fatty acids, including several derived from fish oils (eicosapentaenoic acid, EPA; docosahexaenoic acid, DHA), are discussed later in this chapter.

Cholesterol Cholesterol is vital to human physiology in a variety of ways, so the body needs an adequate supply. Because cholesterol may be manufactured in the body from either fats, carbohydrate, or protein; however, there is apparently little need for us to obtain large amounts, if any, in the foods we eat. Also, because a positive relationship has been established between high blood cholesterol levels and coronary heart disease, reduction of dietary cholesterol has been advocated by a number of health-related associations.

The *2020-2025 Dietary Guidelines for Americans* do not have a quantitative limit for dietary cholesterol. However, the *Guidelines* state that people should eat as little dietary cholesterol as possible, because foods that are higher in cholesterol are often higher in saturated fats, which should be limited in the diet. Thus, even though the older recommendation about eating no more than 300 mg/day of cholesterol no longer exists, by following a diet low in saturated fats, cholesterol intake will likely be between 100 and 300 mg/day (depending on kcal intake).

Table 5.4 indicates the grams of fat and saturated fat and milligrams of cholesterol that may be consumed daily on a diet containing 30 percent of the kcal as fat and less than 300 milligrams of cholesterol. For the 30 percent recommendation, a very simple method to determine the grams of total fat you may consume on a given caloric diet is to simply drop the zero from the daily caloric total and divide by 3—for example,

$$2,100\text{-kcal diet}$$
$$2,100 = 210$$
$$210/3 = 70 \text{ grams total fat}$$

For lower percentages of fat kcal, say 20 percent, simply multiply the daily caloric intake by the percentage desired and divide by 9 to get the grams of fat allowed per day. For example, 20 percent of 2,500 kcal would permit 500 kcal from fat, or about 55 g/day.

Theoretically, high-fat diets may be deleterious to physical performance in several ways. The fat may displace carbohydrate in the diet, may lead to excessive caloric intake and increased body weight, and may cause gastrointestinal distress if consumed as part of a pregame meal. All of these factors could impair physical performance. On the other hand, some investigators have contended that high-fat diets may enhance exercise performance. These issues will be discussed later in this chapter.

TABLE 5.4	Daily allowance for grams of fat and saturated fat, and milligrams of cholesterol*			
Total kcal	Fat kcal	Grams of fat	Grams of saturated fat	mg of cholesterol
1,000	300	33	11	100
1,500	450	50	16	150
2,000	600	66	22	200
2,500	750	83	27	250
3,000	900	100	33	300

*Based on a diet containing 30 percent of kcal as fat with 100 milligrams of cholesterol per 1,000 kcal.

Key Concepts

▶ The three major lipids in human nutrition are triglycerides, cholesterol, and phospholipids.

▶ Triglycerides, which consist of fatty acids and glycerol, account for about 98 percent of the lipids we eat. Fatty acids may be saturated or unsaturated. Unsaturated fatty acids may be monounsaturated or polyunsaturated. Polyunsaturated fatty acids exist in two forms, *cis* and *trans*. Omega-3 fatty acids are also polyunsaturated.

▶ The fat content of foods varies considerably, but generally foods in the fruits, vegetables, and grains groups are good sources of unsaturated fats and are low in total fat, whereas the meat and milk food exchanges contain foods that may have a high total fat and saturated fat content.

▶ Cholesterol is a nonfat substance vital to human metabolism, and although it may be obtained in the diet only from animal foods, the body can produce its own supply from other dietary nutrients such as saturated fats.

▶ The AMDR for dietary fat is 20–35 percent of daily kcal intake. Although some fat is essential in the diet as a source of essential fatty acids (linoleic and alpha-linolenic) and the fat-soluble vitamins (A, D, E, K), these nutrients may be obtained from polyunsaturated fats. The total amount of saturated fats in the diet should be less than 10 percent, preferably 7 percent of daily kcal intake, while monounsaturated and polyunsaturated fats should constitute the majority of the AMDR.

Check for Yourself

▶ Peruse food labels of some of your favorite foods and check the fat content. Pay particular attention to percent Daily Value for total fat, saturated fat, and cholesterol. Check also for *trans* fat content.

Metabolism and Function

In this section we briefly cover the digestion of dietary lipids, their metabolic disposal in the body, interactions with carbohydrate and protein, the major functions of fats in the body, and energy stores of fat.

How does dietary fat get into the body?

The major dietary sources of lipids are the triglycerides, comprising about 98 percent, while the other 2 percent consists mainly of sterols and phospholipids. Most of the dietary triglycerides contain long-chain fatty acids (14 or more carbons). Lipids are insoluble in water, and therefore their digestion and absorption is somewhat more complicated than that of carbohydrates; a broad overview is presented in **figure 5.6**. As lipids enter the small intestine, they stimulate hormonal secretion by the intestine that culminates in the secretion of bile from the gallbladder and lipases from the pancreas into the intestinal lumen. The bile salts serve as emulsifiers, breaking up the lipid droplets into smaller segments that may be hydrolyzed by the lipid enzymes, pancreatic lipases, and cholesterases. In essence, lipids are hydrolyzed into free fatty acids (FFAs), glycerol, cholesterol, and phospholipids, which through an intricate process are then absorbed into the cells of the intestinal mucosa. Here they are combined into a fat droplet called a **chylomicron**, which contains a large amount of triglyceride and smaller amounts of cholesterol, phospholipids, and protein. The chylomicron is one form of a **lipoprotein**, which, by its name, you can see is composed of lipids and protein. A diagram of a lipoprotein is presented in **figure 5.7**. The chylomicron then leaves the intestinal cell and is absorbed by the lacteal in the villi, where it is eventually transported in the lymphatic system to the blood. A schematic of the absorption process is presented in **figure 5.8**.

Medium-chain triglycerides (MCTs) release medium-chain fatty acids (MCFAs) with shorter carbon chain lengths (6–12 carbons), enabling them to be absorbed directly into the blood without being converted into chylomicrons. They are transported directly to the liver. Because of this rapid processing, MCTs have been theorized to possess ergogenic potential, and their efficacy in this regard will be discussed in a later section. Short-chain fatty acids (SCFAs) derived from triglycerides are also absorbed like MCFAs.

What happens to the lipid once it gets in the body?

The digestion of lipids into chylomicrons is slow, and the absorption after a high-fat meal can last several hours. As the chylomicron circulates in the blood, it reacts with various cells in the body, particularly cells in the muscle and adipose tissues. Specific proteins in the outer coat of lipoproteins are known as apolipoproteins. **Apolipoproteins**, or *apoproteins*, increase lipid solubility and enable the various lipoproteins to react with specific receptors in cells throughout the body. The apolipoproteins in the chylomicron interact with an enzyme, lipoprotein lipase, which is produced in the muscle and adipose cells and released to the capillary blood vessels surrounding the cells. The lipoprotein lipase releases fatty acids and glycerol from the chylomicron. The fatty acids are absorbed into the cells by simple diffusion and receptors for some LCFAs, while the glycerol is transported primarily to the liver for conversion to glucose. The remains of the chylomicron, the chylomicron remnant, are transported to the liver for disposal.

In the muscle, the fatty acids may either be used as a source of energy or combine with newly generated glycerol, which is derived as a metabolic by-product of glycolysis, leading to the formation and storage of muscle triglycerides. In exercise science literature, intramuscular triglycerides are often referred to as *intramyocellular triacylglycerol (IMTG)*. In the adipose cell, most of the fatty acids combine with glycerol and are stored as adipose cell triglycerides.

The key organ in the body for the metabolism of most nutrients is the liver. It is a clearinghouse in human metabolism. As blood passes through the liver, its cells take the basic nutrients and convert them into other forms. As mentioned in chapter 4,

FIGURE 5.6 A summary of fat digestion and absorption.

Fats
1 Very minor fat digestion in stomach
2 Bile made by the liver aids fat digestion and absorption
3 Fat digested mainly into monoglycerides and free fatty acids by a lipase enzyme released from the pancreas
4 Absorbed fat is mostly packaged as chylomicrons and transported through lymphatic vessels
5 Less than 5% of fat normally excreted in feces
2 Liver
1 Stomach
3 Pancreas
4 Small intestine
5
Anus

Triglycerides
Protein
Phospholipids
Free cholesterol
Cholesterol bound to fatty acids

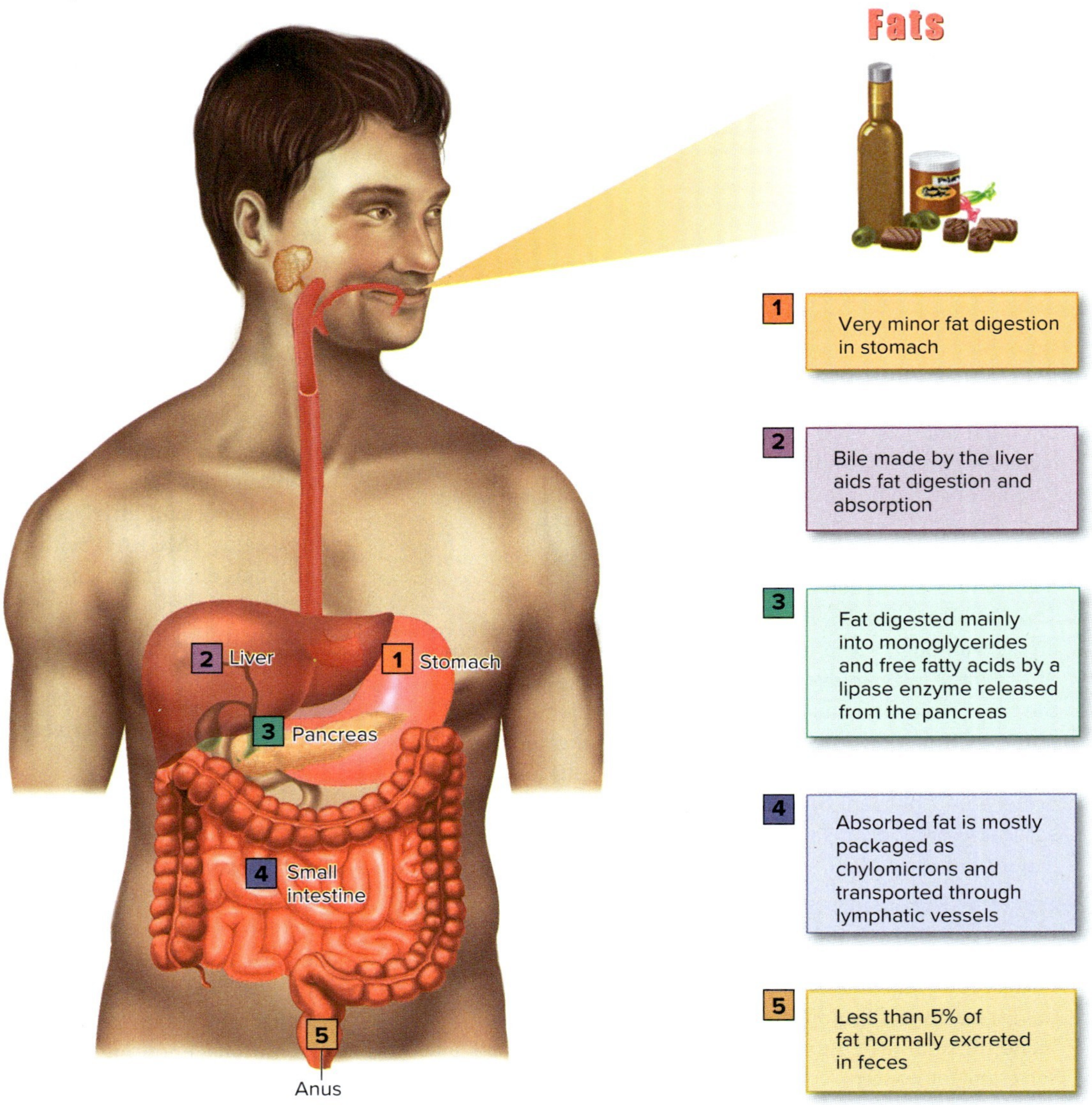

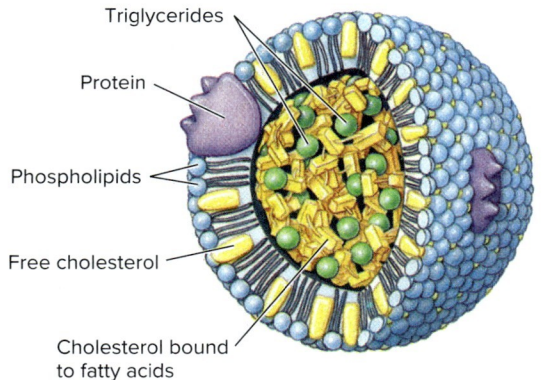

FIGURE 5.7 Schematic of a lipoprotein. Lipoproteins contain a core of triglycerides and cholesterol esters surrounded by a coat of apoproteins, cholesterol, and phospholipids. The proportion of protein, cholesterol, triglycerides, and phospholipids varies among the different types of lipoproteins.

the liver is able to manufacture glucose from a variety of other nutrients, including glycerol. Pertinent to our discussion here is its role in lipid metabolism. As noted previously, glycerol and chylomicron remnants, including phospholipids, are transported to the liver, as are the MCFAs and SCFAs directly from the intestinal tract. Adipose cells are metabolically active in the sense that they are constantly releasing fatty acids for use by the body, including the liver. The major role of the liver is to combine these various components (fatty acids, glycerol, cholesterol, and phospholipids), along with protein, into various forms of lipoproteins.

What are the different types of lipoproteins?

After the chylomicrons have been cleared from the blood, which may take several hours, other lipoproteins constitute approximately 95 percent of the serum lipids. The metabolism of lipoproteins is complex, for they are constantly being synthesized and catabolized

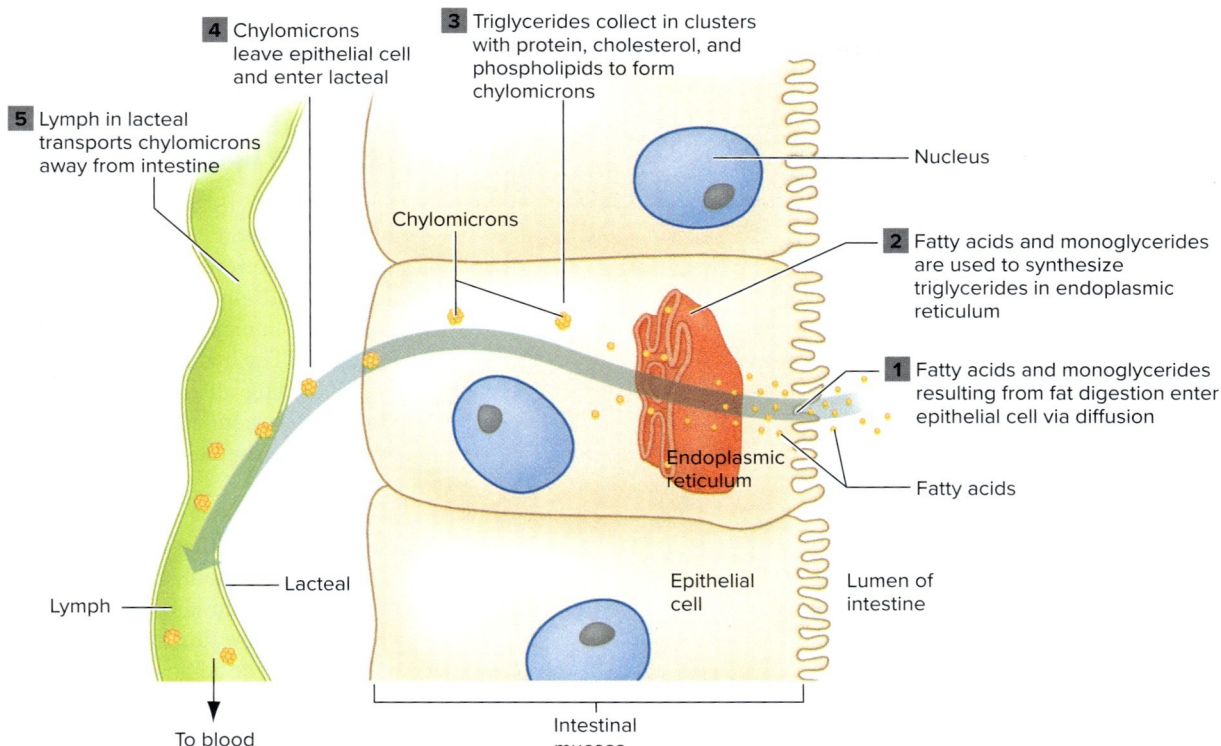

FIGURE 5.8 The absorption of lipids. In the lumen of the intestine, several lipases, assisted by bile salts, digest lipids to various forms of fatty acids, phospholipids, cholesterol, and glycerol, which are absorbed by an intricate process into the epithelial cells of the intestinal mucosa. Here they are combined with protein to form chylomicrons, a form of lipoprotein, which are transported out of the cell and into the lacteal, where the lymph eventually carries them to the blood. Medium-chain triglycerides (MCTs) may be absorbed directly into the blood (not depicted).

by the liver and other body tissues. As a result, there is an exchange of protein and lipid components among the different classes of lipoproteins, which can lead to the conversion of one form into another.

The classification of lipoproteins may be determined by several methods. One of the methods is by the type of apolipoprotein present and its functions. Lipoproteins have a number of different apoproteins, which enable them to react with different tissues. The letters *A, B, C, D,* and *E,* including subdivisions such as *A-I* and *A-II*, are common designations. The second method, most popularly known, is based on the density of the lipoprotein particle. Designations range from high to very low density.

The chylomicron is one form of lipoprotein, but it is relatively short-lived because it is derived from dietary fat intake. For our purposes in this book, the major classifications of lipoproteins along with their suggested composition and function are listed next; a graphical depiction is presented in **figure 5.9**. However, it should be noted that a wide variety of lipoproteins exist, based on their specific lipid and protein content. Additionally, their metabolism and complete functions have not been totally elucidated.

VLDL (very low-density lipoproteins). **VLDL** consist primarily of triglycerides formed in the liver from endogenous sources, whereas chylomicrons contain triglycerides from exogenous sources, that is, the diet. Like chylomicrons, VLDL are transported to the tissues to provide fatty acids and glycerol. The loss of some triglycerides to the liver or tissues produces VLDL remnants, referred to as IDL (intermediate-density lipoprotein) or TRL (triglyceride-rich lipoprotein). These remnants are either taken up by the liver or converted into LDL. Apoprotein B is the major apoprotein associated with both VLDL and IDL.

LDL (low-density lipoproteins). **LDL** contain a high proportion of cholesterol and phospholipids, but little triglycerides. LDL are formed after the VLDL and IDL release most of their stores of triglycerides. LDL size may be important. One form of LDL, a small, dense LDL, with important health implications has been identified. LDL, interacting with cell membrane receptors, deliver cholesterol into body cells. Apolipoprotein B is the major apolipoprotein associated with LDL.

HDL (high-density lipoproteins). **HDL** contain a high proportion of protein (about 45–50 percent), moderate amounts of cholesterol and phospholipids, and very little triglycerides. Various HDL are produced in the liver and intestinal tract. Several subclasses of HDL have been identified, most notably HDL_2 and HDL_3. HDL transport cholesterol from peripheral cells to the liver, known as *reverse cholesterol transport.* Apolipoprotein A is the major apolipoprotein associated with HDL.

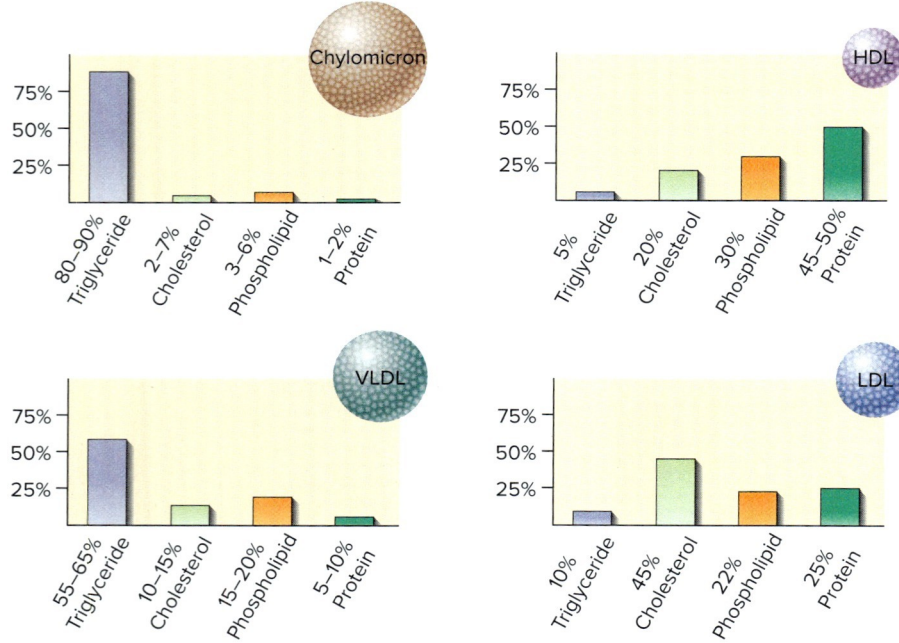

FIGURE 5.9 The approximate content of four different types of lipoproteins.

Lipoprotein (a). **Lipoprotein (a)** is very similar to the LDL, being in the upper LDL density range. The principal apolipoprotein associated with lipoprotein (a) is apolipoprotein (a).

Other apolipoproteins constitute the structure of lipoproteins. For example, apolipoprotein E is formed in the liver and present in all lipoproteins. These lipoproteins are needed to bind with cell membrane receptors.

A simplified schematic of fat metabolism is presented in **figure 5.10**.

Can the body make fat from protein and carbohydrate?

You may recall that glycogen is made up of many individual glucose molecules and is a glucose polymer. In essence, fatty acids are polymers of acetyl CoA, the primary substrate for the Krebs cycle.

As noted in **figure 4.9**, the amino acids of protein may be converted into acetyl CoA, which can then be converted into fat. Carbohydrates also may be converted into fat via acetyl CoA. It is important to understand that the body will take excess amounts of both carbohydrate and protein and convert them into fat when caloric expenditure is less than caloric intake. Thus, in general, it is not necessarily what you eat, but rather how much, that determines whether or not you gain body fat.

It is important to note that although carbohydrates and protein may be converted into fat (primarily fatty acids), fatty acids cannot be converted into carbohydrate. However, keep in mind that glycerol can be converted into carbohydrate. Regarding protein, metabolic by-products of fatty acids can combine with excess nitrogen from protein (if available) to form nonessential amino acids. Fatty acids cannot be converted into protein without this excess nitrogen.

What are the major functions of the body lipids?

The body lipids are derived from the dietary lipids and other carbon sources, namely carbohydrate and protein; however, with the exception of linoleic fatty acid and alpha-linolenic fatty acid, all lipids essential to human metabolism may be produced by the liver. The body lipids serve a variety of functions, including all three purposes of food: They form body structures, help regulate metabolism, and provide a source of energy.

Structure The structure of virtually all cell membranes, including the nerve membranes, consists partly of lipids, notably cholesterol and phospholipids. Lipids form myelin, an important component in the sheath covering nerve fibers. The structural fat deposits in the adipose tissues are used as insulators to conserve body heat and shock absorbers to protect various organs.

Metabolic Regulation Various fatty acids interact with proteins, the major metabolic regulators in the body. Essential fatty acids are found in cell membranes and are involved in various intracellular metabolic pathways, including regulation of gene expression. Cholesterol is a component of several hormones, such as testosterone, estrogen, cortisol, and aldosterone, which have diverse effects in the regulation of human metabolism. The majority of cholesterol in the body is used by the liver to produce bile salts, essential for the digestion of fats. Phospholipids are also instrumental for blood clotting.

Adipose cells produce various substances called **adipokines** (adipocytokines) and release them into the bloodstream. These adipokines may function as hormones, affecting tissues in other parts of the body. Leptin is one such adipokine that we will discuss in chapter 10.

Some derivatives of the omega-6 and omega-3 fatty acids formed by oxidation have some potent biological functions in the body. These derivatives—prostaglandins, prostacyclins, thromboxanes, and leukotrienes—are collectively known as **eicosanoids**. These eicosanoids possess local hormone-like properties that influence a number of physiological functions, including several that may have implications for health or physical performance. Several important eicosanoids are derived from omega-3 fatty acids, and the theorized health and performance implications will be discussed in later sections of this chapter.

Energy Source In general, the function of the majority of the body lipids, the triglycerides, is to provide energy to drive metabolic processes. The majority of the triglycerides in the body are stored in the adipose tissue. They break down to free fatty acids (FFA) and glycerol, which are released into the blood, with the FFA being transported to the tissues and the glycerol going to the

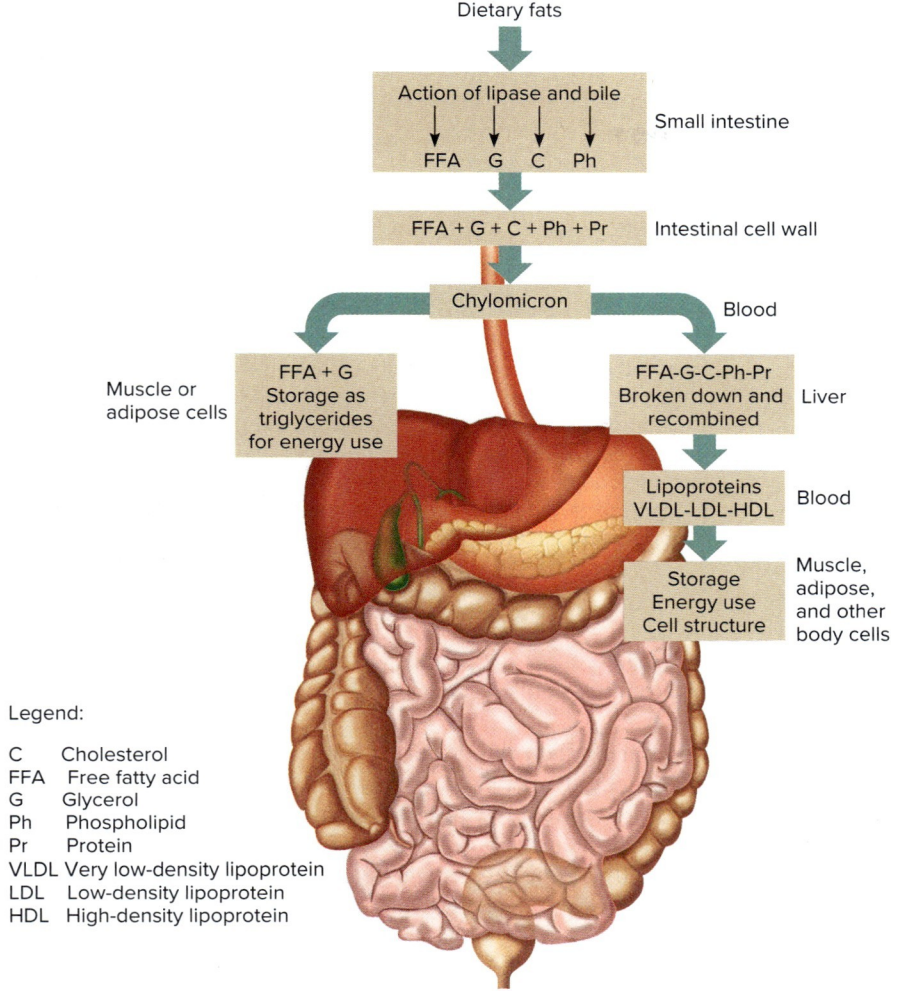

FIGURE 5.10 Simplified diagram of fat metabolism. After digestion, most of the fats are carried in the blood as chylomicrons. Through the metabolic processes in the body, fat may be utilized as a major source of energy, used to help develop cell structure, or stored as a future energy source.

liver. In the tissues, the FFA are reduced to acetyl CoA and enter the Krebs cycle to produce energy via the oxygen system. The glycerol is used by the liver to form other lipids or glucose. This energy-yielding process is illustrated in appendix A, **figure A.4A**.

During rest, nearly 60 percent of the energy supply is provided by the metabolism of fats when the individual consumes a mixed diet, but it may be higher when blood glucose is low, as after an overnight fast. Most of the energy provided is presented to cells as fatty acids, either FFA delivered via the blood or fatty acids from stored intracellular triglycerides. Inside the cell, the metabolism of fatty acids into acetyl CoA (a 2-carbon molecule) is known as **beta-oxidation**, for the beta carbon is the second carbon on the fatty acid chain. The acetyl CoA is then processed through the citric acid (Krebs) cycle and associated electron transport system for production of energy as ATP. At rest, a high-fat meal will increase the proportion of energy derived from fat. In general, the greater the amount of fatty acids available in the plasma, the greater their use as a source of energy. For example, the heart muscle may derive 100 percent of its energy needs from fatty acids after a lipid-rich meal.

Ketones, ketoacids that are metabolic by-products of excess fatty acid metabolism, may also serve as an energy source for body cells. Ketones diffuse from the liver into the blood and are transported to the body tissues, where they can eventually be used as a source of energy. The major ketones are acetoacetic acid, beta-hydroxybutyric acid, and acetone. These ketones usually are produced in small amounts, but when the use of fatty acids as an energy source is high (such as with fasting, high-fat diets, and diabetes) ketone levels in the blood will increase. Ketosis is a metabolic state where blood and urine levels are increased and can be a normal response to energy needs. **Ketoacidosis** is an uncontrolled production of ketones that is not normal and indicates a pathologic state and a medical emergency. This condition is most commonly seen in type 1 diabetes. Untreated ketoacidosis could lead to coma and death.

How much total energy is stored in the body as fat?

The greatest amount of energy stored in the body is fat in the form of triglycerides. Fat is a very efficient, compact means to store energy, for several reasons. First, fat has 9 kcal per gram, more than twice the value of carbohydrate and protein. Also, there is very little water in body fat compared to the 3–4 grams of water stored with each gram of carbohydrate or protein. In essence, based on weight, body fat may be about five to six times as efficient an energy store as carbohydrate and protein. If the average 154-pound man had to carry all the potential energy of his fat stores as carbohydrate, he would weigh nearly 300 pounds.

Most of the triglycerides are stored in the adipose tissues, approximately 80,000–100,000 kcal of energy in the average adult male with normal body fat. The triglycerides within and between the muscle cells may provide approximately 2,500–2,800 kcal, while those in the blood provide only about 70–80 kcal. The free fatty acids (FFA) in the blood total about 7–8 kcal. The liver also contains an appreciable store of triglycerides. Thus, you can see that the human body contains a huge reservoir of energy kcal in the form of fat. A summary is presented in chapter 3, **table 3.3**.

Key Concepts

▶ Fats are transported in the blood primarily as lipoproteins. Lipoproteins may be classified by their density and have various functions. In general, VLDL transport fats to the tissues, LDL transport cholesterol to the tissues, and HDL transport cholesterol from the tissues.

- Dietary lipids may serve the three major functions of nutrients. They may be utilized as an energy source; they may be used as part of body-cell structure; and by-products of fat metabolism, known as eicosanoids, may act as local hormones and affect a variety of metabolic functions.
- The vast majority of dietary fats are stored as triglycerides in the adipose cells, but significant amounts may also be stored in the muscles as intramuscular triglycerides, also known as intramyocellular triacylglycerols.

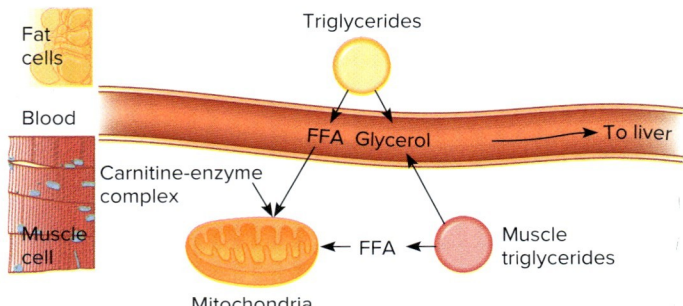

FIGURE 5.11 Fat as an energy source during exercise. Free fatty acids (FFA) are an important energy source during endurance exercise. They may be released by the adipose tissue triglycerides and travel by the blood to the muscle cells, and also may be derived from the muscle cell triglycerides. A carnitine-enzyme complex is needed to transport the FFA into the mitochondria. The glycerol that is released from the triglycerides may be transported to the liver for gluconeogenesis.

Fats and Exercise

Are fats used as an energy source during exercise?

The two major energy sources for the production of ATP during exercise are carbohydrates in the form of muscle glycogen and fats in the form of fatty acids, mainly LCFA. In steady-state exercise, both can be converted into acetyl CoA for subsequent oxidation in the citric acid cycle. In general, a mixture of both fuel sources is used during exercise, although the quantitative values may vary depending on a variety of factors, including the intensity and duration of the exercise bout, the diet, and the training status of the individual. The use of fat as an energy source during endurance exercise, including the marathon, has been the subject of several recent symposia and reviews, including those by Hawley, Jeukendrup, Roepstorff, Spriet, and Spriet and Hargreaves, and several key studies, such as those by Romijn and associates. The following discussion represents some of the key findings from these reviews and studies.

Fat Energy Sources The fatty acids used by the muscle cells during exercise may be derived from a variety of sources, including the plasma triglycerides in the chylomicrons and VLDL, but these sources are considered to be minor, providing less than 10 percent of fat energy.

The two major energy sources are the plasma FFA and the fatty acids derived from intramuscular triglycerides (IMTG). The plasma FFA are in very short supply, so they must be replenished by the vast stores of triglycerides in the adipose tissue. An enzyme in the adipose cells, known as hormone-sensitive lipase (HSL), catabolizes the intracellular triglycerides to FFA and glycerol. The FFA are released into the blood, bound to the protein albumin as a carrier and transported to the muscle cells or other cells. The FFA enter the cell by diffusion and by specific protein receptors (transporters) in the cell membrane. The FFA are activated in the cell cytoplasm, transported into the mitochondria by an enzyme complex containing the amine **carnitine**, metabolized to acetyl CoA by beta-oxidation, and produce energy via the citric acid cycle and the associated electron transport system. The muscle triglycerides may also be metabolized to fatty acids and glycerol by an enzyme similar to HSL, and the fatty acids may be transported into the mitochondria (**figure 5.11**). van Loon indicated that IMTG can function as an important substrate source during exercise and its use is determined by various factors, including exercise intensity and training status.

Use during Exercise During rest, most of the fat energy needs of the body are met by the supply of plasma FFA to the cells. Fatty acids are constantly being mobilized from the adipose tissues to replenish the plasma FFA. Most of the FFA released during rest, about 70 percent, are actually re-esterified back into triglycerides, the remainder being delivered to the body cells for energy.

During exercise, only about 25 percent of these FFA are re-esterified, so this alone provides a substantial increase in FFA delivery to the muscle cells. Additionally, hormones that activate HSL, such as epinephrine, are secreted during exercise, stimulating the breakdown of adipose cell triglycerides and the release of FFA into the blood for transport and entrance to the muscle cell. Spriet noted that fatty acid proteins may be important in regulating the FFA transport into cells, and as Glatz and others reported, muscle contraction activates the fatty transporters in the muscle cell membrane, thus increasing FFA uptake into the muscle cell. Epinephrine also stimulates intramuscular lipases to catabolize muscle triglycerides into FFA. These fatty acids then enter the mitochondria and are degraded to acetyl CoA.

During mild exercise at about 25 percent of VO_2 max, 20 percent or less of the total energy cost is derived from carbohydrate, while the other 80 percent or more comes from fat. Wolfe indicates that exercise-induced lipolysis normally provides FFA at a rate in excess of that needed during exercise. Thus, the plasma FFA provided by the adipose tissue appear to be the major source of fat energy during mild exercise, but their percentage use decreases and that of muscle triglycerides increases as the exercise intensity increases up to about 65 percent VO_2 max. At this point, fats and carbohydrates appear to contribute equally to the energy expenditure, and the plasma FFA and muscle triglycerides contribute equally to the energy derived from fats. Carbohydrate increasingly becomes the predominant fuel as exercise exceeds 65 percent of

TABLE 5.5	Fat energy sources during exercise
Plasma chylomicrons	Not a major source
Plasma VLDL	Not a major source
Plasma FFA	Major source; replenished by adipose cell release of FFA; used in exercise at low to moderate intensity, i.e., 25–65 percent VO_2 max; use decreases as exercise intensity increases toward 65 percent VO_2 max
Muscle FFA	Major source; released from intramuscular triglycerides; low use during mild exercise; used increasingly as exercise intensity increases toward 65 percent VO_2 max

Note: With high-intensity exercise, 65 percent VO_2 max or higher, total fat oxidation falls.

VO_2 max. At high-intensity exercise levels, about 85 percent VO_2 max, the percentage contribution from fats (mostly muscle triglycerides) diminishes to 25 percent or less as muscle glycogen becomes the preferential energy source. These percentages are relevant to trained athletes and will be somewhat lower in untrained individuals. A summary of fat utilization during exercise is presented in **table 5.5**.

Limiting Factors In a review, Lange stated that the human being is far from optimally designed for fat oxidation during exercise, noting that fat oxidation alone can sustain a metabolic rate corresponding to only 50–60 percent of VO_2 max. Although researchers contend that factors limiting fatty acid utilization during high-intensity exercise are largely unknown, several have been suggested. First, inadequate FFA mobilization from adipose tissue may limit FFA delivery to the muscle. Wolfe indicates that fat oxidation is increased significantly at exercise intensities of 85 percent VO_2 max when lipid is infused, but carbohydrate still remains the most significant source of energy. Second, suboptimal intramuscular processes may also limit fat oxidation. For example, Wolfe indicated that the high rate of carbohydrate oxidation during high-intensity exercise may inhibit fatty acid oxidation by limiting transport into the mitochondria, possibly by inhibition of the carnitine-enzyme complex.

In his review, Spriet noted that these two factors may limit fat utilization during high-intensity exercise, but other regulatory factors may be involved as well, such as limited transport of fatty acids into the muscle cell and the optimal muscle triglyceride lipase activity. Frayn, in a review, concluded that there is a problem with delivery of sufficient fatty acids to muscle from adipose tissue only during high-intensity exercise. This limitation may be due to feedback inhibition of lipolysis, possibly due to lactate or catecholamine concentrations. As noted by Spriet, research is needed to determine what downregulates fat use during exercise intensity above 85 percent of VO_2 max.

Dietary Effects As noted in previous chapters, the amount of energy that may be obtained from muscle and liver glycogen is rather limited. Feeding carbohydrate before and during exercise may reduce fat utilization. Spriet and Hargreaves indicated that the reduction in fat metabolism following glucose ingestion appears to be the result of increased plasma insulin levels. Insulin may decrease adipose tissue lipolysis, thus decreasing FFA availability in the plasma. Also, insulin may reduce fatty acid oxidation in the muscle, possibly by inhibiting fatty acid transport into the mitochondria. Although fat oxidation may be reduced, the available carbohydrate would provide a more efficient energy source. In one study, Larson-Meyer and others found that reducing intramuscular lipid levels with a low-fat diet over the course of 3 days had no effect on endurance running performance, as measured by a 2-hour run at 62 percent of VO_2 max followed by a 10-kilometer time trial.

Carbohydrate intake before and during prolonged exercise helps, but within 90–120 minutes or more of high-intensity aerobic exercise, glycogen stores approach very low levels and the body shifts to an increasing usage of FFA, leading to a decrease in the intensity of the exercise. In cases such as prolonged endurance tasks like ultramarathons, FFA may provide nearly 90 percent of the energy in the later stages of the event when muscle glycogen and blood glucose levels are inadequate to sustain higher-intensity exercise.

Use of Ketones Although ketones may be utilized by the muscle, they do not appear to contribute significantly to energy production during exercise.

However, Jeff Volek, a leading expert on high-fat diets and nutritional ketosis, notes that there is a need for longer-term studies on performance. Volek contends that allowing an athlete to keto adapt, or allowing the athlete to spend sufficient time on a low-carbohydrate/high-fat diet to achieve very high fat oxidation rates, could have many health and performance benefits. But, this is controversial and a great deal more research needs to be conducted. **Ketogenic** diets and ketone supplements are discussed later in the chapter.

Does sex influence the use of fats as an energy source during exercise?

maridav/123RF

Sex Women typically possess a greater percentage of body fat than men, and several writers for popular runners' magazines have suggested that women could process this fat more efficiently and thus be more effective in ultramarathon events. Tarnopolsky indicated some theoretical rationale supports this viewpoint, as both the muscle lipid content and the maximal activity of a key enzyme in fat metabolism are higher in females. The greater fat oxidation for women during submaximal endurance exercise compared with men seems to occur partly through a sex hormone–mediated enhancement of lipid-oxidation pathways. In several studies Tarnopolsky and his associates, using the respiratory exchange ratio to evaluate energy substrate utilization following carbohydrate loading, reported that women oxidized significantly more fat and less carbohydrate than men when exercising at 65 or 75 percent VO_2 max. Similar findings were presented in studies headed by Knechtle and Venables. Roepstorff noted that some sport scientists think women may use more IMTG during exercise than men, possibly attributed to sex differences in skeletal muscle HSL regulation.

However, other studies have shown similar utilization of fat as an energy source by men and women during exercise, particularly when matched for their aerobic capacity. Replicating their previous study with males, Romijn and others studied substrate metabolism in well-trained females at 25, 65, and 85 percent of VO_2 max. As with the men, carbohydrate oxidation in women increased progressively with exercise intensity, whereas the highest rate of fat oxidation was during exercise at 65 percent of VO_2 max. They concluded that the patterns of use of carbohydrate and fat during moderate- and high-intensity exercise are similar in trained men and women.

In a recent meta-analytic review of 35 studies, Cano and colleagues reported that, when compared to women, respiratory exchange ratio was significantly higher, indicating greater carbohydrate oxidation, in sedentary and athletic men. How these differences in metabolism influences endurance exercise performance is debatable. For example, in one of their studies, cited previously, Tarnopolsky and others found that although the women oxidized more fat and less carbohydrate than the men, the men improved their performance in the exercise task following carbohydrate loading but the women did not.

What effect does exercise training have on fat metabolism during exercise?

A single bout of exercise influences fat utilization during exercise. Tunstall and others found that a single endurance exercise session led to an increased expression of genes in the skeletal muscle that increase the capacity for fat oxidation, and 9 days of training augmented this effect.

In general, research indicates that trained athletes use more fat than untrained athletes during a standardized exercise task. For example, if you ran an 8-minute mile both before and after a 2-month endurance training program, you would use the same amount of caloric energy each time. However, after training, more of that energy would be derived from fat. Hence, training helps you become a better fat burner, so to speak, which may help spare some of the glycogen in your muscles. Although the exact mechanisms have not been identified, the reviews by Horowitz and Klein and Jeukendrup and his colleagues document multiple factors, as presented in **table 5.6**.

Overall, according to Martin, the increased content and use of muscle triglycerides may be the primary mechanism underlying the greater capacity of trained muscle to oxidize fatty acids during exercise. Increased utilization of fat during exercise is one of the major effects of training experienced by the endurance athlete.

Even though carbohydrate becomes more important as an energy source during high-intensity exercise, Coggan and others found highly trained endurance athletes may be able to use fats more efficiently at exercise intensity levels of 75–80 percent VO_2 max, and in a review Spriet noted that IMTG is an important energy source during prolonged, moderate dynamic exercise up to about 85 percent of VO_2 max in well-trained athletes. The ability to derive a substantial proportion of the energy demands of intensive exercise from fatty acids is extremely important for athletes such as marathoners who may be able to save some of their muscle glycogen for utilization in the later stages of the race. An optimal mixture of fatty acids and glycogen for energy will enable them to sustain their pace, whereas the total depletion of muscle glycogen and subsequent reliance on fatty acids as the sole energy supply would force them to slow down. Thus, it is important for the endurance athlete to become a better "fat burner," and a variety of ergogenic aids have been proposed to enhance this effect.

TABLE 5.6 Possible mechanisms associated with the increased use of fat as an energy source during aerobic endurance exercise following exercise training

- Increased blood flow and capillarization to the muscle, delivering more plasma FFA
- Increased muscle triglyceride content, possibly associated with increased insulin sensitivity. Insulin regulates movement of FFA into muscle cells; exercise training may also increase the activity of lipoprotein lipase or fatty acid transporters at the muscle cell membrane
- Increased sensitivity of both adipose and muscle cells to epinephrine, resulting in increased FFA release to the plasma and within the muscle from triglycerides
- Increased number of fatty acid transporters in the muscle cell membrane to move fatty acids from the plasma into the muscle cell
- Improved ability to use ketones as an energy source
- Increased number and size of mitochondria, and associated oxidative enzymes for processing of activated FFA
- Increased activation of FFA and transport across the mitochondrial membrane
- Increased activity of oxidative enzymes

Key Concepts

▶ One of the major functions of fat is to provide energy. Although fat may be an important energy source for low- to moderate-intensity exercise, it is not the optimal energy source during high-intensity aerobic or anaerobic exercise. Nevertheless, well-trained athletes may use IMTG as an energy source up to 85 percent of VO_2 max.

▶ Some research suggests that when compared to adult males, adult females may use more fat as a source of energy during endurance exercise; however, research with similarly well-trained male and female athletes reveals no differences in fuel utilization endurance exercise.

▶ Endurance exercise training, by enacting multiple mechanisms, enhances the use of fat for energy during aerobic exercise. Endurance athletes are better *fat burners*.

Fats: Ergogenic Aspects

Because exercise training leads to an increased utilization of fatty acids as an energy source and improved performance in prolonged endurance events (theoretically by sparing muscle glycogen), a variety of dietary practices, dietary supplements, and pharmacological agents have been employed in attempts to facilitate this metabolic process during exercise. In a review of the proposed mechanism, Jeukendrup noted that the increased availability and oxidation of fatty acids would generate more acetyl CoA, which in turn would inhibit a cascade of events that would essentially decrease the activity of enzymes involved in carbohydrate breakdown. Such an effect could spare the use of muscle glycogen and enhance endurance performance.

Both acute and chronic dietary strategies have been used to increase the concentration of muscle triglycerides and serum level of FFA, and increase fat oxidation. Such strategies include high-fat diets, fasting, and even infusion of lipids into the bloodstream. Similarly, dietary supplements such as medium-chain triglycerides, lecithin, glycerol, omega-3 fatty acids or fish oil, hydroxycitrate, conjugated linoleic acid, phosphatidylserine, and ketones have been used to increase the supply of oxidizable fats or the rate of fat metabolism. Caffeine also has been theorized to enhance fat oxidation, and its role as an ergogenic aid is covered in chapter 13.

High-fat diets

High-fat diets are popular yet controversial styles of nutritional intake. First, a great deal of research points to diets high in fat as disease promoting; this will be discussed later in the chapter. Second, if one macronutrient is increased, and kcal are held constant, the amount of another macronutrient in the diet must decrease. The beneficial effect of high-carbohydrate diets on exercise performance, as described in the previous chapter, has been well documented, so if an increase in dietary fat results in a decrease in dietary carbohydrate, performance could be compromised. High-fat diets may be used on an acute or chronic basis. Some people refer to acute high-fat diets as **fat loading**.

Acute fat loading involves the intake of fats immediately prior to exercise in the hopes of increasing energy production from fats and sparing of limited carbohydrate stores. Hargreaves notes that elevated plasma FFA levels are, in fact, associated with reduced muscle glycogenolysis, which could help spare muscle glycogen. Because the rate at which FFA are oxidized in the muscle is dependent in part upon their concentration in the blood plasma, several different acute techniques have been tried in attempts to increase plasma FFA levels.

Chronic fat loading involves increasing dietary fat intake for days or weeks prior to endurance exercise. These strategies are designed to increase lipid metabolism and gene expression in skeletal muscle, resulting in an increased ability of the skeletal muscle to use fat as an energy source during exercise. In a review, Roepstorff and others indicated that ingestion of a fat-rich diet induces an increase in intramuscular triglyceride (IMTG) content, primarily in the type I muscle fibers used for oxidative energy production. Spriet and Hargreaves reported that IMTG levels can be increased by 50–80 percent following the consumption of a high-fat diet.

Acute High-Fat Diets   Lipid digestion and absorption are slow, so one strategy is to infuse a lipid solution (such as Intralipid®) directly into the blood along with heparin, a substance that stimulates lipoprotein lipase activity and increases plasma FFA levels. Such a strategy increases fat oxidation and reduces carbohydrate oxidation, which may enhance endurance exercise performance. However, after a lipid solution was used by a national team in the Tour de France, the entire team withdrew from the race allegedly due to adverse reactions. No research has been uncovered that supports this ergogenic technique.

A second strategy is to ingest a high-fat meal prior to exercise performance. For example, in two studies, Okano and others reported no significant differences between a high-fat meal (61 percent fat content) and a control or low-fat meal on performance in a cycling test to exhaustion at 78–80 percent VO_2 max following 2 hours of riding at 60–67 percent VO_2 max. Additionally, Rowlands and Hopkins investigated the effect of a high-fat (85 percent fat energy) meal, as compared to a high-carbohydrate (85 percent carbohydrate energy) meal and high-protein (30 percent protein energy) meal consumed 90 minutes before an endurance cycling test, which involved a 1-hour preload at 55 percent VO_2 max, five 10-minute incremental loads from 55 to 82 percent of peak power, and a 50-kilometer time trial that included several 1-km and 4-km sprints. Subjects consumed a carbohydrate supplement during the cycling protocol. The meal composition had no clear effect on sprint or 50-kilometer cycle performance.

An acute high-fat dietary strategy does not appear to enhance performance and, in fact, may actually impair performance if it contributes to gastrointestinal distress because of the delayed gastric emptying associated with fats. Research has shown that consuming a high-fat diet for 1 or 2 days, another acute approach, may actually impair performance in high-intensity exercise tasks.

One of the principles behind keto-adaptation is that athletes must be consuming high-fat diets for long enough to gain an effect. Thus, the lack of a benefit seen on exercise performance in studies of acute high-fat diets is not surprising.

Chronic High-Fat Diets   Several investigators have challenged the dogma that endurance athletes need high-carbohydrate diets and suggest that endurance performance may benefit from diets containing about 50 percent or more energy from fat. Brown and Cox note that athletes can adapt to such a diet and maintain physical endurance capacity, will increase their muscle triglyceride levels, and may increase the use of fat and decrease use of carbohydrate during exercise. Volek and colleagues studied a group of ultra-marathoners and triathletes who followed either a 25 or 70 percent fat diet for about 20 months. Peak fat oxidation was more than two times greater in the high-fat diet group. Interestingly, muscle glycogen concentrations were not different between the two groups. Unfortunately, endurance performance was not reported in this study, so it is unknown if this metabolic advantage translated into a performance advantage. In general, research has shown that when an individual is placed on a chronic low-carbohydrate and high-fat diet for about a week or more, the body adjusts its metabolism to use fats more efficiently. However, metabolic improvements do not necessarily correlate with or guarantee performance improvements.

Studies showing ergogenic effect Few studies support an ergogenic effect of chronic fat loading.

In one study, Lambert and others placed athletes on a high-fat (70 percent fat kcal) diet or low-fat (12 percent fat kcal) diet for 2 weeks, testing the effects on three performance tests done in consecutive order with 30 minutes rest between each. The three exercise tests included a Wingate high-power test, a high-intensity test to exhaustion at 90 percent VO_2 max, and a moderate-intensity test to exhaustion at 60 percent VO_2 max. Although there were no significant effects between the two diets for performance in the first two exercise tests, the cyclists rode significantly longer on the moderate-intensity test following adaptation to the high-fat diet. The investigators noted that adaptation to the high-fat diet decreased the reliance on carbohydrate as an energy source during exercise and thus suggested the improved performance was due to muscle-glycogen sparing. In a subsequent study, Lambert and others evaluated the effect of a 10-day high-fat diet, prior to 3 days of a carbohydrate-loading regimen, on cycling performance in a trial consisting of 150 minutes at 70 percent VO_2 max followed by a 20-kilometer time trial. Compared to their habitual diet with carbohydrate loading, the high-fat diet enabled the endurance cyclists to increase total fat oxidation and reduce carbohydrate oxidation, as well as significantly improve time trial performance by 1.4 minutes.

Studies showing no ergogenic effect Conversely, other well-controlled research suggests that chronic high-fat diets do not benefit endurance performance. The most often cited study used to argue the beneficial effects of a low-carbohydrate, high-fat diet on endurance performance was published in 1983 by Phinney and colleagues. After 4 weeks of keto-adaptation, there were dramatic increases in fat oxidation (on average 90 g/hour) and decreases in exercise RQ (from 0.83 to 0.72) in five cyclists. Interestingly, pre-exercise muscle glycogen levels at the end of the study were half of what they were at baseline (143 versus 76 mmol/kg). However, there were no changes in VO_2 max and no improvements in endurance performance as a result of the diet. Using well-trained cyclists as subjects, Havemann and others reported no significant differences in 100-kilometer time-trial performance when consuming either a high-fat (68 percent fat) or high-carbohydrate (68 percent carbohydrate) diet for 6 days, followed by 1 day of carbohydrate loading, but the high-fat diet impaired 1-kilometer sprint power. The authors noted that although the high-fat diet increased fat oxidation, it compromised high-intensity sprint performance.

Several studies have shown no effect of chronic fat loading on endurance performance in trained individuals. Brown and Cox, in a randomized study using 32 endurance-trained cyclists, reported no difference in 20-km road time performance over a period of 12 weeks when subjects consumed either a high-fat (47 percent of energy) or a high-carbohydrate (69 percent of energy) diet. Burke and her associates found no beneficial effects in two studies. In the first study they investigated the effects of a high-fat diet for 5 days, followed by a 1-day high-carbohydrate diet, on an endurance cycling protocol involving cycling at 70 percent VO_2 max for 2 hours followed by an intense time trial. Compared to a high-carbohydrate diet, the high-fat diet induced greater utilization of fat with an associated muscle glycogen sparing, but there was no significant improvement in the cycling time trial, even though performance following the high-fat phase was about 3.4 minutes faster. In a subsequent study using a similar protocol, Burke and her associates reported no significant effect on endurance performance.

In a similar vein, Rowlands and Hopkins compared the effects of three different 14-day dietary regimens on 100-kilometer cycling performance. One of the diets was high-carbohydrate, one was high-fat, and one was high-fat before 2.5 days of carbohydrate loading. Both high-fat diets increased fat oxidation during the cycling trial, and although there were no significant differences between

---

Training Table

Should athletes "go keto"?

High-fat, low-carbohydrate, or ketogenic, diets have been around for many years, and occasionally see a resurgence in popularity. Despite the large body of evidence showing that carbohydrate is necessary to fuel athletic performance, some athletes question the potential benefit of a ketogenic diet. First, as described in chapter 3, carbohydrates are a more efficient fuel than fats, which require more oxygen to produce less ATP. If your goal is to complete an endurance exercise event, is it possible to do so on a ketogenic diet? Louise Burke and colleagues at the Australian Institute of Sport answered this question in a comprehensive study of elite race walkers. Volunteers followed a high-carbohydrate diet (60–65 percent CHO, 15–20 percent protein, 20 percent fat), a periodized carbohydrate diet (same as a high-carbohydrate diet, but with different timing, such that some training sessions were with high- or low-carbohydrate availability), and a low-carbohydrate diet (75–80 percent fat, 15–20 percent protein, greater than 50 g/day CHO). After 3 weeks of dietary adaptation and intense training, endurance performance improved on the high-carbohydrate diets, but not in the ketogenic diet group, probably due to decreased economy (more oxygen needed to do the same amount of work). If your goal is success in endurance sports, it appears that carbohydrate is still king!

Source: Burke, L. M., et al. 2017. Low carbohydrate, high fat diet impairs exercise economy and negates the performance benefit from intensified training in elite race walkers. *Journal of Physiology* 595(9):2785–2807.

For more on ketogenic diets and endurance performance, visit: www.mysportscience.com/post/2016/12/26/lchf-diets-and-performance-in-elite-athletes

the three trials, the 100-kilometer performance was approximately 3–4 percent faster with the high-fat diets. Interestingly, power output in the last 5 kilometers of the time trial was significantly greater for the high-fat with carbohydrate-loading trial as compared to the high-carbohydrate trial. The authors noted that although the main effects of the study were not significant, there was some evidence for enhanced ultraendurance cycling with a high-fat diet compared to a high-carbohydrate diet. Carey and others compared the effects of two 6-day diets, either high-carbohydrate or high-fat, on a prolonged cycling task involving 4 hours at 65 percent VO_2 peak followed by a 1-hour time trial. In both diets, a high-carbohydrate diet was consumed on the day prior to testing. Similar to other studies, the high-fat diet increased total fat oxidation and reduced carbohydrate oxidation, but there was no effect on cycling performance. The recent Supernova study conducted by Burke and colleagues is discussed in the *Training Table* in this section. As it is one of the best designed studies on high-fat diets and endurance performance, and showed a performance impairment with this dietary style, it provides strong evidence that ketogenic diets are not the best option for endurance athletes.

There are other concerns associated with high-fat diets that have not been well studied. For example, in a crossover study, Holloway and others showed that a 70 percent fat diet impaired cognitive function, speed, and mood. A decrement in any one of these areas could affect an individual's ability to train and compete, particularly in activities that require an athlete to react to an opponent or conditions.

Reviews concluding no ergogenic effect Trent Stellingwerff, an expert in nutrition and athletic performance, reviewed the data on the effects of ketogenic diets on exercise performance. He noted that out of 21 published studies, 12 showed a performance decrease and 7 showed no benefit of a ketogenic diet. Only two studies are available that show improved performance. Stellingwerff estimates, via the rates of energy expenditure, ATP production, and available fuel stores, that an elite endurance athlete can "complete" a marathon while on a ketogenic diet but that the athlete's time to complete the race will be dramatically slower.

In the most comprehensive review of ketogenic diets and endurance exercise performance in athletes, Louise Burke notes the following:

> In trained athletes on a verified ketogenic diet, performance was not improved in 8 of 9 studies. There was strong evidence of substantial increases in fat oxidation within 3 to 4 weeks and possibly within 5 to 10 days on the diet. Exercise fat use doubled to 1.5 g/min and maximal fat oxidation rates shifted from 45 to 70% of maximal aerobic capacity. However, there was decreased carbohydrate oxidation and a subsequent reduction in performance of higher intensity endurance exercise. It is clear that athletes contemplating a ketogenic diet should balance the risk of impaired higher intensity endurance exercise performance with a greater reliance on fat for fuel.

One of the questions that must be addressed when considering a low-carbohydrate, high-fat diet is not if you can finish the race but how fast you can finish the race. Volek and others correctly point out that a keto-adapted endurance athlete is able to oxidize up to 1.5 grams of fat/minute during an Ironman triathlon, and so would not need to ingest exogenous fuels such as carbohydrate to complete the race. However, Stellingwerff points out that under keto-adapted circumstances and based on the caloric requirements to complete a marathon, an athlete would only be able to complete the race in 3–4 hours. Thus, low-carbohydrate, high-fat diets may allow athletes to perform, but possibly not at their best and far from an elite level. In a *New York Times* article, Louise Burke, an expert on low-carbohydrate, high-fat/ketogenic diets, noted that "sports performance requires metabolic flexibility," which means that athletes need to be able to use more than one system well, not just fat oxidation.

Several things are not known and must be clearly defined for research on low-carbohydrate, high-fat/ketogenic diets to progress:

1. How much dietary fat and dietary carbohydrate constitute a low-carbohydrate, high-fat/ketogenic diet in terms of exercise performance?
2. How long does one have to stay on a low-carbohydrate, high-fat/ketogenic diet for beneficial effects to occur?
3. Is there a biological marker (e.g., blood ketones) that can be used to indicate when one is keto-adapted?
4. Are the effects of a low-carbohydrate, high-fat/ketogenic diet different based on training status, athletic event, body size, or any other variables?
5. Are low-carbohydrate, high-fat/ketogenic diets healthy for extended periods of time?
6. How long does it take to restore muscle glycogen levels and carbohydrate oxidation rates after a long-term ketogenic diet?

High-fat diets and weight loss

The long-term impact of a low-carbohydrate and high-fat (ketogenic) diet on weight loss in those with overweight or obesity is unknown. In a meta-analysis of 13 studies containing almost 1,600 individuals, Bueno and colleagues concluded that low-carbohydrate, high-fat diets result in greater reductions in body mass, triglycerides, and diastolic blood pressure. However, as has been described elsewhere in this text, care must be taken when interpreting the results of a meta-analysis. For instance, only 4 of the studies included lasted longer than 12 months, and weight loss relapse typically happens after this point in time. Importantly, although some cardiovascular risk factors improved, the authors point out that pathological markers such as hepatic lipids, endothelial function, cardiovascular events, and renal function, which are important when describing the safety of a diet, were not assessed. Also, compliance to these diets is an important factor, some studies had participant dropout rates as high as 50 percent, and by the end of the study, dietary carbohydrate intake from participants was higher than allowed in many studies. Finally, average BMI in the sample indicates that subjects had obesity, so it is unclear what the effects of a low-carbohydrate, high-fat diet in healthy-weight people might be.

Training Table

Low-fat or high-fat diet for weight loss?

One of the most frequently asked questions in nutrition is which diet is best to lose weight. Typically, this question focuses on the weight-loss diet being either high in carbohydrate and low in fat (HCLF) or low in carbohydrate and high in fat (LCHF). The real question is, does one style of eating somehow promote more weight loss (diet quality) or does weight loss happen when people eat less (diet quantity) regardless of macronutrient composition? Gardner and colleagues recently provided an answer to this question, in a 12-month weight-loss study with over 600 participants. The results of the study showed that whether one follows a HCLF or LCHF diet, if kcal intake is the same, so is weight loss. For best results, it makes sense to follow the dietary plan that is most sustainable. Consult with your doctor or a registered dietitian/nutritionist to find out what style of eating might be best for you.

Source: Gardner, C. D., et al. 2018. Effect of low-fat vs low-carbohydrate diet on 12-month weight loss in overweight adults and the association with genotype pattern or insulin secretion: The DIETFITS randomized clinical trial. *Journal of the American Medical Association* 319(7):667–79.

For an in-depth analysis of the recent Gardner study, visit: https://examine.com/nutrition/low-fat-vs-low-carb-for-weight-loss/.

The *Training Table* in this section highlights one of the largest and most comprehensive low-carbohydrate weight loss studies ever conducted. The overall message? When it comes to weight loss, kcal matter more than macronutrient distribution. The popularity of low-carbohydrate, high-fat diets continues to be very high, but a great deal more research needs to be conducted regarding the safety and efficacy of this style of eating. For instance, if people lose more weight when restricting carbohydrates, is it because they are eating less due to their displeasure with the diet plan? If so, it would seem that this style of eating may not be effective in the long run. Also, if this style of eating is sustainable, what are the long-term effects on health outcomes such as cancer or other diseases? As an example, a recent study of nearly 50,000 postmenopausal women revealed a significant decrease in deaths after breast cancer in women assigned to a diet with fat intake of 20 percent of total kcal.

Often, when describing the effects of diet on health, we use proxy markers such as serum cholesterol, LDL, or blood pressure to estimate health. However, these markers do not perfectly predict or correlate with health. Ultimately, to answer a question about health, one must consider mortality as an outcome. Noto and colleagues analyzed 17 low-carbohydrate diet studies and found that low-carbohydrate diets were associated with increased risk of mortality but that they did not increase risk of cardiovascular disease mortality.

Does exercising on an empty stomach or fasting improve performance?

A controversial area related to dietary fat intake is the issue of fasting for a limited period of time prior to exercise. Some people contend that this will improve training adaptations and subsequently performance. Indeed, fasting, even an overnight fast, can increase FFA availability and, theoretically, increase fat oxidation, spare glycogen, and improve performance. On the other hand, if fasting reduces muscle glycogen or causes hypoglycemia, it could impair performance. Another popular theory is that training while fasted will encourage greater weight loss either through metabolic adaptations or reductions in appetite. Despite the large amount of available research on the benefits of carbohydrate ingestion before exercise, several research groups have examined exercising after various durations of fasting. The following is a summary.

Gutierrez and others reported decreased aerobic endurance physical working capacity in young men following a 3-day fast. Data from Peter Hespel's lab showed that following a 6-week endurance training program, training in a fasted state increased oxidative enzymes, such as citrate synthase and β-hydroxyacyl coenzyme A dehydrogenase, and intramyocellular lipid breakdown more than training in a fed state. Other work from this group showed that GLUT-4 and proteins involved with fatty acid transport increased when training in the fasted state. However, following the endurance training program, there were no differences in VO$_2$ max between those who trained under fasted or fed conditions. Stannard and colleagues compared 4 weeks of fasted or fed endurance training and showed greater increases in VO$_2$ max in fasted subjects and greater increases in muscle glycogen in fed subjects. There were no differences in citrate synthase and β-hydroxyacyl coenzyme A dehydrogenase between groups. The reader is cautioned that beneficial metabolic changes may not lead to beneficial performance changes.

Some believe that training while fasted improves weight loss more than training in a fed state. Schoenfeld and others examined the effects of 4 weeks of training in a fed or fasted state while on a hypocaloric diet. Both groups decreased body mass and fat mass, but there were no differences between groups. The authors acknowledge that a longer period of time may be needed for any differences between protocols to take effect. In acute studies, both Gonzalez and Deighton showed that there was no advantage of exercising after an overnight fast or after a meal on subsequent appetite or daily energy intake. Headland and others reviewed the effects of intermittent compared with continuous energy restriction on weight loss. The authors noted no differences in weight loss between the two dietary styles. Trepanowksi and others reported no differences in weight loss and cardiovascular disease risk factors between alternate day fasting and continuous energy restriction diets in 100 adults with obesity. Thus, there are no convincing data that athletes should train while in the fasted state for performance or weight loss. However, many Muslim athletes may be training or competing while fasting for religious reasons during Ramadan. Trabelsi and others showed that resistance training was unaffected by fasting during Ramadan but that fasted athletes were more likely to be dehydrated. In a similar study, endurance training while fasting during Ramadan resulted in changes to various metabolic parameters related to hydration and kidney function. Although it is not currently known if sports performance *per se* is negatively impacted by fasting during Ramadan, athletes should be aware that proper hydration might be compromised.

Can the use of medium-chain triglycerides improve endurance performance or body composition?

It has been suggested that medium-chain triglycerides (MCTs) are ergogenic, possibly because they are water soluble, which may confer two advantages: They can be absorbed by the portal circulation and delivered directly to the liver instead of via the chylomicron route in the lymph, and they more readily enter the mitochondria in the muscle cells, as they do not need carnitine. MCTs have been marketed commercially. Research has shown that MCTs do not inhibit gastric emptying, as common fat does, and may be absorbed rapidly in the small intestine. Also, Massicotte and his associates reported that exogenous MCTs are oxidized at a rate comparable to exogenous glucose, being oxidized within the first 30 minutes of exercise. MCT supplements are available as a liquid, which can be added to food, or softgel capsules. MCTs are also added to meal-replacement powders, energy, and protein bars. Recently, possibly due to an interest in low-carbohydrate, high-fat diets or in foodstuffs such as coconut oil, there has been a resurgence in MCT use.

Bueno and colleagues recently completed a meta-analysis of the effects of replacing at least 5 grams of dietary long-chain fatty acids with medium-chain fatty acids in 16 studies that included 399 participants. Reportedly, replacement with MCTs resulted in greater reductions in body mass, body fat, and waist circumference. But the authors stated that their results should be taken with caution until longer-duration and better-quality research trials are completed. Mumme and others also conducted a meta-analysis of the effects of MCTs on weight loss and body composition, studying 13 trials with 749 individuals. MCT ingestion resulted in greater decreases in body mass, body fat, and subcutaneous and visceral fat, but these changes were small. Consuming MCTs resulted in about a 0.51-kilogram extra decrease in body mass over a 10 week period. More importantly, the authors systematically identified several areas of concern with the studies they included in their analysis, such as how well the MCT products were blinded, if data were selectively reported, and if there was a commercial bias. Bueno and colleagues also note that the blinding of the source of the fat in their analysis was rated "low" or "unclear" for every study.

Some early research found that MCT supplementation alone may actually impair endurance exercise performance, whereas an MCT-carbohydrate supplement might be ergogenic. Van Zyl and others, using endurance-trained cyclists, compared the effects of three supplements on an endurance performance task consisting of a 2-hour ride at 60 percent VO_2 max followed by a 40-kilometer performance ride. The three supplements, consumed throughout the performance task, were carbohydrate only, MCTs only, and carbohydrate with MCTs; the MCT dose was about 86 grams. Compared to the carbohydrate supplement, the MCT supplement actually impaired 40-kilometer performance, whereas the combination carbohydrate-MCT supplement improved performance. These investigators suggested that the carbohydrate-MCT supplement improved performance in the 40-kilometer performance ride by decreasing oxidation of muscle glycogen during the preliminary submaximal 2-hour ride, thus sparing the glycogen for the more intense exercise task.

However, most studies have not shown any beneficial effects of MCT supplementation, either alone or combined with carbohydrate, on endurance exercise performance. One such study was conducted by Goedecke and others, from Tim Noakes's laboratory in South Africa. Nine endurance-trained cyclists cycled for 2 hours at 63 percent of VO_2 peak, then completed a 40-kilometer time trial under three conditions: glucose, glucose and low-dose MCTs, and glucose and high-dose MCTs, all in solution. The

Training Table

Coconut oil: Miracle food or dietary fad?

In the media and on the internet, coconut oil is often presented as a "super food," and sales of coconut oil have increased dramatically in recent years. A recent survey revealed that 72 percent of Americans rated coconut oil as a "healthy" food, but far fewer nutrition professionals agree with this perspective. The reality is that coconut oil is more than 80 percent saturated fat, and when directly compared to consumption of mono- and polyunsaturated fats, it increases LDL cholesterol, thus increasing cardiovascular disease risk. "Bulletproof coffee," which consists of coffee, coconut oil, butter, and sometimes additional MCTs, became an "energy promoting and weight-loss" concoction. Coconut oil does not increase thermogenesis more than corn oil, and satiates less than MCT oil, making any weight-loss effects unlikely. MCT supplements, as described in this chapter, are also generally regarded as ineffective at improving exercise performance or stimulating weight loss. There are no data to support an increase in energy levels after consuming bulletproof coffee, except, perhaps, from the caffeine. In addition, a serving of bulletproof coffee contains about 400 kcal per serving, and regular consumption could easily lead to weight gain. Sales of coconut oil have begun to decline to normal levels, and it looks like another dietary fad has passed.

For additional reading on dietary fat, see the American Heart Association presidential advisory on dietary fats and cardiovascular disease: https://circ.ahajournals.org/content/early/2017/06/15/CIR.0000000000000510.

solutions were ingested immediately before and every 10 minutes during exercise. Although MCT ingestion increased serum FFA concentration, there were no beneficial effects on performance as compared to the glucose trial. In a well-designed crossover study, Angus and others compared the effects of carbohydrate to carbohydrate plus MCT on 100-kilometer cycling time trial in eight endurance-trained males. The beverages were provided during the trial at about every 15 minutes and consisted of a 6 percent carbohydrate solution with and without a 4.2 percent MCT solution. They found that compared to the placebo trial, the carbohydrate enhanced 100-kilometer cycling performance, but the addition of MCT did not provide any further performance enhancement. In a similar study with ultraendurance athletes, but with carbohydrate or MCTs provided before an ultradistance ride, Goedecke and others reported that MCT supplementation actually impaired periodic sprint performance within the event. Using muscle biopsies, Horowitz and others have also shown that a carbohydrate-MCT solution does not spare muscle glycogen during high-intensity aerobic exercise.

Using a chronic MCT feeding protocol, Misell and others had 12 trained male endurance runners consume either corn oil (LCT) or MCT (60 grams) daily for 2 weeks. The runners then performed an endurance treadmill test consisting of a 30-minute run at 85 percent VO_2 max followed by a run to exhaustion at 75 percent VO_2 max. The investigators reported that chronic MCT consumption neither enhances endurance performance nor significantly alters exercise metabolism in trained male runners. Louise Burke states that there is no evidence of benefits of chronic MCT ingestion for athletes. Also, the effectiveness of MCT is limited by the inability of athletes to tolerate the amount necessary to have an effect. So if there were any possible beneficial effects on endurance performance, they would likely be outweighed by gastrointestinal distress. The A to Z supplement guide, authored and edited by leading experts in the sport nutrition field, rates MCT research as providing good evidence to support mixed or no effect on performance and no studies or credible evidence for beneficial effects on health. The *Training Table* in this section discusses the topic of coconut oil.

Is the glycerol portion of triglycerides an effective ergogenic aid?

As you may recall, glycerol is one of the by-products of triglyceride breakdown. Burelle and others have noted that exogenous glycerol can be oxidized during prolonged exercise, presumably following conversion into glucose in the liver. Thus, researchers theorized that it could be an efficient energy source during exercise. However, in well-controlled research, glycerol feedings did not prevent either hypoglycemia or muscle glycogen depletion patterns in several prolonged exercise tasks. Apparently the rate at which the human liver converts glycerol to glucose is not rapid enough to be an effective energy source during strenuous prolonged exercise. As noted in chapter 9, glycerol may be used to increase body water stores, including plasma volume, prior to exercise, and has been theorized to be ergogenic for endurance athletes performing under warm environmental conditions.

Omega-3 fatty acid and fish oil supplements

There is a great deal of interest in the health-promoting and ergogenic effects of omega-3 polyunsaturated fatty acids. It is well known that populations who consume diets high in these fatty acids have decreased risk of cardiovascular disease. In particular, eicosapentaenoic (EPA) and docosahexaenoic (DHA) acid have been studied because of their potential anti-inflammatory effect. People who do not regularly consume oily fish probably consume less than 200 mg/day, and fish oil supplements are an easy way to increase omega-3 fatty acid intake.

Omega-3 fatty acids are theorized to be ergogenic not because of their energy content but because they may elicit favorable physiological effects relative to several types of physical performance. One theory is based on the finding that omega-3 fatty acids may be incorporated into the membrane of the red blood cell (RBC), making the RBC less viscous and less resistant to flow. Another theory is based on the role of certain by-products—the eicosanoids mentioned previously—whose production in the body cells is related to omega-3 fatty acid metabolism. In particular, two specific forms of the eicosanoids, prostaglandin E_1 (PGE_1) and prostaglandin I_2 (PGI_2), may elicit a vasodilation effect on the blood vessels. Walser and others noted that 6 weeks of EPA and DHA (total 5 g/day) supplementation enhances blood flow during exercise in healthy individuals. Theoretically, the less viscous RBC and the vasodilative effect should enhance blood flow, facilitating the delivery of blood and oxygen to the muscles during exercise, benefiting the endurance athlete.

Unfortunately, although the ergogenic potential of omega-3 fatty acids is an interesting hypothesis, there are few supportive scientific data. Results from well-controlled, peer-reviewed scientific research indicate that omega-3 fatty acids do not affect energy metabolism during exercise. For example, Bortolotti and others recently reported that supplementation with 7.2 grams of fish oil daily, containing 1.1 g/day eicosapentaenoic acid and 0.7 g/day docosahexaenoic acid, for 14 days exerted no effect on glucose or lipid energy metabolism during 30 minutes of cycling at 50 percent VO_2 max.

Research also indicates that omega-3 fatty acid supplementation has no effect on aerobic endurance performance. Buckley and others supplemented the diets of 25 professional Australian Football League players for 5 weeks in a randomized, double-blind study. The players were matched for performance of a 2,200-meter running time trial and provided either 6 × 1 gram capsules of either sunflower oil (placebo) or DHA-rich fish oil (1.56 grams DHA and 0.36 gram eicosapentaenoic acid). At the end of the 5 weeks of supplementation, the fish oil group had lower serum triglycerides and lower heart rate during submaximal exercise. Buckley and others did not find improvements in endurance exercise performance, as determined by time to exhaustion, or recovery. In a slightly longer study, Peoples and colleagues supplemented the diets of 16 well-trained male cyclists for 8 weeks with either 8 × 1 grams of olive oil (control) or fish oil in a double-blind, parallel design. The fish oil supplementation lowered heart rate during VO_2 peak testing and steady-state submaximal exercise, as well as whole-body oxygen consumption and rate pressure product (systolic blood pressure × heart rate). Time to voluntary fatigue was not influenced by fish oil supplementation.

In their review of 53 studies of omega-3 fatty acid supplementation with doses ranging from 0.06 to 4.9 g/day for up to 24 weeks, Thielecke and Blannin noted that although these supplements may have measurable physiological and metabolic effects, this did not always translate into improved performance.

There are interesting preliminary data that support athletes purposefully eating foods high in omega-3 fatty acids or taking omega-3 supplements. First, cognitive processing improves following omega-3 fatty acid supplements, but it is not known if this will work in healthy young athletes, or if it will benefit sports performance. Second, animal research and some case studies show that omega-3 fatty acid supplementation reduces the damage and cognitive decline associated with concussion. Again, data in athletes are not available. Finally, data from Kevin Tipton and Oliver Witard's labs show that omega-3 fatty acid supplements can increase protein synthesis, and, when combined with protein, carbohydrate, and leucine, decrease muscle soreness. Murphy and McGlory theorize that omega-3 fats may potentiate the anabolic response of skeletal muscle at both resting and postexercise conditions but only in response to suboptimal protein ingestion. Xin and Eshaghi completed a meta-analysis of ten omega-3 supplementation studies and reported reduced blood markers of exercise-induced muscle damage, although data on muscle function and recovery were not reported. In their systematic review, Heileson and Funderburk also supported the idea that omega-3 fatty acid supplementation could enhance recovery from damaging exercises. The recent International Olympic Committee consensus on dietary supplements and review by Rawson and others state that it is unclear if omega-3 fatty acid supplementation should be pursued by athletes, instead of just including fatty fish in the diet as a natural and concentrated source. However, Murphy and McGlory suggested that it might take 8-20 servings of oily fish per week to achieve a dose of omega-3 fatty acids that could elicit ergogenic effects (e.g. about 3 g/day).

Mickleborough noted the following potential safety issues related to fish oil supplementation:

- Supplements could be made from fish that are contaminated with heavy metals. This is theoretical and has not been reported in the literature, but as discussed throughout this textbook, dietary supplement quality control can be a concerning issue.
- Fish oil supplements can cause increased bleeding due to decreased platelet stickiness. This could be an issue, especially at higher doses.
- Digestive problems, such as flatulence and diarrhea, are possible. Ingesting fish oil supplements before or during exercise performance could be ergolytic due to gastrointestinal distress.
- High doses of fish oil may increase LDL-cholesterol.
- Fish oil supplements can decrease blood pressure, which could be a problem for an athlete who is already hypotensive.
- Fish oil supplements can result in a fishy taste in the mouth.

Can carnitine improve performance or weight loss?

Carnitine is a water-soluble, vitamin-like compound that facilitates the transport of long-chain fatty acids into the mitochondria. There are basically two forms of carnitine, L-carnitine and D-carnitine, but other forms are available, such as L-propionylcarnitine. L-carnitine is the physiologically active form in the body, so in the following discussion, carnitine will refer mostly to L-carnitine, but in some studies L-propionylcarnitine has been used.

Dietary Sources Carnitine is not an essential dietary nutrient because it may be formed in the liver from other nutrients—principally two amino acids, lysine and methionine. Also, carnitine is found in substantial amounts in animal foods. Meat products, particularly beef and pork, are good sources of carnitine; much less is found in fish and poultry, and even lower amounts in dairy products. Only minimal amounts of carnitine are found in fruits, vegetables, and grains. For example, for similar weights, beef has about 300 times as much carnitine as bread; 3 ounces of beef contains about 60 mg of carnitine. There is no RDA for carnitine. Most individuals consume enough carnitine in the daily diet, and the body has an effective conservation system. The typical nonvegetarian diet provides about 100–300 mg/day. Carnitine deficiencies are very rare.

Theory as an Ergogenic Aid Carnitine supplementation has been theorized to enhance physical performance and weight loss because of several of its metabolic functions in the muscle cell. Approximately 90 percent of the body supply of carnitine is located in the muscle tissues, where it is part of an enzyme (carnitine palmitoyl transferase) important for transport of long-chain fatty acids into the mitochondria for oxidation. Theoretically, supplemental carnitine might facilitate the transport of LCFAs into the mitochondria for oxidation, which would be an important consideration if the oxidation of fatty acids were limited by their transport into the mitochondria.

Effects on Performance Although these are logical theories and interesting medical applications, the available scientific evidence is somewhat equivocal and in general does not appear to support an ergogenic effect of carnitine supplementation. Major reviews regarding the effect of carnitine supplementation on physical performance have been published, and the following are the key points regarding the ergogenic effects of carnitine supplementation emanating from these reviews and studies:

1. Supplementation will increase plasma levels of carnitine, but much of this will be excreted by the kidneys. Stephens and Greenhaff state that the bioavailability of carnitine is less than 15 percent of a 2- to 6-gram dose. They estimate that it would take about 100 days of supplementation to increase muscle carnitine 10 percent. Several earlier studies showed that oral carnitine supplementation cannot increase muscle carnitine levels. However, Wall and others were able to increase muscle carnitine 21 percent in 168 days when combined with 80 grams of carbohydrate twice daily. These data must be reproduced, and athletes should consider the effects of an extra 640 kcal of sugar per day, particularly if they are attempting to lose weight.

Training Table

Dietary supplements for athletes

One of the most commonly asked questions by athletes is "What supplements should I take?" The International Olympic Committee convened a group of researchers to write a consensus statement and an in-depth series of review articles to answer this question. As mentioned earlier in the text, despite the large number of dietary supplements marketed as ergogenic aids, the number of safe and effective dietary supplements is quite small. For omega-3 fatty acid supplementation, it was noted that it was not known if cognitive processing improvements result in improved athletic performance; there are few data on mild traumatic brain injury (i.e. concussion); reduced muscle damage is not a consistent finding; and it was unclear if supplementation should be pursued by athletes, or if including fatty fish in the diet as a source of omega-3 fatty acids was more prudent.

To read the consensus statement on dietary supplements and the high-performance athlete, visit: http://dx.doi.org/10.1136/bjsports-2018-099027.

5. D-carnitine may be toxic, as it can deplete L-carnitine, leading to a carnitine deficiency. L-carnitine appears to be a safe supplement, but some reviewers recommend no more than 2–5 grams per day, possibly for only 1 month at a time.

Carnitine is generally thought of as an ineffective ergogenic aid, but this is probably because it is difficult to increase muscle carnitine. Until a way to increase muscle carnitine becomes available and more practical, supplementation cannot be recommended, although more research needs to be done.

Can hydroxycitrate (HCA) enhance endurance performance?

imagenavi/Getty Images

Hydroxycitric acid is derived from a tropical fruit and marketed as a dietary supplement, hydroxycitrate (HCA). Kriketos and others noted that as a competitive inhibitor of citrate lyase, HCA has been hypothesized to modify citric acid cycle metabolism to promote fatty acid oxidation. Although some studies with mice suggest that HCA supplementation may enhance endurance performance, human studies do not support such an ergogenic effect.

Kriketos and others, in an excellent crossover study, reported no significant effect of HCA supplementation (3.0 g/day for 3 days) on blood serum energy substrates, fat metabolism, or energy expenditure either during rest or during moderately intense exercise. However, the subjects were sedentary males. Using endurance-trained cyclists as subjects, van Loon and others evaluated the effect of HCA supplementation (3.1 ml/kg body mass of a 19 percent HCA solution) given twice to the cyclists before and during 2 hours of cycling at 50 percent VO_2 max. They concluded that HCA supplementation, even in large quantities, does not increase total fat oxidation in endurance-trained cyclists.

Melinda Manore, an expert in sports nutrition, recently reviewed several common weight-loss supplements. Regarding HCA, she stated that there is no benefit of HCA on weight loss, fat oxidation, or appetite. Additionally, there have been several safety issues raised with HCA, including liver injuries, although it is unknown if this is related to poor quality control or the plant extract itself. Saito, using an animal model, reported that HCA was toxic to and caused atrophy of the testes. In their review, Márquez and others state that most HCA supplementation studies in humans have been small and short in duration (e.g., not longer than 12 weeks). They concluded that there is little evidence to support the effectiveness HCA.

Can conjugated linoleic acid (CLA) enhance exercise performance or weight loss?

Conjugated linoleic acid (CLA) has been marketed as a sports dietary supplement to resistance-trained individuals, mainly as a means to promote weight loss and gain muscle mass. For similar and other reasons, CLA also has been marketed as a means to improve health.

2. The primary theory underlying carnitine supplementation is enhanced fat utilization. If muscle carnitine does not increase with supplementation, no increase in fat oxidation should be expected, and this has been reported. However, in the study by Wall and others in which muscle carnitine did increase, muscle glycogen use decreased 55 percent, muscle lactate was 44 percent lower, and work output improved. Numerous studies have reported no ergogenic effect under a variety of exercise tests conditions, likely because muscle carnitine did not increase in those trials. As the number of studies showing no metabolic or ergogenic effect of carnitine far outweigh the number that have shown an effect, carnitine cannot be recommended until more research is conducted.

3. Preliminary data, from William Kraemer's research group, suggest that combining carnitine with tartrate may have some beneficial effects for individuals engaged in resistance training. Tartrate, a salt, possesses antioxidant properties. Spiering and others reported that carnitine supplementation, either 1- or 2-gram doses for 3 weeks, reduced markers of metabolic stress and perceived muscle soreness following a resistance training workout. Kraemer and others noted that carnitine/tartrate supplementation may increase responses of androgenic receptors that help muscle formation, which they suggest may help promote recovery from resistance exercise. These preliminary findings are interesting and merit additional research, possibly evaluating the effects of carnitine and tartrate separately.

4. Both Asker Jeukendrup and Melinda Manore, two experts on dietary supplements and sports nutrition, have reviewed the purported effects of carnitine on weight loss. Neither finds any evidence that carnitine is effective for weight loss.

Research regarding its ergogenic effect on exercise-trained individuals is very limited. Using resistance-trained athletes as subjects, Kreider and others investigated the effect of CLA supplementation (6 grams prescribed daily dose; 28 days) on body composition and muscular strength, and reported no significant effects on total body mass, fat mass, fat-free mass, or strength as measured by a single maximal repetition in the bench press and leg press. Pinkoski and others, in a crossover design, also found that CLA supplementation (5 grams daily for 7 weeks) resulted in minimal changes in body composition and no changes in the strength tests in males and females involved in resistance training. However, in the first phase of the study, which did not involve a crossover design, male subjects receiving CLA experienced significant increases in bench press strength compared to the placebo group. The crossover phase of the study, as the authors note, is a stronger experimental design.

Macaluso and others recently reviewed seven studies on CLA supplementation and exercise performance and noted that more than half showed no beneficial effect. Onakpoya and colleagues, in their meta-analysis, concluded that CLA supplementation results in a statistically significant, albeit small and possibly clinically irrelevant, decrease in body mass relative to placebo ingestion. Presently, adverse reactions to CLA supplementation are mild, with gastrointestinal distress being most commonly reported. Based on limited research, CLA supplementation does not currently appear to be an effective ergogenic aid for trained individuals, but confirming research is needed.

Can ketone supplements improve endurance performance?

While some have attempted to use low-carbohydrate high-fat diets to increase ketone production, a new approach is ketone supplements. Cox and others found increased serum ketones, and improved cycling time-trial performance (about 2 percent) following ketone diester ingestion. In contrast, Leckey and others showed increased serum ketones, but impaired cycling time-trial performance in world-class cyclists ingesting a ketone diester beverage. Although the effects of ketone supplements on exercise performance is a small body of literature, several reviews are now available. Margolis and O'Fallon, in their systematic review, noted that the effects of ketone supplements are equivocal and that out of 16 performance outcomes, 3 were positive, 10 showed no effect, and 3 were negative. In their systematic review and meta-analysis of eight studies of ketone supplementation, Brooks and others also found that ketone supplements did not improve endurance performance. Finally, in their systematic review and meta-analysis, Valenzuela and colleagues found no effect of ketone supplements on exercise performance. It seems that there is insufficient evidence to recommend ketone supplements at this time. It is difficult to draw conclusions from such a small body of literature, but there are several reasons to be cautious. In the study by Leckey and colleagues, all of the participants reported some form of gastrointestinal distress, including dry retching, nausea, reflux, discomfort, vomiting, and dizziness. Sport nutrition experts Asker Jeukendrup and Luc van Loon report that ketone esters are very expensive and taste badly, which, along with poor gastrointestinal tolerance, will likely limit consumption of enough ketones to make a difference in fueling sports performance.

Primoz Korosec/123RF

What's the bottom line regarding the ergogenic effects of fat burning diets or strategies?

As noted in this section, numerous fat burning strategies have been employed in attempts to enhance prolonged endurance exercise performance. Theoretically, such strategies would increase the oxidation of fat, decrease the utilization of carbohydrate, and thus spare some muscle glycogen for use in the later stages of exercise. As muscle glycogen is a more efficient fuel compared to fats, performance should be enhanced.

However, John Hawley, an international sports science scholar, reviewed all such fat burning strategies and concluded that endurance exercise capacity is not systematically improved with increases in serum FFA availability and fat oxidation, even in some studies with substantial muscle glycogen sparing. He noted that for some reason, exercise capacity is remarkably resistant to change. Other studies suggest that high-fat diets may impair performance in some events, such as high-intensity surges during a race or sprints to the finish. Moreover, individuals may find it difficult to adhere to a high-fat diet.

> ### Key Concepts
> ▸ Fat-loading practices, either acute or chronic, may increase utilization of fat during endurance exercise but do not appear to enhance exercise or sports performance. Chronic high-fat diets may increase the psychological stress of exercise training.
> ▸ Fasting, or exercising on an empty stomach, does not improve performance or weight loss.
> ▸ Medium-chain triglycerides and ketone supplements do not improve and may impair endurance exercise performance.
> ▸ Omega fatty acid supplements may offer some benefits in terms of reducing symptoms of muscle damage, but performance enhancement is unlikely.
> ▸ In general, various dietary strategies and dietary supplements theorized to increase oxidation of fat during exercise and enhance prolonged aerobic endurance performance have not been shown to be effective ergogenic aids.

> **Check for Yourself**
>
> ▶ Check an online store that sells sports supplements that supposedly will help you burn more fat, primarily as a means to enhance endurance performance. Evaluate the supplement fact labels for content and performance claims and compare them to the text discussion. Do you recommend the supplement? Why or why not?

Dietary Fats and Cholesterol: Health Implications

As noted in previous chapters, the etiology of chronic diseases such as cancer and coronary heart disease is complex and involves multiple risk factors. Eliminating or reducing as many risk factors as possible is the best approach to optimize your health. Your diet is one of the most important risk factors that can be modified to promote good health, particularly the amount and composition of fat you eat. Every few years professional societies such as the Academy of Nutrition and Dietetics and the American Heart Association provide reviews of the evidence regarding various dietary practices and effects on cardiovascular disease. In this section, we shall focus on the role of dietary fat in the etiology of coronary heart disease (CHD). Chapters 10 and 11 then discuss the impact of dietary fat intake on obesity and related health conditions.

CHD is still the number one cause of death in industrialized nations. In its recommendations regarding dietary fat, the National Academy of Sciences indicated that although very high-fat diets may predispose to CHD, so, too, may very low-fat diets. For example, Siri and Krauss reported that increased dietary carbohydrates, particularly simple sugars and starches with high glycemic index, can modify the serum lipid profile in ways that may also be conducive to CHD. Lipid expert Penny Kris-Etherton and her associates recently noted that the main message regarding dietary fat is very simple: Avoid diets that are *very low* and *very high* in fat. The guiding principle is that moderation in total fat is the defining benchmark for a contemporary diet that reduces risk of chronic disease. Moreover, within the range of a moderate-fat diet, it is still important to individualize the total fat prescription. Fats are not all equal. Some are better for your health. The key is moderation.

Because the available evidence relating dietary lipids to cardiovascular disease is so compelling, we shall treat this subject in some detail. However, note that the dietary and exercise recommendations advanced later in this chapter for the prevention of cardiovascular disease may also help prevent other chronic diseases, such as obesity and certain forms of cancer.

How does cardiovascular disease develop?

Nearly one out of every two deaths in the United States is due to diseases of the heart and blood vessels. Each year, approximately 1 million Americans die from some form of cardiovascular disease, including coronary heart disease, stroke, hypertensive disease, rheumatic heart disease, and congenital heart disease.

Coronary heart disease is the major disease of the cardiovascular system; of the million deaths noted previously, it is responsible for more than half. Although the total percentage of deaths due to coronary heart disease has been declining in recent years, it is still an epidemic and the number one cause of death among both males and females.

Coronary heart disease (CHD) is also known as **coronary artery disease (CAD)** because obstruction of the blood flow in the coronary arteries is responsible for the pathological effects of the disease. The coronary arteries, which nourish the heart muscle, are illustrated in **figure 5.12**. The major manifestation of CHD is a heart attack, which results from a stoppage of blood flow to parts of the heart muscle. A decreased blood supply, known as **ischemia**, will deprive the heart of needed oxygen. In some individuals, ischemia results in **angina**, a sharp pain in the chest, jaw, or along the inside of the arm indicative of a mild heart attack. Other terms often associated with a heart attack include **coronary thrombosis**, a blockage of a blood vessel by a clot (thrombus); **coronary occlusion**, which simply means blockage; and **myocardial infarction**, death of heart cells that do not get enough oxygen due to the blocked coronary artery. The major cause of blocked arteries is atherosclerosis.

Arteriosclerosis is a term applied to a number of different pathological conditions wherein the arterial walls thicken and lose their elasticity. It is often defined as hardening of the arteries. **Atherosclerosis**, one form of arteriosclerosis, is characterized by the formation of **plaque**, an accumulation of fatty acids, oxidized LDL-cholesterol, macrophages (white blood cells that oxidize LDL), foam cells (macrophages that consume cholesterol), cytokines (immune system mediators of inflammation), cellular debris, fibrin, and calcium on the inner lining of the coronary artery wall. A cap of smooth muscle cells forms around the plaque to prevent contact with the arterial wall. **Figure 5.13** presents a schematic of the content of arterial plaque.

Inflammatory processes precipitated by cytokines are now recognized to play a central role in the pathogenesis of atherosclerosis by interacting with serum cholesterol and serving as an initiating factor in plaque buildup. As the plaque accumulates, the diameter of the artery is diminished, decreasing blood flow to the heart muscle. Foam cells continue to accumulate, becoming a major component of plaque. The foam cells secrete a substance that can weaken the muscle cap. If the muscle cap ruptures, plaque will leak into the bloodstream and trigger the formation of a clot, which partially or completely blocks blood flow to a section of heart muscle, leading to death of cardiac cells due to inadequate oxygen and nutrients. The process is depicted in **figure 5.14**.

Atherosclerosis is a slow, progressive disease that begins in childhood and usually manifests itself later in life. Because of its prevalence in industrialized society, scientists throughout the world have been conducting intensive research to identify the cause or causes of atherosclerosis and coronary heart disease. The actual cause has not yet been completely identified, but considerable evidence has identified factors that may predispose an individual.

As noted previously, a risk factor represents a statistical relationship between two items such as high serum cholesterol and heart attack. This does not mean that a cause-and-effect relationship exists, although such a relationship is often strongly supported by

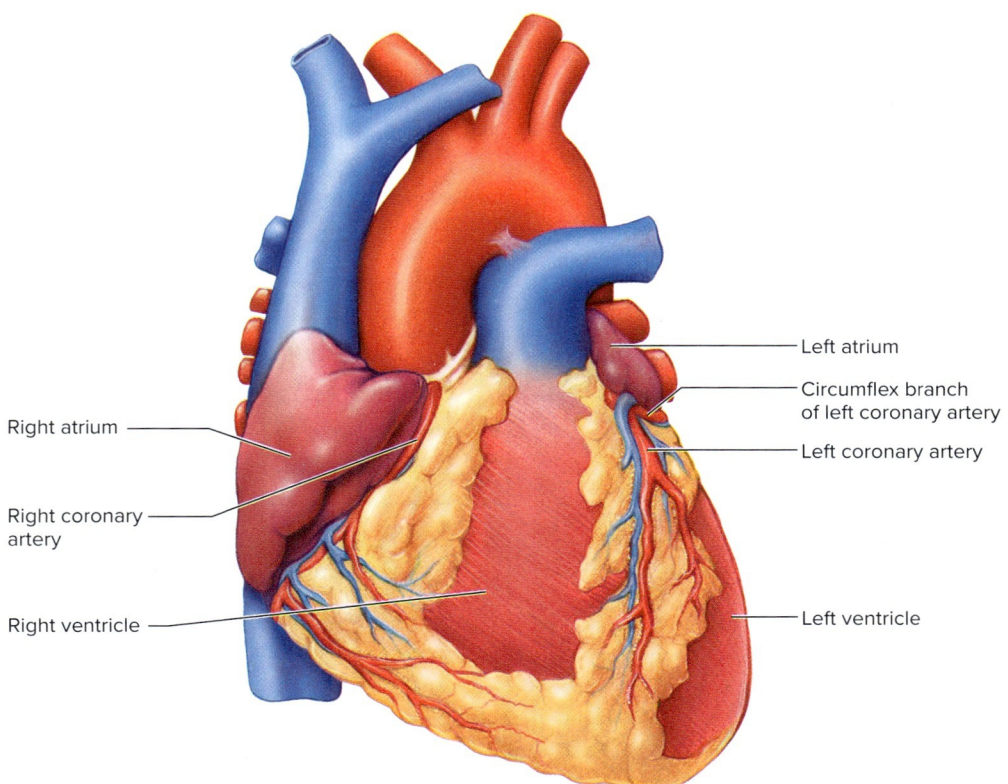

FIGURE 5.12 The heart and the coronary arteries. The heart muscle itself receives its blood supply from the coronary arteries. The main coronary arteries are the left coronary artery, one of its branches named the circumflex, and the right coronary artery. Atherosclerosis of these arteries leads to coronary heart disease.

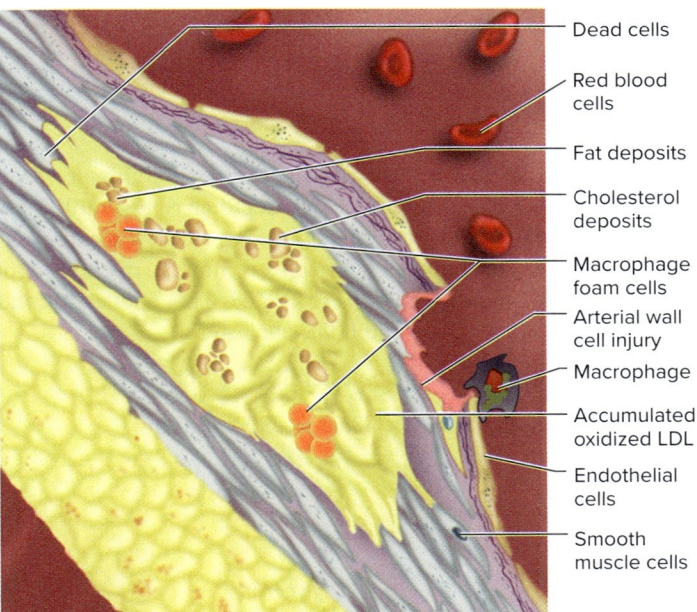

FIGURE 5.13 An enlargement of atherosclerotic plaque. Oxidized LDL-cholesterol, macrophages, foam cells, fibrous material, and other debris collect beneath the endothelial cells lining the coronary artery. The site of the plaque may be initiated by some form of injury to the cell lining, possibly an ulceration as shown.

the available evidence. The three principal risk factors associated with CHD are high blood pressure, high serum cholesterol levels, and cigarette smoking. Several major professional and governmental health organizations also believe that physical inactivity is a fourth principal risk factor. Other interacting risk factors are heredity, diabetes, diet, obesity, age, gender, and stress. Elevated blood levels of homocysteine and C-reactive protein (CRP) have been associated with CHD, and their role as risk factors continues to be studied. Homocysteine is discussed in chapter 7. CRP is a marker for inflammation.

www.heart.org/en/healthy-living/healthy-lifestyle/my-life-check-lifes-simple-7
Use this link to assess your risk for things like heart attacks and stroke and to learn how to improve heart health.

How do the different forms of serum lipids affect the development of atherosclerosis?

In atherosclerosis, the plaque that develops in the arterial walls is composed partly of fats and cholesterol. Hence, high levels of blood lipids (triglycerides and cholesterol) are associated with increased plaque formation. However, as you recall, triglycerides and cholesterol may be transported in the blood in a variety of

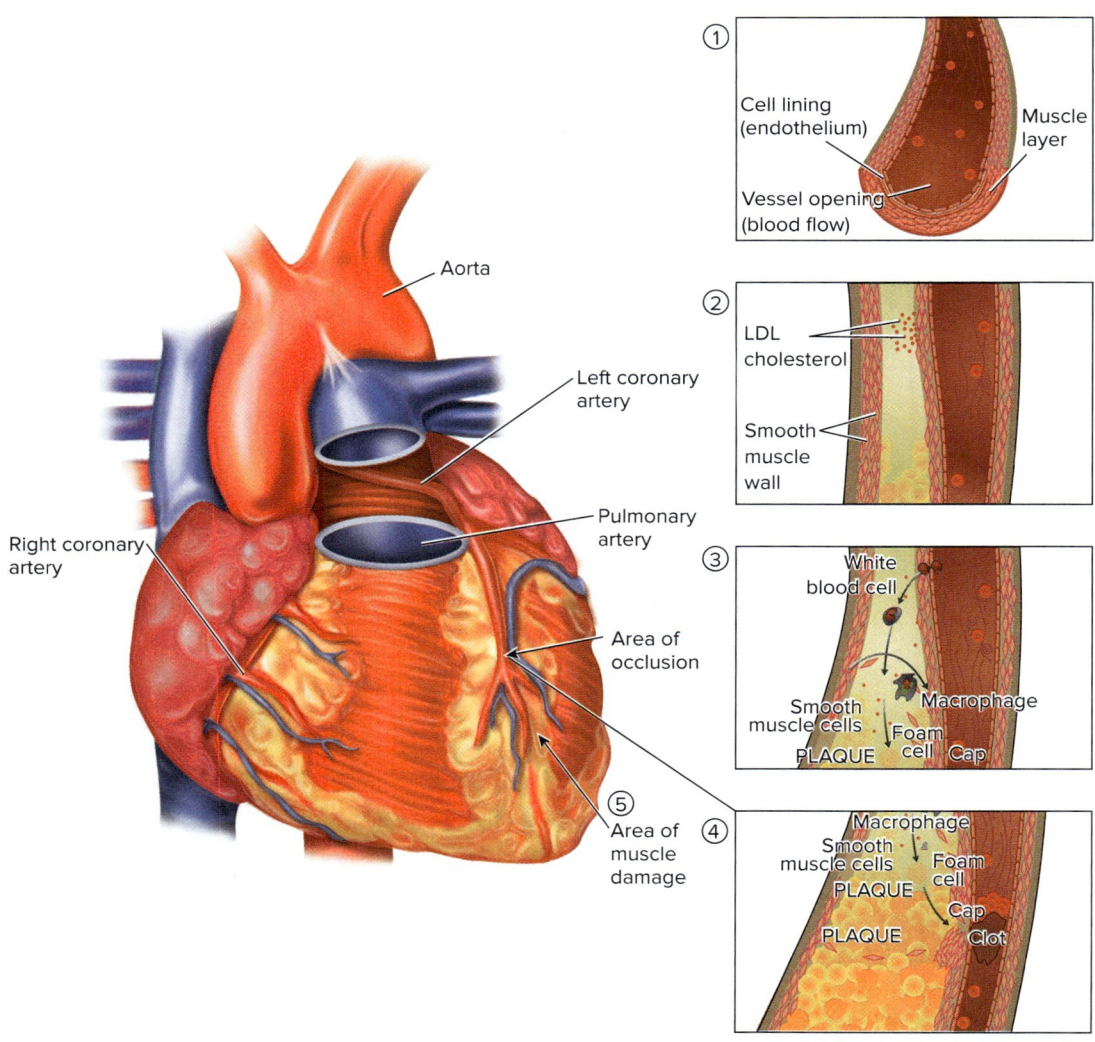

FIGURE 5.14 The developmental process of atherosclerosis and thrombosis. (1) Normal coronary artery. (2) LDL-cholesterol seeps through the endothelium into the smooth muscle wall. (3) Immune system responds, sending macrophages (white blood cells) to ingest the LDL-cholesterol; macrophages become foam cells, a major component of plaque; a smooth muscle cap forms to protect the endothelium. (4) Foam cells accumulate, secreting a substance that weakens the cap; the cap ruptures; plaque leaks into the bloodstream, triggering clot formation. (5) Blocked artery leads to a decrease or cessation of blood flow to a section of heart tissue, leading to heart attack, which may be mild or fatal depending on the severity of the blockage.

ways, but primarily as constituents of lipoproteins. Considerable research has been devoted to identifying those specific lipoproteins and other lipid components that may predispose to CHD, and although there is some debate about the meaningfulness of specific serum lipid profiles, some theories prevail.

The four main serum lipid factors associated with increased risk of atherosclerosis are total cholesterol, LDL-cholesterol, HDL-cholesterol, and triglycerides. The guidelines for blood serum level profile are presented in **table 5.7**.

Serum lipid levels are normally given in milligrams per deciliter (mg/dl), as shown in **table 5.7**. A deciliter is 100 milliliters. However, you may see cholesterol and triglyceride levels expressed as millimole per liter (mmol/L). To convert mmol/L of cholesterol to mg/dl, simply multiply mmol/L by 38.67. This applies to total cholesterol as well as LDL- and HDL-cholesterol.

Example: Total cholesterol = 7.5 mmol/l

7.5 mmol/L × 38.67 = 290 mg/dL

TABLE 5.7	Recommended lipoprotein profile		
Total cholesterol	**LDL-cholesterol**	**HDL-cholesterol**	**Triglycerides**
Less than 200—desirable 200–239—borderline high 240 or above—high	Less than 100—optimal 100–129—near optimal 130–159—borderline high 160–189—high 190 and above—very high	Less than 40—low 60 or above—protective	Less than 150—normal 151–199—borderline high 200–499—high 500 and above—very high

*Fasting levels expressed in mg/dl; testing recommended every 5 years for those over 20.

To convert mmol/L of triglycerides into mg/dl, simply multiply by 88.57.

Example: Serum triglycerides = 1.5 mmol/l

1.5 × 88.57 = 132.9 mg/dl

To convert in the opposite direction, from mg/dl to mmol/L, simply divide mg/dl by the appropriate numerical factor for cholesterol or triglycerides.

Total Blood Cholesterol As noted in **table 5.7**, a cholesterol level below 200 mg/dl is considered to be desirable, between 200 and 239 mg/dl is borderline-high, and above 240 mg/dl is high. However, you should be aware that there is a rather large standard error of measurement involved in some tests of cholesterol, being on the order of 30 mg. What this means is that if your blood cholesterol is reported as 220 mg/dl (borderline high), you may actually have a cholesterol level of 190 mg/dl (desirable) or 250 mg/dl (high) if you vary, respectively, one standard error below or above your actual measurement of 220 mg/dl. For this reason, it may be a good idea to have a second test completed if you are concerned about your total cholesterol level.

LDL-Cholesterol The form by which cholesterol is transported in your blood may also be related to the development of atherosclerosis. In general, a high level of low-density lipoproteins (LDL) is the major risk factor associated with atherosclerosis. A current theory suggests various forms of LDL, such as small, dense LDL and the variant lipoprotein (a), may be more prone to oxidation by macrophages at an injured site in the arterial epithelium, leading to an influx into the cell wall and the formation of plaque. The presence of oxygen-free radicals has been suggested to accelerate this process. Other mechanisms, such as increased clotting ability, may be operative. As noted in **table 5.7**, LDL levels less than 100 mg/dl are optimal, while those above 160 mg/dl pose a high risk and those above 190 mg/dl a very high risk. Although not normally listed in risk factor tables, lipoprotein (a) values greater than 25–30 mg/dl of blood are associated with increased risk of CHD. A high level of IDL (which some consider a form of LDL) is also recognized as a risk factor, as is apolipoprotein B, involved in cholesterol transport to the tissues.

HDL-Cholesterol Conversely, high levels of high-density lipoproteins (HDL), particularly the subfraction HDL_2 and HDL with apolipoprotein A-I, appear to be protective against the development of atherosclerosis, although research is continuing to explore other relationships. Levels of 60 mg/dl or more of HDL appear to be protective, but because HDL varies daily, several measurements over time may be required to obtain an accurate reading. Research suggests that HDL interacts with the arterial epithelium, acting as a scavenger by picking up cholesterol from the arterial wall and transporting it to the liver for removal from the body, known as *reverse cholesterol transport*. HDL may also inhibit LDL oxidation and platelet aggregation. HDL_2 levels are higher in women until menopause and then decrease, with an associated increased risk for CHD.

Triglycerides Jacobson and others noted that elevated triglyceride levels may be a significant independent risk factor for coronary heart disease. Also, it is often associated with increased levels of LDL, particularly the small, dense LDL, and decreased levels of HDL. Current guidelines from the American Heart Association indicate that triglyceride levels below 150 mg/dl are normal, whereas the risk associated with progressively increasing levels goes from borderline high to very high (**table 5.7**). Jacobson and others indicate increasing concern over the increasing rate of hypertriglyceridemia, which is associated with overweight and obesity.

A summary of serum lipid factors associated with increased risk of atherosclerosis is presented in **table 5.8**.

TABLE 5.8 Serum lipid factors associated with increased risk of atherosclerosis

High levels of total cholesterol
High levels of LDL-cholesterol
High levels of dense form of LDL-cholesterol
High levels of IDL-cholesterol
High levels of abnormal lipoprotein, lipoprotein (a)
High levels of apolipoprotein B
High levels of triglycerides
Low levels of HDL-cholesterol
Low levels of HDL_2-cholesterol
Low levels of apolipoprotein A-I

Cholesterol Ratios and Other Tests If your total blood cholesterol is borderline or high, a determination of the LDL and HDL levels may be desirable, for they provide additional information relative to your risk. Based on epidemiological data, several ratios have been developed to assess risk of CHD, with the lower the ratio, the lower the risk.

One common comparison is the ratio of total cholesterol (TC) to the HDL level, or TC/HDL. A ratio of about 4.5 is associated with an average risk for CHD. For example, an individual with a total cholesterol of 200 mg/dl and an HDL of 60 mg/dl would have a ratio of 3.33 (200/60), or a lower risk, while someone with the same total cholesterol but an HDL of 20 mg/dl would have a much higher risk with a ratio of 10 (200/20).

Another comparison is the ratio of LDL to HDL or LDL/HDL. An LDL to HDL ratio of about 3.5 is considered to be an average risk for CHD. Thus, a ratio of 140/60, or 2.3, would be a much lower risk than 140/20, or 7.0.

Additional tests are merited for those at high risk for CHD. Special lipoprotein tests may measure density levels of the various lipoproteins. Tests for other markers of atherosclerosis, such as homocysteine and CRP, will add to the diagnosis. For example, Ridker and others noted that high CRP levels may be more effective predictors of heart attacks than high LDL-cholesterol levels, but the greatest risk is when both are high.

https://static.heart.org/riskcalc/app/index.html#!/baseline-risk
Check your cardiovascular disease risk with this calculator from the American Heart Association and the American College of Cardiology.

Can I reduce my serum lipid levels and possibly reverse atherosclerosis?

Lowering your serum cholesterol, particularly your LDL-cholesterol, is a very effective means to help prevent CHD. LaRosa indicated that for each 1 percent reduction in LDL-cholesterol, there is a corresponding 1 percent reduction in coronary heart disease risk. Several approaches may help you improve your serum lipid levels and reduce the risk of atherosclerosis. A healthy lifestyle is one, and appropriate drug therapy is another. However, even with a healthy lifestyle some individuals may have poor serum lipid profiles. Certain forms of hypercholesteremia are genetic. In the future, gene therapy may be the treatment of choice for such individuals, possibly manipulating genes to decrease LDL and increase HDL.

Drug therapy may be required to reduce serum lipid levels in genetically predisposed individuals, as well as in those with poor diets. Some drugs stimulate liver degradation and excretion of cholesterol, others increase lipoprotein lipase activity or LDL receptor function to decrease serum triglyceride levels, while still others bind with bile salts in the intestines so that they are not reabsorbed; because bile salts are derived from cholesterol, it is effectively excreted from the body. Statins, drugs that inhibit an enzyme (HMG-CoA reductase) that regulates cholesterol, have been particularly effective to reduce serum LDL-cholesterol. For more detail on current and proposed medicinal means to help lower serum lipids, the interested reader is referred to this summary from the American Heart Association: www.heart.org/en/health-topics/cholesterol/prevention-and-treatment-of-high-cholesterol-hyperlipidemia/cholesterol-medications.

What should I eat to modify my serum lipid profile favorably?

One of the first steps is to identify those individuals with high serum cholesterol by various simplified screening techniques, such as the measurement of total cholesterol by small samples of blood obtained through fingertip capillary blood. If this measure is borderline high (200–239 mg/dl) or high (>240 mg/dl), venous blood samples may be taken to determine LDL and HDL levels. If high serum cholesterol levels are detected, dietary modifications and other appropriate lifestyle changes may be recommended. The updated guidelines from the American Heart Association and the American College of Cardiology encourage a healthy diet and regular physical activity as a foundation to prevent cardiovascular disease and related risk factors, such as hyperlipidemia. However, if there is disease, or if lifestyle modifications do not improve risk factors, medications are recommended in addition to lifestyle modification.

What is a heart healthy diet?

The dietary modifications most able to lower blood cholesterol are decreased intake of saturated and trans fats. This is best accomplished by reducing saturated fat intake to less than 6 percent of daily kcal intake. As described in earlier chapters, limiting red meat, whole fat dairy products, and fried foods and cooking with mono- and poly-unsaturated oils are sensible choices. Additionally, diets like the heart-healthy DASH diet emphasize fruits, vegetables, and grains; replace red meats with poultry, fish, and nuts; and limit sodium and sugar sweetened foods and drinks. Achieving recommended physical activity levels, smoking cessation, and appropriate weight loss augment these dietary modifications. See **figure 5.15** for DASH diet heart healthy recommendations. A sensible plan to reduce serum lipid levels is presented in **figure 5.16**, and representative results are shown in **figure 5.17**. The American Heart Association has collected information about cholesterol and heart health here: www.heart.org/en/health-topics/cholesterol

If atherosclerotic cardiovascular disease (ASCVD) is present, the American Heart Association and American College of Cardiology have established the recommendations outlined in **table 5.9**.

A list of the top ten prudent dietary guidelines to reduce high serum lipids or maintain normal serum lipids is highlighted in

FIGURE 5.15 Recommended changes using the DASH eating plan to help lower blood pressure and LDL.
Source: NIH.

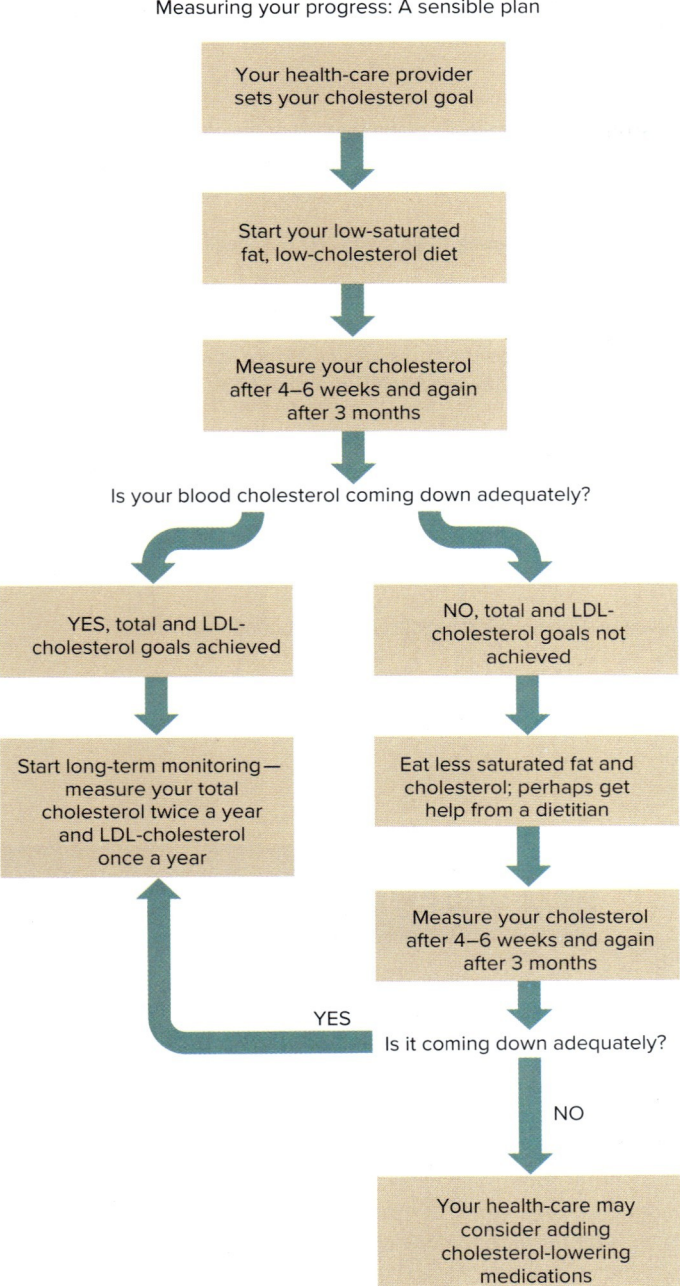

FIGURE 5.16 A sensible plan to monitor the effects of a cholesterol-lowering diet. In individuals highly resistant to dietary modifications, drug therapy may be prescribed.

Source: U.S. Department of Health and Human Services.

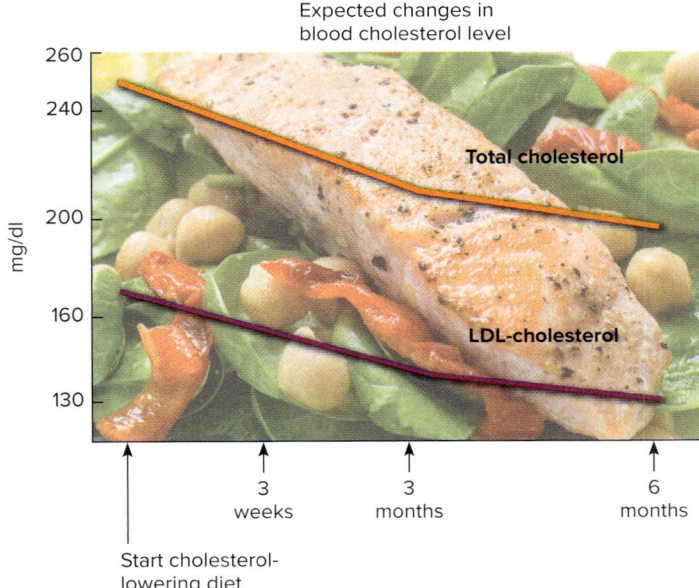

FIGURE 5.17 Representative expected changes in total serum cholesterol and LDL-cholesterol associated with appropriate dietary modifications.

Source: U.S. Department of Health and Human Services.

a *Training Table* in this section. Detailed recommendations on managing serum lipids are as follows:

1. *Adjust caloric intake to achieve and maintain healthy body-weight and composition.* One of the most common causes of high triglyceride levels is too much body fat, particularly in the abdominal region. The health risks of obesity are detailed in chapter 10. In many cases, simply losing body weight or reducing caloric intake will reduce serum lipids.

2. *Reduce the total amount of fats in the diet.* Reducing the total amount of fat will usually reduce the amount of kcal, but nutrient content will actually improve. Reducing total fat intake to 20 percent or lower of total daily kcal, as recommended in some healthy diet plans, will reduce total and LDL-cholesterol even more. However, a very low-fat diet (10 percent fat kcal) may actually cause a negative lipid profile in some individuals by decreasing HDL-cholesterol. To help prevent this, the carbohydrate kcal that replace the fat kcal should *not* be derived from refined carbohydrates but rather from complex carbohydrates containing dietary fiber.

However, for a style of eating to be effective at improving health, it must be sustainable. Gibson and others, in their meta-analysis, demonstrated that individuals on low-carbohydrate, high-fat/ketogenic diets were less hungry, had a decreased desire to eat, and had a greater feeling of satiety/fullness. However, Kevin Hall, an obesity expert, conducted a meta-analysis of 32 isocaloric, controlled feeding studies, and showed that both energy expenditure and fat loss were greater with low fat diets. Johnston and others recently analyzed 48 randomized weight-loss trials with dietary fat intakes ranging from low fat (<20 percent of daily kcal) up to high fat (55 percent of total kcal). They determined that the differences between the diets were too small to be important to affect weight loss; thus, low fat or high fat was not an important factor for determining weight loss. However, long-term diet-induced weight-loss studies are among the most difficult type of research studies to complete, and participants often have poor compliance to the dietary intervention. Because

TABLE 5.9 Guidelines on the Management of Blood Cholesterol

1. In all individuals, emphasize a heart-healthy lifestyle across the life course.
2. In patients with clinical ASCVD, reduce low-density lipoprotein cholesterol (LDL-C) with high-intensity statin therapy or maximally tolerated statin therapy.
3. In very high-risk ASCVD, use an LDL-C threshold of 70 mg/dl to consider the addition of nonstatins to statin therapy. Very high-risk includes a history of multiple major ASCVD events or 1 major ASCVD event and multiple high-risk conditions.
4. In patients with severe primary hypercholesterolemia (LDL-C level ≥190 mg/dl), without calculating 10-year ASCVD risk, begin high-intensity statin therapy.
5. In patients 40–75 years of age with diabetes mellitus and LDL-C ≥70 mg/dl, start moderate-intensity statin therapy without calculating 10-year ASCVD risk.
6. In adults 40–75 years of age evaluated for primary ASCVD prevention, have a clinician–patient risk discussion before starting statin therapy.
7. In adults 40–75 years of age without diabetes mellitus and with LDL-C levels ≥70 mg/dl, at a 10-year ASCVD risk of ≥7.5%, start a moderate-intensity statin if a discussion of treatment options favors statin therapy.
8. In adults 40–75 years of age without diabetes mellitus and 10-year risk of 7.5%–19.9% (intermediate risk), risk-enhancing factors favor the initiation of statin therapy (see no. 7).
9. In adults 40–75 years of age without diabetes mellitus and with LDL-C levels ≥70 mg/dl–189 mg/dl, at a 10-year ASCVD risk of ≥7.5%–19.9%, if a decision about statin therapy is uncertain, consider measuring CAC.
10. Assess adherence and percentage response to LDL-C-lowering medications and lifestyle changes with repeat lipid measurement 4–12 weeks after statin initiation or dose adjustment, repeated every 3–12 months as needed.

Source: 2018 AHA/ACC Guideline on the Management of Blood Cholesterol: Executive Summary: www.ahajournals.org/doi/10.1161/CIR.0000000000000624#d3e1470

weight loss will only occur if the diet is maintained, the authors recommended that the diet that promotes the best adherence is the ideal diet.

3. *Reduce the amount of saturated fat to 5 or 6 percent of dietary kcal.* The American Heart Association decreased the upper limit of saturated fat intake from 10 percent to 5 or 6 percent of total kcal. As a matter of fact, scientists recommend reducing intake of saturated fats as low as possible while consuming a nutritionally adequate diet. The National Academy of Sciences indicated that there is a positive linear trend between total saturated fatty acid intake and total and LDL-cholesterol concentration and increased risk of CHD. Saturated fats may also increase blood clotting, another risk factor for CHD. In two meta-analyses, Howell and others found that consumption of saturated fatty acids was the major dietary determinant of plasma cholesterol response to diet, while Clarke and others indicated that reducing the intake of saturated fat produced the most significant benefits regarding the prevention of CHD.

The effects of dietary fat intake on human health are extremely complex, and not as straight forward as saturated fats are bad and must be avoided. As an example, Volk and colleagues from Jeff Volek's lab placed obese volunteers on 3-week diets, with dietary carbohydrate increasing from 47 g/day to 346 g/day and saturated fat decreasing from 84 g/day to 32 g/day over the study. Dietary saturated fat intake had no effect on the saturated fatty acid content of any plasma lipids. However, as dietary carbohydrate increased, palmitoleic acid, which is associated with poor health outcomes, also increased. This study, which showed no relationship between dietary and plasma saturated fat, but instead showed a potentially negative aspect of dietary carbohydrate in obese patients with metabolic syndrome, highlights the complex relationship between diet

Training Table

Top ten prudent dietary guidelines to reduce high serum lipids or maintain normal serum lipids:

1. Adjust caloric intake to achieve and maintain healthy bodyweight and composition.
2. Consume 10–25 percent of total kcal from fat.
3. Reduce the amount of saturated fat to 5–6 percent of total kcal.
4. Reduce the consumption of *trans* fats in the diet and, comparable to saturated fat, keep dietary intake as low as possible.
5. Substitute monounsaturated fats for saturated fats and simple or refined carbohydrate.
6. Consume adequate amounts of polyunsaturated fatty acids.
7. Limit the amount of dietary cholesterol.
8. If you consume foods with artificial fats, do so in moderation.
9. Reduce intake of refined carbohydrates and increase consumption of plant foods high in complex carbohydrates and dietary fiber.
10. Nibble food throughout the day.

Read detailed, evidence-based analyses of these points on the following pages.

and health. The same team studied the effects of 12 weeks of consuming three whole eggs per day versus a yolk-free substitute while on a low- (25-30 percent of kcal) carbohydrate diet in patients with metabolic syndrome. Although both groups lost weight and improved certain aspects of metabolic health, the group consuming whole eggs improved more in some inflammatory markers, plasma insulin, VLDL, and HDL-cholesterol. This research demonstrates that fat intake interacts with carbohydrate intake, body mass, weight loss, and existing metabolic disease. Clifton and Keogh have stated that replacing dietary saturated fat with carbohydrate will not lower coronary heart disease or cardiovascular disease mortality. However, replacing saturated fat with poly- or monounsaturated fats, or high-quality carbohydrates, will lower coronary heart disease events.

A significant contributor of saturated fat to the American diet is hamburger meat. However, extra lean hamburger may be low in total fat and saturated fat. Note the following comparison for 1 serving (4 ounces) of hamburger (ground beef) containing various percentages of fat:

> Regular hamburger (25 percent) = 331 kcal, 28 grams of fat (10.7 grams saturated fat)
> Lean hamburger (15 percent) = 243 kcal, 17 grams of fat (6.6 grams saturated fat)
> Extra lean hamburger (5 percent) = 155 kcal, 5.6 grams of fat (2.5 grams saturated fat)
> Extra lean hamburger (3 percent) = 136 kcal, 4.5 grams of fat (1.5 grams saturated fat)

Processed meats, such as most luncheon meats, are relatively high in fat. In contrast, fish, chicken, turkey, and very lean cuts of beef (eye of round) and pork (tenderloin) contain much less fat and saturated fat.

All oils contain saturated fats and unsaturated fats. Avoid the tropical oils, such as palm, palm kernel, and coconut, which may be 50-90 percent saturated fat. Use mainly monounsaturated and polyunsaturated oils that have no more than about 2 grams of saturated fat per tablespoon:

- Canola
- Safflower
- Sunflower
- Corn
- Olive
- Sesame
- Soybean
- Peanut

Food labels must list the amount of saturated fat per serving and its percentage of the Daily Value, which provides a sound means to select products with little to no saturated fat.

4. *Reduce the consumption of* trans *fats in the diet.* In a review, Lichtenstein noted that *trans* fats elevate LDL-cholesterol and at relatively high intakes decrease HDL-cholesterol levels. Some data suggest that *trans* fatty acids may also trigger inflammation. The National Academy of Sciences also noted that, like saturated fats, there is a positive linear trend between *trans* fat and CHD. Although some reviewers contend that the adverse effects of *trans* fatty acids are somewhat less than those associated with saturated fatty acids, Mozaffarian and others cited research indicating that they may be as bad, or even worse. They indicate that, on a per-kcal basis, *trans* fats appear to increase the risk of coronary heart disease more than any other macronutrient. An increased risk is seen with low levels of consumption corresponding to 1-3 percent of total energy, which for a 2,000-kcal diet would be approximately 20-60 kcal from *trans* fat. Similarly, Remig and others found that an increase in 2 percent of total energy from *trans* fat increases cardiovascular risk by 23 percent.

The vast majority of *trans* fat consumed by Americans has traditionally been in processed foods, particularly margarine; vegetable shortening; white bread; packaged goods such as cookies, crackers, potato chips, and cakes; and fried fast foods such as french fries. The major component of each of these foods that adds *trans* fatty acids is usually partially hydrogenated vegetable oil. *Trans* fatty acid content per serving is now listed in the Nutrition Facts food label. However, the food label may list 0 grams of *trans* fat if there is less than 0.5 gram per serving. Thus, consuming multiple servings of such products daily may accumulate, totaling several grams or more of *trans* fat. In late 2015, the FDA decided that artificial *trans* fat is not "generally recognized as safe" for use in human food and that manufacturers must remove partially hydrogenated oils from their products by 2018. This is an ongoing process.

5. *Substitute monounsaturated fats for saturated fats and simple or refined carbohydrates.* Consume about 10-15 percent of kcal from monounsaturated fats. Although no DRI have been set for monounsaturated fats, the National Academy of Sciences notes that they may have some benefit in the prevention of chronic disease.

Epidemiological research has indicated that the Mediterranean diet is associated with reduced risk of CHD. Olive oil, the primary source of dietary fat, is a staple of the Mediterranean diet. It is rich in monounsaturated fatty acids, is a good source of various phytochemicals, and contains the antioxidant vitamin E. In separate reviews, Covas and Perez-Jimenez and others highlighted various mechanisms, such as improved serum lipoprotein profile and reduced inflammation, whereby olive oil may reduce the risk of heart disease. However, they noted that the specific mechanisms underlying the beneficial effects of olive oil need further research. Nevertheless, the FDA has approved a *qualified* health claim for olive oil, indicating that eating 2 tablespoons of olive oil daily may, due to its monounsaturated fat content, reduce the risk of CHD. Keep in mind that *qualified* health claims are based on limited, not conclusive, evidence.

Walter Willett, the renowned nutrition research scientist from Harvard University, reported that both epidemiological and metabolic studies suggest that individuals can benefit greatly by adopting elements of the Mediterranean diet. Although olive oil is one of those elements, others include a diet rich in vegetables, whole grains, and seafood, all of which may confer health benefits in the prevention of CHD.

Dietary fats can have powerful effects on metabolic health. One recent example is the study from Wang and colleagues in which obese individuals were placed on a low-fat diet (24 percent of energy from fat), a moderate-fat diet (34 percent of energy from fat), and a moderate-fat diet with one Hass avocado per day (also 34 percent of energy from fat). After 5 weeks, the decrease in LDL was greater in the avocado group than in the other moderate-fat or the low-fat group. Additionally, LDL particle number, small-density LDL-cholesterol, and LDL/HDL ratio were all decreased in the avocado group. Avocados, olive oil, and nuts are rich sources of MUFAs.

The American Heart Association provided its stamp of approval to diets rich in MUFAs, provided saturated fatty acid intake is limited to a minimum and caloric intake is in balance. These two points may be the key to the role of MUFAs, such as olive oil, in helping prevent CHD.

Additionally, the OmniHeart study by Appel and others found that substituting monounsaturated fats, mainly olive oil, canola oil, and nuts, for simple carbohydrates, primarily desserts, may help promote heart health. For example, in a pooled analysis of 25 intervention trials, Sabaté and others found that 67 grams (2.4 ounces) of nuts consumed daily lowered triglyceride and LDL-cholesterol levels. These effects were dose related and more marked in those with higher LDL-cholesterol levels. Thus, consumption of nuts may lower the risk of CHD. The OmniHeart plan provides the following tips to enrich the diet with monounsaturated fats.

- Have a teaspoon per day of olive or canola oil–based margarine on bread at lunch.
- Have 1 or 2 tablespoons of salad dressing made with olive or canola oil and vinegar in salads each day.
- Add a teaspoon of olive or canola oil or margarine in vegetables at dinner.
- Use olive or canola oil to sauté or stir-fry vegetables and add to recipes.
- Have 1 ounce of unsalted nuts rich in monounsaturated fat, such as almonds, peanuts, and pecans, as a snack or add to cereals.

6. *Consume adequate amounts of polyunsaturated fatty acids.* As indicated previously, the National Academy of Sciences developed an AI for both the omega-6 and omega-3 polyunsaturated fatty acids because there may be some health benefits associated with such dietary intakes. Both types of fatty acids help promote healthy skin and are a source of various eicosanoids that may influence health processes. When substituted for saturated fat, polyunsaturated fat may reduce total serum cholesterol, including LDL-cholesterol. In their study of pooled data from 11 American and European studies, Jakobsen and others examined associations between type of fat consumed and risk of CHD. They found that when polyunsaturated fats were substituted for saturated fats, there was a reduced risk of both coronary events and coronary deaths.

One of the most controversial papers published on dietary fat intake and health was the meta-analysis by Chowdhury and colleagues. The limitations of the meta-analysis procedure are described earlier in the text. Essentially, in their analysis of more than 70 studies with over 500,000 people, Chowdhury and others concluded that encouraging high consumption of polyunsaturated fats and low consumption of saturated fats was not supported by evidence, so should not be recommended for cardiovascular health. However, this analysis had some errors that were corrected in a subsequent article. Also, the authors did not include some studies that would not have supported their conclusions. In fact, several leading nutrition researchers called for the article to be retracted, but it was not.

A great deal can be learned from such controversial papers. One important point is if dietary saturated fat is reduced, what is it replaced with? Replacing dietary saturated fat with polyunsaturated fats or carbohydrate could yield different effects on health. One thing that can be learned from this controversial article is to focus not on nutrients (e.g., eat less saturated fat and more polyunsaturated fat) but on whole foods (e.g., eat more avocados and less butter). If this whole-foods approach is used over an entire diet plan, it is likely there will be an overall consumption of health-promoting nutrients and a decrease in anti-health-promoting nutrients.

Essential fatty acids Polyunsaturated fatty acids should constitute about 10 percent of the daily caloric intake, and if foods are selected wisely this should provide adequate amounts of both omega-6 and omega-3 fatty acids. The essential omega-6 linoleic fatty acid is found in various vegetable oils that constitute food products such as margarine, salad dressings, and cooking oils. Several sources are listed in **table 5.10**. The essential omega-3 alpha-linolenic fatty acid is found in green leafy vegetables, canola oil, flaxseed oil, soy products, some nuts, and fish. Some nuts are especially rich in both the omega-3 and omega-6 fatty acids. Although both essential fatty acids may confer separate health benefits, the omega-3 fatty acids are thought to be more important for the prevention of CHD.

Omega-3 fatty acids The three principal omega-3 fatty acids—alpha-linolenic, EPA (eicosapentaenoic acid), and DHA (docosahexaenoic acid)—are believed to reduce the risk of CHD, but EPA and DHA are believed to be more potent. EPA and DHA may be formed in the body from alpha-linolenic acid, but this process appears to be limited. However, both EPA and DHA are found in substantial quantities in various fish oils, as

TABLE 5.10 Alpha-linolenic content of selected oils, nuts, and seeds

Oils, nuts, and seeds	Alpha-linolenic content, grams/tablespoon
Olive oil	0.1
Walnuts, English	0.7
Soybean oil	0.9
Canola oil	1.3
Walnut oil	1.4
Flaxseed	2.2
Flaxseed (linseed) oil	8.5

Source: USDA Nutrient Data Laboratory.

TABLE 5.11 Grams of EPA and DHA in fish per 3-ounce edible fish portion and in fish oils per gram of oil

>1 gram/ 3 ounces	0.5–1.0 gram/ 3 ounces	<0.5 gram/ 3 ounces
Herring	Halibut	Catfish
Oysters, Pacific	Omega-3 concentrate*	Cod
Salmon, Atlantic, farmed	Salmon, sockeye	Cod liver oil*
Salmon, Atlantic, wild	Trout	Crab
Salmon, chinook	Tuna, fresh	Flounder/sole
Sardines	Tuna, white, canned in water	Haddock
		Lobster
		Oysters, Eastern
		Scallops
		Shrimp
		Tuna, light, canned in water

*Note: Omega-3 content in fish oils or supplements is per gram of oil. Check supplement labels for content.

Source: Adapted from USDA Nutrient Data Laboratory. Ranges listed are rough estimates because oil content can vary markedly with species, season, diet, and packaging and cooking methods.

highlighted in **table 5.11**. Eggs from chickens fed a special diet may contain DHA in amounts ranging from 0.05 to 0.15 gram, but the eggs are still high in cholesterol. Fish oil supplements may contain 0.3–0.5 gram.

As mentioned earlier, omega-3 fatty acids have been theorized to be ergogenic because of the production of specific eicosanoids. Omega-3 fatty acids are also being studied for their potential health benefits, which also may be related to specific eicosanoids that are produced. Although the health-related role of omega-3 fatty acids and eicosanoids is complex and has not been totally determined, here is a simple summarization. The cell membrane contains a variety of molecular compounds, including phospholipids and their associated fatty acids. When the diet is high in linoleic acid, one of the main fatty acids in the phospholipids is arachidonic acid, which produces one form of eicosanoids when it is metabolized. When the diet is high in fish oils, EPA and DHA become the major source of eicosanoids, which are different in nature compared to those derived from arachidonic acid. In essence, the different forms of eicosanoids function as local hormones in body cells affecting metabolism and gene expression, and the effects associated with omega-3 fatty acid–derived eicosanoids appear to provide some health benefits. Metcalf and others found that dietary omega-3 fatty acids are rapidly incorporated into the phospholipids of human heart muscle cells, displacing arachidonic acid.

The effects of fish intake on cardiovascular health are less controversial than research on the effects of saturated fat intake on health. Chowdhury found that fish consumption is associated with decreased risk of cerebrovascular disease, such as strokes. This points to the beneficial effects of omega-3 fats on health. Epidemiological research has suggested that populations consuming diets rich in fish products have a lower incidence rate of CHD, and experimental research has suggested a number of possible mechanisms underlying this relationship:

- Reduce serum triglycerides
- Increase HDL-cholesterol
- Prevent clot formation
- Decrease platelet aggregation and stickiness
- Improve vascular tone
- Decrease blood viscosity
- Optimize blood pressure
- Promote anti-inflammatory activity
- Decrease abnormal heart rhythms

A great deal of research has focused on the effects of fish oil or omega-3 fatty acid supplementation on CHD, but the results of recent reviews highlight inconsistent findings among randomized controlled trials. For instance, Weinberg and colleagues recommend two or more servings of fatty fish per week as part of a heart healthy diet for people without atherosclerotic cardiovascular disease. If this dietary recommendation cannot be achieved, they consider supplementation reasonable. They note that omega-3 fatty acid supplementation may benefit cardiovascular health through reductions in triglyceride rich lipoproteins, decreased inflammation, plaque and membrane stabilization, and antithrombotic effects. However, they also point out that clinical trial data are inconsistent in terms of cardiovascular outcomes and that more needs to be known about supplement formulation and dose. The authors suggest that some individuals, perhaps those with hyperlipidemia or atherosclerotic disease, if under the supervision of a qualified health care provider, might benefit from omega-3 supplementation. In their review, Liao and colleagues identify similar inconsistencies in the literature; not every trial shows a beneficial effect. They also recognize the same multiple beneficial effects of omega-3 fatty acid intake, which are supported by many basic science studies. In their analysis, they found that omega-3 fatty acid supplementation resulted in small decreases in blood pressure in hypertensive and normotensive populations. There was no definite benefit in the primary prevention of CVD, but supplementation could augment the effects of statin medication if there was hypertriglyceridemia and hypercholesterolemia. In the secondary prevention of CHD, supplementation could improve outcomes in patients receiving suboptimal medical treatment, but the effects in patients receiving good treatment were not known. Also, the protective effect in secondary prevention was related to the dose and supplementation protocol; more favorable results were evident in higher-dose, longer-term studies. While earlier studies indicated a reduced risk of ventricular arrhythmias, these data have not been reproducible. Finally, in postoperative patients atrial fibrillation could be reduced, but high-dose omega-3 ingestion might increase atrial fibrillation risk. Omega-3 fatty acid intake is only one component of the diet, and as Cundiff and others noted higher EPA and DHA intake is associated with decreased kcal, total fat, saturated fat and increased fiber intake, indicating that some of the benefit of dietary fish or omega-3 fatty acid intake may be from other associated dietary improvements.

Wallin and others reported geographic differences in the effect of fish consumption on type 2 diabetes risk, such that there were small increases (United States), no effect (Europe), or decreases in risk (Asia). This is another example of how a meta-analysis of different types of studies can add to the confusion around dietary fat and health. Szostak-Wegierek and others point out that an increase in omega-6 polyunsaturated fats without regard to omega-3 fats might be problematic in terms of cardiovascular disease. They claim a healthy Mediterranean diet has an omega-6 to omega-3 ratio of 2:1 and that this is not being adhered to in countries that increased omega-6 fat intake. The NIH has concluded that there is supportive, but not conclusive, evidence in the primary prevention of cardiovascular disease, which is the basis for the *qualified* health claim on food labels that consumption of conventional foods containing omega-3 fatty acids may reduce the risk of CHD.

Tufts University, one of the leading international nutrition research universities, noted that the predominance of the medical literature continues to support eating fish for cardiovascular health. Based on the available evidence, Kris-Etherton and others provided some dietary guidelines in the American Heart Association Scientific Statement: Fish Consumption, Fish Oil, Omega-3 Fatty Acids, and Cardiovascular Disease. The key points are:

- Eat fish, particularly fatty fish, at least two times a week. Fatty fish such as mackerel, lake trout, herring, sardines, albacore tuna, and salmon are high in EPA and DHA.
- Eat plant foods rich in the alpha-linolenic acid, an omega-3 fatty acid that may be converted into EPA and DHA in the body.
- Individuals who have high serum triglycerides may benefit from a fish oil supplement of 2–4 grams of EPA and DHA per day. Patients with CHD may benefit but should consult with their physicians.

Although eating more fish and fish oils is a healthful recommendation, a report by the Institute of Medicine, *Seafood Choices: Balancing Benefits and Risks,* raises some caveats. Some types of fish may contain significant amounts of mercury, as methylmercury, which if consumed in excess may harm the nervous system and impair neurodevelopment in the fetus or young children. Sushi is generally made from large blue fin, or *ahi,* tuna, which may contain mercury. Some types of fish, particularly older, larger predatory fish such as shark and farmed fish such as Atlantic salmon, may contain environmental contaminants such as dioxins and polychlorinated biphenyls.

In a major review, Mozaffarian and Rimm concluded that the benefits of fish intake exceed the potential risks. However, along with the AHA, FDA, and Dietary Guideline for Americans, caution is recommended depending on a person's stage in life:

- For women of childbearing age, benefits of modest fish intake, excepting a few selected species, outweigh risks. Pregnant women should avoid intake of shark, swordfish, king mackerel, and tilefish and limit consumption of other fish to no more than 12 ounces per week. The FDA indicates that shrimp, salmon, pollock, and catfish are generally low in mercury. Some avoid canned light tuna may contain as much mercury as white (albacore) tuna, so it could be recommended that women who are pregnant avoid canned tuna entirely. Women of childbearing age should also limit weekly consumption of canned tuna: no more than three 6-ounce cans of light tuna or no more than one can of white tuna.
- Young children should also limit consumption of fish that may be high in mercury. It is recommended that children consume just two servings a week from fish known to be low in mercury. Best choices for children include clam, cod, crab, haddock, salmon, shrimp, tilapia, and others suggested by the FDA. For more information, click this link: www.epa.gov/fish-tech/epa-fda-advice-about-eating-fish-and-shellfish.
- The health effects of low-level methylmercury in adults are not clearly established. Adults should consume a variety of seafood. Select fresh, local seafood where available. The Consumers Union recommends the following safe fish consumption for adult men and women:

Daily	**Several times a week**
Salmon	Flounder
Tilapia	Sole
Pollock	Herring
Sardine	Mackerel
Oyster	Croaker
Shrimp	Scallop
Clam	Crab
Crawfish	

http://seafood.edf.org/ The Environmental Defense Fund provides information on fish that are safe to eat and offers substitutes for overfished choices.

Although consuming fish is the recommended means to obtain EPA and DHA, fish oil supplements are also available. Weber and others noted that the possible higher mercury content of some fish may make the intake of omega-3 fatty acids as capsules the better choice. Mercury accumulates in the muscle of the fish, not the fat tissues that is the source of fish oils. Although some health professionals recommend higher amounts for individuals with high levels of serum triglycerides or heart disease, one recommendation for healthy individuals is 500–1,300 milligrams a day of EPA and DHA combined. This is more liberal than the recommendation from Lee and colleagues in their review. They propose at least 1 gram of long-chain omega-3 fatty acids daily for patients with known

CHD, and 250–500 mg daily for those without disease. The amount of EPA and DHA in a typical fish oil capsule varies, but generally a 1,000-milligram fish oil softgel will contain about 180 mg of EPA and 120 mg of DHA or a total of 300 mg of omega-3 fatty acids. Consuming two to four capsules a day would provide the recommended amount. Some capsules may contain more than 500 mg of omega-3 fatty acids. However, Brunton and Collins note that many types of omega-3 fatty acid dietary supplements are available, but the efficacy, quality, and safety of these products are open to question because they are not regulated by the same standards as pharmaceutical agents. Health professionals recommend that you discuss the benefits and risks of taking fish oil supplements with your doctor. If you do take supplements, do so with food to help absorption.

You may also see omega-3 fatty acids prominently displayed on many food labels, such as cereals, pasta, yogurt, and soy milk. However, check the ingredient list for the source. If it is soybean, canola, or flaxseed oil, the omega-3 fatty acid is alpha-linolenic, which is a good choice but not considered to be as healthful as EPA and DHA.

Conjugated linoleic acid (CLA) Conjugated linoleic acid (CLA), a polyunsaturated omega-6 fatty acid found naturally in small amounts in dairy foods and beef, has been studied for its potential to reduce body fat. Whigham and others conducted a meta-analysis of 18 human studies and concluded that CLA supplementation in a dose of about 3.2 grams daily produces a modest loss of body fat in humans, about 0.1 pound per week. Although this may be meaningful over time, the Consumers Union reports some research suggesting that CLA supplementation may induce some effects, such as impaired blood glucose regulation and inflammation, that might contribute to chronic health problems. Based on these reports, individuals at risk for CHD might be advised not to use CLA as a weight-loss supplement.

7. *Limit the amount of dietary cholesterol.* In recent years, some have contended that dietary cholesterol does not influence serum cholesterol and the development of CHD. For example, Hasler noted that it is now known that there is little if any connection between dietary cholesterol and blood cholesterol levels, and consuming up to one or more eggs per day does not adversely affect blood cholesterol levels. In contrast, in a meta-analysis covering 17 studies that evaluated cholesterol intake for at least 14 days, Weggemans and others noted that the addition of 100 milligrams of dietary cholesterol per day would increase slightly the ratio of total cholesterol to HDL-cholesterol, an adverse effect on the serum cholesterol profile. They concluded that the advice to limit cholesterol intake by reducing consumption of eggs and other cholesterol-rich foods may still be valid.

Limiting cholesterol intake is particularly important for cholesterol responders, those individuals with a genetic predisposition whose body production of cholesterol does not automatically decrease when the dietary intake increases. The average U.S. daily intake is approximately 137 mg/1000 kcal.

8. *If you consume foods with artificial fats, do so in moderation.* The fat substitutes discussed earlier in this chapter are generally recognized as safe and have been approved by the Food and Drug Administration (FDA). Although olestra has been approved by the FDA, some contend that its use may interfere with the absorption of several fat-soluble vitamins and beta-carotene. However, the FDA requires that products containing olestra-type fat substitutes be enriched with fat-soluble vitamins to offset potential losses.

Snack foods containing olestra, particularly when consumed in large amounts, may cause intestinal cramps and loose stools. McRorie and others reported that although olestra-containing potato chips induced a gradual stool softening effect after several days of consumption, when consumed in smaller amounts, such as 20–40 grams, there were no objective measures of diarrhea or increased gastrointestinal symptoms. Nevertheless, the Center for Science in the Public Interest noted that the FDA received more than 18,000 adverse-reaction reports from people who had eaten olestra-containing foods. Sales of such foods have dropped dramatically and olestra is banned in several countries including Canada and the United Kingdom.

The majority of fat replacers, when used in moderation by adults, can be safe and useful adjuncts to lowering the fat content of foods and may play a role in decreasing total dietary energy and fat intake. However, they are effective only if they lower the total caloric content of the food and if the consumer uses these foods as part of a balanced meal plan, such as that promoted in the *2020–2025 Dietary Guidelines for Americans*.

Some fat is needed in the diet to provide the essential fatty acids and fat-soluble vitamins, and this fat should be obtained easily through natural, wholesome foods such as whole grains, fruits, and vegetables. Foods with fat replacers have the potential to help people reduce the total and saturated fat intake, which could help reduce serum lipids and assist in weight loss.

9. *Reduce intake of refined carbohydrates and increase consumption of plant foods high in complex carbohydrates and dietary fiber, particularly water-soluble fiber.* Refined sugar and starches provoke higher triglyceride concentrations more than complex carbohydrates with fiber do. Again, the value of complex carbohydrates in the diet is stressed, particularly high-fiber foods, as a means to help reduce serum cholesterol. Research has suggested that without adequate amounts of fiber, a diet low in saturated fats and cholesterol has only modest effects on lowering CHD risk. Thus, replace high-fat foods with high-fiber foods. Legumes, such as beans, are an excellent source of carbohydrate and water-soluble fiber. Beans also contain protein, and soy protein, as found in products such as tofu, has been shown to reduce cholesterol in men with both normal and high serum cholesterol. Oat products, such as found in oatmeal, may effectively lower serum cholesterol. Increased consumption of fruits and vegetables is recommended as well, for they may provide substantial amounts of the antioxidant vitamins (C, E, and beta-carotene) that may help to prevent undesired oxidations in

the body. Guidelines presented in the preceding chapter are helpful to increase carbohydrate intake, and the role of antioxidant vitamins will be discussed in chapter 7.

Some plant foods, such as almonds and oats, may also contain various sterols and stanols, which are known to reduce serum cholesterol. Devaraj and Jialal indicated that about 2 grams of stanols or sterols per day may lower total and LDL-cholesterol, possibly by interfering with the uptake of both dietary and biliary cholesterol from the intestinal tract. Commercial margarines such as Benecol® and Take Control® contain such plant stanols and sterols. In a meta-analysis of six well-designed studies, Moruisi and others concluded that fat spreads (margarines) providing about 2.5 grams of phytosterols/stanols daily over the course of 1–3 months reduced both total cholesterol and LDL-cholesterol. As noted in chapter 2, a diet rich in plant sterols and stanols may reduce the risk of heart disease. Plant foods also contain several phytochemicals that may reduce serum cholesterol. Many food manufacturers, such as those who produce breakfast cereals and orange juice, are fortifying their products with sterols, stanols, phytochemicals, and other nutrients, creating functional foods designed to reduce serum cholesterol and provide other health benefits as well.

10. *Nibble food throughout the day.* Interestingly, David Jenkins showed a significant reduction in serum LDL-cholesterol if subjects consumed their daily kcal, actually the same food, throughout the day rather than in three concentrated meals at breakfast, lunch, and dinner. In particular, it may be wise to avoid eating a high-fat meal. Nicholls and others reported that a single meal rich in saturated fats may impair blood vessel function and reduce the anti-inflammatory potential of HDL-cholesterol. A single high-fat meal can significantly increase serum triglycerides and decreased blood flow through the heart. Jakulj and others found that a single high-fat meal could increase the blood pressure response to a stressful situation, such as public speaking. All of these factors may increase the short-term risk of a heart attack in susceptible individuals.

In simple, practical terms, what do all of these recommendations mean? Some, such as Lawrence and Harcombe and others, claim that the adverse health effects associated with saturated fat intake are incorrect and that advice on dietary fat consumption and health should never have been introduced to the public. This is likely an oversimplification and can be misleading. It is best to focus on an overall healthy lifestyle that meets recommendations for physical activity, fitness, sleep, body composition, and a diet that is not extreme in terms of the intake of specific nutrients.

You may have noted that alcohol intake was not one of these recommendations. As detailed in chapter 13, low-risk alcohol intake may provide some protection against CHD for those who do drink. However, most health professionals do not recommend that nondrinkers begin to consume alcohol for its potential health benefits because of other health risks associated with drinking in excess.

> For the interested reader, here are two open access, comprehensive, and evidence-based summaries of dietary fat intake and heart health:
>
> Kris-Etherton, P. and Krauss, R. 2020. Public health guidelines should recommend reducing saturated fat consumption as much as possible: YES. *American Journal of Clinical Nutrition* 112:13–18.
> *https://doi.org/10.1093/ajcn/nqaa110*
>
> Weinberg, R. et al. 2022. Cardiovascular impact of nutritional supplementation with omega-3 fatty acids: JACC focus seminar. *Journal of the American College of Cardiology.* 77:593–608.
> *https://doi.org/10.1016/j.jacc.2020.11.060*

Can exercise training also elicit favorable changes in the serum lipid profile?

Physical inactivity, or lack of exercise, has been identified as one of the primary risk factors associated with an increased incidence of atherosclerosis and cardiovascular disease. Hence, exercise programs stressing aerobic endurance-type activities have been advocated as a means of reducing the incidence levels of these conditions, possibly via direct beneficial effects on the heart or blood vessels. However, the precise mechanism whereby exercise may help reduce the morbidity and mortality of CHD has not been identified. Therefore, many authorities believe that the beneficial effect may not be due to exercise itself but rather the possible associated effects, such as reductions in body fat and blood pressure. Although some investigators believe that endurance exercise may have a preventive function independent of these associated effects, it also exerts a significant beneficial influence on the serum lipid profile, which, like blood pressure, is one of the major risk factors.

An acute bout of exercise may reduce risk factors for CHD. For example, Thompson and others recently noted that acute exercise may reduce blood pressure and serum triglycerides, increase HDL-cholesterol, and improve insulin sensitivity and glucose homeostasis. Thus, some of the beneficial effects of exercise on risk factors for CHD, including the serum lipid profile, may be attributed to recent exercise bouts. However, some of these benefits may become long-lasting with a chronic exercise training program.

Chronic exercise training has been shown to favorably affect the serum lipid profile. Hundreds of epidemiological and experimental studies have been conducted over the past several decades to investigate the effects of exercise on serum lipids. Space does not permit a detailed analysis of each, but major reviews of the worldwide literature have been reported by prominent authorities such as Stefanik and Wood, Leon and Sanchez, Durstine, and Williams. Although most of these reviews involved males, similar reviews have evaluated the effects of exercise training on women, such as those by Dowling and the meta-analysis by Kelley and others. These reviews have noted a rather consistent pattern relating exercise and blood lipids, and some of the benefits have been associated with concomitant body weight control. In general, increased levels of exercise are associated with lower plasma levels of triglycerides and higher levels of HDL, as documented in several recent meta-analyses by Kelley and Kelley. However,

Durstine and others note that exercise training seldom alters total and LDL-cholesterol, although some studies have shown small decreases in the later. Moreover, research has shown that exercise may not improve the lipid profile of some individuals, primarily in those with genetic defects. These individuals may receive other health benefits of exercise but may need drug therapy to control elevated serum lipid levels.

In an attempt to quantify the serum lipid changes with the amount of exercise, Durstine and his associates conducted a meta-analysis of well-controlled studies. One of the dose-response findings from their analysis indicated that an exercise training volume of 1,200-2,200 kcal per week is often effective at elevating HDL-C levels from 2 to 8 mg/dl and lowering triglyceride levels by 5-38 mg/dl. Their analysis also suggests that greater increases in HDL-cholesterol can be expected with additional increases in exercise training volume. This amount of physical activity is reasonable and attainable for most individuals and is within the ACSM recommended range for healthy adults. Recently, Fikenzer and colleagues reported that effective endurance training with an intensity of 75-85 percent of maximal heart rate, for 40-50 minutes per session, on 3-4 days/week for 26-40 weeks lowered total cholesterol about 4 percent, LDL about 5 percent, triglycerides about 8 percent, and increased HDL about 4 percent. This analysis proves that endurance training can favorably change blood lipids, but that the beneficial effect is small. Lifetime aerobic exercise appears to be the key, and moderately intense leisure-time activity, such as brisk walking, may elicit beneficial effects in men, women, children, and adolescents. As noted by Kelley and others, walking favorably affects the adult serum lipid profile independent of changes in body composition.

As noted previously, a single high-fat meal may increase the risk of heart attack. However, Katsanos indicates that expending about 500 kcal or more through moderate-intensity exercise within 16 hours before the meal will minimize adverse changes in the lipid profile. As discussed below, this supports the finding that endurance athletes who consume diets rich in fat maintain normal serum lipid profiles.

Although the precise biochemical mechanisms underlying the beneficial effects of exercise on serum lipids have not been identified, researchers have found that in physically trained males and females, activity levels of several enzymes, such as hepatic lipase and lipoprotein lipase, are modified in such a way as to promote a more rapid catabolism of triglycerides and a greater production of HDL. The muscle cell membrane may be modified favorably to become more insulin sensitive, helping clear lipids from the blood into the muscle. Exercise may also favorably modify the serum lipid levels by helping the individual lose body fat or influencing changes in other aspects of his or her lifestyle, such as diet.

Research has revealed that the beneficial effects of exercise training are additive to a diet modified in fat content, such as one reduced in total and saturated fat. A low-fat diet may reduce total cholesterol and LDL-cholesterol but may also undesirably decrease HDL-cholesterol. Exercise may prevent or attenuate the decrease in HDL-cholesterol on such diets, but when combined with omega-3 fatty acid supplementation may actually increase serum HDL-cholesterol, as reported in a study by Thomas and others. Thus, the combination of both dietary modifications and exercise is the recommended approach to modify favorably serum lipid levels.

Research also reveals that highly trained endurance runners who increase their dietary fat to about 40 percent or more of daily caloric intake for 4 weeks do not experience any adverse effects in their blood lipid profiles. Although this type of diet is not recommended on a long-term basis, the review by Brown and Cox illustrates some of the protective effects of exercise training on serum lipid changes associated with short-term increases in dietary fat. They suggest that the strenuous physical training seems to metabolize the increased fat intake for energy and prevents adverse changes in the lipid profile.

www.heart.org/en/health-topics/cholesterol/prevention-and-treatment-of-high-cholesterol-hyperlipidemia/cooking-to-lower-cholesterol Use this site for cooking advice to help lower blood cholesterol.

Key Concepts

▸ High levels of low density forms of lipoproteins (LDL) may predispose certain individuals to coronary heart disease, whereas high-density forms (HDL) may be protective.

▸ In general, a low- to moderate-fat diet is recommended for both health and physical performance. One should consume less high-fat meat and dairy products and more fruits, vegetables, whole-grain products, dietary fiber, lean meats, and skim milk. Fish, including fatty fish such as salmon, is part of a heart-healthy diet.

▸ Diets rich in saturated fats, *trans* fats, and cholesterol may increase the risk of coronary heart disease. In the United States, in general, the recommended dietary intake of total fat is 10–25 percent or less of the total caloric intake, with saturated fats at 5–6 percent of the total. The recommended cholesterol intake is to consume as little as possible while maintaining a nutritionally adequate diet.

▸ Aerobic exercise training increases the ability of the muscles to use fat as an energy source and can be an important adjunct to diet in beneficially modifying the serum lipid profile and reducing body fat, two factors that may reduce the risk of coronary heart disease (CHD). Relative to the serum lipid profile, aerobic exercise training is most effective in reducing serum triglycerides and increasing serum HDL-cholesterol.

Check for Yourself

▸ Major professional health organizations, such as the American Heart Association, the National Cancer Society, and the American Diabetes Association, all have internet sites providing dietary recommendations to help prevent related diseases. Visit websites for several such organizations and evaluate the findings relative to dietary fat. Compare your findings.

APPLICATION EXERCISE

What type of diet is easiest to follow? What are the effects on exercise?

Calculate your total daily energy intake need. Then, using NutritionCalc Plus or a nutrition app, develop two, 3-day isocaloric diets, one with 75 percent of kcal from fat, and one with 75 percent of kcal from carbohydrate. Describe how easy/difficult it would be to follow each diet. Were there any deficiencies in vitamins, minerals, or fiber in either diet? To assess the effects of the two diets on exercise performance, choose an endurance task such as running a 5K, completing a group exercise class, or completing a fitness test on the cardiovascular exercise equipment in a gym, such as a treadmill or elliptical trainer. Use NutritionCalc Plus or an app to track your energy intake, and try to consume 75 percent of your kcal from fat for 3 days. After following a high-fat diet for 3 days, complete the exercise test. Record your time to complete the exercise test or your energy level during a group exercise class. Then, using the same method to track your dietary intake, follow an isocaloric, but high-carbohydrate (75 percent kcal) diet for 3 days. Complete the same exercise test again. Do you feel your performance was affected by the diet? Which one allowed you to perform at your best?

Review Questions—Multiple Choice

1. If a 2,000-kcal diet contains 100 grams of fat, the percentage of fat kcal in the diet is which of the following?
   a. 20 percent
   b. 25 percent
   c. 35 percent
   d. 45 percent
   e. 60 percent
2. Which of the following dietary supplements has been proven to increase fat utilization during exercise, store muscle glycogen, and enhance endurance exercise performance?
   a. carnitine
   b. conjugated linoleic acid
   c. omega-3 fatty acids
   d. hydroxycitrate
   e. medium-chain triglycerides
   f. a and b
   g. none of the above
3. Which of the following is most conducive to the development of atherosclerosis?
   a. a total cholesterol of 190 mg/dl
   b. a low level of very low-density lipoprotein cholesterol
   c. a high-density lipoprotein cholesterol of 70 mg/dl
   d. a low-density lipoprotein cholesterol of 170 mg/dl
   e. a total cholesterol/high-density lipoprotein cholesterol ratio of less than 3.5
4. Which lipid dietary component appears to be most likely to cause an increase in serum cholesterol and the development of atherosclerosis?
   a. saturated fats
   b. polyunsaturated fats
   c. monounsaturated fats
   d. omega-3 fatty acids
   e. phospholipids
5. What compound in the diet cannot be used to form fat if it is consumed in excess?
   a. fat
   b. complex carbohydrate
   c. simple carbohydrate
   d. protein
   e. alcohol
   f. All are capable of forming fat.
6. Which essential fatty acids are needed in the diet?
   a. linoleic and alpha-linolenic
   b. oleic and linoleic
   c. stearic and alpha-linolenic
   d. palmitic and stearic
   e. palmitoleic and stearic
7. Which of the following statements relative to fats is false?
   a. Hydrogenation of fats makes them more saturated.
   b. Saturated fats are found primarily in animal foods.
   c. Vegetable fats are primarily unsaturated fats.
   d. Polyunsaturated fats are theorized to be more healthful than saturated fats.
   e. Saturated fats appear to help lower blood cholesterol levels.
8. Which of the following is not good advice in attempts to reduce *serum* cholesterol?
   a. Limit whole egg consumption to about two to four per week.
   b. Eat fish and white poultry meat in place of saturated fat meat.
   c. Drink skim milk instead of whole milk.
   d. Use butter instead of olive oil.
   e. Eat more fruits, vegetables, and whole-grain products.
9. Fats may be a significant source of energy during exercise of low intensity and long duration. What is the main form of fats used for energy production during low-intensity exercise?
   a. phospholipids derived from the cell membrane
   b. chylomicrons from the liver
   c. free fatty acids from the adipose cells and muscle cells
   d. VLDL from the liver
   e. cholesterol from the kidney
10. Aerobic endurance exercise may have some beneficial effects on the serum lipid profile and help to prevent coronary heart disease (CHD). In particular, what aspects of the serum lipid profile are improved from exercise to help reduce risk of CHD?
    a. lower both total cholesterol and LDL-cholesterol
    b. lower total cholesterol and increase LDL-cholesterol
    c. lower both triglycerides and LDL-cholesterol
    d. lower both triglycerides and HDL-cholesterol
    e. lower triglycerides and increase HDL-cholesterol

Answers to multiple choice questions: 1. d; 2. g; 3. d; 4. a; 5. f; 6. a; 7. e; 8. d; 9. c; 10. e

Critical Thinking Questions

1. List the major classes of dietary fatty acids and discuss their relative importance to cardiovascular health. Include in your discussion specific fatty acids as deemed relevant.
2. Describe the role that the blood lipoproteins play in the etiology of atherosclerosis and cardiovascular disease.
3. What is carnitine and how is it theorized to enhance endurance exercise performance? Does research support the theory?
4. Describe the process of chronic fat loading as a strategy to enhance endurance exercise performance, and provide a synthesis of research findings relative to its efficacy.
5. List at least five dietary strategies that may help reduce the risk of atherosclerosis and cardiovascular disease, including specific foods in the diet.

References

Books

American Institute of Cancer Research. 2007. *Food, Nutrition, Physical Activity, and the Prevention of Cancer: A Global Perspective,* Washington, DC: AICR.

Castell, L. M., Stear, S. J., and Burke, L. M., eds. 2015. *Nutritional Supplements in Sport, Exercise, and Health: An A to Z Guide.* London: Routledge.

Institute of Medicine. 2006. *Seafood Choices: Balancing Benefits and Risks.* Washington, DC: National Academies Press.

National Academy of Sciences. 2005. *Dietary Reference Intakes for Energy, Carbohydrates, Fiber, Fat, Protein and Amino Acids (Macronutrients).* Washington, DC: National Academies Press.

U.S. Department of Health and Human Services Public Health Service. 2010. *Healthy People 2020: National Health Promotion and Disease Prevention Objectives.* Washington, DC: U.S. Government Printing Office.

Reviews and Specific Studies

Achten, J., and Jeukendrup, A. 2004. Optimizing fat oxidation through exercise and diet. *Nutrition* 20:716-27.

American Diabetes Association. 2000. Role of fat replacers in diabetes medical nutrition therapy. *Diabetes Care* 23:S96-97.

American Heart Association. 2006. Diet and lifestyle recommendations revision 2006: A scientific statement from the American Heart Association Nutrition Committee. *Circulation* 114:82-96.

Angus, D., et al. 2000. Effect of carbohydrate or carbohydrate plus medium-chain triglyceride ingestion on cycling time trial performance. *Journal of Applied Physiology* 88:113-19.

Appel, L., et al. 2005. Effects of protein, monounsaturated fat, and carbohydrate intake on blood pressure and serum lipids: Results of the OmniHeart randomized trial. *Journal of the American Medical Association* 294:2455-64.

Barrett, E. C., et al. 2014. Omega-3 fatty acid supplementation as a potential therapeutic aid for the recovery from mild traumatic brain injury/concussion. *Advances in Nutrition* 5:268-77.

Blesso, C., et al. 2013. Effects of carbohydrate restriction and dietary cholesterol provided by eggs on clinical risk factors in metabolic syndrome. *Journal of Clinical Lipidology* 7:463-71.

Blesso, C., et al. 2013. Whole egg consumption improves lipoprotein profiles and insulin sensitivity to a greater extent than yolk-free egg substitute in individuals with metabolic syndrome. *Metabolism* 62:400-10.

Bortolotti, M., et al. 2007. Fish oil supplementation does not alter energy efficiency in healthy males. *Clinical Nutrition* 26:225-30.

Brilla, L., and Landerholm, T. 1990. Effect of fish oil supplementation and exercise on serum lipids and aerobic fitness. *Journal of Sports Medicine & Physical Fitness* 30:173-80.

Brooks, E. 2022. Acute ingestion of ketone monoesters and precursors do not enhance endurance exercise performance: A systematic review and meta-analysis. *International Journal of Sport Nutrition and Exercise Metabolism* 32:214-25.

Brown, G., et al. 1993. Lipid lowering and plaque regression: New insights into prevention of plaque disruption and clinical events in coronary disease. *Circulation* 87:1781-89.

Brown, R., and Cox, C. 2000. High-fat versus high-carbohydrate diets: Effect on exercise capacity and performance of endurance trained cyclists. *New Zealand Journal of Sports Medicine* 28:55-59.

Brown, R., and Cox, C. 2001. Challenging the dogma of dietary carbohydrate requirements for endurance athletes. *American Journal of Medicine & Sports* 3:75-86.

Brunton, S., and Collins, N. 2007. Differentiating prescription omega-3-acid ethyl esters (P-OM3) from dietary-supplement omega-3 fatty acids. *Current Medical Research & Opinion* 23:1139-45.

Buckley, J., et al. 2009. DHA-rich fish oil lowers heart rate during submaximal exercise in elite Australian Rules footballers. *Journal of Science & Medicine in Sport* 12:503-7.

Bueno, N. B., et al. 2015. Dietary medium-chain triacylglycerols versus long-chain triacylglycerols for body composition in adults: Systematic review and meta-analysis of randomized controlled trials. *Journal of the American College of Nutrition* 34:175-83.

Bueno, N., et al. 2013. Very-low-carbohydrate ketogenic diet v. low-fat diet for long-term weight loss: A meta-analysis of randomised controlled trials. *British Journal of Nutrition* 110:1178-87.

Burelle, Y., et al. 2001. Oxidation of [(13)C]-glycerol ingested along with glucose during prolonged exercise. *Journal of Applied Physiology* 90:1685-90.

Burke, L. 2021. Ketogenic low-CHO, high-fat diet: The future of elite endurance sport? *Journal of Physiology* 599:819-43.

Burke, L. M., et al. 2017. Low carbohydrate, high fat diet impairs exercise economy and negates the performance benefit from intensified training in elite race walkers. *Journal of Physiology* 595(9):2785-2807.

Burke, L. M. 2015. Medium-chain triglycerides. In *Nutritional Supplements in Sport, Exercise, and Health. An A-Z Guide,* ed. L. M. Castell, S. J. Stear, and L. M. Burke. New York: Routledge.

Burke, L., et al. 2000. Effect of fat adaptation and carbohydrate restoration on metabolism and performance during prolonged cycling. *Journal of Applied Physiology* 89:2413-21.

Burke, L., et al. 2002. Adaptations to short-term high-fat diet persist during exercise despite high carbohydrate availability. *Medicine & Science in Sports & Exercise* 34:83-91.

Cano A. et al. 2022. Analysis of sex-based differences in energy substrate utilization during moderate-intensity aerobic exercise. *European Journal of Applied Physiology* 122:29-70.

Carey, A., et al. 2001. Effects of fat adaptation and carbohydrate restoration on prolonged endurance exercise. *Journal of Applied Physiology* 91:115-22.

Center for Science in the Public Interest. 1999. Olean times at P & G. *Nutrition Action Health Letter* 26 (8):2.

Center for Science in the Public Interest. 2013. Six reasons to eat less meat. *Nutrition Action Health Letter* (June):1-7

Cheuvront, S. 1999. The zone diet and athletic performance. *Sports Medicine* 27:213-28.

Chowdhury, R., et al. 2012. Association between fish consumption, long chain omega 3 fatty acids, and risk of cerebrovascular disease: Systematic review and meta-analysis. *BMJ* 345:e6698.

Chowdhury, R., et al. 2014. Association of dietary, circulating, and supplement fatty acids with coronary risk: A systematic review and meta-analysis. *Annals of Internal Medicine* 160:398-406.

Clarke, R., et al. 1997. Dietary lipids and blood cholesterol: Quantitative meta-analysis of metabolic ward studies. *British Medical Journal* 314:112-17.

Clifton, P. M., and Keogh, J. B. 2017. A systematic review of the effect of dietary saturated and polyunsaturated fat on heart disease. *Nutrition, Metabolism and Cardiovascular Diseases* 27(12):1060-80.

Coggan, A., et al. 2000. Fat metabolism during high-intensity exercise in endurance-trained and untrained men. *Metabolism* 49:122-28.

Consumers Union. 2002. Chest pain after eating? *Consumer Reports on Health* 14 (7):3.

Consumers Union. 2006. Mercury in tuna. *Consumer Reports* 71 (7):20-21.

Costill, D., et al. 1979. Lipid metabolism in skeletal muscle of endurance trained males and females. *Journal of Applied Physiology* 47:787-91.

Covas, M. 2007. Olive oil and the cardiovascular system. *Pharmacological Research* 55:175-86.

Cox, P., et al. 2016. Nutritional ketosis alters fuel preference and thereby endurance performance in athletes. *Cell Metabolism* 24:256-68.

Cundiff, D., et al. 2007. Relation of omega-3 fatty acid intake to other dietary factors known to reduce coronary heart disease risk. *American Journal of Cardiology* 99:1230-33.

De Bock, K., et al. 2008. Effect of training in the fasted state on metabolic responses during exercise with carbohydrate intake. *Journal of Applied Physiology* 104:1045-55.

Decombaz, J., et al. 1993. Effect of L-carnitine on submaximal exercise metabolism after depletion of muscle glycogen. *Medicine & Science in Sports & Exercise* 25:733-40.

Deighton, K., et al. 2012. Appetite, energy intake and resting metabolic responses to 60 min treadmill running performed in a fasted versus a postprandial state. *Appetite* 58:946-54.

Devaraj, S., and Jialal, I. 2006. The role of dietary supplementation with plant sterols and stanols in the prevention of cardiovascular disease. *Nutrition Reviews* 64:348-54.

Dowling, E. 2001. How exercise affects lipid profiles in women. *Physician and Sportsmedicine* 29(9):45-52.

Durstine, J., et al. 2001. Blood lipid and lipoprotein adaptations to exercise: A quantitative analysis. *Sports Medicine* 31:1033-62.

Fikenzer, K., et al. 2018. Effects of endurance training on serum lipids. *Vascular Pharmacology* 101:9-20.

Franco, O., et al. 2005. Effects of physical activity on life expectancy with cardiovascular disease. *Archives of Internal Medicine* 165:2355-60.

Frayn, K. 2010. Fat as fuel: Emerging understanding of the adipose tissue-skeletal muscle axis. *Acta Physiologica* 199:509-18.

Gardner, C. D., et al. 2018. Effect of low-fat vs low-carbohydrate diet on 12-month weight loss in overweight adults and the association with genotype pattern or insulin secretion: The DIETFITS randomized clinical trial. *Journal of the American Medical Association* 319(7):667-79.

Gibson, A., et al. 2015. Do ketogenic diets really suppress appetite? A systematic review and meta-analysis. *Obesity Reviews* 16:64-76.

Glatz, J., et al. 2002. Exercise and insulin increase muscle fatty acid uptake by recruiting putative fatty acid transporters to the sarcolemma. *Current Opinion in Clinical Nutrition & Metabolic Care* 5:365-70.

Goedecke, J., et al. 1999. Effect of medium-chain triacylglycerol ingested with carbohydrate on metabolism and exercise performance. *International Journal of Sports Medicine* 9:35-47.

Goedecke, J., et al. 2005. The effects of medium-chain triacylglycerol and carbohydrate ingestion on ultra-endurance exercise performance. *International Journal of Sport Nutrition & Exercise Metabolism* 15:15-27.

Gonzalez, J., et al. 2013. Breakfast and exercise contingently affect postprandial metabolism and energy balance in physically active males. *British Journal of Nutrition* 110:721-32.

Gray, P., et al. 2014. Fish oil supplementation reduces markers of oxidative stress but not muscle soreness after eccentric exercise. *International Journal of Sport Nutrition and Exercise Metabolism* 24:206-14.

Grundy, S. 2006. Nutrition in the management of disorders of serum lipids and lipoproteins. In *Modern Nutrition in Health and Disease,* ed. M. Shils, et al. Philadelphia, PA: Lippincott Williams & Wilkins.

Gutierrez, A., et al. 2001. Three days fast in sportsmen decreases physical work capacity but not strength or perception-reaction time. *International Journal of Sport Nutrition & Exercise Metabolism* 11:420-29.

Hall, K. D., and Guo, J. 2017. Obesity energetics: Body weight regulation and the effects of diet composition. *Gastroenterology* 152:1718-27.

Harcombe, Z., et al. 2015. Evidence from randomised controlled trials did not support the introduction of dietary fat guidelines in 1977 and 1983: A systematic review and meta-analysis. *Open Heart* 2015;2: doi:10.1136/openhrt-2014-000196

Hargreaves, M. 2006. The metabolic systems: Carbohydrate metabolism. In *ACSM's Advanced Exercise Physiology,* ed. C. M. Tipton. Philadelphia, PA: Lippincott Williams & Wilkins.

Harris, W. 2004. Are omega-3 fatty acids the most important nutritional modulators of coronary heart disease risk? *Current Atherosclerosis Reports* 6:447–52.

Harris, W. 2007. Omega-3 fatty acids and cardiovascular disease: A case for omega-3 index as a new risk factor. *Pharmacological Research* 55:217–23.

Hasler, C. 2000. The changing face of functional foods. *Journal of the American College of Nutrition* 19:499S–506S.

Havemann, L., et al. 2006. Fat adaptation followed by carbohydrate loading compromises high-intensity sprint performance. *Journal of Applied Physiology* 100:194–202.

Hawley, J. 2002. Effect of increased fat availability on metabolism and exercise capacity. *Medicine & Science in Sports & Exercise* 34:1485–91.

Hawley, J., et al. 2000. Fat metabolism during exercise. In *Nutrition in Sport,* ed. R. Maughan. Oxford: Blackwell Scientific.

Headland, M., et al. 2016. Weight-loss outcomes: A systematic review and meta-analysis of intermittent energy restriction trials lasting a minimum of 6 months. *Nutrients* 8:E354.

Heileson, J., and Funderburk, L. 2020. The effect of fish oil supplementation on the promotion and preservation of lean body mass, strength, and recovery from physiological stress in young, healthy adults: A systematic review. *Nutrition Reviews* 78:1001–14.

Heinonen, O. 1996. Carnitine and physical exercise. *Sports Medicine* 22:109–32.

Holloway, C., et al. 2011. A high-fat diet impairs cardiac high-energy phosphate metabolism and cognitive function in healthy human subjects. *American Journal of Clinical Nutrition* 93:748–55.

Horowitz, J., and Klein, S. 2000. Lipid metabolism during endurance exercise. *American Journal of Clinical Nutrition* 72:558S–63S.

Horowitz, J., et al. 2000. Preexercise medium-chain triglyceride ingestion does not alter muscle glycogen use during exercise. *Journal of Applied Physiology* 88:219–25.

Howell, W., et al. 1997. Plasma lipid and lipoprotein responses to dietary fat and cholesterol: A meta-analysis. *American Journal of Clinical Nutrition* 65:1747–64.

Huffman, D., et al. 2004. Effect of n-3 fatty acids on free tryptophan and exercise fatigue. *European Journal of Applied Physiology* 92:584–91.

Hulbert, A. 2005. Dietary fats and membrane function: Implications for metabolism and disease. *Biological Reviews of the Cambridge Philosophical Society* 80:155–69.

Jacobson, T., et al. 2007. Hypertriglyceridemia and cardiovascular risk reduction. *Clinical Therapy* 29:763–77.

Jakobsen, M., et al. 2009. Major types of dietary fat and risk of coronary heart disease: A pooled analysis of 11 cohort studies. *American Journal of Clinical Nutrition* 89:1425–32.

Jakulj, F., et al. 2007. A high-fat meal increases cardiovascular reactivity to psychological stress in healthy young adults. *Journal of Nutrition* 137:935–39.

Jenkins, D., et al. 1989. Nibbling versus gorging: Metabolic advantages of increased meal frequency. *New England Journal of Medicine* 321:929–34.

Jeukendrup, A. E. 2016. Ketone bodies: Fuel or hype? www.mysportscience.com/single-post/2016/05/16/Ketone-bodies-Fuel-or-hype.

Jeukendrup, A. 2002. Regulation of fat metabolism in skeletal muscle. *Annals of the New York Academy of Sciences* 967:217–35.

Jeukendrup, A. E., et al. 1998. Fat metabolism during exercise: A review. Part II: Regulation of metabolism and the effects of training. *International Journal of Sports Medicine* 19:293–302.

Jeukendrup, A., and Aldred, S. 2004. Fat supplementation, health, and endurance performance. *Nutrition* 20:678–88.

Jeukendrup, A., and Randell, R. 2011. Fat burners: Nutrition supplements that increase fat metabolism. *Obesity Reviews* 12:841–51.

Jeukendrup, A., et al. 1996. Effect of endogenous carbohydrate availability on oral medium-chain triglyceride oxidation during prolonged exercise. *Journal of Applied Physiology* 80:949–54.

Jeukendrup, A., et al. 1998. Effect of medium-chain triacylglycerol and carbohydrate ingestion during exercise on substrate utilization and subsequent cycling performance. *American Journal of Clinical Nutrition* 67: 397–404.

Johnson, E., and Schaefer, E. 2006. Potential role of dietary n-3 fatty acids in the prevention of dementia and macular degeneration. *American Journal of Clinical Nutrition* 84:1494S–98S.

Johnston, B., et al. 2014. Comparison of weight loss among named diet programs in overweight and obese adults: A meta-analysis. *Journal of the American Medical Association* 312:923–33.

Jones, P., and Kubow, S. 2006. Lipids, sterols, and their metabolites. In *Modern Nutrition in Health & Disease,* eds. M. Shils, et al. Philadelphia, PA: Lippincott Williams & Wilkins.

Jouris, K. B., et al. 2011. The effect of omega-3 fatty acid supplementation on the inflammatory response to eccentric strength exercise. *Journal of Sports Science & Medicine* 10:432–38.

Katsanos, C. 2006. Prescribing aerobic exercise for the regulation of postprandial lipid metabolism: Current research and recommendations. *Sports Medicine* 36:547–60.

Kelley, G., and Kelley, K. 2006. Aerobic exercise and HDL_2-C: A meta-analysis of randomized controlled trials. *Atherosclerosis* 184:207–15.

Kelley, G., et al. 2004. Aerobic exercise and lipids and lipoproteins in women: A meta-analysis of randomized controlled trials. *Journal of Women's Health* 13:1148–64.

Kelley, G., et al. 2004. Walking, lipids, and lipoproteins: A meta-analysis of randomized controlled trials. *Preventive Medicine* 38:651–61.

Kinsella, R., et al. 2017. Coconut oil has less satiating properties than medium chain triglyceride oil. *Physiology and Behavior* 179:422–26.

Knechtle, B., et al. 2004. Fat oxidation in men and women endurance athletes in running and cycling. *International Journal of Sports Medicine* 25:38–44.

Kodama, S., et al. 2007. Effect of aerobic exercise training on serum levels of high-density lipoprotein cholesterol: A meta-analysis. *Archives of Internal Medicine* 167:999–1008.

Konig, A., et al. 2005. A quantitative analysis of fish consumption and coronary heart disease mortality. *American Journal of Preventive Medicine* 29:335–46.

Kraemer, W., et al. 2006. Androgenic responses to resistance exercise: Effects of feeding and L-carnitine. *Medicine & Science in Sports & Exercise* 38:1288–96.

Kreider, R., et al. 2002. Effects of conjugated linoleic acid supplementation during resistance training on body composition, bone density, strength and selected hematological markers. *Journal of Strength Conditioning Research* 16:325–34.

Kriketos, A., et al. 1999. (−)-Hydroxycitric acid does not affect energy expenditure and substrate oxidation in adult males in a post-absorptive state. *International Journal of Obesity & Related Metabolic Disorders* 23:867–73.

Kris-Etherton, P., and Krauss, R. 2020. Public health guidelines should recommend reducing saturated fat consumption as much as possible: YES. *American Journal of Clinical Nutrition* 112:13–18.

Kris-Etherton, P., et al. 2002. AHA scientific statement: Fish, fish oils and omega-3 fatty acids. *Circulation* 106:2747–57.

Kris-Etherton, P., et al. 2002. Dietary fat: Assessing the evidence in support of a moderate-fat diet: The benchmark based on lipoprotein metabolism. *Proceedings of the Nutrition Society* 61:287–98.

LaBarrie, J., and St-Onge, M. P. 2017. A coconut oil-rich meal does not enhance thermogenesis compared to corn oil in a randomized trial in obese adolescents. *Insights in Nutrtion and Metabolism* 1:30–36.

Lambert, E., et al. 1994. Enhanced endurance in trained cyclists during moderate intensity exercise following 2 weeks adaptation to a high fat diet. *European Journal of Applied Physiology* 69:287–93.

Lambert, E., et al. 2001. High-fat diet versus habitual diet prior to carbohydrate loading: Effects of exercise metabolism and cycling performance. *International Journal of Sport Nutrition & Exercise Metabolism* 11:209–25.

Lange, K. 2004. Fat metabolism in exercise—With special reference to training and growth hormone administration. *Scandinavian Journal of Medicine & Science in Sports* 14:74–99.

LaRosa, J. 2007. Low-density lipoprotein cholesterol reduction: The end is more important than the means. *American Journal of Cardiology* 100:240–42.

Larsen, T., et al. 2006. Conjugated linoleic acid supplementation for 1 y does not prevent weight or body fat regain. *American Journal of Clinical Nutrition* 83:606–12.

Larson-Meyer, D., et al. 2008. Effect of dietary fat on serum and intramyocellular lipids and running performance. *Medicine & Science in Sports & Exercise* 40:892–902.

Lawrence, G. 2013. Dietary fats and health: Dietary recommendations in the context of scientific evidence. *Advances in Nutrition* 4:294–302.

Leckey, J. J., et al. 2017. Ketone diester ingestion impairs time-trial performance in professional cyclists. *Frontiers in Physiology* 8:806.

Lee, J., et al. 2009. Omega-3 fatty acids: Cardiovascular benefits, sources, and sustainability. *Nature Reviews: Cardiology* 6:753–8.

Leon, A., and Sanchez, O. 2001. Response of blood lipids to exercise training alone or combined with dietary interventions. *Medicine & Science in Sports & Exercise* 33:S502–15.

Liao, J. et al. 2022. The effects of fish oil on cardiovascular diseases: Systematical evaluation and recent advance. *Frontiers in Cardiovascular Medicine* 8:802306.

Lichtenstein, A. 2000. Trans fatty acids and cardiovascular disease risk. *Current Opinion in Lipidology* 11:37–42.

Macaluso, F., et al. 2013. Do fat supplements increase physical performance? *Nutrients* 5:509–24.

Manore, M. 2012. Dietary supplements for improving body composition and reducing body weight: Where is the evidence? *International Journal of Sports Nutrition and Exercise Metabolism* 22:139–54.

Margolis, L., and O'Fallon, K. 2020. Utility of ketone supplementation to enhance physical performance: A systematic review. *Advances in Nutrition* 11:412–19.

Márquez, F., et al. 2012. Evaluation of the safety and efficacy of hydroxycitric acid or *Garcinia cambogia* extracts in humans. *Critical Reviews in Food Science and Nutrition* 52:585–94.

Martin, W. 1997. Effect of endurance training on fatty acid metabolism during whole body exercise. *Medicine & Science in Sports & Exercise* 29:635–39.

Massicotte, D., et al. 1992. Oxidation of exogenous medium-chain free fatty acids during prolonged exercise: Comparison with glucose. *Journal of Applied Physiology* 73:1334–39.

Maughan, R., et al. 2018. IOC consensus statement: Dietary supplements and the high-performance athlete. *International Journal of Sport Nutrition and Exercise Metabolism* 28:104–125.

McGlory, C., et al. 2016. Fish oil supplementation suppresses resistance exercise and feeding-induced increases in anabolic signaling without affecting myofibrillar protein synthesis in young men. *Physiologial Reports* 4:1–11.

McRorie, J., et al. 2000. Effects of olestra and sorbitol consumption on objective measures of diarrhea: Impact of stool viscosity on common gastrointestinal symptoms. *Regulatory Toxicology & Pharmacology* 31:59–67.

Metcalf, R., et al. 2007. Effects of fish-oil supplementation on myocardial fatty acids in humans. *American Journal of Clinical Nutrition* 85:1222–28.

Mickleborough, T. 2008. A nutritional approach to managing exercise-induced asthma. *Exercise & Sport Science Reviews* 36:135–44.

Misell, L., et al. 2001. Chronic medium-chain triacylglycerol consumption and endurance performance in trained runners. *Journal of Sports Medicine & Physical Fitness* 41:210–15.

Mora-Rodriguez, R., and Coyle, E. 2000. Effects of plasma epinephrine on fat metabolism during exercise: Interactions with exercise intensity. *American Journal of Physiology Endocrinology & Metabolism* 278:E669–76.

Mori, T., and Woodman, R. 2006. The independent effects of eicosapentaenoic acid and docosahexaenoic acid on cardiovascular risk factors in humans. *Current Opinion in Clinical Nutrition & Metabolic Care* 9:95–104.

Moruisi, K., et al. 2006. Phytosterols/stanols lower cholesterol concentrations in familial hypercholesterolemic subjects: A systematic review with meta-analysis. *Journal of the American College of Nutrition* 25:41–48.

Mozaffarian, D., and Rimm, E. 2006. Fish intake, contaminants, and human health: Evaluating the risks and the benefits. *Journal of the American Medical Association* 296:1885–99.

Mozaffarian, D., et al. 2006. *Trans* fatty acids and cardiovascular disease. *New England Journal of Medicine* 354:1601–16.

Mumme, K., and Stonehouse, W. 2015. Effects of medium-chain triglycerides on weight loss and body composition: A meta-analysis of randomized controlled trials. *Journal of the Academy of Nutrition and Dietetics* 115:249–63.

Muoio, D., et al. 1994. Effect of dietary fat on metabolic adjustments to maximal VO_2 and endurance in runners. *Medicine & Science in Sports & Exercise* 26:81–88.

Murphy C., and McGlory, C. 2021. Fish oil for healthy aging: Potential applications to masters athletes. *Sports Medicine* 51:S31–S41.

Nicholls, S., et al. 2006. Consumption of saturated fat impairs the anti-inflammatory properties of high-density lipoproteins and endothelial function. *Journal of the American College of Cardiology* 48:715–20.

Noakes T., et al. 2014. Low-carbohydrate diets for athletes: What evidence? *British Journal of Sports Medicine* 48:1077–78.

Noto, H., et al. 2013. Low-carbohydrate diets and all-cause mortality: A systematic review and meta-analysis of observational studies. *PLoS One* 8:e55030.

Okano, G., et al. 1996. Effect of 4 h preexercise high carbohydrate and high fat meal ingestion on endurance performance and metabolism. *International Journal of Sports Medicine* 17:530–34.

Okano, G., et al. 1998. Effect of elevated blood FFA levels on endurance performance after a single fat meal ingestion. *Medicine & Science in Sports & Exercise* 30:763–68.

Onakpoya, I., et al. 2012. The efficacy of long-term conjugated linoleic acid (CLA) supplementation on body composition in overweight and obese individuals: A systematic review and meta-analysis of randomized clinical trials. *European Journal of Nutrition* 51:127–34.

Oostenbrug, G., et al. 1997. Exercise performance, red blood cell deformability, and lipid peroxidation: Effects of fish oil and vitamin E. *Journal of Applied Physiology* 83:746–52.

Peoples, G., et al. 2008. Fish oil reduces heart rate and oxygen consumption during exercise. *Journal of Cardiovascular Pharmacology* 52:540–47.

Perez-Jimenez, F., et al. 2005. International conference on the healthy effect of virgin olive oil. *European Journal of Clinical Investigation* 35:421–24.

Philippe, J. M., et al. 2017. Ketone bodies and exercise performance: The next magic bullet or merely hype? *Sports Medicine* 47:383–91.

Philpott, J. D., et al. 2018. Adding fish oil to whey protein, leucine, and carbohydrate over a six-week supplementation period attenuates muscle soreness following eccentric exercise in competitive soccer players. *International Journal of Sport Nutrition and Exercise Metabolism* 28:26–36.

Phinney, S. D., et al. 1983. The human metabolic response to chronic ketosis without caloric restriction: Preservation of submaximal exercise capability with reduced carbohydrate oxidation. *Metabolism* 32:769–76.

Pinkoski, C., et al. 2006. The effects of conjugated linoleic acid supplementation during resistance training. *Medicine & Science in Sports & Exercise* 38:339–48.

Raastad, T., et al. 1997. Omega-3 fatty acid supplementation does not improve maximal aerobic power, anaerobic threshold and running performance in well-trained soccer players. *Scandinavian Journal of Medicine & Science in Sports* 7:25–31.

Rawson, E. S., et al. 2018. Dietary supplements for health, adaptation, and recovery in athletes. *International Journal of Sport Nutrition and Exercise Metabolism* 19:1–12.

Remig, V., et al. 2010. *Trans* fats in America: A review of their use, consumption, health implications, and regulation. *Journal of the American Dietetic Association* 110: 585–92.

Reynolds, G. 2015. Should athletes eat fat or carbs? *New York Times,* February 25, http://well.blogs.nytimes.com/2015/02/25/should-athletes-eat-fat-or-carbs/?_r=0

Ridker, P., et al. 2002. Comparison of C-reactive protein and low-density -lipoprotein cholesterol levels in the prediction of first cardiovascular events. *New England Journal of Medicine* 347:1557–65.

Roepstorff, C., et al. 2005. Intramuscular triacylglycerol in energy metabolism during exercise in humans. *Exercise & Sport Sciences Reviews* 33:182–88.

Romijn, J., et al. 1993. Regulation of endogenous fat and carbohydrate metabolism in relation to exercise intensity and duration. *American Journal of Physiology* 265:E380–E391.

Romijn, J., et al. 2000. Substrate metabolism during different exercise intensities in endurance-trained women. *Journal of Applied Physiology* 88:1707–14.

Rowan, T. 2017. Low-fat dietary pattern and breast cancer mortality in the Women's Health Initiative randomized controlled trial. *Journal of Clinical Oncology* 35:2919–26.

Rowlands, D., and Hopkins, W. 2002. Effect of high-fat, high-carbohydrate, and high-protein meals on metabolism and performance during endurance cycling. *International Journal of Sport Nutrition & Exercise Metabolism* 12:318–35.

Rowlands, D., and Hopkins, W. 2002. Effects of high-fat and high-carbohydrate diets on metabolism and performance in cycling. *Metabolism* 51:678–90.

Sabaté, J., et al. 2010. Nut consumption and blood lipid levels. *Archives of Internal Medicine* 170:821–27.

Saito, M., et al. 2005. High dose of *Garcinia cambogia* is effective in suppressing fat accumulation in developing male Zucker obese rats, but highly toxic to the testis. *Food Chemistry Toxicology* 43:411–9.

Schoenfeld, B. J., et al. 2014. Body composition changes associated with fasted versus nonfasted aerobic exercise. *Journal of the International Society of Sports Nutrition* 11:54.

Shrier, I. 2005. Mediterranean diet for reducing mortality and easing the metabolic syndrome. *Physician and Sportsmedicine* 33 (5):8–9.

Shulman, D. 2002. Fuel on fat for the long run. *Marathon & Beyond* 6 (5):128–36.

Siri, P., and Krauss, R. 2005. Influence of dietary carbohydrate and fat on LDL and HDL particle distributions. *Current Atherosclerosis Reports* 7:455–59.

Spiering, B., et al. 2007. Responses of criterion variables to different supplemental doses of L-carnitine L-tartrate. *Journal of Strength & Conditioning Research* 21:259–64.

Spriet, L. 2002. Regulation of skeletal muscle fat oxidation during exercise in humans. *Medicine & Science in Sports & Exercise* 34:1477–84.

Spriet, L., and Hargreaves, M. 2006. The metabolic systems: Interaction of lipid and carbohydrate metabolism. In *ACSM's Advanced Exercise Physiology,* ed. C. Tipton. Philadelphia: Lippincott Williams & Wilkins.

Stannard, S. R., et al. 2010. Adaptations to skeletal muscle with endurance exercise training in the acutely fed versus overnight-fasted state. *Journal of Science and Medicine in Sport* 13:645-9.

Stefanik, M., and Wood, P. 1994. Physical activity, lipid and lipoprotein metabolism, and lipid transport. In *Physical Activity, Fitness, and Health*, ed. C. Bouchard, et al. Champaign, IL: Human Kinetics.

Stellingwerff, T., et al. 2006. Decreased PDH activation and glycogenolysis during exercise following fat adaptation with carbohydrate restoration. *American Journal of Physiology. Endocrinology & Metabolism* 290:E380-88.

Stephens, F., and Greenhaff, P. 2015. L-carnitine. In *Nutritional Supplements in Sport, Exercise, and Health. An A-Z Guide*, ed. L. M. Castell, S. J. Stear, and L. M. Burke. New York: Routledge.

Studer, M., et al. 2005. Effect of different antilipidemic agents and diets on mortality: A systematic review. *Archives of Internal Medicine* 165:725-30.

Szostak-Wegierek, D., et al. 2013. The role of dietary fats for preventing cardiovascular disease. A review. *Roczniki Panstwowego Zakladu Higieny* 64:263-9.

Tarnopolsky, M. 2000. Gender differences in metabolism: Nutrition and supplements. *Journal of Science & Medicine in Sport* 3:287-98.

Tarnopolsky, M. 2000. Gender differences in substrate metabolism during endurance exercise. *Canadian Journal of Applied Physiology* 25:312-27.

Tarnopolsky, M. 2008. Sex differences in exercise metabolism and the role of 17-beta estradiol. *Medicine & Science in Sports & Exercise* 40:648-54.

Tarnopolsky, M., et al. 1995. Carbohydrate loading and metabolism during exercise in men and women. *Journal of Applied Physiology* 68:302-8.

Tartibian, B., et al. 2010. The effects of omega-3 supplementation on pulmonary function of young wrestlers during intensive training. *Journal of Sciences & Medicine in Sport* 13:281-86.

Taylor, J., et al. 2006. Conjugated linoleic acid impairs endothelial function. *Arteriosclerosis, Thrombosis & Vascular Biology* 26:307-12.

Thomas, D. T., et al. 2016. American College of Sports Medicine Joint Position Statement. Nutrition and athletic performance. *Medicine and Science in Sports and Exercise* 48:543-68.

Thomas, T., et al. 2004. Effects of omega-3 fatty acid supplementation and exercise on low-density lipoprotein and high-density lipoprotein subfractions. *Metabolism* 53:749-54.

Thompson, P., et al. 2001. The acute versus the chronic response to exercise. *Medicine & Science in Sports & Exercise* 33:S438-45.

Thielecke, F., and Blannin, A. 2020. 3 Omega-3 fatty acids for sport performance – Are they equally beneficial for athletes and amateurs? A narrative review. *Nutrients* 12: 3712.

Timmons, B., et al. 2007. Energy substrate utilization during prolonged exercise with and without carbohydrate intake in pre-adolescent and adolescent girls. *Journal of Applied Physiology* 103:995-1000.

Trabelsi, K., et al. 2012. Effect of resistance training during Ramadan on body composition and markers of renal function, metabolism, inflammation, and immunity in recreational bodybuilders. *International Journal of Sport Nutrition and Exercise Metabolism* 22:267-75.

Trabelsi, K., et al. 2012. Effects of fed- versus fasted-state aerobic training during Ramadan on body composition and some metabolic parameters in physically active men. *International Journal of Sport Nutrition and Exercise Metabolism* 22:11-18.

Trepanowski, J. F. 2017. Effect of alternate-day fasting on weight loss, weight maintenance, and cardioprotection among metabolically healthy obese adults. A randomized clinical trial. *JAMA Internal Medicine* 177:930-38.

Tufts University. 2007. Studies find new omega-3 benefits. *Tufts University Health & Nutrition Letter* 25(5):4-5.

Tufts University. 2010. No dessert for heart checkmark program. *Tufts University Health & Nutrition Letter* 28(2):3.

Tunstall, R., et al. 2002. Exercise training increases lipid metabolism gene expression in human skeletal muscle. *American Journal of Physiology. Endocrinology & Metabolism* 283:E66-72.

Valenzuela, P. et al. 2020. Acute ketone supplementation and exercise performance: A systematic review and meta-analysis of randomized controlled trials. *International Journal of Sports Physiology and Performance* 10:1-11.

van Loon, L. 2004. Use of intramuscular triacylglycerol as a substrate source during exercise in humans. *Journal of Applied Physiology* 97:1170-87.

van Loon, L., et al. 2000. Effects of acute (–)-hydroxycitrate supplementation on substrate metabolism at rest and during exercise in humans. *American Journal of Clinical Nutrition* 72:1445-50.

Van Proeyen, K., et al. 2010. Training in the fasted state improves glucose tolerance during fatrich diet. *Journal of Physiology* 588:4289-302.

Van Proeyen, K., et al. 2011. Beneficial metabolic adaptations due to endurance exercise training in the fasted state. *Journal of Applied Physiology* 110:236-45.

Van Zyl, C., et al. 1996. Effects of medium-chain triglyceride ingestion on fuel metabolism and cycling performance. *Journal of Applied Physiology* 80:2217-25.

Venables, M., et al. 2005. Determinants of fat oxidation during exercise in healthy men and women: A cross-sectional study. *Journal of Applied Physiology* 98:160-67.

Volek, J. S., et al. 2015. Rethinking fat as a fuel for endurance exercise. *European Journal of Sport Science* 15:13-20.

Volk, B., et al. 2014. Effects of step-wise increases in dietary carbohydrate on circulating saturated fatty acids and palmitoleic acid in adults with metabolic syndrome. *PLoS One* 9:e113605.4.

Vukovich, M., et al. 1994. Carnitine supplementation: Effect on muscle carnitine and glycogen content during exercise. *Medicine & Science in Sports & Exercise* 26:1122-29.

Wall, B., et al. 2011. Chronic oral ingestion of L-carnitine and carbohydrate increases muscle carnitine content and alters muscle fuel metabolism during exercise in humans. *Journal of Physiology* 589:963-73.

Wallin, A., et al. 2012. Fish consumption, dietary long-chain n-3 fatty acids, and risk of type 2 diabetes: Systematic review and meta-analysis of prospective studies. *Diabetes Care* 35:918-29.

Walser, P., et al. 2006. Supplementation with omega-3 polyunsaturated fatty acids augments brachial artery dilation and blood flow during forearm contraction. *European Journal of Applied Physiology* 97:347-54.

Wang, L., et al. 2015. Effect of a moderate fat diet with and without avocados on lipoprotein particle number, size and subclasses in overweight and obese adults: A randomized, controlled trial.

Warber, J., et al. 2000. The effects of choline supplementation on physical performance. *International Journal of Sport Nutrition & Exercise Metabolism* 10:170-81.

Weber, H., et al. 2006. Prevention of cardiovascular diseases and highly concentrated *n*-3 polyunsaturated fatty acids (PUFAs). *Herz* 31 (Supplement 3):24-30.

Weggemans, R., et al. 2002. Dietary cholesterol from eggs increases the ratio of total cholesterol to high-density lipoprotein cholesterol in humans: A meta-analysis. *American Journal of Clinical Nutrition* 73:885-91.

Weinberg, R. et al. 2022. Cardiovascular impact of nutritional supplementation with omega-3 fatty acids: JACC focus seminar. *Journal of the American College of Cardiology* 77:593-608.

Whigham, L., et al. 2007. Efficacy of conjugated linoleic acid for reducing fat mass: A meta-analysis in humans. *American Journal of Clinical Nutrition* 85:1203-11.

Willett, W. 2006. The Mediterranean diet: Science and practice. *Public Health & Nutrition* 9:105-10.

Williams, P. 2001. Health effects resulting from exercise versus those from body fat loss. *Medicine & Science in Sports & Exercise* 33:S611-21.

Wolfe, R. 1998. Fat metabolism in exercise. *Advances in Experimental Medicine & Biology* 441:147-56.

Xin, G., and Eshaghi, H. 2021. Effect of omega-3 fatty acids supplementation on indirect blood markers of exercise-induced muscle damage: Systematic review and meta-analysis of randomized controlled trials. *Food Science and Nutrition* 9:6429-42.

Zderic, T., et al. 2004. High-fat diet elevates resting intramuscular triglyceride concentration and whole body lipolysis during exercise. *American Journal of Physiology. Endocrinology & Metabolism* 286:E217-25.

Protein: The Tissue Builder

CHAPTER SIX

KEY TERMS

- alanine
- alpha-ketoacid
- amino acids
- ammonia
- beta-alanine
- carnosine
- complete proteins
- creatine
- deamination
- delayed onset of muscle soreness (DOMS)
- essential (indispensable) amino acids
- glucogenic amino acids
- Beta-Hydroxy Beta-methylbutyrate (HMB)
- human growth hormone (HGH)
- incomplete proteins
- inosine
- ketogenic amino acids
- legumes
- limiting amino acid
- nitrogen balance
- nonessential (dispensable) amino acids
- protein hydrolysate
- protein-sparing effect
- proteinuria
- purines
- sports anemia
- urea

Don Hammond/Design Pics

LEARNING OUTCOMES

After studying this chapter, you should be able to:

1. Distinguish between complete and incomplete protein and identify foods that may be one or the other.
2. Calculate the approximate grams of protein that should be included in your daily diet.
3. Name the nine essential amino acids.
4. Describe the digestion of protein, its metabolic fate and distribution in the body, and its major functions in human metabolism.
5. Explain the role of protein in human energy systems during exercise.
6. Understand the rationale underlying the judgment of some investigators that both strength and endurance athletes need more dietary protein than the amount provided by the RDA. State general recommendations to various athletes based on grams of protein per kilogram body weight.
7. Based on current research findings, describe the dietary strategies involving protein intake (amount, type, and timing) before and/or after exercise that may help provide the substrate and hormonal milieu conducive to muscle tissue anabolism during recovery.
8. Explain the theory underlying the use of protein and protein-related dietary supplements to improve adaptation to exercise or exercise performance, and highlight the major research findings relative to their efficacy.
9. Identify the potential health risks associated with either inadequate or excess dietary intake of protein or individual amino acids.

Introduction

Don Hammond/Design Pics

Protein is one of the most essential nutrients for optimal health. Additionally, it has a wide variety of physiological functions that are essential for optimal physical performance. For example, protein forms the structural basis of muscle tissue, is the major component of most enzymes in the muscle, and can serve as a source of energy during exercise. Because protein is so important to the development and function of muscle tissue, and because most feats of human physical performance involve strenuous muscular activity in one form or another, it is no wonder that protein has persisted throughout the years as the food of the athlete. Indeed, surveys have revealed that many athletes believe that athletic performance is improved by a high-protein diet. Protein and protein-related dietary supplements continue to be among the best selling sports supplements.

Companies that market nutritional supplements for athletes have capitalized on the belief that increased protein consumption is needed for success. It appears that the groups most heavily targeted for protein supplement advertisements are bodybuilders and strength-type athletes, such as weight lifters and gridiron football players. Numerous high-protein products have been developed for these athletes in attempts to exploit the protein-muscle size and strength relationship. In recent years, specific amino acids have been theorized to maximize muscle mass and strength gains and have been advertised extensively on the internet and in magazines. Some advertisements even suggest that certain amino acid mixtures have an effect similar to drugs such as anabolic steroids, which have been used to stimulate muscle development.

Protein supplements are marketed for other types of athletes as well. Although protein is not regarded as a major energy source during exercise, research has suggested that endurance athletes may use some specific amino acids for energy production under certain conditions. Also, combined protein and carbohydrate supplements are marketed as recovery beverages for endurance and other athletes. Specific amino acids continue to be marketed and studied as potential nutrients that can delay the onset of fatigue through their effect on neurotransmitters in the brain.

There is no doubt that an adequate amount of dietary protein and related essential amino acids is required by all individuals. However, some advertisements directed toward athletes imply that protein or amino acid supplements are necessary for optimal performance. While research supports that increased dietary protein intake augments adaptations to resistance training, such as increased muscle mass or strength, questions remain about the exact amount and source of this additional protein. Most investigators recommend that additional dietary protein be derived from natural food sources when possible.

Other dietary supplements related to protein, such as creatine, beta-hydroxy beta-methylbutyrate (HMB), beta-alanine, and gelatin, have become increasingly popular in recent years. In most cases these dietary supplements have been marketed to strength-trained athletes as a means to foster muscle growth and strength development, but some are intended to improve sport performance or speed recovery from injury.

While physically active individuals certainly need more protein in their diet, are whole foods or protein supplements the better choice? Also, are protein-related supplements safe, effective, or necessary for optimal health and performance? The information presented in this chapter will provide answers to these and other questions. Topics to be covered include dietary needs and sources of protein; metabolic fates and functions in the body; effects of exercise on protein metabolism and dietary requirements; safety and efficacy of protein and protein-related supplements; and health aspects of dietary protein.

Dietary Protein

What is protein?

Protein is a complex chemical structure containing carbon, hydrogen, and oxygen—just as carbohydrates and fats do. Protein has one other essential element—nitrogen, which constitutes about 16 percent of most dietary protein. These four elements are combined into a number of different structures called **amino acids**, each one possessing an amino group (NH_2) and an acid group (COOH), with the remainder being different combinations of carbon, hydrogen, oxygen, and in some cases sulfur. There are 20 amino acids, all of which can be combined in a variety of ways to form the proteins necessary for the structure and functions of the human body. The body may also modify the structure of dietary amino acids, such as converting proline to hydroxyproline, to meet its needs. **Figure 6.1** depicts the formula of **alanine**, an amino acid discussed later.

Proteins are created when two amino acids link and form a peptide bond; hence, a dipeptide is formed. As more amino acids are added, a polypeptide is formed. Most proteins are polypeptides, combining up to 300 amino acids. **Figure 6.2** depicts the building of a protein. Protein is contained in both animal and plant foods. Humans obtain their supply of amino acids from these two general sources.

Is there a difference between animal and plant protein?

To answer this question, let us first look at a basic difference between two groups of amino acids. Humans can synthesize some amino acids in their bodies but cannot synthesize others. The nine amino acids that cannot be manufactured in the body are called **essential**, or **indispensable, amino acids** and must be supplied in the diet. Those that may be formed in the body are called **nonessential**, or **dispensable, amino acids**. Six of the dispensable amino acids are conditionally indispensable, which means that they must be obtained through the diet when endogenous synthesis cannot meet metabolic demands, such as in severe catabolic states. Although nutrition scientists prefer the terms *indispensable* and *dispensable,* this text uses the terms *essential* and *nonessential,* because they are most commonly used.

It should be noted that all 20 amino acids are necessary for protein synthesis in the body and must be present simultaneously for optimal maintenance of body growth and function. The use of the terms *essential* and *indispensable* in relation to amino acids is to distinguish those that must be obtained in the diet. **Table 6.1** presents the dietary essential, nonessential, and conditionally essential amino acids.

The National Academy of Sciences indicates that different dietary sources of protein vary widely in their composition and nutritional value. The quality of a source of protein is an expression of its ability to provide the nitrogen and amino acid

TABLE 6.1	The dietary amino acids
Essential amino acids	**Nonessential amino acids**
Histidine	Alanine
Isoleucine*	Arginine**
Leucine*	Asparagine
Lysine	Aspartic acid
Methionine	Cysteine**
Phenylalanine	Glutamic acid
Threonine	Glutamine**
Tryptophan	Glycine**
Valine*	Proline**
	Serine
	Tyrosine**

*Branched-chain amino acids.
**Conditionally essential.

requirements for growth, maintenance, and repair. The key factors are digestibility and the ability to provide the indispensable amino acids. All natural, unprocessed animal and plant foods contain all 20 amino acids. However, the amount of each amino acid in specific foods varies. Over the years a number of different techniques have been used, usually with animals, to assess the quality of protein in selected foods. One of the most widely used is the Protein Digestibility-Corrected Amino Acid Score (PDCAAS), which incorporates real-life variables, including the amino acid content and digestibility of the protein. Scores can range from 1.0 to 0.0, with 1.0 being the highest quality. We need not go into a detailed discussion of all techniques to evaluate protein quality, but essentially they focus on the concept of **nitrogen balance**, the ability of the body to retain nitrogen. In essence, nitrogen balance is protein balance. In positive nitrogen balance the body is retaining protein to adequately support growth and development, whereas in negative nitrogen balance the body is losing protein, with possible impairment in growth and development. The quality of the protein in foods we eat may affect nitrogen balance.

In general, those foods that contain an adequate content of all nine essential amino acids to support both life and growth are

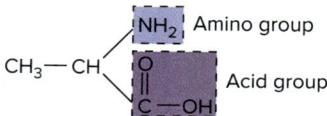

FIGURE 6.1 The chemical structure of alanine, an amino acid. The amino group (NH_2) contains nitrogen, while the acid group is represented by COOH.

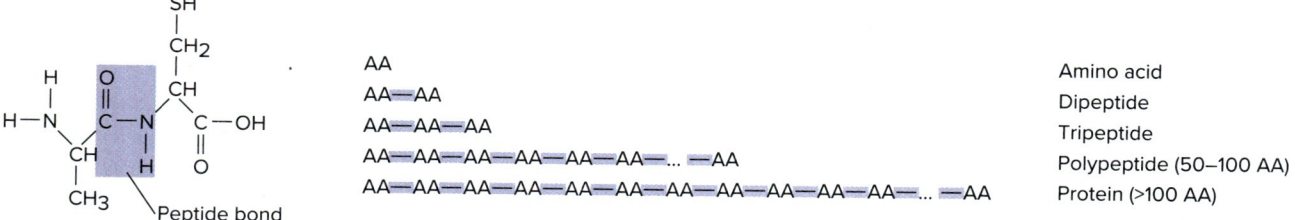

FIGURE 6.2 Formation of peptides and polypeptides from amino acids, with eventual formation of proteins.

known as **complete proteins**, or high-quality proteins, and will have a high PDCAAS score, while those that have a deficiency of one or more essential amino acids and are unable to support life or growth are called **incomplete proteins**, or low-quality proteins, and have a lower score. Relative to human requirements, an essential amino acid that is in limited supply in a particular food is labeled a **limiting amino acid**.

The proteins ingested as animal products are generally regarded to be of a higher quality than those found in plants. This is not to say that an amino acid found in a plant is inferior to the same amino acid found in an animal. They are the same. When we look at the distribution of all the amino acids in the two food sources, however, we can then see two major reasons animal protein is called a high-quality protein, whereas plant protein is of lower quality. The PDCAAS for egg white and meat is, respectively, 1.0 and 0.92, while the score for kidney beans and whole wheat is, respectively, 0.68 and 0.40. The PDCAAS for collagen, a popular dietary supplement, is zero. Collagen supplements are discussed later in the chapter.

Animal protein is a complete protein because it contains each essential amino acid in the proper proportion to human requirements. As noted, all 20 amino acids must be present simultaneously for the body to synthesize them into necessary body proteins. If 1 amino acid is in short supply, protein construction may be blocked. Having the proper amount of animal protein in the diet is a good way to ensure receipt of a balanced supply of amino acids. Some protein supplements marketed to athletes are made from animal protein, including milk, egg, and whey protein.

Plant proteins can provide you with all the protein and amino acids you need for optimal growth and development. However, proteins usually exist in smaller concentrations in plant foods. For example, 2 ounces of fish contain about 14 grams of protein, while 2 ounces of cooked macaroni have only 2 grams; 2 ounces of beans, which are generally regarded to be good sources of protein, have only 5 grams. In addition, most plant proteins have insufficient amounts of 1 or more of the essential amino acids (i.e., limiting amino acids). Grain products are usually deficient in lysine, whereas legumes are low in methionine. An exception to this generality is the protein isolated from soybeans, which when processed properly is comparable to animal protein. As noted in chapter 2, vegetarians who eat plant foods in proper combinations over the course of the day will receive a balanced supply of amino acids. Some populations receive most of their protein from plant sources.

What are some common foods that are good sources of protein?

Animal Foods Animal foods from the protein foods and dairy groups generally have substantial amounts of high-quality protein. One glass of cow's milk or its equivalent contains about 7–8 grams of protein, as does 1 ounce of meat, fish, or poultry. One egg contains 6 grams of high-quality protein, as does 1 serving of Egg Beaters, but the later has half the kcal and no fat or cholesterol.

Plant Foods and Supplements **Legumes**, such as dry beans (black, garbanzo, great northern, kidney, lima, navy, pinto, soybeans), lentils, and peas (black-eyed, split), are relatively good sources of protein. One-half cup contains about 7–9 grams of protein. Nuts contain fair amounts of protein but are high in fat. Fruits, vegetables, and grain products all have some protein, but the content varies; generally speaking, the protein content is low, ranging from less than 1 gram to about 3 grams of protein per serving, although some products may contain more, such as protein-enriched pasta. Some sports drinks and sports bars contain significant amounts of protein. Protein supplements targeted to strength-trained individuals may contain substantial amounts of protein, but may also be expensive.

Table 6.2 and **figure 6.3** present some common foods in each of several food groups, with the number of grams of protein in each. Notice the effect combination-type foods have on protein content: for example, macaroni and cheese versus plain macaroni. Most food labels today list the grams of protein per serving. For plant foods commonly eaten in a vegan diet, see chapter 2.

How much dietary protein do I need?

Humans actually do not need protein *per se* but rather an adequate amount of nitrogen and essential amino acids. However, because all nine essential amino acids and almost all dietary nitrogen are derived from dietary protein, it serves as the basis for our daily requirements. There are three basic ways to characterize protein intake: absolute intake in grams per day, relative intake based on grams per unit body weight (the basis of the RDA), and as percentage of daily energy intake (the basis of the AMDR).

In the United States, the recommended dietary intake of protein is based upon the RDA. The amount of protein necessary in the diet varies in different stages of the life cycle, as may be noted in the Dietary Reference Intakes table for macronutrients. During the early years of life, children manufacture protein tissue during rapid growth stages, with the rate of growth (and thus the protein needs) varying from infancy through late adolescence. In young adulthood, the protein requirement stabilizes. Throughout the life cycle, however, the protein requirement established in the RDA is based upon the body weight of the individual. As a person passes from infancy to adulthood, the protein RDA per unit body weight decreases, but the absolute amount of protein needed by the body as a whole actually increases because of increases in body weight.

Table 6.3 presents the amount of protein needed per kilogram or per pound of body weight for different age groups. A variety of scientific techniques have been used over the years to determine human protein needs, and more recent research has reaffirmed these estimates for adults and children. The values for the first year of life are AI, while the remainder are RDA. The values in this table are dependent upon adequate daily energy intake (i.e., kcal), because a low-energy diet will increase protein needs. To calculate your requirement, simply determine your body weight in kilograms or pounds and multiply by the appropriate figure for your age group. Recall that 1 kilogram is equal to 2.2 pounds. As an example, compute the protein requirement for a 154-pound, or 70-kilogram, average 23-year-old male:

$$0.36 \text{ g protein/pound} \times \text{pounds} = 55.4 \text{ or } 56 \text{ g protein/day}$$
$$0.8 \text{ g protein/kg} \times 70 \text{ kg} = 56 \text{ g protein/day}$$

TABLE 6.2 Protein content in some common foods

Food	Amount	Protein (grams)	Food	Amount	Protein (grams)
Dairy			*Fruits*		
Cow's milk, whole	1 c	8	Banana	1	1
Cow's milk, skim	1 c	8	Orange	1	1
Cheese, cheddar	1 oz	7	Pear	1	1
Yogurt	1 c	8	*Grains*		
Greek yogurt	1 c	22	Bread, wheat	1 slice	3
Protein foods			Bran flakes	1 c	4
Beef, lean	1 oz	8	Doughnuts	1	1
Chicken breast	1 oz	8	Macaroni	½ c	3
Luncheon meat	1 oz	5	Macaroni and cheese	½ c	9
Fish	1 oz	7	Peas, green	½ c	4
Eggs	1	6	Potato, baked	1	3
Navy beans, cooked	½ c	7	Quinoa	1 c	8
Peanuts, roasted	¼ c	9	*Sports drinks and bars*		
Peanut butter	1 tbsp	4	Gatorade® Nutrition Shake	11 oz	20
Vegetables			Power Bar® Protein Plus	1	30
Broccoli	½ c	2	Endurox® R4 Recovery Drink	12 oz	11
Carrots	1	1	*Protein sports supplements*		
			Whey protein concentrate	40 g	25

FIGURE 6.3 Foods high in protein include meats, milk, cheese, eggs, and plants such as wheat and legumes. Some sports bars are also rich in protein.

John Thoeming/McGraw Hill

On a protein-free diet, the average individual loses approximately 0.34 g protein/kg body weight per day, which could be replaced by a similar amount of high-quality egg protein. However, allowances are made in the RDA for the fact that individual protein needs vary, that the biological quality of all dietary protein is not as good as egg protein, and that the efficiency of utilization decreases at higher dietary protein-intake levels. Hence, the RDA is adjusted upward to account for these factors.

TABLE 6.3 Grams of protein needed per kilogram or per pound body weight during the life cycle

Age in years	Grams/kg body weight	Grams/pound body weight
0.0–0.5	1.52	0.69
0.5–1.0	1.10	0.50
1–3	1.10	0.50
4–8	0.95	0.43
9–13	0.95	0.43
14–18	0.85	0.39
19 and up	0.8	0.36

The RDA for protein, as noted, is based upon body weight of the individual at different ages. If you took the recommended energy intake in kcal for each age group, say 2,500 kcal for the average adult male, and calculated the percentage of this value that the RDA for protein supplies, the values approximate 10 percent for each age group. The National Academy of Sciences indicated that the AMDR should not be set below levels for the RDA for protein, which is about 10 percent of energy. Mathematically, 56 grams of protein, at 4 kcal per gram, total 224 kcal, which is about 9 percent of 2,500, or near the lower limit of the AMDR.

The Acceptable Macronutrient Distribution Ranges (AMDR) for individuals were set based on evidence from interventional trials, with support of epidemiological evidence, to suggest a role in the prevention of increased risk of chronic disease and based on ensuring sufficient intakes of essential nutrients. Thus, the AMDR for protein was not set with athletes in mind. The AMDR for protein is 10–35 percent of energy for adults and 5–20 and 10–30 percent for young and older children, respectively. Although beyond the scope of this chapter, there are likely differences in protein needs in several segments of the population beyond athletes and nonathletes or children and adults. For example, as highlighted by Rogeri and colleagues, research shows that older adults can be resistant to the anabolic effects of exercise and protein ingestion and possibly require more dietary protein to maintain muscle mass and strength. In their review, Freitas and Katsanos describe similar blunted responses to exercise and protein ingestion in the muscles of people with obesity.

Energy intake is an important determinant of protein needs. For example, if energy intake is inadequate, as may be the case in those on weight-loss diets or the elderly, dietary protein may be used for energy instead of its core purpose of building tissue. Thus, some individuals may need more or higher-quality protein because their energy intake may be low. The role of dietary protein in healthful and successful weight-loss programs is discussed later in the chapter.

How much of the essential amino acids do I need?

The Academy has established RDAs for the nine essential amino acids. The amounts for adult males age 19 and over are found in **table 6.4**. Slightly larger amounts are recommended for children and adolescents. For the average adult, about 25 percent of the total protein requirement should consist of the essential amino acids; this amounts to about 14–15 grams. Phenylalanine is an essential amino acid, whereas tyrosine is normally listed as a nonessential amino acid. The two are of similar chemical structure so that when substantial quantities of tyrosine are contained in the diet, the need for phenylalanine will decrease somewhat. The same holds true for the essential sulfur-containing amino acid methionine and its chemically related counterpart, cysteine. Individuals who obtain the RDA for protein should have no problem obtaining these recommended values.

What are some dietary guidelines to ensure adequate protein intake?

To answer in one sentence: Eat a wide variety of animal and plant foods. The high-quality, complete proteins are obtained primarily from animal foods. Meat, fish, eggs, poultry, milk, and cheese

TABLE 6.4   RDA for the essential amino acids in an adult male (70 kg)

	RDA (mg/kg)	Total mg
Histidine	14	980
Isoleucine	19	1,260
Leucine	42	2,940
Lysine	38	2,660
Methionine plus cysteine	19	1,260
Phenylalanine plus tyrosine	33	2,310
Threonine	20	1,400
Tryptophan	5	350
Valine	24	1,680
Total		14,840

contain the type and amount of the essential amino acids necessary for maintaining life and promoting growth and development. They are high-nutrient-density foods, particularly if fat content is low to moderate. Because animal protein is of high quality, you do not need as much of it to satisfy your RDA. For example, for a male who needs about 56 grams of protein per day, only 45 grams are needed if it is animal protein. One glass of milk, with 8 grams of protein, will provide almost 20 percent of his protein RDA. Two glasses of milk, one egg, and 3 ounces of lean meat, fish, or poultry will provide 100 percent of his RDA. In addition, a substantial proportion of daily vitamin and mineral needs will also be supplied in these foods. As noted in chapter 5, selection of low-fat foods will enhance the nutrient density by reducing kcal.

Plant foods also may provide good sources of protein. Grain products such as wheat, rice, and corn, as well as soybeans, peas, beans, and nuts, have a substantial protein content. However, most plant foods contain incomplete proteins because they lack a sufficient quantity of some essential amino acids. For this reason, the protein RDA for the adult male is 65 grams per day when plant proteins are the primary source. However, the position stand on vegetarian diets by the Academy of Nutrition and Dietetics, states that if certain plant foods are eaten over the course of a day, such as grains and legumes, they may supply all the essential amino acids necessary for human nutrition and be as complete a protein as animal protein.

Some research has suggested that if the daily dietary protein is obtained through a mixture of animal and plant foods in a ratio of 30:70, that is, 30 percent of the protein from animal foods and 70 from plant foods, the protein quality would be similar to the use of animal foods alone. Mixing animal and plant foods in the same meal is common and is healthful and nutritious. Animal foods provide excellent sources of essential minerals, such as iron, zinc, and calcium, while plant foods provide carbohydrate, dietary fiber, and various phytochemicals.

Key Concepts

- Protein contains nitrogen, an element essential to the formation of 20 different amino acids, the building blocks of all body cells. All 20 amino acids are necessary for protein formation in the body.
- Essential, or indispensable, amino acids cannot be adequately synthesized in the body and thus must be obtained through dietary protein, whereas nonessential, or dispensable, amino acids may be synthesized in the body. Conditionally indispensable amino acids may be synthesized from other amino acids under normal conditions, but their synthesis may be limited under certain conditions when insufficient amounts of their precursors are available.
- The RDA for protein is based upon the body weight of the individual, and the amount needed per unit body weight is greater during childhood and adolescence than during adulthood. The adult RDA is 0.8 gram of protein per kilogram body weight, or 0.36 gram per pound body weight.
- The Acceptable Macronutrient Distribution Range (AMDR) for protein indicates that dietary intake should be no less than 10 percent and no greater than 35 percent of daily energy needs. The RDA and AMDR for daily protein intake are not based on the needs of athletes.
- The human body needs a balanced mixture of essential amino acids, and although animal protein provides all of the essential amino acids in the proper blend, a combination of certain plant proteins, such as grains and legumes, will satisfy this dietary requirement.
- Although animal foods in the meat and milk groups have high protein content, they may also be high in fat. Increasing the proportion of dietary protein intake from plant sources is recommended. Combining animal and plant proteins in one meal, such as milk and cereal or stir-fry vegetables and meat, will increase the protein quality of the meal.

Check for Yourself

- Peruse food labels of some of your favorite foods and check the protein content. Check both animal and plant foods and compare the grams of protein provided by a serving.

Metabolism and Function

What happens to protein in the human body?

Dietary protein consists of long, complex chains of amino acids. In the digestive process, enzymes (proteases) in the stomach and small intestine break the complex protein down into polypeptides and then into individual amino acids. The amino acids are absorbed through the wall of the small intestine, pass into the blood, and then pass to the liver via the portal vein (**figure 6.4**).

The digestion of protein takes several hours, but once the amino acids enter the blood they are cleared within 5–10 minutes. There is a constant interchange of amino acids among the blood, the liver, and the body tissues. The liver is a critical center in amino acid metabolism. It is continually synthesizing a balanced amino acid mixture for the diverse protein requirements of the body. These amino acids are secreted into the blood and carried as free amino acids or as plasma proteins such as albumin. All these functions consume energy, and the thermic effect (TEF) is greater for protein than for carbohydrate and fat.

The most important metabolic fate of the amino acids is the formation of specific proteins, including the structural proteins such as muscle tissue and the functional proteins such as enzymes. Body cells obtain amino acids from the blood, and the genetic apparatus in the cell nucleus directs the synthesis of proteins specific to the cell needs. The body cells may also use some of the nitrogen from the amino acids to form nonprotein nitrogen compounds, such as creatine. For example, the muscle cells will form contractile proteins as well as the enzymes and creatine phosphate necessary for energy production. The body cells will use only the amount of amino acids necessary to meet their protein needs. They cannot store excess amino acids to any significant amount, although the protein formed may be catabolized to release amino acids back to the blood.

Because the human body does not have a mechanism to store excess nitrogen, it cannot store amino acids *per se*. Through the process of **deamination**, the amino group (NH_2) containing the nitrogen is removed from the amino acid, leaving a carbon substrate known as an **alpha-ketoacid**. The excess nitrogen must be excreted from the body. In essence, the liver forms **ammonia** (NH_3) from the excess nitrogen; the ammonia is converted into **urea**, which passes into the blood and is eventually eliminated by the kidneys into the urine.

The alpha-ketoacid that is released may have several fates. For one, this carbon substrate may be oxidized for the release of energy. For another, it may accept another amino group and be reconstituted to an amino acid. It also may be channeled into the metabolic pathways of carbohydrate and fat. The liver is the main organ where this conversion occurs. In essence, some of the amino acids are said to be **glucogenic amino acids**, that is, glucose forming. At various stages of the energy transformations within the liver, the glucogenic amino acids may be converted to glucose. As noted in chapter 4, this process is called *gluconeogenesis*. The **ketogenic amino acids** are metabolized in the liver to acetyl CoA, which may be used for energy production via the Krebs cycle or converted to fat. The glucose and fat produced may be transported to other parts of the body to be used. Thus, although excess protein cannot be stored as amino acids in the body, the energy content is not wasted, for it is converted to either carbohydrate or fat.

Protein turnover represents the process by which all body proteins are being continuously broken down and resynthesized, and it is an ongoing process. The National Academy of Sciences indicates that about 250 grams of body protein turns over daily in an adult, or about 0.5 pound of body mass. **Figure 6.5** presents a summary of the fates of protein in human metabolism.

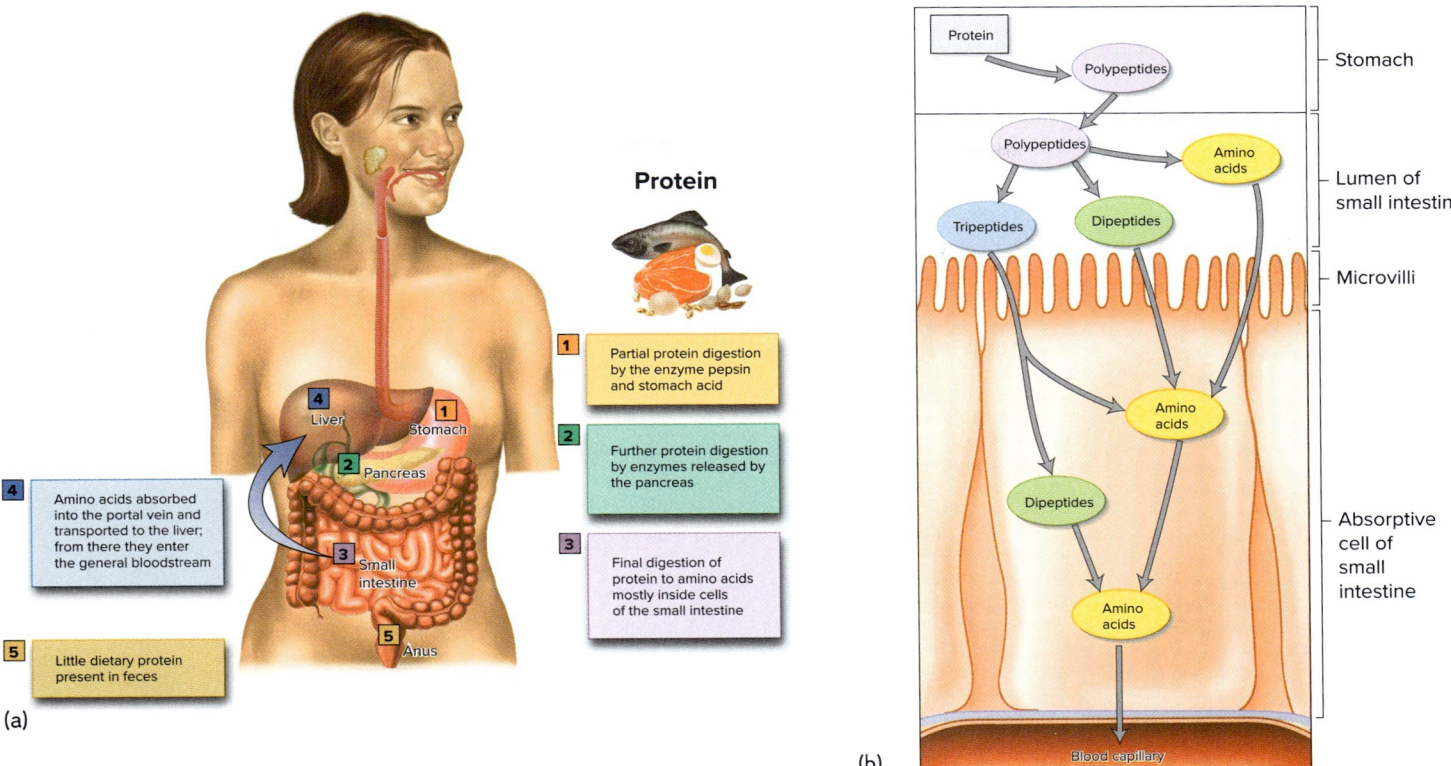

FIGURE 6.4 (a) A summary of protein digestion and absorption. (b) Stomach acid and enzymes contribute to protein digestion. Enzymatic protein digestion begins in the stomach and ends in the absorptive cells of the small intestine, where the last peptides are broken down into single amino acids.

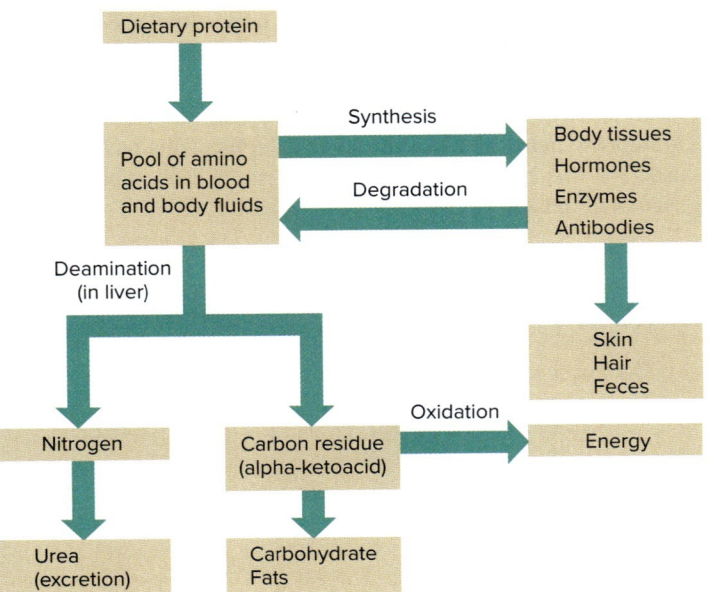

FIGURE 6.5 Simplified diagram of protein metabolism. Following the digestion of dietary proteins, one of the major functions of the amino acids is the synthesis of body tissues, enzymes, hormones, and antibodies. However, protein also is constantly being degraded by the liver. The excess nitrogen is excreted as urea, while the carbon residue may be converted into carbohydrate or fat or used to produce energy.

Can protein be formed from carbohydrates and fats?

Yes, but with some major limitations. Protein has one essential element, nitrogen, which is not possessed by either carbohydrate or fat. However, if the body has an excess of amino acids, the liver may be able to use the nitrogen-containing amino groups from these excess amino acids and combine them with alpha-ketoacids derived from either carbohydrate or fat metabolism. A key alpha-ketoacid from carbohydrate is pyruvic acid, while fat yields acetoacetic acid. The net result is the formation in the body of some of the nonessential amino acids using carbohydrates and fats as part of the building materials. Keep in mind that nitrogen must be present for this to occur, and its source is dietary protein.

What are the major functions of protein in human nutrition?

Dietary protein may be utilized to serve all three major functions of food. Through the action of the individual amino acids, protein serves as the structural basis for the vast majority of body tissues, is essential for regulating metabolism, and can be used as an energy source. In one way or another protein is involved in almost all body functions. Its individual roles are beyond the scope of this text, so the following discussion represents just some of its major functions of importance to health and fitness. **Table 6.5** highlights the major functions of protein in the body.

TABLE 6.5 Summary of the functions of proteins and amino acids in human metabolism

1.	Structural function	Form vital constituents of all cells in the body, such as contractile muscle proteins
2.	Transport function	Transport various substances in the blood, such as the lipoproteins for conveying triglycerides
3.	Enzyme function	Form almost all enzymes in the body to regulate numerous, diverse physiological processes
4.	Hormone and neurotransmitter function	Form various hormones, such as insulin; form various neurotransmitters, or neuropeptides, that function in the central nervous system, such as serotonin
5.	Immune function	Form key components of the immune system, such as antibodies
6.	Acid-base balance function	Buffer acid and alkaline substances in the blood to maintain optimal pH
7.	Fluid balance function	Exert osmotic pressure to maintain optimal fluid balance in body tissues, particularly the blood
8.	Energy function	Provide source of energy to the Krebs cycle when deaminated; excess protein may be converted to glucose or fat for subsequent energy production
9.	Movement function	Provide movement when structural muscle proteins use energy to contract

Protein is the main nutrient used in the formation of all body tissues. This role is extremely important in periods of rapid growth, such as childhood and adolescence. Athletes who attempt to gain muscle tissue also need an adequate dietary supply of protein to create a positive protein balance. Certain amino acids, such as the branched-chain amino acids (BCAA) leucine, isoleucine, and valine, constitute a significant amount of muscle tissue.

Protein is critical in the regulation of human metabolism. It is used in the formation of almost all enzymes, many hormones, and other compounds that control body functions. Insulin, hemoglobin, and the oxidative enzymes in the mitochondria are all proteins that have important roles in regulating metabolism during exercise. Other metabolic roles of protein include the maintenance of water balance and acid-base balance, regulation of the blood clotting process, prevention of infection, and development of immunity to disease. Proteins also serve as carriers for nutrients in the blood, such as the free fatty acids (FFA) and the lipoproteins, and help transport nutrients into the body cells.

Although protein is not a major energy source for humans at rest, it can serve such a function under several conditions. In nutritional energy balance, the priority use of dietary protein is to promote synthesis of body proteins essential for optimal structure and function. However, as noted previously, excess dietary protein may be deaminated and used for energy, or it may be converted to carbohydrate or fat and then enter metabolic pathways for energy production or storage. During periods of starvation or semistarvation, adequate amounts of dietary or endogenous carbohydrates and dietary fats may not be available. Both dietary protein and the body protein stores are used for energy purposes in such a situation, because energy production takes precedence over tissue building in metabolism. Hence, if the active individual desires to maintain lean body mass, it is essential to have not only adequate protein intake but also sufficient carbohydrate kcal in the diet to provide a **protein-sparing effect**. In other words, carbohydrate kcal will be used for energy production, thus sparing utilization of protein as an energy source and allowing it to be used for its more important structural and metabolic functions.

Although body proteins are composed of all 20 amino acids, individual amino acids may have important specific effects in the body. For example, the amino acid glycine is a neurotransmitter inhibitor; tryptophan and tyrosine are important for the formation of several chemical transmitters in the brain; and the branched-chain amino acids (leucine, isoleucine, and valine) are major components of muscle tissue that may provide a source of energy.

Because of the diverse roles of protein and amino acids in the body, athletes have used protein supplements for years in attempts to improve performance. Amino acid supplements have also been used for this purpose. The effectiveness of such supplements is evaluated in later sections.

Key Concept

▸ The major functions of dietary protein are to build and repair tissues and to synthesize hormones, enzymes, and other body compounds essential in human metabolism, but it also may be used as a significant source of energy under certain conditions.

Proteins and Exercise

Protein has always been considered one of the main staples of an athlete's diet. In this section we discuss several topics regarding protein and exercise, including its use as an energy source, possible avenues for protein loss, protein metabolism during recovery from exercise, and dietary requirements and recommendations for athletes.

Are proteins used for energy during exercise?

The average individual consumes about 10 percent or more of daily energy intake from protein. For individuals in protein and energy balance, this protein intake must be balanced through energy expenditure and other body losses. In general, scientists suggest that about 5 percent of daily protein intake may be used directly for energy. Protein in excess of tissue needs may be converted to carbohydrate or fat, which may also be used for energy. Protein may also be lost from the body in various ways, such as creatinine and urea excreted in the urine and sloughed body cells. In one review, Gibala noted that protein is regarded as a minor source of fuel during rest, usually accounting for less than 5 percent of total daily energy expenditure. However, given its nutritional importance, researchers have attempted to determine the effect of exercise on protein balance and needs.

Scientists have used a variety of techniques to study protein metabolism during exercise. Because urea is a by-product of protein metabolism, its concentration in the urine, blood, and sweat has been analyzed. Also, the presence in the urine of a marker for muscle protein breakdown, known as 3-methylhistidine, a modified amino acid, has been studied to evaluate protein catabolism. The nitrogen balance technique consists of precisely measuring nitrogen intake and excretion to determine whether the individual is in positive or negative protein balance. Finally, labeled isotopes of amino acids have been ingested or injected to study their metabolic fate during exercise, not only in the whole body but also in isolated muscle groups.

Using these techniques, investigators have evaluated the use of protein during both resistance exercise and aerobic endurance exercise. Although both types of exercise affect protein metabolism, the use of protein as an energy source appears to be more prevalent in aerobic exercise, particularly when prolonged.

Resistance Exercise As noted by Rennie and Tipton, resistance exercise has little effect on amino acid oxidation. Phillips and colleagues were first to demonstrate that resistance exercise increases both protein synthesis and protein breakdown for about 48 hours post-exercise. Muscle protein net balance remained negative during this time. As McGlory and colleagues note, the combination of resistance exercise and protein ingestion is synergistic, meaning that when resistance exercise is performed before protein ingestion, the combination increases protein synthesis rates above protein breakdown, allowing for muscle hypertrophy. So, although resistance exercise alone does not increase protein oxidation, it can provoke protein breakdown especially during intense resistance exercise, such as with eccentric contractions.

Aerobic Endurance Exercise Poortmans has noted that although protein may be used to produce significant amounts of ATP in the muscle, during aerobic endurance exercise the rate of production is much slower than carbohydrate and fat, the preferred fuels. On the other hand, Rennie and Tipton noted that sustained dynamic exercise stimulates amino acid oxidation, mainly by activating an enzyme (BCAA dehydrogenase, or BCAAD) that oxidizes BCAA and increases ammonia production in proportion to exercise intensity. If the exercise is intense enough, there is a net loss of muscle protein as a result of decreased protein synthesis, increased catabolism, or both. Some of the amino acids are oxidized as fuel, and the rest provide substrates for gluconeogenesis. In this regard, prolonged exercise may be comparable to a state of starvation. As the endurance athlete depletes the endogenous carbohydrate stores, the body catabolizes some of its protein for energy or eventual conversion to glucose. Protein catabolism increases significantly, even when muscle glycogen is depleted by only about 33–55 percent, which can be common in sport or during exercise training.

Mechanisms and By-Products In general, a brief session of exercise lowers the rate of protein synthesis and speeds protein breakdown. The exact mechanisms of protein metabolism during exercise have not been determined, though several mechanisms have been proposed. Parkhouse has reported that exercise, particularly exercise to exhaustion, activates specific proteolytic enzymes in the muscle that degrade the myofibrillar protein. Fitts and Metzger found elevated levels of proteolytic enzymes in fatigued muscle. In a review, Wagenmakers indicated that six amino acids (BCAA; asparagine, aspartate, glutamate) may be metabolized in the muscle, providing the nitrogen needed for the synthesis of ammonia, alanine, and glutamine. During exercise, particularly prolonged aerobic exercise, Graham and MacLean noted significant muscle efflux of ammonia, glutamine, and alanine. These three products carry excess nitrogen from the muscle to other parts of the body, most notably the liver, for recycling or conversion to urea.

Ammonia, a nitrogen by-product of protein catabolism, is an indicator of increased muscle amino acid breakdown. Even though no underlying mechanism has been identified, increasing levels of ammonia in the body have been associated with fatigue, somewhat comparable to the accumulation of lactic acid. One theory is that increased ammonia levels in the muscle may impair oxidative processes, thus decreasing energy production, while another theory suggests that increased plasma ammonia may impair brain functions and induce central fatigue. Because ammonia is formed in the muscle from the amino group, removal of the amino group by alanine or glutamine may help decrease the production of ammonia and delay the onset of fatigue. Glutamine release from the muscle also increases during exercise, and it is an important fuel for various cells in the body, particularly those in the immune system.

Leucine and the Glucose-Alanine Cycle Although a number of amino acids may be used as energy substrate during exercise, the major research effort has focused on the fate of leucine. Wagenmakers noted that the increase in BCAA oxidation during prolonged exercise seems to be specific for leucine only. In essence, the amino group of leucine catabolism eventually combines with pyruvate in the muscle cell to form alanine and leaves the residual alpha-ketoacid. The alpha-ketoacid may enter the Krebs cycle and be used for energy production. The alanine is released into the bloodstream and transported to the liver, where it is converted into glucose. The glucose may then be released into the blood to be used by the central nervous system and may eventually find its way to the contracting muscle to be used as

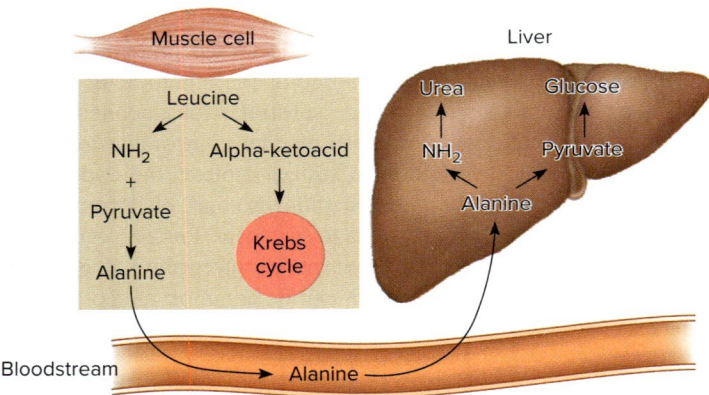

FIGURE 6.6 Glucose-alanine cycle. Alanine may be produced in the muscle tissue from the breakdown of other amino acids, most notably leucine. The alanine is then released into the blood and travels to the liver for eventual conversion to glucose through the process of gluconeogenesis.

an energy source. Alanine appears to be the most important means of transporting the amino group to the liver for excretion as urea. This overall process involving gluconeogenesis, known as the glucose-alanine cycle, is depicted graphically in **figure 6.6**. Some investigators have noted that during the later part of endurance exercise, the blood levels of alanine increase, presumably because more is released from the muscle. However, the estimated glucose production approximates only 4 grams per hour, which might make a limited contribution in mild-intensity exercise but is possibly insignificant during high-intensity exercise when carbohydrate use may approximate 3 grams or more per minute. Additionally, several investigators have reported an increased release of BCAAs from the liver during endurance exercise, with subsequent uptake by the muscle cells.

Protein Use and Importance of Carbohydrate Protein (amino acids) can be utilized during exercise to provide energy directly in the muscle and via glucose produced in the liver, particularly when the body stores of glycogen and glucose are low. In earlier research, Lemon reported that in the later stages of prolonged endurance exercise, protein could contribute up to 15 percent of the total energy cost. However, less protein would be used for energy during the early stages of prolonged exercise when carbohydrate is adequate. Thus, Gibala suggested that the contribution of amino acid oxidation to total energy expenditure is negligible during short-term, intense exercise and accounts for 3-5 percent of the total energy production during prolonged exercise. In his review, Tarnopolsky cited similar figures, indicating that protein oxidation could account for 1-6 percent of the energy cost of aerobic exercise. Tarnopolsky also noted that women oxidize less protein compared with men and show lower leucine oxidation during aerobic exercise.

It is important to note how dietary carbohydrate influences protein as an energy source during exercise. A low-carbohydrate diet leading to decreased muscle glycogen levels will lead to increased dependence on protein as an energy source. However, adequate carbohydrate intake before and during prolonged exercise will help reduce the use of body protein for this purpose, because the presence of adequate muscle glycogen appears to inhibit enzymes that catabolize muscle protein. Scientists from the University of Maastricht in the Netherlands noted that high-carbohydrate diets may have a protein-sparing effect for endurance athletes.

Although the available evidence suggests that metabolism of protein and its use as an energy source are increased during exercise, the magnitude of its contribution may depend on a variety of factors, such as the intensity and duration of exercise and the availability of other fuels, such as carbohydrate, either as stored muscle glycogen or consumed during exercise.

Does exercise increase protein losses in other ways?

Other than loss of protein from oxidation, exercise has been shown to increase protein losses from the body in several other ways.

Urinary Losses Exercise may cause an elevated level of protein in the urine, a condition known as **proteinuria**. This condition has been observed following competition in a wide variety of sports, including running, football, basketball, and handball. Research suggests that the greater the intensity of the exercise, such as a high-intensity 400-meter sprint versus a lower-intensity 3,000-meter run, the greater the loss of protein in the urine. Prolonged aerobic exercise, such as the triathlon, has also increased urinary protein loss. Yaguchi and others suggested that heavy, prolonged exercise, such as a triathlon, may induce some transient kidney damage in some individuals. Such damage may explain the findings of Poortmans, who reported a decreased reabsorption of protein in the kidney tubules following intense exercise. However, the total amount of protein lost in this manner appears to be rather negligible, amounting to less than 3 grams per day.

Sweat Losses Protein also may be lost in the sweat. Several investigators have reported the presence of both amino acids and proteins in exercise-induced sweat, with both sweat rate and sweat nitrogen losses increasing with greater exercise intensities. Again, the losses are relatively minor, on the order of 1 gram per liter of sweat in adult males. This avenue could account for 2-4 grams of protein in an endurance athlete training in a warm environment.

Gastrointestinal Losses As shall be noted in chapter 8, prolonged, intense exercise may also increase gastrointestinal losses of iron, which may be bound to blood proteins. Again, however, these protein losses would be relatively minor.

What happens to protein metabolism during recovery after exercise?

Although net protein breakdown occurs during exercise, in general protein synthesis is believed to predominate during the recovery period. In a review, Tipton and Wolfe noted that whole body protein breakdown is generally reduced following aerobic endurance exercise, while whole body protein synthesis is either increased or unchanged. However, they note that whole body protein synthesis may not represent changes in specific muscle groups. Muscle groups that have been exercised may experience protein synthesis, as they may be especially insulin sensitive after exercise. In another review, Walberg-Rankin notes that although leucine may be oxidized during moderate aerobic exercise, leucine balance returns to normal in 24 hours, indicating a reduced leucine catabolism over this time frame.

Tipton and Wolfe report that resistance exercise induces protein breakdown in the exercised muscles, which may persist in the immediate recovery period. However, over the next 24–48 hours, protein anabolism appears to predominate and the net effect is increased protein synthesis. Wolfe notes that anabolic states appear to be due more to stimulation of synthesis rather than a decrease in breakdown. This is especially so if adequate amino acids are available.

Eccentric exercise, such as lowering weights during resistance exercise or running downhill rapidly, puts tremendous stress on muscle tissue, and often induces muscle soreness in the following days. Tipton and Wolfe note that following eccentric exercise, whole body protein breakdown is increased. Microtears in muscle fibers may impair protein synthesis and delay recovery from such exercise. The muscle fiber microtears are believed to be the causative factor underlying the muscle soreness that, because its onset is usually delayed for 1–2 days, is referred to as **delayed onset of muscle soreness (DOMS).**

What effect does exercise training have upon protein metabolism?

Protein metabolism may also become more efficient as a result of training. Rennie noted that the response of muscle protein turnover to habitual exercise is comparable to other metabolic changes in the muscle associated with exercise training. Although an initial bout of exercise may markedly elevate protein breakdown and synthesis in an untrained individual, the effect would be much less in one who has trained habitually. Tarnopolsky indicated that training induces a decreased *activity* in BCAAD, the enzyme that oxidizes BCAAs, when exercising at a standardized workload of the same absolute intensity.

As mentioned, protein synthesis appears to predominate in the muscles during recovery. With training, or repeated bouts of exercise on a regular basis, changes in the muscle structure and function are additive. Numerous studies have found that after resistance or endurance exercise, protein balance becomes positive. Trained individuals, during rest, have been shown to experience a preferential oxidation of fat and a sparing of protein, as measured by leucine metabolism and the respiratory quotient. The specific exercise task apparently stimulates the DNA in the muscle cell nucleus to increase the synthesis of protein, and the type of protein that is synthesized is specific to the type of exercise. Aerobic exercise stimulates syntheses of mitochondria and oxidative enzymes, which are composed of protein and are necessary for energy production in the oxygen system. Resistance training promotes synthesis of the contractile muscle proteins. These adaptations are the key factors underlying improved performance (**figure 6.7**).

Other than its beneficial effects in increasing structural and functional protein important to resistance or aerobic endurance exercise, training may influence protein metabolism in other ways to help prevent premature fatigue. You may recall from previous chapters that there is substantial research to support the conclusion that aerobic endurance training improves the ability of the muscle cell to use both carbohydrate and fat as energy sources during exercise. Although extensive evidence is not available, in a review, Graham and others noted that following endurance training, enzymes in the muscles appear to develop the potential for increased capacity for oxidation of leucine and the other BCAAs, which are an abundant source of energy in the muscles. Thus, endurance training may increase the capacity of the muscle to derive energy from protein in a fashion similar to the increased utilization of fat, another possible means to spare the use of carbohydrates such as blood glucose, to help protect the main energy source for the brain, and muscle glycogen. Although these changes do not appear to spare the use of muscle protein, they would appear to spare the use of carbohydrate, and thus may help prevent fatigue when carbohydrate levels are decreased during exercise.

Additionally, when exercising at a standardized workload before and after training, training may also decrease the production or accumulation of ammonia. Extrapolating from animal research, some investigators theorize that instead of forming ammonia, the nitrogen is incorporated into other amino acids, such as alanine, for transportation from the muscle to the liver. Theoretically, reduced plasma ammonia levels may be associated with less fatigue.

Training properly may also help prevent muscle injury associated with eccentric exercise. Research shows that physically trained individuals, when compared to untrained individuals, do not experience as much muscle tissue damage during prolonged, eccentric exercise tasks. However, specificity of training is also important, and individuals preparing for a race with a significant downhill component, such as the Boston Marathon, should gradually incorporate increasing intensities of eccentric exercise into their training.

In general, these changes associated with appropriate training appear to represent another means whereby the body adapts to endurance training in an attempt to prevent fatigue. However, some research indicates that excessive training leading to the overtraining syndrome is associated with a persistent decrease in plasma amino acids, particularly glutamine.

The effect of training in producing a positive nitrogen balance or a positive protein balance, and possibly the effect of preventing the overtraining syndrome, depends on an adequate dietary supply of protein and kcal.

Does exercise increase the need for dietary protein?

In their joint position stand, the American College of Sports Medicine, the Academy of Nutrition and Dietetics, and Dietitians of Canada concluded that protein requirements are higher in very

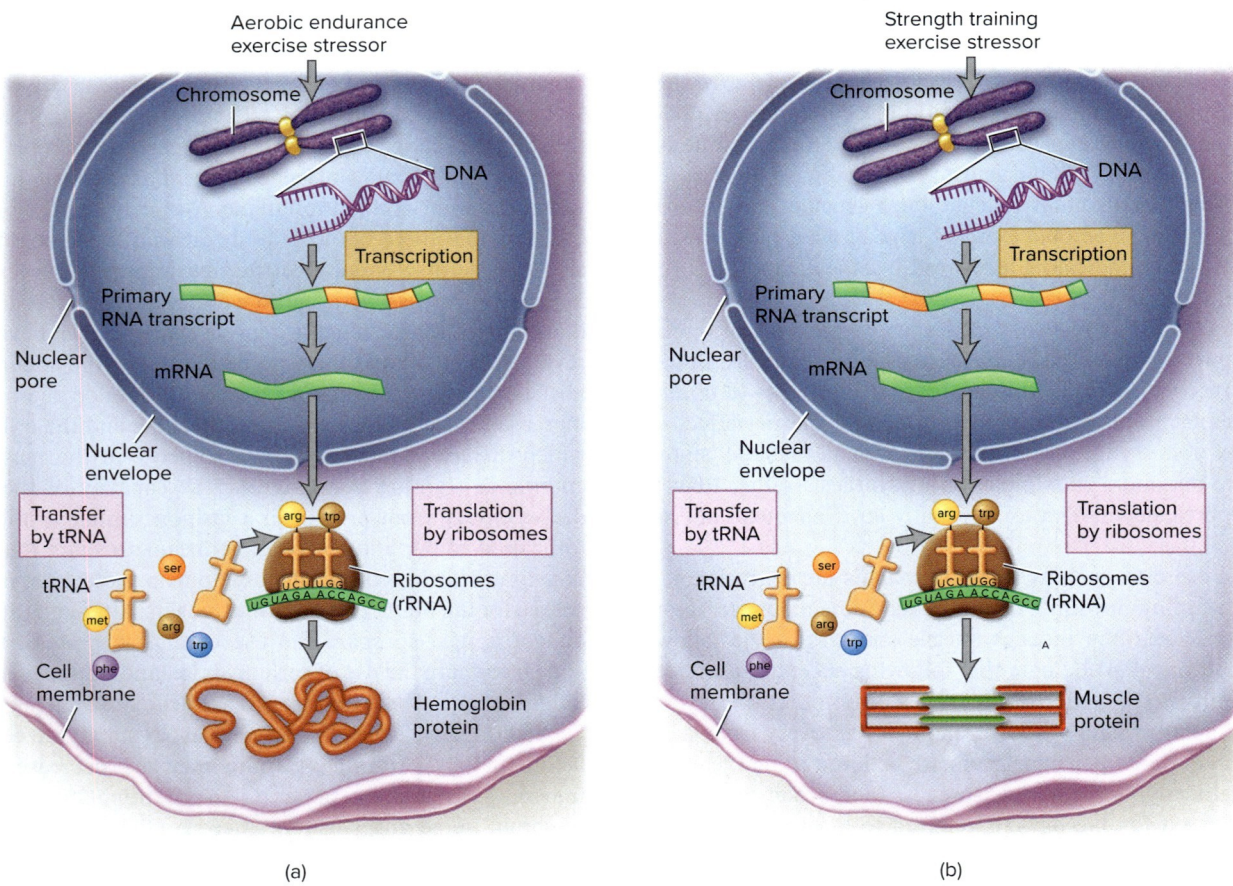

FIGURE 6.7 Adaptations in body tissues following exercise training are specific to the type of exercise. The exercise stressor influences the genes to initiate the formation of protein synthesis through the transcription, transfer, and translation processes. The transcription process provides the template (messenger RNA:mRNA) for the specific protein; the transfer (transfer RNA) process moves specific amino acids to the ribosome; the translation process by the ribosomes (ribosomal RNA) forms the protein. (a) Aerobic exercise training will lead to an increase in serum hemoglobin levels, which transport oxygen to the muscle cells. (b) Resistance training will lead to an increase in skeletal muscle protein.

active individuals and suggested that endurance and resistance athletes need 1.2–2.0 grams per kilogram body weight. In a similar vein, the International Society of Sports Nutrition, in a position stand developed by Campbell and others, concluded that protein intake of 1.4–2.0 grams per kilogram body weight daily may improve body adaptations to exercise training. Stuart Phillips, a protein metabolism expert, points out that protein needs for athletes should be based on maintenance or gain of muscle, bone, and connective tissue, and not simply nitrogen balance. Skeletal muscle, bone, tendon, and ligament protein turnover, which can be 50 times slower than gut or plasma proteins, are likely not well represented in a short-term nitrogen balance study. This section focuses on the optimum protein intake for those individuals seeking to increase or maintain the functional capacity of skeletal muscle/connective tissue, such as athletes.

Strength-Type Activities Our ability to maintain or increase skeletal muscle mass is based on our ability to properly balance two processes, muscle protein synthesis (MPS) and muscle protein breakdown (MPB). Following the ingestion of a high-protein meal (from either food or protein supplement), there is a period of time when blood levels of amino acids increase dramatically. This aminoacidemia stimulates MPS, while the resulting increase in blood insulin (hyperinsulinemia) inhibits MPB. In this scenario, which fluctuates throughout a normal day, a positive protein balance occurs when MPS is greater than MPB, which can stimulate muscle hypertrophy and increase strength. As described by Devries and Phillips, two factors that greatly influence MPS and MPB are resistance training and energy intake. Following a resistance exercise workout, both MPS and MPB are increased, with positive protein balance achieved when protein is consumed following the exercise. When energy intake is restricted, which occurs in athletes as well as the general population, MPS decreases, which can cause a negative protein balance and subsequently a decrease in muscle mass.

A common question by those attempting to increase muscle mass and strength with resistance training is whether or not additional dietary protein above the RDA will cause a further increase in muscle hypertrophy and strength. The answer to this is unequivocally yes. A meta-analysis of 49 studies and over 1,800 participants by Morton and others showed that protein supplementation (up to 1.62 grams of total protein intake/kg of body weight/day) during resistance training further increased gains in strength, fat-free mass, and muscle fiber cross-sectional area. Similarly, the meta-analysis by Cermak and colleagues showed that protein intakes greater than 1.2 grams of protein/kg of body weight/day, during resistance training, further increased strength and muscle fiber cross-sectional area relative to placebo ingestion.

Endurance-Type Activities It is important to reinforce the viewpoint that carbohydrate is the main energy source for endurance-type athletes. Besides its efficiency as a metabolic fuel during exercise, carbohydrate also provides a potent protein-sparing effect. Consuming sufficient carbohydrate will decrease reliance on protein during aerobic endurance exercise, reduce the formation of ammonia, and better maintain normal protein status in the body.

Nevertheless, additional dietary protein has been recommended for endurance athletes because they may utilize protein as an energy source during exercise and may need protein to synthesize oxidative enzymes and mitochondria. Athletes involved in prolonged, intermittent, high-intensity sports, such as soccer, field or ice hockey, lacrosse, and basketball, may also fall into this category. Additionally, extra protein has been recommended to prevent development of anemia during the early stages of training. In brief, **sports anemia** is thought to occur because the body uses dietary protein for these aforementioned purposes at the expense of hemoglobin formation, leading to anemia. Sports anemia also has been associated with iron nutriture and will be discussed in more detail in chapter 8.

Moore and colleagues recently reviewed the dietary protein needs of endurance athletes and noted that muscle remodeling is needed to break down old and damaged proteins and synthesize new proteins. Acute endurance exercise and chronic endurance exercise training cause an increase in MPS, and this adaptive response to endurance training can be enhanced by providing additional dietary protein.

As with strength-trained individuals, various investigators have also estimated the protein requirement of endurance-trained individuals. In a review based on protein balance studies, Lemon recommends that individuals involved in vigorous aerobic endurance exercise consume 1.1–1.4 grams/kg/day; slightly higher amounts, about 1.4–1.7 grams per kilogram body weight, have been recommended for individuals involved in intermittent, high-intensity sports because these include a balance of endurance and power activities.

What are some general recommendations relative to dietary protein intake for athletes?

Tipton and Witard, in a review of protein requirements for athletes, concluded that athletes are frequently put into general categories of strength and endurance athletes within research studies. A difficulty with these categories is they may not be specific enough

Erik Isakson/RubberBall Selects/Alamy Stock PHoto

to examine the protein requirements of many athletes. In fact, there are gaps in our knowledge that do not allow for meaningful recommendations for athletes in sports such as football, rugby, or decathlon. Additionally, Tipton and Witard state that few data exist on protein intakes and performance measures with respect to gender. Nevertheless, recent reviews by some of the most prominent investigators in protein metabolism during exercise provide the basis for some reasonable guidelines. All of the recommendations are consistent with the AMDR for protein.

The following mathematical presentations are possible scenarios for various athletes who wish to maintain or increase protein balance. The protein needs are based upon the RDA, a rough estimate of the amount of protein used during strenuous exercise, and additional protein needs for the individual who wants to gain weight.

Let us look first at the young resistance-trained athlete who wants to gain body weight, preferably in the form of muscle tissue, through a weight-training program. The protein RDA for an adult male is 0.8 gram per kilogram. At moderate activity levels, the average 70-kilogram male would be in protein balance with about 56 grams daily. Based on a recommendation of 1.7 grams per kilogram, he would need about 119 grams daily if involved in a strenuous training program.

One pound of muscle tissue is equal to 454 grams, and its composition is approximately 70 percent water, 7 percent lipids, and 22 percent muscle tissue. Hence, 1 pound of muscle contains about 100 grams of protein (454 × 0.22). If the desired weight gain is 1 pound of lean body mass per week, a reasonable goal, then this young male would need to assimilate an additional 14 grams of protein per day (100 g/7 days) to supply the amount in 1 pound of muscle tissue. A gain of 2 pounds per week, although probably more difficult to accomplish, would require the assimilation of 28 additional grams of protein per day. Let us be liberal and estimate an additional 22 grams of protein per day to cover losses due to exercise. In summary, assuming that a portion of these protein needs are not covered by the safety margin incorporated in the RDA, this young athlete would need approximately 106 grams of protein per day (56 + 28 + 22) to gain 2 pounds of lean body tissue per week, or about 1.5 grams of protein per kilogram body weight. This value falls within the recommended range for resistance-trained athletes.

Endurance athletes are not necessarily interested in gaining weight, but they may need to replenish the protein that may serve as an energy source during training. Running 10 miles per day would expend approximately 1,000 kcal or more. Again, let us be liberal and assume that if 6 percent of this energy cost, or 60 kcal, was derived from protein, then approximately 15 grams of protein (60 kcal/4 kcal per gram of protein) would have to be replaced. With a liberal estimated additional loss of 10 grams of protein in the urine and sweat, the total daily protein requirement for a 70-kilogram young male, again assuming that these additional needs are not accounted for by the safety margin in the RDA, would be 81 grams (56 + 15 + 10), or 1.2 g/kg. Again, this value falls within the recommended range. Here are some general protein intake recommendations for athletes:

1. *Obtain about 15 percent or more of daily energy intake from protein.*

As do most investigators, Peter Lemon indicates that it is possible to get these recommended amounts from natural food sources in the daily diet. For example, the average caloric intake for a moderately active young male averages 2,500–3,000 kcal. This caloric intake may be increased through physical training as more kcal are expended. Thus, caloric intake would be increased to approximately 3,500–4,000 kcal. It is important to note that adequate energy intake, primarily in the form of carbohydrates, will improve protein balance. In essence, an increased energy intake appears to decrease protein requirements somewhat.

If the protein portion of the dietary kcal averaged 12 percent, a general recommended level of protein intake, then the intake of protein would approximate 1.5–1.7 g/kg body weight, which parallels the amounts estimated in the examples cited previously. Currently, the protein content of the average American diet is 12–16 percent, and Walberg-Rankin noted that many athletes consume this much or more in their daily diets. Consuming a diet with a protein content of 15 percent could provide a value of 2.0 grams or more per kilogram body weight, which is an amount commonly consumed by strength-type athletes. These values approach or exceed the higher amounts recommended by some investigators for individuals in training. The calculations are presented in **table 6.6**.

Some athletes may need to increase the percentage of protein in their daily caloric intake. Athletes in weight-control sports, such as wrestlers and gymnasts, may be in greater need of protein because of low caloric intake. Endurance athletes, particularly ultraendurance athletes, are susceptible to overtraining and chronic fatigue. Such athletes have been found to have depressed plasma levels of amino acids, which Kingsbury and others indicated appeared to be associated with inadequate protein intake.

Maintaining muscle mass during weight loss is critical for athletes and nonathletes alike, as muscle mass is a major determinant of basal metabolic rate, is the primary site of glucose disposal, and is necessary to perform sporting activities. Recently, Longland and others showed that the combination of a very high protein diet (2.4 g/kg/day versus 1.2 g/kg/day), resistance training, and a very low kcal weight-loss diet (40 percent less than daily requirements), causes greater increases in lean body mass and decreases in fat mass. Antonio showed that a very high protein diet (3.4 g/kg/day versus 2.3 g/kg/day) during resistance training resulted in a greater decrease in fat mass but with similar increases in fat-free mass. In his review, Phillips noted that a high-protein diet (25–35 percent of energy intake) is more effective at attenuating the loss of lean body mass that occurs on low-kcal diets than a normal protein diet (12–15 percent of energy intake).

Helms and others reached a similar conclusion, noting that 2.3 to 3.1 grams of protein/kg of body mass/day was needed to offset losses in lean body mass. Churchward-Venne and colleagues report that overfeeding energy in the form of protein (about 1.8–3.0 g/kg/day, 15–25 percent of kcal) results in gains in lean body mass even without resistance training. As a note of caution, increasing the protein content of the diet, without increasing the kcal content, could displace another important macronutrient (e.g., carbohydrate). One way that higher-protein diets can encourage weight loss is by promoting satiety (i.e., feeling of fullness). In a review, Leidy and others reported that 1.2–1.6 grams of protein/kg body mass/day improves appetite and body weight management. Thus, although the exact dose is not known, dietary protein intake above recommended levels, and at least at what has been reported to maximize MPS in hard-training athletes, may be important in athletes in weight-control sports or those actively trying to lose weight.

Increasing the protein percentage of daily energy intake may help such athletes obtain adequate protein. For example, a young wrestler attempting to lose weight to compete in a lower weight category might be on a diet of 1,600 kcal per day. With an RDA of 0.85 gram of protein per kilogram body weight, a 60-kilogram (132-pound) wrestler would need 51 grams of protein to meet the RDA and up to 102 grams to meet the highest recommendation for strength-type athletes. To consume 51 grams of protein, the wrestler would need to obtain about 13 percent of daily energy intake from protein, but to obtain 102 grams of protein he would need to obtain about 25 percent of daily energy intake from protein. These amounts are well within the AMDR for protein.

Wise selection of high-quality protein foods will provide adequate amounts through a balanced diet to meet bodily needs during the early and continued stages of training. It is not difficult to increase the protein content of the diet. For example, 8 ounces of roasted, skinless chicken breast and two glasses of cow's skim milk, a total of less than 600 kcal, will provide more than 70 grams of high-quality protein, the RDA for our typical 70-kilogram adolescent and more than half of the 125 grams that may be recommended for such an athlete attempting to gain muscle mass. Perusal of **table 6.2** will help you select high-protein foods.

2. *Consume protein, preferably with carbohydrate, before and after workouts.*

TABLE 6.6	Calculation of grams protein/kg body weight	
Body weight: 70 kg One gram protein = 4 kcal		
Daily caloric intake:	3,500–4,000	3,500–4,000
Percent protein:	15	12
kcal in protein:	525–600	420–480
Grams of protein:	131–150	105–120
Grams protein/kg:	1.9–2.1	1.5–1.7

Eating strategies may also be an important consideration, particularly timing of protein intake. Most research has focused on amino acid or protein ingestion following exercise, but some studies have also studied the effect of pre-exercise protein intake. The feedings usually occurred immediately before or within an hour of exercise. Many studies have used **protein hydrolysate**, a high-protein dietary supplement containing a solution of amino acids and peptides prepared from protein by hydrolysis. Wolfe noted that the protein supplement should contain all the essential amino acids, and many studies also added carbohydrate.

Protein Intake before Exercise In separate reviews, Tipton and Wolfe both noted research studies suggesting that ingestion of free amino acids plus carbohydrates before exercise results in a superior anabolic response to exercise than if ingested after exercise. Lemon indicates that the practice of consuming nutrients prior to exercise may be beneficial by providing fuel and/or minimizing catabolic processes. However, in his review Tipton noted that the difference in anabolic response between pre-exercise and postexercise ingestion of protein is not apparent.

Protein Intake after Exercise Perhaps the two most frequently asked questions by those interested in adding protein to their diet are "what is the best time to eat protein?" and "is one protein better than another?" These questions are often in reference to which time and what type of protein will cause the greatest increase in MPS, muscle hypertrophy, and strength. As reviewed by van Loon, many proteins, including whey, casein, soy, casein hydrolysate, egg, whole milk, and fat-free milk have been shown to improve postexercise protein balance or MPS. However, the goal of an athlete is maximal or optimal MPS. There are a wide variety of protein supplements available for athletes and other consumers to choose from, including whey, casein, egg, soy, and collagen. During the past few years, experts in protein metabolism from the laboratories of Stuart Phillips, Kevin Tipton, and Luc van Loon have published research studies that have helped to answer these two important questions about protein.

The two most commonly consumed protein supplements are the two proteins in milk: casein and whey. Data from the Phillips group have demonstrated that whey protein stimulates MPS more than casein or soy protein, both at rest and postexercise. It makes sense that whey protein causes the largest increase in MPS because (1) whey is rapidly digested, exits the stomach quickly, and results in a large increase in blood amino acids; (2) whey is a complete protein, containing all the essential amino acids; and (3) whey is high in the amino acid leucine with about 3 grams of leucine per 25 grams of whey protein. Conversely, soy protein has lower levels of essential amino acids, branched-chain amino acids, and leucine. Also, soy protein is less bioavailable than whey or casein and when consumed, is more likely to be oxidized or involved with organ protein catabolism and urea synthesis. Casein, although it is digested and absorbed more slowly than soy, has higher levels of essential amino acids, branched-chain amino acids, and leucine than soy. Collagen is an incomplete protein; is low in essential amino acids, branched-chain amino acids, and leucine; and is a poor choice for a protein supplement.

It appears that the branched-chain amino acid leucine is an important factor in the MPS response to a particular type of protein. The powerful effect of leucine on MPS is referred to as the leucine trigger or leucine threshold. Essentially, muscle leucine levels must reach a critical level for protein ingestion to cause the greatest increase in MPS. The exact amount of leucine needed to maximally stimulate MPS is not known, because it varies based on age and physical activity level. However, a range of leucine doses can be suggested based on recent studies. Moore and Witard, in separate studies, achieved the highest rates of MPS following ingestion of 20 grams of whey protein containing 1.7 to 2.4 grams of leucine. In a follow-up study, by Churchward-Venne, MPS was equal following ingestion of 25 grams of whey protein or 6.25 grams of whey protein plus 5 grams of leucine.

Several groups are conducting research in an attempt to determine the optimal dosing pattern of postexercise protein consumption. Atherton has described the "muscle full effect," where MPS returns to baseline values after 2 hours, even if amino acids are still provided. Should protein be ingested as a single bolus, for example, 20 grams, or as several small servings, for example, 5 grams every 15 minutes? Also, how long should postexercise protein ingestion last to get the maximal MPS response? Areta showed that the timing and distribution of whey protein can modify MPS over 12 hours of recovery. Four servings of 20 grams of whey protein, ingested every 3 hours, caused the largest increase in MPS compared to 8 servings of 10 grams ingested every 1.5 hours, and 2 servings of 40 grams ingested every 6 hours.

Plant-Based vs. Animal-Based Proteins An interest in replacing animal-based proteins in the diet with plant-based proteins has led to an increase in research in this area, as well as an increase in the number of plant-based protein supplements. These supplements are made from various sources, including soy, wheat, pea, and potato protein. Pinckaers and others note that the ingestion of some plant-based proteins results in lower MPS when compared to animal-based proteins. For instance, Gorissen and colleagues found no increase in MPS, following the ingestion of 35 grams of wheat protein, but a large increase in MPS following the ingestion of 60 grams. The larger amount of wheat protein increased the leucine content to the same as 35 grams of whey protein, which is considered a very high-quality protein in terms of MPS. This difference is attributed to differences in digestion, amino acid absorption, and amino acid content. As described earlier in the chapter, most plant-based proteins have a low essential amino acid content and/or are deficient in one or more essential amino acids. Leucine, in particular, is considered a driver of MPS, and some plant proteins are lower in leucine than commonly consumed animal proteins. It is possible that some of the differences between plant and animal-based proteins could be accounted for with food combining to improve amino acid content, consuming a larger amount of the plant-based protein, using plant-based protein supplements that combine protein sources, or by fortifying the plant-based source with the specific amino acid known to be deficient. These theories have to be tested scientifically, and we cannot simply assume they will work because it makes sense on paper. As an example, Yang showed that fortification of 20 grams of soy protein with 2.5 grams of leucine did not increase muscle protein synthesis over 20 grams of soy protein alone.

In their meta-analysis and systematic review, Lim and colleagues reported that animal protein ingestion resulted in greater gains in lean mass than plant-based protein, although there were no differences in strength gains. Sixteen articles were analyzed, which offers a large sample population to study; however, the individuals studied varied widely in age (21–64 years), adiposity (obese vs. nonobese), and training status (trained vs. untrained), making interpretation of the results complex. As Freitas and Katsanos have reviewed, it is known that individuals with obesity have decreased MPS. Similarly, older adults often develop an "anabolic resistance" where they also have a blunted response in MPS to protein ingestion or resistance exercise. Much more research is needed in the area of plant-based proteins. Particular attention must be given to evaluating the differences between whole foods and plant- and animal-based isolates and concentrates. Additionally, there are likely differences between populations, including those with obesity, older adults, athletes, and athletes on low-energy diets.

Nutrient Timing/Postexercise Protein Supplements containing both protein and carbohydrate have been marketed to enhance performance when consumed during exercise and also to facilitate recovery and improve subsequent exercise performance when consumed after exercise. Ivy and others compared the effects of adding protein to a carbohydrate solution on aerobic endurance performance. Trained cyclists, consuming 200 milliliters of the solution every 20 minutes, exercised for 3 hours at intensities varying from 45 to 75 percent of VO_2 max and then exercised to exhaustion at 85 percent of VO_2 max. The carbohydrate solution (7.75 percent) significantly improved endurance performance compared to the placebo, but the carbohydrate/protein combination (7.75/1.94 percent) enhanced performance even more than the carbohydrate solution. Saunders, in a review, reported that other researchers have found similar results. However, Saunders did note an important limitation to these positive studies; the protein supplied in the beverage was in addition to the carbohydrate, thus providing more energy (kcal).

Subsequently, other researchers showed no ergogenic effect of protein/carbohydrate supplementation when compared to carbohydrate supplements containing equal energy (kcal). For example, van Essen and others reported that when trained athletes ingested a sports drink during exercise at a rate considered optimal for carbohydrate delivery, protein provided no additional performance benefit during an 80-kilometer cycling trial, an event that simulated "real life" competition. Romano-Ely and others also compared the effects of a protein/carbohydrate drink, which also included antioxidants, to an isocaloric carbohydrate drink on cycling time to exhaustion at 70 percent VO_2 max. The drinks were consumed every 15 minutes during exercise. There was no significant difference between the treatments on performance time. In his review, Gibala indicated that there is no established mechanism by which protein intake during exercise should improve acute endurance performance, and these later studies support this viewpoint.

As noted previously, consuming protein and carbohydrate following exercise provides a milieu conducive to enhanced anabolism and muscle recovery. Some have contended that ingesting carbohydrate plus protein following prolonged exercise may restore exercise capacity more effectively than ingestion of carbohydrate alone. For example, compared to a standard commercial carbohydrate/electrolyte sports drink, Karp and others found that both low-fat chocolate milk and a commercial protein/carbohydrate sports drink, each of which contain about 34 grams of carbohydrate and 9 grams of protein per 8 fluid ounces, enhanced performance in a cycling test to exhaustion 4 hours after a hard interval workout. However, although the subjects consumed the same amount of fluid in each trial, the two protein/carbohydrate drinks contained more than twice as much carbohydrate and more than three times as much caloric energy as the carbohydrate/electrolyte sports drink. When the energy content is similar, consuming protein/carbohydrate meals or supplements during recovery, as compared to carbohydrate alone, has no effect on subsequent exercise performance.

In two separate studies, Betts and others had physically active males complete a 90-minute run at 70 percent of VO_2 max followed by a 4-hour recovery, during which they consumed either a carbohydrate or carbohydrate/protein solution with similar energy (kcal) content. The recovery was followed by a run to exhaustion at either 70 or 85 percent of VO_2 max, but there was no difference between the two recovery diets. Studies by Rowlands and Millard-Stafford, along with their colleagues, have shown that consuming high-protein/carbohydrate meals or fluid supplements following intense exercise training, as compared to comparable carbohydrate feedings, does not enhance performance in intermittent, high-intensity (10 maximal 2.5-minute cycling sprints) or aerobic endurance (5-kilometer track run) the following day. In a systematic review, Lieberman and colleagues assessed the effects of protein plus carbohydrate ingestion on acute and repeated endurance exercise performance in 26 studies. This review included studies of protein and carbohydrate ingestion during an acute exercise bout, and protein and carbohydrate ingestion during and after exercise and the effects on a subsequent bout of exercise. The authors concluded that the addition of protein to a carbohydrate supplement had no effect on endurance exercise performance.

In his review, Saunders also noted that consumption of protein/carbohydrate solutions has been associated with reduced markers of muscle damage and less muscle soreness. Several studies support this viewpoint. In their study cited earlier, which involved a 21-kilometer outdoor run at a set pace followed by a treadmill run to exhaustion at 90 percent of VO_2 max, Millard-Stafford and others reported that the carbohydrate/protein mixture resulted in less muscle soreness. Romano-Ely and others also reported that subjects experienced less muscle soreness on the carbohydrate/protein drink, and blood enzyme tests indicated less muscle tissue damage. Conversely, Green and others found that a carbohydrate-protein drink consumed following an exercise task (downhill running) designed to induce muscle injury and soreness had no effect on quadriceps strength recovery or muscle soreness.

The existence of a postexercise anabolic window and importance of postexercise protein ingestion in general has recently been called into question by Schoenfeld, an expert in muscle hypertrophy. In a meta-analysis, Schoenfeld showed a beneficial effect of postexercise protein consumption on muscle hypertrophy, but this effect was not evident when total protein intake was considered. He contends that the benefit of postexercise protein ingestion was not necessarily due to timing but simply to greater protein consumption. Also, Aragon and Schoenfeld have pointed out that much of the postexercise protein consumption research is conducted with volunteers in a fasted state. This is unlikely to be the case with real athletes. They suggest pre- or postexercise protein ingestion of about 0.4–0.5 grams of protein/kg of lean body mass as a general guideline to create the optimal anabolic effect.

Practically speaking, an inexpensive, nutrient-dense food such as fat-free milk or cottage cheese would be an excellent choice for a high-quality/high essential amino acid, high-leucine food. However, athletes often seek the optimal adaptive training response and may choose to consume a protein supplement at certain times during the day. Additionally, if an athlete is attempting to optimize muscular adaptations to training as well as decrease body fat, a postexercise leucine-enriched, low-dose whey protein supplement might be the logical choice. Finally, the constraints of sport can influence dietary protein intake behaviors. Some athletes compete multiple times per day, which would necessitate a fast-digesting protein, while others travel a great distance to a training facility or competition and may not have access to a refrigerator.

Are protein supplements necessary?

Few researchers would argue that protein supplements are necessary; rather, it is optimal to consume high-quality protein several times per day to support positive protein balance. This can be achieved with high-protein foods, protein supplements, or a combination of the two. However, in any discussion of protein supplements vs. high-protein foods, it is worth noting that humans shouldn't focus on eating individual nutrients, but rather whole foods that supply a variety of essential nutrients. For example, a protein supplement may contain high amounts of essential amino acids. A food such as milk may also contain high amounts of essential amino acids, as well as important vitamins such as vitamin D, important minerals such as calcium, important macronutrients that support human performance such as carbohydrate, and much more. Similarly, a food such as lean beef contains high concentrations of essential amino acids, but also high amounts of the minerals iron and zinc. Also, as demonstrated in **table 6.8,** protein supplements generally cost much more than high-protein foods. Although there is nothing inherently wrong with protein supplements, they should be treated as just that, supplements to a healthy diet.

Protein and Carbohydrate Intake after Exercise Carbohydrate alone may have some effect on protein synthesis following exercise, as it may help to decrease secretion of cortisol, a hormone that promotes protein catabolism. Moreover, Lemon has noted that the insulin response to dietary carbohydrate has been shown to enhance the already elevated protein synthetic rate in muscle following a strength-training session. However, van Loon, Drummond, and others indicate that if adequate protein is available, there is no need for carbohydrates to promote muscle protein synthesis, but also noted that because resistance exercise uses muscle glycogen, the carbohydrate could help replenish muscle glycogen.

Several studies and reviews have suggested that such protein intake soon *after* exercise, possibly combined with carbohydrate, was beneficial. Anton Wagenmakers, an international authority on protein metabolism during exercise, noted that protein provides the amino acids and carbohydrate increases insulin secretion. Wagenmakers noted that the effect of amino acids and insulin on protein synthesis is substantially larger after exercise, suggesting that exercise potentiates the anabolic effect of insulin and amino acids. Levenhagen and others reported that consuming protein immediately after exercise enhanced accretion of whole body and leg protein as compared to protein consumption several hours later. These findings were supported in more recent studies by Koopman and others, who used a protein/carbohydrate solution during exercise and added leucine to a protein/carbohydrate feeding after exercise. Esmarck also found that consumption of protein with carbohydrate soon after exercise also benefits muscle protein synthesis in the elderly as well.

Nutrient Timing and Performance Based on available research findings, the ratio of carbohydrates to protein requires additional investigation, as concluded by the International Society of Sports Nutrition in its position stand on nutrient timing. In this position stand, Kerksick and others conclude that the approach often used is to consume a supplement containing a ratio of 3–4 grams of carbohydrate per gram of protein within 30 minutes following exercise. Although commercial protein/carbohydrate sport drinks and supplements are available, Phillips and others recommend high-quality protein food sources, such as dairy protein, eggs, and lean meat, which provide an abundance of essential amino acids. In their review, Rennie and Tipton note that a solution of mixed amino acids will increase human muscle protein synthesis to about the same extent as complete meals. A turkey breast sandwich on whole-wheat bread consumed with a glass of chocolate skim milk provides a good balance of protein and carbohydrate. Karp and Elliot, along with their associates, indicated that milk, including chocolate milk, may be an effective recovery drink for athletes and provides an alternative to supplements.

It should be noted that although these dietary practices may provide the nutrients necessary for an anabolic response in the muscle, Gibala indicates it remains to be determined if the acute effects of such supplementation eventually lead to greater gains in muscle mass following habitual training. In a study involving a 10-week resistance training program, Cribb and Hayes found that consuming a supplement containing protein, creatine, and glucose (1 g/kg body weight) immediately before and after training, as compared to consuming the same supplement in the morning and evening, resulted in greater increases in lean body mass, muscle cross-section area, and muscular strength. Although this study would appear to support anabolic effects of protein and carbohydrate intake immediately before and after exercise training, the supplement also contained creatine, which as noted later in this chapter may by itself increase muscle mass and strength. Most investigators in this area indicate that more research is needed to determine the optimal combination of protein and timing for the various types of exercise training adaptations.

Protein Intake before Sleep van Loon and colleagues have pioneered research in the area of protein consumption prior to sleep, with the goal of optimizing protein synthesis during the overnight period. Historically, strength and power athletes have instinctively ingested protein during the middle of the night to aid in recovery and encourage muscle hypertrophy. This has not been studied until very recently. van Loon's team showed a 22 percent higher rate of muscle protein synthesis during the overnight period when 40 grams of casein protein was consumed 30 minutes before sleep. In a 12-week resistance training study, the same group showed increased strength and muscle cross sectional area in participants ingesting 27.5 grams of protein each night before sleep compared to those ingesting a placebo. In reviews, Trommelen and van Loon and Snijders and others note that 40 grams of protein ingested prior to sleep is digested, is absorbed, and increases plasma amino acid availability. This causes a robust increase in MPS during overnight sleep, and, when combined with exercise on the same evening, the increased MPS response is further augmented. Overall, presleep protein ingestion, combined with resistance training, further increases gains in muscle mass and strength when compared to no protein supplementation.

3. *Be prudent regarding protein intake.*

Two experts in protein metabolism, Kevin Tipton and Robert Wolfe, noted that given sufficient energy intake, lean body mass can be maintained within a wide range of protein intakes. They note that since a high protein intake is not likely harmful, and since there is a metabolic rationale for the efficacy of an increase in dietary protein if muscle hypertrophy is the goal, a higher protein intake within the context of an athlete's overall dietary requirements may be beneficial. However, they also note that there are few convincing data to indicate that the ingestion of a high amount of protein (2–3 g/kg body weight) is necessary. Based on current literature, they conclude that it may be too simplistic to rely on recommendations of a particular amount of protein per day, because the amount depends on energy intake, type of protein, and timing of intake.

Consuming additional protein within the AMDR is both safe and prudent. **Table 6.7** presents a summary of some prudent daily protein intakes for sedentary and physically active individuals and the *Training Table* in this section provides specific protein intake recommendations for athletes.

Training Table

Protein intake recommendations for athletes:

- Daily protein intake for maintaining or building muscle should be about 1.4–2.0 g/kg body weight/day, or about twice the RDA (0.8 g/kg body weight/day).
- To promote optimal protein synthesis throughout the day, consume about 20–25 grams of protein (0.25–0.4 g/kg body weight/meal) every 3–4 hours as part of each meal.
- Consume high-quality proteins that have high levels of essential amino acids, especially leucine, that are most effective at increasing muscle protein synthesis. When possible, choose whole food proteins instead of protein supplements.
- Consume 20–25 grams of high-quality protein immediately postexercise to maximize protein synthesis.
- Consume 40 grams of high-quality protein prior to sleep.

It has recently been proposed by Dr. Stuart Phillips, a protein metabolism expert, that protein intake values could be higher. Visit www.mysportscience.com/post/2017/10/18/how-much-protein-do-i-need-to-eat-to-build-muscle.

TABLE 6.7 Prudent protein intakes in grams per kilogram body weight for sedentary and physically active individuals

	Grams of protein/kg body weight
Sedentary	0.8
Strength-trained, maintenance	1.2–2.0
Strength-trained, gain muscle mass	1.6–2.0
Endurance-trained	1.2–2.0
Intermittent, high-intensity training	1.4–2.0
Weight-restricted	1.4–2.0

The values presented represent a synthesis of those recommended by leading researchers involved in protein metabolism and exercise. Teenagers should add 10 percent to the calculated values.

To calculate body weight in kilograms, simply multiply your weight in pounds by 0.454. Then, multiply your weight in kilograms by the appropriate value in the grams per kilogram body weight column to determine the range of grams of protein intake per day. Teenagers should increase this amount by 10 percent.

TABLE 6.8 Costs of protein found in various food sources

Source	Serving size	Grams of protein/serving	Cost per serving	Cost per 8 grams of protein
Powdered milk	23 grams	8	$0.13	$0.13
Egg	1	6	$0.10	$0.13
Turkey breast	4 ounces	28	$0.75	$0.21
Skim milk	8 fluid ounces	8	$0.20	$0.20
Protein capsules	8 capsules	8	$1.20	$1.20
MetRx® Bar	3.5-ounce bar	27	$2.50	$0.74
Boost®	8 fluid ounces	10	$1.10	$0.88
Avalanche® Power Drink	16 fluid ounces	40	$3.00	$0.60
Whey Pro	1 scoop	23	$1.01	$0.34

Key Concepts

▶ During aerobic endurance exercise, particularly with low carbohydrate stores in the body, muscle protein may supply nearly 5 percent of the energy kcal.
▶ Although protein catabolism may occur during exercise, protein synthesis predominates in the recovery period. The type of protein synthesized is specific to the type of exercise program, such as resistance (weight, strength) training or aerobic endurance.
▶ Several recognized authorities have recommended a protein intake of 1.6–2.0 g/kg body weight per day for athletes attempting to gain weight, and about 1.2–2.0 g/kg body weight per day for endurance athletes.
▶ Timing of protein intake may be an important consideration. Consuming a protein/carbohydrate combination immediately before or after exercise training may provide a nutritional and hormonal milieu favorable to muscle anabolism and recovery. However, whether such a dietary strategy increases muscle mass or exercise performance is unknown.

Check for Yourself

▶ Using **table 6.7** as a guide, calculate how many grams of protein may be recommended for you. If you have not already done so, keep a record of your daily food intake for several days and determine if you are obtaining sufficient dietary protein.

Protein-Related Supplements

Given the potential importance of protein to optimal physical performance, a wide variety of ergogenic aids associated with protein nutrition have been used in attempts to enhance performance. As highlighted in the *Training Table* in this section, only a small number of these supplements have strong evidence that they have the ability to improve body composition or performance. The three supplements that have strong evidence—creatine, beta-alanine, and beetroot/dietary nitrate—will be discussed first. Other supplements with less or controversial evidence of efficacy will be discussed second, and supplements with little to no evidence of efficacy or data that do not support an ergogenic effect will be discussed last.

Training Table

Protein-related dietary supplements

A group of leading sport nutrition researchers was brought together by the International Olympic Committee to create an evidence-based dietary supplement consensus statement. This effort was spearheaded by sport nutrition expert Ron Maughan, and the consensus was published concurrently in both the *International Journal of Sport Nutrition and Exercise Metabolism* and the *British Journal of Sports Medicine*. In terms of protein-related supplements discussed in this chapter, strong evidence of an ability to improve performance or body composition was only found for protein, creatine, beta-alanine, and nitrate. Limited or low to moderate support were found for glutamine and colostrum, not for muscle building, but for reducing symptoms associated with upper respiratory illness. Visit https://doi.org/10.1136/bjsports-2018-099027 https://doi.org/10.1123/ijsnem.2018-0020.

Strong Evidence of Efficacy

Creatine Monohydrate Creatine, the nutrient, was first discovered in beef and reported in 1832. Creatine is a nonessential nutrient that is primarily ingested through meat (about 1 g/day) and synthesized by the body (about 1 g/day). Vegetarians have decreased blood and muscle creatine, although this does not indicate a deficiency. The role of creatine in energy production, described in chapter 3, was not revealed until the discovery of creatine kinase, ATP, and the creatine kinase reaction in the 1930s. Nonetheless, creatine metabolism was well studied throughout the 1800 and 1900s. In 1926, the first human supplementation trial was published, which demonstrated that oral creatine ingestion resulted in some of the creatine being excreted in the urine, while some was not recovered, meaning it was retained by the body. Many years later, in 1992, Roger Harris and others, using the muscle biopsy technique, showed that oral creatine monohydrate supplementation increased muscle creatine content. These findings have been reproduced many times, and it is generally recognized that oral ingestion of creatine monohydrate can increase skeletal muscle creatine content. As discussed later in this section, there are many different formulations of creatine supplements, but about 99 percent of the safety and efficacy data are on creatine monohydrate. In this text, unless otherwise stated, creatine supplementation refers to creatine monohydrate supplementation. A detailed summary of creatine monohydrate supplementation can be found in the Position Stand of the International Society of Sport Nutrition: https://doi.org/10.1186%2F1550-2783-4-6.

Supplementation Creatine can be ingested using two primary supplementation protocols, high-dose, short-term supplementation and low-dose, longer-term supplementation. To increase muscle creatine and phosphocreatine quickly, consume about 0.3 g/kg of body weight/day for about 5 days. As an example, for a 70 kg athlete, this equals 21 g/day for 5 days. In terms of increasing muscle creatine levels, there should be no difference between food-based or supplemental creatine; however, the ingestion of 20+ grams per day from food sources is unlikely and would have a drastic effect on energy, protein, and fat intake. A good general recommendation that applies to many athletes is to ingest 20 g/day for 5 days to quickly saturate muscle creatine levels. Excess supplemental creatine is excreted in the urine, and this simple protocol, where one does not have to calculate daily dosage based on body weight, will be effective for most sized athletes. Supplementation should be divided into four equal doses spread evenly throughout the day. To maintain elevated muscle creatine stores, a smaller dose of creatine, about 3–5 g/day, should be ingested. Alternatively, smaller doses of creatine (about 2–3 g/day) will increase muscle creatine stores over about a 4-week period. Either protocol, high-dose, short-term or low-dose, longer-term, should increase muscle creatine. The increase in muscle creatine and phosphocreatine following supplementation is somewhat variable and ranges from about 10 to 40 percent, but most everyone experiences an increase. Vegetarians, who have lower muscle creatine due to their low creatine intake, often have the largest muscle creatine increase in response to supplementation. Because muscle creatine uptake is insulin mediated, uptake can be further increased by ingesting creatine monohydrate supplements with carbohydrate, carbohydrate and protein, or following exercise. Athletes and others ingesting creatine wishing to optimize uptake could be advised to ingest creatine following meals or exercise, although many studies that have shown increased muscle creatine following supplementation did not employ these techniques.

Mechanism of Action Increasing muscle creatine through creatine supplementation can be ergogenic through several mechanisms, including directly enhancing performance, improving the results of strength and conditioning programs, and enhancing adaptation to training. Following supplementation both muscle phosphocreatine and glycogen are increased, both of which can increase exercise performance. Essentially, there is more fuel in the muscle, so intense exercise performance could be improved or sustained longer. For example, during maximal sprinting, about 80 percent of ATP is produced through the creatine kinase reaction. Additionally, because phosphocreatine and glycogen resynthesis is increased, creatine supplements could allow for faster recovery times during rest periods and improve performance over repeated bouts of exercise. For example, increased muscle creatine and faster phosphocreatine enhance the performance of repeated bouts of intense exercise, such as cycling, swimming, and running, and also resistance training. Many studies, as reviewed by Rawson and Volek and Lahners and colleagues, have shown that creatine supplementation improves resistance exercise performance. Thus, creatine supplements allow for increased training volume during strength and conditioning workouts, which should transfer into improved sports competition performance. In terms of exercise adaptation, increasing muscle creatine through supplementation increases the expression of several growth factors, satellite cell numbers, and many genes related to muscle function. There is also an increase in intracellular water (i.e. cellular hyper-hydration) that can decrease protein breakdown and increase glycogen, protein, DNA, and RNA synthesis.

Efficacy Creatine supplements have been heavily studied, and several hundred peer-reviewed studies have been published. It has been well described that following creatine supplementation and increased muscle creatine levels, the performance of brief, intense exercise lasting <30 seconds is typically improved. The ergogenic effect appears to be most evident when there are repeated bouts, as benefits may not be measurable until later bouts. The longer the exercise task is beyond 30 seconds, the less likely there is to be an effect of creatine supplements. There are a small number of studies showing increased power output, speed, or decreased time to complete a fixed distance in the 30-second to 5-minute time domain following creatine ingestion. In theory, increased muscle creatine could decrease the oxygen cost of endurance exercise lasting longer than 5 minutes, but this is not supported by research where most studies show no improvements in performance. Creatine supplements have been shown to benefit endurance athletes by increasing power output, speed, or decreasing fatigue and time to complete a fixed distance of sprints embedded within or following an endurance ride.

Even if creatine supplements do not benefit the performance of a specific sport, for example, field hockey or tennis, most sports benefit from strength and conditioning, and creatine improves resistance training performance. Rawson and Volek showed creatine supplementation plus resistance training increases strength and muscle endurance more than resistance training plus placebo ingestion. In meta-analyses, Lahners and others reported increased strength and Branch reported increased lean body mass subsequent to creatine ingestion plus resistance training. It is clear that creatine supplements can directly improve performance in some sports, for example, sprinting. For some athletes, creatine supplements may function more as a training aid, enhancing the quality of strength and conditioning programs, which could then translate into improved sports performance. For a brief, practical summary, see the *Training Table* in this section on creatine supplementation recommendations for athletes. For the interested reader, a more detailed review of creatine supplementation can be found here: www.gssiweb.org/sports-science-exchange/article/the-safety-and-efficacy-of-creatine-monohydrate-supplementation-what-we-have-learned-from-the-past-25-years-of-research.

Training Table

Creatine supplementation recommendations for athletes:

Supplementation

- Take 5 grams four times/day (20 g/day) for 5 days to increase muscle creatine levels.
- Take 3–5 g/day to maintain increased muscle creatine levels.
- Consume creatine after meals and/or exercise, both of which might increase muscle creatine uptake.

Potential benefits

- Improved performance of brief, intense exercise (<30 seconds), especially repeated bouts
- Improved quality of resistance/strength and conditioning workouts
- Possible improved performance of some tasks that rely on the lactic acid or oxygen systems
- Increased muscle glycogen when combined with carbohydrate loading
- Enhanced muscle recovery via decreased muscle damage/inflammation and reduced severity of, or enhanced recovery from, traumatic brain injury

Caution

- Majority of research on safety and efficacy is on creatine monohydrate.
- Small weight gain may impair performance in weight-dependent sports or make it difficult to lose enough body mass to make a weight class limit.

Formulation A number of different formulations and types of creatine supplements are available for sale (e.g., liquid, complexed with other compounds). Typically, these products are marketed as having better absorption than creatine monohydrate, of which more than 99 percent is absorbed. Many of these products are also marketed as being safer than creatine monohydrate, which has an excellent and well-documented safety profile. Typically, these products are also advertised as being more effective than creatine monohydrate. These claims are largely unsubstantiated. The most comprehensive review of different creatine supplement formulations was published by Kreider and others. They note that although products are marketed with claims of better bioavailability, safety, or efficacy, compared to creatine monohydrates, there is little to no evidence to support these claims. Additionally, while creatine monohydrate is well absorbed and tolerated, some of these creatine supplements are pro-creatinine or contain very little creatine at all. At this point in time, given the enormous amount of data on the safety and efficacy of creatine monohydrate, it is unadvisable to recommend a different creatine formulation.

Safety The safety of creatine monohydrate supplementation has been thoroughly reviewed by several experts including Persky and Rawson, Gualano and others, Lopez and colleagues, and Rawson, Clarkson, and Tarnopolsky. Gualano and colleagues note that there are several hundred published studies and millions of exposures to creatine monohydrate supplements, and creatine supplementation maintains an excellent record of safety. Although creatine monohydrate supplements are viewed as safe when taken in recommended doses, misinformation persists about the safety of creatine supplements. The interested reader can refer to these comprehensive reviews, which cover the safety of creatine supplementation on renal, muscular, and thermoregulatory systems. Additionally, the review from Antonio and others, "Common questions and misconceptions about creatine supplementation: what does the scientific evidence really show?" can help clear up some of the most common myths regarding safety. A brief summary of the most discussed safety concerns is summarized here.

Multiple studies (>20) and several reviews are available that demonstrate no decline in renal function following intake of recommended doses of creatine supplements. There are a few case studies related to renal function, but they are confounded by drug/medication use, prior kidney disease, and use of other supplements. Over the past three decades of creatine research and use across various populations, there have been tens of thousands of exposures, yet no link between creatine supplements and renal health has emerged. In their meta-analysis of ten studies, Lopez and colleagues found no evidence to support that creatine supplementation impacts thermoregulation by hindering heat dissipation or negatively altering the fluid balance. In this analysis, multiple controlled experimental trials of athletes exercising in the heat, following creatine supplementation, resulted in no adverse effects. In their review, Rawson, Clarkson, and Tarnopolsky highlighted the fact that creatine does not promote muscle dysfunction. Across several different types of research studies where individuals were supplemented with creatine and then subjected to stressful exercise or where athletes who ingested creatine were tracked across the competitive season, no increases in muscle dysfunction were noted. In fact, in several investigations, decreased muscle damage or better recovery from stressful exercise or fewer cramps, strains, and injuries in competitive athletes were noted.

Brain Function Creatine is synthesized in the brain, where, like in skeletal muscle, it is used for energy production. There are a small number of studies that indicate that creatine supplementation increases brain creatine, although to a smaller extent than in muscle, possibly 5–10 percent. Additionally, as reviewed by Forbes and colleagues and Roschel and colleagues, creatine supplementation appears to improve cognitive processing, although it is unknown how this applies to athletic performance. Beneficial effects of creatine on cognitive processing have been reported in unstressed and stressed (e.g., following strenuous exercise and/or sleep deprivation) individuals, in older and younger populations, and in athletes and nonathletes, which makes the comparison of studies difficult. More work needs to be done in this area, especially as it pertains to (1) finding an optimal supplementation protocol to increase brain creatine, (2) understanding under what conditions supplementation is most beneficial, and (3) conducting studies where cognitive processing and brain creatine are both assessed. Finally, there is a potential for creatine supplements to help reduce the severity of or enhance recovery from traumatic brain injury (concussion). In theory, many of the changes in the brain related to concussion, for example, decreased brain creatine, cell membrane disruption leading to calcium influx, nerve damage, mitochondrial dysfunction, oxidative stress, and inflammation, could benefit from creatine supplementation. Available human data are uncommon, but animal data are supportive of less tissue damage following traumatic- and hypoxic-induced brain injuries in laboratory animals ingesting creatine supplements. Two small studies of children with brain injuries did report that creatine supplementation improved cognition, communication, self-care, personality, and behavior and reduced headaches, dizziness, and fatigue. Although data on creatine supplementation and brain injury are scarce, as reported by Roschel and colleagues, creatine supplements have documented muscular performance benefits, are inexpensive, are widely available, and have a strong safety profile. In this light, and given the serious effects of concussion, some think that the use of creatine as a preventive measure against concussion in high-risk athletes might be sensible.

Medical Applications Creatine supplementation improves muscle and brain function, which can benefit more than young athletes. Gualano and others report that creatine monohydrate supplementation can benefit individuals suffering from a variety of diseases, including myopathies, neurodegenerative disorders, cancer, rheumatic diseases, and type 2 diabetes. The recent review from Harmon and colleagues also summarizes the findings of research that focused on medical applications of creatine supplements. Rawson and Venezia and Candow and Chillibeck have shown that creatine supplementation plus resistance training is beneficial for older adults. Creatine supplementation holds promise as a nutritional medical therapy, and researchers will continue to explore this important topic. The interested reader should refer to the findings presented at the recent Creatine Conference linked at the end of this section.

> Every few years, there is an entire conference dedicated to creatine metabolism and supplementation. Based on the research shared at the most recent conference, an open-access edition of the journal *Nutrients* was dedicated to creatine research. The interested reader can find links to these articles here: www.mdpi.com/journal/nutrients/special_issues/creatine_supplementation and a single PDF of all articles as an e-book here: https://creatineforhealth.com/education/

Beta-Alanine Beta-alanine (β-alanine) is a naturally occurring nonessential amino acid that is synthesized by the body and consumed in the diet, mostly through meat, poultry, and fish. Unlike the normal form of alanine (l-alpha alanine), it is not used in the formation of any major proteins or enzymes. However, beta-alanine can be ingested orally, taken up by muscle cells, and combined with histidine to form a peptide, carnosine. Carnosine is highly concentrated in muscle tissue and is a robust intracellular buffer. Although carnosine only accounts for about 7 percent of the buffering capacity of skeletal muscle, this can be doubled through beta-alanine supplementation. The seminal research on beta-alanine, as was the case with creatine supplementation, was conducted by Roger Harris and colleagues. Since that time, many studies of the effects of beta-alanine supplementation on muscle carnosine content and on exercise performance have been conducted. Although these studies used different methods to test the efficacy of beta-alanine ingestion, enough studies have been conducted for the data to be synthesized and published in several extensive reviews. Drs. Craig Sale and Bryan Saunders are leading authorities on muscle carnosine and beta-alanine supplementation.

Supplementation Harris and others first reported that 3.2–6.4 g/day of beta-alanine supplementation, taken in multiple doses throughout the day, for 4 weeks, results in about a 42–64 percent increase in muscle carnosine. These data have been reproduced in other studies. Saunders and others showed that the increased muscle carnosine remains for up to 24 weeks, as long as supplementation (6.4 g/day) is maintained. It is possible that a lower dose of supplementation can be used to maintain increased muscle carnosine; Stegen and others reported that after beta-alanine loading (3.2 g/day for 46 days), muscle carnosine remained 30–50 percent elevated above baseline levels with the ingestion of 1.2 g/day. More research is needed to find the most precise supplementation protocol.

Efficacy In their comprehensive meta-analysis of 40 studies of 1,461 participants, Saunders and colleagues demonstrated that beta-alanine supplementation improved exercise performance over specific time domains. Short duration (≤0.5 minutes) and longer-term (>10 minutes) exercise performance did not benefit from supplementation. However, exercise performance of moderate duration (0.5–10 minutes) was improved. As the cause of fatigue in exercise lasting less than 60 seconds is unlikely to be acidosis, it appears that very brief and intense tasks may not improve subsequent to beta-alanine supplementation. Sale and Harris rate beta-alanine as a supplement that has "Level IV" or the highest level of evidence available to show the efficacy of a supplement, and the International Olympic Committee includes beta-alanine on the shortlist of effective dietary supplements for the high-performance athlete.

Safety The only known side effect of beta-alanine is paresthesia (flushing), but this appears to have been minimized with timed-release supplements or by ingesting smaller doses more frequently throughout the day. One intriguing possibility is the potential combination of dietary supplements, such as creatine monohydrate, beta-alanine, and sodium bicarbonate, that are effective in improving the performance of brief, intense exercise but that work through different mechanisms. The small number of studies in this area, such as by Tobias and colleagues who combined beta-alanine and sodium bicarbonate supplements, has demonstrated an additive effect.

Dietary Nitrate/Beetroot

It is generally accepted that a diet rich in vegetables is associated with a healthy, long life. Nitrate is found in vegetables, particularly in leafy greens and beetroot, and has been examined for its potential benefits to the cardiovascular system, including exercise performance. The goal, through the ingestion of dietary nitrate, is to increase nitric oxide, which plays an important role in vasodilation, blood flow, blood pressure, mitochondrial respiration, and skeletal muscle contraction. Nitric oxide has a short half-life and must be continuously produced to aid in these processes, which is how dietary nitrate can help. In addition to foods that naturally contain nitrate, there are many beetroot sports bars and drinks with standardized nitrate levels that are marketed toward athletes.

Supplementation Nitrate supplementation can be effective when ingested acutely (single dose) or over multiple days (about 3-7 days). Performance benefits are obtained quickly, approximately 3 hours after ingestion and can be maintained over days if daily nitrate ingestion is continued. A systematic review and meta-analysis by Senefeld and others suggested that dietary nitrate is ergogenic at doses between about 5 and 29 mmol/day but not at low doses of about 2-5 mmol/day. A recent expert panel refined this recommendation agreeing that acute supplementation of 8 to 16 mmol, or chronic supplementation of 4->16 mmol/day, is ergogenic and that acute supplementation with <4 mmol/day is not ergogenic. The Dietary Approaches to Stop Hypertension (DASH) diet contains up to 20 mmol/day of nitrate. Jones states that, at this time, there does not appear to be any benefit in consuming more than 740 milligrams (12 mmol). A nitrate standardized beetroot juice shot contains 400 milligrams of nitrate.

Unlike creatine monohydrate and beta-alanine, where recommended supplement doses are difficult to achieve with food alone, recommended levels of dietary nitrate can more easily be achieved through food sources. This can include beetroot juice standardized for nitrate content, beetroot sports bars, high nitrate vegetables, and vegetable smoothies. In one study, ingestion of nitrate-rich beetroot juice, rocket (arugula) salad beverage, and spinach beverage all increased plasma nitrate and nitrite levels and lowered blood pressure more than sodium nitrate. Although it is sensible to encourage athletes to consume extra vegetables, van der Avoort and others reported larger increases in plasma nitrite and nitrate following beetroot rather than vegetable ingestion. Zhong and colleagues have published a database of the nitrate content of foods. Acute nitrate supplementation up to about 16 mmol is believed to be safe by experts, although less is known about the safety of chronic nitrate supplementation.

Mechanism of Action Nitric oxide is involved with muscle contraction, metabolism, and blood flow and must be produced via an enzymatic pathway requiring arginine and oxygen or the nitrate-nitrite-nitric oxide pathway, which depends on the availability of nitrate. Nitrate is produced endogenously or can be ingested through the diet. Briefly, ingested nitrate is converted to nitrite by bacteria in the mouth, and nitrite enters the circulation and is converted to nitric oxide, especially in areas where oxygen is needed, such as exercising muscle. Antibacterial mouthwash can disrupt the oral microbiota, blunting the conversion to, and increase in plasma nitrite, and subsequently attenuate the anti-hypertensive and performance-enhancing effects. Dr. Andrew Jones is the world's leading authority of dietary nitrate and beetroot supplementation, and the interested reader can learn more about dietary nitrate here: www.gssiweb.org/en/sports-science-exchange/Article/dietary-nitrate-and-exercise-performance-new-strings-to-the-beetroot-bow

Efficacy Larsen and colleagues first reported that sodium nitrate supplementation reduced the oxygen cost of submaximal cycling, while Bailey and others from Dr. Jones' laboratory first showed that beetroot juice ingestion reduced oxygen uptake while cycling at a fixed submaximal power output. Subsequently, many research studies and reviews have concluded that dietary nitrate supplementation, usually from beetroot juice, enhances endurance exercise performance. However, improved performance has not been shown in every study, and highly trained athletes, in particular, may not receive benefits. In an expert consensus on dietary nitrate as an ergogenic aid, Shannon and other experts report that the effects of dietary nitrate appear to be diminished in individuals with higher aerobic fitness ($\dot{V}O_{2peak}$ > 60 ml/kg/minute). Dietary nitrate supplementation may also improve high-intensity sprint exercise performance. In their review, Tan and others noted improved aspects of cycling and running sprint performance as well as resistance and isokinetic exercise performance. Jones notes that the nitrate to nitrite to nitric oxide pathway is favored under conditions of low pH and low oxygen availability, such as during high-intensity exercise when the lactic acid energy system is a larger contributor to energy turnover. Also, type II (fast-twitch) muscle fibers may have a more optimal environment for the reduction of nitrite to nitric oxide and some well-trained athletes may have a high proportion of type II muscle fibers.

Safety A meta-analysis on 16 studies of dietary nitrate and blood pressure by Siervo and colleagues revealed that beetroot juice and inorganic nitrate supplementation significantly reduce blood pressure. While this is great news from a public health perspective, people taking prescribed medication for hypertension should consult with their doctor about nitrate to avoid a potentially dangerous hypotensive episode. Similarly, athletes may be ingesting dietary supplements that already contain nutrients that may alter blood pressure (e.g., preworkout supplements), and the addition of nitrate could cause a dangerous hypotensive response. More research needs to be conducted on the efficacy and safety of dietary nitrate, but current data supports an athletic performance and antihypertensive effect. Athletes should know that some report gastrointestinal discomfort following beetroot ingestion; there is often a temporary pink/purple color change of urine and stools, which is thought to be harmless, and supplementation of nitrite, compared to foods or standardized nitrate products like shots or bars, is discouraged. Ingesting too much nitrite is likely easier than ingesting too much nitrate from dietary or natural food sources or food-based supplemental sources.

Less or Controversial Evidence of Efficacy

Taurine Taurine is sulfur-containing substance synthesized from amino acids, mainly methionine and cysteine, and is found only in animal foods. It is a vitamin-like compound that has multiple functions in the body, including effects on heart contraction, insulin actions, and antioxidant activity that could be of interest to the athlete. Taurine is an ingredient in several energy drinks, such as Red Bull®.

In their review of 19 studies, Kurtz and others reported that both limited and varied outcomes prohibit definitive conclusions regarding the efficacy of taurine on aerobic and anaerobic exercise performance, recovery from training, and/or mitigating muscle damage. In their meta-analysis of ten studies, Waldron and others reported a small to moderate improvement in endurance performance using various supplementation protocols (from 1 gram of taurine in a single dose to 6 grams for 2 weeks). Lawrence Spriet, an expert in taurine supplementation, noted that human skeletal muscle is resistant to large, prolonged increases in plasma taurine, which can prevent muscle taurine uptake. This is the opposite of rodent skeletal muscle, so animal research must be interpreted cautiously. If there are benefits of taurine supplementation on human skeletal muscle, this likely occurs outside the muscle cell. Much more research is needed before taurine transport into the muscle is well understood and also before taurine can be recommended as a supplement.

Gelatin and Collagen One important area of research that remains understudied is the effect of nutritional interventions on connective tissue proteins, such as tendons. In one recent study, Baar and colleagues, as reported in a paper by Shaw and others, showed that the combination of 15 grams of gelatin and 50 milligrams of vitamin C increased collagen production in vitro and amino-terminal propeptide in the blood. These findings indicate increased synthesis of connective tissue proteins, which could prove valuable in preventing injury, reducing time lost to injury, and speeding recovery and return to play following an injury. A small number of studies of the effects of collagen supplementation on osteoarthritis symptoms (five studies) were highlighted in a review by Garcia-Coronado and colleagues, which showed improved scores on the Western Ontario and McMaster Universities Osteoarthritis Index (WOMAC) scale. In their systematic review, Khatri and others found significant improvements in joint pain/recovery from injury and muscle soreness/recovery (seven total studies). Gelatin and collagen are generally considered low-quality proteins and poor choices for muscle building. Conclusions on the effects of these protein-related supplements on connective tissue, joint health, or recovery from injury are premature as only a small number of studies are available and few are in athletes.

HMB (Beta-Hydroxy-Beta-Methylbutyrate) Beta-hydroxy-beta-methylbutyrate (HMB) is not a nutrient per se but a by-product of leucine metabolism in the human body. The body produces about 0.2–0.4 grams of HMB per day depending on dietary leucine intake. HMB is marketed as a dietary supplement in the form of calcium-HMB or HMB-free acid (HMB-FA), mostly to strength-power athletes. HMB supplementation is theorized to increase lean muscle mass, decrease body fat, increase muscle strength, and reduce muscle damage. Although the underlying mechanism is not known, investigators who developed HMB speculate that it may inhibit the breakdown of muscle tissue during strenuous exercise. Initially, in attempts to increase the nutritional quality of animal meat, research on various farm animals indicated that HMB supplementation may increase lean muscle mass and decrease body fat. About 40 human HMB supplementation studies have been published, and several extensive reviews are available. Compared to creatine monohydrate and beta-alanine, which have consistently been shown to be ergogenic, the literature on HMB is much more difficult to interpret. One reason for this is that there are large differences in the methods and populations used to study HMB supplementation. Another reason is that some of the studies focus on muscle damage, while others focus on outcomes like gains in lean body mass or strength.

Several years ago, in a small meta-analysis of nine studies, Nissen and Sharp concluded that HMB supplementation reduced muscle damage and increased strength and lean body mass. However, subsequent studies failed to support these initial conclusions. More recent meta-analyses by Rowlands and Thomson and Sanchez-Martinez and others conclude that HMB supplementation had trivial effects on strength and fat-free mass, especially in trained competitive athletes, and effects on muscle damage were unclear. Molfino and colleagues reviewed the effects of HMB supplementation in young adults from 22 studies and reported that HMB supplementation increased lean body mass and strength in only about half of the studies. Similar findings from the review by Zanchi and colleagues make it difficult to recommend HMB supplementation for young adults seeking to increase lean body mass or strength. The recommended daily dosage of HMB is about 3 grams per day, which appears safe, but there are few safety data available.

Recently, in a controversial study, Wilson and colleagues reported significant but very large gains in lean body mass (7.4 kg) and strength (1-RM bench press + squat + deadlift: 77 kg) in trained males ingesting HMB-FA during resistance training. As these gains rival those of high-dose anabolic steroids combined with resistance training and have been challenged by leading experts, they should be interpreted cautiously. Wilkinson and colleagues described that HMB-FA supplementation stimulated muscle protein synthesis, increased anabolic signaling, and attenuated muscle protein breakdown, which could explain how HMB supplementation might benefit resistance training athletes. Although HMB-FA is absorbed better than calcium HMB, there are very few data available on HMB-FA supplementation, and the findings need to be replicated. Slater concluded that the beneficial effects of HMB supplementation on adaptations to resistance training are small in untrained individuals and negligible in athletes. Much more research needs to be conducted on HMB before it can be recommended.

Supplements with Little to No Evidence of Efficacy, or, Data That Do Not Support an Ergogenic Effect

Inosine Inosine is not an amino acid but is classified as a nucleoside. It is included for discussion here because it is associated with the development of purines, nonprotein nitrogen compounds that have important roles in energy metabolism. On the basis of animal research and studies of blood storage techniques, writers in popular media have theorized that inosine may be an effective ergogenic aid for a variety of athletes. Advertisements have suggested that inosine may improve ATP production in the muscle and thus be of value to strength-type athletes. Additionally, some marketing and media sources suggest inosine enhances oxygen delivery to the muscles, thus being beneficial to aerobic endurance athletes. There are no data to support these claims. Research from Mel Williams' laboratory at Old Dominion University revealed no ergogenic effect following 6 grams of inosine ingestion and possibly a decrement in endurance performance. Similarly, Starling and his colleagues showed no beneficial effect of 5 grams of inosine supplementation for 5 days on anaerobic and endurance exercise performance, but there was a performance decrement on a supramaximal cycling test. Finally, McNaughton and others provided 10 grams of inosine to cyclists for 10 days and tested anaerobic and endurance cycling performance, and again reported no ergogenic effect. Thus, on the basis of the available data, inosine does not appear to be an effective ergogenic aid.

Radius Images/Design Pics/Alamy Stock Photo

Bovine Colostrum Bovine colostrum is the first milk secreted by cows. Standardized preparations of bovine colostrum are available as dietary supplements and marketed to athletes as a way to alter body composition or improve performance. Brinkworth and others indicate that bovine colostrum is a rich source of protein, carbohydrates, vitamins, minerals, and various biologically active components, also including growth factors. Bovine colostrum is purported to increase levels of serum insulin-like growth factor (IGF-1), which could be anabolic. However, many studies have shown no increases in IGF-1 following supplementation. In a recent review, Davison summarized the effects of bovine colostrum on body composition, performance, recovery, gut damage and permeability, immune function, and illness risk. In terms of body composition and performance, although there are some reports of an ergogenic effect, there are few studies, and some have potential confounding factors. Overall, the evidence of an ergogenic effect is minimal. There is stronger evidence for a protective effect on exercise-induced increases in gut permeability and reduced risk of upper respiratory infection, which may be of interest to athletes. Athletes should be concerned that, if bovine colostrum contains prohibited substances, (e.g. IGF-1) it could affect the outcome of doping tests.

Chondroitin and Glucosamine Chondroitin and glucosamine are derived from connective tissue, and each has been marketed as a dietary supplement, either separately or in combination, to help promote healthy joints in individuals who exercise. Although weight-bearing or resistance exercise training has not been shown to cause excessive wear-and-tear on healthy joints and may actually improve joint health, some dietary supplement entrepreneurs may suggest otherwise. In a sense, if a dietary supplement could prevent joint pain and thus promote optimal exercise training, it could be considered ergogenic.

Both chondroitin and glucosamine may be synthesized in the human body from amino acids and other nutrients, and both are found in human cartilage, one of the main components involved in joint health. Glucosamine is believed to help form compounds, such as proteoglycans, that form the structural basis for cartilage, while chondroitin is part of a protein that helps cartilage hold water to give it elasticity and resiliency. Cartilage serves as a kind of shock absorber and prevents bone-to-bone contact. Excessive wear of cartilage leads to osteoarthritis, a painful joint condition. Dietary supplements of chondroitin are made from cattle cartilage, while those of glucosamine are made from shellfish. Different salt forms of supplements are available, such as sulfate and hydrochloride. Theoretically, such supplements will help maintain normal cartilage levels and prevent the development of osteoarthritis.

Numerous studies have investigated the role of chondroitin and/or glucosamine supplementation on symptoms of arthritic pain. However, the results are somewhat equivocal. The National Institutes of Health (NIH) funded a large multicenter clinical study called GAIT (Glucosamine/Chondroitin Arthritis Intervention Trial) designed to provide a clearer picture of the role that these dietary supplements, both separately and in combination, may play in the treatment of osteoarthritis. Nearly 1,600 subjects, with an average age of 59 and experiencing arthritic knee pain, were assigned to receive daily 1,500 milligrams of glucosamine hydrochloride, 1,200 milligrams of chondroitin sulfate, both glucosamine hydrochloride and chondroitin sulfate, 200 milligrams of an anti-inflammatory drug, or placebo for 24 weeks. The overall findings indicated that glucosamine, chondroitin, or the glucosamine/chondroitin combination did not reduce knee pain in patients with osteoarthritis more than a placebo. The NIH study was designed to provide a definitive answer regarding the efficacy of such dietary supplements in reducing arthritic pain, but it did not. In two recent meta-analyses, researchers reached opposite conclusions about the potential benefits of glucosamine and chondroitin supplementation. In their analysis of eight studies, Meng and others concluded that the total Western Ontario and McMaster Universities Arthritis Index (WOMAC) score was better with glucosamine/chondroitin supplementation compared with the placebo group. However, Simental-Mendía and colleagues, in their analysis of ten studies, concluded that there was no effect of the supplements on the total WOMAC score.

It seems that most scientists agree that chondroitin supplements do not reduce joint pain. However, some believe that glucosamine sulfate, as opposed to glucosamine hydrochloride, could be effective. Indeed, in separate reviews, both Vlad and Reginster noted that trials using glucosamine sulfate produced more positive results than studies using glucosamine hydrochloride. Athletes considering supplementation should consider that most studies have been conducted with older people, and data may not be generalizable to hard-training young athletes. Ostojic and colleagues found improved knee flexibility at a single time point but no improvements in knee flexibility, pain, or swelling at multiple time points in 100 athletes (aged 25 years) following 28 days of glucosamine sulfate ingestion (1,500 mg/day). Thus, there are no data that these supplements will prevent the development of joint pain or osteoarthritis in young, healthy athletes. More research needs to be conducted to determine if glucosamine sulfate supplementation is valuable in athletic populations.

Amino Acid Supplements

Providing all essential amino acids during recovery from exercise increases muscle protein synthesis and net protein balance (i.e., protein synthesis > protein breakdown). It would seem most prudent to simply eat a high-protein food following exercise, or if more convenient, a protein supplement in the form of a shake or a food bar. However, the ingestion of groups of amino acids, such as branched-chain amino acids or individual amino acids postexercise or at other times of day, has become very popular. Amino acids are typically marketed as enhancing adaptation to exercise, such as increasing muscle hypertrophy or strength. In some cases, amino acid supplements are marketed as nutrients that can reduce fatigue and improve exercise performance and/or health. But amino acid metabolism is very complex. It depends on a variety of factors, such as the concentration in the blood, competition with other amino acids, feedback control mechanisms, and the presence in the diet of other nutrients. Consumption of specific amino acid mixtures may actually lead to nutritional imbalances, as an overload of one amino acid may inhibit the absorption of others into the body. While research on the effects of dietary and supplemental protein strongly points to increased protein synthesis and improved exercise adaptations, research on supplementation of groups of or individual amino acids discussed in this section is much less convincing.

Branched-Chain Amino Acids (BCAAs) The three BCAAs are leucine, isoleucine, and valine. BCAA supplementation has been theorized to enhance exercise performance in two primary ways: (1) by increasing skeletal muscle protein synthesis or decreasing protein breakdown, and (2) by altering neurotransmitter function and reducing mental fatigue during exercise. Robert Wolfe, a world authority on protein metabolism, notes that a significant increase in muscle protein synthesis requires all essential amino acids to be available. By only ingesting BCAAs, the availability of other essential amino acids will become rate limiting and they are unable to support increased muscle protein synthesis. Not surprisingly, two studies have shown decreased protein synthesis following BCAA supplementation. Dr. Wolfe states that BCAA supplements alone do not support muscle anabolism. In their review, Plotkin and others note that most studies do not support a role for BCAA supplements for muscle strength or hypertrophy, especially when daily protein needs are met. In agreement with Wolfe, these researchers state that building muscle requires all nine essential amino acids, and so, if that condition and the total protein needs are met, additional BCAAs cannot provide additional benefits. A small number of studies, including two recent meta-analyses that analyzed nine clinical trials, show small but inconsistent improvements in some indirect markers of muscle damage following intense exercise. Although interesting, these results are not enough evidence to recommend BCAA supplements for recovery from exercise-induced muscle damage.

Central fatigue hypothesis In the 1980s, Eric Newsholme, a biochemist at Oxford University, proposed the central fatigue hypothesis, postulating that high levels of serum free-tryptophan (fTRP) in conjunction with low levels of BCAA, or a high fTRP:BCAA ratio, may be a major factor in the etiology of fatigue during prolonged endurance exercise. Research with animals has shown that a high fTRP:BCAA ratio may lead to increased production of serotonin. Newsholme suggested that serum BCAA levels eventually decrease in endurance exercise, such as marathon running, because they may be used for energy production. Such an effect would possibly increase the fTRP:BCAA ratio, facilitating the transport of tryptophan into the brain and increasing serotonin production, which could lead to fatigue because increased serotonin levels may depress central nervous system functions. Some, but not all, research supports an increased fTRP:BCAA ratio in humans undergoing endurance exercise, so the theory that BCAAs can improve performance by reducing mental fatigue is plausible. The central fatigue hypothesis is described in chapter 4 and in **figure 4.11**.

Romain Meeusen, an authority on nutrition and the brain, notes that BCAA ingestion does cause a rapid increase in plasma levels and brain uptake of BCAAs. He notes that there is some evidence that BCAA ingestion reduces ratings of perceived exertion (RPE), a measure of how mentally stressful a person perceives a given exercise task to be. However, Meeusen and Decroix point out that BCAA supplementation studies have failed to consistently show improved exercise performance during various exercise challenges, including prolonged exercise to exhaustion, prolonged time-trial performance, incremental exercise, or intermittent shuttle running. It could be postulated that if serum BCAA levels fall during prolonged exercise, then BCAA supplements might be used as an additional energy source during endurance exercise, thereby reducing fatigue. However, Rennie and others indicate that the total contribution of BCAAs to fuel provision during exercise is minor, and most would recognize pre- or during-exercise carbohydrate ingestion would be a better choice to provide additional energy. In terms of improving endurance exercise performance, it appears that the theory behind BCAA supplementation does not translate into real exercise performance.

Glycine Glycine is a nonessential amino acid. Because it is involved in the formation of creatine and phosphocreatine, it could theoretically be an ergogenic aid by serving as a precursor for creatine. Several studies conducted more than a half-century ago suggested a beneficial effect of glycine or gelatin, which is high in glycine, supplementation on various measures of strength, but the experiments were poorly designed. More contemporary research with proper experimental design and relatively large doses of glycine reveals no beneficial effects on physical performance.

Tyrosine Tyrosine is a precursor for the catecholamine hormones and neurotransmitters, specifically epinephrine, norepinephrine, and dopamine. Some have suggested that inadequate production of these hormones or transmitters could compromise optimal physical performance. Thus, as a precursor for the formation of these hormones and neurotransmitters, tyrosine has been suggested to be ergogenic. Research in this area is very limited. Sutton and others, in a well-designed, placebo-controlled, crossover study, had subjects consume tyrosine (150 mg/kg body weight) 30 minutes prior to taking a series of physical performance tests. Although the tyrosine supplementation significantly increased plasma tyrosine levels, there were no significant ergogenic effects on aerobic endurance, anaerobic power, muscle strength, or plasma levels of epinephrine or norepinephrine. In their review, Meeusen and Decroix state that despite a good rationale for its use, evidence of an ergogenic benefit of tyrosine supplementation on exercise performance is limited. They note that tyrosine supplementation appears to prevent declines in some aspects of cognitive processing and mood in individuals subjected to stress; however, this still does not appear to consistently translate into enhanced exercise performance. In multi-nutrient supplements and energy drinks, tyrosine is sometimes combined with other potentially ergogenic substances, such as caffeine, carbohydrate, or taurine. Enhanced exercise performance from the use of these products is likely from the effects of caffeine and carbohydrate.

Glutamine Glutamine, a nonessential amino acid, is the most abundant amino acid in the plasma. Houston indicates that it represents about 60 percent of the body's amino acid pool. Glutamine, like alanine, is synthesized in the muscle tissue, where it is found in high concentrations and is a major means for removing excess amino groups from the muscle, delivering the amino groups to the liver and kidneys for excretion or reuse of excess nitrogen. It is also an important fuel for the immune system. Glutamine supplements have been marketed as muscle builders, performance enhancers, and immune-stimulants. Phillips notes that despite the popularity of glutamine supplements, there is a lack of evidence of benefits related to skeletal muscle, such as strength or recovery. In reviews, Hargreaves and Snow and Nieman, along with Akerström and Pedersen, indicated that there is little support from controlled studies to recommend glutamine supplementation for enhanced immune function and prevention of upper respiratory tract infections. Coquiero and others, in a review of 55 studies, report that glutamine supplements may improve some indirect markers of fatigue but have limited effects on physical performance. As documented previously, when it comes to reducing fatigue and improving endurance exercise performance, consuming adequate daily carbohydrates appears to be a sound dietary strategy for most athletes.

Arginine and Citrulline Arginine, a conditionally essential amino acid, has several functions in human metabolism. One of its most important is to serve as a precursor for nitric oxide (NO) synthesis. NO acts as a vasodilator to increase blood flow. The role of nitric oxide in exercise performance is discussed elsewhere in this chapter, along with dietary nitrate. Citrulline is also an amino acid but is not one of the 20 essential or nonessential amino acids because it is not involved in protein synthesis. However, dietary citrulline is eventually taken up by the kidney and metabolized to generate large amounts of arginine. Hickner and others noted that citrulline supplementation increases plasma arginine levels to a higher level than arginine supplementation.

Ergogenic effects of arginine supplementation are unlikely, and several research teams have found no benefits of supplementation. However, data on citrulline ingestion, although preliminary, are promising. In their review, Gonzalez and Trexler report that oral L-citrulline and citrulline malate supplementation can increase plasma citrulline, arginine, nitrate, and nitrite concentrations. Although blood flow enhancement is a proposed mechanism for the ergogenic potential of L-citrulline, evidence supporting improvements in vasodilation and skeletal muscle tissue perfusion is scarce and inconsistent. They note that several studies have reported that L-citrulline supplementation can enhance exercise performance and recovery but that more work should be done to investigate the effects of both acute and chronic supplementation on markers of blood flow and exercise performance. In their review of citrulline malate supplementation, Gough and colleagues note that current evidence supports that an acute 8-gram dose of citrulline malate may increase muscular endurance-strength performance, although this has not been shown consistently. They identify methodological concerns in some studies, differences in supplementation strategies, and supplement quality control issues as some of the reasons behind unsupportive studies. Of note, they state that there is little evidence to advocate for supplementation in terms of muscular power, maximal strength, recovery of muscular function, or supporting muscular adaptations to training. If drinking beetroot juice or eating green leafy vegetables is more effective at increasing nitric oxide levels, as described in the dietary nitrate section, then the ingestion of citrulline supplements may be unnecessary. The International Olympic Committee does not include citrulline supplements on their list of recommended supplements for high-performance athletes.

Arginine, Lysine, and Ornithine Research has shown that infusing any of a number of amino acids into the blood potentiates the release of human growth hormone (HGH), a polypeptide. HGH is released from the pituitary gland into the bloodstream, affecting all tissues. One of its effects is to stimulate the production of another hormone, insulin-like growth factor-1, that spurs the growth of tissue, including muscle tissue. Some amino acids may also stimulate the release of insulin, an anabolic hormone, from the pancreas. Such effects could be ergogenic for strength-trained individuals, and unfortunately, arginine, lysine, and ornithine have been advertised as being more powerful than anabolic steroids, potent and illegal drugs discussed in chapter 13, that are used by some athletes to increase muscle mass.

As mentioned previously, amino acid metabolism is complex. For example, arginine supplementation may increase HGH secretion at rest but impair it during exercise. In one study, Collier and others found that oral arginine supplementation (7 grams) increases the secretion of HGH but not as much as a bout of resistance exercise. In a review, Kanaley concluded that arginine alone increases HGH levels by about 100 percent, while exercise can increase HGH levels by 300–500 percent. However, when arginine and exercise are combined, the increase is less than seen with exercise alone, suggesting that arginine supplementation does not augment and may actually decrease the HGH response to exercise. There could be other unwanted effects as well. Research by Bucci and his colleagues supported the effect of ornithine on increasing serum HGH levels. But, using dosages of 40, 100, and 170 mg/kg body weight, only the highest dose of ornithine increased HGH levels, which caused intestinal distress (osmotic diarrhea) in many of the subjects. Thus, its use at this effective dose may be impractical.

Early studies that showed the benefits of these amino acids on body composition or strength have been criticized in the literature for using questionable measurement techniques and statistical procedures. In their review, Chromiak and Antonio indicated that oral doses of arginine, lysine, and ornithine that are great enough to induce significant growth hormone release are likely to cause gastrointestinal discomfort. Moreover, they reported no studies finding that such supplementation augments HGH release; nor do any studies support an ergogenic effect to increase muscle mass and strength to a greater extent than resistance training alone.

Tryptophan Although tryptophan is one of the amino acids that may increase the release of HGH, its theoretical ergogenic effect is based upon another function. A neurotransmitter in the brain, serotonin (5-hydroxytryptamine), is derived from tryptophan. This neurotransmitter may induce sleepiness and elicit a mellow mood, and Segura and Ventura hypothesized that it may help to decrease the perception of pain. They postulated that individuals who are most resistant to pain may be able to delay the onset of fatigue and that tryptophan supplementation, which could increase serotonin levels, might improve exercise performance. As has been seen with other individual amino acids in this section, the theory behind an ergogenic effect often does not translate into an improvement in exercise performance in real life. Tryptophan supplementation does not appear to be an effective ergogenic in either short-term or prolonged exercise tasks in exercise-trained individuals, a finding supported in the review by Anton Wagenmakers.

Key Concepts

- Consumption of adequate protein or protein/carbohydrate preparations in the postexercise period may may provide a milieu conducive to muscle protein anabolism. For optimal performance, athletes should ingest meals with adequate protein and carbohydrate, throughout the day, including the postexercise period.
- Central fatigue during prolonged aerobic exercise is hypothesized to occur when BCAA levels are decreased. BCAAs normally compete with free-tryptophan (fTRP) for entry into the brain. fTRP increases serotonin production, which is believed to induce fatigue. Thus, an increased fTRP:BCAA ratio may induce central fatigue. Benefits of BCAAs on central fatigue have not been consistently shown.
- Although several interesting hypotheses have been proposed, individual amino acid supplements are not currently considered to be effective as a means of improving physical performance and may create adverse imbalances in amino acid concentrations and may create adverse imbalances in amino acid concentrations.
- Creatine, beta-alanine, and nitrate supplements are well researched and, when properly used, have been shown to enhance exercise performance under certain conditions. More research is needed on taurine, gelatin/collagen, and HMB supplements before supplementation could be recommended to athletes.
- Other protein-related supplements, such as inosine, bovine colostrum, chondroitin, and glucosamine, BCAAs, and individual amino acid supplements have little to no evidence of efficacy, or data that do not support an ergogenic effect.

Check for Yourself

- Using **table 6.8** as a guide, check an online store that sells sports supplements and check the cost of various protein supplements. Calculate the cost per serving and the average cost per 8 grams of protein. Compare to the table.

Dietary Protein: Health Implications

The Acceptable Macronutrient Distribution Range for individuals has been set for protein based on evidence from interventional trials, with support of epidemiological evidence, to suggest a role in the prevention of increased risk of chronic diseases and based on ensuring sufficient intakes of essential nutrients. As noted previously, the AMDR for protein is 10-35 percent of energy, and slightly lower percentages were set for children (5-20 percent for young and 10-30 percent for older children). There may be several possible adverse health effects from consuming a diet which is consistently outside the AMDR for protein, either too low or too high. Excess intake of individual amino acids may also pose health risks.

Does a deficiency of dietary protein pose any health risks?

A short-term protein deficiency (several days) is not likely to cause any serious health problems, mainly because body metabolism adjusts to conserve its protein stores. However, because protein is the source of the essential amino acids, and because protein-rich foods also contain an abundance of essential vitamins and minerals, a prolonged deficiency could be expected to cause serious health problems. Such is the case in certain parts of the world where protein intake is inadequate for political, economic, or other reasons. Protein-kcal malnutrition is one of the major nutritional problems in the world today, particularly for young children. Infections develop because the immune system, which depends on adequate protein, is weakened. Death is common. For children who survive, physical and mental growth may be permanently retarded. Protein deficiency may also occur in individuals who abuse sound nutritional practices, such as drug addicts, chronic alcoholics, and extreme food faddists, but adults are more likely to recover fully with adequate nutrition.

Older adults, those over 65 years, may be more prone to protein undernutrition because they may eat less protein-rich food and may use protein less efficiently. Churchward-Venne, van Loon, and Phillips, protein experts, all report that older adults need a larger dose of protein to maximally increase MPS than young adults. This is even so in the postexercise period, where older adults may need as much as 40 grams to stimulate MPS. Lesourd indicated that protein undernutrition in the elderly may impair immune function, making them more susceptible to infections. Adequate protein also plays an important role in bone development, thereby influencing peak bone mass. Bonjour and others note that low protein intake can be detrimental for both the acquisition of bone mass during growth and its conservation during adulthood. Low protein intake impairs both the production and action of IGF-I, an essential factor for bone longitudinal growth. Protein intake is especially important during childhood, but is very important during adulthood as well. According to a review by Gaffney-Stomberg and colleagues, between 32 percent and 41 percent of women 50 years of age and older consume less than the current RDA of 0.8 grams of good-quality protein per kilogram body weight per day. This is higher than the 22-38 percent of men in the same age group who do not meet the protein RDA. This is of particular concern given the loss of muscle mass, or *sarcopenia,* seen with aging. Genaro and Martini found that sarcopenia in the elderly is associated with decreased metabolic rate, increased risk of falls and fractures, and thus an increased morbidity and loss of independence.

Individuals who are on a low-protein diet plan, or young athletes who are on very low kcal diets to lose weight for such sports as gymnastics, ballet, or wrestling, may experience periods of protein insufficiency. During this time, the individual may be in negative nitrogen balance; that is, more nitrogen is being excreted from the body than is being ingested. Body tissues such as muscles and hemoglobin may be lost, with a possible reduction in strength and endurance capacity. Adequate protein intake is essential for proper physiological functioning and health, both in the inactive and active individual.

Several major health problems associated with excessive weight loss, both in nonathletes and athletes, are related to both energy and protein balance.

Does excessive protein intake pose any health risks?

The potential dangers of excess protein intake have focused on three primary areas: cardiovascular disease, renal function, and bone health. There is no direct effect of protein on cardiovascular disease; however, some protein foods are high in kcal, high in total fat, and/or high in saturated fat. As described in this text, a healthful diet with adequate energy and protein can be consumed with healthy levels of saturated and total fat. Several decades ago, Dr. Barry Brenner hypothesized that higher-protein diets would lead to declines in kidney function. To this day, as noted in reviews by Devries and colleagues and Van Elswyk and others that there are no data to support the theory that high-protein diets in healthy adults cause a deterioration in renal function. In their review, Wallace and Frankenfeld showed that protein intake above the current RDA is beneficial as a preventative measure for hip fractures and bone density loss. Similarly, Shams-White from the National Osteoporosis provided evidence that there are no adverse effects of higher protein intakes, and high protein intake was associated with positive trends in bone density at most bone sites. Thus, protein and protein supplements can be and should be incorporated into a healthful diet for physically active individuals and competitive athletes.

Does the consumption of individual amino acids pose any health risks?

Amino acids do not exist free in foods we eat but are complexed with other amino acids to form protein. Free amino acids have been manufactured to be given to patients intravenously for adequate protein nutrition. They are also used as food additives to enhance the protein quality of foods deficient in specific amino acids and marketed as dietary supplements. The National Academy of Sciences noted no evidence that amino acids derived from usual or even high intakes of protein from foodstuffs present any health risk. Although no UL has been established for amino acids, the academy indicated that the absence of a UL means that caution is warranted in using any single amino acid at levels significantly above that normally found in food. Extreme consumption of individual amino acid supplements may pose a health risk. As dietary supplements, purity and safety are not guaranteed. In 1989, a serious epidemic of eosinophilia-myalgia syndrome (EMS), a neuromuscular disorder characterized by weakness, fever, edema, rashes, bone pain, and other symptoms, was attributed to an L-tryptophan supplement contaminated during manufacturing. A practical concern is that excessive reliance on free-form amino acids, in comparison to dietary protein, may lead to a diet deficient in key vitamins and minerals that are normally found in protein foods, such as iron and zinc in meat, fish, and poultry.

> **Key Concept**
>
> ▶ Dietary deficiencies, as well as dietary excesses, of protein and amino acids may interfere with optimal physiological efficiency, which may lead to impairment of health status. However, dangers of protein intakes that exceed the RDA are not supported by data.

APPLICATION EXERCISE

Obtain a supply of creatine monohydrate—about 100 grams, enough to provide 20 grams a day for 5 days. Measure your weight accurately in the morning after arising and your normal bathing routine, but before eating breakfast. Consume 20 grams of creatine per day for 5 days, taking it in 4 equal doses of 5 grams at breakfast, at lunchtime, late afternoon, and before bed. Weigh yourself again the morning after taking the last dose. Did your body weight change? Record your weight again next week, and the following three weeks. Did your body weight change again? Compare your findings to the text discussion.

Creatine Monohydrate Trial

	Day 1	Day 6	1 Week Later	2 Weeks Later	3 Weeks Later
Morning Weight					

Review Questions—Multiple Choice

1. A high-quality protein is best described as one that _____:
    a. contains 10 grams of protein per 100 grams of food.
    b. contains all of the essential amino acids in the proper amounts and ratio.
    c. contains all of the nonessential amino acids.
    d. contains adequate amounts of glucose for protein sparing.
    e. contains the amino acids leucine, isoleucine, and valine.

2. Which of the following statements involving the interaction of protein and exercise training is false?
    a. Small amounts of protein may be used as an energy source during endurance exercise but usually account for less than 5 percent of the energy cost of the exercise.
    b. Small amounts of protein may be lost in the urine and sweat during exercise.
    c. Resistance weight-training programs usually result in the development of a positive nitrogen balance in most athletes who are attempting to gain body weight in the form of muscle mass.
    d. Although weight lifters and endurance athletes may need slightly more protein than accounted for by the RDA, such increased protein may be obtained readily and more economically through a planned diet.
    e. Research has shown conclusively that all amino acid supplements and other protein supplements will enhance performance in sports.

3. Which of the following has the least amount of dietary protein?
    a. one ounce of chicken breast
    b. one-half cup of baked beans
    c. one slice of whole wheat bread
    d. one orange
    e. one-half glass of skim milk

4. Which of the following statements relative to protein and exercise is false?
    a. Protein may be catabolized during exercise and used as an energy source, but the contribution is less than 10 percent.
    b. Carbohydrate intake may exert a protein-sparing effect during exercise.
    c. Very low levels of protein intake during training may lead to the development of a condition known as sports anemia.
    d. Research has shown that individuals who are training to gain weight need about 6–8 grams of protein per kilogram body weight.
    e. In general, research has shown that protein supplementation above the RDA will not improve physiological performance capacity during aerobic endurance exercise.

5. In the recommendations for a healthy diet from the National Academy of Sciences, what is the Acceptable Macronutrient Distribution Range for protein as a percent of daily energy intake in kcal?
   a. 15-20
   b. 10-35
   c. 4-6
   d. 12-14
   e. 40-65
6. Which of the following statements relative to protein metabolism is false?
   a. Excess protein may be converted to glucose in the body.
   b. The liver is a critical center for the control of amino acid metabolism.
   c. Essential amino acids can be formed in the liver from carbohydrate and nitrogen from nonessential amino acids.
   d. Excess protein may be converted to fat in the body.
   e. Urea is a waste product of protein metabolism.
7. Which is most likely to be a complete, high-quality protein food?
   a. cheddar cheese
   b. peanut butter
   c. green peas
   d. corn
   e. macaroni
8. Supplementation with some amino acids has been theorized to decrease the formation of serotonin in the brain and possibly help delay the onset of central nervous system fatigue in prolonged aerobic endurance exercise. Which amino acids are theorized to do this?
   a. leucine, isoleucine, and valine
   b. arginine, ornithine, and inosine
   c. tryptophan, arginine, and creatine
   d. inosine, creatine, and alanine
   e. asparagine, aspartic acid, and glutamine
9. If an adult weighed 176 pounds, the RDA for protein would be what, in grams?
   a. 176
   b. 140.8
   c. 80
   d. 64
   e. 309.7
10. Research has suggested that creatine supplementation may enhance performance in which of the following types of physical performance tasks?
    a. an all-out power lift in 1 second
    b. high-intensity exercise lasting 6-30 seconds
    c. 10-kilometer race lasting about 30 minutes
    d. marathon running (26.2 miles)
    e. ultramarathons, such as Ironman-type triathlons

Answers to multiple choice questions:
1. b; 2. e; 3. d; 4. d; 5. b; 6. c; 7. a; 8. a; 9. d; 10. b.

Critical Thinking Questions

1. Differentiate between complete and incomplete proteins as related to essential and nonessential amino acids and indicate several specific foods that are considered to contain either complete or incomplete protein.
2. Describe the process of gluconeogenesis from protein.
3. Explain why some scientists recommend that both strength and endurance athletes may need more dietary protein than the RDA. Provide some recommended values and calculate the recommended grams of protein for a 70-kilogram athlete.
4. Explain the central fatigue hypothesis as related to BCAA supplementation for endurance athletes and summarize the research findings as to the related ergogenic efficacy of BCAA supplementation.
5. Discuss how healthful food-based protein choices can play a role in the prevention of heart disease.

References

Books
Houston, M. 2006. *Biochemistry Primer for Exercise Science.* Champaign, IL: Human Kinetics.

National Academy of Sciences. 2005. *Dietary Reference Intakes for Energy, Carbohydrates, Fiber, Fat, Fatty Acids, Cholesterol, Protein and Amino Acids (Macronutrients).* Washington, DC: National Academies Press.

Williams, M., Kreider, R., and Branch, J. 1999. *Creatine: The Power Supplement.* Champaign, IL: Human Kinetics.

Reviews and Specific Studies
Akerström, T., and Pedersen, B. 2007. Strategies to enhance immune function for marathon runners: What can be done? *Sports Medicine* 37:416-19.

American College of Sports Medicine, Academy of Nutrition and Dietetics, Dietitians of Canada. 2016. Joint position statement: Nutrition and athletic performance. *Medicine & Science in Sports & Exercise* 48:543-68.

Antonio, J. et al. 2021. Common questions and misconceptions about creatine supplementation: What does the evidence really show? *Journal of the International Society of Sports Nutrition* 18.

Antonio, J., et al. 2015. A high protein diet (3.4 g/kg/d) combined with a heavy resistance training program improves body composition in healthy trained men and women—A follow-up investigation. *Journal of the International Society of Sport Nutrition* 12:39.

Appel, L., et al. 2005. Effects of protein, monounsaturated fat, and carbohydrate intake on blood pressure and serum lipids: Results of the OmniHeart randomized trial. *JAMA* 294:2455-64.

Aragon, A., and Schoenfeld, B. 2013. Nutrient timing revisited: Is there a post-exercise anabolic window? *Journal of the International Society of Sport Nutrition.* 10:5.

Areta, J., et al. 2013. Timing and distribution of protein ingestion during prolonged recovery from resistance exercise alters myofibrillar protein synthesis. *Journal of Physiology* 591:2319-31.

Atherton, P., and Smith, K. 2012. Muscle protein synthesis in response to nutrition and exercise. *Journal of Physiology* 590:1049-57.

Bailey, S., et al. 2009. Dietary nitrate supplementation reduces the O_2 cost of low-intensity exercise and enhances tolerance to high-intensity exercise in humans. *Journal of Applied Physiology* 107:1144-55.

Betts, J., et al. 2005. Recovery of endurance running capacity: Effect of carbohydrate-protein mixture. *International Journal of Sport Nutrition & Exercise Metabolism* 15:590-609.

Betts, J., et al. 2007. The influence of carbohydrate and protein ingestion during recovery from prolonged exercise on subsequent endurance performance. *Journal of Sports Sciences* 13:1-12.

Branch, J. 2003. Effect of creatine supplementation on body composition and performance: A meta-analysis. *International Journal of Sport Nutrition & Exercise Metabolism* 13:198-226.

Brinkworth, G., et al. 2002. Oral bovine colostrum supplementation enhances buffer capacity, but not rowing performance in elite female rowers. *International Journal of Sport Nutrition & Exercise Metabolism* 12:349-63.

Brinkworth, G., et al. 2004. Effect of bovine colostrum supplementation on the composition of resistance trained and untrained limbs in healthy young men. *European Journal of Applied Physiology* 91:53-60.

Burke, D., et al. 2003. Effect of creatine and weight training on muscle creatine and performance in vegetarians. *Medicine & Science in Sports & Exercise* 36:1946-55.

Campbell, B., et al. 2007. International Society of Sports Nutrition position stand: Protein and exercise. *Journal of the International Society of Sports Nutrition* 4:8.

Candow, D., and Chilibeck, P. 2007. Effect of creatine supplementation during resistance training on muscle accretion in the elderly. *Journal of Nutrition, Health & Aging* 11(2):185-88.

Cermak, N., et al. Protein supplementation augments the adaptive response of skeletal muscle to resistance-type exercise training: A meta-analysis. *American Journal of Clinical Nutrition* 96:1454-64.

Chromiak, J., and Antonio, J. 2002. Use of amino acids as growth-hormone releasing agents by athletes. *Nutrition* 18:657-61.

Churchward-Venne, T., et al. 2013. Role of protein and amino acids in promoting lean mass accretion with resistance exercise and attenuating lean mass loss during energy deficit in humans. *Amino Acids* 45:231-40.

Churchward-Venne, T., et al. 2016. What is the optimal amount of protein to support post-exercise skeletal muscle reconditioning in the older adult? *Sports Medicine* 46:1205-12.

Collier, S., et al. 2006. Oral arginine attenuates the growth hormone response to resistance exercise. *Journal of Applied Physiology* 101:848-52.

Coqueiro, A., et al. 2019. Glutamine as an anti-fatigue amino acid in sports nutrition. *Nutrients* 11:863.

Cribb, P., and Hayes, A. 2006. Effects of supplement timing and resistance exercise on skeletal muscle hypertrophy. *Medicine & Science in Sports & Exercise* 38:1918-25.

Dalbo, V., et al. 2008. Putting to rest the myth of creatine supplementation leading to muscle cramps and dehydration. *British Journal of Sports Medicine* 42:567-73.

Davison, G. 2021. The use of bovine colostrum in sport and exercise. *Nutrients* 13(6), 1789.

Deldicque, L., et al. 2005. Increased IGF mRNA in human skeletal muscle after creatine supplementation. *Medicine & Science in Sports & Exercise* 37:731-36.

Devries, M. et al. 2018. Changes in kidney function do not differ between healthy adults consuming higher-compared with lower-or normal-protein diets: A systematic review and meta-analysis. *Journal of Nutrition* 148:1760-75.

Devries, M., and Phillips, S. 2014. Creatine supplementation during resistance training in older adults—A meta-analysis. *Medicine & Science in Sports & Exercise* 46:1194-203.

Devries, M., and Phillips, S. 2015. Supplemental protein in support of muscle mass and health: Advantage whey. *Journal of Food Science* 80:A8-A15.

Drummond, M., et al. 2009. Nutritional and contractile regulation of human skeletal muscle protein synthesis and mTORC1 signaling. *Journal of Applied Physiology* 106:1374-84.

Elliot, T., et al. 2006. Milk ingestion stimulates net muscle protein synthesis following resistance exercise. *Medicine & Science in Sports & Exercise* 38:667-74.

Engelhardt, M., et al. 1998. Creatine supplementation in endurance sports. *Medicine and Science in Sports and Exercise* 30:1123-9.

Esmarck, B. 2001. Timing of postexercise protein intake is important for muscle hypertrophy with resistance training in elderly humans. *Journal of Physiology* 535:301-11.

Fitts, R., and Metzger, J. 1993. Mechanisms of muscular fatigue. In *Principles of Exercise Biochemistry,* ed. J. Poortmans. Basel, Switzerland: Karger.

Forbes, S. et al. 2022. Effects of creatine supplementation on brain function and health. *Nutrients* 14:921.

Freitas, E., and Katsanos, C. 2022 (Dys)regulation of protein metabolism in skeletal muscle of humans with obesity. *Frontiers in Physiology* 13:843087.

Gaffney-Stomberg, E., et al. 2009. Increasing dietary protein requirements in elderly people for optimal muscle and bone health. *Journal of the American Geriatrics Society* 57:1073-79.

Garcia-Coronado, J. et al. 2019. Effect of collagen supplementation on osteoarthritis symptoms: A meta-analysis of randomized placebo-controlled trials. *International Journal of Orthopedics* 43:531-38.

Genaro, Pde, S., and Martini, L. 2010. Effect of protein intake on bone and muscle mass in the elderly. *Nutrition Reviews* 68:616-30.

Gibala, M. 2001. Regulation of skeletal muscle amino acid metabolism during exercise. *International Journal of Sport Nutrition & Exercise Metabolism* 11:87-108.

Gibala, M. 2002. Dietary protein, amino acid supplements, and recovery from exercise. *Sports Science Exchange* 15(4):1-4.

Gibala, M. 2007. Protein metabolism and endurance exercise. *Sports Medicine* 37:337-40.

Gonzalez, A., and Trexler, E. 2020. Effects of citrulline supplementation on exercise performance in humans: A review of the current literature. *Journal of Strength and Conditioning Research* 34:1480-95.

Gorissen, S. et al. 2016. Ingestion of wheat protein increases in vivo muscle protein synthesis rates in healthy older men in a randomized trial. *Journal of Nutrition* 146:1651-59.

Gough, L. et al. A critical review of citrulline malate supplementation and exercise performance. *European Journal of Applied Physiology* 121:3283-95.

Graham, T., and MacLean, D. 1998. Ammonia and amino acid metabolism in skeletal muscle: Human, rodent and canine models. *Medicine & Science in Sports & Exercise* 30:34-46.

Graham, T., et al. 1997. Effect of endurance training on ammonia and amino acid metabolism in humans. *Medicine & Science in Sports & Exercise* 29:646-53.

Green, A., et al. 1996. Carbohydrate ingestion augments creatine retention during creatine feeding in humans. *Acta Physiologica Scandinavica* 158:195-202.

Green, M., et al. 2008. Carbohydrate-protein drinks do not enhance recovery from exercise-induced muscle injury. *International Journal of Sport Nutrition & Exercise Metabolism* 18:1-18.

Greenhaff, P., et al. 1994. Effect of oral creatine supplementation on skeletal muscle phosphocreatine resynthesis. *American Journal of Physiology* 266:E725-E730.

Gualano, B., et al. 2012. In sickness and in health: The widespread application of creatine supplementation. *Amino Acids* 43 (2):519-29.

Hargreaves, M., and Snow, R. 2001. Amino acids and endurance exercise. *International Journal of Sport Nutrition & Exercise Metabolism* 11:133-45.

Harmon, K. et al. 2021. The application of creatine supplementation in medical rehabilitation. *Nutrients* 13:1825.

Harris, R., et al. 1992. Elevation of creatine in resting and exercised muscle on normal subjects by creatine supplementation. *Clinical Science* 83:367-74.

Harris, R., et al. 1993. The effect of oral creatine supplementation on running performance during maximal short term exercise in man. *Journal of Physiology* 467:74P.

Harris, R., et al. 2006. The absorption of orally supplied beta-alanine and its effect on muscle carnosine synthesis in human vastus lateralis. *Amino Acids* 30:279-89.

Hartman, J., et al. 2007. Consumption of fat-free fluid milk after resistance exercise promotes greater lean mass accretion than does consumption of soy or carbohydrate in young, novice, male weightlifters. *American Journal of Clinical Nutrition* 86:373-81.

Helms, E., et al. 2014. A systematic review of dietary protein during caloric restriction in resistance trained lean athletes: A case for higher intakes. *International Journal of Sport Nutrition and Exercise Metabolism* 24:127-38.

Hernandez, M., et al. 1996. The protein efficiency ratios of 30:70 mixtures of animal: Vegetable protein are similar or higher than those of the animal foods alone. *Journal of Nutrition* 126:574-81.

Hespel, P., et al. 2006. Dietary supplements for football. *Journal of Sports Sciences* 24:749-61.

Hickner, R., et al. 2006. L-citrulline reduces time to exhaustion and insulin response to a graded exercise test. *Medicine & Science in Sports & Exercise* 38:660-66.

Hobson, R., et al. 2012. Effects of beta-alanine supplementation on exercise performance: A meta-analysis. *Amino Acids* 43:25-37.

Hultman, E., et al. 1996. Muscle creatine loading in men. *Journal of Applied Physiology* 81:232-37.

Ivy, J., et al. 2003. Effect of a carbohydrate-protein supplement on endurance performance during exercise of varying intensity. *International Journal of Sport Nutrition & Exercise Metabolism* 13:382-95.

Jäger, R., et al. 2011. Analysis of the efficacy, safety, and regulatory status of novel forms of creatine. *Amino Acids* 40:1369-83.

Jones, A. 2014. Dietary nitrate supplementation and exercise performance. *Sports Medicine* 44:S35-45.

Jones, A. 2022. Dietary nitrate and exercise performance: New strings to the beetroot bow. *Gatorade Sports Science Exchange* 35:1-5.

Kanaley, J. 2008. Growth hormone, arginine and exercise. *Current Opinion in Clinical Nutrition & Metabolic Care* 11:50-54.

Karp, J., et al. 2006. Chocolate milk as a postexercise recovery aid. *International Journal of Sport Nutrition & Exercise Metabolism* 16:78-91.

Kerksick, C. M., et al. 2017. International Society of Sports Nutrition position stand: Nutrient timing. *Journal of the International Society of Sport Nutrition* 14:33.

Kerksick, C., et al. 2006. The effects of protein and amino acid supplementation on performance and training adaptations during ten weeks of resistance training. *Journal of Strength & Conditioning Research* 20:643-53.

Khatri, M. et al. 2021. The effects of collagen peptide supplementation on body composition, collagen synthesis, and recovery from joint injury and exercise: A systematic review. *Amino Acids* 53:1493-506.

Kilduff, L., et al. 2004. The effects of creatine supplementation on cardiovascular, metabolic, and thermoregulatory responses during exercise in the heat in endurance-trained humans. *International Journal of Sport Nutrition & Exercise Metabolism* 14:443-60.

Kirkendall, D. 2004. Creatine, carbs, and fluids: How important in soccer nutrition? *Sports Science Exchange* 17(3):1-6.

Koopman, E., et al. 2004. Combined ingestion of protein and carbohydrate improves protein balance during ultra-endurance exercise. *American Journal of Physiology-Endocrinology & Metabolism* 287: E712-20.

Koopman, R. 2007. Role of amino acids and peptides in the molecular signaling in skeletal muscle after resistance exercise. *International Journal of Sport Nutrition & Exercise Metabolism* 17:S47-S57.

Koopman, R., et al. 2005. Combined ingestion of protein and free leucine with carbohydrate increases postexercise muscle protein synthesis in vivo in male subjects. *American Journal of Physiology, Endocrinology & Metabolism* 288:E645-53.

Kreider, R. et al. 2022. Bioavailability, efficacy, safety, and regulatory status of creatine and related compounds: A critical review. *Nutrients* 14:1035.

Kreider, R., et al. 2010. ISSN exercise & sport nutrition review: Research and recommendations. *Journal of the International Society of Sports Nutrition* 7:7.

Kurtz, J. et al. 2021. Taurine in sports and exercise. *Journal of the International Society of Sports Nutrition* 18:39.

Lanhers, C., et al. 2015. Creatine supplementation and lower limb strength performance: A systematic review and meta-analyses. *Sports Medicine* 45:1285-1294.

Lanhers, C., et al. 2017. Creatine supplementation and upper limb strength performance: a systematic review and meta-analysis. *Sports Medicine* 47:163-173.

Larsen, F. et al. 2007. Effects of dietary nitrate on oxygen cost during exercise. *Acta Physiologica* 191:59-66.

Leidy, H., et al. 2015. The role of protein in weight loss and maintenance. *American Journal of Clinical Nutrition* 101:1320S-29S.

Lemon, P. 1998. Effects of exercise on dietary protein requirements. *International Journal of Sport Nutrition,* 8:426-47.

Lemon, P. 2000. Effects of exercise on protein metabolism. In *Nutrition in Sport,* ed. R. J. Maughan. Oxford: Blackwell Science.

Lemon, P. 2000. Protein metabolism during exercise. In *Exercise and Sport Science,* eds. W. E. Garrett and D. T. Kirkendall. Philadelphia, PA: Lippincott Williams & Wilkins.

Lemon, P., et al. 1992. Protein requirements and muscle mass/strength changes during intensive training in novice bodybuilders. *Journal of Applied Physiology* 73:767-75.

Lesourd, B. 1995. Protein undernutrition as the major cause of decreased immune function in the elderly: Clinical and functional implications. *Nutrition Reviews* 53:S86-S94.

Levenhagen, D., et al. 2001. Postexercise nutrient intake timing in humans is critical to recovery of leg glucose and protein homeostasis. *American Journal of Physiology. Endocrinology & Metabolism* 280:E982-93.

Levenhagen, D., et al. 2002. Postexercise protein intake enhances whole-body and leg protein accretion in humans. *Medicine & Science in Sports & Exercise* 34:828-37.

Liappis, N., et al. 1979. Quantitative study of free amino acids in human eccrine sweat excreted from the forearms of healthy trained and untrained men during exercise. *European Journal of Applied Physiology* 42:227–34.

Lim, M. et al. 2021. Animal protein versus plant protein in supporting lean mass and muscle strength: A systematic review and meta-analysis of randomized controlled trials. *Nutrients* 13:661.

Longland, T., et al. 2016. Higher compared with lower dietary protein during an energy deficit combined with intense exercise promotes greater lean mass gain and fat mass loss: A randomized trial. *American Journal of Clinical Nutrition* 103:738–46.

Lopez, R., et al. 2009. Does creatine supplementation hinder exercise heat tolerance or hydration status? A systematic review with meta-analyses. *Journal of Athletic Training* 44:215–23.

Lukaszuk, J., et al. 2002. Effect of creatine supplementation and a lactoovovegetarian diet on muscle creatine concentration. *International Journal of Sport Nutrition & Exercise Metabolism* 12:336–48.

Mayhew, D., et al. 2002. Effects of long-term creatine supplementation on liver and kidney functions in American college football players. *International Journal of Sport Nutrition and Exercise Metabolism* 12:453–60.

McGlory, C. 2017. Skeletal muscle and resistance exercise training; the role of protein synthesis in recovery and remodeling. *Journal of Applied Physiology* 122:541–48.

McLellan, T., et al. 2014. Effects of protein in combination with carbohydrate supplements on acute or repeat endurance exercise performance: A systematic review. *Sports Medicine* 44:535–50.

McNaughton, L., et al. 1999. Inosine supplementation has no effect on aerobic or anaerobic cycling performance. *International Journal of Sport Nutrition* 9:333–44.

Meeusen, R., and Wilson, P. 2007. Amino acids and the brain: Do they play a role in "central fatigue"? *International Journal of Sport Nutrition & Exercise Metabolism* 17:S37–S46.

Meeusen, R., et al. 2006. The brain and fatigue: New opportunities for nutritional interventions? *Journal of Sports Sciences* 24:773–82.

Melina, V., et al. 2016. Position of the academy of nutrition and dietetics: Vegetarian diets. *Journal of the Academy of Nutrition and Dietetics* 116:1970–80.

Meng, Z. et al. 2022. Efficacy and safety of the combination of glucosamine and chondroitin for knee osteoarthritis: A systematic review and meta-analysis. *Archives of Orthopedic Trauma and Surgery* Jan 13.

Millard-Stafford, M., et al. 2005. Recovery from run training: Efficacy of a carbohydrate-protein beverage. *International Journal of Sport Nutrition & Exercise Metabolism* 15:610–24.

Miller, B. 2007. Human muscle protein synthesis after physical activity and feeding. *Exercise & Sport Sciences Reviews* 35:50–55.

Mitchell, W., et al. 2015. A dose- rather than delivery profile-dependent mechanism regulates the "muscle-full" effect in response to oral essential amino acid intake in young men. *Journal of Nutrition* 145:207–14.

Molfino, A., et al. 2013. Beta-hydroxy-beta-methylbutyrate supplementation in health and disease: A systematic review of randomized trials. *Amino Acids* 45:1273–92.

Moore, D., et al. 2009. Ingested protein dose response of muscle and albumin protein synthesis after resistance exercise in young men. *American Journal of Clinical Nutrition* 89:161–68.

Moore, D., et al. 2014. Beyond muscle hypertrophy: Why dietary protein is important for endurance athletes. *Applied Physiology Nutrition and Metabolism* 39:987–97.

Morton, R., et al. 2017. A systematic review, meta-analysis and meta-regression of the effect of protein supplementation on resistance training-induced gains in muscle mass and strength in healthy adults. *British Journal of Sports Medicine* July 11. epub.

Newsholme, E., and Blomstrand, E. 1996. The plasma level of some amino acids and physical and mental fatigue. *Experientia* 52:413–15.

Newsholme, E., and Castell, L. 2000. Amino acids, fatigue and immunodepression in exercise. In *Nutrition in Sport*, ed. R. J. Maughan. Oxford: Blackwell Science.

Newsholme, E., et al. 1992. Physical and mental fatigue: Metabolic mechanisms and importance of plasma amino acids. *British Medical Bulletin* 48:477–95.

Nieman, D. 2001. Exercise immunology: Nutrition countermeasures. *Canadian Journal of Applied Physiology* 26:S45–55.

Nissen, S., and Sharp, R. 2003. Effect of dietary supplements on lean mass and strength gains with resistance exercise: A meta-analysis. *Journal of Applied Physiology* 94:651–59.

Olsen, S., et al. 2006. Creatine supplementation augments the increase in satellite cell and myonuclei number in human skeletal muscle induced by strength training. *Journal of Physiology* 573:525–34.

Ostojic, S., et al. 2007. Glucosamine administration in athletes: Effects on recovery of acute knee injury. *Research in Sports Medicine* 15:113–24.

Paddon-Jones, D., et al. 2008. Protein, weight management, and satiety. *American Journal of Clinical Nutrition* 87:1558S–1561S.

Parkhouse, W. 1988. Regulation of skeletal muscle myofibrillar protein degradation: Relationships to fatigue and exercise. *International Journal of Biochemistry* 20:769–75.

Pasiakos, S., et al. 2014. Effects of protein supplements on muscle damage, soreness and recovery of muscle function and physical performance: A systematic review. *Sports Medicine* 44:655–70.

Pasiakos, S., et al. 2015. The effects of protein supplements on muscle mass, strength, and aerobic and anaerobic power in healthy adults: A systematic review. *Sports Medicine* 45:111–31.

Persky, A. M., and Rawson, E. S. 2007. Safety of creatine supplementation in health and disease. In *Creatine and Creatine Kinase in Health and Disease,* eds. Gajja J. Salomons and Markus Wyss. Dordrecht, the Netherlands: Springer.

Phillips, G. 2007. Glutamine: The nonessential amino acid for performance enhancement. *Current Sports Medicine Reports* 6:265–68.

Phillips, S. 2004. Protein requirements and supplementation in strength sports. *Nutrition* 20:689–95.

Phillips, S. 2006. Dietary protein for athletes: From requirements to metabolic advantage. *Applied Physiology, Nutrition & Metabolism* 31:647–54.

Phillips, S. 2013. Defining optimum protein intake for athletes. In *Sports Nutrition. The Encyclopedia of Sports Medicine,* ed. Ronald J. Maughan. Oxford: Wiley-Blackwell.

Phillips, S. 2014. A brief review of higher dietary protein diets in weight loss: A focus on athletes. *Sports Medicine* 44:S149–53.

Phillips, S. et al. 1997. Mixed muscle protein synthesis and breakdown after resistance exercise in humans. *American Journal of Physiology* 273:E99–E107.

Phillips, S., et al. 2007. A critical examination of dietary protein requirements, benefits, and excesses in athletes. *International Journal of Sport Nutrition & Exercise Metabolism* 17:S58–S76.

Piknova, B. et al. 2022. Skeletal muscle nitrate as a regulator of systemic nitric oxide homeostasis. *Exercise and Sport Science Reviews* 50:2–13.

Pinackaers, P. et al. 2021. The anabolic response to plant-based protein ingestion. *Sports Medicine* 51:59–74.

Plotkin, D. et al. 2021. Isolated leucine and branched-chain amino acid supplementation for enhancing muscular strength and hypertrophy: A narrative review. *International Journal of Sport Nutrition and Exercise Metabolism* 31:292–301.

Preen, D., et al. 2003. Creatine supplementation: A comparison of loading and maintenance protocols on creatine uptake by human skeletal muscle. *International Journal of Sport Nutrition & Metabolism* 13:97–111.

Rawson E., et al. 2018. Dietary supplements for health, adaptation, and recovery in athletes. *International Journal of Sport Nutrition & Exercise Metabolism* 28:188–99.

Rawson, E. S. 2018. The safety and efficacy of creatine monohydrate supplementation: What we have learned from the past 25 years of research. *Gatorade Sports Science Exchange* 29:1–6.

Rawson, E. S., and Volek, J. S. 2003. The effects of creatine supplementation and resistance training on muscle strength and weightlifting performance. *Journal of Strength and Conditioning Research* 17(4):822–31.

Rawson, E., and Venezia, A. 2011. Use of creatine in the elderly and evidence for effects on cognitive function in young and old. *Amino Acids* 40:1349–62.

Rawson, E., et al. 2013. Sport-specific nutrition: Practical issues—Strength and power events. In *Sports Nutrition. The Encyclopedia of Sports Medicine,* ed. Ronald J. Maughan. Oxford: Wiley-Blackwell.

Rawson, E., et al. 2017. Perspectives on exertional rhabdomyolysis. *Sports Medicine* 47:33–49.

Reginster, J., et al. 2007. Current role of glucosamine in the treatment of osteoarthritis. *Rheumatology* 46:731–35.

Rennie, M., and Tipton, K. 2000. Protein and amino acid metabolism during and after exercise and the effects of nutrition. *Annual Review of Nutrition* 20:457–83.

Res, P., et al. 2012. Protein ingestion before sleep improves postexercise overnight recovery. *Medicine and Science in Sports and Exercise* 44:1560–9.

Rockwell, J., et al. 2001. Creatine supplementation affects muscle creatine during energy restriction. *Medicine & Science in Sports & Exercise* 33:61–68.

Rodriguez, N., et al. 2007. Dietary protein, endurance exercise, and human skeletal-muscle protein turnover. *Current Opinions in Clinical Nutrition & Metabolic Care* 10:40–45.

Romano-Ely, B., et al. 2006. Effect of an isocaloric carbohydrate-protein-antioxidant drink on cycling performance. *Medicine & Science in Sports & Exercise* 38:1608–16.

Roschel, H. et al. 2021. Creatine supplementation and brain health. *Nutrients* 13:586.

Rowlands, D., and Thomson, J. 2009. Effects of β-hydroxy-β-methylbutyrate supplementation during resistance training on strength, body composition, and muscle damage in trained and untrained young men: A meta-analysis. *Journal of Strength & Conditioning Research* 23(3): 836–46.

Rowlands, D., et al. 2007. Effect of protein-rich feeding on recovery after intense exercise. *International Journal of Sport Nutrition & Exercise Metabolism* 17: 521–43.

Sahlin, K., et al. 1998. Energy supply and muscle fatigue in humans. *Acta Physiologica Scandinavica* 162:261–66.

Sale, C., and Harris, R. 2015. β-alanine and carnosine. In *Nutritional Supplements in Sport, Exercise, and Health. An A-Z Guide,* ed. L. M. Castell, S. J. Stear, and L. M. Burke. New York: Routledge.

Sale, C., et al. 2010. Effect of beta-alanine supplementation on muscle carnosine concentrations and exercise performance. *Amino Acids* 39:321–33.

Sale, C., et al. 2013. Carnosine: From exercise performance to health. *Amino Acids* 44:1477–91.

Sanchez-Martinez, J., et al. 2018. Effects of beta-hydroxy-beta-methyl-butyrate supplementation on strength and body composition in trained and competitive athletes: A meta-analysis of randomized controlled trials. *Journal of Science and Medicine in Sport* 21:727–735.

Saunders, B., et al. 2017. β-alanine supplementation to improve exercise capacity and performance: A systematic review and meta-analysis. *British Journal of Sports Medicine* 51:658–69.

Saunders, M. 2007. Coingestion of carbohydrate-protein during endurance exercise: Influence on performance and recovery. *International Journal of Sport Nutrition & Exercise Metabolism* 17:S87–S103.

Schilling, B., et al. 2001. Creatine supplementation and health variables: A retrospective study. *Medicine & Science in Sports & Exercise* 33:183–88.

Schoenfeld, B, et al. 2013. The effect of protein timing on muscle strength and hypertrophy: A meta-analysis. *Journal of the International Society of Sports Nutrition* 10:53.

Segura, R., and Ventura, J. 1988. Effect of L-tryptophan supplementation on exercise performance. *International Journal of Sports Medicine* 9:301–5.

Senefeld, J. W. et al. 2020. Ergogenic effect of nitrate supplementation: A systematic review and meta-analysis. *Medicine and Science in Sports and Exercise* 52:2250–61.

Shams-White, M. et al. 2017. Dietary protein and bone health: A systematic review and meta-analysis from the National Osteoporosis Foundation. *American Journal of Clinical Nutrition* 105:1528–43.

Shannon, O. et al. 2022. Dietary inorganic nitrate as an ergogenic aid: An expert consensus derived via the modified Delphi Technique. *Sports Medicine* May 23.

Shaw, G., et al. 2017. Vitamin C-enriched gelatin supplementation before intermittent activity augments collagen synthesis. *American Journal of Clinical Nutrition* 105:136–43.

Siervo, M., et al. 2013. Inorganic nitrate and beetroot juice supplementation reduces blood pressure in adults: A systematic review and meta-analysis. *Journal of Nutrition* 143:818–26.

Simental-Mendía, M. et al. 2018. Effect of glucosamine and chondroitin sulfate in symptomatic knee osteoarthritis: A systematic review and meta-analysis of randomized placebo-controlled trials. *Rheumatology International* 38:1413–28.

Skare, O., et al. 2001. Creatine supplementation improves sprint performance in male sprinters. *Scandinavian Journal of Medicine & Science in Sports* 11:96–102.

Slater, G. 2015. β-hydroxy β-methylbutyrate (HMB). *In Nutritional Supplements in Sport, Exercise, and Health. An A-Z Guide,* ed. L. M. Castell, S. J. Stear, and L. M. Burke. New York: Routledge.

Slater, G., and Jenkins, D. 2000. Beta-hydroxy-beta-methylbutyrate (HMB) supplementation and the promotion of muscle growth and strength. *Sports Medicine* 30:105–16.

Snijders, T., et al. 2015. Protein ingestion before sleep increases muscle mass and strength gains during prolonged resistance-type exercise training in healthy young men. *Journal of Nutrition* 145:1178–84.

Snow, R., and Murphy, R. 2003. Factors influencing creatine loading into human skeletal muscle. *Exercise & Sport Sciences Reviews* 31:154–58.

Spriet, L. 2015. Taurine. In *Nutritional Supplements in Sport, Exercise, and Health. An A-Z Guide,* ed. L. M. Castell, S. J. Stear, and L. M. Burke. New York: Routledge.

Spriet, L., and Whitfield, J. 2015. Taurine and skeletal muscle function. *Current Opinion in Clinical Nutrition and Metabolic Care* 8:96–101.

Starling, R., et al. 1996. Effect of inosine supplementation on aerobic and anaerobic cycling performance. *Medicine & Science in Sports & Exercise* 28:1193–98.

Steenge, G. 2000. Protein- and carbohydrate-induced augmentation of whole body creatine retention in humans. *Journal of Applied Physiology* 89:1165–71.

Stegen, S. et al. 2014. β-Alanine dose for maintaining moderately elevated muscle carnosine levels. *Medicine and Science in Sports and Exercise* 46:1426–32.

Sutton, E., et al. 2005. Ingestion of tyrosine: Effects on endurance, muscle strength, and anaerobic performance. *International Journal of Sport Nutrition & Exercise Metabolism* 15:173–85.

Tang, J., et al. 2009. Ingestion of whey hydrolysate, casein, or soy protein isolate: Effects on mixed muscle protein synthesis at rest and following resistance exercise in young men. *Journal of Applied Physiology* 107:987–92.

Tarnopolsky, M. 2004. Protein requirements for endurance athletes. *Nutrition* 20:662–68.

Tarnopolsky, M. 2008. Sex differences in exercise metabolism and the role of 17-beta estradiol. *Medicine & Science in Sports & Exercise* 40:648–54.

Tarnopolsky, M., and Safdar, A. 2008. The potential benefits of creatine and conjugated linoleic acid as adjuncts to resistance training in older adults. *Applied Physiology, Nutrition & Metabolism* 33:213–27.

Tarnopolsky, M., et al. 2001. Creatine-dextrose and protein-dextrose induce similar strength gains during training. *Medicine & Science in Sports & Exercise* 33:2944–52.

Tarnopolsky, M., et al. 2004. Creatine monohydrate enhances strength and body composition in Duchenne muscular dystrophy. *Neurology* 62:1771–77.

Theodorou, A., et al. 2005. Effects of acute creatine loading with or without carbohydrate on repeated bouts of maximal swimming in high-performance swimmers. *Journal of Strength & Conditioning Research* 19:265–69.

Thompson, C., et al. 2017. Influence of dietary nitrate supplementation on physiological and muscle metabolic adaptations to sprint interval training. *Journal of Applied Physiology* 122:642–52.

Tipton, K. 2007. Role of protein and hydrolysates before exercise. *International Journal of Sport Nutrition & Exercise Metabolism* 17:S77–S86.

Tipton, K., and Witard, O. 2007. Protein requirements and recommendations for athletes: Relevance of ivory tower arguments for practical recommendations. *Clinics in Sports Medicine* 26:17–36.

Tipton, K., and Wolfe, R. 1998. Exercise-induced changes in protein metabolism. *Acta Physiologica Scandinavica* 162: 377–87.

Tipton, K., and Wolfe, R. 2004. Protein and amino acids for athletes. *Journal of Sports Sciences* 22:65–79.

Tipton, K., et al. 1999. Postexercise net protein synthesis in human muscle from orally administered amino acids. *American Journal of Physiology* 276:E628–E634.

Tomcik, K., et al. 2018. Effects of creatine and carbohydrate loading on cycling time trial performance. *Medicine & Science in Sports & Exercise* 50:141–50.

Trommelen, J., and van Loon, L. 2016. Pre-sleep protein ingestion to improve the skeletal muscle adaptive response to exercise training. *Nutrients* 8:763.

Van der Avoort, C. et al. 2018. Increasing vegetable intake to obtain the health promoting and ergogenic effects of dietary nitrate. *European Journal of Clinical Nutrition* 72:1485–89.

Van Elswyk, M. et al. 2018. A systematic review of renal health in healthy individuals associated with protein intake above the US Recommended Daily Allowance in randomized controlled trials and observational studies. *Advances in Nutrition* 9:404–18.

van Essen, M., and Gibala, M. 2006. Failure of protein to improve time trial performance when added to a sports drink. *Medicine & Science in Sports & Exercise* 38:1476–83.

van Loon, L. 2007. Application of protein or protein hydrolysates to improve postexercise recovery. *International Journal of Sport Nutrition & Exercise Metabolism* 17:S104–S117.

van Loon, L., et al. 2013. Dietary protein as a trigger for metabolic adaptations. In *Sports Nutrition. The Encyclopedia of Sports Medicine,* ed. Ronald J. Maughan. Oxford: Wiley-Blackwell.

Vandebuerie, F., et al. 1998. Effect of creatine loading on endurance capacity and sprint power in cyclists. *International Journal of Sports Medicine* 19:490–5.

Vanhatalo, A., et al. 2010. Acute and chronic effects of dietary nitrate supplementation on blood pressure and the physiological responses to moderate-intensity and incremental exercise. *American Journal of Physiology & Regulatory, Integrative & Comparative Physiology* 299:R1121–31.

Vlad, S., et al. 2007. Glucosamine for pain in osteoarthritis: Why do trial results differ? *Arthritis & Rheumatism* 56:2267–77.

Volek, J., and Rawson, E. 2004. Scientific basis and practical aspects of creatine supplementation for athletes. *Nutrition* 20:609–14.

Wagenmakers, A. 1999. Amino acid supplements to improve athletic performance. *Current Opinion in Clinical Nutrition & Metabolic Care* 2:539–44.

Wagenmakers, A. 2000. Amino acid metabolism in exercise. In *Nutrition in Sport,* ed. R. J. Maughan. Oxford: Blackwell Science.

Wagenmakers, A. 2006. The metabolic systems: Protein and amino acid metabolism in muscle. In *ACSM's Advanced Exercise Physiology,* ed. C. M. Tipton. Philadelphia, PA: Lippincott Williams & Wilkins.

Walberg-Rankin, J. W. 1999. Role of protein in exercise. *Clinics in Sports Medicine* 18:499–511.

Waldron, M. et al. 2018. The effects of an oral taurine dose and supplementation period on endurance exercise performance in humans: A meta-analysis. *Sports Medicine* 48:1247–53.

Wallace, T. C., and Frankenfeld, C. L. 2017. Dietary protein intake above the current RDA and bone health: A systematic review and meta-analysis. *Journal of the American College of Nutrition* 36:481–96.

Wilkinson, D., et al. 2013. Effects of leucine and its metabolite beta-hydroxy-betamethylbutyrate on human skeletal muscle protein metabolism. *Journal of Physiology* 591:2911–23.

Wilkinson, S., et al. 2007. Consumption of fluid skim milk promotes greater muscle protein accretion after resistance exercise than does consumption of an isonitrogenous and isoenergetic soy-protein beverage. *American Journal of Clinical Nutrition* 85:1031–40.

Williams, M., et al. 1990. Effect of oral inosine supplementation on 3-mile treadmill run performance and VO2 peak. *Medicine & Science in Sports & Exercise* 22:517–22.

Willoughby, D., and Rosene, J. 2001. Effects of oral creatine and resistance training on myosin heavy chain expression. *Medicine & Science in Sports & Exercise* 33:1674–81.

Wilson, J., et al. 2014. The effects of 12 weeks of beta-hydroxy-beta-methylbutyrate free acid supplementation on muscle mass, strength, and power in resistance-trained individuals: A randomized, double-blind, placebo-controlled study. *European Journal of Applied Physiology* 114:1217-27.

Witard, O., et al. 2014. Myofibrillar muscle protein synthesis rates subsequent to a meal in response to increasing doses of whey protein at rest and after resistance exercise. *American Journal of Clinical Nutrition* 99:86-95.

Wolfe, R. 2001. Control of muscle protein breakdown: Effects of activity and nutritional states. *International Journal of Sport Nutrition & Exercise Metabolism* 11:S164-69.

Wolfe, R. 2001. Effects of amino acid intake on anabolic processes. *Canadian Journal of Applied Physiology* 26:S220-27.

Wolfe, R. 2006. Skeletal muscle protein metabolism and resistance exercise. *Journal of Nutrition* 136:525S-28S.

Wolfe, R. 2017. Branched-chain amino acids and muscle protein synthesis in humans: Myth or reality? *Journal of the International Society of Sports Nutrition* 14:30.

Yaguchi, H., et al. 1998. The effect of triathlon on urinary excretion of enzymes and proteins. *International Urology & Nephrology* 30:107-12.

Yoshimura, H. 1970. Anemia during physical training (sports anemia). *Nutrition Review* 28:251-53.

Zanchi, N., et al. 2011. HMB supplementation: Clinical and athletic performance-related effects and mechanisms of action. *Amino Acids* 40:1015-25.

Zhang, M., et al. 2004. Role of taurine supplementation to prevent exercise-induced oxidative stress in healthy young men. *Amino Acids* 26:203-07.

Zhong, L. et al. 2022. A food composition database for assessing nitrate intake from plant-based foods. *Food Chemistry* 394:133411.

Zoeller, R., et al. 2007. Effects of 28 days of beta-alanine and creatine monohydrate supplementation on aerobic power, ventilatory and lactate thresholds, and time to exhaustion. *Amino Acids* 33:505-10.

KEY TERMS

alpha-tocopherol
beta-carotene
bioavailability
biotin
carotenemia
cholecalciferol (vitamin D3)
choline
coenzyme
dietary folate equivalents (DFE)
enzymes
folate (folic acid)
free radicals
gamma-tocopherol
homocysteine
hypervitaminosis
intrinsic factor
menadione
niacin
niacin equivalents (NE)
osteomalacia
pantothenic acid
retinol
retinol activity equivalents (RAE)
retinol equivalents (RE)
riboflavin (vitamin B2)
rickets
thiamin (vitamin B1)
vitamin A
vitamin B6 (pyridoxine)
vitamin B12 (cobalamin)
vitamin C (ascorbic acid)
vitamin D
vitamin E (alpha-tocopherol)
vitamin K
xerophthalmia

Vitamins: Fat-Soluble, Water-Soluble, and Vitamin-Like Compounds

CHAPTER SEVEN

BananaStock/Getty Images

LEARNING OUTCOMES

After studying this chapter, you should be able to:

1. Describe the general characteristics and key functions of vitamins.
2. Identify the fat-soluble vitamins, water-soluble vitamins, and vitamin-like substances.
3. Describe the key functions, food sources, recommended dietary intake (RDA or AI), and health effects of a deficiency or toxicity of each of the vitamins, as well as choline.
4. Explain the potential effects on health and sports performance associated with a deficiency of each essential vitamin, as well as choline.
5. List the antioxidant vitamins and provide recommendations on antioxidant intake for athletes.
6. Summarize the research on how supplementation with each of the vitamins may impact sports performance.

Introduction

BananaStock/Getty Images

Vitamins are a diverse class of 13 known specific nutrients that are involved in almost every metabolic process in the human body. We need only small amounts of vitamins in our daily diet, so they are classified as micronutrients. Nevertheless, they are critical nutrients. Noticeable symptoms of a deficiency may appear in 2–4 weeks for several of the vitamins, and major debilitating diseases may occur with prolonged deficiencies. Vitamin deficiencies are widespread in many developing countries. Indeed, such deficiencies impact a greater number of people in the world than does protein-energy malnutrition. In many parts of the world, enriching grains with certain vitamins and/or fortifying foods with vitamins and minerals may help alleviate many health problems associated with insufficient intake.

Major vitamin deficiencies are rare in industrialized societies because a wide variety of food products are available, many of them fortified with vitamins. For most healthy children and adults, health professionals recommend "food first" as a source of vitamins. However, certain segments of the population may not be obtaining adequate amounts of vitamins from food alone and supplementation may be advantageous. According to the *2020–2025 Dietary Guidelines for Americans*, the vitamins most likely to be insufficient in the diet of Americans are vitamins A, C, D, and E. Vitamin D is considered a "nutrient of public health concern," because low intake is associated with specific health concerns.

In general, most studies reveal that athletes are obtaining adequate vitamins in their diet, probably because of the additional food energy intake associated with the increased energy expenditure of exercise. As well, some athletes take vitamin supplements. For example, Jacobson and others reported that multivitamin/mineral pills were the most commonly used dietary supplement by female NCAA Division I collegiate athletes. However, like members of the general population, many athletes may not be obtaining optimal levels of vitamins from the diet. Moreover, certain athletic groups, particularly those who are on weight-reduction programs to qualify for competition or to enhance performance, may not receive adequate vitamin nutrition. Furthermore, individual athletes in generally well-nourished athletic groups may have a suboptimal vitamin intake.

As noted throughout this chapter, adequate vitamin nutrition is essential for both optimal health and athletic performance. But, if you do not obtain the RDA for a specific vitamin or vitamins, will your health or physical performance suffer? Will vitamin supplements above and beyond the RDA improve your health or performance? A major purpose of this chapter is to provide you with evidence-based recommendations, based on the available research, to help answer these two very general questions.

The first section of this chapter provides some basic facts about the general role of vitamins in the human body. The next two sections cover the fat-soluble and water-soluble vitamins, respectively, with each individual vitamin discussed in terms of its Recommended Dietary Allowance (RDA) or Adequate Intake (AI); good food sources; metabolic functions in the body with particular reference to health and the physically active individual; and the findings of research relative to the impact of deficiencies and supplementation. The fourth section focuses on ergogenic aspects of special vitamin or vitamin-like preparations and provides guidelines for selecting a multivitamin/mineral for those who choose to supplement.

©Tammy Stephenson

247

Basic Facts

What are vitamins and how do they work?

Vitamins are a class of complex organic compounds that are found in small amounts in most foods. They are essential for the optimal functioning of many different physiological processes in the human body but do not provide kcal to the diet. The activity levels of many of these physiological processes are increased greatly during exercise, and an adequate bodily supply of vitamins must be present for these processes to function best.

Coenzyme Functions For the fundamental physiological processes of the body to proceed in an orderly and controlled fashion, a number of complex chemicals known as **enzymes** are necessary to regulate the diverse reactions involved. Hundreds of enzymes have been identified in the human body. Enzymes are necessary to digest our foods, to make our muscles contract, to release the energy stores in our bodies, to help us transport body gases such as carbon dioxide, to help us grow, and to help clot our blood. Because they are capable of inducing changes in other substances without changing themselves, enzymes are also able to serve as catalysts.

Enzymes are chemicals that generally consist of two parts. One part is a protein molecule and to it is attached the second part, a **coenzyme**. For the enzyme to function properly, both parts must be present. The coenzyme often contains a vitamin or some related compound (**figure 7.1**). The enzyme is not used up in the chemical process that it initiates or in which it participates, but enzymes may deteriorate with time. Coenzymes may also be degraded through body metabolism. The B complex vitamins are essential in human nutrition because of their role as coenzymes, and thus a fresh supply of these water-soluble vitamins is constantly needed.

Antioxidant Functions Various oxidative reactions in the body produce substances called free radicals. **Free radicals** are chemical substances that contain a lone, unpaired electron in the outer orbit. The superoxide radical ($O_2^{\cdot -}$) and hydroxyl radical ($OH^{\cdot}$) are true free radicals. Two other related substances, referred to as nonradical oxygen species, are hydrogen peroxide (H_2O_2) and singlet oxygen ($^1O_2$). These substances are known as reactive oxygen species (ROS) and, when nitrogen is involved, are known as reactive oxygen/nitrogen species (RONS).

Free radicals are unstable compounds that possess an unbalanced magnetic field that affects molecular structure and chemical reactions in the body. As well, free radicals may be very reactive with body tissues.

Formation of free radicals, including superoxide anions and hydrogen peroxide, during oxidative processes is essential to normal cellular function, such as gene expression and muscle contractile force. However, although oxidative processes are essential to life, some oxidations may cause cellular damage by oxidation of unsaturated fats in cellular and subcellular membranes. Free radicals may cause such undesirable oxidation, thus damaging DNA, lipids, proteins, and other molecules. The excessive production of free radicals has been associated with an increased risk of several forms of chronic disease, including cardiovascular disease, cancer, and Alzheimer's disease.

Fortunately, although free radicals are formed naturally in the body, body cells produce a number of antioxidant enzymes. Superoxide dismutase, glutathione peroxidase, and catalase are enzymes that help neutralize free radicals and prevent cellular damage. To function properly, these enzymes, often referred to as free radical–scavenging enzymes, must contain certain nutrients such as copper, zinc, and selenium. Comparable to these enzymes, as depicted in **figure 7.2**, vitamins A (as beta-carotene), C, and E also possess antioxidant properties. The effects of these vitamins on health and physical activity have been well-researched and are discussed at appropriate points later in this chapter.

Hormone Functions Although vitamin D exists in vitamin form, it undergoes several conversions in the body and, in its active form, functions as a hormone. After being activated in the liver and kidney, vitamin D circulates in the blood like other hormones and exerts its functions on various tissues to promote bone metabolism. Other vitamins, such as A and K, may be produced in the liver and intestines, respectively, and exert functions in other parts of the body. Some vitamins may be critical in the formation of various hormones—such as the role vitamin C plays in the formation of epinephrine—but are not classified as hormones. Only vitamin D is assigned hormonal status in its active form.

What vitamins are essential to human nutrition?

An *essential vitamin* is one that cannot be synthesized in the body in sufficient quantity, causes deficiency symptoms when dietary intake is inadequate, and alleviates deficiency symptoms when added back

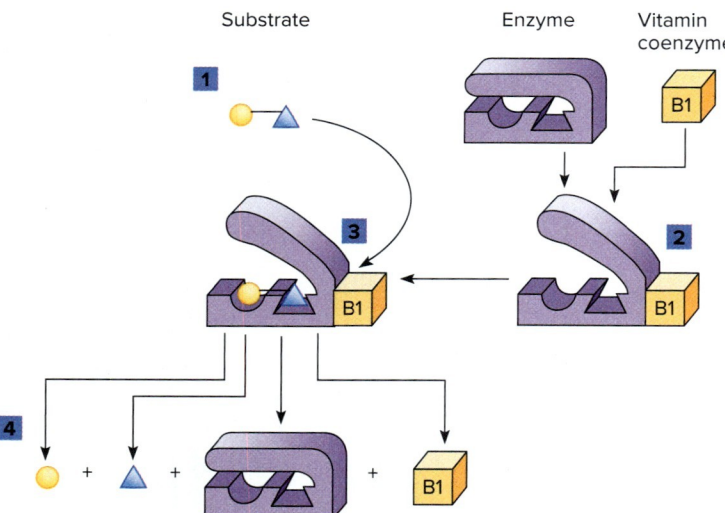

FIGURE 7.1 Role of vitamins as coenzymes. (1) Substrates, such as pyruvate, need enzymes to be converted into more usable compounds. However, many enzymes must be activated before a reaction occurs. Note that the enzyme is in a closed position. (2) An enzyme and a vitamin coenzyme (B1) combine to form an activated complex, in essence opening up the enzyme. (3) The open, activated enzyme accepts the substrate and (4) splits it into two compounds while releasing the enzyme and coenzyme.

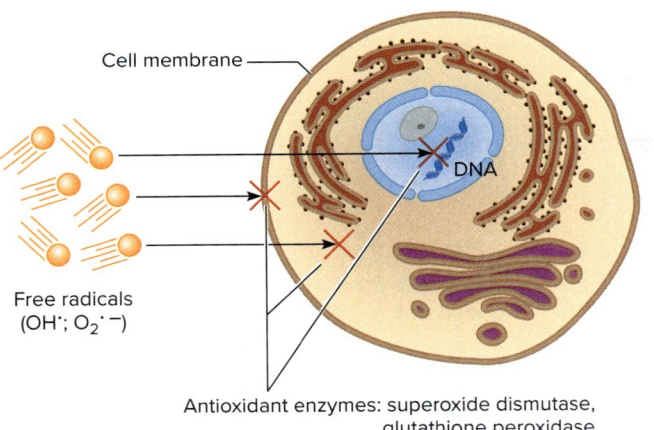

FIGURE 7.2 The antioxidant role of vitamins. To protect against the destructive nature of free radicals, such as hydroxyl radicals and superoxides, cells contain a number of different enzymes (superoxide dismutase, glutathione peroxidase, catalase) to help neutralize the free radicals. By doing so, this helps to prevent disintegration of cell membranes or the genetic material within the cell. In addition, vitamins A (as beta-carotene), C, and E also serve as antioxidants.

to the diet. At present, the human body is known to need an adequate supply of 13 different vitamins. Vitamin-like substances, such as choline, are needed for various physiological functions, but the body can make these substances. The *Training Table* in this section shows the classification of each vitamin and vitamin-like substance.

A well-balanced diet will satisfy the vitamin requirements of most individuals. Four of these vitamins are soluble in fat and are obtained primarily from foods containing fats and oils in our diet, while the other nine, water-soluble, vitamins are distributed rather widely in a variety of foods. Although most vitamins must be obtained from the food we eat, several of them may be formed in the body from other ingested nutrients, by the action of ultraviolet rays from sunlight on our skin, or by the activity of some intestinal bacteria.

Table 7.1 presents an overview of the 13 essential vitamins and choline with commonly used, interchangeable synonyms, major food sources, major functions in the body, and symptoms associated with deficiencies or excessive consumption. RDA and AI values for other age groups may be found in DRI tables available at www.nal.usda.gov/fnic/dri-tables-and-application-reports. The health and physical performance effects of the essential vitamins and selected vitamin-like substances are covered in the following sections.

In general, how do deficiencies or excesses of vitamins influence health or physical performance?

Whether or not a vitamin deficiency affects one's health or physical performance may depend on the magnitude of the deficiency. Four stages of vitamin deficiency associated with the duration of undernourishment and inadequate vitamin intake have been described. These same four stages may apply to mineral deficiency diseases discussed in the next chapter.

1. *A preliminary stage* is associated with inadequate amount or availability of the vitamin in the diet. For example, a drastic change in the diet may influence vitamin **bioavailability** (the amount of a nutrient that the body absorbs), whereas pregnancy may increase the need for several vitamins.
2. *Biochemical deficiency.* In this stage, the body's pool of the vitamin is decreased. For a number of vitamins, biochemical deficiency can be identified by blood or tissue tests. As an example, deficiencies of riboflavin may be detected by the activity of an enzyme in the red blood cells.
3. *Physiological deficiency* is associated with the appearance of unspecific symptoms such as loss of appetite, weakness, or physical fatigue.

    These first three stages are known as latent or marginal vitamin deficiency, or subclinical malnutrition. Whether or not these stages impair physical performance may depend upon the nature of the sport, but weakness or physical fatigue would certainly be counterproductive to optimal performance.
4. *Clinically manifest vitamin deficiency.* In this final stage, specific clinical symptoms are observed. For example, anemia is a clinical symptom associated with a deficiency of several vitamins, such as folic acid, vitamin B6, and/or vitamin B12. Both health and performance would also be adversely affected with a clinically manifest deficiency of vitamin D.

How are vitamin needs determined?

In the past, the recommended intakes for vitamins were established to prevent vitamin-deficiency diseases. However, recommendations today incorporate the role of vitamins in health promotion

Training Table

Vitamins are classified based on their solubility in fat versus water—fat-soluble or water-soluble. Vitamin-like substances are similar to vitamins, but the body can synthesize them.

Fat-soluble vitamins	Water-soluble vitamins	Vitamin-like substances
Vitamin A	B-Complex Vitamins:	Carnitine
Vitamin D	Thiamin (vitamin B1)	Choline
Vitamin E	Riboflavin (vitamin B2)	Inositol
Vitamin K	Niacin (vitamin B3)	Lipoic acid
	Vitamin B6	Taurine
	Vitamin B12	
	Folate	
	Biotin	
	Pantothenic acid	
	Vitamin C	

TABLE 7.1 Essential vitamins*

Vitamin name (other terms)	RDA or AI for adults age 19–50*	Major sources
Fat-soluble vitamins		
Vitamin A (retinol: provitamin carotenoids)	RDA: 900 mcg RAE ♂ 700 mcg RAE ♀ (RAE = retinol activity equivalents)	Retinol in animal foods: liver, fortified cow's milk, cheese; carotenoids in plant foods: carrots, green leafy vegetables, sweet potatoes
Vitamin D (cholecalciferol)	RDA: 600 IU or 15 mcg	Vitamin D-fortified foods like dairy products, fish oils; action of sunlight on the skin
Vitamin E (tocopherol)	RDA: 15 mg alpha-tocopherol	Vegetable oils, products made from vegetable oil, certain fruits and vegetables, wheat germ, nuts and seeds, whole-grain products, and egg yolks
Vitamin K (phylloquinone, menoquinone)	AI: 120 mcg ♂ 90 mcg ♀	Dark green leafy vegetables, canola and soybean oils; formation in the human intestine by bacteria
Water-soluble vitamins		
Thiamin (vitamin B1)	RDA: 1.2 mg ♂ 1.1 mg ♀	Ham, pork, lean meat, whole-grain products, enriched breads and cereals, and legumes
Riboflavin (vitamin B2)	RDA: 1.3 mg ♂ 1.1 mg ♀	Cow's milk and dairy products, meat, eggs, enriched grain products, green leafy vegetables, beans
Niacin (nicotinamide, nicotinic acid, vitamin B3)	RDA: 16 mg ♂ 14 mg ♀	Lean meats, fish, poultry, whole-grain products, beans; may be formed in the body from tryptophan, an essential amino acid
Vitamin B6 (pyridoxal, pyridoxine, and pyridoxamine)	RDA: 1.3 mg	Protein foods: lean meats, fish, poultry, legumes; green leafy vegetables, baked potatoes, bananas
Vitamin B12 (cobalamin; cyanocobalamin)	RDA: 2.4 mcg	Animal foods only: meat, fish, poultry, cow's milk, and eggs
Folate (folic acid)	RDA: 400 mcg (DFE) (DFE = dietary folate equivalents)	Dark green leafy vegetables, legumes, nuts, orange juice, and enriched breads and cereals
Biotin	AI: 30 mcg	Meats, legumes, cow's milk, egg yolk, whole-grain products, and most vegetables
Pantothenic acid	AI: 5 mg	Beef and pork liver, lean meats, cow's milk, eggs, legumes, whole-grain products, and mushrooms
Vitamin C (ascorbic acid)	RDA: 90 mg ♂ 75 mg ♀ (nonsmokers)	Citrus fruits, green leafy vegetables, broccoli, peppers, strawberries, and potatoes
Choline**	AI: 550 mg ♂ 425 mg ♀	Cow's milk, meats, eggs, peanuts; found in most foods as part of cell membranes

(Continued)

TABLE 7.1   Essential vitamins* (Continued)

Major functions in the body	Deficiency symptoms	Symptoms of excessive consumption*
Fat-soluble vitamins		
Maintains epithelial tissue in skin and mucous membranes; forms visual purple for night vision; promotes bone development	Night blindness, intestinal infections, impaired growth, and xerophthalmia	UL is 3 mg/day. Nausea, headache, fatigue, liver and spleen damage, skin peeling, and pain in the joints
Acts as a hormone to increase intestinal absorption of calcium and promote bone and tooth formation	Rare; rickets in children and osteomalacia in adults	UL is 4,000 IU, or 100 mcg/day. Loss of appetite, nausea, irritability, joint pain, calcium deposits in soft tissues such as the kidney
Functions as an antioxidant to protect cell membranes from destruction by oxidation	Extremely rare; disruption of red blood cell membranes; anemia	UL is 1,000 mg/day. General lack of toxicity with doses up to 400 mg. Some reports of headache, fatigue, or diarrhea with megadoses
Essential for blood coagulation processes	Increased bleeding and hemorrhage	No UL set. Possible clot formation (thrombosis), vomiting
Water-soluble vitamins		
Serves as a coenzyme for energy production from carbohydrate; essential for normal functioning of the central nervous system	Poor appetite, apathy, mental depression, pain in calf muscles, and beriberi	No UL set. General lack of toxicity
Functions as a coenzyme involved in energy production from carbohydrates and fats; maintenance of healthy skin	Dermatitis, cracks at the corners of the mouth, sores on the tongue, and damage to the cornea	No UL set. General lack of toxicity
Functions as a coenzyme for the aerobic and anaerobic production of energy from carbohydrate; helps synthesize fat and blocks release of FFA; needed for healthy skin	Loss of appetite, weakness, skin lesions, gastrointestinal problems, and pellagra	UL is 35 mg/day. Nicotinic acid causes headache, nausea, burning and itching skin, flushing of face, and liver damage
Functions as a coenzyme in protein metabolism; necessary for formation of hemoglobin and red blood cells; needed for glycogenolysis and gluconeogenesis	Nervous irritability, convulsions, dermatitis, sores on tongue, and anemia	UL is 100 mg/day. Loss of nerve sensation, impaired gait
Functions as a coenzyme for formation of DNA, RBC development, and maintenance of nerve tissue	Pernicious anemia, nerve damage resulting in paralysis	No UL set. General lack of toxicity
Functions as coenzyme for DNA formation and RBC development	Fatigue, gastrointestinal disorders, diarrhea, anemia, neural tube defects in newborns	UL is 1,000 mcg/day. May prevent detection of pernicious anemia caused by B12 deficiency
Functions as coenzyme in the metabolism of carbohydrates, fats, and protein	Rare; may be caused by excessive intake of raw egg whites: fatigue, nausea, and skin rashes	No UL set. General lack of toxicity
Functions as part of coenzyme A in energy metabolism	Rare; produced only clinically: fatigue, nausea, loss of appetite, and mental depression	No UL set. General lack of toxicity
Forms collagen essential for connective tissue development; aids in absorption of iron; helps form epinephrine; serves as antioxidant	Weakness, rough skin, slow wound healing, bleeding gums, anemia, and scurvy	UL is 2,000 mg/day. Diarrhea, possible kidney stones, and rebound scurvy
Functions as a precursor for lecithin, a phospholipid in cell membranes	Rare; liver damage	UL is 3.5 g/day. May lead to fishy body odor, gastrointestinal distress, and vomiting, low blood pressure

*RDA, AI, and UL values for all age groups may be found at www.nal.usda.gov/fnic/dri-tables-and-application-reports.
**Not classified as a vitamin.

and chronic disease prevention. For example, the original RDA for vitamin C was based on the amount of vitamin C needed to prevent scurvy, the deficiency disease associated with consuming insufficient vitamin C. Current recommendations for vitamin C are based on how much vitamin C is needed to prevent scurvy, but also how much promotes optimal immune health and prevents cancer and other forms of chronic disease.

In general, it is very difficult to obtain excessive amounts of vitamins through the diet to the point that health or physical performance is impaired. Even when supplements are taken, the body may excrete some vitamins, keeping body functions normal. However, overconsumption of some vitamins may induce **hypervitaminosis**, a condition in which a vitamin may function comparable to a drug, not a nutrient, and induce toxic reactions. An excessive intake of vitamin supplements may be a common cause of hypervitaminosis, but some dietary practices, such as overconsuming vitamin-fortified foods, may also lead to excessive vitamin intake. Tolerable Upper Intake Levels (ULs) have been established for most vitamins and are noted in **table 7.1**.

Key Concepts

- Vitamins are complex organic compounds that function in the body in a variety of ways. Some act as coenzymes to help regulate metabolic processes; others are antioxidants that protect cell membranes; and vitamin D is classified as a hormone. Vitamins do not provide kcal, but they do help regulate energy processes in the body.
- There may be four stages in a vitamin deficiency: the preliminary stage, the biochemical deficiency stage, the physiological deficiency stage, and the clinically manifest deficiency stage.
- Recommended intakes for vitamins have been established to determine the amount of the vitamin optimal for health. This includes both the amount of the vitamin necessary to prevent a deficiency disease, but also to prevent chronic disease.

Check for Yourself

- Look at the label for a multivitamin supplement and compare the amount of each vitamin provided to your RDA/AI and UL for that vitamin. How do they compare?

Fat-Soluble Vitamins

The four fat-soluble vitamins are A, D, E, and K. Because they are soluble in fat, but not in water, dietary sources often include foods that have some fat content. The body may contain appreciable stores of each fat-soluble vitamin, and several of them may be manufactured by the body, making deficiencies less common than for most of the water-soluble vitamins. On the other hand, excessive intake may be toxic.

Vitamin A (retinol)

Vitamin A is a fat-soluble, unsaturated alcohol. The physiologically active form of vitamin A is known as **retinol**. The human body is capable of forming retinol from provitamins known as carotenoids, primarily **beta-carotene**. Both preformed vitamin A, or retinol, and carotenoids are found in the foods we eat.

DRI The RDA for vitamin A may be obtained by consuming preformed retinol, beta-carotene and other carotenoids, or a combination of the two. The RDA may be expressed in several ways, including as **retinol equivalents (RE)**, **retinol activity equivalents (RAE)** as a combination of retinol and carotenoids, or as international units. In brief, 1 RAE equals 1 mcg of retinol, or 12 mcg of beta-carotene, or about 3.3 IU. The RDA is 900 mcg RAE, or 3,000 IU, for adult males and 700 mcg RAE, or 2,300 IU, for adult females. The UL for adults is 3 mg RAE/day from retinol, or 10,000 IU. No UL has been established for carotenoids.

Food Sources Preformed vitamin A is found in substantial amounts in some animal foods such as liver, butter, cheese, egg yolks, fish liver oils, and fortified cow's milk. Provitamin A, as beta-carotene, is found in dark-green leafy and yellow-orange vegetables, as well as in some fruits such as oranges, limes, pineapples, prunes, and cantaloupes. One glass of cow's milk provides about 15 percent of the RDA, while one medium carrot will supply nearly 200 percent and a serving of liver (3 oz) a whopping 1,000 percent or more of the RDA.

Pixtal/age fotostock

Major Functions Vitamin A is essential for maintenance of the epithelial cells, those cells covering the outside of the body and lining the body cavities. It is also essential for proper visual function, such as night vision and peripheral vision. Vitamin A has a variety of other physiological roles in the body that are not well understood, although it is considered essential in proper bone development and for maintaining optimal function of the immune system. Beta-carotene functions as an antioxidant and, in this role, may be beneficial in the prevention and treatment of chronic disease.

Deficiency: Health and Physical Performance Vitamin A is stored in the body in relatively large amounts and, therefore, deficiencies are rare. However, an inadequate intake of vitamin A could have serious health implications if prolonged. The gradual loss of night vision is one of the first symptoms of vitamin A deficiency. Other symptoms of mild deficiencies include increased susceptibility to infection and skin lesions. Epidemiological research also has suggested that a deficient intake of beta-carotene could predispose the individual to the development of cancer in the epithelial tissues such as the skin, lungs, breasts, and intestinal lining. Although severe deficiencies are not common in industrialized nations, they do occur in some parts of the world and lead to blindness through destruction of the cornea of the eye, a condition known as **xerophthalmia**. Vitamin A deficiency has been associated with higher mortality rates in children in developing countries, with several million childhood deaths annually. Some, but not all, studies have shown that vitamin A supplementation to such children may decrease the death rate, possibly by strengthening the immune system.

Theoretically, vitamin A deficiency could affect physical performance. Some investigators have suggested that a deficiency may impair the process of gluconeogenesis in the liver, which may be an important consideration for the endurance athlete in the later stages of competition. Others have implied a reduction in the synthesis of muscle protein and impaired vision, which could negatively affect strength athletes or those involved in sports requiring eye alertness. Very little research is available to support these theoretical views.

Supplementation: Health and Physical Performance Vitamin A in supplements can come from retinol (vitamin A palmitate or acetate) or beta-carotene, or both. Check the label; it may or may not list the separate components. The UL for adults is set at 3,000 mcg RAE, or 10,000 IU. The UL applies only to preformed vitamin A, or retinol, not to carotenoids. In general, supplements of vitamin A as retinol are not recommended unless under the guidance of a health professional. Excessive amounts of vitamin A, generally caused by self-medication with megadoses, can cause a condition known as hypervitaminosis A. Symptoms may include weakness, headache, loss of appetite, nausea, pain in the joints, and peeling of skin. Similar symptoms were reported in a young soccer player who took about 100,000 IU daily for 2 months in an attempt to improve performance. The symptoms were relieved when he stopped taking the supplements.

Excess vitamin A may also weaken the bones. Specifically, excess vitamin A stimulates bone resorption and inhibits bone formation, leading to bone loss and contributing to osteoporosis. Feskanich and others found that females with the highest intake of vitamin A as retinol, greater than 3,000 mcg daily, had double the risk of hip fractures compared to females with the lowest intake, less than 1,250 mcg per day.

Excessive intake of vitamin A during pregnancy may be teratogenic, causing deformities in the developing embryo or fetus. Indeed, vitamin A doses as low as four times the RDA can increase a pregnant female's chances of having a baby with birth defects, such as a cleft palate, heart defects, or other problems. Although this amount would not be consumed with a normal diet, it could be obtained by someone who takes a daily supplement, drinks substantial amounts of cow's milk, eats liver, and has several servings of fortified cereals. Additionally, consuming alcohol during pregnancy may worsen these teratogenic effects of vitamin A. Finally, extremely large doses of vitamin A may lead to severe liver damage, especially with concomitant alcohol intake, and may be fatal.

Beta-carotene supplements are not believed to be toxic but may cause **carotenemia**, a condition characterized by yellowing of the skin, due to accumulation of beta-carotene in the fat tissues. The supplements may also cause adverse health effects in some individuals, particularly smokers.

There appears to be little theoretical value in using vitamin A supplementation for ergogenic purposes, and no scientific evidence supports its use as a means to enhance physical performance. However, beta-carotene has been combined with other antioxidants in an attempt to prevent muscle damage during exercise, and this research is discussed later.

Prudent Recommendations In summary, vitamin A supplementation to the diet of the active individual does not have a sound theoretical basis. Moreover, the research conducted with vitamin A and physical performance has shown no beneficial effect. Hence, there appears to be no advantage for the active individual to supplement the diet with vitamin A, particularly not with megadoses that may have undesirable effects. The advisability of beta-carotene supplementation for its antioxidant properties is discussed in the later sections of this chapter dealing with ergogenic and health issues. As shall be noted, individuals should not consume high doses of beta-carotene.

Vitamin D (cholecalciferol)

Vitamin D, a term representing a number of compounds, has been classified as both a fat-soluble vitamin and a hormone. Vitamin D is naturally found in foods, but can also be made when the skin cells are exposed to the sun's UV radiation. The physiologically active form is calcitriol, which is the hormone form of vitamin D. In brief, the ultraviolet rays from sunshine initiate a process that eventually converts a provitamin found in the skin (7-dehydrocholesterol) into **cholecalciferol (vitamin D3)**, a prohormone, which is released into the blood and is eventually converted by the liver and kidneys into the active hormone, calcitriol (1,25-dihydroxycholecalciferol). **Figure 7.3** illustrates the activation of vitamin D from both dietary sources and sunlight.

Dietary supplements may contain vitamin D2 (ergocalciferol) and vitamin D3. For those who take vitamin D supplements, health professionals recommend vitamin D3 because it is more effective in raising the serum marker (25-hydroxyvitamin D) of vitamin D status. Check supplement labels for "Vitamin D as cholecalciferol" or the ingredients list for vitamin D3 or cholecalciferol.

DRI The RDA for vitamin D is given in mcg of cholecalciferol or as IU; 1 mcg of cholecalciferol is the equivalent of 40 IU. Even though sunlight is a major source of vitamin D for some people, the RDA is based on minimal exposure to sunlight. For infants 0–12 months old, the AI is 10 mcg (400 IU); for ages 1 to 70 years, the RDA is 15 mcg (600 IU); and for 70 years and older, the RDA is 20 mcg (800 IU). The UL for those over age 9 is 100 mcg (4,000 IU).

Oleksandra Naumenko/Shutterstock

Food Sources Fatty fish, such as salmon, mackerel, sardines, and catfish, are good sources of vitamin D and may contain about 200–500 IU in 3 ounces. Shitake mushrooms are also a good source, containing about 250 IU in four mushrooms. Small

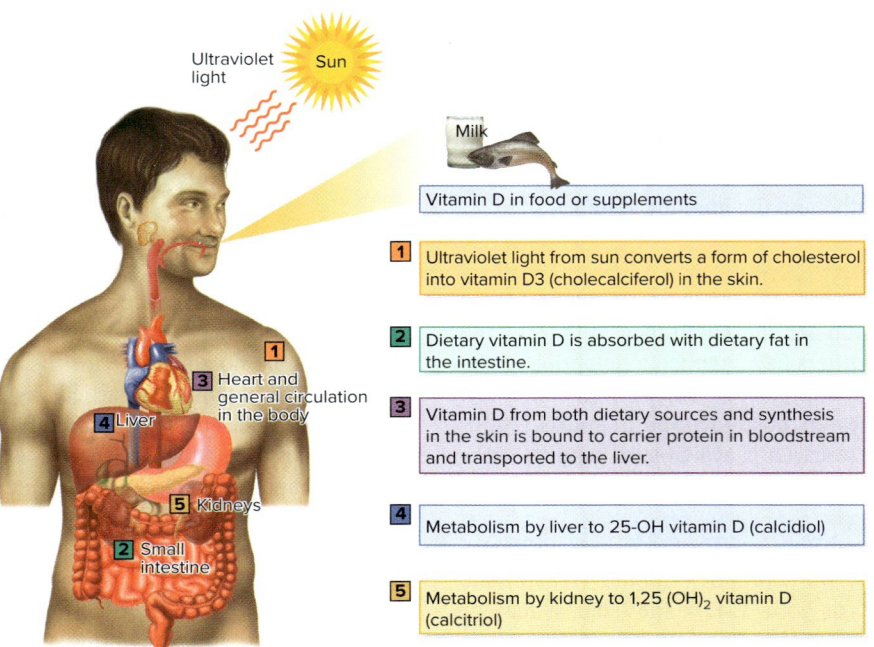

FIGURE 7.3 Activation of vitamin D. Whether synthesized in the skin or obtained from dietary sources, vitamin D ultimately functions as a hormone known as calcitriol.

amounts are also found in egg yolk and butter. Cow's milk is often fortified with vitamin D, making dairy foods a good source of the vitamin. One glass of fortified cow's milk will provide approximately 100 IU of vitamin D, which is about 15 percent of the RDA for children and adults to age 70. Some ready-to-eat cereals and orange juice may also be fortified with vitamin D.

Research suggests that we normally get about 90 percent of our vitamin D from sunlight and the remaining 10 percent from food. In general, a light-skinned person out in the sun in a bathing suit, with no sunscreen, can make 20,000–30,000 IU in 30 minutes. However, African-Americans may need up to ten times as much sunlight as Caucasians to make similar amounts of vitamin D. Clothes block the action of the sun, but the RDA for vitamin D may be obtained by exposing the hands, arms, and face to 10–20 minutes of summer sunshine about two or three times per week. The ultraviolet-B (UV-B) radiation waves promote vitamin D formation in the skin; however, UV-B waves also cause wrinkles and skin cancer. While sunscreen may inhibit vitamin D formation, its use is still recommended for the prevention of skin cancer. Longer periods of exposure are often necessary in the winter, and it may be difficult to obtain adequate vitamin D by sunlight in northern latitudes. Formation of vitamin D from sunlight also decreases in older adults and in individuals with darker skin.

Major Functions Most tissues and cells in the body have receptors for the hormonal form of vitamin D, and between 200 and 2,000 genes are controlled by vitamin D. In particular, vitamin D plays a central role in bone metabolism through its effect on calcium and phosphorus, whose roles in bone metabolism are discussed in the next chapter. It works in conjunction with several other hormones, particularly parathormone (PTH) secreted by the parathyroid gland. Vitamin D aids in the absorption of calcium from the intestinal tract and the kidneys, helping to maintain normal serum calcium levels and proper bone metabolism. Vitamin D also helps regulate phosphorus metabolism, another mineral essential in bone formation. Verhaar and others note that, besides the classical actions of vitamin D in bone metabolism, it also appears to be important for muscle function. Vitamin D status has been associated with chronic, nonskeletal diseases, including cardiovascular disease, hypertension, multiple sclerosis, arthritis, infection, autism, and certain cancers. As well, taking vitamin D supplements has been shown to be protective against mortality and intensive care unit (ICU) admission in patients with COVID-19 infection.

Deficiency: Health and Physical Performance According to the *2020-2025 Dietary Guidelines for Americans*, vitamin D is a "nutrient of public health concern." Indeed, research suggests that nearly half of American adults have low blood levels of vitamin D. These low levels of vitamin D have been related to poor intake of the vitamin as well as insufficient exposure to UV light. More Americans are spending time indoors and in sedentary activities. As well, we are drinking less cow's milk, a primary source of vitamin D, and not obtaining vitamin D from other alternative dietary sources.

A deficiency of vitamin D is known as **rickets** in children and **osteomalacia** in adults. Both rickets and osteomalacia contribute to softening and weakening of the bones. Bones can start to bend and break more easily. As well, those with osteomalacia experience muscle weakness. This muscle weakness has been theoretically linked with impaired calcium metabolism in the muscle, possibly due to inadequate activation of vitamin D receptors in muscle.

There is increasing evidence that some athletes should be concerned with vitamin D status. The position paper of the Academy of Nutrition and Dietetics (AND), Dietitians of Canada, and the American College of Sports Medicine (ACSM) suggests that vitamin D is a "micronutrient of key interest" for athletic performance. According to the position paper, vitamin D status has been related to injury prevention, rehabilitation, improved neuromuscular function, increased type II muscle fiber size, reduced inflammation, decreased risk of stress fracture, and reduced risk of acute respiratory illness. Ogan and Pritchett suggest that vitamin D deficiency and insufficiency in athletes likely mirrors what is observed in the general population.

In a year-long study of 41 athletes (12 indoor and 29 outdoor), Halliday and others reported that 64 percent of the athletes were vitamin D deficient or insufficient in the winter, 12 percent in the fall, and 20 percent in the spring. Indoor sport athletes (e.g., wrestling and basketball) had lower serum levels of 25-hydroxyvitamin D than did outdoor sport athletes (football, cross-country). Low vitamin D status was correlated with upper respiratory infections, colds, influenza, and gastroenteritis. Shindle and others reported that 80 percent of 89 professional American football players were either vitamin D deficient (less than 20 ng/ml) or insufficient (20 to 31.9 ng/ml). African-American players and players who suffered muscle injuries had significantly lower vitamin D levels.

Larson-Meyer and Willis noted vitamin D deficiency or insufficiency in athletes from all over the world. In a cross-sectional study of 950 ethnically diverse athletes, Allison and others reported that 57 percent were vitamin D deficient or severely deficient. Although European and African athletes had higher spine, neck, and hip bone mineral density than their Asian or Middle Eastern counterparts, there was no evidence of osteoporosis. In their review, Cannell and others noted that studies from the 1950s showed improved performance following exposure to ultraviolet light; peak performance is seasonal and is related to serum 25-hydroxyvitamin D levels; and vitamin D increases type II fiber size and number and muscle function in older adults. They concluded that vitamin D may improve performance in vitamin D-deficient athletes, and that peak performance may occur when 25-hydroxyvitamin D levels approach 50 ng/ml. Such findings have been replicated and shared by multiple researchers.

Todd and others commented that vitamin D levels below 30 mmol/L (12 ng/ml) are associated with reduced training volume in athletes. Moran and others noted the influence of vitamin D extends beyond calcium metabolism to include proper skeletal and cardiac muscle function, immune function, and cancer prevention. Vitamin D has a recently discovered role in transcription of a binding protein for insulin-like growth factor-1 (IGF-1), an anabolic mediator of human growth hormone, which is discussed in chapters 6 and 13. Vitamin D receptors (VDRs) are found in the nuclei of many tissues, including muscle and immune cells. Problems with VDR transcription are thought to play a role in muscle weakness.

http://ods.od.nih.gov/factsheets/vitamind/ Find the "Vitamin D Dietary Supplement Fact Sheet" from the NIH Office of Dietary Supplements at this site.

Vitamin D may help inhibit cell proliferation, so a deficiency could lead to increased cell proliferation, a key characteristic in cancer growth. Solar UVB irradiance and/or vitamin D have been found inversely correlated with incidence, mortality, and/or survival rates for breast, colorectal, ovarian, and prostate cancer and Hodgkin's and non-Hodgkin's lymphoma with evidence emerging that nearly two dozen different types of cancer are likely to be vitamin D sensitive.

Vitamin D deficiency is a worldwide problem with serious health consequences. Low levels of serum vitamin D3 appear to increase health risks. Evidence suggests that low levels of vitamin D are associated with higher risk of myocardial infarction in a graded manner, even after controlling for factors known to be associated with coronary artery disease. Some research has found that low vitamin D levels are independently associated with all-cause and cardiovascular mortality. However, a causal relationship has yet to be proved by intervention trials using vitamin D.

Low levels of vitamin D in the blood have also been associated with insulin resistance and increased risk of diabetes. Additionally, vitamin D deficiency may lead to an increased production of renin by the kidney, which could lead to an increased blood pressure. Low levels of vitamin D may also affect mental health. Indeed, evidence suggests that 1,25 dihydroxyvitamin D synthesized in the brain mediates the function of various brain growth factors. In a meta-analysis of 37 studies, the authors concluded that low vitamin D status (less than 50 nmol/liter) is associated with impaired cognitive function and greater risk of Alzheimer's disease.

Supplementation: Health and Physical Performance While much of the early vitamin D supplementation research focused on bone health and related issues, more current research has investigated the effects of vitamin D supplementation on other aspects of health.

Bone health Although some studies have shown that vitamin D and calcium supplementation do not prevent bone fractures, some such studies have limitations. In particular, researchers used the relatively less potent vitamin D2 or a too low dose of D3 (400 IU) that may have limited the increase in vitamin D status necessary to prevent bone fractures. More recent extensive reviews indicate that considerable evidence supports the role of both calcium and vitamin D in protecting the skeleton. For example, in the NIH report on multivitamin/mineral supplements and chronic disease prevention, vitamin D alone did not increase bone mineral density or decrease fracture risk, but it did work in combination with calcium to decrease the risk of fractures in postmenopausal females. Each nutrient is necessary to maximize the benefits of the other in bone health.

Additionally, research suggests vitamin D supplementation may lower the risk of bone fractures in older people by increasing muscular strength, which helps maintain balance and prevent falls. In a meta-analysis, Kong and others reported supplementation with 800–1,000 IU of vitamin D daily was associated with a reduced risk for fracture and falls.

Cancer Vitamin D has an antiproliferative effect that can regulate cell differentiation and function, which may explain how a vitamin

D deficiency can play a role in the pathogenesis of certain diseases, such as cancer. Epidemiological studies have shown a reduced risk of colorectal cancer with increased intake of calcium and vitamin D. Some randomized, controlled studies, such as the report by Wactawshi-Wende and others, revealed no effect of supplementation with 1,000 mg of calcium and 400 IU of vitamin D3, for 7 years, on the development of colorectal cancer in postmenopausal females.

However, Lappe and others, using a supplemental dose of 1,400-1,500 milligrams of calcium plus 1,100 IU of vitamin D3, did find a substantial reduction in all-cancer risk in postmenopausal females. Moreover, in a meta-analysis, Gorham and others reported that a 50 percent lower risk of colorectal cancer was associated with elevated serum vitamin D levels that could be obtained with a daily intake of 1,000-2,000 IU/day of vitamin D3. Research is ongoing to clarify the role of vitamin D3 supplementation as a possible means to help prevent cancer. The supplement dosage appears to be important.

Diabetes Vitamin D may enhance immune cell function to help prevent autoimmune diseases, such as type I diabetes. For example, doses of 2,000 IU and higher daily may have a strong protective effect in children at risk for type I diabetes.

Kidney stones and other adverse health effects Evidence suggests that supplementation with calcium and vitamin D may increase the risk for kidney stones. The elevated serum calcium levels may combine with other substances, such as oxalates or phosphates, to form the kidney stones, which may pass through the urinary tract and cause considerable pain. The calcium may also become incorporated into plaque in the arteries, leading to calcified or hardened plaque. Additionally, hypervitaminosis D may lead to vomiting, diarrhea, weight loss, and loss of muscle tone.

To help promote health, primarily bone health, scientists suggest that a healthy serum D (25-hydroxyvitamin D) concentration should be about 30-60 nanograms per liter (ng/L), which may require 1,000-2,000 IU of vitamin D3 daily. However, many vitamin D scientists are recommending that the UL be increased. In support of this viewpoint, Hathcock and others assessed the risk of vitamin D supplementation and noted the absence of toxicity in trials conducted in healthy adults who used vitamin D3 doses up to 10,000 IU daily, which supports the selection of this value as the UL.

Nutrient status for both vitamin D and calcium tends to be deficient in the adult population of the industrialized nations. Given the health implications of adequate vitamin D and calcium intake, many experts believe that fortification with both nutrients may be appropriate and, based on contemporary diets and sun exposure, probably necessary because many people do not adhere to a regimen of taking supplements daily.

Physical performance Vitamin D supplementation may be related to physical performance. Vitamin D receptors and response elements are located in skeletal muscle and other tissues and regulate over 900 gene variants. In a recent review, Dahlquist and others noted several possible ergogenic mechanisms for vitamin D. Although a specific mechanism remains unclear, the enzymes converting vitamin D3 to 1,25-dihydroxyvitamin D contain heme iron which may influence oxyhemoglobin affinity and improve aerobic capacity. Vitamin D may also improve post-exercise recovery by inhibiting myostatin (a protein which inhibits muscle protein synthesis); increase force and power production via increased calcium kinetics and cross-bridge cycling; and influence muscle growth by augmenting testosterone production and androgen binding.

Ardestani and colleagues reported that serum 25-hydroxyvitamin D levels are correlated with VO_2 max. In their meta-analysis of seven studies, Tomlinson and colleagues observed that vitamin D doses ranging from 4,000 to 8,500 IU/day for 4 weeks to 6 months significantly increased upper and lower body strength in subjects 18-40 years of age. They also noted that baseline vitamin D levels in three of these studies was below levels deemed adequate (less than 50 nmol/l or 20 ng/ml) by the Institute of Medicine. Baseline serum vitamin D levels may affect supplementation efficacy. For example, vitamin D supplementation appears to reduce risk of falls in susceptible populations such as older adults and may also affect cell signaling pathways which improve muscle function. Weight training increases the serum concentration of vitamin D, which is theorized to be related to the increased bone mass developed through exercise. Willis and others reported that little is known about the vitamin D status of athletes but suggested sports nutritionists should advise athletes to obtain adequate amounts, either through safe sun exposure or dietary supplementation with 1,000-2,000 IU vitamin D3 daily. Given data suggesting high prevalence rates of vitamin D deficiency and insufficiency, athletes should consider having their vitamin D levels checked.

Prudent Recommendations Based on the current research, athletes should consume a vitamin D-rich diet to optimize health and athletic performance. Those athletes living and training at higher latitudes and/or who are indoors are at the greatest risk for a vitamin D deficiency. Supplementation may be warranted for those athletes with a history of stress fracture, or bone or joint injury. However, further research is needed to evaluate the potential ergogenic effects of vitamin D supplementation.

For those considering supplementation with vitamin D, experts recommend having blood levels of vitamin D assessed before supplementation is commenced, because indiscriminate supplementation with vitamin D is not advised.

Vitamin E (alpha-tocopherol)

In its natural form, **vitamin E** is a complex family of eight compounds including tocopherols and tocotrienols, each with alpha, beta, gamma, and delta forms. The two major forms are **alpha-tocopherol** and **gamma-tocopherol**. Alpha-tocopherol alone is the basis for the RDA and it is the most common form in the bloodstream and dietary supplements. Gamma-tocopherol is found naturally in foods, including vegetable oil. Compared to alpha-tocopherol, the biological activity of gamma-tocopherol is significantly lower.

DRI The RDA for vitamin E for adults is 15 mg of alpha-tocopherol. One mg alpha-tocopherol is equivalent to about 1.5 IU. The UL for adults is 1,000 mg/day, which is equal to 1,500 IU.

Food Sources The most common dietary sources of vitamin E are polyunsaturated vegetable oils, such as corn, soybean, and safflower oils; 1 tablespoon of vegetable oil contains about 5 mg of vitamin E. The amount of the different tocopherols and tocotrienols varies among different oils. Sunflower seeds, almonds, peanuts, wheat germ, and several vegetables are also good sources of vitamin E. One-quarter cup of sunflower seeds provides about 8 mg of vitamin E, while ½ cup of cooked spinach provides about 2 mg of vitamin E. Meats, refined grain products, and dairy foods are typically poor sources of vitamin E.

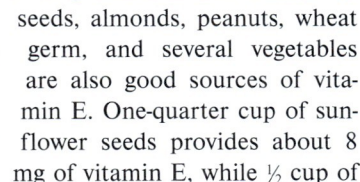
I. Rozenbaum & F. Cirou/PhotoAlto

Vitamin E supplements may contain natural or synthetic sources. The form should be listed on the food label. Unlike other vitamins, synthetic vitamin E (dl-alpha-tocopherol) is not the same as the natural source (d-alpha-tocopherol). Both forms are appropriate, but you need to take about 40–50 percent more of the synthetic form to get the potency of the natural form.

Major Functions The key role of vitamin E is to serve as an antioxidant in the cell membrane. Vitamin E works in concert with other antioxidants, such as vitamin C, to prevent the oxidation of unsaturated fatty acids in cell membrane phospholipids, thereby protecting the cell from damage. It may also help prevent the oxidation of vitamin A. Other claims, extrapolated from research with animals, have suggested that vitamin E may also play a key role in the synthesis of hemoglobin or serve a pro-oxidant effect by activating enzymes in the mitochondria to improve cellular oxygen utilization; however, these claims are not well documented in humans. Some of the vitamin E antioxidant effects are theorized to help prevent the development of several chronic diseases, or muscle tissue damage during exercise.

Deficiency: Health and Physical Performance Because vitamin E is rather widely distributed in foods, and because it is also stored in the body, a true vitamin E deficiency in humans is rare. However, certain individuals with genetic diseases, such as the inability to absorb fat, do experience a deficiency. Others on very low-fat diets may not obtain enough dietary vitamin E. In such cases, anemia may occur because the membranes of the red blood cells (RBCs) are oxidized and release their hemoglobin. Deficiency symptoms noted in animals include nutritional muscular dystrophy and damage to the heart and blood vessels.

Since vitamin E plays a role in preventing damage from free radical oxidation, a deficiency has been theorized to contribute to the development of heart disease and cancer in humans.. For example, vitamin E may help prevent the oxidation of LDL-cholesterol, which, as noted in chapter 5, is associated with atherosclerosis and coronary heart disease. Some studies have suggested that a deficiency will lead to premature aging and decreased fertility.

Although vitamin E deficiency is rare in humans, several researchers have used the data from research with animals and those humans with genetic defects to support the need for supplementation by athletes. They suggest that a vitamin E deficiency may lead to impaired oxygen transport due to RBC damage and to reduced oxidative capacity within the muscle cell. These effects would reduce VO_2 max and lead to a decrease in aerobic endurance capacity.

Supplementation: Health and Physical Performance Vitamin E supplementation, based on the potential antioxidant effects, has been studied extensively as a means of enhancing health and physical performance. Vitamin E has often been combined with other antioxidants to evaluate the effects on disease processes or muscle damage during exercise, and these topics are discussed later in the chapter.

Vitamin E supplementation has also been studied as a means to enhance VO_2 max and aerobic endurance capacity by maintaining red blood cell membrane integrity and polyunsaturated fatty acid content. The resulting improvement in membrane fluidity and deformability could theoretically enhance oxygen delivery. Early studies showed a beneficial effect, particularly at higher altitudes, but the experiments were poorly designed. However, Kobayashi, using a well-designed double-blind, placebo protocol, found that 1,200 IU of vitamin E supplements daily for 6 weeks improved VO_2 max and increased aerobic endurance in sedentary subjects at altitudes of 5,000 and 15,000 feet. More recent studies have supported these findings in athletes.

Exercising in highly polluted areas could also damage the RBC membrane. Since sports competitions may occur in large, urban areas with poor air quality, strategies to prevent RBC damage and peroxidation would be of interest to athletes such as marathon runners. Early evidence from animal and human studies suggested that vitamin E may attenuate the harmful effects of smog exposure. Yet, data from animal models do not clearly document an ergogenic effect of supplemented vitamin E above normal values. For ethical reasons, human data are based on short-term exposure instead of chronic long-term exposure. Given the possible effects of vitamin E on preserving RBC membrane integrity and exercise-induced muscle damage (discussed later in this chapter), Simon-Schnass commented that athletes should consume 100–200 IU daily and that it would be bordering on malpractice not to point out the benefits of such supplementation to athletes.

However, the majority of the recent well-designed studies on athletes, using doses from 400 to 1,200 IU, revealed no significant effect of vitamin E supplementation on variables such as VO_2 max, aerobic endurance at sea level, or marathon or Ironman-distance triathlon performance. One possible reason for this finding is that the plasma level of vitamin E appears to rise significantly during intense exercise. A study of national-class racing cyclists found no improvement in VO_2 max and other cycling performance measures after 5 months of vitamin E supplementation, as compared to a placebo treatment, despite a significant increase in serum vitamin E levels.

The antioxidant effect of vitamin E on exercise-induced oxidative stress has been the subject of several reviews. Nikolaidis and others reviewed the effects of vitamin E or C supplementation on training adaptations and oxidative markers in humans and rat models and concluded that there is no basis for recommendation of vitamin E and/or C supplementation. Although Stepanyan and others found no significant effect of tocopherol supplementation on exercise-induced oxidative stress or muscle damage in their meta-analysis of 20 studies, they noted the complex antioxidant function of tocopherols and that future research should examine additional markers of oxidative stress.

Prudent Recommendations Vitamin E is an important antioxidant, and health professionals indicate that diets rich in vitamin E may be beneficial. While vitamin E supplements were once recommended for potential health benefits, caution is now advised due to potential risks associated with high-dose supplementation. Moreover, as a means of enhancing physical performance, vitamin E supplements are not recommended.

Vitamin K (menadione)

Vitamin K is often called the blood coagulation vitamin or antihemorrhagic vitamin. Vitamin K was discovered in Denmark, and K is derived from *koagulation,* the Danish word for coagulation. Vitamin K includes phylloquinone (in plant foods), menaquinone (in animal foods and the form synthesized by bacteria in the colon), and **menadione** (synthetic form). Of the three forms, phylloquinone has the highest biological activity.

DRI The AI for vitamin K is 120 and 90 mcg/day for adult males and females, respectively. No UL has been established.

Food Sources Vitamin K is found in a variety of plant and animal foods. Good plant sources include vegetable oils (soybean, olive) and green and leafy vegetables such as kale, turnip greens, broccoli, and spinach. Meats and cow's milk contain much lower amounts. Phylloquinone is the major dietary source of vitamin K; 3 ounces of spinach contain 380 mcg, whereas 3 ounces of meat contain less than 1 mcg. The typical mixed American diet provides about 100–150 mcg of vitamin K daily. Additionally, the menaquinone form of vitamin K is also synthesized in the colon by bacteria, so a deficiency is unlikely. The AI is usually met easily by a combination of dietary intake and bacterial synthesis.

jita/iStock/Getty Images

Major Functions Vitamin K is needed for the formation of four compounds that are essential in the blood-clotting process. If vitamin K is not available, blood-clotting factors are inactive and blood clotting is impaired. In addition, vitamin K appears to enhance the function of osteocalcin, a protein that plays an important role in strengthening bones.

Deficiency: Health and Physical Performance A deficiency of vitamin K is uncommon in healthy adults, but may occur with very low–vitamin K diets and in some individuals when antibiotic medications destroy the intestinal bacteria that produce the vitamin. Deficiencies may impair blood clotting and lead to hemorrhage. Low vitamin K intake has also been associated with an increased risk of osteoporotic bone fractures. There are no data indicating deficiency states in physically active individuals or related effects on physical performance.

Supplementation: Health and Physical Performance Recent evidence shows an inverse relationship between vitamin K intake and risk for osteoporosis. In a meta-analysis of 20 studies, Salma and others reported that supplementation with vitamin K for at least 6 months was associated with a decreased risk for fracture. The authors concluded that vitamin K supplementation may counter bone loss disorders, but more research is needed to evaluate optimal dosing and the long-term safety of vitamin K supplementation.

No supplementation studies are available to evaluate the effect of vitamin K on physical performance because it does not appear to play an important role in this regard. However, as related to bone health, Craciun and others reported an increased index of bone formation markers in female elite athletes following vitamin K supplementation (10 mg/day) for 1 month. Important to this study, some of the athletes were vitamin K deficient prior to the supplementation protocol. Conversely, in a 2-year clinical study, Braam and others failed to find any effect of supplementary vitamin K on the rate of bone loss in female endurance athletes. The importance of bone health in young female athletes is discussed further in chapter 8.

The National Academy of Sciences has noted that, unlike vitamins A and D, vitamin K as phylloquinone is not toxic when consumed in large doses. Individuals taking anticoagulant drugs should consult with their health-care provider about diet and supplements. As vitamin K promotes clotting, consuming excess vitamin K in the diet or through supplementation may reduce the effectiveness of the anticoagulant medication.

Prudent Recommendations No available evidence supports vitamin K supplementation as a means to improve the health status of the average individual or to improve performance in athletes. Some research suggests that vitamin K supplementation may help promote bone health by increasing bone mineral content. Individuals desiring to take vitamin K supplements are recommended to do so under the guidance of a physician.

> ### Key Concepts
>
> ▶ The fat-soluble vitamins are A, D, E, and K. Although most vitamins must be obtained from the food we eat, several fat-soluble vitamins may be manufactured in the body. Vitamin A may be produced from dietary beta-carotene, vitamin D from exposure to the sun, and vitamin K from intestinal bacteria.

- Current research suggests that some individuals may need to obtain more vitamin D in their diet. Although obtaining vitamin D through consumption of healthy, nutrient dense foods is ideal, supplements may be recommended for some.
- In general, supplementation with fat-soluble vitamins has not been shown to enhance sports performance.

Water-Soluble Vitamins

As described earlier in this chapter, there are nine water-soluble vitamins, including eight in the vitamin B complex and vitamin C (ascorbic acid). The B complex vitamins include thiamin, riboflavin, niacin, B6, B12, folate, biotin, and pantothenic acid. There are also vitamin-like substances, including choline, that have similar general characteristics to the water-soluble vitamins. Being water soluble, they are not, with a few exceptions, stored to any significant extent in the body. For some of these vitamins, the effects of a deficiency may be noted in as little as 2–4 weeks, often reducing physical performance capacity. Excesses of these vitamins are usually excreted in the urine and are generally considered to be relatively harmless. However, there are some exceptions.

Figure 7.4 provides a broad perspective on the major sites of activity of the water-soluble vitamins, highlighting sites for vitamin E and other antioxidants as well. Most water-soluble vitamins act as components of specific coenzymes necessary for energy metabolism. For example, niacin is a component of the coenzyme nicotinamide adenine dinucleotide (NAD), necessary for over 200 pathways in metabolism.

Thiamin (vitamin B1)

Thiamin, also known as **vitamin B1**, was the first vitamin to be discovered.

DRI The RDA for adult males is 1.2 mg/day and for adult females is 1.1 mg/day. Because very little thiamin is stored in the human body, deficiency symptoms can develop within a few days of insufficient thiamin intake. No UL has been established.

Food Sources Thiamin is widely distributed in both plant and animal tissues. Excellent sources include whole-grain cereals, beans, and pork. One lean pork chop contains more than 50 percent of the daily recommendation for thiamin. Several fortified, ready-to-eat cereals contain 100 percent of the RDA for thiamin, as well as most of the other B vitamins.

Major Functions Thiamin has a central role in the metabolism of carbohydrates. It is part of a coenzyme known as thiamin pyrophosphate (TPP), which is needed to convert pyruvate into acetyl CoA for entrance into the Krebs cycle. Thiamin is essential for the normal functioning of the nervous system and energy derivation from glycogen in the muscles.

Pixtal/age fotostock

Deficiency: Health and Physical Performance Deficiency symptoms may occur within 1 to 3 weeks and include loss of appetite, mental confusion, muscular weakness, and pain in the calf muscles. Prolonged deficiencies lead to beriberi, a serious disease involving damage to the nervous system and the heart.

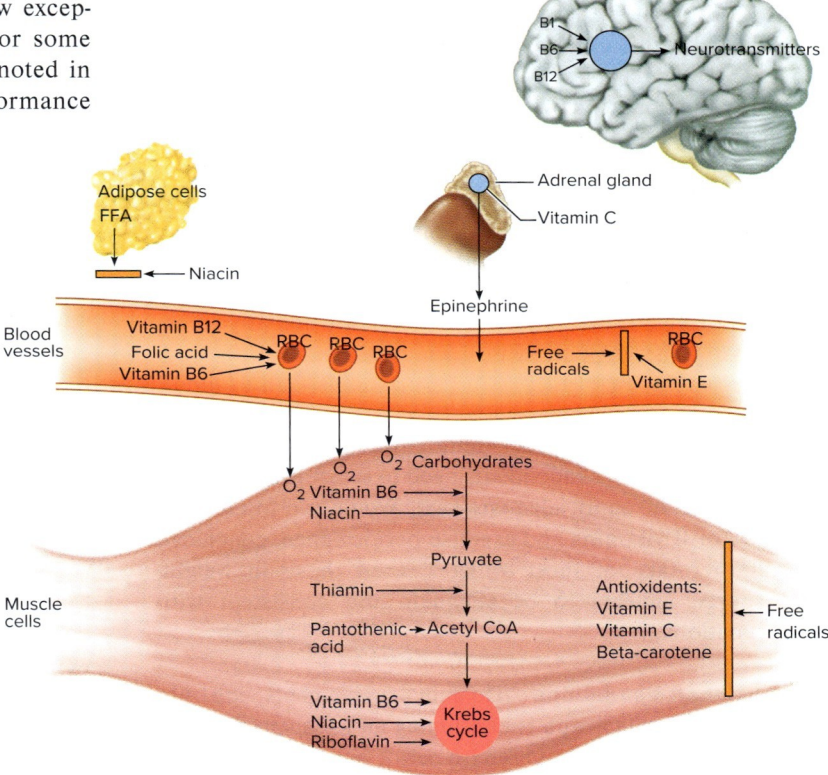

FIGURE 7.4 Roles of vitamins important to sports performance. A number of B vitamins, including thiamin, riboflavin, niacin, B6, and pantothenic acid, are essential for the conversion of carbohydrate into energy for muscular contraction. Vitamin B12 and folic acid are essential for the development of the red blood cells (RBCs), which deliver oxygen to the muscle cells. Vitamin E helps protect the RBC membrane from destruction by free radicals. Vitamin E and other antioxidant vitamins are theorized to prevent free radical damage in muscle cells during exercise. Vitamin C is needed for the formation of epinephrine (adrenaline), a key hormone during strenuous exercise. Niacin may actually block the release of free fatty acids from the adipose tissue, which could be a disadvantage for ultraendurance athletes. Finally, several of the B vitamins are also involved in the formation of neurotransmitters in the brain, which may induce a relaxation effect.

Fortunately, thiamin deficiency is not very common, being most common in alcoholics and those consuming inadequate daily kcal.

Of importance to the athlete, two factors that increase the need for thiamin are exercise and high carbohydrate intake. A deficiency of thiamin could prove to be detrimental to the active individual who might rely on high levels of carbohydrate metabolism for aerobic energy production during exercise, such as endurance athletes. Indeed, some well-controlled research conducted during World War II to evaluate military nutrition needs in combat noted decreased endurance capacity after several weeks of a thiamin-deficient diet. More contemporary research has also investigated the role of thiamin deficiency on exercise performance, but in conjunction with riboflavin and niacin deficiencies. These reports are discussed in the section on vitamin B complex.

Supplementation: Health and Physical Performance   Thiamin supplementation apparently has no health benefits for a well-nourished individual. Toxicity from thiamin is rare as the excess will be excreted in the urine. No UL has been set for thiamin.

No recent research appears to exist relative to the effect of thiamin supplementation upon physical performance, although results from a number of studies conducted more than 50 years ago are available. Following a careful review of these studies, many of which had problems in establishing a proper experimental design, there appears to be no conclusive evidence to support the contention that vitamin B1 intake above and beyond the normal RDA will enhance performance.

As noted previously, physical activity, particularly high-intensity, endurance-type activity, increases the need for both thiamin and energy in the diet. With proper selection of foods, the increased thiamin need may be met by the content in the additional foods eaten. An adequate thiamin intake is one of the reasons physically active individuals need to select foods that are nutrient-dense and supply adequate kcal. Woolf and Manore, in their review of B vitamins and exercise performance, concluded that thiamin intake of most athletes, even those in weight-control sports, appears to be adequate and consistent with recommendations.

Prudent Recommendations   Thiamin supplements are not needed by the individual who is consuming an adequate diet, and they do not appear to enhance exercise performance.

Riboflavin (vitamin B2)

Riboflavin is also known as **vitamin B2**. Like thiamin, riboflavin is a component of the B complex.

DRI   The RDA for riboflavin is 1.3 mg for the adult male and 1.1 mg for the adult female. No UL has been established, because excess riboflavin is rapidly excreted in the urine.

Food Sources   Riboflavin is distributed widely in foods. Major food sources include cow's milk and other dairy products.

Keith Leighton/Alamy Stock Photo

Additional good sources include liver, eggs, dark-green leafy vegetables, wheat germ, yeast, whole-grain products, and enriched breads and cereals. Exposure of riboflavin to light causes the vitamin to degrade quickly. As such, food sources of riboflavin, including cow's milk, should not be packaged or stored in clear glass containers exposed to light.

Major Functions   Riboflavin is a component of two coenzymes necessary for energy metabolism, flavin mononucleotide (FMN) and flavin adenine dinucleotide (FAD). FMN and FAD are also important for fatty acid and folate metabolism.

Deficiency: Health and Physical Performance   Ariboflavinosis, a riboflavin deficiency, is very rare but has been seen in alcoholics and those adhering to various fad diets. Early signs of deficiency include glossitis (an inflammation of the tongue), cracks at the corners of the mouth, and dry, scaly skin at the corners of the nose. Confusion and headaches have also been reported in those experiencing ariboflavinosis.

Although the effect of a riboflavin deficiency on physical performance has not been studied directly, research suggests that physically untrained individuals who initiate an aerobic training program may need a higher intake of riboflavin to synthesize more flavoproteins in the muscles. Haralambie reported a possible deficiency in trained athletes but did not relate the deficiency to metrics of performance. Most recently, Woolf and Manore reported that, although athletes may need more riboflavin than the general population and the current RDA, they can satisfy these increased needs if adequate energy is consumed.

Supplementation: Health and Physical Performance   Riboflavin supplementation apparently has no health benefits for a well-nourished individual. Research on the effects of riboflavin supplementation on physical performance is limited. Tremblay and others, studying elite swimmers, reported that 60 mg of riboflavin daily for 16–20 days did not improve VO_2 max, anaerobic (lactate) threshold, or swim performance.

Prudent Recommendations   Considering the available research data and the absence of riboflavin deficiency in most individuals, one must conclude that riboflavin supplementation will not enhance health or physical performance. Consumption of a well-balanced diet is recommended to optimize riboflavin in the body.

Niacin

Niacin, sometimes identified as vitamin B3, is also known as nicotinic acid, nicotinamide, or the antipellagra vitamin.

DRI Niacin is found naturally in many foods, but it also may be formed in the body from excess amounts of dietary tryptophan, an essential amino acid. Therefore, the RDA is expressed in **niacin equivalents,** or **NE**. One NE equals 1 mg of niacin or 60 mg of tryptophan—1 mg of niacin can be produced from that amount of tryptophan. The RDA for niacin is 16 NE for adult males and 14 NE for adult females. The adult UL for niacin is 35 mg/day.

Food Sources Niacin is typically found in foods that have a high protein content. For example, niacin is most abundant in lean meats, organ meats, fish, poultry, whole-grain cereal products, legumes, and enriched foods. Cow's milk and eggs contain small amounts of niacin, but they contain sufficient tryptophan. One-half of a chicken breast contains more than 60 percent of the RDA of niacin.

Mark Steinmetz/McGraw Hill

Major Functions Niacin serves as a component of two coenzymes concerned with energy processes within the cell. Nicotinamide adenine dinucleotide (NAD) is important in the process of glycolysis and the Krebs cycle, which is the means by which muscle glycogen produces energy both aerobically and anaerobically. NAD is also involved in fatty acid and amino acid metabolism. Nicotinamide adenine dinucleotide phosphate (NADP) is involved in fat metabolism by promoting fat synthesis in the body, which may block release of free fatty acids from adipose cells.

Deficiency: Health and Physical Performance Although niacin deficiency was prevalent in the past, the enrichment of foods with niacin has nearly eliminated this problem. Deficiency symptoms include loss of appetite, skin rashes, mental confusion, lack of energy, and muscular weakness. Serious deficiencies lead to pellagra, a disease characterized by severe dermatitis, diarrhea, and symptoms of mental illness.

In theory, physical performance would be impaired by a niacin deficiency because the production of energy from carbohydrate could be impaired. Both aerobic- and anaerobic-type performances could be affected. However, no research has been uncovered that has directly studied the effects of niacin deficiency alone on exercise performance.

Supplementation: Health and Physical Performance Megadoses of niacin may function as drugs and have been used in attempts to treat several health problems, being relatively ineffective in the treatment of mental disease and somewhat successful in reducing high serum lipid levels. For those individuals who cannot tolerate statin medications, megadoses of niacin in the form of nicotinic acid may reduce elevated LDL levels and improve low HDL levels. By doing so, risk for stroke and heart attack may be reduced in those with abnormal blood lipid levels.

Although niacin is generally considered to be nontoxic, large doses in the form of nicotinic acid may cause flushing, with burning and tingling sensations around the face, neck, and hands occurring within 15–20 minutes after ingestion (i.e., the histamine-like effect). Taken over long periods, niaci may contribute to liver problems such as hepatitis and peptic ulcers. Prescription doses of niacin come in three formulations: immediate-release, sustained-release, and extended-release. Extended-release niacin preparations appear to minimize the flushing effects seen with the immediate-release form and the hepatotoxic effects seen with the sustained-release form.

Because of the role of niacin in energy metabolism, a number of studies have been conducted relative to niacin supplementation and physical performance capacity. However, research has not supported a beneficial effect of niacin supplementation on athletic performance. In fact, niacin supplementation is not recommended for most athletes, particularly those involved in endurance-type exercise such as marathon running, because excessive niacin intake (3–9 g/day) influences fat metabolism, blocking the release of FFA from the adipose tissue. This will decrease the supply of FFA to the muscle, which may lead to an increased dependence on carbohydrate as an energy source during exercise. From there, muscle glycogen, an important energy source during exercise, may be depleted more rapidly. This finding was recently confirmed by Davis and Nelson who reported niacin supplementation impaired exercise performance in a study of untrained male athletes completing aerobic activities.

Prudent Recommendations Unless recommended under the treatment of a health-care professional, niacin supplements are not recommended for a physically active individual consuming a balanced diet. Excessive intake may actually impair certain types of athletic performance and elicit adverse health effects.

Vitamin B6 (pyridoxine)

Vitamin B6 is a collective term for three naturally occurring substances that are all metabolically and functionally related: pyridoxine, pyridoxal, and pyridoxamine. *Pyridoxine* is most often used as a synonym.

DRI The adult RDA for vitamin B6 is 1.3 mg/day for ages 19–50, and then increases to 1.7 mg for males and 1.5 for females above age 51. Slightly different amounts are needed at different stages of the life cycle. Recommendations for vitamin B6 are based on protein intake, so requirements may increase with high-protein diets. As excess B6 intake may cause some health problems, an UL of 100 mg/day has been set for adults.

Purestock/SuperStock

Food Sources Vitamin B6 is widely distributed in foods. The most reliable sources are protein foods such as meats, poultry, fish, wheat germ, whole-grain products, brown rice, and eggs. One-half of a chicken breast contains more than 25 percent of the RDA.

Major Functions In its coenzyme form (primarily pyridoxal phosphate, PLP), vitamin B6 is critically involved in the metabolism of protein, but it is also involved in carbohydrate and fat metabolism. PLP functions with more than 60 enzymes in such processes as the synthesis of dispensable amino acids, the conversion of tryptophan to niacin, the formation of neurotransmitters in the nervous system, and the incorporation of amino acids into body proteins such as hemoglobin, myoglobin, and oxidative enzymes. Vitamin B6 is also involved in the breakdown of muscle glycogen, as well as gluconeogenesis in the liver.

Deficiency: Health and Physical Performance Vitamin B6 deficiency is not considered to be a major health problem in the United States. Deficiency may be seen in those experiencing alcoholism or certain genetic conditions impacting metabolism of vitamin B6. As well, certain medications, including oral contraceptives, may contribute to deficiency. Woolf and Manore reported that studies have found poor vitamin B6 status in some athletes, particularly in endurance athletes. Female athletes on low-energy diets may also have low vitamin B6 intakes. Deficiency symptoms include nausea, impaired immune function, skin disorders, mouth sores, weakness, mental depression, anemia, and epileptic-like convulsions.

Theoretically, a B6 deficiency could adversely affect endurance activities dependent on oxygen, for it is involved in the formation of protein compounds, such as hemoglobin, that are essential to oxidative processes. Its role in carbohydrate metabolism, particularly muscle glycogen utilization, is also important to the endurance athlete. And, vitamin B6's role in the formation of neurotransmitters could be important to athletes engaged in fine motor control sports, such as archery and riflery. In addition, the requirement for B6 increases with protein intake, which may have some implications for athletes who may be on high-protein diets. However, because B6 is typically found in protein-rich food sources, it should be easily obtainable in such a diet.

No research has been uncovered that has directly studied the effect of a B6 deficiency on physical performance. One report did suggest that runners who covered 5-10 miles per day appear to use more B6 than their sedentary counterparts, but these investigators also noted that exercise may actually promote storage of the vitamin in the athlete, thus helping to prevent a deficiency state. Serum levels of B6 actually increase during exercise. Woolf and Manore indicate that exercise increases plasma levels of PLP, which can be converted into an acid and lost in the urine during exercise; thus, exercise may increase the turnover and loss of vitamin B6 from the body.

Supplementation: Health and Physical Performance Vitamin B6 supplementation has been used to treat nausea during pregnancy, mental depression associated with the use of oral contraceptives, and premenstrual syndrome (PMS), but its effectiveness for these purposes has received mixed reviews.

As discussed later, folate supplementation may decrease plasma homocysteine levels, which has been studied as a risk factor for CHD. While not as effective as folate, low doses of B6 (1.6 mg/day) may also provide additional homocysteine-lowering benefits to older adults who have adequate intake of the other B vitamins. Vitamin B6 has also been studied as part of B complex supplementation for prevention of CHD, and the results are presented later in this chapter.

Woolf and Manore suggest that some active individuals, depending on training level, may require 1.5-2.5 times the current RDA for B6 to maintain good B6 status. Several reports relative to the effect of B6 supplementation on physical performance are available. Although the muscle may store B6, Coburn and others noted that B6 supplementation did not markedly increase muscle stores. Moreover, in general, the studies reveal no significant effect on metabolic functions during exercise or the capacity to do more work. One investigator suggested that B6 may actually be detrimental to endurance athletes because it may facilitate the use of muscle glycogen and lead to earlier depletion in prolonged events. However, in a study from the laboratory at Oregon State University, Virk and others reported that B6 supplementation (20 g/day for 9 days) did not influence, either positively or negatively, performance of trained males in an exhaustive aerobic endurance exercise task just under 2 hours in duration.

There appears to be little or no toxicity associated with moderate doses, but high levels of vitamin B6 supplementation can cause peripheral nerve damage, evidenced by problems including loss of natural sensation from the limbs and an impaired gait. Such nerve damage has been seen in individuals who exceed the UL for vitamin B6 for an extended time period.

Prudent Recommendations Vitamin B6 can be consumed in adequate amounts from a well-balanced diet. Vitamin B6 supplementation does not appear to be warranted for the physically active individual and may be associated with some health risks if consumed in large doses for prolonged periods.

Vitamin B12 (cobalamin)

Vitamin B12 (cobalamin) is a part of the B complex and is the most recent vitamin to be discovered.

DRI The adult RDA for vitamin B12 is 2.4 mcg/day. The average mixed American diet provides about 5-15 mcg. No UL has been established for vitamin B12.

Food Sources Vitamin B12 is found in good supply only in animal foods such as meat, fish, poultry, cheese, eggs, and cow's milk. One glass of cow's milk contains nearly 50 percent of the RDA. Vitamin B12 is not found in natural plant foods such as fruits, vegetables, beans, and grains, but vitamin-fortified cereals may be an excellent source. As well, the vitamin is present in microorganisms such as bacteria and yeast, which may be found in some plant foods;

Mitch Hrdlicka/Photodisc/Getty Images

however, the bioavailability of B12 from these sources is uncertain. Although B12 may be produced by microorganisms in the human bowel, the site of production is below the point of absorption.

Vitamin B12 digestion and absorption is unique in that it requires a complex series of steps. When consumed from natural food sources, the vitamin B12 is most often bound to animal protein, which prevents its digestion and absorption. To release the vitamin B12 from the protein, HCl in the stomach is essential. Then, in the small intestine, vitamin B12 binds to **intrinsic factor**, a substance made and secreted from the cells of the stomach. In the ileum, the intrinsic factor–vitamin B12 complex separates and the vitamin is absorbed. Without normal HCl or intrinsic factor secretion, vitamin B12 absorption is impaired.

Major Functions Vitamin B12 is a part of coenzymes present in all body cells and is essential in the synthesis of DNA. As shown in **figure 7.5**, vitamin B12 works closely with folic acid, and both have important roles in the development of red blood cells. Vitamin B12 is also essential for the formation of the protective sheath around nerve fibers (the myelin sheath) and for the metabolism of homocysteine.

Deficiency: Health and Physical Performance Deficiency of vitamin B12 is most common in vegans and those experiencing malabsorption of the vitamin from the small intestine as a result of inflammation of the stomach and/or insufficient production of HCl or intrinsic factor. As described previously, HCl is necessary to break apart the protein–vitamin B12 complex in the stomach, and intrinsic factor is necessary for absorption of vitamin B12 from the ileum. As a person ages, production of intrinsic factor from the cells of the stomach decreases. Gastritis, inflammation of the stomach lining, can reduce both HCl and intrinsic factor synthesis and secretion.

Compared to other water-soluble vitamins, the body also stores a considerable amount of vitamin B12 in the liver, which may last for years; the body contains about 2,500 mcg but loses only about 1 mcg daily. The major symptoms of vitamin B12 deficiency are megaloblastic and pernicious ("deadly") anemia. Long-term vitamin B12 deficiency can also lead to nerve damage, including paralysis.

Because of its role in the formation of RBCs, a deficiency of B12 resulting in pernicious anemia would be theorized to decrease aerobic endurance capacity. No research is available relative to the effect of a vitamin B12 deficiency on performance, but other types of anemia have been shown to impair exercise performance.

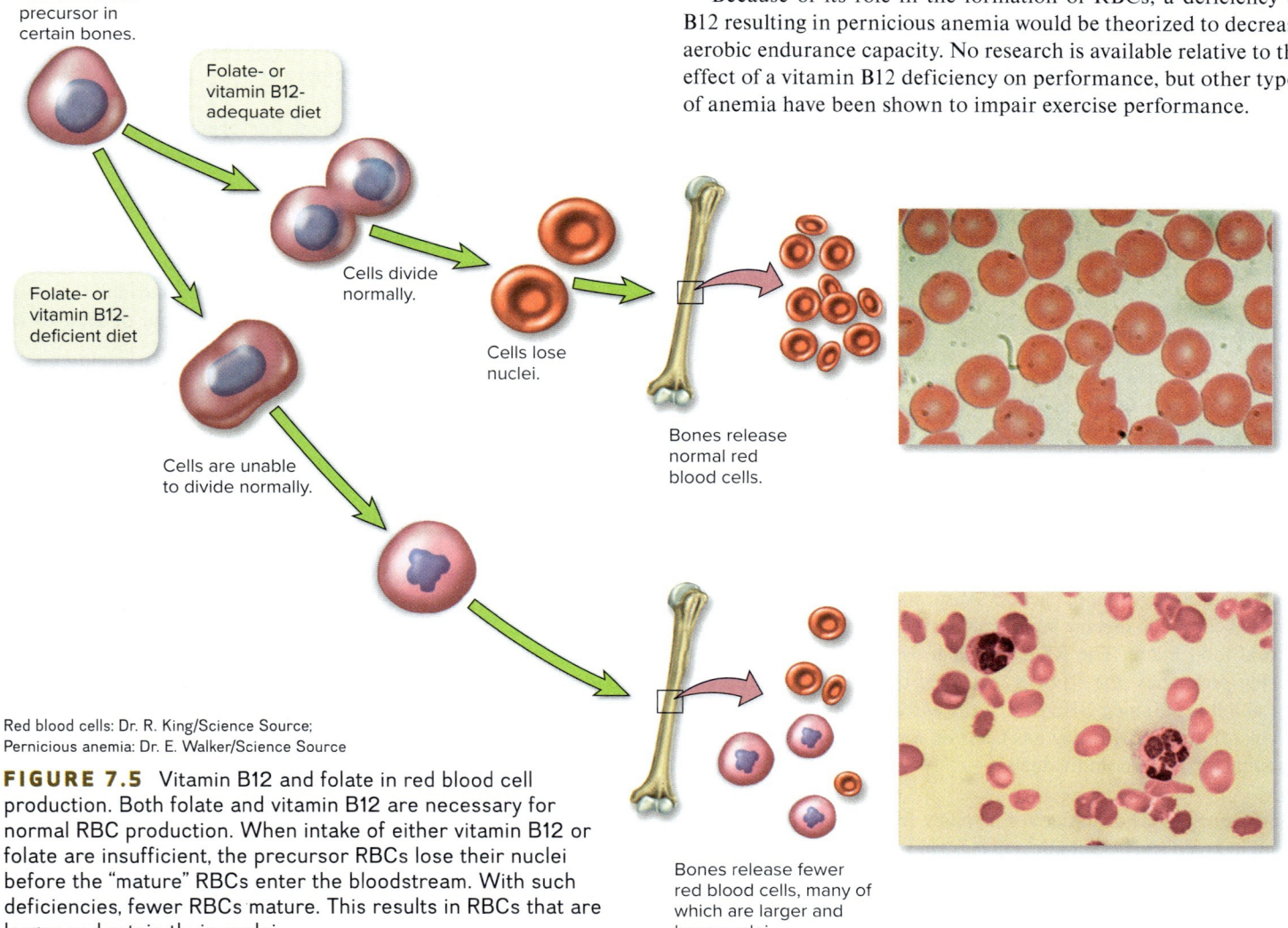

Red blood cells: Dr. R. King/Science Source; Pernicious anemia: Dr. E. Walker/Science Source

FIGURE 7.5 Vitamin B12 and folate in red blood cell production. Both folate and vitamin B12 are necessary for normal RBC production. When intake of either vitamin B12 or folate are insufficient, the precursor RBCs lose their nuclei before the "mature" RBCs enter the bloodstream. With such deficiencies, fewer RBCs mature. This results in RBCs that are larger and retain their nuclei.

Supplementation: Health and Physical Performance Vitamin B12 megadoses may be an effective medical treatment for pernicious anemia, but do not appear to benefit the active individual who eats a balanced diet. However, as noted in the following text, lower-dose supplements may be recommended for some individuals.

Relative to sports, vitamin B12 has been one of the most abused vitamins in the athletic world, with some reports of athletes receiving large amounts by injection just prior to competition. The belief probably exists that, if a little vitamin B12 can prevent anemia, then a lot of it will do something magical to increase performance capacity. However, several well-controlled studies conducted with B12 supplementation reached the general conclusion that it will not help to increase metabolic functions, such as VO_2 max or endurance performance. Multivitamin supplements containing vitamin B12 appear to be safe, but do not seemingly benefit the active individual who is following a balanced diet.

Prudent Recommendations Individuals who consume animal products typically consume adequate amounts of vitamin B12 and do not need vitamin B12 supplements. As well, there is no sound data that those who consume animal foods will benefit from supplementation to enhance their sports performance. However, vegans should consume food products fortified with vitamin B12, such as certain breads and cereals. Additionally, individuals over age 50, including older adults who regularly engage in physical activity, may have decreased intrinsic factor and limited B12 absorption from natural B12 food sources, so they are advised to get about 2.4 mcg of B12 from fortified food products, such as cereals, or supplements. One serving of B12-fortified cereal contains about 1.5 mcg, whereas a multivitamin tailored for older adults usually contains 25 mcg.

Folate (folic acid)

Folate, or **folic acid**, is also a part of the B complex. Folate is found naturally in some foods, but folic acid is a synthetic form found in fortified foods and dietary supplements; collectively they are called *folacin*.

DRI The RDA for folate is given as mcg of **dietary folate equivalents (DFE)**. The DFE accounts for the difference in absorption between natural food sources of folate and folic acid in fortified foods and dietary supplements. Absorption of folic acid from fortified foods or dietary supplements is about 1.7 times that of natural food folate. The RDA for folate is 400 mcg of DFE/day. During pregnancy, the RDA is increased to 600 mcg of DFE per day, and it is 500 mcg DFE during the early stages of lactation. An UL of 1,000 mcg DFE/day has been set for adults, and even lower amounts for younger individuals.

Food Sources Folate derives its name from *foliage* because it is found in green leafy vegetables such as spinach. Other good sources include seafood, poultry, nuts, legumes, whole-grain products, and some fruits. One papaya provides nearly 700 mcg DFE, approximately twice the RDA for the average healthy adult. As well, all cereal and grain products in the United States are fortified with folic acid to reduce the occurrence of neural tube defects. The average fortification is 140 mcg per 100 grams of food product. Studies indicate that such fortification has increased daily dietary folic acid intake by about 200 mcg/day.

Major Functions Folate serves as part of tetrahydrofolic acid (THFA), a coenzyme that plays a critical role in the metabolism of methionine, an essential amino acid. As well, folic acid is critical to the formation of DNA, the genetic material that regulates cell division. As shown in **figure 7.5**, folate is essential for maintaining normal production of RBCs, one of the most rapidly dividing cells in the body. Folate is also critical during the very early stages of pregnancy when cells in the fetus divide rapidly. Researchers indicate during periods of rapid cell division large amounts of folate are needed to make DNA.

Folate is also involved in the metabolism of **homocysteine**, an amino acid derived in a conversion from methionine and considered by some experts to be a possible risk factor for cardiovascular disease. With vitamin B12, folate is necessary for the reformation of methionine from homocysteine.

Deficiency: Health and Physical Performance According to recent data, less than 1 percent of adults and children age 4 years and older in the United States are deficient in folate. Deficiency rates have decreased significantly since 1996 when the mandatory fortification of enriched grain cereals with folic acid was authorized in the United States. Deficiency risk is greatest during times of rapid growth, including pregnancy, infancy, and childhood. A folic acid deficiency may impair DNA formation, or lead to an increase in homocysteine.

Due to its effect on DNA synthesis, one of the major effects of folate deficiency is pernicious anemia, attributed to inadequate RBC regeneration. Anemia could impair delivery of oxygen to tissues and significantly impair performance in aerobic endurance events.

DNA and chromosomal damage caused by folic acid deficiency has been suggested to increase the risk of cancer. For example, in a Tufts University review, a low level of folate in large intestine cells was associated with an increased risk of colon cancer. A recent meta-analysis by Fu and others reports that high folate intake was protective against colon cancer. As well, Ren and others report that higher folate intake was correlated with a lower breast cancer risk in premenopausal females but not in postmenopausal females.

Due to either DNA damage or the adverse neural effects of homocysteine, females who are folate deficient and become pregnant may give birth to children with neural tube defects (NTDs). Such NTDs occur when the neural tube, what ultimately develops into the spinal cord and brain, fails to form correctly during the first few weeks after conception. Spina bifida and anencephaly are the most common NTDs. Such defects may cause paralysis and severe disabling conditions in the child.

Elevated plasma levels of homocysteine are thought to damage the lining of the blood vessels or initiate growth of cells that form the framework of plaque, increasing the risk for several vascular diseases, including coronary heart disease (CHD), stroke, peripheral vascular disease (PVD), and Alzheimer's disease. According to a review by e Silva Ade and da Mota, physical activity may contribute to decreased plasma homocysteine levels, thus potentially protecting against chronic diseases associated with elevated homocysteine. However, as indicated in a meta-analysis by Deminice and colleagues, more research is needed to better understand the impact of physical activity on homocysteine levels both immediately following exercise as well as long-term.

Supplementation: Health and Physical Performance Although folic acid supplementation is effective in preventing NTDs, there is ongoing debate relative to beneficial vascular effects associated with homocysteine lowering by folic acid supplementation. For example, Wang and others conducted a meta-analysis of eight studies and found that folic acid supplementation significantly reduced the risk of a first stroke by 18 percent, particularly when supplemented over the course of 3 years, and reduced homocysteine levels by more than 20 percent. Conversely, Bazzano and others did a meta-analysis of 12 randomized, controlled studies and reported that folic acid supplementation has not been shown to reduce risk of cardiovascular diseases or all-cause mortality among participants with prior history of vascular disease. Recently, Li and others reported that folic acid supplementation resulted in a 10 percent lower risk of stroke and 4 percent lower risk of overall cardiovascular disease.

An UL of 1,000 mcg/day has been established for folic acid from fortified foods or supplements because megadoses could mask a vitamin B12 deficiency by preventing the development of anemia that would otherwise be discovered by a blood test. Unfortunately, folic acid does not prevent nerve damage, so the B12 deficiency may lead to paralysis if not detected. In addition, excessive folate intake may negatively impact other aspects of health, possibly even increasing risk of certain types of cancer. However, more research is needed.

One advocate of vitamin supplementation to athletes has reported that runners need additional folic acid to replace RBCs that may be destroyed in heavy training programs. At this time, there is no evidence available to support this theory, nor are there any data showing that folic acid supplements will benefit physical performance. Only one study has been uncovered related to folate supplementation to athletes. Matter and her colleagues provided folate and iron therapy (5 mg/day for 11 weeks) to female marathon runners who were diagnosed as being folate and iron deficient. While the folate therapy restored serum folate levels to normal, no improvements were noted in VO₂ max, maximum treadmill running time, peak lactate levels, or running speed at the lactate anaerobic threshold.

Prudent Recommendations All individuals should increase their intake of folate-rich foods, particularly fruits and vegetables, to obtain both folate and other health-promoting nutrients. Females of childbearing potential should obtain 400 mcg DFE daily from a supplement and/or dietary sources. While adequate intake of folate is important for optimizing health and preventing chronic disease, too much folic acid from supplements can have negative impacts on health. As such, monitoring daily intake of synthetic folic acid and limiting intake to less than 1,000 mcg DFE is recommended.

Pantothenic acid

Pantothenic acid is a water-soluble B-vitamin found in a wide variety of foods. The Greek word *pantothen* means "everywhere," referring to the presence of pantothenic acid in many different types of foods.

DRI The adult AI for pantothenic acid is 5 mg and no UL has been established.

Pixtal/age fotostock

Food Sources Pantothenic acid is distributed widely in foods. It is found in all natural animal and plant products, but best sources include organ meats, eggs, legumes, yeasts, and whole grains. Grains, including cereal, have often been fortified with pantothenic acid. One cup of mushrooms provides about 3.5 mg pantothenic acid.

Major Functions Pantothenic acid is an essential component of coenzyme A (CoA), which plays a central role in energy metabolism. You may recall that acetyl CoA, a key component of metabolism of carbohydrates, fats, and proteins, is the principal substrate for the Krebs cycle. Pantothenic acid is also involved in gluconeogenesis, the synthesis and breakdown of fatty acids, the modification of proteins, and the synthesis of acetylcholine, a chemical released by the motor neuron to initiate muscle contraction.

Deficiency: Health and Physical Performance Deficiencies of pantothenic acid are quite rare and are most likely to develop in individuals who are deficient in many of the B-vitamins. If a deficiency was to develop, symptoms most commonly include fatigue, muscle cramping, headache, and impaired motor coordination. Because deficiencies of pantothenic acid have not been observed, such effects upon physical performance have not been studied.

Supplementation: Health and Physical Performance Pantothenic acid supplementation apparently has no health benefits for a well-nourished individual. Published research findings do not support an effect of pantothenic acid supplementation on coenzyme A levels or physical performance. In highly trained cyclists, Webster reported no significant effects of pantothenic acid supplementation (1.8 g/day for 7 days, combined with allithiamin) on a 2,000-meter time trial. There were no significant metabolic, physiological, or psychological effects. Wall and others compared 1 week of supplementation with d-pantothenic acid (1.5 mg/day) combined with l-cysteine

(1.5 mg/day) with placebo and found no effect on resting coenzyme A levels; similar declines in postexercise coenzyme A, lactate levels, and energy substrate use; and no difference in work output.

Supplements of pantothenic acid appear to be relatively nontoxic. Large doses of 10–20 grams have been known to cause diarrhea.

Prudent Recommendations Given the fact that pantothenic acid deficiency is rather nonexistent and there is little research to support a beneficial effect on health or physical performance at this time, supplementation is not recommended. A balanced diet should provide adequate pantothenic acid for the healthy, physically active individual.

Biotin

Biotin is the final vitamin in the water-soluble B-vitamin complex.

DRI The adult AI for biotin is 30 mcg and no UL is available.

Digital Vision/age fotostock

Food Sources Good dietary sources of biotin include organ meats, such as liver, egg yolk, legumes, and dark-green leafy vegetables. Bacteria in the colon can also synthesize biotin.

Major Functions As a coenzyme, biotin participates in chemical reactions involving the transfer of carbon dioxide. Biotin is important in the synthesis of glucose and fatty acids. Because biotin is an important coenzyme for gluconeogenesis, it may have some implications relative to endurance performance.

Deficiency: Health and Physical Performance Deficiency of biotin is rare but may occur when the diet contains large amounts of raw egg whites. A protein in the raw egg white binds biotin and prevents its absorption into the body. In such cases, symptoms include loss of appetite, mental depression, dermatitis, and muscle pain. For athletes who consume eggs for their protein content, it may be important to know that cooking the egg white eliminates this problem while providing the same amount of high-quality protein. Keep in mind that raw eggs pose a risk of *Salmonella*, a type of bacteria associated with foodborne illness.

Although a biotin deficiency could impair physical performance, no data are available to support this hypothesis.

Supplementation: Health and Physical Performance At this time, there is no evidence that biotin supplementation improves health or increases physical performance capacity.

Prudent Recommendations According to the available research, biotin supplements are unnecessary for the physically active individual. Consuming a mixed diet that provides adequate amounts of biotin is advised.

Vitamin C (ascorbic acid)

Vitamin C, or **ascorbic acid**, is a water-soluble vitamin widely known for its deficiency disease, scurvy, which was common for centuries in sailors who had little access to fresh fruits and vegetables.

DRI The adult RDA of vitamin C for males is 90 mg/day and for females is 75 mg/day. For adults, the UL has been set at 2,000 mg/day.

Food Sources The best food sources of vitamin C are fruits and vegetables, primarily citrus fruits and the leafy parts of green vegetables. Excellent sources include oranges, grapefruit, broccoli, and salad greens. Other good sources are green peppers, potatoes, strawberries, and tomatoes. Cow's milk, meats, and grain products are typically poor sources of vitamin C.

Nataliia K/Shutterstock

Major Functions Although vitamin C does not directly participate in enzyme-catalyzed conversions of substrate to product, the vitamin modifies mineral ions in the enzymes to make them active. Vitamin C has a number of different functions in the body, some of which have important implications for the physically active individual. **Table 7.2** lists the key functions of vitamin C.

Deficiency: Health and Physical Performance Deficiencies of vitamin C are rare in industrialized countries. The human body (adult) has a pool of vitamin C ranging from 1.5 to 3.0 grams. However, smoking, aspirin use, oral contraceptive use, and stress may increase a person's need for vitamin C. Those who consume inadequate amounts of fruits and vegetables are also at risk for a vitamin C deficiency. The major deficiency disease is scurvy, a disintegration of the connective tissue in the gums, skin, tendons, and cartilage that may develop in a month on a vitamin C–free

TABLE 7.2 Key functions of vitamin C in human health
Powerful antioxidant that protects cells from free radical damage
Synthesis of collagen, necessary for formation and maintenance of cartilage, tendon, and bone
Formation of certain hormones and neurotransmitters, including epinephrine (adrenaline) during exercise
Absorption of iron from the GI tract
Synthesis of red blood cells
Healing of wounds through development of scar tissue
Regulates metabolism of folic acid, cholesterol, and amino acids
Promotes health of the immune system

diet. Typical symptoms of scurvy include bleeding gums, rupture of blood vessels in the skin, impaired wound healing, muscle cramps, and weakness. Anemia also may develop.

Because of the role of vitamin C as a potent antioxidant, the vitamin appears to play an important role in optimizing health. Specifically, some forms of cancer, cardiovascular disease, age-related macular degeneration (AMD) and cataracts, and the common cold have been associated with a deficiency of vitamin C. Epidemiological research suggests that higher dietary intake of fruits and vegetables, often rich sources of vitamin C, is associated with a reduced risk of most types of cancer. Similarly, higher intake of vitamin C–rich fruits and vegetables has also been associated with a reduced risk of cardiovascular disease. Oxidative stress contributes to both AMD and cataracts and, because of that, vitamin C appears to play a role in preventing and/or treating those diseases of the eye. Perhaps most well known to the average consumer, vitamin C may protect against and/or treat the common cold. This topic has been the subject of hundreds of research studies since Linus Pauling first suggested such a relationship in the 1970s. However, the results of trials on dietary and supplemental vitamin C to prevent or treat the common cold have found inconsistent results. For some individuals, particularly older adults, smokers, or those with compromised immune systems, vitamin C may be effective in doses of at least 200 mg/day.

Supplementation: Health and Physical Performance The effect of vitamin C supplementation on physical performance has received considerable attention, mainly because it is one of the vitamins that athletes consume in rather substantial quantities. Both early and contemporary research have shown that vitamin C supplementation improves physical performance in subjects who are vitamin C deficient, but a thorough analysis of these studies supports the general conclusion that vitamin C supplementation does not increase physical performance capacity in subjects who are not vitamin C deficient. Bell and others found that vitamin C supplementation of 500 mg daily for 30 days did not increase VO_2 max or cardiac output in either young or older males. No solid experimental evidence supports the use of megadoses of 5–10 grams that some athletes take.

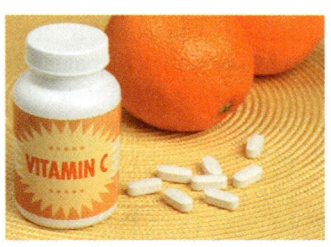
Denise Bush/Getty Images

Because exercise is a stressor, some investigators have recommended that the active individual may need slightly more vitamin C than the RDA, for example, 200–300 mg per day. Although research with runners doing 5–10 miles a day does not support this viewpoint, this amount could easily be obtained by wise selection of foods high in vitamin C content. Vitamin C supplementation may also be beneficial to heat acclimation, a topic that merits additional research with trained athletes.

While exercise normally promotes health benefits, in contrast to moderate levels of physical activity, prolonged and intensive exertion, such as running an ultramarathon, causes numerous changes in the human immune system that may predispose to certain illnesses, such as upper respiratory tract infection (URTI). Large doses of vitamin C have been claimed to strengthen the immune system and prevent URTI. Some early research suggests that the antihistamine effects of vitamin C may decrease the severity of some symptoms of a cold. For example, Peters and others reported that 600 mg of vitamin C supplementation for 21 days prior to an ultramarathon reduced symptoms of respiratory tract infections, while Douglas and others reported that doses greater than 200 mg may also help prevent URTI in heavy physical exercise in cold environments, such as soldiers in subartic weather conditions. However, more recent research revealed that 200 mg of vitamin C daily will lead to full saturation of plasma and white blood cells, which should optimize immune functions associated with vitamin C. It may be possible that smaller amounts of vitamin C, such as 200 mg, provide effects comparable to those seen with larger doses.

In general, vitamin C supplementation does not appear to benefit immune function, even in ultraendurance athletes. David Nieman, an international expert in exercise and immune function, along with his colleagues, reported no effect of 7 days of vitamin C supplementation (1,500 mg/day) on oxidative and immune responses in runners both during and after a competitive ultramarathon race. Two studies by Davison and Gleeson also reported no beneficial effects of vitamin C supplementation, either alone or with carbohydrate, on immune functions following 2.5 hours of cycling at 60 percent VO_2 max. In a review of his own and others' research, Nieman indicated that vitamin C supplementation, as well as combined antioxidant vitamin supplementation, is not an effective countermeasure to exercise-induced immunosuppression.

There is some debate regarding the safety of megadoses of vitamin C. Several investigators have found that excessive amounts of vitamin C, such as 5–10 grams daily, may produce some undesirable side effects such as diarrhea; destruction of vitamin B12 in the diet; excessive excretion of vitamin B6; decreased copper bioavailability; predisposition to gout, creating pain in the joints; and formation of kidney stones from oxalate salts, one of the breakdown products of vitamin C. Although for some individuals increased iron absorption is a beneficial effect of vitamin C, it may be a major health problem for individuals prone to iron-storage disease, discussed in the next chapter. Moreover, one study reported increased markers of oxidative DNA damage in some subjects when vitamin C and iron supplements were taken together. Excess vitamin C may potentiate the oxidative effects of iron. As well, too much vitamin C also may interfere with the correct interpretation of certain blood and urine tests. Finally, several case studies revealed the development of a condition known as rebound scurvy when the individual stopped taking the supplements. The researchers suggested a mechanism whereby the increased activity of an enzyme in the body that destroys excess vitamin C during the supplement stages continued after the supplements were stopped, leading to a deficiency and symptoms of scurvy.

Conversely, others have reported megadoses to be relatively harmless because excessive amounts are excreted by the kidneys. They criticize the research upon which claims of adverse effects are based, noting that some of the conclusions rest on isolated case studies. Others support a middle viewpoint, noting that larger doses may be harmless to many, but certain individuals may be prone to problems, such as those who have a family history of kidney stones. For example, a highly controlled supplementation study by Massey and others reported that large daily intakes of vitamin C (2,000 mg) increased the excretion of urinary oxalate and increased the risk for calcium oxalate kidney stones in 40 percent of study participants. Evidence suggests that oxalates appear in the urine when individuals are supplemented with only 1,000 milligrams of vitamin C per day. These were short-term studies and no observations of kidney stones were detected, but increased oxalates could lead to stone formation over time in those at risk. However, in a longer prospective study involving males between the ages of 40 and 75 with no history of kidney stones, Curhan and others found that after 6 years of follow-up there was no association between vitamin C intake—up to levels of 250–1,500 mg/day—and kidney stone formation. Nevertheless, Gerster notes that those at risk for kidney stones should restrict daily vitamin C intake to about 100 mg. The National Academy of Sciences recommends against the routine use of large supplements. Overall, megadose supplementation with vitamin C remains controversial.

Prudent Recommendations Consuming adequate vitamin C from natural food sources appears to be optimal for health and wellness for both the sedentary and physically active individual. Health experts recommend including adequate amounts of fruits and vegetables in the diet to meet vitamin C needs. While supplementation with vitamin C may be recommended for some cases, further research is needed, and consumers should choose "food first" when it comes to vitamin C.

Vitamin-like compounds: Choline

Choline, an amine, is a water-soluble essential nutrient listed in the DRI report with the B vitamins, but it has not been classified as a vitamin. Commercial choline products are available as lecithin or choline salts, but the actual choline content may vary. Choline is also marketed as a powder with carbohydrate and electrolytes to make a sports drink for athletes.

Pixtal/age fotostock

DRI An AI has been set at 550 and 425 mg/day for male and female adults, respectively. The UL is 3.5 g/day.

Food Sources Choline is found in most foods, particularly as lecithin (phosphotidylcholine) in animal foods and free choline in plants. Meat, fish, and eggs are good animal sources, while plant sources include vegetables, legumes, nuts, seeds, and whole grains.

Major Functions Choline functions as a precursor for lecithin, a phospholipid in cell membranes. Choline is also involved in the formation of acetylcholine, an important neurotransmitter in the central nervous system.

Deficiency: Health and Physical Performance As choline is found in most foods as part of cell membranes, choline deficiency is very rare. The only research available, involving individuals fed a choline-deficient solution, revealed the development of fatty livers and liver damage. Plasma choline levels have been reported to be significantly reduced following exhaustive exercise such as marathon running, and thus a possible reduction in acetylcholine levels in the nervous system may be theorized to be a contributing factor to the development of fatigue. However, exercise performance was not evaluated in these reports. Penry and Manore suggested that choline supplementation might only increase endurance performance in activities that reduce circulating choline levels below normal.

Supplementation: Health and Physical Performance As choline deficiency is very rare, the effect of choline supplementation on health status has received little research attention. Comparatively, its effect on exercise performance has been the focus of several studies in recent years. According to several reports, choline administration accelerates synthesis and release of acetylcholine by neurons. Choline is a common ingredient found in commercially available pre-workout supplements.

Research has shown that choline supplementation, either as choline salts or as lecithin preparations, will increase blood choline levels at rest and during prolonged exercise. Some preliminary field and laboratory research has suggested that increased plasma choline levels are associated with a significantly decreased time to run 20 miles and improved mood states of cyclists 40 minutes after completion of a cycle ergometer ride to exhaustion. In contrast, well-controlled laboratory research has revealed that choline supplementation, although increasing plasma choline levels, exerted no effect on either brief, high-intensity anaerobic cycling tests lasting about 2 minutes or more prolonged aerobic exercise tasks lasting about 70 minutes.

Several experts have recommended more research with choline supplementation, particularly controlled laboratory research involving prolonged aerobic endurance exercise tasks greater than 2 hours' duration. Warber and others provided choline (6 grams of total free choline) to United States Army Rangers during and after 4 hours of strenuous exercise. Although choline supplementation increased plasma choline levels by 128 percent, there were no significant effects on any exercise performance measure, including a treadmill run to exhaustion following the 4-hour treadmill exercise protocol. In a short-term supplementation field study, Buchman and others observed similar results in runners who consumed either a lecithin supplement (2.2 grams total) or placebo 1 day prior to a marathon. The lecithin supplementation increased plasma choline concentration, while the placebo group experienced a significant decrease, but there were no significant differences between marathon running times. At this time, evidence suggests that supplementation with phosphatidylcholine (0.2 g/kg) is more effective than choline in increasing plasma choline levels and may improve performance in activities that have decreased plasma choline concentrations.

Prudent Recommendations Given the evidence that choline deficiency states are very rare and that supplementation does not appear to significantly enhance health or exercise performance, choline supplementation is not recommended. Rather, a balanced diet should provide adequate choline.

> ### Key Concepts
>
> - The water-soluble vitamins consist of those in the B complex and vitamin C. Choline is an example of a vitamin-like substance that is water-soluble, but that is nonessential because it can be produced in the human body.
> - In general, water-soluble vitamins are not toxic in excess. Exceptions include that excess niacin may interfere with proper liver function, and excess vitamin B6 has been associated with neurological problems.
> - Water-soluble vitamin deficiencies are rare in developed countries. However, supplements may be advised for some individuals. Females of childbearing age should consume 400 mcg of folic acid daily to prevent the possibility of birth defects in the newborn. Older adults and those following an exclusively plant-based diet should obtain adequate vitamin B12 through fortified foods and/or vitamin supplements.
> - A water-soluble vitamin deficiency may impair physical performance, usually by interfering with some phase of the energy-producing process. In some cases, impairment may be seen within 2–4 weeks on a deficient diet.
> - Supplementation with water-soluble vitamins has not been shown to enhance sports performance.
> - Vitamin C plays a wide variety of roles in the body, including as a key antioxidant that protects against free radical damage. Diets rich in fruits and vegetables often provide adequate amounts of vitamin C.

Vitamin Supplements: Ergogenic Aspects

Like the general population, the vast majority of athletes receive recommended amounts of most vitamins as part of a mixed daily diet. While some studies report that certain groups of athletes receive less than the RDA/AI for some vitamins, or even have indicators of a biochemical deficiency, others report that biochemical indices of nutritional status are usually within normal limits. Sarah Short of Syracuse University, in her exhaustive review of dietary surveys with athletes, and Larry Armstrong and Carl Maresh in their review, found that vitamin deficiency symptoms rarely are reported. Moreover, at least in developed countries, dietary vitamin intake by athletes appears to be more than required for maximal exercise performance. Nevertheless, elite endurance athletes, such as Tour de France cyclists, and the majority of both high school and college athletes believe that vitamins are essential for success, and it is a matter of fact that many consume vitamin supplements either as nutritional insurance or in the hope of improving performance. For example, a recent study of preadolescent and adolescent Canadian athletes found that 100 percent of those athletes surveyed reported using dietary supplements. The most commonly used supplements were multivitamin-multiminerals, vitamin C, vitamin D, vitamin-enriched water, protein powder, fatty acids, probiotics, and plant extracts. Elite athletes use supplements more than college or high school athletes, and females more often than males. Athletes appear to use supplements more than the general population, and some take high doses that may lead to nutritional problems.

> https://ods.od.nih.gov/factsheets/list-all/ The National Institutes of Health Office of Dietary Supplements provides fact sheets on nearly a hundred different dietary supplements, including vitamins and minerals.

Should physically active individuals take vitamin supplements?

According to a recent review by Rawson and colleagues, some dietary supplements may improve exercise performance, while others may improve overall health, adaptation to exercise, and/or injury recovery. In terms of vitamins, research suggests that supplementation with vitamin D may help athletes train and compete more effectively.

There may be some good reasons for physically active individuals to take vitamin supplements. For example, athletes with relative energy deficiency syndrome in sport (RED-S) are often consuming inadequate nutrients, including vitamins. Research suggests that vitamin depletion, mainly the water-soluble vitamins, can occur rapidly in humans on low-kcal diets and that these vitamins should be replaced daily. Athletes may also need vitamin supplementation if they are subsisting on poor diets, as discussed in the next section. Moreover, some vitamin requirements are increased in pregnant females and older adults, so those who exercise need to consume vitamin-rich diets. Active individuals should consult appropriate health professionals before self-prescribing vitamin supplements, particularly megadoses of individual vitamins.

As is obvious from the evidence presented in this chapter, the athlete who is on a balanced diet likely will not benefit from vitamin supplementation to improve performance. Nevertheless, some interesting hypotheses suggest that antioxidant vitamin supplements may help prevent muscle tissue damage during training, and a variety of special vitamin-like compounds have been marketed specifically for athletes.

Can the antioxidant vitamins prevent fatigue or muscle damage during training?

Aerobic exercise induces oxidative stress in the body, increasing the production of free radicals. Finaud and others indicate that the effects of the free radicals may be positive or negative. On the positive side, Neiss indicates growing evidence that regular aerobic exercise training enhances the functional capacity of the antioxidant network, upgrading the capacity of the natural antioxidant enzymes in the muscles. Reviews by Ji and Powers suggest that *chronic* exercise training increases the activity of superoxide dismutase and glutathione peroxidase in response to free radical generation. These enzymes are important to muscle cell survival during increased oxidative stress. Knez and others reported that the volumes and intensities of exercise associated with ultraendurance training, such as for Ironman-type triathlon competition, induce favorable changes in innate antioxidant defenses against free radical damage, resulting in improved oxidative balance. One possible benefit is improved immune function. On the negative side, excessive exercise-induced oxidative stress may occur if the generation of free radicals overwhelms tissue antioxidative defenses, which may disturb cellular homeostasis and cause fatigue or lipid peroxidation and muscle tissue damage.

Antioxidant supplements have been studied as a means to enhance physical performance. However, as noted previously, individual supplementation with either vitamin C or vitamin E has not been shown to enhance exercise performance, with the possible exception of vitamin E at altitude. Moreover, reviews by Powers and Sen and their colleagues concluded that antioxidants, including antioxidant cocktails containing several vitamins, have not been shown to reliably improve physical performance. For example, Zoppi and others found that supplementation with vitamins C (1,000 mg/day) and E (800 mg/day) to professional soccer players over the course of 3 months of training had no effect on strength, speed, or aerobic capacity. As well, a recent study by Sellitto and others found that antioxidant supplementation actually hindered the body's natural antioxidant systems.

Atalay and others suggested that performance should not be the only criterion to evaluate the success of antioxidant supplementation. Faster recovery and minimization of injury time could also be affected by antioxidant therapy. Although Zoppi and others found no performance enhancement with antioxidant supplementation to the soccer players, supplementation was associated with reduced levels of blood markers for muscle tissue damage.

It has been known for years that certain forms of physical training for sports, particularly intense training, can induce muscle damage and soreness. Eccentric muscle contractions, such as those incurred in the quadriceps muscle when running downhill, may cause mechanical trauma to the muscle and connective tissue, resulting in soreness during the following days. Neiss indicates that, although an *acute* bout of exercise increases the activity of antioxidant enzymes in the body, strenuous exercise may generate reactive oxygen species to a level to overwhelm tissue antioxidant defense systems. The result is oxidative stress. The magnitude of the stress depends on the ability of the tissues to detoxify ROS. Excessive production of free radicals may induce lipid peroxidation, possibly damaging the integrity of cellular and subcellular membranes in the muscles, leading to muscle injury and muscle soreness.

Most of the research with antioxidant supplements to athletes has focused on prevention of muscle tissue damage and soreness. Theoretically, prevention of muscle tissue damage may enable the athlete to train more effectively, the desired result being improvement in competition. Some endurance athletes will train at altitude in attempts to enhance their oxygen-delivery ability and, as noted earlier in this chapter, vitamin E supplements may convey some benefits when exercising at altitude. Additionally, older individuals may be more susceptible to oxidative stress during exercise, for optimal functioning of the free radical–scavenging enzymes appears to decline with the aging process. Millions of older individuals perform aerobic exercise for the related health benefits and often become involved in various forms of athletic competition, including local, national, and international competition for athletes over age 40. Companies are now marketing *super antioxidants* to speed recovery in athletes during sport training. Do they help?

Numerous studies have been conducted to evaluate the effects of antioxidant supplements on exercise-induced muscle damage and, in some studies, on performance. The designs of these studies have varied, including differences in subjects (animals versus humans, age), methods to induce muscle soreness (e.g., downhill running versus level running, resistance exercise), the type and amount of supplement given, and the biochemical markers used to assess muscle damage. The most common supplements used were vitamins E and C and beta-carotene, but coenzyme Q_{10}, selenium, and other substances have also been used. Some studies used "antioxidant cocktails" consisting of approximately 800 IU of vitamin E, 1,000 mg of vitamin C, and 10–30 mg of beta-carotene. The markers of muscle tissue damage include serum enzymes that may leak from the muscle, such as creatine kinase (CK) and lactic acid dehydrogenase (LDH); end products of lipid peroxidation, such as malondialdehyde (MDA); myoglobin leakage from the muscle tissue; and others.

Overall, the results of these studies may be regarded as promising. A number of studies, such as the report by Itoh and others, have shown some beneficial effects of antioxidant supplementation, that is, reduced markers of muscle tissue damage when compared to the placebo treatment. Benefits have been reported for both young and old physically active individuals. Most studies used multiple markers of muscle tissue damage, and in some cases, one marker of muscle damage would be improved by the antioxidant supplements but another would be unaffected. Some studies compared different antioxidants, for example, C versus E, reporting a beneficial effect of one but not the other.

In contrast, more recent studies have reported no significant benefits of antioxidant supplementation to prevent muscle tissue damage during exercise. For example, Mastaloudis and others reported that supplementation with vitamin C (1,000 mg) and vitamin E (300 mg alpha-tocopheryl acetate) before a 50-kilometer ultramarathon had no effect on leg muscle damage or recovery following the race and up to 6 days afterwards. Machefer and

Mikadun/Shutterstock

others investigated the effect of moderate supplementation with vitamin C, vitamin E, and beta-carotene on muscle tissue damage during and following the Marathon des Sables, a 6-day, 156-mile (254-kilometer) ultramarathon race across the Sahara Desert. The supplement elevated serum vitamin levels, but had no effect on markers of muscle tissue damage. Bryer and Goldfarb found that prolonged vitamin C supplementation (3 grams) both before and after eccentric exercise designed to induce muscle soreness produced divergent results on markers of muscle damage and measures of muscle soreness. In general, there were lower levels of creatine kinase 48 hours after testing, but no effect on muscle soreness.

Some studies actually reported adverse effects of supplementation on both exercise performance and health. Nieman and others found that vitamin E supplementation promoted lipid peroxidation and inflammation during an Ironman-distance triathlon. Close and others also reported that consuming 1 gram of ascorbic acid 2 hours before and 14 days after completing a downhill run did not prevent delayed onset of muscle soreness (DOMS). Muscle function was impaired in both the vitamin C and placebo groups, but more so in the supplement group, suggesting that vitamin C supplementation may actually inhibit the recovery of muscle function.

Antioxidant supplementation may also adversely affect cellular adaptations to chronic training. Gomez-Cabrera and others reported that vitamin C supplementation during training may impair training adaptations, possibly by decreasing production of mitochondria. Similar results were observed by Paulsen and others, who assigned subjects to supplementation (1,000 mg vitamin C/day; 345 mg vitamin E/day) and placebo groups followed by 11 weeks of high-intensity interval training. Although VO_2 max improved by 8 percent in both groups, the supplemented group had decreased activity in cell signaling pathways, which normally stimulate increases in mitochondrial respiration and biogenesis. The endurance athlete may wish to exercise caution in consuming high doses of antioxidants.

The role of antioxidants to prevent exercise-induced muscle tissue damage has been the subject of much research. A recent meta-analysis by Kim and others found that dietary vitamin E supplementation lower than 500 IU was beneficial in preventing exercise-induced muscle damage in athletes. While a promising finding, experts agree that more research is needed in this area.

Prudent Recommendations Most experts in this area recommend that physically active individuals obtain antioxidant vitamins naturally from food sources. Increasing consumption of fruits and vegetables will enable athletes to obtain the proposed beneficial amounts of beta-carotene (10–30 mg) and vitamin C (250–1,000 mg). Vitamin E can also be consumed through a variety of food sources, including nuts, sunflower seeds, and healthy oils. There seems no valid reason to recommend antioxidant supplements to most athletes, except in those known to be consuming a low-antioxidant diet for prolonged periods. For health benefits, remember that most research documents beneficial effects when these vitamins, along with other phytochemicals, are obtained through natural foods, primarily fruits and vegetables.

> *https://sportsrd.org/educational-resources-2/educational-resources/*   The Collegiate and Professional Sports Dietitians Association (CPSDA) provides fact sheets on a variety of sports nutrition topics, including dietary supplements.

How effective are the multivitamin supplements marketed for athletes?

Special athletic vitamin packs have been appearing on the market—even in single packets at your local convenience store—that have been advertised as a means for the athlete to increase energy and reach peak performance. Many of these have simply been multivitamin-mineral supplements, while others have been special concoctions that contain plant and herbal ingredients.

Because in human metabolism vitamins often work together, and often in conjunction with minerals, the ergogenic potential of multivitamin-mineral compounds has been studied for half a century. In a review of the older research, Williams reported that, although results of a number of studies suggested ergogenic effects, the experimental designs were usually poorly controlled. In contrast, contemporary research indicates that such supplements, consumed for substantial periods, are not ergogenic for the athlete on a balanced diet. From Timothy Noakes's laboratory in South Africa, Weight conducted a thorough 9-month double-blind, placebo, crossover study. While multivitamin-mineral supplements did raise blood levels of some vitamins, the authors reported that 3 months of supplementation did not improve maximal oxygen uptake, the anaerobic (lactate) threshold, treadmill run time to exhaustion, or running performance in a 15-kilometer time trial. Anita Singh and her colleagues provided either a high-potency multivitamin-mineral supplement or a placebo to 22 healthy, physically active males for 90 days. The vitamin dosages ranged from 300 to 6,000 percent of the RDAs. Although serum levels of many of the vitamins increased, there were no significant effects on physiological variables during a 90-minute run, nor were there any effects on maximal heart rate, VO_2 max, or time to exhaustion. Finally, Richard Telford and his colleagues matched 82 nationally ranked Australian athletes in training at the Australian Institute of Sport and assigned them to either a supplement or placebo treatment. The supplement contained an assortment of vitamins and minerals, ranging from about 100 to 5,000 percent of the RDA. The supplement was taken for approximately 7–8 months, and the subjects were tested on a variety of sport-specific tests (e.g., swim bench) as well as common tests of strength (torque), anaerobic power (400-meter run), and aerobic endurance (12-minute run and VO_2 max). These investigators reported no significant effect of the supplement on any measure of physical performance when compared with athletes whose vitamin and mineral RDA were met by dietary intake.

Vitamins and minerals are also marketed to physically active individuals in liquid forms, such as sports drinks. Although research is limited, Fry and others reported no significant improvement in two tests of anaerobic exercise performance (30-second cycle sprint and one set of squats) following 8 weeks of supplementation with a multivitamin/mineral liquid. Thus, all of the current reputable research refutes an ergogenic effect of multivitamin-mineral supplements in adequately nourished athletes.

Prudent recommendations Multivitamin-mineral supplements may not enhance athletic performance in well-nourished athletes. Those involved in weight-control sports with limited caloric intake might consider taking a simple one-a-day supplement with no more than 100 percent of the RDA for the essential vitamins and minerals. As recently recommended by Wardenaar and colleagues, "Athletes should consider making better food choices and the daily use of a low-dosed multivitamin supplement." The *Training Table* in this section provides prudent recommendations for those taking a multivitamin supplement.

- Read the supplement label carefully and avoid "added ingredients" such as caffeine, herbs, botanicals, and other such substances.

Training Table

Nutrient supplements display a nutrition label that is slightly different from that on foods. The Supplement Facts label must list the ingredient(s), amount(s) per serving, serving size, suggested use, and % Daily Value (if one has been established). This sample label also includes a structure/function claim. Thus, it must also include the FDA warning that these claims have not been evaluated by the agency.

For consumers wanting to take a multivitamin supplement, the following prudent guidelines are recommended for healthy (nonpregnant or lactating) adults:

- Talk with your health-care provider about taking any dietary supplement.
- Check the % daily values for each vitamin. In general, select multivitamins with 100% or less of daily values for each vitamin.
- Purchase vitamin supplements from reputable sources.

Key Concepts

▶ Although research findings regarding the ability of antioxidant vitamin supplements to prevent muscle tissue damage following intense exercise are somewhat equivocal, in general, the benefits are minor. Antioxidant vitamin supplementation has not been shown to enhance sports or exercise performance.

▶ Results from well-controlled research generally indicate that multivitamin/mineral supplements are not effective ergogenic aids.

Check for Yourself

▶ Search online for "sports vitamins." Select at least two of these vitamins and describe the marketing of the supplement product and reported claims to sports performance and/or health. Then, compare that information to what is provided by the NIH Office of Dietary Supplements: https://ods.od.nih.gov/factsheets/list-VitaminsMinerals/.

APPLICATION EXERCISE

Construct a brief, one-page survey regarding vitamin supplement use in athletes. Questions you might include are:

1. Do you take a vitamin supplement?
   _____ Yes _____ No
2. What type of supplement do you take?
   _____ Multivitamin
   _____ Multivitamin/mineral
   _____ Vitamin B complex
   _____ Vitamin D
   _____ Other
3. How often do you take the supplement?
   _____ Two or more times a day
   _____ Daily
   _____ Several times a week
   _____ Once a week
   _____ Several times a month
   _____ Once a month
   _____ Never
4. Why do you take the vitamin supplement?
   _____ To help guarantee good health
   _____ To enhance my sports or exercise performance
   _____ Other

Once your survey is developed, get permission to administer it to some physically active individuals or athletes, such as participants in recreational sports activities or sports at your school, members of a local cycling or running club, or members of a commercial fitness facility. Share the findings with your class.

Review Questions—Multiple Choice

1. Vitamin A toxicity is most likely to occur from _____.
   a. consuming too many dark green and deep orange vegetables
   b. eating liver more than once per week
   c. consuming high-dosage vitamin A supplements
   d. drinking too much vitamin A–fortified cow's milk
   e. eating red meat 2 days per week
2. Which of the following populations is at the greatest risk for a vitamin B12 deficiency?
   a. young adults
   b. endurance athletes
   c. children
   d. vegans
   e. male athletes
3. Which of the following vitamins is most important for promoting optimal bone health?
   a. vitamin C
   b. vitamin B12
   c. vitamin E
   d. riboflavin
   e. vitamin D
4. A deficiency of either of these two vitamins produces a similar anemia:
   a. thiamin and riboflavin
   b. riboflavin and niacin
   c. thiamin and vitamin B2
   d. pantothenic acid and biotin
   e. vitamin B12 and folate
5. If an individual is on a well-balanced diet, which of the following vitamin supplements will increase physical performance at sea level competition?
   a. thiamin
   b. niacin
   c. vitamin C
   d. vitamin E
   e. none of the above
6. Most of the B vitamins function in human metabolism as _____.
   a. coenzymes
   b. hormones
   c. antioxidants
   d. a source of kcal
   e. activators of mineral metabolism
7. Which of the following statements about antioxidant vitamins is true?
   a. Antioxidant vitamins include vitamins A, D, E, and K.
   b. Beta-carotene is a form of vitamin E.
   c. Athletes should take a daily antioxidant supplement to optimize performance.
   d. Animal food sources are the best sources of antioxidant vitamins.
   e. Athletes should be encouraged to consume antioxidant vitamins from natural food sources, including fruits and vegetables.
8. Which of the following are fat-soluble vitamins?
   a. vitamins B, C, D, niacin
   b. vitamin E, niacin, thiamin, riboflavin
   c. vitamins A, D, E, K
   d. vitamins A, B, C, D
   e. vitamins B1, B2, B6, C
9. All of the following are B vitamins except _____.
   a. riboflavin
   b. choline
   c. pantothenic acid
   d. biotin
   e. thiamin
10. The main function of vitamin E in the body is to act as a(n) _____.
    a. antioxidant
    b. superoxide
    c. free radical
    d. hormone
    e. source of energy

Answers to multiple choice questions:
1. c; 2. d; 3. e; 4. e; 5. e; 6. a; 7. e; 8. c; 9. b; 10. a

Critical Thinking Questions

1. Describe at least three specific reasons why vitamin D is important for athletes. Which athletes are at the greatest risk for a vitamin D deficiency and why?
2. For each of the B vitamins, identify the coenzyme form of that B vitamin and include a brief description of its importance in nutrient metabolism.
3. Develop a table that includes the adult RDA/AI and good food sources for each of the vitamins as well as choline.
4. Your friend is a former high school athlete who now works out at the gym six days per week and has come to you with questions about supplementation with vitamins. Your friend drinks several energy drinks per day and eats predominately processed and fast foods. Do you recommend your friend take a multivitamin supplement? Why or why not? What other advice do you have for your friend about vitamins?
5. Dominique is an elite alpine skier who competes internationally. She has been feeling under the weather lately and has considered taking a vitamin C supplement. Do you recommend such a supplement? Why or why not?

References

Academy of Nutrition and Dietetics. 2016. Position of the Academy of Nutrition and Dietetics: Vegetarian diets. *Journal of the Academy of Nutrition and Dietetics* 116(12):1970–80.

Alf, D., et al. 2013. Ubiquinol supplementation enhances peak power production in trained athletes: A double-blind placebo controlled study. *Journal of the International Society of Sports Nutrition* 10:24.

Allison, R. J., et al. 2015. No association between vitamin D deficiency and markers of bone health in athletes. *Medicine and Science in Sports and Exercise* 47 (4):782–88.

Ardestani, A., et al. 2011. Relation of vitamin D level to maximal oxygen uptake in adults. *American Journal of Cardiology* 107:1246–49.

Armstrong, L., and Maresh, C. 1996. Vitamin and mineral supplements as nutritional aids to exercise performance and health. *Nutrition Reviews* 54:S148–S58.

Atalay, M., et al. 2006. Dietary antioxidants for the athlete. *Current Sports Medicine Reports* 5:182–86.

Awasthi, S., et al. 2022. Prevalence of specific micronutrient deficiencies in urban school going children and adolescence of India: A multicenter cross-sectional study. *PLoS One* 17(5):e0267003.

Backx, E., et al. 2017. Seasonal variation in vitamin D status in elite athletes: A longitudinal study. *International Journal of Sport Nutrition and Exercise Metabolism* 27(1):6–10.

Bazzano, L., et al. 2006. Effect of folic acid supplementation on risk of cardiovascular diseases: A meta-analysis of randomized controlled trials. *Journal of the American Medical Association* 296:2720–26.

Beck, K., et al. 2021. Micronutrients and athletic performance: A review. *Food and Chemical Toxicology* 158:112618.

Bell, C., et al. 2005. Ascorbic acid does not affect the age-associated reduction in maximal cardiac output and oxygen consumption in healthy adults. *Journal of Applied Physiology* 98:845–49.

Benardot, D., et al. 2001. Can vitamin supplements improve sport performance? *Sports Science Exchange Roundtable* 12(3):1–4.

Bloomer, R. J., et al. 2012. Impact of oral ubiquinol on blood oxidative stress and exercise performance. *Oxidative Medicine and Cellular Longevity* 465020. doi: 10.1155/2012/465020.

Bonetti, A., et al. 2000. Effect of ubidecarenone oral treatment on aerobic power in middle-aged trained subjects. *Journal of Sports Medicine and Physical Fitness* 40:51–57.

Bonke, D. 1986. Influence of vitamin B_1, B_6 and B_{12} on the control of fine motoric movements. *Bibliotheca Nutritio et Dieta* 38:104–9.

Braakhuis, A. J., and Hopkins, W. G. 2015. Impact of dietary antioxidants in sport performance: A review. *Sports Medicine* 45(7):939–55.

Braam, L., et al. 2003. Factors affecting bone loss in female endurance athletes: A two-year follow-up study. *American Journal of Sports Medicine* 31:889–95.

Brancaccio, M., et al. 2022. The biological role of vitamins in athletes' muscle, heart and microbiota. *International Journal of Environmental Research and Public Health* 19(3):1249.

Brisswalter, J., and Louis, J. 2014. Vitamin supplementation benefits in master athletes. *Sports Medicine* 44 (3):311–18.

Bryer, S., and Goldfarb, A. 2006. Effect of high dose vitamin C supplementation on muscle soreness, damage, function, and oxidative stress to eccentric exercise. *International Journal of Sport Nutrition and Exercise Metabolism* 16:270–80.

Brzezianski, M., et al. 2022. Correlation between the positive effect of vitamin D supplementation and physical performance in young male soccer players. *International Journal of Environmental Research and Public Health* 19(9): 5138.

Buchman, A., et al. 2000. The effect of lecithin supplementation on plasma choline concentrations during a marathon. *Journal of the American College of Nutrition* 19:768–70.

Cannell, J., et al. 2009. Athletic performance and vitamin D. *Medicine and Science in Sports and Exercise* 41:1102–10.

Capo, X., et al. 2016. Effects of almond- and olive oil-based docohexaenoic- and vitamin E-enriched beverage dietary supplementation on inflammation associated to exercise and age. *Nutrients* 8(10):E619.

Chung, M., et al. 2011. Vitamin D with or without calcium supplementation for prevention of cancer and fractures: An updated meta-analysis for the U.S. Preventative Services Task Force. *Annals of Internal Medicine* 155(12):827–38.

Close, G., et al. 2006. Ascorbic acid supplementation does not attenuate post-exercise muscle soreness following muscle-damaging exercise but may delay the recovery process. *British Journal of Nutrition* 95:976–81.

Coburn, S., et al. 1990. Effect of vitamin B_6 intake on the vitamin content of human muscle. *FASEB Journal* 4:A365.

Craciun, A., et al. 1998. Improved bone metabolism in female elite athletes after vitamin K supplementation. *International Journal of Sports Medicine* 19:479–84.

Curhan, G., et al. 1996. A prospective study of the intake of vitamin C and B_6, and the risk of kidney stones in men. *Journal of Urology* 155:1847–51.

Dahlquist, D.T., et al. 2015. Plausible ergogenic effects of vitamin D on athletic performance and recovery. *Journal of the International Society of Sports Nutrition* 12:33 doi:10.1186/s12970-015-0093-8.

Davis, R., and Nelson, A. 2021. Niacin supplementation impairs exercise performance. *International Journal of Vitamin and Nutrition Research* October 26, 2021. Online ahead of print.

Davison, G., and Gleeson, M. 2005. Influence of acute vitamin C and/or carbohydrate ingestion on hormonal, cytokine, and immune responses to prolonged exercise. *International Journal of Sport Nutrition and Exercise Metabolism* 15:465–79.

Davison, G., and Gleeson, M. 2006. The effect of 2 weeks vitamin C supplementation on immunoendocrine responses to 2.5 h cycling exercise in man. *European Journal of Applied Physiology* 97:454–61.

De la Puenta Yague, M., et al. 2021. Role of vitamin D in athletes and their performance: Current concepts and new trends. *Nutrients* 12(2):579.

Deminice, R., et al. 2016. The effects of acute exercise and exercise training on plasma homocysteine: A meta-analysis. *PLoS One* 11(3):e0151653.

Dhalla, N. S., et al. 2013. Mechanisms of the beneficial effects of vitamin B6 and pyridoxal 5-phosphate on cardiac performance in ischemic heart disease. *Clinical Chemistry and Laboratory Medicine* 51 (3):535–43.

Doyle, M., et al. 1997. Allithiamine ingestion does not enhance isokinetic parameters of muscle performance. *International Journal of Sport Nutrition* 7:39–47.

Dysken, M. W., et al. 2014. Effect of vitamin E and memantine on functional decline in Alzheimer disease: The TEAM-AD VA cooperative randomized trial. *Journal of the American Medical Association* 31 (1):33–44.

e Silva Ade, S., and da Mota, M. P. 2014. Effects of physical activity and training programs on plasma homocysteine levels: A systematic review. *Amino Acids* 46(8):1795–804.

Englehart, M., et al. 2002. Dietary intake of antioxidants and risk of Alzheimer disease. *Journal of the American Medical Association* 287:3223–29.

Feskanich, D., et al. 2002. Vitamin A intake and hip fractures among postmenopausal women. *Journal of the American Medical Association* 287:47–54.

Finaud, J., et al. 2006. Oxidative stress: Relationship with exercise and training. *Sports Medicine* 36:327–48.

Fortmann, S. P., et al. 2013. Vitamin and mineral supplements in the primary prevention of cardiovascular disease and cancer: An updated systematic evidence review for the U.S. Preventive Services Task Force. *Annals of Internal Medicine* 159:824–34.

Franzke, B., et al. 2019. Fat soluble vitamins in institutionalized elderly and the effect of exercise, nutrition and cognitive training on their status—the Vienna Active Aging Study (VAAS): A randomized controlled trial. *Nutrients* 11(6):1333.

Fry, A., et al. 2006. Effect of a liquid multivitamin/mineral supplement on anaerobic exercise performance. *Research in Sports Medicine* 14:53–64.

Fu, H., et al. 2022. Folate intake and risk of colorectal cancer: A systematic review and up-to-date meta-analysis of prospective studies. *European Journal of Cancer Prevention* May 11, 2022. Online ahead of print.

Gerster, H. 1997. No contribution of ascorbic acid to renal calcium oxalate stones. *Annals of Nutrition and Metabolism* 41:269–82.

Gomez-Cabrera, M., et al. 2008. Oral administration of vitamin C decreases muscle mitochondrial biogenesis and hampers training-induced adaptions in endurance performance. *American Journal of Clinical Nutrition* 87:142–49.

Gorham, E., et al. 2007. Optimal vitamin D status for colorectal cancer prevention: A quantitative meta-analysis. *American Journal of Preventive Medicine* 32:210–6.

Gül, I., et al. 2011. Oxidative stress and antioxidant defense in plasma after repeated bouts of supramaximal exercise: The effect of coenzyme Q_{10}. *Journal of Sports Medicine and Physical Fitness* 51:305–12.

Halliday, T., et al. 2011. Vitamin D status relative to diet, lifestyle, injury, and illness in college athletes. *Medicine and Science in Sports and Exercise* 43:335–43.

Haralambie, G. 1976. Vitamin B_2 status in athletes and the influence of riboflavin administration on neuromuscular irritability. *Nutrition and Metabolism* 20:1–8.

Hathcock, J., et al. 2007. Risk assessment for vitamin D. *American Journal of Clinical Nutrition* 85:6–18.

Heath, E., et al. 1993. Effect of nicotinic acid on respiratory exchange ratio and substrate levels during exercise. *Medicine and Science in Sport and Exercise* 25:1018–23.

Higgins, M., et al. 2020. Antioxidants and exercise performance: With a focus on vitamin E and C supplementation. *International Journal of Environmental Research and Public Health* 17(22):8452.

Hosseini, B., et al. 2022. Effects of vitamin D supplementation on COVID-19 related outcomes: A systematic review and meta-analysis. *Nutrients* 14(10):2134.

Itoh, H., et al. 2000. Vitamin E supplementation attenuates leakage of enzymes following 6 successive days of running training. *International Journal of Sport Medicine* 21:369–74.

Jacobson, B., et al. 2001. Nutrition practices and knowledge of college varsity athletes: A follow-up. *Journal of Strength and Conditioning Research* 15:63–68.

Janssen, J., et al. 2021. The effect of a single bout of exercise on vitamin B2 status is not different between high- and low-fit females. *Nutrients* 13(11):4097.

Janousek, J., et al. 2022. Vitamin D: Sources, physiological role, biokinetics, deficiency, therapeutic use, toxicity, and overview of analytical methods for detection of vitamin D and its metabolites. *Critical Reviews in Clinical Laboratory Sciences* May 16, 2022. Online ahead of print.

Jastrzebska, M., et al. 2016. Effect of vitamin D supplementation on training adaptation in well-trained soccer players. *Journal of Strength and Conditioning Research* 30(9):2648–55.

Ji, Y., et al. 2013. Vitamin B supplementation, homocysteine levels, and the risk of cerebrovascular disease: A meta-analysis. *Neurology* 81 (15):1298–307.

Joubert, L., and Manore, M. 2006. Exercise, nutrition, and homocysteine. *International Journal of Sport Nutrition and Exercise Metabolism* 16:341–61.

Keum, N., and Giovannucci, E. 2014. Vitamin D supplements and cancer incidence and mortality: A meta-analysis. *British Journal of Cancer* 111(5):967–80.

Kim, M., et al. 2022. Can low-dose of dietary vitamin E supplementation reduce exercise-induced muscle damage and oxidative stress? A meta-analysis of randomized controlled trials. *Nutrients* 14(8):1599.

Kimmick, G., et al. 1997. Vitamin E and breast cancer: A review. *Nutrition and Cancer* 27:109–17.

Knez, W., et al. 2007. Oxidative stress in half and full Ironman triathletes. *Medicine and Science in Sports and Exercise* 39:283–88.

Kobayashi, Y. 1974. Effect of vitamin E on aerobic work performance in man during acute exposure to hypoxic hypoxia. Unpublished doctoral dissertation. University of New Mexico.

Kolka, M., and Stephenson, L. 1990. Skin blood flow during exercise after niacin ingestion. *FASEB Journal* 4:A279.

Krumbach, C., et al. 1999. A report of vitamin and mineral supplement use among university athletes in a Division I institution. *International Journal of Sport Nutrition* 9:416–25.

Laaksonen, R., et al. 1995. Ubiquinone supplementation and exercise capacity in trained young and older men. *European Journal of Applied Physiology* 72:95–100.

Lappe, J., et al. 2007. Vitamin D and calcium supplementation reduces cancer risk: Results of a randomized trial. *American Journal of Clinical Nutrition* 85:1586–91.

Larson-Meyer, D., and Willis, K. 2010. Vitamin D and athletes. *Current Sports Medicine Reports* 9:220–6.

Lawson, K., et al. 2007. Multivitamin use and risk of prostate cancer in the National Institutes of Health-AARP Diet and Health Study. *Journal of the National Cancer Institute* 99:754–64.

Lee, I., et al. 2005. Vitamin E in the primary prevention of cardiovascular disease and cancer: The Women's Health Study: A randomized controlled trial. *Journal of the American Medical Association* 294:56–65.

Leonard, S., and Leklem, J. 2000. Plasma B-6 vitamer changes following a 50-km ultra-marathon. *International Journal of Sport Nutrition and Exercise Metabolism* 10:302–14.

Li, Y., et al. 2016. Folic acid supplementation and the risk of cardiovascular diseases: A meta-analysis of randomized controlled trials. *Journal of the American Heart Association* 5(8):e003768.

Machefer, G., et al. 2007. Nutritional and plasmatic antioxidant vitamins status of ultra endurance athletes. *Journal of the American College of Nutrition* 26(4):311–6.

Madden, R. F., et al. 2017. Evaluation of dietary intakes and supplement use in paralympic athletes. *Nutrients* 9(11):E1266.

Manore, M. 1994. Vitamin B_6 and exercise. *International Journal of Sport Nutrition* 4:89–103.

Manson, J. E., and Bassuk, S. S. 2018. Vitamin and mineral supplements: What clinicians need to know. *Journal of the American Medical Association* 319(9):859.

Marawan, A., et al. 2019. Association between serum vitamin D levels and cardiorespiratory fitness in the adult population of the USA. *European Journal of Preventative Cardiology* 26(7):750.

Maroon, J. C., et al. 2015. Vitamin D profile in National Football League players. *American Journal of Sports Medicine* 43(5):1241–5.

Massey, L., et al. 2005. Ascorbate increases human oxaluria and kidney stone risk. *Journal of Nutrition* 135:1673–77.

Mastaloudis, A., et al. 2006. Antioxidants did not prevent muscle damage in response to an ultramarathon run. *Medicine and Science in Sports and Exercise* 38:72–80.

Matter, M., et al. 1987. The effect of iron and folate therapy on maximal exercise performance in female marathon runners with iron and folate deficiency. *Clinical Science* 72:415–22.

McGinley, C., et al. 2009. Does antioxidant vitamin supplementation protect against muscle damage? *Sports Medicine* 39:1011–32.

Mehran, N., et al. 2016. Prevalence of vitamin D insufficiency in professional hockey players. *Orthopaedic Journal of Sports Medicine* 4(12):2325967116677512.

Moran, D. S., et al. 2013. Vitamin D and physical performance. *Sports Medicine* 43:601–11.

Munoz, D., et al. 2017. Oxidative stress, lipid peroxidation indexes and antioxidant vitamins in long and middle distance athletes during a sport season. *Journal of Sports Medicine and Physical Fitness* 58(12):1713.

National Institutes of Health. 2006. National Institutes of Health State-of-the-Science conference statement: Multivitamin/mineral supplements and chronic disease prevention. *Annals of Internal Medicine* 145:364–71.

Neiss, A. 2005. Generation and disposal of reactive oxygen and nitrogen species. In *Molecular and Cellular Exercise Physiology*, ed. F. Mooren and K. Völker. Champaign, IL: Human Kinetics.

Nieman, D. 2001. Exercise immunology: Nutritional countermeasures. *Canadian Journal of Applied Physiology* 26:S45–55.

Nieman, D., et al. 2002. Influence of vitamin C supplementation on oxidative and immune changes after an ultramarathon. *Journal of Applied Physiology* 92:1970–77.

Nieman, D., et al. 2004. Vitamin E and immunity after the Kona Triathlon World Championship. *Medicine and Science in Sports and Exercise* 36:1328–35.

Nieman, D., et al. 2007. Quercetin reduces illness, but not immune perturbations after intensive exercise. *Medicine and Science in Sports and Exercise* 39:1561–69.

Nikolaidis, M. G., et al. 2012. Does vitamin C and E supplementation impair the favorable adaptations of regular exercise? *Oxidative Medicine and Cellular Longevity* 2012:707941 doi:10.1155/2012/707941.

Ogan, D., and Pritchett, K. 2013. Vitamin D and the athlete: Risks, recommendations, and benefits. *Nutrients* 5 (6):1856–68.

Owens, D. J., et al. 2017. Efficacy of high-dose vitamin D supplements for elite athletes. *Medicine and Science in Sports and Exercise* 49(2):349–56.

Owens, D. J., et al. 2018. Vitamin D and the athlete: Current perspectives and new challenges. *Sports Medicine* 48(Suppl 1):3.

Parnell, J. A., et al. 2016. Dietary intakes and supplement use in pre-adolescent and adolescent Canadian athletes. *Nutrients* 8(9):E526.

Paulsen, G., et al. 2014. Vitamin C and E supplementation hampers cellular adaptation to endurance training in humans: A double-blind, randomized, controlled trial. *Journal of Physiology* 592(8):1887–901.

Peake, J. 2003. Vitamin C: Effects of exercise and requirements with training. *International Journal of Sport Nutrition and Exercise Metabolism* 13:125–51.

Peklaj, E., et al. 2022. Is RED-S in athletes just another face of malnutrition? *Clinical Nutrition ESPEN* 48:298.

Pelletier, D. M., et al. 2013. Effects of quercetin supplementation on endurance performance and maximal oxygen consumption: A meta-analysis. *International Journal of Sport Nutrition and Exercise Metabolism* 23(1):73–82.

Penry, J., and Manore, M. 2008. Choline: An important micronutrient for maximal endurance-exercise performance. *International Journal of Sport Nutrition and Exercise Metabolism* 18:191–203.

Peters, E., et al. 1993. Vitamin C supplementation reduces the incidence of postrace symptoms of upper-respiratory-tract infection in ultramarathon runners. *American Journal of Clinical Nutrition* 57:170–74.

Pingitore, A., et al. 2015. Exercise and oxidative stress: Potential effects on antioxidant strategies in sports. *Nutrition* 31(7-8):916–22.

Powers, S., et al. 2004. Dietary antioxidants and exercise. *Journal of Sports Sciences* 22:81–94.

Rafnsson, S., et al. 2013. Antioxidant nutrients and age-related cognitive decline: A systematic review of population-based cohort studies. *European Journal of Nutrition* 52:1553–67.

Rawson, E. S., et al. 2018. Dietary supplements for health, adaptation, and recovery in athletes. *International Journal of Sports Nutrition and Exercise Metabolism* 28:188–99.

Ren, X., et al. 2020. Association of folate intake and plasma folate level with the risk of breast cancer: A dose-response meta-analysis of observational studies. *Aging* 12(21):21355.

Rock, C. 2007. Multivitamin-multimineral supplements: Who uses them? *American Journal of Clinical Nutrition* 85:277S-79S.

Rokitzki, L., et al. 1994. α-tocopherol supplementation in racing cyclists during extreme endurance training. *International Journal of Sport Nutrition* 4:253-64.

Rokitzki, L., et al. 1994. Acute changes in vitamin B6 status in endurance athletes before and after a marathon. *International Journal of Sport Nutrition* 4:154-65.

Sebastian, R., et al. 2007. Older adults who use vitamin/mineral supplements differ from nonusers in nutrient intake adequacy and dietary attitudes. *Journal of the American Dietetic Association* 107:1322-32.

Sellitto, C., et al. 2022. Antioxidant supplementation hinders the role of exercise training as a natural activator of SIRT1. *Nutrients* 14(10):2092.

Shindle, M., et al. 2011. Vitamin D status in a professional American football team. *Proceedings from the AOSSM Annual Meeting* ID 46-9849.

Shreshta, S., et al. 2022. Is routine vitamin A supplementation still justified for children in Nepal? Trial synthesis findings applied to Nepal national mortality estimates. *PLoS One* 17(5):e0268507.

Simon-Schnass, I. 1993. Vitamin requirements for increased physical activity: Vitamin E. In *Nutrition and Fitness for Athletes,* ed. A. Simopoulos and K. Pavlou. Basel, Switzerland: Karger.

Simon-Schnass, I., and Pabst, H. 1988. Influence of vitamin E on physical performance. *International Journal for Vitamin and Nutrition Research* 58:49-54.

Sindhughosa, D., et al. 2022. Additional treatment of vitamin D for improvement of insulin resistance in non-alcoholic fatty liver disease patients: A systematic review and meta-analysis. *Scientific Reports* 12(1):7716.

Singh, A., et al. 1992. Chronic multivitamin-mineral supplementation does not enhance physical performance. *Medicine and Science in Sport and Exercise* 24:726-32.

Snider, I., et al. 1992. Effects of coenzyme athletic performance system as an ergogenic aid on endurance performance to exhaustion. *International Journal of Sport Nutrition* 2:272-86.

Sobol, J., and Marquart, L. 1994. Vitamin/mineral supplement use among athletes: A review of the literature. *International Journal of Sport Nutrition* 4:320-34.

Stepanyan, V., et al. 2014. Effects of vitamin E supplementation on exercise-induced oxidative stress: A meta-analysis. *Applied Physiology, Nutrition, and Metabolism* 39(9):1029-37.

Stratton, M., et al. 2022. The influence of caffeinated and non-caffeinated multi-ingredient pre-workout supplements on resistance exercise performance and subjective outcomes. *International Journal of the International Society of Sports Nutrition* 19(1):126.

Suzuki, M., and Itokawa, Y. 1996. Effects of thiamine supplementation on exercise-induced fatigue. *Metabolism in Brain Diseases* 11:95-106.

Takanami, Y., et al. 2000. Vitamin E supplementation and endurance exercise: Are there benefits? *Sports Medicine* 29:73-83.

Telford, R., et al. 1992. The effect of 7 to 8 months of vitamin/mineral supplementation on athletic performance. *International Journal of Sport Nutrition* 2:135-53.

Tiidus, P., and Houston, M. 1995. Vitamin E status and response to exercise training. *Sports Medicine* 20:12-23.

Todd, J. J., et al. 2015. Vitamin D: Recent advances and implications for athletes. *Sports Medicine* 45:213-29.

Tomlinson, P. B., et al. 2014. Effects of vitamin D supplementation on upper and lower body muscle strength levels in healthy individuals. A systematic review with meta-analysis. *Journal of Science and Medicine in Sport* 18(5):575-80.

Tremblay, A., et al. 1984. The effects of a riboflavin supplementation on the nutritional status and performance of elite swimmers. *Nutrition Research* 4:201.

van der Beek, E. 1991. Vitamin supplementation and physical exercise performance. *Journal of Sports Sciences* 92:77-79.

Verhaar, H., et al. 2000. Muscle strength, functional mobility and vitamin D in older women. *Aging* 12:455-60.

Virk, R., et al. 1999. Effect of vitamin B-6 supplementation on fuels, catecholamines, and amino acids during exercise in men. *Medicine and Science in Sports and Exercise* 31:400-8.

Volpe, S. 2007. Micronutrient requirements for athletes. *Clinics in Sports Medicine* 26:119-30.

Wactawshi-Wende, J., et al. 2006. Calcium plus vitamin D supplementation and the risk of colorectal cancer. *New England Journal of Medicine* 354:684-96.

Wall, B. T., et al. 2012. Acute pantothenic acid and cysteine supplementation does not affect muscle coenzyme A content, fuel selection, or exercise performance in healthy humans. *Journal of Applied Physiology* 112 (2):272-78.

Wang, X., et al. 2007. Efficacy of folic acid supplementation in stroke prevention: A meta-analysis. *Lancet* 369:1876-82.

Warber, J., et al. 2000. The effects of choline supplementation on physical performance. *International Journal of Sport Nutrition and Exercise Metabolism* 10:170-81.

Wardenaar, F., et al. 2017. Micronutrient intakes in 553 Dutch elite and sub-elite athletes: Prevalence of low and high intakes in users and non-users of dietary supplements. *Nutrients* 9(2):E14.

Wardenaar, F. C., et al. 2017. Nutritional supplement use by Dutch elite and sub-elite athletes: Does receiving dietary counseling make a difference? *International Journal of Sports Nutrition and Exercise* 27(1):32-42.

Watson, T., et al. 2005. Antioxidant restriction and oxidative stress in short-duration exhaustive exercise. *Medicine and Science in Sports and Exercise* 37:63-71.

Webster, M., et al. 1997. The effect of a thiamin derivative on exercise performance. *European Journal of Applied Physiology* 75:520-24.

Weight, L., et al. 1988. Vitamin and mineral supplementation: Effect on the running performance of trained athletes. *American Journal of Clinical Nutrition* 47:192-95.

Williams, M. 1989. Vitamin supplementation and athletic performance. *International Journal for Vitamin and Nutrition Research,* Supplement 30:161-91.

Williams, M. 2011. Sports supplements: Quercetin. *ACSM Health & Fitness Journal.* 15(5):17-20.

Williams, S., et al. 2006. Antioxidant requirements of endurance athletes: Implications for health. *Nutrition Reviews* 64:93-108.

Willis, K., et al. 2008. Should we be concerned about the vitamin D status of athletes? *International Journal of Sport Nutrition and Exercise Metabolism* 18:204-24.

Woolf, K., and Manore, M. 2006. B-vitamins and exercise: Does exercise alter requirements? *International Journal of Sport Nutrition and Exercise Metabolism* 16:453-84.

Yavari, A., et al. 2015. Exercise-induced oxidative stress and dietary antioxidants. *Asian Journal of Sports Medicine* 6(1):e24898.

Zoppi, C., et al. 2006. Vitamin C and E supplementation effects in professional soccer players under regular training. *Journal of the International Society of Sports Nutrition* 3:37-44.

KEY TERMS

athletic amenorrhea
electrolytes
female athlete triad
ferritins
hematuria
heme iron
hemochromatosis
hemolysis
hepcidin
ions
iron-deficiency anemia
iron deficiency without anemia
major minerals
metalloenzymes
mineral
nonheme iron
osteoporosis
peak bone mass
secondary amenorrhea
trabecular bone
trace minerals

Minerals: The Inorganic Regulators

CHAPTER EIGHT

Erik Isakson/Blend Images LLC

LEARNING OUTCOMES

After studying this chapter, you should be able to:

1. Contrast major minerals, trace minerals, and minerals that may possibly be essential and identify which minerals fit into each category.

2. Name the major and trace minerals, state the RDA or AI for each, describe key functions, and identify several foods rich in each mineral.

3. Explain the role of calcium in bone metabolism and identify the factors that may contribute to bone health and those that may contribute to osteoporosis.

4. Describe the importance of phosphorus to health and the theory as to how phosphate salt supplementation may enhance sports performance, and highlight the research findings regarding its ergogenic efficacy.

5. Provide a list of heme and nonheme sources of iron, describe which is more readily absorbed, and explain factors that influence iron bioavailability.

6. Explain the importance of copper, chromium, selenium, and zinc in human health and exercise performance.

7. Summarize why health professionals may recommend mineral supplements under certain circumstances to improve the health of some individuals.

Introduction

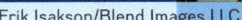
Erik Isakson/Blend Images LLC

You may recall the periodic table of the elements hanging on the wall in your high school or college chemistry class. At latest count there were 118 known elements, 78 of them occurring naturally and the remainder being synthetic. Many of the natural elements, including a wide variety of minerals, are essential to human bodily structure and function. For example, calcium and phosphorus are main constituents of bone, while iron, magnesium, zinc, and other minerals are cofactors of hundreds of enzymes regulating numerous body functions.

Much research attention is currently being devoted to the role of minerals in health and disease. For example, using the RDA as a basis for comparison, national surveys among the general population have revealed that either an inadequate dietary intake of some minerals or an excessive dietary intake of others in certain small segments of the population may be contributing to several health problems. Laboratory studies using either animals or humans as subjects have explored the roles of both deficiencies and excesses of various minerals, including both essential and nonessential elements, on human health and disease processes.

An increasing number of research studies have been conducted with athletes to evaluate the effect of mineral nutrition on physical performance and the converse—the effect of exercise on mineral metabolism. Because some minerals function similarly to vitamins, a deficiency state could adversely affect performance. Moreover, exercise in itself may be a contributing factor to mineral deficiencies or impaired mineral metabolism in some types of athletes. Additionally, several mineral supplements have been marketed specifically for physically active individuals.

This chapter is especially important to all females and young athletes because it addresses two major dietary concerns: obtaining sufficient calcium and obtaining sufficient iron. These key minerals are of particular importance to females who participate in sports or are otherwise physically active. The female athlete triad and relative energy deficiency in sport (RED-S) are introduced in this chapter with the major focus on osteoporosis because of its relationship to calcium metabolism. An expanded discussion of disordered eating and eating disorders is presented in chapter 10. Female endurance athletes also need to obtain adequate dietary iron intake because of its important role in the oxygen energy system.

The major purpose of this chapter is to summarize the available data relative to the effect of mineral nutrition on physical performance and health. The first section discusses some basic facts about the general role of those minerals that are essential to human nutrition. The second and third sections cover, respectively, major minerals and trace minerals that are important to human health and physical activity performance. In these two sections, each of the minerals is discussed in terms of its dietary requirement, good food sources, metabolic functions in the body with particular reference to the physically active individual, an evaluation of the research pertaining to the effects of deficiencies or supplementation on health or exercise performance, and prudent recommendations. The last section summarizes dietary mineral nutrition guidelines for those who exercise for health or sport.

Basic Facts

What are minerals, and what is their importance to humans?

A **mineral** is an inorganic element found in nature. Minerals are found in the soil and are eventually incorporated in growing plants. Most animals get their mineral nutrition from the plants they eat, whereas humans obtain their supply from both plant and animal food. Drinking water may also be a good source of several minerals, including fluoride. As minerals are excreted daily from the body in sweat, urine, or feces, they must regularly be replaced.

Growth and Development Many minerals are used as the building blocks for body tissues, such as bones, teeth, muscles, and other organic structures. In particular, calcium and phosphorus are important to bone health. As well, iron is an important component of hemoglobin, which is needed for optimal oxygen transport during aerobic endurance exercise.

Metabolic Regulation A number of minerals are involved in regulation of metabolic processes. Many are components of enzymes known as **metalloenzymes**, such as the cytochrome enzymes in the mitochondria that facilitate ATP production. Others, such as zinc and copper, are part of the natural antioxidant enzymes discussed in chapter 7. Still others exist as **ions**, or **electrolytes**, which are small particles carrying electrical charges. They are important components or activators of several enzymes and hormones. Specific to athletic performance, minerals are involved in muscle contraction, normal heart rhythm, nerve impulse conduction, oxygen transport, oxidative phosphorylation, enzyme activation, immune function antioxidant activity, bone health, acid-base balance of the blood, and maintenance of body water supplies. **Figure 8.1** provides a broad overview of key mineral functions in the body.

Energy Metabolism Minerals are comparable to vitamins relative to energy production in humans. Although minerals may play a significant role in the generation of energy via their metabolic functions, they do not provide energy (kcal) themselves.

Inadequate mineral nutrition has been associated with impairment of normal physiological functions, as well as with a variety of human diseases, including anemia, high blood pressure, obesity, diabetes, cancer, tooth decay, and osteoporosis. However, excessive intake of minerals may also contribute to significant health risks.

Thus, proper dietary intake of essential minerals is necessary for optimal health and physical performance.

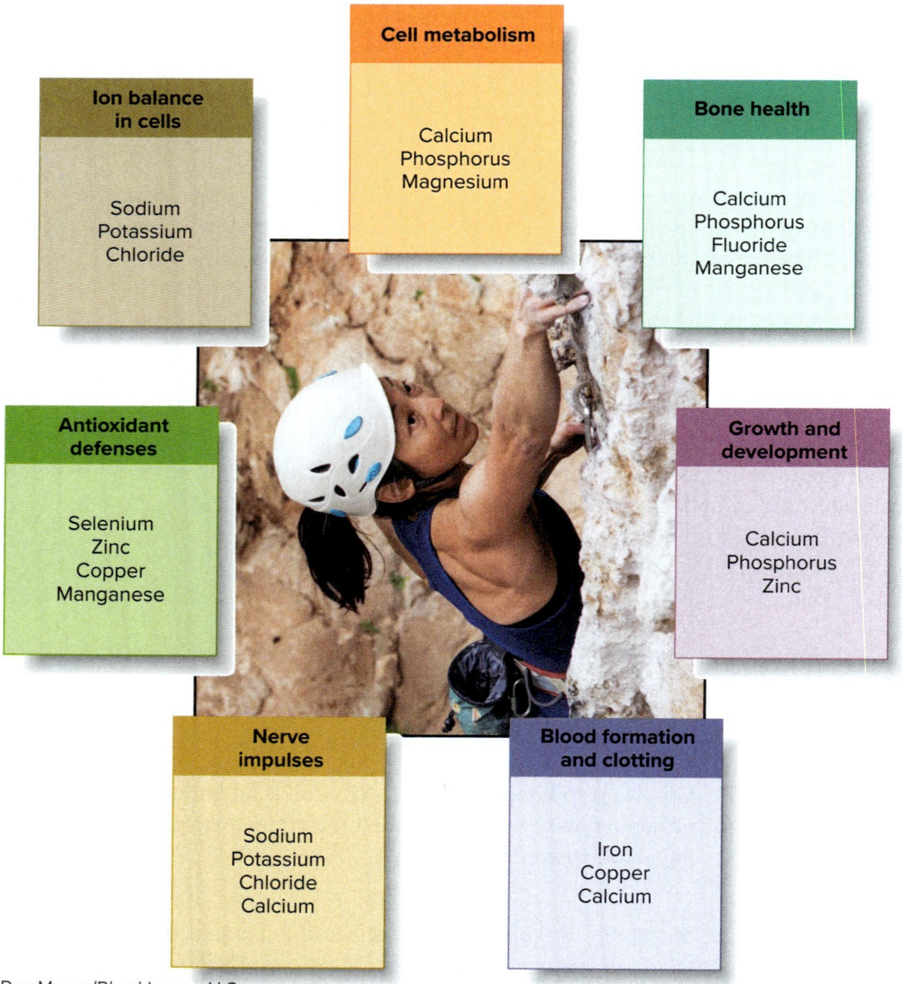

What minerals are essential to human nutrition?

Of all the elements in the periodic table, only 25 are currently known to be, or presumed to be, essential in humans. Five of these elements (hydrogen, oxygen, carbon, sulfur, nitrogen) make up the carbohydrate, fat, and protein that we eat and the water we drink. They are the elements found in body water, protein, fat, and carbohydrates stores, which constitute slightly more than 96 percent of a person's body weight. The remaining minerals compose less than 4 percent of an average person's body weight, but are equally important.

Essential minerals are classified as either major or trace minerals. The classification is based on the amount of mineral required by the body, not on the importance of the mineral to health. A major mineral is one that the human body requires in amounts of 100 mg or more daily. Comparatively, trace minerals are needed in amounts less than 100 mg daily. A third category of minerals are those minerals that may be essential to human health, but further research is needed. Like other minerals, the body only requires small amounts of these possibly essential minerals, including nickel and arsenic. **Table 8.1** lists those minerals that are major minerals, trace minerals, and those that are possibly essential.

In general, how do deficiencies or excesses of minerals influence health or physical performance?

Similar to vitamin deficiencies, mineral deficiencies may occur in several stages. The first three stages (preliminary, biochemical deficiency, and physiological deficiency) may be termed subclinical malnutrition and may or may not have significant effects on health or physical performance. In the clinically manifest deficiency state, however, health and performance most likely will suffer.

The interaction of exercise and mineral nutrition may pose some special health problems, as we shall see in later sections of this chapter. In regard to the preliminary stage of a deficiency, some athletes may reduce their mineral intake as they shift toward a low-kcal diet. Changes in food selection may also be important, for the bioavailability of many minerals is markedly influenced by the form in which they are consumed. In general, most minerals are poorly absorbed from the GI tract. For example, because only 10 percent of dietary iron from the average American diet is absorbed, the RDA for iron is ten times the amount actually needed by the body. Moreover, mineral absorption may be inhibited by certain compounds in foods, and supplementation with one mineral may impair the absorption of another. In athletes, factors that lower intake and

FIGURE 8.1 Minerals contribute to many functions in the body. Mineral deficiencies therefore lead to a variety of health problems and may impair physical performance.

Cell metabolism: Calcium, Phosphorus, Magnesium

Ion balance in cells: Sodium, Potassium, Chloride

Bone health: Calcium, Phosphorus, Fluoride, Manganese

Antioxidant defenses: Selenium, Zinc, Copper, Manganese

Growth and development: Calcium, Phosphorus, Zinc

Nerve impulses: Sodium, Potassium, Chloride, Calcium

Blood formation and clotting: Iron, Copper, Calcium

Don Mason/Blend Images LLC

TABLE 8.1 Classifying minerals as major minerals, trace minerals, or minerals that may possibly be essential

Major minerals	Trace minerals	Minerals that may possibly be essential
Calcium (Ca)	Chromium (Cr)	Arsenic (As)
Chloride (Cl$^-$)	Copper (Cu)	Boron (B)
Magnesium (Mg)	Fluoride (F$^-$)*	Lithium (Li)
Phosphorus (P)	Iodine (I)	Nickel (Ni)
Potassium (K)	Iron (Fe)	Silicon (Si)
Sodium (Na)	Manganese (Mn)	Vanadium (V)
Sulfur (S)	Molybdenum (Mo)	
	Selenium (Se)	
	Zinc (Zn)	

*Fluoride is technically not essential; however, the mineral contributes to strengthening teeth and bones.

absorption may be compounded because certain physical activities may raise some mineral requirements. Greater amounts of certain minerals may be needed for the synthesis of new tissues associated with physical training, or to replace losses often observed during and following intense exercise training via sweat, urine, and feces.

Sports dietitians are becoming concerned that the presence of these factors during the preliminary stage of a mineral deficiency could lead to the subsequent stages of subclinical malnutrition, or even to a clinical deficiency. According to recent dietary surveys, calcium, iron, potassium, and zinc are the minerals most likely to be deficient in the diet of American adults. For adult females in particular, average intakes of calcium and iron are below recommended levels. Athletes with inadequate dietary intake and biochemical deficiencies of several minerals are predominately athletes involved in weight-control sports such as gymnastics, dancing, figure skating, and wrestling. Female athletes in particular could be at risk for mineral insufficiency due to inadequate dietary intake, menstruation, and inflammatory responses to heavy physical activity. Experts disagree about the potential adverse effects of such dietary or biochemical deficiencies, but certain physiological and clinically manifest mineral deficiencies are known to impair performance.

The human body possesses a very effective control system for some minerals. When a deficiency occurs, the body absorbs more of the mineral from the food in the intestine and excretes less via routes such as the urine. When an excess is consumed, the opposite is true; less is absorbed and more is excreted. However, the body has a limited ability to excrete certain minerals, so excessive consumption may override these natural control systems and cause a number of health problems, even in relatively small dosages.

Additionally, a few minerals not important to human nutrition, such as lead, mercury, cadmium, and arsenic as well as some industrial forms of silicon and chromium, may be extremely toxic to the human body. For example, as discussed in chapter 5, mercury in polluted waterways may accumulate in fish, which if eaten may damage the nervous system. According to the *2020-2025 Dietary Guidelines for Americans,* young children and pregnant/lactating females should choose seafood lowest in methylmercury and possibly eat less seafood than recommended in a healthy US-style dietary pattern.

The following sections summarize the role of select major and trace minerals in health and exercise performance.

Key Concepts

▸ Minerals perform two of the three major functions of nutrients in food: the formation of several body tissues and the regulation of numerous physiological processes. Minerals are not a source of energy (kcal) but, like vitamins, they are essential for energy metabolism.

▸ Minerals are classified as major or trace minerals based on the human body's requirement for such minerals. Both major and trace minerals are essential to health. A third class of minerals includes those that play potentially essential roles. Further research is needed on these minerals.

▸ Mineral deficiencies or excess may have adverse effects on both health and physical performance.

Check for Yourself

▸ Based on your age and sex. look up your RDA/AI and UL for the major minerals. Develop a table with these recommendations.

Major Minerals

The seven **major minerals** are calcium, phosphorus, magnesium, potassium, sodium, chloride, and sulfur. In general, the human body maintains a proper balance of these minerals through precise hormonal control mechanisms, but deficiencies or excesses may occur and disturb normal physiological functions, thus impairing health or physical performance. Sulfur, an integral component of several amino acids and vitamins, is not discussed here as its functions are associated with those nutrients. Because potassium, sodium, and chloride are the major electrolytes in sweat, and in some sports drinks, they are covered in the following chapter dealing with water and temperature regulation. **Table 8.2** provides an overview of the major minerals.

Calcium (Ca)

Calcium, a silver-white metallic element, is the most abundant mineral in the body, representing almost 2 percent of the body weight. According to the *2020-2025 Dietary Guidelines for Americans*, calcium is considered a "nutrient of public health concern."

TABLE 8.2 Overview of the major minerals

Major mineral	RDA/AI and UL for adults age 19–50	Major food sources	Key functions in the body	Deficiency symptoms	Toxicity symptoms
Calcium	RDA: 1,000 mg UL: 2,500 mg	Dairy products including cow's milk, cheese, yogurt and ice cream; egg yolk; dark green leafy vegetables; canned fish; soy foods; calcium-fortified food products	Bone formation; enzyme activation; nerve impulse transmission; muscle contraction; cell membrane potential	Osteoporosis; rickets; impaired muscle contraction; muscle cramps	Constipation; inhibition of absorption of other minerals; kidney stones, calcification of soft tissues
Chloride	AI: 2,300 mg UL: 3,600 mg	Processed foods including salty snacks; table salt; seaweed; olives	Fluid balance; maintenance of acid-base balance; production of stomach acid (HCl); nerve impulse transmission	Convulsions (in infants)	Hypertension
Magnesium	RDA: 400–420 mg ♂ 310–320 mg ♀ UL: 350 mg from medications and not food sources	Green vegetables (e.g., spinach); legumes; dairy foods; nuts; whole-grain products; wheat bran	Protein synthesis; metalloenzyme; 2,3-DPG formation; glucose metabolism; smooth muscle contraction; bone component	Rare. Muscle weakness; apathy; muscle twitching; muscle cramps; cardiac arrhythmias	Nausea; vomiting; diarrhea
Phosphorus	RDA: 700 mg UL: 4,000 mg	Dairy products including cow's milk, cheese, yogurt and ice cream; hummus; meat; soft drinks; fish; eggs; sesame seeds; whole-grain products	Bone formation; acid-base balance; cell membrane structure; B vitamin activation; organic compound component (e.g., ATP-PCr)	Rare. Deficiency symptoms parallel calcium deficiency; muscular weakness	Rare. Impaired calcium metabolism; poor bone mineralization; GI distress from phosphate salts
Potassium	AI: 3,400 mg for ♂ 2,600 mg for ♀ UL: Not determined	Fruits and vegetables; dairy foods; legumes; meat; whole grains	Fluid balance; maintenance of acid-base balance; nerve impulse transmission	Muscle cramping; irregular heartbeat	No UL has been determined
Sodium	AI: 1,500 mg UL: 2,300 mg	Table salt; processed foods such as pretzels, chips, and other snack foods; lunch meats; sauces; condiments	Fluid balance; maintenance of acid-base balance; nerve impulse transmission; muscle contraction; transport of substances into cells	Muscle cramping	Hypertension in salt-sensitive individuals; increase loss of calcium in urine
Sulfur	None	Protein-rich foods including fish, poultry, meat, eggs, dairy foods, and cheese; legumes; nuts	Part of organic compounds, including certain amino acids (cysteine and methionine) and vitamins (biotin and thiamin)	None reported	Unlikely from dietary sources

DRI The RDA for calcium is 1,000 mg/day for adults age 19–50. For females age 51 and older and males over age 70, the RDA increases to 1,200 mg/day. Comparatively, for children age 9–18, the RDA is actually slightly higher at 1,300 mg/day to account for increased bone mass as a result of growth and development. The UL for adults is 2,000–2,500 mg/day.

Food Sources Calcium content is highest in dairy products. One 8-ounce glass of cow's milk, which contains about 300 mg of calcium, supplies about one-third of the RDA for adults age 19–50. Other equivalent dairy foods are 1½ ounces of cheese, 1 cup of yogurt, and 1¾ cups of ice cream. In the United States, dairy products supply approximately 75 percent of daily calcium intake for

Image Source/Glow Images

adults. Other quality sources are fish with small bones, such as sardines and canned salmon, dark-green leafy vegetables (particularly broccoli, kale, and turnip greens), calcium-set tofu, legumes, and nuts. For individuals with lactose intolerance, the use of yogurt, lactase enzymes, or smaller portions of cow's milk may be helpful. Calcium is also used as a preservative in some foods, such as breads, which may provide small amounts. Additionally, some food products, such as fruit juice and cereals, are fortified with calcium; 1 serving of calcium-fortified orange juice may contain 300–350 milligrams, while some brands of fortified cereal contain 100 percent of the DV for calcium.

There are several factors that influence the bioavailability of calcium. For the average adult, only about one-third of the calcium consumed in the diet is absorbed. Absorption rates for calcium are greater in infants and children as well as during pregnancy. The *Training Table* in this section summarizes key factors influencing calcium absorption and, thus, bioavailability.

Training Table

Calcium absorption occurs throughout the intestines with the majority of absorption occurring in the duodenum. There are several factors that impact calcium absorption.

Factors that **INCREASE** calcium absorption:

- Adequate vitamin D (calcitriol)
- Adequate HCl secretion as part of gastric juices
- Lactose (in infants)

Factors that **DECREASE** calcium absorption:

- Deficiency of vitamin D
- Reduced HCl secretion as part of gastric juices
- Consumption of phytic and oxalic acids
- High-fiber diet
- High-phosphorus diet
- Fat malabsorption
- Chronic diarrhea (for example, as a result of IBS)

Major Functions The vast majority of body calcium, about 98 percent, is found in the skeleton, where it gives strength by the formation of salts such as calcium phosphate. One percent is used for tooth formation; the remainder, which exists in an ionic state or in combination with certain proteins, exerts considerable influence over human metabolism. Intracellular calcium ions (Ca^{2+}) are involved in all types of muscle contraction, including that of the heart, skeletal muscle, and smooth muscle found in blood vessels such as the arteries. Calcium activates a number of enzymes; in this capacity it plays a central role in both the synthesis and breakdown of muscle glycogen and liver glycogen. Calcium also helps regulate nerve impulse transmission, blood clotting, and secretion of hormones. It should be noted that the skeletal content of calcium is not inert. The physiological functions of calcium, such as nerve cell transmission, take precedence over formation of bone tissue. If the diet is low in calcium for a short time, the body can mobilize some from the skeleton through the action of hormones, such as parathyroid hormone and calcitriol (hormonal form of vitamin D), to maintain an adequate amount in ionic form. Given the large skeletal reserves, impaired metabolic processes associated with calcium deficiency are almost nonexistent.

Deficiency: Health and Physical Performance Calcium balance in the human body is rather complex. **Figure 8.2** depicts the fate of an intake of 1,000 mg of calcium. Only 300 mg (about 30 percent) is absorbed, while the remaining 700 mg is excreted in the feces. The calcium that is absorbed into the blood interacts with the current body stores, the net result being the excretion of 300 mg through the intestines, kidneys, and sweat to balance the amount originally absorbed. Calcium deficiency may develop from inadequate dietary intake or increased excretion.

Inadequate dietary intake of calcium is the major cause of calcium deficiency. According to recent reports, the average intake of calcium in adults in the United States is 857 mg/day for females and 1,084 mg/day for males.

Those individuals at the greatest risk for inadequate calcium intake include middle-aged and older females (50 years and older), adolescent girls (9–18 years), and older males (70 years and older). Other groups of people who may be at risk for calcium inadequacy include:

- Individuals with lactose intolerance or an allergy to cow's milk. Those with such conditions may eliminate cow's milk from the diet, a major source of dietary calcium.
- Individuals with an eating disorder, including anorexia nervosa. Eating disorders, which may interfere with calcium metabolism, are discussed in chapter 10.
- Some individuals following a predominately plant-based diet. Those who consume no dairy products, particularly vegans, are at greatest risk.
- Athletes exhibiting signs of relative energy deficiency in sport (RED-S).
- Additionally, studies indicate that strenuous exercise may increase calcium losses in the sweat.

The major health problems associated with impaired calcium metabolism involve diseases of the bones. A number of factors are involved in the formation, or mineralization, of bone tissue, including mechanical stresses such as exercise; hormones such as parathyroid hormone, calcitonin, vitamin D (calcitriol), and estrogen; and dietary calcium. An imbalance in any one of these factors could lead to bone demineralization, resulting in the development of rickets in children and osteoporosis in adults. Rickets is most often associated with vitamin D deficiency and is discussed in chapter 7.

Osteoporosis, thinning and weakening of the bones related to loss of calcium stores, is a debilitating disease that affects more than 50 million people in the United States. Approximately

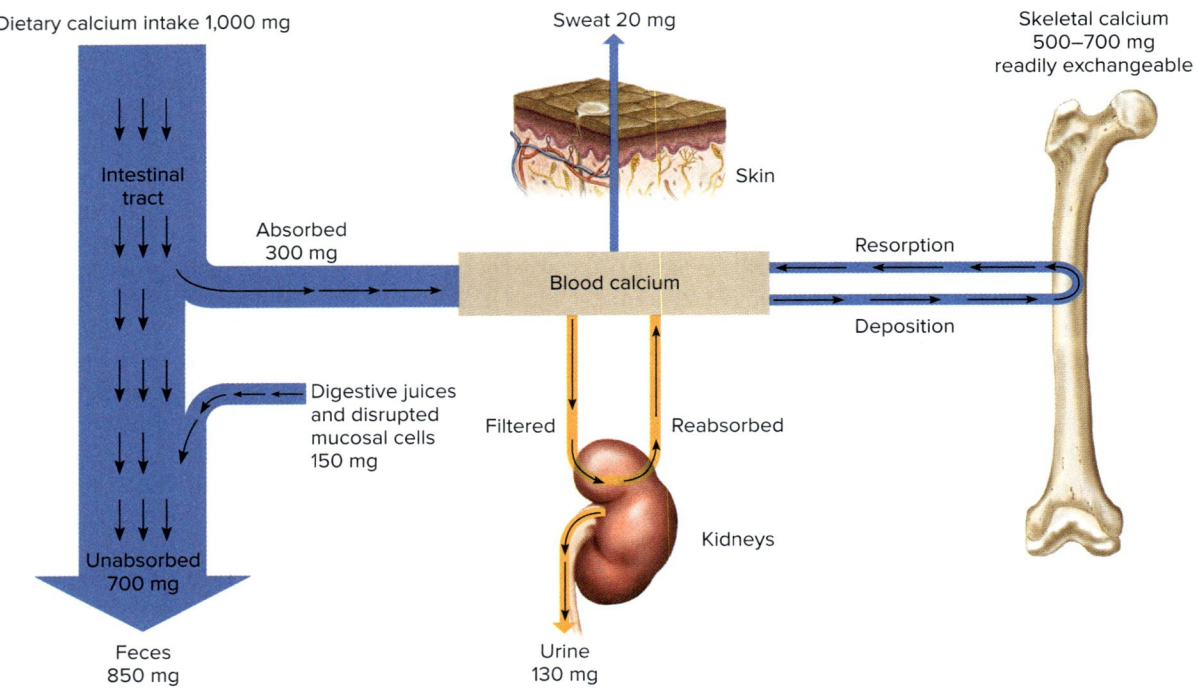

FIGURE 8.2 Calcium balance in an adult who requires 1,000 mg daily. On an intake of 1,000 mg, only about 30 percent, or 300 mg, are absorbed into the body, the remaining 700 mg being excreted in the feces. To maintain calcium balance, 300 mg are excreted, including an additional 150 mg in the feces, 130 mg through the kidneys to the urine, and 20 mg in sweat.

10 million American adults over the age of 50 have osteoporosis, and an additional 43 million are at risk due to low bone mass. Just in the United States, over 1.5 million people experience an osteoporosis-related fracture each year. Such fractures are particularly common in older adult females. Osteoporosis is a multi-factorial disease with potential contributions from genetic, endocrine, exercise, and nutritional factors, with the last including calcium, vitamin D, fluoride, magnesium, and other trace elements. The condition is age- and sex-related, being most prevalent in white, postmenopausal females. However, as noted by the National Institutes of Health, osteoporosis occurs in all populations and at all ages and has significant physical, psychosocial, and financial consequences.

> www.bonehealthandosteoporosis.org/ Visit the Bone Health and Osteoporosis Foundation website for more information about osteoporosis, including information on diagnosis tools and educational materials for patients.

A positive family history, or heredity, and low levels of estrogen are the two primary risk factors in females. European and Asian females are at higher risk for osteoporosis than females of African ancestry. Following menopause, estrogen production is diminished. Estrogen is a hormone essential for optimal calcium balance in females. Bone receptors for estrogen have been identified, indicating an active role in bone metabolism. In general, reduced levels of estrogen lead to negative calcium balance and a rapid onset of bone demineralization. This softening of the bones predisposes individuals to fractures, particularly in the spine, the end of the radius in the forearm, and the neck of the femur at the hip joint, as illustrated in **figure 8.3**. These fractures

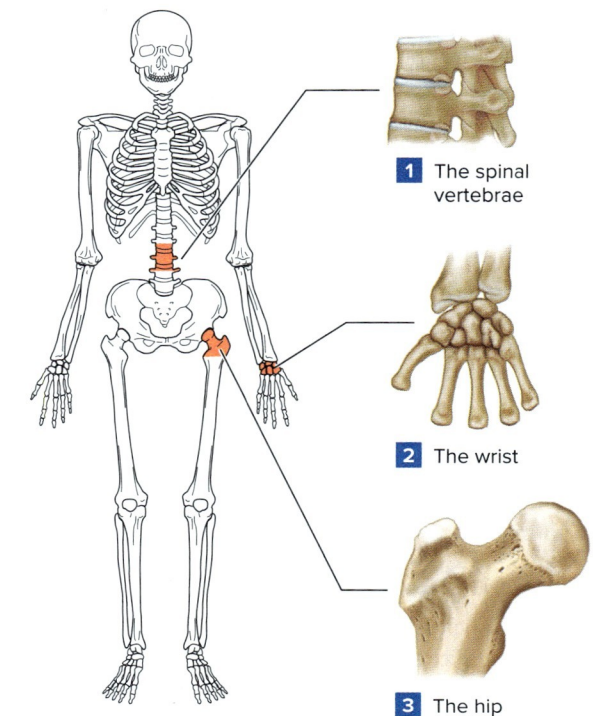

FIGURE 8.3 Three principal sites of osteoporosis fractures.

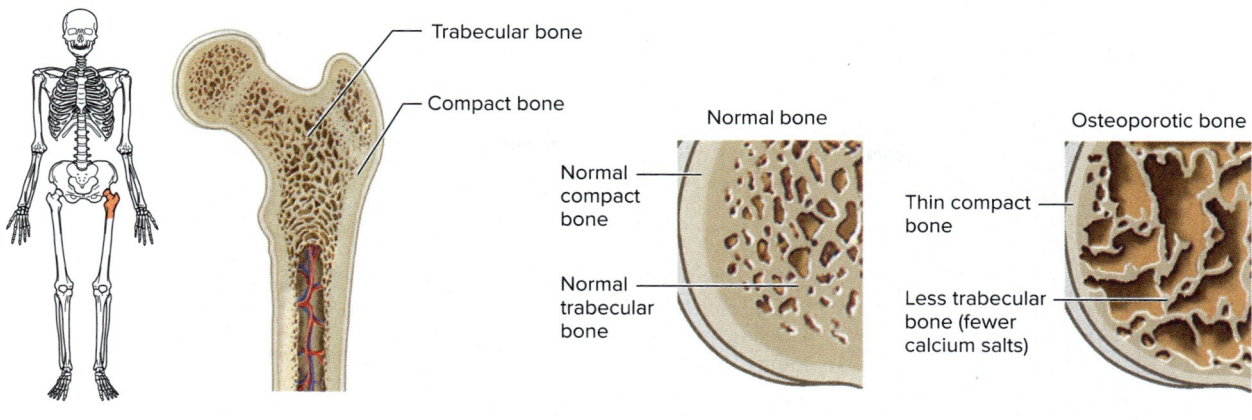

FIGURE 8.4 In osteoporosis, impaired calcium metabolism may decrease the external compact bone thickness and the strength of the internal trabecular bone lattice network.

may be completely debilitating to the older individual. The spinal fracture is more common because the vertebrae are composed of **trabecular bone**, a spongy type of bone more susceptible to calcium loss than the more dense compact bone. However, both types of bone may be lost during osteoporosis, as depicted in **figure 8.4**.

Osteoporosis is known as the "silent killer." The disease itself causes no pain, but following a serious bone fracture nearly a third or more of older adults die due to accompanying illnesses within a year. Health professionals recommend that females age 65 and over, and others with risk factors for osteoporosis, routinely have measurements of bone mineral density (BMD).

Although heredity and estrogen status are strong risk factors for osteoporosis, some lifestyle factors may impair optimal bone metabolism. Both physical inactivity and inadequate dietary intake of calcium are risk factors for osteoporosis, and so are cigarette smoking, stress, and use of various medications. **Table 8.3** highlights the risk factors for osteoporosis. All these factors may influence **peak bone mass**, the highest bone mass in young adulthood. Modifying your lifestyle to increase your peak bone mass is like putting money in the bank for later in life. However, although the idea that the growing years are an opportune time to optimize bone mass and strength certainly has merit, the importance of these gains depends largely on their permanence. Thus, you need to continue to practice good dietary and exercise habits to maintain optimal bone health.

Relative to physical performance, the effect of a calcium deficiency depends on whether calcium levels are low in the blood or in the bones. Serum calcium levels are usually regulated by several hormones in the average individual. The body can adapt to low dietary intake by increasing the rate of absorption from the intestines and decreasing the rate of excretion by the kidneys. Because the skeleton is a large reservoir of body calcium, low serum levels are rare. When they do occur, it usually is because of hormonal imbalances rather than dietary deficiencies.

Fortunately, serious deficiencies of serum calcium are rare in athletes because hormones may extract calcium from the bone as needed. Nevertheless, as shall be noted, bone health is one of the major concerns in the female athlete triad and with RED-S.

Supplementation: Health and Physical Performance Considerable research has focused on the role of calcium supplementation, often in concert with vitamin D supplementation, on bone health. While each of these micronutrients is necessary for the full expression of the effect of the other, and while their actions are independent, their effects on skeletal health are complementary. In addition, exercise may play a role in enhancing bone health.

TABLE 8.3	Risk factors for osteoporosis
Heredity	Positive family history
Race	White or Asian
Sex	Female
Menstrual status	Postmenopausal; amenorrheic
Age	Older adults
Exercise	Physical inactivity; bed rest
Diet	Inadequate calcium; inadequate vitamin D; excessive coffee; excessive alcohol
Tobacco	Cigarette smoking
Alcohol	Excessive use
Stress	Excessive stress; anxiety
Medications	Certain medications increase calcium losses
Hormonal status	Low estrogen; low testosterone

Bone health As noted previously, bone health is an important consideration, particularly as one ages. Although genetic factors are involved in optimal bone health, so are lifestyle factors, including diet and exercise. The following discussion focuses on the role of supplements, exercise, and drug therapy, as well as a brief discussion of osteoporosis in sports.

Calcium supplements Calcium supplements come in a variety of forms, such as calcium carbonate, calcium citrate, calcium lactate, and calcium gluconate and are found in certain antacids, such as Tums®. Calcium from most supplements is absorbed as well as from cow's milk, but a few calcium salts, including calcium citrate/malate and calcium ascorbate, have superior absorbability. Be sure to check the label for the calcium content per tablet, which may range from 50 to 600 mg depending on the brand.

Supplementation with calcium, along with vitamin D, may be necessary in persons not achieving the recommended dietary intake. Specifically, calcium and vitamin D supplements may protect and even improve bone mineral density as well as reduce fracture risk in postmenopausal females. However, the benefits of supplementation with calcium and vitamin D need to be weighed against the potential negative effects of calcium supplementation on cardiovascular events, kidney stones, and other health problems.

For those who take a calcium supplement, it may be wise to take a supplement containing about 200 mg of calcium with snacks and/or small meals three times a day, rather than one supplement with 600 mg, as it appears that more calcium is absorbed when the intake is spread throughout the day. Moreover, when the supplement is combined with meals, gastric acidity and slower transit time in the gut promote calcium absorption. A daily total supplement of 600 mg calcium, combined with a dietary intake of 500–600 mg, should provide adequate calcium nutrition for most individuals. Multivitamin/mineral supplements normally contain about 200 milligrams of calcium, or 20 percent of the RDA. However, three tablets are needed to provide 600 milligrams of calcium and may not be the best means to obtain calcium, because the three tablets may provide excess amounts of other nutrients, such as vitamin A.

Hypercalcemia essentially never occurs from ingestion of natural food sources, but rather with supplement use. Although supplements up to 600 mg per day do not appear to pose much danger, excessive amounts may contribute to various health problems, as discussed later. Moreover, excessive dietary calcium or calcium supplements may interfere with the absorption of other key minerals, notably iron and zinc. The calcium RDA for some age groups (1,000–1,300 milligrams) may require supplementation for some individuals, but the National Academy of Sciences recommends against supplementation to a total much above the RDA. The UL is 2,500 mg, which may be exceeded if one takes supplements and consumes too many calcium-fortified foods. Meal replacement powders and energy-dense protein bars, popular dietary supplements in strength-power athletes, often have 1,000 mg of calcium per serving. An athlete who consumes three of these products per day for extra kcal and protein can very easily exceed the RDA for calcium.

Mark Dierker/McGraw Hill

Physical Activity Exercise places a mechanical stress on the bone, stimulating mechanical receptors that facilitate the deposition of calcium salts and an increase in bone mass and density. In a review of exercise regimens to increase bone strength, Turner and Robling indicated that mechanical loading through exercise will add bone, and a small amount of added new bone results in dramatic increases in bone strength because the location of the new bone through exercise is where mechanical stresses in the bone will cause fracture. However, they note that not all exercises are equally effective. Dynamic exercise, as compared to static exercise, creates a unique stress on the bone to stimulate bone strength gains. Dynamic, high-impact exercise, such as running and jumping, may stimulate bone growth more than low-impact exercise. Kato and others found that young college females who did ten maximal vertical jumps for 3 days a week over the course of 6 months increased bone mineral density in the hip and spinal bones. Studies find that female athletes in high-impact sports have greater bone mineral density than athletes in low-impact sports.

Age is an important consideration in optimizing bone mass with exercise, and youth is the age of opportunity. Turner and Robling also indicate that, although exercise has clear benefits for the skeleton, engaging in exercise during skeletal growth is clearly more osteogenic than exercise during skeletal maturity. Periosteal expansion occurs predominantly during growth, and consequently, the childhood and adolescent years provide a window of opportunity to enhance periosteal growth with exercise. As Bloomfield notes, you should get bone in your bone bank by age 30. Exercise is an important means of doing so with experts recommending that you should engage in multiple, brief bouts of activity during the day, focusing on weight-bearing activities and a variety of movements.

Although adult bone health is dependent on maximal attainment of peak bone mass in youth and the prevention of bone loss during young adulthood, whether or not exercise prevents bone loss after menopause appears to be debatable. Miller and others, in a review of 13 studies, concluded that the findings provide support for regular aerobic activity in postmenopausal females as a means to offset age-related declines in bone mineral density. Additionally, in a meta-analysis of 18 randomized, controlled trials, Bonaiuti and others concluded that aerobics, weight-bearing, and resistance exercises are all effective in increasing the bone mineral density of the spine in postmenopausal females, and walking is also effective on the hip.

However, others suggest that exercise provides minimal, if any, benefits. The American College of Sports Medicine (ACSM), in a position statement on physical activity and bone health, indicated that physical activity for optimal bone health is important across the age spectrum, yet currently there is no strong evidence that even vigorous physical activity attenuates the menopause-related loss of bone mineral in females. These findings were recently confirmed in a review by Fonseca and others, noting that most exercise intervention studies reveal either no effect or only minor benefits of exercise programs in improving bone mineral density in osteoporotic patients. However, they suggest that exercise interventions in individuals with osteoporosis may be beneficial by improving other determinants of bone strength, such as collagen properties and osteocyte density.

The *Training Table* feature in this section provides recommendations relative to exercise and the prevention of osteoporosis.

Training Table

In addition to adequate dietary intake of calcium and vitamin D, regular physical activity may also be beneficial in the prevention of osteoporosis. While supplementation with calcium may be necessary in certain individuals, such as postmenopausal females, the following prudent recommendations for exercise can also improve bone health:

1. Choose dynamic exercises, such as resistance training and high-impact, weight-bearing aerobic exercises to stimulate development of bone.
2. Youth participation in regular physical activity is particularly beneficial for optimizing bone mass during the peak developmental years.
3. While exercise may enhance bone mass slightly in adulthood, a primary benefit of exercise is preventing further loss of bone that occurs with inactivity.

Walking, jogging, hiking, practicing yoga or pilates, playing tennis, and dancing are all examples of weight-bearing exercises that may be beneficial to keeping bones strong and healthy. In addition, strength and balance exercises can help protect individuals from falls, which are a primary cause of broken bones, particularly in older adults.

jo Crebbin/Shutterstock

Osteoporosis in sports Although osteoporosis occurs primarily in older individuals, disturbed calcium metabolism in young athletes is of concern, particularly in endurance athletes and those involved in weight-control sports. The **female athlete triad**—low energy availability, *amenorrhea*, low bone mineral density—has been reported in numerous studies. More recently, the **male athlete triad**—low energy availability, low bone mineral density, and low testosterone—has been an important topic of study in performance nutrition.

While the exact cause has not been identified, the underlying behavior appears to be disordered eating, resulting in low energy availability. Athletes who attempt to lose body weight in order to improve their appearance or competitive ability in sports may modify their diets, decreasing energy and protein intake. They may also exercise excessively to burn energy. Restrictive diets and excessive exercise regimens may affect hormone status in various ways, including disturbed functioning of the hypothalamus and pituitary gland. These two glands significantly influence overall hormone status in the body, including female and male reproductive hormones. Decreased levels of body fat may also lead to a reduced production of estrone, a form of estrogen. **Secondary amenorrhea** (cessation of menses for prolonged periods) is a classic sign of disturbed hormonal status associated with disordered eating in postpubertal females, as seen in patients with anorexia nervosa. When observed in athletic females, secondary amenorrhea is often referred to as **athletic amenorrhea** and it may involve oligomenorrhea, or intermittent periods of amenorrhea. In young female athletes, athletic amenorrhea is often associated with osteoporosis.

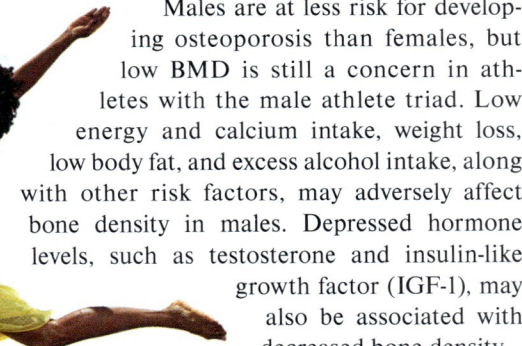

Males are at less risk for developing osteoporosis than females, but low BMD is still a concern in athletes with the male athlete triad. Low energy and calcium intake, weight loss, low body fat, and excess alcohol intake, along with other risk factors, may adversely affect bone density in males. Depressed hormone levels, such as testosterone and insulin-like growth factor (IGF-1), may also be associated with decreased bone density.

As discussed in chapter 10, the ACSM and Academy of Nutrition and Dietetics (AND) indicate that medical attention is necessary for

Erik Isakson/Blend Images LLC

athletes with disordered eating and/or amenorrhea, noting that treatment of the triad often requires intervention via a team approach, including a physician, a registered dietitian nutritionist (RDN), a psychologist, and the support of family, friends, teammates, and coaches.

Cardiovascular disease Some experts have expressed concern about the potential adverse effects of excess calcium intake, particularly from supplements, on cardiovascular health. According to experts, calcium may be involved in cardiovascular disease development through multiple pathways. One concern is calcification of blood vessels, which could impair blood flow in the coronary arteries. A meta-analysis of calcium supplements by Reid found a 27–31 percent increase in the risk of myocardial infarction and a 12–20 percent increase in the risk of stroke, and concluded that the health risks of calcium supplements appear to outweigh any skeletal benefits. However, Waldman and others indicate that in general the studies of calcium use and cardiovascular disease published to date have had important limitations, such as small sample size and limited follow-up, and the findings should be interpreted with caution. In their review, Heaney and others noted a number of issues in studies that could interfere with the interpretation of the results. Lewis and others also noted that the design of the meta-analyses showing increased risk of myocardial infarction with calcium supplements has raised some concerns. For example, such design issues included "nonadherence to the clinical protocol, multiple endpoint testing, and failure to correctly adjust for endpoint ascertainment."

Studies and reviews by other scientists suggest calcium supplementation has no adverse effects on cardiovascular health. Paik and others, in the Nurse's Health Study over the course of 24 years, reported that their findings do not support the hypothesis that calcium supplement intake increases cardiovascular disease risk in females. Lewis and others conducted their own meta-analysis using only randomized, controlled trials. They concluded that the current evidence does not support the hypothesis that calcium supplementation with or without vitamin D increases cardiovascular disease risk in older adult females. Heaney and others concluded that the current evidence regarding the hypothesized relationship between calcium supplement use and increased cardiovascular disease risk is insufficient to warrant a change in the Institute of Medicine recommendations, which advocate use of supplements to promote optimal bone health in individuals.

The current state of affairs relative to the role of calcium supplementation in the development of cardiovascular disease is probably best expressed by Rautiainen and others. In their review, they note the results of research trials suggest calcium supplementation has no effect on cardiovascular disease development, but current research findings also do not allow a definitive conclusion to be drawn. They conclude that only large-scale, randomized trials designed to investigate the effects of calcium supplementation on cardiovascular disease events as the primary end point, as well as short-term trials investigating the effect on coronary disease biomarkers, can provide a definitive answer.

Obesity and weight control Calcium supplementation has been theorized to promote weight loss. This is related to the fact that low-calcium diets cause an increase in calcitriol, the vitamin D hormone, which can stimulate adipose cell calcium influx, promoting fat accumulation. On a related note, higher-calcium diets may inhibit the formation of body fat.

Zemel and others, in several studies, have shown that supplementing a reduced-kcal diet with calcium—in particular, calcium-rich dairy products—led to greater weight loss compared to placebo groups. However, most other studies comparable to the design of those of Zemel reported no beneficial effects of high-calcium diets or supplements on weight loss.

The results of two reviews are somewhat contradictory. De Oliveira Freitas and others indicated the results of most studies suggested that calcium ingestion may improve body composition. Conversely, Soares and others analyzed the effects of 15 randomized, controlled trials on calcium with or without vitamin D and indicated the evidence does not consistently support the contention that calcium accelerates weight or fat loss in obesity.

However, Tremblay and Gilbert suggest dairy calcium may play a role in body weight control, possibly accentuating the impact of an intervention supporting weight loss. They suggest that calcium/dairy supplementation may promote weight loss in several ways: (1) by promoting fecal fat loss and oxidation, (2) by favoring a decrease in energy intake, and (3) by facilitating appetite control. More details on interventions to treat overweight and obesity are presented in chapter 11.

Kidney stones Kidney stones can form in the urine and cause pain in the urinary tract if they are too large to pass. There are several forms of kidney stones, the most common being calcium oxalate formed from calcium and oxalate in the urine.

Possible causes of kidney stones include inadequate fluid intake, too much protein intake, too much salt, or too much sugar, especially fructose. Experts note that kidney stones are not caused by *dietary* calcium. Rather, although kidney stones contain calcium oxalate, it is the oxalate that may be the problem. Calcium intake to about 1,200 milligrams daily may reduce the risk of kidney stones, as the calcium can bind oxalate in the intestines and lead to its excretion, thus decreasing the amount of oxalate in the body.

Sports performance Research regarding the effect of calcium supplementation on physical performance is limited. Given its theoretical role in fat metabolism in weight-loss programs, White and others investigated the acute effect of calcium on fat metabolism and exercise performance in trained female runners. The subjects consumed either a high-calcium (500 mg) or low-calcium (80 mg) drink 60 minutes prior to a 90-minute run at 70 percent VO_2 max, which was followed by a 10-kilometer treadmill time trial. The high-calcium drink did not affect carbohydrate or fat metabolism during the 90-minute run or 10-kilometer run time. In a similar study, Gonzalez and others investigated the effect of 2 weeks of calcium supplementation (1,400 milligrams daily) on carbohydrate and lipid metabolism during a cycling-based exercise test. However, the supplement did not affect either carbohydrate or fat utilization rates during the exercise task.

Calcium may be combined with other salts, such as lactate, to form calcium lactate, which could be used to buffer lactic acid during exercise. Painelli and others studied the effect of calcium lactate on intense anaerobic exercise, and although there were increases in bicarbonate and the alkalinity of the blood, performance on intense Wingate cycling tests was not improved. Detailed coverage of the use of buffering agents, such as sodium bicarbonate, is presented in chapter 13.

Low serum calcium levels may also impair neuromuscular functions, but such conditions are rare because the body will draw from calcium reserves in the skeleton to restore serum levels. In a review, Kunstel concluded that increased physical activity alone does not necessarily demand an increased intake of dietary calcium. Thus, acute or chronic calcium supplementation is not recommended as a means to enhance sports performance. However, calcium supplementation may be useful to help maintain bone mass in some female and male athletes, particularly those who do not consume adequate calcium from foods.

Prudent Recommendations Adequate daily calcium intake and weight-bearing exercise are important during the developmental years of childhood and adolescence to maximize peak bone mass, and such practices should be continued throughout adult life. Calcium supplementation may be recommended to attain the RDA in individuals who do not obtain a sufficient amount through their normal diet. The possible positive benefits of calcium supplementation on bone health appear to outweigh the few risks.

Phosphorus (P)

Phosphorus is the second most abundant mineral in the body after calcium. Like calcium, a significant amount of the phosphorus in the human body, about 85 percent, is found as part of teeth and bones.

DRI The adult RDA is 700 mg/day for both males and females. Higher amounts are needed during periods of rapid skeletal growth, particularly between the ages of 9 and 18 years. The UL for adults is 4 g/day.

Food Sources Phosphorus is distributed widely in foods. Indeed, all natural foods contain phosphorus as inorganic phosphate salts or organic molecules. Excellent sources include dairy foods such as cow's milk and cheese, seafood, meat, eggs, nuts, seeds, grain products, and a wide variety of vegetables. Phosphate is also a common food additive, and soft drinks often have a relatively high phosphate content. In some foods, phosphorus is also a part of phytate, which may diminish the absorption of minerals like calcium, iron, zinc, and copper by forming insoluble phosphate salts in the intestine. However, this is not a major problem with the typical diet in the United States.

Julia Sudnitskaya/Shutterstock

Most Americans consume about twice the RDA for phosphorus and, as noted previously, too little calcium. The recommended calcium: phosphorus ratio is about 1:1, that is, equal amounts of each. Consuming too much dietary phosphorus, which may occur because of phosphate additives in food, may impair calcium metabolism and increase a person's risk for osteoporosis. Too much phosphorus may also stimulate the release of parathyroid hormone. A calcium:phosphorus ratio of up to 1:2 may be compatible with bone health; however, ratios of 1:4 may be associated with osteoporosis. Similar to too much phosphorus impacting the bioavailability of calcium, too much calcium can also impair phosphorus absorption and eventually lead to phosphorus insufficiency and osteoporosis. A high dietary calcium:phosphorus ratio can occur with the use of calcium supplements and calcium-fortified foods.

Major Functions Phosphorus is a critically important element in every cell in the body. In the human body, phosphorus occurs only as the salt phosphate, which exists as inorganic phosphate or is coupled with other minerals or organic compounds. Phosphates are extremely important in human metabolism. About 80–90 percent of the phosphorus in the body combines to form calcium phosphate, which is used for the development of bones and teeth. As with calcium, the bones represent a sizable store of phosphate salts. Other phosphate salts, such as sodium phosphate, are involved in acid-base balance.

The remainder of the body phosphates are found in a variety of organic forms, including the phospholipids, which help form cell membranes, and DNA. Several other organic phosphates are of prime importance to the active individual. For example, organic phosphates are essential to the normal function of most of the B vitamins involved in the energy processes within the cell. They are also part of the high-energy compounds found in the muscle cell, such as ATP and PCr, which are needed for muscle contraction. Glucose also must be phosphorylated in order to proceed through glycolysis. Organic phosphates also are a part of a compound in the RBC known as 2,3-BPG (2,3-biphosphoglycerate), which facilitates the release of oxygen to the muscle tissues.

Deficiency: Health and Physical Performance Because phosphorus is distributed so widely in foods, and because hormonal control is very effective, deficiencies are rare. Symptoms of a phosphorus deficiency parallel those of calcium deficiency, such as loss of bone density, resulting in rickets or osteomalacia. Muscle weakness, poor appetite, fatigue, and weight loss may also occur with a phosphorus deficiency. Extreme muscular exercise may increase phosphorus excretion in the urine, but has not been reported to cause a deficiency state. While phosphorus deficiency could theoretically impair physical performance, it has not been the subject of study because such deficiencies are rare.

Supplementation: Health and Physical Performance Although various health problems, most notably rickets and osteoporosis, could occur with a phosphorus deficiency, supplementation for health-related benefits has not received research attention because deficiency states are rare.

However, phosphate salts supplementation has been studied as a means of enhancing exercise performance. Sodium phosphate and potassium phosphate were reported to relieve fatigue in German soldiers during World War I. Other research in Germany during the 1930s suggested that phosphate salts could improve physical performance. More than 70 years ago, one reviewer discredited much of this early research, but he did note that phosphates probably could increase human work output when consumed in quantities exceeding the amounts found in the normal diet. Indeed, they are still marketed online and in some stores selling sports supplements. Although phosphate salt supplementation may influence various physiological processes associated with physical performance, its effects to increase 2,3-BPG and related oxygen dynamics during aerobic endurance exercise have received most research attention. Bremner and others found that a 7-day phosphate loading protocol would increase erythrocyte phosphate pools and 2,3-BPG.

The results of contemporary research relative to the ergogenic effect of phosphate supplementation are inconsistent, some showing positive performance-enhancing effects and others reporting no significant improvement in performance. The efficacy of phosphate supplementation as an ergogenic aid appears to be related to the specific type of phosphate, dosing, timing, length of supplementation, and individual variations. For example, Cade and others studied the effect of 4 grams (1 gram four times daily) on the performance of highly trained runners. The phosphate salts increased the serum 2,3-BPG, which related closely to an increase in VO_2 max. The amount of lactate produce at a standard exercise workload decreased and was accompanied by a lower rating of perceived exertion (RPE), suggestive of lower psychological stress. Similarly, Kreider and colleagues found that 4 grams of trisodium phosphate for 6 days produced a significant increase in VO_2 max in highly trained cross-country runners. However, no changes in 2,3-BPG were found. In a similar protocol, Stewart and others reported that 3.6 grams of sodium phosphate for 3 days increased 2,3-BPG levels and increased VO_2 max in trained cyclists. Exercise time to exhaustion in cycling also increased by nearly 16 percent following supplementation.

In contrast, the results of other well-designed studies have not shown any performance-enhancing effects of phosphate salt supplementation. For example, Kreider and colleagues found that 4 grams of trisodium phosphate for 6 days had no effect on a 5-mile competitive run in highly trained cross-country runners. Similarly, Mannix and team report that, although phosphate supplementation did increase 2,3-BPG levels, it did not improve cardiovascular function or oxygen efficiency in subjects exercising at 60 percent of VO_2 max. Similar results have been found by others evaluating the efficacy of phosphate supplementation in athletes.

In a review 30 years ago, Tremblay and others addressed methodological differences between studies that could contribute to the equivocal results reported at that time. In a more recent review, Buck and others, although suggesting sodium phosphate may have positive benefits for sporting performance, also indicated not all studies are supportive. Possible differences between study protocols may explain the difference, including the type of phosphate used, the dose, the dosing protocol, and the fitness level of the subjects. Additional research may be warranted.

As you may recall, adenosine triphosphate (ATP) is the immediate source of energy for muscle contraction. Although some entrepreneurs have marketed ATP supplements for athletes, there is no available evidence that they enhance physical performance. Several studies have evaluated the effect of ATP supplementation on exercise performance but reported no significant effects. Arts and others studied the bioavailability of oral ATP supplements and found they do not appear in the blood, suggesting that such preparations are not bioavailable to the tissues and may help explain why studies show no ergogenic effect.

Creatine phosphate is used to replenish ATP rapidly as a component of the ATP-PCr energy system. As noted in chapter 6, research with oral creatine supplementation has suggested some ergogenic effects on muscular strength and muscle mass with resistance training. While creatine phosphate has been studied in a medical case study and has shown some beneficial effects on muscle mass during rehabilitation, creatine supplementation also will suffice as there is an adequate amount of phosphate available in the body to form creatine phosphate.

Excesses of phosphorus in the body are excreted by the kidneys. Phosphorus excess *per se* does not appear to pose any problems, with the exception of individuals who have limited kidney function. Subjects consuming phosphate supplements may experience gastrointestinal distress, which may be alleviated by mixing the salts in a liquid and consuming with a meal. Excessive amounts of phosphate over time may impair calcium metabolism and balance.

Prudent Recommendations Given that the average daily intake of phosphorus is approximately twice the RDA, it may be reasonable to reduce intake of dietary sources high in phosphates, many of which are found in highly processed foods that usually contain food additives. Check the food label for terms that contain the word *phosphate*, such as dicalcium phosphate, hexametaphosphate, and sodium tripolyphosphate. Eating a diet rich in unprocessed foods, such as fresh fruits and vegetables, may reduce dietary intake of phosphate.

For athletes, some evidence suggests that sodium phosphate salt supplementation may enhance aerobic endurance performance. An accepted protocol appears to be a total of 4 grams of trisodium phosphate per day, consumed in 1-gram portions with food and drink, for 5-6 days. If you decide to experiment with phosphate supplementation, do so in training before using it in conjunction with competition. Given the association with possible adverse health effects, this procedure should be used sparingly and under the guidance of a health professional. Currently, use of phosphate salts for sports competition is not prohibited by the World Anti-Doping Agency.

Magnesium (Mg)

Magnesium is the sixth most abundant mineral found in the body. The mineral plays an important role in energy metabolism, muscle activity, and heart health.

DRI The adult RDA for magnesium is 400-420 mg/day for males and 310-320 mg/day for females. The UL of 350 mg for magnesium, which is greater than the RDA for females, applies only to pharmacological forms of magnesium, such as is found in supplements and foods fortified with magnesium. There are no restrictions in obtaining magnesium via natural food sources.

Food Sources Magnesium is widely distributed in foods, particularly nuts, seafood, green leafy vegetables, other fruits and vegetables, black beans, and whole-grain products. One cup of cooked spinach provides approximately 160 mg of magnesium, which is about half of the adult female RDA for the mineral. Areas with hard water may contain up to 20 mg of magnesium per liter, and some bottled waters may contain more than 100 mg per liter. The type of magnesium found in most multivitamins/multiminerals is magnesium oxide, which is poorly absorbed from the digestive tract.

Major Functions The body stores about 50-60 percent of its magnesium in the skeletal system, which may serve as a reserve during short periods of dietary deficiency. Magnesium influences bone metabolism and helps prevent bone fragility. Only about 1 percent is in the extracellular fluid, but the remainder is found in soft tissues such as muscle, where it is a component of numerous enzymes. Interestingly, magnesium is involved in more than 300 essential metabolic reactions, many of which are important to the physically active individual, including neuromuscular, cardiovascular, and hormonal functions. For example, magnesium is involved in metabolic processes related to ATP production and use, including the glycolytic pathway, the citric acid cycle, and ATPase. As you may recall, ATP is the primary source of energy for muscle contraction and numerous other metabolic processes. Magnesium helps regulate the synthesis of protein and other compounds, such as 2,3-BPG, which may be essential for optimal oxygen metabolism. Magnesium also helps block some of the actions of calcium in the body, such as contraction in both the skeletal and smooth muscles.

Deficiency: Health and Physical Performance Magnesium deficiencies in the United States are rare in the healthy adult population, even though many Americans do not consume recommended amounts of the mineral in their diet. This is because absorption rates increase in the GI tract when the body is lacking the mineral. For example, normal absorption rates are 40-60 percent, but can be has high as 80 percent when magnesium is lacking in the body.

Those at the greatest risk for a magnesium deficiency include those with poorly controlled diabetes, those taking certain medications (such as diuretics), and those with alcohol use disorder. Risk for deficiency also increases as one ages due to the increased urinary excretion of the mineral and reduced absorption through the digestive process. Symptoms of a deficiency include muscle weakness, muscle twitching, muscle cramping, apathy, and, in some cases, cardiac arrhythmias. Chronic deficiency of magnesium has been associated with a variety of chronic health conditions, including type 2 diabetes, metabolic syndrome, hypertension, cardiovascular disease, osteoporosis, migraines, asthma, alcohol-related liver disease, and colon cancer.

Relative to exercise or sports performance, a magnesium deficiency could certainly impair performance, as some symptoms are muscle weakness, tremor, and cramps. In addition, overtraining may lead to a decrease in exercise performance, and in a joint consensus statement on prevention, diagnosis, and treatment of the overtraining syndrome by the European College of Sport Science and the ACSM, Meeusen and others indicated nutrient deficiencies, including magnesium, should be considered.

Dietary magnesium intake in athletes may vary. For example, Lukaski reported that dietary survey findings indicated that most male athletes equaled or exceeded the RDA for magnesium. However, Czaja and others reported that elite male Polish athletes consumed low levels of magnesium and recommended supplementation. Lukaski also reported many female athletes were obtaining only about 60-65 percent of the RDA, while athletes in weight-control sports were getting only 30-35 percent. Over time, athletes with such low intakes may incur a magnesium deficiency.

Exercise appears to influence magnesium metabolism in various ways. Deuster has noted that one of the most common research observations is a decrease in plasma levels of magnesium following exercise. It is thought that magnesium enters the tissues in response to exercise-related requirements, for example, of the muscle tissue for energy metabolism and the adipose tissue for lipolysis. Exercise may also increase magnesium losses via urine and sweat. Although the reported sweat losses of 4-15 mg per liter are relatively small in comparison to body stores and daily intake, Nielsen and Lukaski stated that strenuous exercise apparently increases urinary and sweat losses that may increase magnesium requirements by 10-20 percent. Keep in mind that, to replace 15 milligrams of magnesium lost from the body, one would need to consume about 40 milligrams at a 40 percent rate of absorption. That is almost 10 percent of the male RDA. On the basis of some of these findings, several reports have recommended that individuals undergoing prolonged, intensive physical training should increase their daily intake of magnesium, but these recommendations are within range of the RDA. The extra kcal consumed when energy expenditure increases during exercise should provide the additional magnesium.

Supplementation: Health and Physical Performance Several reviews have suggested that magnesium supplementation may help reduce blood pressure, a major risk factor for heart disease. Kass and others, in a meta-analysis of 22 studies, concluded that magnesium supplementation, with an average dose of 410 milligrams daily, appears to achieve a small but clinically significant reduction in both systolic (3-4 mmHg) and diastolic (2-3 mmHg) blood pressure. In addition, Houston reported that magnesium intake of 500 to 1,000 milligrams per day may reduce both systolic (5.6 mmHg) and diastolic (2.8 mmHg) blood pressure, but clinical studies show a wide range of reduction, with some showing no reduction. However, Houston suggests the combination of increased intake of magnesium and potassium, coupled with reduced sodium intake, is more effective in reducing blood pressure. The DASH diet, discussed in chapter 9 as a means to lower blood pressure, is rich in magnesium and potassium and low in sodium and is one of the most recommended healthful diets, as noted throughout this text.

The effect of magnesium supplementation on exercise performance has received limited research attention over the years, but several reviews are available. In 1988, McDonald and Keen indicated that they are not aware of any data showing a positive effect of magnesium supplementation on exercise performance in individuals who are in adequate magnesium status. In 2000, Newhouse and Finstad conducted a meta-analysis of human supplementation studies involving magnesium supplementation and exercise performance, and they concluded that the strength of the evidence favors those studies finding no effect of magnesium supplementation on any form of exercise performance, including aerobic, anaerobic-lactic acid, and strength activities. In a 2006 review, Nielsen and Lukaski also concluded that magnesium supplementation of physically active individuals with adequate magnesium status has not been shown to enhance physical performance. However, if a magnesium deficiency is corrected, exercise performance may be improved. For example, Lukaski noted that some earlier studies have shown that magnesium supplementation improved strength and cardiorespiratory function in healthy persons and athletes but also noted that it is unclear as to whether these observations related to improvement of an impaired nutritional status or a pharmacological effect. More recently, Setaro and others investigated the effect of magnesium supplementation (350 milligrams daily for 4 weeks) in volleyball players on a variety of exercise performance variables, including VO₂ max, isokinetic strength tests, and plyometric jump tests. They noted significant increases in two of the plyometric tests and suggested that magnesium supplementation may improve alactic anaerobic metabolism, even though the players were not magnesium deficient.

At present, evidence suggests that exercise performance may be compromised as a result of magnesium deficiency. Animal research suggests that magnesium supplementation may improve energy efficiency during sports performance. While such evidence is not available through human trials, magnesium supplementation has been associated with improved parameters in both aerobic and anaerobic types of exercise. Large-scale intervention trials are needed to better understand the role of magnesium supplementation in athletes.

As noted previously, a UL of 350 mg has been established for magnesium supplements or fortified foods. Mildly excessive intakes of magnesium may cause nausea, vomiting, and diarrhea. In individuals with kidney disorders who cannot excrete the excess, increased serum levels of magnesium may lead to coma and death.

Prudent Recommendations Physically active individuals should obtain adequate magnesium from a balanced diet containing foods rich in magnesium. The DASH diet is recommended. For those trying to lose body weight for competition, dietary magnesium may be compromised. In particular, female athletes are at risk for magnesium deficiency. In cases of suspected dietary insufficiency, magnesium supplementation may be recommended via supplements or fortified foods. The amount of magnesium in such products may vary. Pure magnesium supplements may contain about 200–500 milligrams; multivitamin-mineral supplements contain varying amounts, about 50 milligrams or more; fortified foods also contain varying amounts. Check the Supplement Facts and Nutrition Facts labels for magnesium content. For adults, the recommended amount from such sources should not exceed 350 milligrams daily. Much lower amounts are recommended for children. Ultimately, there is no need for larger doses, as they may be associated with some adverse effects.

> ### Key Concepts
>
> ▶ Major minerals are those needed in greater amounts in the diet and stored in greater amounts in the body. The seven major minerals are calcium, chloride, magnesium, phosphorus, potassium, sodium, and sulfur. **Table 8.2** summarizes the major minerals.
> ▶ Calcium intake is important for normal bone formation, but also for enzyme activation, nerve impulse transmission, muscle contraction, and other essential physiological functions in the body. Children and adolescents should obtain adequate calcium to increase their peak bone mass. Adults need calcium to help prevent losses during aging. Two keys to the prevention of osteoporosis are weight-bearing or resistance exercise and adequate calcium and vitamin D in the diet.
> ▶ Phosphorus is the second most abundant mineral and, like calcium, is found predominately stored in the human body as part of teeth and bones. Phosphate salts have been used for more than 80 years in attempts to improve athletic performance. In general, recent studies support an ergogenic effect of phosphate salt supplementation, but confirmation by additional research is needed.
> ▶ Magnesium has essential roles in many aspects of health, including participating in over 300 different chemical reactions in the body, participating in macronutrient metabolism, and being essential for smooth muscle cell contraction and nerve function. Magnesium deficiency may be associated with various health conditions, such as diabetes, hypertension, and heart disease. As well, a magnesium deficiency may contribute to impaired physical performance. Magnesium supplementation may be beneficial for some athletes, but additional research is needed.

Trace Minerals

As described earlier, **trace minerals** are those needed in quantities less than 100 mg per day. For several, the body needs only extremely minute amounts, such as a few micrograms per day. The term *ultratrace* is applied to these minerals. The trace minerals include chromium, fluoride, copper, iodine, iron, manganese, molybdenum, selenium, and zinc. **Table 8.4** provides an overview of the trace minerals, including dietary sources, recommended daily intakes, key functions, and symptoms associated with deficiencies and toxicities.

The remainder of this section will then provide details on key trace minerals believed to be of particular importance to athletic performance and health. We will start with a discussion of iron, followed by a discussion of copper, zinc, chromium, and selenium.

Iron (Fe)

Iron is a metallic element that exists in two general forms, ferrous (Fe^{2+}) and ferric (Fe^{3+}). According to the *2020–2025 Dietary Guidelines for Americans,* iron is considered a "nutrient of public health concern" for certain populations, including young children, females capable of becoming pregnant, and females who are pregnant.

DRI Depending upon age and sex, the average individual needs to replace about 1.0–1.5 mg of iron that is lost from the body daily. However, because the bioavailability of iron is very low, with only about 10 percent of food iron being absorbed, the RDA is ten times the need. Currently, the RDA for adult males is 8 mg/day and for adult females age 19–50 is 18 mg/day. Slightly different amounts are needed by other age groups and may be found in the DRI tables available at www.nal.usda.gov/fnic/dri-tables-and-application-reports. For example, pregnant females need 27 mg/day, whereas postmenopausal females need only 8 mg/day. The UL range is 40–45 mg/day.

TABLE 8.4 Overview of the trace minerals

Major mineral	RDA/AI and UL for adults age 19–50	Major food sources	Key functions in the body	Deficiency symptoms	Toxicity symptoms
Chromium	AI: 30–35 mcg♂ 20–25 mcg♀ UL: Not determined	Whole-grain products; organ meats; egg yolks; pork; oysters; nuts; fruits and vegetables, including broccoli	Enhances insulin function as part of glucose tolerance factor	Glucose intolerance; impaired lipid metabolism	Unknown, but currently under scientific investigation
Copper	RDA: 900 mcg UL: 10,000 mcg	Organ meats, including liver; meat; fish; poultry; shellfish; legumes; whole-grain products; cocoa	Proper use of iron and hemoglobin in the body; metalloenzyme involved in connective tissue formation; part of antioxidant enzymes	Anemia; impaired immune function; poor growth	Nausea; vomiting; liver damage; abnormal nervous system function
Fluoride	AI: 4 mg♂ 3 mg♀ UL: 10 mg	Fluoridated water, tea, coffee, seaweed	Enhances bone formation; protects tooth enamel against decay and cavities	Dental erosion — dental caries or cavities	Dental fluorosis—abnormal change to appearance of tooth enamel; Skeletal fluorosis—joint stiffness, bone pain
Iodine	RDA: 150 mcg UL: 1,100 mg	Iodized salt; saltwater fish; seaweed	Component of thyroid hormones—controls rate of cellular metabolism	Goiter; hypothyroidism (low levels of thyroid hormones); fatigue; difficulty concentrating; decreased metabolic rate; cretinism (in infants)	Reduced function of thyroid gland
Iron	RDA: 8 mg♂ 18 mg♀ UL: 45 mg	Meat; fish; poultry; shellfish, especially oysters; whole-grain products; enriched-grain products; fortified cereals; green leafy vegetables; legumes	Hemoglobin and myoglobin formation; electron transport; essential in oxidative processes; immune system function	Anemia; fatigue; impaired temperature regulation; decreased resistance to infection	GI upset; hemochromatosis; liver damage; death
Manganese	AI: 2.3 mg♂ 1.8 mg♀ UL: 11 mg	Green leafy vegetables; whole-grain products; nuts; tea	Cofactor for a variety of enzymes including those involved in carbohydrate metabolism, bone and cartilage formation, and antioxidant activity	None reported	Abnormal nervous system function
Molybdenum	RDA: 45 mcg UL: 2,000 mcg	Legumes; whole-grain products; nuts; green leafy vegetables	Cofactor for four enzymes, including sulfite oxidase	None reported	Unlikely from dietary sources

(Continued)

TABLE 8.4 Overview of the trace minerals (Continued)

Major mineral	RDA/AI and UL for adults age 19–50	Major food sources	Key functions in the body	Deficiency symptoms	Toxicity symptoms
Selenium	RDA: 55 mcg UL: 400 mcg	Brazil nuts; oysters; fish; poultry; eggs; whole-grain products	Cofactor for glutathione peroxidase, an antioxidant enzyme	Cardiac muscle damage; muscle pain and weakness	Nausea; vomiting; abdominal pain; hair loss; weakness; liver damage
Zinc	RDA: 11 mg♂ 8 mg♀ UL: 40 mg	Oysters; seafood; meat; legumes; whole-grain products; dairy products, including yogurt	Cofactor for many enzymes involved in energy metabolism, protein synthesis, immune function, etc.	Depressed immune function; impaired wound healing; poor appetite; failure to grow; skin rash; hair loss	Nausea; vomiting; impaired immune function; abnormal blood lipid levels; impaired copper absorption

Food Sources Dietary iron comes in two forms. **Heme iron** is associated with hemoglobin and myoglobin and is found only in animal foods, such as meat, chicken, and fish. About 35–55 percent of the iron found in meat is heme iron, the percentage being somewhat higher in beef as compared to chicken and fish. **Nonheme iron** is found in both animal and plant foods. About 20–70 percent of the iron in animal foods and 100 percent in plant foods is in the nonheme form. Heme iron has greater bioavailability than nonheme iron. Indeed, about 15–35 percent of heme iron is absorbed from the intestines compared to only 2–20 percent for nonheme iron. In addition to being heme versus nonheme iron, the percent absorbed depends on the iron status of the individual. Those with higher levels of body iron stores will absorb less, whereas those with lower levels will absorb more.

Olga Nayashkova/Shutterstock

Foodandmore/Stockcreations/123RF

Excellent animal sources of dietary iron include liver, lean meats, oysters, clams, and dark poultry meat. One ounce of lean meat provides about 1 mg of heme iron. Good sources of nonheme iron include dried fruits such as apricots, prunes, and raisins; vegetables such as broccoli and peas; legumes; and whole-grain products. Six dried apricot halves or ½ cup of beans provides about 3 mg of nonheme iron, and some breakfast cereals are fortified with nonheme iron to provide 100 percent of the RDA. Cooking in iron pots or skillets also contributes some iron to the diet. With a balanced diet, about 6 mg of iron is provided in every 1,000 kcal ingested.

There are many factors that influence the absorption and bioavailability of iron. **Table 8.5** summarizes key factors that increase and decrease iron bioavailability. For those who are deficient in iron, considering these factors is important to optimizing iron bioavailability.

Major Functions Iron is an essential element for almost all living organisms, as it participates in a wide variety of metabolic processes. A major function of iron in the body is the formation of compounds essential to the transportation and utilization of oxygen. The vast majority is used to form hemoglobin, a protein-iron compound in the RBC that transports oxygen from the lungs to the body tissues. Other iron compounds include myoglobin, the cytochromes, and several Krebs-cycle metalloenzymes, which help use oxygen at the cellular level. The remainder of the body iron is stored in the tissues, principally as protein compounds called **ferritins**. The iron in the blood, serum ferritin,

TABLE 8.5 Factors influencing iron bioavailability

Increases iron bioavailability	Decreases iron bioavailability
Consumption of heme iron or nonheme iron with a source of heme iron	Consumption of excessive calcium with iron-containing foods
Leavening of bread	Medications that reduce HCl secretion (acidity) in stomach
Soaking beans or grains prior to cooking	Oxalic acid (found in vegetables, such as spinach)
Consumption of vitamin C	Phytic acid (found in grains)
	Soy protein–rich foods (such as tofu, edamame)
	Tannins (found in tea)

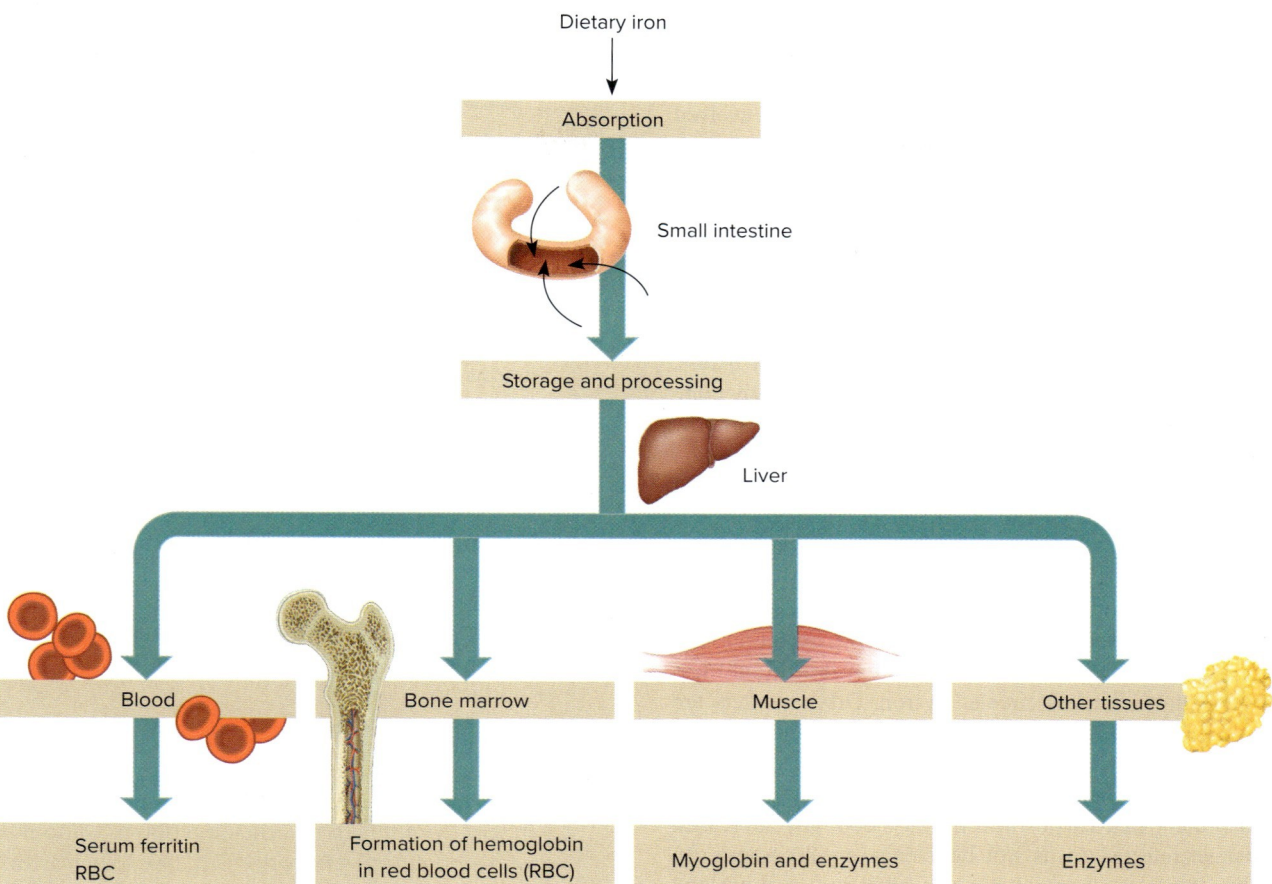

FIGURE 8.5 Simplified diagram of iron metabolism in humans. After digestion and absorption, iron is used in the formation of hemoglobin, myoglobin, and certain cellular enzymes, all of which are essential for transportation of oxygen in the body.

is used as an index of the body iron stores, as are a number of other markers such as transferrin, protoporphyrin, and hemoglobin. Other major storage sites include the liver, spleen, and bone marrow. When ferritin levels become excessive in the liver, the iron is stored as hemosiderin, an insoluble form. Approximately 30 percent of the body iron is in storage form, while the remaining 70 percent is involved in oxygen metabolism. Because iron is so critical to oxygen use in humans, it is essential that those individuals engaged in aerobic endurance-type exercises have an adequate dietary intake. **Figure 8.5** provides a brief outline of iron metabolism in humans.

As noted below, a deficiency or excess of iron in the body may lead to serious health problems. Fortunately, humans have an important iron regulatory hormone, **hepcidin**, that helps regulate body iron stores. According to Zhao and others, the main function of hepcidin is to decrease iron concentrations in the blood by various means. Nemeth and Ganz indicate that when serum iron levels are elevated, the liver will synthesize more hepcidin to help inhibit iron absorption from the intestines and suppress release into the blood. In contrast, when serum iron stores are depressed, synthesis of hepcidin is decreased, leading to increased iron absorption.

Deficiency: Health and Physical Performance Iron deficiency is one of the leading risk factors for disability and death worldwide. According to the World Health Organization (WHO), more than 2 billion people, particularly females and children in developing countries, are iron deficient. A deficiency in children may alter the energy metabolism of the brain, impairing brain development in early life with associated learning and motor defects. In adults, an iron deficiency can contribute to a reduced capacity to work, ultimately leading to serious economic challenges. In many cases, iron deficiency is exacerbated by infectious diseases such as malaria, HIV/AIDS, and worm infections.

The body normally loses very little iron through such routes as the skin, gastrointestinal tract, hair, and sweat. Typically, about 8 mg of dietary iron daily will replace these losses. Females also lose some additional iron in the blood flow during menstruation. Therefore, females need about 15–18 mg of dietary iron per day to replace their total losses. Adolescent males need about 11 mg, as they are increasing muscle tissue and blood volume during this rapid period of growth. With 6 mg of iron per 1,000 kcal, the adult male typically has no problem meeting his requirement of 8 mg per day. With a normal intake of 2,900 kcal, he will receive 17.4 mg. With 2,200 kcal, the average intake for females, only 13.2 mg of iron would be provided. This is somewhat short of the 15–18 mg needed.

The main factor underlying iron deficiency in Western diets is inadequate dietary intake. The typical American diet has evolved so that many individuals consume iron-poor diets, such as snack foods, white bread, and sugar-sweetened beverages. Yet, most females still have normal hemoglobin and serum ferritin status. Because the normal loss of iron from the body is relatively low, and because excessive amounts in the body may be harmful, the intestine limits the amount absorbed from the diet. In contrast, when an individual becomes iron deficient, the intestine may increase the amount of dietary iron absorbed to above 30 percent. Nevertheless, iron deficiency and iron-deficiency anemia are still relatively common in adolescent girls and females of childbearing age. In the United States, approximately 10–19 percent of females between the ages of 12 and 49 are iron deficient. Comparatively, only 1 percent of males in this age range experience the same deficiency.

Deficiency stages Iron deficiency occurs in stages. The first stage involves depletion of the bone marrow stores and a decrease in serum ferritin. It is referred to as the stage of iron depletion. The second stage involves a further decrease in serum ferritin and less iron in the hemoglobin, or less circulating iron. Other markers are used to evaluate iron stores in this stage, including free erythrocyte protoporphyrin (FEP), which is used to form hemoglobin. FEP in the blood increases when adequate iron is not available. Serum transferrin, a protein that carries iron in the blood, also increases. This is the stage referred to as iron-deficiency erythropoiesis. In these first two stages the hemoglobin concentration in the blood is still normal. Collectively, these stages are often alluded to as **iron deficiency without anemia**. Reduced serum ferritin is the key indicator of iron depletion, or iron deficiency without anemia, with levels lower than 30 nanograms per milliliter. The third stage consists of a very low level of serum ferritin and decreased hemoglobin concentration, or **iron-deficiency anemia**. The *Training Table* feature in this section summarizes key signs and symptoms of iron-deficiency anemia, globally one of the most common nutrient deficiencies.

Deficiency in athletes Because iron is so critical to the oxygen energy system, it is essential for endurance athletes to have adequate iron in the diet to maintain optimal body supplies. Buratti and others note that low body iron levels can cause anemia and thus limit the delivery of oxygen to exercising muscle, but they also indicate iron deficiency may also have adverse effects on oxidative metabolism within the muscle.

Recent reviews suggest certain groups of athletes may be at risk for iron deficiency. Kong and others indicated that exercise-induced iron-deficiency anemia is notably high in athletic populations, particularly those with heavy training loads. This could include some male athletes, such as gymnasts, wrestlers, and distance runners. Relative to distance runners, Zourdos and others noted that, among competitive athletes, marathoners are at greater risk to develop anemia and other clinical syndromes that may be associated with inadequate dietary intake of iron. In particular, McClung and others indicated female athletes are at risk, noting recent studies that have documented poor iron status

Training Table

Iron-deficiency anemia occurs when the body lacks enough iron to make the hemoglobin that is necessary for the production of normal red blood cells. This deficiency results in red blood cells that are small (microcytic) and pale (hypochromic). Approximately 30 percent of the world population has anemia; most cases are a result of iron deficiency.

Commons signs and symptoms of iron-deficiency anemia include:

- Fatigue and weakness
- Paleness
- Shortness of breath
- Abnormally shaped fingernails (brittle, cupped)
- Irritability
- Poor appetite
- Difficulty concentrating (e.g., studying for exams)
- Feeling cold
- Headache
- Hair loss

Individuals experiencing any of these symptoms should visit their health-care provider for further evaluation and diagnosis. Do not take an iron supplement without speaking with a health-care provider first, because iron toxicity can be quite harmful to the body.

and associated declines in both cognitive and physical performance. In a study of elite female soccer players 6 months before the FIFA Women's World Cup, Landahl and others reported that 57 percent had iron deficiency and 29 percent had iron-deficiency anemia. Recently, Attwell and colleagues explored contributing factors to high rates of iron deficiency in dancers and the impact on training, rehearsal, and performance.

The normal hemoglobin level is 14–16 grams per deciliter (100 ml) of blood for males and 12–14 grams for females. As noted previously, males have been classified as anemic with less than 13 grams, whereas values less than 12 grams have been used as the criterion for anemia in females. Randy Eichner, a hematologist involved in sports medicine, poses the interesting question of whether an athlete whose usual hemoglobin is 16 grams per deciliter is anemic if his level decreases to 14 grams.

Most sport scientists indicate it is important to monitor the diet and iron status of certain athletes. In its position statement relative to nutrition for the athlete, the ACSM, AND, and Dietitians of Canada professional organizations recommended periodic screening of iron status for some athletes. In the United States some, but not all, universities screen for iron deficiency in female athletes.

Causes of deficiency in athletes There may be a number of causes for the low iron or hemoglobin levels found in some athletes, including inadequate dietary intake, heavy menstrual cycles, and various exercise protocols, training at high altitudes, foot-strike hemolysis, and injury. As well, intense training can contribute to the loss of iron through sweat, urine, and feces.

Inadequate dietary intake Inadequate dietary intake of iron has been identified as one key factor underlying decreased performance in athletes, such as swimmers and marathon runners. In fact, iron is one of the few nutrients recommended as a supplement by the position paper of the ACSM and AND.

Heavy menstrual cycles Increased loss of blood through menstruation could eventually lead to an iron deficiency. In such cases, consultation with a health-care provider could be necessary to determine the cause and appropriate treatment.

Various exercise protocols Exercise may contribute to iron deficiency in a variety of ways. For example, strenuous endurance exercise training has been associated with iron deficiency. Sim and others reported that intense exercise training may induce increases in inflammatory cytokines, such as interleukin-6. This may then elevate hepcidin levels, leading to a decrease in iron absorption. They note hepcidin levels have been shown to be significantly elevated after prolonged exercise training, such as for a marathon, in female athletes. Decreased iron absorption and metabolism could lead to anemia.

Losses through urine, feces, and sweat Some types of exercise may increase iron loss though the urine, feces, and sweat. Distance runners may experience **hematuria**, the presence of hemoglobin or myoglobin in the urine. Repetitive impact of the feet with the ground may rupture blood cells as they pass through the foot vascular system, a process known as **hemolysis**, releasing hemoglobin for circulation to the kidneys and excretion in the urine. As well, prolonged running may lead to ruptured muscle cells, releasing iron-containing myoglobin, which may have the same fate.

Intense, prolonged exercise has also been shown to cause inflammation and bleeding in the GI tract. Additionally, some athletes may also use aspirin or other anti-inflammatory drugs as a pain reliever during intense training, and use of such medications has been associated with GI bleeding. Bleeding in the gastrointestinal tract leads to a loss of iron in the feces. Additionally, athletes may lose some iron during heavy sweating. Iron sweat losses may vary, one approximation being about 0.18–0.20 milligrams per liter of sweat. Although not a substantial amount of iron loss, copious sweat losses could accumulate to 0.4 milligrams, which would necessitate 4 milligrams of dietary iron to replace based on a 10 percent absorption rate.

Training at high altitudes Athletes who begin to train at altitude will need to ensure adequate dietary iron intake, because the increased production of red blood cells at altitude will draw upon the body reserves.

Sports anemia As would be expected, the problem of concern to the endurance athlete is the development of iron-deficiency anemia. A number of studies have shown that anemia causes a significant reduction in the ability to perform prolonged, high-level exercise.

As mentioned in Chapter 6, one form of anemia associated with endurance training is sports anemia. Sports anemia is actually not a true anemia. Although the hemoglobin concentration is toward the lower end of the normal range, the other indices of iron status are normal. Whether sports anemia is a beneficial physiological response to endurance exercise or a condition that will hinder performance is not known. Short-term sports anemia appears to develop in some individuals during the early phases of training or when the magnitude of training increases drastically. One of the effects of endurance training is to increase both the plasma volume and the number of RBC. However, the plasma expansion appears to be greater, so there is a dilution of the RBC and a lowering of the hemoglobin concentration. This effect is believed to be beneficial to the athlete, however, because it reduces the viscosity, or thickness, of the blood and allows it to flow more easily. This adaptation also enhances thermoregulation. In many athletes, the hemoglobin concentration returns to normal after the first month or so of training. Long-term sports anemia is often seen in highly trained endurance athletes. One theory proposes that the production of RBC by the bone marrow is decreased in endurance athletes because the RBC become so efficient in releasing oxygen to the tissues. The authors of this theory suggest that sports anemia is not due to poor iron status. Moreover, in a review relating hematological factors to aerobic power, Gledhill and his associates indicated that an increase in blood volume can compensate for a moderate reduction in hemoglobin concentration.

Eichner noted that the term *sports anemia* indicates a pseudoanemia, or a false anemia in athletes who are aerobically fit. Many experts believe the term is misleading and its use should be discouraged because athletes who develop anemia do so not because of exercise, but for the same reasons that nonathletes do, primarily inadequate dietary iron.

Andrew Rich/E+/Getty Images

Iron deficiency without anemia As noted above, an iron deficiency in the tissues may have adverse effects on exercise performance. Thus, an athlete may have an iron deficiency but not to the extent of being anemic. The effect of iron deficiency without anemia on exercise performance has been studied over the course of 30 years. In a review, Rowland noted that whether iron deficiency without anemia decreases performance in humans is debatable. He indicated that research evidence in animals is clear cut, indicating that depletion of iron stores without a decrease in hemoglobin diminishes exercise performance. Decreased iron stores in the tissues diminished intracellular metabolic capacity. However, Rowland notes that research in humans has not provided sufficient evidence to support the effect seen in animals, indicating that research with human subjects suffers from numerous methodological weaknesses. From the available data, he concludes that it is difficult to mount a compelling argument that an iron deficiency without anemia impairs endurance performance in athletes.

Supplementation: Health and Physical Performance The importance of iron to oxygen transport and endurance capacity and the possibility that many athletes, particularly females, may be iron deficient have led a number of investigators in the area of sports nutrition to recommend that more athletes take dietary iron or supplements. Others discourage the indiscriminate use of iron supplements by athletes.

Does iron supplementation improve physical performance? Many studies have been conducted in attempts to answer this question, and the answer appears to be dependent upon the iron status of the individual.

Iron-deficiency anemia If the individual suffers from iron-deficiency anemia, iron therapy could help correct this condition and concomitantly improve health status and exercise performance capacity. In a review, Lomagno and others reported improvement in various aspects of mood, memory, and intellectual ability after iron supplementation. In a meta-analysis, Avni and others found that iron replacement therapy in cardiac patients improved quality-of-life measures as well as performance in a 6-minute walking test. In their review, Rodenberg and Gustafson concluded that athletes who are found to be anemic secondary to iron deficiency do benefit and show improved performance with appropriate iron supplementation.

Iron deficiency without anemia The effect of iron supplementation in iron-deficient, nonanemic individuals has been studied for possible effects on both health and physical performance.

Relative to health, Lobera-Jáuregu notes that iron deficiency without anemia may cause cognitive disturbances, mainly in attention span, intelligence, and sensory perception, but also notes that despite methodological differences among studies, there is some evidence that iron supplementation improves cognitive functions. Supplementation may have some other health benefits as well, particularly for females during pregnancy, as plasma and red blood cell production increases, possibly causing a deficiency. During pregnancy, the RDA for iron increases by 50 percent.

The effect of iron supplementation on physical performance capacity in iron-deficient but nonanemic individuals is somewhat controversial. Although iron supplementation may improve markers of iron status, the effect on exercise performance is debatable. In their review, Rodenberg and Gustafson indicated that supplementation does not appear to be justified solely to improve performance. Conversely, two separate reviews by DellaValle and Burden and others concluded that iron supplementation would improve iron status in female athletes. DellaValle also concluded supplementation may improve measures of physical performance, while Burden and others reported a moderate effect on improving hemoglobin and aerobic capacity as measured by VO_2 max. Additional research is needed to help resolve these contradictory findings.

Oxygen transport Iron supplementation offers no benefits to individuals with normal hemoglobin and iron status. Some well-controlled research on highly active, nonanemic females with normal iron status has shown no effect of iron supplementation on hemoglobin concentration or exercise performance.

However, endurance athletes with normal hemoglobin status who attempt to increase their red blood cells (RBC) and hemoglobin levels may benefit from iron supplementation. Increased hemoglobin levels increase the ability to transport oxygen to the muscles, with the goal of enhancing performance. Previously, athletes used blood doping techniques (reinfusion of one's own blood previously drawn or from a blood-matched donor) or injection of recombinant erythropoietin (rEPO), a hormone that stimulates RBC and hemoglobin production. However, both blood doping and rEPO have been prohibited by the World Anti-Doping Agency, not only because they may provide an unfair advantage but also because their use may be lethal.

The technique of "Live high, train low" may be an effective alternative. When one ascends to altitude, such as 2,500 meters (8,225 feet) or so, the atmospheric oxygen pressure decreases, leading to lower oxygen pressure in the blood. The body immediately begins to adapt. The kidneys produce natural EPO, which stimulates the bone marrow to produce more RBC. Over time, the RBC and hemoglobin concentration are elevated. This is the benefit of "Live high." However, given the decreased oxygen pressure, athletes may not train as intensely at altitude, and thus may not train optimally, as they could at sea level, the "Train low" component. Some athletes may reside in locations where it is possible to live high and train low by driving an hour or so to a lower altitude. Such is not the case for most athletes, so scientists have constructed houses at sea level whose inside atmosphere has been manipulated to resemble one of high altitude. Thus, the athlete can "Live high" in the house, and produce EPO naturally, but can also step outside and "Train low" at sea-level atmosphere. Additionally, tents used as sleeping chambers are available commercially.

DreamPictures/Jensen Walker/Blend Images LLC

Research supports the efficacy of a "Live high, train low" protocol to enhance aerobic endurance performance. Wilber and others suggest that the optimal dose is to live at a natural elevation of 2,000–2,500 meters for at least 4 weeks with more than 22 hours a day at that altitude. Although living at natural altitude is effective, Mazzeo and Fulco contended that the results from using alternative means, such as altitude houses or sleeping tents, are equivocal.

For athletes who use this strategy, iron supplementation may be necessary to provide substrate for hemoglobin synthesis. Mazzeo and Fulco note that this is true particularly for females who come to altitude with inadequate or borderline iron stores. Iron supplementation prior to and during their stay at altitude will increase hematocrit similarly to males. In a study conducted by Hannon and others, the researchers compared males and females with and without iron supplementation before, during, and after 60 days at altitude (4,300 meters). Males and females with iron supplementation had similar increases in hematocrit levels (% over sea level baseline) compared to females without iron supplementation. Thus, they recommend that, prior to coming to altitude, individuals should ensure that they have adequate iron stores, especially for athletes who plan on training or competing at high elevations. Females may benefit from iron supplementation during altitude acclimatization.

Excess Iron Iron is both an essential nutrient and potentially harmful to cells when in excess. If you plan to take iron supplements, you should have your serum ferritin checked, because some danger is associated with iron supplements if they lead to excessive iron in the body. Prolonged consumption of large amounts can cause a disturbance in iron metabolism in susceptible individuals. Iron then tends to accumulate in the liver as hemosiderin, which in excess can cause **hemochromatosis**, the most serious health consequence of excessive iron intake. This condition causes cirrhosis and may lead to the ultimate destruction of the liver. In addition to dietary factors contributing to hemochromatosis, approximately one million people in the United States have hereditary hemochromatosis, a genetic disease.

Experts believe that iron overload can contribute to the development of additional health problems. Other than the liver, iron can also accumulate in other body organs, such as the heart and pancreas. Too much iron in the heart can cause irregular heartbeats and contribute to heart failure. Additionally, excessive iron may be fatal to young children; more than 30 deaths occur each year from overdoses of iron obtained by eating large amounts of candy-flavored vitamin tablets with iron.

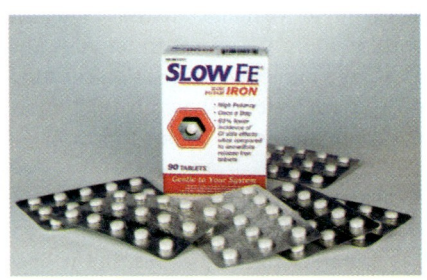
David A. Tietz/Editorial Image, LLC

Prudent Recommendations In summary, it would be wise for developing adolescents to be aware of the iron content in their diets. This concern is especially important to endurance athletes, although it would appear that the extra kcal they eat to meet the additional energy requirements of training would provide the necessary iron. Eating foods rich in vitamin C with nonheme iron-containing foods and using iron cookware also will increase iron bioavailability.

Female athletes at the greatest risk for iron deficiency are often advised to be screened at the beginning of and during the training season using tests for hemoglobin and serum ferritin, and appropriate dietary and/or supplementation recommendations should be made to those with compromised iron status. Iron supplementation by commercial preparations may be recommended for certain individuals who have, or who are at high risk of having, low serum ferritin levels. This includes female distance runners, some vegetarian athletes, those who experience heavy menstrual blood flow, athletes who initiate altitude training, and athletes who are on restricted caloric intake. Over-the-counter multivitamin/mineral preparations vary in iron content. Some contain none, such as those marketed to males and postmenopausal females, while others contain 18 milligrams of iron, which is 100 percent of the RDA for adult females and 120 percent for adolescent girls. One tablet a day may be advisable for these individuals, and it should be consumed on an empty stomach to minimize adverse effects of some foods on absorption. For females who have iron deficiency without anemia who want a rapid restoration of serum iron, injections or infusions may be necessary.

It is important to reemphasize that iron supplementation should not be done indiscriminately, but preferably only after determination of one's iron status. Increased iron stores are a common finding in elite athletes who have used long-term iron supplementation, putting the athletes at an increased risk of developing iron overload–related diseases. Health professionals indicate that iron supplements should be given to athletes only by prescription, primarily only in cases of iron-deficiency anemia.

> www.ods.od.nih.gov/factsheets/Iron-HealthProfessional/ The NIH Office of Dietary Supplements provides health fact sheets for minerals, including iron. Use the supplement fact sheet for additional information on iron.
>
> www.genome.gov/Genetic-Disorders/Hereditary-Hemochromatosis The NIH National Human Genome Research Institute shares practical information about hemochromatosis.

Copper (Cu)

Copper is a vital mineral that is a component of several enzymes.

DRI The RDA for copper is 900 mcg/day for adults age 19–50, but amounts vary for other age groups. The UL for adults is 10,000 mcg/day.

Ken Weinrich/Shutterstock

Food Sources Copper is widely distributed in foods and is found in high amounts in seafood, meats, nuts, beans, whole-grain products, and foods containing chocolate. Copper also may be found in drinking water, particularly soft water, which leaches it from copper pipes.

Major Functions Copper serves a prominent role as a cofactor for metalloenzymes, known as cuproenzymes, most of which are involved in oxidation processes. Copper also works closely with iron in oxygen metabolism. It is needed for the absorption of iron from the intestinal tract, helps in the formation of hemoglobin, and is involved in the activity of a specific cytochrome (cytochromes C oxidase), an oxidative enzyme in the mitochondria. Copper is also a component of ceruloplasmin, a glycoprotein in the plasma, and is in superoxide dismutase (SOD), an enzyme that functions as an antioxidant to quench free radicals.

Deficiency: Health and Physical Performance Copper deficiency is rare in humans. A genetic disorder, Menkes syndrome, may impair copper metabolism, eliciting deficiency symptoms in the newborn with numerous developmental and neurological defects. In many cases, affected children live only a few years. Other genetic disorders may impair copper metabolism and cause deficiency. Individuals consuming large doses of zinc supplements or antacids may also experience copper deficiency. The major deficiency symptom is anemia, but osteoporosis, neurological defects, and heart disease may also develop.

Available surveys indicate that most athletes consume ample amounts of copper. The effects of exercise or exercise training on serum copper levels are variable, with studies showing increases, decreases, or no changes. Several studies have reported decreases in serum copper in athletes involved in prolonged training or after an endurance exercise task. The authors theorized that the decreased levels were due to sweat or fecal losses. However, no deficiency symptoms were noted. In his review, Lukaski indicated that physical training does increase the copper-containing SOD, and normal body stores of copper are apparently adequate to support the increase of this antioxidant enzyme.

Supplementation: Health and Physical Performance Dietary supplements contain varying amounts of copper, but most provide the RDA for copper, 900 mcg/day. Fortified foods may also contain copper. Although copper (2 mg) is used in a supplement being studied to help prevent age-related macular degeneration, it is not because it plays a role in eye health but rather to help counter the effect of the high zinc content that could reduce copper absorption. Supplements are not recommended because excessive copper intake, even 5–10 milligrams, may cause nausea and vomiting. The UL is 10 mg/day. The National Academy of Sciences reports that toxicity from dietary sources is rare. However, evidence suggests that copper poisoning can result from excess intake via supplements.

In recent years, scientists have been exploring the role of copper in the etiology of Alzheimer's disease. Some individuals have a genetic disorder predisposing them to increased concentrations of copper in various body tissues, including the brain. This condition is known as Wilson disease and has numerous symptoms, including some comparable to Alzheimer's. Although Wilson disease is hereditary, researchers suggest a diet-gene interplay may be involved. In this regard, Brewer hypothesizes that the dietary factor is inorganic copper in drinking water, leached from copper plumbing, the use of which coincides with the epidemic of Alzheimer's. Kaler reported that symptoms associated with Wilson disease are reduced by copper chelation therapy, a method used to bind and excrete metals from the body.

Prudent Recommendations Consuming a balanced diet should provide adequate amounts of dietary copper. Copper supplements are not recommended for either health or enhancement of sports performance.

Zinc (Zn)

Zinc is a blue-white metal that is an essential nutrient for humans and widespread in a variety of foods.

DRI The RDA for zinc is 11 mg/day for adult males and 8 mg/day for adult females. The UL for adults is 40 mg/day.

Carolyn Taylor Photography/Stockbyte/Getty Images

Food Sources Good sources of zinc are meat, cow's milk, and seafood, particularly oysters. Three ounces of meat contain approximately 30–50 percent of the RDA, whereas only one oyster will provide more than 70 percent. Whole-grain products and legumes also contain significant amounts of zinc, but the phytate and fiber content will slightly decrease its bioavailability. Zinc is lost in the milling process of wheat, but some breakfast cereals are fortified to 25–100 percent of the RDA. In general, if you receive enough protein in the diet, you will obtain the RDA for zinc. About 20–50 percent of dietary zinc is absorbed and daily intake of zinc is recommended.

Major Functions Zinc is found in virtually all tissues in the body. Total body zinc is approximately 1.5 grams in females and 2.5 grams in males. Zinc has a large number of physiological functions, which may be related to its key role in the activity of more than 300 enzymes. **Table 8.6** summarizes the major functions of zinc in human metabolism and health.

Deficiency: Health and Physical Performance Plasma zinc concentration is normally used as a marker for zinc status, but it does not reflect zinc status in the cells. According to experts, nearly half of the world's population is at risk for zinc deficiency, mainly in Asia and Africa. Livingstone notes, in general, that adaptive mechanisms enable the body to maintain normal total body zinc status over a wide range of intakes, but deficiency can occur because of reduced absorption or increased gastrointestinal losses. In developing countries in

TABLE 8.6	Major functions of zinc
Promotion of immune system functions	
Promotion of eye health	
Promotion of wound healing	
Production of energy via the lactic acid energy system	
Synthesis of DNA, protein, and insulin	
Support of cellular and body growth	
Promotion of bone formation	
Promotion of red blood cell production	
Regulation of gene expression	
Optimization of the senses of taste and smell	

Asia, Africa, and other parts of the world, the diet consists primarily of plant-based foods, which are rich in phytates. As you may recall, phytates decrease zinc absorption. Additionally, most daily losses of zinc are via the gastrointestinal tract. Diarrhea may be prevalent in less developed countries, leading to increased gastrointestinal losses of zinc. Other possible mechanisms of zinc loss from the body include menstruation and sweat losses.

Zinc deficiency poses serious health risks, particularly in the young and in pregnant females, as it may lead to impaired growth and development. Worldwide, zinc deficiency contributes to about 4 percent of child morbidity and mortality.

Overt zinc deficiency is uncommon in North America. However, certain individuals may be at risk and need to include zinc-rich foods in their diets, including vegetarians, individuals with gastrointestinal disorders such as Crohn's disease, and pregnant and lactating females. Those with alcohol use disorder may also have low zinc status because alcohol decreases zinc absorption and increases excretion of the mineral.

Epidermal, gastrointestinal, central nervous, immune, skeletal, and reproductive systems are the organs most affected clinically by zinc deficiency. Some common symptoms of zinc deficiency include failure to grow properly, impaired wound healing, and depressed appetite, but other symptoms such as weight loss, taste abnormalities, mental depression, and impotence may occur. Such symptoms may be associated with a variety of factors, so authorities indicate a medical examination is necessary to determine the presence of a zinc deficiency.

Most research indicates that athletes who obtain sufficient dietary kcal generally meet the RDA for zinc, but some athletes may be at risk. In particular, young athletes in sports that stress weight loss for optimal performance or competition, such as gymnastics, wrestling, and cross-country running, may adopt very low-kcal diets, often low in animal protein, and thus may not obtain sufficient dietary zinc. Athletes following a predominately plant-based diet may also be at risk due to decreased zinc absorption associated with the high phytate content of plant foods. Sweat losses of zinc may also contribute a zinc deficiency, with one study reporting zinc losses in sweat accounted for approximately 8 percent to 9 percent of the daily RDA. However, in reviews by two experts in mineral nutrition, both Lane and Lukaski noted that, in general, there is no evidence that exercise causes a poor zinc status or that a marginal deficiency impairs performance.

Supplementation: Health and Physical Performance Zinc supplementation has been studied in relation to a number of health conditions, including asthma and chronic obstructive pulmonary diseases (COPD). Das and others indicated that food fortification with zinc may be a cost-effective strategy to overcome zinc deficiency in developing countries. In this regard, King and Cousins noted that zinc supplementation has been associated with physical growth of children in various population groups and, in some studies, impressive reductions in childhood morbidity and mortality. In 2013, Penny reported that zinc supplements could reduce diarrhea mortality in children age 12–59 months by an estimated 23 percent. Mayo-Wilson and others, in an analysis of 80 studies, concluded that the benefits of preventive zinc supplementation may outweigh any potentially adverse effects in areas where risk of zinc deficiency is high.

Relative to eye health, zinc is found in high concentrations in the retina, and diets rich in zinc have been theorized to help decrease the risk of age-related macular degeneration (AMD), a major cause of visual problems and blindness in the United States. In a recent study, Cunningham and colleagues describe the beneficial role of zinc in the treatment of AMD. As well, the Age-Related Eye Disease Study (AREDS) reported that zinc may be recommended as a supplement when combined with other nutrients, particularly copper and vitamins C and E. The National Eye Institute provides recommendations for AREDS formulation for prevention and treatment of AMD. Visit www.nei.nih.gov/learn-about-eye-health/eye-conditions-and-diseases/age-related-macular-degeneration for those recommendations.

Zinc supplementation has been advocated as a means of enhancing immune functions, particularly treatment of the common cold, a health problem that could impair exercise training. Many clinical studies have evaluated the effect of zinc lozenges as a means to reduce the duration of symptoms of the common cold, but a review by Nieman and a meta-analysis by Jackson and others concluded that evidence supporting its effectiveness in this regard is still lacking. King and Cousins also note that the results of studies relative to zinc supplementation, mainly zinc lozenges, and the common cold are mixed. Additionally, in a review of nutrition support to maintain proper immune status during intense exercise training written by a prominent exercise immunologist, Gleeson noted that, although an adequate intake of dietary zinc is particularly important in the maintenance of immune function, it is safe to say with reasonable confidence that supplementation with zinc is unlikely to boost immunity or reduce infection risk in athletes. Recently, the relationship between zinc status and COVID-19 outcomes has been investigated, with research suggesting that adequate zinc status can be beneficial in protecting individuals with COVID-19 against more severe disease progression.

Beletate and others indicated that zinc plays a key role in the synthesis and action of insulin, both physiologically and in diabetes mellitus. Therefore, hypothetically, zinc supplementation may help prevent type 2 diabetes. However, in their review, research was very limited and they concluded that there is currently not enough evidence to suggest the use of zinc supplementation in the prevention of type 2 diabetes mellitus.

Given the many metabolic functions of zinc, it is unusual that only limited research has evaluated its effects on exercise performance. Several studies have been conducted and used several different tests of exercise performance to evaluate the effect of zinc supplementation, but the results revealed significant benefits of supplementation on some exercise tests but not on others. In his review, Lukaski noted that study designs limit our ability to provide recommendations regarding zinc supplementation to athletes. Zinc supplementation studies are needed, particularly using well-trained elite athletes.

While taking small amounts of zinc from supplement sources does not appear to pose any major health problems, zinc toxicity can occur in both acute and chronic forms. The acute adverse effects of high zinc intake (about 500 milligrams) include nausea, vomiting, loss of appetite, abdominal cramps, diarrhea, and headaches. Chronic effects associated with intakes of 150–450 milligrams of zinc per day include low copper status, altered iron function, and reduced immune functions. The doses of zinc used in the AREDS study (80 mg/day for about 6 years) have been associated with a significant increase in hospitalizations for genitourinary causes. Thus, long-term use of zinc doses greater than the UL may increase the risk of adverse health effects. However, the National Eye Institute notes that the UL does not apply to individuals receiving zinc for medical treatment, but such individuals should be under the care of a physician who monitors them for adverse health effects.

Prudent Recommendations   On the basis of available evidence, zinc supplementation is not warranted for most individuals, including athletes. Foods rich in zinc should be selected to replace the increased kcal expended through exercise. However, athletes such as wrestlers and others incurring weight losses, as well as older endurance athletes whose immune system normally declines, should be exceptionally aware of high-zinc foods. Zinc-fortified foods, such as some grains, may provide the RDA for zinc. Be sure to check food labels as several servings of fortified foods daily could provide enough zinc to exceed the UL. If a supplement is recommended, it should not exceed the RDA.

Chromium (Cr)

Chromium is a very hard metal that exists in various oxidative states, with chromium III and chromium IV being the most important to human health. Chromium VI, hexavalent chromium, is an industrial waste product that, when inhaled or consumed, has been associated with an increased risk of cancer.

DRI   The AI for chromium is 20–25 mcg/day for females and 30–35 mcg/day for males. No UL has been set.

nicolebranan/E+/Getty Images

Food Sources   Good sources of chromium include whole-grain products, organ meats, egg yolks, pork, oysters, nuts, and fruits and vegetables. Beer also contains some chromium. Broccoli is a rich source of chromium with one-half cup providing approximately 25 percent of the AI for chromium. While whole wheat bread is a good source of chromium, processed white bread contains only about half as much. As with some other minerals, the amount of chromium in plant foods is related to the chromium content of the soil where the crop is grown. Chromium is poorly absorbed from the intestinal tract; less than 1 percent is absorbed when intakes are within the AI range. At lower dietary intakes, absorption is somewhat increased.

Major Functions   Chromium is particularly important in optimizing insulin activity, ultimately playing an important role in carbohydrate and lipid metabolism. Specifically, chromium is considered to be an essential component of the glucose-tolerance factor (GTF) associated with insulin in the proper metabolism of blood glucose. By enhancing the action of insulin, allowing glucose to enter cells, chromium as a part of GTF is important for carbohydrate, fat, and protein metabolism. Chromium helps maintain blood glucose levels, promotes glycogen formation in the muscles, and helps promote muscle tissue synthesis.

Deficiency: Health and Physical Performance   Chromium deficiency is rare as well-balanced diets typically provide more than adequate amounts of the mineral. Because of its role in GTF, deficiencies can contribute to impaired glucose tolerance, which may increase one's risk for type 2 diabetes. As well, chromium deficiencies have been associated with abnormal blood lipid levels, primarily elevated total cholesterol and triglyceride levels.

The major problem in assessing the effect of chromium deficiency on health or physical performance is determining if a deficiency exists. Chromium status is difficult to determine because various measures, such as blood and urine tests, do not reflect body stores and no specific biochemical marker to reliably access chromium status in humans has been found.

Nevertheless, given its potential role in human metabolism, a chromium deficiency could impair insulin activity and both health and physical performance could be adversely affected. Impaired insulin sensitivity could lead to diabetes and body weight gain, accompanied by their associated health risks, particularly cardiovascular disease. Impaired insulin sensitivity could also impair physical performance, possibly by interfering with carbohydrate and protein metabolism. Decreased muscle glycogen could impair performance in endurance athletes, while decreased uptake of amino acids into the muscle could impair muscle growth in strength athletes.

Lefavi speculated that athletes may incur a negative chromium balance under three conditions. One, increased intensity and duration of exercise may increase urinary excretion of chromium. Two, athletes who consume substantial amounts of carbohydrates may need more chromium to process glucose. And three, athletes who lose weight for competition may decrease dietary intake of chromium.

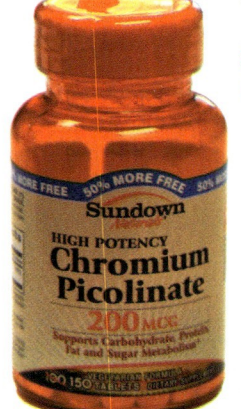
Mark Dierker/McGraw Hill

Supplementation: Health and Physical Performance   Given the potential role of chromium in glucose metabolism, supplementation has been studied for possible benefits to blood sugar and glucose tolerance, most notably in individuals with type 2 diabetes. In a meta-analysis of 41 studies, Balk and others noted that chromium supplementation significantly improved blood glucose control among patients with diabetes, but future studies that address the limitations in the current evidence are needed before definitive claims can be made about the effect of chromium supplementation. Several recent studies on the relationship between chromium supplementation and diabetes risk and treatment have found that chromium appears to be beneficial. These potential benefits are summarized in recent reviews by Asbaghi and others and Zhao and colleagues. Again, further research is needed.

www.diabetes.org The American Diabetes Association provides patient education materials that summarize the current research and recommendations related to chromium supplementation in those with diabetes. Visit the ADA website and search "chromium."

Theoretically, chromium supplementation might benefit the endurance athlete by improving insulin sensitivity and carbohydrate metabolism during exercise. Also, because chromium may enhance the anabolic effect of insulin, it may increase amino acid uptake into the muscle and modify the body composition, increasing muscle mass and decreasing body fat. Given the potential commercial application of this later theoretical possibility to both athletes and the general population, most of the research to date has focused on the effect of chromium supplementation on body composition, but several recent studies have evaluated its effect on strength. Most studies have used chromium picolinate.

Although chromium supplementation may provide beneficial effects for some individuals with diabetes, its major popularity was related to marketing to the public as a supplement for body weight control, mainly by promoting fat loss. Chromium supplements were also marketed to physically active individuals as a means to promote gains in muscle mass. Such chromium supplements became so popular that sales were second only to calcium among mineral supplements. In particular, chromium picolinate was a best-selling supplement. Picolinate is a natural derivative of tryptophan, an amino acid, and apparently facilitates the absorption of chromium into the body. The popularity of chromium picolinate may have been associated with some research in the late 1980s.

In a review of several of his own studies involving chromium picolinate supplementation to male college athletes and students involved in weight training, Evans suggested chromium picolinate supplementation enhanced body composition by decreasing body fat and increasing lean body mass. Following the publicity associated with this report, chromium picolinate was billed in certain muscle magazines as the alternative to anabolic steroids. The advertisers suggested that chromium's insulin-like effects may elicit significant anabolic hormone effects in the body. Advertisements also appeared in magazines targeted for the general population, suggesting that chromium picolinate would facilitate the loss of body fat.

However, well-controlled research does not support these advertisement claims. Numerous studies have concluded that supplementation with chromium picolinate and other forms of chromium, usually in conjunction with an exercise training program, had no effect on body composition or various measures of physical performance when compared to the placebo condition. Several of the studies used research protocols similar to those of Evans. The following is a summary of those findings:

- No effect on body composition or strength in untrained males after 12 weeks of resistance training
- No effect on lean body mass, body fat percentage, or strength in college football players after 9 weeks of training
- No effect on body weight and composition in nonexercising females over the course of 12 weeks
- No effect on body fat or lean muscle mass in males after 8 weeks of training
- No effects on body fat or lean muscle mass in individuals engaging in aerobic exercise for 16 weeks
- No effect on body composition, resting metabolic rate, or blood glucose in females with obesity in a walking exercise program for 12 weeks
- No effect on body composition or strength of adult females after 12 weeks of resistance training
- No effect on body composition or neuromuscular or metabolic performance in highly trained NCAA Division I wrestlers during 14 weeks of training
- No effect on body composition and muscular strength in female softball players during 6 weeks of resistance training
- No effect on prolonged, intermittent, high-intensity exercise in physically active males following ingestion of a sports drink with added chromium

Relative to weight loss, recent reviews and meta-analyses do not support claims of major body weight or body fat losses following supplementation with chromium. Onakpoya and others, in a meta-analysis of 11 studies, did show a statistically significant difference in weight loss favoring chromium supplementation over placebo. However, although statistically significant, the effect was very small, only a loss of about 1 pound, which the investigators noted had uncertain clinical significance. In their review, Tian and others found no current, reliable evidence to provide firm decisions about the efficacy of chromium picolinate supplements in reducing body weight in individuals with overweight or obesity. In his review, Vincent has a stronger viewpoint, indicating chromium has been conclusively shown not to have beneficial effects on body mass or composition in human studies.

Overall, these findings indicate chromium supplementation has little to no effect on body composition in sedentary individuals, individuals engaged in various forms of exercise training, or athletes engaged in training for their sport. Chromium supplementation also does not appear to benefit various forms of exercise performance.

Relative to safety, no UL has been established for chromium, mainly because few serious adverse effects have been linked with high intakes of dietary chromium. The Institute of Medicine reported that no consistent, frequent adverse events were evident from human studies with chromium supplementation. They also noted there is a lack of information on the long-term effects of chronic chromium picolinate supplementation at the recommended doses, and recommend additional research to resolve any uncertainties.

Prudent Recommendations Consuming a well-balanced diet should provide adequate chromium. For those who are glucose intolerant or who have prediabetes or diabetes, chromium supplementation may be beneficial, but should be discussed with a healthcare provider. In general, chromium supplementation will not help you lose weight or gain muscle. If you insist on taking a chromium supplement, it should not exceed a total of 200 micrograms daily taken in three separate doses. Chromium might best be taken as part of an inexpensive multivitamin-mineral tablet containing the other essential vitamins and minerals. The typical multivitamin/multimineral tablet contains about 120 micrograms of chromium. Keep in mind, however, that the best sources of chromium are whole grains, fruits, and vegetables.

Selenium (Se)

Selenium is a chemical element resembling sulfur that is found naturally in many foods and, while originally thought to be a toxin, is now known to be an essential nutrient.

DRI The RDA for selenium is 55 mcg/day for males and females, with lower amounts for children. The UL has been set at 400 mcg/day for adults, but somewhat lower levels for children and adolescents.

Food Sources Most selenium found in nature is a part of proteins known as selenoproteins. One of the richest sources of dietary selenium is Brazil nuts; a 1-ounce serving contains over 500 mcg. Other rich sources of selenium are organ meats such as kidney and liver, seafood such as tuna, and other meats. Plants such as cereals, grains, fruits, and vegetables grown in soil abundant in selenium are also good sources. The U.S. Geological Survey provides a map showing the typical soil selenium content throughout the United States. Visit https://mrdata.usgs.gov/geochem/doc/averages/se/usa.html to view that map and notice the significant range in soil selenium content.

Nishihama/Shutterstock

Major Functions Selenium is actually part of selenocysteine, an amino acid incorporated in about 25 selenoproteins, many of which are involved in oxidoreductase activity involving the transfer of electrons from one molecule to another. In particular, selenium is involved in the activity of glutathione peroxidase, an antioxidant enzyme that helps protect cells, such as the membranes of red blood cells, from potentially damaging oxidation. Selenoproteins may serve as biomarkers of several diseases, such as diabetes and several forms of cancer.

Deficiency: Health and Physical Performance Selenium deficiency in industrialized countries, including the United States, is rare. However, deficiency diseases may be noted in geographical areas where the selenium content in the soil is low. Keshan disease, which is associated with impaired heart function, was evident in parts of China because the primary sources of food are plants grown locally in selenium-depleted soil. The disease has virtually disappeared from China with improved economic and living conditions. In particular, the government initiated a program of selenium fortification in foods.

Given selenium's role in antioxidant activity in the body, a selenium deficiency has been hypothesized to be related to the development of various chronic diseases. For example, decreased prevention of DNA damage could lead to cancer, and a diminished ability to prevent the oxidation of LDL-cholesterol could lead to heart disease. Some researchers suggest a link between selenoproteins and glucose metabolism, possibly associated with type 2 diabetes. Other potential health problems associated with selenium deficiency include cognitive decline with aging and impaired thyroid gland function. Yet, as noted previously, selenium deficiency is rare in developed countries.

For the athlete, selenium deficiency may impair antioxidant functions during intense exercise, possibly leading to muscle tissue or mitochondrial damage, thus impairing physical performance. Currently, there are no data to support this notion.

Supplementation: Health and Physical Performance Intervention trials in China have shown that selenium supplementation may help prevent Keshan disease. However, can selenium supplementation help prevent cardiovascular disease, cancer, or diabetes in Western populations with better nutritional status?

Relative to cardiovascular disease, selenium supplementation may help prevent LDL oxidation, one of its major risk factors. A recent meta-analysis by Ju and colleagues reported that selenium supplementation resulted in decreased serum C-reactive protein levels and increased glutathione peroxidase levels, suggesting a beneficial effect on oxidative stress and inflammation. Additional research is needed to better understand the contribution of selenium from food and supplemental sources on cardiovascular health.

Some earlier epidemiological studies had indicated that the higher the body selenium levels, the lower the risk of prostate cancer. At one time the National Institutes of Health suggested that selenium may reduce risk for prostate cancer. However, Thompson notes that in a major clinical trial, the Selenium and Vitamin E Cancer Prevention Trial (SELECT), selenium supplementation had no effect on prostate cancer risk. Sunde indicates the study was stopped prematurely because a monitoring committee found an increased risk of prostate cancer in males assigned to the vitamin E supplementation group, but not vitamin E and selenium. In two separate reviews, Vinceti and others concluded selenium supplementation has no beneficial effect on cancer risk, more specifically on risk of prostate cancer. They also indicate some trials suggest harmful effects of selenium exposure.

As related to physical performance, the effects of selenium supplementation by itself has received only limited research attention. Although antioxidant supplements have not universally been shown to prevent peroxidation of lipids in cell membranes and other cell structures, some studies by Tessier and his associates have shown that selenium supplementation will enhance glutathione peroxidase status and reduce lipid peroxidation during prolonged aerobic exercise. Although these findings are intriguing, selenium supplementation did not improve actual physical performance, as evaluated by VO_2 max or running performance of an aerobic/anaerobic nature. In a subsequent study, Margaritis and others reported that selenium supplementation (180 mcg/day) had no effect on muscle antioxidant capacity or exercise performance during 10 weeks of endurance training.

Some health food supplements contain more than 700 times the RDA for selenium per serving, and episodes of acute selenium poisoning have occurred mainly by consuming such products. Symptoms may include nausea, diarrhea, fatigue, brittle fingernails, loss of hair and nails, garlicky body odor, and peripheral neuropathy.

Prudent Recommendations Adequate selenium may be obtained through a well-balanced diet providing adequate amounts of grain products. In the United States and Canada, most grains are produced in the upper Great Plains, where the soil is rich in selenium. Selenium in foods is present in an organic form, which may be more effectively used by the body than inorganic selenium salt supplements. Selenium supplements do not appear to enhance exercise training or performance. If you decide to take a selenium supplement, note that most experts agree such a supplement, or a multivitamin-mineral supplement, should not exceed 200 micrograms. Some experts recommend caution, possibly limiting selenium to 100 micrograms a day.

Key Concepts

- The trace minerals include chromium, fluoride, copper, iodine, iron, manganese, molybdenum, selenium, and zinc.
- Trace minerals are involved in a wide variety of physiological functions in humans and a trace mineral deficiency may lead to adverse health effects. Supplementation with trace minerals typically has no beneficial effects on health or physical performance except in cases where a deficiency is corrected.
- Iron deficiency, particularly among females and young children, is a major public health concern in the United States. Consuming foods rich in iron, particularly heme iron, can protect against deficiency.
- Iron status is important to aerobic endurance athletes because insufficient hemoglobin or other factors associated with iron deficiency may impair performance. Athletes may have poor iron status due to inadequate dietary iron and increased losses of iron in the urine, sweat, and feces.
- Iron supplementation may improve performance in individuals with iron-deficiency anemia, but not in individuals with normal iron status. Although individuals who have iron deficiency without anemia may experience less favorable responses to aerobic training, findings from studies are equivocal as to the efficacy of iron supplementation to improve performance, and investigators recommend additional research. Athletes who train at altitude may consider iron supplementation.
- Zinc deficiency has been shown to impair the growth process in children, so it may be a problem for young athletes who incur heavy sweat losses and are on low-kcal diets, such as wrestlers, dancers, endurance runners, or gymnasts.
- Chromium is a component of glucose tolerance factor and, as such, may enhance insulin function. Chromium supplements have been marketed to increase muscle mass and decrease body fat, but research does not support those claims.
- Selenium is a cofactor for glutathione peroxidase, an antioxidant enzyme. The selenium content of plant foods varies based on where the plant was grown due to differences in selenium soil content.

Mineral Supplements: Exercise and Health

Perusal of the internet reveals numerous advertisements for mineral supplements and their effects on health or sports performance. For example, various mineral supplements are marketed to improve bone health, brain health, and sexual health, and one company markets a mineral supplement designed to provide anabolic mineral support for elite-level athletes. As the foregoing discussion indicates, individuals who have a mineral deficiency may experience improved health or physical performance if that deficiency is corrected. In general, however, supplementation has little effect on the individual whose mineral status is adequate. This last section summarizes some key points relative to mineral nutrition, focusing on the need for supplementation.

Does exercise increase my need for minerals?

Exercise may induce mineral losses from the body by several mechanisms. Many minerals appear to be mobilized into the circulation during exercise, probably being released from body stores in the muscles or elsewhere. As they circulate, some may be removed by the kidneys and excreted in the urine, whereas others may appear in the sweat, particularly in a warm environment. Losses from the gastrointestinal tract may also occur during exercise, although the mechanism is not totally understood.

Other body changes mediated by exercise may influence mineral requirements. The female athlete who develops secondary amenorrhea may need additional calcium, as might the male endurance athlete in whom trabecular bone mass is decreased. The need for iron in the female athlete may decrease somewhat with the cessation of menses in secondary amenorrhea.

Because of potential mineral losses, some sports nutritionists have suggested that mineral supplementation be considered for athletes, particularly those with poor dietary habits. However, although supplementation may be helpful for some, the first concern should be to educate the athlete about obtaining adequate mineral nutrition through dietary means.

Can I obtain the minerals I need through my diet?

As many dietary surveys have shown, many Americans are not obtaining the RDA or AI for a variety of minerals, including iron, zinc, calcium, potassium, and chromium. Similar dietary deficiencies have been noted in surveys with athletes, but mainly with athletes who participate in activities such as wrestling, distance running, dance, and gymnastics, where weight control is a concern.

In general, as with all other nutrients, a balanced diet is essential. Select a wide variety of foods from all the food groups and make nutrient-dense selections. The bioavailability of minerals varies but consumption of phytates, oxalates, and tannins can decrease mineral bioavailability. As well, minerals often compete for absorption from the intestine and, therefore, excessive intake of one mineral can negatively impact absorption of another. Balance is important in optimizing health.

A basic principle of mineral nutrition is to eat natural foods that are rich in calcium, potassium, and iron, nutrients of "public health concern" in the United States. If you select a diet to provide your RDA for these three minerals, you should receive adequate amounts of the other major and trace minerals at the same time.

Are mineral megadoses or some nonessential minerals harmful?

One of the generally accepted facts relative to mineral nutrition in the healthy individual is that the levels associated with toxicity can normally be obtained only through the use of supplements or fortified foods, not through natural dietary sources. Because they hope to improve health or physical performance, many individuals purchase supplements containing minerals. However, surveys indicate that the most common preparations purchased contain the RDA or DV, or less, which should pose no health problems to the healthy individual. Unfortunately, as indicated in the last chapter, many individuals self-prescribe and may consume more than the recommended daily dosage. Although the toxicity and possible health problems associated with excessive intake of several minerals, such as calcium, iron, zinc, and copper, are fairly well documented, the level of safety for intake of a variety of other minerals, particularly some of the trace minerals suggested to be therapeutic in nature, has been more difficult to document. Nevertheless, the National Academy of Sciences has noted that all trace minerals are toxic if consumed at high doses for a long enough time.

Of increasing concern is the potential role of several metals in the etiology of Alzheimer's disease. Ayton and others have proposed a metal hypothesis relative to the development of Alzheimer's disease, primarily involving iron, copper, and zinc. Such minerals may affect brain proteins, possibly leading to neurodegeneration, the hallmark of Alzheimer's. Obtaining adequate amounts of iron and zinc and avoiding excess intake of copper may be important to preventing and treating the disease.

Several nonessential minerals may be consumed inadvertently and cause significant health problems, even in small amounts. For example, lead can displace other minerals, such as calcium and zinc, in various enzymes and thus interfere with intracellular processes involving protein and gene expression. As well, hexavalent chromium, also known as industrial chromium, is regarded as a carcinogen. Of recent concern is mercury, which may be found in foods that we normally think of as healthy—fish!

Methylmercury, an industrial waste product, has been dumped into the seas, where it may be consumed by fish. As noted in chapter 5, high levels of mercury may accumulate more in larger, older, predatory fish, such as shark, swordfish, king mackerel, and tilefish. Tuna may contain somewhat less mercury. However, some FDA data suggest that some light tuna, which normally has lower levels of mercury than white (Albacore) tuna, may have as much as or more.

Houston indicates that mercury may displace other minerals, such as zinc and copper, thus reducing the effectiveness of various metalloenzymes, including antioxidant enzymes, inducing numerous pathological effects. In particular, too much mercury may damage the nervous system, especially the brain during its formative years prior to birth and the first 7 years of life. Thus, the FDA advises females who are or who can become pregnant and small children not to eat any shark, swordfish, king mackerel, or tilefish.

There are several "possibly essential" minerals that are offered as part of dietary supplements, but for which there is little research on their efficacy or safety. The *Training Table* in this section provides information on boron and vanadium, two such possibly essential minerals.

Should physically active individuals take mineral supplements?

In general, the answer to this question for most athletes is *no*—for several reasons.

- First, contrary to advertising claims of mineral-supplement manufacturers, you can obtain adequate mineral nutrition from the diet if you adhere to some of the guidelines presented throughout this chapter.
- Second, although some athletes may not be obtaining the recommended amounts of several minerals, such as zinc and calcium, mineral deficiencies to the point of impairing physical performance are rare. Very limited data are available on this topic; current evidence suggests that physical performance is

Training Table

A quick search at the supplement shop might yield supplements with some lesser known minerals, including boron and vanadium. These minerals are classified as "possibly essential." For many of these supplements, there is limited research on the ergogenic and health benefits, yet consumers are still interested.

What is boron?

Boron is a metalloid, an element with properties associated with both metals and nonmetals. Although boron is an essential nutrient for plants, no RDA or AI has been established for humans. However, some scientists suggest that it is of nutritional and clinical importance and most likely is an essential nutrient for humans.

Boron may form compounds known as boroesters, which may have beneficial effects on bone growth, may reduce arthritic symptoms, may enhance central nervous system function, and may reduce risk for some types of cancers. The metalloid may also have a role in estrogen and testosterone metabolism. Limited research data are available relative to the ergogenic efficacy of boron supplementation; available research suggests that boron supplements do not appear to enhance exercise performance.

What is vanadium?

Vanadium is a light gray metallic element found in food as vanadyl sulfate. Vanadium can also exist in other forms, such as vanadium pentoxide, which is a toxic industrial pollutant. Research with animals indicates vanadium may be involved in several enzymatic reactions in the body, including carbohydrate and lipid metabolism. The metallic element may also potentiate the effects of insulin, but the underlying mechanism remains unknown.

While a deficiency of vanadium has not been noted, if vanadium does have an insulin-like effect, a deficiency could potentially impair glucose metabolism. Limited research has been conducted using vanadyl salts as a means to improve body composition in athletes. The research suggests that vanadyl salt supplementation is not recommended for the average healthy adult, nor for athletes, as such supplementation has not been found to improve body composition or exercise performance. Moreover, excess amounts may be toxic.

not affected even if serum levels of most minerals are low. An exception may be low levels of serum iron, for, as noted previously, supplementation, although controversial, has been helpful to some athletes.

- Third, many minerals may be harmful when taken in excess. As noted throughout this chapter, the absorption rate for most minerals is relatively low. Also, a high dietary intake of several minerals that are easily absorbed increases their excretion rate by the kidney. Thus, a low absorption or high excretion rate prevents the accumulation of excess amounts of minerals in the body, which may interfere with normal metabolism. However, large supplemental doses may overload the body and cause numerous health problems and, as noted for several minerals, may be fatal.

Nevertheless, it is recognized that certain athletes may not be obtaining adequate mineral nutrition from their diets and may benefit from supplementation. As noted previously, athletes who are attempting to lose weight for performance are at most risk for developing a mineral deficiency. Because many of the dietary surveys of these athletes have reported intakes lower than the RDA for iron and calcium, it may be assumed that their diets are also low in other trace minerals.

If there is concern for the nutritional status of the athlete, the ideal situation would be to consult a sports nutritionist or nutritionally oriented physician. Unfortunately, this approach does not appear to be common among athletes who may be in need of nutritional counseling, although the situation is improving. For example, many collegiate and professional sports teams now have access to sport dietitians with expertise in sport physiology, many of whom are members of the Collegiate and Professional Sport Dietitians Association (CPSDA). Visit www.sportsrd.org to learn more about CPSDA and for educational resources on sports nutrition.

For athletes who cannot or will not seek professional advice, it may be prudent to recommend a one-a-day vitamin-mineral supplement to those who are known to have poor nutritional habits. The tablet should contain no more than 50–100 percent of the RDA for any mineral. Additionally, the point should be made to the athlete that the supplement is being recommended to help prevent a deficiency, not for any ergogenic purposes. As noted in chapter 7, large doses of multivitamin-mineral supplements taken over prolonged periods of time have not been shown to enhance physical performance. In the meantime, efforts should be undertaken to educate the athlete concerning sound nutritional practices.

For those considering mineral supplementation for health reasons, the DRI developed by the National Academy of Sciences reflect a paradigm in which the determination of nutrient requirements includes consideration of the total health effects of nutrients, not just their roles in preventing deficiency pathology. For example, the RDA's for calcium and vitamin D as a possible means to prevent osteoporosis reflect this paradigm. While much research is needed before concrete recommendations may be made relative to mineral supplementation and purported health benefits, the recommendations presented in chapters 7 and 8 may be useful guidelines for healthy sedentary and physically active individuals to use in the meantime.

Key Concepts

- Health professionals recommend that individuals obtain their mineral needs from a well-balanced diet. A diet that provides the RDA's for iron and calcium, as well as kcal from a balanced selection of foods throughout the different food groups, will provide adequate amounts of both the major and trace minerals.
- Mineral supplements may be recommended for some individuals as a means to improve health or sports performance, but excessive intake is not recommended because of potential health problems.

Check for Yourself

- Use dietary analysis software, such as NutritionCalc Plus, to track your food intake for 1–3 days. Then, evaluate your daily intake of major and trace minerals. How does your intake compare to your recommended intake? Provide practical recommendations to optimize your mineral intake.

APPLICATION EXERCISE

Choose one of the major or trace minerals that is marketed for potential ergogenic benefits in athletes. Visit a store that sells supplements or search online for supplements providing that mineral. Take a picture or screenshot of the supplement bottle, including the supplement facts label and any marketing for the supplement.

Then, develop an informational handout for someone interested in taking that supplement to optimize physical activity performance. Compare what is on the supplement label to what you have learned about the supplement in the textbook as well as online at the NIH Office of Dietary Supplements website www.ods.od.nih.gov/factsheets/list-all/. Search for the relevant fact sheet on the mineral at the NIH website.

As part of your informational handout, include the image(s) of the supplement container, reported structure/function claims, nutrients and amounts, and suggested uses. How does the information provided on the supplement bottle compare to what is provided by the NIH Office of Dietary Supplements? What recommendation would you make to an athlete interested in taking the supplement?

Review Questions—Multiple Choice

1. Which of the following is not considered a major mineral?
   a. phosphorus
   b. iron
   c. sodium
   d. sulfur
   e. potassium

2. A deficiency of iron is most common in which of the following groups?
   a. older females
   b. older males
   c. young adult females
   d. young adult males
   e. male gymnasts

3. To help prevent the development of osteoporosis in later life, females should consume adequate quantities of which nutrient during the years in which they are developing peak bone mass?
   a. boron
   b. iron
   c. selenium
   d. sulfur
   e. calcium

4. Which mineral is a component of glucose tolerance factor, which enhances insulin function and may be beneficial in preventing type 2 diabetes?
   a. phosphorus
   b. copper
   c. chromium
   d. magnesium
   e. selenium

5. Excessive intake of iron can lead to a condition called hemachromatosis, potentially damaging which organ in the body?
   a. arterial walls
   b. kidney
   c. heart
   d. liver
   e. lungs

6. The mineral _____ acts as a component of glutothione peroxidase, an important antioxidant vitamin.
   a. selenium
   b. boron
   c. phosphorus
   d. chloride
   e. fluoride

7. Which of the following contributes to an increase in iron bioavailability?
   a. eating a diet rich in phytates and oxalates
   b. taking an antacid to reduce stomach acid secretion
   c. consuming a diet rich in calcium
   d. consuming vitamin C
   e. consuming polyphenols, such as found in tea

8. Which of the following is not a major function of zinc?
   a. supporting cellular and body growth
   b. promotion of red blood cell production
   c. regulation of gene expression
   d. promotion of wound healing
   e. antioxidant activity to prevent against chronic disease

9. Which of the following is the best source of heme iron?
   a. steak
   b. whole-grain cereal
   c. low-fat cow's milk
   d. broccoli
   e. oysters

10. Which of the following statements concerning magnesium is false?
    a. Magnesium is found in green leafy vegetables, such as spinach.
    b. Magnesium is important for protein synthesis.
    c. Deficiency results in cretinism.
    d. Absorption rates increase in times of greater need for the mineral.
    e. Deficiency may impair athletic performance.

Answers to multiple choice questions:
1. b; 2. c; 3. e; 4. c; 5. d; 6. a; 7. d; 8. e; 9. a; 10. c

Critical Thinking Questions

1. Contrast major minerals and trace minerals. Select one of the major minerals and provide a summary of the mineral, including functions in human health, food sources, and potential health or ergogenic effects.

2. Identify at least five key functions of zinc in the human body and at least three rich food sources of zinc. Do you recommend supplementation for athletes? Why or why not?

3. Describe the modifiable and nonmodifiable risk factors for osteoporosis. What recommendations would you provide to an athlete at risk for osteoporosis? Provide specific dietary recommendations as they relate to minerals.

4. Explain which athletes are at the greatest risk for iron-deficiency anemia and what recommendations you might make to these athletes to prevent and/or treat the anemia. Contrast dietary sources of heme and nonheme iron. What are factors that increase the bioavailability of iron?

5. You have a friend who watched a video on social media that chromium supplements might be beneficial to sports performance. Provide an overview of chromium to your friend, including theoretical ergogenic effects, sources, and whether or not you recommend supplementation.

References

Abbott, A., et al. 2022. Part II: Risk factors for stress fractures in female military recruits. *Military Medicine* February 27, 2022. Online ahead of print.

Alaunyte, I., et al. 2015. Iron and the female athlete: A review of dietary treatment methods for improving iron status and exercise performance. *Journal of the International Society of Sports Nutrition* 12:38.

American College of Sports Medicine. 2004. Physical activity and bone health. *Medicine & Science in Sports & Exercise* 36:1985–96.

Arts, J., et al. 2012. Adenosine 59-triphosphate (ATP) supplements are not orally bioavailable: A randomized, placebo-controlled cross-over trial in healthy humans. *Journal of the International Society of Sports Nutrition* 9 (1):16. doi: 10.1186/1550-2783-9-16.

Asbaghi, O., et al. 2021. Effects of chromium supplementation on blood pressure, body mass index, liver function enzymes and malondialdehyde in patients with type 2 diabetes: A systematic review and dose-response meta-analysis of randomized controlled trials. *Complementary Therapies in Medicine* 60:102755.

Attwell, C., et al. 2022. Dietary iron and the elite dancer. *Nutrients* 14(9):1936.

Aune, D., et al. 2015. Dairy products, calcium, and prostate cancer risk: A systematic review and meta-analysis of cohort studies. *American Journal of Clinical Nutrition* 101:87–117.

Avni, T., et al. 2012. Iron supplementation for the treatment of chronic heart failure and iron deficiency: Systematic review and meta-analysis. *European Journal of Heart Failure* 14:423–9.

Ayton, S., et al. 2015. Biometals and their therapeutic implications in Alzheimer's disease. *Neurotherapeutics* 12:109–20.

Balk, E., et al. 2007. Effect of chromium supplementation on glucose metabolism and lipids: A systematic review of randomized controlled trials. *Diabetes Care* 30:2154–63.

Beletate, V., et al. 2007. Zinc supplementation for the prevention of type 2 diabetes mellitus. *Cochrane Database of Systematic Reviews* 24:CD005525.

Bielik, V., and Kolisek, M. 2021. Bioaccessibility and bioavailability of minerals in relation to a healthy gut microbiome. *International Journal of Molecular Sciences* 22(13):6803.

Blaszczykm, U., and Duda-Chodak, A. 2013. Magnesium: Its role in nutrition and carcinogenesis. *Roczniki Państwowego Zaktadu Higieny* 64:165–71.

Bloomfield, S. 2001. Optimizing bone health: Impact of nutrition, exercise, and hormones. *Sports Science Exchange* 14 (3):1–4.

Bonaiuti, D., et al. 2002. Exercise for preventing and treating osteoporosis in postmenopausal women. *Cochrane Database of System Reviews* 3:CD000333.

Bremner, K., et al. 2002. The effect of phosphate loading on erythrocyte 2,3-bisphophoglycerate levels. *Clinical Chimica Acta* 323:111–14.

Brewer, C. P., et al. 2015. Effect of sodium phosphate supplementation on repeated high-intensity cycling efforts. *Journal of Sports Science* 33(11):1109–16.

Brewer, G. 2012. Copper excess, zinc deficiency, and cognition loss in Alzheimer's disease. *Biofactors* 38:107–13.

Brewer, G., and Kaur, S. 2013. Zinc deficiency and zinc therapy efficacy with reduction of serum free copper in Alzheimer's disease. *International Journal of Alzheimers Disease* 2013:586365.

Bristow, S., et al. 2013. Calcium supplements and cancer risk: A meta-analysis of randomised controlled trials. *British Journal of Nutrition* 110:1384-93.

Buck, C., et al. 2013. Sodium phosphate as an ergogenic aid. *Sports Medicine* 43:425-35.

Buck, C. L, et al. 2015. Effects of sodium phosphate and beetroot juice supplementation on repeated-sprint ability in females. *European Journal of Applied Physiology* 115(10):2205-13.

Buratti, P., et al. 2015. Recent advances in iron metabolism: Relevance for health, exercise, and performance. *Medicine & Science in Sports & Exercise* 47:1596-604.

Burden, R., et al. 2014. Is iron treatment beneficial in iron-deficient but non-anaemic (IDNA) endurance athletes? A meta-analysis. *British Journal of Sports Medicine* 49(21):1389.

Burk, R. 2002. Selenium, an antioxidant nutrient. *Nutrition in Clinical Care* 5:75-79.

Cade, R., et al. 1984. Effects of phosphate loading on 2,3-diphosphoglycerate and maximal oxygen uptake. *Medicine & Science in Sports & Exercise* 16:263-68.

Castell, L., et al. 2015. *Nutritional Supplements in Sport, Exercise and Health: An A-Z Guide*. New York: Routledge Publishing.

Chiodini, I., and Bolland, M. 2018. Calcium supplementation in osteoporosis: Useful or harmful? *European Journal of Endocrinology* 178(4):D13.

Cunningham, F., et al. 2021. A potential new role for zinc in age-related macular degeneration through regulation of endothelial fenestration. *International Journal of Molecular Science* 22(21):11974.

Czaja, J., et al. 2011. Evaluation for magnesium and vitamin B6 supplementation among Polish elite athletes. *Roczniki Państwowego Zaktadu Higieny* 62:413-8.

Das, J., et al. 2013. Systematic review of zinc fortification trials. *Annals of Nutrition & Metabolism* 62 (Supplement 1):44-56.

de Oliveira Freitas, D., et al. 2012. Calcium ingestion and obesity control. *Nutricion Hospitalaria* 27:1758-71.

DellaValle, D. 2013. Iron supplementation for female athletes: Effects on iron status and performance outcomes. *Current Sports Medicine Reports* 12:234-9.

Derom, M., et al. 2013. Magnesium and depression: A systematic review. *Nutritional Neuroscience* 16:191-206.

Desbrow, B., et al. 2014. Sports Dietitians Australia position statement: Sports nutrition for the adolescent athlete. *International Journal of Sport Nutrition and Exercise Metabolism* 24:570-84.

De Souza, M., et al. 2022. The path towards progress: A critical review to advance the science of the female and male athlete triad and related energy deficiency in sport. *Sports Medicine* 52(1):13.

Deuster, P. 1989. Magnesium in sports medicine. *Journal of the American College of Nutrition* 8:462.

Embden, G.E., et al. 1921. Über Steigerung der Leistungsfahigkeit durch Phosphatzufuhr. *Zeitschrift für Physikalische Chemie* 113:67.

Eichner, E. 1992. Sports anemia, iron supplements, and blood doping. *Medicine & Science in Sports & Exercise* 24:S315-S18.

Eichner, E. 2001. Anemia and blood boosting. *Sports Science Exchange* 14 (2):1-4.

Eichner, E. 2021. Athletes, anemia, and iron redox. *Current Sports Medicine Reports* 20(7):335.

European Food Safety Authority. 2014. Scientific opinion on Dietary Reference Values of chromium. *EFSA Journal* 12:3845-70.

Evans, G. 1989. The effect of chromium picolinate on insulin controlled parameters in humans. *International Journal of Biosocial and Medical Research* 11:163-80.

Fahrenholtz, I., et al. 2022. Risk of low energy availability, disordered eating, exercise addiction, and food intolerances in female endurance athletes. *Frontiers in Sport and Active Living* May 3, 2022. Online ahead of print.

Fazelian, S., et al. 2017. Chromium supplementation and polycystic ovarian syndrome: A systematic review and meta-analysis. *Journal of Trace Elements in Medicine and Biology* 42:92-6.

Fonseca, H., et al. 2014. Bone quality: The determinants of bone strength and fragility. *Sports Medicine* 44:37-53.

Gledhill, N., et al. 1999. Haemoglobin, blood volume, cardiac function, and aerobic power. *Canadian Journal of Applied Physiology* 24:54-65.

Gleeson, M. 2013. Nutritional support to maintain proper immune status during intense training. *Nestle Nutrition Institute Workshop Series* 75:85-97.

Gonzalez, J., et al. 2014. The influence of calcium supplementation on substrate metabolism during exercise in humans: A randomized controlled trial. *European Journal of Clinical Nutrition* 68:712-8.

Ishibashi, A., et al. 2022. Iron metabolism following twice a day endurance exercise in female long-distance runners. *Nutrients* 14(9):1907.

Hannon, J.P., et al. 1969. Effects of altitude acclimatization on blood composition of women. *Journal of Applied Physiology* 26(5):540-7.

Hansen, T. H., et al. 2018. Bone turnover, calcium homeostasis, and vitamin D status in Danish vegans. *European Journal of Clinical Nutrition* 72(7):1046.

Heaney, R., et al. 2012. A review of calcium supplements and cardiovascular disease risk. *Advances in Nutrition* 3:763-71.

Heine-Bröring, R., et al. 2015. Dietary supplement use and colorectal cancer risk: A systematic review and meta-analysis of prospective cohort studies. *International Journal of Cancer* 136(10):2388-401.

Houston, M. 2011. The role of magnesium in hypertension and cardiovascular disease. *Journal of Clinical Hypertension* 13:843-7.

Jentjens, R. L, and Jeukendrup, A. E. 2002. Effect of acute and short-term administration of vandayl sulphate on insulin sensitivity in healthy active humans. *International Journal of Sport Nutrition and Exercise Metabolism* 12(4):470-9.

Jackson, J., et al. 2000. Zinc and the common cold: A meta-analysis revisited. *Journal of Nutrition* 130:1512S-5S.

Ju, W., et al. 2017. The effect of selenium supplementation on coronary heart disease: A systematic review and meta-analysis of randomized controlled trials. *Journal of Trace Elements in Medicine and Biology* 44:8-16.

Kaler, S. 2013. Inborn errors of copper metabolism. *Handbook of Clinical Neurology* 113:1745-54.

Kass, L., et al. 2012. Effect of magnesium supplementation on blood pressure: A meta-analysis. *European Journal of Clinical Nutrition* 66:411-8.

Kass, L. S., and Poeira, F. 2015. The effect of acute vs chronic magnesium supplementation on exercise and recovery on resistance exercise, blood pressure and total peripheral resistance on normotensive adults. *Journal of the International Society of Sports Nutrition* 12:19.

Kato, T., et al. 2006. Effect of low-repetition jump training on bone mineral density in young women. *Journal of Applied Physiology* 100(3):839-43.

King, J., and Cousins, R. 2014. Zinc. In *Modern Nutrition in Health and Disease*, ed. A. C. Ross, et al. Philadelphia:Wolters Kluwer/Lippincott Williams & Wilkins.

Kipec, B. J., et al. 2016. Effects of sodium phosphate and caffeine ingestion on repeated-sprint ability in male athletes. *Journal of Science and Medicine in Sport* 19(3):272-6.

Kong, W., et al. 2014. Hepcidin and sports anemia. *Cell & Bioscience* 4:19.

Kraemer, W., et al. 1995. Effects of multibuffer supplementation on acid-base balance and 2,3-diphosphoglycerate following repetitive anaerobic exercise. *International Journal of Sport Nutrition* 5:300-14.

Kreider, R., et al. 1990. Effects of phosphate loading on oxygen uptake, ventilatory anaerobic threshold, and run performance. *Medicine & Science in Sports & Exercise* 22:250-56.

Kreider, R., et al. 1992. Effects of phosphate loading on metabolic and myocardial responses to maximal and endurance exercise. *International Journal of Sport Nutrition* 2:20-47.

Kreider, R. 1999. Dietary supplements and the promotion of muscle growth with resistance exercise. *Sports Medicine* 27:97-110.

Kunstel, K. 2005. Calcium requirements for the athlete. *Current Sports Medicine Reports* 4:203-6.

Landahl, G., et al. 2005. Iron deficiency and anemia: A common problem in female elite soccer players. *International Journal of Sport Nutrition and Exercise Metabolism* 15:689-94.

Lane, H. 1989. Some trace elements related to physical activity: Zinc, copper, selenium, chromium, and iodine. In *Nutrition in Exercise and Sport,* ed. J. Hickson and I. Wolinsky. Boca Raton, FL: CRC Press.

Lefavi, R. 1992. Efficacy of chromium supplementation in athletes: Emphasis on anabolism. *International Journal of Sport Nutrition* 2:111-22.

Lewis, J., et al. 2015. The effects of calcium supplementation on verified coronary heart disease hospitalization and death in postmenopausal women: A collaborative meta-analysis of randomized controlled trials. *Journal of Bone Mineral Research* 30:165-75.

Liu, X., et al. 2022. The role of zinc in the pathogenesis of lung disease. *Nutrients* 14(10):2115.

Livingstone, C. 2015. Zinc: Physiology, deficiency, and parenteral nutrition. *Nutrition in Clinical Practice* 30: 371-82.

Lobera-Jáuregu, J. 2014. Iron deficiency and cognitive functions. *Journal of Neuropsychiatric Disease and Treatment* 10:2087-95.

Lomagno, K., et al. 2014. Increasing iron and zinc in pre-menopausal women and its effects on mood and cognition: A systematic review. *Nutrients* 14:5117-41.

Lukaski, H. 1995. Micronutrients (magnesium, zinc, and copper): Are mineral supplements needed for athletes? *International Journal of Sport Nutrition* 5:S74-S83.

Lukaski, H. 1999. Chromium as a supplement. *Annual Reviews in Nutrition* 19:279-302.

Lukaski, H. 2001. Magnesium, zinc, and chromium nutrition and athletic performance. *Canadian Journal of Applied Physiology* 26:S13-22.

Lukaski, H. 2004. Vitamin and mineral status: Effects on physical performance. *Nutrition* 20:632-44.

Lukaski, H., et al. 1991. Altered metabolic response of iron deficient women during graded, maximal exercises. *European Journal of Applied Physiology* 63:140-45.

Lukaski, H., et al. 1996. Chromium supplementation and resistance training: Effects on body composition, strength, and trace element status of men. *American Journal of Clinical Nutrition* 63:954-65.

Lukaski, H., et al. 2007. Chromium picolinate supplementation in women: Effects on body weight, composition, and iron status. *Nutrition* 23:187-95.

Mannix, E., et al. 1990. Oxygen delivery and cardiac output during exercise following oral phosphate-glucose. *Medicine & Science in Sports & Exercise* 22:341-47.

Margaritis, I., et al. 1997. Effects of endurance training on skeletal muscle oxidative capacities with and without selenium supplementation. *Journal of Trace Elements in Medicine and Biology* 11:37-43.

Mayo-Wilson, E., et al. 2014. Preventive zinc supplementation for children, and the effect of additional iron: A systematic review and meta-analysis. *BMJ Open* 4 (6):e004647.

Mazzeo, R., and Fulco, C. 2006. Physiological systems and their responses to conditions of hypoxia. In *ACSM's Advanced Exercise Physiology,* ed. C. Tipton. Philadelphia: Lippincott Williams & Wilkins.

McClung, J., et al. 2014. Female athletes: A population at risk of vitamin and mineral deficiencies affecting health and performance. *Journal of Trace Elements in Medicine and Biology* 28:388-92.

McDonald, R., and Keen, C. 1988. Iron, zinc and magnesium nutrition and athletic performance. *Sports Medicine* 5:171-84.

Meeusen, R., et al. 2013. Prevention, diagnosis, and treatment of the overtraining syndrome: Joint consensus statement of the European College of Sport Science and the American College of Sports Medicine. *Medicine & Science in Sports & Exercise* 45:186-205.

Meng, L., et al. 2022. Are micronutrient levels and supplements casually associated with risk of Alzheimer's disease? A two-sample Mendelian randomization analysis. *Food and Function* June 2, 2022. Online ahead of print.

Micheletti, A., et al. 2001. Zinc status in athletes: Relation to diet and exercise. *Sports Medicine* 31:577-82.

Miller, L., et al. 2004. Bone mineral density in postmenopausal women. *Physician and Sportsmedicine* 32 (2):18-24.

Nemeth, E., and Ganz, T. 2006. Regulation of iron metabolism by hepcidin. *Annual Review of Nutrition* 26:323-42.

Newhouse, I., and Finstad, E. 2000. The effects of magnesium supplementation on exercise performance. *Clinical Journal of Sport Medicine* 10:195-200.

Nicoll, R., et al. 2022. The role of micronutrients in the pathogenesis of alcohol-related liver disease. *Alcohol and Alcoholism* 57(3):275.

Nielsen, F. 2014. Update on human health effects of boron. *Journal of Trace Elements in Medicine and Biology* 28:383-9.

Nielsen, F., and Lukaski, H. 2006. Update on the relationship between magnesium and exercise. *Magnesium Research* 19:180-89.

Nieman, D. 2001. Exercise immunology: Nutritional countermeasures. *Canadian Journal of Applied Physiology* 26:S45-S55.

Onakpoya, I., et al. 2013. Chromium supplementation in overweight and obesity: A systematic review and meta-analysis of randomized clinical trials. *Obesity Reviews* 14:496-507.

Paik, J., et al. 2014. Calcium supplement intake and risk of cardiovascular disease in women. *Osteoporosis International* 25:2047-56.

Painelli, V., et al. 2014. The effects of two different doses of calcium lactate on blood pH, bicarbonate, and repeated high-intensity exercise performance. *International Journal of Sport Nutrition and Exercise Metabolism* 24:286-95.

Parks, R. B., et al. 2017. Iron deficiency and anemia among collegiate athletes: A retrospective chart review. *Medicine and Science in Sports and Exercise* 49(8):1711-5.

Pedlar, C. R., et al. 2017. Iron balance and iron supplementation for the female athletes: A practical approach. *European Journal of Sports Science* 27:1-11.

Penny, M. 2013. Zinc supplementation in public health. *Annals of Nutrition and Metabolism* 62 (Supplement 1):31-42.

Pelczynska, M., et al. 2022. The role of magnesium in the pathogenesis of metabolic disorders. *Nutrients* 14(9):1714.

Rautiainen, S., et al. 2013. The role of calcium in the prevention of cardiovascular disease—A review of observational studies and randomized clinical trials. *Current Atherosclerosis Reports* 15(11):362.

Reid, I. 2013. Cardiovascular effects of calcium supplements. *Nutrients* 5:2522-9.

Renata, R., et al. 2022. Immunomodulatory role of microelements in COVID-19 outcome: A relationship with nutritional status. *Biological Trace Element Research* June 6, 2022. Online ahead of print.

Rodenberg, R., and Gustafson, S. 2007. Iron as an ergogenic aid: Ironclad evidence? *Current Sports Medicine Reports* 6:258-64.

Roman, M., et al. 2014. Selenium biochemistry and its role for human health. *Metallomics* 6:25-54.

Rossi, K. A. 2017. Nutritional aspects of the female athlete. *Clinics in Sports Medicine* 36(4):627-53.

Rowland, T. 2012. Iron deficiency in athletes. *American Journal of Lifestyle Medicine* 6:319-27.

Rowland, T., et al. 1988. The effect of iron therapy on the exercise capacity of non-anemic iron-deficient adolescent runners. *American Journal of Diseases in Children* 142:165-69.

Setaro, L., et al. 2014. Magnesium status and the physical performance of volleyball players: Effects of magnesium supplementation. *Journal of Sport Sciences* 32:438-45.

Sim, M., et al. 2014. Iron regulation in athletes: Exploring the menstrual cycle and effects of different exercise modalities on hepcidin production. *International Journal of Sport Nutrition and Exercise Metabolism* 24:177-87.

Soares, M., et al. 2011. Calcium and vitamin D for obesity: A review of randomized controlled trials. *European Journal of Clinical Nutrition* 65:994-1004.

Speich, M., et al. 2001. Minerals, trace elements and related biological variables in athletes and during physical activity. *Clinical Chimica Acta* 312:1-11.

Stanzione, J., et al. 2022. A systematic intervention for low iron status in collegiate distance runners. *Nutrition and Health* May 9, 2022. Online ahead of print.

Stewart, I., et al. 1990. Phosphate loading and the effects on VO_2 max in trained cyclists. *Research Quarterly for Exercise and Sport* 61:80-84.

Straub, D. 2007. Calcium supplementation in clinical practice: A review of forms, doses, and indications. *Nutrition in Clinical Practice* 22:286-96.

Sunde, R. 2014. Selenium. In *Modern Nutrition in Health and Disease,* ed. A. C. Ross et al. Philadelphia: Wolters Kluwer/Lippincott Williams & Wilkins.

Tabizi, R., et al. 2017. The effects of selenium supplementation on glucose metabolism and lipid profiles among patients with metabolic diseases: A systematic review and meta-analysis of randomized controlled trials. *Hormone and Metabolism Research* 49(11):826-30.

Tessier, F., et al. 1995. Selenium and training effects on the glutathione system and aerobic performance. *Medicine & Science in Sports & Exercise* 27:390-96.

Thompson, I., et al. 2014. Prevention of prostate cancer: Outcomes of clinical trials and future opportunities. *American Society of Clinical Oncology Educational Book* e76-e80.

Tian, H., et al. 2013. Chromium picolinate supplementation for overweight or obese adults. *Cochrane Database of Systematic Reviews* (November 29) 11:CD010063.

Tremblay, A., and Gilbert, J. 2011. Human obesity: Is insufficient calcium/dairy intake part of the problem? *Journal of the American College of Nutrition* 30:449S-53S.

Tremblay, M., et al. 1994. Ergogenic effects of phosphate loading: Physiological fact or methodological fiction? *Canadian Journal of Applied Physiology* 19:1-11.

Turner, C., and Robling, A. 2003. Designing exercise regimens to increase bone strength. *Exercise and Sport Sciences Reviews* 31:45-50.

Varahra, A., et al. 2018. Exercise to improve functional outcomes in persons with osteoporosis: A systematic review and meta-analysis. *Osteoporosis International* 29(2):265.

Vincent, J. 2014. Is chromium pharmacologically relevant? *Journal of Trace Elements in Medicine and Biology* 28:397-405.

Vincent, J. 2015. Is the pharmacological mode of action of chromium(iii) as a second messenger? *Biological Trace Element Research* 166: 7-12.

Vinceti, M., et al. 2001. Adverse effects of selenium in humans. *Reviews on Environmental Health* 16:233-51.

Vinceti, M., et al. 2013. Friend or foe? The current epidemiologic evidence on selenium and human cancer risk. *Journal of Environmental Science and Health. Part C, Environmental Carcinogenesis & Ecotoxicology Reviews* 31:305-41.

Vinceti, M., et al. 2018. Selenium for preventing cancer. *Cochrane Database of Systematic Reviews* (January 29) 1:CD005195.

Vinceti, M., et al. 2014. Selenium neurotoxicity in humans: Bridging laboratory and epidemiologic studies. *Toxicology Letters* 230:295-303.

Vishwanathan, R., et al. 2013. A systematic review on zinc for the prevention and treatment of age-related macular degeneration. *Investigative Ophthalmology & Visual Science* 54:3985-98.

Waldman, T., et al. 2015. Calcium supplements and cardiovascular disease: A review. *American Journal of Lifestyle Medicine* 9(4):298.

Wang N., et al. 2017. Supplementation of micronutrient selenium in metabolic diseases: Its role as an antioxidant. *Oxidative Medicine and Cellular Longevity* 2017: 7478523.

White, K., et al. 2006. The acute effects of dietary calcium intake on fat metabolism during exercise and endurance exercise performance. *International Journal of Sport Nutrition and Exercise Metabolism* 16:565-79.

Wilber, R., et al. 2007. Effect of hypoxic "dose" on physiological responses and sealevel performance. *Medicine & Science in Sports & Exercise* 39:1590-99.

Zemel, M., et al. 2005. Effects of calcium and dairy on body composition and weight loss in African-American adults. *Obesity Research* 13:1218-25.

Zhang, Y., et al. 2017. Can magnesium enhance exercise performance? *Nutrients* 9(9):E946.

Zhao, F., et al. 2022. Effect of chromium supplementation on blood glucose and lipid levels in patients with Type 2 diabetes mellitus: A systematic review and meta-analysis. *Biological Trace Element Research* 200(2):516.

Zhao, N., et al. 2013. Iron regulation by hepcidin. *Journal of Clinical Investigations* 123:2337-43.

Zhao, J. G., et al. 2017. Association between calcium or vitamin D supplementation and fracture incidence in community-dwelling older adults: A systematic review and meta-analysis. *Journal of the American Medical Association* 318(24):2466-82.

Zourdos, M., et al. 2015. A brief review: The implications of iron supplementation for marathon runners on health and performance. *Journal of Strength & Conditioning Research* 29:559-65.

Design element: Training Table (orange) ©mphillips007/Getty Images

Water, Electrolytes, and Temperature Regulation

CHAPTER NINE

Greg Epperson/Shutterstock

LEARNING OUTCOMES

After studying this chapter, you should be able to:

1. Identify the principal water compartments in the body, describe the general function of each, and explain how your body maintains overall water balance.
2. Provide recommendations for foods that are high or low in sodium and potassium, and explain the physiological responses of your body to restore normal serum sodium levels following a high salt intake.
3. List the four components of environmental heat stress recorded by the wet-bulb globe temperature (WBGT), and explain how each may affect the heat balance equation during exercise under hot, humid environmental conditions.
4. Describe how exercise in the heat may impair endurance performance as compared to exercise in a cooler environment, and explain the physiological responses your body would make to promote heat loss.
5. Outline the key guidelines for consuming water, electrolytes, and carbohydrate before, during, and after exercise under warm or hot environmental conditions, and offer general recommendations to athletes participating in such events.
6. Describe the theory underlying the use of glycerol as an ergogenic aid, and understand the current research findings regarding its efficacy in enhancing exercise performance.
7. Explain various strategies to reduce the risk of heat illness while exercising in a hot environment, and also be able to identify the various heat illnesses along with their causes, clinical findings, and appropriate treatment.
8. Describe American Heart Association categories of high blood pressure and associated health risks, and describe the role that diet and exercise may play in its prevention and treatment.

KEY TERMS

acclimatization
aldosterone
antidiuretic hormone (ADH)
carbohydrate-electrolyte solutions (CES)
compensated heat stress
conduction
convection
core temperature
DASH diet
dehydration
dipsogenic drive
electrolyte
essential hypertension
euhydration
evaporation
exercise-associated collapse
exercise-associated muscle cramps
exertional heat stroke
extracellular water
heat-balance equation
heat exhaustion
heat index
heat stroke
heat syncope
high blood pressure
homeostasis
hyperhydration
hyperkalemia
hypertension
hyperthermia
hypohydration
hypokalemia
hyponatremia
hypothermia
insensible perspiration
intercellular water
intracellular water
intravascular water
metabolic water
normohydration
osmolality
osmotic demyelination
postexercise hypotension
radiation
shell temperature
sodium loading
specific heat
sweat gland fatigue
tonicity
uncompensated heat stress
WBGT index

313

Introduction

Greg Epperson/Shutterstock

Water (H_2O) is the most important nutrient and, except for minerals, the simplest nutrient. The average individual can survive several weeks without food but only a week without water. Rapid losses of water and electrolytes through dehydration or diarrhea can be fatal in an even shorter period of time. The essence of water to life was keenly understood by primitive humankind many centuries prior to our current (and evolving) understanding of the nutrients described in chapters 4–8. As described by Smith, in 1855 John Snow demonstrated the importance of clean drinking water in cholera prevention by observing a death rate in London households' supplied drinking water by the Southwark and Vauxhall Company (downstream from fecal waste discharge into the Thames River) that was 8.5 times that of households supplied by the Lambeth Company (upstream of waste discharge). Clean drinking water is described by Rush as being taken for granted in developed countries. Water is the solvent for the body's chemical processes and accounts for most of our body mass. Our ability to derive energy from nutrients in food requires water for the chemical reaction known as hydrolysis. In separate reviews, Kenefick and Cheuvront and Maughan note the importance of water intake, availability, and adequate hydration for regulation of body temperature in both physically active and vulnerable populations, respectively.

Electrolytes, part of the mineral class of nutrients discussed in the previous chapter, have a myriad of physiological roles, some of which are muscle contraction, neural action potentials, cofactors in energy metabolism, and activation of enzymes. Although electrolyte balance is normally tightly regulated by the body and adequate electrolyte intake is available in a healthy diet, a significant loss of electrolytes accompanies the loss of water by profuse sweating, vomiting, or diarrhea. According to Suh and others, deaths in children under 5 years of age have dropped significantly since 1990 due in part to the standard World Health Organization (WHO) oral rehydration therapy (ORT) solution (sodium, potassium, chloride, and glucose) to rapidly replace lost fluid and electrolytes. However, diarrhea remains one of the leading causes of childhood deaths.

In addition to health, proper fluid replacement and electrolyte balance are important for performance in recreational/competitive sports and in the workplace. As described by Pandolf and others, research conducted by the United States Army Research Institute of Environmental Medicine has focused on preparing military personnel to function in extreme environmental conditions. Much of this research has focused on risk factors associated with exercise under warm or hot conditions, strategies to acclimatize personnel, and the development of policies to reduce morbidity and mortality from heat illness. Scientists and coaches also recognized the need for sports ORT to offset impaired performance and fatigue from fluid and electrolyte losses through sweating in runners, cyclists, football players, and other athletes. The formulation of Gatorade® in 1965 by the University of Florida's Dr. Robert Cade was the first entry in the "sports drink" (carbohydrate-electrolyte solution or CES) industry, which thrives today. Optimal CES composition, palatability, and factors promoting absorption continue to receive research attention.

Physically active individuals are aware that a hot and/or humid environment can impair performance. Environmental and metabolic heat results in loss of body water and electrolytes, which compromises the body's ability to dissipate heat and regulate core temperature. At best, loss of fluid and electrolytes will impair performance. Inability to regulate core temperature can lead to heat illness, the most extreme form of which can be fatal. This chapter will discuss the roles of water and the key electrolytes sodium, chloride, and potassium in fluid regulation and performance; the effects of water and electrolyte loss on performance; strategies for proper hydration before exercise and proper rehydration during and after exercise; the recognition and prevention of heat illnesses; and the role of fluid/electrolyte balance, diet, and exercise in high blood pressure (hypertension) prevention and management.

Water

How much water do you need per day?

The U.S. National Academies of Sciences, Engineering, and Medicine Institute of Medicine (IOM) has established the daily adequate intake (AI) values for water intake of 3.7 liters and 2.7 liters (3.9 and 2.9 quarts) for healthy males and females, respectively. Lower AI values have been established for teenagers and children. Using 2011-2016 National Health and Nutrition Examination Survey (NHANES) data, Vieux et al. reported average daily water intake of 2.95 liters and 2.5 liters for males and females, with only 40 percent of the sample meeting IOM AI. The actual requirement for a given individual depends on factors such as body mass, health status, age, environmental conditions, and physical activity. Water requirements may be increased substantially during exercise, particularly under warm environmental conditions, and will be discussed later in this chapter. Valtin noted that thirst is a good guide to help maintain normal body-water balance and questioned the anecdotal recommendation to drink at least eight 8-oz (237 milliliter) glasses of water per day. Vivanti identified problems with generalized equations for estimating water intake in a health-care setting. As will become apparent in this chapter, water requirements are highly variable between and within individuals as a result of a variety of factors such as health status, environmental conditions, physical activity, acclimatization, and diet.

Body water balance is maintained when the output of body fluids is matched by the input of water. Water intake comes from many sources. In their analysis of 2011-2016 NHANES data, Vieux and others reported that total daily water (2,718 ml/day) came from tap and bottled water (1,066 ml/day, 39 percent), 15 other beverages such as coffee tea, alcohol, and milk (1,036 ml/day, 38 percent), and food moisture (618 ml/day, 23 percent). A small amount of water is lost in the feces and through the exhaled air in breathing. **Insensible perspiration** on the skin, which is not visible, is almost pure water and accounts for about 30 percent of body-water losses. Sweat on the skin is noticeable and increases considerably during exercise and/or hot environmental conditions. Urinary output is the main avenue for water loss. It may increase somewhat through the use of diuretics, including alcohol and caffeine. Although Stookey reported water losses of about 1 ml/mg of caffeine, and 10 ml/g of alcohol, other research has challenged the assumption that caffeine is a significant diuretic. In their meta-analysis, Zhang and others concluded that caffeine does not promote excessive fluid loss at rest and has no diuretic effect during exercise. Armstrong and others reported no differences in hydration status in subjects following 0, 3, and 6 mg/kg doses of caffeine each day for 4 days. Using a crossover design, Silva and colleagues reported no effects of caffeine ingestion (5 mg/kg/day for 4 days) on water volume and distribution or daily water ingestion in 30 nonsmoking males who were low users of caffeine compared to placebo. Consumption of a high-protein diet produces urea, which has to be excreted by the kidneys and may increase urine output.

Fluid intake of beverages, such as water, soda, milk, juice, coffee, and tea, provides about 80 percent of total water needed to replenish losses. Valtin noted that we obtain significant amounts of fluid from caffeinated and alcoholic beverages. The diuretic effect of each may be offset by the amount of fluid in the beverage and more fluid is probably gained than lost from beer, coffee, and other beverages. As previously noted by Vieux and others, about 20 percent of our daily total water intake comes from the foods we eat. Solid foods also contribute as a water source in two different ways. First, food contains water in varying amounts; certain foods such as lettuce, celery, melons, and most fruits contain about 80-90 percent water; meats and seafood contain about 60-70 percent water; even bread, an apparently dry food, contains 36 percent water. Second, the metabolism of carbohydrate, fat, and protein for energy produces water. Fat, carbohydrate, and protein all produce water when broken down for energy. As discussed in chapter 3, when glucose is metabolized to produce energy, with one of the by-products being **metabolic water**:

$$C_6H_{12}O_6 + 6O_2 \rightarrow \text{Energy} + 6CO_2 + 6H_2O$$

Figure 9.1 summarizes the daily water loss and intake for the maintenance of water balance for an adult female. In general, amounts would be greater for an adult male. As shall be seen later, however, these amounts may change drastically under certain conditions.

What else is in the water we drink?

The tap water we drink, although generally safe, is not pure. Solutes entering drinking water from natural geological formations include minerals such as calcium, magnesium, fluoride, iron, sodium, and zinc. Fluoride is beneficial in preventing dental caries and is added to the municipal water supply, but too much can stain teeth and adversely affect bone health. Some minerals, such as excess sodium, arsenic, and lead, may lead to various health problems, whereas others, such as calcium and magnesium, may be beneficial. Exposure to

John Lund/Drew Kelly/Blend Images LLC

lead in water can result in behavioral and learning problems, lowered IQ, and stunted growth in children. Dignam and others report that while blood lead levels in children have dropped by 94 percent from 1976 to 2016, 500,000 children ≤5 years of age still have blood lead levels ≥5 µg/dl, as established by the Centers for Disease Control and Prevention. In 2014, the discovery of lead in the water supply of Flint, Michigan, was a prominent news story. Of greater concern to public health in economically developed nations are human-made solutes in the water supply. Murray and others identified three broad classes of "emerging contaminants" as industrial chemicals, pesticides, and pharmaceuticals/personal care products. Under the Safe Drinking Water Act, the Environmental Protection Agency (EPA) provides a list of contaminants and their maximum contaminant level (MCL) at www.epa.gov/dwstandardsregulations. Most, but not all, municipal water treatment facilities conform to these standards. The EPA also provides guidance for special populations such as pregnant women, infants and children, and the elderly regarding exposure to contaminants in drinking water.

If you are concerned about your tap water, water filters added to your tap may help remove unwanted substances, such as chlorine or

FIGURE 9.1 Estimate of water balance—intake versus output—in a woman. We primarily maintain our volume of body fluids by adjusting water output to intake. As you can see, most water comes from the liquids we consume. Some comes from the moisture in more solid foods, and the remainder is manufactured during metabolism. Water output includes losses from the lungs, urine, skin, and feces.

Drinking water: vm/Getty Images; Standing on treadmill: Purestock/SuperStock

chlorinated by-products. Some water filters are designed to eliminate lead and mercury, and others can even trap parasites. Many types of water filters are on the market, so, if interested, have your water analyzed and then seek an appropriate filter to help purify your tap water. Water filling stations can be found in many public buildings and college campuses. To get information on the quality of your water supply, you may contact your local water utility and ask for the latest water quality or Consumer Confidence Report, or contact the EPA website.

http://water.epa.gov/drink/local/ The Environmental Protection Agency provides information on the quality of local drinking water.

www.fda.gov/food/buy-store-serve-safe-food/fda-regulates-safety-bottled-water-beverages-including-flavored-water-and-nutrient-added-water The Food and Drug Administration provides information on bottled water.

www.bottledwater.org/ The International Bottled Water Association website.

Bottled water is the current rage. According to the International Bottled Water Association (IBWA, www.bottledwater.org/types-of-water-bottled), bottled water was the largest beverage category by sales volume in the United States, with total retail sales of 15 billion gallons and retail sales of $36.25 million in 2020, a 3.8 percent increase over 2019. The Food and Drug Administration (FDA) regulates bottled water as stringently as the EPA regulates tap water. Artesian water is drawn from a well that taps a confined aquifer. Mineral water comes from a protected underground source and must naturally contain ≥250 parts per million in minerals from a geological underground source. Spring water flows naturally from an underground source. Purified water is produced by distillation, deionization, or some comparable process. Sparkling water contains carbon dioxide in the same concentration after treatment as before. Vitamin, herbal, nutraceutical, fitness, and oxygen waters, the last of which will be discussed later in this chapter, are also in the marketplace. Proponents extol the benefits of specialized water products such as "alkaline" water to improve health and attenuate disease risk. For example, there is no research evidence for claims that alkaline water improves metabolism, increases energy, slows aging, improves digestion, or reduces bone loss (www.webmd.com/diet/what-is-alkaline-water). About 85 percent of bottled water manufacturers belong to the IBWA, which sets even tougher standards for its members than the FDA. Consumers should be aware that, according to Food and Water Watch (www.foodandwaterwatch.org/wp-content/uploads/2021/04/fs_1803_bottled-water-web.pdf), the percentage of bottled water originating from municipal tap water increased from 52 percent in 2009 to 64 percent in 2014. For example, the two best-selling bottled waters in the United States, Aquafina and Dasani®, are purified municipal water. Moreover, surveys have shown that most bottled waters do not contain fluoride. Lalumandier and Ayers reported that only 5 percent of 57 samples of bottled water contained fluoride within the recommended range. In addition to checking bottled water for mineral content, consumers should also be aware of the possibility of contamination. Mason and others reported that 93 percent of a sample consisting of 11 different bottled water brands purchased in nine different countries had measurable microplastic particle contamination. Bottled water is regulated for quality and safety by the FDA, according to the Federal Food, Drug, and Cosmetic Act, according to standards at least as stringent as the regulation of municipal tap water by the Environmental Protection Agency.

Bottled water can be expensive. Bling H_2O is marketed at an average of $55 per bottle. One bottled water product, Acqua Di Cristallo Tributo a Modigliani, costs $72,000 for 750 ml (25.4 oz), a price due mainly to the hand-made bottle covered in 24-carat gold. According to Business Insider (www.businessinsider.com/bottled-water-costs-2000x-more-than-tap-2013-7), bottled water costs 2,000 times more per gallon than tap water and twice as much as a gallon of gasoline.

Where is water stored in the body?

As illustrated in **figure 9.2**, water is stored in several body compartments but moves constantly between compartments. The reference 70-kilogram (154-lb) man is 60 percent water, or 42 liters. The average female is approximately 50 percent water. More water is found in muscle and nonfat tissue than in fat tissue, so water content can vary from 40 percent in obese to 70 percent in very lean individuals. In other words, an individual with obesity generally has lower body-water content than a normal-weight individual. In the average male, approximately 60–65 percent (25–28 liters) of total body water is **intracellular water**, while the extracellular water compartment

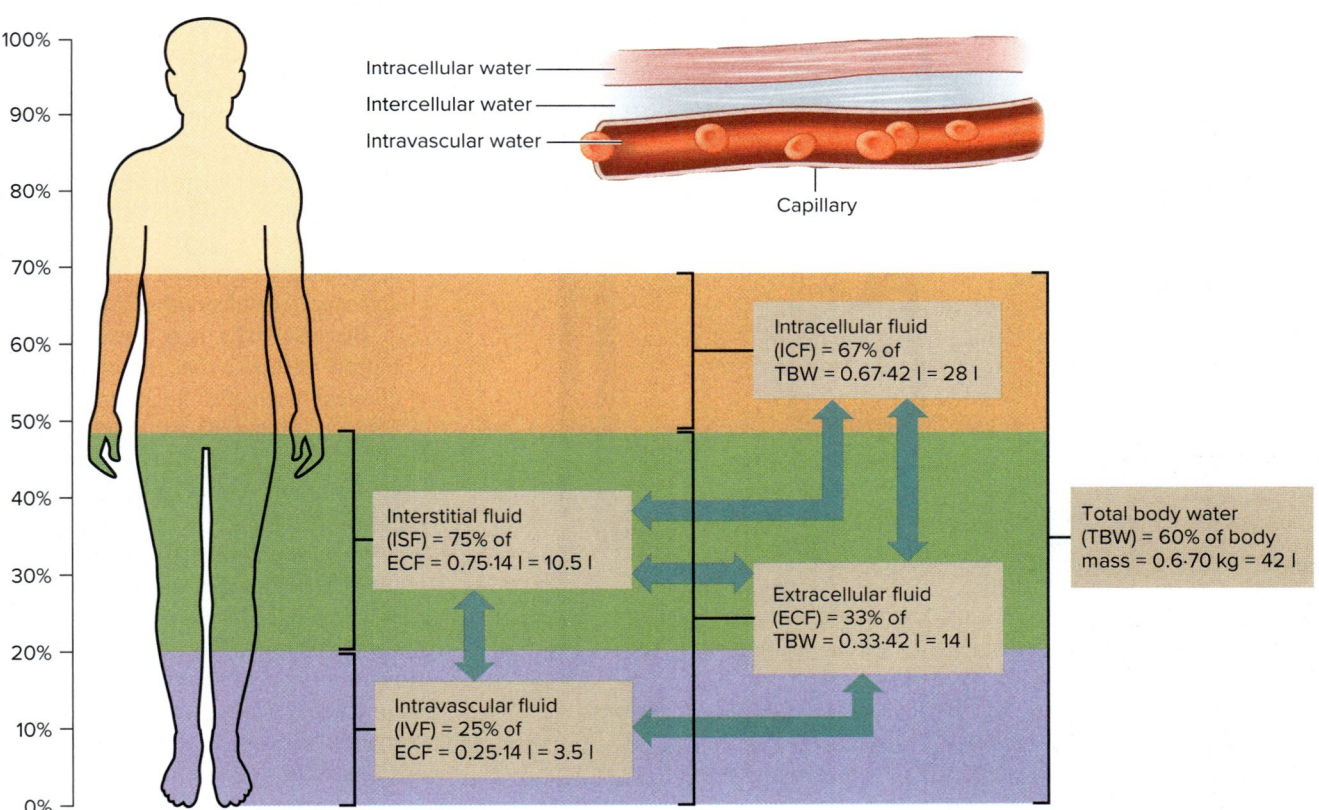

FIGURE 9.2 Body-water compartments in the 70-kg reference male. As illustrated by the blue arrows, there is a constant interchange among the different body-water compartments depending on osmolality between compartments. The water inside the body cells, the intracellular water, is important for cell functions. The extracellular space includes intercellular, intravascular, and miscellaneous (e.g., cerebrospinal fluid). The intracellular water constitutes about 60–65 percent of the total body water, while the extracellular water constitutes the remaining 35–40 percent. In the blood, some of the water is intracellular in the blood cells, while the remainder is extracellular in the plasma. Decreases in blood volume (intravascular water) may adversely affect endurance capacity.

makes up the remaining 35–40 percent (14–17 liters). The **extracellular water** is further subdivided into the **intercellular** (interstitial) **water** between or surrounding the cells (8–9 liters), the **intravascular water** within the blood vessels (3–4 liters), and miscellaneous water compartments such as the cerebrospinal fluid (1–3 liters). Water compartments are measured using the indicator dilution method, where a known quantity of an appropriate tracer that cannot leave the compartment is injected into the space. For example, deuterium oxide is a common tracer to measure total body water. Other tracers are required to measure water volume in specific areas. After equilibration, a compartment sample is analyzed for the concentration of the tracer. Compartment volume (ml) is calculated as injected tracer mass (mg) ÷ sample tracer concentration (mg/ml).

Water is held in the body in conjunction with protein, carbohydrate, and electrolytes. Protein in the muscles, blood, and other tissues helps bind water to those tissues. Plasma proteins, notably albumin, bind approximately 15 milliliters of water per gram and account for colloid oncotic pressure, which pulls water into the vascular space. As discussed in chapter 4, muscle glycogen has considerable amounts of water bound to it (about 3 grams of water per gram of glycogen), which may prove to be an advantage when exercising in the heat. According to Hoyt and Honig, normal metabolic water production is between 250 and 350 ml/day. The sodium in the extracellular fluid, including sodium in the vascular system, also attracts water.

Proper water and electrolyte balance within these compartments is of extreme importance to the athletic individual. Fluid shifts such as decreases in blood volume and cellular dehydration, both of which may develop during exercise in the heat, could contribute to the onset of fatigue or heat illness.

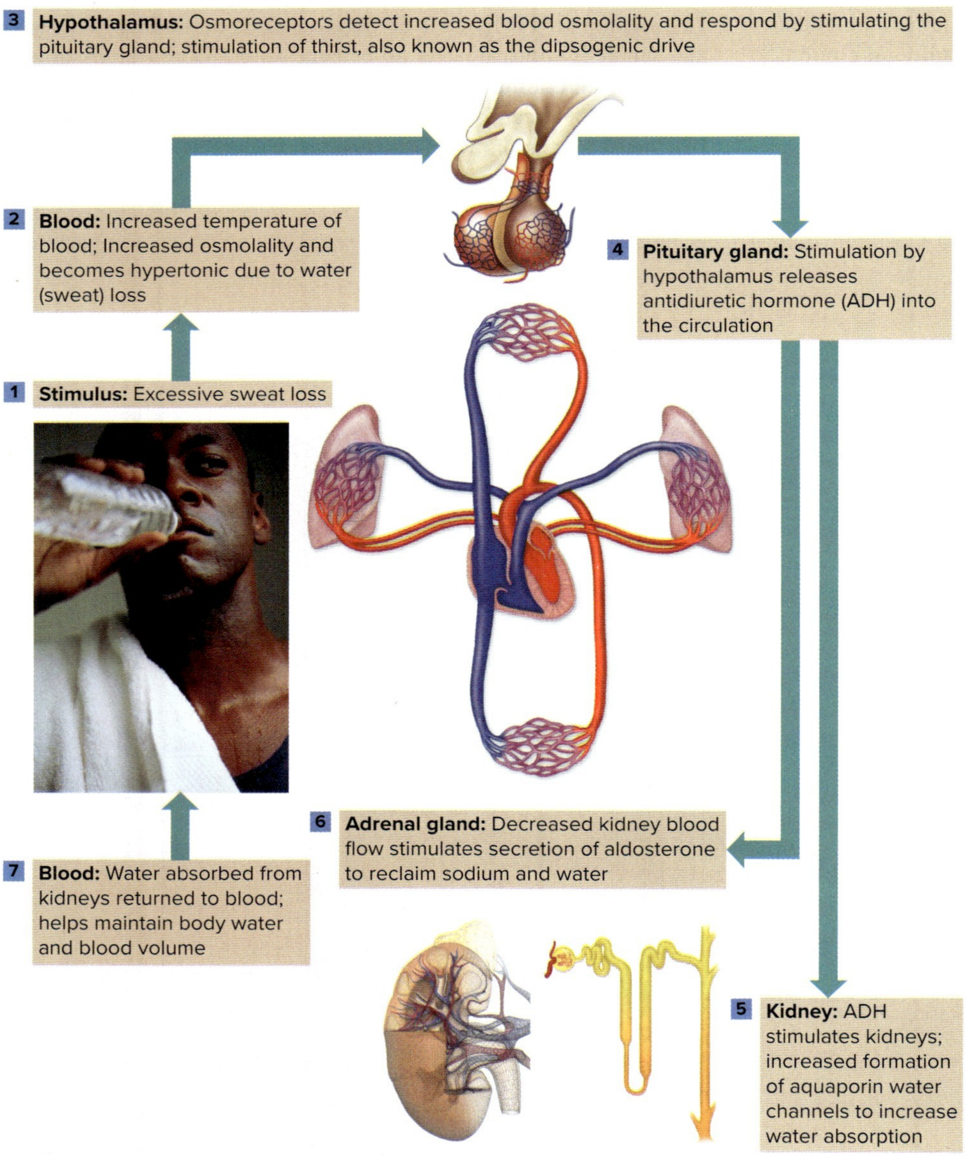

How is body water regulated?

Johnson notes that maintaining body water and sodium balance is so critical to health that the central nervous system has developed specific neural patterns fostering an appetite for both water and salt. Body water is maintained at a normal level through kidney function. Normal body-water content is called **normohydration**, or **euhydration**. **Dehydration**, the loss of body water, results in a state of **hypohydration**, or low body-water content. **Hyperhydration** represents a condition in which the body retains excess body fluids. Normal kidneys function very effectively to eliminate excess water during hyperhydration and conserve water during hypohydration.

Because water is so essential to life, it is indeed fortunate that the body possesses an efficient mechanism to maintain proper water balance. **Homeostasis** is the term used to describe the maintenance of a normal internal environment so that the body has the proper distribution and use of water, electrolytes, hormones, and other substances essential for life processes. Homeostatic mechanisms are essentially feedback loops consisting of commands from the CNS to fluid and thermoregulatory organs (efferent outflow) in response to communication from the tissue (afferent input). A detailed discussion is beyond the scope of this text, but several feedback loops to regulate water balance, electrolytes, and temperature are illustrated in **figure 9.3**. If these feedback devices are functioning properly, the body usually has no problem in maintaining the normal physical and chemical composition of its fluid compartments.

FIGURE 9.3 One feedback mechanism for homeostatic control of body water and blood volume. Other feedback mechanisms operate concurrently. For example, the hypothalamus also stimulates the thirst response to increase fluid intake.

[1] **Stimulus:** Excessive sweat loss

[2] **Blood:** Increased temperature of blood; Increased osmolality and becomes hypertonic due to water (sweat) loss

[3] **Hypothalamus:** Osmoreceptors detect increased blood osmolality and respond by stimulating the pituitary gland; stimulation of thirst, also known as the dipsogenic drive

[4] **Pituitary gland:** Stimulation by hypothalamus releases antidiuretic hormone (ADH) into the circulation

[5] **Kidney:** ADH stimulates kidneys; increased formation of aquaporin water channels to increase water absorption

[6] **Adrenal gland:** Decreased kidney blood flow stimulates secretion of aldosterone to reclaim sodium and water

[7] **Blood:** Water absorbed from kidneys returned to blood; helps maintain body water and blood volume

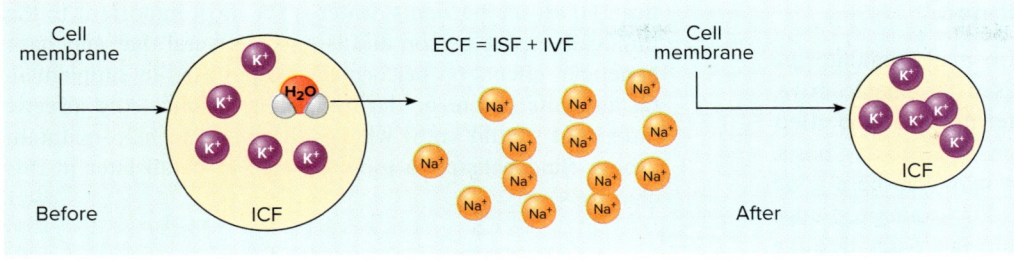
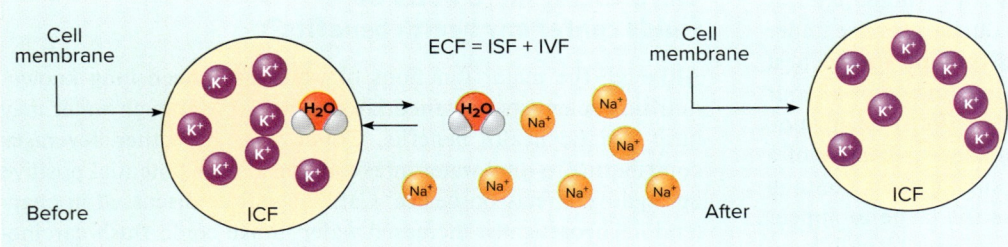
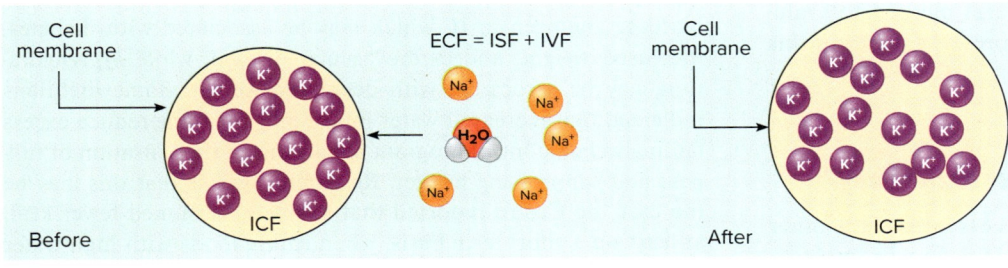

FIGURE 9.4 Osmosis and tonicity. (a) When the extracellular fluid (ECF) contains more electrolytes or other osmotic substances, it is hypertonic to the intracellular fluid (ICF) and water will flow from the ICF space to the ECF space (interstitial fluid [ISF] + intravascular fluid [IVF]) and the cell decreases in volume. In other words, water flows to the area of greater osmotic pressure. (b) When the solute concentration is the same in the cell and outside the cell, there is no net gain or loss of water from either space; therefore, there is no change in cell volume. (c) When the ICF contains more electrolytes or higher osmotic pressure than the ECF, water will flow into the ICF space from the ECF space, and the cell increases in volume.

The main feedback device for the control of body water is the osmolality of the various body fluids. **Osmolality** refers to the amount, or concentration, of dissolved substances, known as *solute,* in a solution. In the body, a number of different substances affect osmolality, including glucose, protein, and several electrolytes, most notably sodium. These substances are dissolved in the body water. One mole of a nonionic substance, such as glucose, dissolved in a liter of water is 1 osmole. One millimole (1/1,000 mole) is 1 milliosmole. However, a mole of a substance that can dissociate into two ions, such as sodium chloride, is equivalent to 2 osmoles. One millimole of sodium chloride would be 2 milliosmoles (mOsm). Osmolality is expressed in mOsm/kg of body mass. In normally hydrated conditions, osmolality is 275–295 mOsm/kg.

A term often used in conjunction with osmolality is **tonicity**, which means tension or pressure. When two solutions have the same osmotic pressure they are said to be *isosmotic* or, more commonly, *isotonic*. Iso means "same." When two solutions with different solute concentrations are compared, the one with the higher osmotic pressure is called hypertonic and the other is hypotonic.

When two solutions with different solute concentrations are separated by a permeable membrane, as in the human body between the fluid compartments, a potential pressure difference may develop between the solutions that will allow for water movement. This pressure is known as *osmotic pressure.* Water moves through cell membrane proteins, known as aquaporin water channels, from the hypotonic solution (low solute concentration and high water content) to the hypertonic solution (high solute concentration and low water content). In essence, high solute concentrations create high osmotic pressures and tend to draw water into their compartments. **Figure 9.4** depicts this mechanism between the blood and the body cells.

To briefly illustrate the feedback mechanism for control of body water, let us look at what happens when you become dehydrated due to excessive body-water losses or lowered water intake. The blood becomes more concentrated, or hypertonic. Because maintenance of a normal blood volume is of prime importance, the blood tends to draw water from the body cells. Certain cells in the hypothalamus, called *osmoreceptors,* are sensitive to changes in osmotic pressure. These cells react to the more concentrated body fluids by stimulating the release of **antidiuretic hormone (ADH)**, also known as arginine vasopressin, from the posterior pituitary gland. The ADH travels by the blood to the kidneys and directs them to reabsorb more water. Hence, urinary output of water is diminished considerably. **Figure 9.3** illustrates several feedback loops that regulate body water and blood volume. During hyperhydration, which would produce a hypotonic condition in the body fluids, a reverse process would occur, leading to increased water excretion.

Because maintenance of euhydration is critical for health and physical performance, it is important to note that ADH is only one of several hormones that help to regulate body-water balance. Other hormones are involved in the maintenance of sodium and potassium, which also affect body-water levels, and the role of one is discussed in the next section on electrolytes.

How do I know if I am adequately hydrated?

Osmoreceptors and other mechanisms also may stimulate the sensation of thirst, also known as increased **dipsogenic drive,** which is usually a good guide to body-water needs and is effective in restoring body water to normal on a day-to-day basis. Some sports scientists contend that thirst may also be a good guide to hydration status during exercise. For example, Fudge and others studied elite Kenyan runners during a 5-day training period and found that they remained well hydrated day-to-day with *ad libitum* fluid intake. However, as shall be noted later, it may be advisable to start consuming fluids during exercise under warm conditions before you become thirsty. One of the best guides to indicate a state of normohydration is the color of your urine. In general, it should be a clear, pale yellow. A deeply colored urine, usually excreted in small amounts, is indicative of a state of hypohydration. However, vitamin supplements containing riboflavin (vitamin B2) may also cause the urine to appear yellow and suggest a state of dehydration when the individual is euhydrated. Maughan and Shirreffs indicate that change in body weight is a reliable and easy short-term method to evaluate hydration status. Some guidelines are presented later in this chapter.

What are the major functions of water in the body?

Water is the solvent for life and therefore is essential if the other nutrients are to function properly within the human body. It has a number of diverse functions that may be summarized as follows:

1. Water provides the essential building material for cell protoplasm, the fundamental component of all living matter.
2. Because water cannot be compressed, it protects key body tissues such as the spinal cord and brain.
3. Water is essential in the control of the osmotic pressure in the body, or the maintenance of a proper balance between water, interstitial and intravascular proteins, and the electrolytes. Any major changes in the electrolyte concentration may adversely affect cellular function. A serious departure from normal osmotic pressure cannot be tolerated by the body for long.
4. Water is the main constituent of blood, the major transportation mechanism in the body for conveying oxygen, nutrients, hormones, and other compounds to the cells for their use, and for carrying waste products of metabolism away from the cells to organs such as the lungs and kidneys for excretion from the body.
5. Water is essential for the proper functioning of our senses. Sound waves are transmitted by fluid in the inner ear. Fluid in the eye is involved in the reflection of light for proper vision. For the taste and smelling senses to function, the foods and odors must be dissolved in water.
6. Of primary importance to the active individual is the role that water plays in the regulation of body temperature. Water is the major constituent of sweat, and through its evaporation from the surface of the skin, it can help dissipate excess body heat. Of all the nutrients, water is the most important to the physically active person and is one of several that may have beneficial effects on performance when used in supplemental amounts before or during exercise. Hence, the athletic individual should know what is necessary to help maintain proper fluid balance, a topic covered in detail later in this chapter.

Can drinking more water or fluids confer any health benefits?

Although the major functions of water have been long known, nutrition scientists now theorize that drinking enough water may have specific health benefits. Coffee, tea, and other beverages contributing to total water intake may also have potential positive or negative effects on health. Caffeine will be discussed in chapter 13. Theoretically, increased water intake could flush carcinogens from the urinary tract and colon. Thornton comments that *hypo*hydration-related alterations in renin–angiotensin function increase angiogensin II, which may be associated with diabetes, increased weight, and cardiovascular disease, while *hyper*hydration may be associated with decreased weight. Some dietitians indicated that increased water intake may help one reduce excess fat in a weight-control program by increasing the sensation of fullness and suppressing hunger. Research suggests that this may be the case. de Castro reported that subjects consumed fewer kcal, at least on a short-term basis, when eating foods with high water content. As discussed in chapter 11, high-volume, low-energy foods may be an integral part of a weight-control diet. In their review, El-Sharkawy and associates note considerable evidence of associations between fluid imbalances and a number of urological, cardiovascular, neurological, metabolic, gastrointestinal, and dental diseases. Although you can consume water in a variety of beverages, such as juices, soda, or coffee, scientists recommend water itself. Water is inexpensive and is free of added sugars and saturated fats.

However, most of us do obtain our daily water needs from beverages other than water. Some beverages, such as pure fruit juices, may provide some health benefits associated with their vitamin, mineral, and phytonutrients content. Other beverages have been suggested to pose health threats, such as alcoholic beverages, which are discussed in chapter 13. In response to a link between overweight and obesity and consumption of high-kcal beverages, Popkin and other health professionals formed a Beverage Guidance Panel in 2006 to inform consumers on the relative health and nutritional benefits and risks of various beverage categories. In general, the panel recommended preference for beverages with limited kilocalories and no added sugars and concluded that drinking water was ranked as the preferred beverage to fulfill daily water needs. Ferretti and Mariani reported positive associations between affordability and consumption of sugar-sweetened beverages and prevalence rates of overweight and obesity in five countries in each region of the World Health Organization. As this issue deals with weight control, additional information will be presented in the next two chapters.

> ### Key Concepts
>
> ▸ The U.S. National Academies of Sciences, Engineering, and Medicine Institute of Medicine (IOM) has established the daily adequate intake of water as 3.7 liters and 2.7 liters (3.9 and 2.9 quarts), respectively, for a healthy male and female.
> ▸ The actual daily water requirement for a given individual will depend on health status, environmental conditions, amount of physical activity, acclimatization levels, diet, and other factors.
> ▸ Sources of daily water intake are tap/bottled water, various other fluids (coffee, tea, cola, fruit juices, milk, etc.), water in solid food, and metabolic water production.
> ▸ Normal water levels in the various body-fluid compartments are maintained by a feedback mechanism involving specific receptors for osmotic pressure, the antidiuretic hormone (ADH), and the kidneys.
> ▸ Water has a number of functions in the body. One of its most important benefits for people who exercise is the control of body temperature.
> ▸ Plain water is an effective and inexpensive means to help maintain fluid balance in the body. Some beverages, such as pure fruit juices, may provide healthful nutrients, whereas others, such as alcoholic and sugar-sweetened beverages, may pose a health risk.
>
> ### Check for Yourself
>
> ▸ Use a measuring cup or download an app to measure the amount of fluids you drink for a day. Compare your daily intake to the recommendations listed in this section.

Electrolytes

What is an electrolyte?

An **electrolyte** is a substance that, in solution, conducts an electrical current. The solution itself may be referred to as an electrolyte solution. Acids, bases, and salts are common electrolytes, and they usually dissociate into ions, particles carrying either a positive (cation) or a negative (anion) electrical charge. The major electrolytes in the body fluids are sodium, potassium, chloride, bicarbonate, sulfate, magnesium, and calcium. Electrolytes can act at the cell membrane and generate electrical current, such as in a nerve impulse. They can also function in other ways such as activating enzymes to control a variety of metabolic activities in the cell. Some of the important metabolic functions of calcium, phosphorus, and magnesium were discussed in Chapter 8. In this chapter the focus is on sodium, chloride, and potassium because of their presence in sports drinks, popular beverages used to replace fluid losses in physically active people.

The concentration of all elements in the body may be expressed in a variety of ways, such as milligrams per unit volume, millimoles, and milliequivalents. The equivalencies for sodium, chloride, and potassium will be provided, as you may often see these various terms in the literature.

In the following sections, the functions of each of these electrolytes in the human body will be discussed, followed by the interaction of these electrolytes with exercise in hot/humid environmental conditions and their role in the etiology of high blood pressure.

Sodium (Na)

Sodium is a mineral element also known as natrium, from which the symbol *Na* is derived. It is one of the principal positive ions, or electrolytes, in the body fluids. The gram atomic weight (1 mole) of sodium is 23. One millimole (and one milliequivalent) of sodium is 23 milligrams. One millimole (and one milliequivalent) of sodium chloride (NaCl, or table salt) is 58.5 milligrams, containing 23 milligrams of sodium and 35.5 milligrams of chloride.

Chronic Disease Risk Reduction (CDRR) Sodium Levels

The U.S. National Academies of Sciences, Engineering, and Medicine, Institute of Medicine, has established daily dietary sodium limits to the following age-related CDRR levels: $\leq 1,200$ milligrams for ages 1–3; $\leq 1,500$ milligrams for ages 4–8; $\leq 1,800$ milligrams for ages 9–13; and $\leq 2,300$ milligrams for age ≥ 14. CDRR limits, included in the *2020-2025 Dietary Guidelines for Americans*, are consistent with dietary patterns to lower the risk of hypertension and cardiovascular disease. Since common table salt (NaCl, sodium chloride) is about 40 percent sodium, only about 2.3 grams (2,300 milligrams) is needed to supply the minimum sodium requirement. According to *2020-2025 Dietary Guidelines for Americans*, 97 and 86 percent of males and nonpregnant/nonlactating females aged 2 to >60 have exceeding CDRR levels, with average daily sodium intakes of 3,520 and 2,785 milligrams, respectively. There is no evidence that higher sodium intakes confer any additional health benefits. However, these recommendations do not take into account the large sodium losses that may occur during exercise-induced sweating, as will be discussed later in the chapter.

Food Sources

Sodium is distributed widely in nature but is found in rather small amounts in most natural foods. However, significant amounts of salt, and hence sodium, are usually added from the salt shaker for flavor. One teaspoon of salt contains about 2,000 milligrams of sodium. Moreover, processing techniques add significant amounts of salt and sodium to food. For example, turnip greens cooked with salt has 890 percent more sodium than without salt. In general, natural foods are low in sodium, whereas processed foods are relatively high.

Table 9.1 highlights the sodium content in selected foods within the major food groups and some popular restaurant fast foods. Fresh foods generally have lower sodium content than processed foods. Canned soups typically contain high amounts of sodium per serving, but lower sodium brands are available. Checking food labels is the best means to control sodium intake. Food labels must list the sodium content in milligrams and in percent of the Daily Value and may carry claims such as "sodium free" if the product meets certain restrictions as shown in **table 9.2**. Most Americans obtain about

TABLE 9.1 Sodium content of selected food items

Food group	Item	Serving	Sodium (mg)
Dairy*	Milk, low-fat, fluid, 1%, protein fortified, with added vitamin A and vitamin D	1.0 cup	143
	Cheese, cottage, creamed, large or small curd	4.0 oz	356
	Cream, fluid, light (coffee cream or table cream)	1.0 fl oz	22
	Cheese, provolone	1.0 cup, diced	960
	Cheese, swiss	1.0 cup, diced	247
	Cheese, cream	1.0 tbsp	46
	Yogurt, plain, low fat	1.0 container (6 oz)	119
Vegetables*	Turnip greens and turnips, frozen, cooked, boiled, drained, with salt	1.0 cup	416
	Turnip greens, cooked, boiled, drained, without salt	1.0 cup, chopped	42
	Potatoes, french fried, cottage-cut, salt not added in processing, frozen, as purchased	10.0 strips	21
	Potatoes, baked, skin, without salt	1.0 skin	12
Fruits*	Peaches, canned, light syrup pack, solids and liquids	1.0 cup, halves or slices	13
	Apples, frozen, unsweetened, unheated (includes foods for USDA's Food Distribution Program)	1.0 cup slices	5
	Fruit cocktail (peach and pineapple and pear and grape and cherry), canned, extra light syrup, solids, and liquids	0.5 cup	5
	Oranges, raw, with peel	1.0 cup	3
Grains*	Bread, reduced-calorie, white	1.0 oz	136
	Bread, reduced-calorie, wheat	1.0 oz	94
	Cereals ready-to-eat, granola, homemade	1.0 cup	32
	Cereals ready-to-eat, Quaker Oatmeal Squares, Golden Maple	1.0 cup	194
Protein foods*	Beans, white, canned, no added fat	0.5 cup (130 g)	345
	Beef, ground, 90% lean meat/10% fat, patty, cooked, broiled	3.0 oz	58
	Bologna, chicken, turkey, pork	1.0 serving	258
	Bologna, pork and turkey, lite	1.0 serving 2 oz	401
	Chicken breast, oven-roasted, fat-free, sliced	1.0 serving 2 slices	457
	Fish, tuna, skipjack, fresh, cooked, dry heat	3.0 oz	40
	Fish, tuna, white, canned in oil, drained solids	3.0 oz	337
Fats and oils*	Margarine-like, margarine-butter blend, soybean oil and butter	1.0 tbsp	101
	Butter, salted	1.0 pat (1″ sq, 1/3″ high)	32
Canned foods*	Soup, vegetable beef, microwavable, ready-to-serve, single brand	1.0 serving	1,098
	Soup, chicken gumbo, canned, condensed	0.5 cup (4 fl oz)	873
	Soup, chicken, canned, chunky, ready-to-serve	1.0 cup	867
	Chicken noodle soup, classic Campbell's chunky	1.0 cup	790
	Soup, cream of chicken, canned, condensed, reduced sodium	0.5 cup	443
	Soup, tomato, canned, condensed, reduced sodium	1.0 serving 0.5 cup	27
Condiments*	Mustard (Burman's Dijon)	1 tsp (5 g)	120
	Ketchup (Burman's Tomato)	1 tbsp (17 g)	160
	Salad dressing (Ken's Steak House Balsamic Vinaigrette)	2 tbsp (31 g)	240
Restaurant fast foods[†]	Arby's classic roast beef sandwich	1	970
	Burger King Whopper with cheese	1	1,339
	McDonald's Big Mac sandwich	1	950
	Pizza Hut large (6″) personal pepperoni pan pizza	1	1,410
	Subway turkey breast 6 inch sandwich	1	810

As you can see in this table, the sodium content of foods can vary greatly. In general, canned and processed foods have much higher sodium content than do fresh foods. Eat fresh fruits, vegetables, and whole grain products whenever possible and prepare them with little or no salt. Avoid highly salted foods. Look for "sodium free" or "low sodium" labels when shopping for canned foods.

*Source: www.nal.usda.gov/fnic/nutrient-lists-standard-reference-legacy-2018
[†]Source: https://fastfoodnutrition.org

TABLE 9.2	Nutrition Facts label terms for sodium*
Salt/Sodium-Free Less than 5 milligrams per serving	
Very Low Sodium 35 milligrams or less per serving	
Low Sodium 140 milligrams or less per serving	
Reduced-Sodium At least 25 percent less than the regular product	
Light in Sodium or Lightly Salted At least 50 percent less than the regular product **No Salt Added or Unsalted** No salt is added during processing—but these products may not be salt/sodium-free unless stated	

*Food labels must list the milligrams of sodium and the percent of the Daily Value, which is 2,300 mg.
Source: www.fda.gov/food/nutrition-education-resources-materials/sodium-your-diet

75–80 percent of their sodium intake from processed and restaurant foods. As of May 7, 2018, certain restaurants and other retail establishments must comply with US Food and Drug Administration labeling guidelines to assist consumers in making better nutrition choices, including information about sodium content.

Cooking your own food can help reduce salt intake. With some canned vegetables, draining and rinsing the product with fresh water removes some of the sodium. Herbs and other spices can add flavor and be used to replace salt added to home-prepared meals. Salt substitutes are available, such as Morton® Salt Substitute, which contains only potassium and is sodium free. Light salts are also available, such as Morton® Lite Salt, that contain less than 50 percent sodium. Most table salt has added iodine, which is important for normal thyroid function as discussed in chapter 8. Otherwise, sea salt and table salt have similar nutritional values (www.webmd.com/food-recipes/difference-between-sea-salt-and-table-salt). Both are possible sources of excess sodium intake, which may contribute to increased blood pressure and heart disease risk as noted by Aburto and others.

Major Functions Sodium is an important element in a number of body functions. As the principal electrolyte in the extracellular fluids, it primarily helps maintain normal body-fluid balance and osmotic pressure. In this regard it is essential in the control of normal blood pressure through its effect on the blood volume. The role of sodium in the etiology of high blood pressure is discussed in a later section.

In conjunction with several other electrolytes, sodium is critical for nerve impulse transmission and muscle contraction. It is also a component of several compounds, such as sodium bicarbonate, that help maintain normal acid-base balance and, as noted in chapter 13, may be an effective ergogenic aid. An overview of sodium is presented in **table 9.3**.

Deficiency and Excess Regulation of blood pressure is essential to life. Sodium is essential to maintain normal blood volume and pressure. Stanhewicz and Kenney indicate that thirst, sodium concentrations, and fluid balance are highly regulated by interdependent behavioral and neuroendocrine mechanisms, including a drive to ingest salt known as "sodium appetite." The human body has developed an effective regulatory feedback mechanism allowing for a wide range of dietary sodium intake. The hypothalamus helps regulate sodium as well as water balance in the body. If the sodium concentration decreases in the blood, a series of complex reactions leads to the secretion of **aldosterone**, a hormone produced in the adrenal gland, which stimulates the kidneys to retain more sodium. In contrast, excesses of serum sodium will lead to decreased aldosterone secretion and increased excretion of sodium by the kidneys in the urine. Other hormones, notably ADH via its effect on water absorption in the kidneys, help maintain normal sodium equilibrium in the body fluids. During exercise, particularly intense exercise, sodium concentration increases in the blood, which helps to maintain blood volume. Exercise also leads to increased secretion of ADH and aldosterone, which helps conserve body water and sodium.

Because this regulatory mechanism is so effective, deficiency states due to inadequate dietary intake of sodium are not common. Indeed, our natural appetite for salt assures adequate sodium intake and sodium balance over time. Nevertheless, excessive losses of sodium from the body, usually induced by prolonged sweating while exercising in the heat, may lead to short-term deficiencies that may be debilitating to the athletic individual. The importance of sodium in thirst stimulation and complete volume replacement in all water compartments following exercise is described by Stachenfeld in her review. These problems are discussed later in this chapter in the sections on fluid and electrolyte replacement and health aspects.

Chloride (Cl)

Chloride is the major anion in the extracellular fluids. The gram atomic weight (one mole) of chloride is 35.5. One millimole (one milliequivalent) of chloride is 35.5 milligrams.

DRI The chloride AI is 2,300 milligrams (3,800 milligrams of salt) for individuals ages 9–50 and is lower for both younger and older individuals as described in **table 9.3**. The chloride upper limit is 3,600 milligrams (6,000 milligrams of salt).

Food Sources Chloride is distributed in a variety of foods. Its dietary intake is closely associated with that of sodium, notably in the form of common table salt, which is 60 percent chloride.

Major Functions Chloride ions work with sodium in the regulation of body-water balance and in the formation of electrical potentials across cell membranes. An example of the later is the conduction of nerve impulses. Chloride is part of hydrochloric acid (HCl) in the stomach, which is necessary for certain digestive processes, and is also involved with the transport of CO_2 from tissues in an exchange with bicarbonate across the red blood cell membrane—a process known as the "chloride shift."

TABLE 9.3 Major electrolytes: sodium, chloride, and potassium*

Major electrolyte	Adequate intake†	Major functions in the body	Deficiency symptoms	Symptoms of excess consumption
Sodium	110 mg (ages 0–6 months) 370 mg (ages 7–12 months) 800 mg (ages 1–3 years) 1,000 mg (ages 4–8 years) 1,200 mg (ages 9–13 years) 1,500 mg (ages ≥14 years)	Primary positive ion in extra-cellular fluid; nerve impulse conduction; muscle contraction; acid-base balance; blood volume homeostasis	Hyponatremia; muscle cramps; nausea; vomiting; loss of appetite; dizziness; seizures; shock; coma	Hypertension (high blood pressure) in susceptible individuals
Chloride	1,500 mg (ages 1–3 years) 1,900 mg (ages 4–8 years) 2,300 mg (ages 9–50 years) 2,000 mg (ages 51–70 years) 1,800 mg (ages ≥71 years)	Primary negative ion in extra-cellular fluid; nerve impulse conduction; hydrochloric acid formation in stomach	Rare; may be caused by excess vomiting and loss of hydrochloric acid; convulsions	Hypertension, in conjunction with excess sodium
Potassium	400 mg (ages 0–6 months) 860 mg (ages 7–12 months) 2,000 mg (ages 1–3 years) 2,300 mg (ages 4–8 years) 2,300 mg (females ages 9–18 years) 2,600 mg (females ages 19–51+ years) 2,500 mg (males ages 9–13 years) 3,000 mg (males ages 14–18 years) 3,400 mg (males 19–≥50 years)	Primary positive ion in intracellular fluid; same functions as sodium, but intra-cellular; glucose transport into cell	Hypokalemia; loss of appetite; muscle cramps; apathy; irregular heartbeat	Hyperkalemia; inhibited heart function

*Food sources for sodium and potassium may be found in tables 9.1 and 9.4, respectively; food sources for chloride are similar to those for sodium.
†Chronic Disease Risk Reduction (CDRR) Sodium Levels are ≤1,200 mg for ages 1–3; ≤1,500 mg for ages 4–8; ≤1,800 mg for ages 9–13; and ≤2,300 mg for age ≥14. AI and CDRR values are from the National Academies of Science Institute of Medicine and the 2020–2025 Dietary Guidelines for Americans.

Deficiency Under normal circumstances chloride deficiency is rather rare. Chloride is normally excreted from the body in urine. However, the loss of stomach HCl through excessive vomiting can significantly decrease chloride. Because the losses of sodium and chloride in sweat are directly proportional, the symptoms of chloride loss during excessive dehydration through sweating parallel those of sodium loss. The effects of sweat electrolyte losses and replacement on physical performance and health are covered in later sections of this chapter. An overview of chloride is presented in **table 9.3**.

Potassium (K)

Potassium is a mineral element also known as kalium, from which the symbol K is derived. It is the primary intracellular cation. The gram atomic weight (one mole) of potassium is 39. One millimole (one milliequivalent) of potassium is 39 milligrams.

DRI Potassium AIs are 3,400 and 2,600 milligrams for adult males and females ages 9–50 and are lower for younger individuals, as described in **table 9.3**. During pregnancy and lactation, recommended AIs are 2,600 and 2,500 for ages 14–18 and 2,900 and 2,800 for ages 19–50. The DV for potassium is 4,700 mg for adults and children age ≥4 years. Note that even though the DV exceeds that AI, potassium supplements are not recommended.

Food Sources Potassium is found in most foods and is especially abundant in bananas, citrus fruits, fresh vegetables, milk, meat, and fish. **Table 9.4** provides some data on the potassium content of several common foods in the major food groups.

Major Functions As the major electrolyte inside the body cells, potassium works in close association with sodium and chloride in the maintenance of body fluids and in the generation of electrical impulses in the nerves and the muscles, including the heart muscle. During depolarization and repolarization, potassium and sodium move through ion-specific cell membrane channels. Resting membrane potential is reestablished by the transfer of sodium out of and potassium into the cell in a 3:2 ratio by Na^+-K^+-ATPase activity. Potassium also plays a role in energy metabolism, glucose transport, and glycogen storage.

Deficiency and Excess Potassium balance, like sodium balance, is regulated by aldosterone but in a reverse way. A high serum potassium level stimulates the release of aldosterone from the adrenal cortex, leading to an increased excretion of potassium by the kidneys into the urine. A decrease in serum potassium levels elicits a drop in aldosterone secretion and hence a greater conservation of potassium by the kidneys. Because a potassium imbalance in the body may have serious health consequences, potassium regulation is quite precise. Deficiencies or excessive accumulation is extremely rare under normal circumstances.

TABLE 9.4 Potassium content in some common foods in the major Food Groups

Food Group	Amount	Milligrams of potassium (% DV)
Dairy		
Milk, low fat (1 percent)	1 cup	366 (8)
Yogurt, Greek, plain, nonfat	6 oz	240 (5)
Cheese, cheddar	1 oz	22 (<1)
Protein foods		
Chicken breast, boneless, grilled	3 oz	332 (7)
Black beans	5 oz	480 (10)
Fish, salmon, atlantic, farmed, cooked	3 oz	326 (7)
Grains		
Bread, whole wheat	1 slice	81 (2)
Cereal, Honey Nut, Cheerios	1 cup	150 (3)
Fruits		
Raisins	½ cup	618 (13)
Banana	1 medium	422 (9)
Orange juice	1 cup	496 (11)
Apple	1 medium 3" diameter	195 (4)
Vegetables		
Squash, acorn, mashed	1 cup	644 (14)
Potato, baked, no skin	1 medium	610 (13)
Broccoli, cooked, chopped	1 cup	229 (5)
Carrot, raw	1 medium, 7" long/1.25" diameter	250 (5)

Source: https://ods.od.nih.gov/factsheets/Potassium-HealthProfessional
Banana ©McGraw-Hill Education

Although potassium deficiencies are rare, causes include diarrhea, vomiting, excessive use of alcohol, certain prescribed medications such as diuretics and some antibiotics, extreme fasting diets, and diabetic ketoacidosis. In such cases **hypokalemia**, or low serum potassium levels, could lead to muscular weakness and even cardiac arrest due to a decreased ability to generate nerve impulses and an irregular heartbeat.

Excessive body potassium stores also are not very common. Causes include kidney disease, certain prescribed medications such as angiotensin converting enzyme inhibitors and β-blockers, dehydration, abuse of potassium supplements, and diabetes. **Hyperkalemia**, or excessive potassium in the blood, may disturb electrical impulses, causing cardiac arrhythmias and possible death. It may result when potassium ingestion overwhelms the aldosterone regulatory system discussed above and in the previous chapter. John and others reported two case studies of near-fatal hyperkalemia, one from a salt substitute and the other from a muscle-building supplement. For these reasons, individuals should never take potassium supplements in large doses without the consent of a physician. An overview of potassium is presented in **table 9.3**.

In theory, a potassium deficiency could adversely affect physical performance capacity. However, given the potential risks associated with excess potassium supplementation, there is very little research evaluating its ergogenic effects. The role of potassium in the etiology of high blood pressure has also been studied. The results of this research are presented in later sections of this chapter.

Key Concepts

▶ Chronic Disease Risk Reduction (CDRR) Sodium Levels established by the National Academies of Science Institute of Medicine are ≤1,200 milligrams for ages 1–3; ≤1,500 milligrams for ages 4–8; ≤1,800 milligrams for ages 9–13; and ≤2,300 milligrams for age ≥14. The AI value for individuals ≥14 years of age is 1,500 milligrams.

▶ Sodium and chloride perform vital functions such as generating electrical impulses for contraction of muscles, including the heart. Sodium is also very important in the regulation of blood pressure.

▶ Potassium and sodium are important in depolarization and repolarization of muscle fibers and neurons.

▶ Potassium works with sodium and chloride in the regulation of blood pressure and normal neural functions, and it participates in other metabolic functions such as storage of muscle glycogen.

▶ Sodium, chloride, and potassium concentrations in the body are precisely regulated. Deficiencies or excesses are rare but can and do occur without replacement of daily losses or excessive supplementation, respectively. Electrolyte deficiencies or excess may contribute to serious health problems.

Check for Yourself

▶ Go to the supermarket or look online and compare the cost of a serving of various bottled waters, the sodium content of various brands of soup, and the contents of various sports drinks.

Regulation of Body Temperature

What is the normal body temperature?

The temperature of different body parts may vary considerably. The skin may be very cold but the body internally is much warmer. Body temperature refers to the internal, or **core temperature**, and not the external shell temperature. **Shell temperature**, which represents the temperature of the skin and the tissues directly under it, varies considerably depending upon the surrounding environmental temperature. A recent review by Cheuvront and Kenefick describes the importance of a wide gradient between core and shell temperature for optimal regulation of body temperature and the adverse effect of dehydration on heat dissipation and performance.

Normal resting body temperature (98.6°F, 37°C) may range from 97°F to 99°F (36.1°C–37.2°C). Resting rectal temperature is normally about 0.5–1.0°F higher than the oral temperature. Although body temperature is measured using oral, rectal, esophageal, tympanic, axillary, and infrared thermographic (used in COVID-19 screening) methodologies, recent reviews by Masserole and Mekjavic and their colleagues conclude that core temperature is underestimated by many common methodologies. Rectal and esophageal methodologies using thermocouples or a swallowed capsule with wireless telemetry are considered valid measurements of core temperature. The latter methodology allows researchers to study core temperatures in field settings. Shell temperatures may be measured by adhesive thermometer pads attached to the skin.

Humans can survive a range of core temperatures for a short time, but optimal physiological functioning usually occurs within a range of 97–104°F (36.1–40.0°C). A variety of factors may affect body temperature. Here we are concerned with the effect exercise has on the core temperature and how the body adjusts to help maintain heat balance.

What are the major factors that influence body temperature?

Humans are warm-blooded animals and are able to maintain a constant body temperature under varying environmental temperatures. To do this, the body must constantly make adjustments to either gain or lose heat.

Approximately 80 percent of metabolism is in the form of thermal (heat) energy. The basal metabolic heat production is provided through normal burning (oxidation) of the three basic foodstuffs in the body—carbohydrate, fat, and protein. A higher basal metabolic rate, infectious diseases, shivering, and exercise are several factors that might increase heat production.

The human body also has a variety of means to lose heat. Heat loss is governed by four physical means—conduction, convection, radiation, and evaporation.

Conduction—heat is transferred from the body by direct physical contact, such as when you sit on a cold seat.

Convection—heat is transferred by movement of air or water over the body.

Radiation—heat energy radiates from the body into the surrounding air.

Evaporation—heat is lost from the body when it is used to convert sweat to a vapor, known as the heat of vaporization. Sweating is the body's most important heat loss mechanism during hot weather exercise. The lungs also help to dissipate heat through evaporation.

During rest and under normal environmental temperatures, body heat is transported from the core to the shell by way of conduction and convection, the blood being the main carrier of the heat. The vast majority of the heat escapes from the body by radiation and convection, with a smaller amount being carried away by the evaporation of insensible perspiration. A cooler environment, increased air movement such as a cool wind, increased blood circulation to the skin, or an increased radiation surface would facilitate heat loss.

In contrast, under certain environmental conditions, such as exercising in the sunlight on a hot day, some of these processes may be reversed, with the body gaining heat instead of losing it. For example, radiant energy from the sun could add heat to the body.

The well-known **heat-balance equation** may be used to illustrate these interrelationships:

$$H = M \pm W \pm C \pm R - E$$

where H = heat balance, M = resting metabolic rate, W = work done (exercise), C = conduction and convection, R = radiation, E = evaporation. Note that C and R are means by which the body can either gain or lose heat, while E is a heat loss mechanism.

If any of these factors governing heat production or heat loss is not balanced by an opposite reaction, heat balance will be lost and the body will deviate from its normal value. During exercise, W increases heat production. Hence, compensating adjustments in C, R, and E must be made to dissipate the extra heat. **Figure 9.5** illustrates heat stress factors and mechanisms of heat loss during exercise.

How does the body regulate its own temperature?

Body temperature is controlled by the autonomic division of the central nervous system. The hypothalamus is an important structure in the brain that is involved in the control of a wide variety of physiological functions, including body temperature. It is thought to function as the thermostat does in your house. If your house gets too cold, the heat comes on; if it gets too warm, the air conditioning system starts. The human body makes similar adjustments.

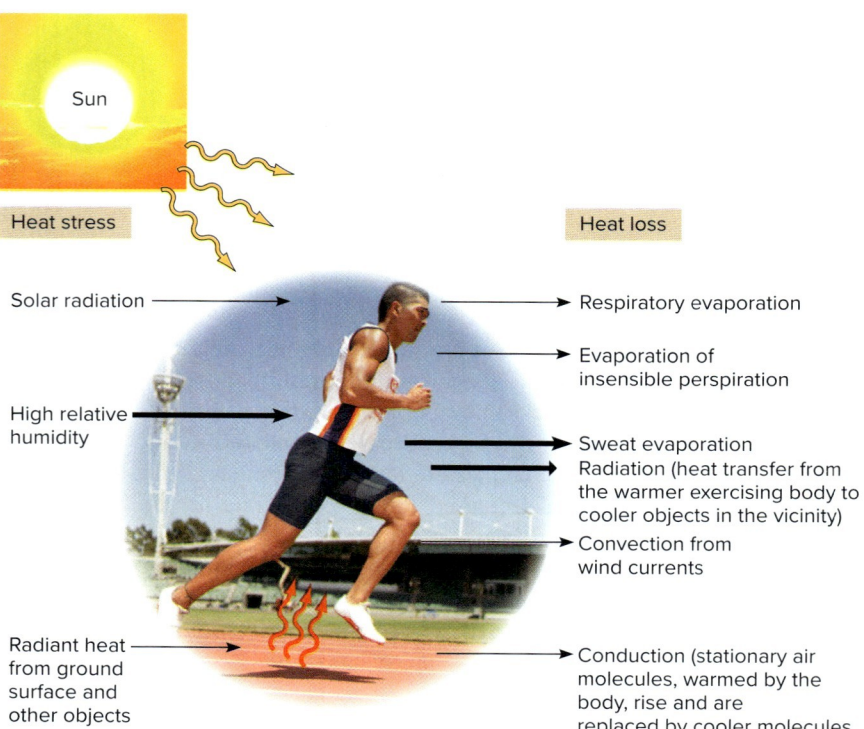

Sunset: Galyna Andrushko/Shutterstock; Runner: Tom Merton/Getty Images

FIGURE 9.5 Sources of heat gain and heat loss to the body during exercise. High relative humidity with high heat is the greatest heat stressor. Evaporative heat loss from sweating is the most effective heat loss mechanism. See text for details.

The temperature-regulating center in the hypothalamus receives input from several sources. First, receptors in the skin can detect temperature changes and send afferent input to the hypothalamus. Second, the temperature of the blood can directly affect the hypothalamus as it flows through that structure.

In general, if the skin receptors detect a warmer temperature or the blood temperature rises, the body will make adjustments in an attempt to lose heat. Two major adjustments may occur. First, the blood will be channeled closer to the skin so that the heat from within may get closer to the outside and radiate away more easily. Second, sweating will begin and evaporation of the sweat will carry heat away from the body.

If the skin receptors detect a colder temperature or the blood temperature is lowered, then the body will react to conserve heat or increase heat production. First, the blood will be shunted away from the skin to the central core of the body. This decreases heat loss by radiation and helps keep the vital organs at the proper temperature. Second, shivering may begin. Shivering is nothing more than the contraction of muscles, which produces extra heat by increasing the metabolic rate. Thermoregulatory feedback loops related to increased blood temperature and dehydration are illustrated in **figure 9.3**. **Figure 9.5** illustrates heat exchange and loss mechanisms to compensate for environmental heat stress. **Figure 9.6** is a simplified schematic of body temperature control.

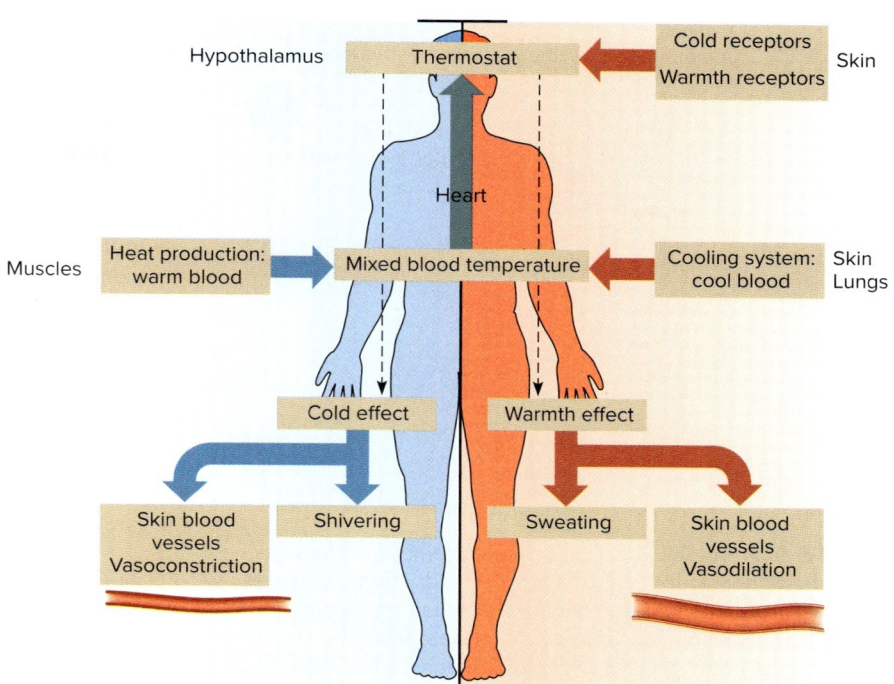

FIGURE 9.6 Simplified schematic of body temperature control. The temperature of the blood returning from the muscles and the skin stimulates the temperature regulation center (thermostat) in the hypothalamus, as do nerve impulses from the warmth and cold receptors in the skin. An overall cold effect will elicit a constriction of the blood vessels near the body surface and muscular shivering, thus helping to conserve body heat. An overall warmth effect will elicit a dilation of blood vessels near the skin and sweating, thus increasing the loss of body heat.

The hypothalamus is usually very effective in controlling body temperature. However, certain conditions may threaten temperature control. For example, an individual who falls into cold water will lose body heat rapidly, for water is an excellent conductor of heat. Such a situation may lead to **hypothermia** (low body temperature) and a rapid loss of temperature control. Hypothermia may also develop in slower runners during the later part of a road race under cold, wet, and windy environmental conditions, when heat is lost more rapidly than it is produced through exercise. Muscular incoordination and mental confusion are early signs of hypothermia.

On the other hand, the most prevalent threat to the athletic individual is **hyperthermia**, or the increased body temperature that occurs with exercise in a warm or hot environment. Hyperthermia, one of the major factors limiting physical performance and one of the most dangerous, will be discussed later in this chapter.

What environmental conditions may predispose an athletic individual to hyperthermia?

Four environmental factors interact to determine the heat stress imposed on an active individual:

1. **Air temperature.** Caution should be advised when the air temperature is 80°F (27°C) or above. However, if the relative humidity and solar radiation are high, lower air temperatures may pose a risk of heat stress during exercise.

2. **Relative humidity.** Evaporation of sweat is the body's main cooling system during exercise. Higher relative humidity values at any temperature reflect greater water content, which impairs the ability of sweat on the skin to vaporize and cool the body. In subjects performing low-intensity exercise under relative humidity conditions ranging from 40 to 85 percent, Moyen and others reported decreased sweat evaporation and heat loss with higher humidity. With humidity levels from 90 to 100 percent, heat loss via evaporation nears zero. Some note that caution should be used when the relative humidity exceeds 50–60 percent, especially when accompanied by warmer temperatures.

3. **Air movement.** Still air limits heat carried away by convection. Even a small breeze may help keep body temperature near normal by moving heat away from the skin surface.

4. **Radiation.** Radiant heat from the sun and other warmer objects in the vicinity may create an additional heat load.

Some useful guidelines have been developed taking these four factors into consideration. The wet-bulb globe temperature (WBGT) thermometer, illustrated in **figure 9.7**, measures all four. Small hand-held WBGT thermometers are available. The dry-bulb thermometer (DB) measures air temperature, the globe thermometer (G) measures radiant heat, and the wet-bulb thermometer (WB) evaluates relative humidity and air movement as they influence air temperature. The **WBGT index** is computed as follows:

$$\text{WBGT index} = 0.7\ \text{WB} + 0.2\ \text{G} + 0.1\ \text{DB}$$

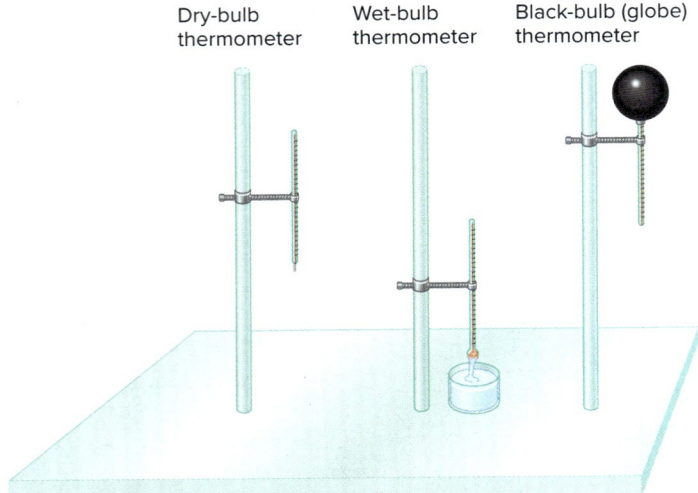

FIGURE 9.7 A typical setup for measurement of the wet-bulb globe temperature (WBGT) index. The dry bulb measures air temperature, the wet bulb indirectly measures humidity, and the black bulb measures the radiant heat from the sun. Computerized commercial devices that measure the WBGT rapidly are also available.

For example, if the WB reads 70, the G is 100, and the DB is 80, then the WBGT = $(0.7 \times 70) + (0.2 \times 100) + (0.1 \times 80) = 77°F$. Relative humidity and wind convection account for 70 percent of WGBT heat stress because these factors determine the efficacy of evaporative heat loss from sweating in regulating the core temperature.

Another indicator of heat stress is the **heat index** (figure 9.8), which combines the air temperature and relative humidity to determine the apparent temperature, or how hot it feels. Figure 9.8 also contains some temperature levels predisposing to heat disorders based on the heat index. Other models such as the revised Predicted Heat Strain model validated by Malchaire incorporate core temperature, skin temperature, and sweat rate; effects of movement; and thermal properties of clothing.

The American College of Sports Medicine (ACSM) has published a position statement with guidelines for the prevention of heat illness during distance exercise training and competition. These guidelines are discussed in the last section of this chapter.

How does exercise affect body temperature?

As noted in chapter 3, exercise increases the metabolic rate and the production of energy. Under a normal mechanical efficiency ratio of 20-25 percent, the remaining 75-80 percent of energy is released as

Heat index

Relative humidity (%) \ Air temperature (°F)	70°	75°	80°	85°	90°	95°	100°	105°	110°
100	72°	80°	91°	108°					
90	71°	79°	88°	102°	122°				
80	71°	78°	86°	97°	113°	136°			
70	70°	77°	85°	93°	106°	124°	144°		
60	69°	76°	82°	90°	100°	114°	132°	149°	
50	70°	75°	81°	88°	96°	107°	120°	135°	150°
40	68°	74°	79°	86°	93°	101°	110°	123°	137°
30	67°	73°	78°	84°	90°	96°	104°	113°	123°
20	66°	72°	77°	82°	87°	93°	99°	105°	112°
10	65°	70°	75°	80°	85°	90°	95°	100°	105°
0	64°	69°	73°	78°	83°	87°	91°	95°	99°

Heat index	Heat disorders possible with prolonged exposure and/or physical activity
80° – 89°	Fatigue
90° – 104°	Sunstroke, heat cramps, and heat exhaustion
105° – 129°	Sunstroke, heat cramps, or heat exhaustion likely and heat stroke possible
130° or higher	Heat stroke/sunstroke highly likely

NOTE: Direct sunshine increases the heat index by up to 15°F.

FIGURE 9.8 Possible heat disorders in runners and other high-risk groups based on the heat index (air temperature and relative humidity versus apparent temperature).

heat. The total amount of heat produced in the body depends on the intensity and duration of the exercise. Exercise performed at a higher intensity and/or longer duration produces more metabolic heat.

A hypothetical example of the change in core temperature for a 70 kilograms runner during rest and exercise *without* evaporative heat loss from insensible perspiration, pulmonary water vapor and/or sweating and the same conditions *with* evaporative heat loss is presented in the following *Training Table*. Humans have a mechanical efficiency of approximately 20 percent, so the remaining 80 percent is metabolic heat that must be dissipated to prevent a dangerous rise in core temperature. In this example, metabolic heat is 59 and 720 kcal for an hour of resting and running, respectively. The **specific heat** of body tissues is 0.83 kcal per kg or 58 kcal in this example. Without sweating, core temperature would increase 1°C (59 ÷ 58) during rest and 12.4°C (720 ÷ 58) during exercise. One would not survive a core temperature of 49.4°C (37°C + 12.4°C; 120°F). For each milliliter of evaporated sweat, 0.6 kcal of heat is dissipated. Evaporative heat loss from insensible perspiration and pulmonary water vapor maintains core temperature at rest. Evaporation of 67 percent of 1,500 ml of exercise sweat results in a 2°C increase in core temperature (39°C, 102.2°F). The average core temperature during exercise, even during moderately warm temperatures, may reach about 102.2-104.0°F (39-40°C). This is because of the body's cooling system.

How is body heat dissipated during exercise?

During exercise in a cold or cool environment, body heat is lost mainly through radiation and convection via the air movement around the body. Some evaporation of sweat and evaporative heat loss from the lungs may also contribute to maintenance of heat balance.

Training Table

Hypothetical example of temperature regulation at rest and during intense exercise *without* and *with* evaporative heat loss from sweating. These calculations do *not* include effects of radiation, conduction, or convection, which may contribute to heat loss *or* heat gain. See the text for explanation.

Row letter	Row description	Calculation	No evaporative heat loss from sweating		Evaporative heat loss from insensible perspiration and pulmonary water vapor (rest) and sweating (exercise)	
			Rest	Running (7.5 miles in 60 min)	Rest	Running (7.5 miles in 60 min)
A	Core temperature at the beginning of the hour (°C)		37	37	37	37
B	kcal/hour		74	900	74	900
C	Metabolic heat (80%)	= B × 0.8	59	720	59	720
D	Specific heat (tissue heat storage capacity=0.83 kcal per kg)		0.83	0.83	0.83	0.83
E	Total heat storage capacity	= 70 kg × D	58	58	58	58
F	Sweat rate (ml sweat per min)		0	0	1.5	25
G	Sweat (or insensible perspiration/pulmonary water volume) per hour	= F × 60 min	0	0	95	1,500
H	Percent evaporated		0	0	100%	67%
I	Evaporated volume	= H × 0.75	0	0	95	1,005
J	Capacity for evaporative cooling (0.6 kcal per ml sweat evaporated) dissipated heat kcal per hour	= I × 0.6 kcal per ml evaporated sweat	0	0	57	603
K	Increase in temperature	= (C − J) ÷ E	1.0	12.4	0.0	2.0
L	Core temperature at the end of the hour (°C)	= A + K	38.0	49.4	37.0	39.0
Comment				Fatal	No change	Live to run another day

However, when the environmental temperature rises, the evaporation of sweat becomes the main means of controlling an excessive rise in the core temperature. For example, evaporation of sweat may account for about 20 percent of total heat loss when exercising in an ambient temperature of 50°F (10°C) but increases to about 45 percent at 68°F (20°C) and 70 percent at 86°F (30°C). Although variable, the maximal evaporation rate is about 30 milliliters of sweat per minute, or 1.8 liters per hour. However, greater sweat rates may occur when sweat drops off the skin without vaporizing. Only sweat that evaporates has a cooling effect. One liter of sweat, if perfectly evaporated, will dissipate about 580 kcal of heat. In the previous example, the evaporation of 1.24 liters of sweat (720/580) would prevent a dangerous rise in the core temperature. However, the evaporation of sweat from the body is not perfect, as sweat can drip off the body and not carry away body heat, so more than 1.24 liters may be lost. If we assume that 2.0 liters were lost, then this individual would have lost 4.4 lbs (2.2 lbs per kg) of body fluids during the 1-hour run. It should be noted that sweat rates may vary considerably within and between individuals. Baker reported an intra-individual variability of 5–7 percent in whole-body sweat rates with inter-individual rates ranging from 0.5 to 2 or more liters/hour. Prolonged exercise decreases sweat rate sensitivity and sweat capacity. This phenomenon, reported by Wyndham over 50 years ago, occurs in acclimatized and unacclimatized individuals and is known as **sweat gland fatigue.**

Under most warm environmental circumstances, the evaporative mechanisms and the body's natural warning signals are able to keep the core temperature during exercise below 104°F (40°C) and prevent heat injuries. However, an excessive rise in the core temperature, above 104°F, or excessive fluid and electrolyte losses may lead to diminished performance or serious thermal injury in some individuals. Heat illnesses will be discussed in the last section of this chapter.

Key Concepts

- Our core temperature is about 98.6°F (37°C). One degree Celsius (C) equals 1.8 degrees Fahrenheit (F). Convert C to F by the formula [(C × 1.8°) + 32°]. Convert F to C by the formula [(F − 32°)/1.8°].
- Core temperature is regulated by the hypothalamus in response to increases or decreases in the blood temperature and osmolality and afferent input from thermal receptors in the skin in response to exercise and hot/humid or cold environmental conditions.
- High environmental temperatures, high relative humidity, or radiant heat from the sun can impose a severe heat stress on those who exercise under such conditions.
- Exercise can produce significant amounts of heat, but the body temperature usually can be regulated quite effectively by activation of heat-loss mechanisms. Body heat is lost mainly by radiation, conduction, and convection, but evaporation of sweat is a major avenue of heat loss during exercise in the heat.

Check for Yourself

▶ Calculate how much heat you would generate if you ran for 60 minutes expending energy at a rate of 12.25 kcal/minute. Assume that you were running at 20 percent mechanical efficiency. How much sweat would you have to evaporate to keep your body temperature at the same level? Assume that you do not lose any heat from other avenues such as radiation and convection.

Exercise Performance in the Heat: Effect of Environmental Temperature and Fluid and Electrolyte Losses

Athletes train and compete in all types of weather conditions, as do many individuals who exercise for fitness and health. Not all types of physical performance are impaired when performed under warm or hot environmental conditions, but some are. Cheuvront and Kennefick commented that a dehydration threshold of greater than 2 percent is supported by the literature for impaired aerobic performance, but no such criterion or mechanism exists for dehydration-related impaired performance in strength, power, and speed events. The major concern is performance in prolonged exercise and whether or not the core body temperature is maintained. Sawka and Young describe a balance between heat production and heat loss as **compensated heat stress** where core temperature and exercise intensity are maintained. In contrast, during **uncompensated heat stress,** heat production exceeds heat loss with an increase in core temperature followed by impaired exercise performance. Sawka and others point out that impaired aerobic performance during exercise in the heat is due to dehydration and the resulting inability to maintain a significant core-skin temperature gradient to dissipate body heat. Environmental heat stress itself may contribute to impaired performance, but so can fluid and electrolyte losses over time.

How does environmental heat affect physical performance?

Performance in more prolonged aerobic endurance activities is normally worse in warm compared to cooler temperatures. In their review, Nybo and others noted that the cause of fatigue during prolonged exercise in the heat has not been clearly established, but changes in brain function, alterations in dopaminergic activity, inhibitory afferent feedback, blood circulation, skeletal muscle function, hyperthermia, and dehydration, either separately or collectively, could impair performance. Much of the research evaluating the effect of heat on endurance performance has been conducted with runners and cyclists. Montain and others analyzed data of elite runners from 140 race-years of major marathons and found that as environmental temperature increased, so did finishing times. Slower runners suffered even greater performance decrements in warmer weather. Sawka and Young note that marathon running performance declines by about

1 minute for each 1°C increase in air temperature beyond 8-15°C (each 1.8°F increase in air temperature beyond 46-59°F). Junge and others reported an average 15 percent decline in cycling power output with wind speed and acclimatization exerting significant modulating influences on the degree of performance impairment.

Environmental heat may affect exercise performance in the following ways, which are discussed in the next sections:

- Central neural fatigue caused by increased brain temperature
- Cardiovascular strain caused by changes in blood circulation
- Muscle metabolism changes caused by increased muscle temperature
- Dehydration caused by excessive sweat losses

Central Neural Fatigue The brain appears to play an important role in the development of fatigue during exercise in the heat. Nybo and others indicated that the main factor adversely affecting muscle tissue activation appears to be elevated brain temperature. Subjects exercising in the heat seem to reach the point of voluntary fatigue at similar and consistent core body temperatures despite various experimental manipulations. In essence, the elevated brain temperature impairs central arousal of voluntary activation of muscle. Marino discussed two neuroprotective mechanisms of CNS protection during exercise in the heat. The "critical limiting temperature" mechanism involves a decrease in voluntary activation of exercising muscles due to elevated brain temperature, leading to termination of exercise. In the "selective cooling" mechanism, brain temperature is regulated by countercurrent heat exchange between warmer arterial blood entering and cooler venous blood leaving the brain. Tucker and others reported reductions in both neuromuscular stimulation and power output during exercise in the heat prior to abnormal increase in rectal temperature, heart rate, or perception of effort, suggesting that the brain apparently anticipates heat stress and reduces heat production (by decreasing muscle contraction) accordingly. In contrast, Périard and others found no difference in 20-second maximal voluntary contraction activation and force production following 40-km time trials under cool and hot conditions, despite a significantly greater rectal temperature in the hot condition. Central fatigue is discussed in more detail in chapter 3.

Cardiovascular Strain Other factors may also inhibit performance in the heat, such as high levels of cardiovascular strain. For example, in a 5-kilometer race the runner will be performing at a rather high metabolic rate and thus will be producing heat rapidly. To prevent hyperthermia, blood flow to the skin will increase so as to dissipate heat to the environment. This shifting of blood to the skin will result in a smaller proportion of blood, and hence oxygen, being delivered to the active musculature. As illustrated in **figure 9.9**, decreased plasma volume from sweating will decrease central blood volume. The inability of increased heart rate to offset decreases in diastolic filling and stroke volume will ultimately result in decreased cardiac output and delivery of oxygen to exercising muscle.

Muscle Metabolism Yaspelkis and others reported Jeukendrup noted that exercising in the heat shifts energy metabolism toward increased carbohydrate use and decreased fat use with greater

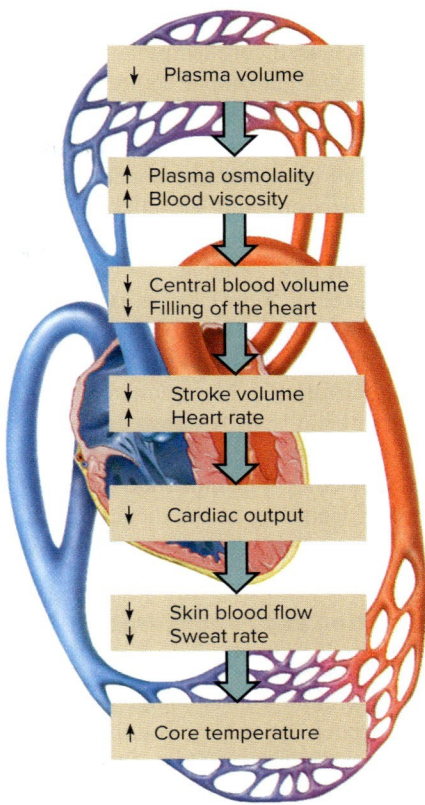

FIGURE 9.9 Some physiological effects of dehydration. The decreased blood volume and increased core temperature may contribute to premature fatigue and heat illness.

lactate accumulation. Increased intramuscular temperature and increased epinephrine due to increased sympatho-adrenal stress response could increase the rate of muscle glycolysis. A more rapid depletion of muscle glycogen could impair prolonged endurance performance. Young as well as Kenefick and others have reported greater lactate accumulation during exercise in a warm environment compared to the same work rate performed in a cooler environment. Yaspelkis and others reported similar lactate increases with no increase in muscle glycogen utilization. It is possible that increased lactate could be associated with decreased clearance by the liver due to vasoconstriction. Depletion of muscle glycogen as a cause of fatigue is discussed in chapter 4.

The loss of intracellular water due to dehydration also stimulates glycolysis. Regulation of intracellular volume is critical to homeostasis. According to Lang, an increase in the concentration of intracellular molecules such as glucose-6-phosphate (a metabolite in glycolysis) is an attempt to attract water into the cell **(figure 9.3)**. Racinais and Oska indicate that increased muscle temperature greater than 1°C adversely affects central and peripheral neural drive and increases muscle protein degradation, which would also impair performance. Supporting this viewpoint, Bongers and others noted that precooling the body prior to or during exercise, such as by taking a cold shower or bath to lower the body temperature, may be beneficial for endurance exercise tasks up to 30-40 minutes. Body cooling techniques as an ergogenic aid are discussed later in this chapter.

If circulatory adjustments and efforts to minimize dehydration are not adequate, increased core temperature and accelerated glycogen use may lead to fatigue, impaired performance, and possibly heat-related illness, which will be discussed later in this chapter.

Dehydration Dehydration may also impair exercise performance. In their review, Cheuvront and Kenefick report a dehydration threshold of 2 percent for impaired endurance exercise performance due to volume loss. Although the 5-kilometer runner will sweat heavily, the duration of the event is usually short, so an excessive loss of body fluids does not occur. However, in more prolonged events, athletes may suffer the problems noted previously plus the adverse effects of dehydration. Marathoners may lose 5 percent or more of their body weight (mostly water) during a race, which may not only deteriorate performance but have serious health consequences as well. Goulet recommends fluid intake during prolonged (>1 hour) exercise according to the dictates of thirst in order to maintain exercise-induced weight loss at no more than 3 percent.

How do dehydration and hypohydration affect physical performance?

The dehydration literature consists largely of studies conducted under tightly controlled laboratory conditions, but the ultimate application of control of water balance during exercise is in the field (i.e., the actual practices of athletes in training and in competition). Stachenfeld notes that field research provides fundamental questions to be examined in the laboratory. The effect of dehydration on physical performance has been studied from two different viewpoints. Voluntary dehydration is often used by athletes such as wrestlers and boxers to qualify for lower weight classes prior to competition. In other athletes, dehydration occurs involuntarily during training or competition as the body attempts to maintain temperature homeostasis.

Although dehydration and hypohydration are used interchangeably, dehydration is a process leading to the outcome of hypohydration, which may affect numerous physiological processes, leading to impaired physical performance. Sawka and others noted that hypohydration may lead to decreases in both intracellular and extracellular fluid volumes (particularly blood volume) with associated decreases in stroke volume and cardiac output. Hypohydration also reduces both skin blood flow and sweating rate which can increase body heat storage. Kenefick and others found that hypohydration induced an earlier onset of the lactate threshold during exercise, an adverse effect relative to aerobic endurance performance. Hypohydration could also lead to electrolyte imbalances in the muscle, with subsequent adverse effects. Despite these well-documented adverse effects on performance, Cheuvront and others acknowledge the difficulty in quantitative assessment of dehydration due to volemic and/or osmotic regulatory mechanisms.

Voluntary Dehydration Voluntary dehydration techniques used by wrestlers have included exercise-induced sweating, thermal-induced sweating such as the use of saunas, diuretics to increase urine losses, and decreased intake of fluids and food.

Much of the research with voluntary dehydration has been conducted with wrestlers. Evaluation criteria have emphasized factors such as strength, power, local muscular endurance, and performance of anaerobic exercise tasks designed to mimic wrestling. In their review, Kraft and others report the level of dehydration (3–4 percent) impairing anaerobic performance is mode dependent and depends on factors such as dehydration mode, level, and work/rest interval for repeated anaerobic performance. In its position stand on fluid replacement, the ACSM indicated that dehydration of 3–5 percent of body weight does not degrade either anaerobic performance or muscular strength.

In contrast, separate studies by Shoffstall, Judelson, and Pallarés and their respective colleagues have reported initial declines in various strength and power tasks (e.g., bench press, back squat, countermovement vertical jump) in a dehydrated state, followed by various degrees of recovery following partial rehydration. Judelson and others reported greater back squat performance following partial rehydration of 2.5 percent dehydration compared to 5.0 percent dehydration. The adverse effects on strength are not consistent, but anaerobic muscular endurance tasks lasting longer than 20–30 seconds have been impaired when subjects were hypohydrated. For example, Montain and others reported that a 4 percent decrease in body weight decreased knee-extension endurance by 15 percent. Suggested mechanisms of impairment include loss of potassium from the muscle, higher muscle temperatures during exercise, and decreased ability of the central nervous system to stimulate the musculature. It should also be noted that there is no evidence that hypohydration improves performance in these exercise tasks. Investigators recommend more research.

Involuntary Dehydration Involuntary dehydration is most common during prolonged physical activity. Dehydration may occur during exercise in cold or temperate environments, but the ACSM, in its position stand on fluid replacement, indicated that dehydration (3 percent body weight) has marginal influence on degrading aerobic exercise performance when exercising in colder environments. However, the adverse effects of involuntary dehydration are most severe on aerobic endurance performance when exercising in warm, humid environmental conditions. The following *Training Table* presents the major highlights of the ACSM position stand on fluid replacement relative to dehydration and prolonged endurance exercise performance.

Sawka and others have suggested that the deterioration in aerobic endurance performance appears to be related to adverse effects on cardiovascular functions and temperature regulation. Reduction in the plasma volume may reduce cardiac output and blood flow to the skin and the muscles. Reductions in skin blood flow have been shown to lower the sweat rate and raise the core temperature. In a meta-analysis of literature, Montain and others noted that hypohydration decreased cardiac output, and the greater the intensity of the exercise, the greater the decrease. The effects of dehydration on cardiovascular dynamics are depicted in **figure 9.9**.

One of the key points of the ACSM position stand is the effect an individual's unique biological characteristics and the exercise task may play regarding hydration status and exercise performance, and some research suggests that highly trained endurance

Training Table

Key highlights of the American College of Sports Medicine position stand on exercise and fluid replacement:

- Dehydration increases physiological strain and perceived effort to perform the same exercise task and is accentuated in warm or hot weather.
- Dehydration can degrade aerobic exercise performance, especially in warm or hot weather.
- The greater the dehydration level, the greater the physiological strain and aerobic exercise performance impairment.
- The critical water deficit and magnitude of exercise performance degradation are related to the heat stress, exercise task, and the individual's unique biological characteristics.

Andersen Ross/Blend Images LLC

runners may be able to better tolerate some, but not all, of the adverse effects of dehydration. Several researchers have offered evidence that a certain amount of exercise-induced dehydration may not adversely affect certain aspects of endurance performance. Noakes contends that there is no evidence that athletes who drink according to thirst are at any significant disadvantage from the 3–5 percent level of dehydration that they may develop. Although Armstrong and others observed increased heart rate and rectal temperature, which are concurrent with reduced stroke volume and cardiac output, would impair performance over the course of prolonged exercise, they reported that dehydration ≤5.7 percent had no adverse effect on running economy in highly trained collegiate distance runners during 10 minutes of running at 70 or 85 percent VO_2 max. Byrne and others suggested that the effects of body weight loss (used as the measure for dehydration via sweat loss) on temperature regulation may vary in actual outdoor race competition under warm conditions, as compared to controlled laboratory conditions. They reported that core temperature after running a half-marathon was not affected by the level of dehydration, which ranged from 1.62 to 4.0 liters in 18 nonelite runners. However, the authors did not appear to evaluate the effect of weight loss on performance time, as they had no measures of running speed during the race. In a meta-analysis of 15 studies of cycling performance comparing exercise-induced dehydration of < 2 and ≥ 2 percent, Goulet reported that ≥2 percent dehydration adversely affected constant power output tasks, but not time trial ("real world") tasks. He concluded that the 2 percent criterion of exercise-induced dehydration leading to impaired performance may not apply to field studies of time trial performance and that athletes should rely on thirst for fluid replacement.

The ACSM also noted that dehydration might degrade mental/cognitive performance, which may be caused by adverse effects of hyperthermia on mental processes. Separate studies conducted by Baker and others reported that dehydration may impair vigilance in dynamic sports environments, such as basketball, leading to increased errors of omission and commission and impaired reaction time. In one study, skilled basketball players were dehydrated by 1, 2, 3, and 4 percent prior to taking a test mimicking basketball skills in a fast-paced game. The players experienced a progressive deterioration in performance as dehydration progressed from 1 to 4 percent, but performance was not significantly impaired until dehydration reached 2 percent. In another study on dehydration in basketball players, Dougherty and others also reported impaired sprint times and shooting percentage in 12–15-year-old skilled basketball players who were dehydrated by 2 percent prior to a simulated 48-minute basketball game. Edwards and colleagues studied the impact of dehydration on sport-specific activities in male soccer players who completed no fluid intake, fluid intake, and mouth-rinse treatments in counterbalanced order. After a 45-minute soccer match, body mass was reduced by 2.14, 2.4, and 0.7 percent, respectively, following mouth-rinse, no-fluid, and fluid treatments. Although dehydration had no impact on mental concentration tests, sport-specific soccer performance was decreased by 13–15 percent when fluid was not ingested. Adan reported that 2 percent dehydration adversely affects attention, psychomotor skills, and immediate memory, but not long-term/working memory and executive functions. Cheuvront and Kennedick comment that impaired cognition as related to dehydration is relatively small, primarily related to discomfort, and is task dependent. Additional research would appear to be warranted to explore the effects of dehydration on other mental aspects of sports performance.

Dehydration may also be a major factor in the onset of gastrointestinal (GI) distress. GI symptoms associated with dehydration include nausea, vomiting, bloating, GI cramps, flatulence, diarrhea, and GI bleeding, many of which could impair performance if severe enough. In their review, de Oliveira and others reported that 30–50 percent of athletes experience gastrointestinal distress that is physiological, mechanical, and/or nutritional in nature and that our understanding of the underlying etiology and efficacy of possible intervention strategies is far from complete.

The ACSM indicated that dehydration is also a risk factor for various heat illnesses, which are covered in a later section of this chapter.

How fast may an individual become dehydrated while exercising?

In her review, Baker indicated that whole-body sweat rate typically ranges from 0.5 to 2.0 liters (1.1–4.4 lbs) per hour but can exceed 3 liters (>6.6 lbs) per hour in approximately 2 percent of athletes. Most athletes may lose somewhat less, maybe 2–3 liters, when exercising strenuously in the heat, but even then it will not take long to incur a 2–3 percent decrease in body weight. A 150-pound runner producing 2 liters of sweat/hour would have a 3 percent loss of body mass (4.4/150 = 0.03; 0.03 × 100 = 3 percent), which could cause premature fatigue. According to Godek and colleagues, sweat rate is positively related to both body mass and surface area in American football players. Linemen may lose up to 10 kg (22 lbs) over a day with multiple daily workouts. Training intensity also plays a role in rate of dehydration. Duffield and colleagues reported greater sweat rates and electrolyte losses in professional male soccer players following "high-intensity" and "game simulation" conditions compared to "low-intensity" conditions. They concluded that individualized rehydration strategies should be employed based on training intensity and other factors.

There are some gender and age differences in sweating, sweat composition, and fluid balance. According to Kaciuba-Uscilko and Grucza, smaller sweat rates in females are offset by greater evaporative efficiency compared to males after adjustment for body surface area, body composition, and exercise capacity. Smith and Havenith reported greater regional and gross sweat rates of aerobically trained males compared to females at both 60 and 75 percent of VO_2 max. Eijsvogels and others reported lower fluid intake, higher fluid losses, and higher blood sodium levels in males compared to females in response to prolonged exercise of the same duration and intensity. Meyer and others indicate that children sweat somewhat less than adults and have differences in sweat solute concentrations, but they still may reach hypohydration levels comparable to adults. Young tennis players may lose 1–2 liters of sweat per hour in tournament play, and some older adolescents as much as 3 liters per hour. Excessive dehydration may impair not only one's physical performance but possibly also one's health, as discussed later in this chapter.

How can I determine my sweat rate?

The rate of sweating varies among different individuals, so some may be more prone to dehydration than others. As Cheuvront and others state, dehydration can be difficult to assess and the average individual will not have access to clinical markers such as plasma and urine osmolality and indicator dilution techniques to measure volumes of different water compartments. Although there are a number of methods to evaluate sweat loss and hydration status, body weight change is reliable and easy to do. However, Maughan and others note that although sweat rate and hydration status are often estimated from body weight loss, several sources of error may give rise to misleading results. For example, substantial respiratory water losses may occur during intense exercise in dry environments. Other factors such as water production from the oxidation of energy substrate and body mass loss from fat oxidation may result in a maintenance of the total body water despite a loss of body weight. As a result, body weight loss may not always be a reliable marker of changes in hydration status. However, Cheuvront and others note that body weight loss adjusted for fluid intake and urine losses is primarily a function of sweat losses. Urine is darker in the dehydrated state. Another test of possible volume depletion as described by Cheuvront and others is a simple 20+ beats/minute increase in heart rate from the sitting to the standing position.

The calculations of total sweat volume and sweat rate in the following example are based on methodology from the Gatorade Sports Science Institute. The exerciser's pre-exercise body weight, postexercise body weight, consumed fluid volume, and urine production (if any) must be measured. You may use the following examples as a guide to calculate your own sweat rate during exercise (see the Application Exercise at the end of this chapter). The sweat rate for athlete A is calculated in the metric measurement system and for athlete B in the English system. According to the joint Academy of Nutrition and Dietetics, Dietitians of Canada, and the American College of Sports Medicine's position statement on nutrition and performance, hydration during exercise should limit the total body fluid deficit to a body weight loss of less than 2 percent. Athlete A, but not athlete B, ingested sufficient fluids to accomplish this goal.

Training Table

Calculation of sweat volume and sweat rate

		Calculations	Athlete A 2.0% dehydration Metric	Athlete B 3.5% dehydration English
A	Pre-exercise body weight (hydrated, early AM, post void)		70 kg	185 lbs
B	Post exercise body weight		68.7 kg	178.5 lbs
C	Change in body weight	=B − A	−1.3 kg	−6.5 lbs
		=C × 1,000 g/kg	1,300 g (ml)*	
		=C × 16 oz/lb		104 oz[†]
D	Percent change in body weight	=C ÷ A × 100	−1.9%	−3.5%
E	Drink volume during exercise		500 ml	16 oz
F	Urine production		100 ml	3 oz
G	Sweat loss	=D + E − F	1,700 ml	117 oz
H	Exercise duration (minutes)		60 minutes	90 minutes
I	Sweat rate	=G ÷ H	28.3 ml/minute	1.30 oz/minute

*1 kg = 1,000 g ~ 1 ml; [†]1 lb = 16 oz.

What is the composition of sweat?

The human body contains two different types of sweat glands. Apocrine sweat glands, located in hairy areas of the body such as the armpits, secrete an oily mixture to decrease friction and are the source of odor associated with sweating. Eccrine sweat glands, about 2–3 million over the surface of the body, are primarily involved in temperature regulation.

Sweat is mostly water (about 99 percent), but a number of major electrolytes and other nutrients may be found in varying amounts. Sweat is hypotonic in comparison to the fluids in the body. This means that the concentration of electrolytes is lower in sweat than in the body fluids.

The composition of sweat may vary somewhat from individual to individual and will even be different in the same individual when acclimatized to the heat, as contrasted to the unacclimatized state. The major differences are the concentrations of the solid matter in the sweat, the electrolytes or salts. During 10 days of heat acclimatization in adult males and females, Klous and others observed that sodium, chloride, and lactate levels were lower in the sweat in days 3–6 preceding increased back and arm sweat rates on days 7–8. There was no change in the level of potassium in sweat. Meyer and others reported higher sweat lactate and ammonia concentrations and lower sweat pH in boys and girls compared to adults following two 20-minute bouts of exercise at 50 percent of VO_2 max.

The major electrolytes found in sweat are sodium and chloride, as sweat is derived from the extracellular fluids, such as the plasma and intercellular fluids, which are high in these electrolytes. You may actually note the formation of dried salt on your skin or clothing after prolonged sweating. Lara and others report that salt concentration in runners after a marathon is variable but averages 42.9 mmol/liter for sodium and 32 mmol/liter for chloride with lower values in females than in males.

Other minerals lost in small amounts include potassium, magnesium, calcium, iron, copper, and zinc. As noted in chapter 8, certain athletes, especially those who lose large amounts of sweat, may need to increase their dietary intake of certain trace minerals, such as iron and zinc, to replace losses during exercise.

Small quantities of nitrogen, amino acids, and some of the water-soluble vitamins also are present in sweat, but these amounts are easily restored by consuming a balanced diet.

Is excessive sweating likely to create an electrolyte deficiency?

The likelihood for excessive sweating to create an electrolyte imbalance depends on electrolyte balance during exercise and post exercise recovery on a day-to-day basis.

The concentration of electrolytes in the blood during exercise with excessive sweating has been studied under laboratory conditions, as well as immediately after endurance events such as the Ironman triathlon and a marathon run. In general, exercise increases the concentration of several electrolytes in the blood. Sodium and potassium concentrations are elevated. The sodium increase may be due to greater body-water loss than sodium loss, so a concentration effect occurs. The potassium may leak from the muscle tissue to the blood, thereby increasing the blood concentration of this ion. Calcium ion concentration remains relatively unchanged during exercise. Magnesium levels usually fall, possibly because the active muscle cells and other tissues need this ion during exercise and it passes from the blood into the tissues. Thus, during acute, prolonged bouts of exercise, even in marathon running, it appears that an electrolyte deficiency will not occur. Meyer and others found increased sodium and chloride and decreased potassium in the sweat of young adult males and females compared to pubescent boys and girls. There were no gender differences between age groups. They concluded that maturation may be a factor in selecting an "optimal" fluid-electrolyte replacement beverage.

This is not to say that electrolyte replacement is not important. As we shall see in the next section, an electrolyte imbalance may occur in the body during extremely prolonged endurance events, such as ultramarathoning and Ironman-type triathlons, if proper fluid replacement techniques are not used. Moreover, what happens during the recovery period after excessive sweating may contribute to an electrolyte deficiency. In their study of marathoners, Lara and others observed that "salty sweaters" (all males with forearm sweat sodium > 60 mmol/liter) may benefit from special sodium intake to replace high sweat sodium losses. If sodium, chloride, and potassium sweat losses are not replaced daily, an electrolyte deficiency may occur over time. The next section deals with the need for water and electrolyte replacement.

Key Concepts

- Both hyperthermia and dehydration may impair endurance capacity through mechanisms contributing to central nervous fatigue, cardiovascular strain, and impaired energy metabolism.
- Sweat consists mainly of water and some minerals, primarily sodium and chloride. It is hypotonic compared to the body fluids.
- In the literature, a decrease in body weight of ≥ 2 percent due to dehydration is a common, but not universally agreed upon, threshold for impaired endurance performance. More severe dehydration is generally associated with greater decrements in performance.
- According to a joint ACSM/AND/DC position statement, hydration during exercise should limit the total body fluid deficit to a body weight loss of less than 2 percent.
- Strength, power, and other anaerobic tasks are somewhat less susceptible to dehydration-related performance decrements.

Exercise in the Heat: Fluid, Carbohydrate, and Electrolyte Replacement

Evidence-based guidelines for maintaining hydration and electrolyte balance before, during, and following exercise have evolved as the body of literature on fluid and electrolyte replacement has grown in recent years. American College of Sports Medicine (ACSM) and International Olympic Committee (IOC) guidelines from 1996 were criticized by Noakes and other researchers because the ACSM recommendation to drink "as much as tolerable" during exercise may contribute to excess fluid intake and hyponatremia (a serious condition to be discussed later in the chapter) and because of the dearth of appropriately controlled, randomized, prospective studies providing the foundation for the IOC recommendations. Updated ACSM guidelines in 2007 represent a synthesis of the best research available, provide recommendations that are considered to be prudent, are more likely to help delay the onset of premature fatigue during prolonged exercise in the heat, may help in the prevention of exercise-associated heat illness, and are less likely to cause other exercise-associated health problems. More recently, the 2017 joint ACSM, Academy of Nutrition and Dietetics, and Dietitians of Canada's position statement on nutrition and athletic performance includes guidelines for hydration and electrolyte balance before, during, and after exercise.

Which is most important to replace during exercise in the heat—water, carbohydrate, or electrolytes?

In the 1960s Robert Cade, a scientist-physician working at the University of Florida, developed an oral fluid replacement for athletes that was designed to restore some of the nutrients lost in sweat. This product was eventually marketed as Gatorade® (Gator is the nickname for University of Florida athletes) and was the first glucose-electrolyte solution (GES, now referred to as **carbohydrate-electrolyte solutions,** or **CES**) to appear as a sports drink in the athletic marketplace. Endurance athletes and individuals who are engaged in prolonged (over 1 hour) activity, especially in a hot, humid environment, will derive the greatest benefit from CES consumption.

CES were the first commercial fluid-replacement preparations designed to replace both fluid and carbohydrate. Today, Gatorade has many competitors in the CES market. Other than water, the major ingredients in these solutions are carbohydrates, usually in various combinations of glucose, glucose polymers, sucrose, or fructose and some of the major electrolytes. As noted in chapter 4, sports drinks containing multiple carbohydrates, such as glucose, fructose, sucrose, and maltodextrins (glucose polymer), may be a good choice. Jeukendrup recommended that carbohydrate intake be individualized based on the intensity and duration of the exercise task and stated that multiple carbohydrates increase intestinal absorption and oxidation. The sugar content ranges from about 5 to 10 percent depending on the brand. The caloric values range from about 6 to 12 kcal per ounce. The major electrolytes include sodium, chloride, potassium, and phosphorus. These ions are found in varying amounts in different brands. Some brands may also include a variety of other substances, including vitamins (B vitamins and C), minerals (calcium and magnesium), branch-chain amino acids (BCAA), drugs (caffeine), herbals (ginseng), and artificial coloring and flavoring. Standard sports drinks should not be confused with "Energy," "Sports Energy," or "sports shots" drinks in the marketplace, which contain considerably more carbohydrate, high levels of caffeine, and numerous other ingredients. In their review, Higgins and others call for greater regulatory oversight and changes in the marketing of energy drinks because of adverse events linked to energy drink consumption. In 2018, the European Specialist Sports Nutrition Alliance called for differentiation between "sports drinks" (i.e., CES) legitimately used by endurance athletes and highly caffeinated "energy drinks" with ingredients that may adversely affect children and young adults (www.nutraingredients.com/Article/2018/07/11/Sports-drinks-versus-energy-drinks-The-two-must-be-differentiated-argues-ESSNA). The FDA has also provided guidance for industry regarding highly caffeinated beverages including energy drinks. Other beverages that appear to be sports drinks may contain minimal carbohydrate content. Nutrition Facts labels on sports drinks will provide you with the actual content, including source of carbohydrates. The contents of selected ingredients for several CES are presented in **table 9.5**. As noted by Baker and Jeukendrup, fluid-electrolyte replacement is important to individuals dehydrated from vomiting or diarrheal disease as well as individuals exposed to exercise and/or environmental stress. Therefore, the optimal content of fluid-electrolyte replacement beverages depends on many factors, not the least of which is the reason for fluid and/or electrolyte loss.

Each of the components of CES may be important to the athlete, depending on the circumstances. When dehydration or hyperthermia is the major threat to performance, water replacement is the primary consideration. In prolonged endurance events, where muscle glycogen and blood glucose are the primary energy sources, carbohydrate replacement, as noted in chapter 4, may help improve performance. In very prolonged exercise in the heat with heavy sweat losses, such as ultramarathons, electrolyte replacement may be essential to prevent heat injury. Although the beneficial effects of carbohydrate intake during exercise were covered in chapter 4, the role of carbohydrate as a component of the CES is stressed in this chapter.

The following questions focus on the importance and mechanisms of water, carbohydrate, and electrolyte replacement for the individual incurring sweat losses while exercising under heat stress conditions.

What are some sound guidelines for maintaining water (fluid) balance during exercise?

InsideOutPix/age fotostock

Proper hydration is probably the most important nutritional strategy an athlete can use in training and competition. Adequate hydration before and during exercise will help decrease fluid loss, reduce cardiovascular strain, enhance performance, and prevent some heat illnesses. Athletes have used several strategies to help prevent hypohydration and excessive increases in body temperature associated with certain types of sports competition. Depending on the sport, three commonly used practices are skin wetting, hyperhydration, and rehydration. Another procedure, body cooling, is discussed in the section on ergogenic aids.

Skin Wetting Skin wetting techniques, such as sponging the head and torso with cold water or using a water spray, have been shown to decrease sweat loss. This could be an important

TABLE 9.5 Fluid-replacement and high-carbohydrate* beverage comparison chart per 8-oz serving

Beverage	Carbohydrate ingredient	Carbohydrate (% concentration grams/100 ml)	Carbohydrate (grams)	Sodium (mg)	Potassium (mg)
Gatorade Thirst Quencher (Gatorade Company Lemon-Lime)	Sucrose, glucose, fructose	6	14	108	32
Gatorade Endurance Formula	Sucrose, glucose, fructose	6	15	208	94
Accelerade® (Pacific Health Laboratories Mountain Berry)	Sucrose, trehalose (disaccharide), fructose	6	14	141	64
PowerAde® (The Coca-Cola Company)	High-fructose corn syrup	6	15	100	24
Lucozade Sport (Suntory)	Glucose; maltodextrin	6	15	Trace	Trace
Cytomax Performance Plus Cytosport Tropical Fruit	alpha-l-polylactate	5.5	13	55	30
Coca-Cola®	High-fructose corn syrup, sucrose	11	26	30	0
Diet soft drinks	None	0	0	0–25	Low
Tropicana® Orange juice (100% juice)	Fructose, sucrose	11	26	0	450
Water	None	0	0	Low	Low
Gatorade Endurance Formula–Orange	Maltodextrin, glucose, fructose	6	15	200	94
Coconut water, unsweetened	Sucrose, glucose, fructose	3.8	8.9	252	600
Beverage	Carbohydrate ingredient	Carbohydrate (% concentration, g/100 ml)	Carbohydrate (g)	Sodium (mg)	Potassium (mg)
Pedialyte® Sport Powder Pack—lemon lime	Anhydrous dextrose, short-chain fructooligosaccharides	2	5	325	300
Gatorade G2	Sugar, sulacrose	2	4.6	107	31
Propel	Sucrose syrup, sucralose	1	2	5	Trace

*Compiled from nutrition facts for Gatorade, Coca Cola, and Tropicana products; some products are in powdered form to be mixed with water. For information about coconut water, visit: https://tools.myfooddata.com/nutrition-facts/170174/wt1.

consideration in a long run, as body-water supplies may be depleted less rapidly. These techniques also cool the skin and offer an immediate sense of psychological relief from the heat stress, which may help to improve performance. However, skin wetting techniques as in athletic competition have not been shown to cause any major reductions in core temperature or cardiovascular responses. Bassett and others reported no change in rectal temperature, heart rate, perceived exertion, sweat loss, or plasma volume change following skin wetting in runners completing 120-minute treadmill runs under two humidity conditions. Decreases in skin temperature and skin blood flow were observed, presumably due to vasoconstriction. Some researchers have theorized that skin wetting techniques may be potentially harmful. The psychological sense of relief may encourage athletes to accelerate their pace, increasing heat production without providing for control of the body temperature. If the core temperature increases, heat illness may occur. Although some scientists suggest that skin wetting is not beneficial, many endurance athletes claim that it helps. There is little research on skin wetting. Bassett's study was published almost 35 years ago. Additional research appears to be warranted.

Hyperhydration Hyperhydration, also known as superhydration, is simply an increase in body fluids by the voluntary ingestion of water or other beverages. It is an attempt to ensure that the body-water level is high before exercising in a hot environment. In a recent review, Périard and others concluded that hyperhydration results in a slight decrease in core temperature with no effect on sweat rate compared to euhydration. The efficacy of hyperhydration may be affected by the hyperhydration solution. Water may upregulate the renin-aldosterone-angiotensin system to excrete excess water as urine whereas an osmotic solution may promote greater fluid retention. Hyperhydration may increase pre-exercise blood volume, leading to a smaller decrease in exercise blood volume (i.e., attenuated hypohydration and less cardiovascular strain) with greater blood flow to active the muscle and skin. The increased pre-exercise hydration volume may also serve as a heat sink. Sawka and others concluded that hyperhydration may delay hypohydration but provides no clear thermoregulatory or performance advantage. However Morris and others reported that sodium-induced hyperhydration decreased exercise dehydration and increased water consumption, retention, and cycle performance compared to placebo or no treatment.

Given hyperhydration's potential benefits, the American College of Sports Medicine recommends that it be used prior to exercise in heat stress environments. Cold water or a CES may be used to hyperhydrate, although the carbohydrates and electrolytes in CES may be helpful in greater retention of fluids, as previously noted by Périard and others. The ACSM guidelines relative to hyperhydration are presented later in this section.

Most research in this area has focused on glycerol supplementation, which may help retain more water with hyperhydration, an effect that has been theorized to improve endurance performance. The proposed ergogenic effects of glycerol-induced hyperhydration are discussed later in this chapter.

Rehydration Of the various techniques used, research has shown that rehydration is the most effective to enhance performance. Rehydration techniques have been used to replenish fluid loss associated with both voluntary and involuntary dehydration in sports such as wrestling and distance running, respectively.

One research approach to evaluate the effects of rehydration involves the sport of wrestling, in which athletes dehydrate to qualify for a lower weight class and then attempt to rehydrate rapidly prior to competition. In this approach, subjects performed some exercise or mental task, such as a measure of strength, power, anaerobic endurance, or cognitive function, were then dehydrated and tested again, and finally were rehydrated and tested one more time to see if rehydration could improve performance back to the predehydration level. The results of such research are mixed. Some studies have shown dehydration may not impair strength, power, or local muscular endurance. Thus, rehydration would not improve performance as measured by these criteria beyond that usually seen in euhydration. For example, McKenna and Gillum reported that anaerobic power output was not impaired in wrestlers following a rapid weight-loss protocol resulting in 3 percent dehydration, or after subsequent glycerol-induced rehydration. Choma and associates found that negative mood following dehydration was reversed following rehydration. In contrast, Timpmann and others reported reduced upper-body intermittent sprint performance and a negative affective state in wrestlers, following a 5 percent rapid weight-loss protocol.

Performance and affect were not improved by sodium citrate rehydration compared to placebo.

In the aftermath of the 1997 deaths of three collegiate wrestlers, the practice of aggressive dehydration to achieve a lower weight class is no longer allowed by the National Collegiate Athletic Association (NCAA). For the past 20 years, the NCAA has mandated a preseason minimum wrestling weight (MWW) class certification based on body composition assessment (5 percent body fat) in a hydrated state (urine specific gravity < 1.02 g/ml). Similar preseason MWW certification exists at the high school level based on 7 percent body fat.

A second approach in studying rehydration is to have subjects ingest fluids during prolonged endurance exercise, particularly in warm environments. Rehydration has been shown to minimize the rise in core temperature, to reduce stress on the cardiovascular system by minimizing the decrease in blood volume, and to help maintain an optimal race pace for a longer period. This beneficial effect is usually attributed to decreased dehydration and the maintenance of a better water balance in the blood and other fluid compartments. Rehydration techniques, both with water alone and with CES solutions, have been shown to improve performance in exercise tasks of 1 hour or more in the heat.

If fluid replacement is to be effective, water has to be absorbed into the circulating blood so that the reduction in blood volume and sweat production that occurs during prolonged endurance exercise will be minimized. Research in which water was labeled with radionuclides showed that water ingested during exercise may appear in plasma and sweat within 10-20 minutes. However, the amount of the ingested fluid that enters the circulation to benefit the athlete depends on two factors: gastric emptying and intestinal absorption.

The ACSM position stand on exercise and fluid replacement stresses individualized rehydration schedules using water or carbohydrate-electrolyte solutions for more prolonged exercise in order to minimize dehydration to less than 2 percent of body mass loss. More specific recommendations from this position stand are presented later in this section.

What factors influence gastric emptying and intestinal absorption?

In a later section, we will discuss factors, such as palatability, that may influence how much fluid is consumed during exercise. For any fluid to be of benefit during exercise, it must first empty from the stomach and then be absorbed into the bloodstream from the intestines.

Gastric Emptying A number of factors may influence the gastric emptying rate, including volume, solute or caloric density, osmolality, drink temperature, exercise intensity, mode of exercise, and dehydration.

Volume is one of the most important factors affecting gastric emptying. Leiper noted that greater rates of gastric emptying occur with ingested volumes in the 600-800 ml range. However, larger volumes consumed during exercise may slow gastric emptying and also cause discomfort to the athlete because of abdominal distention.

CES energy content, osmolality, and temperature also affect gastric emptying. Leiper noted that gastric emptying is slower with carbohydrate solutions exceeding 6 percent and that energy density exerts a greater influence than osmolality on gastric emptying. CES with higher energy content may slow gastric emptying, but could also increase the rate of energy delivery to the duodenum for absorption. Solutions with carbohydrate concentrations above 10 percent may impair gastric emptying, possibly due to increased osmolality. Polymer solutions may empty more rapidly than monosaccharides solutions, but only at higher energy densities. De Oliveira and others recommend CES with multiple transportable monosaccharide (e.g., 8-10 percent glucose/fructose) solutions to minimize a delay in gastric emptying and subsequently maximize absorption. They also comment that the gut may be "trained" to tolerate higher carbohydrate-containing CES and that CES mouth rinsing may be ergogenic for endurance tasks of less than 1 hour. Rowlands and others reported significantly faster fluid absorption (measured by uptake of deuterium oxide tracer) following consumption of a hypotonic sports drink compared to isotonic and hypertonic sports drinks and a noncaloric control drink in cyclists consuming these drinks (2 liters: 250 milliliters each 15 minutes) over a 2-hour period, followed by an incremental test to exhaustion. To summarize, a 6-8 percent carbohydrate solution may provide the athlete optimal intake of water and carbohydrate, and is in accord with both the ACSM and ACSM/AND/DC position stands on fluid replacement during exercise. In general, cold fluids empty rapidly and may help cool the body core. In a meta-analysis, Burdon and colleagues reported increased subject preference for, greater consumption of, and attenuated dehydration following consumption of cold (0-10°C) or cool (10-22°C) beverages compared to warmer beverages.

Gastric emptying is affected by the intensity and mode of exercise as well as timing of rehydration. Horner and others concluded absorbed volume was greater during low-moderate intensity exercise compared to high-intensity exercise. Intense exercise greater than 70 percent VO_2 max and high-intensity interval appears to exert an inhibitory effect. They also reported decreased absorbed volume in cycling compared to running with greatest volume absorbed during walking exercise; greater absorption in hydration before versus during exercise; and negative effects of increasing carbohydrate content, osmolality, and exercise duration on gastric emptying.

Ryan and others reported that hypohydration to approximately 3 percent of body weight does not impair gastric emptying. De Oliveira and others also note that high-intensity exercise and dehydration can decrease mesenteric blood flow, esophageal sphincter tone, and peristalsis, all of which can contribute to decreased gastric emptying and general gastrointestinal distress.

Intestinal Absorption
Intestinal absorption has not been studied as extensively as gastric emptying. As highlighted in **figure 9.10(a)**, glucose and sodium interact in the intestinal wall. Glucose stimulates sodium absorption, and sodium is necessary for glucose absorption. When glucose and sodium are absorbed, these solutes tend to pull fluid with them via an osmotic effect, thus facilitating the absorption of water from the intestine into the circulation. Gisolfi and others reported no effect of varying sodium levels (0-50 mEq/l) on fluid absorption and plasma volume. They concluded that intestinal epithelial cells probably contain sufficient sodium to facilitate carbohydrate transport. As previously noted by Stachenfeld, CES sodium content stimulates the dipsogenic (thirst) drive to promote greater consumption and more complete rehydration.

Leiper notes the intestinal absorption of water and carbohydrates is directly related to gastric emptying volume and is inversely related to osmolality of the ingested solution. Sodium facilitates glucose absorption through a sodium-glucose epithelial co-transport. Glucose polymer solutions (maltodextrin) have a lower osmolality and may be better absorbed than glucose monosaccharide solutions of the same caloric value. Fructose uptake occurs with a different transporter than glucose, is slower, and is less effective in promoting water uptake. In their review, Shi and Passe report that water absorption is influenced by osmolality as well as carbohydrate concentration and type (multiple versus single saccharides and/or polymers). Water is absorbed by passive diffusion through the permeable duodenum (proximal segment), while the jejunum (middle segment) has more carbohydrate and electrolyte (Na^+) transporters which promotes water absorption via an osmotic gradient.

As discussed in chapter 4, excess carbohydrate in the intestine may cause a reverse osmotic effect, as depicted in **figure 9.10(b)**. Highly concentrated sugars in the intestine draw water from the blood, leading to gastrointestinal distress with symptoms such as abdominal cramping and diarrhea. Although de Oliveira and others comment that CES osmolality, high carbohydrate concentration, and acidity can contribute to reverse osmosis and gastrointestinal distress, they also note the possibility of gut "trainability" to tolerate solutions of higher osmolality and carbohydrate content, which could result in greater exogenous carbohydrate absorption and oxidation.

Whether exercise impairs intestinal absorption is controversial. High-intensity exercise may compromise blood flow to the intestine, which might impair absorption. In an early review, Gisolfi cited studies showing that exercise either reduced or had no effect on intestinal absorption and noted some of the methodological difficulties in studying this problem. In a more recent review, Shi and Passe reported greater water absorption during all exercise (range = 30-78 percent of VO_2 max) trials compared to resting trials, but they acknowledged there was no examination of an exercise intensity dose-response regarding water absorption.

It should be noted that individual differences in both gastric emptying and intestinal absorption may be significant. In reviewing studies of gastric emptying, Costill noted some subjects could empty 80-90 percent of the ingested solution in 15-20 minutes, whereas others emptied only 10 percent. As noted previously, some subjects may also develop diarrhea caused by ineffective intestinal absorption of fluids. Training to drink during exercise is recommended as a possible means of enhancing tolerance to consuming larger amounts of fluids. As suggested by Jeukendrup, endurance athletes should ingest fluids during training in order to "train" the stomach to handle fluids during competition.

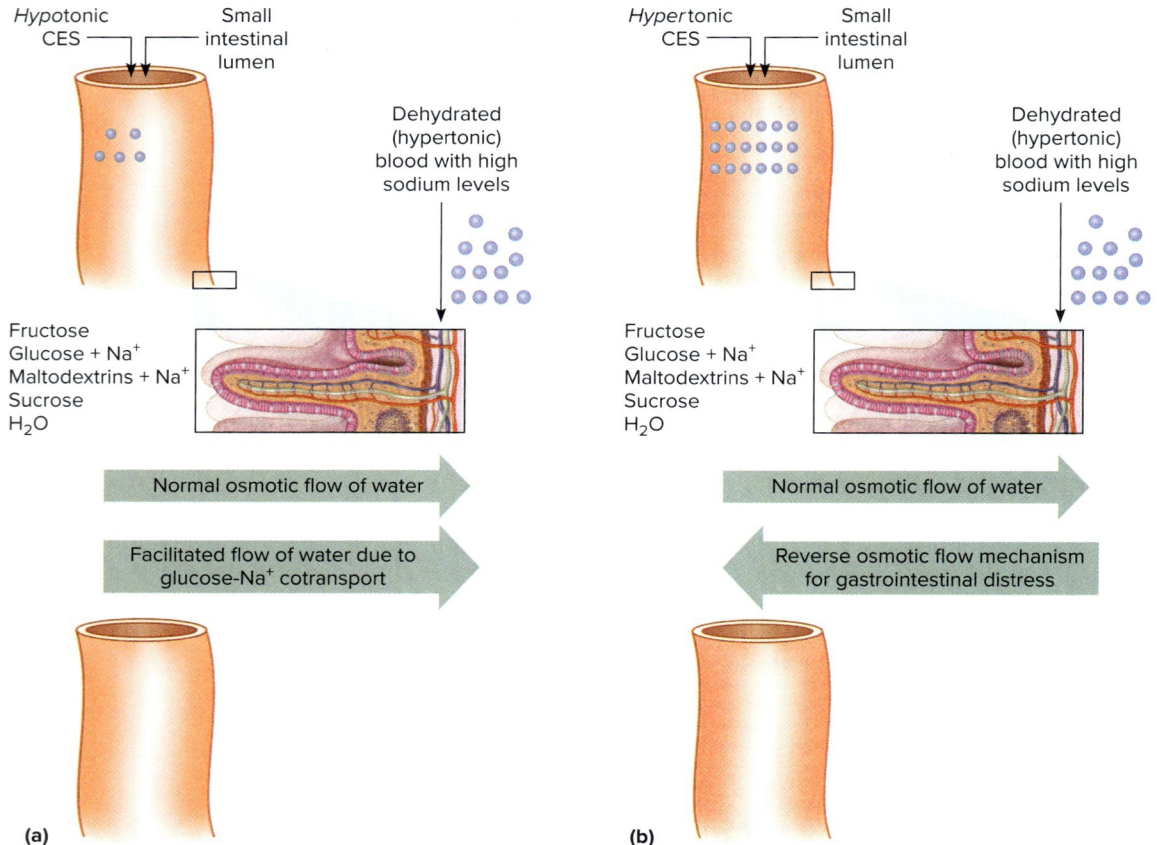

FIGURE 9.10 (a) Water normally diffuses from the intestine to the circulation via osmosis. Glucose and sodium in a CES enhance osmosis, as shown by the larger arrow. Rehydration + exogenous CHO absorption = adequate thermoregulation + sustained performance. (b) A hypertonic solution may actually reverse osmosis, moving fluid from the circulatory system to the intestines, possibly leading to gastrointestinal distress symptoms such as diarrhea. Reverse osmosis impairing rehydration + less exogenous CHO absorption = suboptimal thermoregulation + impaired performance.

How should carbohydrate be replaced during exercise in the heat?

The value of carbohydrate intake during exercise as a means to improve performance was detailed in chapter 4, primarily in relation to performance in a cool environment. Carbohydrate is the primary fuel for high-intensity endurance exercise for a duration of an hour or longer. Logan-Sprenger and others observed accelerated muscle glycogenolysis and total body carbohydrate oxidation at dehydration levels less than 1 percent. Glycogen depletion contributes to fatigue. Therefore, carbohydrate intake may be useful to improve performance during such exercise tasks, but if temperature regulation is of prime importance, water replacement should receive top priority. Hence, one of the goals of researchers has been to develop a fluid that will help replace carbohydrate during exercise in the heat without affecting water absorption. As discussed previously, carbohydrate-electrolyte solutions have been developed for this purpose.

An optimal amount of carbohydrate in solution may maintain body temperature as effectively as water and may enhance performance during prolonged exercise. Water and carbohydrate complement each other to improve physical performance. Fritzsche and others compared the effects of water, 6 percent carbohydrate solution (water + carbohydrates), carbohydrates only, and placebo on power output in endurance-trained cyclists exercising for 2 hours in a hot environment. Water + carbohydrate was more effective in attenuating the decline in power output compared to water only, carbohydrate only, and placebo treatments, with no difference between carbohydrate only and placebo. Similar findings have been reported by Below and others in cycling performance and Millard-Stafford in running performance of approximately 1 hour duration.

Scores of studies have compared the effectiveness of different carbohydrate combinations and concentrations in enhancing physical performance during prolonged endurance tasks. Most of this research is discussed in chapter 4. The following are the pertinent general findings relative to CES intake during prolonged exercise under warm environmental conditions.

According to Leiper, gastric emptying is similar with water and carbohydrate solutions of up to 2.5 percent and generally decreases in carbohydrate solutions exceeding 6 percent. Such solutions have no adverse effect on temperature regulation or sweat rate compared to water and help maintain plasma volume, plasma glucose, and liver glycogen.

As previously noted, a CES with multiple transportable monosaccharides and glucose polymers generally results in greater carbohydrate absorption due to reliance on multiple intestinal epithelial cell transport proteins. In their review of 14 studies, Rowlands and others reported that ingesting a beverage containing a fructose:glucose ratio of 0.5–1.0:1.0 with maltodextrin glucose polymer at 1.3–2.4 g/minute is more effective in improving endurance performance in a variety of tasks compared to an isocaloric glucose/maltodextrin beverage. Fructose is absorbed in the intestine by facilitated GLUT-2 and GLUT-5 transport instead of by SGLT-1 co-transport.

Although higher concentrations of carbohydrates deliver more glucose to the intestine, solutions higher than 10–12 percent may significantly delay gastric emptying, decrease intestinal absorption, and cause gastrointestinal distress, as noted previously. High concentrations of fructose in some fruit juices or juice blends may be particularly debilitating. However, higher carbohydrate concentrations may "train" the gut by increasing the density and activity of sodium-dependent glucose-1 intestinal co-transport (SGLT-1), as noted by Jeukendrup and de Oliveira and others. In ultra-endurance cycling, where competitors may ride 20 hours or more a day, such high carbohydrate concentrations may be necessary to meet the high energy demands.

In summary, there are no disadvantages to the consumption of beverages containing recommended amounts of carbohydrate and electrolytes. The ACSM position stand on fluid replacement during exercise includes some guidelines on the composition of fluids to be consumed, including carbohydrate concentration.

Table 9.6 calculates the amount of fluid you must consume, for a given concentration, to obtain 30–100 grams of carbohydrate.

Additional strategies for carbohydrate intake before and after exercise are presented in chapter 4.

Training Table

The following calculations solve for the amount of fluid of a known concentration (g carbohydrate/100 ml) of a carbohydrate–electrolyte solution required to deliver total exogenous carbohydrate (g) given the desired rate of carbohydrate consumption (g/kg/hour).

Problem	How much of a 6 percent carbohydrate–electrolyte solution (CES) would a 132 lb. (60 kg) female need to ingest over a 2-hour time trial in order to deliver exogenous carbohydrate at a rate of 1.25 g/kg/hour? How much should she consume every 15 minutes?
Solution	Total exogenous carbohydrate = 60 kg × 1.25 g/kg × 2 hr = 150 g
	Total volume of 6% CES to consume over 2-hr = Total carbohydrate mass ÷ carbohydrate concentration = 150 g ÷ 6 g/100 ml = 150 ÷ 0.06 = 2,500 ml (42.3 oz.)
	15 min hydration rate = 2,500 ml ÷ 120 min = 20.8 ml/min × 15 min = 312.5 ml (10.7 oz) every 15 min

How should electrolytes be replaced during or following exercise?

Because the major solid component of sweat consists of electrolytes, considerable research has been conducted relative to the need for replacement of these lost nutrients, primarily sodium and

TABLE 9.6 Fluid consumption (milliliters) at a given percent carbohydrate concentration to obtain desired grams of carbohydrate

Percent concentration (%)	Total grams of carbohydrate in solution (g/100 ml)							
	30	40	50	60	70	80	90	100
2	1,500	2,000	2,500	3,000	3,500	4,000	4,500	5,000
4	750	1,000	1,250	1,500	1,750	2,000	2,250	2,500
6	500	667	833	1,000	1,167	1,333	1,500	1,667
8	375	500	625	750	875	1,000	1,125	1,250
10	300	400	500	600	700	800	900	1,000
12	250	333	417	500	583	667	750	833
14	214	286	357	429	500	571	643	714
16	188	250	313	375	438	500	563	625
18	167	222	278	333	389	444	500	556
20	150	200	250	300	350	400	450	500

Values are calculated as total grams of carbohydrate ÷ percent solution. For example, total fluid to deliver 60 grams as a 6% (6 g/100 ml) solution = 60 ÷ 0.06 = 1,000 ml.

potassium. We shall look at this question from two points of view, one dealing with the need for replacement during exercise and the other involving daily replacement.

During Exercise Because sweat is hypotonic to the body fluids, the concentration of electrolytes in the blood and other body fluids actually increases during exercise and makes the body fluids hypertonic. During moderately prolonged exercise, water rehydration maintains electrolyte balance. Electrolyte supplementation is unnecessary. Sanders and others examined fluid shifts in three 90-minute cycle exercise bouts at 60 percent VO_2 max while consuming no fluid, water (1.2 l), and 100 mmol saline (40 mmol sodium) solutions (1.2 l). Plasma volume was similarly maintained by water and saline, leading the authors to conclude that sodium replacement provides little benefit to athletes rehydrating approximately 50 percent of lost volume during exercise. Several studies have reported that even during strenuous prolonged exercise with high levels of sweat losses, like marathon running for several hours, water alone is the recommended fluid replacement to help maintain electrolyte balance, although added carbohydrate may provide some needed energy.

However, electrolyte replacement, particularly sodium, may be necessary for some athletes participating in very prolonged bouts of physical activity, such as marathons, ultramarathons, Ironman-type triathlons, or tennis tournaments where one might play off and on all day. A number of medical case studies following such events have reported complications resulting from an electrolyte imbalance in the blood, which is the topic of a subsequent question.

Daily Replacement In general, heavy daily sweat losses do not lead to an electrolyte deficiency. If body levels of sodium and potassium begin to decrease, the kidneys begin to reabsorb more of these minerals and less are excreted in the urine. Evans and others conclude that water, combined with a balanced diet, will adequately maintain proper body electrolyte levels from day to day, even when an individual is exercising and is losing large amounts of sweat.

However, if electrolytes are not adequately replaced because of poor dietary intake, a deficit may occur over 4–7 days of very hard training, especially in hot environmental conditions where fluid losses will tend to be high. Evans and others comment that added sodium in a rehydration beverage and fluids increases extracellular fluid osmolality and volume. Adding salt to meals may also help promote rehydration. The sodium is needed in the body to help retain water and maintain normal osmotic pressures. Stachenfeld states that plain water preferentially hydrates plasma volume over interstitial and intracellular volumes and that sodium ingestion during and after exercise maintains the dipsogenic (thirst) drive for more complete rehydration.

A good method of checking on the adequacy of fluid replenishment on a day-to-day basis is to measure body weight in the morning. It should be nearly the same every day. If you weigh several pounds less from one day to the next, it is likely that you are hypohydrated. Conversely, if you weigh several pounds more, you may be overhydrated.

What is hyponatremia and what causes it during exercise?

Hyponatremia is a condition of subnormal levels of sodium in the blood. Also known as *water intoxication,* it can occur at rest simply by consuming too much water. Hyponatremia can also occur following prolonged exercise, in which case it may be known as *exercise-associated hyponatremia (EAH).* The statement by Hew-Butler and others following the Third International Exercise-Associated Hyponatremia Consensus Development Conference in 2015 defined EAH as a serum sodium concentration below the normal reference range or less than 135 mmol/liter (135 mEq/liter).

Training Table

Symptoms of hyponatremia

Mild cases (serum [sodium] <130 mmol/liter):

- Bloating
- Puffiness of hands and feet
- Nausea
- Vomiting
- Headache

Severe cases (serum [sodium] <120 mmol/l): possible massive brain swelling, which may be associated with:

- Seizures
- Coma
- Respiratory arrest
- Permanent brain damage
- Death

Knechtle and others report EAH prevalence rates of 8 percent for marathon (26.2 miles, 42.2 kilometers) races, ≤3 percent for ultramarathon races of distances up to 100 kilometers, and >20 percent for 100-mile races. They noted higher EAH prevalence rates in running sports, among females, and in the United States compared to Europe due to warmer ambient conditions. According to Rogers and Hew-Butler, EAH has also been documented in hikers, climbers, trekkers, and cold-climate endurance athletes. Treatment of individuals with symptomatic hyponatremia is a medical emergency, and transportation to a hospital is essential. Infusion of hypertonic solutions may be necessary.

Various risk factors have been identified that predispose individuals to development of EAH in marathons and other endurance events, including the following:

- Excessive drinking of fluids before, during, and after the event
- Considerable weight gain over the course of the event
- Slower finishers
- Females
- Low body weight
- Heat-unacclimatized, poorly trained competitors
- High sweat sodium losses

- Novice participant
- Nonsteroidal anti-inflammatory drug (NSAID) use, altered kidney functions to excrete fluids

Hyponatremia may be caused by water dilution, excess sodium losses, or both. The EAH consensus conference committee statement by Hew-Butler and others indicated that dilutional hyponatremia, caused by an increase in total body water relative to the amount of total body sodium, is the current etiology of EAH. The ACSM, in its position stand on fluid replacement, indicates that fluid consumption exceeding sweating rate is the primary factor leading to exercise-associated hyponatraemia. Additionally, factors that normally control body water balance, mainly hormones such as arginine vasopressin and the kidney, may malfunction. Hew-Butler and others comment that use of nonsteroidal anti-inflammatory drugs (NSAIDs) may enhance the effects of arginine vasopressin in absorbing water in the nephron, but that more research is need to establish NSAID use as a potential risk factor for EAH. Seal and Kavouras note that overconsumption of hypotonic fluids leading to weight gain, abnormal antidiuretic hormone secretion, longer exercise time (e.g., slower competitors), and smaller body mass (e.g., females) may contribute to exercise-associated hyponatremia.

Many of these factors may interact to contribute to the development of EAH. For example, females may be of lower body weight; generally run slower marathon times than males; and may be more conscientious about consuming fluids, given the old adage to "drink as much as you can." As a result, they have more time to drink more fluids and gain weight during the competition. The weight gain is water, which dilutes the serum sodium concentration. The 2007 ACSM position stand on exercise and fluid replacement recommends managing fluid loss during prolonged exercise (>60 minutes) through an individually developed schedule of rehydration with fluids, electrolytes, and carbohydrates to prevent excessive dehydration (>2 percent body weight loss), which can contribute to decreased performance and increased risk of heat-related illness.

Sports drinks are constituted to be palatable so that athletes will drink more. Peacock and others reported a 46 percent greater *ad libitum* fluid ingestion with CES compared to water. Sports drinks do contain some sodium but are hypotonic solutions. According to Sports Dietitians Australia, the sodium content in CES products ranges from 10 to 25 mEq (230-575 milligrams) per liter. Twerenbold and others reported that consuming a 30 mEq (680 mg/l) sodium solution minimized hyponatremia in female endurance athletes running for 4 hours while two athletes in the water control trial developed hyponatremia (<130 mmol serum sodium). However, Hoffman and others found no difference in sodium intakes between hyponatremic and nonhyponatremic participants in an ultramarathon (161 km) run in hyperthermic (up to 39°C) conditions. Anastasiou and others compared consumption of water, mineral water, and sports drinks containing 19.9 and 36.2 mmol (460 and 833 mg) of sodium consumed at a rate equal to body mass change and reported that a sodium solution of 19.9 mmol/liter was effective in preventing possible hyponatremia when fluids are consumed at the rate of weight loss during exercise in the heat. To summarize, excessive fluid consumption, even of sodium-containing CES, may still lead to hyponatremia, as noted by Hew-Butler and others in the Third EAH Consensus Statement on EAH. However, rehydration with an appropriate volume of a sports drink containing electrolytes may decrease the severity of EAH.

Treatment for hyponatremia varies. Buffington and Abreo recommend that for mild symptomatic hyponatremia, drinking of hypotonic fluids should be restricted until the athlete is urinating. Hypertonic solutions may be provided if the athlete can drink fluids. For severe hyponatremia, intravenous hypertonic (e.g., 3 percent) sodium chloride solutions will speed recovery and improve outcomes. As discussed by Sterns, **osmotic demyelination,** destruction of the myelin sheath of brain cells in the brainstem, can occur if low sodium levels are increased too quickly. Athletes who do not recover rapidly should be sent to the nearest medical emergency facility.

Individual differences may dictate who may be prone to developing hyponatremia during prolonged exercise, but given the current evidence, it appears that athletes involved in ultraendurance events should consume adequate salt in their diet the days before competition to help assure normal serum sodium levels and consume fluids with added sodium during the event. More research is needed to help refine current recommendations. In the meantime, experiment with salty solutions during practice. You can carry some fluids with you in competition, and in others you may have personal beverages located at specific aid stations on the course.

www.sportsdietitians.com.au/wp-content/uploads/2015/04/Sports-Drinks.pdf Sports drink fact sheet from Sports Dietitians Australia

Are salt tablets or potassium supplements necessary?

In general, the use of salt tablets to replace lost electrolytes, primarily sodium, is not necessary. As previously noted, a nutrient-dense diet as recommended in the *2020-2025 Dietary Guidelines for Americans* will replace daily electrolyte lost from sweating.

The concentrations of salt in sweat may vary. As previously noted by Lara and others, individuals who have a high amount of salt in their sweat are sometimes referred to as "salty sweaters." Maughan and Shirreffs suggest that wearing a black t-shirt during exercise and looking for salt stains when the sweat has evaporated may provide a rough self-assessment of sodium loss. The average sweat salt concentration may be about 3.2 grams of salt (1.3 grams sodium) per liter, although there are reports as high as 4.5 grams (1.8 grams sodium) per liter in unacclimatized individuals, and as low as 1.75 grams (0.7 grams sodium) per liter in the heat-acclimatized individual. If an athlete lost about 8-9 pounds of body fluids during an exercise period, a total of 4 liters of fluid (about 4 quarts) would be lost because a liter weighs 2.2 pounds. Four liters of sweat would contain, at the most, 7.2 grams of sodium in the unacclimatized individual, but less than 3 grams in one who was acclimatized. Because the average meal may contain 2-3 grams of sodium if well salted, three meals a day could offer 6-9 grams, about enough to just cover the losses in the sweat. However, sodium is lost through other means, primarily in the urine; thus, a slight increase in sodium intake may be reasonable for the unacclimatized athlete. In the recent Sports Dietitians Australia position stand, McCubbin and others comment that during heat acclimatization, 20–40 mg sodium/kg body weight with 10 ml fluid/kg body weight ingested 1–2 hours before exercise is effective in fluid retention. For a 70 kilogram individual, sodium ingestion of 40 mg/kg (2.8 grams) would exceed the Chronic Disease Risk Reduction intake of 2.3 grams recommended by the National Academy of Sciences. However, that recommendation is based on the sedentary individual, not an athlete losing copious amounts of sodium during a period of acclimatization. Stachenfeld notes that acute high sodium intake following exercise may cause a transient increase in blood pressure, but not sustained hypertension in healthy individuals. However, once an athlete is acclimatized to the heat, sodium intake may be reduced to normal.

Common salt tablets contain only sodium and chloride. They are not necessary to replace lost sodium but may be recommended for unacclimatized athletes who do not replace sodium through normal dietary means in the early stages of an acclimatization program. Salt tablets should be taken only if the athlete loses substantial amounts of weight via sweat losses during a workout. Checking the body weight before and after a workout provides a good estimate of sweat loss. If we switch to the English system, 1 quart of sweat equals 2 pounds; 1/2 quart, or a pint, is 1 pound. One recommendation is that salt tablets should be taken only if the athlete needs to drink more than 4 quarts of fluid per day to replace that lost during sweating: that is, an 8-pound weight loss. The general rule is to take two salt tablets with each additional quart of fluid beyond the 4 quarts. This would be equal to 1 gram of sodium (the average tablet has 1/2 gram of sodium) per quart. Another way to look at it is to take 1 pint of water with every salt tablet. The use of salt tablets should be discontinued after the athlete is acclimatized, usually about 10–14 days.

Potassium supplements are not recommended for several reasons. First, Costill and others reported that a deficiency of potassium is rare, even with large sweat losses and a diet low in potassium. Second, as noted previously, excessive potassium may result in a fatal disturbance in the electrical rhythm of the heart. The moderate use of substitutes, such as potassium chloride for common table salt, may be helpful in assuring potassium replacement, but investigators recommended particular attention to the diet, citing citrus fruits and bananas as two of the many foods high in potassium. For example, a large glass of orange juice will replace the potassium lost in 2 liters of sweat.

What are some prudent guidelines relative to fluid replacement while exercising under warm or hot environmental conditions?

In sports nutrition, no other area has received as much research attention as the objective of determining the optimal formulation of a carbohydrate electrolyte solution for individuals doing prolonged exercise under warm or hot environmental conditions. This may be because water and carbohydrates are two nutrients that may enhance performance in such events, and water and electrolytes may also help to prevent heat-related illnesses. As discussed previously in relation to the need for fluid, carbohydrate, or electrolytes, a number of factors—in particular, the intensity and duration of the exercise task, the prevailing environmental conditions, and individual differences in sweat rate, gastric emptying, and intestinal absorption—may influence the desired composition of the sports drink. Given these considerations, many of the leading investigators in exercise-hydration research indicate that there is no agreement on the optimal formulation of an oral rehydration solution that would suit the needs of all individuals who engage in a variety of prolonged exercise tasks. Indeed, as noted previously, the ACSM identified the individual's unique biological characteristics as a factor affecting hydration status and exercise performance. As noted by Baker and Jeukendrup, the source of fluid loss (e.g., sweat, diarrhea, vomiting, other) is a factor in determining the optimal composition of a fluid-replacement solution.

Through their concerted research efforts over the years, the many sports scientists previously cited in this chapter have provided a sound basis to promote prudent recommendations regarding fluid replacement before, during, and after exercise. The latest guidelines on fluid replacement for exercise have been published by the ACSM, which serve as the basis for these prudent recommendations.

> https://www.acsm.org/education-resources/pronouncements-scientific-communications/position-stands Access the American College of Sports Medicine 2007 position stand on exercise and fluid replacement, which is located in the list of ACSM Position Stands and Joint Position Statements.

Before Competition and Practice The goal of the ACSM guidelines is to start in a state of euhydration with normal plasma electrolyte levels. Unfortunately, not all athletes come to practice adequately hydrated. Vukasinović-Vesić and others observed pregame dehydration based on urine osmolality in 75 percent of young (19 ± 1 year) elite European male basketball players with progressively greater dehydration (loss of 0.9 ± 0.7 kg body weight) during the game. Volpe and others reported dehydration in 66 percent of National Collegiate Athletic Association Division I athletes prior to practice, with a greater percentage of dehydrated men compared to women. Here are some key points:

1. Athletes should be adequately hydrated the day before competition. Minimize consumption of alcoholic beverages the night before competition, for they may lead to hypohydration in the morning.
2. Drink slowly about 5–7 ml/kg (0.08–0.11 ounce/pound) body weight at least 4 hours prior to exercise.
   - Fluid palatability (temperature, sodium, flavoring) will enhance fluid intake. Peacock and others reported that CES ingestion significantly increased pleasure ratings, ingested volume, and plasma glucose concentrations compared to water.
   - If the exercise task is to be prolonged, carbohydrate may be added. A concentration of 6–8 percent is advisable, but concentrations of 20 percent and higher have been used by some individuals without adverse effects.
   - Beverages with sodium (20–50 mEq/l) and/or salty foods or snacks will help stimulate thirst and retain fluids.

3. If no urine is produced, or urine is dark or highly concentrated, drink another 3–5 ml/kg body weight about 2 hours prior to exercise. Your urine should be clear, pale yellow before competition or practice.
4. Do not excessively overhydrate, which may increase the risk of dilutional hyponatremia if fluids are aggressively replaced during and after exercise.

Training Table

The Academy of Nutrition and Dietetics, Dietitians of Canada, and the American College of Sports Medicine recommendations pertaining to fluid intake and carbohydrate intake, some of which is in the form of CES:

- Before exercise: Fluid ingestion to optimize hydration and glycogen status
- During exercise: Fluid ingestion to minimize dehydration/hypohydration, which impairs performance (≥2 percent loss of body weight). Replace sweat loss and attenuate plasma volume loss for optimal thermoregulation
- During exercise: Avoid overly aggressive hydration to prevent hyponatremia
- During exercise: Carbohydrate ingestion (≥30–60 g/hour, some in the form of CES) during exercise > 60 minutes in duration
- During exercise: Mouth rinsing with CES may also improve performance
- Following exercise: Rehydration of ~1.25–1.5 liters of fluid for every 1 kilogram of body weight lost. Rapid rehydration for repetitive activity within 24 hours; otherwise more leisurely pace of rehydration. CHO intake of 1.0–1.2 g/kg/hour for 4–6 hours to replace muscle/liver glycogen

Source: Position of the Academy of Nutrition and Dietetics, Dietitians of Canada, and the American College of Sports Medicine: Nutrition and athletic performance. 2016. *Journal of the Academy of Nutrition and Dietetics* 116:501–28.

---

www.hydrationcheck.com/pocket_chart.php   Provides a Dehydration Urine Color Chart to help estimate your level of dehydration.

During Competition and Practice According to Garth and Burke, there is little research on self-selected hydration by elite and nonelite athletes during practice and competition. Varied practices exist both across and within competitive events. Many factors, some beyond the athlete's control, such as official fluid stations in running events, cast doubt on the existence of true *ad libitum* fluid intake during competition. The goal of the ACSM guidelines is to prevent excessive dehydration (>2 percent body weight loss from water deficit). The amount and rate of fluid replacement depend on individual sweating rate, exercise duration, and opportunities to rehydrate. Here are some key considerations:

1. Individuals should monitor body weight changes during training/competition sessions to estimate fluid losses during a particular exercise task.
2. Although some believe that thirst is the best stimulus for fluid ingestion in situations where fluid is readily available, the ACSM believes it is important to start rehydrating early in endurance events because thirst does not develop until about 1–2 percent of body weight has been lost. The ACSM guidelines recommend a possible starting point for marathon runners is to drink *ad libitum* about 0.4 (smaller runner) to 0.8 (larger runner) liter of fluids per hour. Consuming 0.4 liter (about 14 ounces) per hour could be accomplished by drinking 3–4 ounces every 15 minutes, while drinking about 7 ounces every 15 minutes would provide 0.8 l/hour. These amounts may be adjusted to individual preferences and to increase carbohydrate content, as discussed later.

   Although Church and others reported that deuterium oxide (D_2O) can appear in sweat in as little as 3 minutes after ingestion, it is very difficult to consume enough fluids to replace those lost. Costill found that 50 or more milliliters per minute of fluid may be lost though sweating (3 l/hour), but only 20–30 milliliters per minute may be absorbed from the intestines. Although some dehydration will occur, rehydration will help maintain circulatory stability and heat balance, thereby delaying deterioration of endurance capacity. Cheuvront and Haymes reported that replacing 60–70 percent of sweat losses helps to maintain thermoregulatory responses during hot and warm weather conditions. Holland and others concluded that fluid consumption rates of 0.15–0.34 ml/kg body mass/minute reduced high-intensity 1-hour cycling performances. However, moderate-intensity performances of 1–2 hours and 2+ hours were improved by 2 and 3+ percent, respectively, with fluid consumption rates of 0.15–0.27 ml/kg body mass/minute. By calculating your typical sweat losses, as discussed in the *Training Table* earlier in this chapter, you may be able to estimate how much fluid you should consume per hour.
3. Cold water is effective when carbohydrate intake is of little or no concern, for example, in endurance events lasting less than 50–60 minutes. Sports drinks with 6–8 percent carbohydrates and normal electrolyte content may also be consumed but, in general, provide no advantages over water alone.
4. The composition of the fluid is considered important for prolonged endurance events. Carbohydrates and electrolytes may provide some advantages, and these components may be in the drink or nonfluid sources such as gels or energy bars. Palatability and the presence of other ingredients may also be important considerations.
   - Carbohydrate provides energy for longer-duration events. If carbohydrate is desired in the drink, the concentration should not be excessive. A 6–10 percent concentration is recommended. Concentrations greater than 10–12 percent may retard gastric emptying and contribute to gastrointestinal distress. Use a sports drink containing multiple sources

of carbohydrate, including glucose, sucrose, fructose, and maltodextrins. Such a mixture may enhance absorption and utilization of the exogenous carbohydrate, possibly up to 1.2-1.7 g/minute. Check the Nutrition Facts label for ingredients.

The ACSM recommends that athletes consume enough fluid to provide about 30-80 grams of carbohydrate per hour. On average, sports drinks containing 6-8 percent carbohydrate provide about 2 grams of carbohydrate for every ounce of fluid consumed. Thus, you need to drink about 15-16 ounces of a sports drink per hour to obtain about 30 grams of carbohydrate, or about 1 liter per hour to obtain 60 grams. See **table 9.6** for guidelines.

It may be difficult for some athletes, such as marathon runners, to consume the amount of fluid during exercise to obtain the recommended grams of carbohydrate. Many runners do not consume a liter per hour, which is needed to provide 60 grams from a 6 percent sports drink. However, consuming other sources of carbohydrate with the sports drink, such as sports gels or sports beans, can provide the additional grams of carbohydrate.

Athletes in prolonged, intermittent, high-intensity sports, such as soccer, may use various rehydration procedures. Clarke and others compared the effects of isovolumic and isocaloric total amounts of a CES provided to soccer players at 0 and 45 minutes compared to 1, 15, 30, 45, 60, and 75 minutes of a 90-minute soccer specific protocol. There were no differences in metabolic responses during the soccer protocol, and sprint power was not different, suggesting that the two rehydration scheduled were equally effective.

- The fluid should contain small amounts of electrolytes, particularly sodium and potassium, to help replace lost electrolytes. The ACSM recommends about 20-30 mEq of sodium and 2-5 mEq of potassium, which are amounts present in many commercial sports drinks. However, for athletes involved in very prolonged endurance events under warm environmental conditions, some recommend a range of about 700-1,150 milligrams of sodium per liter (approximately 30-50 mmol per liter) and 120-225 milligrams of potassium per liter (approximately 3-6 mmol per liter). Some commercial sports drinks contain electrolytes comparable to this range. For example, Gatorade Lemon-Lime Endurance Formula contains 620 milligrams of sodium and 280 milligrams of potassium per 12-ounce (360 ml) container, which is more than 1,724 milligrams of sodium (78 mmol) and 750 milligrams of potassium (21 mmol) per liter.
- The fluid should be palatable and not interfere with normal gastrointestinal functions. As previously noted by Peacock, voluntary intake of fluids increases when they are tasty. Being cold and sweet enhances palatability. Burdon and others noted 50 percent greater consumption by volume of cool/cold versus warm beverages. Leiper commented that beverage carbonation has little effect on beverage consumption volume and no effect on gastric emptying at rest or during exercise. Rowlands and others note that the neurophysiological effects of beverage sweetness are complex, but the addition of artificial sweeteners does not appear to affect carbohydrate absorption and performance. However, Zachwieja and others reported that certain flavorings, such as citrate, could cause gastrointestinal distress and impair gastric emptying.
- Caffeine is found in some sports drinks and may help sustain performance. Detailed information on caffeine and exercise performance is presented in chapter 13. In brief, however, caffeine supplementation appears to be an effective ergogenic aid for aerobic endurance performance, and its use is currently not prohibited by the World Anti-Doping Agency (WADA). Moreover, caffeinated sports drinks maintain hydration and metabolic and thermoregulatory functions as well as standard sports drinks.
- Some sports drinks include protein, but they do not appear to enhance performance more than typical CESs. Stearns and others concluded that time to exhaustion was greater for CES + protein trials compared to isocarbohydrate CES without protein trials. The ergogenic effect of protein may be simply due to added fuel (kcal), a conclusion underscored by their finding of no difference in time to exhaustion between isocaloric trials with and without protein. As noted in chapter 6, other studies have not reported performance enhancement associated with protein-containing CES when compared to typical CES. Review chapter 6 for details on protein.

After Competition and Practice The goal of the ACSM guidelines is to fully replace any fluid and electrolyte deficit. Replacement may have to be rapid, such as in preparation for a subsequent exercise endeavor on the same day. For example, tennis players may compete in two or more events daily with a short recovery period, while some athletes may also train twice daily. In other situations, fluid replacement may be more leisurely if time permits.

1. If time is short to the next exercise session, aggressive rehydration is important.
   - Drink 1.5 liters of fluid for every kilogram of body weight loss, or about 1.5 pints for each pound loss. The additional fluid is needed to compensate for increased urine output.
   - Consume adequate carbohydrates and electrolytes as well. Fruit juices and sports drinks are helpful when you need to replenish both fluids and carbohydrates. Pretzels and other salty snacks may provide sodium, as well as carbohydrate. This may be especially important for competition. Osterberg and others found that the inclusion of carbohydrate (3-12 percent) to an electrolyte beverage enhanced the retention of fluid following exercise-induced dehydration. In a study comparing ingestion of 220 and 55 grams of ^{13}C-labeled glucose over the course of 4 hours between two bouts of treadmill exercise, Bilson and others reported increased glycogen synthase activity between exercise bouts, greater total carbohydrate oxidation, greater $^{13}CO_2$ production, and lower fat oxidation during

bout two, following 220 grams of carbohydrate supplementation. Exercise duration was similar between 220- and 55-gram treatments and was limited to increased core temperature and not substrate availability.
- Some preliminary research suggests that consuming a sports drink with protein may help. Li and others compared the effects of high- and low-carbohydrate CES and CES with three levels of whey protein on rehydration during 4 hours following exercise (60 minutes at 65 percent VO_{2max}), eliciting 2.15 ± 0.05 percent body weight loss. Total hydration volume for all five solutions was 150 percent body weight loss. CES combined with high (33 g/l) and medium (22 g/l) whey protein significantly decreased urine production and increased fluid retention compared to the other solutions.

2. If recovery time permits (24 hours), normal meal and water intake will restore euhydration, provided sodium intake is adequate. Stachenfeld notes that sodium is necessary in the recovery period to restore fluid balance.
   - A diet rich in wholesome, natural foods adhering to healthy eating practices will help replenish needed electrolytes.
   - Sodium replacement is important, particularly if your sweat contains significant amounts. Check your skin and clothing for traces of dried salt after exercise; if white streaks occur, you may be losing substantial amounts of sodium during exercise.
   - Extra salt may be added to meals when sodium losses are high.
   - Drink fluids with added sodium or consume salty foods or snacks.

3. Minimize alcohol intake. In the ACSM, Academy of Nutrition and Dietetics, and Dietitians of Canada joint position statement on nutrition and performance, athletes are encouraged to minimize or avoid alcohol use when rapid recovery and tissue repair are critical.

In Training

1. Maughan and Meyer stress the importance of proper hydration during training in order to maintain high training loads. Jeukendrup describes "training" the gut to accommodate more fluid during competition by practicing fluid consumption during training. Some research indicates that consuming CES during training, particularly high-intensity training, may result in a more effective workout, which could lead to better performance in competition. By experimenting with various formulations while you train, particularly during training comparable to the intensity and duration experienced in competition, you will be able to determine what fluids work specifically for you. Consuming fluids during training may help you overcome some of the factors that inhibit fluid intake during exercise, such as the uncomfortable sensation of fluid in the stomach, and lead to increased fluid intake during competition. This is especially important for older athletes, who tend to drink less than their younger counterparts.

2. If you sip sports drinks throughout the day to stay hydrated, be sure to practice sound dental hygiene. As noted in chapter 4, some research has shown that sports drinks may exert significant eroding effects on dental enamel.

A summary of guidelines for fluid, carbohydrate, and electrolyte replacement during exercise under warm environmental conditions is presented in **table 9.7**. In brief, the ACSM guidelines present three key points to prevent hypohydration during exercise. Hydrate well before the exercise task; drink according to your personal needs; and rehydrate rapidly in preparation for subsequent exercise bouts on the same day, or more leisurely if recovery time permits. The reviews and studies presented in this section support the efficacy of commercial carbohydrate-electrolyte beverages in improving endurance performance compared to a placebo treatment. They can also effectively supplement water and dietary electrolyte intake for postexercise rehydration.

Key Concepts

- Hyperhydration before exercise is important, but rehydration during and following exercise is the most important nutritional consideration when exercising in the heat.
- Rehydration with cold water is effective in moderating body temperature during exercise in the heat, but carbohydrate solutions may be equally effective and provide a source of energy for more prolonged endurance exercise.
- An effective rehydration solution is one that optimizes gastric emptying and intestinal absorption of fluid.
- Electrolyte replacement generally is not needed during exercise but may be helpful during very prolonged exercise tasks. Water alone, in combination with a balanced diet, including adequate sodium and potassium, will adequately restore normal electrolyte levels in the body on a day-to-day basis.
- Current research suggests that 6–10 percent solutions of glucose, fructose, sucrose, glucose polymers, or combinations of these different carbohydrates, may be effective for athletes who need carbohydrate replacement during exercise.
- Research has shown that combinations of carbohydrates (e.g., glucose, fructose, small glucose polymers) in CES may increase the oxidation rate of ingested carbohydrates.
- Excessive fluid consumption during very prolonged exercise, coupled with inadequate salt intake, may contribute to hyponatremia, a potentially dangerous condition.
- One ounce of a typical 6–8 percent carbohydrate-electrolyte solution (CES) contains about 2 grams of carbohydrate. Drinking 20 ounces of a CES per hour during exercise will provide about 40 grams of carbohydrate.
- Individuals who desire to rehydrate rapidly following exercise should consume about 120–150 percent of the fluid lost in order to account for continued urine production. Added salt will help the body retain the ingested fluids.

TABLE 9.7 Fluid intake guidelines before, during, and after exercise in warm or hot environmental conditions

Before competition and practice
The goal of the ACSM guidelines is to start in a state of euhydration with normal plasma electrolyte levels.

- Drink slowly about 5–7 ml/kg (0.08–0.11 ounce/pound) body weight at least 4 hours prior to exercise.
- Drink another 3–5 ml/kg body weight about 2 hours prior to exercise if no urine is produced or the urine is dark or highly concentrated. Your urine should be clear, pale yellow before competition or practice.
- Drink water (≤60 minutes of exercise) or carbohydrate-electrolyte solutions (CES) (>60 minutes of exercise).
- Drink beverages with carbohydrate (6–8 percent) to help increase body stores of glucose and glycogen for use in prolonged exercise bouts.
- Drink beverages with sodium (20–50 mEq/l) and/or consume salty foods or snacks to help increase body stores of sodium and water for prolonged exercise.
- Do not drink excessively, which may increase the risk of dilutional hyponatremia if fluids are aggressively replaced during and after exercise.

During competition and practice
The goal of the ACSM guidelines is to prevent excessive dehydration (>2% body weight loss from water deficit).

- Determine your sweat loss for a given intensity and duration of exercise in the heat to estimate a schedule for fluid intake to minimize dehydration according to the example in this section of the chapter.
- In general, smaller athletes may consume 14 ounces/hour (3–4 ounces every 15 minutes) while larger athletes may consume 28 ounces/hour (7 ounces every 15 minutes). Fluids may also be consumed *ad libitum,* or as other time schedules, rules and conditions permit. Athletes can adjust the amounts according to personal needs.
- Drink cold water when carbohydrate intake is of little or no concern, such as in endurance events of less than 50–60 minutes. CES may be consumed during such events if preferred but provide no advantages over water alone.
- Drink fluids with carbohydrates for longer-duration events.
  - Select a CES with a 6–8 percent concentration.
  - Use a CES containing multiple sources of carbohydrate, including glucose, sucrose, fructose, and maltodextrins.
  - Consume enough fluid to provide about 30–80 grams of carbohydrate per hour. One ounce of a CES provides about 2 grams of carbohydrate.
  - Use sports gels or sports beans to provide additional carbohydrate if the necessary fluid intake would be unreasonable. Sports gels and beans may provide about 25–30 grams of carbohydrate per serving.
- Drink fluids with small amounts of electrolytes, particularly sodium and potassium. Many CES contain about 20–30 mEq of sodium and 2–5 mEq of potassium, which amounts to about 110–166 mg of sodium and 19–47 mg of potassium in an 8-ounce serving.

After competition and practice
The goal of the ACSM guidelines is to fully replace any fluid and electrolyte deficit.

- Rapid replacement
  - Drink 1.5 liters of fluid for every kilogram of body weight loss, or about 1.5 pints for each pound loss.
  - Consume about 1.0–1.5 grams of carbohydrate per kilogram body weight (about 0.5 to 0.7 gram per pound body weight) each hour for 3–4 hours. For a 60-kg athlete, this would represent about 60–90 grams of carbohydrate per hour.
  - Consume adequate sodium. Salty carbohydrate snacks, such as pretzels, may provide both sodium and carbohydrate.
- Leisurely replacement (24-hour recovery)
  - Eat a diet rich in wholesome, natural foods adhering to healthy eating practices to help replenish needed electrolytes.
  - Extra salt may be added to meals when sodium losses are high.
  - Drink fluids with added sodium or consume salty foods or snacks.

These guidelines have been adapted from the position stand on fluid replacement developed by the American College of Sports Medicine. The guidelines are appropriate for athletes competing or training for endurance or high-intensity, intermittent sports, such as 10-kilometer races, marathons and ultramarathons, endurance cycling races, Olympic- to Ironman-distance triathlons, soccer and field hockey games, and tennis matches. These guidelines are approximations and may be modified based on individual preferences derived through personal experience in both training and competition. See text for additional information.

Ergogenic Aspects

If preventing or correcting a nutrient deficiency is seen as an ergogenic technique, then certainly water could be construed to be an ergogenic aid. Compared to taking in no fluid during exercise, rehydration has been shown to enhance temperature regulation or exercise performance by optimizing hydration status. However, some athletes have attempted to lose body water for ergogenic purposes. Although we have seen that hypohydration generally does not improve performance, and indeed may actually impair performance in endurance-type events, certain athletes such as high jumpers may use drugs like diuretics to lose weight rapidly without losses in power. Research has shown that diuretic-induced weight losses may improve vertical jumping ability because the athlete can develop the same power to move a lower body weight. Detailed coverage of these drugs is beyond the scope of this text. Moreover, the use of certain diuretics and other masking agents is banned by the World Anti-Doping Agency and the National Collegiate Athletic Association.

Over the years, athletes have attempted to modify body water stores or body temperature using various ergogenic techniques, including supplementation with various nutrients, in attempts to enhance sports performance.

Does oxygen water enhance exercise performance?

Oxygen water, or water oxygenated before bottling, has been marketed to physically active individuals. One brand claims to be a *performance water,* suggesting that tests show that it positively affects cardiovascular and muscular performance and endurance. Fleming and others recently reported increased lactate clearance, but no performance improvement in trained runners before, during, and after a 5,000-meter run in trainer runners consuming an enhanced oxygen solution compared to a placebo. They suggested enhanced liver clearance of lactate via the Cori cycle as a possible mechanism for their observation, but they found no evidence of enhanced oxygen saturation to support this possibility. However, most peer-reviewed research does not support marketing claims that oxygen water enhances energy and boosts athletic performance. Wing-Gaia and others reported no differences in blood oxygenation, performance, or other physiological response in recreational cyclists exercising under hypoxic conditions following consumption of oxygen versus regular water. In a brief review, Piantadosi notes that the amount of oxygen dissolved in a bottle of water is about the same as that found in a single breath. Oxygen is relatively insoluble in water. Since only 0.3 milliliters of the normal arterial oxygen content of 20 ml oxygen/100 ml blood is physically dissolved in the blood, it is unlikely that oxygen dissolved in water is absorbed by the intestine into the bloodstream. Piantadosi concludes that oxygenated water fails to meet quantitative analysis for dissolved oxygen claims by marketers and lacks evidence for substantial intestinal oxygen absorption. In summary, the small body of current research does not support an ergogenic effect of oxygenated water.

Do precooling techniques help reduce body temperature and enhance performance during exercise in the heat?

There are at least two reasons why precooling may be an effective ergogenic strategy for some athletes competing in the heat. Sawka and Young indicated that cooling the skin will decrease skin blood flow, so theoretically more blood could be shuttled to the muscles during exercise. According to Stevens and others, precooling can decrease skin temperature which would increase the thermal gradient between the core and the shell for enhanced heat dissipation. Other mechanisms include reduced cardiovascular strain and improvements in heat sensation and central nervous system function.

Precooling has been examined in several meta-reviews. Rodrigues and others concluded that external (e.g., ice water, cold packs, cooling clothes), internal methods (e.g., ingestion of ice water), and mixed cooling methods have significantly increased distance covered, mean power output, time to exhaustion, work, and mean peak torque, and decreased completion time in competitive, unacclimatized athletes. They reported the most effective general strategy to be a combination of external and internal methods, followed by cold-water immersion. Bongers and others also reported similar effect sizes on exercise performance for pre-cooling and percooling (during exercise). Precooling resulted in a lower core temperature compared to control conditions. Multiple techniques and ice vests were the most effective strategies for precooling and percooling, respectively. Wegmann and colleagues reported the greatest precooling effect for hot conditions, in subjects with high aerobic capacity, and for endurance compared to sprint performance. The best methods were cold drinks and cooling packs. The precooling effect on time-trial performance, the most relevant task for an ergogenic effect, was 3.7 percent. Bongers and others comment that scheduling of athletic events in hot, humid environments will continue to drive research to develop strategies for attenuating the rise in core temperature during hyperthermic exercise.

Does sodium loading enhance endurance performance?

The adverse effects of dehydration of 2 percent of body mass or more on endurance performance are well-documented. An electrolyte deficiency could impair physical performance, but supplements above and beyond normal electrolyte nutrition have not been recommended for ergogenic purposes. However, sodium concentration is one of the main determinants of water retention and blood volume. Decreased blood volume loss during hyperthermic exercise increases cardiovascular strain and impairs performance. In separate studies of male runners and female cyclists by Sims and others, subjects consumed high (164 mmol/liter) and low (10 mmol/liter) sodium solutions before exercise at 70 percent of VO_2 max in a warm (32°C) environment. In both studies, the high-sodium trial increased plasma volume before exercise, reduced ratings of perceived exertion during exercise, cardiovascular strain, and increased exercise performance compared to the low-sodium trial. In their review, Mora-Rodrigues and Hamouti note that ingestion of a sodium solution of up to 164 mEq/liter, a practice known as **sodium loading**, can expand pre-exercise plasma volume and improve subsequent endurance performance.

These data are supportive of an ergogenic effect of sodium loading *before* exercising in a hot environment. As mentioned in chapter 13, sodium bicarbonate supplementation has been shown to enhance endurance performance in some studies when provided before exercise, which could be related to the sodium content in the bicarbonate preparation. However, consuming sodium tablets *during* exercise does not appear to enhance performance. In a study of more than 400 triathletes, Hew-Butler and others reported no ergogenic effect from consuming additional sodium in tablet form (a total of 3.6 grams more than the placebo and control groups) on finishing time in Ironman competition.

Does glycerol supplementation enhance endurance performance during exercise under warm environmental conditions?

As noted in chapter 5, glycerol is an alcohol that combines with fatty acids to form a triglyceride. Years ago, interest in glycerol as an ergogenic aid focused on its role as a substrate for gluconeogenesis in the liver or in energy metabolism. However, Massicotte and others reported that only a small portion, about 4 percent, of glycerol consumed during exercise is converted to glucose for oxidation or directly oxidized for energy. Research findings have not supported an ergogenic effect of glycerol when used for its energy content.

More recently, glycerol supplementation has been studied in attempts to enhance endurance performance in warm environments. Glycerol has been combined with water during hyperhydration prior to exercise to study its potential ergogenic effects on performance in prolonged endurance events, often under warm environmental conditions. Van Rosendal and others suggest consumptions of 1.2 g glycerol/kg BW in 26 ml/kg BW of fluid 1 hour before exercise for optimal hydration; 0.125 g glycerol/kg BW equal to 5 ml/kg BW during exercise to delay dehydration; and postexercise consumption of 1.0 g glycerol/kg BW in each 1.5 liters of fluid to accelerate the replacement of plasma volume. Glycerol supplementation has also been based on lean body mass or total body water and combined with carbohydrates. Theoretically, glycerol-induced hyperhydration will increase osmotic pressure in the body fluids, helping to retain more total body water and possibly increase the plasma volume, factors that could enhance temperature regulation and exercise performance.

Glycerol-induced hyperhydration protocols have normally been compared to water-induced hyperhydration techniques. Research results generally favor the efficacy of glycerol in improved hydration status. Van Rosendal and colleagues reported the greatest fluid retention following 4 percent dehydration with oral glycerol with intravenous infusion of saline + sports drink compared to either oral sports drink + water, oral glycerol + sports drink + water, or intravenous saline + sports drink + water.

The research is equivocal regarding the effects of glycerol-induced hyperhydration on thermoregulation and performance. Coutts and others compared the effects of glycerol hyperhydration and a placebo on competitive Olympic distance triathlon performance in a counterbalanced crossover field study. The glycerol hyperhydration induced a significantly greater plasma expansion than the placebo, and the increase in the triathlon completion time between the hot and warm conditions was significantly less than the placebo trial. The authors suggested that glycerol hyperhydration prior to triathlon competition in high ambient temperatures may provide some protection against the negative performance effect of competing in the heat. However, the WBGT temperature for day 1 was significantly warmer (30.5°C) than for day 2 (25.4°C), which may have confounded the results. Kavouras and others have reported improved cycling performance following glycerol-induced hyperhydration and increased plasma volume, but no other thermoregulatory benefit following glycerol-induced hyperhydration. Van Rosendal and colleagues reported no difference in cycle performance in the heat (40-kilometer time trial) between any of the previously described rehydration strategies involving oral glycerol + water and oral or intravenous saline + sports drink. Hillman and others reported reduced postexercise oxidative stress in cyclists following glycerol and water preexercise hyperhydration treatments compared to euhydration, but there was no difference between any of the hydration conditions in thermoregulatory response or performance. In its position stand on fluid replacement, the ACSM indicated that although hyperhydration, including use of glycerol, does not provide any thermoregulatory advantages, it can delay the onset of dehydration, which may be responsible for any small performance benefits that are occasionally reported.

The ergogenic effect of glycerol-induced hyperhydration has been the subject of three recent reviews, and even the conclusions of these reviewers are somewhat equivocal. Some reviewers concluded that glycerol hyperhydration is an effective ergogenic for endurance athletes exercising in the heat, whereas others concluded that it is ineffective and should be avoided. Van Rosendal and others note favorable thermoregulatory and cardiovascular effects of glycerol hyperhydration, including higher sweat rates and reduced heart rates, core temperatures, thermal sensation, thirst sensation, and perceived exertion. Performance improvements such as increased endurance time to exhaustion; increased work and power, and improved triathlon performance have been reported by some, but not all researchers. There is little evidence of a glycerol effect on sport-specific skill and agility tasks. Goulet and others analyzed 14 studies comparing glycerol hyperhydration with water hyperhydration. However, only 4 studies met the criteria for comparing the two treatments on endurance performance. The meta-analysis indicated that glycerol hyperhydration significantly improved performance by 2.6 percent, but due to the limited research available, more research is needed before definitive conclusions can be drawn as to the ergogenic effects of glycerol hyperhydration. Ergogenic effects of glycerol-induced hyperhydration appear to be more apparent in cycling studies such as that of Kavouras and others. Hyperhydration may be ergogenic for cyclists, who need not be too concerned with the additional body weight associated with water retention, but runners may be at a slight disadvantage because the potential benefits of hyperhydration may not counteract the potential adverse effects of the extra weight, which may result in impaired performance. However, Van Rosendal and Coombes noted no adverse effect of extra body mass of glycerol hyperhydration on submaximal running economy.

In summary, while glycerol and fluid ingestion results in hyperhydration, the documented benefits to exercise performance remain inconsistent. Concerns that glycerol hyperhydration might be a strategy to mask the use of other banned agents resulted in the addition of glycerol to the World Anti-Doping Agency's (WADA) banned list in 2010. Effective January 1, 2018, the WADA removed glycerol from the banned list based on subsequent research concluding that glycerol-induced plasma volume expansion has a minimal masking effect.

> ### Key Concepts
>
> ▸ There is no evidence from well-controlled research that oxygen water has an ergogenic effect.
> ▸ Various precooling strategies such as wearing an ice vest and cool water immersion have been shown to improve performance.
> ▸ Sodium loading *before* hyperthermic exercise may increase plasma volume and improve endurance performance. However, there is no evidence for an ergogenic effect of sodium loading *during* exercise.
> ▸ Glycerol supplementation appears to enhance hyperhydration, increase blood volume, and produce favorable physiological responses during exercise. However, research findings are equivocal regarding its effects as a means of improving endurance performance. Glycerol hypohydration, if abused, may also be associated with several adverse health effects.

Health Aspects: Heat Illness

Heat illness, as the name implies, involves various health problems associated with environmental heat stress. As noted previously, excessive dehydration may impair physical performance, and individuals who overhydrate in attempts to prevent dehydration during exercise may experience exercise-associated hyponatremia, which could have serious health consequences. Dehydration and loss of electrolytes may also cause health problems during exercise, some more serious than others.

However, high environmental heat stress poses one of the most serious health threats to athletes and others who exercise, and such individuals should use caution when exercising in the heat. In a case report by Boddu and others, a healthy 35-year-old female suffered cardiac arrest in a hot yoga class.

Should I exercise in the heat?

The American College of Sports Medicine position stand on exertional heat illness during training and competition presents guidelines targeted to sports medicine personnel, such as athletic trainers, and other sport administrators who should be aware of environmental heat conditions that suggest modification or cancellation of competition or practice. However, individuals may also use these guidelines to determine when to modify exercise protocols in the heat. The guidelines are based on the WBGT, which may not be readily available to most individuals. However, local television stations or various websites usually can provide a heat index, which is a good approximation of the WBGT when exercising in the shade. Exercising in the sun adds to the heat index. **Table 9.8** presents a modification of the ACSM guidelines.

> *www.acsm.org/education-resources/pronouncements-scientific-communications/position-stands* Access the ACSM position stand "Exertional Heat Illness during Training and Competition" which is located in the list of ACSM Position Stands and Joint Position Statements.
>
> *www.wunderground.com* Various websites may provide you with the temperature, humidity, wind, and possibility of sunshine. They also generally provide a heat index, indicating that although the air temperature is only 85°F, it may feel like 95°F due to humidity.

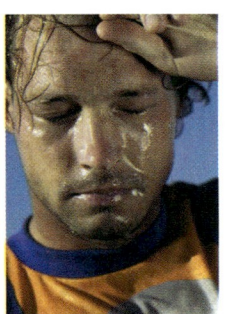

PNC/Brand X Pictures/PunchStock/Getty Images

What are the potential health hazards of excessive heat stress imposed on the body?

One of the most serious threats to the performance and health of the physically active individual is heat illness, which is often referred to as exertional heat illness or exercise-associated heat illness when it occurs with exercise. Basu and colleagues reported strong positive associations (percent increased risk per 10°F) between ambient temperature and morbidities such as central and cerebral vascular disease, diabetes, hypotension, renal failure, and intestinal infection in addition to heat illness and dehydration. Vaidyanathan and others report an average of 702 heat-related deaths in the United States (415 with heat as the underlying cause and 287 as a contributing cause) per year from 2004 to 2018. Using data from the Centers for Disease Control Environmental Public Health Tracking Network and the National Center for Health Statistics from 2005 to 2010, Fetcher-Leggett and others reported a significantly greater relative risk for heat-illness-related emergency visits in non-metropolitan compared to large metropolitan areas in all US climate regions. These differences affect the design and implementation of strategies for the prevention and intervention of heat-related illnesses. Any athlete who exercises in a warm environment is susceptible to heat injury, but the increasing popularity of road racing has generated concern for runners who are not prepared for strenuous exercise in the heat, or who participate in races that are poorly organized in regard to preventing and treating heat injuries. Marathons are popular events, with some major races having tens of thousands of runners. Even well-organized events may experience problems in providing for the needs of runners when the environmental heat stress becomes excessive, such as during an unexpected heat wave in races that normally have cooler weather.

The individual who exercises unwisely under conditions of environmental heat stress may experience one or several of a variety of heat injuries. Three factors may contribute to these injuries: increased core temperature, loss of body fluids, and loss of electrolytes. However, other factors may also be involved, as Noakes noted that several of the heat illnesses, such as muscle cramps, also occur during exercise in cold environments.

Figure 9.11 represents a simple flow chart of heat disorders. When a combination of exercise and environmental heat stress is imposed on the body, vasodilation and sweating increase as the body tries to cool

TABLE 9.8 American College of Sports Medicine guidelines for modifying or canceling competition or training to help prevent heat illness

WBGT (°F)	WBGT (°C)	Continuous activity and competition	Training and noncontinuous activity
<50–65	(10–18.3)	Generally safe	Normal activity
65.1–72.0	(18.4–22.2)	Risk of heat illness begins to rise *High-risk:* Should be monitored or not compete	*Low-risk:* Normal activity *High-risk:* Increase rest: exercise ratio and monitor fluid intake
72.1–78.0	(22.3–25.6)	Risk for all competitors is increased	*Low-risk:* Normal activity and monitor fluid intake *High-risk:* Increase the rest: exercise ratio and decrease total duration of activity
78.1–82.0	(25.7–27.8)	*High-risk:* Risk is high	*Low-risk:* Normal activity and monitor fluid intake *High-risk:* Increase the rest: exercise ratio and decrease intensity and total duration of activity
82.1–86.0	(27.9–30.0)	Cancel for those at risk of exertional heat stroke	*Low-risk:* Plan intense or prolonged exercise with discretion* *High-risk:* Increase the rest:exercise ratio to 1:1 and decrease intensity and total duration of activity*
86.1–90.0	(30.1–32.2)		*Low-risk:* Limit intense exercise and total daily exposure to heat and humidity *High-risk:* Cancel or stop practice and competition
>90	(>32.3)		Cancel exercise when uncompensable heat stress exists for all athletes*

Low-risk: Individuals acclimatized to the heat for 3 weeks; high fitness level.
High-risk: Individuals nonacclimatized to the heat; unfit; using certain medications; dehydrated; recent illness; previous heat illness, particularly exertional heat stroke.
*Differences of local climate and individual heat acclimatization status may allow activity at higher levels than outlined in the table. Athletes and coaches should consult with sports medicine staff and be cautious when exercising in extreme heat conditions.
The WBGT is the wet-bulb globe thermometer temperature. Commercial devices are available to quickly and accurately measure the WBGT and should be used to help assess environmental heat stress and modify training or competition as recommended.

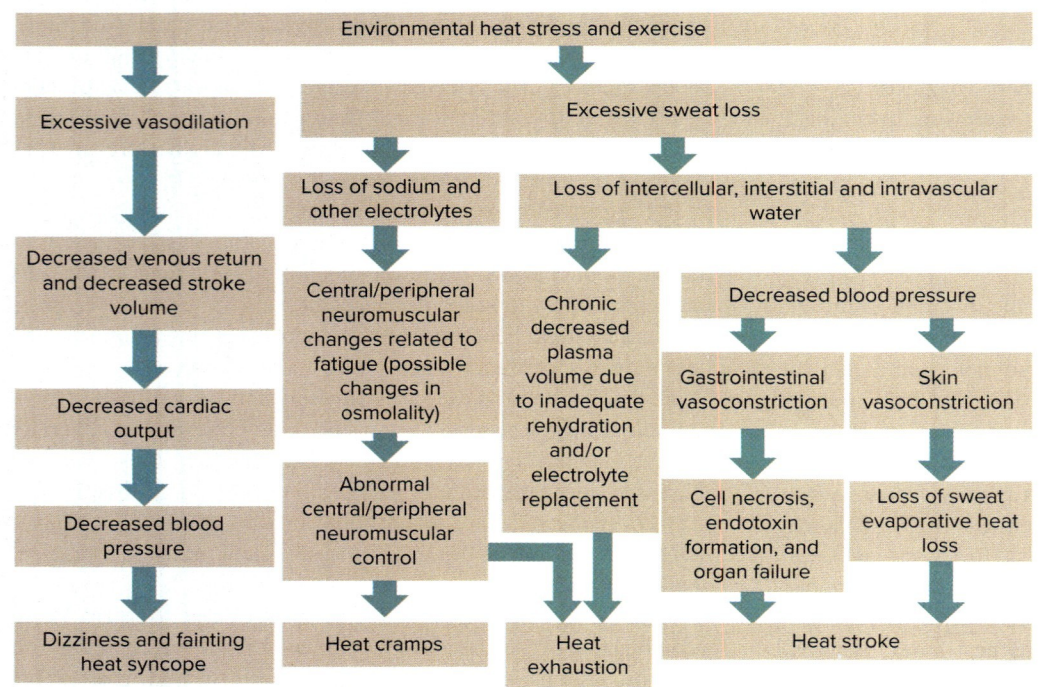

FIGURE 9.11 Basic flow chart for heat illnesses. The combination of environmental heat and exercise may cause an excessive vasodilation or pooling of blood. These conditions may decrease blood return to the heart and brain, causing dizziness and fainting. Excessive loss of sweat may cause significant losses of body water and electrolytes, leading to various heat illnesses. See text for details.

itself. When these two adjustments begin to falter, problems develop. Circulation attempts to regulate both body temperature and blood pressure at the same time, and when stressed excessively, control of blood pressure wins and body temperature regulation is impaired. In addition, if the exercise metabolic load is very great, heat injuries may develop independent of circulatory and sweating inadequacies. For example, in their 2015 review, Sawka and others describe a pathological pathway of heat stress, beginning with lack of gut blood flow and resulting in cell necrosis; the release of endotoxins and mitochondrial DNA fragments; and cascade of vascular and immunological pathologies that result in multiorgan failure and damage.

Heat Syncope Excessive vasodilation may contribute to circulatory instability. The blood vessels expand and have a much greater capacity. A decrease in blood volume with corresponding decreases in cardiac output and blood pressure may result in dizziness and fainting. This condition is called **heat syncope** and is usually associated with heat exhaustion, as discussed later. A newer term, **exercise-associated collapse**, has been introduced, of which simple fainting may be a mild form, whereas more severe forms may include heat stroke or hyponatremia.

Kenefick and Sawka also note that fainting after a race may be due to the reduced muscle pump activity. When the runner stops, venous pooling in the legs may reduce blood return to the heart and, subsequently, to the brain. Runners are advised to keep the legs moving at the completion of the race. If dizzy, lie down with feet elevated.

Heat Cramps The ACSM indicates that exercise-associated muscle cramping can occur with exhaustive work in any temperature range but appears to be more prevalent in hot and humid conditions (**heat cramps**).

Eichner notes that although not all cramps are alike, he indicates three lines of evidence suggesting that heat cramping is caused by "salty sweating," specifically by the triad of salt loss, fluid loss, and muscle fatigue. First, historically, heat cramping in industrial workers is alleviated by salt. Second, field studies of athletes show that heat-crampers tend to be salty sweaters. Stofan and others reported that American football players who were prone to muscle cramps averaged more than twice the amount of sodium loss than those not prone to cramping. Third, intravenous saline can reverse heat cramping, and more salt in the diet and in sports drinks can help prevent heat cramping. Eichner concludes that for heat cramping, the solution is saline.

To help prevent cramps, Bergeron recommends that at the first sign of subtle muscle twitching, which usually is about 20–30 minutes before full-blown cramps, the athlete should consume a salt solution, such as half a teaspoon of salt in 16 ounces of a sports drink. The athlete should then continue to drink small amounts of a similar solution (the same amount of salt in 32 ounces of a sports drink) at regular intervals for the remainder of the exercise session.

Several electrolyte products have been developed for athletes who are prone to muscle cramps. Products such as EnduroLytes® Extreme and Gatorade Endurance GatorLYTES® contain sodium, potassium, calcium, and magnesium, all of which may be helpful in the prevention of muscle cramping. However, Sulzer and others reported that serum electrolytes did not appear to be associated with muscle cramps in triathletes prone to muscle cramping. A review by Giuriato and others concluded that a neuromuscular hypothesis for **exercise-associated muscle cramps** prevails in the literature over a dehydration/electrolyte imbalance hypothesis. Specifically, action potentials in the soma of the motor neuron are influenced by imbalances between increased excitatory muscle spindle activity from Ia afferent afferents and decreased inhibitory input from Golgi tendon organ Ib afferent input. Although electrolyte deficiencies and neuromuscular factors are reported to be the most discussed mechanisms for cramps, Miller and others commented that the variety of existing prevention strategies and interventions underscore the uncertainty regarding any single mechanism. Nevertheless, the ACSM notes that muscle cramping usually responds to rest and replacement of fluid and salt (sodium).

Heat Exhaustion Heat exhaustion is a common heat illness during exercise. Dehydration is the main risk factor for exercise-associated heat exhaustion. Inadequate salt replacement over the course of several days may also be a contributing factor, as blood volume may decrease. A high body mass index (BMI) also increases the risk. Heat exhaustion is a cause of heat syncope. Fatigue and weakness are key signs of heat exhaustion, which may be associated with dizziness and fainting. Other signs and symptoms include rapid pulse, headache, nausea, vomiting, unsteady walk, muscle cramps, chills or goose bumps. The rectal temperature is usually less than 104°F (40°C), which the ACSM notes may be the only discernable difference between severe heat exhaustion and exertional heat stroke in on-site evaluations. Heat exhaustion may incapacitate the individual for a few hours but will generally resolve with body cooling and fluid intake.

Heat Stroke Heat stroke may occur under heat stress conditions even when resting, particularly in older individuals. When heat stroke occurs during exercise, it is known as **exertional heat stroke**, as illustrated in **figure 9.12**. The ACSM indicates that exertional heat stroke can affect seemingly healthy athletes even when the environment is relatively cool. However, it occurs more during exercise in heat stress environments, and dehydration is a major risk factor. The following are risk factors for exertional heat stroke according to a review by Leon and Bouchama and a commentary by Krohn and others on heat illness in American football players:

- High temperature and/or humidity (WBGT greater than 28°C)
- Strenuous exercise
- Dehydration/inadequate rehydration
- Inadequate heat acclimatization
- Lack of fitness
- Obesity
- Clothing that limits evaporation of sweat (e.g., American football pads)
- Genetic factors
  - Sickle cell trait
  - Problems with genes coding for formation of heat shock proteins which are upregulated during heat acclimatization and protect against heat-related damage to tissues

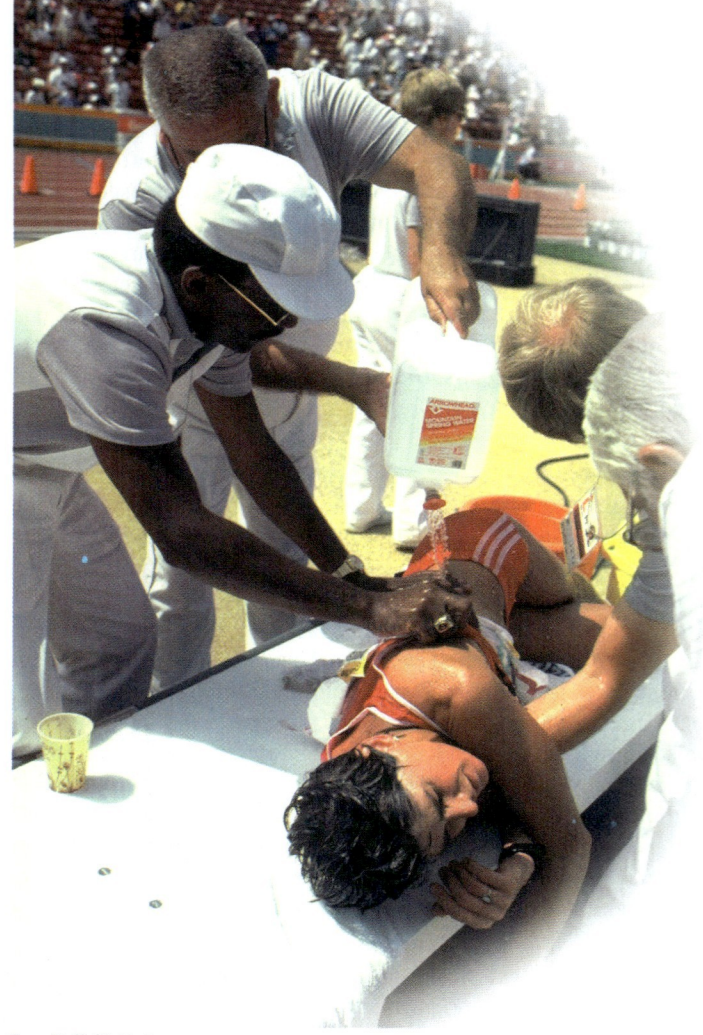

FIGURE 9.12 Proper acclimatization and adequate hydration before, during, and after competition are important factors in preventing heat illness.

- Medications
  - Antihypertensive medications (e.g., diuretics, β-blockers)
  - Stimulants, which increase heat production (e.g., amphetamines, ephedra, thyroid agonists, others)
  - Anticholinergic medications, which may decrease sweat production (e.g., antihistamines, others)
  - Nonsteroidal anti-inflammatory drugs (NSAIDs)
  - Others (e.g., Ritalin, Ecstasy)
- Fever
- Sleep deprivation

Although inconvenient, Ronneberg and others reported that rectal temperature was very effective in diagnosing hyperthermic runners. As reported in a review by Mazerolle and others, oral temperature is not a valid method to assess changing core temperature because differences in measurement exceeded 0.27°C (0.5°F), with even greater discrepancies at highest rectal temperatures observed in exertional heat illnesses. The ACSM defines exertional heat stroke as a rectal temperature greater than 40°C (104°F) accompanied by symptoms or signs of organ system failure, most frequently central nervous system changes such as confusion, disorientation, agitation, aggressiveness, blank stare, apathy, irrational behavior, staggering gait, delirium, convulsions, unresponsiveness, or coma. Other signs of possible heat stroke include weak and rapid pulse, vomiting, involuntary bowel movement, and hyperventilation. Roberts stated that the skin appearance is usually ashen and sweaty, can be either warm or cool to touch, and is rarely pink, hot, and dry as described in textbooks.

Heat stroke is the most dangerous heat injury, as it may be fatal. Several deaths of professional and collegiate athletes have been associated with such circumstances, and one involved the use of ephedrine, which as noted in chapter 13 can stimulate metabolism and heat production and predispose to heat stress. As noted by Krohn and colleagues, at least 33 high school and college American football players died from exertional heat stroke between 2001 and 2015. Lack of fitness, timing of preseason conditioning, and protective padding, which reduces surface area for optimal evaporative sweat heat loss, increase susceptibility of football players to heat-related illness. Exertional heat stroke may lead to rhabdomyolysis, damaged muscle tissue that leaks its contents into the blood, eventually leading to kidney damage and possible death.

What are the symptoms and treatment of heat injuries?

As noted, the symptoms of impending heat injury are variable. Among those reported are weakness, feeling of chills, pilo-erection (goose pimples) on the chest and upper arms, tingling arms, nausea, headache, faintness, disorientation, muscle cramping, and cessation of sweating. Continuing to exercise in a warm environment when experiencing any of these symptoms may lead to heat injury. **Table 9.9** presents the major heat injuries along with principal causes, clinical findings, and treatment. In general, treatment of heat syncope, heat cramps, and heat exhaustion involves resting (preferably lying down), cooling the body if overheated, and drinking fluids, preferably with sodium. Most individuals recover fairly rapidly but should be monitored until in a safe environment.

The ACSM indicates that early recognition and rapid cooling can reduce both the morbidity and mortality associated with exertional heat stroke. Of the 63 studies reviewed by Douma and others, cold water (≤17°C) immersion was the most effective treatment to rapidly lower core temperature during heat stroke. They found no difference in the rate of core temperature reduction between cold (14-17°C/57.2-62.6°F), colder (8-12°C/48.2-53.6°F), and ice (1-5°C/33.8-41°F) water immersion. If water immersion is not available, rapidly rotating ice-water towels to the trunk, extremities, and head, combined with ice packing of the neck, axillae, and groin, may be very effective.

According to Asplund and others, exercise-associated collapse, a phenomenon that is occasionally observed in endurance athletes, is usually, but not always, the result of low blood pressure due to blood pooling in the legs after cessation of exercise. This condition is benign and resolves upon rest. However, it is important for the clinician to check vital signs (especially rectal temperature if heat stroke is suspected), assess fluids status (dehydrated versus fluid

TABLE 9.9 Heat injuries: causes, clinical findings, and treatment

Heat injury	Causes	Clinical findings	Treatment*
Heat syncope (exercise-associated collapse)	Excessive vasodilation: pooling of blood in the skin	Fainting Weakness Fatigue	Place on back in cool environment; give cool fluids
Heat cramps	Excessive loss of electrolytes in sweat; inadequate salt intake	Muscle cramps	Rest in cool environment; oral ingestion of salt drinks; salt foods daily; medical treatment in severe cases
Heat exhaustion	Excessive loss of water and salt; inadequate fluid and salt intake	Fatigue Nausea Cool, pale, moist skin Weakness Dizziness Chills Rectal temperature lower than 104°F (40°C)	Rest in cool environment; replace fluids and salt by mouth; medical treatment if serious
Heat stroke (exercise-associated heat stroke)	Excessive body temperature	Headache Confusion Blank stare Disorientation Unconsciousness Rectal temperature greater than 104°F (40°C)	Cool body immediately to 102°F (38.9°C), preferably with cold-water immersion; if not, cool body areas with ice packs, ice, or cold water; give cool drinks with glucose if conscious; administer intravenous fluids if available; get medical help immediately
Hyponatremia (exercise-associated hyponatremia)	Excessive fluid intake	Confusion Lethargy Agitation Coma ** Need to determine by serum sodium level lower than 135 mmol/l	Mild cases may be treated with hypertonic fluids; more severe cases may need intravenous fluids and possible hospitalization

*Begin treatment as soon as possible. In cases of heat stroke, begin immediately.
**Symptoms of hyponatremia may be similar to other heat illnesses; determination must be made by measurement of serum sodium levels less than 135 mmol/l. Providing hypotonic fluids to individuals with hyponatremia could exacerbate the condition.

overload), and perform laboratory tests when needed in order to rule out hyperthermia or hyponatremia. Providing hypotonic fluids to individuals with hyponatremia will exacerbate the condition because they are already overhydrated.

www.acsm.org/education-resources/pronouncements-scientific-communications/position-stands Access the ACSM position stand "Exertional Heat Illness during Training and Competition" which is located in the list of ACSM Position Stands and Joint Position Statements.

Do some individuals have problems tolerating exercise in the heat?

A number of predisposing factors have been associated with heat injury, including gender, level of physical fitness, age, body composition, previous history of heat injury, and degree of acclimatization.

Poor Physical Fitness The ACSM position on exertional heat illness in training and competition identifies poor physical fitness as a key risk factor. In general, the better the physical fitness, the better tolerance to a given heat stress. Alele and others reported longer run times indicative of lower levels of aerobic fitness to be predictor of exertional heat illness in male and female members of the armed forces. In a comparison of trained and untrained subjects completing an 80-minute fixed-intensity exertional-heat stress test, Ogden and others reported greater intestinal epithelial damage and increased presence of intestinal microbial DNA in plasma of untrained subjects. These findings are consistent with decreased splanchnic blood flow, increased intestinal permeability, and endotoxin formation known to occur with hyperthermia. However, Stacey and others note that both poor and adequate fitness can be related to heat illness of the United Kingdom military personnel. Highly trained athletes may experience heat illnesses when using unsafe training practices. As previously noted, NCAA Division I wrestlers have died from heat stroke or related causes in attempts to reduce body weight for competition through exercise-induced sweating in a hot environment.

Gender Early research findings that females did not tolerate exercise in the heat as well as males were flawed by the use of the same standard work rate, usually a higher percentage of $VO_{2\,max}$, for females. Later and better-designed research matching males and females on fitness, adiposity, acclimatization, body size, and using the same percentage of $VO_{2\,max}$ report no difference in tolerance to exercise in the heat between the genders.

There are differences between males and females in thermoregulatory response. Females have a lower sweat rate than males. This may be a disadvantage to females in hot-dry environments, but an advantage in hot-humid conditions, because females may be less likely to dehydrate. Females have a larger body surface area to mass ratio which increases surface area for evaporative heat loss from sweating. Charkoudian and Stachenfeld note that high estrogen levels favor heat dissipation, while high progesterone levels are associated with higher temperature. Although higher progesterone levels in the luteal phase are associated with a slightly higher core temperature, the menstrual cycle has minimal effects on tolerance to exercise-heat stress. Hyponatremia occurs more in women than men, but this appears to be a function of overconsumption of fluids by females instead of gender *per se*.

Age Children may have problems exercising in the heat. Lower sweat rates have been consistently reported in children despite greater sweat gland density and a larger body surface area-to-mass ratio, which theoretically favors heat transfer. Gomes and others state that young children have "sweat gland immaturity," characterized by a lower release of catecholamines and androgens during exercise, which may partially explain a lower sweat rate in children compared to adults. Notley and others note that various well-established differences exist between children and young adults in body composition, surface-to-mass ratio, cardiovascular function, cardiorespiratory fitness, and economy of motion that potentially impact thermoregulatory function in children, but current literature has not made appropriate adjustments for body weight and child-adult differences in body size. According to the current policy statement from the American Academy of Pediatrics (AAP), exertional heat illness in exercising children and adolescents is due to preventable factors such as inadequate hydration, excessive exertion, inadequate recovery, and attire that retards evaporative heat loss. The statement concludes that "youth do not have less effective thermoregulatory ability, insufficient cardiovascular capacity or lower physical exertion tolerance compared with adults during exercise in the heat when adequate hydration is maintained." Using 2012–2013 to 2016–2017 data from the National High School Sports–related Injury Surveillance System, Kerr and others reported 300 cases of exertional heat illness (EHI, incidence rate = 0.13 cases/10,000 athlete exposures). Almost half of these (44.3 percent) occurred in American football. The EHI incidence rate was higher for football players than for all other sports (0.52 vs. 0.04 cases/10,000 athlete exposures). The AAP's 2011 guidelines for preventing heat illnesses in children are in accord with recommendations provided in the next section. The ACSM's guidelines in **table 9.10** may be used to restrain physical activities for children under conditions of heat stress.

At high levels of heat stress, tolerance to the heat is decreased in older individuals, possibly because they experience decreased blood flow to the skin and sweat less. Reduced heat tolerance in the elderly may also be related to fitness levels. However, more and more people have become and remain physically active throughout middle age and advanced years. Kenney and others note that increased cardiac output during hot weather activity to provide both increased skin and muscle blood flow may be a challenge to left ventricular function in the older individual. However, regulation of body temperature in well-trained and heat-acclimated older athletes is comparable to that of younger athletes. Therefore, the risk of heat-related illness is not significantly greater in healthy older men and women who maintain a high degree of aerobic fitness compared to young adults. Older adults have a decreased dipsogenic (thirst) response and may not rehydrate as quickly as younger adults during and following hyperthermic exercise. As a result, older adults should focus on adequate fluid intake when exercising in the heat.

TABLE 9.10 WBGT temperature guidelines to modify practice sessions for exercising children

WBGT (°F)	WBGT (°C)	Restraints on activities
<75.0	<24.0	All activities allowed, but watch for symptoms of heat illnesses in prolonged events
75.0–78.6	24.0–25.9	Have longer rest periods in the shade; enforce fluid intake every 15 minutes
79.0–84.0	26.0–29.0	Stop activity for unacclimatized and high-risk children; limit activities of others; cancel long-distance races and cut the duration of other activities
>85.0	>29.0	Cancel all athletic activities

Modifications of the recommendations on climatic heat stress in exercising children and adolescents proposed by the American Academy of Pediatrics Committee on Sports Medicine and Fitness.

Obesity Large individuals with obesity have smaller body surface area-to-mass ratios, which decrease the surface area for sweat evaporative heat loss. Such individuals are also usually less fit with lower economies of motion, resulting in more metabolic heat production. High amounts of visceral and subcutaneous fat decrease heat conductance from the core to the shell. While this is advantageous during exposure to ambient cold, it is detrimental during hot and humid conditions. As a result, they may be more susceptible to exertional heat illness. In American football, offensive linemen generally have the highest levels of body fat, lowest body surface area-to-mass ratios, and lowest surface areas for heat loss through sweat evaporation. Further reduction of surface area by wearing protective padding increases their susceptibility for heat illness, especially during pre-season and early-season practice. In their review, Alele and others also noted a body mass index of ≥ 26 kg/m^2, generally but not always due to increased fatness, is a predictor of exertional heat illness in male members of the armed forces.

Previous Heat Illness Previous heat injury may result in less tolerance to exercise in the heat. Many individuals do regain heat tolerance 8–12 weeks after heat injury, while others appear to lose some ability for the circulatory system to adjust to heat stress. It has been proposed that temperature-regulating centers, such as the cerebellum and the preoptic area of the hypothalamus (POAH) which processes afferent input of increased core temperature, may suffer irreversible damage. If so, heat transfer from the core to the skin would be impaired with a rapid increase in body temperature. Leon and Bouchama note damage to the cerebellum in autopsied human hyperthermia casualties, but no histological evidence of POAH damage. They call for more sensitive biomarkers of tissue damage.

Lack of Acclimatization One of the more important factors determining an individual's response to exercise in the heat is degree of acclimatization, which is discussed on the following pages. Although some individuals may be susceptible to heat illnesses, all individuals who exercise under warm or hot environmental conditions may benefit from the following recommendations.

How can I reduce the hazards associated with exercise in a hot environment?

Recently, Racinais and others provided recommendations for heat acclimatization, proper hydration, cooling strategies, and recommendations for event organizers. The following list reflects these recommendations, which if followed, will reduce considerably your chances of suffering heat injury.

1. Check the temperature and humidity conditions before exercising. Even if the dry temperature is only 65–75°F, high humidity will increase the heat stress. Warm, humid conditions cause fatigue sooner, so slow your pace or shorten your exercise session. As noted previously, local news stations and several websites may provide information on heat stress conditions.
2. Exercise in the cool of the morning or evening to avoid the heat of the day.
3. Exercise in the shade, if possible, to avoid radiation from the sun. If you run in the sun, wear an appropriate sunscreen to prevent sun damage to the skin.
4. Wear sports clothing designed for exercise in the heat. Materials with fibers that increase surface area, wick moisture away from the skin by capillary action, and dry quickly will facilitate evaporative heat loss. Clothing should be loose to allow air circulation, white or a light color to reflect radiant heat, and porous to permit evaporation. Do not wear a hat if running in the shade, but wear a loose hat if running in the sun.
5. If you are running and there is a breeze, plan your route so that you are running into the wind during the last part of your run. The breeze increases heat loss by convection and evaporation at the time you need it the most.
6. Hyperhydrate if you plan to perform prolonged, strenuous exercise in the heat. In general, drink about 16 ounces of fluid 30–60 minutes prior to exercising.
7. Drink cold fluids periodically. For a long training run, plan your route so that it passes some watering holes, such as gas stations or other sources of water. Alternatively, you may purchase a water-bottle belt or backpack to carry your own water supply. Take frequent water breaks, consuming about 6–8 ounces of water every 15 minutes or so. As a rough gauge, one mouthful of water is approximately 1 ounce of fluid.
8. Replenish your water daily. Keep a record of your body weight. For each pound you lose, drink about 20–24 ounces of fluid. Your body weight should be back to normal before your next workout.
9. Replenish lost electrolytes (salt) if you have sweated excessively. Put a little extra salt on your meals and eat foods high in potassium, such as bananas and citrus fruits.
10. Avoid excessive intake of protein, as extra heat is produced in the body when protein is metabolized. This may contribute slightly to the heat stress.
11. Avoid dietary supplements containing ephedrine. Ephedrine is a potent stimulant that can increase metabolism and heat production, leading to an increased body temperature during exercise in warm environmental conditions and predisposing to heat illness.
12. If you drink caffeinated beverages, check your responses. Current research indicates that caffeinated beverages may not affect hydration status or temperature regulation during exercise. However, some individuals may respond differently to caffeine intake. Caffeine can increase heat production at rest, which could raise the body temperature before exercise.
13. Because alcohol is a diuretic, excess amounts should be avoided the night before competition or prolonged exercise in the heat.
14. If you are sedentary, overweight, or aged, you are less likely to tolerate exercise in the heat and should therefore use extra caution.
15. Be aware of the signs and symptoms of heat exhaustion and heat stroke, as well as the treatment for each. Chills, goose pimples, tingling arms, dizziness, weakness, fatigue, mental disorientation, nausea, and headaches are some symptoms that may signify the onset of heat illness. Stop activity, get to a cool place, and consume some cool fluids.
16. Do not exercise if you have been ill or have had a fever within the last few days.
17. Check your medications. Some medications, such as antihistamines used to treat cold symptoms, may block sweat production. Drugs used to treat high blood pressure, such as beta-blockers and calcium-channel blockers, may impair skin blood flow and decrease heat loss from the body.
18. If you plan to compete in a sport held under hot environmental conditions, you must become acclimatized to exercise in the heat.

www.gssiweb.com The Gatorade Sports Science Institute (GSSI) provides very useful information on a wide variety of topics in sports nutrition, especially information relative to proper hydration practices to help enhance performance and prevent heat illness when exercising in the heat. Under the Sport Science Exchange tab, relevant information can be located under the Hydration and Thermoregulation option.

How can I become acclimatized to exercise in the heat?

It is a well-established fact that acclimatization to the heat will help increase performance in warm environments as compared with an unacclimatized state. Simply living in a hot environment confers a small amount of acclimatization. Physical training, in and of itself, provides a significant amount of acclimatization, possibly up to 50 percent of that which can be expected and increases body-water levels. However, neither of these two adjustments, either singly or together, can prevent the deterioration of exercise performance in the heat by an unacclimatized individual. Thus, a period of active acclimatization is necessary to optimize performance when exercising in the heat.

The technique of acclimatization is relatively simple. Simply cut back on the intensity or duration of your normal activity when the hot weather begins. Do not avoid exercise in the heat completely, but after an initial reduction in your activity level, increase it gradually. For example, if you were running 5 miles a day, cut your distance back to 2-3 miles in the heat; if you need to do 5 a day, do the remaining miles in the evening. Eventually build up to 3, 4, and 5 miles. As illustrated in **figure 9.13**, the acclimatization process usually takes about 10-14 days to complete. Falk and Dotan comment that children acclimatize to the heat at a slower rate than adults, but children and young athletes are as capable as adults to adapting to environmental change.

An individual living in a cooler location who wishes to compete in a warmer, humid location (or to compete in a winter event in a warm climate) can exercise indoors and/or wear extra layers of clothes to help prevent evaporation and build a hot, humid microclimate around the body. Although this technique can provide a degree of acclimatization, it should only be used in cool weather and not under hot conditions. Individuals using these techniques should be aware of the previously discussed symptoms of impending heat illness.

Repeated exposure to a high core temperature during exercise helps the body make the following important adjustments during acclimatization to the heat. The following *Training Table* lists some of the benefits of heat acclimatization that will help improve performance and reduce the risk of heat stroke and heat exhaustion.

In essence, as illustrated in **figure 9.13**, these changes increase the ability of the body to dissipate heat with less stress on the cardiovascular system. The end result is a more effective body-temperature control and improved performance when exercising in the heat. These adaptations may be maintained by exercising in the heat several days per week but are lost in about 7-10 days in a cool environment. If you are interested in learning more about acclimatization and exercise in hot weather, examine the reviews by Nybo and others and Racinais and others.

Training Table

Selected benefits of acclimatization

- Total body water increases considerably, which usually includes increased plasma volume. This occurs because the blood vessels conserve more protein and sodium, which tend to hold water.
- The increased blood volume allows the heart to pump more blood per beat, so the stress on the heart is reduced.
- When volume increases, more blood flows to the muscles and skin. The muscles receive more oxygen and skin cooling increases, improving endurance performance.
- Less muscle glycogen may be used as an energy source at a given rate of exercise, sparing this energy source in endurance events.
- The sweat glands hypertrophy and secrete about 30 percent more sweat, allowing for greater evaporative heat loss.
- Body salt losses decrease. The amount of salt in the sweat decreases considerably; evaporation becomes more efficient and electrolytes are conserved. Sweat losses of calcium, magnesium, copper, zinc, and iron also decrease with acclimatization.
- Sweating starts at a lower core temperature, leading to earlier cooling.
- The core temperature will not rise as high or as rapidly as in the unacclimatized state.
- The psychological feeling of stress is reduced at a given exercise rate.

Key Concepts

▶ Heat injuries, of which heat stroke is potentially the most dangerous, may be due to increased body core temperature, loss of body fluids, or loss of electrolytes. Some individuals, such as the obese, are more susceptible to heat injury.

▶ The general treatment for heat-stress illnesses is to rest, drink cool liquids, and cool the body. Rapid body cooling is essential in cases of heat stroke.

▶ If you exercise in the heat, you should be aware of signs of impending heat injury, such as chills, dizziness, and weakness. You should also be aware of methods to reduce heat gain to the body and methods to facilitate heat loss.

▶ Acclimatization to exercise in the heat takes about 10–14 days, but endurance capacity in the heat is still limited somewhat even when one is fully acclimated.

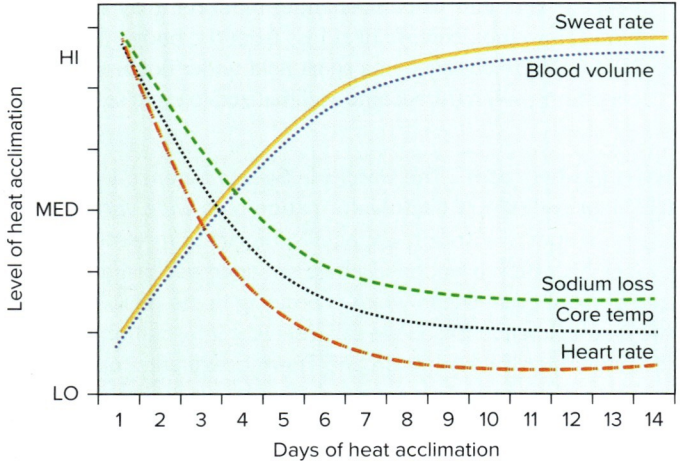

FIGURE 9.13 Changes with acclimatization. Acclimatization to the heat for 7–14 days will lead to an increase in the blood volume and the ability to sweat. For a standardized exercise task in the heat, these changes will lead to a lower heart rate, less sodium loss, and a lower core temperature. These changes will lead to improved exercise performance in the heat.

Health Aspects: High Blood Pressure

What is high blood pressure, or hypertension?

Arterial blood pressure is the product of cardiac output (flow) and the resistance to blood flow, or total peripheral resistance. It is the force that is exerted by the blood against the arterial wall. Blood pressure is an important variable in physiology and overall health. It is usually measured by a sphygmomanometer in millimeters of mercury (mmHg) as illustrated in **figure 9.14**. Blood pressure is variable relative to the working and resting phases of the cardiac cycle. Blood pressure readings are given in two numbers. The higher number (e.g., 120 mmHg) represents the *systolic* phase, when the heart is pumping blood through the arteries. The lower number (e.g., 80 mmHg) represents the *diastolic* phase, when the heart is resting between beats and blood is flowing back into it.

A blood pressure measurement of <120 mmHg and <80 mmHg would be considered "normal" or "normotensive" blood pressure. **High blood pressure,** also known as **hypertension** (hyper = high; tension = pressure), is known as a silent disease. In the past, physicians have used elevations in diastolic blood pressure as the basis for a diagnosis of hypertension. In 2017, a joint American College of Cardiology (ACC) and American Heart Association (AHA) Task Force published a report authored by Welton and others of new blood pressure classifications for adults using systolic and diastolic values. This system is presented in **table 9.11**. Using 2015-2018 data from the National Health and Nutrition Examination Survey (NHANES) and the 2017 ACC/AHA guidelines, the Centers for Disease Control report that 47.3 percent of Americans (116 million) have hypertension. Bell and others have estimated that the prevalence rates of elevated blood pressure (previously called prehypertension under 2004 American Academy of Pediatrics guidelines) in children and adolescents increased from 14.8 percent to 16.3 percent based on the 2017 ACC/AHA categories. According to Mills and others, the total global prevalence rate of hypertension was 31.1 percent of adults (1.39 billion) in 2010, with a higher prevalence in low- to middle income countries (31.5 percent, 1.04 billion people) than in high-income countries (28.5 percent, 349 million people).

Using 2013-2016 National Health and Nutrition Examination Survey data, Virani and others reported that of hypertensive US adults, 35.3 percent are unaware of their disease, 46.6 percent are not receiving treatment, and 75.3 percent do not have their disease under control. High blood pressure is dangerous for several reasons. The heart must work much harder to pump the extra blood volume or to overcome the peripheral vascular resistance. This normally leads to an enlarged heart. Over time the increase in heart size becomes excessive and the efficiency of the heart actually decreases, making it more prone to a heart attack. Second, high blood pressure may directly damage the arterial walls. It is thought to be a major contributing factor in the development of atherosclerosis and a predisposing factor to coronary disease and stroke. High blood pressure is itself a disease and is involved in the etiology of other diseases. It is one of the primary risk factors for heart disease and stroke.

TABLE 9.11 Classification of blood pressure for adults age 18 and older

Blood pressure category	Systolic and Diastolic blood pressure values (mmHg)
Normal	< 120 *and* < 80
Elevated	120–139 *and* < 80
Stage 1 hypertension	130–139 *or* 80–89
Stage 2 hypertension	140–180 *or* 90–120
Hypertensive crisis	> 180 *or* > 120

www.heart.org In the search window, type High Blood Pressure Risk Assessment for links to more information.

How is high blood pressure treated?

General symptoms of hypertension include headaches, dizziness, and fatigue, but these symptoms may not be recognized as symptoms of high blood pressure because they can be caused by a multitude of other factors. Although a great deal of research about the cause of hypertension has been conducted, the exact cause is unknown in about 90 percent of all cases. In these cases, the condition is known as **essential hypertension,** which cannot be cured. Poulter and others comment that genetics account for approximately 30 percent of blood pressure variance. According to Ehret and others, there are 66 genetic loci associated with altered vascular endothelial cell function that affect regulation of vascular tone, blood pressure, and tissue damage in multiple organs. Gene therapy is unlikely to be available in the near future, so individuals must manage current essential hypertension with diet, exercise, and medication. A variety of drugs are available to treat hypertension, including diuretics, beta-blockers, angiotensin-converting enzyme (ACE) inhibitors, and calcium-channel blockers. If your blood pressure is elevated, you and your doctor need to take aggressive action. If drug therapy is recommended, you may need to experiment, with your physician, as to type and dose of medicine to use. This is especially important for athletic individuals. Schleich and others suggest using drugs that are less likely to have adverse effects on exercise performance. For example, diuretics and beta-blockers may impair aerobic endurance performance. Physicians should also be aware that the WADA prohibits the use of beta-blockers for competition in archery, shooting, and some other sports. Diuretics are also banned because of their role as masking agents for performance-enhancing agents. The American College of Sports Medicine position stand on exercise and hypertension suggests angiotensin-converting enzyme (ACE) inhibitors, angiotensin II receptor blockers, or calcium-channel blockers as antihypertensive medications with the lowest adverse impact on exercise capacity.

Other than impairment of aerobic endurance capacity, blood pressure drugs may cause other adverse health effects. Thus, a

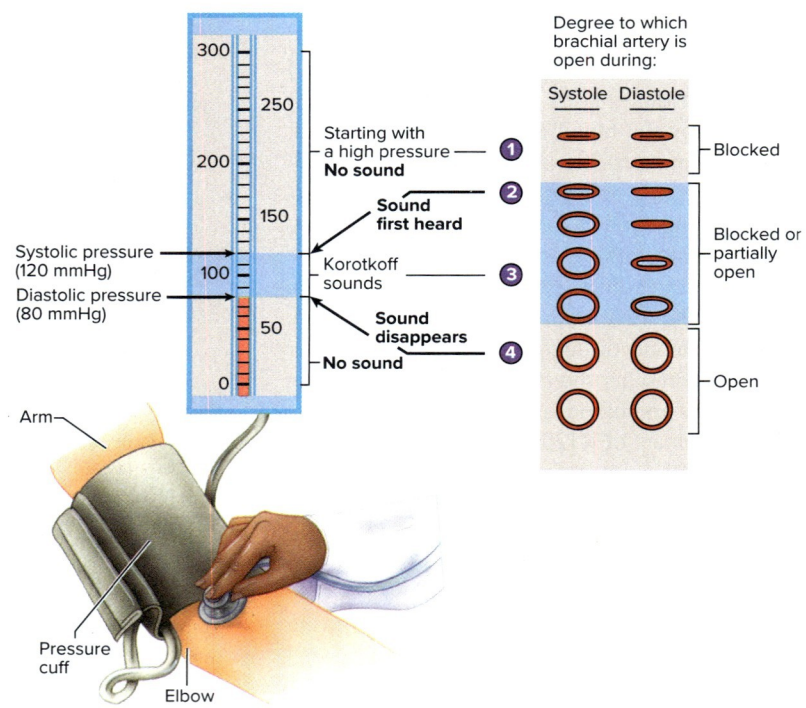

FIGURE 9.14 Blood pressure measurement.

nonpharmacological approach is often a first choice of treatment in cases of mild to moderate hypertension. According to Rabi and others, the Canadian Hypertension Education Program emphasizes the point that lifestyle modifications are the cornerstone of antihypertensive therapy.

What dietary modifications may help reduce or prevent hypertension?

How much and what you eat may influence your blood pressure. The following are the key points to help reduce or prevent hypertension.

1. *Achieve and maintain a healthy body weight.* Numerous studies have shown that reducing body weight, even as little as 10 pounds, will reduce blood pressure in overweight, hypertensive individuals. Maintaining a healthy body weight may be one of the most effective preventive measures. Thus, restriction of caloric intake to either lose or maintain body weight may be a helpful dietary strategy. In a review, Hanevold notes that the length and magnitude of salt intake increases salt sensitivity, which is associated with an increased risk of obesity and hypertension. Prevalence rates of hypertension in young adults who were salt sensitive as children are higher than in young adults who were not salt sensitive as children. Furthermore, decreased obesity can lessen salt sensitivity. Healthful methods of losing excess body fat are presented in chapter 11.

2. *Reduce or moderate sodium intake.* This remains one of the most controversial recommendations, but it may be a prudent behavior for most individuals. A 2013 review of the dietary sodium literature by the Institute of Medicine (IOM) provides no support for any universal recommendation for a particular level of sodium intake. The IOM report stated there was insufficient evidence to conclude that dietary sodium intake of <2,300 mg/day either increases or decreases disease risk. Graudal and others reported a U-shaped association between sodium intake and all-cause and cardiovascular disease mortality, with a lower mortality in "low-usual" (115–165 mmol/day = 2,645–3,795 mg/day) and "high-usual" sodium intake (166–215 mmol/day = 3,818–4,945 mg/day) compared to low (<115 mmol/day) and high sodium intake (>215 mmol/day). The IOM reported considerable methodological variations in the literature regarding the measurement of sodium intake or estimation from urinary excretion, which make it difficult to compare studies. Graudal's meta-analysis did not include any randomized, controlled trials.

In contrast, others recommend a significant reduction in current sodium intake in industrialized and developing societies. Powles and others reported 2010 global mean per capita sodium intake, estimated through urinary excretion, at 3,950 mg/day, almost twice the World Health Organization recommendation of 2,000 mg/day. Using three modeling approaches over 10 years involving, respectively, a gradual 40 percent reduction in sodium intake, an immediate 40 percent reduction to a population mean intake of 2,200 mg/day, and an immediate decrease in mean intake to 1,500 mg/day, Coxson and others projected that 28,000–500,000 deaths might be averted. He and others attributed significant decreases in stroke and ischemic heart disease mortality, decreased blood pressure, and decreased serum cholesterol in part to a 1.4 g/day decrease in sodium intake in England from 2003 to 2011.

Most health governmental agencies and professional groups promote dietary sodium restriction. Penner and others noted that one in four Canadians has hypertension, and with the lifetime risk of developing hypertension being more than 90 percent in an average life span, the need for a population-based approach to reducing hypertension is clear. Thus, the 2020 Canadian Hypertension Education Program, as documented by Rabi and others, recommended restricting sodium intake toward 2,000 mg/day in adults to prevent hypertension and to reduce blood pressure in hypertensive adults. The AI for Americans is 1,500 mg/day and the Chronic Disease Risk Reduction level is 2,300 mg/day for individuals ≥14 years of age. The American Medical Association (AMA) has urged the U.S. Food and Drug Administration (FDA) to remove salt from the GRAS (generally recognized as safe) list and require high-salt foods to carry a distinctive label, such as pictures of salt shakers bearing the word *High*. Although some question the wisdom in recommending salt restriction in all individuals, many health professionals suggest that this is a good policy.

An effort by most normotensive individuals to reduce salt intake is seen by many experts as good public health policy. Although most individuals possess physiological control systems that effectively maintain a proper balance of sodium in the body, many individuals are sodium-sensitive and their blood pressure may increase with excessive consumption of salt. Sodium may accumulate in the body, possibly due to a defect in aldosterone-mediated excretion, and hold fluids, particularly blood, which could increase blood pressure. In a recent meta-analysis of studies assessing sodium urinary excretion as a marker of sodium intake, Milajerdi and others reported a direct nonlinear association between sodium intake and cardiovascular disease mortality, which increased steeply beyond a sodium intake of 2,400 mg/day. The association between sodium intake and all-cause mortality was not significant. As discussed in a review by Richardson and others, low renin-aldosterone response, reduced potassium intake, and other humoral and genetic perturbations may predispose certain individuals (e.g., non-Hispanic blacks and Hispanics) to salt sensitivity, requiring individualized dietary and pharmacological intervention. That being so, and because many individuals do not know their blood pressure, millions of Americans may benefit from the recommendation to moderate salt intake. In a review of prospective studies, D'Elia and others reported a significant increase in gastric cancer in "high" and "moderate" salt consumers compared to "low" salt consumers.

The current prudent medical recommendation for dietary prevention or treatment of hypertension is to decrease sodium consumption simply by eating a wide variety of foods in their natural state. Avoid highly salted foods, restrict intake of processed foods, and hide your salt shakers. Individuals who sweat during training most likely do not need to be concerned with the sodium content in sports drinks. Over the course of a month, Roberts reported no change in blood pressure in normotensive individuals who worked strenuously outside and consumed about 4.5 liters of a typical CES daily.

It should be noted that decreasing salt and sodium intake poses some practical difficulties. Most salt we eat comes from the packaged foods we buy. Even minimally processed foods, such as milk and bread, may contain significant amounts of sodium. Salt is also high in most restaurant foods. Some single servings of fast food have more than 1,000 milligrams of sodium, some up to nearly 5,000 milligrams. Thus, one might have to buy most foods in their natural state and prepare and cook them at home. For those with hypertension, it may be a challenge—but worth it.

3. *Consume a diet rich in fruits, vegetables, and low-fat, protein-rich foods and with reduced saturated and total fat.* The **DASH** (Dietary Approaches to Stop Hypertension) **diet** emphasizes fruits and vegetables, nuts, low-fat dairy foods, fish, and chicken instead of red meat, low-sugar and low-refined carbohydrate foods, and reduced saturated and total fat. In a review of meta-analyses and other reviews, Chiavaroli and others concluded the DASH diet was associated with decreased incident cardiovascular disease, coronary heart disease, stroke, and systolic and diastolic blood pressure. The DASH eating plan for a 2,000-kcal diet is presented in **table 9.12**. The number of servings may be modified to meet other caloric requirements. The DASH diet is rich in potassium, magnesium, calcium, and fiber, which have been suggested to help prevent high blood pressure. According to Houston, foods that are high in potassium (e.g., raisins); nitrates, which are converted to nitrous oxide, a vasodilator (e.g., beets); omega-3 fatty acids and magnesium (e.g., walnuts); plant and lean animal protein; and low-fat fluid dairy products can help lower blood pressure.

> www.nhlbi.nih.gov/health-topics/dash-eating-plan For a detailed DASH eating plan, visit this website and type DASH Eating Plan in the search box. This site also provides other detailed information on high blood pressure.

Appel and others developed the Optimal Macro-Nutrient Intake to Prevent Heart Disease (OmniHeart) diet by replacing 10 percent of carbohydrates in the DASH diet with *good* protein (plants, chicken, and fish) and *good* fat (nuts, olive, and canola oil). Miller and others reported a 12.9 mmHg decrease in systolic blood pressure of hypertensive subjects following the traditional DASH carbohydrate intake compared to slightly greater decreases following partial substitution of

DASH carbohydrates with plant and lean animal protein (16.1 mmHg) and unsaturated fat (15.8 mmHg). They concluded that there is considerable flexibility regarding the macronutrient mix of a heart healthy diet. Chen and others found that a reduction in sugar-sweetened beverage intake over an 18-month period resulted in reductions in both systolic and diastolic blood pressures. A reduction in sugar-sweetened beverage intake is one of the goals of the *2020-2025 Dietary Guidelines for Americans*.

The DASH and OmniHeart diets are healthy, nutrient-dense, and are rich in fruits and vegetables, healthy protein, and phytonutrients and restrict bad fats, sweets, and salt. Several studies have reported that both diets can reduce other risk factors in addition to blood pressure. For example, Miller

TABLE 9.12 The DASH eating plan (servings per day based on a 2,000-kcal diet)

Food group	Servings	Serving sizes	Examples and notes	Significance to the DASH eating plan
Grains	6–8 per day	1 slice bread 1 oz dry cereal ½ cup cooked rice, pasta, or cereal	Whole wheat bread, English muffin, pita bread, bagel, cereals, oatmeal, unsalted pretzels, popcorn	Major sources of energy and fiber
Vegetables	4–5 per day	1 cup raw leafy vegetable ½ cup cooked vegetable ½ cup vegetable juice	Tomatoes, potatoes, carrots, green peas, broccoli, kale, spinach, lima beans, sweet potatoes	Rich sources of potassium, magnesium, and fiber
Fruits	4–5 per day	½ cup fruit juice 1 medium fruit ¼ cup dried fruit ½ cup fresh, frozen, or canned fruit	Apricots, bananas, dates, grapes, oranges, orange juice, melons, peaches, pineapples, raisins, strawberries	Important sources of potassium, magnesium, and fiber
Low-fat or fat-free milk and milk products	2–3 per day	1 cup milk 1 cup yogurt 1½ oz cheese	Fat-free (skim) milk, fat-free or low-fat regular or frozen yogurt, low-fat and fat-free cheese	Major sources of calcium and protein
Plant-based dairy alternatives	2–3	1–1½ cups	Soy, almond, coconut, rice, nut, or oat beverages Dairy-free ice cream and yogurt Vegan cheese	Good sources of vitamins and minerals Good sources of mono- and polyunsaturated fats No cholesterol
Lean meats, poultry, and fish	6 or less per day	1 oz cooked meats, poultry, or fish 1 egg	Select only lean; trim away visible fats; broil, roast, or boil, instead of frying; remove skin from poultry	Rich sources of protein and magnesium
Nuts, seeds, and dry beans	4–5 per week	⅓ cup or 1½ oz nuts 2 Tbsp or ½ oz seeds ½ cup cooked dry beans or peas 2 Tbsp peanut butter	Almonds, mixed nuts, peanuts, walnuts, sunflower seeds, kidney beans	Rich sources of energy, magnesium, potassium, protein, and fiber
Fats and oils	2–3 per day	1 tsp soft margarine 1 Tbsp low-fat mayonnaise 2 Tbsp light salad dressing 1 tsp vegetable oil	Soft margarine, low-fat mayonnaise, light salad dressing, vegetable oil (olive, corn, canola)	DASH has 27 percent of Kcal as fat, including fat in or added to foods 1 Tbsp fat-free dressing equals 0 servings
Sweets and added sugars	5 or less per week	1 Tbsp sugar 1 Tbsp jelly or jam ½ cup sorbet, gelatin 1 cup lemonade	Sugar, jelly, jam, jelly beans, hard candy, sorbet, ices	

Source: Adapted from The DASH Eating Plan. National Heart, Lung, and Blood Institute. For more details, go to www.nhlbi.nih.gov/files/docs/public/heart/dash_brief.pdf.

and others reported mean decreases of 19.8–23.6 mg/100 ml in low-density lipoprotein cholesterol in hyperlipidemic subjects. In a 24-year prospective, cohort study, Fung and others found that adherence to a DASH-style diet was associated with a lower risk of CHD and stroke in middle-aged women. According to Azadbakht and others, the DASH diet can likely increase HDL-cholesterol, and lower triglycerides, fasting blood glucose, and body weight.

The DASH or OmniHeart diet and reduced intake of sodium to 1,500 milligrams per day is recommended by many health professionals. Sacks and others randomly assigned individuals consuming a DASH-style diet to one of three levels of sodium intake per day. Significant reductions were observed in both systolic and diastolic blood pressure with the greatest decrease in the lowest sodium intake of 1,500 milligrams per day.

4. *Moderate alcohol consumption.* As noted in chapter 4, and discussed extensively in chapter 13, for those who consume alcohol, moderate consumption may actually confer some health benefits, particularly the prevention of cardiovascular disease. Day and Rudd report dose–response increase in relative risks of hypertension defined as systolic ≥ 140 mmHg and/or diastolic ≥ 90 mmHg of 1.7 for 50 g ethanol/day and 2.5 at 100 g/day compared to 7–14 g/day (0.5–1 drink).

5. *Be cautious with dietary supplements.* Stone and others note that a DASH-like diet includes many sources of potassium, which may improve blood pressure by various mechanisms including promoting vasodilation through improved vascular endothelial cell function, decreasing vasoconstriction caused by catecholamine release, increased Na^+-K^+-ATPase activity, and increasing sodium excretion. In a review of mostly cross-sectional studies, Perez and Chang concluded the sodium-to-potassium ratio is superior to either electrolyte alone in modulating blood pressure in hypertensive subjects. In their 2015 report of evidence-based nutrition to manage hypertension, Lennon and others recommended increased dietary intake of potassium, calcium, and magnesium and evaluated the available evidence regarding the efficacy of these nutrients as fair.

In reviews of other supplements, Rasmussen and others reported that Coenzyme Q10, fish oil, garlic, and vitamin C are associated with lower blood pressure, while ephedra, Siberian ginseng, and licorice are associated with increased blood pressure. Foods rich in inorganic nitrate are associated with lower blood pressure (BP) due to the vasodilating effects of nitric oxide. In a meta-analysis of 16 studies, Bonilla Ocampo and others reported evidence that supplementation with inorganic nitrate and beetroot juice, which is high in nitrate content, might lower blood pressure in different populations through the formation of nitric oxide. Ephedra and ginseng will be discussed in chapter 13. Given controversies and potential health risks surrounding these supplements, individuals considering *any* supplement purported to lower blood pressure should do so *only* under the guidance of their physician. There may be adverse interactions between a supplement and a prescribed medication. Furthermore, no supplement should ever replace a prescribed antihypertensive medication. More well-designed research is needed to better understand the effects of dietary and over-the-counter supplements on blood pressure.

All of the previously listed recommendations are in accord with the Prudent Healthy Diet, which was discussed in chapters 1 and 2. The more of these recommendations you follow, the better. In their meta-analysis of nonpharmacologic interventions to control hypertension, Fu and others found high-quality research evidence that a DASH diet was superior to other nonpharmacologic interventions such as exercise, low-sodium and high-potassium salt, comprehensive lifestyle modification, breathing control, and meditation, which all had moderate- to high-quality research evidence.

Can exercise help prevent or treat hypertension?

Regular mild- to moderate-intensity aerobic exercise, such as jogging, brisk walking, swimming, cycling, and aerobic dancing, has also been recommended to reduce high blood pressure. The exercise need not be continuous. In their chapter on hypertension and exercise training, Moraes-Silva and others noted that low-to-moderate intensity aerobic exercise is effective in preventing and treating hypertension and that the positive effects of resistance and isometric training on muscle and bone should complement the efficacy of aerobic training in hypertension treatment and management. They conclude by stating that the optimal control of hypertension includes a healthy diet, exercise, and, if necessary, pharmacological treatment.

Because exercise may be an effective means of losing excess body fat, it may exert a beneficial effect on blood pressure through this avenue. However, the exact role or mechanism of exercise as an independent factor in lowering blood pressure has not been totally resolved. Dempney and others report that increased excitatory and decreased inhibitory neural input to the hypothalamic paraventricular nucleus (PVN) can increase sympathetic outflow associated with hypertension. They discuss the efficacy of exercise training in decreasing hypertension by decreasing sympathetic outflow from the PVN, oxidative stress, and renin–angiotensin system activity.

Some individuals, called nonresponders, will not experience a decrease in blood pressure with exercise training. Nevertheless, most health professionals find the available information sufficient to justify an aerobic exercise program as a useful adjunct for the treatment of high blood pressure.

Most research has focused on the chronic effect of exercise training on blood pressure. Significant reductions in both diastolic and systolic blood pressure have been reported, more so in hypertensive than normotensive individuals. In their meta-analysis, Cornelissen and Smart reported reductions in systolic and diastolic pressures after endurance ($-3.5/-2.5$ mmHg); dynamic resistance ($-1.8/-3.2$ mmHg); and isometric ($-10.9/-6.2$ mmHg) training regimens. Interestingly, across all studies, combined training elicited a significant decrease in diastolic (-2.2 mmHg), but not systolic blood pressure. Endurance exercise elicited the greatest blood pressure changes in hypertensive subjects (−8.3/−5.2 mmHg) compared to normotensive subjects (−0.75/−1.1 mmHg), whereas dynamic resistance training was most effective among prehypertensive subjects (−4.0/−5.2 mmHg) compared to hyper- or normotensive subjects. In another meta-analysis, Lv and others reported comparable reductions in systolic and diastolic blood pressures (−7.5/−4.5 mmHg) to significantly lower the relative risk of major cardiovascular events, including myocardial infarction, stroke, and end-stage renal disease.

In the American College of Sports Medicine position stand on exercise and hypertension, exercise is the cornerstone therapy for the primary prevention, treatment, and control of hypertension. These are some of the major points of the position stand:

- Exercise programs that involve endurance activities, such as walking, jogging, running, or cycling, coupled with resistance training, help to prevent the development of hypertension and to lower blood pressure in adults.
- Exercise should be done daily for 30 minutes or more at a moderate level.
- A higher level of physical activity and fitness resulting from long-term exercise training has a protective effect against hypertension; fitter people with hypertension will have lower blood pressure than those who are less fit.
- Even a single-session (acute) exercise bout provides an immediate reduction in blood pressure, which can last for a major portion of the day. This phenomenon is known as **postexercise hypotension.** Brito and others suggest that lower postexercise blood pressures following acute exercise may accumulate over time to result in a chronic adaptation. Promoting the benefits of lowering blood pressure through single bouts of exercise may help motivate people to exercise. In a review comparing 27 acute aerobic and 5 acute resistance exercise studies, Anunciação and Polito reported that aerobic exercise elicited postexercise hypotension of greater magnitude and of longer duration.

Special considerations for exercise with hypertension include the following:

- Individuals with controlled hypertension and no cardiovascular or kidney disease may participate in an exercise program.
- Overweight adults should use exercise to lose weight.
- People on medications, such as beta-blockers, should be cautious of developing heat illness when exercising. As noted by the ACSM, calcium-channel blockers, ACE inhibitors, and angiotensin II receptor blockers are less likely to affect exercise tolerance.
- Adults with hypertension should extend the cool-down period of the workout; some medications may cause blood pressure to lower too much after abruptly ending exercise.

Hypertensive individuals who have concerns about exercising should consult with their physicians about the mode and intensity of exercise. There is considerable evidence that moderate-intensity aerobic exercise may lower resting blood pressure and elicit a lower exercise blood pressure response. There are some concerns that high-intensity anaerobic exercise and activities requiring intense straining, lifting, or hanging, such as isometric exercises, weight lifting, or pull-ups, might be inappropriate for some individuals. However, Cornelissen and Smart reported a significantly larger decrease in systolic blood pressure in five isometric studies compared to endurance, dynamic resistance, or combined training studies. In their meta-analysis, Marçal and others examined studies comparing high-intensity interval exercise (HIIE) and moderate intensity constant exercise (MICE) on postexercise blood pressure. Although there was no difference in 30- and 60-minute postexercise blood pressure following HIIE compared to MICE, significantly lower post-HIIE blood pressures compared to MICE were observed during subsequent daylight hours of monitoring ambulatory blood pressure. The use of handheld weights might be contraindicated for some hypertensive individuals due to the pressor (elevated blood pressure and heart rate) response that can occur with upper-body exercise. Hypertensive individuals should consult with their physicians regarding exercise indications and contraindications.

In summary, the more healthful behaviors one adopts, the greater will be the reduction in blood pressure. The JNCDET noted that a healthy lifestyle may be sufficient to avoid pharmacological therapy for some patients and is a valuable adjunct to drug therapy for most. The JNCDET quantified the potential blood pressure–lowering effects of various health behaviors, as follows:

- Weight reduction (5–20 mmHg/10 kg)
- DASH eating plan (8–14 mmHg)
- Dietary sodium reduction (2–8 mmHg)
- Increased physical activity (4–9 mmHg)
- Moderation of alcohol consumption (2–4 mmHg)

Key Concepts

▶ Lifestyle practices to help prevent or treat high blood pressure include an optimal body weight, aerobic exercise, moderation in salt and sodium intake, moderation in alcohol consumption, and increased intake of fruits, vegetables, whole grains, and low-fat, high-protein foods.

Check for Yourself

▶ Have your blood pressure checked at rest. If possible, have it checked immediately after both an aerobic-type and a resistance-type exercise.

APPLICATION EXERCISE

Determine Your Body Temperature Response and Sweating Rate

Prior to exercise:
1. Accurately measure your body temperature using an oral thermometer. Keep your mouth closed tightly during the measurement.
2. Weigh yourself exactly in your pre-exercise attire (dry, light clothes without shoes).

During exercise:
3. Exercise for 30–60 minutes, preferably outdoors under warm ambient conditions.
4. Make a mental note of the minute during your exercise session that you begin to sweat.

After exercise:
5. Immediately record your heart rate response to your exercise session.
6. Immediately record your body temperature as described in step #1.
7. Record exactly the amount of fluid you consumed *during* exercise in ounces.
8. Record the volume of any urine produced.
9. Immediately towel completely dry.
10. Change into an identical dry set of clothes and weigh yourself without shoes.
11. Calculate your body temperature increase as follows: temperature change = postexercise temperature − pre-exercise temperature.

Sweat Rate

A	Pre-exercise weight (nearest 0.25 lb)
B	Postexercise weight (nearest 0.25 lb)
C	Change in weight = A − B
D	Convert C to ounces (1 lb = 16 ounces)
E	Fluid consumed during exercise (ounces)
F	Urine produced (if any, else 0 ounces)
G	Sweat loss = C + E − F
H	Sweat rate/minute = G ÷ minutes of exercise
I	Sweat rate/hour = H × 60

12. Calculate your sweat loss as follows: sweat loss = preweight − postweight + fluid volume − urine volume.
13. Calculate your sweat rate per minute as follows: sweat rate/minute = sweat loss ÷ exercise time.
14. Calculate your sweat rate per hour as follows: sweat rate/minute × 60.

If feasible, repeat this experiment after acclimatization to exercise in the heat on a day with similar heat and humidity. Compare the pre- and postacclimatization responses, such as when you begin sweating, sweat volume, sweat saltiness, and exercise heart rate.

The final figure provides you with a guide to replenish fluids per hour. You need not fully replenish what you lose per hour, but replacing about 60 percent or more will help you prevent excessive dehydration.

Body temperature response
A. Enter your pre-exercise body temperature in degrees Fahrenheit or Celsius. _____
B. Enter your postexercise body temperature in degrees Fahrenheit or Celsius. _____
C. Subtract B from A. _____

The final figure represents your core body temperature increase for the intensity and duration of exercise and for the given environmental conditions (air temperature, humidity, solar radiation).

Review Questions—Multiple Choice

1. Which of the following is the critical level of dehydration associated with a decline in endurance performance?
   a. 1 percent
   b. 2 percent
   c. 3 percent
   d. 4 percent
   e. 5 percent
2. All of the following are aspects of acclimatization to heat and humidity *except*:
   a. sweat gland hypertrophy
   b. sweating that begins at a lower core temperature
   c. greater rate of glycogen utilization
   d. greater production of more dilute sweat
   e. increased plasma volume
3. Which of the following would most benefit from the use of a carbohydrate-electrolyte solution?
   a. recreational exerciser
   b. Little League baseball player
   c. sprinter
   d. marathon runner
4. Which aspect of rehydration during exercise does *not* increase gastric emptying *and* intestinal absorption of water, electrolytes, and carbohydrate?
   a. drinking a glucose-only carbohydrate solution
   b. drinking a cold fluid
   c. drinking large volumes up to 700 ml
   d. drinking a hypotonic solution
5. An athlete exercises at an intensity of 3 liters of oxygen per minute (5 kcal/l) for 40 minutes with a sweat rate of 20 ml/minute. Seventy-five (75) percent of this sweat evaporates. Each evaporated ml of sweat dissipates 0.6°C. The athlete

weighs 80 kg and has a mechanical efficiency of 20 percent with a specific heat of 0.83 kcal/kg of body tissue. Assume that radiation, conduction, and convection do not add or remove heat. What is the increase in body temperature from rest?
   a. 5.8
   b. 4.8
   c. 3.8
   d. 2.8
   e. 1.8
6. The most effective heat loss mechanism is:
   a. conduction
   b. convection
   c. radiation
   d. evaporation
   e. condensation
7. Select the correct information about blood pressure and hypertension.
   a. The cause of 90 percent of cases of hypertension is unknown.
   b. Determinants of blood pressure are volume (flow) and vascular resistance.
   c. Exercise and diet are effective in managing hypertension in many individuals.
   d. Hypertensive individuals should limit sodium intake and alcohol consumption.
   e. All are correct.
8. Which of the following represents Stage 1 hypertension?
   a. SBP = 130-139 mmHg *or* DBP = 80-89 mmHg
   b. SBP = 140-180 mmHg *and* DBP = 90-120 mmHg
   c. SBP = 120-139 mmHg *and* DBP < 80 mmHg
   d. SBP = <120 mmHg *and* <DBP = 80 mmHg
   e. SBP = 140-180 mmHg *or* DBP = 90-120 mmHg
9. Which of the following is *not* a major ingredient of a typical carbohydrate-electrolyte solution?
   a. water
   b. calcium
   c. sodium
   d. multiple monosaccharides
   e. potassium
10. Which thermoregulatory aid was banned by the World Anti-Doping Agency (WADA) in 2010 as a potential masking agent for other illegal performance-enhancing aids and was removed from the WADA banned list in 2018?
   a. sodium loading
   b. precooling
   c. glycerol
   d. oxygen water
   e. carbohydrate-electrolyte solutions.

Answers to multiple choice questions:
1. b; 2. c; 3. d; 4. a; 5. e; 6. d; 7. e; 8. a; 9. b; 10. c;

Critical Thinking Questions

1. Discuss physiological changes in blood, circulation, sweat glands, the anterior and posterior pituitary gland, the hypothalamus, and the kidney that occur during acclimatization to hot and humid weather.
2. Name the four components of heat stress that are recorded by the wet-bulb globe temperature (WBGT) thermometer, and discuss how each factor may contribute to heat stress during exercise under warm environmental conditions.
3. Compare and contrast CES products "A" and "B" regarding gastric emptying and intestinal absorption leading to optimal pre-exercise hydration, exercise rehydration, and postexercise rehydration:
   CES A: a hypotonic, 6 percent fructose/glucose/maltodextrin (glucose polymer) solution
   CES B: a hypertonic 10 percent glucose solution
4. List and discuss five strategies to help reduce the hazards associated with exercise in a hot environment.
5. What is high blood pressure? Why is it dangerous to your health? What lifestyle behaviors may help in its prevention or treatment?

References

Books

Food and Drug Administration, U.S. Department of Health and Human Services. Voluntary Sodium reduction goals: Target mean and upper bound concentrations for sodium in commercially processed, packaged, and prepared foods: Guidance for industry. 2016. *Federal Register* 81(106): 1-18. Thursday, June 2, 2016.

Institute of Medicine. 2005. *Dietary Reference Intakes for Water, Potassium, Sodium, Chloride, and Sulfate.* Washington, DC: The National Academies Press. https://doi.org/10.17226/10925.

Moraes-Silva I.C., et al. 2017. Hypertension and Exercise Training: Evidence from Clinical Studies. In: Xiao J. (eds) *Exercise for Cardiovascular Disease Prevention and Treatment. Advances in Experimental Medicine and Biology,* vol 1000. Springer, Singapore. https://doi.org/10.1007/978-981-10-4304-8_5.

National Academies of Sciences, Engineering, and Medicine. 2019. *Dietary Reference Intakes for Sodium and Potassium.* Washington, DC: The National Academies Press. https://doi.org/10.17226/25353.

U.S. Department of Agriculture and U.S. Department of Health and Human Services. *Dietary Guidelines for Americans, 2020-2025.* 9th Edition. December 2020. Available at DietaryGuidelines.gov.

World Anti-Doping Agency International Standard Prohibited List, January 2017. www.wada-ama.org. Accessed May 24 2017.

Reviews and Specific Studies

Aburto, N.J., et al. 2013. Effect of lower sodium intake on health: Systematic review and meta-analyses. *British Medical Journal* 346:f1326. doi: 10.1136/bmj.f1326. PMID: 23558163; PMCID: PMC4816261.

Adan, A. 2012. Cognitive performance and dehydration. *Journal of the American College of Nutrition* 31(2):71-78.

Alele, F., et al. 2020. Systematic review of gender differences in the epidemiology and risk factors of exertional heat illness and heat tolerance in the armed forces. *British Medical Journal Open* 10:e031825. doi:10.1136/bmjopen-2019-031825.

American Academy of Pediatrics. 2011. Policy Statement—Climatic heat stress and exercising children and adolescents. *Pediatrics* 128(3):e741-7. doi: 10.1542/peds.2011-1664.

American College of Sports Medicine. 2004. American College of Sports Medicine position stand: Exercise and hypertension. *Medicine & Science in Sports & Exercise* 36:533-53.

American College of Sports Medicine. 2005. Youth football: Heat stress and injury risk. *Medicine & Science in Sports & Exercise* 37:1421-30.

American College of Sports Medicine. 2007. American College of Sports Medicine Position Stand: Exertional heat illness during training and competition. *Medicine & Science in Sports & Exercise* 39:556-72.

American College of Sports Medicine. 2007. Exercise and fluid replacement. *Medicine & Science in Sports & Exercise* 39:377-90.

American College of Sports Medicine. 2016. Nutrition and Athletic Performance. Joint Position Statement. Academy of Nutrition and Dietetics, Dietitians of Canada, and the American College of Sports Medicine. *Medicine and Science in Sports and Exercise* 48(3):543-68.

Anastasiou, C., et al. 2009. Sodium replacement and plasma sodium drop during exercise in the heat when fluid intake matches fluid loss. *Journal of Athletic Training* 44:117-23.

Anunciação, P. G., and Polito, M. D. 2011. Review on post-exercise hypotension in hypertensive individuals. *Arquivos Brasileiros de Cardiologia* 96 (5):e100-9.

Appel, L., et al. 2005. Effects of protein, monounsaturated fat, and carbohydrate intake on blood pressure and serum lipids: Results of the OmniHeart randomized trial. *Journal of the American Medical Association* 294:2455-64.

Armstrong, L., et al. 2005. Fluid, electrolyte, and renal indices of hydration during 11 days of controlled caffeine consumption. *International Journal of Sport Nutrition and Exercise Metabolism* 15:252-65.

Armstrong, L., et al. 2006. No effect of 5% hypohydration on running economy of competitive runners at 23 degrees C. *Medicine & Science in Sports & Exercise* 38:1762-69.

Asplund, C. A., et al. 2011. Exercise-associated collapse: An evidence-based review and primer for clinicians. *British Journal of Sports Medicine* 45(14):1157-62.

Azadbakht, L., et al. 2005. Beneficial effects of a Dietary Approaches to Stop Hypertension eating plan on features of the metabolic syndrome. *Diabetes Care* 28:2823-31.

Baker, L., et al. 2005. Sex differences in voluntary fluid intake by older adults during exercise. *Medicine & Science in Sports & Exercise* 37:789-96.

Baker, L., et al. 2007. Dehydration impairs vigilance-related attention in male basketball players. *Medicine & Science in Sports & Exercise* 39:976-83.

Baker, L., et al. 2007. Progressive dehydration causes a progressive decline in basketball skill performance. *Medicine & Science in Sports & Exercise* 39:1114-23.

Baker, L. B., and Jeukendrup, A. E. 2014. Optimal composition of fluid-replacement beverages. *Comprehensive Physiology* 4(2):575-620.

Baker, L. B. 2017. Sweating rate and sweat sodium concentration in athletes: A review of methodology and intra/interindividual variability. *Sports Medicine* 47(S1):S111-S128.

Bassett, D. R., et al. 1987. Thermoregulatory responses to skin wetting during prolonged treadmill running. *Medicine and Science in Sport and Exercise* 19 (1):28-32.

Basu, R., et al. 2012. The effect of high ambient temperature on emergency room visits. *Epidemiology* 23 (6):813-20.

Bell, C.S. et al. 2019. Prevalence of hypertension in children. *Hypertension* 73(1):148-152. doi: 10.1161/hypertensionaha.118.11673. PMID: 30571555; PMCID: PMC6291260.

Below, P., et al. 1995. Fluid and carbohydrate ingestion independently improve performance during 1 h of intense exercise. *Medicine & Science in Sports & Exercise* 27:200-210.

Bergeron, M. F. 2008. Muscle cramps during exercise–Is it fatigue or electrolyte deficit? *Current Sports Medicine Reports* 7(4):S50-S55.

Bilzon, J. L., et al. 2002. Influence of glucose ingestion by humans during recovery from exercise on substrate utilisation during subsequent exercise in a warm environment. *European Journal of Applied Physiology* 87 (4-5):318-326. doi: 10.1007/s00421-002-0614-4.

Boddu, P., et al. 2016. Sudden cardiac arrest from heat stroke: Hidden dangers of hot yoga. *American Journal of Medicine* 129(8):e129-30. doi: 10.1016/j.amjmed.2016.03.030.

Bongers, C. C. W. G., et al. 2015. Precooling and percooling (cooling during exercise) both improve performance in the heat: A meta-analytical review. *British Journal of Sports Medicine* 49(6):377-84.

Bonilla Ocampo, D.A., et al. 2018. Dietary nitrate from beetroot juice for hypertension: A systematic review. *Biomolecules* 8(4):134. doi: 10.3390/biom8040134. PMID: 30400267; PMCID: PMC6316347.

Brito, L.C., et al. 2018. Postexercise hypotension as a clinical tool: A "single brick" in the wall. *Journal of the American Society of Hypertension* 12(12):e59-e64. doi: 10.1016/j.jash.2018.10.006.

Buffington, M.A., and Abreo, K. 2016. Hyponatremia: A Review. *Journal of Intensive Care Medicine* 31(4):223-236. doi: 10.1177/0885066614566794.

Burdon, C. A., et al. 2012. Influence of beverage temperature on palatability and fluid ingestion during endurance exercise: A systematic review. *International Journal of Sport Nutrition and Exercise Metabolism* 22:199-211.

Byrne, C., et al. 2006. Continuous thermoregulatory responses to mass-participation distance running in heat. *Medicine & Science in Sports & Exercise* 38:803-10.

Centers for Disease Control and Prevention (CDC). 2021. *Hypertension cascade: Hypertension prevalence, treatment and control estimates among US adults aged 18 years and older applying the criteria from the American College of Cardiology and American Heart Association's 2017 Hypertension Guideline–NHANES 2015-2018.* Atlanta, GA: US Department of Health and Human Services.

Charkoudian, N., and Stachenfeld, N. 2016. Sex hormone effects on autonomic mechanisms of thermoregulation in humans. *Autonomic Neuroscience* 196:75-80. doi: 10.1016/j.autneu.2015.11.004.

Chen, L., et al. 2010. Reducing consumption of sugar-sweetened beverages is associated with reduced blood pressure: A prospective study among United States adults. *Circulation* 121:2398-406.

Cheuvront, S. N., and Kenefick, R. W. 2014. Dehydration: Physiology, assessment and performance effects. *Comprehensive Physiology* 4 (1):257-85. doi: 10.1002/cphy.c130017.

Cheuvront, S. N., et al. 2013. Physiologic basis for understanding quantitative dehydration assessment. *American Journal of Clinical Nutrition* 97:455-462.

Cheuvront, S., and Haymes, E. 2001. Ad libitum fluid intakes and thermoregulatory responses of female distance runners in three environments. *Journal of Sport Sciences* 19:845-54.

Choma, C. W., et al. 1998. Impact of rapid weight loss on cognitive function in collegiate wrestlers. *Medicine and Science in Sports and Exercise* 30(5):746-49.

Church, A., et al. 2017. Transition duration of ingested deuterium oxide to eccrine sweat during exercise in the heat. *Journal of*

Thermal Biology 63:88–91. doi: 10.1016/j.jtherbio.2016.11.018.

Clarke, N., et al. 2005. Strategies for hydration and energy provision during soccer-specific exercise. *International Journal of Sport Nutrition and Exercise Metabolism* 15:625–40.

Cornelissen, V. A., and Smart, N .A. 2013. Exercise training for blood pressure: A systematic review and meta-analysis. *Journal of the American Heart Association* 2:e004473. doi: 10.1161/jaha.112.004473.

Costill, D. 1990. Gastric emptying of fluids during exercise. In *Perspectives in Exercise Science and Sports Medicine. Fluid Homeostasis during Exercise,* ed. C. Gisolfi and D. Lamb. Indianapolis, IN: Benchmark.

Costill, D., et al. 1982. Dietary potassium and heavy exercise: Effects on muscle water and electrolytes. *American Journal of Clinical Nutrition* 36:266–75.

Coutts, A., et al. 2002. The effect of glycerol hyperhydration on Olympic distance triathlon performance in high ambient temperatures. *International Journal of Sport Nutrition and Exercise Metabolism* 12:105–19.

Coxson, P. G., et al. 2013. Mortality benefits from US population-wide reduction in sodium consumption. Projections from 3 modeling approaches. *Hypertension* 61:564–70.

D'Elia, L., et al. 2012. Habitual salt intake and risk of gastric cancer: A meta-analysis of prospective studies. *Clinical Nutrition* 31(4):489–98. doi: 10.1016/j.clnu.2012.01.003.

Dampney, R.A., et al. 2018. Regulation of sympathetic vasomotor activity by the hypothalamic paraventricular nucleus in normotensive and hypertensive states. *American Journal of Physiology. Heart and Circulatory Physiology* 315(5):H1200–H1214. doi: 10.1152/ajpheart.00216.2018. PMID: 30095973; PMCID: PMC6297824.

Day, E., and Rudd, J.H.F. 2019. Alcohol use disorders and the heart. *Addiction* 114(9):1670–78. doi: 10.1111/add.14703. PMID: 31309639; PMCID: PMC6771559.

de Castro, J. 2005. Stomach filling may mediate the influence of dietary energy density on the food intake of free-living humans. *Physiology & Behavior* 86:32–45.

de Oliveira, E. P., et al. 2014. Gastrointestinal complaints during exercise: Prevalence, etiology, and nutritional recommendations. *Sports Medicine* 44 (Supplement 1):S79–S85.

Dignam, T. et al. 2019. Control of lead sources in the United States, 1970–2017: Public health progress and current challenges to eliminating lead exposure. *Journal of Public Health Management and Practice* Jan/Feb;25 Suppl 1, Lead Poisoning Prevention (Suppl 1 Lead Poisoning Prevention):S13–22. doi: 10.1097/PHH.0000000000000889. PMID: 30507765; PMCID: PMC6522252.

Dougherty, K., et al. 2006. Two percent dehydration impairs and six percent carbohydrate drink improves boy's basketball skills. *Medicine & Science in Sports & Exercise* 38:1650–58.

Douma, M.J., et al. 2020. First aid task force of the International Liaison Committee on Resuscitation. First aid cooling techniques for heat stroke and exertional hyperthermia: A systematic review and meta-analysis. *Resuscitation* 1(148):173–90. doi: 10.1016/j.resuscitation.2020.01.007.

Duffield, R., et al. 2012. Hydration, sweat and thermoregulatory responses to professional football training in the heat. *Journal of Sports Science* 30 (10):957–65.

Edwards, A., et al. 2007. Influence of moderate dehydration on soccer performance: Physiological responses to 45 min of outdoor match-play and the immediate subsequent performance of sport-specific and mental concentration tests. *British Journal of Sports Medicine* 41:385–91.

Ehret, G.B., et al. 2016. The genetics of blood pressure regulation and its target organs from association studies in 342,415 individuals. *Nature Genetics* 48(10):1171–84. doi: 10.1038/ng.3667.

Eichner, R. 2007. The role of sodium in "heat cramping." *Sports Medicine* 37:368–70.

Eijsvogels, T. M., et al. 2013. Sex differences in fluid balance responses during prolonged exercise. *Scandinavian Journal of Medicine and Science in Sports* 23 (2):198–206.

El-Sharkawy, A. M., et al. 2015. Acute and chronic effects of hydration status on health. *Nutrition Reviews* 73(S2):97–109.

European Food Safety Authority. 2011. Scientific opinion on the substantiation of health claims related to carbohydrate-electrolyte solutions and reduction in rated perceived exertion/effort during exercise, enhancement of water absorption during exercise, and maintenance of endurance performance pursuant to Article 13(1) of Regulation (EC) No 1924/20061. *European Food Safety Authority Journal* 9 (6):2211–40.

Evans, G. H., et al. 2017. Optimizing the restoration and maintenance of fluid balance after exercise-induced dehydration. *Journal of Applied Physiology* 122: 945–51.

Falk, B., and Dotan. R. 2011. Temperature regulation and elite young athletes. *Medicine and Sport Science* 56:126–49. doi: 10.1159/000320645.

Fechter-Leggett, E.D., et al. 2016. Heat stress illness emergency department visits in national environmental public health tracking states, 2005–2010. *Journal of Community Health* 41(1):57–69. doi: 10.1007/s10900-015-0064-7. PMID: 26205070; PMCID: PMC4715715.

Ferretti, F., and Mariani, M. 2019. Sugar-sweetened beverage affordability and the prevalence of overweight and obesity in a cross section of countries. *Global Health* Apr 18;15(1):30. doi: 10.1186/s12992-019-0474-x. PMID: 30999931; PMCID: PMC6472017.

Fleming, N., et al. 2017. Ingestion of oxygenated water enhances lactate clearance kinetics in trained runners. *Journal of the International Society of Sports Nutrition* Mar 29;14:9. doi: 10.1186/s12970-017-0166-y. PMID: 28360825; PMCID: PMC5371271.

Food and Drug Administration. 2008. Guidance for industry: A labeling guide for restaurants and other retail establishments selling away-from-home foods. (April).

Fritzsche, R., et al. 2000. Water and carbohydrate ingestion during prolonged exercise increase maximal neuromuscular power. *Journal of Applied Physiology* 88:730–37.

Fu, J., et al. 2020. Nonpharmacologic interventions for reducing blood pressure in adults with prehypertension to established hypertension. *Journal of the American Heart Association* 9(19):e016804. doi: 10.1161/JAHA.120.016804. PMID: 32975166; PMCID: PMC7792371.

Fudge, B., et al. 2008. Elite Kenyan endurance runners are hydrated day-to-day with *ad libitum* fluid intake. *Medicine & Science in Sports & Exercise* 40:1171–79.

Fung, T., et al. 2008. Adherence to a DASH-style diet and risk of coronary heart disease and stroke in women. *Archives of Internal Medicine* 168:713–20.

Garth, A. K., and Burke, L. M. 2013. What do athletes drink during competitive sporting activities? *Sports Medicine* 43 (7):539–64.

Gillespie, C. D., and Hurvitz, K. A. 2013. Prevalence of hypertension and controlled hypertension—United States, 2007–2010. *Morbidity and Mortality Weekly Report* 62(03):144–48. www.cdc.gov/mmwr/.

Gisolfi, C. 1996. Fluid balance for optimal performance. *Nutrition Reviews* 54: S159–S168.

Gisolfi, C., et al. 2001. Intestinal fluid absorption during exercise: Role of sport drink osmolality and [Na^1]. *Medicine & Science in Sports & Exercise* 33:907–15.

Godek, S. F., et al. 2010. Sweat rates, sweat sodium concentrations, and sodium losses in 3 groups of professional football players. *Journal of Athletic Training* 45(4):364–71.

Gomes, L. H. L. S., et al. 2013. Thermoregulatory responses of children exercising in a hot environment. *Revista Paulista de Pediatria* 31 (1):104–10.

Goulet, E.D.B. 2013. Effect of exercise-induced dehydration on endurance performance: Evaluating the impact of exercise protocols on outcomes using a meta-analytic procedure. *British Journal of Sports Medicine* 47:679–86.

Goulet, E., et al. 2007. A meta-analysis of the effects of glycerol-induced hyperhydration on fluid retention and endurance performance. *International Journal of Sport Nutrition and Exercise Metabolism* 17:391–410.

Graudal, N., et al. 2014. Compared with usual sodium intake, low- and excessive-sodium diets are associated with increased mortality: A meta-analysis. *American Journal of Hypertension* (April 26). [Epub ahead of print] doi:10.1093/ajh/hpu028

Giuriato, G., et al. 2018. Muscle cramps: A comparison of the two-leading hypothesis. *Journal of Electromyography and Kinesiology* 41:89–95. doi: 10.1016/j.jelekin.2018.05.006.

Hanevold, C.D. 2021. Salt sensitivity of blood pressure in childhood and adolescence. *Pediatric Nephrology* Jul 29. doi: 10.1007/s00467-021-05178-6. Epub ahead of print. PMID: 34327584.

Hausswirth, C. et al. 2012. Post-exercise cooling interventions and the effects on exercise-induced heat stress in a temperate environment. *Applied Physiology Nutrition and Metabolism* 37:965–975.

He, F. J., et al. 2014. Salt reduction in England from 2003 to 2011: Its relationship to blood pressure, stroke, and ischaemic heart disease mortality. *British Medical Journal Open* 4 (4):e004549. doi: 10.1136/bmjopen-2013-004549.

Hew-Butler, T., et al. 2003. The incidence, risk factors, and clinical manifestations of hyponatremia in marathon runners. *Clinical Journal of Sports Medicine* 13:41–47.

Hew-Butler, T., et al. 2015. Statement of the third international exercise-associated hyponatremia consensus development conference. *Clinical Journal of Sports Medicine* 25:303–20.

Hew-Butler, T., et al. 2006. Sodium supplementation is not required to maintain serum sodium concentrations during an Ironman triathlon. *British Journal of Sports Medicine* 40:255–59.

Higgins, J.P., et al. 2018. Energy drinks: A contemporary issues paper. *Current Sports Medicine Reports* 17(2):65–72. doi: 10.1249/JSR.0000000000000454. PMID: 29420350.

Hillman, A. R., et al. 2013. A comparison of hyperhydration versus ad libitum fluid intake strategies on measures of oxidative stress, thermoregulation, and performance. *Research in Sports Medicine* 21 (4):305–17. doi: 10.1080/15438627.2013.825796.

Hoffman, M. D., et al. 2015. Sodium intake during an ultramarathon does not prevent muscle cramping, dehydration, hyponatremia, or nausea. *Sports Medicine - Open* 1:39. doi 10.1186/s40798-015-0040-x.

Holland, J. J., et al. 2017. The influence of drinking fluid on endurance cycling performance: A meta-analysis. *Sports Medicine* doi: 10.1007/s40279-017-0739-6. [Epub ahead of print].

Horner, K. M., et al. 2015. Acute exercise and gastric emptying: A meta-analysis and implications for appetite control. *Sports Medicine* 45:659–78.

Houston, M. 2014. The role of nutrition and nutraceutical supplements in the treatment of hypertension. *World Journal of Cardiology* 6(2):38–66.

Hoyt, R.W., and Honig, A. 1996. Environmental influences on body fluid balance during exercise: Altitude. In: Buskirk, E.R., Puhl, S.M., Eds. *Body fluid balance: Exercise and sport.* Boca Raton, FL: CRC Press. pp. 183–196.

Institute of Medicine. 2013. *Sodium Intake in Populations: Assessment of Evidence.* Washington, DC: National Academies Press.

Jeukendrup, A. E. 2007. Fueling during exercise. www.nestlenutrition-institute.org/resources/library/Secured/sport Focus/Documents/Publication00266/6_Fueling%20During%20Exercise.pdf.

Jeukendrup, A. E. 2014. A step towards personalized sports nutrition: Carbohydrate intake during exercise. *Sports Medicine* 44 (Suppl 1):S25–33.

Jeukendrup, A. E. 2003. Modulation of carbohydrate and fat utilization by diet, exercise and environment. *Biochemical Society Transactions* 31(Pt 6):1270–73.

Jeukendrup, A. E. 2017. Training the gut for athletes. *Sports Medicine* 47(S1):101–10.

John, S. K., et al. 2011. Life-threatening hyperkalemia from nutritional supplements: Uncommon or undiagnosed? *American Journal of Emergency Medicine* 29 (9):1237.e1-2. doi: 1016/j,ajem.2010.08.029.

Johnson, A. 2007. The sensory psychobiology of thirst and salt appetite. *Medicine & Science in Sports & Exercise* 39:1388–400.

Joint National Committee on Prevention, Detection, Evaluation and Treatment of High Blood Pressure. 2003. The Seventh Report of the Joint National Committee on Prevention, Detection, Evaluation and Treatment of High Blood Pressure. *Hypertension* 42:1206–52.

Judelson, D., et al. 2007. Effect of hydration state on strength, power, and resistance exercise performance. *Medicine & Science in Exercise & Sports* 39:1817–24.

Junge, N., et al. 2016. Prolonged self-paced exercise in the heat-environmental factors affecting performance. *Temperature* 3(4):539–48.

Kaciuba-Uscilko, H., and Grucza, R. 2001 Gender differences in thermoregulation. *Current Opinions in Clinical Nutrition and Metabolic Care.* 4(6):533–36. doi: 10.1097/00075197-200111000-00012. PMID: 11706289.

Kavouras, S., et al. 2006. Rehydration with glycerol: Endocrine, cardiovascular, and thermoregulatory responses during exercise in the heat. *Journal of Applied Physiology* 100:442–50.

Kenefick, R. W., and Cheuvront, S. N. 2012. Hydration for recreational sport and physical activity. *Nutrition Reviews* 70:S137–42.

Kenefick, R., and Sawka, M. 2007. Heat exhaustion and dehydration as causes of marathon collapse. *Sports Medicine* 37:378–81.

Kenefick, R., et al. 2002. Hypohydration adversely affects lactate threshold in endurance athletes. *Journal of Strength and Conditioning Research* 16:38–43.

Kenney, W.L., et al. 2021. Temperature regulation during exercise in the heat: Insights for the aging athlete. *Journal of Science and Medicine in Sport* 24(8):739–46. doi: 10.1016/j.jsams.2020.12.007.

Kerr, Z.Y., et al. 2020. The epidemiology and management of exertional heat illnesses in high school sports during the 2012/2013–2016/2017 academic years. *Journal of Sport Rehabilitation* 29(3):332–38. doi: 10.1123/jsr.2018-0364.

Klous, L., et al. 2020. Sweat rate and sweat composition during heat acclimation. *Journal of Thermal Biology* Oct;93:102697. doi: 10.1016/j.jtherbio.2020.102697. Epub 2020 Aug 26. PMID: 33077118.

Knechtle, B., et al. 2019. Exercise-associated hyponatremia in endurance and ultra-endurance performance-aspects of sex, race location, ambient temperature, sports discipline, and length of performance: A narrative review. *Medicina (Kaunas)* 55(9):537. doi: 10.3390/medicina55090537. PMID: 31455034; PMCID: PMC6780610.

Kraft, J. A., et al. 2012. The influence of hydration on anaerobic performance: A review. *Research Quarterly for Exercise and Sport* 83(2):282–92.

Krohn, A. R., et al. 2015. Heat illness in football: Current concepts. *Current Sports Medicine Reports* 14(6):463–71.

Lalumandier, J., and Ayers, L. 2000. Flouride and bacterial content of bottled water vs tap water. *Archives of Family Medicine* 9:246–50.

Lang, F. 2011. Effect of cell hydration on metabolism. Nestlé Nutrition Institute Workshop Series 69:115–26; discussion 126–130. doi: 10.1159/000329290.

Lara, B., et al. 2016. Interindividual variability in sweat electrolyte concentration in marathoners. *Journal of the International Society of Sports Nutrition* 13:31. doi: 10.1186/s12970-016-0141-z.

Lefferts, L. 1990. Water: Treat it right. *Nutrition Action Health Letter* 17:5–7.

Leiper, J. B. 2015. Fate of ingested fluids: Factors affecting gastric emptying and intestinal absorption of beverages in humans. *Nutrition Reviews* 73(S2):57-72.

Lennon, S.L., et al. 2017. Evidence analysis library evidence-based nutrition practice guideline for the management of hypertension in adults. *Journal of the Academy of Nutrition and Dietetics* 117(9):1445-58.e17. doi: 10.1016/j.jand.2017.04.008.

Leon, L. R., and Bouchama, A. 2015. Heat stroke. *Comprehensive Physiology* 5(2):611-47.

Li, L., et al. 2018. Effects of whey protein in carbohydrate-electrolyte drinks on post-exercise rehydration. *European Journal of Sport Science* 18(5):685-94. doi: 10.1080/17461391.2018.1442499.

Logan-Sprenger, H. M., et al. 2012. Effects of dehydration during cycling in skeletal muscle metabolism in females. *Medicine and Science in Sports and Exercise* 44(10):1949-57.

Lv, J., et al. 2012. Effects of intensive blood pressure lowering on cardiovascular and renal outcomes: A systematic review and meta-analysis. *Public Library of Science Medicine* 9(8): e1001293. doi:10.1371/journal.pmed.1001293

Malchaire, J.B.M. 2006. Occupational heat stress assessment by the predicted heat strain model. *Industrial Health* 44(3):380-7. doi: 10.2486/indhealth.44.380. PMID: 16922181.

Marçal, I.R., et al. 2021. Post-exercise hypotension following a single bout of high intensity interval exercise vs. a single bout of moderate intensity continuous exercise in adults with or without hypertension: A systematic review and meta-analysis of randomized clinical trials. *Frontiers in Physiology* 12:675289. doi: 10.3389/fphys.2021.675289. PMID: 34262474; PMCID: PMC8274970.

Marino, F. E. 2011. The critical limiting temperature and selective brain cooling: Neuroprotection during exercise? *International Journal of Hyperthermia* 27 (6):582-590.

Mason, S.A., et al. 2018. Synthetic polymer contamination in bottled water. *Frontiers in Chemistry* 11 September https://doi.org/10.3389/fchem.2018.00407.

Massicotte, D., et al. 2006. Metabolic fate of a large amount of ^{13}C-glycerol ingested during prolonged exercise. *European Journal of Applied Physiology* 96:322-29.

Maughan, R. J. 2012. Hydration, morbidity, and mortality in vulnerable populations. *Nutrition Reviews* 70:S152-55.

Maughan, R. J., and Meyer, N. L. 2012. Hydration during intense exercise training. 76th Nestlé Nutrition Institute Workshop, Oxford.

Maughan, R., and Shirreffs, S. 2010. Development of hydration strategies to optimize performance for athletes in high-intensity sports and in sports with repeated intense efforts. *Scandinavian Journal of Medicine & Science in Sports* 20 (Suppl 2):59-69.

Maughan, R., et al. 2005. Fluid and electrolyte balance in elite male football (soccer) players training in a cool environment. *Journal of Sports Sciences* 23:72-79.

Maughan, R., et al. 2007. Errors in the estimation of hydration status from changes in body mass. *Journal of Sports Sciences* 25:797-804.

Mazerolle, S. M., et al. 2011. Is oral temperature an accurate measurement of deep body temperature? A systematic review. *Journal of Athletic Training* 46 (5):566-73.

McCubbin, A.J., et al. 2020. Sports Dietitians Australia position statement: Nutrition for exercise in hot environments. *International Journal of Sports Nutrition and Exercise Metabolism* 30(1):83-98. doi: 10.1123/ijsnem.2019-0300. PMID: 31891914.

McKenna, Z. J., and Gillum, T. L. 2017. Effects of exercise induced dehydration and glycerol rehydration on anaerobic power in male collegiate wrestlers. *Journal of Strength and Conditioning Research* 31(11):2965-68. doi: 10.1519/JSC.0000000000001871. PMID: 28240714.

Mekjavic, I.B., and Tipton, M.J. 2020. Myths and methodologies: Degrees of freedom—limitations of infrared thermographic screening for Covid-19 and other infections. *Experimental Physiology* Dec 28. doi: 10.1113/EP089260.

Meyer, F., et al. 2007. Effect of age and gender on sweat lactate and ammonia concentrations during exercise in the heat. *Brazilian Journal of Medical and Biological Research* 40:135-43.

Milajerdi, A., et al. 2019. Dose-response association of dietary sodium intake with all-cause and cardiovascular mortality: A systematic review and meta-analysis of prospective studies. *Public Health Nutrition* 22(2):295-306. doi: 10.1017/S1368980018002112.

Millard-Stafford, M. 1997. Water versus carbohydrate-electrolyte ingestion before and during a 15-km run in the heat. *International Journal of Sport Nutrition* 7:26-38.

Miller, E. R., et al. 2006. The effects of macronutrients on blood pressure and lipids: An overview of the DASH and OmniHeart Trials. *Current Atherosclerosis Reports* 8(6):46-51.

Miller, K. C., et al. 2010. Exercise-associated muscle cramps: Causes, treatment, and prevention. *Sports Health* 2(4):279-83.

Mills, K.T., et al. 2020. The global epidemiology of hypertension. *Nature Reviews. Nephrology* 16(4):223-37. doi: 10.1038/s41581-019-0244-2.

Montain, S., et al. 1998. Hypohydration effects on skeletal muscle performance and metabolism: A 31P-MRS study. *Journal of Applied Physiology* 84:1889-94.

Montain, S., et al. 1998. Thermal and cardiovascular strain from hypohydration: Influence of exercise intensity. *International Journal of Sports Medicine* 19:87-91.

Montain, S., et al. 2007. Marathon performance in thermally stressing conditions. *Sports Medicine* 37:320-23.

Mora-Rodrigues, R., and Hamouti, N. 2012. Salt and fluid loading: Effects on blood volume and exercise performance. *Medicine and Sport Science* 59:113-19.

Morbidity and Mortality Weekly Report (MMWR) QuickStats: Number of Heat-Related Deaths by Sex 1999-2010. National Vital Statistics System, United States 61 (36);729 September 14, 2012. www.cdc.gov/mmwr/.

Morris, D. M., et al. 2015. Acute sodium ingestion before exercise increases voluntary water consumption resulting in preexercise hyperhydration and improvement in exercise performance in the heat. *International Journal of Sport Nutrition and Exercise Metabolism* 25:456-62.

Moyen, N. E., et al. 2014. Increasing relative humidity impacts low-intensity exercise in the heat. *Aviation Space and Environmental Medicine* 85:112-19.

Murray, K. E., et al. 2010. Prioritizing research for trace pollutants and emerging contaminants in the freshwater environment. *Environmental Pollution* 158:3462-71.

National Collegiate Athletic Association. National NCAA Wrestling Weight Management Program. http://www.sectiononewrestling.com/ncaa_wrestling_weight_management_policy.pdf. Accessed September 1, 2021.

Noakes, T. 2007. Drinking guidelines for exercise: What evidence is there that athletes should drink "as much as tolerable," "to replace the weight lost during exercise" or "ad libitum?" *Journal of Sports Sciences* 25:781-96.

Noakes, T. 2007. Hydration in the marathon. *Sports Medicine* 37:463-66.

Notley, S.R., et al. 2020. Exercise thermoregulation in prepubertal children: A brief methodological review. *Medicine and Science in Sports and Exercise* 52(11):2412-22.

Nybo, L., et al. 2014. Performance in the heat-physiological factors of importance for hyperthermia-induced fatigue. *Comprehensive Physiology* 4(2):657-89.

Office of Dietary Supplement Programs. Center for Food Safety and Applied Nutrition. Food and Drug Administration. 2018. Highly concentrated caffeine in dietary supplements: Guidance for industry. www.fda.gov/media/112363/download

Ogden, H.B., et al. 2020. Influence of aerobic fitness on gastrointestinal barrier integrity and microbial translocation following a fixed-intensity military exertional heat stress test. *European Journal of Applied Physiology* 120(10):2325-37. doi: 10.1007/s00421-020-04455-w.

Osterberg, K., et al. 2010. Carbohydrate exerts a mild influence on fluid retention following exercise-induced dehydration. *Journal of Applied Physiology* 108:245-50.

Pallarés, J.G., et al. 2016. Muscle contraction velocity, strength and power output changes following different degrees of hypohydration in competitive olympic combat sports. *Journal of the International Society of Sports Nutrition* Mar 8;13:10. doi: 10.1186/s12970-016-0121-3. PMID: 26957952; PMCID: PMC4782333.

Pandolf, K. B., et al. 2011. United States Army Research Institute of Environmental Medicine: Warfighter research focusing on the past 25 years. *Advances in Physiology Education* 35:353-60.

Peacock, O. J., et al. 2012. Voluntary drinking behavior, fluid balance, and psychological affect when ingesting water or a carbohydrate-electrolyte solution during exercise. *Appetite* 58:56-63.

Penner, S., et al. 2007. Dietary sodium and cardiovascular outcomes: a rational approach. *Canadian Journal of Cardiology* 23:567-72.

Perez, V., and Chang, E. T. 2014. Sodium-to-potassium ratio and blood pressure, hypertension, and related factors. *Advances in Nutrition* 5(6):712-41.

Piantadosi, C. A. 2006. "Oxygenated" water and athletic performance. *British Journal of Sports Medicine* 40(9):740-41.

Popkin, B., et al. 2006. A new proposed guidance system for beverage consumption in the United States. *American Journal of Clinical Nutrition* 83:529-42.

Poulter, N. L., et al. 2015. Hypertension. *Lancet* 386: 801-12.

Powles, J., et al. 2013. Global, regional and national sodium intakes in 1990 and 2012; a systematic analysis of 24 h urinary sodium excretion and dietary surveys worldwide. *British Medical Journal Open* 3 (12):e003733. doi: 10.1136/bmjopen-2013-003733.

Périard, J. D., et al. 2011. Neuromuscular function following prolonged intense self-paced exercise in hot climatic conditions. *European Journal of Applied Physiology* 111:1561-69.

Périard, J.D., et al. 2021. Exercise under heat stress: Thermoregulation, hydration, performance implications, and mitigation strategies. *Physiological Reviews* 101(4):1873-79. doi: 10.1152/physrev.00038.2020. Epub 2021 Apr 8. PMID: 33829868.

Quader, Z. S., et al. 2017. Sodium intake among persons aged ≥2 years—United States, 2013-2014. *Morbidity and Mortality Weekly Report* March 31, 66(12):324-38.

Rabi, D.M., et al. 2020. Hypertension Canada's 2020 comprehensive guidelines for the prevention, diagnosis, risk assessment, and treatment of hypertension in adults and children. *Canadian Journal of Cardiology* 36(5):596-624. doi: 10.1016/j.cjca.2020.02.086. PMID: 32389335.

Racinais, S., and Oksa, J. 2010. Temperature and neuromuscular function. *Scandinavian Journal of Medicine and Science in Sports* 20(S3):1-18.

Racinais, S., et al. 2015. Consensus recommendations on training and competing in the heat. *Scandinavian Journal of Medicine and Science in Sports* 25(S1): 6-19.

Rasmussen, C. B., et al. 2012. Dietary supplements and hypertension: Potential benefits and precautions. *Journal of Clinical Hypertension* 14(7): 467-71.

Richardson, S. I., et al. 2013. Salt sensitivity: A review with a focus on non-Hispanic blacks and Hispanics. *Journal of the American Society of Hypertension* 7 (2):170-79.

Roberts, D. 2006. Blood pressure response to 1-month electrolyte-carbohydrate beverage consumption. *Journal of Occupational and Environmental Hygiene* 3:131-36.

Roberts, W. 2007. Exercise-associated collapse care matrix in the marathon. *Sports Medicine* 37:431-33.

Rodríguez, M.Á., et al. 2020. A matter of degrees: A systematic review of the ergogenic effect of pre-cooling in highly trained athletes. *International Journal of Environmental Research and Public Health* 7(8):2952. doi: 10.3390/ijerph17082952. PMID: 32344616; PMCID: PMC7215649.

Rogers, I., and Hew-Butler, T. 2009. Exercise-associated hyponatremia: Overzealous fluid consumption. *Wilderness and Environmental Medicine* 20:139-43.

Ronneberg, K., et al. 2008. Temporal artery temperature measurements do not detect hyperthermic marathon runners. *Medicine & Science in Sports & Exercise* 40:1373-75.

Rowlands, D. S., et al. 2011. Unilateral fluid absorption and effects on peak power after ingestion of commercially available hypotonic, isotonic, and hypertonic sports drinks. *International Journal of Sports Nutrition and Exercise Metabolism* 21 (6):480-91.

Rowlands, D. S. 2015. Fructose-glucose composite carbohydrates and endurance performance: Critical review and future perspectives. *Sports Medicine* 45:1561-76.

Rush, E. C. 2013. Water: Neglected, unappreciated and under researched. *British Journal of Clinical Nutrition* 67:492-95.

Ryan, A., et al. 1998. Effect of hypohydration on gastric emptying and intestinal absorption during exercise. *Journal of Applied Physiology* 84:1581-88.

Sacks, F., et al. 2001. Effects on blood pressure of reduced dietary sodium and the Dietary Approaches to Stop Hypertension (DASH) diet. *New England Journal of Medicine* 344:3-10.

Sanders, B., et al. 1999. Water and electrolyte shifts with partial fluid replacement during exercise. *European Journal of Applied Physiology* 80:318-23.

Sawka, M.N., et al. 1984. Influence of hydration level and body fluids on exercise performance in the heat. *Journal of the American Medical Association* 252(9):1165-69. PMID: 6471340.

Sawka, M. N., et al. 2012. High skin temperature and hypohydration impair aerobic performance. *Experimental Physiology* 97 (3):327-32.

Sawka, M. N., et al. 2015. Hypohydration and human performance: Impact of environment and physiological mechanisms. *Sports Medicine* 45(S1):S51-60.

Sawka, M., and Young, A. 2006. Physiological systems and their responses to conditions of heat and cold. In *ACSM's Advanced Exercise Physiology,* ed. C. Tipton. Philadelphia: Lippincott Williams & Wilkins.

Schleich, K.T., et al. 2016. Hypertension in athletes and active populations. *Current Hypertension Reports* 18(11):77. doi: 10.1007/s11906-016-0685-y. PMID: 27739019.

Schoffstall, J., et al. 2001. Effects of dehydration and rehydration on the one-repetition maximum bench press of weight-trained males. *Journal of Strength and Conditioning Research* 15:102-8.

Seal, A.D., and Kavouras, S.A. 2021. A review of risk factors and prevention strategies for exercise associated hyponatremia. *Autonomic Neuroscience* 238:102930. doi: 10.1016/j.autneu.2021.102930.

Shi, X., and Passe, D. H. 2010. Water and solute absorption from carbohydrate-electrolyte solutions in the human proximal small intestine: A review and statistical analysis. *International Journal of Sport Nutrition and Exercise Metabolism* 20:427-42.

Silva, A. M., et al. 2012. Total body water and its compartments are not affected by ingesting a moderate dose of caffeine in healthy young adult males. *Applied Physiology Nutrition and Metabolism* 36 (6):626-32.

Sims, S., et al. 2007. Preexercise sodium loading aids fluid balance and endurance for women exercising in the heat. *Journal of Applied Physiology* 103:534-41.

Sims, S., et al. 2007. Sodium loading aids fluid balance and reduces physiological strain of trained men exercising in the heat. *Medicine & Science in Sports & Exercise* 39:123-30.

Smith, C. J., and Havenith, G. 2012. Body mapping of sweating patterns in athletes: A sex comparison. *Medicine and Science in Sports and Exercise* 44 (12):2350-61.

Smith, G.D. 2002. Commentary: Behind the Broad Street pump: aetiology, epidemiology and prevention of cholera in mid-19th century Britain. *International Journal of Epidemiology* 31:920-932.

Stacey, M. J., et al. 2015. Susceptibility to exertional heat illness and hospitalisation risk in UK military personnel. *British Medical Journal Open Sport and Exercise Medicine*. 1:000055. doi:10.1136/bmjsem-2015-000055.

Stachenfeld, N. S. 2008. Acute effects of sodium ingestion on thirst and cardiovascular function. *Current Sports Medicine Reports* 74 (4):S7-S13.

Stachenfeld, N. S. 2014. The interrelationship of research in the laboratory and the field to assess hydration status and determine mechanisms involved in water regulation during physical activity. *Sports Medicine* 44(S1):S97-S104.

Stanhewicz, A. E., and Kenney, W. L. 2015. Determinants of water and sodium intake and output. *Nutrition Reviews* 73 (S2):73-82.

Stearns, R. L., et al. 2010. Effects of ingesting protein in combination with carbohydrate during exercise on endurance performance: A systematic review with meta-analysis. *Journal of Strength and Conditioning Research* 24(8): 2192-2202.

Sterns, R.H. 2018. Treatment of severe hyponatremia. *Clinical Journal of the American Society of Nephrology* 13(4):641-49. doi: 10.2215/CJN.10440917.

Stevens, C.J., et al. 2017. Cooling during exercise: An overlooked strategy for enhancing endurance performance in the heat. *Sports Medicine* 47(5):829-41. doi: 10.1007/s40279-016-0625-7. PMID: 27670904.

Stofan, J., et al. 2005. Sweat and sodium losses in NCAA football players: A precursor to heat cramps. *International Journal of Sport Nutrition and Exercise Metabolism* 15:641-52.

Stone, M.S., et al. 2016. Potassium intake, bioavailability, hypertension, and glucose control. *Nutrients* 22;8(7):444. doi: 10.3390/nu8070444. PMID: 27455317; PMCID: PMC4963920.

Stookey, J. 1999. The diuretic effects of alcohol and caffeine and total water intake misclassification. *European Journal of Epidemiology* 15:181-88.

Suh, J. S., et al. 2010. Recent advances in oral rehydration therapy (ORT). *Electrolyte Blood Press* 8:82-86.

Sulzer, N., et al. 2005. Serum electrolytes in Ironman triathletes with exercise-associated muscle cramping. *Medicine & Science in Sports & Exercise* 37:1081-85.

Thornton, S.N. 2016. Increased hydration can be associated with weight loss. *Frontiers in Nutrition* Jun 10;3:18. doi: 10.3389/fnut.2016.00018. PMID: 27376070; PMCID: PMC4901052.

Timpmann, S., et al. 2012. Dietary sodium citrate supplementation enhances rehydration and recovery from rapid body mass loss in trained wrestlers. *Applied Physiology, Nutrition, and Metabolism* 37(6):1028-37.

Tucker, R., et al. 2004. Impaired exercise performance in the heat is associated with an anticipatory reduction in skeletal muscle recruitment. *Pflugers Archive* 448:422-30.

Twerenbold, R., et al. 2003. Effects of different sodium concentrations in replacement fluids during prolonged exercise in women. *British Journal of Sports Medicine* 37:300-3.

Vaidyanathan, A., et al. 2020. Heat-related deaths—United States, 2004-2018. *MMWR Morbidity and Mortality Weekly Report* 69:729-34. doi: http://dx.doi.org/10.15585/mmwr.mm6924a1.

Valtin, H. 2002. "Drink at least eight glasses of water a day." Really? Is there scientific evidence for "8 × 8?" *American Journal of Physiology. Regulatory Integrative and Comparative Physiology* 283:R993-1004.

van Rosendal, S. P., and Coombes, J. S. 2012. Glycerol use in hyperhydration and rehydration: Scientific update. *Medicine and Sport Science* 59:104-12.

van Rosendal, S. P., et al. 2012. Performance benefits of rehydration with intravenous fluid and oral glycerol. *Medicine and Science in Sports and Exercise* 44 (9):1780-90.

van Rosendal, S.P., et al. 2010. Guidelines for glycerol use in hyperhydration and rehydration associated with exercise. *Sports Medicine* 40(2):113-29. doi: 10.2165/11530760-000000000-00000. PMID: 20092365.

Vieux, F., et al. 2020. Trends in tap and bottled water consumption among children and adults in the United States: Analyses of NHANES 2011-16 data. *Nutrition Journal* 19(1):10. doi: 10.1186/s12937-020-0523-6. PMID: 31996207; PMCID: PMC6990513.

Virani, S.S., et al. 2020. American Heart Association Council on Epidemiology and Prevention Statistics Committee and Stroke Statistics Subcommittee. Heart disease and stroke statistics—2020 update: A report from the American Heart Association. *Circulation* 141(9):e139-e596. doi: 10.1161/CIR.0000000000000757.

Vivanti, A. P. 2012. Origins for the estimations of water requirements in adults. *European Journal of Clinical Nutrition* 66(120):1282-89.

Volpe, S., et al. 2009. Estimation of prepractice hydration status of National Collegiate Athletic Association Division I athletes. *Journal of Athletic Training* 44:624-29.

von Fraunhofer, J., and Rogers, M. 2005. Effects of sports drinks and other beverages on dental enamel. *General Dentistry* 53:28-31.

Vukasinović-Vesić, M., et al. 2015. Sweat rate and fluid intake in young elite basketball players on the FIBA Europe U20 Championship. *Vojnosanitetski Pregled* 72(12):1063-68. doi: 10.2298/vsp140408073v. PMID: 26898028.

Wegmann, M., et al. 2012. Pre-cooling and sports performance: A meta-analytic review. *Sports Medicine* 42 (7):545-64.

Whelton, P.K. et al. 2018. 2017 ACC/AHA/AAPA/ABC/ACPM/AGS/APhA/ASH/ASPC/NMA/PCNA Guideline for the prevention, detection, evaluation, and management of high blood pressure in adults: A report of the American College of Cardiology/American Heart Association Task Force on clinical practice guidelines. *Hypertension* 71(6):1269-324. doi: 10.1161/HYP.0000000000000066.

White, J., et al. 1998. Fluid replacement needs of well-trained male and female athletes during indoor and outdoor steady state running. *Journal of Science and Medicine in Sport* 1:131-42.

Wing-Gaia, S., et al. 2005. Effects of purified oxygenated water on exercise performance during acute hypoxic exposure. *International Journal of Sport Nutrition and Exercise Metabolism* 15:680-88.

Wyndham, C.H., et al. 1966. Fatigue of the sweat gland response. *Journal of Applied Physiology* 21(1):107-10.

Yaspelkis, B., et al. 1993. Carbohydrate metabolism during exercise in hot and thermoneutral environments. *International Journal of Sports Medicine* 14:13-19.

Young, A. 1990. Energy substrate utilization during exercise in extreme environments. *Exercise and Sport Sciences Reviews* 18:65-118.

Zachwieja, J., et al. 1992. The effects of a carbonated carbohydrate drink on gastric emptying, gastrointestinal distress, and exercise performance. *International Journal of Sport Nutrition* 2:239-50.

Zhang, Y., et al. 2015. Caffeine and diuresis during rest and exercise: a meta-analysis. *Journal of Science and Medicine in Sport* 18:569-74.

Design element: Training Table (orange) ©mphillips007/Getty Images

Body Weight and Composition for Health and Sport

CHAPTER TEN

LEARNING OUTCOMES

After studying this chapter, you should be able to:

1. List different components that constitute human body composition.
2. Describe techniques used to assess body composition and discuss the general uses and limitations of such techniques.
3. Identify body mass index and body fat values associated with underweight, overweight, and degrees of obesity.
4. Explain the mechanisms whereby the human body regulates body weight, including the role of the central nervous system and feedback from peripheral organs and tissues.
5. List and explain the various genetic and environmental factors that may affect the normal regulation of body weight, particularly factors that may predispose to the development of overweight and obesity.
6. Outline the health problems that are associated with obesity in both adults and children.
7. Describe the health problems associated with excessive weight loss involving the use of prescription medications for weight loss, very-low kcal diets, and/or bariatric surgery.
8. Discuss and provide examples of how body weight and body composition impact sport-specific exercise performance.
9. Contrast disordered eating and eating disorders and describe characteristics of common eating disorders.
10. Explain the male and female athlete triads and describe common treatment recommendations.

KEY TERMS

activity-stat hypothesis
adaptive thermogenesis
adipocyte
adipokines
adiposopathy (sick fat)
air displacement plethysmography (ADP)
aminostatic theory
android-type obesity
anorexia athletica (AA)
anorexia nervosa (AN)
anorexiant
apoptosis
appestat
binge eating disorder (BED)
bioelectrical impedance analysis (BIA)
body image
body mass index (BMI)
brown adipose tissue (BAT)
bulimia nervosa (BN)
cellulite
Diabesity™
differential susceptibility
disordered eating
dual energy X-ray absorptiometry (DXA, DEXA)
dual intervention point model
eating disorders
essential fat
fat-free mass
female athlete triad
general model
ghrelin
glucostatic theory
gynoid-type obesity
hunger center
hydrostatic weighing
hyperplasia
hypertrophy
hypokalemia
lean body mass
leptin
lipostatic theory
male athlete triad
metabolic syndrome
metabolically healthy obesity (MHO)
neuropeptide Y (NPY)
nonexercise activity thermogenesis (NEAT)
morbid obesity
obesity
population attributable risk
relative energy deficiency in sport (RED-S)
regional fat distribution
rhabdomyolysis
sarcopenia
satiety center
set-point theory
settling-point theory
skinfold technique
storage fat
subcutaneous fat
very low-kcal diets (VLCD)
visceral fat
waist circumference
weight cycling

Introduction

Ingram Publishing/Alamy Stock Photo

The human body is a remarkable machine. Energy systems, which were discussed in chapter 3, capture a portion of the chemical energy primarily in carbohydrates and fats (and protein), as discussed in chapters 4–6, for ATP synthesis. Energy balance reflects, quite literally, a balance between the dietary energy one consumes and the metabolic and thermal energy produced by the body throughout the day. In order to maintain a given body weight, energy input (diet) must balance energy output (metabolism). If there is a long-term imbalance between energy input and energy output, body weight will either increase or decrease. A surplus of 50 kcal per day adds up to 18,250 kcal (over 5 pounds) a year. Conversely, a deficit of 50 kcal per day will result in an estimated 5-pound weight loss over a year.

In their review, Rounsefell and others reported direct associations between high levels of social media exposure and body image dissatisfaction. They detected several themes including awareness by young adults of social media's encouragement for users to engage in body image comparisons and the pursuit of and validation of a perceived ideal body image by social media. A primary reason for dissatisfaction with body image is a perception of being overweight. Using National Health and Nutrition Examination Survey data, Martin and others reported that 56.4 percent of woman and 41.7 percent of males tried to lose weight between 2013 and 2016. According to Research and Markets (www.researchandmarkets.com/reports/5523820/weight-lossweight-management-obesity-market#src-pos-3), $78 billion was spent in the USA on weight loss on 2019, with a 21 percent decline in 2020 due to the COVID-19 pandemic. Market projections point to a strong rebound in 2021. GrandView Research.com (www.grandviewresearch.com/industry-analysis/weight-management-market) reported a 2016 global weight management market including dietary products, equipment, and services of $217 billion. The U.S. Federal Trade Commission investigates many fraudulent or deceptive claims for weight-loss products and services each year. In 2014, the highest court in the European Union ruled that obesity could be considered a disability but did not require protection under antidiscrimination laws for individuals with obesity. Weight-loss strategies will be discussed in detail in chapter 11.

Overweight and obesity, commonly defined as respective BMI values of 25.0–29.9 kg/m^2 and ≥30.0 kg/m^2, affect males and females of all ages and all racial/ethnic groups, and from all areas of the world. According to the World Health Organization (www.who.int/news-room/fact-sheets/detail/obesity-and-overweight), 39 million children under 5 years of age and 340 million children 5–19 years of age in the world had either overweight or obesity in 2020. Almost 2 billion adults 18 or older had overweight in 2016, including 650 who had obesity. Hales and others reported a US adult obesity prevalence rate of 42.2 percent in 2017–2018, with similar rates between males and females and age groups. In the same time period, 14.4 million US children between 2 and 19 years of age (19.3 percent) had obesity with pediatric prevalence rates of 25.6 percent in Hispanics, 24.2 percent in non-Hispanic Blacks, 16.1 percent in non-Hispanic Whites, and 8.7 percent in non-Hispanic Asians.

As we shall see, being overweight or having obesity may contribute to the development of numerous health problems. In their review, Kohut and others conclude that comorbidities associated with childhood obesity previously found predominantly in adults may persist into adulthood and result in a decrease in lifespan. Cawley and others reported significant and progressive increases in annual medical care costs in individuals who have obesity (68.4 percent for BMI of 30–34.9 kg/m^2) to 234 percent for BMI ≥ 40 kg/m^2) compared to normal weight individuals. These increased costs occur in outpatient care, inpatient care, and prescription drugs. In 2016, the total medical cost of obesity in adults was an estimated $260.6 billion in 2016 or about 8.5 percent of total estimated health care costs of $3.1 trillion according to Dieleman and others. Some contend that obesity should be targeted for reduction as aggressively as cigarette smoking has been.

Excess body weight may affect physical performance as well. For some athletes, simply being a little overweight may prove to be detrimental to physical performance because it costs energy to move the extra body mass. In contrast, increased body weight, provided it is of the right composition, may be advantageous to other athletes.

At the other end of the body-weight continuum, excessive weight loss, particularly in combination with reductions in lean body mass, may have an impact on health and physical performance. According to Galmiche and others, the prevalence rate of eating disorder based on accurate diagnostic (*Diagnostic and Statistical Manual of Mental Disorders-V*) criteria has increased from 3.5 percent from 2000–2006 to 7.8 percent in 2013–2018. Although losing excess body fat may improve performance in some sports, excessive weight losses may have a negative impact on both health and athletic performance.

The preceding discussion has focused on extremes in body *weight*. It is important to stress that body weight is not synonymous with optimal body *composition* for health and performance. The major focus of this chapter is on the basic nature of body composition and its effect on health and physical performance. The following two chapters focus on strategies to achieve and maintain a body weight and body composition that best supports both health as well as athletic performance.

Body Weight and Composition

What is the ideal body weight?

Most individuals have a perception of an *"ideal"* body weight for standing height, age, and sex. Many research efforts have attempted to find an ideal body weight for optimal health and prevention of chronic disease. Actuarial data collected by life insurance companies during the past century have been compiled into "normal" or "desirable" ranges of body weight for a given height and age. These height-weight charts, such as the Metropolitan Life Height/Weight Charts appearing in many health-care provider offices, were designed to represent the "ideal" body weight to support health and longevity.

However, there appears to be no sound evidence to suggest a specific ideal weight for a given individual. Instead of *ideal* body weight for health and performance, a more appropriate concept is *healthy* body weight, which is grounded in the concept of healthy body composition. Many factors will influence the body weight for a given individual, not the least of which is the measurement error in various body composition assessment techniques, a topic discussed in this chapter. The major focus of this chapter is to discuss general guidelines that have been proposed for body composition and body weight relative to health and physical performance.

Although height-weight charts have been used to screen for normal body weight in the past, the body mass index (BMI) is the current standard, which is also based on a height–weight relationship. The **body mass index (BMI)**, also known as *Quetelet's Index,* is a weight:height ratio. Using the metric system, the formula is body weight in kilograms ÷ (height in meters)^2. In English units, the formula is [(body weight in pounds) × 705] ÷ (height in inches)^2. The following is an example:

200-pound (90.9-kg), 71-inch (1.8-m) male = $90.9 \div 1.80^2$
$= 28$ kg/m^2
$200 \times 705 \div 71^2 = 28$ kg/m^2

In general, a BMI range of 18.5–24.9 is considered to be "normal" for adults. Adult BMI categories for normal weight, overweight, obesity, and disease risk are presented in **table 10.1**. We will discuss waist circumference, patterns of fat deposition, and risk of disease later in the chapter. You may calculate your BMI using method A in appendix B or at the Centers for Disease Control and Prevention (CDC) website listed in the following text box.

> www.cdc.gov/healthyweight/index.html Click "Assessing Your Weight" on the left to find other web pages providing information, including "Adult BMI Calculator" and "Child and Teen BMI Calculator."

Calculating the BMI for children and teens requires the use of both age and gender and is more complex than BMI calculation for adults. The BMI is used as a screening tool to identify possible weight problems for children, and both the Centers for Disease Control and Prevention and the American Academy of Pediatrics (AAP) recommend the use of BMI in children beginning at 2 years old. The caption in **table 10.2** includes a link to CDC BMI-for-age for males and females that can be consulted to provide a percentile for children and adolescents between 2 and 20 years of age. The resulting percentile ranking places the child in one of the following five categories according to the Pediatric Obesity Algorithm in **table 10.2**.

What are the values and limitations of the BMI?

In relation to determining whether an individual possesses normal body weight for a given height, the BMI may be a useful screening device for health problems. As we will discuss in later sections of the chapter, high BMI values may be associated with overweight or obesity and various metabolic diseases such as diabetes mellitus, cardiovascular disease, and **metabolic syndrome** a group of symptoms including abnormal cholesterol levels, insulin resistance, and hypertension. At the other end of the continuum, low BMI values may also be the result of hormonal imbalances, malnutrition, or eating disorders. BMI is also used in large epidemiological studies instead of more sophisticated body composition assessment techniques as an apparent indicator of increased prevalence of overweight/obesity in a population over time. Assuming the existence of standardized measurements of body weight and height, such studies can detect associations between apparent increases in body fatness (increased BMI) and health. The use of BMI as an indicator of body composition is based on the assumption that increased BMI over time is due to increased body weight (the numerator), which in most individuals is most likely the result of increased adiposity over time. Since the mid-1980s, the Centers for Disease

TABLE 10.1 Classification of disease risk based on body mass index and waist circumference

		Disease risk relative to normal weight and waist circumference*	
		Normal risk visceral fat deposition*	High risk visceral fat deposition*
Category	BMI (kg/m^2)	Males ≤102 cm (40 inches); Females ≤88 cm (35 inches)	Males >102 cm (40 inches); Females >88 cm (35 inches)
Underweight	<18.5	—	—
Normal	18.5–24.9	—	—
Overweight	25.0–29.9	Increased	High
Obesity I	30.0–34.9	High	Very high
Obesity II	35.0–39.9	Very high	Very high
Obesity III	≥40	Extremely high	Extremely high

*Research suggests that waist sizes greater than 37 inches in males and 31.5 inches in females may increase health risks when accompanied by other conditions, such as high blood pressure. Increased risk associated with high waist circumference may occur even in individuals with a "normal" BMI.
Source: National Heart, Lung, and Blood Institute of the National Institutes of Health. www.nhlbi.nih.gov/health/educational/lose_wt/BMI/bmi_dis.htm.

Control and Prevention has been tracking the prevalence of obesity (operationally defined as BMI ≥30) in the United States through the Behavioral Risk Factor Surveillance Survey.

> **www.cdc.gov/obesity/** Click "Adult Overweight and Obesity," then "Adult Obesity Facts," and then "Obesity Prevalence Facts." What was the prevalence of self-reported obesity in your state and inr your self-identified racial group in 2018–2020?

TABLE 10.2 Body mass index percentile categories for children and adolescents ages 2–20 years

Weight status category	Percentile range
Underweight	Less than the 5th percentile
Healthy weight	5th percentile to less than the 85th percentile
Overweight	85th to less than the 95th percentile
Obesity	95–99th percentile or BMI > 30
Severe obesity	BMI > 120th% of the 95th percentile or BMI > 35 kg/m^2

Source: Pediatric Obesity Algorithms® ©Obesity Medicine Association www.obesitymedicine.org/wp-content/uploads/2019/07/Pediatric-Obesity-Algorithm-2018-2020.pdf
CDC Growth Charts for Children and Adolescents are available at www.cdc.gov/growthcharts/clinical_charts.htm

As a caveat, however, BMI actually reveals nothing about body composition. The BMI value does not represent percent body fat. As illustrated in **figure 10.1,** two individuals may be exactly the same height and weight and have the same BMI, but the distribution of their body weight might be so different that one individual could be considered as having obesity while the other might be considered very muscular. In a comparison of BMI and more sophisticated body composition assessment techniques (discussed later) in college athletes and nonathletes, Ode and others reported that BMI incorrectly classified normal-fat athletes as overweight because of generally larger muscle mass present in male and female athletes. Conversely, many females who were not athletes were classified as normal weight by the BMI but actually had higher than recommended levels of body fat. Those who do not exercise, and yet are thin, may have excessive amounts of internal fat and thus may be thin on the outside but fat on the inside. BMI should be used cautiously when classifying body composition in collegiate student-athletes and other muscular individuals as well as individuals who are sedentary.

Although BMI is not perfect and may be inappropriate for use with very muscular individuals, it can be a good guide that the average person may use to think about a healthier body weight. However, other methods are needed to evaluate actual body composition.

What is the composition of the body?

The human body contains many of the elements of the earth, 25 of which appear to be essential for normal physiological functioning. The elements carbon, hydrogen, oxygen, and nitrogen account for 96 percent of the human body's mass and provide the structural basis for body protein, carbohydrate, fat, and water. The remaining

Individual A **Individual B**

FIGURE 10.1 Body mass index (BMI), although a general indicator of overweight and obesity, does not assess body composition in any given individual. The individuals above have the same height, weight, and BMI (23.9 kg/m²), but individual A appears to be more muscular with a low body-fat percentage than individual B.

4 percent of the body is composed of minerals, primarily calcium and phosphorus in the bones; other macrominerals such as chlorine, sulfur, sodium, and magnesium; and the microminerals iron, cobalt, zinc, iodine, selenium, fluorine, manganese, molybdenum, chromium, and others that were discussed in chapter 8.

Because body composition may have a significant impact on health and physical performance, scientists have developed a variety of techniques to measure various body components. Wang and others noted that depending on the purpose, body composition may be evaluated at five levels: atomic, molecular, cellular, tissue-system, and whole-body levels. Of major interest to body composition scientists are four major body components: total body fat, fat-free mass, bone mineral, and body water. The location of body fat deposits, whether under the skin (**subcutaneous fat**) or deep in the body (**visceral fat**), is related to health status. As will be discussed later in the chapter, large deposits of visceral fat are associated with diabetes mellitus, cardiovascular disease, and metabolic syndrome. Each body component has a different density. According to Archimedes' principle, density, also known as specific gravity, equals mass divided by volume. In body composition, density is usually expressed as grams per milliliter (g/ml) or grams per cubic centimeter (g/cc³). The standard for comparison is water, which has a density of essentially 1.0, or 1 g/ml, the exact density depending on the water temperature. Corresponding densities for the other components are approximately 1.3–1.4 for bone, 1.1 for fat-free protein tissue, and 0.9 for fat. The total body density value may be used to estimate the body-fat percentage, with a higher density representing a greater amount of fat-free mass and a lower amount of body fat. The total human body density may range from approximately 1.010 g/ml (estimated 40 percent fat) to 1.090 g/ml (estimated 4 percent fat). Although there is little interindividual variance in the densities of fat and water, bone and fat-free densities may vary considerably between individuals and as a function of health status. As a result, the use of constant densities for these components is a source of error in attempts to estimate body-fat percentage from total-body density.

Depending on the purpose, body composition is usually analyzed as two, three, or all four components. The two components most commonly measured are total body fat and fat-free mass. Bone mineral content and body water may be measured with more elaborate techniques. Wang and others also introduced a six-component model, which in addition to these four components adds measurement of soft-tissue minerals and glycogen.

Total Body Fat The total body fat in the body consists of both essential fat and storage fat. **Essential fat** is necessary for proper functioning of certain body structures such as the brain, nerve tissue, bone marrow, heart tissue, and cell membranes. Essential fat in adult males represents about 3 percent of the body weight. Adult females also have an additional 9–12 percent of essential fat associated with normal reproductive function. This additional sex-specific fat gives females a total of 12–15 percent essential fat, although this amount may vary considerably among individuals. **Storage fat** is simply a depot for excess energy, the quantity of which may vary considerably between and within individuals due to factors discussed later in this chapter.

Some storage fat is found around body organs for protection, but about 70–80 percent of total body fat is found just under the skin and is known as subcutaneous fat. According to Bass and others, **cellulite** results from hardening of the fibrous septa, collagen-containing connective tissue forming chambers of subcutaneous fat. Peripheral **adipocytes**, (fat cells) in these chambers expand with water and protrude, giving rise to a dimpled "orange peel" appearance. A higher prevalence of cellulite in females is due to sex-related differences in the skin, fat, and fibrous septa arrangement, suggesting a possible role for estrogen. Other storage fat is located deep in the body, particularly in the abdominal area. This deep fat is referred to as visceral fat, which as noted later is associated with increased health risks.

Fat-Free Mass **Fat-free mass** primarily consists of protein and water, with smaller amounts of minerals and glycogen. The tissue of skeletal muscles is the main component of fat-free mass, but the heart, liver, kidneys, and other organs are also included. A more common term often used interchangeably with *fat-free mass* is **lean body mass**, but technically lean body mass includes essential fat. In the simplistic two-component model of body composition assessment, total body mass is the sum of fat-free mass (or lean body mass) and fat mass. For example, a 200-pound sedentary person with an estimated 30 percent fat is assumed to have a fat mass of 60 pounds (200 × 0.3) and a corresponding 70 percent, or 140 pounds (200 × 0.7), of fat-free mass.

Bone Mineral Bone gives structure to the body, but it is also involved in a variety of metabolic processes. Bone consists of about 50 percent water and 50 percent solid matter, including protein (collagen) and minerals, primarily calcium and phosphorus. Although total bone weight, including water and protein, may be 12–15 percent of the total body weight, the mineral content is only 3–4 percent of total body weight.

Body Water As was discussed in chapter 9, the average adult body weight is approximately 60 percent water, the remaining 40 percent consisting of dry weight materials that exist in this internal water environment. Some tissues, like the blood, have a high water content, whereas others, like adipose tissue, are relatively dry. The fat-free mass is about 70 percent water, while adipose fat tissue is less than 10 percent. Under normal conditions the water concentration of a given tissue is well regulated relative to its needs. The percentage of the body weight that may be attributed to a given type of body tissue includes its normal water content.

Factors Affecting the Components of Body Composition Body composition may be influenced by a number of factors such as age, sex, diet, and level of physical activity. Age effects are significant during the developmental years as muscle and other body tissues are being formed. There are some minor differences in body composition between boys and girls up to the age of puberty, but at this age the differences become fairly great. In general, girls deposit more fat beginning with puberty, whereas boys develop more muscle tissue. Factors contributing to a loss of muscle mass during adulthood include a decline in physical activity and **sarcopenia,** the age-related loss of muscle mass. Physical inactivity also contributes to a positive energy balance leading to increased body fat storage. As was discussed in chapter 8, loss of bone mineral content (osteopenia) occurs with age, with high prevalence rates of osteoporosis in some populations. Body composition can be affected by *short-term* dietary changes such as water restriction and starvation, but the main effects are due to *long-term* overeating leading to increased body fat stores. Physical activity may also be very influential, with a sound exercise program helping to build muscle and lose fat. Strategies to combat increased body fat and weight will be discussed in the next chapter. As a result of these and other factors, the amount of fat, lean tissue, bone, and water can vary significantly between and within individuals. A lean 154-pound male may be compartmentalized into 60 percent (92 pounds) water and 40 percent (62 pounds) solid matter subdivided into fat (14 percent, 22 pounds); protein (22 percent, 34 pounds); and bone minerals (4 percent; 6 pounds).

What techniques are available to measure body composition and how accurate are they?

The measurement of body fat has become very popular in recent years. Many high school and university athletic departments routinely analyze the body composition of their athletes in attempts to predict an optimal weight for competition. In some sports, body composition assessment is mandated by various state or national sport associations. For example, minimum wrestling weights are calculated in high school and college wrestlers prior to the competitive season based on estimated body fat percentages of 7 and 5 percent, respectively. Fitness and wellness centers also usually include a body-fat analysis as one of their services. Unfortunately, some of the individuals who analyze body composition in these situations are unaware of the limitations of the tests they employ.

The only direct, accurate method of analyzing body composition is by chemical extraction of all fat from body tissues, which is obviously not appropriate with living humans. Thus, a variety of indirect methods have been developed to assess body composition. Some are relatively simple, such as visual observation by an experienced judge, and others are rather complex, such as nuclear magnetic resonance imaging, using multimillion-dollar machines. Indirect methods are used to measure body fat, lean body mass, bone mineral content, and body water. Some techniques are also used to measure fat in specific locations of the body.

All indirect assessment techniques discussed in this section that are employed *in vivo* (i.e., on the living, breathing individual)—even technologically sophisticated ones—ultimately provide only estimates of various components of body composition, including body fat. These estimates vary in precision and are prone to error depending on the technique. Such errors usually are expressed statistically as standard errors of measurement or estimate, which can be used to show the accuracy of the body-fat measurement. For example, a wrestler undergoes preseason body composition assessment. A skinfold technique predicts 7 percent body fat and has a standard error of measurement (SEM) of 3 percent. If the same wrestler is measured repeatedly with no change in body composition, the technique will generate different estimates. Approximately 68 percent of these actual measurements would be within 1 SEM of the average value, so there is a 68 percent level of confidence that the actual body fat percentage of this wrestler is between 4 and 10 percent (±1 SEM). This is actually an example of a *doubly* indirect technique, since an *estimate* of body density is used to *estimate* body fat percentage. Body composition assessment should only be considered as providing a range, which may include the actual body fat percentage at a certain level of confidence, not as a precise measurement.

According to Müller and others, contemporary gold standards for body composition assessment include four-component models, dual energy X-ray absorptiometry (DEXA), whole body computer tomography (CT), and magnetic resonance imaging (MRI). Three- and four-component models that combine measures of body density with body water and mineral content dramatically reduce the errors associated with the traditional simplistic two-component model. Practical lab-based applications (hydrostatic weighing, air-displacement plethysmography [Bod Pod]) and field-based assessment methods (skinfolds, bioelectrical impedance analysis [BIA]) are based on two-component models. Although a number of body composition measurement techniques are highlighted in **table 10.3**, only the more commonly used or promising techniques will be discussed in this chapter.

The validities of various body composition assessment techniques have been the focus of various reviews. Heyward discusses the sources of error and advantages of various body composition techniques in different age and ethnic groups. Prediction equations should have "good" SEE values defined as ≤ 3.5 percent. Ackland and colleagues provide an excellent review of body composition assessment and application in training for weight-class (e.g., wrestling), aesthetic (e.g., gymnastics), and gravitational (e.g., jumping, cycling, and running) sports. Fosbøl and Zerahn review contemporary assessment methods, assumptions underlying the use of each method, sources of error, and variables to consider in selection of a method for a given individual.

The selected body composition technique will depend on the availability of resources, the laboratory-versus field-based nature of assessment, and other factors. Regardless of the chosen method, the following general guidelines should be followed.

1. The measurement error will be smaller if the same method is used to assess changes in an individual throughout the period of assessment.
2. Skinfold measurements may have a low inter-measurer reliability due to differences in grasping technique and anatomical site location. Measurements should be taken by the same trained technician throughout the assessment period.
3. Population-specific equations have been developed from homogeneous subjects based on characteristics such as gender, age, ethnicity, and athletic group. The individual being assessed should belong to the population upon which the equation was developed. Heyward and Wagner's book provides prediction formulas for various athletic populations.

Hydrostatic Weighing

One of the most common research techniques for determining body density is **hydrostatic weighing**, also known as underwater weighing or *hydrodensitometry*. Hydrostatic weighing is a volumetric technique based on Archimedes' principle that a body immersed in a fluid is acted upon by a buoyancy force that is directly related to the volume of water displaced (**figure 10.2**). Because fat is less dense and bone and muscle tissue

David Madison/Getty Images

FIGURE 10.2 Underwater weighing is one of the more common laboratory means for determining body composition. However, all current techniques for estimating percent body fat are subject to error. See text for discussion.

are more dense than water, a given weight of fat will displace a larger volume of water and exhibit a greater buoyant effect than the corresponding weight of bone and muscle tissue. Body density is calculated by the equation below and can be converted to estimated body fat percentage by several equations such as the one developed by Siri.

$$\text{Body density (BD)} = \frac{\text{Body Mass}_{Air}}{\left\{\frac{(\text{Body Mass}_{Air} - \text{Body Mass}_{Water})}{\text{Specific Gravity}_{Water}}\right\} - (\text{Residual Volume} + 0.1)}$$

$$\text{Siri estimated \%fat} = (495 \div \text{BD}) - 450$$

Hydrostatic weighing was once considered the "gold standard" in body composition analysis. However, Wagner and Heyward noted this method should not be regarded as such because it is an indirect two-component model with the previously described weaknesses.

TABLE 10.3 Methods used to determine body composition using the two (2)- or three (3)-component models

Technique	Number of components	Description	Availability	Approximate standard error of measurement and categories according to Heyward	Approximate cost
Air Displacement Plethysmography (BOD POD)	2	Whole-body plethysmograph measures air displacement and calculates body density based on Archimedes' Principle; comparable to water displacement protocol used in underwater weighing; requires measurement or estimation of thoracic gas volume	Low	~2–3% Very Good	$40–$60
Anthropometry	2	Measures body segment girths to predict body fat	High	~4% Fairly Good	$10–$20
Bioelectrical impedance analysis (BIA)	2	Measures resistance to electrical current to predict body-water content, lean body mass, and body fat	Moderate	~3.5–4% Good	$30–$40
Computed tomography (CT)	3	X-ray scanning technique to image body tissues; useful in determining subcutaneous and deep fat to predict body-fat percentage; used to calculate bone mass	Low	~2–3% Excellent	>$3,000
Dual energy X-ray absorptiometry (DEXA, DXA)	3	X-ray technique at two energy levels to image body fat; used to calculate bone mass	Low	~2–3% Excellent	$200–$300
Dual photon absorptiometry (DPA)	3	Beam of photons passes through tissues, differentiating soft tissues and bone tissues; used to predict body fat and calculate bone mass	Low	NA	NA
Infrared interactance	2	Infrared light passes through tissues, and interaction with tissue components used to predict body fat	High	~4.5% Fair	$25–$50
Magnetic resonance imaging (MRI)	3	Magnetic-field and radiofrequency waves are used to image body tissues similar to CT scan; very useful for imaging deep abdominal fat	Low	~2–3% Excellent	>$2,000
Neutron activation analysis	3	Beam of neutrons passes through the tissues, permitting analysis of nitrogen and other mineral content in the body; used to predict lean body mass	Low	NA	NA
Skinfold thicknesses	2	Measures subcutaneous fat folds to predict body-fat content and lean body mass	High	~3–4%	$20–$40
Total body potassium	2	Measures total body potassium, the main intracellular ion, to predict lean body mass and body fat	Low	NA	NA
Total body water (hydrometry)	2	Measures total body water by isotope dilution techniques to predict lean body mass and body fat	Low	~2–3%	NA
Ultrasound	2	High-frequency ultrasound waves pass through tissues to image subcutaneous fat and predict body-fat content	Low	~2% Excellent	NA
Underwater weighing (hydrodensitometry)	2	Underwater-weighing technique based on Archimedes' principle to predict body density, body fat, and lean body mass	Moderate	~2–3% Very Good	$30–75%

Sources: Ackland et al. 2012; Heyward 1998; Schubert et al. 2019.

For example, the assumption that the density of the fat-free mass is 1.10 g/ml may not be valid for all individuals, such as athletes and older persons. The standard error is about 2–3 percent. The hydrostatic-weighing technique is rather time-consuming and difficult for some individuals. The quality of the data is affected by the subject's comfort level with the water medium, so nonswimmers may have difficulty with the technique. The technique also requires the measurement or estimation of residual lung volume, which is the volume of air still remaining in the thorax after a maximal expiration. A standard volume for trapped intestinal gas is also used for calculating the total body density. Both volumes are sources of error in the technique. Other techniques (e.g., air displacement plethysmography, discussed next) have been developed for either research purposes or practical applications. The interested reader is referred to the book by Heyward and Wagner.

Air Displacement Plethysmography (ADP) Another volumetric technique is **air displacement plethysmography (ADP)**, sometimes referred to as body plethysmography. Subjects enter a dual-chamber plethysmograph designed to measure the amount of air they displace, somewhat comparable to the water displacement technique of underwater weighing. One commercial product available is called the Bod Pod (**figure 10.3**). It is somewhat portable, easy to operate, requires little time, and eliminates the necessity of going underwater, several clear advantages compared to underwater weighing. ADP may be more valid and reliable than hydrodensitometry for certain individuals, such as those who fear underwater submersion, but it has similar limitations. Fields and others noted that the ADP and underwater weighing agree within 1 percent of body fat for adults and children, but when compared to multicomponent models ADP generally underestimated body fat by 2–3 percent. Weyers and others noted no significant differences between ADP and DEXA methods for body fat percentage, fat mass, and fat-free mass changes in males and females who are classified as overweight. Tucker and others concluded that ADP is a reliable methodology, but more than two trials may be required to detect small changes in body composition.

Skinfolds The **skinfold technique** is designed to measure the thickness of subcutaneous fat in millimeters at specific anatomical sites (**figure 10.4**). Skinfold measurements are commonly used for nonresearch body composition assessment. The skinfold thickness values are inserted into an appropriate equation to estimate total body density, which in turn is used in another equation to estimate body fat percentage. Some formulas also have been developed for specific athletic groups. To improve the accuracy of this technique, skinfold measures should be obtained from a variety of body sites because using a single skinfold site may be unrepresentative of total storage fat. The test also should be administered with an acceptable pair of skinfold calipers by an experienced tester. Ultrasound techniques are also available to assess skinfold thicknesses, but these are more expensive than calipers. Gomes and others reported moderate to high agreement between skinfolds, ultrasound and DEXA methodologies in body composition assessment in athletes. However, more consistent results were observed with ultrasound. However, correlations vary according to anatomical site and gender. In a study of NCAA Division I athletes, Wagner and others reported better agreement between skinfold, ultrasound, and air-displacement plethysmography (Bod Pod) methods for male compared to female athletes. For all subjects, a higher intertechnician reliability was observed for ultrasound than skinfold measurement.

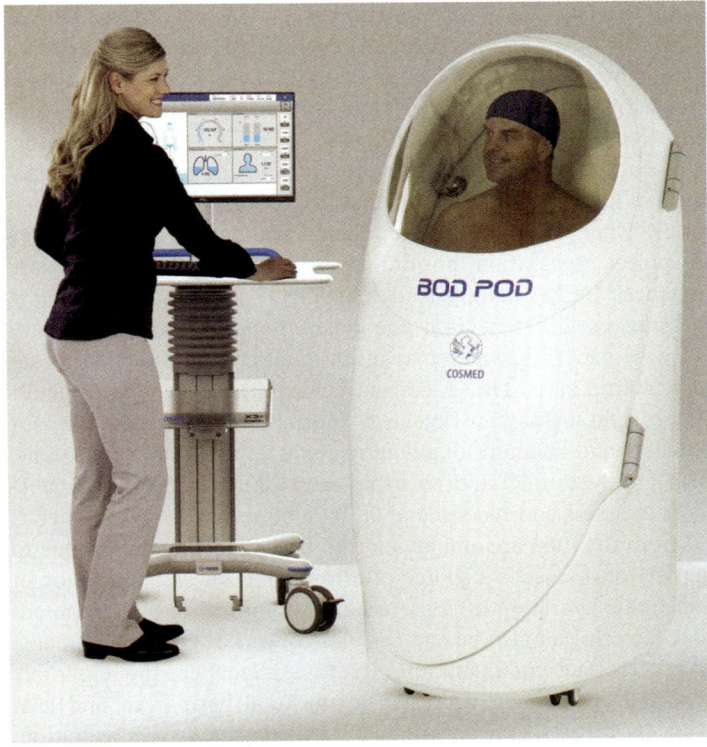

Cosmed USA, Inc.

FIGURE 10.3 The Bod Pod. An application of total body plethysmography, an air displacement technique, to evaluate body composition.

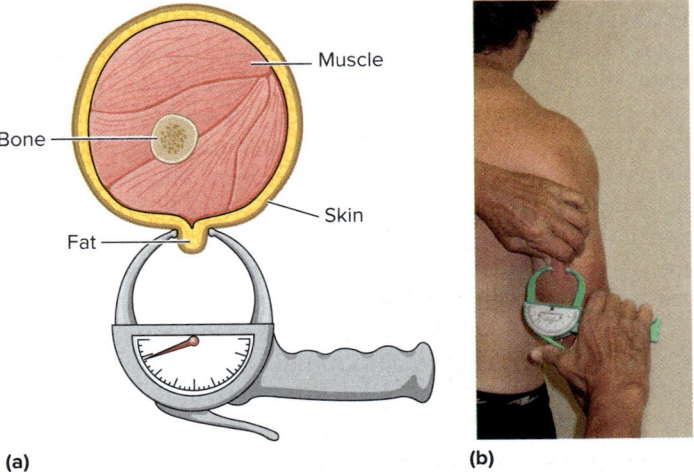

Triceps skinfold measurement.: Connie Mueller/McGraw Hill

FIGURE 10.4 (a) Schematic drawing showing the skinfold of fat that is pinched up away from the underlying muscle tissue. (b) Measurement of the triceps skinfold. Precise location: a vertical fold on the posterior aspect of the right arm halfway between the acromial process of the shoulder and olecranon process of the elbow.

Skinfold equations to estimate body density are usually based on total body density measured by hydrostatic weighing. As a result, the standard error for the skinfold technique is about 3–4 percent and reflects error in both measurement of the skinfold thickness and the hydrostatic weighing used to generate the prediction equation. According to Heyward and Wagner, the skinfold technique is one of the best practical methods to measure body composition in children, adults, and athletes, but should not be used in individuals with obesity. Factors such as gender, age, physical activity, and ethnicity may affect proportions of water, mineral, and protein in fat-free mass and the resulting density of the fat-free component in a two-component assessment of body composition. Therefore, population-specific equations should be used to estimate body fat percentage from skinfold measurements. Although the skinfold technique is widely used in the National Collegiate Athletic Association (NCAA) minimum wrestling weight protocol, Loenneke and others note the body density equation is based on Caucasian adolescent wrestlers and that more research is needed to validate the protocol for college-aged wrestlers generally and African-Americans specifically. Appendix B includes commonly used generalized equations to estimate body density from three gender-specific skinfold sites and age, based on the work of Jackson and Pollock (males), and Jackson, Pollock, and Ward (females), respectively, with subsequent conversion of estimated body density to estimated body fat percentage using the Siri equation. O'Connor and others have developed equations for non-Hispanic White, Hispanic, and African-American adults with standard errors of estimate between 3 and 4 percent.

Bioelectrical Impedance Analysis (BIA)

A more expensive, practical field technique is **bioelectrical impedance analysis (BIA)** illustrated in **figure 10.5**. BIA is based on the principle of resistance to an electrical current that is applied to the body. Lower recorded resistances are associated with a greater water content, a higher body density, and lower body fat. Based on conductivity as a function of total body water, far-free mass (FFM) is predicted based on an assumed FFM hydration of 73 percent, with subsequent prediction of fat mass difference between total mass and predicted FFM. Early research with BIA revealed large standard errors in predicting lean body mass, so it was not considered to be very valid. Despite growing popularity and evidence of the clinical value of BIA, concerns remain regarding BIA as an indirect body composition assessment methodology. According to Buffa and others, BIA measurements that are normalized for subject height and cross-sectional area (known as specific vector BIA) can increase the sensitivity of BIA to detect changes in two-component (fat mass versus fat-free mass) body composition in adults and elderly individuals. Moon noted that most BIA instruments do not have athlete-specific equations and recommended the development of total body water-based 3- and 4-component equations for athletes. In her review, Ward describes three issues regarding the use of BIA: (1) standardization of BIA methodology, (2) emphasis on the clinical relevance of BIA assessment rather than statistical differences between BIA and a "criterion" (e.g., DEXA) methodology, and (3) assumptions regarding BIA as a body composition assessment methodology that remain within the limits of impedance technology. In its position stand on

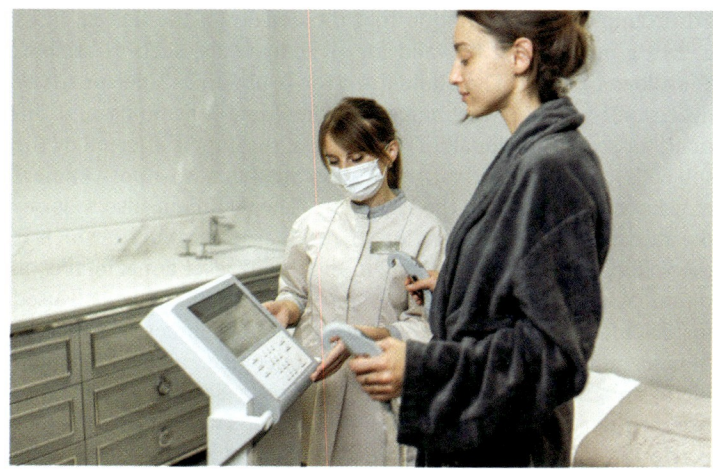

RossHelen/Shutterstock

FIGURE 10.5 Bioelectrical impedance analysis. A low-voltage current is optimally conducted through a body with a high fat-free mass (high water content) while high-fat mass (low water content) offers resistance to the current. Electrodes are placed at the hands and feet.

nutrition for the athlete, the Academy of Nutrition and Dietetics, Dietitians of Canada, and American College of Sports Medicine indicated that the prediction accuracy of BIA is similar to that of skinfold assessment, but BIA may be preferable because it does not require the technical skill associated with skinfold measurements.

Dual Energy X-Ray Absorptiometry (DXA, DEXA)

DEXA, shown in figure 10.6, is a three-component method that assesses bone mineral content, fat mass (e.g., lipid, subcutaneous and visceral adipose tissue, triacylglycerol, and phospholipid bilayer) and lipid-free soft tissue (i.e., lean mass) by the attenuation of tissue-specific X-ray energy. According to Messina and others, the assessment of fat mass and lean body mass by DEXA is highly correlated with CT and MRI imaging techniques. Standardized methodology in subject preparation, positioning, and postscan analysis of data enhances DEXA validity and reliability. Different DEXA manufacturers use different X-ray energy levels and different proprietary software packages, which may result in different proportions of fat and fat-free mass. DEXA-derived fat mass allows for the calculation of a fat mass index (kg fat/m^2, normal < 6 for males and 9 for females) and an android/gynoid regional fat mass ratio analogous, respectively, to BMI and waist/hip ratios. A recent DEXA capability is to assess android visceral fat by subtracting the subcutaneous fat from the total abdominal android fat. Subject hydration due to activity or disease is a source of error as DEXA assumes a nonfat soft tissue hydration of 73 percent. Postprandial digestive content and exercise can also be sources of error in DEXA measurements in their effect on hydration. Rodriguez-Sanchez and Galloway reported 2.3 percent decreases in the total body mass and lean mass in athletes in a dehydrated state compared to euhydration. Bone and others recently reported significant increases in DEXA estimates of lean mass with increased water content that accompanies glycogen supercompensation (discussed in chapter 4) and creatine supplementation (discussed in chapter 6).

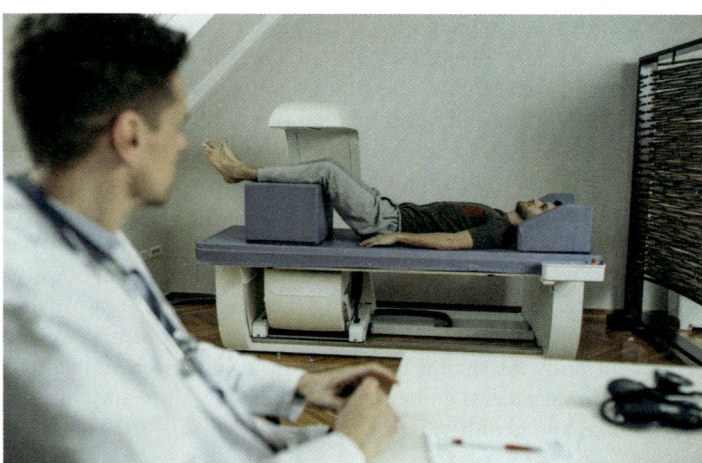

LStockStudio/Shutterstock

FIGURE 10.6 Dual energy X-ray absorptiometry (DEXA). This method measures body fat by releasing small doses of radiation through the body that a detector then quantifies as fat, lean tissue, or bone. The scanner arm moves from head to toe and in doing so can determine body fat as well as bone density. DEXA is currently considered the most accurate method for determining body fat as long as the person is able to fit on the table and/or under the arm of the instrument. The radiation dose is minimal.

Near-Infrared Interactance Another device marketed commercially is based on near-infrared light interactance. Infrared light passes through the tissues, and its interaction with tissue components is used to predict body fat. Wagner and Heyward reported a higher standard errors of measurement with near-infrared interactance of greater than 3.5 percent. In separate studies, Moon and colleagues concluded that near-infrared interactance was an acceptable field method for college-aged females (Futrex 6100/XL® Total Error [TE] = 2.7 percent compared to a 3-component model) but not for male high school athletes (Futrex 5000 TE = 10.4 percent compared with hydrostatic weighing). However, Harbin and others reported lower estimated body fat percentages from 3D infrared scanner (Styku MYBODEE™) in college-aged males and females compared to hydrostatic weighing, bioelectrical impedance, skinfold measurements, and circumference measurements. Decreased precision was observed with increasing adiposity, leading the researchers to conclude that the infrared interactance requires a correction factor in the algorithm before being considered a valid body fat estimation methodology.

Anthropometry Anthropometry, or measurement of body parts, is an inexpensive, practical method to assess body composition. Body measurements include circumferences such as the neck and abdomen, and bone diameters such as the hip, shoulders, elbow, and wrist. Although circumference and/or bone diameter measurements may be incorporated into various equations to predict body fat and lean body mass, Schuna and others reported the circumference-based equation used by the U.S. Department of Defense underestimated changes in percent fat and fat mass and overestimated changes in fat-free mass over 9 months compared to air-displacement plethysmography (Bod Pod) in Army ROTC cadets.

Circumference measurements of the abdomen, hips, buttocks, thigh, and other body parts may indicate **regional fat distribution**. As discussed later in the chapter, the pattern of fat distribution may be an important indicator of major health problems such as obesity, diabetes, and metabolic syndrome. The principal measure of regional fat distribution is the **waist circumference**, the girth measured by a flexible tape at the narrowest section of the waist as seen from the front. The waist circumference is a good screening technique for regional fat distribution, but it does not provide an accurate measure of deep visceral fat, such as provided by CT or MRI techniques.

Multicomponent Model The multicomponent model uses several methods, such as hydrodensitometry, total body water, and DEXA to reduce the errors associated with any single method and to provide information on body fat, body water, bone mass, and lean body mass. Lohman and Going recommend use of a multicomponent model including waist circumference, selected skinfolds, and DEXA when assessing body composition in children and youths. Wagner and Heyward note that the multicomponent model is now regarded as the "gold standard" in body composition assessment and should be used when feasible. Lara and others reported that the average of two- (fat mass + fat-free mass), three- (water + fat mass + fat-free dry tissue), and four-component (water + bone minerals + fat mass + protein + carbohydrates) models improved the accuracy of changes in fat mass in subjects with overweight or obesity compared to any single two-, three-, or four-component approach. While such accuracy is desirable, expensive laboratory-based three- and four-component techniques are more applicable in research, but less so in field assessment of body composition.

What problems may be associated with rigid adherence to body fat percentages in sport?

Table 10.3 lists most of the methods used to estimate body composition. Historically, hydrostatic weighing has been the criterion method by which other techniques have been validated. More recently, DEXA or a four-component model has been used as the criterion. As has been previously discussed, different techniques have sources of error, which contribute to different standard errors of measurement and resulting differences in estimated body fat percentages. For example, Pourhassan and others compared body composition assessment by a four-component model, air displacement plethysmography (ADP), total body water (TBW) measurement, DEXA, and magnetic resonance imaging (MRI) over 2 to 4 years in normal-weight and overweight adult subjects. Compared to the four-component model, DEXA, TBW, and ADP underestimated fat mass gain and overestimated fat-free mass. The researchers attributed this bias to the erroneous assumption in two-component models of a constant hydration state of 73.2 percent for fat-free mass.

The error associated with different techniques can be particularly problematic for athletes such as wrestlers and others who compete by weight class. Devrim-Lanpir and others compared several two-, three-, four-, and seven-site skinfold equations to estimate body density with the criterion air displacement plethysmography (BOD POD) in male and female Olympic wrestlers and reported that only two equations achieved adequate high precision and low bias in body fat estimation. In their review, Nana and others discuss the sources of methodological and technical error in DEXA, generally considered a criterion method, in body composition assessment of athletes. As previously discussed, Loenneke and colleagues comment that much of the research literature for the NCAA minimum weight (MW) protocol is based on studies of high school wrestlers, which may not accurately represent the body composition status of collegiate wrestlers. Aspects of the NCAA MW policy requiring additional research attention include the effect of overhydration (low urine specific gravity); use of skinfold calipers from different manufacturers; and use of the Brożek equation to estimate fat percentage from estimated body density, which may erroneously increase MW in African-American wrestlers.

Given the problems with assumptions underlying the various methods of body composition determination, body-fat percentage estimates are only approximations. The rigid use of body-fat percentages in weight-control sports, such as gymnastics, dancing, cheerleading, and wrestling, may lead to excessive weight loss. For example, if a wrestler who has 8 percent fat is required by his coach to reach 5 percent fat, the wrestler may already be at 5 percent fat because the skinfold technique has a 3–4 percent standard error of measurement. Losing extra weight may be very difficult for the wrestler because he is already near minimal body-fat levels. In his attempt to heed the coach's mandate, the wrestler may lose lean muscle mass, which will place him at a competitive disadvantage. In young athletes, such practices may also lead to disordered eating and, possibly, clinical eating disorders.

How much should I weigh or how much body fat should I have?

The answer to this question depends on many factors and whether you are concerned primarily about appearance, health, or physical performance. For example, Campisi and others assessed body fat in adult males and females by air displacement plethysmography (Bod Pod) and also asked subjects to provide perceived and desired body fat values. Perceived and desired body fat values were lower than assessed body fat in females, while the opposite was true for males. Ultimately, you are the best judge of what is your individual ideal body weight and body composition. Factors to consider include how do you feel—do you have the energy to do the things you want to do? For athletes, considering how performance is impacted by body weight and body composition is important, keeping in mind this can be different for each individual. There is no one "ideal" that is right for everyone.

Photodisc/Getty Images

The effect of body weight and body fat on physical performance is discussed in a later section, although some general guidelines are presented here. The effect of body weight and fat on health has received considerable research attention. By medical definition, **obesity** is simply an accumulation of excess fat in the adipose tissue. Obesity is also referred to as a disease or disorder and is the most common nutritional health problem in North America. The actual measurement and determination of clinical obesity is a controversial issue. Several approaches have been used to define the point at which a person is classified as having clinical obesity.

Unfortunately, our present level of knowledge does not provide us with the ability to predict precisely what the optimal weight or percent body fat should be for health in any given individual. However, some general guidelines have been developed by various professional and health organizations.

Using body mass index to calculate a target body weight BMI, discussed earlier in the chapter, may be used as a crude technique to calculate a target body weight, as shown in appendix B and various websites. According to guidelines from the National Academy of Sciences and the National Institutes of Health, a person with an "ideal" body weight will have a BMI of ≥ 18.5 up to ≤ 24.9. The BMI equation can be rearranged to solve for a target weight as follows:

weight in kg = desired BMI, a value from 18.5 to 24.9 $\times$ height measured in meters^2

or

weight in lbs = desired BMI, a value from 18.5 to 24.9 $\times$ height measured in inches$^2 \div 705$

> www.nhlbi.nih.gov/health/educational/lose_wt/BMI/bmicalc.htm
> Calculate your BMI using the following National Heart, Lung, and Blood Institute website:

Higher BMI values are usually, but not always, associated with overweight and obesity, as was described in **table 10.1**. As previously discussed, BMI values are based on height and weight and should not be confused with body-fat percentages. Most Americans are focused on weight maintenance or weight loss, strategies of which are covered in chapter 11. At the other extreme, low BMI values may be indicative of starvation. According to Shetty and James, the United Nations Food and Agriculture Organization has proposed the following BMI categories as indicative of use of a BMI less than 18.5 as a criterion for chronic energy deficiency (CED). CED grades and associated BMI values (17.0−<18.5=Grade I; 16.0–16.9=Grade II; <16.0=Grade III) are based on a Physical Activity Level (PAL=measured energy expenditure or intake÷measured basal metabolic rate) of less than 1.4. The PAL concept is discussed in chapters 3 and 11.

The health and performance of some individuals will actually benefit from weight gain consisting of fat-free mass, strategies of which will be discussed in chapter 12.

Body-Fat Percentage As noted previously, the use of the BMI height–weight relationship does not evaluate body composition.

A low BMI may be a symptom of a serious disease, not a cause. Muscular individuals may have a high BMI and not have obesity. For health purposes, the body requires a minimal level of fat, called essential fat, found in cell membranes, bone marrow, and nervous tissue and considered to be approximately 3 percent for males and 12–15 percent for females. Several authorities have included additional levels of storage fat and suggested that minimal levels of total body fat for health range from 5 to 10 percent for males and 15 to 18 percent for females. Depending on gender and sport, levels of body fat in the range of 5 to 10 percent may be recommended for distance runners, gymnasts, dancers, and wrestlers. However, some athletes have performed very successfully even though their body-fat percentage was higher than the recommended values.

Different levels of body fat percentages have been cited as criteria for obesity in males and females. In its book on Dietary Reference Intake for macronutrients, the National Academy of Sciences identified criteria for obesity as 25 percent body fat for males and 37 percent for females, with values over 31 percent in males and 42 percent body fat in females being indicative of clinical obesity. According to Bays and others, common American Council on Exercise body fat percentage cutpoints for obesity of ≥25 percent for males and ≥32 percent for females have not been scientifically validated. For example, Okorodudu and others reported the commonly used BMI cutpoint of ≥30 kg/m^2 to detect obesity as defined by excess adiposity had a high specificity (0.97) but a low sensitivity (0.42) as it failed to detect over half of cases with excess body fat.

Body-fat percentage categories from essential to very poor (overweight/obesity) for males and females by age are presented in **table 10.4**. Several points should be kept in mind when using such tables. All body-fat prediction methods contain a source of error. The essential, very lean, and excellent categories pertain to athletes competing in weight-control sports or sports where excess body fat may be a disadvantage. As will be discussed later, obesity generally increases health risks, but some individuals with higher body-fat percentages may not develop obesity-related health problems if they are otherwise physically fit and consume a healthful diet. Finally, as will also be discussed later in this chapter, the location of the fat in the body may have significant health consequences.

Waist Circumference It may not be how much fat you have that affects your health but where that fat is located. A waist circumference ≥35 inches in females and ≥40 inches in males reflects

TABLE 10.4 Body-fat categories by gender and age

	Males					
	Age Group					
Category	20–29	30–39	40–49	50–59	60–69	70+
Essential			3.0			
Very lean	≤6.6	≤10.3	≤12.9	≤14.8	≤16.2	≤15.5
Excellent	7.9–10.5	12.4–14.9	15.0–17.5	17.0–19.4	18.0–20.2	17.5–20.1
Good	11.5–14.8	15.9–18.4	18.5–20.8	20.2–22.3	21.0–23.0	21.0–22.9
Fair	15.8–18.6	19.2–21.6	21.4–23.5	23.0–24.9	23.6–25.6	23.7–25.3
Poor	19.7–23.3	22.4–25.1	24.2–26.6	25.6–28.1	26.4–28.8	25.8–28.4
Very poor	≥24.9	≥26.4	≥27.8	≥29.2	≥29.8	≥29.4
Category	Females					
Essential			12.0–15.0			
Very lean	≤14.0	≤13.9	≤15.2	≤16.9	≤17.7	≤16.4
Excellent	15.1–16.8	15.5–17.5	16.8–19.5	19.1–22.3	20.2–23.3	18.3–22.5
Good	17.6–19.8	18.3–21.0	20.6–23.7	23.6–26.7	24.6–27.5	23.7–26.6
Fair	20.6–23.4	22.0–24.8	24.6–27.5	27.6–30.1	28.3–30.8	27.6–30.5
Poor	24.2–28.2	25.8–29.6	28.4–31.9	30.8–33.9	31.5–34.4	31.0–34.0
Very poor	≥30.5	≥31.5	≥33.4	≥35.0	≥35.6	≥35.3

Modified from *American College of Sports Medicine's Guidelines for Exercise Testing and Prescription*, 10th Edition, and The Cooper Institute, Dallas, TX.

a metabolically dangerous visceral pattern of fat deposition. Excessive waist circumference is a component of the metabolic syndrome. This simple-to-measure variable is a well-documented risk factor for cardiovascular and metabolic diseases according to the American College of Sports Medicine and other organizations. The health implications of regional fat distribution are discussed in greater detail later in the chapter, but you may use method C in appendix B to calculate your waist circumference and evaluate associated health risks. However, as discussed later, lower waist circumferences may be associated with increased risks if accompanied by other conditions, such as high blood pressure. **Table 10.1** highlights the risk of chronic disease associated with the BMI and waist size.

Waist–Hip Ratio The waist–hip ratio (WH) is another metric used by health professionals for assessing abdominal obesity and metabolic risk. In addition to measuring waist circumference, hip circumference is the widest visible portion of the buttocks above the gluteal fold. The ratio is calculated as Waist ÷ Hip. According to a 2008 World Health Organization report, WH ratios of ≥ 0.9 and ≥ 0.85 for males and females, respectively, are markers for abdominal obesity and are strongly associated with increased metabolic risk, cardiovascular disease, some cancers, and overall mortality.

Key Concepts

- Body mass index (BMI) does not measure body composition, but it may be useful as a screening device to determine whether one has overweight or obesity. *Overweight* and *obesity* are not synonymous terms. It is possible to be overweight but not have obesity, or at a "normal" weight with too much body fat.
- The body consists of four components: body fat, protein, minerals, and water. However, for practical purposes, body composition may be classified as consisting of two components: fat-free weight, which is about 70 percent water, and body fat.
- All techniques that are currently used to measure body composition, primarily body fat, are indirect and prone to error by <2 to >5 percent depending on the method.
- Our present level of knowledge does not provide us with the ability to predict precisely what the optimal body composition should be for health or physical performance. However, BMI and body-fat levels higher than normal are associated with increased health risks.

Check for Yourself

- Using appendix B, calculate both your body mass index (BMI) and your waist circumference. Compare your findings to the rating scale in table 10.1.

Regulation of Body Weight and Composition

How does the human body normally control its own weight?

An individual may eat more than a ton of food—nearly a million kcal—a year and yet not gain 1 pound of body weight. For this to occur, the body must possess an intricate regulatory system that helps to balance energy intake and output both on a short-term and a long-term basis. The regulation of human energy balance is complex, involving numerous feedback loops to help control energy balance. Although we do not know all the exact physiological mechanisms whereby body weight is maintained relatively constant over short or long periods, research suggests that a variety of specific interactions between the brain and peripheral tissues may be involved in the control of both energy intake and energy expenditure.

Energy Intake According to Blundell, appetite and body weight are regulated by two driving influences. A *tonic energy* drive to eat is influenced by fat-free mass, tissues with high metabolic rate such as the heart and brain, and resting energy expenditure. An *episodic drive* is influenced by environment and culture. Energy balance (energy intake and energy expenditure) is regulated by complex central nervous system circuitry, specifically in the brain and hypothalamus, and complex afferent input from the gastrointestinal tissue; blood levels of glucose, fatty acids, and protein; hormones; and cytokines in order to either stimulate hunger or create a sense of satiety. These interactions provide short-term (daily) and long-term (years) regulation of appetite and body weight. However, this regulatory balance can be overridden by an array of social, environmental, genetic, developmental, neural, inflammatory, and endocrine factors leading to overweight/obesity or eating disorders.

In their review, Matafome and Seiça discuss homeostatic neurons and pathways in the hypothalamus and brainstem that are **orexigenic** (appetite stimulating) and **anorexigenic** (appetite inhibiting). There are nonhomeostatic neurons in the mesolimbic and corticolimbic area of the brain secreting dopamine, which control reward and pleasure. These neurons are involved in short-term and long-term regulation of energy intake and body weight. A very simplified schematic of these regulatory pathways is presented in **figure 10.7**. In general, short-term (meal-to-meal) regulation of energy intake is governed by brainstem structures, which receive input from stomach fullness and peptides produced in the gut following a meal. The brainstem also controls the motor aspects of eating, such as chewing and swallowing. The hypothalamus contains neural areas promoting eating and satiety, respectively, to regulate food intake on a long-term basis by modulating brainstem sensitivity to the above-described input. This area of the hypothalamus is called the **appestat**, with certain neurons referred to as a **hunger center** and other neurons referred to as a **satiety center**. Specific neural receptors within the appestat monitor various stimuli that may increase or decrease the appetite to control energy intake. Stimulation of dopaminergic

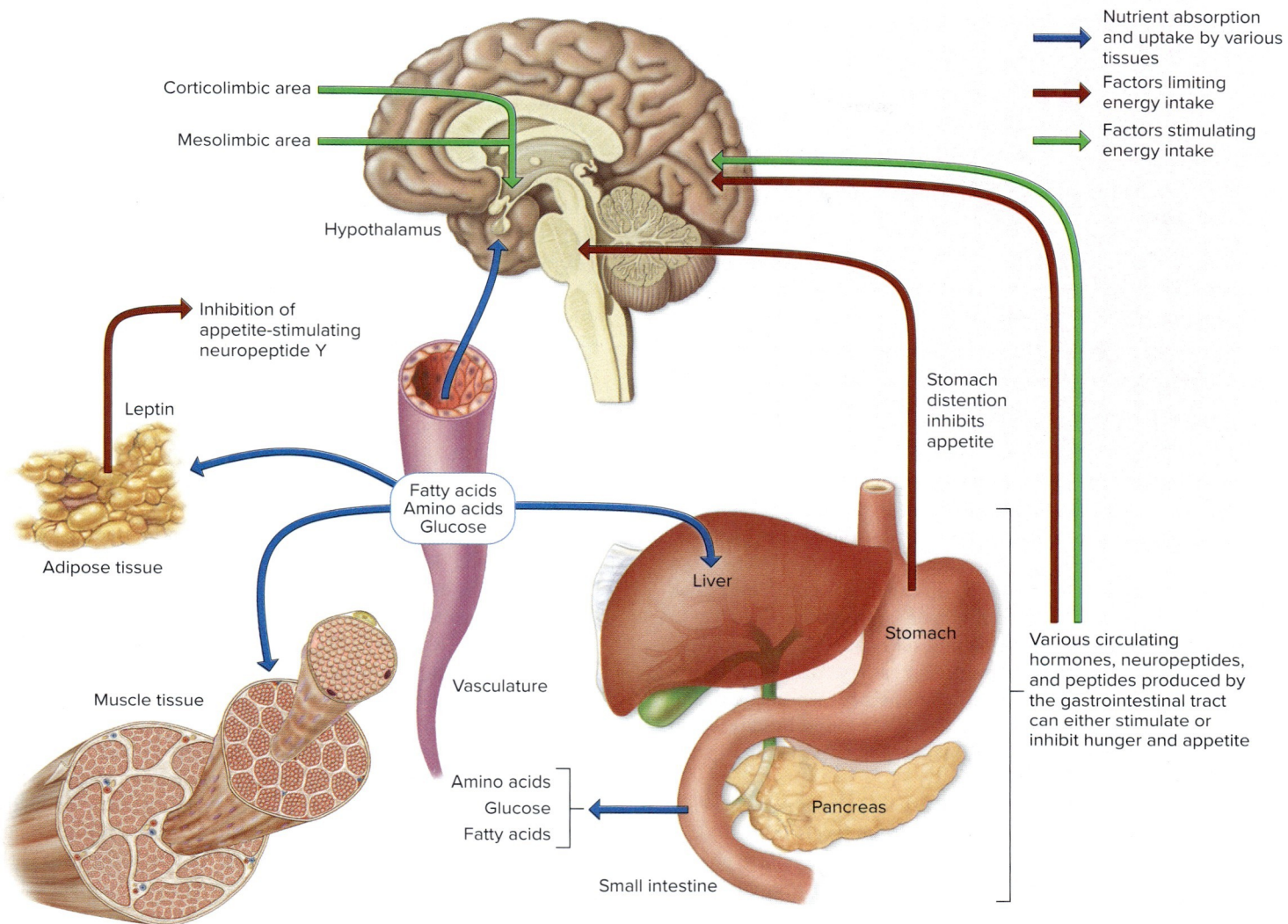

FIGURE 10.7 Simplified schematic of energy balance and appetite regulation. Absorption and uptake of nutrients by various tissues influence glucose, fatty acid, and amino acid levels, which are sensed by the hypothalamus and other brain structures. Stomach fullness and distension inhibits appetite. Some hormones, neuropeptides, and gut peptides inhibit appetite (e.g., cholecystokinin, obestatin). Leptin inhibits production of appetite-stimulating neuropeptide Y by the hypothalamus. Other hormones, neuropeptides, and gut peptides stimulate appetite (e.g., ghrelin). The taste and texture of added sugar and fat can upregulate the dopamine-secreting mesolimbic and corticolimbic areas of the brain that are associated with pleasure and reward, which can then override normal control of energy intake. Detailed coverage of the many brain structures, peptides, neurotransmitters, and stimuli directly or indirectly regulating energy balance is beyond the scope of this text.

receptors in the ventral tegmental area of the midbrain is associated with the addictive response to high-fat and high-sugar foods observed in rat models and in humans. The hypothalamus also has neural receptors that function as a thermostat to either stimulate energy (heat) production or loss in the regulation of body temperature.

Stimuli influencing the appetite include signals from the stomach, intestine, muscles, fat depots, pancreas, liver, and other body tissues and organs; blood levels of various nutrients; metabolites; hormones; and environmental cues. These stimuli are integrated and interpreted by the hypothalamus and directed to other brain centers and body organs to help maintain energy balance. Some factors may function to control body weight on a short-term basis, while others may exert long-term effects. The following stimuli may be involved in body-weight control in one way or another.

- *Senses.* Stimulation of several senses such as sight, sound, and smell may influence neural or hormonal activity to stimulate or depress our appetites, even before food is ingested. The sense of taste also has a significant impact on appetite and energy intake. In her review, Ventura discusses the first 1,000 days of pregnancy and early childhood as a critical window of opportunity to develop healthy lifelong eating behaviors. Amniotic fluid and breast milk transmit flavors from the mother's diet to the developing fetus, neonate, and the young child. Repeated exposures to flavors from a healthy diet *in utero* and during breastfeeding, followed by repeated exposures of healthy foods of different textures during weaning, provide the foundation for eating habits during childhood that may extend through life. Strategies to promote healthy eating in pregnancy and in infants and toddlers are discussed in the *2020-2025 Dietary Guidelines for Americans.*

- *Stomach fullness.* An empty or full stomach may influence mechanical stretch receptors in the stomach walls that provide vagal feedback to the central nervous system. An empty stomach may stimulate the hunger center by various neural pathways, whereas a full stomach may stimulate the satiety center. The stomach may also release hormone-like substances that stimulate or diminish hunger. Although postmeal stretch provides feedback to avoid overeating, Brunstrom also stresses the importance of learned decisions about portion/meal size before a meal (i.e., meal planning based on expected satiety) over postmeal stomach stretch as a control mechanism for food intake. In their meta-analysis on chewing and lubrication as components of oral food processing in promoting satiety, Krop and others reported significant inverse associations between chewing and self-reported hunger and food intake. They found little research on lubrication, which could influence perceived texture in the regulation of food intake.
- *Blood nutrient levels.* Receptors in the hypothalamus, liver, or elsewhere may be able to monitor nutrient levels in the blood. In regard to this, three theories center on the three energy nutrients. The **glucostatic theory**, originally proposed in the 1950s by Jean Mayer, suggested that a decrease in blood glucose stimulates glucose-sensitive neurons in the hypothalamus and increases appetite, whereas an increased blood glucose level decreases appetite. A recent review by Chaput and Tremblay reported links between hypoglycemia, impaired glycemic control, excess energy intake, and increased body weight including obesity, thus lending support to this theory. According to the "selfish brain" theory of glucose homeostasis, described by Peters, the brain regulates "competent" use of glucose in a healthy diet while "incompetent" brain glucose uptake leads to excess body glucose as contributing factors to obesity and type 2 diabetes. The **lipostatic theory** of regulatory afferent feedback from stored fat is based on research in the 1950s identifying *ob/ob* obese mice, which lacked the gene expressing a factor controlling appetite, and *db/db* obese mice, which lacked the receptor gene for this factor. This research ultimately lead to the 1994 discovery of leptin, discussed later in this chapter, by Zhang and others. The **aminostatic theory** is a similar mechanism for regulation of amino acids, or protein. In essence, by-products of carbohydrate, fat, and protein metabolism may influence neurotransmitter production influencing appetite, such as serotonin and norepinephrine, in the hypothalamus.
- *Body temperature.* A thermostat in the hypothalamus may respond to changes in body temperature and influence the feeding center. For example, an increase in body temperature inhibits the appetite.
- *Hormones, cytokines, and neuropeptides.* A number of different hormones, cytokines, and neuropeptides (neurotransmitters) in the body have been shown to affect feeding behavior, including but not limited to insulin, serotonin, norepinephrine, leptin, ghrelin, cortisol, cholecystokinin, glucose-dependent insulinotropic peptide, glucagon-like peptide 1, insulin-like peptide 5, peptide YY, neuropeptide Y, oxyntomodulin, and thyroxine. As discussed later, some hormones may function on a short-term basis to help control meal size, whereas other hormones may be involved in long-term regulation of body weight.
- *Ultra-processed foods.* Recent research suggests that a diet high in ultra-processed food may promote excess energy intake. Hall and others provided diets consisting of ultra-processed and unprocessed diets (e.g., similar to recommendations from *2020-2025 Dietary Guidelines for Americans*) to 20 weight-stable but overweight (BMI = 27 ± 1.5 kg/m^2) adults. Diets were matched on presented kcal, sugar, fat, fiber, sodium, and other macronutrients. Subjects consumed each diet for 2 weeks in counterbalanced order. Energy intake while consuming the ultraprocessed diet was 508 ± 106 kcal greater with higher intake of carbohydrates and fat but not protein compared to the unprocessed diet. Although the study was not designed to examine mechanisms of these differences, they noted that ultra-processed foods are higher in sugar, fat, and sodium, but lower in protein and fiber. While on ultraprocessed and unprocessed diets, subjects gained and lost 0.9 kilograms, respectively. The short-term study was conducted in a metabolic ward, which does not reflect exposure to "real-world" factors associated with increased energy intake and/or decreased energy expenditure. Future studies with larger samples and longer exposures are needed to replicate these findings and gain insight into possible explanatory mechanisms.

In his review, Woods described the increased knowledge base of food intake regulatory mechanisms since Mayer's glucostatic theory to the current view of an extremely complex, plastic (as opposed to hardwired), interwoven system of neural circuitry involving many brain areas that receive input from an ever-growing list of recently discovered peptides, neurotransmitters, and hormones. Our understanding of feeding regulatory mechanisms will no doubt increase in the future.

Energy Expenditure The other side of the energy-balance equation is energy expenditure, or metabolism. Although exercise is one way to increase energy expenditure, the vast majority of the energy that is expended by the body on a daily basis is accounted for by the basal energy expenditure (BEE) or resting energy expenditure (REE), which is slightly higher than BEE due to prior eating and physical activity, as was discussed in chapter 3. Changes in the REE may be involved in the regulation of body weight. Several mechanisms of body-weight regulation have been proposed.

- *Brown adipose tissue.* **Brown adipose tissue (BAT)** differs from white adipose tissue (WAT) that comprises most fat tissue. WAT cells have a large single lipid droplet with fewer and smaller mitochondria, while BAT cells have more cytoplasm with multiple smaller lipid droplets and multiple mitochondria which accounts for the brown color. The key characteristic of BAT is mitochondrial uncoupling proteins (UCP-1), proton pores in the inner mitochondrial membrane that allow hydrogen ions to "leak" from the intermembrane space to the matrix independently of the proton gradient created by the electron transport chain that is harnessed by ATP synthase for ATP formation. BAT has a high rate of metabolism and releases energy in the form of heat without ATP production. UCP-1 activation and heat production occur primarily through β_3-adrenergic receptor stimulation. This activity is referred to as *nonshivering thermogenesis*. BAT may also contribute to diet-induced thermogenesis (DIT), which was discussed in chapter 3. BAT thermogenic activity is much greater in rodents and hibernating animals than in humans. It is found in small amounts around the neck, back, and chest areas of humans (≤ 1 percent of body fat).

  Constitutive (cBAT) cells are developed during embryogenesis while recruitable (rBAT) cells appear postnatally in WAT and skeletal muscle. BAT differentiation and metabolism are regulated by neuroendocrine signals from the brain, thyroid, skeletal muscle, pancreas, WAT, liver, and heart. As discussed by Sepa-Kishi and Ceddia, subcutaneous WAT can undergo "browning" in response to exercise training, which offsets downregulation of UCP-1 thermogenic activity in more central BAT. Other factors associated with WAT browning include lipolysis, cold exposure, various tissue growth factors, the release of the myokines irisin and musclin from skeletal muscle, and upregulation of cellular pathways that promote mitochondrial biogenesis. Mattson suggested that individuals with low levels of brown adipose tissue are prone to obesity, insulin resistance, and cardiovascular disease, whereas those with higher levels maintain lower body weights and exhibit superior health as they age. According to Jeremic and others, BAT exerts a cardio protective effect by releasing healthy adipokines, nerve growth factors, and other anti-inflammatory substances; promoting vascular health, and combating insulin resistance. Higher BAT levels are observed in lean individuals. BAT decreases with aging. The role of brown adipose tissue in the etiology and treatment of obesity, diabetes, and metabolic syndrome as well as clearance of triacylglycerol-rich chylomicrons and very low-density lipoprotein is the subject of ongoing research.

- *White adipose tissue and muscle tissue.* UCPs are also found in other tissues, such as muscle tissue and WAT, where thermogenesis without ATP production can occur with high energy intake. The activation of several transcriptional factors can increase expression of UCP-1 in WAT. As discussed previously, exercise and other neural, hormonal, and other factors may promote the "browning" (also known as "beiging") of WAT (**figure 10.8**). According to Rodríguez and others, there is considerable regulatory communication between adipokines and myokines secreted by fat and muscle tissue, respectively. Irisin, which is produced by the cleavage of a protein found in muscle and adipose tissue, can induce thermogenesis and the browning of WAT and is regulated by other adipokines and myokines.

- *Hormones.* Levels of hormones from the thyroid and adrenal glands may rise or fall and affect energy metabolism accordingly. The thyroid hormones triiodothyronine (T_3) and thyroxine (T_4) may be involved in the stimulation of BAT, which has a high expression of enzymes catalyzing the conversion of T_4 into the more active T_3. Hormones such as epinephrine from the adrenal gland also may increase the activity of certain enzymes, resulting in increased energy expenditure. Decreases in such hormonal activity may depress energy metabolism. Other hormones may stimulate or depress thermogenesis in adipose or muscle tissues.

- *Nonexercise activity thermogenesis (NEAT).* **Nonexercise activity thermogenesis (NEAT)** Chung and others commented that NEAT may be a potential mechanism to significantly increase the total daily energy expenditure in those with obesity who have little or no daily exercise energy expenditure. According to Villablanca and others, supplementation of leisure and occupational activities with NEAT could increase energy expenditure by as much as 2,000 kcal/day depending on body mass and the type/quantity of such activity. Factors affecting NEAT are not completely understood. Separate reviews by Butterick and Teske and their respective colleagues have discussed the role of various neuropeptides such as orexin A, cholecystokinin, corticotropin-releasing hormone, neuromedin U, neuropeptide Y, leptin, agouti-related protein, and ghrelin in increasing spontaneous physical activity. NEAT may also be promoted by genetics, development, environment, and interactions between these factors that are unique to the individual. Increasing NEAT could have a profound impact on prevalence rates of obesity, diabetes, cardiovascular disease, the metabolic syndrome, and other morbidities.

Feedback Control of Energy Intake and Expenditure The human body has developed a number of short-term and long-term *feedback systems* to regulate most physiological processes, including energy balance and body weight. Short-term (daily) control mechanisms may either decrease or increase food intake. The stomach expands during a meal, with afferent input from stretch receptors in the stomach wall to the hypothalamus to suppress food intake. As previously noted in separate reviews by Butterick and Teske, ingested nutrients alter the release of a variety of peptides from the stomach, intestines, and pancreas, which regulate energy intake. Cholecystokinen, pancreatic glucagon, obestatin, and amylin are gut peptides which are released rapidly with eating and have short actions leading to meal termination. A partial list of hormones and peptides affecting energy balance are listed in **table 10.5**. Body stores of carbohydrate, protein, and fat are also regulated on a short-term basis. The human body has a limited capacity to store excess carbohydrate and protein, so changes in blood glucose and amino acid levels help regulate carbohydrate and protein intake. Although the human body possesses a high capacity to store fat, blood lipids and other factors also help maintain body-fat balance on a short-term basis. Other short-term mechanisms increase food intake. **Ghrelin** is a peptide

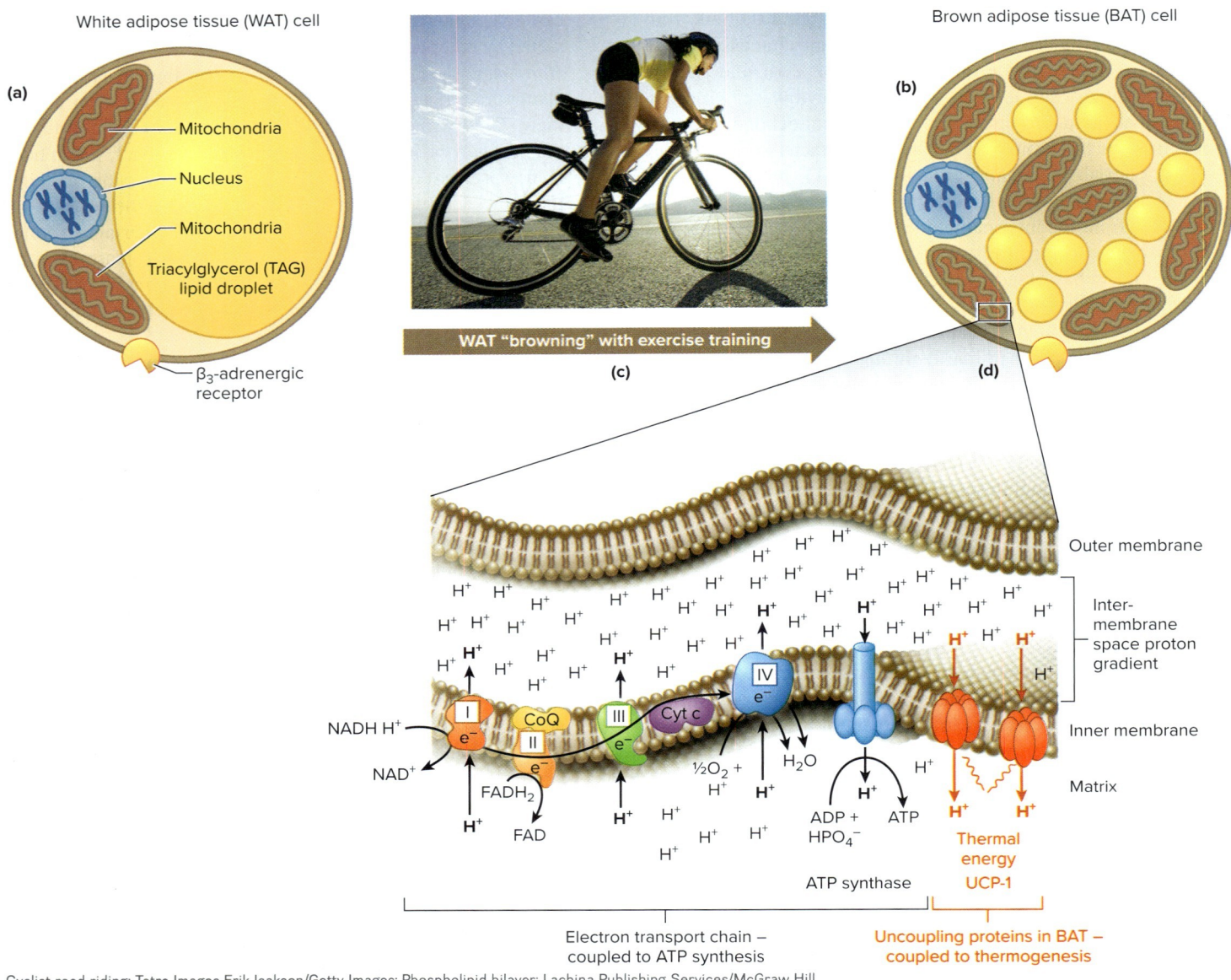

FIGURE 10.8 (a) White adipose tissue (WAT) cells have a single triacylglycerol (TAG) rich droplet, less cytoplasm, and fewer mitochondria. (b) Brown adipose tissue (BAT) cells have multiple smaller TAG droplets and more mitochondria, which accounts for its brown color. (c) Exercise can promote the "browning" (a.k.a., "beigeing") of WAT into BAT in a mechanism that includes the stimulation of β_3-adrenergic receptors. (d) BAT has mitochondrial uncoupling proteins (e.g., UCP-1), which promote thermogenesis instead of ATP synthesis.

hormone released by the stomach, mainly before mealtime when the stomach is relatively empty. Ghrelin acts on the hypothalamus to stimulate the appetite. Yanagi and others note that ghrelin may also affect dopaminergic neurons in the brain, possibly increasing secretion of dopamine, which induces sensations of pleasure that may be involved in long-term regulation of body weight, as discussed later.

The **set-point theory** of weight control is a long-term proposed feedback mechanism primarily developed from rat models and applied to energy intake and expenditure in human behavior. This theory proposes that the body is programmed to be a certain weight, or set point, something comparable to a set body temperature (37°C or 98.6°F). Based on animal studies, Sullivan and others indicate that the maternal nutrition status during the perinatal period, or just before and after birth, may be critical in establishing the offspring's body-weight set point. The arcuate nucleus of the hypothalamus contains neurons that can sense an array of metabolites associated with either negative or positive energy balance to conserve or dissipate energy accordingly. **Leptin**, a peptide hormone, is encoded and produced by the *OB* gene in the adipose cells. It is secreted as an adipokine by adipose tissue and exerts long-term effects on energy balance by suppressing food intake,

TABLE 10.5 Effects of selected hormones and neuropeptides on appetite and energy balance*

Hormone or neuropeptide	Secretion site	Effect on appetite
Amylin	Pancreas	Inhibits
Cholecystokinin	Gastrointestinal tract	Inhibits
Glucagon	Pancreas	Inhibits
Glucagon-like peptide 1	Gastrointestinal tract	Inhibits
Insulin	Pancreas	Inhibits
Leptin	Adipose tissue	Inhibits
Obestatin	Gastrointestinal tract	Inhibits
Oxyntomodulin	Gastrointestinal tract	Inhibits
Pancreatic polypeptide	Pancreas	Inhibits
Peptide YY	Gastrointestinal tract	Inhibits
Proopiomelanocortin derivatives	Pituitary gland	Inhibits
Agouti-related peptide	Brain	Stimulates
Cortisol	Adrenal cortex	Stimulates
Ghrelin	Gastrointestinal tract	Stimulates
Neuropeptide Y	Hypothalamus	Stimulates
Thyroxine; Triiodothyronine	Thyroid gland	Stimulates

*This is not a complete list of neuroendocrine regulators of energy balance. Some, but not all, of the above hormones and neuropeptides are discussed in the text.

decreasing appetite, and inducing weight loss. Leptin production and secretion are greater when fat stores are high and lower when fat stores are low. Blood-borne leptin inhibits the production of **neuropeptide Y (NPY)**, which is known to stimulate the appetite, increase energy intake, and reduce resting energy expenditure (REE). Feedback control of NPY by leptin suppresses hunger and voluntary food intake. Greater fat stores result in greater leptin secretion, which exerts an anorexigenic (appetite-inhibiting) effect and inhibits the orexigenic (appetite-stimulating) effect of NPY. A decrease in NPY formation in the hypothalamus may also increase REE by stimulating thermogenesis in adipose tissue and muscle. Likewise, decreased leptin secretion results in increased NPY formation by the hypothalamus.

The hypothalamus may also regulate energy homeostasis through sensing other metabolites. In their review, Stark and others discuss an integrated approach to nutritional and hormonal feedback in energy homeostasis instead of isolating the effect of one nutrient, hormone, or neuropeptide. As an example, carnitine, discussed in chapter 5 and synthesized from the amino acid lysine and methionine, is essential for long-chain fatty acids to enter the mitochondria to undergo β-oxidation to produce acetyl CoA. Since acetyl CoA is common to the catabolism of carbohydrate, fat, and protein, hypothalamic carnitine sensing could represent a convergence point in energy homeostasis.

Deviation from the set point results in metabolic adjustments known as **adaptive thermogenesis**, which either increases or decreases heat production. According to Wallace and Fordahl, hedonic (pleasure- and reward-seeking) signals from high saturated fat diets stimulate dopamine-secreting neurons in certain areas of the brain, as illustrated in **figure 10.7**, which can influence normal hypothalamic homeostatic control and endocrine, cytokine, and peptide signals regulating hunger and satiety. As a result, homeostatic control may be compromised due to greater reinforcement for food intake associated with hedonic behavior. Over time, less effective feedback control of energy intake and expenditure by peptides would contribute to weight gain and obesity.

Ghrelin may also be involved in long-term control of body weight. As noted by separate reviews by Yanagi, Rui, and their respective colleagues, ghrelin may also affect parts of the brainstem that contain neurons secreting dopamine, which stimulates the mesolimbic and corticolimbic pleasure and reward brain centers (**figure 10.7**). As a result, ghrelin not only stimulates the appetite but may also help establish strong memories between eating certain foods and sensations of pleasure. According to Sclafani and Ackroff, dopamine reward circuits in the brain are involved in the conditioning of carbohydrate and fat intake and may possibly lead to *hypereating* and long-term weight gain. However, less is known about gut/brain pathways regulating protein intake.

Less research has been conducted relative to feedback control of physical activity. According to the **activity-stat hypothesis**, increased or decreased physical activity results in a compensatory change regulated by a center in the brain that functions to increase or decrease physical activity in order to maintain a stable level of activity or energy expenditure over time. Increasing or decreasing the daily amounts of NEAT may be related to the activity-stat hypothesis. As previously discussed and illustrated in **figure 10.7**, several brain structures are thought to regulate NEAT via afferent stimulation by various central and peripheral peptides. If substantiated, an activity-stat center could support the set-point theory of body-weight control. Although the set point is a theory, it may help explain why some people maintain a normal body weight throughout life but, when disrupted, may lead to an excessive gain or loss of body mass. It is important to note that although the set-point theory is based on subconscious control mechanisms underlying energy intake and energy expenditure, the forebrain can consciously override these subconscious mechanisms and increase or decrease body mass if necessary. In their review of 28 studies of differing designs and methodologies which examined this hypothesis, Gomersall and others reported little evidence for physical activity compensation and therefore no clear evidence of an activity-stat center. They proposed several design and methodological recommendations for future research.

How is fat deposited in the body?

The actual deposition of fat in the human body may occur in two ways: **hyperplasia** (an increase in the number of adipocytes, or fat cells) or **hypertrophy** (an increase in the amount of fat in each cell). Earlier research appeared to support the theory that hyperplasia is a major cause of childhood obesity, whereas hypertrophy is the primary cause in adulthood. It is now generally accepted that adipocyte hyperplasia occurs throughout life and that obesity involves an interaction between the increase in both fat cell size and fat cell number in response to normal cell turnover and to the need for additional fat mass stores that arises when caloric intake exceeds nutritional requirements. Lee and others describe adipogenesis as a complex and incompletely understood coordination between regulatory adipogenic transcriptional factors (e.g., peroxisome proliferator-activated receptor gamma; CCAAT/enhancer-binding protein α) and epigenetic modifications (e.g., histone and DNA methylation/demethylation; histone acetylation/deacetylation) to either promote or inhibit differentiation of precursor stem cells into adipocytes. Existing fat cells apparently have a maximal size potential (about 1 microgram of fat per cell), which when exceeded stimulates the formation of new fat cells or the accumulation of fat in preadipocytes, small cells in the adipose tissue that have the potential to become adipocytes. A genetic predisposition for adipocyte hyperplasia would facilitate the development of obesity. However, individuals without this genetic predisposition may still develop obesity with a positive energy balance stored as fat because of adipocyte hypertrophy, leading to adipocyte hyperplasia. Engin comments that death of hypertrophic adipocytes may be due to programmed necrotic pathways and may be a stimulus for adipocyte hyperplasia. In their review, Arner and Spalding noted that extreme weight loss decreases the size but not the number of adipocytes. Smaller adipocytes secrete less leptin, which attenuates the feedback regulation of appetite-stimulating neuropeptide Y. The development of pharmacological agents that inhibit adipogenesis without adversely affecting normal metabolism (fat storage and insulin sensitivity) may be a future strategy in reducing the prevalence of obesity.

What are the contributing factors to obesity?

Energy processes in the human body are governed by the laws of thermodynamics. If the human body consumes less energy in the form of food kcal than it expends in metabolic processes, then a negative energy balance will likely occur and the individual will likely lose body weight. Conversely, a greater caloric intake in comparison to energy expenditure will likely result in a positive energy balance and potentially a gain in body weight.

Although the first law of thermodynamics provides the basic answer as to how weight gain may occur, it does not provide any insight relative to the specific mechanisms. Albuquerque and others comment that obesity is generally accepted as the result of many genetic, environmental, and lifestyle factors and their various interactions, several of which are discussed in the following text.

Genetic and epigenetic factors According to Butler, heritability is a strong factor in the etiology of obesity, particularly **morbid obesity**. Individuals with Prader-Willi syndrome, a rare genetic disease affecting hypothalamic regulation of energy intake and physical activity, have morbid obesity because of the disrupted pathway between leptin binding on receptors on the arcuate hypothalamic nucleus and proteolysis of the large peptide proopiomelanocortin [POMC] at the paraventricular hypothalamic nucleus. Mutations of the POMC gene, the gene cleaving POMC into smaller active neuropeptides, and/or genes for other involved proteins are the reason for the altered hypothalamic pathway. At least 370 identified genes play a role in obesity, with concordance rates in studies of monozygotic and dizygotic twins ranging from 70 to 90 and 35 to 45 percent. In these studies such as ongoing Finnish Twin Cohort studies reviewed by Naukkarinen and others, twins who were raised separately have increased body composition similarities with their biological parents compared to their adoptive parents.

Research into the genetics of obesity has been progressing at a rapid pace since the mapping of a mutation in 2007 in the fat mass and obesity-associated (FTO) gene. Ongoing prospective, multigenerational investigations, and genome-wide association studies (GRAS) have identified many genetic mutations that are associated with obesity. The influence of such mutations in FTO and other genes on the prevalence of obesity in diverse populations has been the focus of considerable research as well as separate reviews by Yeo (2014) and Fall and Ingelsson (2014). Claussnitzer and others manipulated the expression of a variant in the FTO gene contributing to obesity in humans and mice and observed that inhibition increased beige adipocyte differentiation and thermogenesis, while overexpression decreased beige adipocyte differentiation and thermogenesis and increased white adipocyte differentiation and lipid storage. Mutations of genes expressing thermogenic uncoupling proteins in brown fat (UCP-1) and white fat and muscle (UCP-2) may also decrease resting energy expenditure. Altered lipid and carbohydrate metabolism, appetite control, oxidative damage, and thrombolytic (blood clotting) activity are also implicated in obesity-related genetic mutations. Chaput and colleagues in the Quebec Family Study have identified genetic changes affecting intracellular signaling pathways that are associated with fatness as well as the inability to control food intake. In their meta-analysis, Tang and colleagues reported significant associations between obesity and a genetic mutation leading to leptin resistance and ineffective feedback regulation of NPY, which stimulates the appetite.

Hetherington and Cecil indicate that common obesity is polygenic, involving complex gene–gene and gene–environment interactions that increase energy intake and involve neural pathways regulating addiction and reward behaviors which contribute to overeating and obesity. Rohde and others comment that although epigenetic modifications from environmental exposures and behavior may increase risk in individuals predisposed to obesity, our understanding of the specific mechanisms is insufficient to develop personalized prevention or treatment strategies. Prehistoric humans conserved energy in times of plenty against the inevitable times of hardship. The term thrifty gene has been associated with this efficiency of energy conservation, which served our ancestors well. However, this gene does not serve modern humans, who generally have an uninterrupted abundance of food and rely on the same physiology as prehistoric humans to store excess energy. Lutter and Nestler describe the selfish gene, a more accurate reflection of contemporary energy homeostasis and the obesity epidemic, as dominance by the dopaminergic pleasure centers in the brain over the hypothalamus in regulating food intake and increasing the desire to consume foods that are highly palatable, thereby unbalancing the regulation of energy intake. Kälin and others noted that obesity-associated inflammation may result in impaired functioning of hypothalamic neurons regulating energy intake that are not easily reversed.

In summary, research suggests that genes affecting energy balance in the body appear to be very efficient as a means to promote energy intake and weight gain but relatively inefficient in promoting energy expenditure and weight loss. Hetherington and Cecil indicate that common obesity is polygenic, involving complex gene–gene and gene–environment interactions that increase energy intake and involve neural pathways regulating addiction and reward behaviors which contribute to overeating and obesity. Kälin and others noted that obesity-associated inflammation may result in impaired functioning of hypothalamic neurons regulating energy intake that is not easily reversed. Genetic factors related to the development of obesity are listed in **table 10.6**.

TABLE 10.6 Genetic factors that have been implicated in the development of obesity

Predisposed taste for sweet, high-fat foods
Impaired function of hormones such as insulin and cortisol
Decreased levels of human growth hormone
Low plasma leptin concentrations
Leptin resistance
Inability of nutrients or hormones in the blood to suppress the appetite control center
Greater number of white fat cells (adipocyte hyperplasia)
Lower levels of brown fat, and lower "beige" or "brown" differentiation of white adipose tissue
Enhanced metabolic efficiency in storing fat
Lower resting energy expenditure (REE)
Decreased thermic effect of food (TEF)
Lower rate of fat oxidation
Lower levels of nonexercise activity thermogenesis (NEAT)
Lower levels of energy expenditure during light exercise

Environmental Factors In addition to genetic heritability, environmental factors also contribute to obesity. A considerable body of genetic and epidemiological research has examined the gene × environment interaction where individuals with genetic polymorphisms associated with obesity who are also exposed to an "obesogenic environment," including primarily physical inactivity and energy dense (high fat/high sugar) diet, have an increased risk of obesity. According to Rohde and others, although such research has reported that genetic predisposition to obesity can be compensated for by positive lifestyle behavioral changes (e.g., physical activity, healthier diet), a full understanding of the gene × environment interaction remains incomplete. Consuming a high-kcal diet, either high or low in fat, may lead to excess caloric intake and weight gain. Hruby and Hu note that socioeconomic and environmental factors such as income, education, the built environment, gut microbiota, viruses, and social networks, in addition to individual lifestyle factors such as diet and activity, contribute to obesity. The amount of food eaten is strongly influenced by factors such as portion size, food visibility, and the ease of obtaining food. The following discussion highlights the variety of environmental factors that may contribute to excessive weight gain. The main contributors are excessive caloric intake and decreased physical activity, both of which may be influenced by other environmental factors.

High-fat, high-kcal foods Butler and Eckel describe eating as a motivated behavior regulated by the brain, which monitors energy balance with an evolutionary bias toward consumption of an energy surplus. As previously discussed, this mechanism served humans well during past times of food scarcity, but it contributes to obesity and its comorbidities today. Nonhomeostatic hedonic (e.g., pleasure/reward-seeking) influences have the potential to encourage overconsumption of a palatable, energy-dense diet. According to the first law of thermodynamics, an increase in caloric intake compared to expenditure will lead to a weight gain. However, researchers suggest that dietary fat is the main culprit leading to obesity. The National Academy of Sciences concluded that higher fat intakes are accompanied by increased energy intake and therefore increased risk for weight gain in populations already disposed to being overweight and obesity, such as North America and other developed/developing countries. Obesity researchers have identified at least three reasons for these associations. First, the high palatability of dietary fat elicits a previously-mentioned hedonic response in many individuals and may encourage overconsumption. Although dietary fat contains more kcal per gram, the body responds more slowly to dietary fat, resulting in the ingestion of too much fat (and kcal) from a high-fat meal. Samra notes that fats, kcal for kcal, are less effective in suppressing subsequent food intake than carbohydrates. De Vadder and others describe dietary protein as playing a "satiety role" in regulating energy intake by providing input from the intestine to the brain. Second, dietary fat may be stored as fat more efficiently compared to carbohydrate and protein. The conversion of carbohydrate or protein to fat requires three to four times as much energy compared to dietary fat storage. Third, it is thought that a chronic high-fat diet may alter hypothalamic response to factors, such as leptin, that normally suppress appetite. For example, as previously discussed, leptin resistance impairs feedback control of NPY, an orexigenic peptide that stimulates energy intake and body-fat deposition. Butler and Eckel comment that high-fat diet-induced neural inflammation can contribute to an imbalance between homeostatic and nonhomeostatic regulation of energy intake.

Regular consumption of energy-dense foods at fast-food restaurants (FFRs) and sugar-sweetened beverages (SSBs) may also contribute overweight and obesity. In a review of studies with strong methodologies and precise measurement of SSBs, Bucher Della Torre and others concluded a significant association between SSB consumption and obesity in children and adolescents. Lee and others reported public housing neighborhoods had more FFRs with cheap SSBs and "super-size" menu items compared to middle-income neighborhoods. In a study of almost 7,500 California public schools from 2000 to 2010, Sanchez-Vaznaugh and others reported increased numbers of FFRs serving inexpensive energy-dense foods in low-income schools and neighborhoods compared to decreased numbers of FFRs in more affluent neighborhoods and schools. Reviews of large, well-controlled studies by Nago and Lachat and their respective colleagues reported significant associations of frequent "out-of-home eating," specifically at FFRs, with increased BMI, overweight/obesity, waist circumference, and a lower intake of micronutrients (e.g., vitamin C, calcium, and iron). As noted by Weihrauch-Blüher and Wiegand, high consumption of SSB and FFR food leading to weight gain and insulin resistance significantly increases the risk of adult cardiovascular disease and type 2 diabetes. Ronit and Jensen concluded that food industry self-regulation is weak and that some degree of external regulation might improve food industry effectiveness in marketing and labeling to combat obesity. The *2020–2025 Dietary Guidelines for Americans* recommends a diet of low-energy-density and high-nutrient density for all age groups. The interested reader is referred to reviews cited in this section.

Low-fat, large-portion-size, high-kcal foods Dietary fat is not the only explanation for obesity. Most reviews conclude that liquid kcal such as SSB is also involved in weight gain. Walker and others noted that while the prevalence of obesity in the United States has dramatically increased over the past 35 years, daily energy derived from dietary fat has actually decreased.

There are two reasons for this inconsistency. First, "fat-free" foods may be loaded with simple sugars and actually contain more kcal per serving. According to data in the *2020–2025 Dietary Guidelines for Americans*, added sugars should total no more than 10 percent of daily caloric intake. However, depending on age, 57 to 78 percent of females and 54 to 80 percent of males from 2 to over 60 years of age exceed this limit. Based on a 2,000 kcal/day diet, added sugar should be no more than 200 kcal. Welsh and others note that most Americans exceed the recommended limit of kcal from total added sugar per day in SSB alone. A 12-ounce sugar-sweetened soda contains the equivalent of 10 teaspoons of table sugar or about 150 kcal. Malik and others concluded that a greater consumption of SSBs is associated with weight gain in children and adults. Consuming SSBs *per se* may not be a cause of obesity, but it may be one avenue to exceeding recommended caloric intake. An additional intake of 150 kcal in added sugars would increase body weight by over 15.5 pounds over the course of a year, an example of creeping obesity, as discussed later. Restricting advertising to children, limiting school sales, and increasing the availability of smaller portion sizes are beverage industry initiatives to reduce SSB consumption.

Second, portion sizes have increased dramatically over the years. Supersized portions with significantly increased caloric content are aggressively marketed everywhere. The average soft drink is more than 50 percent larger than in past years and contains more sugar and kcal. Consumption of a 20-ounce bottle (unit) of soda will add 100 kcal to energy intake compared to the 12-ounce bottle. According to Steenhuis and Poelman, "unit bias," "visual cues," "bite size," and "value for money" are some of the researched mechanisms explaining large portion size selection. There is no compensation for consumption of larger portions with decreased subsequent intake, so the net effect is increased energy intake.

Physical inactivity and NEAT Excess energy intake is not the sole explanation for overweight and obesity. Although technological advances generally improve life, sedentary entertainment/leisure activities and labor-saving devices lower energy expenditure and can contribute to overweight, obesity, and other chronic diseases. According to the laws of thermodynamics, the total energy in a system is constant, consists of potential and kinetic energies, and is neither created nor destroyed. A change in potential energy is matched by a corresponding opposite change in kinetic energy. As discussed in chapter 3, the metabolic pathways conserve some of the potential energy in the macronutrients in the form of ATP, with the remainder lost as metabolic heat. Surplus dietary potential energy not transferred to ATP or dissipated as heat is stored as fat, glycogen, and/or amino acids. As discussed in separate reviews by Villablanca, Butterick, Teske and their respective colleagues, obesity resistance observed in animal models and in human research is related to a complex interaction between peripheral and central peptides and neurotransmitters and brain structures to regulate spontaneous physical activity and nonexercise activity thermogenesis (NEAT).

Villablanca and others content that the variability in total daily energy expenditure is attributed mainly to NEAT. Technology has decreased the amount of energy expended via NEAT each day, such as walking, standing, and even fidgeting, which may lead to an energy imbalance and weight gain.

Sedentary activities, such as playing video games or working on the computer, decrease NEAT, which may contribute to the development of obesity. Villablanca and others reported that NEAT may vary between individuals by as much as 2,000 kcal per day. In particular, an excessive amount of sitting has been an increasing concern. Swinburn and Shelly noted that excessive television viewing may pose a double threat for weight gain, not only because of the prolonged sitting but also because advertisements for energy-dense foods may increase total caloric consumption. Biswas and others concluded that sedentary behavior had a greater effect on all-cause mortality in individuals with low fitness compared to those with high fitness. Tremblay and others reported a dose-response relationship between sedentary behavior and adverse health outcomes in children age 5-17 years. More than 2 hours per day of screen time was associated with decreased fitness, increased fatness, lowered self-esteem, and lower academic performance. The term *active couch potato* has been used by Owen and others to classify individuals who may have compromised metabolic health if they sit for prolonged periods, even if they meet public health guidelines on physical activity. Bouchard and others noted the need for national and international guidelines for levels of fitness and sedentary behavior in addition to those in place for physical activity.

However, recent reviews conclude that a weaker association exists between adult sedentary behavior and obesity. In a meta-analysis of 23 studies, Campbell and others reported a small but significant increase in waist circumference for each 1 hour/day increase in sedentary behavior from baseline to 5 years of follow-up. They also reported a significant odds ratio of change in BMI to overweight or obesity categories in the highest sedentary category compared to the lowest category at baseline. They concluded that sedentary behavior had little effect on overweight and obesity and that increased cardio-metabolic risk due to sedentary behavior appears to be due to mechanisms other than overweight and obesity. In a review of existing reviews, Biddle and others also concluded that any association between adult sedentary behavior and obesity does not appear to be causal. Both groups noted the need for more prospective studies using standardized methodologies to quantify sedentary behavior, body weight, and body composition.

Other environmental factors Other environmental factors may predispose to weight gain.

Sleep and emotional stress Inadequate sleep may alter hormonal response to energy intake and control. In their review, Ogilvie and Patel describe cross-sectional and longitudinal associations between poor sleep and obesity. A potential two-way causality exits where poor sleep may be associated with poor diet and low levels of physical activity while obesity is a recognized risk factor for sleep apnea, which adversely affects the quality of sleep.

In their review, McNeil and others noted significant epidemiological and experimental evidence of a role of sleep restriction in higher BMI, body fat, and insulin resistance due in part to increased cortisol levels that impair insulin-mediated glucose uptake. Physical and psychological stress provokes the "fight-or-flight" response with an up-regulation of the hypothalamic-pituitary-adrenal (HPA) axis. Although Incollingo Rodriguez and others reported some inconsistency in the literature, in general, they observed an association between up-regulation of the HPA axis, which results in increased cortisol secretion, and greater abdominal fat accumulation. Cortisol, a glucocorticoid hormone secreted by the adrenal cortex in response to HPA upregulation, suppresses immune function, promotes gluconeogenesis, and decreases insulin sensitivity. Individuals with obesity have elevated levels of the enzyme converting inactive cortisone into active cortisol, which promotes lipogenesis and fat deposition, particularly in the visceral/abdominal area. Zimberg and others noted that sleep restriction is also associated with increased ghrelin levels and decreased leptin levels, which promotes increased energy intake. Decreased energy expenditure is a plausible result of fatigue associated with sleep restriction.

Personal relationships Personal relationships may contribute to weight gain. In a 32-year study of weight gain in 12,000 subjects, Christakis and Fowler reported the chances of obesity increased by 57, 40, and 37 percent, respectively, if a friend, sibling, or spouse had obesity. They concluded that obesity appears to be spread through social ties. According to Powell and others, complex interpersonal relationships between significant others, friends and family, and peers, as well as interpersonal networks formed by casual and incidental contact, culture, and geographical proximity, strongly influence overweight or obesity through the interrelated processes of *social contagion* (mirroring and aspiring behaviors); *social capital* (belonging, social support), and *homophily* ("like-attracts-like"). Although the focus of the literature is on overweight/obesity, it is possible that similar complex relationships exist among friends or family who maintain a healthy body weight.

In a review of global studies, Liberali and others reported significant association of an "obesogenic pattern diet," including, but not limited to, red meat, whole dairy, fatty foods, and sugar, with obesity compared to a "nonobesogenic pattern diet." They concluded that a diet with a lower percentage of obesogenic foods should effectively reduce the risk of obesity. In two separate studies, Salvy and others reported greater food consumption by overweight youths in the company of an overweight friend compared to a normal weight friend and a significant maternal influence in young (5-7 years) children in healthy food choices compared to peers. Older (13-15) girls but not boys were more influenced by peers compared to their mothers in making healthy dietary choices. Their data suggest age and gender differences in the effect of peer influence of dietary choices.

Drugs and environmental chemicals Certain drugs may increase body weight. Individuals with asthma or pulmonary disease may use corticosteroids containing cortisone, which may be converted to cortisol and cause weight gain, as previously discussed. Two important factors may be the use of alcohol and nicotine. Alcohol is rich in kcal. At 7 kcal/g, it is comparable to a gram of fat, which contains 9 kcal. The National Academy of Sciences has noted that if the energy derived from alcohol is not utilized, the excess is stored as fat. Traversy and Chaput note that heavy drinking is linked to obesity, but that light to moderate drinking is not, possibly due to generally healthier lifestyles which may offset weight gain. Nevertheless, alcohol intake does not suppress the appetite or fat intake. As a result, the kcal in alcohol are additive and may lead to an energy surplus and weight gain.

Nicotine in cigarettes is a stimulant that may inhibit the appetite and increase REE. Tian and others reported a greater weight gain among quitters (9 lbs) compared to continuing smokers (3.2 lbs) in 35 prospective studies over an average follow-up of almost 6 years. Teenagers and young adults may start smoking to prevent weight gain, but smoking is a well-documented primary risk factor for heart and lung disease and cancer. Healthier and more effective ways to maintain weight will be the focus of chapter 12.

Emerging evidence implicates certain environmental (endocrine-disrupting) chemicals in the development of obesity, diabetes, and the metabolic syndrome discussed later in this chapter. Phthalates, used to soften polyvinylchloride (PVC) in the manufacture of many products, is a thyroid antagonist and stimulates nuclear receptors that play a role in fat storage. Phthalates have been associated with central obesity. Polybrominated biphenyl esters, a flame retardant used in many products, also decrease thyroid function, lipolysis, and insulin-mediated glucose oxidation. Dithiocarbamates found in cosmetics inhibit the enzyme-converting cortisol to cortisone. As previously discussed in upregulation of the HPA axis, increased cortisol concentrations could increase energy intake and abdominal fat deposition. Other chemicals such as bisphenol A and phytoestrogens are also associated with obesity. According to a recent review by Heindel and Blumberg, exposure to environmental chemical "obesogens" can increase obesity risk through genetic and epigenetic mechanisms and interaction with other environmental, developmental, and biological factors. Some obesogens (e.g., bisphenol A, tributyltin, BPA, diethylhexyl, dibutyl phthalates) may cause epigenetic modifications resulting in transgenerational risk of obesity and thus require a multigenerational management approach.

Viruses Recent attention has focused on the effect of viruses on obesity. In their review, Ponterio and Gnessi commented that adenovirus AD-36 (AD-36) causes obesity in animals and is associated with obesity in humans. They call for recognition of AD-36 as a risk factor and note the potential for targeting AD-36 in combating obesity. According to Hur and others, adenovirus infections may decrease immune function, suppress norepinephrine and leptin secretion, and increase leptin resistance, thereby increasing appetite. da Silva Fernandez and others reported significant associations between AD-36 and body weight, obesity, and metabolic markers associated with obesity (e.g., insulin resistance, elevated blood lipoproteins in 31 of 37 reviewed studies.

Built environment The *built environment* is defined as all infrastructure designed by humans for use by humans. Availability (or lack thereof) of parks, recreation centers, sidewalks, and trails impacts the perceived convenience of physical activity as a strategy to combat obesity. Convenience is an important factor to consider when developing an exercise program for weight control. Residents of neighborhoods in close proximity to retail, dining, and entertainment venues may find walking or bicycling to be more convenient than residents of large "urban sprawl" developments. In a meta-analysis of 36 studies across eight countries, Chandrabose and others reported significant associations between objectively measured walkability and decreased obesity, hypertension, and type 2 diabetes. In a review, Dixon and others reported increased diet quality in areas with greater grocery store availability/proximity, more physical activity in areas with greater walkability (less urban sprawl), and lower body weight in areas with greater availability of parks. Stronger partnerships between health-care professionals, urban planners, civil engineers, and others may result in future built environments that encourage more physical activity.

Another aspect of the built environment is the effect of ambient temperature control on thermogenesis, which was discussed in chapter 9. Moellering and Smith comment that ambient temperature contributes significantly to energy intake and expenditure. Prolonged exposure to cold or hot ambient environments would tend to curb weight gain. However, heating, ventilating, and air conditioning (HVAC) technology has minimized time individuals are exposed to ambient extremes. The thermogenic-neutral environment generated by HVAC climate control does not increase basal energy expenditure or curb appetite and may be an important causal cofactor in the worldwide increase in obesity. The Centers for Disease Control and Prevention has provided ten strategies for communities to increase physical activity to combat weight gain, which may be found at the following website.

> www.cdc.gov/nccdphp/dnpao/state-local-programs/built-environment-assessment/ The CDC provides recommendations for communities to consider when creating or modifying environments to make it easier for individuals living in the community to be more physically active.

Gut microbiota The gastrointestinal tract is home to a diverse colony of bacteria that was established in early childhood and remains relatively stable up to late adulthood. According to Villanueva-Millán and others, some factors influencing the individual's unique microbiota composition are method of birth delivery, sanitary conditions, breastfeeding (human milk) versus formula feeding, antibiotic exposure, age, diet, health status, geographic location, and stress. The microbiota play important roles in nutrient absorption, energy balance, and health. For example, bacteria can hydrolyze indigestible polysaccharides into carbohydrates for fermentation into short-chain fatty acids (SCFAs), which are absorbed by the small intestine. SCFA absorption increases satiety, reduces appetite, and promotes a healthy weight. Disruption of microbiota diversity or ratios (e.g., greater Firmicutes/Bacteroidetes) by antibiotics or stressors may be associated with increased intestinal permeability, inflammation, and increased metabolic risk, including obesity. According to Gupta and others, brain–gut communication can alter normal homeostatic and nonhomeostatic (hedonic) regulatory balance of food intake toward a hedonic dominance, which may lead to food addition. Future research will establish if the microbiota/obesity relationship is correlational or causal in nature.

Interaction of Genetics and Environment According to Dalle Molle and others, **differential susceptibility** of a genetic variant is defined as an increased likelihood of an adverse outcome (e.g., obesity) in a negative environment, but a resilience to the outcome in a positive environment. In other words, physiological responses to the environment such as up-regulation of the HPA axis and brain pleasure/reward centers may modify genetic predispositions, either for better or worse.

Environmental stress may induce epigenetic modifications, such as DNA methylation and histone acetylation, that alter gene transcription. The "Dutch hunger winter studies" described by Lumey and others are the focus of ongoing prospective research. Briefly, during World War II, the Nazis blockaded food supplies to occupied lands in the Netherlands from October 1944 until the lands were liberated in May 1945. According to the Barker Hypothesis described by Edwards, maternal kcal deprivation may promote the expression of a "thrifty gene" phenotype, which serves the child well in the kcal-restricted *in utero* condition. However, later exposure to abundant, energy-dense food and inadequate physical activity may create the "selfish gene" phenotype described by Lutter and Nestler. In 1944–45, Dutch children conceived before the blockade who experienced *in utero* malnutrition later in gestation were born with low birth weight (LBW) and gained weight at an accelerated rate when exposed to adequate childhood nutrition. Children exposed *in utero* to maternal kcal deprivation in the first trimester experienced epigenetic modifications, which increased the risk of metabolic risk of obesity, heart disease, and insulin resistance in adulthood. Interestingly, there is evidence of a multigenerational effect as the progeny of these mothers also have LBW babies.

Twin studies, such as the Finnish discordant (obesity, no obesity) monozygotic studies reviewed by Naukkarinen and colleagues, provide unique control of genetics, sex, age, and early childhood experiences. The twin without obesity provides support for physical activity and other factors in offsetting increased risk of insulin resistance, fatty deposits in the liver, atherosclerosis, inflammation, and down regulation of oxidative enzyme activity. On the other hand, the genetically identical but physically inactive mate with obesity may risk a "vicious cycle" of worsening pathological changes leading to less activity and greater obesity.

Kral and Rauh describe eating and sucking rate, neophobia (avoidance of new food), preference for protein, and disinhibition (excess eating due to loss of self-control) as heritable eating behaviors which can be modified by environmental influences, such as parent modeling and healthy food choices in the home. Such studies provide valuable insight into beneficial and detrimental environmental effects on the genomics of obesity, but as Waterland notes, our understanding of epigenetic regulators of energy balance is far from complete. The *2020-2025 Dietary Guidelines for Americans* recommends repeatedly exposing young children from 6 to 23 months to nutrient-dense foods with various flavors and textures to foster acceptance of new healthy foods.

Many individuals who have maintained a normal body weight during childhood and adolescence begin to put on weight gradually in young adulthood (age 20–40). For young college students, the first year may involve increased psychological stress, which may adversely affect various lifestyle behaviors, such as dietary habits and physical activity, and lead to weight gain. The proverbial "freshman 15"—a weight gain of 15 lbs in the freshman year—appears to be an exaggeration by the popular media. A meta-analysis of 22 studies including 5,549 students by Vadeboncoeur and others concluded that 60.9 percent of the subjects gained an average of 7.5 lbs (3.38 kilograms) over the entire year. Although weight gain during the freshman year is not as extreme as once suggested, it may be the first step in the phenomenon called creeping obesity, illustrated in **figure 10.9** and caused by increased caloric intake, decreased levels of physical activity, or a combination of the two over adulthood.

Can the set point change?

Researchers continue to refine models of obesity in order to account for known principles of feedback control, genetic and epigenetic adaptations, and social/environmental influences. Speakman and

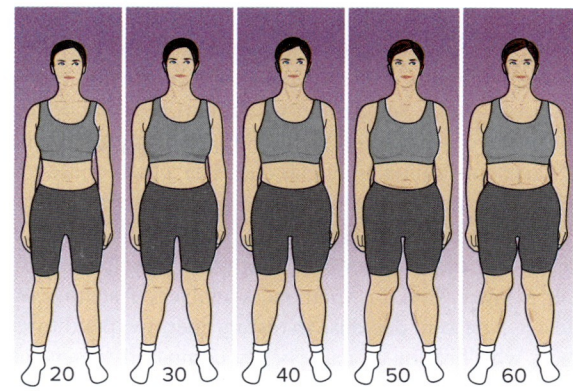
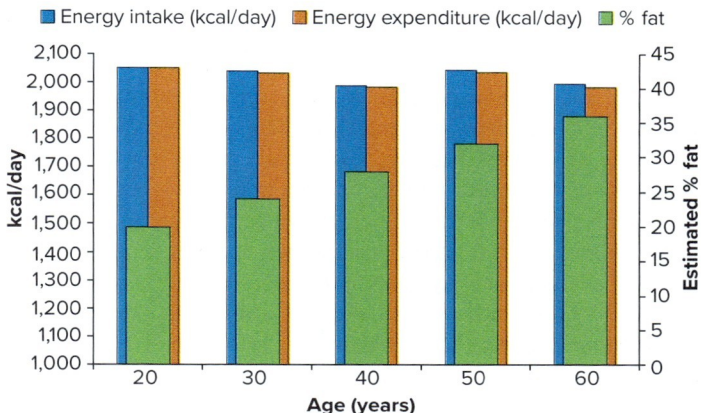

FIGURE 10.9 Adult-onset obesity is often referred to as creeping obesity. Energy balance is the difference between energy intake and energy expenditure. In this hypothetical example, a surplus of as little as ~9 kcal/day through decreased expenditure and/or increased intake can contribute to slow, steady gain in weight and body fat. Note the increases in estimated fat percentage and BMI over the decades of adult life. Of course, a greater surplus results in a greater rate of weight (and fat) gain. If her daily surplus were 200 kcal, she would gain over 20 lbs in 1 year.

others compared various models of energy balance regulation and obesity in order to reconcile physiological, environmental, and epigenetic factors. The set-point (contemporary lipostatic) theory of a defended body weight is well-grounded in physiological feedback control, but it does not explain social and environmental influences contributing to the obesity epidemic. In the **settling-point theory**, the set point may be increased (or decreased) in response to a change in fat mass, which is the direct result of increased (or decreased) intake and to increased (or decreased) expenditure. This model accounts for social and environmental changes over time (e.g., creeping obesity) but does not explain genetic factors, brain regulation, or peptides involved in feedback control of energy intake or expenditure. A **general model** combines aspects of the set-point and settling-point theories in describing regulation of intake as functions of "uncompensated" (social/environmental) and "compensated" (genetics and feedback control) factors. Finally, the **dual intervention point model** describes "competition" between physiological control factors that defend an upper point of weight or fat mass and environmental factors favoring either weight gain or loss. During weight gain, environmental factors promoting increased intake "win out," with physiological control factors forced to defend a higher upper point. Genetics and epigenetics influence this new equilibrium. As an example, Schulz and Chaudhari describe ongoing prospective studies of two groups of the Pima Indian nation, one in the United States consuming a high-fat diet with a high obesity rate and a genetically similar normal-weight group in Mexico consuming a grain- and vegetable-based diet.

A weight-loss diet may lower the set point to help maintain body weight at a lower level. In their meta-analysis, Astrup and others reported a 3–5 percent decrease in resting energy expenditure (REE) in subjects who previously had obesity compared to subjects without obesity. This phenomenon, known as adaptive thermogenesis, is counterproductive to weight loss and actually favors weight regain.

However, Melby and others comment that while factors such as increased hunger, decreased satiety, increased orexigenic peptides, decreased anorexigenic peptides, lower resting metabolic rate due to the loss of fat-free mass, and increased economy of movement can contribute to adaptive thermogenesis, weight regain is not inevitable even in an environment with easy access to palatable, energy-dense food. A nutrient-dense, high-fiber, higher-protein diet, vigilance in monitoring weight, increased physical activity, and decreased time spent in sedentary activity can contribute to long-term weight loss. Nevertheless, low success rates of short- and long-term weight-loss interventions suggest that the body tends to defend a higher set point. Therefore, prevention of obesity may be the key strategy.

Why is prevention of childhood obesity so important?

As noted previously, there is an obesity epidemic in the United States which affects children as well as adults. According to Centers for Disease Control and Prevention, the overall 2017–2018 prevalence rate for obesity in children and adolescents 2–19 years of age was 19.3 percent, with higher rates in Hispanic and non-Hispanic Black children and lower rates in non-Hispanic Asian children (www.cdc.gov/obesity/data/childhood.html). Once an individual has obesity, treatment can be challenging and requires a multi-faceted approach. Waterland notes that neural circuits predisposing to obesity are not easily abolished. Most scientists agree that prevention of obesity is of prime importance, particularly in childhood. In their meta-analysis of 21 studies across various ethnic groups, Liu and others reported significant associations between the fat mass and obesity-associated (FTO) gene mutation and obesity in children. Heritable traits such as eating/sucking rate, preference/avoidance of certain foods, and disinhibition can predispose a child toward obesity. As previously discussed, epigenetic alterations occurring *in utero* can result in transgenerational obesity and related morbidities. Environmental influences such as viruses and chemical exposure can disrupt endocrine function and develop neural circuits that are not easily altered with interventions later in life. In their meta-analysis of over 200,000 subjects across 15 prospective studies, Simmonds and others reported that 55 percent of children with obesity and 80 percent of adolescents with obesity will have obesity as adults. Therefore, prevention at any stage of childhood and adolescence is an important consideration, and as with adults, lifestyle factors, including dietary factors and physical activity, are important.

The role of diet and physical activity in the development of obesity during childhood is not totally clear. Children have little control over their own dietary choices. Care-giver modeling of physical inactivity and/or the purchase/preparation of high-kcal/low-nutrient-density foods can predispose the child toward overweight/obesity. Of the two, a diet with poor nutrient density appears to be the major problem. Foods high in sugar and fat are abundant. As previously mentioned, Lee and others reported disproportionately greater fast-food marketing and availability to youths in low-income compared to middle-class neighborhoods.

Although nine of 13 reviews and meta-analyses in a review by Keller and Bucher Della Torre reported a positive relationship between sugar-sweetened beverage (SSB) consumption and childhood obesity, they concluded the evidence for a relationship is inconsistent and called for higher quality future reviews and studies. Using 1999–2018 National Health and Nutrition Examination Survey data from almost 34,000 children and adolescents, Wang and others reported an increase of ultra-processed food consumption (e.g., ready-to-heat, ready-to-eat, sweet snacks) from 61.4 to 67.0 percent of the total daily energy intake and a corresponding decrease of unprocessed or minimally processed foods from 28.8 to 23.5 percent of the total daily energy intake. Greater adverse effects were observed in children and adolescents from lower income families. Carvalho and colleagues reported a significantly greater frequency of fast-food consumption in children with three or more obesity-related comorbidities compared to those with fewer than three comorbidities. In 2013, the Standing Committee on Childhood Obesity Prevention, Food and Nutrition Board of the National Academy of Sciences' Institute of Medicine sponsored a workshop to discuss challenges and opportunities in the advertisement and marketing of food to children.

The interactions between screen time, diet quality, and physical inactivity may also contribute to childhood obesity. In their review of 13 studies including almost 62,000 children, Avery and others reported that eating while watching television is associated with poor diet quality, specifically, greater fried food and SSB consumption and less consumption of fruits and vegetables. Boulos and others note that excessive television watching (and other sedentary "screen time" activities) may increase the risk of obesity by serving as a substitute for physical activity; increasing exposure to advertisement for high-fat/high-sugar/low-nutrient food; and by providing a stimulus for "mindless eating." Saunders and colleagues comment that increased total time spent in any sedentary activity, including prolonged sitting, is a risk factor for cardio-metabolic disease in children. In a recent American Heart Association scientific statement, Barnett and others note that ever increasing portability of screen-based devices and increased online content may increase food consumption and cardiometabolic risk in children. They suggest that televisions and screen-based devices should be absent from bedrooms and during meals.

In a review of prospective studies on obesity in children age 5–18, Pate and others reported a consistent association with low physical activity but also noted the need for more well-designed, large-scale, long-term prospective studies using appropriate methodology to examine the role of other potential factors, such as sedentary behavior; diet/nutrition; and family, peer, and community support in efforts to combat childhood obesity. Unfortunately, a meta-analysis of 27 school-based obesity prevention programs by Hung and others concluded that such programs have not been effective in reducing BMI or skinfold thicknesses in children. According to Niemeier and others, parental participation is a significant aspect of interventions to reduce BMI in children with obesity. Salvy and others comment that peers can also play an important in role in supporting friends with overweight or obesity, particularly with regard to increasing physical activity levels. However, Rageliené and Grønhøj concluded that peers and siblings generally have a negative influence in the form of high energy-dense and low nutrient-dense choices or no significant influence. However, they note that social interaction factors that may explain these influences are poorly understood. As Hills and others note, habitual physical activity established during the early years may provide the greatest likelihood of impact on health and longevity. It appears that the battle to reduce the prevalence of childhood obesity will be fought on multiple fronts.

Fat is easy to gain, and hard to lose is the key underlying concept cited by scientists who stress the importance of prevention in the battle against obesity. Although prevention of excess weight gain is important at all stages of life, childhood and adolescence are key times to adopt healthier eating and physical activity practices that promote a healthy body weight. Several websites provide information and program activities to help fight childhood obesity.

> *www.newmovesonline.com* A school-based physical education program for girls only to promote increased physical activity and healthier eating.

Key Concepts

▸ The brain's hypothalamus appears to be the central control mechanism in appetite regulation, which involves a complex interaction of physiological and psychological factors. Neural and hormonal feedback from the senses, gastrointestinal tract, adipose cells, and other body organs help to drive hunger or signal satiety.

▸ Although the ultimate cause of obesity is a positive energy balance, the underlying cause is not known but probably involves the interaction of many genetic, epigenetic, and environmental factors.

▸ Strategies to prevent unhealthy weight gain are recommended as, once excess body fat accumulates, it can be difficult to lose and then maintain that loss over time. Childhood and adolescence are key periods of life for prevention.

Check for Yourself

▸ As you go about your normal activities for a day, pay particular attention to environmental conditions that are conducive to consuming high-kcal foods and discouraging physical activity. Compare your results with those of your classmates.

Weight Gain, Obesity, and Health

Although the World Health Organization (WHO) recognized obesity as a disease in 1948, James noted that only within the past 20 years has its global role in disease and escalating medical costs been recognized. According to WHO data from 2016 (www.who.int/news-room/fact-sheets/detail/obesity-and-overweight), 340 million children and adolescents aged 5–19 met criteria for overweight or obesity while prevalence rates for adult overweight and obesity were 39 and 13 percent, respectively. In 2020, 39 million children under the age of 5 met overweight or obesity criteria. The obesity prevalence rate has almost tripled since 1975 and is increasing in every WHO region. The US prevalence rate for obesity was 42.4 percent in 2017 to 2018 (www.cdc.gov/obesity/data/adult.html).

> *www.who.int/mediacentre/factsheets/fs311/en/* This World Health Organization link provides information on worldwide prevalence rates for overweight and obesity.
>
> *www.cdc.gov/obesity/data/adult.html* This Centers for Disease Control and Prevention link provides information on U.S. prevalence rates for overweight and obesity.

The increased health risk of fat deposition in the abdominal region has been recognized for well over 50 years. It is also known that adipose tissue is not simply an inert storage bin for triacylglycerol but rather metabolically active tissue involved in a variety of physiological functions, including hormone production and immune response. **Adiposopathy (or sick fat)** is associated with adipocyte dysfunction, including alterations in adipogenesis, which adversely affect fat storage; insulin resistance in liver and muscle tissue; upregulation of proinflammatory and downregulation of anti-inflammatory **adipokines** (cell signaling pro secreted by adipose tissue); dyslipidemia; and atherosclerosis directly and through high associations with hypertension and diabetes. Adipocytes also contain aromatase, the rate-limiting enzyme in estrosynthesis, and other hormones in the estrogen biosynthetic pathway. The conversion of androgens to estrogens may contribute to visceral fat accumulation and metabolic disease. Adiposopathy is associated with a cluster of symptoms known as the metabolic syndrome including visceral fat deposition, insulin resistance, hypertension, and abnormal triacylglycerol levels and blood lipoprotein profile, which are discussed later in this chapter.

More recent research has focused on adipose tissue differentiation into metabolically healthy obesity (MHO) and metabolically unhealthy obesity (MUO) phenotypes. Although there is not complete agreement on criteria, Phillips noted that MHO includes low visceral/high subcutaneous fat obesity; small adipocytes which secrete protective anti-inflammatory adipokines; and normal blood pressure, insusensitivity, liver function, and lipid profiles. There is also controversy regarding the protective nature of MHO. In their review, Tsatsoulis and Stavroula note the prognostic value of MHO is uncertain and that it can evolve into MUO over time. In their meta-analysis, Zheng and others concluded that the MHO phenotype was associated with a higher risk of cardiovascular events but no difference in all-cause mortality compared to healthy normal-weight subjects. In general, obesity appears to increase health risks and premature death, though the effect may vary in any given individual.

What health problems are associated with overweight and obesity?

Scientific reviews have identified several health problems and risk factors associated with obesity which are listed in **table 10.7**. These health problems may be associated with the strain of the increased body mass, the metabolic effects of adipokines or estrogen, psychological processes, and/or interactions with several of these factors.

Cawley and others reported an aggregate medical cost of obesity of $260 billion in 2016. Compared to normal-weight individuals, medical costs increased linearly by 68.4 and 233.6 percent, respectively, in individuals with obesity class 1 (BMI ≥ 30.0-34.9 kg/m^2) to class 3 obesity (BMI ≥ 40.0 kg/m^2). A Milken Institute report authored by Waters and Graf concluded that direct healthcare costs for cancers, cardiovascular, cognitive, gastrointestinal, metabolic, neuromuscular, pulmonary, and renal diseases associated with overweight (BMI = 25.0-29.9 kg/m^2) and obesity were $480 billion, with an additional $1.254 trillion in lost economic productivity in 2016. The total cost equals almost 10 percent of the US gross domestic product. Waters and Graf estimate that **population attributable risk**, the percent of cases where overweight and obesity contributes to increased risk, ranges from 2 to 30 percent for overweight and 4 to 51 percent for obesity depending on the disease.

Cardiovascular Disease and Related Health Risks The primary health condition associated with excess body fat is coronary heart disease (CHD). Hubert and others were among the first to report obesity as an independent risk factor for CHD in a 26-year follow-up of 5,209 original Framingham cohort members. Obesity is now well recognized as a risk factor for CHD by many professional groups and societies. However, this relationship has recently been complicated by the previously discussed and somewhat controversial concept of metabolically healthy obesity and clinical observations that mild to moderate obesity may be associated with improved survival in heart failure patients. The latter phenomenon, discussed by Clark and colleagues, is called the "obesity paradox" and may be explained by mechanisms including but not limited to cardiorespiratory fitness, greater anti-inflammatory actions of adipokines such as adiponectin, and greater mass to counteract the catabolic effects of heart failure. The obesity paradox should be viewed in the context of obesity prevention to prevent the development of heart failure and cardiovascular disease in the first place.

Reviews by Bays and and Frühbeck and their respective colleagues indicate that adipokines may directly influence homeostasis in heart blood vessels by influencing the function of endothelial cells, arterial smooth muscle cells, and macrophages in the vessel wall. Some of these effects may be beneficial, while others may be harmful to heart health. Adiponectin has potent anti-inflammatory and antiatherogenic properties, whereas other adipokines, such as tissue necrosis factor-alpha (TNF-α), resistin, and visfatin, may promote inflammation and clotting, two factors contributing to atherosclerosis. Adiponectin levels are usually lower in individuals with obesity, while TNF and other atherogenic adipokines may be elevated. According to Frühbeck and others, the adiponectin:leptin ratio is higher in normal weight and lower in overweight and obesity and therefore may be a useful index to assess metabolic risk associated with obesity.

Excess body fat also increases the risk of developing high blood pressure, hypercholesterolemia, and diabetes, all of which are risk factors leading to the development of CHD. In particular, obesity increases the risk of type 2 diabetes, whose rate has increased dramatically over the past decade in concert with the parallel increase in obesity. The term **diabesity** was coined by *Shape Up America*, a non-profit organization promoting a healthy body weight, to describe the adverse effects of the two highly associated conditions. Ng and others comment that diabesity accelerates the atherosclerotic process and results in adverse cardiac remodeling which can lead to heart failure. Adipose tissue secretes an adipokine called resistin, which is reported to induce insulin resistance, linking diabetes to obesity. The greater the amount of adipose tissue, the greater potential for diabetes. Bays and others report a strong relationship between BMI and prevalence of type 2 diabetes. Having below-normal/normal-weight, having overweight, and having obesity account for 17.5, 31.6, and 50.9 percent of cases, respectively. The detrimental effects of obesity relative to the development of chronic disease occur when it persists for 10 years or more. Lawrence and others reported a 95 percent increase in prevalence rate of type 2 diabetes in children aged 10-14 years and adolescents aged 15-19 years from 2001 to 2017 with a greater increase in racial and ethnic minorities compares to white youth. In a review of 15 studies, Owen and others reported considerable variability in estimates across studies but concluded that BMI was directly related to CHD risk later in life. The Centers for Disease Control and Prevention predicts that obesity rates may triple by 2050.

TABLE 10.7 Selected health problems and risk factors associated with obesity

Asthma	Insulin resistance
Cancer (various types)	Low back pain
Cardiovascular disease	Low-grade inflammation
Cognitive decline, dementia, and Alzheimer's disease	Anxiety and depression
Type 2 diabetes	Osteoarthritis and knee pain
Dyslipidemia	Respiratory dysfunction
Gallstones	Sleep apnea
Gastrointestinal reflux disease	Social isolation (avoiding social situations)
Gout	Stroke
Hypertension	Vertebral disk herniation

Cancers Obesity has been regarded as a risk factor for colorectal, esophageal, pancreatic, liver, kidney, thyroid, postmenopausal breast, prostate, ovarian, uterine, and other cancers. Avgerinos and others comment that the attributable risk of obesity to cancers is 11.9 and 13.1 percent in males and females, respectively. In US adults, Islami and others reported that excess body weight contributed to between 3.9 and 6.0 percent of cancers in males and between 7.1 and 11.4 percent of cancers in females.

> www.aicr.org/research/third-expert-report/ Access the World Cancer Research Fund/American Institute for Cancer Research. Diet, Nutrition, Physical Activity, and the Prevention of Cancer: A Summary of the Third Expert Report 2018. Click "View Complete Report" to access a 116 page summary.

Obesity may influence the development of cancers in various ways. Fat tissue leads to an overproduction of estrogen and other steroid hormones. Increased serum estrogen levels are associated with an increased incidence of breast cancer in females. Increased breast cancer risk is also associated with decreased anti-inflammatory adipokine adiponectin, and increased concentrations of leptin and the pro-inflammatory adipokines resistin and visfatin. In their meta-analysis of 93 studies, Gui and others reported lower adiponectin and higher leptin and TNF-α concentrations in females with BMI >25 kg/m^2. However, the role of obesity may be modified by other risk factors. Renehan and others re-examined data from their 2008 meta-analysis data and reported that smoking and use of hormone replacement therapy attenuated associations between BMI and postmenopausal breast, endometrial, ovarian, esophageal, lung, pancreatic, and prostate cancers.

According to Uyar and Sanlier, increases in secretions from visceral adipose tissue promoting inflammation and/or cell proliferation play a role in the development of colorectal cancer. Adipocyte hypertrophy increases insulin-like growth factor-1 (IGF-1), IGF-1 binding proteins, insulin resistance, vascular endothelial growth factors, ghrelin, leptin, and proinflammatory adipokines (interleukin-6, tissue necrosis factor-α, and resistin) and decreases the anti-inflammatory adipokine adiponectin. This milieu can create an imbalance of increased cell proliferation, increased angiogenesis, and decreased **apoptosis** (programmed cell death) to support tumor growth which is a precursor to development of colorectal and other cancers. Obesity may also have a local effect, contributing to gastrointestinal reflux in which stomach acids irritate cells and induce cancer in the lower esophagus. Other unknown factors may also be involved.

Alzheimer's Disease and Other Dementia Mental decline in the elderly has been linked to altered adipokine secretion that occurs during midlife obesity. According to Kacířová and others, obesity and diabetes could be linked to Alzheimer's disease and other age-related cognitive impairments by proinflammatory cytokines such as tissue necrosis factor-α and resistin produced by white adipose tissue. Ishii and Iadecola note that decreased adiponectin and increased leptin may increase susceptibility to dementia and Alzheimer's disease (AD). In their meta-analysis, Pedditizi and others observed a positive association between midlife (before age 65) obesity and incident dementia, but a negative association on older (>65 years) individuals. In other words, decreases in body mass and adiposity may induce changes that promote vascular dysfunction and predispose the elderly to dementia and AD. Loef and Walach modeled significantly higher dementia prevalence estimates for the years 2030 and 2050 in the United States and China than current forecasts that do not include midlife obesity.

Maternal Health During pregnancy, adequate weight gain is important to the health of both the mother and fetus, but excess maternal weight gain may adversely affect pregnancy outcomes and fetal health. In a meta-analysis of 38 studies, Aune and colleagues reported modest but significant associations between maternal BMI and the following pregnancy outcomes: fetal death, stillbirth, perinatal (5 months pre- to 1 month post-birth) death, neonatal death, and infant death.

In the National Academy of Sciences Institute of Medicine (IOM) update on guidelines for weight gain during pregnancy edited by Rasmussen and Yaktine, females today are more likely to have overweight or obesity before becoming pregnant and are more likely to gain weight above recommendations during pregnancy as compared to the past several decades. According to the IOM, the rate of weight gain and total weight gain are based on pre-pregnancy BMI. Recommendations range from approximately 1 pound/week and 28–40 total pounds during pregnancy for underweight females (BMI <18 kg/m^2) to no more than 0.5 pound/week and 11–20 total pounds during pregnancy for females with obesity (BMI ≥30.0 kg/m^2). These guidelines have been developed to best support the health of both mother and child.

Mortality, Morbidity, and Quality of Life Hippocrates (ca. 400 B.C.) recognized that "persons who are naturally fat are apt to die sooner than slender individuals." Flegal and others conducted a meta-analysis of 97 studies including 2.88 million individuals and 270,000 deaths. Using normal-weight BMI (18.5–24.9 kg/m^2) as the referent group, they reported the following hazards ratios between all-cause mortality and BMI categories: 0.94 lower (overweight BMI = 25–29.9 kg/m^2), 0.95 no difference (grade 1 obesity BMI = 30–34.9 kg/m^2), and 1.29 higher (grade 2 obesity BMI = 35–39.9 kg/m^2 and grade 3 obesity BMI ≥40 kg/m^2). According to the World Health Organization (ourworldindata.org/obesity), 8 percent of all global deaths in 2017 were the result of obesity compared to 4.5 percent in 1990. In general, individuals with a healthy BMI and body composition are more likely to live longer, particularly when other lifestyle factors contributing to healthy aging are also considered.

Being overweight, and in particular having obesity, is associated with premature death, but excess fat also contributes to morbidity, or the increase in various diseases such as type 2 diabetes and its associated conditions. Those who are overweight and have such health problems may experience significant impairments in their quality of life.

Epidemiological research clearly indicates that having obesity increases one's health risks. Moreover, the location of the fat in the body is also an important contributing factor underlying these health risks, as will be discussed in the next section.

How does the location of fat in the body affect health?

Classifications of different types of obesity based on regional fat distribution have been proposed, the most popular differentiation being the android versus the gynoid types. **Android-type obesity** is characterized by accumulation in the abdominal region—particularly the intra-abdominal region—of deep, visceral fat but also of subcutaneous fat. Android-type obesity is also known by other terms, such as abdominal, central, upper body, or lower trunk obesity, and is sometimes referred to as apple-shape obesity. In mice, upregulation of the enzyme converting cortisone into cortisol results in overproduction of cortisol, which tends to facilitate the deposition of fat in the abdominal region. **Gynoid-type obesity** is characterized by fat accumulation in the gluteal-femoral region—the hips, buttocks, and thighs. It is also known as lower body obesity and is often referred to as pear-shape obesity (**figure 10.10**). In a cross-sectional study of 625 healthy adult Danish monozygotic and dizygotic twin pairs, Hasselbalch and others reported significant between-twin correlations for overall fat, fat located in peripheral and trunk areas, waist circumference, hip circumference, and leptin. These correlations ranged from 0.49 to 0.94 in males and from 0.48 to 0.98 in females. They concluded that fatness is controlled by shared and different genetic components which underscores the importance of distinguishing between phenotypes in identifying genes responsible for obesity. Waist circumference measurement may be an appropriate screening technique for android-type or gynoid-type obesity. The procedure for determining waist circumference is described in appendix B and the increase in health risk associated with an increased circumference and various BMI levels is presented in **table 10.1**.

Android-type obesity is increasingly being recognized as causing a greater health risk than obesity itself. It appears that android-type fat cells, particularly the deep visceral fat cells, possess dissimilar biochemical functions. These visceral fat cells are large and highly metabolically active, possibly due to differences in the activity of lipoprotein lipase and higher cortisol levels. Pi-Sunyer indicated that android-type obesity is a cardiovascular health risk factor associated with a release of proinflammatory adipokines such as resistin and TNF-a; a decrease in the anti-inflammatory adipokine adiponectin; and an increase in plasminogen activator inhibitor 1, which impairs fibrinolytic (clot-dissolving) action. Resistin is also theorized to be involved in the development of insulin resistance.

Epidemiological data have shown that android-type obesity is associated with a cluster of metabolic disorders, which Reaven labeled "syndrome X" and later renamed metabolic syndrome, symptoms of which are listed in **tables 10.8 and 10.9**. As mentioned previously, some have labeled abdominal fat adiposopathy or *sick fat*.

According to Saklayen, decreased physical activity and increased consumption of a high energy-dense low nutrient-dense diet have resulted in a global epidemic of the metabolic syndrome, defined by the World Health Organization as a cluster of symptoms including abdominal obesity, insulin resistance, hypertension, and

TABLE 10.8 Symptoms of the metabolic syndrome

- Hyperinsulinemia
- Insulin resistance
- Impaired glucose tolerance
- Hypertriglyceridemia
- Increased small, dense LDL-cholesterol particles, which are highly associated with cardiovascular disease
- Decreased HDL-cholesterol
- Hypertension
- Increased plasma fibrinogen and clotting activity
- Android pattern of fat deposition (see table 10.1 and figure 10.10)

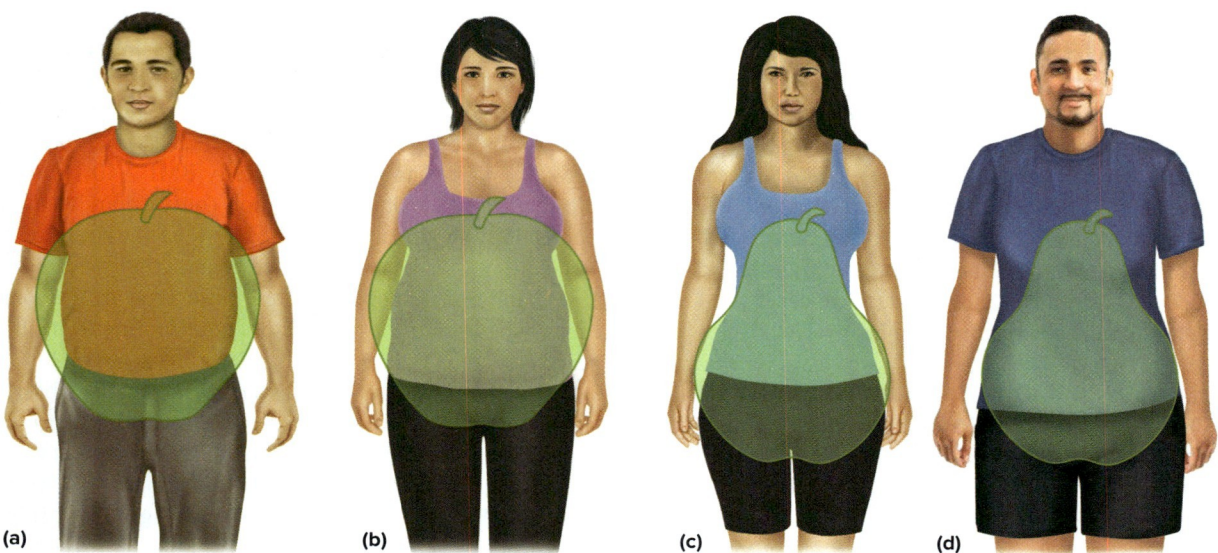

FIGURE 10.10 Patterns of fat deposition. (a) The android, or "apple-shape," pattern is associated with diabetes, coronary heart disease, hypertension, dyslipidemia, metabolic syndrome, and other chronic diseases. (b) Although more common in males, females can also present with the higher-risk apple-shape pattern. (c) The gynoid, or "pear-shape," pattern, more common in females, is generally associated with a lower risk of chronic disease than the apple shape.

TABLE 10.9 International criteria for components of the metabolic syndrome

Criterion	Males	Females
Elevated waist circumference according to population- and country-specific definitions (U.S. criteria in inches [cm])	>40 [102]	>34.6 [88]
Subnormal HDL-cholesterol (mg/dl)	<40	<50
Elevated serum triglycerides (mg/dl)	>150	
Elevated blood pressure (mmHg)	>130/85	
Elevated fasting blood sugar (mg/dl)	>100	

Source: Alberti, K. G. M. M., et al. 2009. Harmonizing the metabolic syndrome. A joint interim statement of the International Diabetes Federation Task Force on Epidemiology and Prevention; National Heart, Lung, and Blood Institute; American Heart Association; World Heart Federation; International Atherosclerosis Society; and International Association for the Study of Obesity. *Circulation* 120:1640–45.

hyperlipidemia. Although the metabolic syndrome is well established as a link between visceral fat, atherogenesis, and insulin resistance, other pathological mechanisms such as increased oxidative stress, altered hypothalamic-pituitary-adrenal axis activity, renin-angiotenson-aldosterone function, and glucocorticoid function may also be involved.

Using 2011-2016 National Health and Nutrition Examination survey, Gaston and others reported metabolic syndrome prevalence rates in US adolescents 12-19 years of age that ranged from 2 percent for females to 11 percent for males depending on the metabolic syndrome risk factor criteria. Males have higher blood pressure, triacylglycerol, fasting blood glucose, and low HDL-cholesterol compared to females. The most prevalent risk factor in males and females was borderline/low HDL-cholesterol. Hirode and Wong analyzed NHANES data from the same period and reported nonsignificant increases in metabolic syndrome prevalence rates (32.5, 34.6, and 36.9 percent) from 2011-2012, 2013-2014, to 2015-2016. However, significant increases occurred in females, young (20-39 year old) adults, non-Hispanic Asians, and Hispanics. From 1999-2000 to 2015-2016, Friar and others reported increases in average waist circumference and BMI for males (99.1-102.1 cm; 27.8-29.1 kg/m^2) and females (92.2-98.0 cm; 28.2-29.6 kg/m^2). In other words, the *average* US adult male and female meets the international waist circumference criterion for the metabolic syndrome described in **table 10.9**. Visceral adiposity continues to be a serious public health concern.

Reaven stressed the importance of treating the associated cardiovascular risk factors, such as high blood pressure. The metabolic syndrome has stimulated drug research leading to development of single drugs that simultaneously treat multiple components of the metabolic syndrome, such as high blood pressure and cholesterol levels.

Differences in diagnostic criteria for components of the metabolic syndrome may partially account for different metabolic syndrome prevalence rates such as those reported by Gaston and others. However, Saklayen comments such differences are minor. In 2009, Alberti and others developed the consensus international clinical criteria for the diagnosis of the metabolic syndrome presented in **table 10.9**.

Although the sex-specific waist circumferences noted in **tables 10.1, 10.8, and 10.9** have been associated with increased health risks, some experts have recommended lower waist circumference criteria (37 inches, 94 cm for males and 31.5 inches, 80 cm for females). Wang and others reported an increased risk for type 2 diabetes in men with waist circumferences greater than 37 inches.

Studies show that android-type obesity is a major risk factor for mortality in both males and females. In the Nurses' Health Study involving nearly 45,000 females over 16 years, Zhang and others reported increased mortality from all causes, cardiovascular disease, and cancer as waist circumference increased. Even in females who were classified as normal-weight, the risk of cardiovascular disease mortality was 3.02 times greater with increased waist circumferences (>88 cm) compared to females who were classified as normal-weight without visceral fat. As indicated in **table 10.1**, waist circumference is positively associated with higher disease risk within all categories of BMI for middle-aged males and females, even those with a recommended BMI level. Jacobs and others concluded that increased mortality risk related to excess body fat is mainly due to abdominal adiposity. In a clinical study of over 2,200 French patients with acute myocardial infarction, Zeller and others reported an inverse relationship between BMI and mortality (the previously described obesity paradox). However, an increase in mortality was observed in males and females with high waist circumference but low to moderate BMI.

Compared to android fat, gynoid fat responds less readily to diet/exercise intervention but appears to pose a lower health risk. In their review, Lee and others describe several physiological differences between visceral and subcutaneous fat depots, which are consistent with sex- and age-related differences in fat deposition.

Does having obesity increase health risks in youth?

As previously mentioned, the World Health Organization estimates that 340 million children and adolescents aged 5-19 met criteria for overweight or obesity in 2016 (www.who.int/news-room/fact-sheets/detail/obesity-and-overweight). Permanent alterations of neural pathways regulating energy intake and expenditure may result from overconsumption of high-energy foods and physical inactivity during childhood. In their review, Kansra and others comment that to obesity in children and adolescents is related to psychological, cardiovascular, dermatological, renal, gastrointestinal, neurological, orthopedic, dental, pulmonary, endocrinological, and reproductive comorbidities.

Children with obesity may also be ridiculed and bullied by peers, adults, and even parents. As a result, they may also suffer

Witthaya/Getty Images

from psychological comorbidities such as anxiety, depression, low self-esteem, negative self-image, attention deficit hyperactivity disorder, delayed social development, and even serious psychological illness. Since personality traits are established in childhood and adolescence, excessive body fat may contribute more greatly to social-emotional problems at this time than during adult-onset obesity. In their review, Smith and others discussed developmental and metal health factors associated with having obesity as a child. Exposure to high maternal stress *in utero* (high cortisol, corticotropin-releasing hormone) are related to high BMI in the first two years. In early childhood years and adolescence, high levels of stress, low socioeconomic status, abuse, high fast-food consumption; emotional eating; low physical activity; increased screen time; poor sleep; bullying and being bullied; perceived bias from peers, parents, clinicians, school personnel; and other forms of social stigma and ostracism are all related to risk for obesity. An unsupportive family environment including, but not limited to, absence and/or depression/stress/anxiety in a parent(s) can contribute to negative reactivity (a quick response to a situation with a negative affect), poor emotional self-regulation, and poor impulse control which are also associated with unhealthy weight gain.

Franks and others indicated that childhood obesity will have very serious long-term health effects through midlife. Researchers are looking for clues, such as genetic, metabolic, or anthropometric markers, that can predict which children will have obesity so that appropriate interventions may be enacted earlier in life, when chances of success to prevent obesity may be greater.

Does losing excess body fat reduce health risks and improve health status?

Numerous studies document the beneficial effects of weight loss by reducing the risk factors for disease, including several major diseases—cardiovascular disease, cancer, and type 2 diabetes. In an American Heart Association policy statement on worksite health screening, Arena and others list ideal weight (BMI 18.5–24.9) and a DASH-type diet as two of seven key metrics. The 2018 World Cancer Research Fund/American Institute of Cancer Research Report on Diet, Nutrition, Physical Activity, and Cancer listed proper weight control as one of its key recommendations to help prevent a variety of cancers. Aucott reported that intentional weight loss reduces the risk of developing type 2 diabetes in the long term. Individuals with type 2 diabetes who lose weight often have reduced clinical symptoms and mortality risk. As noted in chapter 2, consuming a diet that supports a healthy body weight and body composition is an important component to dietary planning.

As previously noted, excess visceral fat is a key component of the metabolic syndrome and, therefore, a logical target for interventions to reduce risk of cardiovascular disease and type 2 diabetes. Separate studies by Normandin and Irving and their respective colleagues have shown that weight loss via dieting and exercise will preferentially decrease fat from the abdominal area and, concomitantly, reduce several of the metabolic syndrome risk factors, such as high blood pressure and serum levels of glucose and triglycerides. Reducing abdominal fat may reduce the production of pathological adipokines, resulting in decreased inflammation and other factors contributing to disease. Romero-Corral and others reported that modest weight gain in subjects who were previously classified as normal-weight, resulted in impaired vascular endothelial function without increasing blood pressure, but that vascular endothelial function returns to normal following weight loss.

As shall be noted in chapter 11, lifestyle changes, including nutritional factors, physical activity, and stress management are important for individuals desiring to attain a healthy weight. For weight loss, caloric balance is important, as weight loss will occur if the total caloric intake, regardless of the macronutrient content of the diet, is less than caloric expenditure. However, as noted in previous chapters, some overall dietary plans, such as the DASH and OmniHeart diets, as well as certain foods, such as fruits, vegetables, and fish, may provide health benefits beyond those associated with weight loss. Chapters 4, 5, and 6 highlight healthful carbohydrates, fats, and protein, and chapter 9 discusses the DASH and OmniHeart diets.

Does being physically fit negate the adverse health effects associated with being overweight?

In general, research findings documenting increased health risks associated with obesity have been derived from the general population and have been based primarily on BMI values. Although BMI may be useful in determining health risks in the general population because a high value is usually associated with excess body fat, it does not apply to muscular individuals. A high BMI may be associated with the development of various chronic diseases if it is due to obesity and related risk factors such as high serum cholesterol levels and high blood pressure. However, BMI does not evaluate body composition or provide an indication of body fat composition in all individuals.

The BMI also does not measure fitness, and a physically active lifestyle may be an important moderating factor in the association between fatness and health.

High levels of cardiorespiratory fitness have been shown to be protective against premature mortality. Ricketts and others

Adam Kaz/Getty Images

followed over 31,000 aerobically fit and unfit males (>20 percentile vs. ≤20 percentile based on treadmill test time) classified as normal weight, overweight, or having obesity who had normal or increased waist circumferences within weight categories. Over 440,000 person-years of follow-up, 1,399 deaths occurred in the cohort. For most of the BMI and waist circumference categories, higher mortality rates were observed in unfit males with increased waist circumference compared with fit males with increased waist circumference, suggesting that inclusion of cardiorespiratory fitness improves mortality risk assessment. In a cross-sectional study of females, Farrell and others reported similar findings that higher levels of cardiorespiratory fitness were associated with lower incidence rates of all-cause mortality across all levels of adiposity.

However, existing comorbidities may confound any protection offered by increased fitness in individuals with overweight or obesity. Tarp and others followed over 77,000 males and females who were classified as normal weight, overweight or having obesity and with low, medium or high cardiorespiratory fitness for a median of 7.7 (0.2–8.2) years. During follow-up, 1,731 deaths occurred, mostly from cancer and cardiovascular disease. Compared to fit subjects who were at a normal weight by gender, a decrease in mortality was observed with increased cardiovascular fitness. However, when subjects with baseline cardiovascular disease and cancer were excluded, the difference in mortality rate in males who were either fit or unfit with obesity decreased but remained significant in males and was insignificant in females. Some suggest that being overweight does not increase mortality unless it is associated with adverse effects on blood pressure, glucose tolerance, serum cholesterol levels, and other risk factors for chronic diseases. Physical activity may help to counteract such adverse effects. However, the issue of whether or not fitness may counteract the adverse health effects of fatness remains controversial.

The Health at Every Size (HAES® asdah.org/health-at-every-size-haes-approach) movement began in the 1960s in order to promote weight-neutral dietary and physical activity strategies for health improvement in overweight and obese individuals. In their commentary on incorporation of the HAES approach into the public health response to obesity, Penney and Kirk comment on the movement's effectiveness in reducing stigma related to body size, promoting body acceptance regardless of size, and participating in health-related physical activity. They also note that further research is necessary to determine if HAES is the optimal approach for individuals with Class II (BMI=34.0–39.9 kg/m^2) or III (≥40 kg/m^2) obesity for whom even modest weight loss may improve metabolic health. Kim and Park state that exercise without weight loss can improve insulin sensitivity and decrease insulin resistance in adults with obesity. Although there are reports of similar effects in children and adolescents, more research is needed on the optimal training volumes and modes to reduce insulin resistance in adolescents with obesity. According to Klip and others, exercise improves muscle glucose uptake by increasing GLUT-4 receptor migration to the muscle cell membrane in an insulin-independent manner and by opposing mechanisms that impair insulin signaling and lead to insulin resistance.

An acute exercise bout can increase insulin sensitivity for up to 16 hours afterwards. Chronic exercise training potentiates the effect of exercise on insulin sensitivity through multiple adaptations in glucose transport and metabolism. As previously mentioned, metabolically healthy obesity discussed by Tastsoulis and others is associated with less visceral obesity and lower metabolic risk including greater insulin sensitivity. While exercise training may not help some individuals lose weight, it may result in significant health improvements.

Although fitness is important to counteract some of the adverse health effects of fatness, losing excess body fat may provide additional health benefits. In their meta-analysis, Kramer and others reported that metabolically healthy individuals with obesity had a small, but significantly higher relative risk for all-cause mortality and cardiovascular events (1.24) compared to metabolically healthy normal-weight individuals.

The combination of exercise and weight loss has been shown to produce numerous health effects, including decreased blood pressure, improved serum lipid status, reduced blood glucose, increased self-esteem, and improved psychological health. In their meta-analysis of 54 randomized controlled trials including over 30,000 subjects, Ma and others reported that weight loss interventions significantly decreased all-cause mortality.

Barry and others examined ten longitudinal studies (average follow-up = 14.3 ± 2.6 years) on all-cause mortality that included measures of cardiorespiratory fitness and BMI. Subjects who were physically fit had a similar mortality risk regardless of apparent fatness. In a prospective cohort of male physicians, Kenchaiah and others reported that hazards ratios for heart failure were somewhat lower in physically active lean, overweight, and obesity groups compared to their inactive counterparts.

Weinstein and others evaluated the interaction of physical activity and body mass index on the risk of coronary heart disease (CHD) in a prospective study of nearly 39,000 females over nearly 11 years. The lowest risk for CHD was in females who were classified as normal weight and physically active. Being overweight and physically inactive increased the risk, and females who had obesity and did not engage in regular physical activity had the highest risk. They concluded that although the risk of CHD associated with elevated body mass index is considerably reduced by increased physical activity levels, the risk is not completely eliminated, reinforcing the importance of being lean and physically active. In other words, being fit does not reverse all of the increased health risks associated with excess adiposity. The results of these studies in part support the HAES® position of weight neutrality in health improvement.

www.niddk.nih.gov/health-information Under Health Information, select Weight Management, then Staying Active at Any Size under Featured Topics.

Key Concepts

▸ Excessive body fat is associated with a variety of chronic diseases and impaired health conditions, including coronary heart disease, type 2 diabetes, high blood pressure, sleep apnea, and arthritis.

- Although having obesity increases health risks, the deposition of fat in the abdominal area exacerbates these risks. Referred to as android-type, apple-type, or male-type obesity, increased abdominal fat is associated with the metabolic syndrome and risk factors for heart disease and type 2 diabetes.
- For children and adolescents, having obesity and abnormally elevated body fat increases risk for type 2 diabetes, metabolic syndrome, and other forms of chronic disease. As well, obesity may also increase risk for anxiety and depression.
- Engaging in regular physical activity is important to not only supporting a healthy body weight and body composition, but also in preventing chronic disease.

Check for Yourself

- Conduct an observational study. Go to a public building (e.g., mall) with at least two floors and an open lobby that will allow you to observe the people, the stairs, and the elevator/escalator. Spend an hour observing the passersby and recording the following: (1) gender; (2) age [child, teen, young, middle-age, older adult]; (3) apparent weight status [normal weight, overweight, with obesity] using the "eyeball" method; and, for those who are fully ambulatory (4) the mode of transit from the ground floor to the upper floor(s). Are there associations between these variables? Search the internet for *The Weight of the Nation: Part 3*. After watching the documentary, consider the following questions: What factors influence the prevalence rates of childhood obesity? What can be done to improve the health of children? What roles do schools and the food industry play in childhood health and obesity?

Excessive Weight Loss and Health

As has been previously discussed, clinically meaningful reductions in metabolic risk may result from losing even a small amount of excess body fat and achieving a healthier body weight. However, many individuals attempt to lose weight for other reasons. For example, male and female athletes, such as distance runners, gymnasts, wrestlers, jockeys, and dancers, practice weight control to improve their performance. Losing body weight for improved performance may also provide some health benefits. For example, Williams theorized that the elevated HDL-cholesterol concentrations of long-distance runners are primarily a result of reduced adiposity. However, excessive weight loss may actually lead to deterioration of health.

What health problems are associated with improper weight-loss programs and practices?

As shall be noted in chapter 11, a well-designed diet and proper exercise are the cornerstones of a sound weight-control program. However, some individuals may establish unrealistically low body-weight goals, which may lead to inappropriate weight-control behaviors such as complete starvation, self-induced vomiting, or the improper or excessive use of drugs, diet pills, laxatives, and diuretics. Although rapid weight loss may be achieved with these techniques, prolonged use may result in serious medical disorders or even death. The following discussion highlights some of the areas of concern in which weight-loss practices may be harmful if abused. The use, efficacy, and potential health risks of dietary supplements for weight loss are discussed in chapter 11.

Dehydration Dehydration may be induced by exercise, exposure to the heat (as with a sauna), or the use of diuretics and laxatives. The effect of dehydration on health, particularly in relation to heat illnesses, was discussed in chapter 9. Laxatives and diuretics, which are illegally used by some athletes to lose weight and to mask the use of illegal supplements, may cause **hypokalemia**, or loss of plasma potassium. Hypokalemia may lead to electrolyte abnormalities; cardiac irregularity; disturbed kidney and neuromuscular function; and in rare case muscle tissue breakdown, known as **rhabdomyolysis**. In a case study, Pfisterer and others reported hypokalemic paralysis in a professional bodybuilder who used triamterene/hydrochlorothiazide, a potent diuretic combination, in order to increase muscle definition prior to a competition. Although this patient responded favorably to gradual potassium replacement after three days of hospitalization, dehydration strategies that alter electrolyte balance can be fatal. In addition to rapid weight-loss agents, diuretics are used as masking agents for other illegal substances and are banned in training and in competition by the World Anti-Doping Agency (WADA). Abuilar-Navarro and others examined WADA test data from 2014 to 2017 and reported greater use of diuretics and masking agents in boxing, wrestling, and martial art—all sports competing by weight class—as well as shooting and gymnastics compared in other sports.

Medications to Treat Overweight and Obesity Lifestyle interventions such as increased physical activity and following a well-balanced diet have not been completely effective in controlling obesity and its comorbidities. Therefore, various prescription medications have been developed to stimulate weight loss and generally are prescribed for individuals with obesity and the greatest risk for associated chronic disease, not for those individuals who want to lose a few pounds. Nevertheless, it appears that individuals across the weight spectrum, including physically active individuals and athletes, are using drugs for weight-control purposes. Some drugs are available without prescription, whereas others should be used only under medical supervision.

Antiobesity drugs listed in **table 10.10** may help reduce body weight by the following mechanisms of action, most of which function as an **anorexiant** which decreases appetite: (1) central nervous system stimulation; (2) increasing energy expenditure (thermogenesis) in the peripheral tissues; (3) inhibiting gastric and intestinal lipase activity from blocking fat absorption; (4) inhibiting the reuptake of serotonin and/or norepinephrine; (5) glucagon-like peptide 1 agonist activity, which lowers blood glucose and may decrease weight as a secondary outcome; (6) combined antidepressant/opioid receptor blocking action; (7) combined β-1 blockade and

TABLE 10.10 Weight loss drugs approved by the US Food and Drug Administration (FDA)

Generic name	Brand name(s)	Rx or OTC	Year of FDA approval	Mode of administration	Mechanism of action
Amphetamine/ methamphetamine	Adzenys XR-ODT, Desoxyn, Dyanavel XR, Evekeo	Rx	1947	Oral	Central nervous system (CNS) stimulant; anorexiant (appetite suppressant)
Diethylproprion	Anorex-SR, Nobesine, Prefamone, Regenon, Tepanil, Tenuate	Rx	1959	Oral	CNS stimulant; anorexiant
Phentermine	Adipex, Adipex-P, Atti-Plex P, Fastin, Ionamin, Lomaria, Phentercot, Phentride, Pro-Fast	Rx	1959	Oral	CNS stimulant; anorexiant
Benzphetamine	Didrex	Rx	1973	Oral	CNS stimulant; anorexiant
Fenfluramine	Fintepla	Rx	1973	Oral	Selective serotonin reuptake inhibitor, component of Fen-phen which was withdrawn from US market in 1997
*Fenfluramine + Prentermine	Redux® ("Fen-Phen")	Rx	1990s; Withdrawn in 1997 due to association with mitral valve disease and pulmonary hypertension	Oral	Serotonin/norepinephrine reuptake inhibitor
*Sibutramine	Meridia®	Rx	1997; Withdrawn in 2010 due to association with increased heart attack risk	Oral	Serotonin/norepinephrine re-uptake inhibitor; anorexiant
Topiramate	Topamax	Rx	1998	Oral	Carbonic anhydrase inhibitor for seizures and migraines. Weight loss is may be a secondary outcome.
Orlistat	Xenical®	Rx	1999	Oral	Inhibitor of gastric and intestinal lipases
Methylphenidate	Methylin, Ritalin	Rx	2002	Oral	CNS stimulant; anorexiant
Orlistat	Alli®	OTC	2007	Oral	Inhibitor of gastric and intestinal lipases
Desvenlafaxine	Pristiq, Khedezla	Rx	2008	Oral	Serotonic norepinephrine reuptake inhibitor; anorexiant
Phendimetrazine	Bontril, Adipost, Anorex-SR, Appecon, Melfiat, Obezine, Phendiet, Plegine, Prelu-2, Statobex, Fendique ER	Rx	2012	Oral	CNS stimulant; anorexiant
Phentremine + Topiramine	Qsymia®	Rx	2012	Oral	CNS stimulant; anorexiant

(Continued)

TABLE 10.10 Weight loss drugs approved by the US Food and Drug Administration (FDA) *(Continued)*

*Lorcaserin	Belviq®	Rx	2012 (withdrawn in 2020 due to increased cancer risk)	Subcutaneous injection	Anorexiant; stimulates the release of alpha-melanocortin-stimulating hormone (alpha-MSH), which acts on melanocortin-4 receptors in the hypothalamic paraventricular nucleus
Liraglutide	Saxenda®, Victoza®	Rx	2014	Subcutaneous injection	Glucagon-like peptide 1 agonist for blood glucose control. Weight loss may be a secondary outcome.
Bupropion + Naltrexone	Contrave®	Rx	2014	Oral	Antidepressant; opioid blocker
Dulaglutide	Trulicity	Rx	2014	Subcutaneous injection	Glucagon-like peptide 1 agonist for blood glucose control. Weight loss may be a secondary outcome.
Semaglutide	Ozempic, Rybelsus	Rx	2017	Subcutaneous injection, oral	Glucagon-like peptide 1 agonist for blood glucose control. Weight loss may be a secondary outcome.
Setmelanotide	Incivree	Rx	2020	Subcutaneous injection	Genetic pro-opiomelanocortin (POMC) deficiency, proprotein subtilisin/kexin type 1 (PCSK1) deficiency, and leptin receptor (LEPR) deficiency
Collagenase Clostridium Histolycium	Xiaflex, Xiapex, Qwo	Rx	2010	Subcutaneous injection	FDA approval for treatment of Dupuytren's contracture and Peyronie's disease, also used to disrupt collagen in cellulite
Semaglutide	Wegovy	Rx	2021	Subcutaneous injection	Glucagon-like peptide 1 agonist for blood glucose control, approved by FDA for weight control
Tesofensine	Saniona	Rx	2021	Oral	Combination of a norepinephrine, dopamine, and serotonin reuptake inhibitor and metoprolol, a β-1 blocking agent for treatment of hypothalamic obesity

Sources: www.drugs.com/condition/obesity.html; www.fda.gov
*Withdrawn from the US market by the FDA due to health concerns.

inhibition of norepinephrine, dopamine, and serotonin reuptake; (8) stimulating the release of alpha-melanocortin-stimulating hormone, which affects melanocortin-4 receptors in the hypothalamic paraventricular nucleus; and (9) stimulating hypothalamic control over appetite in certain genetic defect conditions. In addition, certain medications approved by the Food and Drug Administration (FDA) for other disease conditions that have been prescribed to disrupt collagen in cellulite.

Since the late 1990s, three weight-loss drugs previously approved for use in the United States have been taken off the market by the FDA. In 1997, Fenfluramine+Prentermine (Redux®, a.k.a., "Fen-Phen"), a serotonin/norepinephrine reuptake inhibitor, was linked to mitral valve disease and pulmonary hypertension. Silbutramine (Meridia®), also a serotonin and norepinephrine reuptake inhibitor, was linked to an increased risk for heart attack and was removed in 2010. Lorcaserin (Belviq®), an appetite suppressor that stimulates melanocortin-4 receptors in the hypothalamic paraventricular nucleus, was linked to increased cancer risk and removed in 2020. In the European market, Rimonabant (Acomplia) promoted weight loss by enhancing cannabinoid receptor activity in the central nervous system, but concerns about psychological side effects resulted in its removal in 2009.

Narayanaswami and Dwoskin describe new antiobesity drugs that are in various stages of clinical development and have the following mechanisms of action: (1) simultaneous serotonin, norepinephrine, and dopamine inhibitors; (2) hypothalamic pro-opiomelanocortin neuron agonists, (3) glucagon-like peptide 1 agonists; (4) gastric and pancreatic lipase inhibitors; (5) combination lipogenesis antagonists/lipolysis agonists; (6) appetite suppressants; and (7) sodium/glucose cotransport inhibitors. In their review, Rebello and Greenway describe ongoing areas of research in the development of future pharmacological agents which may combat obesity by inhibiting sodium/glucose cotransport 1/2 activity, upregulating lipid and energy metabolism; inhibiting protein translation; activating energy consumption, developing Peptide YY analogs stimulating satiety, and/or inhibiting fat absorption. Of particular interest is pharmacological research to selectively stimulate β_3-adrenergic receptors to induce the "browning" of white adipose tissue, as previously illustrated in **figure 10.8**. Some scientists contend that obesity is an incurable disease, somewhat comparable to type 1 diabetes. Type 1 diabetics need to take insulin daily to control their disease. It is hoped that continued research into the neural mechanisms underlying the etiology of obesity will eventually provide the database for pharmaceutical companies to develop more effective and safe drugs to prevent and treat obesity. For example, obestatin, which is derived from the same prohormone as ghrelin, depresses the appetite and may serve as the basis for developing an effective and safe weight-loss drug.

Although weight-loss drugs may be effective, the lost weight is regained upon cessation of use of the drug if the lifestyle is not changed. The FDA recommends that weight-control drugs be used only on a short-term basis in conjunction with an education program stressing proper diet, exercise, and behavior modification. Kushner noted that clinicians should counsel patients on diet, physical activity, and behavior modification as the foundation for weight loss, followed by pharmacology, and finally by bariatric surgery if necessary.

Very Low-Calorie Diets Starvation-type diets may involve either complete fasting or **very low-calorie diets (VLCD)** (<800 kcal/day), often referred to as modified fasts. These diet plans are most often used with inpatient programs in hospitals. VLCDs generally include no more than 800 kcal/day, usually in liquid form, and include high percentages of carbohydrates and protein and a low fat percentage. According to Jensen and others, guidelines by the American Heart Association, the American College of Cardiology, and The Obesity Society state that VLCD, under proper medical supervision, are generally regarded as safe and have been effective in inducing rapid weight loss in individuals with obesity. Jensen and others state that VLDCs result in 14–21 kilogram (31–46 lbs) over 11–14 weeks. The efficacy and safety of VLCD may depend on the duration.

Although VLCD may be safe and effective for promoting short-term weight loss, evidence of successful post-VLCD long-term weight maintenance is scant. Apparently little or no harm is caused by a 1- or 2-day fast, and some authorities have reported that a healthy man or woman can fast completely for 2 weeks with no permanent ill effects. In their review, Alhamdan and others concluded that alternate-day fasting increased fat-free mass preservation and fat mass loss compared to VLCD and therefore might be a preferable alternative. In their meta-analysis, Johansson and others reported a mean weight loss during the VLCD or low-kcal diet (<1,200 kcal/day) of 12.3 kilograms (27 lbs) over a median period of 8 weeks. Moreover, greater post-VLDL there was weight maintenance with meal replacement, antiobesity drugs, and high-protein diets compared to control groups with no effect of exercise and other dietary supplements.

Jensen and others note that VLCDs should only be used after a medical exam and under medical supervision. Contraindications include fatigue, weakness, headaches, nausea, constipation, loss of libido, kidney stones, gallbladder disease, decreased HDL-cholesterol, impaired phagocytic function of the white blood cells, inflammation of the intestines and pancreas, decreases in blood volume, decreases in heart muscle tissue, low blood pressure, cardiac arrhythmias, and even death. Muscogiuri and others recommend VLCD programs consisting of three stages. The active stage (600–800 kcal/day) includes three ketogenic phases progressing to more natural low-fat protein during which the participant achieves 80 percent of target weight loss. The re-education low calorie phase (800–1500 kcal/day) consists of three phases during which the goal is maintenance of weight loss and promotion of a healthy diet and diet. The maintenance stage integrates the participant into a diet of high nutrient density and low energy density and a healthy lifestyle including physical activity to maintain weight loss.

Weight Cycling Weight cycling, also known as the "yo-yo" syndrome, is a controversial practice of intermittent weight loss followed by weight gain with unknown long-term consequences. Experts have speculated that the practice may ultimately be counterproductive because of a net weight gain consisting of increased fat mass and loss of lean body mass. During the VLCD phase, the neural circuitry regulating energy balance may respond to caloric deprivation by decreasing both diet-induced thermogenesis and resting energy expenditure (REE), as well as enhancing food storage efficiency. Severe caloric deprivation may also result in catabolism of muscle protein to generate amino acids for conversion into glucose by the liver for central nervous system metabolism. When normal (or greater) caloric intake resumes, energy-conservation mechanisms may continue to function for some time, with storage of the extra kcal as fat instead of being used to replace the lost tissue protein. Moreover, resumption of normal dietary habits may lead to binge eating due to alterations in neural pathways controlling eating.

A study by Bosy-Westphal and others provides some support for these mechanisms. They subjected adults to a weight cycle (13-week low-kcal diet with a 6-month follow-up). Subjects who regained weight had a lower REE compared to those who maintained weight loss. Females had a higher regain of fat in the extremities while males had a lower regain in both extremity and visceral fat.

In 1994, the National Task Force on the Prevention and Treatment of Obesity concluded that although weight cycling did not affect morbidity and mortality, the long-term health effects are unknown. Casazza and others also concluded there is inconsistent evidence for increased mortality associated with weight cycling. In their review, Bosy-Westphal and others argue that a more complete understanding of the effects of weight cycling must include differences in metabolic risk between visceral and subcutaneous fat depots, differences in metabolic rate among various tissues of the lean body mass, and relative changes in fat and lean mass compartments during weight gain and loss. However, Montani and others suggested fluctuations in blood pressure, lipids, and insulin sensitivity may be due to overshooting during weight regain and cause additional cardiovascular stress. They also noted that near-normal weight individuals may actually be more susceptible to increased risk associated with weight-cycling behavior than individuals who have had obesity for many years.

Another potential health-related consequence of weight cycling is chronic inflammation, a known pathology of cardiovascular disease and type 2 diabetes. Strohaker and McFarlin comment that weight cycling may alter adipose tissue to increase secretion of proinflammatory cytokines.

Some investigators believe the issue of weight cycling and health is still open for debate. In a large study, Field and others reported that weight cycling was associated with a higher prevalence of binge eating, lower levels of physical activity, and greater weight gain. It is generally agreed that any perceived risks of weight cycling should not deter an individual with obesity from attempts to lose weight, but the best strategy is a lifelong commitment to healthy behavior, diet, and exercise. There is a clear need for further research on the effects of weight cycling on behavior, metabolism, and health.

Young Athletes   One of the major medical concerns is the effect that severe weight restriction over a longer period may have on children who are still in the growth and development stages of life. Young athletes are at a critical age as far as nutritional needs are concerned, but the importance of making weight for certain sports may outweigh consideration of a balanced diet, adequate fluid intake, and a minimum caloric requirement. As a result, young athletes may receive inadequate intakes of protein, zinc, iron, calcium, and other nutrients, described in earlier chapters, as essential for normal growth and development. Although numerous studies have revealed nutrient deficiencies and pathogenic weight-control behaviors in young athletes such as wrestlers and ballet dancers, there are very few data on the long-range effects of such practices. In an extensive review in 2013, Malina and others concluded that gymnasts, while of shorter average stature, are still within a normal range, with no evidence of impairment of attained adult stature. They also noted that insufficient information exists in the literature regarding an association between intense training and altered endocrine function. The short stature associated with successful, persistent gymnasts may be the result of self-selection or simply being selected by coaches based on stature. The American College of Sports Medicine also concluded that participation in high school wrestling, which normally involves weight cycling, does not adversely affect normal growth patterns. Nevertheless, there has been increasing concern over the development of chronic dietary problems in children, adolescents, and young adults, which could have serious adverse health consequences. According to Desbrow and others, energy deficiency during high volume training is common in young athletes. In order to excel in training and competition and avoid a myriad of potential growth and development disorders related to inadequate nutrition, they recommend ingestion of adequate energy intake to avoid energy deficiency; adequate vitamin/mineral intake; daily protein intake (1.5 g/kg) to offset oxidized protein and support protein synthesis related to training and normal tissue growth; and adherence to hydration guidelines described in chapter 9. In short, the young athlete should form a life-long healthy relationship with nutrient-dense foods.

Paul Bradbury/Tony Talle/Caia Images/Alamy Stock Photo

> ### Key Concepts
>
> ▸ Weight-loss prescription drugs currently approved for long-term use in the United States include orlistat (Xenical®), phentermine-topiramate (Qsymia®), naltrexone HCl/bupropion HCl (Contrave®), liraglutide (Saxenda®) semaglutide (Ozempic, Rybelsus), setmelanotide (Incivree), semaglutide (Wegovy), and tesofensine (Saniona).
> ▸ Drugs and very low-calorie diets may be very effective for weight loss under medical supervision, but they may be associated with a variety of health risks if not used properly.
> ▸ Excessive and rapid weight loss may result in health problems ranging from mild to severe and may have a negative impact on physical performance.

Body Composition and Physical Performance

Modifying body weight and composition is considered by some to be an ergogenic aid. Over the years athletes have used a variety of techniques, including surgery such as liposuction and body sculpting and drugs such as anabolic steroids and human growth hormone (discussed in chapter 13), but most have used diet and specific exercise programs. Some examples of how body-weight change may affect sports performance are presented in the following section. Healthy diet and exercise programs to lose excess body fat or gain muscle mass are discussed in detail in chapters 11 and 12.

What effect does excess body weight have on physical performance?

Although extra body weight might increase stability in contact sports such as football, ice hockey, and sumo wrestling, this advantage may be neutralized if the individual loses a corresponding amount of speed. In rare instances, such as long-distance swimming in cold water, extra body fat may be helpful for its insulation and buoyancy effects. Bosch and others reported that mean BMI values for all positions in 467 NCAA Division I football players met BMI criteria for overweight or having obesity. However, offensive and defensive linemen were the only players without a healthy body fat percentage (13–20 percent) and low visceral fat mass (<500 grams) as measured by DEXA. Steffes and others reported that offensive and defensive linemen had greater risk for metabolic syndrome compared to other positions. As will be discussed in Chapter 12, healthy increases in body weight for sports competition should maximize muscle mass and minimize body-fat gains.

In contrast, there are many sports in which excess body weight may be disadvantageous. Whenever the body has to be moved rapidly or efficiently, excess weight in the form of body fat serves only as a burden. Body-fat percentage is extremely low in sprinters, jumpers, distance runners, dancers and gymnasts.

According to basic principles of physics, body fat in excess of the amount necessary for optimal functioning will impair physical performance. Body fat increases the mass, or inertia, of the individual but does not contribute directly to energy production, so excess fat will detract from performance in events in which the body must be moved. According to the laws of physics and all other factors, including force production, remaining the same, a high jumper who gains 5 pounds of fat would clear a lower height due to the resulting smaller displacement of her center of gravity. Similarly, extra mass on a distance runner could add a considerable energy cost. Adding body fat would slow the running pace. In essence, the body becomes a less efficient machine when it must transport extra weight that has no useful purpose. Losing excess fat will not influence the *absolute* VO_2 max in l/minute, but will increase *relative* VO_2 max which is expressed in milliliters of oxygen per kilogram body weight per minute (ml/kg/minute). A 70-kilogram male runner with 15 percent fat and a total body VO_2 max of 4.9 liters/minute has a weight-relative VO_2 max of 70 ml/kg/minute. A decrease in body fat to 10 percent would reduce his body weight to 66.1 kilograms (based on a two-compartment model) and increase his relative VO_2 max to 74 ml/kg/minute based on the same absolute VO_2 max of 4.9 l/minute (4900 ml/min). The hypothetical ergogenic effect of a competitive 70-kilogram (154-pound) runner losing 1 lb of fat tissue on a 10-kilometer (6.2-mile) race performance is illustrated in **table 10.11**. In theory, the loss of fat mass might result in a 2-second-per-mile increase in race pace with no change in absolute VO_2 max. These beneficial effects are most relevant for those athletes who have excess body fat to lose, and may not be applicable to those who are already at their optimal body weight. The loss of too much weight including fat-free mass would probably adversely affect athletic performance.

For a number of reasons, it is difficult to predict with certainty a precise percentage of body fat for a given athlete that will result in optimal performance. Nevertheless, studies with elite athletes have given us some general guidelines. Depending on the type of athlete, body fat percentages in males and females range from 5 to 20 percent and 10 to 28 percent, respectively, as shown in **table 10.12**. Lower values (5-10 percent) are generally observed in sprinters, long-distance runners, wrestlers, gymnasts, basketball players, soccer players, swimmers, bodybuilders, and football backs while baseball players, football linemen, tennis players, weight lifters, shot-putters, and discus throwers have somewhat higher values (11–25+ percent). Several authorities have suggested that female athletes should carry no more than 20 percent fat, whereas others note that it should be below 15 percent. Although these are some general guidelines, it should be noted that body-fat percentage is only one of many factors that may influence physical performance, and athletes may perform very well even though their body fat is above these levels. However, everything else being equal, excess body fat is a disadvantage when energy efficiency of body movement in sport is an important consideration.

Excess body fat may also increase injury risk and be related to cognitive impairment. In their review of 28 studies of pediatric obesity and athletic injuries, Confroy and others concluded that athletes

TABLE 10.11 Hypothetical ergogenic effect of one pound of fat loss on 10-km (6.2-mile) race performance

		Loss of 1 lb (0.454 kg) of fat	
Mass (kg)	70		69.55
Absolute VO_2 max (ml/min)	4,900		4,900
Relative VO_2 max (ml/kg/min)	70		70.46
10 km best time velocity (miles/hour)	10.91		10.98
10 km best time velocity (meters/min)	292.36		294.27
10 km estimated VO_2 (ml/kg/min)*	61.97		62.35
10 km VO_2 ÷ VO_2 max (%)	88.5%		88.5%
10 km pace (min/mile [sec/mile])	5.50 [330]		5.46 [328]
10 km best time (minutes [min:sec])	34.1 [34:06]		33.88 [33:53]

*Horizontal running VO_2 = 3.5 ml/kg/min + (0.2 ml O_2 × meters/min).
Source: *ACSM Guidelines for Exercise Testing and Prescription, 10th ed.* 2018. Philadelphia, PA: Wolters Kluwer, p. 152.

TABLE 10.12 Body fat percent ranges for male and female athletes

Athletic group	Males	Females
Endurance (cyclists, distance runners, rowers, triathletes, cross-country skiers)	5–15%	10–24%
Gynmasts	5–12%	10–16%
Racquet sports (racketball, tennis)	8–16%	15–24%
Team sports (football, soccer, baseball, hockey, volleyball, basketball)	6–19%	12–27%
Track and field (sprinters, jumpers, shot putters)	7–20%	10–28%
Wrestlers	5–16%	NA
Weightlifters, bodybuilders	5–16%	10–15%

Source: Modified from Jeukendrup, A. and Gleeson, M. *Sport Nutrition*. 2nd Edition. Champaign, IL: Human Kinetics, Inc. 2009.

with higher BMI have decreased general physical conditioning and significantly higher risk of musculoskeletal, exertional heat illness and other orthopedic injuries compared to normal weight athletes. Yard and Comstock noted athletes with obesity sustained a larger proportion of knee injuries than athletes without obesity. It is possible that the increased body mass created more torque on the knee and exceeded the injury threshold of ligaments and tendons. Fedor and Gunstad reported significant associations between high BMI and cognitive impairment in NCAA Division I athletes.

Coaches and practitioners should exercise caution in advising athletes to lose body weight to achieve an arbitrary predetermined goal for the following reasons. As discussed earlier, estimation of body fat percentage will have a 2–4 percent or higher standard error of measurement depending on the assessment technique. The athlete's current body fat percentage may already be near or at the arbitrarily low goal level.

Does excessive weight loss impair physical performance?

In 1998, the National Collegiate Athletic Association adopted a rule mandating testing to establish a minimum wrestling weight for each wrestler. Oppliger and others reported that the NCAA weight-management program appears effective in reducing unhealthy weight-cutting behaviors, such as rapid weight losses and gains associated with competition, and promoting competitive equity. However, some studies found that a significant percent of the high school, college, and international-style wrestlers still used potentially harmful weight-loss methods.

Weight-reduction programs used by wrestlers and other athletes have been condemned by sports medicine groups, not only for health reasons but also because these practices may impair physical performance. In its position stand on weight loss in wrestlers, the American College of Sports Medicine noted that the practice of weight cutting involving food restriction, fluid deprivation, and dehydration could not only affect physical health and growth and development but also impair competitive performance. This impairment in performance may be attributed to decreased blood volume, decreased testosterone levels, impaired cardiovascular function, decreased ability to regulate body temperature, hypoglycemia, or depletion of muscle and liver glycogen stores. However, the ultimate effect on performance may be dependent on the technique used—dehydration or starvation—and the time over which the weight is lost.

The effect of rapid weight loss by voluntary dehydration on physical performance was discussed in chapter 9. In general, events characterized by power, strength, and speed may not be adversely affected by short-term dehydration, whereas performance in aerobic and anaerobic endurance events is likely to deteriorate, particularly if exercising under warm environmental conditions.

Starvation and semistarvation studies have been conducted over periods ranging from 1 day to 1 year. Short-term starvation, involving rapid weight loss, may impair physical performance if blood glucose and muscle glycogen levels are lowered substantially. Although strength and VO_2 max generally are not affected by acute starvation, studies using a 24-hour fast have shown that anaerobic and aerobic endurance performance will suffer if dependent on muscle glycogen or normal blood glucose levels. Long-term semistarvation may lead to significant losses of lean muscle tissue and decreased performance in almost all fitness components. Gradual weight loss and a diet with adequate carbohydrate and protein appear to be the keys to maintaining optimal physiperformance during weight loss. Abaïdia and others examined the effects of fasting from dawn to sunset during Ramadan, the ninth month of the Islamic calendar, as required by devout healthy Muslin teenagers and adults. Across 11 studies, they reported decreased Wingate test peak and mean power and sprint performance during Ramadan. Aerobic performance, strength, jump height, Wingate test fatigue index, and Wingate test total work were unaffected by intermittent Ramadan fasting. Garthe and others reported a slow weekly rate of weight loss (0.7 percent; ~1 pound) increased lean body mass by 2.1 percent compared to no change in a fast weekly rate of weight loss (1.4 percent; ~2 pounds). The athlete who avoids the loss of lean body mass would be less likely to experience a decline in performance during weight loss. Although Koral and Dosseville reported no difference between rapid weight loss (2–6 percent of body mass over 4 weeks) and control judo athletes in squat jump, countermovement jump, or sport-specific movements in judo athletes, research studies in general have concluded that rapid weight loss over the course of several days to a week may impair physical performance, particularly prolonged, sport-specific anaerobic performance. In their meta-analysis of nine studies on the effects of total energy deficit (kcal/day deficit x training duration of 7 to 64 days) on lower-body strength and power in soldiers during military training maneuvers, Murphy and others reported that total deficits during training of $\leq -19,109$ kcal were associated with a change of ≤ -3.3 percent in body mass. Total deficits between $-39,243$ to $-59,377$ kcal were associated with losses of 8, 7, and 10 percent, respectively, in body mass, lower-body power, and strength. In some semistarvation studin which fewer than 1,000 kcal were consumed daily, vigorous exercise programs were maintained even though the subjects were losing substantial amounts of body weight. In summary, the key point is to prevent hypoglycemia, dehydration, and excessive loss of lean muscle mass.

It is difficult to predict the specific body weight at which physical performance will begin to deteriorate for a given individual. For those athletes who are on a weight-loss program, it may be wise to monitor performance through certain standardized tests appropriate for their sport. Some examples include basic fitness tests with measures of strength, local muscular endurance, and cardiovascular endurance. A decrease in performance may be indicative that the weight loss is excessive. Personality changes, excessive tiredness, weakness, and lack of enthusiasm may also be telltale clues.

Weight losses have the potential to either improve or diminish performance. The interested reader is referred to the 2013 International Olympic Committee review and position statement by Sundgot-Borgen and others on minimizing health risks in weight-sensitive sports. The key is to lose weight properly, primarily body fat. The basic guidelines for the development of such a weight-control program to improve physical performance, or health, are presented in chapter 11.

> ### Key Concepts
> - Although the average body-fat percentages for young males and females are, respectively, 15–20 percent and 23–30 percent, those involved in certain types of athletic competition may be advised to reduce those levels.
> - Excessive loss of body weight may impair sports performance. For athletes in sports where weight loss may enhance performance, basic specific fitness tests may be used periodically to ascertain that performance is maintained or improved.

Disordered Eating and Eating Disorders

Eating behaviors vary among individuals and may be influenced by a number of factors, including ethnicity, religion, economics, availability, personal preference, and others. Some eating behaviors may be regarded as "abnormal" with potential negative impacts on a person's health. The following discussion uses the criteria for such eating behaviors as presented in the fifth edition of *Diagnostic and Statistical Manual of Mental Disorders* (*DSM-V*) published by the American Psychiatric Association (APA).

Disordered eating reflects atypical eating behaviors such as avoiding certain foods and food groups, following fad diets, skipping meals, and eating inconsistently throughout the day. Disordered eating patterns often occur in response to a stress and/or a desire to lose weight quickly, such as an athlete before competition. In general, disordered eating habits occur sporadically. For example, a college student might be under a lot of stress during final exams week and their dietary intake becomes sporadic. Once exams are over and stress levels decrease, that individual will likely then return to more "normal" dietary habits. In some cases, disordered eating habits continue and progress to become an eating disorder.

The APA defines eating disorders as "abnormal eating habits that can threaten a person's health or even life." Currently, the APA recognizes three clinical eating disorders in *DSM-V*: anorexia nervosa (AN), bulimia nervosa (BN), and binge eating disorder (BED). As well, *DSM-V describes* other specified feeding or eating disorder and unspecified feeding or eating disorders. According to Udo and Grilo, lifetime prevalence estimates for AN, BN, and BED from a sample of over 36,000 US adults were 0.8, 0.28, and 0.85 percent, respectively. Important to note with this area of research is that eating disorders are often under-reported or not reported. As such, rates are likely higher than currently reported in available research. The following sections summarize the major types of eating disorders, including signs and symptoms typically associated with each condition.

What is anorexia nervosa?

Anorexia nervosa (AN) is a complex disorder that is not completely understood but is thought to be a sign of other psychological problems. Rask-Andersen and others report that AN is a complex, multifactorial disease with high heritability, including genetic influences on mental disorders, hunger regulatory systems, reward systems, and systems regulating energy metabolism and sex hormones. Bulik and others noted that AN was more prevalent between identical twins as compared to fraternal twins, and they indicated that about 56 percent of AN causality is associated with genetics. Sodersen and others suggest that brain mechanisms for reward are stimulated when food intake is reduced, thus sustaining disordered eating behaviors. Research has suggested that individuals with AN often exhibit characteristics that underlie compulsive personality disorders, and the prevalence of AN is greatest in groups that abuse psychoactive substances. The diagnostic criteria for AN, as developed by the APA, are summarized in **table 10.13**.

TABLE 10.13 Symptoms of anorexia nervosa

- Refusal to maintain the body weight over a minimally normal weight for age and height. A weight loss leading to the maintenance of body weight of less than 85% of that expected, including the expected weight gain during the period of growth.
- An intense fear of gaining weight or becoming fat, even though underweight.
- A disturbance in the way one's body weight or shape is perceived.

Source: American Psychiatric Association. 2013. *Diagnostic and Statistical Manual of Mental Disorders-V* (5th ed.). Washington, DC: American Psychiatric Association.

Young females, usually under the age of 25, account for about 85–90 percent of AN cases. As noted earlier, there appears to be a strong genetic predisposition to AN.

Multiple factors underlie the development of AN. Perfectionist tendencies, low-esteem with poor body image, and self-criticism are personality traits that appear to be associated with AN. Upper-middle socioeconomic status, an overprotective home environment, and a culture that idealizes thinness are also associated with AN. Susceptible individuals desire to have control over something in their lives. Individuals with AN are more likely to be depressed, to be anxious, and to suffer obsessive-compulsive disorders. Aoki and others noted that between 40 and 80 percent of individuals with AN engage in excessive exercise to lose weight. Although most individuals with AN are female, males with AN possess similar characteristics.

Treatment of AN, as well as other eating disorders, involves an interdisciplinary team consisting of professionals from medical, nursing, nutritional, and mental health disciplines. In its most recent position stand on treatment of eating disorders by Hackert and others, the Academy of Nutrition and Dietetics emphasized a team approach, including nutritional counseling by a registered dietitian nutritionist (RD/RDN). Psychotherapeutic techniques, such as cognitive behavioral therapy, are designed to foster body acceptance. Intervention programs may use self-help or group approaches and may be prolonged for several years. Medical consequences of AN can be very serious, including hormonal imbalances, anemia, decreased heart muscle mass, heartbeat arrhythmias attributed to electrolyte imbalances, and even death. Amenorrhea, the absence of three consecutive menstrual cycles, is also common in females with AN and was previously used in AN diagnosis. However, amenorrhea is also observed in other conditions, including the female athlete triad, discussed later in this chapter, and is not a diagnostic criterion for AN in *DSM-V*.

Success in treating AN is limited. The dropout rate from AN treatment programs is high. In their review, Bailey and others found that relapse prevention strategies such as cognitive behavioral therapy, psycho-education, and use of selective serotonin reuptake inhibitors were addressed in only six of 64 AN treatment trials. More research is needed to identify effective strategies to prevent AN relapse. Bodell and Mayer reported that females with a lower percent body fat were more likely to experience a poorer long-term outcome. However, El Ghoch and others reported that higher BMI values in AN subjects at discharge from inpatient treatment were associated with normal weight maintenance one year following discharge. They found no association between body fat percentage or distribution and one-year weight maintenance. In their meta-analysis, Arcelus and others reported a weighted mortality rate of AN of 5.1 deaths per 1,000 person years.

What is bulimia nervosa?

The term **bulimia nervosa (BN)** means *morbid hunger*, and the disorder involves a loss of control over the impulse to binge. The bulimic individual repeatedly ingests large quantities of food within a discrete period of time, such as 2 hours, but follows this by self-induced vomiting and other measures to avoid weight gain. This

TABLE 10.14 *DSM-V* criteria for bulimia nervosa

- Recurrent episodes of binge eating, at least one per week for 3 months
- Lack of control over eating during the binge
- Regular use of self-induced vomiting, laxatives, diuretics, fasting, or excessive exercise to control body weight
- Persistent concern with body weight and body shape

Source: American Psychiatric Association. 2013. *Diagnostic and Statistical Manual of Mental Disorders–V* (5th ed.). Washington, DC: American Psychiatric Association.

is known as the binge-purge syndrome. Individuals with bulimia may use other techniques, such as fasting or excessive exercise, to compensate for the binge. Criteria for BN according to DSM-V are listed in **table 10.14**.

According to Udo and Grilo, the lifetime prevalence for BN from a sample of over 36,000 US adults is estimated to be 0.28 percent. As with AN, certain characteristics may underlie the development of BN, including the desire for thinness and chronic low self-esteem.

According to the Mayo Clinic, dehydration, irregularity in heart rhythm, tooth decay, erosion of dental enamel, gum disease, amenorrhea, digestive problems, and substance abuse are associated with BN. Many of these complications are the result of self-induced purging or vomiting. Anxiety and depression are also associated with BN. As with individuals with AN, those with BN are in need of psychological counseling by qualified medical professionals. Prozac (fluoxetine), an antidepressant drug that helps increase serotonin levels, has been approved for medically supervised use in treating BN. McElroy and others report that Prozac helps decrease the core symptoms of BN and associated psychological factors, at least on a short-term basis. Behavioral treatment may also be effective.

Although treatment of BN appears to produce better results than treatments available for AN, the outcome is still unsatisfactory for many patients. Arcelus and others reported a weighted annual mortality rate of BN of 1.7 deaths per 1,000 person years.

What is binge eating disorder?

Binge eating disorder (BED) is a third category of eating disorder. Individuals with BED often eat an unusually large amount of food and feel out of control during the binge. Behaviors associate with BED according to DSM-V are listed in **table 10.15**.

TABLE 10.15 Behaviors associated with binge eating disorder

- At least one episode/week over 3 months of consuming a large amount of food in a short time
- Eating until uncomfortably full
- Eating when not hungry
- Eating alone because of embarrassment
- Feeling disgusted, depressed, or guilty after eating

Source: American Psychiatric Association. 2013. *Diagnostic and Statistical Manual of Mental Disorders–V* (5th ed.). Washington, DC: American Psychiatric Association.

These behaviors are also common to BN, but the individual with BED does not purge. The causes of BED are similar to those of BN. Health consequences include weight gain and obesity, along with increases in CHD risk factors, cancer, and other health problems previously discussed. In their study of mortality rates from 1995 to 2010 in almost 2,500 Finnish subjects (95% female) with eating disorders, Suokas and others reported a hazard ratio (HR) for all-cause mortality of 1.77 (95% CI 0.60–5.27) from BED compared to almost 10,000 normal controls. Treatment of BED is comparable to that for BN, which may involve medication such as antidepressants. In 2015, the U.S. Food and Drug Administration expanded the use of Vyvanse® (lisdexamfetamine dimesylate), previously approved to treat adult attention deficit hyperactivity disorder (ADHD), to also treat BED in adults. Topiramate (Topamax), an anticonvulsant, and antidepressants may also be effective in reducing BED symptoms. Treatment for obesity may also be involved.

What are examples of "other specified" or "unspecified" feeding or eating disorders?

In addition to AN, BN, and BED, the *DSM-5* also includes a discussion of "other specified" feeding or eating disorders that do not meet the full criteria for AN, BN, and/or BED but can be diagnosed by a qualified health-care provider. Such disorders include atypical anorexia nervosa, binge-eating disorder of low frequency and/or limited duration, bulimia nervosa of low frequency and/or limited duration, night eating syndrome, and purging disorder. A final category outlined in *DSM-5* is "unspecified" feeding or eating disorders. These conditions currently do not have specific criteria for diagnosis but are nonetheless important to be aware of given their rising prevalence and potential impact on physical and mental health. These conditions include diabulimia, drunkorexia, muscle dysmorphia, and orthorexia nervosa. **Table 10.16** provides a summary of select "other specified" or "unspecified" feeding or eating disorders.

> www.nationaleatingdisorders.org/ The National Eating Disorders Association (NEDA) provides individuals and their families and friends with information and support, including a short screening tool for eating disorders.
>
> www.apa.org/topics/eating-disorders The American Psychological Association (APA) provides evidence-based content on eating disorders and disordered eating. Treatment options are provided, including a "Find a Psychologist" search tool to find a qualified eating disorder specialist in your area.

What eating disorders are most commonly associated with sports?

Depending on the nature of the sport, the loss of excess body weight may enable an athlete to compete in a lower weight class or improve appearance and/or biomechanics and enhance the

TABLE 10.16 Selected "Other Specified" or "Unspecified" Feeding or Eating Disorders

Other Specified Feeding or Eating Disorders	
Atypical anorexia nervosa	Anorexia nervosa symptoms except patient remains at or above normal weight
Binge eating disorder of low frequency and/or limited duration	BED symptoms except for lower frequency and or for less than 3 months in duration
Bulimia nervosa of low frequency and/or limited duration	BN symptoms except for lower frequency and or for less than 3 months in duration
Night eating syndrome	Late-night eating associated with insomnia, depressed mood, and altered circadian rhythms
Purging disorder	Purging in the absence of binge eating
Unspecified feeding or eating disorders	
Diabulimia	Individuals with type 1 diabetes skip insulin injections or use less insulin than prescribed in an effort to control body weight. Leads to poor diabetes control.
Drunkorexia	Avoidance of food and/or purging behaviors associated with alcohol consumption. More prevalent in young adults.
Muscle dysmorphia	Altered body image characterized by individual's perception of muscular weakness and/or lack of muscle definition. More prevalent in males.
Orthorexia nervosa	Pathological obsession with ritualized food preparation and eating patterns; adherence to a restrictive diet; and avoidance of foods perceived to be unhealthy or impure.

Source: American Psychiatric Association. 2013. *Diagnostic and Statistical Manual of Mental Disorders-V* (5th ed.). Washington, DC: American Psychiatric Association.

potential for success. Wrestlers, gymnasts, dancers, cheerleaders, weightlifters, bodybuilders, figure skaters, divers, distance runners, and other athletes may use weight loss as an ergogenic aid. To lose weight, athletes in these sports, particularly athletes with perfectionistic tendencies, may exhibit some, but not all, of the characteristics associated with clinical eating disorders.

Anorexia athletica (AA) is a term used to describe athletes who become overly concerned with their weight, primarily in the context of enhancing athletic performance, and exhibit some of the diagnostic criteria associated with other diagnosable eating disorders.

Some signs and symptoms of AA as discussed in the review by Sudi and others and the study by Sundgot-Borgen and Torstveit are listed in **table 10.17**. Not all the listed signs and symptoms are necessarily indicative of AA. As an example, Sundgot-Borgen and Torstveit commented that perfectionism and high achievement orientation are desirable traits for athletic success.

Although male athletes, particularly those in bodybuilding, distance running, and wrestling, may be at risk for disordered eating, most studies have focused on female athletes in weight-control aesthetic sports such as gymnastics and dance, as well as female distance runners. Bratland-Sandra and Sundgot-Borgen report prevalence rates of disordered eating and eating disorders in the literature of 0–19 percent in male and 6–45 percent in female athletes.

According to a review by Knapp and others, subclinical eating disorder prevalence estimates in female athletes are much higher than in the general population, as high as 50 percent at the high school level and ranging from 20 to 62 percent at the collegiate level. In a review of three studies using three separate screening tools to identify eating disorders in female athletes, Wanger and others concluded that all three instruments successfully identified eating disorder cases but varied in sensitivity and specificity. Sundgot-Borgen and Torstveit reported significantly higher eating disorder prevalence rates in athletes (13.5 percent) compared to control subjects (4.6 percent), with a 42 percent prevalence rate in female aesthetic athletes. Sudi reports that symptoms of AA generally diminish or disappear at the end of the athlete's career. However, short-term behaviors meant to control weight for training and athletic competition may develop into long-term medical problems.

Female and Male Athlete Triads The **female athlete triad**—low energy availability, amenorrhea, and osteoporosis—was introduced in chapter 8, with a focus on calcium balance. According to studies by Hoch and colleagues and Torstveit and Sundgot-Borgen, the female athlete triad has been reported in athletes and nonathletes. The triad may have serious health consequences, particularly premature osteoporosis. In their 2014 female athlete triad coalition consensus statement, De Sousa and others noted that the three components of the female athlete triad are interrelated and have adverse effects on health and performance.

While the female athlete triad has been identified and studied for several decades, in recent years, experts have described the **male athlete triad**. Components of the male athlete triad include low energy availability, reproductive suppression (reduced testosterone and abnormal sperm), and poor bone health. Risk factors for the male athlete triad include body dissatisfaction, believing that losing weight will improve athletic performance, extreme training loads, frequent injuries, and history of depression and/or anxiety.

In 2014, International Olympic Committee consensus statement by Mountjoy and others proposed replacing "female athlete triad" with "**relative energy deficiency in sport (RED-S)**" based on the rationale that RED-S is a more comprehensive term for a constellation of health problems in both females and males for which low energy intake and/or excessive energy expenditure is the root cause. The triad has been expanded to include males in the context of insufficient energy availability due to low intake and/or excessive energy expenditure by athletes during training. In a recent review, De Souza and others discuss strengths and limitations of the female and male athlete triad and RED-S. The triad is based on well-established interrelationships between energy availability, reproductive function, and bone health from three decades of research while the more recent RED-S model suggesting that energy deficiency is linked to broader health consequences requires more research. The following discussion highlights the key aspects of the female and male athlete triads.

TABLE 10.17 Signs and symptoms of anorexia athletica

- Disturbance in body image characterized by an intense fear of gaining weight or becoming fat, even if underweight
- Motivation to lose weight is based mainly on performance (concern based on body shape or appearance is generally in the context of comparison to a more successful athlete)
- At least 5% less than expected normal weight for age and height
- Lower body weight is achieved by reduction in energy intake, increased training volume, or both, which is voluntary and possibly encouraged by a coach, trainer, and/or parent
- No medical disorder to explain the leanness
- Weight cycling due to variations in caloric intake, training volume, and resulting energy balance
- Binge eating
- Possible use of purging methods (e.g., self-induced vomiting, laxatives, and/or diuretics)
- Gastrointestinal complaints (possibly related to purging methods)
- Delayed puberty
- Amenorrhea or irregular menstrual cycles
- High achievement orientation
- Obsessive-compulsive tendencies
- Perfectionism

Source: Sundgot-Borgen and Torstveit (2004); Sudi, et al. (2004).

---

www.acsm.org Access the ACSM position stand "The Female Athlete Triad" which is located in the list of ACSM Position Stands and Joint Position Statements.

www.femaleandmaleathletetriad.org/professionals/position-stands/ The Female and Male Athlete Triad Coalition is an international consortium that provides educational materials while advocating for a culture supporting healthy athletic environments. The "Education Center" provides links to position statements on the topic.

Disordered eating with energy deficiency Triad experts such as De Souza, Javed, and Sundgot-Borgen contend that disordered eating, either decreased energy intake, and poor food selection or excessive exercise energy expenditure that is not balanced by energy intake may be contributing factors to the triad. The ACSM position stand stresses that the key factor in the triad is low energy availability. Inadequate caloric intake may be inadvertent, intentional, or psychopathological. Whatever the reason, the experts note that most adverse effects of the triad appear to occur when energy intake availability is below 30 kcal/kg of fat-free mass per day, which may lead to impaired reproductive and skeletal health.

Reproductive Dysfunction Reproductive dysfunction occurs in both the male and female athlete triads. In males, reproductive suppression as a result of low energy intake include low testosterone levels, abnormal sperm production, and reduced sex drive. In females, reproductive suppression results in menstrual disturbances ranging from irregular periods to amenorrhea. *Amenorrhea* is the absence of a menstrual period. Primary amenorrhea represents a delay in the onset of puberty and the menarche (first menstrual period), whereas secondary amenorrhea represents an interruption of the normal menstrual cycle, with an interval of 3-6 months or more between periods. Both types of amenorrhea may occur in athletes due to low energy availability, which may be associated with a number of factors, such as exercise intensity and training practices, body weight and composition, disordered eating behaviors, and physical and emotional stress. Collectively, the interaction of low-energy availability and other factors may disturb the hypothalamic-pituitary-gonadal (HPG) axis. The HPG axis is an endocrine feedback loop in which gonadotropin-releasing hormone (GnRH) secreted by the hypothalamus stimulates the subsequent release of follicle-stimulating hormone (FSH) and luteinizing hormone (LH) by the anterior pituitary, which in turn regulates the secretion of estrogen and progesterone by the ovaries. Interruption of the normal HPG axis may interfere with the reproductive cycle and contribute to cessation of menstruation. The ACSM indicates that amenorrhea may be related to infertility and other health problems.

Low Bone Density and Osteoporosis Premature osteoporosis is one of the most serious consequences of the female and male athlete triads. In females, one of the major consequences in disruption of the normal HPG axis may be decreased amounts of estrogen from the ovaries. In addition, hormones released by the adrenal glands may be converted by fat tissue into one form of estrogen. Less estrogen may be produced in individuals with less body fat. As noted in chapter 8, estrogen helps regulate bone metabolism and increase bone mass. Additionally, dietary restrictions will reduce the intake of nutrients important for bone health, such as calcium and protein. Thus, disordered eating predisposes the athlete to osteoporosis. Bone mineral density decreases as the number of menstrual cycles missed increases. Stress fractures occur more commonly in females with low bone mineral density, and the relative risk for stress fractures is two to four times greater in amenorrheic than eumenorrheic athletes. Barrack and others reported that stress fracture risk associated with the female athlete triad increased from 15 to 20 percent for a single risk factor to 30 to 50 percent for combined risk factors. Moreover, the loss of bone mineral density in young athletes may not be reversible. In separate reviews, Javed and De Souza and colleagues commented that suppressed estrogen secretion may also decrease nitric oxide production, which could impair endothelial function and increase cardiovascular disease risk.

How can eating disorders be prevented, detected, and treated in athletes?

Prevention of eating disorders in athletes of all ages should receive more attention. Several investigators have suggested that special attention should be devoted to young athletes, particularly those involved in sports such as gymnastics and ballet, because they may meet the age, sex, and socioeconomic status criteria that may predispose them to AN. Professional organizations agree that, for prevention and early intervention, education of athletes, parents, coaches, trainers, judges, and administrators is a priority. In the National Athletic Trainers' Association position statement, Bonci and others listed 50 recommendations for preventing, detecting, and managing disordered eating in athletes, including predisposing risk factors, clinical features and behavioral warning signs, screening methods, therapeutic interventions, special considerations for adolescent athletes, and prevention strategies. Athletic trainers, team physicians, sports dietitians, and others involved in the health care of athletes should be aware of these recommendations. Athletes should be assessed for the triad at the preparticipation physical and/or annual health screening exam, and whenever an athlete presents with any of the triad's clinical conditions. Table 10.18 summarizes key risk factors for disordered eating or eating disorders.

The National Collegiate Athletic Association (NCAA) provides brochures and other materials to help develop awareness among athletes of the potential health risks associated with these disorders. A summary of some possible warning signs developed by the NCAA is presented in **table 10.18**. Additional information on the female and male athlete triads, eating disorders, or

TABLE 10.18 Key risk factors for disordered eating and eating disorders

- Body dissatisfaction and placing a high degree of importance on an "ideal" body type and shape
- History of frequent dieting
- History of excessive exercising
- Close relative with a history of an eating disorder or another mental illness
- Being bullied or teased
- Poor self-image
- Low self-esteem
- Perfectionist personality
- Coming from a dysfunctional family
- Limited social network and social isolation

disordered eating is provided by the NCAA for athletes, parents, coaches, administrators, athletic trainers, clinicians, and others. The ACSM further notes that sport administrators should also consider rule changes to discourage unhealthy weight-loss practices, similar to what has been done in the sport of wrestling at both the high school and college levels.

In this regard, coaches and others should be aware of the limitations associated with body composition measurement. Unfortunately, imprecise measurement of body composition may be a predisposing factor to the development of eating disorders in athletes. As noted previously, prediction of body fat may vary considerably, so an athlete assessed at 10 percent body fat by one method or prediction equation may be assessed at 15–20 percent by others. If a coach believes that an athlete should achieve a set body-fat percentage (e.g., 8 percent), it would be wise to use a variety of techniques to predict body-fat percentages and use the lowest value predicted. This might be the safest approach to help prevent an excessive target loss of body fat that might lead to disordered eating.

> https://www.ncaa.org/sports/2014/3/4/disordered-eating-in-student-athletes-understanding-the-basics-and-what-we-can-do-about-it.aspx Access these National Collegiate Athletic Association websites for information on eating disorders, disordered eating, and the female athlete triad.
>
> www.girlsontherun.org/ The Girls on the Run program provides resources and programming to support the physical and emotional health of young females.

Treatment Treatment for disordered eating and eating disorders requires a team-based approach from trained professionals including a psychologist, primary care provider (i.e. physician, physician assistant, nurse), and registered dietitian nutritionist. For athletes with eating disorders, treatment professionals need additional training and expertise to understand the importance of sport to an elite athlete and to make a treatment plan accordingly. Many student health and/or counseling units at colleges and universities are staffed by professionals available to all students, including student-athletes. For some student-athletes, they prefer not to discuss potential disordered eating habits and/or eating disorders with their coaches or teammates and feel more comfortable seeking treatment from college resources.

Treatment options for eating disorders range from outpatient therapy to inpatient hospitalization or residential care in a clinic specializing in eating disorders. Cognitive behavioral therapy is a primary mode of treatment for eating disorders. This psychological treatment involves therapeutic approaches addressing unhealthy emotions and behaviors, replacing them with more realistic thoughts. Family therapy and medications may also be beneficial in eating disorder treatment. In recent years, research has explored the potential effects of alternative therapies to complement more traditional medically supervised therapies. The *Training Table* in this section summarizes some of the alternative therapies.

Ultimately, recovery from an eating disorder can take time and often requires a multi-tiered treatment approach. Treatment of eating disorders in athletes can be particularly challenging for the reasons previously described in this section of the chapter. However, current treatment options have a high success rate—seeking treatment is the key.

Training Table

In recent years, alternative therapies to complement traditional medically supervised therapies for eating disorders have been studied with promising results. Again, alternative therapies do not replace traditional therapies but rather complement well-established treatment options such as cognitive-behavioral therapy. Some alternative therapies for eating disorders include:

- Meditation and yoga—mindfulness practice to focus on the present moment, staying centered and grounded
- Acupuncture—insertion of very thin needles through the skin at specific locations on the body
- Phototherapy—exposure to certain wavelengths of light through a specialized lamp
- Biofeedback—use of heart rate and neuro-feedback to decrease anxiety
- Equine therapy—therapy engages patients with horses to support self-discovery and an increased sense of purpose

Prostock-studio/Shutterstock

Key Concepts

▸ Disordered eating habits are generally short-term and a result of stress. Such habits include skipping meals, avoiding certain foods and/or food groups, and following fad diets. In comparison, eating disorders are diagnosable psychological conditions.
▸ According to the fifth edition of the *Diagnostic and Statistical Manual of Mental Disorders (DSM-5)*, anorexia nervosa, bulimia nervosa, and binge eating disorders are diagnosable eating disorders. *DSM-5* also describes other specified or unspecified feeding or eating disorders. These include muscle dysmorphia, orthorexia nervosa, night eating syndrome, and diabulimia.
▸ The female athlete triad is characterized by low energy availability, amenorrhea, and osteoporosis, while the male athlete triad is characterized by low energy availability, reproductive suppression, and poor bone health.
▸ Treatment for eating disorders requires specialized care by trained health professionals. Cognitive behavioral therapy is a primary move of therapy. Alternatives therapies to support traditional medically supervised treatment programs include meditation, acupuncture, phototherapy, and biofeedback.

APPLICATION EXERCISE

Case Study

A wrestling coach would like his 152 pound male athlete to make the 145-pound weight class for the upcoming season, which is scheduled to begin in one month, and seeks your professional advice on the feasibility of this plan. The high school league mandates minimum wrestling weight (MWW) determination based on a minimum 7 percent fat. The table includes body composition assessment data (estimated fat percentages) using four different methods.

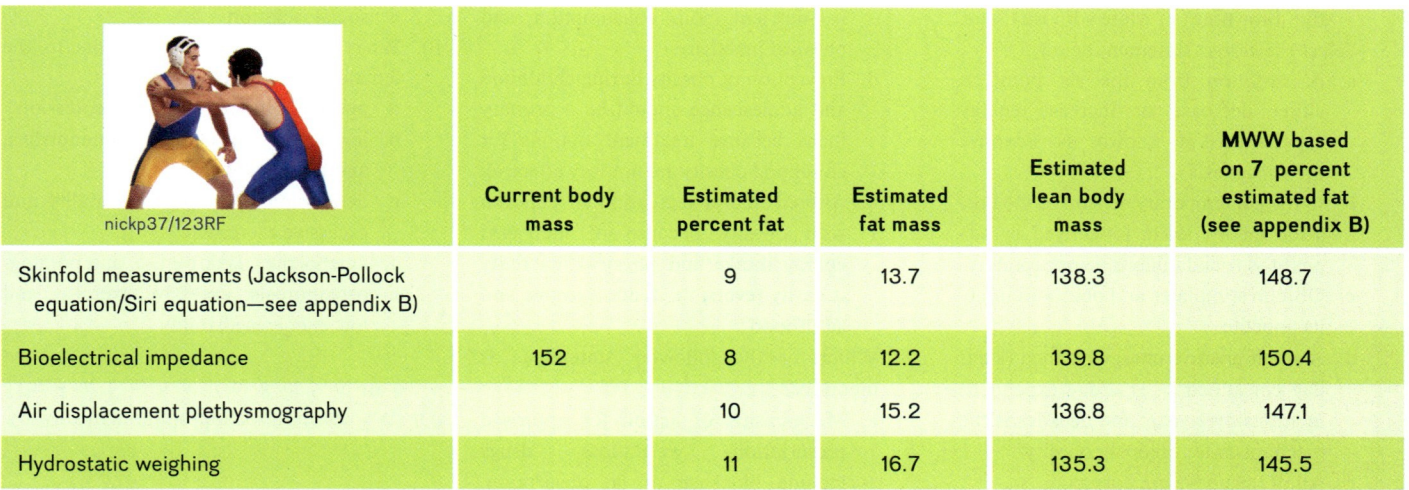
nickp37/123RF

	Current body mass	Estimated percent fat	Estimated fat mass	Estimated lean body mass	MWW based on 7 percent estimated fat (see appendix B)
Skinfold measurements (Jackson-Pollock equation/Siri equation—see appendix B)		9	13.7	138.3	148.7
Bioelectrical impedance	152	8	12.2	139.8	150.4
Air displacement plethysmography		10	15.2	136.8	147.1
Hydrostatic weighing		11	16.7	135.3	145.5

Discuss issues affecting the feasibility of this plan with the coach, athlete, and family members. Include in this discussion the effects of preseason conditioning on lean body mass, the standard errors of measurement of the various body composition assessment methodologies, and the potential risks of rapid weight loss.

Review Questions—Multiple Choice

1. What BMI range is associated with being overweight?
   a. <18.5 kg/m^2
   b. 18.5–24.9 kg/m^2
   c. 25.0–29.9 kg/m^2
   d. 30.0–34.9 kg/m^2
   e. ≥ 35.0 kg/m^2

2. A body composition assessment method has a standard error of measurement (SEM) of 3 percent. If an estimated body fat percentage obtained by this method is 22 percent, there is a 68 percent probability that the true body fat is in the range of _____ percent.
   a. 16–28
   b. 17–27
   c. 18–26
   d. 19–25
   e. 20–24

3. Select the *correct* statement below.
   a. An individual with a BMI of 32 kg/m^2 and an estimated body fat percentage of 15 should be considered as having obesity.

b. A multi (3- or 4-) component model for body composition assessment is more accurate than a 2-component model.
   c. In terms of health risk, it does not matter what an individual's waist circumference is as long as the BMI is less than 30 kg/m^2.
   d. The gynoid pattern of fat deposition is associated with the metabolic syndrome.
   e. The historical "gold standard" for body composition assessment is skinfold measurement.
4. Select the *incorrect* statement below.
   a. Neuropeptide Y suppresses the appetite.
   b. The hypothalamus is involved in regulation of energy intake.
   c. Neurons that secrete dopamine from the brain's pleasure/reward centers can influence energy intake.
   d. Brown adipose tissue contains more mitochondria with uncoupling proteins which cause thermogenesis.
   e. Exercise and other stimuli can result in the "browning" of white adipose tissue.
5. Select the *correct* statement below.
   a. A deviation from the set point to either decrease or increase energy expenditure is known as adaptive thermogenesis.
   b. In the contemporary version of the lipostatic theory, leptin is secreted by adipose tissue and inhibits neuropeptide Y.
   c. Ghrelin stimulates and obestatin inhibits appetite.
   d. A small positive energy balance (kcal/day consumed > kcal/day expended) in adults over years and decades is the mechanism for creeping obesity.
   e. All of the above are correct.

6. Factors contributing to obesity include all of the following *except*:
   a. small serving portions
   b. low-fat/high sugar/high kcal intake
   c. high-fat/high kcal food intake
   d. decreased nonexercise activity thermogenesis (NEAT)
   e. the built environment
7. Which of the following is *incorrect* regarding childhood obesity?
   a. A child who is physically inactive is more likely to have obesity. A child with obesity may be bullied and less likely to engage in regular physical activity. Therefore, physical inactivity can be both a cause and a consequence of obesity.
   b. Excessive TV watching and other "screen time" activities decrease physical activity and increase exposure to advertisements for high-fat/high-sugar/low-nutrient food.
   c. Food intake patterns of children with obesity are the result of genetics, parental modeling of high-energy/low-nutrient foods consumption, and physical inactivity.
   d. Prevention of obesity during childhood and adolescence should be a primary focus because treatment options for childhood obesity are not very effective.
   e. Altered neural regulation in children with obesity because of increased energy intake and physical inactivity is easily reversible in adolescence and adulthood.
8. Which of the following statements is incorrect?
   a. Mechanisms of action for approved prescription weight-loss drugs include blockage of fat absorption, sympathomimetic amines to suppress the appetite, inhibition of serotonin reuptake, and modulation of brain pleasure/reward pathways.
   b. In order to be effective in the long term, very low-kcal diets (VLCDs) for for individuals with obesity should be medically monitored and used in conjunction with lifestyle changes in diet and activity.
   c. Weight cycling is a controversial practice, and experts agree that the perceived risks for weight cycling outweigh the benefits of weight loss.
   d. Fat is easy to gain and hard to lose.
   e. Acute weight loss by dehydration can result in hypokalemia and rhabdomyolosis.
9. Which of the following is *not* a risk factor for eating disorders?
   a. perfectionist personality
   b. being bullied or teased
   c. history of excessive exercise
   d. high self-esteem
   e. social isolation
10. What are the three components to the female athlete triad?
    a. purging, osteoporosis, and depression
    b. low energy availability, amenorrhea, and osteoporosis.
    c. depression, low energy availability, and excessive exercise
    d. amenorrhea, laxative use, and purging
    e. osteoporosis, excessive exercise, and low energy availability

Answers to multiple choice questions: 1. c; 2. d; 3. b; 4. a; 5. e; 6. a; 7. e; 8. c; 9. d; 10. b

Critical Thinking Questions

1. List and describe five different techniques used to evaluate body composition, and highlight at least one advantage and disadvantage of each.
2. Discuss the set-point theory of body-weight control and relate it to body-fat levels, leptin, and neuropeptide Y.
3. List the various genetic and environmental factors that may be involved in the etiology of obesity. Highlight two of each and explain their possible role.
4. Explain the metabolic syndrome and associated health risks.
5. List three components of the female athlete triad and three components of the male athlete triad. Discuss the risk factors associated with both male and female athlete triads and treatment options.

References

Books

American College of Sports Medicine. 2021. *Guidelines for Exercise Testing and Prescription* (11th ed.). Philadelphia, PA: Wolters Kluwer.

American Psychiatric Association. 2013. *Diagnostic and statistical manual of mental disorders* (5th ed.). Washington, DC: American Psychiatric Association.

Bays, H.E., et al. 2019. Obesity Algorithm eBook, presented by the Obesity Medicine Association. www.obesityalgorithm.org. https://obesitymedicine.org/obesity-algorithm (Accessed October 12, 2021).

Edwards, M. 2017. The Barker hypothesis. In: Preedy, V. and Patel, V. (eds.) *Handbook of Famine, Starvation, and Nutrient Deprivation*. Cham: Springer. doi.org/10.1007/978-3-319-40007-5_71-1.

Heyward, V., and Wagner, D. 2004. *Applied Body Composition Assessment*. Champaign, IL: Human Kinetics.

Jeukendrup, A., and Gleeson, M. 2009. *Sport Nutrition*. 2nd Edition. Champaign, IL: Human Kinetics, Inc.

National Academy of Sciences. 2002. *Dietary Reference Intakes for Energy, Carbohydrates, Fiber, Fat, Protein and Amino Acids (Macronutrients)*. Washington, DC: National Academies Press.

National Research Council. 1989. *Diet and Health: Implications for Reducing Chronic Disease Risk*. Washington, DC: National Academies Press.

National Research Council. 2009. *Weight Gain During Pregnancy: Reexamining the Guidelines*. Rasmussen, K. M., and Yaktine, A. L., eds. Washington, DC: The National Academies Press. https://doi.org/10.17226/12584.

Samra R. A. 2010. Fats and satiety. Chapter 15. *In Fat Detection: Taste, Texture, and Post Ingestive Effects*. Montmayeur, J. P., and le Coutre J., eds. Boca Raton, FL: CRC Press/Taylor & Francis.

U.S. Department of Health and Human Services Public Health Service. 2010. *Healthy People 2020*. Washington, DC: U.S. Government Printing Office.

Waters, H., and Graf, M. 2018. *America's Obesity Crisis. The Health and Economic Costs of Excess Weight*. Washington, DC: Milken Institute. milkeninstitute.org/report/americas-obesity-crisis-health-and-economic-costs-excess-weight

World Anti-Doping Agency. International Standard Prohibited List, January 2017. www.wada-ama.org. Accessed May 24, 2017.

World Cancer Research Fund/American Institute for Cancer Research. 2018. *Diet, Nutrition, Physical Activity, and Cancer: a Global Perspective*. Continuous Update Project Expert Report.

World Health Organization. 2008. Waist Circumference and Waist–Hip Ratio: Report of a WHO Expert Consultation. p. 27.

Reviews and Specific Studies

Abaïdia, A. E., et al. 2020. Effects of Ramadan fasting on physical performance: A systematic review with meta-analysis. *Sports Medicine* 50(5):1009–26. doi: 10.1007/s40279-020-01257-0. PMID: 31960369.

Aguilar-Navarro, M, et al. 2020. Sport-specific use of doping substances: Analysis of world anti-doping agency doping control tests between 2014 and 2017. *Substance Use and Misuse* 55(8):1361–69. doi: 10.1080/10826084.2020.1741640.

Ackland, T. R., et al. 2012. Current status of body composition assessment in sport: Review and position statement on behalf of the *ad hoc* research working group on body composition health and performance, under the auspices of the I.O.C. Medical Commission. *Sports Medicine* 42 (3):227–49.

Alberti, K. G. M. M., et al. 2009. Harmonizing the metabolic syndrome. A joint interim statement of the International Diabetes Federation Task Force on Epidemiology and Prevention; National Heart, Lung, and Blood Institute; American Heart Association; World Heart Federation; International Atherosclerosis Society; and International Association for the Study of Obesity. *Circulation* 120:1640–45.

Albuquerque, D., et al. 2015. Current review of genetics of human obesity: From molecular mechanisms to an evolutionary perspective. *Molecular Genetics and Genomics* 290:1191–1221.

Alhamdan, B. A., et al. 2016. Alternate-day versus daily energy restriction diets: Which is more effective for weight loss? A systematic review and meta-analysis. *Obesity Science and Practice* 2(3):293–302.

American College of Sports Medicine. 1996. American College of Sports Medicine position stand: Weight loss in wrestlers. *Medicine & Science in Sports & Exercise* 28 (6):ix–xii.

American College of Sports Medicine. 2007. The female athlete triad: Position stand. *Medicine & Science in Sports & Exercise* 39:1867–82.

American College of Sports Medicine. 2009. ACSM position stand on the appropriate intervention strategies for weight loss and prevention of weight regain for adults. *Medicine & Science in Sports & Exercise* 41:459–71.

American Dietetic Association, Dietitians of Canada, American College of Sports Medicine, and Rodriguez, N., et al. 2009. American College of Sports Medicine position stand: Nutrition and athletic performance. *Medicine & Science in Sports & Exercise* 41:709–31.

American Dietetic Association. 2011. Position of the American Dietetic Association: Nutrition intervention in the treatment of anorexia nervosa, bulimia nervosa, and eating disorders not otherwise specified (EDNOS). *Journal of the American Dietetic Association* 111:1236–41.

Aoki, C., et al. 2014. α4βδ-GABA$_A$RS in the hippocampal CA1 as a biomarker for resilience to activity-based anorexia. *Neuroscience* 265:108–123.

Arcelus, J., et al. 2011. Mortality rates in patients with anorexia nervosa and other eating disorders. A meta-analysis of 36 studies. *Archives of General Psychiatry* 68(7):724–31. doi: 10.1001/archgenpsychiatry.2011.74. PMID: 21727255.

Arena, R., et al. 2014. The role of worksite health screening: A policy statement from the American Heart Association. *Circulation* 130:719–34.

Arner, P., and Spalding, K. L. 2010. Fat cell turnover in humans. *Biochemical and Biophysical Research Communications* 396:101–4.

Astrup, A., et al. 1999. Meta-analysis of resting metabolic rate in formerly obese subjects. *American Journal of Clinical Nutrition* 69:1117–22.

Aucott, L. 2008. Influences of weight loss on long-term diabetes outcomes. *Proceeding of the Nutrition Society* 67:54–59.

Aune, D., et al. 2014. Maternal body mass index and the risk of fetal death, stillbirth, and infant death: A systematic review and meta-analysis. *Journal of the American Medical Association* 311(15):1536–46.

Avery, A., et al. 2017. Associations between children's diet quality and watching television during meal or snack consumption: A systematic review. *Maternal and Child Nutrition* e12428. doi: 10.1111/mcn.12428.

Avgerinos, K. I., et al. 2019. Obesity and cancer risk: Emerging biological mechanisms and perspectives. *Metabolism Clinical and Experimental* 92:121-135. doi: 10.1016/j.metabol.2018.11.001.

Bailey, A. P., et al. 2014. Mapping the evidence for the prevention and treatment of eating disorders in young people. *Journal of Eating Disorders* 2:5. doi:10.1186/2050-2974-2-5.

Barker, D.J. 1995. Fetal origins of coronary heart disease. BMJ Clinical Research Edition 311(6998): 171-174. doi.org/10.1136/bmj.311.6998.171

Barrack, M. T., et al. 2014. Higher incidence of bone stress injuries with increasing female athlete triad-related risk factors: A prospective multisite study of exercising girls and women. *American Journal of Sports Medicine* 42(4):949-58.

Barnett, T.A., et al. 2018. American Heart Association Obesity Committee of the Council on Lifestyle and Cardiometabolic Health; Council on Cardiovascular Disease in the Young; and Stroke Council. Sedentary Behaviors in Today's Youth: Approaches to the Prevention and Management of Childhood Obesity: A Scientific Statement From the American Heart Association. *Circulation* 138(11):e142-e159. doi: 10.1161/CIR.0000000000000591. PMID: 30354382.

Barry, V. W., et al. 2014. Fitness vs. fatness on all-cause mortality: A meta-analysis. *Progress in Cardiovascular Diseases* 56:382-90.

Bass, L. S., et al. 2020. Insights into the pathophysiology of cellulite: A review. *Dermatologic Surgery* 46 Suppl 1(1):S77-S85. doi: 10.1097/DSS.0000000000002388. PMID: 32976174; PMCID: PMC7515470.

Bays, H. E., et al. 2013. Obesity, adiposity, and dyslipidemia: A consensus statement from the National Lipid Association. *Journal of Clinical Lipidology* 7 (4):304-83.

Biddle, S. J. H., et al. 2017. Screen time, other sedentary behaviours, and obesity risk in adults: A review of reviews. *Current Obesity Reports* 6(2):134-47. doi: 10.1007/s13679-017-0256-9.

Biswas, A., et al. 2015. Sedentary time and its association with risk for disease incidence, mortality, and hospitalization in adults: A systematic review and meta-analysis. *Annals of Internal Medicine* 162(2):123-32.

Blundell, J. E., et al. 2020. The drive to eat in homo sapiens: Energy expenditure drives energy intake. *Physiology and Behavior* 219:112846. doi: 10.1016/j.physbeh.2020.112846.

Bodell, L. P., and Mayer, L. E. S. 2011. Percent body fat is a risk factor for relapse in anorexia nervosa: A replication study. *International Journal of Eating Disorders* 44:118-23.

Bonci, C., et al. 2008. National Athletic Trainers' Association position statement: Preventing, detecting, and managing disordered eating in athletes. *Journal of Athletic Training* 43:80-108.

Bone, J. L., et al. 2017. Manipulation of muscle creatine and glycogen changes dual X-ray absorptiometry estimates of body composition. *Medicine and Science in Sports and Exercise* 49(5):1029-35.

Bosch, T. A., et al. 2019. Body composition and bone mineral density of division 1 collegiate football players: A consortium of college athlete research study. *Journal of Strength and Conditioning Research* 33(5):1339-46. doi: 10.1519/JSC.0000000000001888. PMID: 28277428; PMCID: PMC5591058.

Bosy-Westphal, A., et al. 2015. Deep body composition phenotyping during weight cycling: Relevance to metabolic efficiency and metabolic risk. *Obesity Reviews* 16 (Suppl 1):36-44. doi: 10.1111/obr.12254. PMID: 25614202.

Bosy-Westphal, A., et al. 2013. Effect of weight loss and regain on adipose tissue distribution, composition of lean mass and resting energy expenditure in young overweight and obese adults. *International Journal of Obesity* 37 (10):1371-77.

Bouchard, C., et al. 1990. The response to long-term overfeeding in identical twins. *New England Journal of Medicine* 322:1477-82.

Bouchard, C., et al. 2015. Less sitting, more physical activity, or higher fitness? *Mayo Clinic Proceedings* 90(11):1533-40.

Boulos, R., et al. 2012. ObesiTV: How television is influencing the obesity epidemic. *Physiology and Behavior* 107:146-53.

Bratland-Sandra, S., and Sundgot-Borgen, J. 2013. Eating disorders in athletes: Overview of prevalence, risk factors and recommendations for prevention and treatment. *European Journal of Sport Science* 13(5):499-508.

Brożek, J., et al. 1963. Densitometric analysis of body composition: Revision of some quantitative assumptions. *Annals of the New York Academy of Sciences* 110:113-40.

Brunstrom, J. M. 2014. Mind over platter: Premeal planning and the control of meal size in humans. *International Journal of Obesity* 38:S9-S12.

Bucher Della Torre, S., et al. 2016. Sugar-sweetened beverages and obesity risk in children and adolescents: A systematic analysis on how methodological quality may influence conclusions. *Journal of the Academy of Nutrition and Dietetics* 116:638-59.

Buffa, R., et al. 2014. Bioelectrical impedance vector analysis (BIVA) for the assessment of two-compartment body composition. *European Journal of Clinical Nutrition* 68:1234-40.

Bulik, C., et al. 2006. Prevalence, heritability, and prospective risk factors for anorexia nervosa. *Archives of General Psychiatry* 63:305-12.

Butler, M. G. 2016. Single gene and syndromic causes of obesity: Illustrative examples. *Progress in Molecular and Biological Translation Science* 140:1-45. doi: 10.1016/bs.pmbts.2015.12.003.

Butler, M. J., and Eckel, L. A. 2018. Eating as a motivated behavior: Modulatory effect of high fat diets on energy homeostasis, reward processing and neuroinflammation. *Integrative Zoology* 13(6):673-86. doi: 10.1111/1749-4877.12340. PMID: 29851251; PMCID: PMC6393162.

Butterick, T. A., et al. 2013. Orexin: Pathways to obesity resistance? *Reviews in Endocrine and Metabolic Disorders* 14 (4):357-64.

Campbell, S. D. I., et al. 2018. Sedentary behavior and body weight and composition in adults: A systematic review and meta-analysis of prospective studies. *Sports Medicine* 48(3):585-95. doi: 10.1007/s40279-017-0828-6. PMID: 29189928.

Campisi, J., et al. 2015. Sex and age-related differences in perceived, desired and measured percentage body fat among adults. *Journal of Human Nutrition and Dietetics* 28(5):486-92.

Carbone, S., et al. 2019. Obesity paradox in cardiovascular disease: Where do we stand? *Vascular Health and Risk Management* 15:89-100. doi: 10.2147/VHRM.S168946. PMID: 31118651; PMCID: PMC6503652.

Carvalho, R., et al. 2008. Clinical profile of the overweight child in the new millennium. *Clinical Pediatrics* 47 (5):476-82.

Casazza, K., et al. 2015. Weighing the evidence of common beliefs in obesity research. *Critical Reviews in Food Science and Nutrition* 55(14):2014-53.

Cawley, J., et al. 2021. Direct medical costs of obesity in the United States and the most populous states. *Journal of Managed Care and Specialty Pharmacy* 27(3):354–66. doi: 10.18553/jmcp.2021.20410.

Chandrabose, M., et al. 2019. Built environment and cardio-metabolic health: Systematic review and meta-analysis of longitudinal studies. *Obesity Reviews* 20(1):41–54. doi: 10.1111/obr.12759.

Chaput, J., et al. 2007. Currently available drugs for the treatment of obesity: Sibutramine and orlistat. *Mini Reviews in Medicinal Chemistry* 7:3–10.

Chaput, J.P., and Tremblay, A. 2009. The glucostatic theory of appetite control and the risk of obesity and diabetes. *International Journal of Obesity (London)* 33(1):46–53. doi: 10.1038/ijo.2008.221.

Christakis, N., and Fowler, J. 2007. The spread of obesity in a large social network over 32 years. *New England Journal of Medicine* 357:404–7.

Claussnitzer, M. et al. 2015. FTO obesity variant circuitry and adipocyte browning in humans. *New England Journal of Medicine* doi: 10.1056/NEJMoa1502214.

Confroy, K., et al. 2021. Pediatric obesity and sports medicine: A narrative review and clinical recommendations. *Clinical Journal of Sport Medicine* 31(6):e484–e498. doi: 10.1097/JSM.0000000000000839. PMID: 32852300.

da Silva Fernandes, J., et al. 2021. Adenovirus 36 prevalence and association with human obesity: A systematic review. *International Journal of Obesity (London)* 45(6):1342–56. doi: 10.1038/s41366-021-00805-6.

Dalle Molle, R., et al. 2017. Gene and environment interaction: Is the differential susceptibility hypothesis relevant for obesity? *Neuroscience and Biobehavioral Reviews* 73:326–39.

Desbrow, B., et al. 2019. Nutrition for special populations: Young, female, and masters athletes. *International Journal of Sport Nutrition and Exercise Metabolism* 29(2):220–27. doi: 10.1123/ijsnem.2018-0269.

De Souza, M. J., et al. 2014. 2014 female athlete triad coalition consensus statement on treatment and return to play of the female athlete triad. *British Journal of Sports Medicine* 48:289–308.

De Souza, M. J., et al. 2021. The path towards progress: A critical review to advance the science of the female and male athlete triad and relative energy deficiency in sport. *Sports Medicine* Oct 19. doi: 10.1007/s40279-021-01568-w.

De Vadder, F., et al. 2013. Satiety and the role of mu-opoid receptors in the portal vein. *Current Opinion in Pharmacology* 13:959–63.

Devrim-Lanpir, A., et al. 2021. Which body density equations calculate body fat percentage better in Olympic wrestlers?—Comparison study with air displacement plethysmography. *Life (Basel, Switzerland)* 11(7):707. doi: 10.3390/life11070707. PMID: 34357079; PMCID: PMC8306702

Dham, S., and Banerji, M. 2006. The brain-gut axis in regulation of appetite and obesity. *Pediatric Endocrinology Reviews* 3 (Supplement 4):544–54.

Dieleman, J. L., et al. 2020. US health care spending by payer and health condition, 1996–2016. *Journal of the American Medical Association* 323(9):863–84. doi: 10.1001/jama.2020.0734. PMID: 32125402; PMCID: PMC7054840.

Dixon, B. N., et al. 2021. Associations between the built environment and dietary intake, physical activity, and obesity: A scoping review of reviews. *Obesity Reviews* 22(4):e13171. doi: 10.1111/obr.13171.

El Ghoch, M., et al. 2016. Body mass index, body fat and risk factor of relapse in anorexia nervosa. *European Journal of Clinical Nutrition* 70(2):194–8. doi: 10.1038/ejcn.2015.164.

Elliot, D. L., et al. 2008. Long-term outcomes of the ATHENA (Athletes Targeting Healthy Exercise & Nutrition Alternatives) program for female high school athletes. *Journal of Alcohol and Drug Education* 52(2):73–92.

Engin, A. 2017. Fat cell and fatty acid turnover in obesity. *Advances in Experimental Medicine and Biology* 960:135–60. doi: 10.1007/978-3-319-48382-5_6. PMID: 28585198.

Fall, T., and Ingelsson, E. 2014. Genome-wide association studies of obesity and metabolic syndrome. *Molecular and Cellular Endocrinology* 382(1):740–57.

Fallon, E. A., et al. 2014. Prevalence of body dissatisfaction among a United States adult sample. *Eating Behaviors* 15(1):151–8. doi: 10.1016/j.eatbeh.2013.11.007.

Farrell, S., et al. 2010. Cardiorespiratory fitness, adiposity, and all-cause mortality in women. *Medicine & Science in Sports & Exercise* 42:2006–12.

Federal Trade Commission. 2014. FTC testifies before Senate commerce subcommittee on agency efforts to combat fraudulent and deceptive claims for weight-loss products. www.ftc.gov/news-events/press-release/2014/06.

Fedor, A., and Gunstad, J. 2013. Higher BMI is associated with reduced cognitive performance in Division I athletes. *Obesity Facts* 6:185–92.

Field, A., et al. 2004. Association of weight change, weight control practices, and weight cycling among women in the Nurses' Health Study II. *International Journal of Obesity and Related Metabolic Disorders* 28:1134–42.

Fields, D., et al. 2002. Body-composition assessment via air-displacement plethysmography in adults and children: A review. *American Journal of Clinical Nutrition* 75:453–67.

Flegal, K. M., et al. 2013. Association of all-cause mortality with overweight and obesity using standard body mass index categories: A systematic review and meta-analysis. *Journal of the American Medical Association* 309(1):71–82.

Fosbøl, M. Ø., and Zerahn, B. 2015. Contemporary methods of body composition measurement. *Critical Physiology and Functional Imaging* 35(2):81–97.

Franks, P. W., et al. 2010. Childhood obesity, other cardiovascular risk factors, and premature death. *New England Journal of Medicine* 362(6):485–93.

Fryar, C. D., et al. 2018. Mean body weight, height, waist circumference, and body mass index among adults: United States, 1999–2000 through 2015–2016. *National Health Statistics Reports* No. 122. Hyattsville, MD: National Center for Health Statistics.

Galmiche, M., et al. 2019. Prevalence of eating disorders over the 2000–2018 period: A systematic literature review. *American Journal of Clinical Nutrition* 109(5):1402–13. https://doi.org/10.1093/ajcn/nqy342" doi.org/10.1093/ajcn/nqy342

Garthe, I., et al. 2011. Effect of two different weight-loss rates on body composition and strength and power-related performance in elite athletes. *International Journal of Sports Nutrition and Exercise Metabolism* 21(2):97–104.

Gaston, S. A., et al. 2019. Abdominal obesity, metabolic dysfunction, and metabolic syndrome in U.S. adolescents: National Health and Nutrition Examination Survey 2011–2016. *Annals of Epidemiology* 30:30–36. doi: 10.1016/j.annepidem.2018.11.009.

Gerson, L., and Braun, B. 2006. Effect of high cardiorespiratory fitness and high body fat on insulin resistance. *Medicine & Science in Sports & Exercise* 38:1709–15.

Gomersall, S. R., et al. 2013. The ActivityStat hypothesis: The concept, the evidence and the methodologies. *Sports Medicine* 43(2):135–49.

Gomes, A. C., et al. 2020. Body composition assessment in athletes: Comparison of a novel ultrasound technique to traditional skinfold measures and criterion DXA measure. *Journal of Science and Medicine in Sport* 23:1006-10.

Gropper, S. S., et al. 2012. Changes in body weight, composition, and shape: A 4-year study of college students. *Applied Physiology, Nutrition, and Metabolism* 37(6):1118-23.

Gui, Y., et al. 2017. The association between obesity related adipokines and risk of breast cancer: A systematic review and meta-analysis. *Oncotarget* 8(43):75389-99. doi: 10.18632/oncotarget.17853.

Gupta, A., et al. 2020. Brain-gut-microbiome interactions in obesity and food addiction. *Nature Reviews. Gastroenterology and Hepatology* 17(11):655-72. doi: 10.1038/s41575-020-0341-5.

Hackert, A. N., et al. 2020. Academy of Nutrition and Dietetics: Revised 2020 Standards of Practice and Standards of Professional Performance for Registered Dietitian Nutritionists (Competent, Proficient, and Expert) in eating disorders. *Journal of the Academy of Nutrition and Dietetics* 120(11):1902-19.e54. doi: 10.1016/j.jand.2020.07.014. PMID: 33099403.

Hasselbalch, A. L., et al. 2008. Common genetic components of obesity traits and serum leptin. *Obesity (Silver Spring)* 16(12):2723-9. doi: 10.1038/oby.2008.440.

Heindel, J. J. and Blumberg, B. 2019. Environmental obesogens: Mechanisms and controversies. *Annual Review of Pharmacology and Toxicology* 59:89-106. doi: 10.1146/annurev-pharmtox-010818-021304. Epub 2018 Jul 25. PMID: 30044726; PMCID: PMC6559802.

Hetherington, M., and Cecil, J. 2010. Gene-environment interactions in obesity. *Forum on Nutrition* 63:195-203.

Heyward, V. H. 1998. Practical body composition for children, adults, and older adults. *International Journal of Sports Nutrition* 8:285-307.

Hills, A., et al. 2007. The contribution of physical activity and sedentary behaviours to the growth and development of children and adolescents: Implications for overweight and obesity. *Sports Medicine* 37:533-45.

Hirode, G., and Wong, R. J. 2020. Trends in the prevalence of metabolic syndrome in the United States, 2011-2016. *Journal of the American Medical Association* 323(24):2526-28. doi: 10.1001/jama.2020.4501. PMID: 32573660; PMCID: PMC7312413.

Hoch, A., et al. 2009. Prevalence of the female athlete triad in high school athletes and sedentary students. *Clinical Journal of Sport Medicine* 19:421-8.

Hruby, A., and Hu, F. B. 2015 The epidemiology of obesity: A big picture. *PharmacoEconomics* 33:673-89.

Hubert, H. B., et al. 1983. Obesity as an independent risk factor for cardiovascular disease: A 26-year follow-up of participants in the Framingham Heart Study. *Circulation* 67 (5):968-77.

Hung, L. S., et al. 2015. A meta-analysis of school-based obesity prevention programs demonstrates limited efficacy of decreasing childhood obesity. *Nutrition Research* 35(3):229-40.

Hur, S. J., et al. 2013. Effect of adenovirus and influenza virus infection on obesity. *Life Sciences* 93 (16):531-35.

Incollingo Rodriguez, A. C., et al. 2015. Hypothalamic-pituitary-adrenal axis dysregulation and cortisol activity in obesity: A systematic review. *Psychoneuroendocrinology* 62:301-18.

Irving, B.A., et al. 2008. Effect of exercise training intensity on abdominal visceral fat and body composition. *Medicine and Science in Sports and Exercise* 40(11):1863-72.

Ishii, M., and Iadecola, C. 2016. Adipocyte-derived factors in age-related dementia and their contribution to vascular and Alzheimer pathology. *Biochimica et Biophysica Acta* 1862(5):966-74.

Islami, F., et al. 2019. Proportion of cancer cases attributable to excess body weight by US State, 2011-2015. *Journal of the American Medical Association Oncology* 5(3):384-92. doi: 10.1001/jamaoncol.2018.5639. PMID: 30589925; PMCID: PMC6521676.

Jackson, A., and Pollock, M. 1978. Generalized equations for predicting body density of men. *British Journal of Nutrition* 40:497-504.

Jackson, A., et al. 1980. Generalized equations for predicting body density of women. *Medicine & Science in Sports & Exercise* 12:175-82.

Jacobs, E., et al. 2010. Waist circumference and all-cause mortality in a large US cohort. *Archives of Internal Medicine* 170:1293-301.

James, W., 2008. WHO recognition of the global obesity epidemic. *International Journal of Obesity* 32 (Supplement 7):S120-6.

Javed, A., et al. 2013. Female athlete triad and its components: Toward improved screening and management. *Mayo Clinic Proceedings* 88 (9):996-1009.

Jensen, M. D., et al. 2014. 2013 AHA/ACC/TOS guideline for the management of overweight and obesity in adults. A report of the American College of Cardiology/American Heart Association Task Force on Practice Guidelines and The Obesity Society. *Circulation* 129:S102-38.

Jeremic, N., et al. 2017. Browning of white fat: Novel insight into factors, mechanisms, and therapeutics. *Journal of Cellular Physiology* 232:61-68.

Johansson, K., et al. 2014. Effects of anti-obesity drugs, diet, and exercise on weight loss maintenance after a very-low-calorie diet or low-calorie diet: A systematic review and meta-analysis of randomized controlled trials. *American Journal of Clinical Nutrition* 99:14-23.

Kacířová, M., et al. 2020. Inflammation: Major denominator of obesity, Type 2 diabetes and Alzheimer's disease-like pathology? *Clinical Science (London)* 134(5):547-70. doi: 10.1042/CS20191313. PMID: 32167154.

Kälin, S., et al. 2015. Hypothalamic innate immune reaction in obesity. *Nature Reviews. Endocrinology* 11(6):339-51.

Kansra, A. R., et al. 2021. Childhood and adolescent obesity: A review. *Frontiers in Pediatrics* 8:581461. doi: 10.3389/fped.2020.581461. PMID: 33511092; PMCID: PMC7835259.

Keel, P. K., and Striegel-Moore, R. H. 2009. The validity and clinical utility of purging disorder. *International Journal of Eating Disorders* 42:706-19.

Keller, A., and Bucher Della Torre, S. 2015. Sugar-sweetened beverages and obesity among children and adolescents: A review of systematic literature reviews. *Childhood Obesity* 11(4):338-46.

Kenchaiah, S., et al. 2009. Body mass index and vigorous physical activity and the risk of heart failure among men. *Circulation* 119:44-52.

Kim, Y., and Park, H. 2013. Does regular exercise without weight loss reduce insulin resistance in children and adolescents? *International Journal of Endocrinology* 2013:402592. doi: 10.1155/2013/402592.

Klip, A., et al. 2019. Thirty sweet years of GLUT4. *Journal of Biological Chemistry* 294(30):11369-81. doi: 10.1074/jbc.REV119.008351.

Knapp, J., et al. 2014. Eating disorders in female athletes: Use of screening tools.

Current Sports Medicine Reports 13 (4):214-18.

Kohut, T., et al. 2019. Update on childhood/adolescent obesity and its sequela. *Current Opinions in Pediatrics* 31(5):645-53. doi: 10.1097/MOP.0000000000000786. PMID: 31145127.

Koral, J., and Dosseville, F. 2009. Combination of gradual and rapid weight loss: Effects on physical performance and psychological state of elite judo athletes. *Journal of Sports Sciences* 27:115-20.

Kral, T., and Rauh, E. 2010. Eating behaviors of children in the context of their family environment. *Physiology & Behavior* 100:567-73.

Kramer, C. K., et al. 2013. Are metabolically healthy overweight and obesity benign conditions? A systematic review and meta-analysis. *Annals of Internal Medicine* 159 (11):758-69.

Krop, E. M., et al. 2018. Influence of oral processing on appetite and food intake—A systematic review and meta-analysis. *Appetite* 125:253-69. doi: 10.1016/j.appet.2018.01.018.

Kushner, R. F. 2014. Weight loss strategies for treatment of obesity. *Progress in Cardiovascular Diseases* 56(4): 465-72.

Lachat, C. 2012. Eating out of home and its association with dietary intake: A systematic review of the evidence. *Obesity Research* 13 (4):329-46.

Lara, J. et al. 2014. Accuracy of aggregate 2- and 3-component models of body composition relative to 4-component for the measurement of changes in fat mass during weight loss in overweight and obese subjects. *Applied Physiology Nutrition and Metabolism* 39 (8):871-79.

Lawrence, J. M., et al. 2021. SEARCH for diabetes in youth study group. Trends in prevalence of type 1 and type 2 diabetes in children and adolescents in the US, 2001-2017. *Journal of the American Medical Association* 326(8):717-27. doi: 10.1001/jama.2021.11165.

Lee, J. E., et al. 2019. Transcriptional and epigenomic regulation of adipogenesis. *Molecular and Cellular Biology* 39(11):e00601-18. doi: 10.1128/MCB.00601-18. PMID: 30936246.

Lee, M. J., et al. 2013. Adipose tissue heterogeneity: Implications of depot differences in adipose tissue for obesity complications. *Molecular Aspects of Medicine* 34 (1):1-11.

Lee, R. E. et al. 2014. Obesogenic and youth oriented restaurant marketing in public housing neighborhoods. *American Journal of Health Behavior* 38(2):218-24.

Lee, Y. 2009. The role of genes in the current obesity epidemic. *Annals Academy of Medicine Singapore* 38:45-47.

Liberali, R., et al. 2020. Dietary patterns and childhood obesity risk: A systematic review. *Childhood Obesity* 16(2):70-85. doi: 10.1089/chi.2019.0059.

Liu, C., et al. 2013. FTO gene variant and risk of overweight and obesity among children and adolescents: A systematic review and meta-analysis. *Public Library of Science One* 8 (11):e82133. doi: 10,1371/journal.pone.0082133.

Loef, M., and Walach, H. 2013. Midlife obesity and dementia: Meta-analysis and adjusted forecast of dementia prevalence in the United States and China. *Obesity* 21 (1):E51-E55. doi: 10.1002/oby.2037.

Loenneke, J. P., et al. 2011. Validity of the current NCAA minimum weight protocol: A brief review. *Annals of Nutrition and Metabolism* 58:245-49.

Lohman, T., and Going, S. 2006. Body composition assessment for development of an international growth standard for preadolescent and adolescent children. *Food and Nutrition Bulletin* 27:S314-25.

Lumey, L. H., et al. 2007. Cohort profile: The Dutch Hunger Winter families study. *International Journal of Epidemiology* 36(6):1196-204. doi: 10.1093/ije/dym126.

Lutter, M., and Nestler, E. 2009. Homeostatic and hedonic signals interact in the regulation of food intake. *Journal of Nutrition* 139:629-32.

Ma, C., et al. 2017. Effects of weight loss interventions for adults who are obese on mortality, cardiovascular disease, and cancer: Systematic review and meta-analysis. *British Medical Journal* 359:j4849. doi: 10.1136/bmj.j4849. PMID: 29138133; PMCID: PMC5682593.

Malik, V. S., et al. 2013. Sugar-sweetened beverages and weight gain in children and adults: A systematic review and meta-analysis. *American Journal of Clinical Nutrition* 98:1084-102.

Malina, R. M., et al. 2013. Role of intensive training in the growth and maturation of artistic gymnasts. *Sports Medicine* 43:783-802.

Martin, C. B., et al. 2018. Attempts to lose weight among adults in the United States, 2013-2016. NCHS Data Brief, no 313. Hyattsville, MD: National Center for Health Statistics.

Matafome, P., and Seiça R. 2017. The role of brain in energy balance. *Advances in Neurobiology* 19:33-48. doi: 10.1007/978-3-319-63260-5_2. PMID: 28933060.

Mattson, M. 2010. Perspective: Does brown fat protect against diseases of aging? *Ageing Research Reviews* 9:69-76.

Mayer, J. 1996. Glucostatic mechanism of regulation of food intake. *Obesity Research* 4(5):493-96.

McElroy, S. L., et al. 2015. Psychopharmacologic treatment of eating disorders: Emerging findings. *Current Psychiatry Reports* 17:35. doi 10.1007/s11920-015-0573-1.

McNeil, J., et al. 2013. Inadequate sleep as a contributor to obesity and type 2 diabetes. *Canadian Journal of Diabetes* 37 (2):103-8.

Melby, C. L., et al. 2017. Attenuating the biologic drive for weight regain following weight loss: Must what goes down always go back up? *Nutrients* 9(5):468. doi: 10.3390/nu9050468. PMID: 28481261; PMCID: PMC5452198.

Messina, C., et al. 2020. Body composition with dual energy X-ray absorptiometry: From basics to new tools. *Quantitative Imaging in Medicine and Surgery* 10(8):1687-98. doi: 10.21037/qims.2020.03.02. PMID: 32742961; PMCID: PMC7378094.

Moellering, D. R., and Smith, Jr., D. L. 2012. Ambient temperature and obesity. *Current Obesity Reports* 1(1):26-34. doi: 10.1007/s13679-011-0002-7. PMID: 24707450; PMCID: PMC3975627.

Montani, J. P., et al. 2015. Dieting and weight cycling as risk factors for cardiometabolic diseases: Who is really at risk? *Obesity Reviews* S1:7-18. doi: 10.1111/obr.12251.

Moon, J. R., et al. 2008. Validity of the BOD POD for assessing body composition in athletic high school boys. *Journal of Strength and Conditioning Research* 22(1):263-68.

Moon, J. R., et al. 2007. Percent body fat estimations in college women using field and laboratory methods: A three-compartment model approach. *Journal of the International Society of Sports Nutrition* 7:4-16.

Moon, J. R. 2013. Body composition in athletes and sports nutrition: and examination of the bioimpedance analysis technique. *European Journal of Clinical Nutrition* 67:554-59.

Moreno, L. A. et al. 2010. Trends of dietary habits in adolescents. *Critical Reviews in Food Science and Nutrition* 50(2):106-112.

Mountjoy, M., et al. 2014. The IOC consensus statement: Beyond the female athlete triad—Relative energy deficiency in sport (RED-S). *British Journal of Sports Medicine* 48:491-97.

Müller, M. J., et al. 2016. Application of standards and models in body composition analysis. *Proceedings of the Nutrition Society* 75:181-87.

Murphy, N. E., et al. 2018. Threshold of energy deficit and lower-body performance declines in military personnel: A meta-regression. *Sports Medicine* 48(9):2169-78. doi: 10.1007/s40279-018-0945-x. PMID: 29949108.

Muscogiuri, G., et al. 2019. The management of very low-calorie ketogenic diet in obesity outpatient clinic: A practical guide. *Journal of Translational Medicine* 17(1):356. doi: 10.1186/s12967-019-2104-z. PMID: 31665015; PMCID: PMC6820992.

Nago, E. S., et al. 2014. Association of out-of-home eating with anthropometric changes: A systematic review of prospective studies. *Critical Reviews in Food Science and Nutrition* 54 (9):1103-16.

Nana, A., et al. 2015. Methodology review: Using dual-energy X-ray absorptiometry (DXA) for the assessment of body composition in athletes and active people. *International Journal of Sport Nutrition and Exercise Metabolism* 25(2):198-215. doi: 10.1123/ijsnem.2013-0228.

Narayanaswami, V., and Dwoskin, L. P. 2017. Obesity: Current and potential pharmacotherapeutics and targets. *Pharmacology and Therapeutics* 170:116-47.

National Center for Health Statistics. Health, United States, 2015: With Special Feature on Racial and Ethnic Health Disparities. Hyattsville, MD. 2016. Table 53, p. 200.

National Institutes of Health. NHLBI. North American Association for the Study of Obesity. 2000. *The Practical Guide: Identification, Evaluation, and Treatment of Overweight and Obesity in Adults.* NIH Publication Number 00-4084.

National Task Force on the Prevention and Treatment of Obesity. 1994. Weight cycling. *Journal of the American Medical Association* 272:1196-202.

Naukkarinen, J., et al. 2012. Causes and consequences of obesity: The contribution of recent twin studies. *International Journal of Obesity* 36:1017-24.

Ng, A.C.T., et al. 2021. Diabesity: The combined burden of obesity and diabetes on heart disease and the role of imaging. *Nature Reviews Cardiology* 18(4):291-304. doi: 10.1038/s41569-020-00465-5.

Niemeier, B. S., et al. 2012. Parent participation in weight-related health interventions for children and adolescents: A systematic review and meta-analysis. *Preventive Medicine* 55:3-13.

Normandin, E., et al. 2017. Effect of Resistance Training and Caloric Restriction on the Metabolic Syndrome. *Medicine and Science in Sports and Exercise* 49(3):413-419.

O'Connor, D., et al. 2010. Generalized equations for estimating DXA percent body fat of diverse young women and men: The TIGER study. *Medicine & Science in Sports & Exercise* 42:1959-65.

Ode, J., et al. 2007. Body mass index as a predictor of percent fat in college athletes and nonathletes. *Medicine & Science in Sports & Exercise* 39:403-9.

Ogilvie, R. P., and Patel, S. R. 2017. The epidemiology of sleep and obesity. *Sleep Health* 3(5):383-88. doi: 10.1016/j.sleh.2017.07.013.

Okorodudu, D. O., et al. 2010. Diagnostic performance of body mass index to identify obesity as defined by body adiposity: A systematic review and meta-analysis. *International Journal of Obesity (London)* 34(5):791-9. doi: 10.1038/ijo.2010.5.

Oppliger, R., et al. 2006. NCAA rule change improves weight loss among National Championship wrestlers. *Medicine & Science in Sports & Exercise* 38:963-70.

Owen, C. G., et al. 2009. Is body mass index before middle age related to coronary heart disease risk in later life? Evidence from observational studies. *International Journal of Obesity* 33 (8):866-77.

Owen, N., et al. 2010. Too much sitting: The population health science of sedentary behavior. *Exercise and Sport Sciences Reviews* 38:105-13.

Pate, R. R., et al. 2013. Factors associated with development of excessive fatness in children and adolescents: A review of prospective studies. *Obesity Reviews* 14 (8):645-58.

Pearcey, S. M., and de Castro, J. M. 2002. Food intake and meal patterns of weight-stable and weight-gaining persons. *American Journal of Clinical Nutrition* 76(1):107-12.

Pedditizi, E., et al. 2016. The risk of overweight/obesity in mid-life and late life for the development of dementia: A systematic review and meta-analysis of longitudinal studies. *Age and Ageing* 45(1):14-21.

Pereira, M., et al. 2005. Fast-food habits, weight gain, and insulin resistance (the CARDIA study): 15-year prospective analysis. *The Lancet* 365:36-42.

Perry, K., et al. 2015. The role of social networks in the development of overweight and obesity among adults: A scoping review. *BioMed Central Public Health* 15:996. doi 10.1186/s12889-015-2314-0.

Peters, A. 2011. The selfish brain: Competition for energy resources. *American Journal of Human Biology* 23:29-34.

Pfisterer, N., et al. 2020. Severe acquired hypokalemic paralysis in a bodybuilder after self-medication with triamterene/hydrochlorothiazide. *Clinical Journal of Sport Medicine* 30(5):e172-e174. doi: 10.1097/JSM.0000000000000763. PMID: 31770156.

Phillips, C. M. 2017. Metabolically healthy obesity across the life course: Epidemiology, determinants, and implications. *Annals of the New York Academy of Sciences* 1391(1):85-100.

Pi-Sunyer, F. 2006. The relation of adipose tissue to cardiometabolic risk. *Clinical Cornerstone* 8 (Supplement 4):S14-23.

Pi-Sunyer, F. 2007. How effective are lifestyle changes in the prevention of type 2 diabetes mellitus? *Nutrition Reviews* 65:101-10.

Ponterio, E., and Gnessi, L. 2015. Adenovirus 36 and obesity: An overview. *Viruses* 7:3719-40. doi:10.3390/v7072787.

Pourhassan, M., et al. 2013. Impact of body-composition methodology on the composition of weight loss and weight gain. *European Journal of Clinical Nutrition* 67:446-54.

Ragelienė, T. and Grønhøj, A. 2020. The influence of peers' and siblings' on children's and adolescents' healthy eating behavior. A systematic literature review. *Appetite* 2020 148:104592. doi: 10.1016/j.appet.2020.104592.

Rask-Andersen, M., et al. 2010. Molecular mechanisms underlying anorexia nervosa: Focus on human gene association studies and systems controlling food intake. *Brain Research Reviews* 62:147-64.

Reaven, G. 2006. Metabolic syndrome: Definition, relationship to insulin resistance, and clinical utility. In *Modern Nutrition in Health and Disease,* ed. M. Shils et al. Philadelphia: Lippincott Williams & Wilkins.

Reaven, G. 2006. The metabolic syndrome: Is this diagnosis necessary? *American Journal of Clinical Nutrition* 83:1237-47.

Rebello, C. J., and Greenway, F. L. 2020. Obesity medications in development.

Renehan, A. G., et al. 2010. Interpreting the epidemiological evidence linking obesity and cancer: A framework for population-attributable risk estimations in Europe. *European Journal of Cancer* 46:2581–92.

Ricketts, T. A., et al. 2016. Addition of cardiorespiratory fitness within an obesity risk classification model identifies men at increased risk of all-cause mortality. *American Journal of Medicine* 129(5):536.e13–20. doi: 10.1016/j.amjmed.2015.11.015.

Rodriguez, A., et al. 2017. Crosstalk between adipokines and myokines in fat browning. *Acta Physiologica* 219(2):362–81.

Rodriguez-Sanchez, N., and Galloway, S. D. R. 2015. Errors in dual energy x-ray absorptiometry of body composition induced by hypohydration. *International Journal of Sports Nutrition and Exercise Metabolism* 25(1):60–68.

Rohde, K., et al. 2019. Genetics and epigenetics in obesity. *Metabolism Clinical and Experimental* 92:37–50. doi: 10.1016/j.metabol.2018.10.007.

Romero-Corral, A., et al. 2010. Modest visceral fat gain causes endothelial dysfunction in healthy humans. *Journal of the American College of Cardiology* 56:662–66.

Ronit, K., and Jensen, J. D. 2014. Obesity and industry self-regulation of food and beverage marketing: A literature review. *European Journal of Clinical Nutrition* 68(7):753–59.

Rounsefell, K., et al. 2020. Social media, body image and food choices in healthy young adults: A mixed methods systematic review. *Nutrition and Dietetics* 77(1):19–40. doi: 10.1111/1747-0080.12581. PMID: 31583837; PMCID: PMC7384161.

Saklayen, M. G. 2018. The global epidemic of the metabolic syndrome. *Current Hypertension Reports* 20(2):12. doi: 10.1007/s11906-018-0812-z. PMID: 29480368; PMCID: PMC5866840.

Salvy, S. J., et al. 2011. Influence of parents and friends on children's and adolescents' food intake and food selection. *American Journal of Clinical Nutrition* 93 (1): 87–92.

Salvy, S., et al. 2009. The presence of friends increases food intake in youth. *American Journal of Clinical Nutrition* 90:282–7.

Sanchez-Vaznaugh, E. V., et al. 2019. Changes in fast food outlet availability near schools: Unequal patterns by income, race/ethnicity, and urbanicity. *American Journal of Preventive Medicine* 57(3):338–45. doi: 10.1016/j.amepre.2019.04.023.

Saunders, T. J., et al. 2014. Sedentary behaviours as an emerging risk factor for cardiometabolic doseases in children and youth. *Canadian Journal of Diabetes* 38:53–61.

Schulz, L. O., and Chaudhari, L. S. 2015. High-risk populations: The Pimas of Arizona and Mexico. *Current Obesity Reports* 4:92–98.

Schuna, J. M., et al. 2013. The evaluation of a circumference-based prediction equation to assess body composition changes in men. *International Journal of Exercise Science* 6(3):188–98.

Sclafani, A., and Ackroff, K. 2012. Role of gut nutrient sensing in stimulating appetite and conditioning food preferences. *American Journal of Physiology Regulatory and Integrative Physiology* 302 (10):R1119–33.

Sepa-Kishi, D. M., and Ceddia, R. B. 2016. Exercise-mediated effects on white and brown adipose tissue plasticity and metabolism. *Exercise and Sport Sciences Reviews* 44(1):37–44.

Shetty, P. S., and James, W. P. T. 1994. Body mass index—A measure of chronic energy deficiency in adults. Food and Agriculture Organization of the United Nations.

Simmonds, M., et al. 2016. Predicting adult obesity from childhood obesity: A systematic review and meta-analysis. *Obesity Reviews* 17(2):95–107.

Siri, W. E. 1961. Body composition from fluid space and density. In *Techniques for Measuring Body Composition,* ed. J. Brożek and A. Hanschel. Washington, DC: National Academy of Science.

Smith, J. D., et al. 2020. Prevention and management of childhood obesity and its psychological and health comorbidities. *Annual Review of Clinical Psychology* 16:351–78. doi: 10.1146/annurev-clinpsy-100219-060201.

Sodersen, P., et al. 2006. Understanding eating disorders. *Hormones and Behavior* 50:572–78.

Speakman, J. R., et al. 2011. Set points, settling points and some alternative models: Theoretical options to understand how genes and environments combine to regulate body adiposity. *Disease Models and Mechanisms* 4:733–45.

Standing Committee on Childhood Obesity Prevention; Food and Nutrition Board; Institute of Medicine, 2013. *Challenges and opportunities for change in food marketing to children and youth: Workshop summary.* Washington, DC: National Academies Press (US), May 14. PMID: 24872975.

Stark, R., et al. 2015. Hypothalamic carnitine metabolism integrates nutrient and hormonal feedback to regulate energy homeostasis. *Molecular and Cellular Endocrinology* 418(1):9–16.

Steenhuis, I., and Poelman, M. 2017. Portion size: Latest developments and interventions. *Current Obesity Reports* 6:10–17.

Steffes, G. D., et al. 2013. Prevalence of metabolic syndrome risk factors in high school and NCAA Division I football players. *Journal of Strength and Conditioning Research* 27 (7):1749–57.

Stein, D., et al. 2014. Psychosocial perspectives and the issue of prevention in childhood obesity. *Frontiers in Public Health* 2:104. doi: 10.3389/fpubh.2014.00104.

Stobbe, M. 2010. Obesity's impact on costs bigger than thought. *Associated Press,* October 17.

Strohaker, K., and McFarlin, B. K. 2010. Influence of obesity, physical inactivity, and weight cycling on chronic inflammation. *Frontiers in Bioscience* (Elite Edition) 2:98–104.

Stunkard, A., et al. 1990. The body-mass index of twins who have been reared apart. *New England Journal of Medicine* 322:1483–87.

Sudi, K., et al. 2004. Anorexia athletica. *Nutrition* 20:657–61.

Sullivan, E., et al. 2011. Perinatal exposure to high-fat diet programs energy balance, metabolism and behavior in adulthood. *Neuroendocrinology* 93:1–8.

Sundgot-Borgen, J., and Torstveit, M. K. 2004. Prevalence of eating disorders in elite athletes is higher than in the general population. *Clinical Journal of Sports Medicine* 14:25–32.

Sundgot-Borgen, J., et al. 2013. How to minimize the health risks to athletes who compete in weight-sensitive sports review and position statement on behalf of the ad hoc research working group on body composition, health and performance, under the auspices of the IOC Medical Commission. *British Journal of Sports Medicine* 47:1012–22.

Suokas, J. T., et al. 2013. Mortality in eating disorders: A follow-up study of adult eating disorder patients treated in tertiary care, 1995–2010. *Psychiatry Research* 210(3):1101–6. doi: 10.1016/j.psychres.2013.07.042.

Swinburn, B., and Shelly, A. 2008. Effects of TV time and other sedentary pursuits. *International Journal of Obesity* 32 (Supplement 7):S132–6.

Tang, L., et al. 2014. Meta-analysis between 18 candidate genetic markers for overweight/obesity. *Diagnostic Pathology* 9:56.

Tatangelo, G., et al. 2016. A systematic review of body dissatisfaction and sociocultural messages related to the body among preschool children. *Body Image* 18:86-95.

Teske, J. A., et al. 2008. Neuropeptidergic mediators of spontaneous physical activity and non-exercise activity thermogenesis. *Neuroendocrinology* 87(2):71-90.

Tian, J., et al. 2015. The association between quitting smoking and weight gain: A systemic review and meta-analysis of prospective cohort studies. *Obesity Reviews* 16(10):883-901.

Torstveit, M., and Sundgot-Borgen, J. 2005. The female athlete triad exists in both elite athletes and controls. *Medicine & Science in Sports & Exercise* 37:1449-59.

Torstveit, M., and Sundgot-Borgen, J. 2005. Participation in leanness sports but not training volume is associated with menstrual dysfunction: A national survey of 1276 elite athletes and controls. *British Journal of Sports Medicine* 39:141-47.

Traversy, G., and Chaput, J. P. 2015. Alcohol consumption and obesity: An update. *Current Obesity Reports* 4(1):122-30.

Tremblay, M. S., et al. 2011. Systematic review of sedentary behaviour and health indicators in school-aged children and youth. *International Journal of Behavioral Nutrition and Physical Activity* 8:98. doi: 10.1186/1479-5868-8-98.

Tsatsoulis, A. and Paschou, S. A. 2020. Metabolically healthy obesity: Criteria, epidemiology, controversies, and consequences. *Current Obesity Reports* 9(2):109-20. doi: 10.1007/s13679-020-00375-0. PMID: 32301039.

Tucker, L. A., et al. 2014. Test-retest reliability of the Bod Pod: The effect of multiple assessments. *Perceptual and Motor Skills* 118(2):563-70.

Udo, T. and Grilo, C. M. 2018. Prevalence and correlates of DSM-5-defined eating disorders in a nationally representative sample of U.S. adults. *Biological Psychiatry* 84(5):345-54. doi: 10.1016/j.biopsych.2018.03.014. PMID: 29859631; PMCID: PMC6097933.

Uyar, G.O. and Sanlier, N. 2019. Association of adipokines and insulin, which have a role in obesity, with colorectal cancer. *Eurasian Journal of Medicine* 51(2):191-95. doi: 10.5152/eurasianjmed.2018.18089. PMID: 31258362; PMCID: PMC6592445.

Vadeboncoeur, C., et al. 2015. A meta-analysis of weight gain in first year university students: Is freshman 15 a myth? *BioMed Central Obesity* 2:22. doi 10.1186/s40608-015-0051-7.

Ventura, A. K. 2017. Does breastfeeding shape food preferences? Links to obesity. *Annals of Nutrition and Metabolism* 70(Suppl 3):8-15. doi: 10.1159/000478757.

Villablanca, P. A., et al. 2015. Nonexercise activity thermogenesis in obesity management. *Mayo Clinic Proceedings* 90(4):509-19.

Villanueva-Millán, M. J., et al. 2015. Gut microbiota: A key player in health and disease. A review focused on obesity. *Journal of Physiology and Biochemistry* 71(3):509-25.

Wagner, D. R., et al. 2016. Validity and reliability of A-Mode ultrasound for body composition assessment of NCAA Division I athletes. *Public Library of Science ONE* 11(4): e0153146. doi:10.1371/journal.pone.0153146.

Wagner, D., and Heyward, V. 1999. Techniques of body composition assessment: A review of laboratory and field methods. *Research Quarterly for Exercise and Sport* 70:135-49.

Wagner, A. J., et al. 2016. The diagnostic accuracy of screening tools to detect eating disorders in female athletes. *Journal of Sport Rehabilitation* 25(4):395-98. doi: 10.1123/jsr.2014-0337.

Walker, T. B., et al. 2014. Lessons from the war on dietary fat. *Journal of the American College of Nutrition* 33 (4):347-51.

Wallace, C. W. and Fordahl, S.C. 2021. Obesity and dietary fat influence dopamine neurotransmission: Exploring the convergence of metabolic state, physiological stress, and inflammation on dopaminergic control of food intake. *Nutrition Research Review* 28:1-16. doi: 10.1017/S0954422421000196.

Wang, L., et al. 1992. The five-level model: A new approach to organizing body-composition research. *American Journal of Clinical Nutrition* 56:19-28.

Wang, L, et al. 2021. Trends in consumption of ultra-processed foods among US youths aged 2-19 years, 1999-2018. *Journal of the American Medical Association* 326(6):519-30. doi: 10.1001/jama.2021.10238. PMID: 34374722; PMCID: PMC8356071.

Wang, Y., et al. 2005. Comparison of abdominal adiposity and overall obesity in predicting risk of type 2 diabetes among men. *American Journal of Clinical Nutrition* 81:555-63.

Wang, Z., et al. 1998. Six-compartment body composition model: Inter-method comparisons of total body fat measurement. *International Journal of Obesity and Related Metabolic Disorders* 22:329-37.

Ward, L. C., et al. 2019. Bioelectrical impedance analysis for body composition assessment: Reflections on accuracy, clinical utility, and standardisation. *European Journal of Clinical Nutrition* 73(2):194-99. doi: 10.1038/s41430-018-0335-3.

Waterland, R. 2014. Epigenetic mechanisms affecting regulation of energy balance: Many questions, few answers. *Annual Review of Nutrition* 34:337-55.

Webster, B., and Barr, S. 1993. Body composition analysis of female adolescent athletes: Comparing six regression equations. *Medicine & Science in Sports & Exercise* 25:648-53.

Weihrauch-Blüher, S. and Wiegand, S. 2018. Risk factors and implications of childhood obesity. *Current Obesity Reports* 7(4):254-59. doi: 10.1007/s13679-018-0320-0. PMID: 30315490.

Weinstein, A., et al. 2008. The joint effects of physical activity and body mass index on coronary heart disease risk in women. *Archives of Internal Medicine* 168:884-90.

Welsh, J. A., et al. 2013. The sugar-sweetened beverage wars: Public health and the role of the beverage industry. *Current Opinion in Endocrinology, Diabetes, and Obesity* 20 (5):401-6.

Weyers, A. M., et al. 2002. Comparison of methods for assessing body composition changes during weight loss. *Medicine and Science in Sports and Exercise* 34(3):497-502.

Williams, P. T. 2001. Health effects resulting from exercise versus those from body fat loss. *Medicine and Science in Sports and Exercise* 33(6):S611-S621.

Woods, S. C. 2013. Metabolic signals and food intake. Forty years of progress. *Appetite* 71:440-44.

Yanagi, S., et al. 2018. The homeostatic force of ghrelin. *Cell Metabolism* 27(4):786-804. doi: 10.1016/j.cmet.2018.02.008.

Yard, E., and Comstock, D. 2011. Injury patterns by body mass index in US high school athletes. *Journal of Physical Activity and Health* 8:182-91.

Yeo, G. S. H. 2014. The role of the FTO (fat mass and obesity related) locus in regulating body size and composition. *Molecular and Cellular Endocrinology.* doi: 10.1016/j.mce.2014.09.012.

Zeller, M., et al. 2008. Relation between body mass index, waist circumference, and death after acute myocardial infarction. *Circulation* 118:482–90.

Zhang, C., et al. 2008. Abdominal obesity and the risk of all-cause, cardiovascular, and cancer mortality: Sixteen years of follow-up in US women. *Circulation* 117:1658–67.

Zhang, J., et al. 2005. Obestatin, a peptide encoded by the ghrelin gene, opposes effects on food intake. *Science* 310:996–99.

Zhang, Y., et al. 1994. Positional cloning of the mouse obese gene and its human homologue. *Nature* 372:425–32.

Zheng, R., et al. 2016. The long-term prognosis of cardiovascular disease and all-cause mortality for metabolically healthy obesity: A systematic review and meta-analysis. *Journal of Epidemiology and Community Health* 70(10):1024–31.

Zimberg, I. Z., et al. 2012. Short sleep duration and obesity: Mechanisms and future perspectives. *Cell Biochemistry and Function* 30 (6):524–29.

KEY TERMS

behavioral group therapy
calorie (kcal) deficit
cardiorespiratory fitness
energy availability
high-intensity interval training (HIIT)
intermittent fasting
intuitive eating
long-haul concept
low energy availability
mindful eating
mindfulness
motivational interviewing
over-use injuries

sarcopenic obesity
ultra-processed foods (UPFs)
USDA 85–15 guide
weight bias
weight cycling
weight-loss programs

Weight Maintenance and Loss through Proper Nutrition and Exercise

CHAPTER ELEVEN

Mara Zemgaliete/Magone/123RF

LEARNING OUTCOMES

After studying this chapter, you should be able to:

1. Describe rates of overweight and obesity in the United States and across the globe and summarize key lifestyle factors contributing to rising rates overweight of obesity.
2. Define weight bias and describe strategies health-care providers can employ to minimize weight bias in health-care facilities.
3. Explain how energy needs are determined for individuals with overweight and obesity with the goal of losing weight.
4. Summarize generally dietary recommendations to support weight loss.
5. Describe mindfulness and general characteristics of mindful eating. Summarize how mindfulness practices may support both mental and physical health.
6. Provide strategies appropriate for individuals with overweight or obesity to increase physical activity levels to support weight loss and prevention of weight regain.
7. Explain the characteristics of safe and effective weight-loss programs and describe examples of evidence-based weight-loss programs.
8. Explain the importance of exercise and pre- and post operative physical activity recommendations for individuals with obesity undergoing bariatric surgery.
9. Describe potential adverse effects of weight loss and weight cycling in athletes, including low energy availability and long-term effects related to obesity risk and health.

Introduction

Mara Zemgaliete/Magone/123RF

As described in chapter 10, overweight and obesity are significant health concerns across the globe. Worldwide, in 2020, 39 million children under 5 years of age and 340 million children 5–19 years of age had either overweight or obesity. As well, nearly 2 billion adults across the globe had obesity. In the United States, the Centers for Disease Control and Prevention (CDC) reports that 42 percent of adults had obesity between 2017 and 2020. Obesity rates in the US increased over 10 percent in the past 20 years, with rates of severe obesity increasing from 4.7 to 9.2 percent between 1999 and 2000 and 2017 and 2020.

Obesity rates are influenced by a variety of factors, including age, sex, ethnicity, race, socioeconomic status, and where a person lives. Figure 11.1 shows a map of adult obesity rates in the United States based on the state where a person lives. In 2020, Colorado, the District of Columbia, Hawaii, and Massachusetts had the lowest rates of obesity compared to Alabama, Louisiana, Mississippi, and West Virginia, which had the highest rates of obesity. The CDC also provides

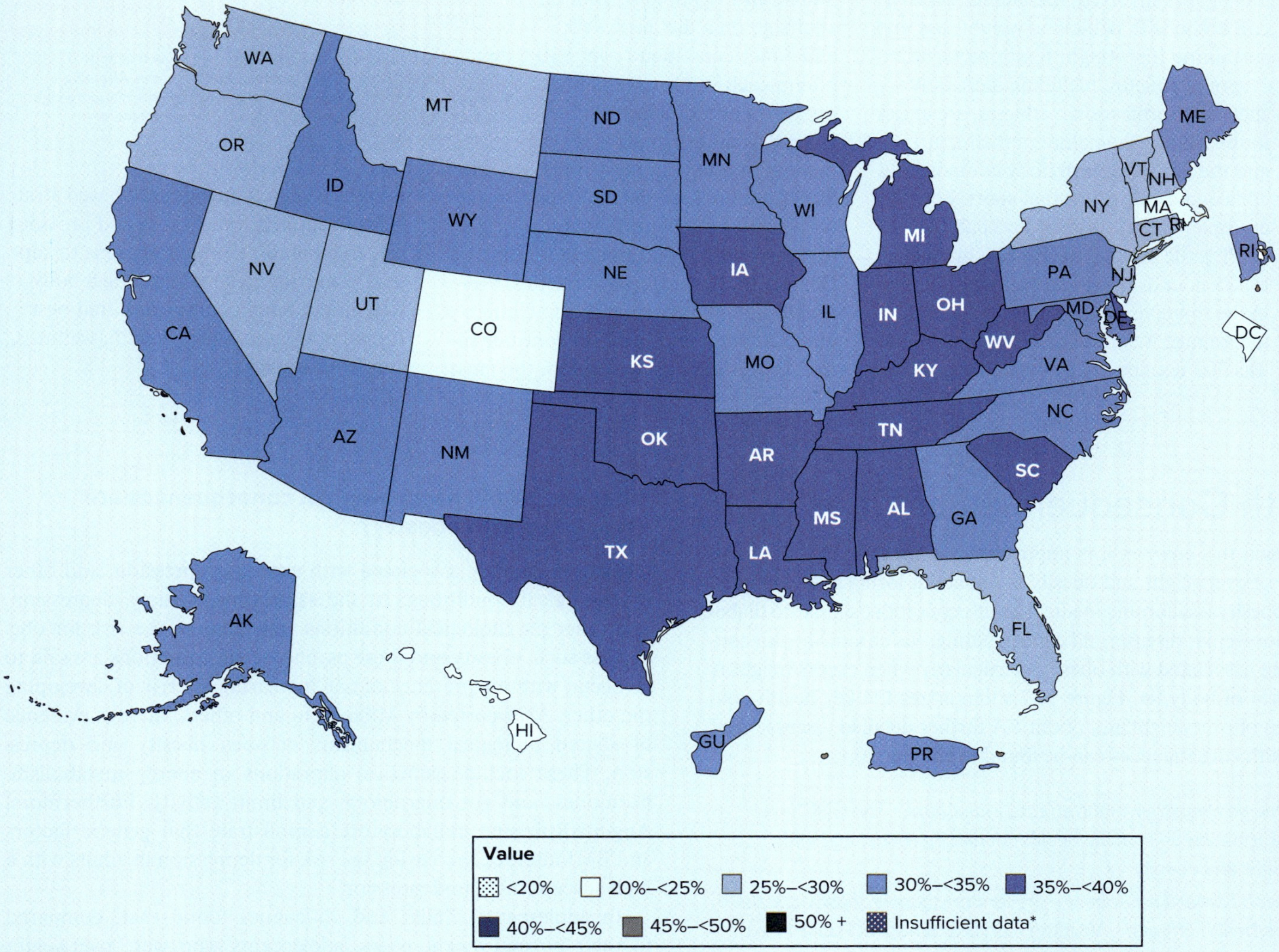

FIGURE 11.1 This map of the United States shows the self-reported obesity rate based on the Behavioral Risk Factor Surveillance System (BRFSS), 2020.
Source: Behavioral Risk factors Survivallance System. U.S. Department of Health & Human Services.

maps showing obesity rates by race and ethnicity, age, sex, education level, and more. Visit www.cdc.gov/obesity/data/prevalence-maps.html to access the adult obesity prevalence maps.

Chapter 10 provided an overview of the complexity of body weight and composition in humans and the numerous factors that influence an individual's body weight and composition. Such factors include biological factors such as genetics and hormone levels, as well as environmental factors, including access to purchase quality foods and beverages and safe places to exercise. Table 11.1 provides a summary of key factors influencing a person's body weight and body composition. As you review this list, keep in mind the complexity of overweight and obesity and, because of the numerous factors involved, the challenges that can come with achieving weight loss and sustaining that weight loss long term.

Some athletes, including incredibly successful professional athletes, are at risk for obesity by the nature of their sport and the body type and body composition considered "ideal" for that sport. For example, in the National Football League (NFL), the average weight of a lineman is 325 pounds. In student-athletes playing in the top division of collegiate football, the average weight of an offensive lineman was 302 pounds in 2014 and 309 pounds in 2021. Across sports, athletes who gain weight early in life are more likely to experience adverse health effects, such as heart disease and sleep apnea, later in life. For many athletes with a higher body mass index (BMI) during adolescence and young adulthood, losing weight and increasing lean body mass (LBM) following competition can be challenging.

Given the high rates of overweight and obesity, the diet industry is a multi-billion-dollar industry. According to recent surveys, 45 percent of people across the globe report that they are currently trying to lose weight. In the United States, an estimated 45 million adults go on a diet each year and Americans spend over $30 billion each year on weight-loss programs and products. On any given day, approximately 17 percent of adults in the United States are on a special diet (see www.cdc.gov/nchs/products/databriefs/db389.htm for more details). Despite these staggering statistics, overweight and obesity rates in the United States and across the world continue to rise. In this chapter we explore the most current research on evidence-based strategies to support weight loss and provide research-based recommendations to support achieving and maintaining a body weight and body composition that best supports an individual's health, wellness, and fitness goals.

TABLE 11.1 Key factors influencing body weight and body composition

- Family history and genes
- Biological factors, including certain hormones
- Race or ethnicity
- Age
- Sex
- Eating and physical habits
- Where you live, work, play, and worship
- Family habits and culture
- Not enough sleep
- Certain medical conditions and medications
- Stress

Source: National Institute of Diabetes and Digestive and Kidney Diseases (NIDDK). Updated 2022. Factors Affecting Weight and Health. www.niddk.nih.gov/health-information/weight-management/adult-overweight-obesity/factors-affecting-weight-health

Health Consequences of Obesity

As we begin this chapter, it is important to emphasize that the management of overweight and obesity is not about losing a few pounds. Rather, obesity is a chronic medical health condition, similar to diabetes, cardiovascular disease, and cancer. Numerous chronic health conditions are associated with obesity, particularly when excess weight is in the form of body fat. Figure 11.2 summarizes the key health consequences of overweight and obesity. A further detailed discussion of these health consequences can be found in chapter 10.

> www.cdc.gov/healthyweight/effects/index.html   The CDC provides information on specific health conditions associated with overweight and obesity.
>
> www.hsph.harvard.edu/obesity-prevention-source/obesity-consequences/health-effects/   Visit the Harvard School of Public Health website to learn more about the relationship between body weight and physical and mental health conditions.

What social and psychological consequences are associated with obesity?

Obesity is strongly associated with anxiety, depression, and other psychological conditions. At the same time, anxiety, depression, and other psychological conditions can increase the risk for obesity. As such, obesity and these psychological conditions are said to co-occur, with the presence of one increasing the risk of developing the other. As reported by Milaneschi and others, there is evidence of shared biological mechanisms between obesity and depression. These include genetics; alterations in energy metabolism, hormones, and gut microbiota; and brain activity. Furthermore, Anguita-Ruiz and collaborators demonstrate that genetic factors and BMI interact, increasing the risk for depression in adults with a high susceptibility to depression.

In adolescents, Zeiler and colleagues found that, compared to their normal-weight peers, adolescents who were overweight had a greater risk for anxiety and depression. As well, the adolescents who were overweight were at a greater risk for an eating

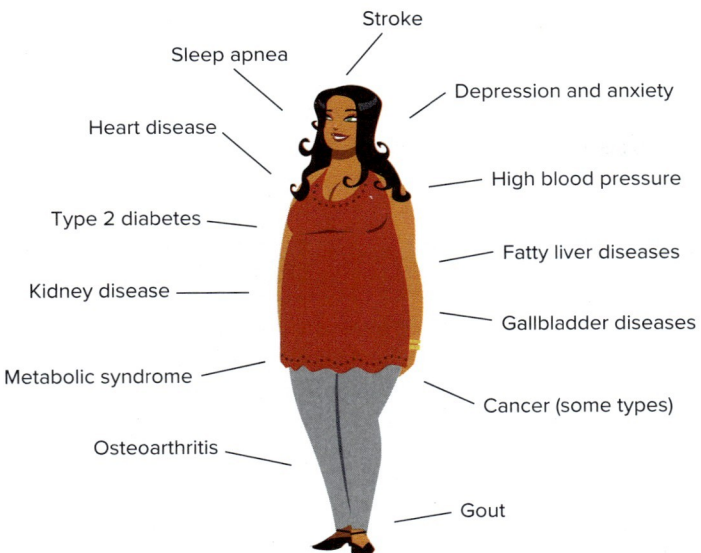

FIGURE 11.2 Overweight and obesity are associated with several chronic health problems.

Source: National Institute of Diabetes and Digestive and Kidney Diseases (NIDDK). Updated 2022. Health Risks of Overweight and Obesity. www.niddk.nih.gov/health-information/weight-management/adult-overweight-obesity/health-risks

disorder and reported lower overall well-being. In a recent meta-analysis, Rao and others reported that adolescents with obesity have a significantly higher risk for major depression compared to adolescents who do not have overweight or obesity. The American Academy of Pediatrics recommends regular screening for depression in all adolescents, including those with obesity.

Prevention and treatment of overweight and obesity can have numerous positive impacts on health and well-being. In addition to reducing the risk for chronic disease, improvements in body weight and body composition can also support a positive quality of life and reduce the risk for serious psychological conditions.

What is weight bias?

The Obesity Action Coalition defines **weight bias** as "negative attitudes, beliefs, judgments, stereotypes, and discriminatory actions aimed at individuals simply because of their weight." Weight bias can occur in individuals of all body shapes and sizes. While weight bias occurs across sectors of society, weight bias among health-care providers significantly impacts the quality of care for patients. Many health-care providers treat their patients with obesity different from patients considered "healthy" by BMI and appearance. This matters because weight bias often results in less time spent with a patient, judgement, and differences in treatment plans. Patients who experience weight bias are less likely to keep up with regular health-care appointments and seek care when symptoms first present. As well, as reported by Phelan and others, individuals with overweight or obesity who have experienced stigmatizing situations with a health-care provider are more likely to switch physicians. All of these impacts of weight stigma in healthcare may contribute to an increased risk for chronic disease, poor quality of life, and increased mortality.

---

www.obesityaction.org/action-through-advocacy/weight-bias/ The Obesity Advocacy Coalition is a nonprofit organization focused on raising awareness of the issues of weight bias and discrimination.

www.obesity.org/wp-content/uploads/2020/05/TOS-Reasons-for-obesity-infographic-2015.pdf The Obesity Society (Professionals Collaborating to Overcome Obesity) shares the most evidence-based information on obesity. This includes an infographic summarizing potential contributors to obesity.

In recent years, the training of physicians, nurses, registered dietitian nutritionists (RDNs), and other health-care providers has changed to include the discussion of weight bias and strategies to prevent this bias. Medical students are receiving more education on nutrition, physical activity, and other factors influencing mental and physical health, including obesity. Health-care providers are receiving additional training on the complexity of obesity and the importance of evaluating each patient as an individual. Patient counseling strategies are now included in the training of most health-care professionals—it is not just the content a provider is sharing with their patient but how that information is being relayed. The *Training Table* in this section provides recommendations for reducing weight stigma in healthcare.

Training Table

Weight bias contributes to reduced quality of care for patients. Stigmatizing and judging a person because of their weight and body type is not productive. For those working in health-care, the following recommendations are provided to support care for your patients of all sizes.

- Remember that you are working with a person, not a medical chart with a bunch of numbers.
- Get to know your patient, using first-person language.
- Show empathy and care.
- Work with your patient to set behavioral and health goals rather than goals focused specifically on weight.
- Separate health and weight, keeping in mind they are not the same.
- Do not assume that your patient's body weight is an indication of lifestyle factors, such as eating habits, sleep, stress, or physical activity. Obesity is complex.
- Provide chairs, gowns, and medical equipment that are appropriate for people of all sizes. Offer a private room with a scale for measuring weight.
- Choose your words carefully, including when charting. For example, use "excess or unhealthy weight" instead of "obese or fat" and "patient living with obesity" instead of "obese patient."

Tetra Images/Getty Images

What are evidence-based strategies to support weight loss?

As has been established, treatment of overweight and obesity is complex and should be approached as other forms of chronic disease would be approached. This includes an individualized plan of action that is realistic and achievable. Table 11.2 provides an overview of strategies that have been found to be successful in supporting weight loss, improvements in body composition, and overall health and wellness. Chapter 10 provided an overview of medical treatments, including medications, devices, and bariatric surgery, for the treatment of overweight and obesity. This chapter will focus on lifestyle factors, including healthy eating strategies and physical activity recommendations. The last section of the chapter introduces mindful or intuitive eating, strategies that allow a person to be more aware of their eating habits, hunger cues, and feelings about food.

Key Concepts

- Overweight and obesity are associated with many chronic diseases, including heart disease, type 2 diabetes, hypertension, certain times of cancer, and anxiety/depression. For individuals who are overweight or have obesity, losing weight and improving body composition can result in a reduced risk for chronic diseases.
- Individuals with obesity are more likely to have anxiety, depression, and other psychological conditions. At the same time, anxiety and depression can contribute to unhealthy weight gain.
- Weight bias occurs frequently and impacts the quality of care and health outcomes when it happens in a medical facility. Health-care providers should receive training to reduce weight bias when caring for patients.
- Treatment of overweight and obesity is complex, and there is no one treatment that works for everyone. Lifestyle and medical interventions can be successful in the treatment of overweight and obesity.

TABLE 11.2 Evidence-based strategies to support weight loss and maintenance of that weight loss in individuals with overweight or obesity

Lifestyle interventions

Follow a well-balanced and nutrient-dense diet that aligns with your food preferences, access to food, culture, traditions, and other personal factors.
- Eat a variety of fruits, vegetables, whole grains, lean sources of quality protein, healthy fats, and low-fat dairy foods (or dairy alternatives).
- Control portion sizes.
- Limit intake of added sugars and saturated fat.
- Focus on taste and enjoy the food you eat.

Engage in regular physical activity, meeting the recommendations of the *Physical Activity Guidelines for Americans, 2018*.
- Start slow – walking or swimming is a great start!
- Find activities you enjoy and that will work with your schedule and lifestyle.
- Invite your friends or family to join you, providing a social network to help stick with the physical activity.
- If you have health conditions, such as heart disease or type 2 diabetes, talk to your health-care provider before beginning any physical activity program.

Ensure you are getting enough quality sleep.

Determine strategies to manage stress, such as physical activity and meditation. Talk to a qualified health professional, if needed.

Practice mindfulness in your daily life, including mindful eating. For example, take your time when eating, sit at a table, avoid screen time while eating, and eat smaller portion sizes.

Take time to enjoy nature with outdoor activities.

Medical interventions

Weight-loss programs

Medications to support weight loss

Bariatric surgery (medical procedures that support weight loss by making changes to the GI tract)

Nutrition and Weight Loss

Chapters 1 and 2 provided general information on nutrition and planning a nutrient-dense and balanced diet to support health and wellness. With regard to nutrition for weight loss, we will see that the recommendations are consistent with general principles of healthy eating. By following the recommendations of the *2020-2025 Dietary Guidelines for Americans*, a person can successfully lose weight while also establishing habits to support health and wellness in the long term. As described by Hall and Kahan, long-term management of weight loss is difficult due to the complex interactions between a person's biology, behavior, and the obesogenic environment of the United States. Experts report that approximately 80–95 percent of individuals who have lost weight regain that weight within 2 years. For this reason, providing realistic and practical recommendations as part of weight-loss diets is important to the sustainability of the diet.

According to the CDC, for those with overweight and obesity, losing weight gradually and steadily is recommended. Losing 1-2 pounds per week has been shown to be a successful weight-loss strategy for not only losing weight but keeping the weight off. Again, losing weight is often not "easy," and it takes commitment, motivation, and support. The following sections summarize key nutrition recommendations to support healthy weight loss.

www.cdc.gov/healthyweight/losing_weight/index.html Losing weight is not easy and requires a commitment to ongoing changes in nutrition and healthy eating. The CDC provides a step-by-step guide to assist those with the goals of weight loss and better health.

How is kcal intake related to weight loss?

You may have heard before that the caloric value of 1 pound of body fat is approximately 3,500 kcal. This is based on 1 pound being equal to 454 grams. Adipose tissue is approximately 86 percent stored fat. Protein, minerals, and water account for the remaining mass. The kcal content of 1 pound of body fat is then calculated based on 0.86×454 g $\times 9$ kcal per gram of fat = 3,514 kcal.

As shown in figure 11.3, the caloric concept of weight control is relatively simple. If you consume more kcal than you expend, you will have a positive energy balance and gain weight. If you expend more than you take in, you will have a negative energy balance and lose weight. To maintain your body weight, caloric input and output must be equal. As far as we know, human energy systems are governed by the same law of physics that rules all energy transformations. The First Law of Thermodynamics is as pertinent to us in the conservation and expenditure of our energy sources as it is to any other machine. Because a kcal is a unit of energy and because energy can neither be created nor destroyed, those kcal that we eat must either be expended in some way or conserved in the body. No substantial evidence is available to disprove the caloric theory. It is still a basis for body-weight control.

Total body weight is made up of different components, those notable in weight control being body water, protein in the fat-free mass, small amounts of carbohydrate, and fat stores. Changes in these components may bring about daily body-weight fluctuations of 3–5 pounds, which would appear to be contrary to the caloric concept because protein and carbohydrate contain only 4 kcal per gram and water contains no kcal. You may gain water weight by consuming a high-salt diet for a day or by menstrual cycle changes. You may lose 5 pounds in an hour, but it will be mostly water weight lost through sweating. Starvation techniques may lead to rapid weight loss, but some of the weight loss will be in glycogen stores, body-protein stores such as muscle mass, and the water associated with glycogen and protein stores. In programs to lose body weight, we usually desire to lose excess body fat, and certain dietary and exercise techniques may help to maximize fat losses while minimizing protein losses. The results of safe weight-loss programs are observed over weeks to months, not hours or days.

How Many Kcal Do I Need to Eat to Lose Weight? Energy needs to support weight loss are highly variable and individualized based on factors including age, sex, physical activity levels, genetics, and more. While there are formulas used to estimate kcal needs, it is important to remember that these are just estimates and that individual needs may vary. However, such formulas do provide a starting point for determining energy needs. Refer to chapter 3 for recommendations on calculating energy needs, including the Estimated Energy Requirement (EER). When using dietary analysis software, such as NutritionCalc Plus (NCP), you will be prompted to enter personal information about your height, weight, physical activity, and weight goals. Caloric intake recommendations are then based on this personalized information.

As described previously, to achieve weight loss, a **calorie (kcal) deficit** is needed. While previous recommendations for adults with overweight or obesity were to reduce kcal intake by 500 kcal per day to lose 1 pound per week, more current guidelines recommend a daily deficit of approximately 300–500 kcals daily. Consuming fewer kcals is not always "better" as lower energy intakes can be associated with reductions in basal metabolic rate (BMR), changes in hormone levels, increased hunger, and feelings of fatigue. When a person feels more tired, they are less likely to engage in physical activity and prepare and consume well-balanced meals.

Keep in mind these recommendations are for adults. Children and adolescents who are overweight or have obesity have different kcal needs as they are still growing, and, therefore, reducing kcal intake too much could compromise normal growth and development. Specialized clinics and health-care providers with expertise in weight loss in children and adolescents should be consulted prior to starting a weight-loss diet in this population. The recently published position paper of the Academy of Nutrition and Dietetics on "Prevention of Pediatric Overweight and Obesity" provides additional guidance on obesity prevention and interventions in the pediatric population.

FIGURE 11.3 Weight control is based on energy balance and governed by the Laws of Thermodynamics. Too much food input or too little exercise output can result in a positive energy balance or weight gain, which can adversely affect health. Decreased food intake or increased physical activity can result in a negative caloric balance or weight loss, which can also adversely affect health.

https://www.nutrition.gov/topics/healthy-living-and-weight/weight-management-youth The United States Department of Agriculture (USDA) provides interactive and skill-building tools to support weight management and weight loss in children and adolescents.

https://kidshealth.org/en/teens/lose-weight-safely.html Dr. Mary Gavin provides recommendations for teenagers to lose weight safely.

What are general dietary recommendations to support weight loss?

Image Source/Glow Images

Chapters 1 and 2 provide an overview of dietary recommendations to support optimal health and wellness, including body weight and body composition. For a detailed review of the *2020-2025 Dietary Guidelines for Americans*, refer to chapter 2 of this textbook. Recall that the *Dietary Guidelines* encourage the consumption of a variety of nutrient-dense food sources from each of the food groups – vegetables, fruits, grains, dairy (or dairy alternatives), and protein. As shown in figure 11.4, the guidelines also share the **USDA 85-15 guide**, meaning that 85 percent of kcal consumed daily should meet food group recommendations healthfully from nutrient-dense foods. That then leaves 15 percent of daily kcal that are available to be consumed from other food sources. For example, desserts, sugar-sweetened beverages, and chips would all fall into the 15 percent category. The premise behind the 85-15 guide is to focus on moderation while not being overly restrictive. For individuals with overweight or obesity, following the 85-15 guide can support the consumption of a well-balanced diet that is still filling and enjoyable.

How does consumption of ultra-processed foods impact body weight?

In recent years, several studies have investigated the relationship between the consumption of **ultra-processed foods (UPFs)** and health. UPFs most likely contain added ingredients such as added sugars, salt, saturated fat, artificial colors, and/or preservatives. Examples of foods considered UPF are fast foods, packaged cookies and cakes, commercial breads, ready-to-eat breakfast cereals, salty snack foods, hotdogs and cold cuts, and frozen meals. Studies have found that eating more UPFs may be associated with overconsumption of kcal, likely due to the palatability of these products. For example, Hall and others conducted an inpatient randomized controlled trial with 20 weight-stable adults. The participants received either ultra-processed or unprocessed diets for 2 weeks, followed by the alternate diet for 2 weeks. Participants could consume as much, or as little, as they wanted during each diet. The researchers found that energy, carbohydrate, and fat intake were greater during the ultra processed diet as compared to the unprocessed diet. As well, research by Zhang and others has shown higher intake of UPFs increases a person's risk for obesity and other forms of chronic disease. However, a recent article by Astrup and Monteiro questions the classification of UPFs, drawing attention to several examples of foods that are considered ultra-processed, yet are nutrient-dense and support healthy eating. Based on the conflicting research on UPFs and obesity, those following a diet to reduce weight should read labels carefully. When possible, try to avoid or limit intake of UPFs and focus on minimally processed foods such as fresh fruits, vegetables, and whole grains.

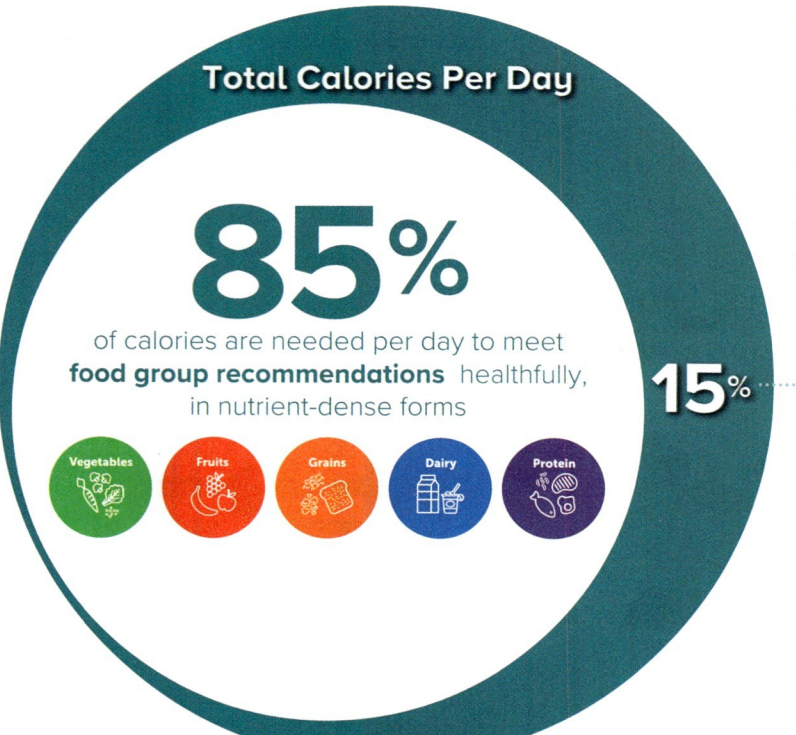

FIGURE 11.4 The USDA provides an 85–15 guide with regard to kcal intake. By following the 85–15 guide, an individual with the goal of weight loss will limit intake of foods high in added sugars and saturated fat.

USDA 2020-2025 Dietary Guidelines for Americans. Pp 37, Chapter 1: Nutrition and Health Across the Lifespan

www.health.harvard.edu/blog/what-are-ultra-processed-foods-and-are-they-bad-for-our-health-2020010918605 Learn more about ultra-processed food and health from the Harvard School of Public Health.

www.fda.gov/food/nutrition-education-resources-materials/calories-menu The U.S. Food and Drug Administration (FDA) provides information on how to make informed decisions when eating out.

Limit Intake of Added Sugars and Saturated Fat The *2020–2025 Dietary Guidelines for Americans* recommend that, for those individuals with a goal of weight loss, added sugars and saturated fat are limited in the diet as a strategy to reduce kcal intake. According to the guidelines, a healthy dietary pattern limits added sugars to less than 10 percent of daily kcal and saturated fats to less than 10 percent of daily kcal. Primary sources of added sugars in the diet of the typical adult in the United States include sugar-sweetened beverages, desserts, sweet snacks, candy, and sweetened coffee and tea. The average daily intake of added sugars is 266 kcal.

As discussed in detail in chapter 5, not all fats are created equal. Different types and sources of fat have different impacts on health – in some cases positive and in others contributing to increased risk for chronic disease. According to the *2020–2025 Dietary Guidelines for Americans*, saturated fat should be limited to less than 10 percent of daily kcal. Saturated fats can be replaced with unsaturated fats, including monounsaturated and omega-3 fatty acids. For individuals with overweight or obesity, changes to meet this recommendation include choosing lean cuts of meat and consuming in moderation, drinking low-fat cow's milk instead of whole milk, and replacing snacks high in saturated fats with nuts, seeds, or avocados. Keep in mind that all fats provide 9 kcal per gram; as such, all sources of fat should be consumed in moderation as part of a weight-loss diet.

Olga Nayashkova/Shutterstock

What are practical recommendations for following a diet to support weight loss?

Following a diet to support weight loss does not mean starving oneself or skipping out on all the foods and beverages a person enjoys. Sustainable weight-loss diets are those that are realistic and take into account a person's food preferences, culture, traditions, access to food, and schedule. The *Training Table* in this section provides practical guidelines to support healthy eating and weight loss.

Training Table

Following a diet to support weight loss can be filling and delicious. If you are not sure where to start, RDNs provide an excellent source of knowledge and can provide individualized support. Find an RDN by visiting www.eatright.org/food/resources/learn-more-about-rdns/find-an-rdn-anywhere-you-need-one. General guidelines to consider when following a diet to support weight loss:

- Calories matter in weight loss. Determine how many kcal best support your body's needs while also contributing to a healthy weight loss of 1–2 pounds per week.
- Pay attention to portion sizes. For some people, weighing and measuring out foods and beverages supports portion control.
- Focus on consuming a variety of vegetables and fruits that you enjoy. Consider vegetables and fruits as snacks.
- Choose whole grains, such as oats, brown rice, and whole grain bread.
- Consume a variety of lean sources of protein from animal and/or plant food sources.
- Select oils that support health, including olive and canola oil, seeds, and nuts.

Key Concepts

▶ When providing nutrition recommendations to support weight loss, it is important to consider not only weight loss but also weight maintenance. In the United States, over 80 percent of individuals who lose weight regain that weight within 2 years.

▶ A kcal deficit is important to successful weight loss. In general, to lose 1 pound of weight per week, kcal intake should be reduced by 300–500 kcal daily. There are individual variations in what is an ideal caloric deficit to support long-term weight loss.

▶ The *2020–2025 Dietary Guidelines for Americans* provide recommendations to support healthy body weight and body composition. These guidelines offer a useful foundation for planning a weight-loss diet.

▶ In general, a diet that limits the intake of ultra-processed foods, added sugars, and saturated fat supports weight loss. Treatment of overweight and obesity is complex, and there is no one treatment that works for everyone. Lifestyle and medical interventions can be successful in the treatment of overweight and obesity.

Check for Yourself

▶ Using dietary analysis software, such as NutritionCalc Plus, to determine your daily kcal needs as calculated by the analysis program. Do you agree with the recommendation of this program? Why or why not?

Mindful Eating Practices

Mindfulness practices are now being recognized as key to supporting a person's overall health and wellness. The Oxford dictionary defines **mindfulness** as "a mental state achieved by focusing one's awareness on the present moment, while calmly acknowledging and accepting one's feelings, thoughts, and bodily sensations." Specific approaches to support mindfulness include yoga and meditation. Everyday approaches to mindfulness involve observing your surroundings, focusing on breath, practicing gratitude, accepting yourself, and living in the moment. Clarke and others report that, according to data from the National Health Interview Survey, the number of adults in the United States practicing yoga and meditation has increased in recent years. In 2017, approximately 14.3 percent of adults in the United States practiced yoga and 14.2 percent used meditation over a 12-month period. As compared to males, females were more likely to practice yoga and meditation.

While mindfulness practices are historically embedded in Buddhist traditions, these practices have gained popularity in the past decade. Mindfulness and meditation are not a religion. Today, nearly 80 percent of medical schools in the United States offer training (required or elective) on mindfulness. Students in health-care professions, including nursing, exercise science, dietetics, dentistry, and medicine, are learning mindfulness skills to support both the care of their future patients and their own physical and mental health. Several studies have found a positive relationship between mindful eating and mental well-being in health-care providers. Choi and Lee reported that a lower mindfulness eating score was associated with obesity and higher occupational stress in nurses.

Much of the available research on the potential health benefits of meditation and mindfulness is still preliminary and/or not scientifically rigorous. More well-controlled and large clinical trials are needed to further explore this relationship. However, research suggests that practicing mindfulness may offer health benefits and improve quality of life. Table 11.3 provides information on the potential health benefits of practicing meditation and mindfulness.

> www.nccih.nih.gov/ The National Center for Complementary and Integrative Health (NCCIH) provides education and research support related to complementary health products and practices. The NCCIH website provides information for consumers and health-care providers on topics related to mindfulness and more.

rdonar/Shutterstock

TABLE 11.3 Potential health benefits of practicing meditation and mindfulness

Reduced stress and decreased risk for anxiety and depression
Reduced blood pressure in people with health conditions such as hypertension, diabetes, or cancer
Decreased acute and chronic pain and improved capacity to tolerate pain
Improved sleep quality
Reduction of post traumatic stress disorders (PTSD), when integrated with traditional therapies
Reduced psychological distress in individuals being treated for breast cancer
Improved eating behaviors and weight loss in adults with overweight or obesity

Source: National Center for Complementary and Integrative Health. Updated 2022. Meditation and Mindfulness: What you need to know. www.nccih.nih.gov/health/meditation-and-mindfulness-what-you-need-to-know#:~:text=A%202019%20analysis%20of%2029,symptoms%20of%20anxiety%20and%20depression.

What is mindful eating?

One area of mindfulness receiving attention is the concept of **mindful eating**. Through mindful eating, a person becomes more aware of food and their experiences consuming foods and meals. While the purpose of mindful eating is not specific to weight loss, practicing mindful eating techniques often support healthy eating habits that then contribute to improvements in health as well as body weight and body composition. According to the Center for Mindful Eating, mindful eating is defined as:

- Allowing yourself to become aware of the positive and nurturing opportunities that are available through food selection and preparation by respecting your own inner wisdom.
- Using all of your senses in choosing to eat food that is both satisfying to you and nourishing to your body.
- Acknowledging responses to food (likes, dislikes, neutral) without judgement.
- Becoming aware of physical hunger and satiety cues to guide your decisions to begin and end eating.

To learn more about these principles of mindful eating, visit the Center for Mindful Eating website at www.thecenterformindfuleating.org/Principles-Mindful-Eating/. Resources are provided for both consumers as well as health-care providers.

Table 11.4 summarizes some of the general characteristics of mindful eating. Rather than implementing all of these recommendations at once, choosing one or two to start with can set a person up for long-term success with mindful eating. Figure 11.5 shares practical recommendations for modifying current habits to those that are more mindful.

How is Mindful Eating Different than Intuitive Eating? While mindful eating and intuitive eating are similar, **intuitive eating** is an anti-diet, weight-inclusive concept. Intuitive eating aligns closely with the *Health at Every Size* movement that was discussed in chapter 10. The principles of intuitive eating were developed in 1995 by RDNs Evelyn Tribole and Elyse Resch. Table 11.5 summarizes the ten principles of intuitive eating, as outlined and updated by Tribole

TABLE 11.4 General characteristics of mindful eating

Eat slowly, taking smaller bites and more time between bites.
Take smaller portions, using smaller plates, bowls, cups, and utensils.
Use hunger cues to make decisions on when to eat. Avoid skipping meals and coming to a meal ravenously hungry.
Enjoy your food, utilizing all of your senses.
Chew food slowly to savor all of the flavors.
If possible, prepare your own food.
Take time to appreciate your food, including the farmers and others responsible for growing and producing the food.
Enjoy meals with others, taking time to engage in conversation.
To avoid distractions, avoid screen time while eating. Rather, put devices in another room or area.
Eat while sitting at a table, rather than in the car, on the couch, or in bed.
Have nutrient-dense food and beverage options readily available.

TABLE 11.5 Ten Key Principles of Intuitive Eating

Reject the diet mentality
Honor your hunger
Make peace with food
Challenge the food police
Discover the satisfaction factor
Feel your fullness
Cope with your emotions with kindness
Respect your body
Movement – feel the difference
Honor your health – gentle nutrition

Source: The Original Intuitive Eating Pros. Updated 2022. 10 principles of intuitive eating. www.intuitiveeating.org/10-principles-of-intuitive-eating/

https://www.intuitiveeating.org/10-principles-of-intuitive-eating/ The Original Intuitive Eating Pros® provide educational resources on intuitive eating. You can also search the site to find RDNs trained and certified in intuitive eating.

and Resch. Like mindful eating, there has been significant research conducted on the health effects of intuitive eating. A meta-analysis by Linardon and others indicates there is strong evidence between intuitive eating and body image, self-esteem, and well-being.

Eating at computer or work desk, not taking a break

Eating processed foods out of to-go boxes

Eating as quickly as possible on-the-go

Eating meals alone

Taking a break and eating meals at a table

Preparing a meal at home

Using all senses to really savor food

Sharing meals with friends, family, coworkers, and others

Businesswoman: Westend61/Getty Images; Grandfather: monkeybusinessimages/iStock/Getty Images; multiethnic young: fizkes/Shutterstock; Cooking lesson: fizkes/Shutterstock; Cooking lesson: tommaso79/Shutterstock; Woman smelling: Antoine Arraou/Alamy Stock Photo; Depressed woman: Dragana991/Getty Images; Happy senior: AlessandroBiascioli/Shutterstock

FIGURE 11.5 Many activities that have become normalized as part of daily life, such as eating while working at the computer or eating quick and processed foods out of to-go containers or wrappers, contradict the recommendations for mindful eating. This figure demonstrates how some common mindless eating habits can be made more mindful.

Food Addiction and Overconsumption of Palatable Foods

De Ridder and Gillebaart explore the relationship between overconsumption (i.e., overeating), overweight and obesity, and negative attitudes toward foods. They suggest that mindful eating practices, such as consuming "normal levels" of food, are associated with social and psychosocial well-being, as well as improvements in body weight and body composition. Overconsumption of food, also called overeating, has been characterized by some as a substance use disorder. In the case of overeating, a person continues to overconsume palatable yet energy-dense foods despite the negative consequences (i.e. unhealthy weight gain). In their review on the topic, O'Connor and Kenny describe the similarities and differences between compulsive drug use and food addiction or overconsumption of foods in individuals with obesity.

Mindful Eating, Weight Loss, and Health Outcomes

When a person eats mindfully, they are aware of physical hunger and satiety cues, which then guide when a person decides to being and end eating. By eating slowly, taking smaller bites, and truly savoring foods, a person with obesity can reduce kcal intake and focus on the consumption of nutrient-dense foods. Hayashi and others share in a review of intervention studies that mindful eating may be beneficial in improving markers related to cardiometabolic function, including blood glucose, blood pressure, and markers of inflammation.

For children and adolescents with obesity, mindfulness interventions are being integrated as part of a comprehensive treatment plan. While research in this area is still limited, early findings suggest that mindful eating practices in children and adolescents with obesity support weight loss, decrease rates of anxiety and depression, and improve blood glucose and lipid levels. According to Keck-Kester and colleagues, "mindfulness interventions are a reasonable addition to a holistic treatment plan of children with obesity."

How can mindfulness practices support both mental and physical health?

As shared by the CDC, 11.3 percent of adults in the United States report having regular feelings of worry, nervousness, or anxiety. As well, approximately 5 percent of adults experience regular feelings of depression. Mental disorders are the primary diagnosis for 56 million visits to health-care providers annually, and nearly 5 million visits are made to emergency departments each year with mental disorders, behavioral, and neurodevelopment as the primary diagnosis. In 2020, over 20 percent of adults in the United States had received mental health treatment in the past 12 months, 16.5 percent had taken prescription medication for their mental health, and 10.1 percent had been in counseling or therapy with a licensed mental health professional.

As discussed throughout this chapter, overweight and obesity are complex disease conditions with inter related mental and physical components. For this reason, successful treatment for overweight and obesity often goes beyond diet and physical activity. In a meta-analysis, Magallares and Pais-Ribeiro quantify and describe the relationship between mental health and obesity. They report that females with obesity are more likely to have mental health issues as compared to females at a "normal" weight. In contrast, males with obesity actually have better mental health than males at a "normal" weight. Tronieri and others further explore this relationship, noting that sex differences exist in relation to obesity and mental health. They report that females are more likely to report weight stigma having a greater impact on their mental health. Males with obesity were more likely to experience anxiety than males without obesity.

Further research is needed to further explore the complex relationships between obesity and mental health. Based on the available evidence at this time, there is an opportunity to develop innovative treatment options to manage obesity and mental health issues concurrently. The *Training Table* in this section shares the potential benefits of mindfulness practices on mental health.

> www.cdc.gov/nchs/fastats/mental-health.htm Data and information on mental health can be found on the CDC website.

Training Table

Stress and anxiety levels are high. For this reason, many people have started to practice mindfulness as part of their daily lives. This not only includes more structured practices such as yoga and meditation but also everyday choices. Below are some of the specific benefits that have been associated with mindfulness practice.

- Meditation for as little as 10 minutes a day can help to reduce stress levels.
- Practicing mindfulness can help to improve mental focus and "quiet your mind" that may be pulled in many different directions.
- Mindfulness can help to minimize bias, both toward others and toward one's self (i.e., negative self-talk).
- Mindfulness practices, when combined with traditional therapies, can further support treatment for anxiety and depression.
- Meditation increases compassion and may be beneficial in improving personal relationships and supporting a positive work environment.

Visit www.mindful.org/ for more information on mindfulness and suggestions for strategies to start a mindfulness practice to support your health and wellness.

Obesity and Eating Disorders

As described by multiple researchers, there are strong associations between obesity and eating disorders, including binge-eating disorder. The relationship is complex and involves numerous factors from a biological, social, and community-level standpoint. Recent research by Breton and

jenifoto/123RF

others has evaluated the overlapping mechanisms related to the immunoinflammatory responses that are seen in both obesity as well as eating disorders. They have found that the underlying alterations in the immunoinflammatory response increase risk for both obesity and eating disorders. As well, Colleluori and colleagues theorize a relationship between early life stress, neural development, and obesity. They hypothesize that the link is with oxytocin, a neurohormone that regulates energy balance while also impacting eating behaviors and emotions.

When developing weight-loss diets and programs, it is important to recognize the relationship between such diets/programs and the risk for eating disorders, including binge-eating disorder. Tylka and others provide a framework for better understanding the differences between weight-inclusive (viewing health and well-being as complex with efforts focused on improving health access) and weight-normative (emphasis on weight and weight loss) approaches to treating obesity. Important is that one is not "better" than the other and a combination of the two approaches likely best supports healthy and sustainable weight loss.

Key Concepts

- Mindful eating practices such as eating slowly, savoring foods, and being aware of hunger and satiety cues, may be beneficial in the management of overweight and obesity. Mindful eating has been associated with positive changes in both mental and physical health.
- Food addiction can contribute to overconsumption of palatable foods, likely leading to unhealthy weight gain. There are complex biological mechanisms at play with regard to overeating.
- A complex relationship exists between obesity and mental health. A person with obesity is more likely to experience anxiety, depression, and other mental health conditions. At the same time, anxiety, depression, and other mental health conditions can contribute to obesity.
- Individuals with obesity may concurrently have an eating disorder, most commonly binge-eating disorder. For this reason, treatment plans for obesity should be developed with care and intention for the individual.

Check for Yourself

- Consider your own dietary habits. Are your habits consistent with mindful eating? Describe why or why not and, if not, provide recommendations for how to incorporate more mindful eating practices into your daily living.

Physical Activity and Weight Loss

Humans are meticulously designed for physical activity, and yet our modern mechanical age has eliminated many of the opportunities that our ancient ancestors had to incorporate moderate physical activity as a natural part of daily living. The regulation of our food intake has not adapted to the highly mechanized conditions in today's society. As discussed in chapter 10, physical inactivity is one of the major contributing factors to obesity as well as other forms of chronic disease. As described throughout this textbook, physical inactivity is a major health concern. In the United States, one in four adults reports that "in the past month, they did not do any physical activity outside of their regular job." Similar to the map showing obesity rates by state (see figure 11.1), maps have been developed for physical inactivity (see figure 11.6). Notice the similarities between the two maps, emphasizing the importance of physical activity in achieving a healthy body weight and composition.

www.cdc.gov/physicalactivity/index.html The CDC provides information about physical activity, including data and statistics as well as individual, community, and worksite strategies to support active living.

How is physical activity related to obesity?

In the United States, the Centers for Disease Control and Prevention (CDC) report that about 50 percent of adults meet recommendations for aerobic physical activity. In high school students, only one in four students meets aerobic physical activity recommendations. Physical inactivity contributes to significant health risks, resulting in an annual health-care cost of $117 billion in the United States.

Physical activity has been shown to be an effective treatment in the prevention and management of obesity. Engaging in exercise can delay the onset of chronic health conditions associated with obesity, such as type 2 diabetes, hypertension, and cardiovascular disease. As well, physical activity improves cardiovascular fitness and can stimulate anti-inflammatory processes. Calcaterra and others report that physical activity in children and adolescents with obesity can limit the negative effects of obesity on health. Similarly, Verduci and collaborators describe the promotion of physical activity and reduction of sedentary behaviors as key to integrative approaches to addressing obesity.

As described in a review and meta-analysis by Nitscheke and others, physical activity interventions are positively associated with the likelihood of achieving at least 5 percent weight loss in adults with overweight and obesity. The National Weight Control Registry reports that, of participants who have successfully lost weight and maintained that weight loss, 94 percent had increased their physical activity, with walking being the most frequently reported form of activity.

Prevalence of Self-Reported Physical Inactivity* Among US Adults by State and Territory, BRFSS, 2017–2020

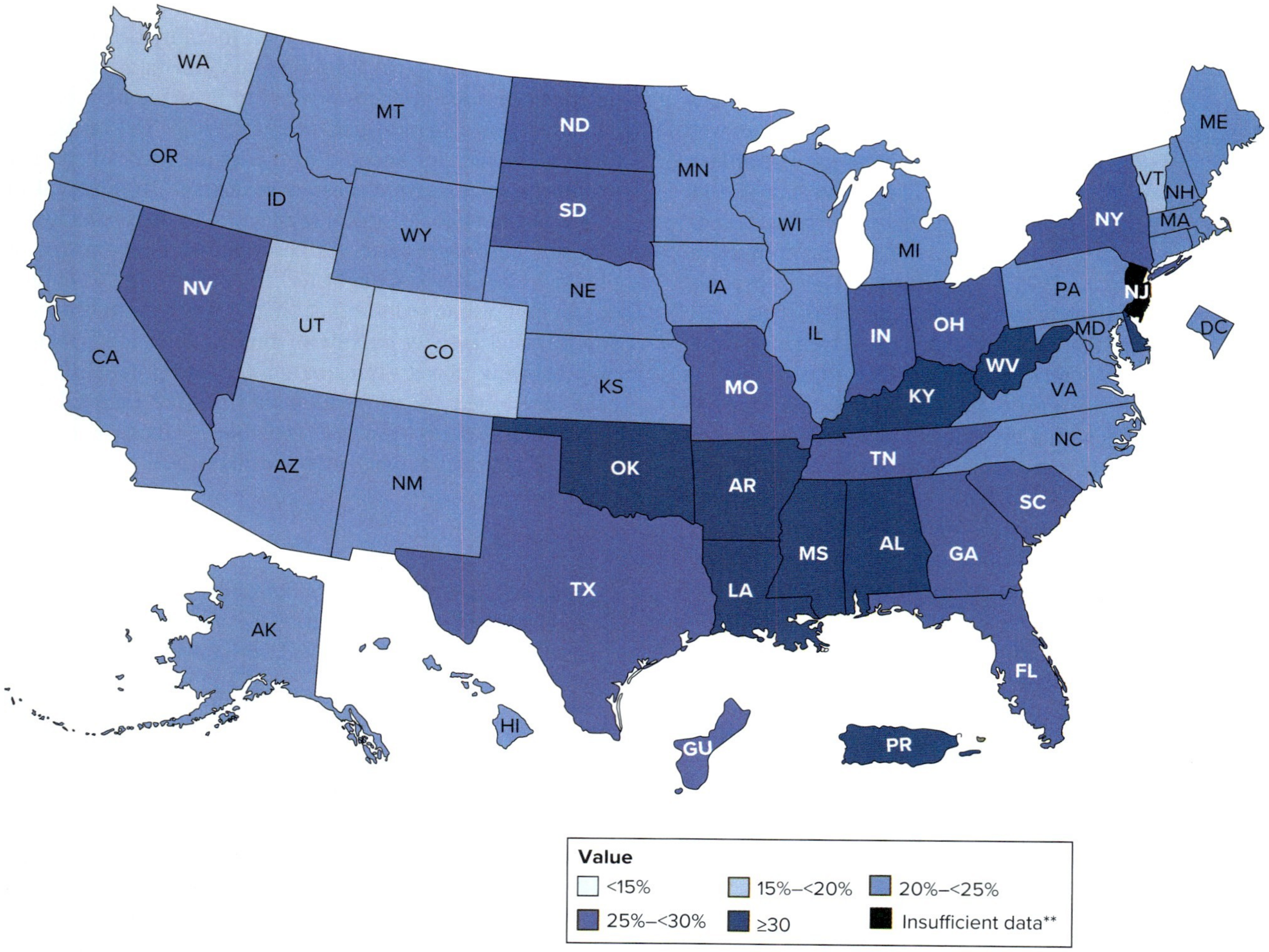

Value
- <15%
- 15%–<20%
- 20%–<25%
- 25%–<30%
- ≥30
- Insufficient data**

* Respondents were classified as physically inactive if they responded "no" to the following question: "During the past month, other than your regular job, did you participate in any physical activities or exercises such as running, calisthenics, golf, gardening, or walking for exercise?"
** Sample size <50, the relative standard error (dividing the standard error by the prevalence) ≥30%, or no data in at least 1 year.
www.cdc.gov/physicalactivity/data/inactivity-prevalence-maps/index.html#Overall

FIGURE 11.6 This map of the United States shows the prevalence of physical inactivity based on the Behavioral Risk Factor Surveillance System (BRFSS), 2020.

What are strategies to increase physical activity?

As outlined in chapter 1, the *Physical Activity Guidelines for Americans, 2018*, recommends adults engage in at least 150 minutes of moderate-intensity physical activity per week to support health. For individuals whose goal is to lose significant weight (more than 5 percent of body weight) and/or those who are trying to prevent weight regain, the guidelines recommend engaging in at least 300 minutes per week of moderate-intensity physical activity. While muscle-strengthening activities do not have the same impact on weight loss and maintenance as aerobic activities, such physical activity is nonetheless beneficial for individuals with overweight or obesity. Strength training activities support increases in LBM and reductions in body fat during weight loss. As well, resistance training can be particularly beneficial in supporting weight maintenance. Figure 11.7 summarizes the key recommendations of the *Physical Activity Guidelines for Americans, 2018*.

The American College of Sports Medicine (ACSM) position stand by Donnelly and other experts in the field provides appropriate physical activity recommendations to support weight loss and prevention of weight regain in adults with obesity. The key guidelines provided in the ACSM position stand are presented in table 11.6. The following sections provide specific recommendations with regard to the mode of physical activity (aerobic and resistance training) as well as community-level opportunities to support physical activity as part of everyday life activities.

Alex Potemkin/Getty Images

Adults need a mix of physical activity to stay healthy.

Moderate-intensity aerobic activity*
Anything that gets your heart beating faster counts.

at least **150 minutes a week**

AND

Muscle-strengthening activity
Do activities that make your muscles work harder than usual.

at least **2 days a week**

* If you prefer vigorous-intensity aerobic activity (like running), aim for at least **75 minutes a week**.

If that's more than you can do right now, **do what you can.** Even 5 minutes of physical activity has real health benefits.

Walk. Run. Dance. Play. What's your move?

Physical Activity Guidelines for Americans. https://health.gov/sites/default/files/2019-09/Physical_Activity_Guidelines_2nd_edition.pdf

FIGURE 11.7 Key recommendations for adults from the *Physical Activity Guidelines for Americans, 2018*

TABLE 11.6 Key recommendations from the ACSM position stand on physical activity and weight loss

Moderate-intensity physical activity between 150 and 250 minutes per week with no changes to diet will provide only modest weight loss.
Moderate-intensity physical activity between 150 and 250 minutes per week will prevent weight gain.
Greater amounts of physical activity (over 250 minutes per week) have been associated with clinically significant weight loss.
Moderate-intensity physical activity between 150 and 250 minutes per week with moderate (but not severe) diet restriction will improve weight loss.
After weight loss, moderate-intensity physical activity of over 250 minutes per week prevents weight regain after weight loss.

Source: Donnelly, J., et al. 2009. American College of Sports Medicine Position Stand. Appropriate physical activity intervention strategies for weight loss and prevention of weight regain for adults. *Medicine and Sciences in Sports and Exercise* 41(2):459.

https://health.gov/sites/default/files/2019-09/Physical_Activity_Guidelines_2nd_edition.pdf Access the second edition of the *Physical Activity Guidelines for Americans*, 2018 for specific recommendations in support of developing physical activity programs for individuals of all ages.

www.acsm.org/ The American College of Sports Medicine provides valuable resources on a range of topics related to physical activity, health, and sports performance. This includes recommendations for physical activity in individuals with overweight or obesity. The ACSM podcast, "*The Sports Medicine Checkup*," includes episodes with experts in the field and is available at www.acsm.org/education-resources/acsm-podcast.

Resistance Exercise Training Resistance-training, or weight-training, programs are detailed in chapter 12 in relation to gaining body weight as muscle mass, but such programs may also be very helpful during weight-loss programs. One possibility, as suggested by the ACSM, is that increased strength through resistance training may lead to a more active lifestyle in individuals who are overweight or have obesity and are sedentary, thus leading to health benefits that may include weight loss and prevention of weight regain. Another possibility is the maintenance of resting energy expenditure (REE) during weight loss. According to Hansen and colleagues, the more significant role of resistance training on body composition during weight loss is in maintaining or increasing fat-free mass rather than through energy expenditure leading to loss of fat mass. Recall that protein tissue, primarily muscle, may be lost along with body fat during a weight-reduction program. However, resistance training may stimulate muscular development and help prevent significant decreases in LBM. Such an effect may also help prevent decreases in the REE. Additionally, as is noted in chapter 12, the typical resistance-training workout does not burn many kcal, mainly because of frequent recovery periods. However, dynamic circuit-type resistance-training programs may also be used to burn additional kcal.

High-intensity interval training (HIIT) is a form of resistance exercise training whereby an individual exercises at high intensity for short durations of time. In their review on the topic, Batrakoulis and Fatouros report that HIIT can not only support weight loss and improvements to body composition but can also be beneficial to psychological health markers such as adherence and exercise enjoyment. Khalafi and Symonds investigated the relations between HIIT and liver fat content in adults with overweight or obesity. They found that HIIT supported reductions in liver fat, a marker of metabolic dysfunction in the body. As such, for those who enjoy HIIT, this form of training, when combined with diet, may be beneficial in supporting weight loss and improving body composition in individuals with overweight or obesity.

Aerobic Exercise Training The primary function of aerobic exercise in a weight-control program is to increase the level of energy expenditure and help tip the caloric equation so that energy output is greater than energy input. As mentioned in chapter 3, the metabolic rate may be increased tremendously during aerobic exercise. For example, while the average person may expend only 60–70 kcal per hour during rest, this value may approach 1,000 kcal per hour during a sustained, high-level activity such as rapid walking, running, swimming, or bicycling. Athletes involved in extreme endurance bicycling and running events have been reported to consume between 6,000 and 13,000 kcal per day.

For individuals with overweight or obesity, the same amount of aerobic weight-bearing exercise will cost the person more kcal than an individual who is considered "normal" weight. Because an individual with overweight or obesity has more weight to move, they will expend more energy and lose more body fat in the long run. For example, the energy cost of jogging 1 mile would be about 70 kcal for a 100-pound individual and about 140 kcal for someone twice that weight. Figure 11.8 depicts this concept for individuals with different body weights who are walking.

Some people may be discouraged to learn that 1 pound of fat contains about 3,500 kcal. Since approximately 100 kcal is expended in walking or running 1 mile, one must walk or jog 35 miles to lose that pound of fat. As a result, they hesitate to become physically active because they mistakenly believe exercise expends a few kcal. However, individuals must consider the **long-haul concept** of weight control. The extra weight did not appear overnight; nor will it disappear overnight. Walking or jogging 2 miles a day will increase caloric expenditure by 6,000 kcal a month (about 1.7 pounds lost), provided the individual does not compensate by consuming more kcal. At this rate, a person may lose 10 pounds in 6 months.

In addition to the direct effect of increased energy output during exercise, exercise has been theorized to facilitate weight loss by other means. As noted in chapter 3, aerobic exercise may increase the REE during the period immediately following the exercise bout and may increase the thermic effect of food (TEF) if you exercise after eating a meal. Unfortunately, the magnitude of this increased REE or TEF is relatively minor and not considered to be of any practical importance in a weight-loss program. Related to the TEF, it does not matter if you exercise before or after a light meal.

During aerobic exercise, the body mobilizes its fat cells to supply energy to the muscle cells. Hence, body-fat stores are reduced. In the long run, this change in body composition may actually favor a slight increase in the REE because muscle tissue is more active metabolically than fat tissue. With diet alone and no physical activity, some lean muscle mass may be lost. According to Cava and others, exercise combined with dietary restrictions is better at sustaining fat-free mass while decreasing fat mass than either exercise or diet alone. Figure 11.9 demonstrates how exercise may support the loss of fat mass.

Community Design to Support Active Living Safety and convenience are often two important barriers to an individual engaging in regular physical activity. For this reason, many organizations, including the CDC and local and state governments, fund projects that promote physical activity through improved community design. The *Training Table* in this section identifies some of the specific efforts being undertaken to connect people from where they live to where they need to go.

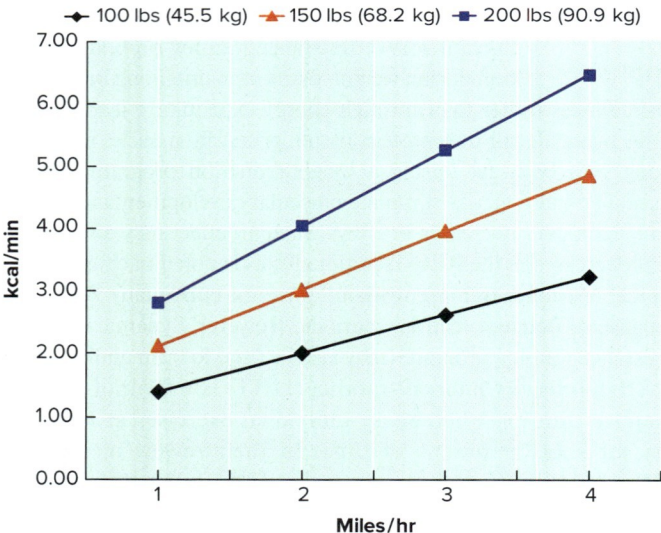

FIGURE 11.8 Effect of speed (mph) and gross body weight (lbs) on energy expenditure (kcal per minute) of walking. An individual with a larger body mass will have a greater expenditure of kcal for any given speed of walking. The same would be true for running and other physical activities in which the body must be moved by foot. Data points in this figure are calculated using the ACSM metabolic equation for steady-state walking exercise.

Source: American College of Sports Medicine. 2018. ACSM's Guidelines for Exercise Testing and Prescription, 10th ed. Philadelphia. PA Wolters Kulwer.

 Training Table

The CDC provides funding to states, communities, and national organizations to support community planning and development that supports active living. Some of the specific efforts of such community design include:

- Making active transportation (i.e., walking, biking, or wheelchair rolling) more feasible and safe. Through land use and zoning policies, schools, workplaces, shops, and parks can be connected closer to an individual's home.

to start incorporating physical activity into their daily living. The CDC provides practical recommendations and tools on "Getting Started with Physical Activity" at www.cdc.gov/healthyweight/physical_activity/getting_started.html. As well, chapter 1 of this textbook provides general guidelines for initiating physical activity and chapter 12 provides more specific details relative to resistance training. Table 11.7 provides general guidelines for getting started with physical activity. These recommendations are suitable for the general population, including those with overweight or obesity.

Can a Person Over-exercise? Like many aspects of lifestyle choices, moderation is the key when it comes to physical activity. While regular aerobic and resistance training has been found to support health and reduce the risk for chronic disease, excessive exercise may have negative impacts on the body. For example, Hoenig and others report that rates of bone stress injuries, including stress fractures, are much greater with excessive exercise. These are described as over-use injuries. Over-use injuries occur from tissue damage, often resulting from repetitive movements. In addition to stress fractures, shoulder impingement and lateral epicondylitis (tennis elbow) are common **over-use injuries**. Growing children and adolescents are particularly prone to over-use injuries, but they can occur at any age. The *Training Table* in this section provides recommendations for preventing over-use injuries. Specifically related to physical activity to support weight loss and maintenance in individuals with overweight or obesity, it is important to provide education on over-use injuries. Over-use injuries can quickly derail a well-intended physical activity program and have negative impacts on the success of a weight-loss program.

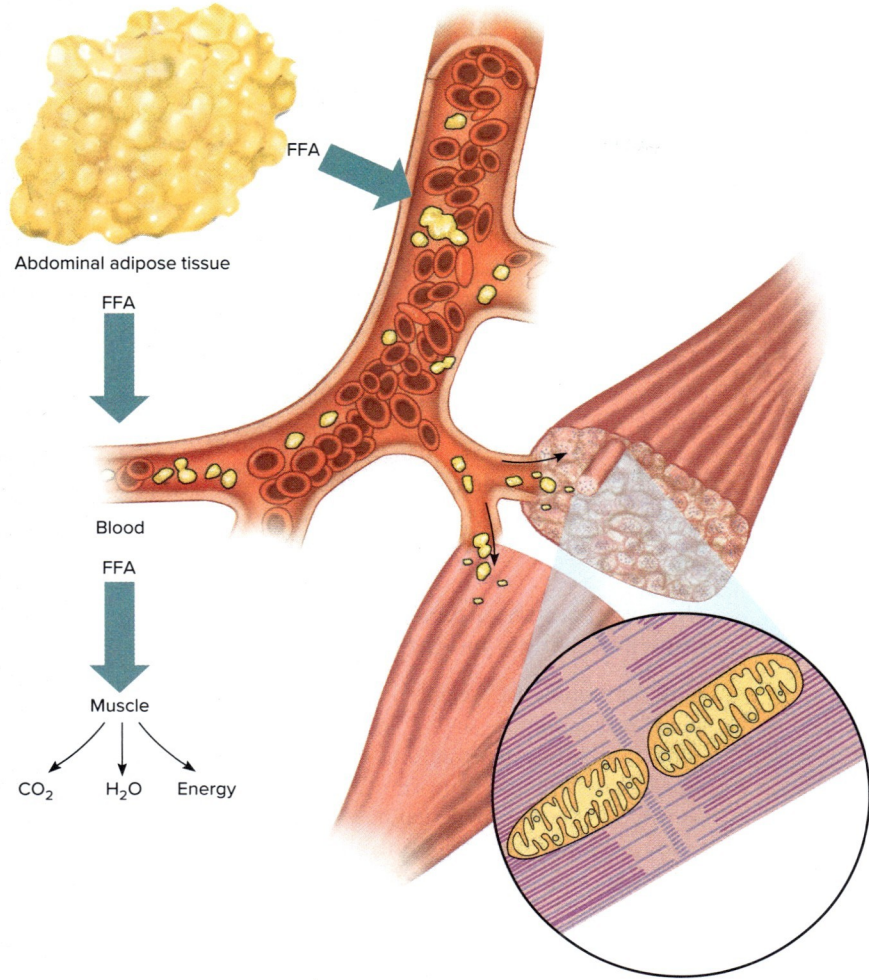

FIGURE 11.9 Exercise helps release fat (free fatty acids) from the adipose tissue, particularly abdominal fat cells. The fat then travels by way of the bloodstream to the muscles, where the free fatty acids are oxidized to provide the energy for exercise. Thus, exercise is an effective means of reducing body fat.

- Supporting safe programs that allow older adults and children to walk, bike, or wheelchair roll to reach their destinations.
- Through implanting policies, making streets safe for people of all ages and abilities to walk, bike, wheelchair roll, or take public transit.
- Supporting planning that connects active transportation and public transit planning.

For more information about community design, visit the CDC website at www.cdc.gov/chronicdisease/resources/publications/factsheets/physical-activity.htm.

How does someone get started with a physical activity program?

There are numerous online resources available from credible sources to assist someone who currently lives a sedentary lifestyle

 Training Table

Over-use injuries occur in individuals and athletes of all levels. As one begins an exercise program, it is particularly important to be aware of opportunities to reduce a person's risk for over-use injury. Such recommendations include:

- Allow the body adequate time to rest and recover. Do not overdo it.
- Limit repetitive movements, such as the number of repetitions with a specific workout routine, like lifting weights.
- Learn and practice correct techniques and exercise with proper equipment.
- Gradually increase exercise to meet goals.

TABLE 11.7	Recommendations for initiating a physical activity program.
	If you have health conditions, such as heart disease or type 2 diabetes, talk to your health-care provider before beginning any physical activity program.
	Look for ways to minimize the amount of time spent in sedentary activities, such as sitting at the computer. If work or school requires a lot of sitting, set a timer to move around at least every hour.
	Start slow – walking or swimming is a great start!
	Find activities you enjoy and that will work with your schedule and lifestyle. Schedule physical activity at the times of day when you feel most energized.
	Invite your friends or family to join you, providing a social network to help stick with the physical activity.
	Make a plan and track your physical activity using an electronic fitness tracker or simply keeping a physical activity diary in a journal.

Source: Centers for Disease Control and Prevention. Updated 2022. Getting started with physical activity. www.cdc.gov/healthyweight/physical_activity/getting_started.html

Key Concepts

- Physical activity is an important component of programs designed to support weight loss in individuals with overweight or obesity. Reducing sedentary activities and participating in physical activity supports weight loss as well as preventing weight regain following loss.
- The *Physical Activity Guidelines for Americans, 2018*, recommends adults engage in at least 150 minutes of moderate-intensity physical activity per week to support health. For individuals whose goal is to lose significant weight (more than 5 percent of body weight) and/or those who are trying to prevent weight regain, the guidelines recommend engaging in at least 300 minutes per week of moderate-intensity physical activity.
- Both resistance training and aerobic training offer benefits to individuals following a weight-loss program. Aerobic training is particularly important for increasing kcal (energy) expenditure, and resistance training reduces fat mass while increasing lean body (fat-free) mass.
- For those just starting a physical activity program, the section provides recommendations for initiating such a program. Moderation is important as someone starts exercising — too much exercise too quickly can contribute to over-use injuries.

Check for Yourself

- Track your physical activity and sedentary behaviors (i.e. screen time, sitting in class) over the course of a week. If available, use a fitness tracker and check the reports daily. Are you meeting physical activity recommendations to support optimal health and wellness? If not, what could you do differently?

Weight-Loss Programs

Every year, approximately 45 million adults in the United States go on a weight-loss diet. Thousands of weight-loss programs, devices, specialized meals and meal programs, and other treatments exist, and Americans spend over $30 billion on these products. As of 2022, the diet industry is worth over $70 billion! Despite the prevalence and popularity of weight-loss diets, research shows that about 95 percent of diets fail. On average, adults in the United States will start a new fad diet twice a year, most often abandoning the diet within 2 weeks.

Rick Gomez/Getty Images

What factors should be considered when selecting a weight-loss program?

With thousands of weight-loss programs available, finding the **weight-loss program** that is right for you can be challenging. Important before starting any weight-loss program is to talk to your health-care provider, who can provide specific recommendations based on your personal health history, family history, medication use, and weight and health goals. Ultimately there is no one "best" weight-loss program, but rather weight-loss programs that work best for individuals. When evaluating weight-loss programs, finding a program that is likely to meet your individual needs is essential. Before selecting a weight-loss program, consider what has worked, or not, for you in the past, as well as your personal preferences in terms of support (online, in person, individual, group, etc.). As well, budget and health insurance coverage become important considerations, with some weight-loss programs being expensive. Table 11.8 reviews the characteristics of safe and effective weight-loss programs, and figure 11.10 summarizes what to look for, and what to avoid, when selecting a weight-loss program.

> *www.eatright.org/* The Academy of Nutrition and Dietetics consumer-based website provides articles on a variety of topics, including weight loss. Search "weight loss" for recently published articles related to weight loss.
>
> *www.cdc.gov/healthyweight/tools/index.html* The CDC website on Healthy Weight, Nutrition, and Physical Activity provides links to a variety of resources to support weight loss.

What types of weight-loss programs are available?

Weight-loss programs come in many different forms, with new programs being developed and marketed each year. There is always

TABLE 11.8 Recommendations to consider when selecting a weight-loss program

Characteristics of safe and effective weight-loss programs:

Program includes behavioral (lifestyle) guidance on how to develop and stick with habits that support health and weight loss, such as diet and physical activity

Program provides information on the importance of getting enough quality sleep and managing stress

Program shares both the benefits and drawbacks associated with weight-loss medications

Program provides ongoing feedback, monitoring and support (in person, by phone, online, or a combination)

Program recommends slow and steady weight-loss goals (usually 1–2 pounds per week, but weight loss may be faster at the start of the program)

Program includes a plan for keeping the weight off (goal setting, self-checks, food journal, counseling support, etc.)

Use caution if weight-loss programs make the following promises:

Lose weight without diet or exercise

Lose weight while eating as much as you want of all of your favorite foods

Lose 30 pounds in 30 days

Lose weight in the problem areas of your body

Source: National Institute of Diabetes and Digestive and Kidney Diseases. Updated 2022. Choosing a safe and successful weight-loss program. www.niddk.nih.gov/health-information/weight-management/choosing-a-safe-successful-weight-loss-program

something new on the market. Figure 11.11 shows some examples of popular weight-loss programs.

In-person Weight-loss Programs Many individuals with overweight or obesity prefer an individualized, in-person approach. RDNs are licensed and trained experts who provide individuals with weight-loss guidance specific to their health history, personal goals, motivation level, and more. As the name implies, it is tailored to the individual. A meta-analysis by Nitschke and others reports that RDNs play key roles in facilitating positive lifestyle behaviors and supporting weight loss for adults with overweight or obesity.

Motivational interviewing is one technique that is utilized as a part of in-person individualized weight-loss programs. With this technique, health-care providers are trained on a counseling approach for supporting behavior changes that may be beneficial in the management of overweight and obesity. Michalopoulo and others conducted a meta-analysis comparing the effectiveness of behavioral-based programs incorporating motivational interviewing and those that do not. They found no significant differences and concluded that, given the training and time involved with motivational interviewing, it may not be worthwhile as an added component to weight-loss programs.

Online or App-based Weight-loss Programs The number of people using online and app-based weight-loss programs has grown exponentially in the past decade. Much of this growth is due to advances in technology and the plethora of information that is now available on common electronic devices including cell phones, tablets, and computers. As well, the COVID-19 pandemic changed how health-care is provided to patients, with more visiting being done virtually through e-Health platforms. During the pandemic, many weight-loss programs transitioned to fully virtual, including

What NOT to look for in a weight-loss program

- Promises quick results
- Is overly restrictive with only certain foods and beverages that you can eat and a long list of "cannot eat" options
- Provides only generic, rather than individualized, meal plans
- All materials are provided up-front with no access to talk with a trained professional about your goals and progress
- Focuses only on diet, with no discussion of other lifestyle factors

What to look for in quality and successful weight-loss programs

- Recommends slow and steady weight loss
- Focuses on nutrient-dense food and beverage items, moderation, and emphasis on what you can eat (rather than avoid)
- Meal plans are individualized based on your food preferences, access to certain foods, culture, etc.
- Trained professionals are readily available to talk by online chat, text, online video, or phone
- Behavioral changes, such as physical activity, sleep, and stress management are emphasized

Plate: Natalia Klenova/Shutterstock; Fresh salad: Tatjana Baibakova/Shutterstock

FIGURE 11.10 When selecting a healthy and successful weight-loss program, consumers should be selective and consider many aspects of the program. This figure summarizes what to look for, and what to avoid, when selecting a weight-loss program.

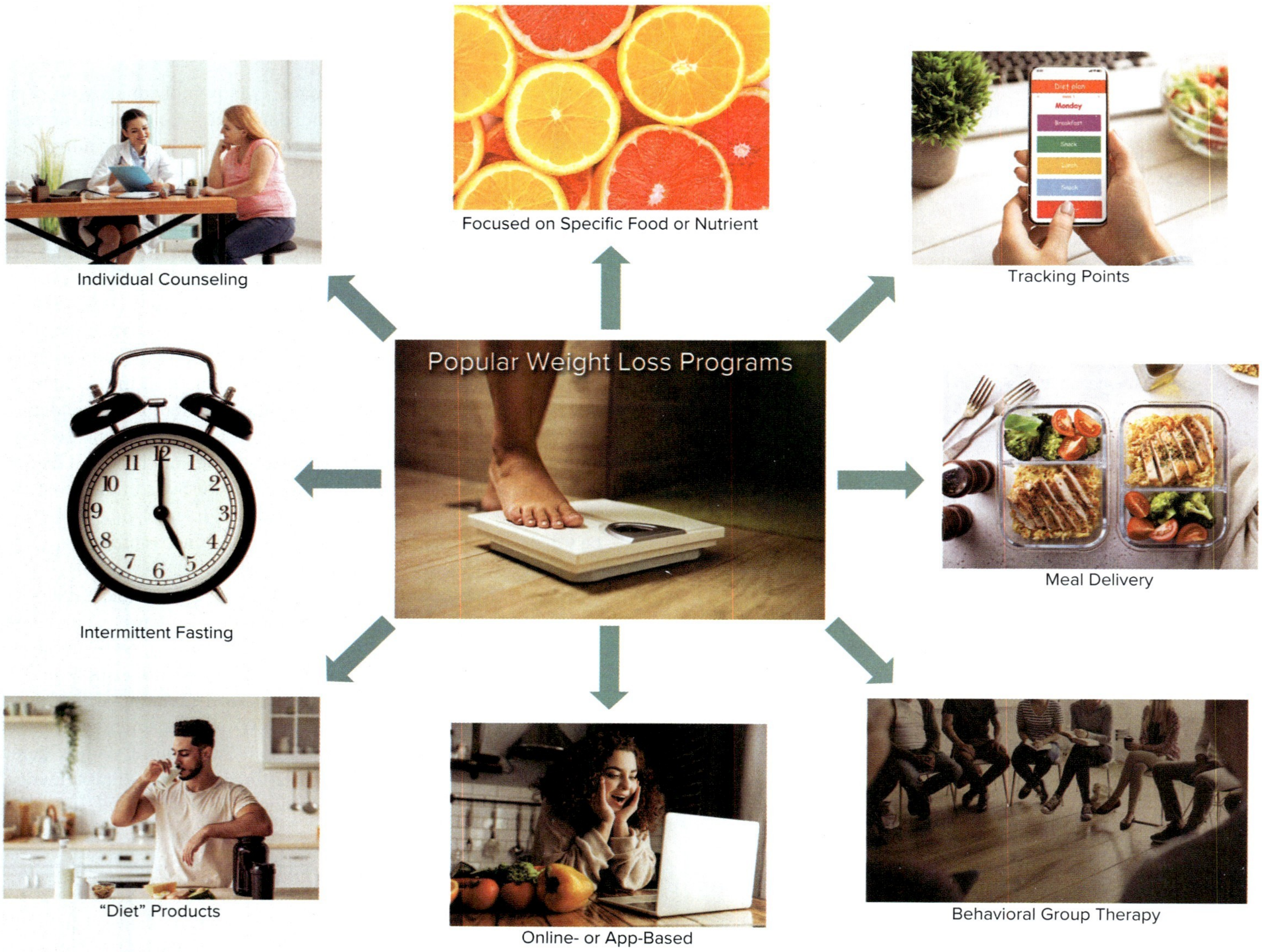

Young female: Africa Studio/Shutterstock; Orange and grapefrui: Len44ik/Shutterstock; Weightloss and healthcare: Prostock-studio/Shutterstock; Alarm clock: Gemenacom/Shutterstock; Woman weighing: Rostislav Sedlacek/Shutterstock; Healthy meal: Elena Veselova/fahrwasser/123RF; Healthy nutrition: Prostock-studio/Shutterstock; Image of health: Dean Drobot/Shutterstock; People sitting: Rawpixel.com/Shutterstock

FIGURE 11.11 In the United States, the diet industry is a multi-billion-dollar industry. There are many different types of weight-loss programs and approaches available. This figure summarizes some of the popular weight-loss programs.

visits with RDNs and other health-care providers. A review by Wright and others reported that e-Health is a suitable option for patients with obesity, particularly when a person has limited or no access to health-care teams and/or requires additional support.

A review and meta-analysis by Mamalaki and others found that web- and app-based weight-loss interventions resulted in more weight regain compared to in-person interventions. However, this finding should be interpreted with caution as others have found conflicting results. For example, Antoun and colleagues report that, while smartphone apps have a role in weight management and preventing weight regain, they are most effectively utilized when combined with human-based behavioral interventions. Given advances in technology, it is likely that web- and app-based weight-loss programs will continue to evolve and become even more popular in health-care. The *Training Table* in this section provides recommendations for individuals looking for online weight-loss programs.

Behavioral Group Therapy

Many programs designed to support weight loss are offered in group settings. Group therapy can be a cost-effective option for individuals seeking a weight-loss program and are less resource-intensive for health-care providers and facilities.

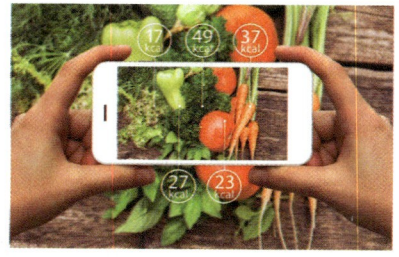

Scharfsinn/Shutterstock

Training Table

For many people seeking out weight-loss programs, particularly those with busy schedules where regular in-person sessions will be challenging to make, they are looking for online programs. Many weight-loss programs now offer fully online options or a combination of in-person and online. When selecting an online weight-loss program, experts recommend asking the following questions?

- Does the program offer organized weekly lessons tailored to your personal goals?
- Does the program offer a plan for you to track your progress in making lifestyle changes, including those related to nutrition, physical activity, sleep, and stress management?
- Does the program include support from a qualified expert who is familiar with your specific goals?
- Does the program provide you with regular feedback on your results and progress toward meeting goals?
- Does the program offer social interactions and support from others in the group through online discussion board platforms or online meet-ups?

Additional guidance on selecting safe and successful online weight-loss programs can be found at www.niddk.nih.gov/health-information/weight-management/choosing-a-safe-successful-weight-loss-program.

TABLE 11.9 Evidence-based interventions to support healthy body weight in children and adolescents

Age	Evidence-based Interventions
2–5 years	Interventions in the home and family, healthcare and community settings
6–12 years	Interventions in school Regulated screen time
All children and teens	Nutrition with physical activity Multicomponent, multilevel, or multisetting interventions Policies to improve access to healthy foods

Source: Hoelscher, D., et al. 2022. Prevention of Pediatric Overweight and Obesity: Position of the Academy of Nutrition and Dietetics Based on an Umbrella Review of Systematic Reviews. *Journal of the Academy of Nutrition and Dietetics* 122(2):410.

The position paper from the Academy of Nutrition and Dietetics (AND) provides evidence-based recommendations for weight-loss interventions that are most likely to be successful based on a child's developmental age. Table 11.9 summarizes key interventions based on a child's age.

Intermittent fasting, high-protein diets, high-fat (keto) diets . . . so many options! Is there one that is the "best" for weight loss?

Many programs designed to support weight loss are focused on specific food groups, food items, and/or nutrients. For example, the "Grapefruit Diet" originated in the 1930s. Those following this diet ate only grapefruit, orange, toast, vegetables, and eggs for a total of 500 kcal daily. People followed this diet for 18 days. Now, in the twenty-first century, people are still following the "Grapefruit Diet" in hopes that this "miracle food" will result in significant weight loss. Do people following this diet often lose quick weight? Yes, but only as a result of the significant cut in kcal intake to only about 500 kcal daily. Experts agree that it is not necessarily the grapefruit but rather kcal deficit that can support short-term weight loss. And, for most people, following the Grapefruit Diet long term is not sustainable.

As of 2022, some of the more common weight-loss programs are focused on intermittent fasting, high-protein, and high-fat (keto) diets. Each of these weight-loss diets will be summarized in the following section. For the latest research on a specific weight-loss diet, visit https://pubmed.ncbi.nlm.nih.gov/.

Intermittent Fasting Intermittent fasting, as the name implies, is a dietary approach that involves consuming foods and beverages at only set times. This is then followed by a fasting period, typically with kcal-free beverages, such as water, being consumed during that period. Interestingly, there are different types of intermittent fasting, each with unique goals and potential impacts on health. Table 11.10 summarizes three main forms of intermittent fasting, but other slightly modified forms are also followed.

As well, social interactions and group support and accountability can support weight loss. Karisson and colleagues conducted a study evaluating the effects of a low-energy diet combined with **behavioral group therapy** in adults with obesity. After 6 months, they reported substantial weight loss as well as significant improvements in health-related quality of life, including physical, psychosocial, and mental health. The researchers suggest that improvements in psychosocial health were likely a result of both the weight loss and the group behavioral therapy. Similar findings were reported by Borek and others, who found that interventions delivered in groups were effective in promoting significant weight loss at 12 months.

Are there specific weight-loss programs for children and adolescents?

When developing and implementing weight-loss programs, it is important to consider that the needs of children and adolescents are vastly different from that of adults. For example, most children do not have autonomy over food that is available in their home or at school. Rather, involving families, schools, and other caregivers in the conversation and weight-loss program is important. As well, children and many adolescents are still growing and restricting nutrient and kcal intake could negatively impact normal growth.

TABLE 11.10 Three main types of intermittent fasting

Roland Magnusson/Shutterstock

	Main characteristics of this form of intermittent fasting
Alternate-day fasting	With alternate-day fasting, a person rotates days of eating and days of fasting. On fasting days, a person consumes no foods or beverages with calories (kcal-free drinks such as black coffee, tea, and water can be consumed), while on feeding days a person eats whatever they want. On eating days, following a nutrient-dense diet is recommended.
Modified fasting	With modified fasting, rather than consuming 0 kcals on fasting days, a person restricts kcal intake to 20–25 percent of daily needs. Other forms of modified fasting limit kcals to 500 per day on fasting days. Another common modified fasting protocol involves the 5:2 fast. In the 5:2 fast, a person eats/feeds 5 days of the week and then fasts for 2 days of the week.
Time-restricted fasting	In time-restricted fasting protocols, consumption of kcals is limited to waking hours. Ideally, a person fasts for 8–12 hours per 24-hour period. This fasting can occur during sleep. For some people, the fasting time is expanded such that a person fasts for 16 or more hours per 24-hour period.

Source: Gordon, B. Reviewed April 2021. What is intermittent fasting? www.eatright.org/health/wellness/fad-diets/what-is-intermittent-fasting

While there have been several studies conducted on the impact of intermittent fasting on weight loss and risk for chronic disease, the results are inconclusive. This is likely due to the variability in how intermittent fasting is defined and the "rules" associated with various forms of fasting. As well, there are also possible risks associated with intermittent fasting, including nutrient deficiencies, dehydration, and low energy levels that impact daily activities. Females who are pregnant or breastfeeding, individuals with diabetes, and persons with a history of disordered eating and/or eating disorders are discouraged from following an intermittent fasting plan due to potential negative effects associated with the fasting period.

High-protein Diets and Weight Loss

As the name implies, a high-protein diet is one that supplies amounts of protein greater than recommended by the Acceptable Macronutrient Density Ranges (AMDRs). Based on the AMDR for protein, adults should consume 10–35 percent of their total daily kcals from protein. There is no "standard" protein recommendation associated with "high" protein diets, which makes evaluation of the effects of such a diet on weight loss challenging. In a meta-analysis by Hansen and others, they evaluated research studies with protein intake of 18–59 percent of total kcals and found that diets rich in protein may have a moderate beneficial impact on weight loss. The weight-loss effects were most pronounced in individuals with obesity who also had prediabetes and/or specific genetic risk factors. Similarly, Kohanmoo and others report in their meta-analysis that acute ingestion of protein may suppress appetite and impact hormones that regulate hunger and satiety. However, long-term trials on weight-loss and high-protein diets are inconclusive. Additional information about high-protein diets can be found in chapter 6.

High-fat (keto) Diets and Weight Loss

Ketogenic diets have been successfully utilized as a therapy for drug-resistant epilepsy for over 100 years. As described in more detail in chapter 5, a ketogenic diet is one that is very low in carbohydrates, moderate in protein, and high in fat. A "classic" ketogenic diet (4:1) provides 90 percent of kcal from fat, 6–8 percent of kcal from protein, and 2–4 percent of kcal from carbohydrates.

Numerous studies have been conducted on the effects of ketogenic diets on weight loss with conflicting results. A review by Golabek and Regulska-Ilow reports that very low-kcal ketogenic diets "may be an effective possibly safe and patient-motivating component of a long-term weight loss plan" but that more long-term studies are needed. Basolo and others explain that a low-kcal ketogenic diet may result in a short-term increase in energy expenditure due to increased fatty acid breakdown and activity of ketogenic diets. When glucose is still available for utilization, there is increased liver oxygen consumption to support gluconeogenesis and increases in triglyceride-fatty acid recycling.

The long-term safety of a ketogenic diet is an area that requires further investigation. As reported by Bostock and others, common side effects of following a ketogenic diet include fatigue, headache, nausea, low blood glucose levels, and constipation. More serious side effects, including pancreatitis and hypertriglyceridemia, have also been reported. As well, following a ketogenic diet for a long period of time will result in nutrient deficiencies, which

SewCream/Shutterstock

may increase the risk for acute and chronic disease. For these reasons, Crosby and others report that, as a diet for weight loss, the risks associated with following a classic, "true," ketogenic diet outweigh any potential benefits.

Key Concepts

- In the United States, 45 million adults are on a weight-loss diet each year. The diet industry is worth over $70 billion. Given the financial gains possible in the weight-loss industry, thousands of unique programs, devices, diet plans, and more exist.
- Consumers should critically evaluate weight-loss programs and determine which best meets their goals and needs. There is no one "ideal" weight-loss program as weight-loss success is quite individualized.
- In-person and online weight-loss programs can support weight loss. Motivational interviewing is one technique utilized in weight-loss interventions. For many people, behavioral group therapy contributes to weight loss and maintenance of that weight loss.
- Children and adolescents have unique needs with regard to weight-loss programs, and health-care professionals with training and experience working with children and adolescents can best support weight-loss efforts in this population.
- Intermittent fasting, high-protein diets, and low-fat (keto) diets are popular as weight-loss diets. Study results are conflicting on the effects of each of these diets on short- and long-term weight loss. More research is needed.

Check for Yourself

- Search online for at least five popular weight-loss diets. Summarize the characteristics of each of the diets and critically evaluate each as a weight-loss program. Would you recommend any of these for a friend or family member with overweight or obesity? If so, describe why. If not, why not?

Weight Loss and Athletic Performance

As described throughout this chapter, overweight and obesity are significant health concerns of the twenty-first century. In this section, we summarize the impacts of weight loss on athletic performance. We start the section by exploring how some treatment options for obesity, including bariatric surgery, can negatively impact fitness levels and how to counteract those potential detrimental effects on physical activity. The second part of this section then explores the impact of weight loss on sport-specific performance. In some instances, weight loss can actually impair exercise performance, including in elite athletes.

What are the recommendations for exercise before and after bariatric surgery?

Bariatric surgery is a major surgery that can result in significant improvements in body weight and body composition while reducing risk and/or treating chronic disease. Despite these important benefits, two potential side effects of bariatric surgery included reduced **cardiorespiratory fitness** and **sarcopenic obesity**. As described in chapter 10, sarcopenic obesity is associated with excess fat mass, low muscle mass, and low physical function. There is an opportunity to utilize specially designed exercise programs both before and after bariatric surgery to support clinical outcomes following the surgery while also reducing the risk for sarcopenic obesity and improving cardiorespiratory fitness.

New Africa/Shutterstock

Ibacache-Saavedra and others conducted a review and meta-analysis to investigate the effects of bariatric surgery on cardiorespiratory fitness postsurgery. Based on the limited evidence available, they found that the weight loss induced by bariatric surgery, while important from a health standpoint, did result in reduced cardiorespiratory fitness. They describe this finding as an opportunity for health-care providers and researchers to develop therapies to prevent the loss of cardiorespiratory fitness to further optimize outcomes following bariatric surgery. One such intervention that has been studied is a preoperative exercise intervention. In a recent meta-analysis, Durey and colleagues found that preoperative exercise resulted in increases in cardiorespiratory fitness both before surgery and at follow-up time points. While the exercise intervention did not result in a significant difference in weight loss, the short- and long-term benefits to fitness in individuals with obesity are an important finding.

In addition to preoperative exercise, other studies have investigated the effects of postoperative exercise on weight loss, cardiorespiratory fitness, and other markers of health. Bellicha and others report that postoperative exercise programs positively impacted bone mineral density as well as weight maintenance following weight loss. Postoperative exercise programs can improve physical fitness and lead to small additional weight and fat loss following bariatric surgery. As summarized by Vierira and others, when resistance training is included as part of the postoperative exercise program, individuals may also have improvements in muscle strength, which are associated with improved metrics related to sarcopenic obesity, functional capacity, and mortality risk.

Based on the available evidence, the American Society for Metabolic and Bariatric Surgery (ASMBS) has developed recommendations for exercise in patients with obesity both before and after bariatric surgery. The *Training Table* in this section summarizes those recommendations.

Training Table

For individuals with obesity who will be having bariatric surgery, pre- and postoperative care, including lifestyle interventions, supports optimal outcomes following the surgery. Before beginning any exercise program, it is important to consult your health-care provider to discuss specific exercise recommendations and any contraindications to any types of physical activity. The ASMBS recommendations for physical activity include:

- Mild exercise, including aerobic conditioning and light resistance training, 20 minutes per day, 3–4 days per week prior to surgery.
- Postoperative exercise recommendations are consistent with general recommendations for adults to support a healthful lifestyle. This includes being physically active as part of activities of daily living as well as at least 150 minutes of moderate-intensity physical activity per week. Muscle-strengthening activities targeting major muscle groups should be completed at least twice per week.

By staying active both before and after surgery, an individual undergoing bariatric surgery can best support their health as well as recovery following surgery.

---

https://asmbs.org/resources/physical-activity The American Society for Metabolic and Bariatric Surgery (ASMBS) provides resources on a variety of topics related to bariatric surgery. This includes recommendations for physical activity.

www.obesityaction.org/ Visit the Obesity Action Coalition website for information on obesity treatments, including bariatric surgery. Search "Physical Activity" for detailed recommendations for patients pre- and postbariatric surgery. This includes appropriate exercise modifications for individuals with obesity as well as during the postoperative care period when caution must be practiced with physical activity.

How does weight loss impact sport-specific performance in athletes?

For some athletes, weight loss can have a negative impact on both health and sport-specific performance. For example, as described by Smith and others, in the sport of bouldering (a form of sport climbing), a low body weight to power ratio is considered ideal. Many bouldering athletes restrict kcal and carbohydrate intake to achieve low body weight, which can have a detrimental effect on performance given the sport is high intensity and requires carbohydrate availability for both mental focus and physical performance. For this reason, an athlete must balance out the potential benefits of having a lower body weight with the negative consequences of following a kcal- and carbohydrate-restrictive diet.

LStockStudio/Shutterstock

Similar to the example of bouldering athletes, such negative health and sport-specific impacts can occur as a result of weight loss in many different sports. As such, it is important to consider both the potential pros and cons of weight loss in athletes, particularly during the competitive season. While many sports have an "ideal" body type for the sport or even specific positions in the sport, it is imperative to recall that we each are individuals and what may be "ideal" for one person is not for another. Setting standards for BMI and body composition in athletes can be detrimental to the athlete, ultimately having a negative impact on both health as well as athletic performance.

Energy Availability in Athletes For elite athletes, either losing weight intentionally (i.e., reducing kcal intake) or unintentionally (i.e., burning kcals through exercise and not refueling with adequate energy) impacts **energy availability**. Energy availability (EA) is calculated from the following equation:

$$EA = \text{Food Energy (kcal) Intake} - \text{Exercise Energy (kcal) Expenditure}$$

When food energy intake is less than exercise energy expenditure, this will likely result in weight loss but also adverse effects associated with **low energy availability** (low EA). With a low EA, the body systems adjust, which leads to changes in hormone levels and metabolic and sport-specific changes. Table 11.11 outlines some of the

TABLE 11.11 Common symptoms of low energy availability in athletes

Chronic fatigue
Recurring illness and/or infections
Inability to gain or build muscle or strength
Poor performance
Repeated injuries, such as stress fractures
Irritability
Training hard but not improving performance
Gastrointestinal problems, such as constipation
Weight loss

Source: National Collegiate Athletic Association (NCAA). Updated 2022. Fact Sheet: Energy Availability. https://ncaaorg.s3.amazonaws.com/ssi/nutrition/SSI_EnergyAvailabilityFactSheet.pdf

common symptoms that an athlete has low EA, and the *Training Table* in this section provides recommendations on managing EA in athletes.

Training Table

Low EA not only impacts sports performance and health in the short term, but there are also serious long-term effects, including osteoporosis and fertility complications. For athletes experiencing low EA, the following recommendations are provided:

- Plan out your meals and snacks, ensuring that you are eating regularly throughout the day, particularly during times with high training loads.
- To ensure you are eating enough kcals, consider high-kcal foods and beverages such as nuts and nut butter, energy shakes, and more.
- If possible, reduce your energy expenditure. While you may not be able to reduce your sport-specific training, consider other energy expenditures during the day, such as walking between classes (is biking an option instead) and training you may do on the side.
- If weight loss is a goal, talk with your health-care provider, trainers, RDN, and other professionals to develop a realistic plan for weight loss. Maybe weight loss in-season is not the best option and may actually impair performance.
- If increases to LBM is a goal, trying to gain muscle mass during the season may be difficult and result in low EA. Again, such changes to LBM might be best achieved during the off-season.

https://ncaaorg.s3.amazonaws.com/ssi/nutrition/SSI_EnergyAvailabilityFactSheet.pdf The National Collegiate Athletic Association (NCAA) fact sheet on low energy availability provides practical guidelines and recommendations for student-athletes as well as the trainers, RDNs, coaches, and others working with the student-athlete.

www.eatright.org/fitness/exercise/exercise-nutrition/6-healthy-ways-to-manage-weight-for-sports The Academy of Nutrition and Dietetics provides practical guidelines for athletes to manage their weight for sport and fitness.

Does weight cycling increase risk for obesity later in life?

As described in chapter 10, some athletes participate in extreme measures that result in rapid weight loss prior to major events and competitions. This rapid weight loss is then often followed by weight regain, resulting in **weight cycling**. Boxers, mixed martial arts fighters, weight lifters, jockeys, dancers, and gymnasts are particularly prone to using weight cycling to meet certain weight categories or subjective appearance markers in their sport. Saarni and colleagues conducted a study evaluating the effects of weight cycling in athletes as young adults and how that impacted weight in middle age. They found that individuals who practiced weight cycling as young athletes were more likely to have obesity than athletes who did not participate in weight cycling as young adults. The researchers conclude that the chronic dieting associated with weight cycling may predispose these athletes to obesity later in life.

In addition to potentially increasing the risk for obesity later in life, weight cycling practices have also been evaluated with regard to their impact on cardiometabolic function and other health metrics. For example, Morehen and others report on findings from a 5-year analysis of the impact of weight cycling in professional boxers on risk for obesity and cardiometabolic disease. Their results suggest that weight cycling practices can contribute to future obesity and cardiometabolic disease risk. Miles-Chan and Isacco provide an informative review discussing the research that has been conducted on this topic; at this time, that research on long-term effects is lacking.

Individuals working with athletes in sports where weight cycling is common should be aware not only of the short-term impacts of weight cycling on parameters of health but also of future risk for obesity and related forms of chronic disease.

Blend Images/Image Source

Key Concepts

- Specialized designed fitness plans implemented both before and after bariatric surgery can support weight loss and cardiorespiratory fitness while reducing the risk for sarcopenic obesity following surgery.
- In some athletes, losing weight can have negative impacts on both health and sport-specific performance. Low energy availability occurs when an athlete's kcal intake is less than kcal expenditure.
- Weight cycling occurs when an athlete rapidly loses weight before a competition or major event and then regains that weight following the competition. Such weight cycling in young adults can contribute to an increased risk for obesity later in life.

APPLICATION EXERCISE

Friends Aaliyah and Jordan recently met at behavioral group therapy for weight loss. Both Aaliyah and Jordan have obesity and are highly motivated to lose weight, increase LBM, and have more energy for their busy schedules.

1. What is the recommended rate of weight loss, and what recommendations do you have in terms of kcal intake to support that rate of weight loss?
2. Provide at least five specific dietary recommendations that support healthy weight loss.
3. Aaliyah and Jordan are interested in learning more about mindful eating. Describe mindful eating, including at least five specific characteristics of eating mindfully.
4. Explain to Aaliyah and Jordan the importance of physical activity as part of their weight loss journey. What recommendations do you have in terms of a physical activity plan?

Shutterstock

Review Questions—Multiple Choice

1. Each of the following is a health effect associated with overweight and obesity except _____.
   a. metabolic syndrome
   b. sleep apnea
   c. high blood pressure
   d. type 1 diabetes
   e. depression and anxiety
2. Which of the following is considered an effective, evidence-based recommendation to support weight loss?
   a. Limit kcal intake to no more than 800 kcal daily.
   b. Losing weight is only about kcal; the source of those kcals does not matter.
   c. The 85–15 guide can support weight loss, limiting the intake of foods high in added sugars, saturated fat, and sodium.
   d. Ultra-processed foods are a healthy addition to a weight-loss diet.
   e. Avoid eating grains as carbohydrates are the cause of obesity.
3. According to the *2020–2025 Dietary Guidelines for Americans*, a healthy weight-loss diet is one that _____.
   a. limits added sugars
   b. includes fresh fruits and vegetables
   c. is low in saturated fat
   d. emphasizes lean sources of quality protein
   e. All of the above are correct.
4. Mindful eating involves _____.
   a. taking time to enjoy food
   b. chewing quickly
   c. setting a timer to eat meals or snacks every 2 hours
   d. eating while watching podcasts on meditation
   e. avoiding meals with others, rather focusing on eating alone
5. Individuals with obesity are at risk for disordered eating and eating disorders. The eating disorder most common in adults with overweight or obesity is _____.
   a. bulimia nervosa
   b. muscle dysmorphia
   c. anorexia nervosa
   d. binge-eating disorder
   e. orthorexia
6. According to the *Physical Activity Guidelines for Americans, 2018*, adults with obesity should engage in moderate-intensity physical activity _____.
   a. daily for at least 1 hour
   b. at least 150 minutes a week
   c. for 30 minutes every day of the week
   d. at least 300 minutes a week
   e. only as part of activities of daily living (Those with obesity should not exercise at a moderate- or high intensity.)
7. For an individual with overweight or obesity, resistance training can be beneficial as part of a weight-loss program. One of the most important effects of resistance training in individuals with obesity is _____.
   a. burning a significant number of kcal to support rapid weight loss
   b. increasing cardiometabolic fitness
   c. improvements to body composition, increasing lean body mass
   d. significantly increasing kcal expenditure
   e. increasing nutrient absorption
8. Which of the following is *not* a characteristic of a safe and effective weight-loss program?
   a. Program provides information on the importance of getting enough quality sleep and managing stress.
   b. Program includes a plan to keep the weight off.
   c. Program provides ongoing feedback, monitoring, and support.
   d. Program encourages rapid weight loss.
   e. Program includes lifestyle guidance, such as related to diet and exercise.
9. _____ involves rotating days of eating and days of fasting. On fasting days, a person consumes no foods or beverages with kcal and on feeding days a person

is encouraged to eat whatever they want while following a nutrient-dense diet.
   a. Alternate-day fasting
   b. Modified fasting
   c. Restrictive fasting
   d. Time-restricted fasting
   e. 5:1 fasting

10. Low energy availability occurs in an athlete when _____.
    a. the athlete does not consume enough fresh foods
    b. consumption of protein is below recommended levels
    c. energy intake is less than energy expenditure
    d. the athlete consumes a plant-based diet
    e. calcium and vitamin D levels are insufficient

Answers to multiple choice questions: 1. d; 2. c; 3. e; 4. a; 5. d; 6. b; 7. c; 8. d; 9. a; 10. c.

Critical Thinking Questions

1. What is weight bias, and what are the impacts on health and wellness for individuals experiencing weight bias with their health-care provider? Provide at least five recommendations for healthcare providers to reduce weight bias in their practice.
2. Compare and contrasts the pros and cons of in-person versus online weight-loss programs. Is one "better" than the other? Why or why not?
3. Explain why weight-loss programs for children and adolescents should not be the same as for adults. What are the key characteristics of weight-loss programs for children and adolescents?
4. A family member has obesity and is scheduled to have bariatric surgery in 2 months. Describe to this family member the importance of appropriately planned physical activity both before and after the surgery. Be specific.
5. You have the opportunity to talk with a community planner as part of the development of several new subdivisions in the community. What recommendations would you make to the community planner to support the health of those who will be living in the community? What can be done in community development to reduce rates of overweight and obesity in the community?

References

Anguita-Ruiz, A., et al. 2022. Body mass index interacts with a genetic-risk score for depression increasing the risk of the disease in high-susceptibility individuals. *Translational Psychiatry* 12(1):30.

Antoun, J., et al. 2022. The effectiveness of combining nonmobile interventions with the use of smartphone apps with various features for weight loss: systematic review and meta-analysis. *JMIR Mhealth Uhealth* 10(4):e35479.

Astrup, A., and Monteiro, C. 2022. Does the concept of "ultra-processed foods" help inform dietary guidelines, beyond conventional classification systems? NO. *American Journal of Clinical Nutrition* June 7, 2022. Online ahead of print.

Batrakoulis, A., and Fatouros, I. 2022. Psychological adaptations to high-intensity interval training in overweight and obese adults: A topic review. *Sports* 10(5):64.

Bellicha, A., et al. 2021. Effect of exercise training before and after bariatric surgery: A systematic review and meta-analysis. *Obesity Reviews* Suppl. 4: e13296.

Borek, A., et al. 2018. Group-based diet and physical activity weight-loss interventions: A systematic review and meta-analysis of randomized controlled trials. *Applied Psychology, Health, and Well Being* 10(1):62.

Bostock, E., et al. 2020. Consumer reports of "keto flu" associated with ketogenic diet. *Frontiers in Nutrition* 7:20.

Breton, E., et al. 2022. Immunoinflammatory processes: Overlapping mechanisms between obesity and eating disorders? *Neuroscience and Behavioral Reviews* May 17, 2022. Online ahead of print.

Calcaterra, V., et al. 2022. Use of physical activity and exercise to reduce inflammation in children and adolescents with obesity. *International Journal of Environmental Research and Public Health* 19(11):6908.

Cava, E., et al. 2017. Preserving healthy muscle during weight loss. *Advances in Nutrition* 8(3):511.

Choi, S., and Lee, H. 2020. Associations of mindful eating with dietary intake pattern, occupational stress, and mental well-being among clinical nurses. *Perspectives in Psychiatric Care* 56(2):355.

Clarke, T., et al. 2018. Use of yoga, meditation, and chiropractors among U.S. adults aged 18 and over. *NCHS Data Brief* No. 325, November 2018.

Colleluori, G., et al. 2022. Early life stress, brain development, and obesity risk: Is oxytocin the missing link? *Cells* 11(4):623.

Crosby, L., et al. 2021. Ketogenic diets and chronic disease: Weighing the benefits against the risks. *Frontiers in Nutrition* 8:702802.

Da Luz, F., et al. 2020. The treatment of binge eating disorder with cognitive behavior therapy and other therapies: An overview and clinical considerations. *Obesity Reviews* December 17, 2020. Online ahead of print.

De Ridder, D., and Gillebaart, M. 2022. How food overconsumption has hijacked our notions about eating as a pleasurable activity. *Current Opinions in Psychology* 46:101324.

Di Giacomo, E., et al. 2022. Disentangling binge eating disorder and food addiction: A systematic review and meta-analysis. *Eating and Weight Disorders* January 18, 2022. Online ahead of print.

Donnelly, J., et al. 2009. American College of Sports Medicine Position Stand. Appropriate physical activity intervention strategies for weight loss and prevention of weight regain for adults. *Medicine and Sciences in Sports and Exercise* 41(2):459.

Durey, B., et al. 2022. The effect of preoperative exercise intervention on patient outcomes following bariatric surgery: A systematic review and meta-analysis. *Obesity Surgery* 32(1):160.

Essel, K., et al. 2022. Discovering the roots: A qualitative analysis of medical students exploring their unconscious obesity bias. *Teaching and Learning in Medicine* March 3, 2022. Online ahead of print.

Golabek, K., and Regulska-Ilow, B. 2022. Possible nonneurological health benefits of ketogenic diet: Review of scientific reports over the past decade. *Journal of Obesity* 2022:7531518.

Hall, K., and Kahan, S. 2018. Maintenance of lost weight and long-term management of obesity. *Medical Clinics of North America* 102(1):183.

Hall, K., et al. 2019. Ultra-processed diets cause excess calorie intake and weight gain: An inpatient randomized controlled trial of ad libitum food intake. *Cell Metabolism* 30(1):67.

Hansen, D., et al. 2007. The effects of exercise training on fat-mass loss in obese patients during energy intake restriction. *Sports Medicine* 37(1):31.

Hansen, T. et al. 2021. Are dietary proteins the key to successful body weight management? A systematic review and meta-analysis of studies assessing body weight outcomes after interventions with increased dietary protein. *Nutrients* 13(9):3193.

Hayashi, L., et al. 2021. Intuitive and mindful eating to improve physiological health parameters: A short narrative review of intervention studies. *Journal of Complementary and Integrative Medicine* December 16, 2021. Online ahead of print.

Hoelscher, D., et al. 2022. Prevention of pediatric overweight and obesity: Position of the academy of nutrition and dietetics based on an umbrella review of systematic reviews. *Journal of the Academy of Nutrition and Dietetics* 122(2):410.

Hoenig, T., et al. 2022. Bone stress injuries. *Nature Reviews* Disease Primer 8(1):26.

Huang, S., et al. 2022. Association of magnitude of weight loss and weight variability with mortality and major cardiovascular events among individuals with type 2 diabetes mellitus: A systemic review and meta-analysis. *Cardiovascular Diabetology* 21(1):78.

Ibacache-Saavedra, P., et al. 2022. Effects of bariatric surgery on cardiorespiratory fitness: A systematic review and meta-analysis. *Obesity Reviews* 23(3):e13408.

Karlsson, J. et al. 2020. Effects on body weight, eating behavior, and quality of life of a low-energy diet combined with behavioral group treatment of persons with class II or III obesity: A 2-year pilot study. *Obesity Science and Practice* 28(7):4.

Keck-Kester, T., et al. 2021. Do mindfulness interventions improve obesity rates in children and adolescents: A review of the evidence. *Diabetes, Metabolic Syndrome and Obesity: Targets and Therapy* 14:4621.

Kohanmoo, A., et al. 2020. Effect of short- and long-term protein consumption on appetite and appetite-regulating gastrointestinal hormones: A systematic review and meta-analysis of randomized controlled trials. *Physiology and Behavior* 226:113123.

Linardon, J., et al. 2021. Intuitive eating and its psychological correlates: A meta-analysis. *International Journal of Eating Disorders* 54(7):1073.

Magallares, A., and Pais-Ribeiro, J. 2014. Mental health and obesity: A meta-analysis. *Applied Research in Quality of Life* 9:295.

Mamalaki, E., et al. 2022. The effectiveness of technology-based interventions for weight loss maintenance: A systematic review of randomized controlled trials with meta-analysis. *Obesity Reviews* June 10, 2022. Online ahead of print.

Michalopoulou, M., et al. 2022. Effectiveness of motivational interviewing in managing overweight and obesity: A systematic review and meta-analysis. *Annals of Internal Medicine* March 29, 2022. Online ahead of print.

Milaneschi, Y., et al. 2019. Depression and obesity: evidence of shared biological mechanisms. *Molecular Psychiatry* 24(1):18.

Miles-Chan, J., and Isacco, L. 2021. Weight cycling practices in sport: a risk factor for later obesity? *Obesity Reviews* 22(Suppl 2):e13188.

Morehen, J., et al. 2021. A 5-year analysis of weight cycling practices in a male world champion professional boxer: Potential implications for obesity and cardiometabolic disease. *International Journal of Sports Nutrition and Exercise Metabolism* 31(6):507.

Nelson, J. 2017. Mindful eating: The art of presence while you eat. *Diabetes Spectrum* 30(3):171.

Nitschke, E., et al. 2022. Impact of nutrition and physical activity interventions provided by nutrition and exercise practitioners for the adult general population: A systematic review and meta-analysis. *Nutrients* 14(9):1729.

O'Connor, R., and Kenny, P. 2022. Utility of "substance use disorder" as a heuristic for understanding overeating and obesity. *Progress in Neuropsychopharmacology and Biology in Psychiatry* May 27, 2022. Online ahead of print.

Phelan, S., et al. 2021. A model of weight-based stigma in health care and utilization outcomes: Evidence from the learning health systems network. *Obesity Science and Practice* 8(2):139.

Rao, W., et al. 2020. Obesity increases risk of depression in children and adolescents: Results from a systematic review and meta-analysis. *Journal of Affective Disorders* 267:78.

Raynor, H., and Champagne, C. 2016. Position of the Academy of Nutrition and Dietetics: Interventions for the treatment of overweight and obesity in adults. *Journal of the Academy of Nutrition and Dietetics* 116(1):129.

Rubin, R. 2019. Addressing medicine's bias against patients who are overweight. *Journal of the American Medical Association* 321(10):925.

Saarni, S., et al. 2006. Weight cycling in athletes and subsequent weight gain in middleage. *International Journal of Obesity* 30(11):1639.

Shook, N., et al. 2022. The search for scientific meaning in mindfulness research: Insights from a scoping review. *PLoS One* 17(5):e0264924.

Smith, E., et al. 2017. Nutritional considerations for bouldering. *International Journal of Sports Nutrition and Exercise Metabolism* 27(4):314.

Tronieri, J., et al. 2017. Sex differences in obesity and mental health. *Current Psychiatry Reports* 19(6):29.

Tylka, T., et al. 2014. The weight-inclusive versus weight-normative approach to health: Evaluating the evidence for prioritizing well-being over weight loss. *Journal of Obesity* 2014:983495.

Verduci, E., et al. 2022. Integrated approaches to combatting childhood obesity. *Annals of Nutrition and Metabolism* June 9, 2022. Online ahead of print.

Vieira, F., et al. 2022. Effect of physical exercise on muscle strength in adults following bariatric surgery: A systematic review and meta-analysis of different muscle strength assessment tests. *PLoS One* 17(6):e0269699.

Wright, C., et al. 2021. Are eHealth interventions for adults who are scheduled for or have undergone bariatric surgery as effective as usual care? A systematic review. *Surgery for Obesity and Related Diseases* 17(12):2065.

Zeiler, M., et al. 2021. Psychopathological symptoms and well-being in overweight and underweight adolescents: A network analysis. *Nutrients* 13(11):4096.

Zhang, Y., and Giovannucci, E. 2022. Ultra-processed foods and health: a comprehensive review. *Critical Reviews in Food Science and Nutrition* June 6, 2022. Online ahead of print.

Design element: Training Table (orange) ©mphillips007/Getty Images

Gaining Lean Body Mass through Proper Nutrition and Exercise

KEY TERMS
circuit aerobics
circuit weight training
concentric contraction
eccentric contraction
isokinetic contraction
isometric contraction
isotonic contraction
macrocycle
mesocycle
microcycle
muscle hyperplasia
muscle hypertrophy
muscle tissue accretion
periodization
principle of exercise sequence
principle of overload
principle of progressive resistance exercise (PRE)
principle of recuperation
principle of specificity
repetition maximum (RM)
sarcopenia
strength-endurance continuum
Valsalva phenomenon

CHAPTER TWELVE

Liquidlibrary/Getty Images

LEARNING OUTCOMES
After studying this chapter, you should be able to:

1. Describe the steps an individual might take to gain body weight, mainly as muscle mass.

2. Plan a diet for an individual who desires to gain muscle mass in concert with a resistance-training program, focusing on recommended caloric intake and foods compatible with the Prudent Healthy Diet.

3. Identify dietary supplements used by physically active individuals to stimulate muscle building and body-fat loss, and list those, if any, that may be effective.

4. List and explain the principles of resistance training.

5. Explain the differences between resistance-training programs for muscular hypertrophy, muscular strength and power, and muscular endurance.

6. Design a total-body resistance-training program for an individual who desires to gain body weight as muscle mass.

7. Identify the potential health benefits of resistance exercise and compare them to the health benefits associated with aerobic endurance exercise, noting similarities and differences.

Introduction

As noted in chapter 11, there are basically three main reasons individuals attempt to lose excess body weight—to improve appearance, health, and/or athletic performance. Some individuals may also wish to gain weight for the same three reasons and may use resistance training, also known as weight training or strength training, as a means to stimulate weight gain.

For those who wish to improve appearance, resistance training will increase muscularity, a desired physical attribute among many males. Resistance training is becoming increasingly popular. According to Healthy People 2030 objectives, 51.1 percent of US adolescents and 27.6 percent of US adults participated in muscle-strengthening activities 2+ and 3+ days per week, respectively, in 2017 and 2018. The goal is to achieve participation rates of 56.1 and 32.1 percent, respectively, by 2030. The *Healthy People 2030* data search website is presented at the end of this introduction.

Gaining weight, particularly muscle mass stimulated by resistance training, may also be associated with some health benefits. An increased muscularity that improves physical appearance and body image may help elevate self-esteem, contributing to positive psychological health. Additionally, resistance training, done alone even without weight gain, is recommended for several other health benefits, such as increased bone mineral density. The American Heart Association and the American College of Sports Medicine, in their reports on physical activity and health, recommend resistance exercise as an effective means to promote overall good health. Resistance training is particularly recommended for older adults to help prevent the muscle wasting, and associated health problems, seen with aging, as documented by Nelson, Raymond, and Williams and their respective colleagues. Using 2018–2018 National Health and Nutrition Examination Survey data, Fryar and others reported that 1.6 percent of Americans 20 years of age and older are underweight. In his popular book, Reuben has provided sound medical advice for such Americans who need to gain weight for a variety of medical and cosmetic reasons. Other websites describing healthy weight gain strategies from selected reputable sources are listed at the end of this introduction.

Increased body weight from increased muscle mass may be associated with improvements in strength, power, and decreased risk of injury. These factors are important in a wide variety of sports. Enhanced muscularity also may influence performance in judged aesthetic sports, such as diving and gymnastics. Most colleges and universities, as well as many high schools, have strength-training programs for their student-athletes. At the elite level, sport-specific resistance-training programs are tailored to the individual athlete.

No matter what the reason for gaining body weight, you should be concerned about where the extra pounds will be stored. The energy-balance equation works as well for gaining weight as it does for losing weight, but excess body fat in general will not improve physical appearance, health, or athletic performance. On the contrary, it may detract from all three. In order to optimally increase body weight for improved health, aesthetics, or performance, one must concentrate on increasing fat-free mass, particularly muscle tissue, with little or no increase in body fat.

Reidy and Rasmussen note that resistance exercise and ingestion of high-quality protein including adequate amounts of the branched-chain amino acid leucine are two independent and major stimuli of muscle protein synthesis and overall muscle growth. Numerous related approaches have been employed to increase muscle protein synthesis, including research on "nutrient timing" and the ingestion of protein in close proximity before and/or after resistance exercise.

Specialized exercise equipment or exercise techniques are advertised as the most effective methods available to build muscles. Protein supplements have been a favorite among weight lifters for years, but numerous other dietary supplements currently on the market are advertised to produce an anabolic, or muscle-building, effect. As discussed in chapter 13, some individuals use illegal drugs to gain weight for enhanced performance or appearance. Although resistance training may confer significant health benefits, excessive weight lifting may be a symptom of a male-dominated psychological condition called muscle dysmorphia (MD), also known as the Adonis complex. Murray and others noted that symptoms of MD such as low self-esteem, body dissatisfaction, perfectionism, and obsessive-compulsive behavior are similar to those observed in eating disorders. Skemp and others reported that competitive weight-training athletes scored higher than noncompetitive athletes in four of six muscle dysmorphia inventory subscales. According to Rohman, anabolic-androgenic steroid (AAS) use and MD are related, but it is unclear if AAS use precedes, coexists with, or is an outcome of MD.

Like weight-loss programs, weight-gaining programs may be safe and effective or they can be potentially harmful to your health. It may be difficult for some individuals to gain muscle mass for a variety of genetic, environmental, and epigenetic reasons. The purpose of this chapter is to present basic information on the type of diet and exercise program that is most likely to be effective as a means to put on weight without compromising your health.

Although some basic information regarding advanced resistance-training programs will be provided, detailed coverage of such programs is beyond the scope of this text. References to advanced resistance-training programs will be provided for the interested reader.

---

https://health.gov/healthypeople/objectives-and-data/browse-objectives/physical-activity This page includes baseline status and Healthy People 2030 objectives for healthy physical activity behaviors in children, adolescents, and adults, as well as the impact of physical activity on quality-of-life in health outcomes such as arthritis, cancers, heart disease and stroke, cognitive health in the elderly, and overweight and obesity.

The following websites from the Mayo Clinic, American Academy of Family Physicians, and the Academy of Nutrition and Dietetics offer sound weight gain advice.

www.mayoclinic.org/healthy-lifestyle/nutrition-and-healthy-eating/expert-answers/underweight/faq-20058429

familydoctor.org/healthy-ways-to-gain-weight-if-youre-underweight/

www.eatright.org/health/wellness/your-overall-health/healthy-weight-gain

Basic Considerations

Why are some individuals underweight?

Being significantly under a healthy body weight may be due to several factors. Heredity may be an important factor, as genetic factors predispose some individuals to leanness. For example, a lean body frame or high basal metabolic rate may have been acquired from your parents. Gastrointestinal problems, infections, or cancer could adversely affect food intake and digestion, so a physician should be consulted to rule out nutritional problems caused by disease, hormonal imbalance, or inadequate absorption of nutrients. Social pressures, such as the strong desire of a teenage girl to have a slender body, could also lead to undernutrition. An extreme example of this is anorexia nervosa, which was discussed in chapter 10. Emotional problems also may affect food intake. In many cases, food intake is increased during periods of emotional crisis, but the appetite may also be depressed in some individuals for long periods. Economic hardship may reduce food purchasing power, so some individuals simply may sacrifice food intake for other life necessities. According to a US Department of Agriculture Economic Research Service report by Coleman-Jensen and others, 10.5 percent of American households were food insecure at least some time in 2020. In the 2021 American Heart Association guidelines to improve cardiovascular health, Lichtenstein and others include food and nutrition insecurity as a challenge to a heart-healthy dietary pattern.

Being considerably underweight, such as a body mass index below 18.5 kg/m^2, may be considered a symptom of malnutrition or undernutrition. It is important to determine the cause before prescribing a treatment. Our concern here is the individual who does not have any of these medical, psychological, social, or economic problems but who simply cannot create a positive energy balance because of excess energy expenditure or insufficient energy intake. Energy intake has to be increased, and the output has to be modified somewhat.

What steps should I take if I want to gain weight?

The following guidelines may help you develop an effective program to maximize your gains in muscle mass and keep body-fat increases relatively low.

1. Have an acceptable purpose for the weight gain. The desire for an improved physical appearance and body image may be reason enough. For athletes, increased muscle mass may be important for a variety of sports, particularly if strength and power are improved. However, you do not want to gain weight at the expense of speed if speed is important to your sport.
2. Calculate your average energy needs daily, as discussed in chapter 11. For a weight-gain program, you may wish to use several techniques to estimate your daily energy needs and select the highest value.
3. Keep a 3- to 7-day record of what you normally eat. See chapter 11 for guidelines to determine your average daily caloric intake. If the obtained value is less than your energy needs calculated under item 2 in this list, this may be a reason you are not gaining weight.
4. Check your living habits. Do you get enough rest and sleep? If not, you may be burning more energy than the estimate in item 2 of this list. Smoking increases your metabolic rate almost 10 percent and may account for approximately 200 kcal per day. Caffeine also increases the metabolic rate for several hours. Getting enough rest and sleep and eliminating smoking and caffeine will help decrease your energy output.
5. Set a reasonable goal within a certain time period. Weight gain may be rapid at first but then tapers off as you near your genetic potential. In their review, Brook and others note that gains in muscle mass are more rapid in the first 4–6 weeks of resistance training, with significant attenuation as training progresses despite increased intensity. In general, about 0.5–1 pound per week is a sound approach for a novice, but weight gaining is difficult for some individuals and may occur at a slower rate. Specific goals may also include muscular hypertrophy in various parts of the body.

6. Start a resistance-training exercise program. This type of exercise program will serve as a stimulus to build muscle tissue. According to Gonzalez and others, resistance training increases anabolic hormonal secretions and mechanically stimulates intramuscular essential cell signaling proteins, both of which stimulate protein synthesis and muscle growth. As noted in separate reviews by Brook and Gonzalez and their respective colleagues, the time course for optimal hypertrophy and optimal resistance-training loads are not completely understood. General guidelines for developing a resistance-training program are presented later in this chapter.

7. As discussed in the *2020-2025 Dietary Guidelines for Americans* and in chapter 2, dietary protein intake should come from high-quality plant and animal sources within the Acceptable Macronutrient Distribution Range of 10-35 percent of total kilocalories. Energy intake should support energy expenditure and allow for muscle repair and growth. A properly designed diet should include adequate kcal and protein and not violate the principles of healthful nutrition. In his review, Phillips noted that resistance training and consumption of high-quality, rapidly digestible protein with high leucine content (0.25-0.30 g/kg body mass/meal) exert independent effects on muscle protein synthesis, with a greater rate of growth due to synergistic effects of training and dietary protein.

8. Use a measuring tape to take body circumferences before and during your weight-gaining program. Be sure you measure at the same points about once a week. Those body parts measured should include the neck, upper and lower arm, chest, abdomen, hips, thigh, and calf. This is to ensure that body-weight gains are proportionately distributed. You should look for good gains in the chest and limbs. The abdominal and hip girth increase should be kept low because that is where fat is more likely to be stored. If available, skinfold calipers may be used to measure subcutaneous fat skinfolds at multiple sites over the body. Fat skinfold thicknesses should remain the same or decrease to ensure that the weight gain is muscle rather than fat.

In summary, adequate rest; sufficient caloric intake; adequate intake of high-quality, highly digestible protein; and a proper resistance-training program may be very effective as a means to gain the right kind of body weight.

> ### Key Concepts
>
> ▸ There may be a variety of reasons an individual is underweight, and the cause should be determined before a treatment is recommended.
> ▸ For those who want to gain weight, a weekly increase of 0.5-1.0 pound is a sound approach, but the desired weight gain should be muscle tissue and not body fat. In essence, adequate rest and sleep; increased caloric intake; adequate intake of high-quality, highly digestible protein; and a proper resistance-training program should be effective in helping to increase lean body mass.

Nutritional Considerations

In the last chapter, we discussed nutritional considerations for losing body weight, particularly body fat. In general, the recommendation consists of a healthy diet but with reduced caloric intake. In this chapter, nutritional considerations to increase lean body mass as muscle include a healthy diet with sufficient energy intake from within the Acceptable Macronutrient Distribution Ranges for carbohydrates, fats, and protein to support net muscle growth.

Gaining weight as muscle mass may be difficult for some. Resistance training, discussed later in this chapter, is important to help stimulate muscle growth so that extra kcal are used to develop muscle, not fat. As noted, gaining about 0.5-1.0 pound per week is a reasonable goal, although some may be able to gain more, while some may gain less.

How many kcal are needed to form 1 pound of muscle?

Muscle tissue consists of about 70 percent water and 22 percent protein, and the remainder is fat, carbohydrate, and minerals. Because the vast majority of muscle tissue is water, which has no caloric value, the total caloric value (carbohydrate + fat + protein) is only about 700-800 kcal per pound. However, extra energy is needed to help synthesize the muscle tissue.

It is not known exactly how many additional kcal are necessary to form 1 pound of muscle tissue in humans. The exact mix of kcal is also unknown, but protein and carbohydrate ingestion appears to facilitate muscle protein synthesis. Dietary protein should have a high Protein Digestibility-Corrected Amino Acid Score (PDCAAS), discussed in chapter 6, which reflects protein quality, digestibility, and absorption. Phillips noted that ingestion of high-quality protein with high leucine content (e.g., whey) may trigger muscle protein synthesis and attenuate postexercise breakdown. Concurrent carbohydrate ingestion may stimulate insulin-mediated amino acid uptake to support muscle protein synthesis. Lambert and others recommended a mix of 25-30, 55-60, and 15-20 percent of total kcal for protein, carbohydrates, and fats, respectively, for bodybuilders attempting to increase muscle mass. They suggested including some saturated fat and monounsaturated fat because high polyunsaturated fat intake has been reported to decrease testosterone secretion.

As noted by Slater and others, muscle tissue is 75 percent water, 20 percent connective and contractile protein (~360 kcal/lb), and 5 percent fat (~160 kcal/lb), glycogen (~18 kcal/lb), phosphates, ions, and other metabolites. According to the National Academy of Sciences, muscle tissue synthesis requires 5.0-5.65 kcal/g or between 2,270 and 2,565 (=5.0-5.65 kcal/g × 454 g), assuming that all these kilocalories are supporting new muscle protein synthesis. Other energy surplus recommendations for a pound of increased muscle tissue, as reviewed by Slater and others, ranging from as little as 163 kcal/day to over 800 kcal/day, depend on body composition, training status and volume, and other factors.

Such recommendations may not account for all energy requirements for muscle growth, including but not limited to initial body composition; training status, mode, and volume; amount and timing of dietary protein intake; increased postexercise protein turnover; energy balance status; and energy required for skeletal muscle remodeling, increased needs of other energy-requiring tissues, increases in diet-induced thermogenesis, resting metabolic rate, exercise energy expenditure, and nonexercise activity thermogenesis. Although the energy in one pound (454 grams) of muscle tissue totals approximately 550 kcal, the energy required for one pound of **muscle tissue accretion** (i.e., muscle growth due to protein synthesis) is subject to interindividual variation and remains unknown. Slater and others recommend a conservative energy surplus with high-quality protein and resistance training, combined with regular assessment of body composition to minimize gains in fat mass and individualize dietary intake. A range of 2,300–3,500 additional kcal in a well-balanced diet appears to provide sufficient energy for total daily energy expenditure, including training, postexercise repair, and protein synthesis. In this chapter, 3,500 kcal/week (500 kcal/day) will be considered as the weekly excess energy intake, which, when combined with resistance training, will support an increase in 1 pound of muscle tissue.

How can I determine the amount of kcal I need daily to gain 1 pound per week?

First, review the techniques presented in chapter 11 to determine the number of kcal needed simply to maintain your current body weight. Then add the kcal that you expend during exercise and the additional amount needed to synthesize the muscle tissue. Consider a 20-year-old male college basketball guard who weighs 170 lbs (77.3 kilograms) and is 76" (1.93 meters) tall (BMI = 20.7 kg/m^2). He has maintained a low level of physical activity (physical activity coefficient = 1.1) since the end of the previous season. He and his conditioning coach recognize that an increase in skeletal muscle mass prior to the season will improve his future basketball prospects. **Table 12.1** presents a plan implemented during the summer prior to the next season of increased physical activity including progressive resistance exercise and aerobic exercise designed by his conditioning coach and a diet designed by the team dietitian including high-quality protein and increased energy to support skeletal muscle growth. You may modify the figures according to your own needs. You can use NutriCalc Plus or another reputable dietary analysis program to plan a weight-gain diet as well as a weight-loss diet. Other methods presented in chapter 11 may also be used. Remember, a weight gain of 0.5–1.0 pound muscle mass per week is a reasonable goal during the early stages of resistance training. If you are maintaining your weight with your current dietary intake, adding 500 kcal/day may also be an acceptable approach.

Increased caloric intake is the key dietary principle, along with adequate protein intake, to gain mass during resistance training, as noted in reviews by Aragon and others; Bosse and Dixon; and Phillips. However, Spendlove and others concluded that research on dietary intake of competitive bodybuilders commented is dated, of limited quality, and called for more contemporary and better controlled research.

Is protein supplementation necessary during a weight-gaining program?

One of the most researched areas in sports nutrition is the recommended protein intake for individuals who are attempting to increase muscle mass with resistance training. Topics of interest include amount, quality, timing, coingestion with carbohydrate, and cost.

Amount From a mathematical viewpoint, adding a pound of muscle protein per week does not require a substantial increase in daily protein intake. As previously noted, muscle tissue is about 22 percent protein and 70 percent water, with the remainder composed of carbohydrate and fat and trace amounts of vitamins and minerals. One pound of muscle is equal to 454 grams, but only about 91 grams (0.20 × 454 grams) is protein. If we divide 91 grams by 7 days, we would need approximately 13 grams of protein per day above our normal protein requirements to support the growth of 1 pound of muscle per week—if we are in protein balance. This amount of dietary protein could be obtained in such small amounts of food as 2 glasses of milk; 2 ounces of meat, fish, or poultry; 2 eggs; or various combinations (**figure 12.1**). However, an increase in dietary protein limited to that found in 1 pound of new muscle tissue (20 percent of 454 grams) is likely an underestimation of the

TABLE 12.1	Estimated kilocalorie intake needed for a 170-pound, 76" tall, sedentary, 20-year-old low-active (Physical Activity Coefficient = 1.1) male basketball player to gain 1 pound per week	
	Energy expenditure	**kcal/day**
*Current energy needs	EER = 662−(9.53 × Age in years) + [PA Coefficient × {(15.91 × weight in kg) + (539.6 × height in meters)}]	2,970
	= 662 − (9.53 × 20) + [1.1 × {(15.91 × 77.3) + (539.6 × 1.93)}]	
Resistance training	=200 kcal/sessions × 5 sessions/week ÷ 7 days/week	143
Aerobic training	=200 kcal/sessions × 4 sessions/week ÷ 7 days/week	114
Muscle tissue synthesis	=3,500 kcal/pound ÷ 7 days/week	500
Total		3,727

*Source: National Academy of Sciences. Food and Nutrition Board. Institute of Medicine. 2005. Dietary Reference Intakes for Energy, Carbohydrate, Fiber, Fat, Protein and Amino Acids. Washington, DC: National Academies Press.

FIGURE 12.1 To add a pound of muscle tissue per week, you need to consume approximately 500 additional kcal and additional daily intake of high-quality protein to support muscle tissue growth (~13–15 grams) and other reasons for increased protein discussed in the text. A weight-training program is an essential part of a muscle-building program. One glass of skim milk, three slices of whole wheat bread, and two hard-boiled egg whites provide the necessary kcal and about 23 grams of protein.

total energy and protein requirement for muscle protein synthesis because of factors previously discussed.

As you may recall, the Acceptable Macronutrient Distribution Range (AMDR) for protein is 10–35 percent of daily caloric intake. As noted in **table 12.1,** a low-active 77.3 kilograms (170-pound) male needs an estimated dietary intake of 3,727 kcal to add a pound of muscle per week. If he consumed a diet containing 15–20 percent of energy from protein, his daily protein intake would approximate 140 (3,727 × 0.15 ÷ 4 kcal/g) to 186 (3,727 × 0.20 ÷ 4 kcal/g) grams or 1.8–2.4 g/kg body mass.

As noted in chapter 6, experts disagree on the necessity of protein over the RDA of 0.8–1.0 g/kg for individuals seeking to gain muscle mass through resistance training. Some contend the RDA includes a safety factor providing adequate protein for muscle hypertrophy. The American College of Sports Medicine/Academy of Nutrition and Dietetics/Dietitians of Canada joint position statement, authored by Thomas and others, recommends 1.2–2.0 g of protein/kg/day for adaptation, repair, remodeling, and protein turnover, with higher intakes during periods of intense training or reduced energy intake. In their review of protein intake and exercise, Jäger and others note the higher (>2–3 g/kg/day) protein intake may be necessary in order to retain lean tissue mass in competitive resistance-trained subjects during periods of hypocaloric intake and may promote loss of fat. The RDA for our 70-kilogram male is 56 grams (0.8 × 70 kilograms) daily. An additional 14 grams totals 70 grams, is less than would be consumed on a diet with 12–15 percent protein. If he consumed 1.6 protein g/kg, as recommended by some, his daily intake would total 112 grams, still within the 12–15 percent range. Protein intake could be increased even more and still remain within the AMDR of 10–35 percent.

As was discussed in chapter 11, increased protein intake may help preserve existing lean body mass during weight loss and weight maintenance. Healthy weight gain consists of increasing lean body mass without increasing fat mass. In their meta-analysis, Clifton and colleagues concluded that higher protein diets (≥5 percent above baseline intake of protein at 12 months) exerted a threefold greater reduction in fat mass compared to a normal (<5 percent over baseline) protein diet. Leidy and others noted that long-term adherence to higher protein intake of 1.2 and 1.6 g protein/kg/day with 25–30 g protein/meal improves appetite control and weight management. Reidy and Rasmussen commented that protein supplementation with acute resistance training upregulates cell signaling mechanisms responsible for protein synthesis in a dose-response manner. However, this effect is attenuated with long-term supplementation and resistance training.

Timing Various aspects of the timing of protein intake have been suggested to stimulate muscle protein synthesis. The general research design employed by studies examining the efficacy of protein timing on muscle protein synthesis has been to assign some subjects to consume the protein supplement immediately before and/or after exercise, while other subjects have consumed the same total supplement at different times of the day, either several hours after the workout or in the morning and evening. The ingestion of protein in close proximity to the end of acute exercise has been supported in a review by Phillips. According to this recommendation, upregulated cell signaling pathways increase messenger-RNA expression to stimulate muscle protein synthesis and inhibit postexercise protein breakdown. As a result, the ingested protein supports net muscle protein accretion and hypertrophy. Beelen and others suggested that 20 grams of intact protein, or about 9 grams of essential amino acids, may maximize muscle protein synthesis rates during the first hours of postexercise recovery. Aragon and Schoenfeld also commented that a postexercise protein intake for optimal muscle growth and repair may be less important if nutrient intake occurs within 1–2 hours of exercise.

Although consuming protein within a short time of completing a resistance-training workout may potentially maximize muscle hypertrophy, and possibly strength and power, some have contended there would be little difference if the total amount of protein consumed throughout the day were similar, in essence hypothesizing that timing is not an important consideration. Witard and others suggested that a postexercise "anabolic window of opportunity" of 24 or more hours may exist, which is consistent with the suggestion by Beelan and others that ingesting small amounts of dietary protein five or six times daily for a total of about 100–120 grams might also support maximal muscle protein synthesis rates throughout the day. In their position stand, Kerksick and others recommend evenly spaced protein feedings every 3 hours during the day and ingestion of 10 grams of essential amino acids (EAA; approximately 10 grams) as part of a daily protein intake of approximately 20–40 grams to maximally stimulate muscle protein synthesis.

There is research support for both viewpoints in both young and older subjects. Protein intake at breakfast as a percent of total daily intake may also affect muscle growth. Aoyama and others assigned older Japanese women high- and low-breakfast protein intake groups with similar total dietary protein intakes. Groups were similar in height, body mass, fat mass, body mass index (BMI), physical function, dietary intake, and physical activity. Skeletal muscle index (kg/m^2) was higher, and muscle mass (kilograms) tended to be higher in the high-breakfast protein intake group. In a similar study of young males consuming the same (1.3–1.4 g/kg body weight) total daily dietary protein and completing high intensity (75–80 percent 1-RM) progressive resistance training, Yasuda and others reported that the total lean tissue mass measured by DEXA tended to be higher in a high-protein breakfast group (0.33 g/kg) compared to a low-protein breakfast group (0.12 g/kg). Cribb and Hayes reported that a creatine/protein/carbohydrate supplement consumed immediately before and after a resistance-training workout, as contrasted to consuming the same supplement in the morning and evening, resulted in greater increases in lean body mass and strength testing.

In contrast, Hoffman and others concluded that the time of protein-supplement ingestion during a 10-week training program did not make any difference in strength, power, or body-composition changes in resistance-trained men. Verdijk and others also concluded that timed protein supplementation immediately before and after exercise did not further augment the increase in skeletal muscle mass and strength after prolonged resistance-type exercise training in healthy elderly men who habitually consume adequate amounts of dietary protein. Jäger and others also noted that the anabolic effect of exercise is at least 24 hours and that an optimal time period for protein ingestion relative to resistance exercise is likely a matter of individual preference or tolerance. According to Bauer and colleagues, there is insufficient evidence for recommendations about the timing of protein intake in older individuals.

In their review, Witard and others also concluded that muscle protein synthesis appears to be equally responsive to pre- and post-exercise supplementation, mixed (carbohydrate/protein) supplements taken together compared to separate ingestion, and daytime compared to nighttime supplementation. According to Pasiakos and others, nitrogen balance, resistance-trained versus untrained status, and variations in protein intake during training are potential confounding factors in research on the timing of protein supplementation. Given the available research, consuming protein after resistance training may be beneficial and does not appear to be detrimental to muscle protein synthesis. The key objective appears to be adequate daily protein intake to support new muscle tissue growth.

Protein Quality (PDCAAS) and Quantity Current research suggests that the quality of the ingested protein may also be an important consideration. Phillips notes that milk proteins, specifically whey and casein proteins, are of the highest quality, with whey protein being better able to support muscle protein synthesis. In their meta-analysis of nine studies, Naclerio and Larumbe-Zabala concluded that whey protein, either alone or as part of a multi-ingredient supplement, added lean body mass gain and strength in resistance-trained subjects compared to either isocaloric carbohydrate or nonwhey protein supplementation. Stokes and others conclude that the ingestion of high-quality protein (~0.3 g/kg/meal) at 3-hour intervals totaling ~20 g protein/day is sufficient to maximally stimulate muscle protein stimulus. Excessive protein intake saturates muscle protein synthesis, leading to increased catabolism of excess amino acids and increased urea production. Witard and others commented that chronic leucine supplementation will not necessarily increase long-term muscle protein synthesis capability, but the addition of leucine may improve the muscle protein synthesis ability of a dietary protein source that is of inferior quality (low PDCAAS) or quantity. According to the *2020–2025 Dietary Guidelines for Americans*, protein should come from a wide variety of plant and animal sources, including beans, peas, lentils, nuts, seeds, and soy products. Game meats can also be excellent sources of low-fat protein. For example, 100 g (3.5 oz.) servings of roasted deer tenderloin and baked wild turkey breast contain low kcal and fat and high protein as shown in the links below.

> *www.fdc.nal.usda.gov/* Type "Game meat, deer, tenderloin, separable lean only, cooked, broiled" in FoodData Central Search Window, enter, click SR Legacy Foods tab, and click link to access nutritional information: 100 g (3.5 oz.) of broiled lean deer tenderloin contains 149 kcal; 29.9 grams of protein; 2.35 grams of total fat
>
> *myfitnesspal.com/food/calories/wild-game-wild-turkey-breast-896933231* 3.5 oz. of wild turkey breast contains 165 kcal; 26 grams of protein; 1.1 grams of total fat

Carbohydrate As noted in chapter 4, consuming carbohydrate after exercise is recommended to replace muscle glycogen, a key energy source for exercise. Phillips noted that carbohydrate alone is not sufficient for muscle protein synthesis but ingesting a protein and carbohydrate mixture immediately following resistance exercise or aerobic endurance exercise may be recommended. Both the carbohydrate and the protein will stimulate insulin secretion, which will help move glucose and amino acids into the muscle to facilitate recovery and support muscle protein synthesis. However, Pasiakos and others commented in their review that carbohydrate added to protein supplements will not stimulate additional changes in lean mass and muscle strength during resistance training.

Several studies have shown that chocolate milk is an effective postexercise recovery fluid. As shown in **figure 12.2,** an 8-ounce glass of low-fat chocolate milk contains 32 grams of carbohydrate and 8 grams of protein (a 4:1 ratio) and is an affordable postexercise recovery beverage. Ferguson-Stegall and others reported that chocolate milk administered immediately and 2 hours following 1.5 hours of cycling exercise at 70 percent of VO_2 max significantly increased

Iconotec/Glowiamges

FIGURE 12.2 Chocolate milk may be an excellent source of nutrients before or after a resistance-training workout. One glass provides about 32 grams of carbohydrate and 8 grams of protein, a 4:1 ratio. Chocolate milk also provides important minerals and vitamins.

cell signaling for protein synthesis compared to isocaloric carbohydrate or placebo supplements. In their review, Pritchett and Pritchett comment that ingestion of 1.0–1.5 grams of chocolate milk/kg body weight immediately after and 2 hours after exercise may reduce muscle damage. Approximately 82 percent of cow's milk is casein protein and about 18 percent is whey protein. Some of the carbohydrate in chocolate milk is lactose, but most is glucose in the added chocolate syrup. Lactose-free chocolate milk is available for those with lactose intolerance. Josse found that fat-free milk is also an effective drink to support favorable body-composition changes with resistance training.

Cost The cost of a standard amount of high-quality protein can vary considerably as noted in chapter 6, table 6.8. In fall 2021, a 4-ounce (114 g) serving of turkey breast containing 28 grams of protein may cost approximately $0.86 ($3.45/pound, 3¢/g protein). In contrast, Boost® (Nestlé Nutrition) High Protein drink (8 oz., 237 milliliters) containing 20 grams of protein may cost $1.50 or more (7.5¢/g protein).

Are dietary supplements necessary during a weight-gaining program?

Dietary supplements appear to be very popular among athletes and others attempting to increase muscle mass and strength. For example, in a survey of only five magazines targeted to bodybuilding athletes, Grunewald and Bailey reported more than 800 performance claims made for 624 commercially available supplements. Although this survey was conducted almost 30 years ago, such marketing practices continue unabated today. Numerous products are marketed not only to increase muscle mass, strength, and power but also to help prevent injuries during resistance training. Harty and others reviewed a number of strategies to attenuate exercise-induced muscle damage from strength training including many fruits and fruit derivatives, vegetables and plant derivatives, herbs and herbal derivatives, amino acids and proteins, vitamins, and other nutritional supplements. They also caution that attempts to modulate inflammation and oxidative stress associated with muscle damage may attenuate normal training adaptations.

Although there may be some truth underlying the alleged performance-enhancing mechanisms of these supplements, online advertisement claims of efficacy of many supplements are not substantiated by well-designed research. The effects of many of these supplements, such as specific individual amino acids, vitamins, minerals, and other individual nutrients, have been discussed in previous chapters and in general have not been found to be effective as performance-enhancing agents. Research has found that many are ineffective, whereas other supplements are inadequately researched.

Protein and creatine supplements may help increase muscle mass. Pasiakos and colleagues reported that protein may promote muscle hypertrophy and power with increased duration, frequency, and volume of resistance training. Stokes and others noted that whey and other high-quality proteins with high leucine content may increase muscle growth and repair. Creatine monohydrate, discussed in chapter 6, also appears to increase muscle mass. In separate meta-analyses, Lanhers and others concluded that creatine supplementation was effective in various populations in improving both upper- and lower-limb performance. Early increases in muscle mass may be due to water retention, but increased resistance-training capacity may ultimately increase muscle mass. Creatine may help augment the anabolic effects of resistance training in the elderly, increasing muscle mass and strength, according to reviews by Candow and by Rawson and Venezia. In their meta-analysis, Devries and Phillips reported that creatine and resistance training increased total mass and fat-free mass in adults over 60 years of age more than resistance training alone. The importance of resistance training to help prevent or reduce loss of muscle mass in the elderly will be covered in the next section.

Beta-hydroxy-beta-methylbutyrate (HMB), a metabolite of leucine catabolism, has also been reported to promote muscle tissue growth. As noted in the International Society of Sports Nutrition position stand by Wilson and others, proposed mechanisms of HMB include up-regulating cell signaling pathways which promote muscle protein synthesis and other pathways which decrease postexercise muscle breakdown. In their meta-analysis of seven studies, Wu and others reported a significant difference in muscle mass (+0.325 kilograms), but no change in fat mass, in older (≥65 years) adults who received HMB compared to control subjects. However, separate meta-analyses by Rowlands and Thompson and Sanchez-Martinez and others concluded that HMB produced insignificant or only small effects sizes.

Several other dietary supplements marketed to athletes are patterned after anabolic-androgenic steroids. Some herbal products, such as *Tribulus terrestris,* are marketed for their supposed anabolic potential. In general, as noted in chapter 13, research indicates that such supplements do not increase muscle mass and may cause adverse health effects.

Most experts provide the following general advice to those trying to gain muscle mass:

- Be consistent with your fitness routine and focus on quality workouts followed by appropriate rest and recovery.
- Consume a nutrient-dense diet that provides adequate complex carbohydrates, quality protein, and healthy fats as recommended by the *2020-2025 Dietary Guidelines for Americans.*
- Don't rely on dietary supplements.

If you do use supplements, be sure to consult with an expert, such as a sports-oriented registered dietitan or physician, not the clerk at a health food store.

ods.od.nih.gov/factsheets   Access the National Institutes of Health's Office of Dietary Supplements website. Click Exercise and Athletic Performance on the list of Dietary Supplement Fact Sheets.

What is an example of a balanced diet that will help me gain weight?

As discussed in chapter 11, weight loss to reduce the metabolic risk associated with overweight and obesity is the primary societal and public health concern. Healthy weight gain, the focus of this chapter, is discussed in the *2020-2025 Dietary Guidelines for Americans* only in the context of normal and healthy gains during pregnancy and not specifically in the context of increased skeletal muscle mass which is the focus of this chapter. Nevertheless, these guidelines include suggested energy intakes for active adults of 2,600-3,200 and 2,400-2,600 kcal (males 19-50 years and 60-≥76 years, respectively) and 2,200-2,400 and 2,000-2,200 kcal (females 19-50 years, 60-≥76 years, respectively), with lower and higher energy intakes for smaller and larger individuals. As previously discussed by Stokes and others, additional energy and protein is also required to support the energy-intensive process of skeletal muscle accretion with the objective of increasing fat-free mass instead of fat mass.

According to the *2020-2025 Dietary Guidelines for Americans,* dietary protein sources contributing to a dietary pattern include lean meats, poultry, and eggs; seafood; beans, peas, and lentils; and nuts, seeds, and soy products. The following general suggestions that were introduced in chapters 1 and 2 may be helpful for those trying to gain weight as muscle. The focus is on healthy dietary sources of carbohydrates, protein, and fats.

Dairy—Drink 1 percent or 2 percent milk instead of skim milk, which will add 15-30 kcal per glass. Chocolate milk provides both carbohydrate and high-quality protein. If you want to decrease kcal intake from fat, fat-free chocolate milk is available. Prepare milk shakes with dry milk powder and supplement with fruit. Add low-fat cheeses to sandwiches or snacks. Eat yogurt supplemented with fruit.

Protein foods including meats, seafood, nuts seeds, beans, soy products—Increase your intake of lean meats, poultry, and fish, which are also sources of high-quality protein. Legumes such as beans and dried peas are high in protein, carbohydrate, and kcal and low in fat. According to the *2020-2025 Dietary Guidelines for Americans,* beans, peas, and lentils have nutrient profiles that are similar to protein foods and can be considered as either vegetables or protein foods for menu planning. Use nuts, seeds, and limited amounts of peanut butter for snacks.

Starchy vegetables—Increase your consumption of whole-grain products. Pasta, rice, and potatoes can be nutritious sources of kcal, but be sure to make high-fiber selections. Breads and muffins can be supplemented with fruits and nuts. Whole-grain breakfast cereals can provide substantial kcal and even make a tasty dessert or snack with added fruit. Starchy vegetables are high in complex carbohydrates but also contain about 15 percent of its kcal as protein.

Grains—A healthy diet profile includes choosing whole grain products at least 50 percent of the time and limiting the intake of refined grain products. For example, choose brown instead of white rice and whole wheat bread over white bread. Choose nutrient-dense cereals with less added sugar.

Fruits—Add fruit to other foods. Although fruit juices can be high in nutrients, be aware that they are also high in sugar kcal, so consider using them sparingly. Dried fruits such as apricots, pineapple, dates, and raisins are high in kcal and make excellent snacks.

Vegetables: dark green; red and orange; beans, peas, lentils; other—Use fresh vegetables like broccoli and cauliflower as snacks with melted low-fat cheese or a nutritious dip.

Lipids/oils—Limit, but do not eliminate the intake of saturated fats. Substitute monounsaturated fats for saturated fats. Nuts, seeds, olive, and canola oils are good sources of monounsaturated fats. High polyunsaturated fat intake has been reported to decrease testosterone secretion and its anabolic effects, as previously noted by Lambert and others.

Beverages—Milk is nutritious and high in kcal. Those who drink alcohol should obtain only limited amounts of kcal in this way. Some liquid supplements are available commercially and may contain 300-400 kcal with substantial amounts of protein. However, check the label for fat and sugar content.

Snacks—Eat three balanced meals per day supplemented with two or three snacks. Dried fruits, nuts, and seeds are excellent snacks. Some of the high-kcal, high-protein, high-nutrient liquid meals, and sports bars on the market also make good snacks.

Table 12.2 presents an example of a high-protein, high-kcal 24-hour diet plan to provide increased energy and protein intake to support resistance and aerobic training requirements and synthesis of additional muscle protein at a theoretical rate of 1 pound per week for the basketball player described in **table 12.1**. It consists of three main meals and three snacks and totals 3,731 kcal, with 199 protein grams (21 percent of total kcal), 114 fat grams (27 percent

TABLE 12.2 A sample high-kcal 24-hr meal plan diet with sufficient protein intake combined with progressive resistance training to support muscle synthesis in a hypothetical 170-pound, 76˝ tall, sedentary, 20-year-old basketball player.

Meal	Food item	Protein grams	Fat grams	Carbohydrate grams	Protein calculated kcal	Fat calculated kcal	Carbohydrate calculated kcal	*Total calculated kcal
Breakfast	†2% milk (8 oz.)	8.3	5	12	33.2	45	48	126
	Eggs (two poached, 100 g)	12.6	9.5	0.4	50.4	85.5	1.6	138
	Ham (two slices extra lean)	7.3	1.2	1.1	29.2	10.8	4.4	44
	Whole wheat toast (two slices, 62 g)	6.0	4	28	24	36	112	172
	Orange juice (8 oz. 1 cup, 248 g)	1.7	0.5	25.8	6.96	4.5	103.2	115
	Apricot, raw (1 cup, sliced, 165 g)	2.3	0.65	18.3	9.24	5.85	73.2	88
AM snack	Concord grape jelly (1 tablespoon)	0.0	0	13	0	0	52	52
	Whole wheat bread (two slices, 62 g)	6.0	4	28	24	36	112	172
	Natural peanut butter (two tablespoons	8.0	16	7	32	144	28	204
	2% milk (8 oz.)	8.3	5	12	33.04	45	48	126
Lunch	†2% milk (12 oz.)	12.5	7.5	18	50	67.5	72	190
	Chicken breast (4 oz. 113 g) boneless, grilled	25.4	3.5	0.5	101.6	31.5	2	135
	Whole wheat bread (two slices, 62 g)	6.0	4	28	24	36	112	172
	†Yogurt, plain, low fat (8 oz., 227 g)	11.9	3.5	16	47.6	31.5	64	143
	Banana (1 large, 8-8¾" long 145 g)	1.5	0.5	29	6	4.5	116	127
PM snack	Raisins (½ cup 80 g)	2.6	0.2	60	10.56	1.8	240	252
	†2% milk (8 oz.)	8.3	5	12	33.04	45	48	126
Dinner	†2% milk (12 oz.)	12.5	7.5	18	50	67.5	72	190
	Salmon (5 oz., 142 g), Atlantic wild, grilled	28.0	9	0	112	81	0	193
	Whole wheat bread (two slices, 62 g)	6.0	4	28	24	36	112	172
	Apple pie (1 slice = 16% of pie, 125 g)	3.0	17	43	12	153	172	337
	Peas (1 cup, 160 g), green, cooked, boiled, drained, without salt	8.6	0.354	25	34.32	3.186	100	138
	Sweet potato (1 medium, cooked, boiled without skin)	2.1	0.21	26.7	8.4	1.89	106.8	117
Evening snack	Peach (1 large 2.75", 175 g, yellow, raw)	1.6	0.44	16.7	6.4	3.96	66.8	77
	2% milk (8 oz.)	8.3	5	12	33.04	45	48	126
Column totals		199	114	479	795	1,022	1,914	3,731
Percent of total kcal					21	27	51	100

*Calculated using grams and Atwater factors for protein (4 kcal/g), fat (9 kcal/g), and carbohydrate (4 kcal/g).
†Soy yogurt and fortified soy beverages, which have similar nutritional content as milk, may be used as well.
Source: United States Department of Agriculture Food Central Database. fdc.nal.usda.gov/

of total kcal), and 479 carbohydrate grams (51 percent of total kcal). Adequate carbohydrate intake increases insulin secretion, which facilitates muscle tissue uptake of glucose and amino acids to promote muscle protein synthesis. The percentages of protein (21), fat (27), and carbohydrates (51) are within the Acceptable Macronutrient Distribution Ranges as established by the Institute of Medicine of the National Academy of Sciences. Carbohydrate also spares the use of protein as an energy source. Alternative foods of similar nutritional density and energy content may be substituted as necessary or desired. This suggested diet provides the necessary nutrients, kcal, and protein essential for increased development of muscle mass, and yet fewer than 30 percent of the kcal are derived from fat. The total number of kcal can be adjusted to meet individual needs. You may also plan your weight-gain diet using NutriCalc Plus or another dietary analysis program. As skeletal muscle accretion occurs during preseason conditioning, these estimates must be periodically adjusted to reflect the increased skeletal muscle mass.

Meal plans may be adjusted to the energy needs of individual athletes. For example, Lambert and others note that competitive bodybuilders should consume a diet that contains about 55-60 percent carbohydrate, 25-30 percent protein, and 15-20 percent fat for both the off-season and precontest phases. However, during the off-season, the diet should be slightly hyperenergetic (approximately 15 percent increase in energy intake) and during the precontest phase the diet should be hypoenergetic (approximately 15 percent decrease in energy intake). As noted previously, for 6-12 weeks prior to competition, bodybuilders attempt to retain muscle mass and reduce body fat to low levels, so they need to be in negative energy balance to facilitate oxidation of body fat. Lambert and others recommend that the diet contain 30 percent protein at this time to help prevent loss of muscle mass and possibly also provide a thermic effect to help burn fat.

For more detailed meal plans, the interested reader is referred to *Nutrient Timing* by Ivy and Portman.

Would such a high-kcal diet be ill advised for some individuals?

As noted in chapter 5, one of the general recommendations for an improved diet is to reduce the consumption of fats, particularly saturated fats. Unfortunately, many high-kcal diets are also high in fats. If there is a history of heart disease in the family or if an individual is known to have high blood lipid levels, then high-fat diets may be contraindicated. Individuals with kidney problems also may have difficulty processing high-protein diets because of the increased need to excrete urea. Any person initiating such a weight-gaining program as advised here should consult with his or her physician.

Selection of food for a weight-gaining diet, if done wisely, can satisfy the criteria for healthful nutrition. Foods high in complex carbohydrates with moderate amounts of protein and a moderate fat content are able to provide substantial amounts of kcal and nutrients yet minimize health risks that have been associated with the typical American and Canadian diet. To gain weight wisely, you need to continue to eat healthful foods, but just more of them.

Key Concepts

▸ The individual attempting to gain body weight should obtain necessary high-quality protein for muscle synthesis through a well-balanced diet, rather than by consuming expensive protein supplements.

▸ Although creatine supplementation may help increase muscle mass during resistance training, most dietary supplements marketed to strength-trained individuals are not effective or have not been evaluated by scientific research.

▸ Individuals can achieve healthy weight gain by adding skeletal muscle mass through a healthy dietary pattern recommended by the *2020–2025 Dietary Guidelines for Americans* including additional protein and calories within Acceptable Macronutrient Distribution Ranges combined with progressive resistance training.

Check for Yourself

▸ Using the information presented in the text, calculate the additional number of kcal you would need in your daily diet to accumulate a weight gain of 0.5 pound per week. What foods could you add to your daily diet to provide these kcal?

Exercise Considerations

Chapter 11 focused on aerobic exercise and other physical activities to promote energy expenditure for the loss of excess fat and maintenance of a healthy weight, supplemented by resistance training to prevent the decrease in resting energy expenditure that can occur with a loss of lean body mass. This chapter focuses on resistance training, sometimes called weight training or strength training, as a means to increase lean body mass and body weight. Before we discuss the principles underlying the design of a proper resistance-training program, let us introduce some basic terminology.

Repetition means the number of times you do a specific exercise. *Intensity* is determined by the weight, or resistance, that is lifted. A term used to describe the interrelationship between repetitions and intensity in weight training is **repetition maximum (RM)**. For example, if an individual can bench-press 150 pounds one time, but not twice, s/he has performed a one repetition maximum, or 1 RM. The resistance in a weightlifting exercise can be expressed as a percentage of 1 RM. As illustrated in **figure 12.3,** resistances to develop increased strength and muscle mass are higher percentages of 1 RM (e.g., 70-80 percent) while lower percentages (e.g., 40-50 percent) are used to develop muscular endurance. The bench press resistance at 80 percent of a 1 RM of 150 pounds would be 120 pounds. If the individual above can bench press 120 pounds five, but not six, times, 120 pounds can also be identified as the individual's 5 RM. A set is any particular number of repetitions, such as five or ten. The total volume of work accomplished for a given exercise in a workout is the product of sets, repetitions, and resistance (=sets × repetitions × resistance).

For example, three bench press sets of five repetitions, each with a resistance of 120 pounds, equals a bench press volume of 1,800 pounds (3 × 5 × 120). The total volume of work accomplished in the entire workout is the sum of the volumes for all resistance exercise performed in the session. Volumes for eight resistance exercises and for the entire session are illustrated in **table 12.3**. The recovery period can refer to the rest intervals between sets in a single workout as well as the rest interval between workouts during the week.

What are the primary purposes of resistance training?

Ronnie Kaufman/Blend Images

As you probably know, there is an inverse relationship between the amount of weight you can lift and the number of repetitions you can do. If your 1 RM in the bench press is 150 pounds, you can do more repetitions with 100 pounds than you can with 140. The **strength-endurance continuum** is a training concept that focuses on the interrelationship between resistance and repetitions. As depicted in **figure 12.3**, to train for muscular strength you must combine high resistance with a low number of repetitions. Conversely, to train for muscular endurance, you must combine a low resistance with a high number of repetitions.

Resistance-training programs may be designed to train all three of the human energy systems. The ATP-PCr energy system predominates in strength and power activities, the lactic acid energy system is primarily involved in anaerobic endurance, and the oxygen system is involved in aerobic endurance activities. Thus, resistance-training programs may be developed for various purposes. One purpose may be to improve health, as discussed below. Another purpose may be to enhance sports performance,

Lifting weight: Jack Mann/Photodisc/Getty Images; Lifting barbell: Comstock/Stockbyte/Getty Images

FIGURE 12.3 The strength-endurance continuum. To gain strength, you need to train on the strength end of the continuum; to gain endurance, you need to train on the endurance end of the continuum.

TABLE 12.3 Volumes for eight resistance exercises, total volumes for separate upper-body and back/lower body sessions, and weekly volume in a hypothetical 170-pound, 76″ tall, sedentary, 20-year-old basketball player.

Body area	Chest	Shoulder	Front arm	Back arm	Back	Thigh	Calf	Quadriceps	Total volume (lbs.) for *each* upper-body resistance exercise session (sessions 1 and 3)	Total volume (lbs.) for *each* back/lower-body resistance exercise session (sessions 2 and 4)	Total weekly volume (lbs.) for all four sessions
	Upper body-weekly sessions 1 and 3				Back/lower body-weekly sessions 2 and 4						
Resistance exercise	Bench press	Lateral raise	Bicep curls	Triceps extension	Lat pulldown	Half squat	Heel raise	Leg press			
Figure in chapter	12.6	12.9	12.11	12.12	12.7A	12.8	12.10				
1-RM (lbs., 170 lb. basketball player)	225	40	60	130	200	350	250	400			
Resistance (75% of 1-RM)	169	30	45	98	150	263	188	300			
Repetitions	5	5	5	5	5	5	5	5			
Sets	3	3	3	3	3	3	3	3			
Total volume (lbs.) (= resistance × repetitions × sets) for each resistance exercise	2,531	450	675	1,463	2,250	3,938	2,813	4,500	5,119	13,500	37,238

including improved strength and power for such sports as weight lifting and aesthetic appearance for sports such as bodybuilding.

The American College of Sports Medicine (ACSM) and the American Heart Association (AHA), both separately and collectively, have provided recommendations for participation in resistance-training programs. The recommendations from the ACSM/AHA committee, chaired by Haskell, focus on resistance training as a component of an overall exercise program to improve muscular strength and endurance in healthy young adults, whereas those from the ACSM/AHA committee, chaired by Nelson, focus on programs for the elderly. The AHA scientific statement, developed by Williams and others, provides recommendations on resistance training for individuals with and without cardiovascular disease. The ACSM also developed a set of recommendations dealing with progression, or the gradual increase in overload placed on the body during training, and is more appropriate for the individual training to maximize muscular size and strength for bodybuilding or sports competition. Overall, these four sets of recommendations highlight the following purposes of resistance training, along with the recommended type of program.

Training for Muscular Hypertrophy A high-volume resistance-training program consisting of multiple-sets; a relatively short recovery interval between sets; and a resistance in the range of 6–12 RM will emphasize muscle hypertrophy. In a 2017 meta-analysis, Schoenfeld and others reported a trend for greater muscle hypertrophy with higher (8.3 percent hypertrophy; >60 percent of 1 RM) compared to lower (7.0 percent hypertrophy; ≤60 percent of 1 RM) resistances. They concluded that a range of resistances can be effective to achieve muscle hypertrophy.

Training for Strength and Power General recommendations to maximize strength include multiple sets with fewer repetitions and a high resistance in the range of 4–6 RM. In a meta-analysis, Schoenfeld reported significantly greater improvements in 1 RM with higher (35.3 percent increase using >60 percent of 1 RM) compared to lower (28.0 percent hypertrophy using ≤60 percent of 1 RM) resistances. Power will be improved by incorporating lighter resistance sets (30–60 percent of 1 RM) lifted with a fast contraction velocity.

Training for Local Muscular Endurance Multiple sets with more repetitions and light to moderate loads, such as 15 or more repetitions at 40–60 percent of 1 RM, are recommended. Use a short recovery period between sets.

Training for Health-Related Benefits Single sets are sufficient, approximating 8–12 RM. Include a variety of exercises that stress the major muscle groups of the body.

What are the basic principles of resistance training?

Although the design and progression of a resistance training program will vary depending on the individual's goals and objectives, the underlying principles are the same. For example, athletes training to maximize muscle mass, strength, and power for their sport will probably engage in a more rigorous training program compared to someone doing resistance training for health benefits. The following discussion will highlight recommendations to gain muscle mass.

As noted in chapter 1, the following principles are not restricted to resistance training but apply to all forms of exercise training. For example, intensity of exercise is simply another way of phrasing the overload principle.

Overload The **principle of overload** is the most important principle in all resistance-training programs. The use of weights places a greater than normal stress on the muscle cell. This overload stress stimulates the muscle to grow—to become stronger—in effect to overcome the increased resistance imposed by the weights (**figure 12.4**).

Erik Isakson/Blend Images

FIGURE 12.4 Lifting heavy weights illustrates the principle of overload in action with weight training. If improvement in strength is to continue, weights must be increased.

Overload of a muscle requires an increase in the total volume of work by increasing the resistance, the number of repetitions, and/or the number of sets. In their meta-analysis of studies on males, Ralston and others concluded that high- or medium-weekly set resistance training more effectively increased muscle strength than low-weekly set training in advanced weightlifters while medium-weekly set resistance training was more effective in novice and intermediate weightlifters.

Using 60–80 percent of 1 RM should provide a resistance that is approximately in the 8–10 RM range. For example, if your bench press 1 RM is 150 pounds, you should be able to do approximately 8 RM with 80 percent of that value, or 120 pounds (0.80 × 150).

> www.shapesense.com/fitness-exercise/calculators/1rm-calculator.aspx Use a resistance you can lift more than one but not more than ten times to estimate your 1 RM for any resistance exercise.

Progression
As the muscle continues to get stronger during your training program, you must increase the amount of resistance—the overload—to continue to get the proper stimulus for sustained muscle growth. This is known as the **principle of progressive resistance exercise (PRE)**, another basic principle of resistance training.

The ACSM provides guidelines on progression as the individual advances in weight-lifting skill and strength, from novice to intermediate and advanced. Rayston and others reported a set dose–response in strength gains for both multijoint and isolation exercises in intermediate weightlifters. Pre- to poststrength gains were greater with high-weekly set number (i.e., a higher training volume) compared to low-weekly set number (i.e., a lower training volume). Following a learning period, a recommended program for beginners is three to five sets using loads corresponding to 8–12 RM. The first step is to determine the maximum amount of weight that you can lift for 8 repetitions. If you can do more than 8 repetitions, the weight is too light and you need to add more poundage. As you get stronger during the succeeding weeks, you will be able to lift the original weight more easily. When you can perform 12 repetitions, add more weight to force you back down to 8 repetitions. This is the progressive resistance principle. Over several months' time, the weight will probably have to be increased several times as you continue to get stronger. Such a transition is illustrated in **figure 12.5**. The ACSM recommends that an intermediate or advanced lifter should emphasize loads of 1–6 RM. The lifter should progress to a higher resistance when able to perform one to two more repetitions than the upper limit of the range. For example, if an initial load of 120 pounds is a 6 RM for a particular resistance exercise, the resistance should be increased when 120 pounds can be lifted eight times.

Specificity
The **principle of specificity** is a broad training principle with many implications for resistance training, including specificity for various sports movements, strength gains, endurance gains, and body-weight gains. Frost and others suggest that to facilitate the greatest improvements to athletic performance, the resistance-training program employed by an athlete must be adapted to meet the specific demands of the sport. Free weights and other mass-based approaches are well-established training stimuli to increase mass, strength, and rate of force development, but these approaches may be less effective in increasing force in sport-specific movement patterns and velocities. For example, a swimmer who wants to gain strength and endurance for a stroke should attempt to find a resistance weight-training program that exercises the specific muscles in a way as close as possible to the form used in that stroke. If you want to gain muscle mass in a certain part of the body, those muscles must be exercised. Strength coaches can help athletes develop resistance-training programs for specific sports.

Week	1	4	7	10	13	16
Weight (lbs)	20	20	25	25	30	30
Repetitions	8	12	8	12	8	12
Sets	4	4	4	4	4	4

John Lund/Marc Romanelli/Blend Images

FIGURE 12.5 The principle of progressive resistance exercise (PRE) states that as you get stronger, you need to progressively increase the resistance to continue to gain strength and muscle. In this example, the individual increases the resistance when she can complete 12 repetitions with a given weight but then does only eight repetitions with the increased weight.

Exercise Sequence
Your exercise routine should be based upon the **principle of exercise sequence**. This means that if you have ten exercises in your routine, they should be arranged in logical order so that fatigue does not limit your lifting ability. For example, the first exercise in a sequence of ten might stress the biceps muscle, the second the abdominals, the third the quadriceps, and so forth. After you perform one full set of each of the ten exercises, you then do a complete second set, followed by the third set. This approach may be best for beginners and is the sequence for the resistance exercises presented in this chapter.

Another popular option is to do three sets of the same exercise with a rest between sets; then do three sets of the second exercise, and so on. This approach may be a little more fatiguing because you are using the same muscle group in three successive sets, but it appears to be very effective. The time spent in recovery will also lengthen the total time for the workout. The ACSM also provides some guidelines on exercise sequence, recommending the following:

- Do multiple-joint exercises before single-joint exercises.
- Do large-muscle-group exercises before small-muscle-group exercises.
- Do higher-intensity exercises before lower-intensity exercises.
- Include concentric (lifting *against* gravity) and eccentric (lowering *with* gravity) muscle actions. In a meta-analysis of 15 resistance-training studies, Schoenfeld and others reported a trend for a greater hypertrophic response with eccentric (10 percent) compared to concentric (6.8 percent) contractions. They concluded that both are important for inducing muscle hypertrophy.

Recuperation Resistance training, if done properly to achieve the greatest gains, imposes a rather severe stress on the muscles, requiring a period of recovery both during the workout and between workouts. During anaerobic exercise such as high-intensity weightlifting (e.g., a set of a 5-RM resistance), high-energy intramuscular phosphates ATP and phosphocreatine are rapidly depleted. However, these high-energy phosphates are reconstituted during 2–3 minutes of rest (e.g., postexercise recovery prior to the next set). This is the **principle of recuperation.** When performing multiple sets of the same resistance exercise, a period of recovery should occur between sets in order to maintain the desired total session volume and quality of the workout. In their review of 35 studies, de Salles and others concluded that rest intervals of 3–5 minutes resulted in a greater number of repetitions of resistances between 50 and 90 percent of 1 RM and produced greater training volumes, power production, and improvement in strength. They also noted that moderate-intensity sets performed with shorter (30- to 60-second) rest intervals may increase the secretion of growth hormone and insulin-like growth factors which may facilitate the development of muscle hypertrophy. Grgic and others concluded that resistance-trained individuals seeking an increase in muscle strength benefit from a rest interval of at least 2 minutes between sets while shorter rest intervals are effective to increase strength in novice weightlifters. In a meta-analysis of 21 studies, Schoenfeld and others reported greater strength gains from high-load (>60 percent of 1 RM) compared to low-load (≤60 percent of 1 RM) resistance training with similar gains in hypertrophy. They concluded that muscle hypertrophy can occur in response to a range of training loads. If the goal is muscular endurance, the ACSM recommends that rest periods be shortened to less than 90 seconds.

According to Witard and others, muscle protein synthesis can occur 24–48 hours after a single bout of heavy resistance exercise. The novice weightlifter should generally engage in resistance training three times a week with a rest or recuperation day between sessions to allow adequate time for muscle repair and growth. The ACSM indicates that advanced lifters could train 4–5 days per week. For health benefits, the ACSM and AHA recommended resistance exercise at least twice a week on nonconsecutive days.

Periodization Periodization is a training concept pioneered in the 1970s by the Russian physiologist Leonid Matveyev. It is based on the alarm, resistance, and recovery or exhaustion stages of Hans Selye's general adaptation theory of how the body adapts to stress. Periodization divides training into time cycles of various durations determined by the individual. A **microcycle** is a short period of time, typically 1 week. A **mesocycle** is an intermediate period of several months or more, including multiple microcycles, and generally beginning with a high-volume/low-intensity phase preceding a progressive increase in intensity with lower volume and ending with tapering in preparation for competition. A **macrocycle** may be a year or more in duration and include multiple mesocycles. The classical linear periodization model includes successive mesocycles of increasing intensity and decreasing volume with variations in resistance, repetitions, and sets to emphasize endurance, hypertrophy, strength, and power. The undulating, or nonlinear, model includes greater variation in volume and intensity within a mesocycle. Periodization is more applicable to competitive athletes than to beginning resistance exercisers. As noted in chapter 3, excessive exercise may predispose athletes to overreaching and overtraining, which may contribute to impaired performance. A more detailed discussion of periodization is beyond the scope of this text. The interested reader is referred to the meta-analysis by Harries and others; texts by Bompa and Buzzichelli, and Stone and others; and the ACSM position statement on resistance-training progression models.

With the exception of periodization, these general principles should serve as guidelines during the beginning phase of your resistance-training program and should be used to guide your progress during the first 3 months of the basic resistance-training program described next. If you become serious about resistance training, additional reading is advised. Personal trainers may also be helpful.

What is an example of a resistance-training program that may help me to gain body weight as lean muscle mass?

If your goal is to gain significant amounts of muscle mass, you may wish to exercise on the strength end of the strength-endurance continuum, using six to ten different exercises that stress the major muscle groups of the body. Wernbom and others commented that as little as one set of 8–12 repetitions may increase muscle thickness. However, de Sallis and others concluded that moderate-intensity sets performed with shorter (30- to 60-second) rest intervals may stimulate muscle hypertrophy. Schoenfeld concluded that high-load training stimulates a greater increase in muscle strength than low-load training but that muscle hypertrophy can occur in response to a range of training loads. Controlled concentric (1–3 seconds) and eccentric (2–4 seconds) muscle actions throughout the range and planes of motion will result in maximal recruitment of muscle fibers. The use of a split routine (e.g., exercising upper- and lower-body muscles on different days) incorporates the previously discussed principle of recuperation. The model 8–12 repetition program presented in **figures 12.6** through **12.14B** employs the PRE concept, starting with an 8-repetition resistance. After progressing

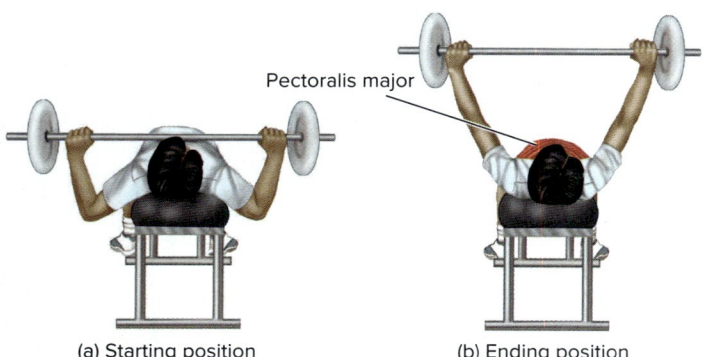

(a) Starting position   (b) Ending position

Chest	
Exercise	Bench press
Chest muscles	Pectoralis major
Other muscles	Deltoid, triceps
Sets	3–5
Repetitions	8–12, PRE concept
Safety	Have spotter stand behind bar to assist as fatigue sets in.
Equipment	Bench with support for weight, or two spotters to hand weight to you.
Description	Lie supine on bench. Use wide grip for chest development. Secure bar and lower *slowly* to chest. Press bar straight up to full extension. Do not arch back.

Kelly Redinger/Design Pics

FIGURE 12.6 The bench press primarily develops the pectoralis major muscle group in the chest. It also develops the deltoids in the shoulder and the triceps at the back of the arm.

Note: For safety reasons, a spotter should be present to assist in case of difficulty. The spotter is not depicted for purposes of illustration clarity.

to 12 repetitions, the resistance should be increased to allow for 8 repetitions (**figure 12.5**).

Numerous resistance-training exercises are available to stress the major muscle groups in the body. The following steps will provide the basis for a safe and effective resistance-training program for the adolescent and adult beginner.

1. Learn the proper technique for each exercise with a light weight, possibly only the bar itself, for 2 weeks. Do 8–12 repetitions of each exercise to develop form. Do not strain during this initial learning phase. Concentrate on lowering the weight slowly.
2. For each exercise, determine the maximum weight that you can lift for 8 repetitions after the 2-week learning phase.
3. A weekly record form, similar to the one presented in **table 12.4**, should be used to keep track of your progress.
4. Do one set of the 9 exercises shown in **figures 12.6–12.14B**. The sequence of exercises should be
   a. Bench press: chest muscles
   b. Lat machine pulldown or bent-arm pullover: back muscles
   c. Half squat: thigh muscles
   d. Standing lateral raise: shoulder muscles
   e. Heel raise: calf muscles
   f. Standing curl: front upper arm muscles
   g. Seated overhead press: back upper arm muscles
   h. Curl-up: abdominal muscles
   i. Horizontal leg press: thigh, buttock muscles
5. Because the exercise sequence is designed to stress different muscle groups in order, less recuperation is necessary between exercises.
6. Do three to five complete sets. You may wish to rest 2–3 minutes between sets.
7. Exercise 3 days per week; in each succeeding day try to do as many repetitions as possible for each exercise in each set. When you can do 12 repetitions each after a month or so, add more weight so you can do only 8 repetitions.
8. Repeat step 7 as you progressively increase your strength.

Because barbells and dumbbells appear to be the most common means of doing resistance training, this is the method utilized. However, other apparatus, such as the Nautilus®, Hammer Strength®, resistance and bands, as well as Pilates and other techniques, can also be used effectively to gain fat-free mass and strength. Most of the exercises described here using barbells or dumbbells have similar counterparts on other apparatus.

Note that muscles seldom operate alone, and that most resistance-training exercises stress more than one muscle group. Thus, keep in mind that although an exercise may be listed specifically for the chest muscles, it may also stress the arm and shoulder muscles. The exercises described in this section generally stress more than one body area, although their main effect is on the area noted.

It is important to note that your muscle contracts during both the up and down phase of weight lifting. As noted later, some training methods are based on this concept. When you lower a weight, your active muscle is actually contracting to help decrease the force of gravity. Lowering a weight slowly increases the time your muscle must contract. Raising the weight takes more force as you are working against gravity.

These nine exercises stress most of the major muscle groups in the body and thus provide an adequate stimulus for gaining body weight and strength through an increase in muscle mass.

TABLE 12.4 Weekly record for resistance-training program of nine exercises

Body area		Chest	Back	Thigh	Shoulder	Calf	Front arm	Back arm	Abdomen	Thigh, buttocks
Resistance exercise		Bench press	Lateral pulldown	Half squat	Lateral raise	Heel raise	Curls	Seated overhead press	Curlups	Horizontal Leg Press
Figures		12.6 A–B	12.7 A–C	12.8 A–B	12.9 A–B	12.10 A–B	12.11 A–B	12.12 A–B	12.13 A–B	12.14 A–B
Date	Set	Resistance=	Resistance=	Resistance=	Resistance=	Resistance=	Resistance=	Resistance=	Resistance=	Resistance=
	1									
	2									
	3									
	1									
	2									
	3									
	1									
	2									
	3									
	1									
	2									
	3									

Literally hundreds of different resistance-training exercises and techniques to train are available. If you become interested in diversifying your program, consult an authoritative source specific to resistance training. Several may be found in the reference list at the end of this chapter. For example, nearly 30 different types of programs are presented in the classic text by Fleck and Kraemer. Some professional organizations, such as the National Strength and Conditioning Association, also provide excellent resistance-training exercises.

Individuality in responses to resistance-training programs is an important consideration. According to Hayes and others, strength protocols elicit greater increases in the male anabolic hormone testosterone and greater gains in hypertrophy and strength in some individuals, whereas others respond preferentially to power or strength/endurance protocols. Other factors that may affect hypertrophy and strength gains between individuals include, but are not limited to, daily variations in testosterone and cortisol levels, which stimulate muscle synthesis and breakdown, respectively, and the testosterone/cortisol ratio. Testosterone will be discussed in chapter 13.

Another important factor is the time of day for resistance training. It is well established, as noted by Chtourou and Souissi, that anaerobic power and strength performances generally peak in the late afternoon, with correspondingly greater posttraining adaptations. According to Hayes and others, greater increases in strength and hypertrophy generally observed in the later afternoon may be related to a higher testosterone/cortisol ratio, which optimizes muscle protein synthesis related to increased testosterone secretion, with a concurrent decrease in the cortisol level, which is associated with proteolysis. However, adaptations are also greater at the time of day when training normally occurs as compared to other times of day. Although the mechanisms explaining these time-of-day adaptations are not completely understood, the novice weight lifter should strive for consistency in time-of-day scheduling if afternoon training is not feasible.

www.nsca.com/Videos For detailed illustrations of a wide variety of resistance-training exercises in video format, click on Exercise Technique under Videos on the right of the page. The muscles involved in each exercise and safety tips are also provided. Some of the videos demonstrate lifts that should be performed only by intermediate and advanced weight lifters.

www.nia.nih.gov/health/four-types-exercise-can-improve-your-health-and-physical-ability Information from the National Institute on Aging for safely improving strength in older adults. Click "Strength" under "Exercise and Physical Activity."

Are there any safety concerns associated with resistance training?

Several health problems discussed later in this chapter may be contraindications for participation in a strenuous resistance-training program. However, resistance training is generally regarded as a relatively safe sport, particularly if appropriate safety precautions are taken. The following guidelines should be incorporated into all resistance-training programs.

1. *Learn to breathe properly.* During the most strenuous part of the exercise, you are likely to hold your breath. This is a natural response to increase intrathoracic pressure, which stabilizes the spine and provides a stable base for muscle contraction. Usually, the breath hold is short and no problems occur. However, if prolonged, it may increase the chance of suffering a hernia if there is a weak area of the abdominal musculature.

   Also associated with prolonged breath holding is a response known as the **Valsalva phenomenon** (Valsalva maneuver), which may lead to a blackout. As you reach a sticking point in your lift and strain to overcome it, you normally hold your breath. This causes your glottis to close over your windpipe and the pressure in your chest and abdominal area to rise rapidly. The pressure creates resistance to blood flow, reducing the return of blood to the heart, and eventually leading to decreased blood flow to the brain and a possible blackout. Additionally, the Valsalva maneuver exaggerates the increase in blood pressure during resistance exercises, and although a brief Valsalva maneuver is unavoidable when doing near-maximal exercises, its effect may be minimized by proper breathing.

   A recommended breathing pattern for beginning weight lifters that will help minimize these adverse effects is to breathe out while lifting the weight, especially through the sticking point, and breathe in while lowering it. You should breathe through both your mouth and nose while exercising. Practice proper breathing when you learn new resistance-training exercises.

2. *Use spotters.* When using free weights, use spotters when doing exercises that are potentially dangerous, such as the bench press. If you are doing a bench press alone and reach a sticking point in your lift, the Valsalva phenomenon may lead to serious consequences if you lose control of the weight directly above your head. The use of machines such as Nautilus helps eliminate the need for spotters.

3. *Use safety equipment.* If using free weights, place lock collars on the bar ends so that the plates do not fall off and cause injury to the feet. Again, the use of machines eliminates this safety hazard. However, do not attempt to change weight plates on machines while they are being used. Your fingers may get caught between the weights.

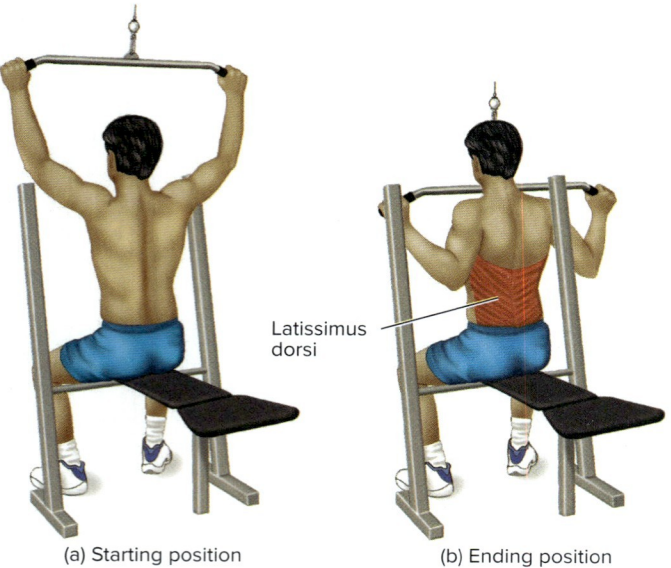

(a) Starting position   (b) Ending position

Blend Images/Alamy Stock Photo

Back	
Exercise	Lat machine pulldown
Back muscles	Latissimus dorsi
Other muscles	Biceps, pectoralis major
Sets	3–5
Repetitions	8–12, PRE concept
Safety	A very safe exercise
Equipment	Lat machine
Description	From seated or kneeling position, take a wide grip at arm's length on the bar overhead. Pull bar down until it reaches your chest. Return slowly to starting position.

FIGURE 12.7A The lat machine pull-down trains the latissimus dorsi in the back and side of the upper body, and it develops the biceps on the front of the upper arm and the pectoralis major in the chest.

Note: If a lat machine is not available, the bent-arm pullover may be substituted.

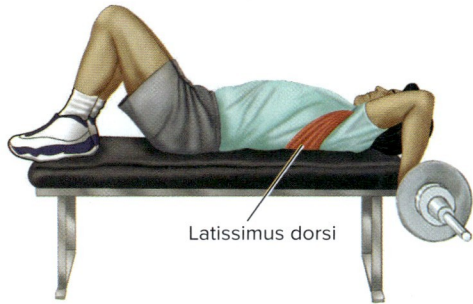

(a) Starting position

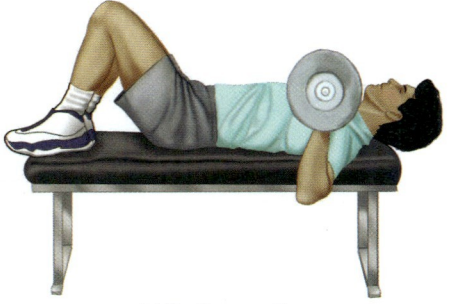

(b) Ending position

Nicholas Piccillo/Shutterstock

4. *Warm up.* Warm up with proper stretching exercises. Gently stretch the muscles to be used during exercise. Slow, static methods are recommended for cold muscles. In separate studies by Kokkonen and others, 10 weeks of static stretching for 40 minutes/day 3 days/week significantly improved knee extension and flexion 1 RM and that novice weight trainers completing weight training with static stretching had significant increases in leg press and knee extension strength compared to weight training without static stretching. Stretching may facilitate a return to resistance training following injury as well as enhance early gains in strength. However, research is equivocal regarding the effect of static stretching during the rest interval between sets in the weightlifting session on flexibility, strength, and hypertrophy. Evangelista and others reported that benefits of interset static stretching were insufficient to conclude a superior effect on muscle hypertrophy compared to traditional strength training. Similarly, Nakamura and others concluded that interset static stretching did not impact muscle hypertrophy in untrained young males. In their review of ten studies, Nunes and others concluded that there is no evidence of significant effects of low-intensity stretching on skeletal muscle architecture and hypertrophy. There is scant evidence that high-intensity stretching while the muscle is loaded or between contractions may induce muscle hypertrophy.

5. *Use proper technique.* Use light weights to learn the proper technique of a given exercise so that you do not strain yourself if you do the exercise incorrectly. Learn to lift smoothly without jerking motions and using the full range of motion. When the proper technique is mastered, the weights may be increased. Using proper technique ensures that the desired muscle group is being exercised.

Back	
Alternate exercise	Bent-arm pullover
Back muscles	Latissimus dorsi
Other muscles	Pectoralis major
Sets	3–5
Repetitions	8–12, PRE concept
Safety	Do not arch back. Start with light weights when learning the technique.
Equipment	Bench
Description	Lie supine on bench, entire back in contact with the bench, feet on the bench, knees bent. Hold weight on chest with elbows bent. Swing weight over head, just brushing hair, and lower as far as possible without taking back off the bench. Keeping elbows in, return the weight to the chest.

FIGURE 12.7B The bent-arm pullover trains the latissimus dorsi and develops the pectoralis major.

Note: For safety reasons, a spotter should be present to assist in case of difficulty. The spotter is not depicted for purposes of illustration clarity.

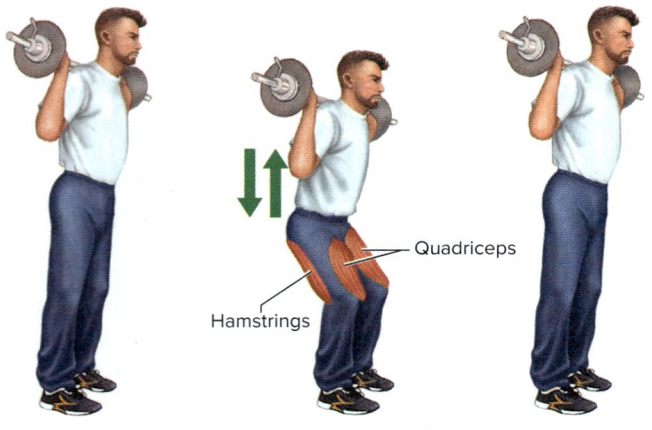

Thigh	
Exercise	Half squat or parallel squat
Thigh muscles	Quadriceps (front), hamstrings (back)
Other muscles	Gluteus maximus
Sets	3–5
Repetitions	8–12, PRE concept
Safety	Have two spotters to assist if using free weights. Keep back straight. Drop weight behind you if you lose balance. Do not squat more than halfway down.
Equipment	Squat rack if available. Pad the bar with towels if necessary.
Description	In standing position, take bar from squat rack or spotters and rest on the shoulders behind the head. Squat until thighs are parallel to ground or until buttocks touch a chair at this parallel position. Do not squat beyond halfway. Keep back as straight as possible. Return to standing position, but do not lock your knees. This will maximize stress on your thighs.

studioloco/Shutterstock

FIGURE 12.8 The half squat or parallel squat develops the quadriceps muscle group on the front of the thigh and the hamstrings on the back of the thigh.

Note: For safety reasons, spotters should be present to assist in case of difficulty. The spotters are not depicted for purposes of illustration clarity.

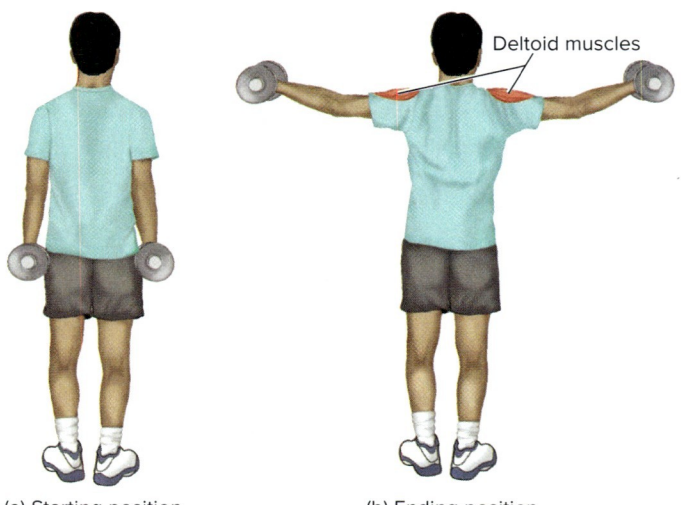

Shoulders	
Exercise	Standing lateral raise
Shoulder muscles	Deltoid
Other muscles	Trapezius
Sets	3–5
Repetitions	8–12, PRE concept
Safety	Do not arch back.
Equipment	Dumbbells
Description	Stand with dumbbells in hands at sides. With palms down, raise straight arms sideways to shoulder level. Bend elbows slightly. Return slowly to starting position.

MDV Edwards/Shutterstock

FIGURE 12.9 The standing lateral raise primarily develops the deltoid muscles in the shoulder. The trapezius in the upper back and neck area is also trained.

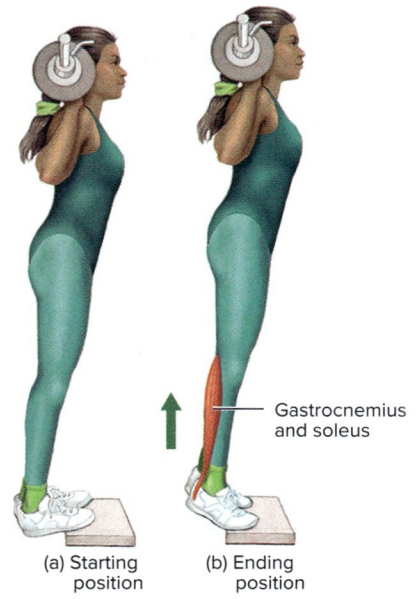

(a) Starting position (b) Ending position

Gastrocnemius and soleus

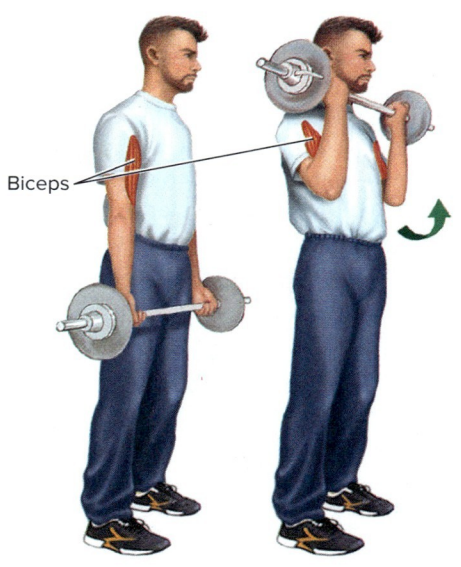

Biceps

(a) Starting position (b) Ending position

Calf	
Exercise	Heel raise
Calf muscles	Gastrocnemius, soleus
Other muscles	Deep calf muscles
Sets	3–5
Repetitions	8–12, PRE concept
Safety	Have two spotters if you use free weights.
Equipment	Squat rack if available. Pad the bar with a towel if necessary.
Description	Place bar on back of shoulders as in squat exercise. Raise up on your toes as high as possible and then return to standing position. Place the toes on a board so heels can drop down lower than normal. Point toes in, out, and straight ahead during different sets to work the muscles from different angles.

Mary Rice/Shutterstock

FIGURE 12.10 The heel raise develops the two major calf muscles—the gastrocnemius and the soleus.

Note: For safety reasons spotters should be present to assist in case of difficulty. The spotters are not depicted for purposes of illustration clarity.

Front of arm	
Exercise	Standing curl
Arm muscle	Biceps
Other muscles	Several elbow flexors
Sets	3–5
Repetitions	8–12, PRE concept
Safety	Do not arch back. Place back against wall to control arching motion.
Equipment	Curl bar if available
Description	Stand with weight held in front of body, palms forward. Place back against wall. Bend the elbows and bring the weight to the chest. Lower it slowly.

Tetra Images, LLC/Alamy Stock Photo

FIGURE 12.11 The standing curl strengthens the biceps muscle in the front of the upper arm as well as several other muscles in the region that bend the elbow.

6. *Protect your lower back.* Avoid exercises that may cause or aggravate low back problems. Try to prevent an excessive forward motion or stress in the lower back region. **Figure 12.15** illustrates some positions that should be avoided.
7. *Lower weights slowly.* If the resistance is lowered rapidly, the muscles have to contract rapidly to maintain control and slow the angular velocity back to the starting position. This necessitates the development of a large amount of force that may tear some muscle fibers or connective tissue and cause muscle soreness.

How does the body gain weight with a resistance-training program?

It is well established that the mechanical stress of high-intensity resistance training tends to increase muscle size. Secretion of anabolic hormones such as testosterone, growth hormone, insulin-like growth factors, and insulin, as well as upregulation of certain cell signaling pathways, increase muscle protein synthesis. Muscle growth can also be facilitated by down-regulation of other cell signaling pathways which promote muscle protein breakdown as noted in reviews by Yoshiuda and Delafontaine, as well as Rodriguez and others. Vingren and others note that testosterone is considered the major promoter of muscle growth and subsequent increase in muscle strength in response to resistance training in men. However, an increase in testosterone following acute resistance exercise has been reported in some but not all studies. According to Hooper and others, increased testosterone secretion and uptake by muscle is thought to be related to androgen receptor upregulation and activation of anabolic cell signaling but is not fully understood. They also note that increased strength and hypertrophy can occur in the absence of acute or chronic increases in testosterone secretion. Human growth hormone and insulin-like growth factors are also involved. These hormones are discussed in more detail in chapter 13. According to Zanou and Gailly, the activation of myogenic stem cells, also known as satellite cells, increases regulatory proteins controlling the formation of the contractile myofilament myosin (**figure 12.16**) as well as growth factors that play a key role in the repair of muscle damage. Stimulation of DNA in skeletal muscle nuclei by resistance training increases RNA and results in protein synthesis. Ribosomes are the sites of translation of mRNA into new protein. As discussed by Chaillou, ribosomal biogenesis and specialization are regulated by complex factors, including but not limited to anabolic and catabolic cell signaling pathways, ribosome location within the cell, cell type, tissue type, ribosomal protein modification and turnover, physiological stimuli, and circadian rhythm.

As noted by Gonzalez and others, the cellular mechanisms which link the mechanical stress of resistance training with muscle growth

(a) Starting position   (b) Ending position

leezsnow/E+/Getty Images

Back of arm	
Exercise	Seated overhead press (triceps extension)
Arm muscle	Triceps
Other muscles	Trapezius, deltoids
Sets	3–5
Repetitions	8–12, PRE concept
Safety	Do not arch back excessively. Have spotter available as fatigue sets in.
Equipment	Bench or chair
Description	Sit on bench with weight held behind the head near the neck. Hands should be close together, elbows bent. Keep elbows in. Straighten elbows and press weight over head to arm's length. Lower weight slowly to starting position.

FIGURE 12.12 The seated overhead press primarily develops the triceps muscle on the back of the upper arm. The exercise also trains the trapezius in the upper back and neck and the deltoids in the shoulder.

Note: For safety reasons spotters should be present to assist in case of difficulty. The spotters are not depicted for purposes of illustration clarity.

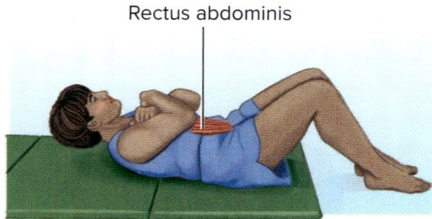
(a) Starting position

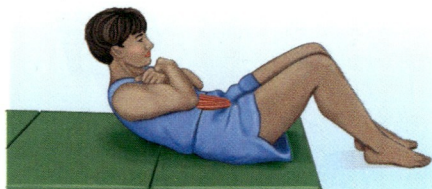
(b) Ending position

Abdominal area	
Exercise	Curl-up
Abdominal muscles	Rectus abdominis
Other muscles	Oblique abdominis muscles
Sets	3–5
Repetitions	8–12, PRE concept
Safety	Develop sufficient abdominal strength before using weights with this exercise. Do not arch back when exercising.
Equipment	Free-weight plates; incline sit-up bench if available
Description	Lie on back, knees bent with heels close to buttocks. Hands should hold weights on chest. Curl up about a third to halfway. Return to starting position slowly.

ESB Basic/Shutterstock

FIGURE 12.13 The curl-up trains the rectus abdominis and the oblique abdominis muscles.

Note: This exercise may be done without weights but with an increased number of repetitions.

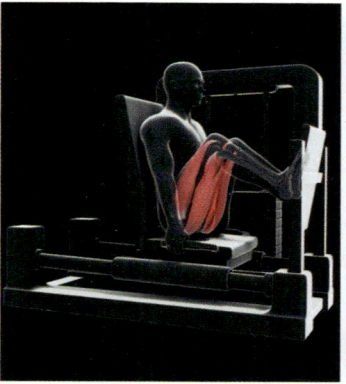 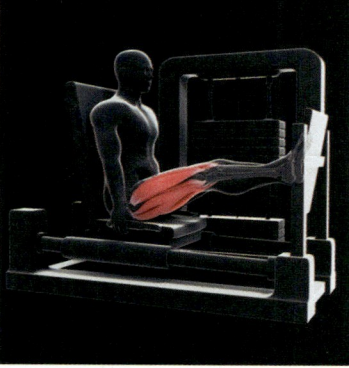
(a) Starting position   (b) Ending position

Thigh/Quadriceps	
Exercise	Horizontal Leg Press
Thigh muscles	Quadriceps
Other muscles	Gluteus maximus, Gluteus medius, Gluteus minimus, Tensor fasciae latae (hips); hamstrings (back of thigh); Gastrocnemius (calves)
Sets	3–5
Repetitions	8–12, PRE concept
Safety	Develop sufficient thigh and knee extension strength before using weights with this exercise. Breathe during the exercise.
Equipment	Universal-type multi-station
Description	Place feet on the platform approximately 1–1½ feet apart. Adjust seat until you have approximately 90° of knee flexion. While exhaling, push with the heels of your feet and use your quadriceps to extend your legs without locking your knees.

(a): MadiGraphic/Shutterstock; (b): Capuski/E+/Getty Images

FIGURE 12.14 The horizontal leg press trains the quadricels, gluteus maximus, gluteus medius, gluteus minimus, tensor fasciae latae, hamstrings, and gastrocnemius muscle groups of the lower extremity.

FIGURE 12.15 Avoid exercises or body positions that place excessive stress on the low-back region. Poor form in exercises like (a) the bench press and (b) the curl exaggerates the lumbar curve. Be sure to keep the lower back as flat as possible. Exercises similar to (c), the bent-over row, place tremendous forces on the lower back because the weight or resistance is so far in front of the body.

through increased anabolic hormone secretion and cell signaling are not completely understood. At the muscle fiber organizational level (**figure 12.16**), muscle growth could be explained by **muscle hypertrophy**, an increase in muscle fiber *size*; **muscle hyperplasia**, an increase in muscle fiber *number*; or the combined effects of hypertrophy and hyperplasia. Although there is some research support for hyperplasia, current evidence strongly favors muscle hypertrophy as the dominant adaptation to explain muscle growth. The incorporation of more protein can result in more myofibrils in each muscle fiber. In their review, Folland and Williams noted that the primary factor contributing to increased overall muscle size is the increase in the size and number of the myofibrils. The amount of connective tissue around each muscle fiber and around each bundle of muscle may also increase and thicken. The fiber may increase its content of enzymes and energy storage, particularly ATP and glycogen. Increased muscle glycogen, along with increased muscle protein, binds additional water, which contributes to an increased body weight. As previously discussed, water accounts for 75 percent of skeletal muscle, so greater water content contributes to hypertrophy of muscle fibers.

In addition to the effects on muscle, some but not all studies indicate that resistance training may stimulate bone formation, possibly due to increased muscle tension on the bone. If so, this may account for a small increase in body weight. In a meta-analysis of six studies of the effects of high-load (>70 percent of 1 RM) and low-load (≤70 percent of 1 RM) resistance training on bone mineral density (BMD) in middle-aged and older (≥45 years) males and females, Souza and others reported significant increases in femoral neck density following high-load resistance training in subjects with normal baseline BMD and in response to training up to 6 months in duration.

Resistance training may be an effective means to increase muscle size and mass regardless of gender and age. Burd and others concluded that while young males and females have similar relative changes in muscle hypertrophy following resistance training, older women have a lower capacity for muscle protein synthesis and decreased hypertrophy in response to resistance training. Decreased secretion of testosterone that occurs with aging may attenuate the capacity for muscle hypertrophy as noted by Vingren and others. However, Peterson and others concluded from their meta-analysis of 49 training studies that resistance training programs, particularly with higher-volume programs, can increase lean body mass in aging adults.

Is any one type of resistance-training program or equipment more effective than others for gaining body weight?

There are a variety of methods for resistance training. **Isometric contractions** involve a muscle contraction against an immovable object, such as pushing against a thick concrete wall. However, if you succeed in moving the object, then you are doing an isotonic exercise. **Isotonic contractions** are of two types. As mentioned on page 471, a **concentric contraction** shortens the muscle as the resistance is lifted against gravity. During an **eccentric contraction**, the muscle lengthens with tension while the resistance is lowered with the assistance of gravity. The muscle tension during the eccentric contraction controls the rate of descent. For example, the biceps muscle contracts concentrically in the up phase of a pull-up and eccentrically in the down phase as the body is lowered under control to the starting point. Finally, **isokinetic contractions** use machines or other devices to regulate the speed at which you can shorten your muscles. For example, you may try to move your arm as fast as possible, but you will be able to move only as fast as the setting on the isokinetic machine. Isokinetic exercise is also known

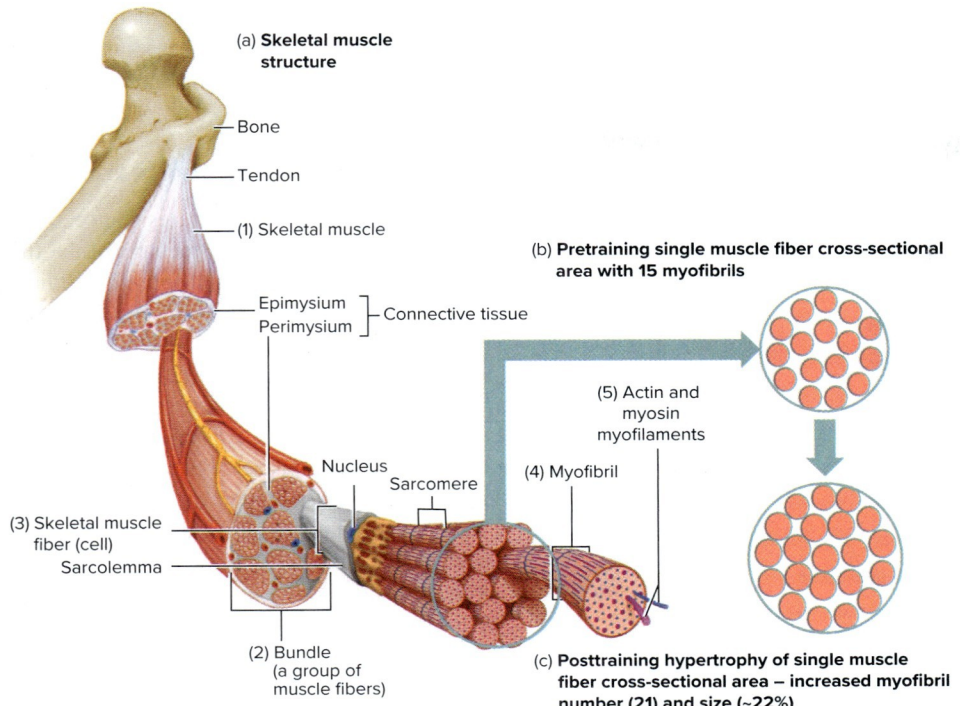

FIGURE 12.16 (a, 1–5) Skeletal muscle structure. The whole muscle is composed of separate bundles of individual muscle fibers. Each fiber is composed of numerous myofibrils, each of which contains thin protein filaments arranged so that they can slide by each other to cause muscle shortening or lengthening. Several layers of connective tissue surround the muscle fibers, bundles, and whole muscles, which eventually band together to form the tendon. (b) Preresistance training single muscle fiber cross-sectional area. (c) Postresistance training single muscle fiber cross-sectional area illustrating hypertrophy. See text for explanation.

as accommodating-resistance exercise because the resistance automatically adjusts to the force exerted, thus controlling the speed of movement. Maximal torques for commercial isokinetic systems (e.g., Biodex System 4 Pro™) are 680 and 540 Newton-meters for concentric and eccentric muscle contractions, respectively. Most commercial isokinetic systems have a maximum *concentric* contraction velocity of 500°/sec.

Several different resistance-training products are available, such as Atlantis®, Nautilus®, Hammer Strength®, Cybex®, Hydra-Gym, Soloflex®, and other similar machines. Depending upon the model, they are designed to utilize one or more of the training methods cited previously.

Current research suggests that the various training methods are comparable in their ability to produce gains in muscle size and strength. For example, in a comprehensive review comparing the effects of dynamic exercise (including free weights and weight machines), accommodating resistance (isokinetic and semi-isokinetic devices), and isometric resistance, Wernbom and others concluded that there is insufficient evidence for the superiority of any mode and/or type of muscle action over other modes and types of training.

Both concentric and eccentric contractions appear to be important for inducing muscle hypertrophy. In their review, Franchi and others concluded that eccentric and concentric resistance training, when matched for maximum load or total work volume, result in similar muscle hypertrophy in young healthy populations. They also noted that concentric and eccentric contractions stimulate different cellular responses leading to increased muscle size. As stated earlier, Schoenfeld and others reported a trend for a greater hypertrophic response with eccentric compared to concentric contractions. Roig and others, in a meta-analysis of 20 studies, concluded that when eccentric exercise was performed at higher intensities compared with concentric training, total strength and eccentric strength increased more significantly. They suggested that the superiority of eccentric training to increase muscle strength and mass appears to be related to the higher loads developed during eccentric contractions. However, such training programs should be used with caution at least until the muscle adapts to a program of progression in intensity because high-intensity eccentric exercise may be more likely to cause muscle tissue damage and muscle soreness. Such programs would appear to be of interest primarily to highly trained athletes.

All methods may be effective in increasing body weight, provided the basic principles of resistance training, particularly the principle of overload, are followed. The ACSM recommends that resistance training programs include both concentric and eccentric exercise, which are incorporated into most machines and free weights. If you use machines, be sure to exercise all major muscle groups. Free weights are relatively inexpensive and can be used for a wide variety of exercises. They also may be constructed at home, using pipe or solid broomstick handles for the bar and different-sized tin cans filled with cement for the weights. Regardless of equipment or system, it is essential to use proper lifting and breathing techniques. In addition, free weightlifting in community and home settings should be performed with spotters, balancing of weight plates on bars, and use of clips to secure plates.

There may be some specific training programs that are better suited for specific purposes, such as specific sports performance enhancement or injury rehabilitation. These topics are beyond the scope of this text, so the interested reader is referred to more detailed resources, such as the texts cited in the reference list at the end of this chapter.

If exercise burns kcal, won't I lose weight on a resistance-training program?

Although exercise does cost kcal, the amount expended during resistance training is relatively small compared to more active aerobic exercise. Resistance training can be a high-intensity exercise, but the time spent actually lifting during a typical workout is usually

1 hour = 600–800 kcal   1 hour = 150–200 kcal

Woman running: Daxiao Productions/Shutterstock; Woman lifting weights: John Lund/Drew Kelly/Blend Images LLC

FIGURE 12.17 All modes of exercise increase caloric expenditure. However, an hour of regular weight training expends only about one-third to one-fourth as many kcal as vigorous aerobic activity. Combining aerobic exercises with weight training (circuit aerobics) helps burn more kcal than weight training alone. It also provides cardiovascular health benefits while increasing muscular strength and endurance.

short, therefore limiting the number of kcal used. For example, in an hour workout, only about 15 minutes may be involved in actual exercise, the remaining time being recovery between each exercise. Based on metabolic data collected in research studies, the average-sized male uses about 200 kcal in a typical workout, while the average-sized female uses about 150 (**figure 12.17**). Resistance training has been reported to have a small, but significant effect on postexercise energy expenditure. In a study of untrained, overweight young adults, Heden and others reported similar (~5 percent) elevations in resting energy expenditure 24, 48, and 47 hours following one-set and three-set acute bouts of resistance exercise (10 exercises, 10 RM) compared to pre-exercise baseline. Jamurtas and others reported increases in resting energy expenditure with increased fat oxidation in trained males 10 and 24 hours following a weight-lifting session (resistance at 70–75 percent of 1 RM) compared to running exercise (72 percent of VO_2 max). Ormsbee and others reported an increase in energy expenditure and fat oxidation following a 45-minute intense resistance-training workout (10 percent, 10 kcal/hour) compared to a nonexercise day. Greer and others compared isocaloric high-intensity interval aerobic exercise sessions with and without resistance training in low to moderately active males. The researchers did not counterbalance the treatment order. They reported small but significant increases in oxygen consumption at 12 and 21 hours following aerobic exercise plus resistance training, compared to steady-state aerobic exercises. However, they reported no effect on fat oxidation. In summary, additional lean tissue from resistance training increases resting energy expenditure. This was discussed in chapter 11 in the context of weight loss and maintenance. Although increased postexercise fat oxidation following resistance exercise with or without aerobic exercise has not been observed in every study, such an outcome would be desirable for healthy body composition during weight gain.

Are there any contraindications to resistance training?

Several health conditions may be aggravated by resistance training, primarily by the increased pressures that occur within the body when straining to lift heavy weights while breath holding. Because the blood pressure can increase rapidly and excessively during weight lifting, to 300 mmHg or higher, individuals with resting blood pressures of 140 mmHg or higher (systolic) or 90 mmHg or higher (diastolic) should refrain from heavy lifting, for they may be exposed to an increased risk of blood vessel rupture and a possible stroke. Lifting with the arms and straining exercises also increase the stress on the heart and thus should be avoided by individuals who have heart problems, such as arrhythmias. Individuals with a hernia (a weakness in the musculature of the abdominal wall) also should refrain from strenuous weight lifting because the increased pressure may cause a rupture. Improper lifting technique may contribute to muscle strain, cause pain in the shoulder and other joints, and aggravate existing low back problems. Some individuals may suffer peripheral nerve injuries, such as carpal tunnel syndrome, involving weakness and numbness in the hand due to compressed nerves from improper lifting techniques. Individuals with these types of health problems should seek medical advice and be cleared by a physician before initiating or continuing with a resistance-training program.

There has been some concern about the advisability of prepubescent youth lifting weights. However, a 2020 American Academy of Pediatrics (AAP) position statement by Strickler and others, as well as a consensus statement endorsed by several international scholarly organizations authored by Lloyd and others, has assessed the safety of resistance-training programs for children and adolescents. These position statements can be summarized by the following ten conclusions.

- Resistance training for youth should be designed and supervised by qualified, knowledgeable adults.
- Resistance training programming for youth should consider the level of training, age, motor skill competency, and proficiency in technique.
- Resistance training by youth should primarily focus on technique at an appropriate volume and in a safe and enjoyable environment.
- Resistance training by youth should include movement patterns in addition to traditional adult-like weightlifting activities.
- Research and science support youth participation in appropriately designed and supervised resistance training with a low risk of proximate injury.
- Resistance training can reduce the risk of sports-related injuries and be of benefit in rehabilitation from injury.
- Resistance training can be part of an effective elementary and secondary physical education curriculum.
- Resistance training by youth can facilitate a lifelong appreciation of and participation in physical activity and promote psychosocial health.

- Gains in strength from resistance training programs for youth are primarily the result of neurological adaptations (e.g., increased motor unit recruitment) with a lesser contribution of hypertrophic adaptations.
- Certain medical conditions require the approval of a pediatrician prior to commencing a resistance training program.

The key to these benefits is a properly designed and supervised program. Gradual progression is important, as it is with adults. The AAP recommends that both preadolescents and adolescents avoid power lifting, bodybuilding, and maximal lifts until they reach physical and skeletal maturity. Caution should be used with young athletes who have preexisting hypertension or other cardiovascular problems, as strength training may aggravate such conditions.

Are there any health benefits associated with resistance training?

Although the effects of resistance training on muscle mass, body weight, and strength are well documented, the health benefits of resistance training have been the focus of relatively recent research. Some of the health benefits are associated with increases in lean body mass and strength. Additionally, de Salles and others, in a review of 17 studies, concluded that resistance training could increase the secretion of beneficial cytokines, such as adiponectin, and decrease the secretion of harmful cytokines, such as tumor necrosis factor-alpha. Both an increase in muscle mass and beneficial cytokines could increase insulin sensitivity and protect against diabetes. Direct effects of resistance exercise on other body systems, such as the neuromuscular and skeletal systems, may produce health benefits. There are multiple health-related benefits of resistance training in diabetic individuals, as discussed in the American College of Sports Medicine/American Diabetes Association position statement on exercise and diabetes by Colberg and others. The conclusions of reviews such as those listed in **table 12.5** are contrary to the belief that resistance training does not confer any health benefits comparable to aerobic exercise. It is also notable that many cardiac rehabilitation programs now incorporate resistance-training exercises.

Nevertheless, it still appears to be prudent health behavior to incorporate some aerobic exercise into your lifestyle, even when trying to gain body weight. Although the ACSM and AHA have incorporated resistance training into their recommended exercise program for healthy adults, it is designed to complement aerobic exercise, not to substitute for it. The American Academy of Pediatrics also noted that if long-term health benefits for children and adolescents are the goal, then strength training should be combined with an aerobic-training program. Aerobic activity generates more caloric expenditure as previously discussed, but excessive expenditure is not necessary to derive benefits from aerobic training. For example, running 2–3 miles about 4 days/week would provide adequate aerobic conditioning for an estimated expenditure of only 200–300 kcal/day depending on body size, economy

TABLE 12.5 Selected health effects of resistance training

Health indicator	Change	Recent reviews or meta-analyses
Activities of daily living	Improvement in ability to perform activities of daily living	Papa, et al. 2017
Blood pressure	Improvement in systolic and diastolic pressures	Cornelissen, et al. 2012; Lemes, et al. 2016; Strasser, et al. 2012
Bone mineral density	Increase	Antoniak, et al. 2017
Chronic pain	Increase in strength of deep trunk muscles; core muscles alleviate chronic low back pain	Chang, et al. 2015
Falls risk	Reduced risk; improvement in walking endurance, balance, stability, functional mobility, gait speed, and stair climbing	Papa, et al. 2017
Glycemic control	Improvement in insulin sensitivity	Lee, et al. 2017; Strasser, et al. 2010
Lipid profile	Limited research on resistance training compared to aerobic exercise; recommendations of 50–85% of 1 RM resistance training to complement aerobic exercise training	Mann, et al. 2014
Pre- and postorthopedic surgery	Stronger evidence for resistance training to facilitate hip replacement rehabilitation compared to knee replacement	Skoffer, et al. 2015
Sarcopenia (age-related loss of muscle tissue)	Attenuation of the rate and magnitude of loss of muscle mass with aging	Papa, et al. 2017
Self-concept and self-efficacy	Improvement in body satisfaction, appearance evaluation, social physique anxiety	SantaBarbara, et al. 2017

of movement, and other factors. This 200- to 300-kcal expenditure could be replaced easily by consuming two glasses of orange juice or similar small amounts of food.

Combining resistance and aerobics exercise may provide additional health benefits. For example, as noted by Mann and others, resistance training can supplement—and possibly enhance—the lipid-lowering effects of aerobic exercise.

Can I combine aerobic and resistance-training exercises into one program?

Although the principles underlying the development of an aerobic-training program and a resistance-training program are similar, the purposes of the two programs are rather different. An aerobic exercise program is designed to improve the efficiency of the cardiovascular system. The basic purpose of a resistance-training program is to increase muscle size, strength, and body weight.

One form of resistance training that has been used to provide some moderate benefits to the cardiovascular system is **circuit weight training**, a method in which the individual moves rapidly from one exercise to the next. Generally, this type of program uses lighter weights with greater numbers of repetitions, thus increasing the aerobic component of training. In their meta-analysis, Muñoz-Martínez and others concluded that circuit weight training (14–30 sessions over 6–12 weeks using resistances of 60–90 percent of 1 RM and session durations of 20–30 minutes) can improve VO_2 max. Upper-body 1 RM was optimally improved by circuit weight training in normal-weight subjects with resistances of ≤ 60 percent of 1 RM and session durations of ≤ 60 minutes.

A newer version of this method is **circuit aerobics**. Circuit aerobics may be done in a variety of ways, but basically it involves an integration of aerobic and resistance-training exercises. It is actually a form of interval aerobic training, but instead of resting or doing a lower level of aerobic activity during the recovery interval, you do resistance-training exercises. Circuit aerobics may offer multiple health and performance benefits, such as improved cardiovascular fitness, increased caloric expenditure for loss of body fat, improved muscular strength and endurance, and increased muscle tone in body areas not normally stressed by aerobic exercise alone. In their review, Romero-Arenas and others concluded that circuit resistance training may improve cardiovascular fitness, muscle strength, body composition, bone mineral density, and risk factors for age-related diseases in elderly subjects.

However, if the main purpose of your resistance-training program is to gain body weight as muscle mass, then you need to train near the strength end of the strength-endurance continuum. Some researchers suggest that a high intensity and volume of aerobic training may be detrimental for athletes seeking maximal gains in lean body mass, strength, and especially power. The rationale for this recommendation is a potential "competition" between the cell signaling pathways that increase aerobic fitness and those that promote hypertrophy. As noted by Elliott and others, overreaching and overtraining could result in adverse physiological outcomes such as catabolic hormonal profile, inappropriate neuromuscular adaptations, and/or ineffective motor learning environment. A high volume of aerobic training with a concurrent emphasis on increased lean body mass, strength, and power, which could be detrimental to the power athlete. However, recent research suggests that moderate intensity aerobic training and resistance training can coexist with each other. Wang and others reported that resistance training can enhance mRNA activity of key mitochondrial proteins, while Kasior and others reported that endurance training can upregulate cell signaling pathways that increase muscle fiber size. In their meta-analysis, Murlasits and others concluded that resistance training should precede aerobic training for optimal lower-body strength improvement. Aerobic fitness improvement does not appear to be affected by the within-session order of aerobic and resistance training.

physiolifenutrition.com/pages/growing-stronger-strength-training-for-older-adults-tufts-university This Tufts University website presents an evidence-based exercise program designed to increase muscle strength, maintain bone health, and improve balance, coordination, and mobility in older adults.

Key Concepts

▶ A basic principle underlying all resistance-training programs is the overload principle, which simply means the muscle should be stressed beyond normal daily levels.

▶ Progressive resistance is also a basic principle of resistance training, for as you get stronger through use of the overload principle, you must progressively increase the resistance.

▶ To increase muscle mass and body weight, you should exercise near the strength end of the strength-endurance continuum.

▶ Your resistance-training program should exercise all major muscles groups in the body.

▶ A variety of methods and equipment are available for resistance training, but research suggests that they are equally effective as a means of gaining strength and muscle mass if the basic principles of resistance training are followed.

▶ Resistance training is generally regarded as a safe form of exercise, but it may be contraindicated in some individuals, for example, those with high blood pressure and hernias.

▶ Although resistance-training programs may confer some significant health benefits, it is also highly recommended that one add an aerobic exercise program to help condition the cardiovascular system.

Check for Yourself

▶ Do this after 2–3 weeks of resistance training. Using free weights or an appropriate weight-training machine, such as Nautilus, determine your one repetition maximum (1 RM) for a given exercise, such as the bench press. If using free weights, do not attempt this exercise without a spotter. Start out with a light weight that you know you can easily lift one time, and then gradually add on amounts until you reach your limit. You should need only three or four attempts. Record the weight of the last successful attempt.

APPLICATION EXERCISE

Bob weighs 150 pounds and has an estimated 15 percent body fat. He consults with you about designing a program to gain healthy weight over a 17-week program including moderate aerobic exercise and resistance training of all major muscle groups. You advise Bob to increase intake of healthy kcal and protein to promote an increase in muscle tissue at a rate of 1 pound per week. The table below includes progressive resistance exercise and 1-RM data for the bench press and as well as body composition data.

Using the two-component model of body composition assessment discussed in chapter 10, calculate fat tissue weight and lean tissue weight and comment on the efficacy of Bob's program in increasing healthy weight at 17 weeks.

Weeks	1	3	5	7	9	11	13	15	17
Bench press resistance	125	125	125	138	138	138	151	151	151
% increase in resistance				10			10		
Repetitions/Set	8	10	12	8	10	12	8	10	12
Sets/week	3	3	3	3	3	3	3	3	3
Bench press 1 RM	149		164			181			199
Body weight	150	152	154	156	158	160	162	164	166
Estimated % fat	15.0	15.0	15.0	14.5	14.5	14.5	14.0	14.0	14.0
Fat tissue weight	22.5	22.8	23.1	22.6	22.9	23.2	22.7	23.0	23.2
Lean tissue weight	127.5	129.2	130.9	133.4	135.1	136.8	139.3	141.0	142.8

Review Questions—Multiple Choice

1. All of the following aspects of protein intake to increase lean body weight are correct *except*:
    a. High-quality protein sources are recommended.
    b. The recommended protein intake in g/kg of weight is somewhat more compared to what is needed by a sedentary individual, but this amount is easily accommodated by the protein AMDR of 10–35 percent of daily kcal.
    c. Protein consumed after resistance exercise may increase muscle protein synthesis.
    d. Commercial protein supplements are required to increase lean body weight.
    e. Carbohydrate and protein ingestion may facilitate insulin-mediated uptake of amino acids.

2. An additional 14 g/day of protein may support the gain of approximately 1 pound per week during resistance training. Which of the following will provide this extra protein?
    a. 1 eight-ounce glass of skim milk
    b. 2 ounces of chicken
    c. 2 eight-ounce glasses of orange juice
    d. 1 medium banana
    e. 1 ounce of cheese

3. Important factors to consider when increasing intake of kcal and protein for healthy weight gain include which of the following?
    a. Avoid protein sources which also have high saturated fat.
    b. Avoid increasing fat intake if you have elevated blood lipids.
    c. Discuss your plans to increase protein intake with your physician if you have kidney problems.
    d. Choose foods high in complex carbohydrates with moderate amounts of protein and a moderate fat content.
    e. All of the above.

4. Your bench press 1 RM is 150 pounds. All of the following are correct regarding resistance training *except*:
    a. The total bench press weekly volume using 80 percent of 1 RM for 10 repetition/set, 3 sets/session, and 3 sessions/week is 10,800 pounds.
    b. Training to increase muscle hypertrophy is facilitated by using multiple sets of concentric and eccentric contractions of a moderate load with relatively short rest intervals.
    c. A 5-RM resistance should be used for muscular endurance, while a 12-RM should be used for muscular strength.
    d. Power is improved by incorporating sets of lower resistance loads at higher contraction velocities with higher resistance sets.
    e. Single sets using resistances lifted 8–12 times can improve health-related fitness.

5. Which of the following is not related to a principle of progressive resistance exercise (PRE) training?
   a. Sequence small muscle exercises before large muscle exercises.
   b. In an 8–12 repetition PRE approach, once you can accomplish 12 repetitions, increase the resistance to force you back to 8 repetitions.
   c. An athlete should strive to use resistance-training exercises that most closely mimic the movement patterns and demands of the sport.
   d. Proper rest intervals between sessions is important for the novice weight lifter to allow for muscle repair and growth.
   e. Do multiple-joint exercises before single-joint exercises and higher-intensity exercises before lower-intensity exercises.
6. Which of the following are safety concerns for effective resistance-training exercise?
   a. Use proper breathing/avoidance of the Valsalva maneuver.
   b. Use proper technique to protect the lower back.
   c. Lower weights slowly.
   d. Use spotters and safety equipment.
   e. All of the above
7. Which of the following is not generally considered to contribute to increased muscle size with resistance training?
   a. increased protein uptake to increase myofibril size and number
   b. increased ATP and glycogen
   c. increased number of muscle fibers
   d. increased connective tissue in muscle
   e. increased water content
8. _____ training includes _____ contractions where the muscle shortens while lifting the resistance *against* gravity and _____ contractions which generate tension while the muscle lengthens to control the rate of descent of the resistance *with* gravity.
   a. Isometric/eccentric/concentric
   b. Isotonic/concentric/eccentric
   c. Isokinetic/eccentric/concentric
   d. Isokinetic/accommodating resistance/eccentric
   e. Isotonic/eccentric/concentric
9. All of the following dietary recommendations would be appropriate for someone trying to gain lean body weight *except*:
   a. Wait at least 4 hours after exercise before eating some protein and carbohydrate.
   b. Increase the intake of high-kcal, high-nutrient foods.
   c. Keep the intake of saturated fats to a minimum.
   d. Eat three balanced meals a day, supplemented with snacks.
   e. Supplement the diet with high-kcal, high-nutrient liquids.
10. All the following are health benefits of proper resistance training *except*:
    a. decreased symptoms of the metabolic syndrome (lower blood pressure; increased insulin sensitivity; improved lipid profile)
    b. increased lean mass in older individuals to counter sarcopenia
    c. improved strength and gait to lower falls risk in older individuals
    d. lower bone mineral density
    e. improved mood image and psychological health

Answers to multiple choice questions: 1. d; 2. b; 3. e; 4. c; 5. a; 6. e; 7. c; 8. b; 9. a; 10. d

Critical Thinking Questions

1. Explain the strength-endurance continuum as a training concept.
2. List the five basic principles of resistance training and provide an example of each.
3. Discuss the physiological means whereby resistance training leads to increases in muscle growth.
4. Describe at least five of the potential health benefits associated with resistance training.
5. Discuss the importance of protein in a weight-gaining diet and provide some recommendations for amounts of protein and types of protein-rich foods in the diet.

References

Books

Bompa, T. O., and Buzzichelli, C. 2015. *Periodization Training for Sports*. 3rd ed. Champaign, IL: Human Kinetics.

Fleck, S., and Kraemer, W. 2014. *Designing Resistance Training Programs*. 4th ed. Champaign, IL: Human Kinetics.

Ivy, J., and Portman, R. 2004. *Nutrient Timing: The Future of Sports Nutrition*. North Bergen, NJ: Basic Health Publications.

National Academy of Sciences. Food and Nutrition Board. Institute of Medicine. 2005. *Dietary Reference Intakes for Energy, Carbohydrate, Fiber, Fat, Protein and Amino Acids*. Washington, DC: National Academies Press.

Reuben, D. 1996. *Dr. David Reuben's Quick Weight-Gain Program*. New York: Crown.

Stone, M., et al. 2007. *Principles and Practices of Resistance Training*. Champaign, IL: Human Kinetics.

U.S. Department of Health and Human Services. 2010. *Healthy People 2020*. Washington, DC: U.S. Government Printing Office.

Reviews and Specific Studies

American College of Sports Medicine. 2009. American College of Sports Medicine position stand: Progression models in resistance training for healthy adults. *Medicine & Science in Sports & Exercise* 41:687–708.

Antoniak, A. E., et al. 2017. The effect of combined resistance exercise training and vitamin D3 supplementation on musculoskeletal health and function in older adults: A systematic review and meta-analysis. *British Medical Journal Open* 7:e014619. doi:10.1136/bmjopen-2016-014619.

Aoyama, S., et al. 2021. Distribution of dietary protein intake in daily meals influences skeletal muscle hypertrophy via the muscle clock. Cell Reports 36(1):109336. doi: 10.1016/j.celrep.2021.109336. PMID: 34233179.

Aragon, A. A., and Schoenfeld, B. J. 2013. Nutrient timing revisited: Is there a post-exercise anabolic window? *Journal of the International Society of Sports Nutrition* 10(1):5. doi: 10.1186/1550-2783-10-5.

Aragon, A. A., et al. 2017. International Society of Sports Nutrition position stand: Diets and body composition. *Journal of the International Society of Sports Nutrition* 14:16. doi:10.1186/s12970-017-0174-y.

Bauer, J., et al. 2013. Evidence-based recommendations for optimal dietary protein intake in older people: A position paper from the PROT-AGE study group. *Journal of the American Medical Directors Association* 14 (8):542–59.

Beelen, M., et al. 2010. Nutritional strategies to promote postexercise recovery. *International Journal of Sport Nutrition and Exercise Metabolism* 20:515–32.

Bloomer, R. 2007. The role of nutritional supplements in the prevention and treatment of resistance exercise-induced skeletal muscle injury. *Sports Medicine* 37:519–32.

Bosse, J. D., and Dixon, B. M. 2012. Dietary protein to maximize resistance training: A review and examination of protein spread and change theories. *Journal of the International Society of Sports Nutrition* 9(1):42. doi: 10.1186/1550-2783-9-42.

Brook, M. S., et al. 2016. The metabolic and temporal basis of muscle hypertrophy in response to resistance exercise. *European Journal of Sport Science* 16(6):633–44.

Burd, N. A., et al. 2009. Exercise training and protein metabolism: Influences of contraction, protein intake, and sex-based differences. *Journal of Applied Physiology* 106(5):1692–701.

Candow, D. 2011. Sarcopenia: Current theories and the potential beneficial effect of creatine application strategies. *Biogerontology* 12:273–81.

Chaillou, T. 2019. Ribosome specialization and its potential role in the control of protein translation and skeletal muscle size. *Journal of Applied Physiology* (1985) 127(2):599–607. doi: 10.1152/japplphysiol.00946.2018.

Chang, W. D., et al. 2015. Core strength training for patients with chronic low back pain. *Journal of Physical Therapy Science* 27(3):619–22.

Chtourou, H., and Souissi, N. 2012. The effect of training at a specific time of day: A review. *Journal of Strength and Conditioning Research* 26(7):1984–2005.

Clifton, P. M., et al. 2014. Long term weight maintenance after advice to consume low carbohydrate, higher protein diets—A systematic review and meta analysis. *Nutrition, Metabolism and Cardiovascular Diseases* 24(3):224–35.

Colberg, S.R., et al. 2010. Exercise and type 2 diabetes. American College of Sports Medicine and the American Diabetes Association: joint position statement. *Medicine and Science in Sports and Exercise* 42(12):2282-303. doi: 10.1249/MSS.0b013e3181eeb61c. PMID: 21084931.

Coleman-Jensen, A. et al. 2021. *Household Food Security in the United States in 2020.* ERR-298. U.S. Department of Agriculture, Economic Research Service.

Cornelissen, V. A., et al. 2012. Impact of resistance training on blood pressure and other cardiovascular risk factors: A meta-analysis of randomized, controlled trials. *Hypertension* 58:950–58.

Cribb, P., and Hayes, A. 2006. Effect of supplement timing and resistance training on skeletal muscle hypertrophy. *Medicine & Science in Sports & Exercise* 38:1918–25.

de Salles, B. F., et al. 2009. Rest interval between sets in strength training. *Sports Medicine* 39(9):765–77.

de Salles, B., et al. 2010. Effects of resistance training on cytokines. *International Journal of Sports Medicine* 31:441–50.

Devries, M. C,. and Phillips, S. M. 2014. Creatine supplementation during resistance training in older adults—A meta-analysis. *Medicine and Science in Sports and Exercise* 46(6):1194–1203.

Elliott, M., et al. 2007. Power athletes and distance training: Physiological and biomechanical rationale for change. *Sports Medicine* 37:47–57.

Evangelista, A. L., et al. 2019. Interset stretching vs. traditional strength training: Effects on muscle strength and size in untrained individuals. *Journal of Strength and Conditioning Research* 33(S1):S159–S166. doi: 10.1519/JSC.0000000000003036. PMID: 30688865.

Ferguson-Stegall, L., et al. 2011. Postexercise carbohydrate-protein supplementation improves subsequent exercise performance and intracellular signaling for protein synthesis. *Journal of Strength and Conditioning Research* 25(5):1210–24.

Folland, J., and Williams, A. 2007. The adaptations to strength training: Morphological and neurological contributions to increased strength. *Sports Medicine* 37:145–68.

Franchi, M. V., et al. 2017. Skeletal muscle remodeling in response to eccentric vs. concentric loading: Morphological, molecular, and metabolic adaptations. *Frontiers in Physiology* 8:447. doi: 10.3389/fphys.2017.00447.

Frost, D., et al. 2010. A biomechanical evaluation of resistance: Fundamental concepts for training and sports performance. *Sports Medicine* 40:303–26.

Fryar, C. D., et al. 2020. Prevalence of underweight among adults aged 20 and over: United States, 1960–1962 through 2017–2018. NCHS Health E-Stats. www.cdc.gov/nchs/data/hestat/underweight-adult-17-18/ESTAT-Underweight-Adult-H.pdf.

Gonzalez, A. M., et al. 2016. Intramuscular anabolic signaling and endocrine response following resistance exercise: Implications for muscle hypertrophy. *Sports Medicine* 46:671–85.

Greer, B. K., et al. 2015. EPOC comparison between isocaloric bouts of steady-state aerobic, intermittent aerobic, and resistance training. *Research Quarterly for Exercise and Sport* 86(2):190–95.

Grgic, J., et al. 2018. Effects of rest interval duration in resistance training on measures of muscular strength: A systematic review. *Sports Medicine* 48(1):137–51. doi: 10.1007/s40279-017-0788-x. PMID: 28933024.

Grunewald, K., and Bailey, R. 1993. Commercially marketed supplements for bodybuilding athletes. *Sports Medicine* 15:90–103.

Harries, S. K., et al. 2015. Systematic review and meta-analysis of linear and undulating periodized resistance training programs on muscular strength. *Journal of Strength and Conditioning Research*. 29(4):1113–25.

Harty, P. S., et al. 2019. Nutritional and supplementation strategies to prevent and attenuate exercise-induced muscle damage: A brief review. *Sports Medicine Open* 5(1):1. doi: 10.1186/s40798-018-0176-6. PMID: 30617517; PMCID: PMC6323061.

Haskell, W., et al. 2007. Physical activity and public health: Updated recommendation for adults from the American College of Sports Medicine and the American Heart

Association. *Medicine & Science in Sports & Exercise* 39:1423-34.

Hayes, L., et al. 2010. Interactions of cortisol, testosterone, and resistance training: Influence of circadian rhythms. *Chronobiology International* 27:675-705.

Heden, T., et al. 2011. One-set resistance training elevates energy expenditure for 72 h similar to three sets. European *Journal of Applied Physiology* 111(3):477-84. doi: 10.1007/s00421-010-1666-5.

Hoffman, J., et al. 2009. Effect of protein-supplement timing on strength, power, and body-composition changes in resistance-trained men. *International Journal of Sport Nutrition and Exercise Metabolism* 19:172-85.

Hooper, D. R., et al. 2017. Endocrinological roles for testosterone in resistance exercise responses and adaptations. *Sports Medicine* 47(9):1709-20. doi: 10.1007/s40279-017-0698-y. PMID: 28224307.

Jamurtas, A. Z., et al. 2004. The effects of a single bout of exercise on resting energy expenditure and respiratory exchange ratio. *European Journal of Applied Physiology* 92: 393-98.

Josse, A., et al. 2010. Body composition and strength changes in women with milk and resistance exercise. *Medicine & Science in Sports & Exercise* 42:1122-30.

Kazior, Z., et al. 2016. Endurance exercise enhances the effect of strength training on muscle fiber size and protein expression of Akt and mTOR. *PLoS One* 11(2):e0149082. doi: 10.1371/journal.pone.0149082. PMID: 26885978; PMCID: PMC4757413.

Kokkonen, J., et al. 2007. Chronic static stretching improves exercise performance. *Medicine & Science in Sports & Exercise* 39:1825-31.

Kokkonen, J., et al. 2010. Early-phase resistance training strength gains in novice lifters are enhanced by doing static stretching. *Journal of Strength and Conditioning Research* 24(2):502-6.

Lambert, C., et al. 2004. Macronutrient considerations for the sport of bodybuilding. *Sports Medicine* 34:317-27.

Lanhers, C., et al. 2015. Creatine supplementation and lower limb strength performance: A systematic review and meta-analyses. *Sports Medicine* 45(9):1285-1294.

Lanhers, C., et al. 2017. Creatine supplementation and upper limb strength performance: A systematic review and meta-analysis. *Sports Medicine* 47(1):163-73.

Lee, J., et al. 2017. Resistance training for glycemic control, muscular strength, and lean body mass in old type 2 diabetic patients: A meta-analysis. *Diabetes Therapy* 8:459-73.

Leidy, H. J., et al. 2015. The role of protein in weight loss and maintenance. *American Journal of Clinical Nutrition* 101(Suppl):1320S-1329S.

Lemes, Í. R., et al. 2016. Resistance training reduces systolic blood pressure in metabolic syndrome: A systematic review and meta-analysis of randomised controlled trials. *British Journal of Sports Medicine* 50(23):1438-42.

Lichtenstein, A. H., et al. 2021. 2021 dietary guidance to improve cardiovascular health: A scientific statement from the American Heart Association. *Circulation* 144(23):e472-e487. doi: 10.1161/CIR.0000000000001031.

Lloyd, R. S., et al. 2014. Position statement on youth resistance training: The 2014 International Consensus. *British Journal of Sports Medicine* 48(7):498-505.

Mann, S., et al. 2014. Differential effects of aerobic exercise, resistance training and combined exercise modalities on cholesterol and the lipid profile: Review, synthesis and recommendations. *Sports Medicine* 44(2):211-21.

Muñoz-Martínez, F. A., et al. 2017. Effectiveness of resistance circuit-based training for maximum oxygen uptake and upper-body one-repetition maximum improvements: A systematic review and meta-analysis. *Sports Medicine* doi: 10.1007/s40279-017-0773-4. [Epub ahead of print].

Murlasits, Z., et al. 2017. The physiological effects of concurrent strength and endurance training sequence: A systematic review and meta-analysis. *Journal of Sports Sciences* 36(11):1212-19. doi: 10.1080/02640414.2017.1364405.

Murray, S. B., et al. 2010. Muscle dysmorphia and the DSM-V conundrum: Where does it belong? A review paper. *International Journal of Eating Disorders* 43(6):483-91.

Naclerio, F., and Larumbe-Zabala, E. 2016. Effects of whey protein alone or as part of a multi-ingredient formulation on strength, fat-free mass, or lean body mass in resistance-trained individuals: A meta-analysis. *Sports Medicine* 46(1):125-37.

Nakamura, M., et al. 2021. Effects of adding inter-set static stretching to flywheel resistance training on flexibility, muscular strength, and regional hypertrophy in young men. *International Journal of Environmental Research and Public Health* 18(7):3770. doi: 10.3390/ijerph18073770. PMID: 33916599; PMCID: PMC8038434.

Nelson, M., et al. 2007. Physical activity and public health in older adults: Recommendation from the American College of Sports Medicine and the American Heart Association. *Medicine & Science in Sports & Exercise* 39:1435-45.

Nunes, J. P., et al. 2020. Does stretch training induce muscle hypertrophy in humans? A review of the literature. *Clinical Physiology and Functional Imaging* 40(3):148-56. doi: 10.1111/cpf.12622.

Ormsbee, M., et al. 2007. Fat metabolism and acute resistance exercise in trained men. *Journal of Applied Physiology* 102:1767-72.

Papa, E. V., et al. 2017. Resistance training for activity limitations in older adults with skeletal muscle function deficits: A systematic review. *Clinical Interventions in Aging* 12:955-61.

Pasiakos, S. M., et al. 2015. The effects of protein supplements on muscle mass, strength, and aerobic and anaerobic power in healthy adults: A systematic review. *Sports Medicine* 45(1):111-31.

Peterson, M., et al. 2011. Influence of resistance exercise on lean body mass in aging adults: A meta-analysis. *Medicine & Science in Sports & Exercise* 43:249-58.

Phillips, S. 2009. Physiologic and molecular bases of muscle hypertrophy and atrophy: Impact of resistance exercise on human skeletal muscle (protein and exercise dose effects). *Applied Physiology, Nutrition, and Metabolism* 34:403-10.

Phillips, S. M. 2012. Dietary protein requirements and adaptive advantages in athletes. *British Journal of Nutrition* 108:S158-67.

Pritchett, K., and Pritchett, R. 2012. Chocolate milk: A post-exercise recovery beverage for endurance sports. *Medicine and Sports Science* 59:127-34.

Ralston, G. W., et al. 2017. The effect of weekly set volume on strength gain: A meta-analysis. *Sports Medicine* 47(12):2585-601. doi: 10.1007/s40279-017-0762-7. PMID: 28755103; PMCID: PMC5684266.

Rawson, E., and Venezia, A. 2011. Use of creatine in the elderly and evidence for effects on cognitive function in young and old. *Amino Acids* 40:1349-62.

Raymond, M. J., et al. 2013. Systematic review of high-intensity progressive resistance strength training of the lower limb compared with other intensities of strength training in older adults. *Archives of Physical Medicine and Rehabilitation* 94:1458-72.

Reidy, P. T., and Rasmussen, B. B. 2016. Role of ingested amino acids and protein in the promotion of resistance exercise-induced muscle protein anabolism. *Journal of Nutrition* 146:155-83.

Rodriguez, J., et al. 2014. Myostatin and the skeletal muscle atrophy and hypertrophy signaling pathways. *Cellular and Molecular Life Sciences* 71(22):4361-71.

Rohman, L. 2009. The relationship between anabolic androgenic steroids and muscle dysmorphia: A review. *Eating Disorders* 17(3):187-99.

Roig, M., et al. 2009. The effects of eccentric versus concentric resistance training on muscle strength and mass in healthy adults: A systematic review with meta-analysis. *British Journal of Sports Medicine* 43:556-68.

Romero-Arenas, S., et al. 2013. Impact of resistance circuit training on neuromuscular, cardio-respiratory and body composition adaptations in the elderly. *Aging and Disease* 4 (5):256-63.

Rowlands, D., and Thomson, J. 2009. Effects of beta-hydroxy-beta-methylbutyrate supplementation during resistance training on strength, body composition, and muscle damage in trained and untrained young men: A meta-analysis. *Journal of Strength & Conditioning Research* 23(3):836-46.

Sanchez-Martinez, J., et al. 2018. Effects of beta-hydroxy-beta-methylbutyrate supplementation on strength and body composition in trained and competitive athletes: A meta-analysis of randomized controlled trials. *Journal of Science and Medicine in Sport* 21:717-35.

SantaBarbara, N. J., et al. 2017. A systematic review of the effects of resistance training on body image. *Journal of Strength and Conditioning Research* doi: 10.1519/JSC.0000000000002135. [Epub ahead of print].

Schoenfeld, B. J. 2010. The mechanisms of muscle hypertrophy and their application to resistance training. *Journal of Strength and Conditioning Research* 24(10):2857-72.

Schoenfeld, B. J., et al. 2017. Hypertrophic effects of concentric vs. eccentric muscle actions: A systematic review and meta-analysis. *Journal of Strength and Conditioning Research* 31(9):2599-2608.

Schoenfeld, B. J., et al. 2017. Strength and hypertrophy adaptations between low- vs. high-load resistance training: A systematic review and meta-analysis. *Journal of Strength and Conditioning Research* 31(12):3508-23. doi: 10.1519/JSC.0000000000002200. PMID: 28834797.

Skemp, K. M., et al. 2013. Muscle dysmorphia: Risk may be influenced by goals of the weightlifter. *Journal of Strength and Conditioning Research* 27(9):2427-32.

Skoffer, B., et al. 2015. Progressive resistance training before and after total hip and knee arthroplasty: A systematic review. *Clinical Rehabilitation* 29(1):14-29.

Slater, G.J., et al. 2019. Is an Energy Surplus Required to Maximize Skeletal Muscle Hypertrophy Associated With Resistance Training. *Frontiers in Nutrition* 6:131. doi: 10.3389/fnut.2019.00131. PMID: 31482093; PMCID: PMC6710320.

Souza, D., et al. 2020. High and low-load resistance training produce similar effects on bone mineral density of middle-aged and older people: A systematic review with meta-analysis of randomized clinical trials. *Experimental Gerontology* 138:110973. doi: 10.1016/j.exger.2020.110973.

Spendlove, J., et al. 2015. Dietary Intake of Competitive Bodybuilders. *Sports Medicine* 45(7):1041-63. doi: 10.1007/s40279-015-0329-4. PMID: 25926019.

Stokes, T, et al. 2018. Recent Perspectives Regarding the Role of Dietary Protein for the Promotion of Muscle Hypertrophy with Resistance Exercise Training. *Nutrients* 10(2):180. doi: 10.3390/nu10020180. PMID: 29414855; PMCID: PMC5852756.

Strasser, B., et al. 2010. Resistance training in the treatment of the metabolic syndrome: A systematic review and meta-analysis of the effect of resistance training on metabolic clustering in patients with abnormal glucose metabolism. *Sports Medicine* 40:397-415.

Strasser, B., et al. 2012. Resistance training, visceral obesity and inflammatory response: A review of the evidence. *Obesity Reviews* 13:578-91.

Stricker, P. R., et al. 2020. American Academy of Pediatrics Council on Sports Medicine and Fitness. Resistance training for children and adolescents. *Pediatrics* 145(6):e20201011.

Thomas, D. T., et al. 2016. American College of Sports Medicine Joint Position Statement. Nutrition and Athletic Performance. *Medicine and Science in Sports and Exercise* 48(3):543-568.

Verdijk, L., et al. 2009. Protein supplementation before and after exercise does not further augment skeletal muscle hypertrophy after resistance training in elderly men. *American Journal of Clinical Nutrition* 89:608-16.

Vingren, J., et al. 2010. Testosterone physiology in resistance exercise and training: The up-stream regulatory elements. *Sports Medicine* 40:1037-53.

Wang, L., et al. 2011. Resistance exercise enhances the molecular signaling of mitochondrial biogenesis induced by endurance exercise in human skeletal muscle. *Journal of Applied Physiology* (1985) 111(5):1335-44. doi: 10.1152/japplphysiol.00086.2011.

Wernbom, M., et al. 2007. The influence of frequency, intensity, volume and mode of strength training on whole muscle cross-sectional area in humans. *Sports Medicine* 37:225-64.

Williams, M., et al. 2007. Resistance exercise in individuals with and without cardiovascular disease: 2007 update: A scientific statement from the American Heart Association Council on Clinical Cardiology and Council on Nutrition, Physical Activity, and Metabolism. *Circulation* 116:572-84.

Wilson, J. M., et al. 2013. International Society of Sports Nutrition position stand: Beta-hydroxy-beta-methylbutyrate (HMB). *Journal of the International Society of Sports Nutrition* 10(1):6. doi: 10.1186/1550-2783-10-6. PMID: 23374455; PMCID: PMC3568064.

Witard, O. C., et al. 2016. Protein considerations for optimising skeletal muscle mass in healthy young and older adults. *Nutrients* 8, 181; doi:10.3390/nu8040181.

Wu, H., et al. 2015. Effect of beta-hydroxy-beta-methylbutyrate supplementation on muscle loss in older adults: A systematic review and meta-analysis. *Archives of Gerontology and Geriatrics* 61(2):168-75.

Yasuda, J., et al. 2020. Evenly distributed protein intake over 3 meals augments resistance exercise-induced muscle hypertrophy in healthy young men. *Journal of Nutrition* 150(7):1845-51. doi: 10.1093/jn/nxaa101. PMID: 32321161; PMCID: PMC7330467.

Yoshida, T., and Delafontaine, P. 2020. Mechanisms of IGF-1-mediated regulation of skeletal muscle hypertrophy and atrophy. *Cells* 9(9):1970. doi: 10.3390/cells9091970. PMID: 32858949; PMCID: PMC7564605.

Zanou, N., and Gailly, P. 2013. Skeletal muscle hypertrophy and regeneration: Interplay between the myogenic regulatory factors (MRFs) and insulin-like growth factors (IGFs) pathways. *Cellular and Molecular Life Sciences* 70 (21):4117-30.

Design element: Training Table (orange) ©mphillips007/Getty Images

Nutritional Supplements and Ergogenic Aids

CHAPTER THIRTEEN

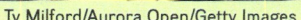
Ty Milford/Aurora Open/Getty Images

KEY TERMS

alcoholism
alcohol use disorders (AUD)
anabolic-androgenic steroids (AAS)
androstenedione
apoptosis
beta-alanine
blood alcohol concentration (BAC)
caffeine
cirrhosis
ciwujia
dehydroepiandrosterone (DHEA)
ephedra
ephedrine
ergogenic aids
ergolytic
ethanol
fetal alcohol effects (FAE)
fetal alcohol syndrome (FAS)
ginseng
immunomodulatory agent
ma huang
nutrient timing preworkout supplements
proof
salsolinol
sarcopenia
sodium bicarbonate

LEARNING OUTCOMES

After studying this chapter, you should be able to:

1. Describe the metabolic, physiological, and psychological effects of alcohol in the body, and evaluate its efficacy as an ergogenic acid.

2. Explain the possible beneficial and detrimental effects of alcohol consumption on health.

3. List and explain the several theories whereby caffeine supplementation has been and is currently proposed to be an effective ergogenic aid, and summarize its effect on exercise performance.

4. Explain the possible beneficial and detrimental health-related effects of caffeine in the body, and cite current recommendations for coffee consumption.

5. Understand the potential health problems associated with dietary supplements containing stimulants such as ephedra.

6. Describe the theory underlying the use of sodium bicarbonate and other buffering agents as an ergogenic aid, and understand the current research findings regarding its the efficacy of such agents to enhance exercise performance.

7. Identify drugs and related dietary supplements used by physically active individuals to stimulate muscle building, and summarize the effects on exercise performance and potential health risks associated with their use.

8. Explain the theory as to how ginseng may enhance sports performance, and highlight the research findings regarding its ergogenic efficacy.

9. List those drugs or dietary supplements discussed in this chapter that are prohibited by the World Antidoping Agency (WADA) in-training and/or in competition.

10. Describe the four evidence levels for efficacy, safety, and permissibility of dietary supplements as ergogenic aids according to the Australian Institute of Sport and provide examples of supplements/ergogenic aids at each level.

Introduction

Ty Milford/Aurora Open/Getty Images

Peak athletic performance requires genetic endowment, optimal nutrition, a strong work ethic, and training techniques to develop specific physiological, psychological, and biomechanical attributes. Over the years, athletes have attempted to go beyond training to gain a competitive edge by using a variety of **ergogenic aids.** *The roots for the word "ergogenic" are "ergon" ("work") and "genic" ("production" or "generation"). Ergogenic aids in athletics refer to a variety of dietary ingredients, supplements, drugs, and training strategies purported to increase performance in training and competition.*

The most historic and popular international sporting event is the Olympic Games. The ancient Olympics were held in Greece from 776 B.C. until being canceled in A.D. 393 by Roman Emperor Theodosius I because they were regarded as a pagan ritual. In 1896, the modern Olympics resumed in Athens, Greece. Throughout history, successful Olympic athletes have achieved fame and fortune.

Most elite Grecian athletes had personal trainers, called *paidotribes,* to plan their exercise program and their diet to prepare them for competition. In a sense, paidotribes were the first sports scientists and sports nutritionists. Several famous Greek scientists, including Galen and Pythagoras, were paidotribes who advocated specific, but different, diets for their athletes; Galen promoted beans as a staple of the athlete's diet, whereas Pythagoras prohibited them. During these early Olympic Games, athletes consumed various plants such as mushrooms containing hallucinogens and *strychnos nux vamica,* the source of strychnine, for their potential ergogenic effects.

As in the ancient Olympics, present-day athletes competing in dozens of sports have been reported to use drugs and dietary supplements in attempts to obtain that competitive edge. Some plant extracts, or phytochemicals, may be marketed as dietary supplements designed to imitate ergogenic drugs. Others, particularly herbals, may be used for their pharmaceutical effects by practitioners of alternative medicine.

In this chapter, we will examine alcohol, caffeine, and other dietary supplements that could, depending on many factors, affect exercise performance and health. Alcohol and caffeine have been studied for their ergogenic effects for more than 100 years and have been used by athletes as a means to enhance performance since the early 1900s. Each has also been studied extensively for possible health effects, both positive and negative.

The dietary supplement ma huang contains ephedra, or ephedrine, a stimulant theorized to enhance exercise performance, particularly when combined with caffeine. Ephedrine has also been studied for its potential health effects.

Sodium bicarbonate, or baking soda, is used in many food products and has been studied for 80 years as a means to enhance performance by buffering acidity. It also has been marketed as part of a dietary supplement for athletes. More recently, beta-alanine has also the focus of research on its effectiveness as a buffering agent.

Several anabolic hormones, and related steroid drugs, are used to increase muscle mass and have been popular with strength/power athletes for more than 50 years. However, such drugs have been illegal in Olympic competition since 1975. These drugs are highly regulated by the Anabolic Steroid Control Act of 1990. Their use is limited to legitimate medical purposes. As a result, companies have marketed prohormones, or precursors to these hormones, as dietary supplements to avoid drug regulations. Prohormones are now also classified as controlled drugs and their use is illegal in sports. Use of these anabolic agents may also pose serious health risks.

Finally, several herbals and other phytochemicals, most notably ginseng, are used in alternative medicine for various purposes, one being enhancement of physical performance. Collectively, with the exception of alcohol and anabolic hormones, most of these performance-enhancing substances are marketed as dietary supplements, or sports supplements when targeted to athletes. The sports supplement industry has become a multibillion-dollar business.

Some of the dietary supplements discussed in this chapter have been banned by the World Anti-Doping Agency (WADA) for use in sports competition, and others are being monitored by WADA. Prevalence estimates of the use of legal and illegal supplements by athletes vary in the literature. According to Garthe and Maughan, prevalence rates for athletes using supplements range from 40 to 100 percent and depend on the sport, competitive level, and the definition of supplements. In a survey of athletes who participated in two major competitions held in 2011, Ulrich and others reported past-year doping prevalence rates of between 30 and 45 percent, while de Hon and others used questionnaires and models of biological parameters to estimate intentional doping in 14–39 percent of adult elite athletes. These rates are much higher than what was detected by random testing.

www.wada-ama.org/sites/default/files/resources/files/2022list_final_en_0.pdf The World Anti-Doping Agency (WADA) provides the complete list of drugs and doping techniques whose use is prohibited or monitored in sports competition. Click on the Prohibited List. The list is updated annually and is effecting on January 1 of every year.

Alcohol: Ergogenic Effects and Health Implications

The alcohol produced for human consumption is ethyl alcohol, or **ethanol**. Ethanol may be classified as a psychoactive drug, a toxin, or a nutrient.

The use of alcohol as a means to enhance exercise or sports performance has a long history. Ancient Greek athletes drank wine or brandy prior to competition, thinking that these alcoholic beverages enhanced performance. In more modern times, Olympic marathon runners in the Paris (1900) and London (1908) games drank brandy or cognac to enhance performance, while in the Paris Olympics in 1924 wine was served at fluid-replacement stations in the marathon. In 1939, Boje noted that in cases of extreme athletic exertion or in events of brief maximal effort, alcohol has been given to athletes to serve as a stimulant by releasing inhibitions and lessening the sense of fatigue.

The use of alcohol as a social, psychoactive drug also has a long history, and its effects on human health have been studied extensively. Historically, most research has focused on the numerous adverse health effects of excessive alcohol consumption, but recent research has identified some potential health benefits of light to moderate alcohol consumption. According to Costanzo and others, the "J-curve" relationship of alcohol consumption, where considerable epidemiological research has reported moderate alcohol consumption to be associated with beneficial health effects and lower mortality compared to abstinence or heavy drinking, continues to be both controversial and scrutinized for bias and confounding influences.

What is the alcohol and nutrient content of typical alcoholic beverages?

Alcohol is a transparent, colorless liquid derived from the fermentation of sugars in fruits, vegetables, and grains. Although classified legally as a drug, alcohol is a component of many common beverages served throughout the world. In the United States, alcohol is consumed mainly as a natural ingredient of beer, wine, and liquors. As illustrated in **figure 13.1**, typical alcohol contents by volume are 4-5 percent for beer; 12-14 percent for wine; and 40-45 percent for distilled spirits such as whiskey, vodka, and rum. Based on these typical percentages, similar amounts of alcohol are found in a 4-oz glass of wine, a 12-oz bottle or can of beer, and a 1.25-oz shot of liquor. Other products containing alcohol are wine coolers and energy drinks, which contain 5-7 percent and 10-12 percent alcohol, respectively. Energy drinks will be discussed later in this chapter. The term **proof** is a measure of the alcohol content in a beverage and is double the percentage. It is very important to remember that the alcohol content can vary significantly among products within beer, wine, and liquor categories. As examples, certain craft and microbrews may contain 10 or more percent alcohol and some wines are fortified to 18-24 percent. An 86-proof bottle of whiskey contains 43 percent alcohol, but a 151-proof bottle of Caribbean rum contains over 75 percent alcohol.

Such beverages would provide significantly more alcohol per standard drink. Technically, alcohol may be classified as a nutrient because it provides energy, one of the major functions of food. Alcohol contains about 7 kcal per gram, almost twice the value of an equal amount of carbohydrate or protein. Beer and wine also contain some carbohydrate, a source of additional kcal. In general, a bottle of regular beer has about 150 kcal, while a 4-oz glass of wine or a shot glass of liquor contains about 100 kcal. **Table 13.1** provides an approximate analysis of the caloric content of common alcoholic beverages and nonalcoholic beer.

In general, the alcohol kcal found in beer, wine, and liquor are empty kcal. Although wine and beer contain trace amounts of protein, vitamins, minerals, and phytochemicals, liquor is void of any nutrient value.

What is the metabolic rate of alcohol clearance in the body?

About 20 percent of the alcohol ingested may be absorbed by the stomach. The remainder passes on to the intestine for absorption. Absorption is rapid, particularly if the digestive tract is empty. The alcohol enters the blood and is distributed to the various tissues, being diluted by the water content of the body. A small portion of the alcohol, about 3-10 percent, is excreted from the body through the breath, urine, or sweat, but the majority is metabolized by the liver, the organ that metabolizes other drugs. As the blood circulates, the liver of an average adult male will metabolize about 1/3 ounce (8-10 grams) of alcohol per hour, or somewhat less than the amount of alcohol in one drink.

Although alcohol is derived from the fermentation of carbohydrates, it is metabolized in the body as fat. The liver helps convert the metabolic by-products of alcohol into fatty acids, which may be stored in the liver or transported into the blood. Alcohol is metabolized by the liver to acetaldehyde or pyruvate by enzymatic action of alcohol dehydrogenase. Pyruvate can be reduced to

1 glass of wine	1 can or bottle of beer	1 shot glass with distilled spirits
4 oz of table wine	12 oz of beer	1.25 oz of whiskey or other hard liquor
12% alcohol by volume	4% alcohol by volume	40% alcohol by volume, or 80 proof
4 × 0.12 = 0.48 oz of ethanol per serving	12 × 0.04 = 0.48 oz of ethanol per serving	1.25 × 0.40 = 0.50 oz of ethanol per serving

Bordeaux wine glass: PLAINVIEW/E+/Getty Images; Beer bottle: broken3/Getty Images; Shot glass: gresei/Getty Images

FIGURE 13.1 Alcohol equivalencies in typical beverages.

TABLE 13.1 Caloric content of commonly consumed alcoholic beverages

Beverage	Amount	Carbohydrate Grams	Carbohydrate kcal	Alcohol Grams	Alcohol kcal	Total kcal
Beer, regular	12 ounces	13	52	13	91	150
Beer, light	12 ounces	7	28	11	77	109
Beer, nonalcoholic	12 ounces	12	48	1	7	55
Beer, alcohol-free	12 ounces	12	48	0	0	48
Wine, table	4 ounces	4	16	12	84	100
Liquor, 80-proof	1.25 ounces	0	0	14	98	100
Energy drink, alcoholic	12 ounces	32	128	15	105	233

The small discrepancies in the calculation of total kcal for beer and liquor may be attributed to a small protein content in beer and trace amounts of carbohydrate in liquor.

form lactate or as a substrate in gluconeogenesis or in transamination reactions. Catalase in peroxisomes and cytochrome P450 in the endoplasmic reticulum can also convert alcohol to acetaldehyde with cytochrome P450 being responsible for most alcohol metabolism during heavy consumption. Acetaldehyde is then converted to acetate in a reaction catalyzed by aldehyde dehydrogenase. Acetate, pyruvate, and lactate may be utilized for energy and converted into carbon dioxide and water. A schematic of alcohol metabolism is presented in **figure 13.2**.

As noted, the liver of an average male can metabolize only about 1/3 ounce (approximately 8–10 grams) of alcohol, or less than one drink, per hour. The rate is lower in smaller individuals and higher in larger individuals. Thus, consumption of alcohol at a rate greater than one drink per hour will result in an accumulation of alcohol in the blood which is measured as the **blood alcohol concentration (BAC)** in grams per 100 milliliters of blood which is equivalent to grams percent. The ingested alcohol is diluted throughout the total body water, both inside and outside body cells, including the blood. For the average male, one drink will result in a BAC of about 0.025, or 0.025 gram (25 milligrams) per 100 milliliters of blood whereas four drinks in an hour would lead to a BAC of a little less than 0.10 (100 mg/100 ml of blood) because a small amount would be metabolized by the liver. However, BAC concentrations resulting from the same amount of drinks may vary widely among individuals, due to food intake, gender, differences in body weight and body fat, tolerance due to a history of alcohol consumption, genetics, and differences in activities of the enzymes, particularly alcohol dehydrogenase, in **figure 13.2** that metabolize alcohol. The following website may be used to calculate BAC.

> www.calculator.net/bac-calculator.html Calculate a person's BAC based on their gender, body weight, number of drinks consumed, and amount of time over which drinks were consumed.

Is alcohol an effective ergogenic aid?

For more than a century, athletes have consumed alcohol just prior to or during competition in attempts to improve performance. Potential ergogenic mechanisms attributed to alcohol include altered energy metabolism and improvements in physiological processes and psychological factors. These claims are examined in the following sections.

Use as an energy source Although alcohol contains a relatively large number of kcal and its metabolic pathways in the body are short, the available evidence suggests that it is not utilized to any significant extent during exercise. First, the major sources of energy for exercise are carbohydrates and fats, which are in ample supply in most individuals. Alcohol may help form fats, but there is no evidence that it can substitute for other fat sources in the body. Triglyceride synthesis from fatty acids formed in the liver by alcohol consumption and subsequent hydrolysis would be an insignificant source of circulating free fatty acids as energy substrate during prolonged exercise compared to fatty acid mobilization from hydrolysis of triglycerides in the lipid droplet stored in adipocytes and in/near active muscle. By-products of alcohol metabolism released by the liver into the blood and entering skeletal muscles appear to be of little importance as an energy substrate. Energy from alcohol catabolism would be an uneconomical energy substrate because the amount of oxygen needed to release the kcal from alcohol is greater than for an equivalent amount of carbohydrate and fat. Finally, the relatively slow rate at which the liver metabolizes alcohol limits its usefulness as an energy source during exercise, particularly in an individual working at a high level of intensity. In summary, alcohol is not a key energy source during exercise, and even if it were, it would not offer any advantages over natural supplies of carbohydrate and fat. Research with carbon labeling revealed that alcohol did not significantly modify endogenous carbohydrate and fat utilization during exercise.

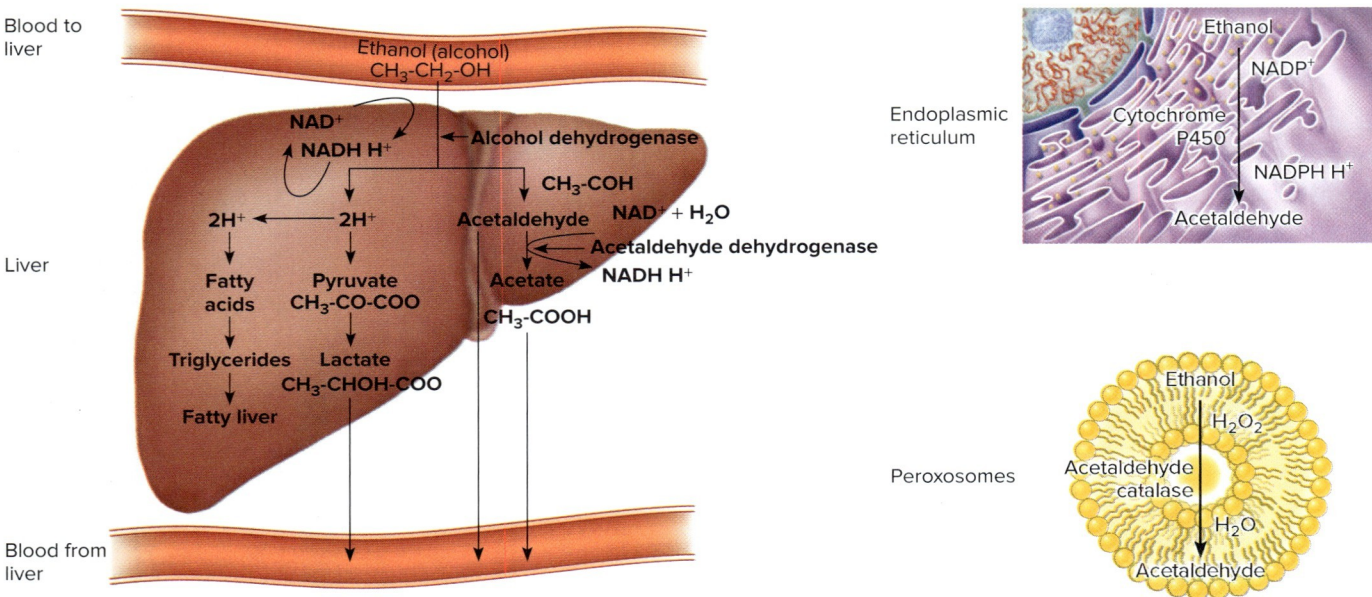

FIGURE 13.2 Simplified metabolic pathways of ethanol (alcohol) in the liver. Acetaldehyde, the immediate downstream metabolite from alcohol oxidation, is formed by enzymatic action of alcohol dehydrogenase, calalase, and cytochrome P450. Hydrogen ions are removed from ethanol as it is converted into acetaldehyde, which may be released into the blood for transport to other tissues. The excess hydrogen ions may combine with fatty acids to form triglycerides or with pyruvate to form lactate. Excessive accumulation of triglycerides may lead to the development of a fatty liver and eventually to cirrhosis.

Effect on exercise metabolism and performance Although alcohol has been used by athletes as a purported ergogenic aid for centuries, peer-reviewed research began in the 1970s. The effects of alcohol on endurance performance were the focus of early research by the late Melvin H. Williams, the founding author of this textbook and to whom it is dedicated. In 1975, Delgado and others reported impaired left ventricular function following small doses of alcohol, which would be a mechanism for **ergolytic** or impaired endurance. Three early studies by McNaughton and Preece (1986), Houmard and others (1987), and Kendrick and others (1993) reported dose-dependent impairments in run performance, trends toward impaired performance, or no effects, respectively, following alcohol ingestion. Later studies by Wang and others (1995) and LeCoutre and Schultz (2009) reported dose-dependent impairments in heart rate, blood pressure, oxygen consumption, blood lactate, power output, and/or perceived exertion during cycling exercise following alcohol ingestion.

More recent research since 2012 suggests that alcohol can also impair anaerobic power and neuromuscular function. Barnes and others reported that alcohol ingestion (1 g/kg), combined with exercise-induced muscle damage, decreased maximal voluntary isometric force production of the quadriceps for up to 60 hours, possibly due to decreased neural activity. Haugvad and others observed that although alcohol did not delay postresistance-exercise muscle recovery, decreases in testosterone/cortisol ratios resulting from decreased testosterone secretion and/or increased cortisol secretion would potentially adversely affect skeletal muscle repair. In their review, Bianco and others noted that alcohol might inhibit protein synthesis cell signaling pathways associated with muscle hypertrophy.

Alcohol consumption may have detrimental effects on energy substrate availability and cofactors in energy metabolism. Burke and others reported that postexercise alcohol consumption might displace carbohydrate kcal, which would adversely affect glycogen synthesis and replacement. Lawler and Cialdella comment that alcohol consumption can inhibit the expression of glycogen synthase, glucose transport (GLUT-4) proteins, and gluconeogenesis, which would adversely affect hepatic and muscle glycogen storage and potentially blood glucose levels. According to El Sayed and others, alcohol consumption decreases the use of glucose and amino acids by skeletal muscles, adversely affects energy supply, and impairs energy metabolism during exercise. Alcohol reduces gluconeogenesis by the liver and glucose uptake by the legs during the later stages of exercise. In prolonged exercise, such as marathons, these effects could lead to an earlier onset of hypoglycemia or muscle glycogen depletion and a subsequent decrease in performance. Subramanya and others reported that chronic ingestion of alcohol inhibited intestinal epithelial cell absorption of thiamine (vitamin B_1), which was discussed in chapter 8 as an important cofactor in energy metabolism. In theory, this could impair prolonged high-intensity endurance performance because thiamine is a critical cofactor with the pyruvate dehydrogenase complex in the aerobic metabolism of carbohydrates.

Alcohol consumption also has implications for dehydration and subsequent rehydration following hyperthermic exercise. The American College of Sports Medicine (ACSM) position stand on fluid replacement states that alcohol consumption increases urine output by inhibiting the release of antidiuretic hormone, which delays full rehydration. Shirreffs and Maughan commented that higher alcohol content reduces the percent retention of ingested fluid. Drinks containing 4 percent alcohol or more, such as beer, tend to delay the restoration of blood and plasma volume compared to carbohydrate-electrolyte solutions or water. Consumption of a significant amount of alcohol in the dehydrated state is problematic for at least two reasons. First, as previously stated, increased urine output delays rehydration attempts, leading to impaired thermoregulation and performance in subsequent competition and training, especially in warm or hot weather. Second, Hobson and Maughan reported that the dehydrated individual would have a higher BAC for a given amount of ingested alcohol due to a lower total water volume. As discussed later, a higher BAC could have adverse consequences. The ACSM position stand for the prevention of cold-weather injuries during exercise implicates alcohol as one of several factors that impair thermoregulation. Graham noted that alcohol consumed before cold-weather exercise could cause vasodilatation, decrease the core temperature, and lower blood glucose levels. These changes could impair performance and possibly contribute to hypothermia.

Based on the available research, it appears that alcohol intake before or during aerobic and anaerobic endurance exercise is not ergogenic. In their review, Pesta and others described alcohol as an ergolytic agent that impairs athletic performance.

Effect on psychological processes Alcohol's well-documented effect as a central nervous system depressant would initially suggest it is not a logical ergogenic aid. However, one of the effects of alcohol is a feeling of euphoria. In their review, Font and others describe the involvement of alcohol and its metabolite, acetaldehyde (**figure 13.2**), in the release of beta-endorphin, an analgesic (which dulls pain), and dopamine, a neurotransmitter associated with the pleasure center of the brain. Peana and others note that **salsolinol**, a condensation product of acetaldehyde and the neurotransmitter dopamine, is the focus of recent research on alcohol abuse and addiction. Interactions between these alcohol metabolites may explain suppression of inhibition, increased self-confidence, reduced anxiety, and a perceived decrease in sensitivity to pain that are sometimes attributed to alcohol consumption. Some of these effects could benefit performance in some tasks. For example, small amounts of alcohol may depress parts of the brain normally inhibiting behavior, which could result in a transitory sensation of excitement and a paradoxical stimulatory effect.

Although these effects may occur, research does not support the use of alcohol in sports involving psychological processes such as perceptual-motor abilities, which involve the perception of a stimulus, integration of this stimulus by the brain, and an appropriate motor response (movement). The evidence overwhelmingly supports the conclusion that alcohol adversely affects psychomotor performance skills, such as reaction time, balance, hand/eye coordination, and visual perception. These are important in events with rapidly changing stimuli, such as tennis.

As previously stated, an ergogenic effect of alcohol is generally not supported by available research. However, a reduction in anxiety and hand tremor following alcohol consumption could enhance performance in sports requiring eye-hand coordination and fine motor control such as riflery, pistol shooting, dart throwing, and archery.

In separate but similar studies by Reilly and others, holding time and a smoother release in archery and accuracy in dart throwing improved with moderate (0.02 BAC) ethanol doses compared to placebo, sober, or higher (0.05 BAC) doses. They also reported a decreased pre-release tremor following both ethanol doses. No performance data were reported for archery. The effect of alcohol on fine motor skills merits additional research.

Social drinking and sports Based on limited research, there is general agreement that light social drinking will not impair performance the next day. In females consuming from zero to six bottles of beer, Kruisselbrink and others reported decreased performance on a four-choice reaction-time test after consuming six beers, but no change in grip strength, aerobic power, or levels of blood glucose and lactate the next morning. Shaw and others reported an 11 percent decrease in time to cycle ergometer test exhaustion following consumption of 1.09 g ethanol/kg fat-free mass compared to a water placebo but no difference in performance in vertical jump test, isometric mid-thigh pulls, and biceps curls test. They concluded that anaerobic capacity and/or high aerobic power may be adversely affected by previous-day alcohol consumption. These studies suggest that alcohol hangover may impair some aspects of next-day performance.

Permissibility The use of alcohol *in competition* by Olympic athletes had been banned previously by the IOC, but because wine and beer are commonly consumed as a part of many traditional European meals, it was removed from the banned list prior to the 1972 Olympics. The World Anti-Doping Agency (WADA) previously prohibited the use of alcohol *in competition* for archery, air, automobile, and powerboating sports. A blood alcohol content of 0.10 g/liter or higher was the doping threshold. However, alcohol was removed from the WADA prohibited list as of January 1, 2018. Protocols have been established by International Federations governing air, automobile, powerboating sports and archery—sports where alcohol use may have performance and public safety consequences—to continue monitoring of alcohol doping/use with assistance from the WADA as necessary.

In their review, Burke and Maughan noted that although alcohol consumption is not prohibited for use by athletes *out of competition*, there are no data supporting an ergogenic effect, and some data to suggest an ergolytic effect. Athletes who drink socially should do so in moderation, and possibly abstain 24 hours prior to a prolonged endurance contest.

What effect can drinking alcohol have upon my health?

Consumption of alcoholic beverages is a popular pastime worldwide. People drink mainly for social reasons, but when and how much they drink may have a significant impact on their health and the health of others. As Costanzo and others note, the "J-curve" relationship whereby "moderate" alcohol consumption has been reported to impart health benefits compared to either abstinence or "heavy" alcohol consumption continues to be remains controversial and is the topic of continued epidemiological research.

Moderate alcohol consumption is defined by the *2020-2025 Dietary Guidelines for Americans* as up to one drink per day for women and up to two drinks per day for men. In its 2018 Global Status Report on Alcohol and Health, the World Health Organization (WHO) stated that alcohol abuse resulted in the death of 3 million people (5.3 percent) in 2016. Causes of death attributable to alcohol consumption include injury (28.7 percent), digestive diseases (21.3 percent), cardiovascular diseases (19 percent), infectious diseases (12.9 percent), and cancers (12.6 percent). The highest and lowest prevalence rates for alcohol dependence and harmful use for both males and females occur in the European and Eastern Mediterranean WHO regions.

Although men and women of all ages may incur health problems with alcohol use, there are potential age- and gender-related differences in health risks. Women have less body water in which to dilute a standard ingested amount of alcohol and have lower levels of a form of alcohol dehydrogenase in the stomach that would decrease the rate of alcohol metabolism. These differences could result in a higher BAC for females, leading to increased short- and long-term health risk. According to the WHO 2018 *Global Status Report on Alcohol and Health,* alcohol-attributable causes of death and disability differ between men and women. The top three gender-specific causes of alcohol-attributable deaths were unintentional injuries, digestive diseases, and communicable diseases in men and cardiovascular diseases, digestive diseases, and unintentional injuries in women. In their review, Kalla and Figueredo commented that there is little research on the effects of alcohol use by the largest-growing U.S. age group, males and females over 65 years of age. Specific issues include interactions between alcohol and pharmacological agents commonly prescribed to older individuals and pathological changes associated with age-related diseases.

Negative Effects Alcohol affects all cells in the body, and many of these effects may have significant health implications. According to the 2018 WHO *Global Status Report on Alcohol and Health,* alcohol is causally related to more than 230 different medical conditions and is a major challenge to public health. Alcohol and its metabolite, acetaldehyde, can have direct toxic effects on cells, possibly damaging DNA. Alcohol may adversely affect functions of major body organs, particularly the liver and brain, and other metabolic functions important to good health.

Many of the adverse health effects of alcohol consumption are associated with heavy, or binge, drinking. Binge drinking in men is defined as consuming five or more drinks in one occasion, while the corresponding amount for women is four or more drinks.

Liver disease The liver is the only organ in the body that metabolizes alcohol, and alcohol may affect liver function in several ways. It may interfere with the metabolism of other drugs, increasing the effect of some and lessening the effects of others. Even with a balanced diet high in protein, consuming six drinks a day for less than a month has been shown to cause significant accumulation of fat in the liver. If continued for 5 years or more, the liver cells degenerate.

Eventually, the damaged liver cells are replaced by nonfunctioning scar tissue, a condition known as **cirrhosis**. As liver function deteriorates, fat, carbohydrate, and protein metabolism are not regulated properly. This has possible pathological consequences for other body organs such as the kidney, pancreas, and heart.

In their meta-analysis of nine studies including over 2.6 million participants with over 5,500 cases of liver cirrhosis, Roerecke and others reported increased risk with even low consumption of alcohol ($\leq$1 drink/day) among women compared to long-term abstainers but not men. The risk for men increased beyond 1 drink/day. Coffee may include an ingredient that protects against cirrhosis. In their meta-analysis of 1,990 cirrhosis cases in 432,133 participants across nine studies, Kennedy and others reported a dose-response between coffee consumption and reduced risk of cirrhosis. For example, an additional two cups of coffee reduced the risk of cirrhosis by almost 50 percent. Resveratrol, an antioxidant in grape skin and red wine, has been reported by researchers to reduce oxidative stress in rat liver. Silva and others commented that the resveratrol may inhibit alcohol-induced oxidative stress and liver cell injury by serving as an antioxidant to scavenge reactive oxygen species; upregulating activities of antioxidant enzymes such as catalase, superoxide dismutase, and glutathione peroxidase; increasing cell signaling pathways regulating lipid metabolism; decreasing inflammation, and inhibiting cell **apoptosis,** which is the normal or programmed death of cells. As a result, some scientists theorize that wine may be the least damaging alcoholic beverage. However, liver damage is one of the most consistent adverse effects of excess alcohol intake. Aside from alcohol abstinence, treatments are limited but there are recent reviews on the use of corticosteroids, probiotics, and peroxisome proliferator-activated receptor agonists, transcriptional factors which regulate fatty acid metabolism and inflammation of liver tissue. These strategies are discussed respectively by Stickel and others, Lee and Suk, and Meng and others.

Psychological problems Many of the adverse health effects of alcohol consumption are associated with disturbed mental functions. As previously mentioned, a small amount of alcohol often provides pleasurable effects. However, alcohol generally acts as a depressant and its effects on the brain are dose-dependent. The effects occur in a hierarchical fashion related to the development of the brain. In general, alcohol initially affects the higher brain centers. Increasing doses impair lower brain functions. This hierarchy of brain functions, from higher levels to lower levels, and some of the functions affected by alcohol are summarized in **table 13.2**.

Driving while impaired A blood alcohol content (BAC) of 0.06–0.09 may impair judgment, fine motor ability, and coordination—three factors that are extremely important in the safe operation of an automobile and other modes of transportation. A BAC of 0.08 or higher is the threshold for *driving under the influence* (*DUI*) in all 50 states. DUI has serious social and personal consequences, including death. According to the National Highway Traffic Safety Administration's National Center for Statistics and Analysis, a BAC of $\geq$ 0.01 percent was a factor in 34 percent of the 36,560 US traffic fatalities occurring in 2018. The economic cost of alcohol-impaired automobile accidents including but not limited to lost productivity, medical and legal expenses, and property damage

TABLE 13.2 General and specific effects of increasing blood alcohol content

Number of drinks consumed in 2 hours*	Blood alcohol content**	General hierarchy of effects on brain function	Specific effects
2–3	0.02–0.04	-Euphoria, Impaired thinking and reasoning (e.g., lower inhibition)	Reduced tension, relaxed feeling, relief from daily stress
4–5	0.06–0.09	-Impaired thinking and reasoning (e.g., greater loss of inhibition) -Impaired perceptual-motor responses (e.g., reaction time)	Legally drunk (0.08) in all states; impaired judgment, a high feeling, impaired fine motor ability and coordination
6–8	0.11–0.16	-Impaired fine motor coordination (e.g., muscles of speech)	Slurred speech, impaired gross motor coordination, staggering gait
9–12	0.18–0.25	-Impaired gross motor coordination (e.g., impaired walking) -Impaired vision (e.g., double vision)	Loss of control of voluntary activity, erratic behavior, impaired vision
13–18	0.27–0.39	-Impaired alertness (e.g., sleep, coma)	Stupor, total loss of coordination
≥19	>0.40	-Respiratory control: (e.g., respiratory failure, death)	Coma, depression of respiratory centers, death

*One drink = 12 ounces regular beer
　　　　　　4 ounces wine
　　　　　　1.25 ounces liquor

**BAC based on body weight of 160 pounds (72.6 kg). The BAC will increase proportionally for individuals weighing less (such as a 120-pound female) and will decrease proportionally for individuals weighing more (such as a 200-pound football player). For example, four to five drinks in 2 hours could lead to a BAC of 0.08–0.12 in a 120-pound individual.

was more than $44 billion in 2010, the most recent for which these data are available. As the saying goes, "Don't drink and drive!"

www.nhtsa.gov/research-data/fatality-analysis-reporting-system-fars At the National Highway Traffic Safety Administration's Fatality Analysis Reporting System website, type Alcohol-Impaired Driving in the search window on the next page to access various summary reports.

In recent years, alcoholic energy drinks have become increasingly popular among the young. Some believe that the caffeine and other possible stimulants in these drinks may counteract the depressant effects of alcohol. Marczinski and others reported greater tolerance for alcohol mixed with a high-caffeine energy drink compared to alcohol alone. This could result in failure to recognize driving impairment, subsequent DUI, and increased threat to public safety. Energy drinks are discussed in more detail in the section on caffeine.

Aggressive behavior An extensive body of literature supports a positive association between acute alcohol consumption and aggressive behaviors. Kuypers and others report a causal link between alcohol intoxication and aggression in placebo-controlled studies, Higher levels of aggression occur at a dose of ≥0.75 g/kg, but various factors may influence this cause-and-effect relationship. In their review, Beck and Heinz note that alcohol is a contributing factor in half of the violent crimes and sexual assaults worldwide and is second only to depression in the psychiatric diagnosis for suicide. The quantity of alcohol consumed also is related to aggression and fatal nontraffic outcomes. In their review, Lees and others note that adolescents who binge drink or are heavy drinkers have decreased cognitive function including poorer learning and psychomotor speed; abnormal neural activity during executive functioning, attention control and reward seeking; decreases in gray matter volume; and lower increases in white matter volume compared to nondrinkers. Galaif and others noted that suicide and alcohol use are related in young individuals. The suicide rate for those between 15 and 19 years of age is 9.7 per 100,000, while half of 12th graders have been intoxicated at least once.

Attwood and Munafò commented that alcohol may disrupt signal processing between the frontal lobe and amygdala, brain centers involved in normal decision making and emotional reactions. Males are more prone to alcohol-related aggressive behavior than females. In susceptible individuals, such disruption may cue an increase in aggression and lead to learned aggressive behavior. Specific outcomes include disinhibition of suppressed behavior, heightened reaction to potentially aggressive situations, loss of empathy, and reduced recognition of submissive cues from others.

Alcohol consumption in college students and athletes Alcohol consumption is commonly considered as a "rite of passage" into adulthood. Carter and others noted that high prevalence rates for alcohol abuse in young adults are largely explained by multiple factors such as Greek and dorm life, athletics, and spring break trips. International comparisons of drinking behaviors in young adults are difficult due to differences in legal drinking ages, cultural attitudes, and in higher education systems. Merrill and Carey noted that "pregaming" and drinking games are dangerous alcohol-related behaviors common in college students. Wilkerson and others also reported high rates of disordered eating, excessive physical activity, and binge drinking, a constellation of behaviors associated with "drunkorexia," which was discussed in chapter 10. According to the National Institute of Alcohol Abuse and Alcoholism's 2019 National Survey on Drug Use and Health (www.niaaa.nih.gov/publications/brochures-and-fact-sheets/alcohol-facts-and-statistics), prevalence rates for drinking, binge drinking (females: ≥4 drinks in 2 hours; males: ≥5 drinks in 2 hours), and heavy alcohol use (females: 3 drinks/day or >7 drinks/week; males: 4 drinks/day or >14 drinks/week) among college students were 52.5, 33.0, and 8.2 percent, respectively. Gruenewald and others concluded that many problems among college students, including poor academic performance, are associated with drinking relatively small amounts of alcohol (2–4 drinks).

Many physically active individuals, including competitive athletes, consume alcohol on a regular basis. In a survey of NCAA male athletes, Buckman and others noted that alcohol was the most widely used drug, with 82 percent of respondents reporting use within the last 12 months. Ford commented that higher binge drinking among college athletes versus nonathletes may be related to differences in perceived social norms, peer influence, and athletes' "work hard, play hard" attitude. Sports participation may, in some way, be associated with increased alcohol use. Kwan and others noted that a positive relationship between alcohol use and sports participation was reported in 14 of 17 studies in their review. The consumption of alcohol in the celebration of victory and the commiseration of defeat appears to be deeply ingrained in sports culture.

Cardiovascular disease According to Minzer and Costanzo and their respective colleagues, purported beneficial effects of moderate consumption (≤1 and ≤2 drinks/day for females and males, respectively) such as decreases in blood pressure and oxidative stress and increased high-density lipoprotein cholesterol continue to be topics of debate. However, associations between excessive alcohol consumption and hypertension, type 2 diabetes, elevated triglycerides, programmed cell death, and fibrosis in the heart; increased sympathetic response and renin-angiotensin-aldosterone system function; decreased myocardial contractile muscle protein synthesis; and impaired vascular endothelial function are well established. **Table 13.3** lists several alcohol-related cardiovascular diseases and symptoms as well as several reviews for the interested reader.

Cancer *In vitro* (i.e., in a test tube) research has shown that damage to DNA caused by alcohol and its metabolite acetaldehyde is similar to that elicited by carcinogens. This DNA damage may occur at an alcohol concentration equivalent to one to two drinks. Based on these findings, the Environmental Protective Agency has concluded that acetaldehyde is a probable human carcinogen.

TABLE 13.3 Selected cardiovascular diseases and symptoms that are associated with excessive alcohol consumption

Disease or symptom	Reviews
Alcoholic dilated cardiomyopathy (impaired heart contractility and function)	George and Figueredo, 2011; Gardner and Mouton, 2015
Hypertension (increased blood pressure)	Kawano, 2010; Gardner and Mouton, 2015
Increased risk of ischemic (lack of blood flow) and hemorrhagic (bleeding) stroke	Cargiulo, 2007.
Increased risk of certain heart rate arrhythmias, such as atrial fibrillation, which can lead to clot formation and a heart attack or stroke	Tonelo, et al., 2013; George and Figueredo, 2010.
Sudden cardiac death	George and Figueredo, 2010.

According to the 2018 World Cancer Research Fund (WCRF)/American Institute for Cancer Research (AICR) report, there is strong evidence that alcohol ingestion in any amount increases the risk of mouth, pharynx, larynx, esophageal, and pre- and postmenopausal breast cancers. Two or more drinks (≥30 grams/day) increases the risk of colorectal cancer while three or more drinks (≥45 grams/day) increases the risk of stomach and liver cancers. Interestingly, the report notes that up to two drinks (≥30 grams/day) *decreases* the risk of kidney cancer.

The relationship between alcohol use and breast cancer is controversial. In a meta-analysis of 60 studies, Choi and others reported dose-response increases of 4, 9, and 13 percent in breast cancer risk with consumption of 0.5, 1 (12.5 grams of alcohol), and 2 drinks/day.

Scoccianti and others noted that younger age at the start of drinking may also increase breast cancer risk by early and longer exposure of mammary tissue to the carcinogenic effects of acetaldehyde. In addition to potential DNA damage, alcohol ingestion may also increase estrogen levels, a factor that increases breast cancer risk. Alcohol also increases estrogen and progesterone receptor expression, which can in turn increase breast cancer risk. According to the National Institute of Alcohol Abuse and Alcoholism, women with significant family history for breast cancer or those on estrogen replacement therapy should discuss their alcohol intake with their health-care provider in order to weigh potential increases in breast cancer risk against potential decreases in cardiovascular disease risk.

Fetal alcohol spectrum disorders Women who drink should abstain during pregnancy because even moderate consumption of alcohol, or even a single drinking binge, may affect DNA in the embryo and fetus. Mandal and others concluded that ethanol is a classic teratogen capable of inducing a wide range of developmental abnormalities, particularly via epigenetic reprogramming in the fetus by DNA methylation. In a meta-analysis of 33 clinical studies, Steane and others concluded that prenatal alcohol exposure increased the risk of placental abruption (separation of the placenta from the uterine wall); decreased placental weight and alterations in placental vasculature, DNA methylation, and gene expression. Given these effects, alcohol may cause health problems in the newborn which are identified as *fetal alcohol spectrum disorder (FASD)*. FASD includes an array of physical, behavioral, and/or cognitive abnormalities in a child's development that are associated with maternal alcohol consumption during pregnancy. Diagnostic criteria for FASD are described by Hoyme and others.

Fetal alcohol syndrome (FAS) is the most severe of the FASD. The Centers for Disease Control and Prevention National Center on Birth Defects and Developmental Disabilities indicates that FAS is one of the leading known causes of intellectual disabilities and birth defects. According to a Centers for Disease Control and Prevention report by Fox and others, the prevalence estimate for FAS in several U.S. states was 0.3 cases per 1,000 live births in 2010. However, studies using in-person assessment suggest prevalence rates as high as 6-9 cases per 1,000 children. Some experts estimate that physical exams may reveal prevalence rates of the full range of FASD as high as 1-5 percent. The child may experience retardation in growth and mental development as well as facial birth defects (**figure 13.3**). In their meta-analysis, Lucas and others reported a significant association between prenatal alcohol exposure or FASD and impairment in gross motor skills such as balance, coordination, and ball-handling skills.

www.cdc.gov/ncbddd/fasd/data.html Access the Centers for Disease Control and Prevention website on Fetal Alcohol Spectrum Disorders.

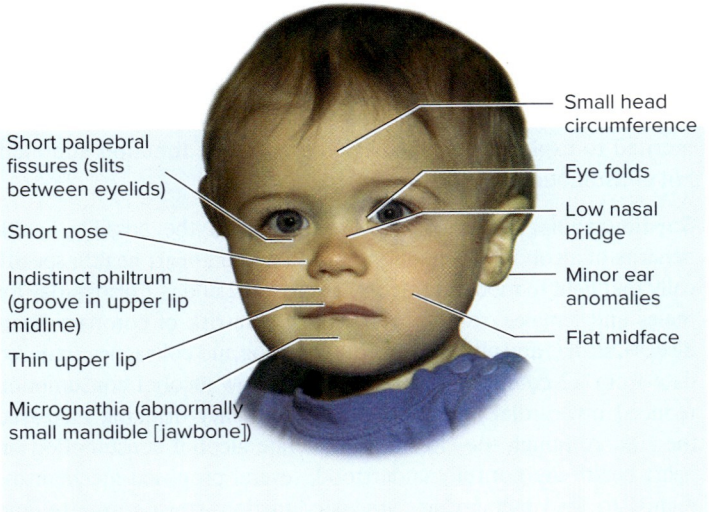

Realistic Reflections

FIGURE 13.3 Common facial characteristics of children with fetal alcohol syndrome (FAS).

Fetal alcohol effects (FAE) may be observed in children when full-blown FAS is not present. Children with FAE are easily distracted and have poor attention spans but do not have the facial features of FAS. Both FAS and FAE are associated with learning disorders in children. According to a National Academies of Sciences' Institute of Medicine document edited by Stratton and others, fetal alcohol spectrum disorders (FASD) is an umbrella term including fetal alcohol syndrome (FAS), *partial fetal alcohol syndrome* (pFAS), *alcohol-related neurodevelopmental disorder* (ARND), and *alcohol-related birth defects* (ARBD). There is conflicting research regarding pregnancy outcomes associated with "light drinking." Kelly and others reported no increase in risk for emotional or cognitive problems in children of mothers who engaged in ≤2 drinks/week during pregnancy compared to children of nondrinkers. Coathup and others reported that ≤1 drink/day during pregnancy was associated with higher intakes of fruit, vegetables, whole grains, and fish and lower intake of processed meats, a pattern generally consistent with *2020-2025 Dietary Guidelines for Americans* during pregnancy. However, Hamilton and others reported that rats exposed to moderate prenatal alcohol demonstrated altered behavior suggesting adverse effects on frontal cortex function that persisted into adulthood. The National Institute of Alcohol Abuse and Alcoholism states there is no known safe level of alcohol consumption in pregnancy. According to the *2020-2025 Dietary Guidelines for Americans*, women who are or may be pregnant should not drink alcohol.

Obesity Alcohol is a significant source of kcal, about 7 per gram, somewhat comparable to the caloric content of fat. Research has indicated that if small amounts of alcohol (5 percent of daily caloric intake) are interchanged for an equivalent caloric intake from carbohydrates, there is no effect on daily energy expenditure. In other words, alcohol kcal themselves will not increase body fat as long as total daily caloric intake matches daily caloric expenditure. In general, the NIAAA indicates that the relationship between moderate alcohol consumption and obesity remains inconclusive. In their study, Beulens and others reported that moderate alcohol consumption (40 grams daily for 4 weeks) was not associated with increased adiposity or increased weight circumference.

However, Traversy and Chaput noted that although light to moderate alcohol intake generally does not promote increased adiposity, heavy alcohol consumption is consistently associated with weight gain. The association between alcohol consumption and weight gain is modulated by factors including but not limited to gender, genetics, physical activity, diet, sleep, and psychosocial issues. Several physiological effects of alcohol intake, which are listed in **table 13.4,** may alter energy balance and contribute to obesity.

Alcohol dependence Alcohol abuse, or excessive alcohol consumption, is a major drug problem in the United States. The American Psychiatric Association's *Diagnostic Statistical Manual-V* incorporates previous alcohol abuse and alcohol dependence diagnoses into a single diagnosis called alcohol use disorders (AUD). Mild, moderate, and severe AUD cases are diagnosed based on 2-3, 4-5, and ≥ 6 symptoms, respectively, out of 11 experienced within the last 12 months. According to the NIAAA National Survey on Drug Use and Health, 5.3 percent of U.S. citizens ≥12 years of age (14.5 million total; 9.9 million males, 6.8 percent; 5.5 million

TABLE 13.4 Physiological effects of alcohol consumption which may be related to weight gain

- Alcohol kcal are generally additive (e.g., alcohol does not reduce dietary kcal).
- Alcohol can stimulate appetite (e.g., cultural/learned association between alcohol and food consumption).
- Alcohol inhibits certain appetite-suppressing hormones (e.g., leptin, glucagon-like peptide 1).
- Alcohol increases the effects of certain central neurotransmitters that stimulate appetite (e.g., gamma-aminobutyric acid, opioids) and decreases effects of others that inhibit appetite (e.g., serotonin).
- Alcohol can inhibit fat oxidation (e.g., fat sparing and accumulation over the long term).
- Alcohol binge drinking and binge eating are related.

Source: Traversy and Chaput, 2015.

females, 3.9 percent) had AUD in 2019. Treatment was sought by only 7.2 percent of those with AUD. Alcohol dependency is more commonly known as **alcoholism**.

The etiology of alcoholism is unknown but probably is related to a variety of physiological, psychological, and sociological factors. Söderpalm and Ericson indicate that alcohol activates the dopamine system, an important part of the brain reward system, and the positive psychological effects may reinforce the desire to continue to consume alcohol. Morozova and others note that no single genetic locus explains alcohol dependence. Studies have reported many complex genetic and epigenetic factors that potentially influence an individual's susceptibility to alcohol dependence. According to the National Council on Alcoholism and Drug Dependence (NCADD), there is no single definition for alcoholism. It may be related to a variety of behaviors. A deficiency of B complex vitamins, particularly thiamine (B_1), may contribute to many of the neuropsychiatric problems seen in alcoholism. The number of behaviors exhibited by the drinker may be related to various stages in the progression toward alcoholism.

www.niaaa.nih.gov/publications/brochures-and-fact-sheets/alcohol-facts-and-statistics   Click here to see results on the consequences of alcohol abuse from the most recent National Survey on Drug Use and Health from the National Institute of Alcohol Abuse and Alcoholism.

ncadd.org/get-help/take-the-test   The NCADD has developed an assessment of behaviors that may signal AUD. Access the questionnaire by clicking Take the Alcohol Test at the lower left of the page. A link to a teenager version can be accessed at the lower right side.

www.testandcalc.com/etc/tests/audit.asp   This website offers a shorter, online version known as the Alcohol Use Disorders Identification Test (AUDIT).

pubs.niaaa.nih.gov/publications/dsmfactsheet/dsmfact.pdf   This link lists the 11 symptoms used to diagnose AUD according to DSM-V and also compares changes in alcohol-related diagnoses between DSM-IV and DSM-V.

Positive Effects   On the positive side, most recent epidemiological research and reviews have shown that light to moderate consumption of alcohol is associated with lessened mortality. In their meta-analysis of 31 studies, Ronksley and others reported drinkers consuming 2.5–14.9 grams of alcohol/day had the lowest relative risk of all-cause mortality (0.87) compared to nonuser referent group (1.0), with a significant increase in risk (1.3) in heavy drinkers (>60 g/day). However, the frequently reported "J-curve" relationship alcohol use and various morbidities and mortalities in epidemiological research remains controversial as previously noted by Minzer and Costanzo and their respective colleagues. Consumption of even low amounts of alcohol may increase risk of breast cancer risk and alcohol should be completely avoided during pregnancy.

Alzheimer's disease, other dementia, and cognitive function   In a review of systematic reviews and meta-analyses, Rehm and others concluded an association exists between light to moderate alcohol consumption in middle aged adults and reduced dementia diagnosis and mortality which falls short of a cause-and-effect relationship. Heavy alcohol consumption defined as >60 and >40 g/day for males and females, respectively was associated with increased dementia risk. Proposed mechanisms include reduced cardiovascular risk, elevated levels of high density lipoprotein cholesterol, the reduced platelet-aggregating effects of alcohol, decreased inflammation, and the memory-enhancing effects of acetylcholine on the hippocampus. Piazza-Gardner and others reported no effect of alcohol on Alzheimer's disease in 9 of 19 reviewed studies. However, there were inconsistencies in measured alcohol consumption in 7 studies reporting moderate alcohol as protective against Alzheimer's disease. Sachdeva and others commented that alcohol-related dementia may be the result of structural and functional brain damage leading to impaired visuospatial function, memory, and executive tasks. They noted that this condition is currently difficult to diagnose, but likely to increase in the future given increased alcohol consumption in young individuals. According to Rehm and others, inconsistencies across studies of cognitive function include but are not limited to self-reported alcohol consumption, lack of standardization in dose and levels of alcohol consumption, lack of control of confounding influences, inclusion of former drinkers with lifelong abstainers in control groups. Given these discordant findings, additional research is merited to explore the links between cognitive function and alcohol consumption.

Cardiovascular Disease   Most research on the possible health benefits of alcohol consumption has focused on heart health, specifically that light to moderate alcohol intake (≤2 and ≤1 drinks/day for males and females, respectively) reduces the risk of coronary heart disease, stroke, and all-cause mortality. Leong and colleagues analyzed data from 52 countries and reported that low alcohol consumption reduced myocardial infarction risk, while heavy drinking increased the risk. Although the effects of moderate alcohol consumption on heart health are not fully understood, several proposed mechanisms related to lipid metabolism, vascular function, stress response, and insulin sensitivity are listed in **table 13.5**.

Some investigators have theorized that the consumption of certain types of alcoholic beverages, most notably red wine, is responsible for

TABLE 13.5 Proposed mechanisms for health benefits of light to moderate alcohol consumption on cardiovascular health

Mechanism	Reference
Increased relaxation/stress reduction	Jones, et al., 2013
Decreased platelet aggregability (clotting ability) by increasing the activity of a clot-dissolving enzyme in the blood to reduce risk of myocardial infarction and stroke	Kasuda, 2006
Increased brain blood flow, which reduces the risk of stroke	Marxen, et al., 2014
Increased insulin sensitivity to reduce diabetes, a risk factor for cardiovascular disease	Koppes, et al., 2005
Increased levels of HDL-cholesterol (particularly the smaller, dense, protein-rich HDL_3), which protects against the development of cardiovascular disease	Brinton, 2012
Favorably affects phospholipids of endothelial cells to improve vascular function	Cahill and Redmond, 2012

the reported health benefits. Polyphenols from moderate wine and beer consumption may protect against the atherosclerotic process. Specific mechanisms attributed to polyphenols include reductions in low density lipoprotein cholesterol oxidation, platelet aggregation, and biomarkers of inflammation, as well as improved vasodilation, blood flow, and vascular function. As previously mentioned, resveratrol has received the most research attention. According to Salehi and others, resveratrol has potential cardioprotective, anticarcinogenic, antiviral, neuroprotective, anti-inflammatory, and antioxidant properties with no major adverse side effects. However, factors such as different isoforms (*cis* vs. the more bioactive *trans*) and poor solubility may limit its bioavailability and the resulting therapeutic potential. There is less support for these effects in distilled spirits, which contain fewer polyphenols. Compared to wine, beer contains more protein and B vitamins, is rich in flavonoids, and has an equivalent antioxidant content, but of different specific antioxidants derived from barley and hops. Stackelberg and others reported a lower risk of abdominal aortic aneurysm with low to moderate consumption of either wine or beer, but not spirits, in Swedish men and women. In their review, Chiva-Blanch and Badimon conclude that low-to-moderate alcohol consumption may benefit cardiovascular function based on findings from epidemiological research. However, self-reporting of alcohol consumption and lack of standardization of consumed quantity may impact this association. Questions that remain include, but are not limited to, specific quantity; differences between types of alcohol; and differences between WHO regions, age, and gender with regard to a cardioprotective effect. Furthermore, any benefits must be weighed against risks such as genetic predispositions to alcohol dependence and certain cancers, pregnancy, and other individual factors.

These factors may be important, but Naimi and others contend that individuals who drink moderately may also practice other lifestyle behaviors that reduce the risk for CHD. For example, Piazza-Gardner and Barry noted a positive relationship between physical activity and alcohol consumption in studies of youth, college-aged students, and the general population. Furthermore, Mukamal and others reported that moderate alcohol intake, up to two drinks daily, was associated with lower risk of experiencing a heart attack even in men already at low risk for heart disease on the basis of body mass index, physical activity, smoking, and diet. However, Williams reported that blood pressure increased in association with the amount of alcohol intake regardless of running level in male runners. There appear to be some positive health effects associated with light to moderate alcohol intake, but it should be reemphasized that heavier drinking is another matter.

Others have challenged the health benefits of alcohol from another perspective. Although lifetime abstainers are the ideal comparison group for the health effects of alcohol use, nonusers in some studies have included individuals who decreased or stopped drinking due to illnesses. According to Fillmore and others, such individuals who are known as *sick quitters* may represent a systematic error in studies supporting an association between moderate alcohol consumption and reduced CHD risk. In other words, their mortality was due to the underlying illness rather than lack of alcohol use. On the other hand, Mukamal and others reported heart benefits from moderate alcohol intake in only healthy men who exercised, ate a good diet, were not obese, and did not smoke. Rehm and others recommend using irregular, light lifetime drinkers with abstainers as the comparison group in future studies. In over 480,000 registrants in the UK biobank, Lankester and others reported positive associations between an additional alcohol drink/day and systolic blood pressure, hemorrhagic stroke, and atrial fibrillation while accounting for gene variants including alcohol dehydrogenase isoform rs1229984 which is known to be associated with lower levels of alcohol dependence. They concluded there is no evidence that any level of alcohol consumption provides cardioprotection and suggested that moderate alcohol use should not be promoted as part of a healthy lifestyle. This finding calls into questions the cardioprotective effects of moderate alcohol consumption.

Health authorities generally recommend either abstinence or prudent consumption in order to minimize the previously discussed health risks associated with alcohol abuse. Clearly, abstinence is the best policy for pregnant women or those operating a motor vehicle. *Low-risk drinking* is an emerging term to represent light to moderate alcohol intake. The National Institute on Alcohol Abuse and Alcoholism indicates that low-risk drinking, along with a balanced diet, should not pose any health problem to the average, healthy individual. Moderate drinking is defined in the United States as ≤2 drinks/day for males and ≤1 drinks/day for females. However, recommendations vary by country. According to the European Union's National Institute for Health and Welfare, a standard drink in member countries ranges from 8 to 20 grams of alcohol with recommended upper limits of consumption ranging from 20 to 48 and 10 to 32 g/day for males and females, respectively. A standard drink in Canada contains 13.45 grams of alcohol with recommended upper consumption limits of 15 and 10 drinks/week for males and females, respectively. Numerous individual differences, such as age, genetics, and metabolic rate, may affect the response to alcohol so moderate alcohol use is not synonymous with *healthy alcohol use*. Bagnardi and others reported significant associations between light alcohol consumption, defined as ≤12 g/day (≤1 drink/day), and oral/pharyngeal cancer, esophageal squamous-cell carcinoma, and breast cancer. Since higher levels of consumption increase health risk, men and women should not exceed four and three drinks/day, respectively, on any day. You should consult with your physician if you are considering drinking for its possible health benefits.

In June 2018, the National Institutes of Health terminated the National Institute of Alcohol Abuse and Alcoholism's (NIAAA) $100 million Moderate Alcohol and Cardiovascular Health (MACH) Trial due to scientific integrity violations which included solicitation of funds from the alcohol industry, design flaws, and other conflicts-of-interest committed by NIAAA personnel. This action was taken amid concerns that results might maximize beneficial effects and minimize harmful effects of alcohol.

Kidney cancer. Three meta-analyses agree that light or moderate alcohol intake can be associated with a reduced risk of renal center. In an analysis of pooled data from 12 studies, Lee and others reported a negative association between ingestion of ≤30 grams of alcohol (2 drinks)/day and risk of renal cell carcinoma. Song and others also reported a decreased risk of renal cancer associated with one drink (~12.5 g)/day with no additional benefit from greater daily consumption of alcohol. In similar findings, Belloco and others reported a decreased risk of renal cancer in light (0.01–12.49 g/day) and moderate (12.5–49.9 g/day) drinkers but not in heavy drinkers (≥50 g/day), compared to nondrinkers. The 2018 WRRF/AICR Continuous Update Project concluded that alcohol consumption of ≤30 g/day probably protects against cancer. Proposed mechanisms are uncertain but may include improved lipoprotein profile (e.g., higher high-density lipoprotein levels), higher levels of the anti-inflammatory cytokine adiponectin, and possibly the diuretic effect of alcohol.

> **Check for Yourself**
>
> ▶ Visit a local beer/wine store that carries a wide variety of products, including microbrews and fortified wines. Check the labels for percentage alcohol content, listing those from lowest to highest. Calculate how much alcohol would be in a standard drink from each.

> *www.niaaa.nih.gov* This website provides detailed information on a wide variety of alcohol-related topics. For example, if you want to decrease the amount of alcohol you drink, click on Publications and Multimedia, then Brochures and Fact Sheets, then Rethinking Drinking.

Caffeine: Ergogenic Effects and Health Implications

Coffee is one of the most widely consumed beverages throughout the world. The coffee bean, a plant product, contains caffeine, which has been theorized to enhance exercise performance. Besides caffeine, coffee also contains numerous other biologically active phytonutrients, such as antioxidants, and has been studied for its possible beneficial or adverse effects on health.

What is caffeine, and in what food products is it found?

Caffeine is an odorless, bitter, white alkaloid that appears naturally in many plants. Technically, caffeine may be classified as a food ingredient, a dietary supplement, or a drug.

As a food ingredient, caffeine is found in many of the foods and beverages that we consume every day, not only coffee but tea, colas, caffeinated waters, juices, energy drinks, sports drinks and sports bars, preworkout supplements, and chocolate. Some approximate amounts in the beverages we consume are 80–135 milligrams in an 8 ounce (237 milliliter) cup of brewed coffee, 40–60 milligrams in a cup of tea, 35–45 milligrams in a can of cola, and 80–120 milligrams or more in an 8-ounce serving of an energy drink. Using NHANES data from 2007 to 2012, Lieberman and others reported an average per capita caffeine intake of 164 mg/day with 70 percent of daily consumption occurring before 12:00 noon and the greatest consumption by middle-aged, non-Hispanic white individuals. Caffeine is also found in various dietary supplements, such as kola nuts and guarana, and even in some over-the-counter stimulant supplements targeted to athletes. Most recently, caffeine has been marketed as *performance candy*, such as Jolt® Caffeine-Energy Gum and Buzz Bites®, a chocolate chew candy. Such products may be construed to be sports supplements.

Caffeine is also legally classified as a drug and has some powerful physiological effects on the human body. A normal therapeutic dose of caffeine may range from 100 to 300 milligrams. As noted in **table 13.6,** some food products and supplements provide a therapeutic dose and meet the standards for classification as a drug. Indeed, caffeine has been identified as the most popular social drug in the United States (**figure 13.4**).

> **Key Concepts**
>
> ▶ One drink of alcohol contains approximately 13–14 grams of alcohol, or about 1/2 ounce. One drink is typically the equivalent of 12 ounces of beer, 4 ounces of wine, or 1.25 ounces of 40-proof whiskey. However, the alcohol content in some beverages may be substantially greater.
> ▶ Alcohol is not an effective ergogenic aid; in fact, it may actually be ergolytic, that is, impair performance.
> ▶ Consumption of alcohol in moderation appears to cause no major health problems for the normal, healthy adult and may actually confer some health benefits. However, alcohol may be contraindicated for some, such as women during pregnancy. The effect of alcohol on breast cancer remains controversial. Heavy drinking is associated with numerous health problems.

TABLE 13.6 Caffeine content in selected products

Product	Serving size	Caffeine (milligrams)
Coffee, brewed	8-ounce cup	96
Coffee, instant	1 packet (2 grams)	63
Coffee, decaffeinated	6-ounce cup (179 grams)	2
Starbucks® Featured Dark Roast Coffee	Grande 16-ounce cup	260
Starbuck's® Nitro Cold Brew Dark Caramel	9.6-ounce can	115
Tea, black or green	8-ounce cup	47
Hot cocoa	8-ounce cup	5
Coca-Cola®	12-ounce can	32
Diet Coke®	12-ounce can	42
Mountain Dew Amp	16-ounce can	142
Gatorade® Bolt24 Energize	16.9-ounce bottle	75
Red Bull Energy Drink	16-ounce serving	143
5 Hour Energy® Extra Strength	1.93 ounce bottle	230
Central Nervous System Stimulant (OTC)	1 Vivarin® tablet	200
Guarana	5 grams	250
Preworkout supplement (e.g., Cellucor® C4® Original)	1 scoop (6 grams)	150

Note: Check labels of over-the-counter stimulants and dietary supplements for caffeine content.
Sources: US Department of Agriculture FoodData Central fdc.nal.usda.gov/fdc; CaffeineInformer www.caffeineinformer.com/the-caffeine-database

What effects does caffeine have on the body that may benefit exercise performance?

One of the primary effects of caffeine is to block the neurotransmitter adenosine, and thus influence a wide variety of metabolic processes throughout the body. Additionally, caffeine may stimulate the adrenal gland to release epinephrine into the circulation. The combination of caffeine and epinephrine stimulates the central nervous system; potentiates muscle contraction; and increases the rate of muscle and liver glycogen breakdown, the release of free fatty acids (FFAs) from adipose tissue, the use of muscle triglycerides, and resting blood fatty acid concentration.

Early research on the ergogenic mechanism of caffeine on prolonged endurance performance focused on increased fat oxidation which spared the use of muscle glycogen. This hypothesis was tested by pioneers such as Costill and others over the years. In a

Shannon Fagan /Image Source

FIGURE 13.4 Caffeine is the most popular social drug in the United States. Up to 85 percent of all U.S. adults consume caffeine in one form or another, mainly as coffee.

recent meta-analysis of 19 placebo-controlled crossover designs, Collado-Mateo and others reported evidence of increased fat oxidation (increased VO_2, decreased respiratory exchange ratio) following a caffeine dose of ≥ 3 mg/kg body mass. However, increased fat oxidation was higher in untrained subjects and attenuated in recreationally trained individuals and endurance athletes. Graham and others conclude there is little evidence to support the fat oxidation/glycogen sparing ergogenic mechanism in competitive endurance athletes.

www.caffeineinformer.com/ This website contains an extensive database of drinks and foods containing caffeine, as well as a calculator to determine safe daily caffeine intake.

Guest and others note that caffeine's effect on the central nervous system is the primary mechanism to improve performance. By competing with adenosine in binding to adenosine A_1 and A_{2A} receptors, caffeine may increase CNS neurotransmitter release and motor unit firing rate, decrease perceived exertion, and suppress pain related in increase release of dopamine. Most but not all individuals experience heightened alertness, improved mood, and increased cognitive function with caffeine consumption. In their meta-analysis, Doherty and Smith concluded that the effect of caffeine on RPE (5.6 percent decrease) was associated with improved exercise performance (11.2 percent increase), which explained approximately 29 percent of the improvement in performance. This reduction in the psychological perception of effort during exercise appears to be an important factor underlying caffeine's ergogenic effect.

Other peripheral mechanisms include increases in myofibrillar calcium availability for increased force and reduced fatigue during muscle contraction and in substrate availability for metabolism. In her review, Kalmar indicated that caffeine may influence muscle performance by stimulating the nervous system at various points along the α-motor neuron to the muscle. Research by Tarnopolsky and Cupido, using electrical stimulation of the muscle, supports the hypothesis that some of the ergogenic effect of caffeine in endurance exercise performance occurs directly at the skeletal muscle level. In research involving electrical stimulation of *in vitro* isolated mouse muscle, Tallis and others reported a greater effect of caffeine on muscle force production in oxidative muscle fibers compared to glycolytic fibers.

Overall, caffeine may influence central and peripheral metabolic processes, as well as psychological processes, to help delay the onset of fatigue and has been theorized to enhance performance in many types of exercise, including endurance, strength, speed, and power.

Does caffeine enhance exercise performance?

Hundreds of studies have been conducted to test the ergogenic effectiveness of caffeine. Considerable differences exist in the experimental designs of caffeine studies in such aspects as caffeine delivery system (caffeine in food, e.g., coffee, capsule, or powder form); caffeine dosage (3–15 mg/kg body weight); the type of exercise task (power, strength, reaction time, short-term endurance, prolonged endurance); the intensity of the exercise (submaximal exercise, maximal exercise); the training status of the subject (trained, untrained); the pre-exercise diet (high-carbohydrate, mixed); the subjects' caffeine status (habitual user, abstainer); individual variability (responder, nonresponder); and other ingested ingredients (e.g., preworkout supplements). These differences complicate interpretation of the results. Additionally, some investigators have combined caffeine with other related stimulants, such as ephedrine, which is discussed later in the chapter. Ergogenic effects of caffeine on selected types of performance tasks are summarized in **table 13.7**.

TABLE 13.7 Effect of caffeine on selected types of performance tasks

Type of task	Ergogenic effect
Aerobic power (prolonged, high intensity; various modes)	• Increased central and peripheral neural activity; delayed fatigue; decreased perception of effort • Increased oxidation rates of glycogen and exogenous carbohydrate
Tasks requiring alertness	• Increased attention, reaction time, and focus (low [≤3 mg/kg] doses only; higher doses may be ergolytic)
Lower-body neuromuscular endurance	• Attenuated decline in α-motor neuron firing rate, which decreases the decline in muscle force production • Increased intramuscular calcium kinetics in excitation contraction-coupling relaxation
Anaerobic power (sustained 1–3 minute; repetitive tasks; intermittent anaerobic activity over a prolonged duration, e.g., soccer match)	• Attenuated decline in α-motor neuron firing rate which decreases the decline in muscle force production • Increased intramuscular calcium kinetics in excitation contraction-coupling relaxation

Use by Athletes and Other Active Individuals

Caffeine is a popular ergogenic aid among athletes, the military, and other active individuals. According to Chaudhary and others, military personnel consume between 212 and 285 mg of caffeine/day, primarily in energy drinks (<30 years of age) and coffee (≥30 years of age) for improved cognitive function and performance, especially during sleep deprivation. From 1984 to 2004, caffeine was banned by the International Olympic Committee, the WADA, and other monitoring organizations. In 1987, a urine caffeine concentration of >12 mcg/ml was set as evidence of doping. Caffeine was removed from the WADA banned list in 2004 but has remained on the in-competition Monitoring Program list. Aguilar-Navarro and others reported slight increases in median levels (mcg/ml) of caffeine in 7488 urine samples of Spanish athletes from 2004 (0.7) to 2008 (0.7) to 2015 (0.9). However, percentages of samples >12 mcg/ml remain well below 1 percent which suggests moderate use, and not abuse, of caffeine.

Effect on Psychomotor Responses

Caffeine may affect a number of psychomotor responses that could enhance performance in some sports. In their meta-analysis, Irwin and others concluded that caffeine doses of <80–600 mg may counteract impaired performance associated with sleep deprivation by improving response time, accuracy on attention tests, executive function, reaction time, response time and accuracy on information processing tasks, and lateral and longitudinal measures of vehicular control on driving tests. Maintaining cognitive function in many endurance activities may enhance performance. In the International Society of Sports Nutrition position statement on caffeine and exercise performance, Guest and others concluded that caffeine can enhance vigilance during bouts of extended exhaustive exercise, as well as periods of sustained sleep deprivation. Spriet commented that low doses of caffeine (<3 mg/kg) ingested before and during exercise may be ergogenic in a variety of continuous and intermittent tasks, primarily due to improvement of vigilance, alertness, and cognitive function.

In their review, Hespel and others noted that optimal caffeine doses associated with improved visual information processing (1–2 mg/kg) are lower than doses which reported to improve endurance performance (4–5 mg/kg). As a result, a higher dose taken by a soccer goalkeeper may impair visual processing and performance. Caffeine doses ≥5.5 mg/kg may also increase anxiety and nervousness in some individuals and impair performance, especially in tasks requiring fine motor control. The effects of caffeine on memory and higher executive functioning are less clear.

Effect on Aerobic Endurance Performance

Most research on the ergogenic effects of caffeine has focused on aerobic endurance performance (**figure 13.5**). The efficacy of caffeine as an ergogenic

FIGURE 13.5 Caffeine may enhance performance in a wide variety of exercise endeavors, particularly events involving aerobic endurance.

aid is supported by several reviews, meta-analyses, and position statements by sports-related associations. As previously noted, Doherty and Smith reported in their meta-analysis that caffeine decreased perceived exertion by 5.6 percent while it improved exercise performance by 11.2 percent. In a meta-analysis restricted to studies including an endurance time-trial, which is more applicable to sport, Ganio and others reported a 3.2 percent improvement in performance following caffeine ingestion compared to the placebo. Similarly, Schubert and Astorino identified an average 1.1 percent improvement in running performance across seven studies. Higgins and others reported that coffee providing 3–8 mg/kg of caffeine improved performance times to exhaustion and completion by 24 and 3 percent, respectively. Glaister and Gissane reported increases in minute ventilation and levels of blood lactate and glucose, and decreases in perceived exertion across 26 studies of submaximal (5–30 minutes at 60–85 percent of VO_2 max) exercise following caffeine supplementation compared to placebo. Finally, Southward and others reported significant effects of caffeine supplementation (3–6 mg/kg) compared to placebo on mean power output and endurance time trial performance across 44 studies.

In the International Society of Sports Nutrition position stand on caffeine and exercise performance, Guest and others concluded that caffeine is ergogenic for various modes of sustained high-intensity endurance exercise and has been shown to be highly effective for time-trial performance, as noted earlier.

The American College of Sports Medicine's position statement on fluid replacement during exercise, authored by Sawka and others, also noted that caffeine intake may help sustain performance. Although there are individual differences in response to caffeine, the available evidence supports its ergogenicity and popularity among endurance athletes.

Effect on High-Intensity Anaerobic Exercise Although the effect of caffeine on endurance performance is well documented, much less research has been conducted on the ergogenic potential of caffeine on anaerobic performance. For purposes of this discussion, high-intensity anaerobic exercise involves maximum exercise for 10–180 seconds, or intermittent high-intensity exercise in sports such as soccer, lacrosse, and field hockey. According to Davis and Green, untrained subjects and flawed designs have possibly contributed to nonsignificant findings in older studies, but more recent studies have featured trained subjects and more appropriate sport-specific research designs. They concluded that caffeine is highly ergogenic for speed-endurance exercise ranging in duration from 60 to 180 seconds. In a meta-analysis of 16 studies investigating the effects of caffeine supplementation (1–6 mg/kg) administered ~60 minutes prior to a Wingate test, a common 30-second all-out cycle test of anaerobic power, Grgic reported significant increases of 3 and 4 percent (standardized mean differences of 0.18 and 0,27) in mean and peak power output, respectively, compared to placebo. In another review, Astorino and Roberson reported significant improvements following caffeine ingestion in 11 of 17 studies of team sports exercise and power-based sports. An ergogenic effect was most commonly observed in elite athletes who do not regularly ingest caffeine. Possible ergogenic mechanisms include competitive adenosine receptor blocker to prevent a decline in motor unit firing rate and sustain the force of muscle contraction; increased arousal; and decreased pain perception.

In the International Society of Sports Nutrition position stand on caffeine and exercise performance, Guest and others commented that energy drinks and preworkout supplements containing caffeine have been reported to improve performance in anaerobic sport-specific basketball, soccer, rugby, field hockey, and combat sports skills. In summary, caffeine appears to be an effective ergogenic for high-intensity, intermittent exercise, particularly within a period of prolonged duration, such as a soccer match.

Effect on Muscular Strength and Endurance The effects of caffeine and resistance performance have been the focus of considerable recent research, with over 50 new studies and several meta-analyses between 2018 and 2021. Astorino and Roberson commented that 6 of 11 studies show that caffeine elicits significant benefits in resistance training. In their meta-analysis, Warren and others concluded that, overall, caffeine ingestion results in a beneficial effect of about 7 percent on maximal voluntary contraction strength but primarily in the knee extensors and not in other muscle groups. Grgic and others have conducted several meta-analyses of various aspects of caffeine and resistance exercise. In a meta-analysis of 10 studies, Grgic and Pickering reported that caffeine increased isokinetic strength, primarily in knee extensor muscles and at greater angular velocities. In a meta-analysis of eight studies limited to women, Grgic and Del Coso reported similar effects in muscular strength and endurance, as previously reported in men. Caffeine (1.4–6 mg/kg) resulted in significant increases in muscle strength (1-RM) and endurance (standardized mean differences of 0.18 and 0.25, respectively) compared to placebo. Caffeine significantly improved upper-body but not lower-body strength and endurance.

In a summary of recent research, Grgic concluded that caffeine may increase isotonic, isometric, and isokinetic strength; improve muscular endurance, velocity, and power; and enhance adaptations to resistance training. The probable mechanism is adenosine A_{2A} receptor blockade. A dose of 2–3 mg/kg appears to produce an ergogenic effect comparable to higher (~6 mg/kg) doses. The International Society of Sports Nutrition position statement on caffeine and exercise performance by Guest and others concurs with this viewpoint, concluding that caffeine may be an effective ergogenic aid to increase velocity and power in resistance exercise.

Effect on Carbohydrate and Fat Oxidation Early caffeine research focused on increased fatty acid mobilization and oxidation and the subsequent "sparing" of glycogen as an ergogenic mechanism. As discussed in chapter 4, carbohydrate is the main energy source for high intensity aerobic endurance exercise. The effect of caffeine combined with carbohydrate on carbohydrate oxidation and endurance performance has been the focus of several studies and reviews. Gant and others reported that a caffeinated carbohydrate-electrolyte solution (CES) improved performance in a 90-minute intermittent shuttle-running trial designed to mimic a soccer match compared to a standard CES, suggesting that a caffeine/carbohydrate combination may be useful in prolonged intermittent, high-intensity sports. In their meta-analysis, Conger and others concluded that supplementation with carbohydrate and caffeine provides a small, but significant effect to improve endurance exercise performance when compared to carbohydrate alone. However, caffeine alone was more ergogenic than caffeine with exogenous carbohydrate. Graham and others pooled data from several muscle biopsy studies which directly measured fat and carbohydrate oxidation and concluded there is no evidence of enhanced fat oxidation or glycogen sparing. As previously discussed, Collado-Mateo and others reported greater evidence of increased fat oxidation (increased VO_2, decreased respiratory exchange ratio) following a caffeine dose of ≥3 mg/kg body mass in untrained subjects compared to recreationally trained individuals and endurance athletes. In summary, recent research using contemporary methodologies demonstrates that although caffeine can increase carbohydrate utilization and fat mobilization, there is little evidence of increased fat oxidation with glycogen sparing.

Effect of Caffeine Status Early research suggested that habitual caffeine use might decrease its ergogenic effects, possibly by decreasing circulating epinephrine levels and circulating free fatty acids during exercise. James and Rogers concluded that ergogenic effects of caffeine are largely explained by terminating the symptoms of caffeine withdrawal. Many studies have included a caffeine withdrawal period of at least 2–4 days to counter this decreased sensitivity. Ganio and others recommended abstaining from caffeine for at least 7 days before use to optimize an ergogenic effect.

However, more recent research casts doubt on decreased ergogenicity with habitual caffeine use. Irwin and others observed that a 3-mg/kg caffeine dose significantly improves exercise performance irrespective of whether a 4-day withdrawal period is imposed on habitual caffeine users. Gonçalves and others recently reported similar improvements in cycling time-trial performance in low (58 ± 29 mg/day), moderate (143 ± 25 mg/day), and high (351 ± 139 mg/day) caffeine users following 6 mg/kg caffeine supplementation. Clark and Richardson also found similar improvements in 5-km cycle time trial performance in low and high caffeine consumers following a 3 mg/kg caffeine dose compared to placebo. de Sallis Painelli and others reported significant improvements in vertical jump and strength endurance performance following acute caffeine supplementation (6 mg/kg) in low (20±11 mg/day), moderate (88±33 mg/day), and high habitual caffeine consumers (281±167 mg/day). In their review, Filip and others noted inconsistent criteria for habitual caffeine use. As will be discussed later, an individual's sensitivity to caffeine may affect its ergogenicity.

Effect of Delivery System Another factor influencing caffeine's effectiveness may be how it is consumed. In addition to caffeine in coffee or in tablets, Guest and others comment that significant improvements in aerobic and/or anaerobic performance have been reported following use of caffeinated chewing gum, mouth rinses, gels/chews/bars, energy drinks, and in combination with other ingredients, e.g., preworkout supplements. Caffeine supplementation has also been studies in aerosol and inspired powder forms. In an early study, Graham and others compared the effects of consuming the same dose of caffeine in coffee or as a capsule in water and found that although the plasma epinephrine increased with the coffee, the increase was significantly greater with the caffeine capsule. Additionally, the caffeine capsule was the only treatment that improved treadmill run time to exhaustion at 85 percent VO_2 max. They suggested that some component or components in coffee may moderate the effect of caffeine. However, Liquori and others reported higher peak salivary caffeine levels and more rapid time to peak absorption in coffee or cola compared to capsule. Other studies have reported significant ergogenic effects when coffee was used to deliver caffeine. For those who enjoy their coffee the morning of an event, McLellan and Bell reported that consuming coffee 30 minutes prior to taking capsulated caffeine did not negate its beneficial ergogenic effect on cycling endurance performance. Thus, drinking coffee would not seem to impair the potential ergogenic effect of caffeine tablets.

Effect of Exercise in the Heat Caffeine has been classified as a diuretic and it stimulates metabolism. Theoretically, increased water losses and an elevated metabolism before competition could impair exercise performance under warm, humid environmental conditions, possibly because of retarded sweat losses and excessive increases in body temperature. However, research refutes any ergolytic effects of caffeine related to impairments in thermoregulation or fluid balance. Del Coso and others reported that caffeine, whether consumed alone or in combination with water or a sports drink, did not alter heat production, forearm skin blood flow, or sweat rate and did not impair heat dissipation when exercising for 120 minutes in a hot environment. Ganio and colleagues found that a 6-mg/kg caffeine dose increased performance under cool or warm conditions compared to a placebo, with no difference in rectal temperature.

These findings are supported by two meta-analyses. Armstrong and others concluded that in contrast to popular beliefs, caffeine consumption does not result in hypohydration, water-electrolyte imbalances, hyperthermia, or impaired exercise-heat tolerance. Zhang and others concluded that although females were more susceptible to the diuretic effects of caffeine compared to males, increases in urine production following caffeine consumption are offset by the antidiuretic effect of exercise. Position statements of the American College of Sports Medicine and International Society of Sport Nutrition conclude that moderate caffeine consumption will not markedly change urine output, induce dieresis during exercise, or harmfully change fluid balance leading to impaired performance.

Athletes will not incur detrimental fluid-electrolyte imbalances if they consume caffeinated beverages in moderation and eat a typical diet. Additionally, Armstrong and others reported that consuming caffeine, either 3 or 6 milligrams per kilogram body weight for 5 days, did not cause hypohydration, and they questioned the widely accepted notion that caffeine consumption acts chronically as a diuretic.

The Placebo Effect As mentioned in chapter 1, subjects in a study may experience a placebo effect if they think they have received an effective ergogenic, such as caffeine. Two studies report evidence of a placebo effect. Beedie and others informed cyclists they would receive placebo, 4.5 mg/kg, and 9 mg/kg caffeine doses in a double-blind counterbalanced manner before a 10-km trial. Instead, all subjects received only the placebo for all conditions. When subjects were asked which treatment they received, those who thought they had received caffeine had improved time-trial performance in a dose-response manner (high dose > moderate dose). In a study of 40-kilometer time trial performance in cyclists, Foad and others examined the effects of four experimental conditions involving the interaction of caffeine (C) and placebo (P) and being informed they had (Y) or had not (N) received caffeine (CY, PY, CN, PN), each administered twice interspersed with 6 baseline trials for a total of 14 trials. They reported both pharmacological ergogenic effects of caffeine, possible beneficial psychological placebo effects, and possible negative placebo (nocebo) effects, i.e., impaired performance when subjects were correctly informed they had received no caffeine. As noted by Guest and others, such studies demonstrate the importance of adequate blinding in placebo-controlled designs.

Effect of Dosage: Permissibility Because caffeine is a stimulant drug, its use as an ergogenic in sports has been regulated by the International Olympic Committee (IOC) and the WADA. The International Society of Sports Nutrition notes that caffeine is effective for enhancing sports performance in trained athletes when consumed in low to moderate dosages, about 3 to 6 mg/kg, and overall does not result in further enhancement in performance when consumed in higher dosages, such as greater than 9 mg/kg. However, the amount of caffeine Olympic and international-class athletes have been permitted to use has varied over the years. The IOC banned the use of caffeine as a drug prior to the 1972 Olympics, removed it from the doping list from 1972 to 1982, banned the use of large amounts (8–10 mg/kg) for the 1984 Olympics (possibly for its ergogenic effects), and under the recommendation of the WADA removed it from the banned list effective January 1, 2004. The WADA felt that the doping list should be adjusted to reflect changing times. The removal of caffeine from the prohibited list may be a reflection of the increased prevalence of caffeinated beverages, such as specialty coffees, fortified colas, energy drinks, and even sports drinks. These drinks, which may be larger and contain more caffeine, may be consumed in quantity by athletes. As noted earlier, caffeine is also found in a wide variety of other food products.

Currently, the WADA has caffeine in its monitoring list, meaning that caffeine levels in international-class athletes are tested and if caffeine abuse increases, it may be returned to the prohibited list. As previously noted, Del Coso and others reported that many athletes use caffeine but only 0.6 percent had urine caffeine levels above the old limit of 12 mcg/ml. Some athletic governing organizations, such as the National Collegiate Athletic Association (NCAA), may test urine samples for caffeine concentration, and a level of 15 mcg/ml (approximately 9 mg/kg) is considered to be evidence of doping. However, such a large dose is unnecessary based on reports from multiple studies that approximately 5 mg/kg is an effective ergogenic dose.

Individuality Although studies have generally not reported a decrease in performance following caffeine ingestion, there are interindividual differences in response to caffeine. Adverse reactions can result in impaired performance in some subjects. According to Southward and others, much of the inter-individual variance in response to caffeine can be attributed to two genes. CYP1A2 codes for cytochrome P450 which is responsible for caffeine metabolism. ADORA2A codes for the adenosine A_{2A} receptor to which caffeine competitively binds as its primary ergogenic mechanism. Pirastu and others reported that coffee consumption has been shown to downregulate the expression of caffeine metabolism genes. Therefore, individual characteristics should be considered when administering caffeine for performance enhancement.

Caffeine appears to be an effective ergogenic aid in doses that are both safe and legal. However, some athletes believe that taking caffeine may be considered unethical because it is an artificial means of enhancing performance. Given its safety and legality, the decision to use caffeine as a performance enhancer rests with the ethical standards of the individual athlete. Although combining ephedrine with caffeine (as discussed in the next section) may increase the ergogenic effect of caffeine alone, ephedrine use may be illegal—so its use to increase sports performance is unethical.

Self-Experimentation If you are considering using caffeine as a potential ergogenic aid, it is wise to experiment with its use in training prior to use in competition. You might start by taking 200–400 milligrams of caffeine not to exceed a dose of 5 mg/kg about an hour prior to some of your workouts. For example, if you are a distance runner, do your long runs periodically with and without the coffee or other caffeine source, and judge for yourself if it works for you. To make it a more valid case study, have someone randomly give you, blinded, a placebo (vitamin capsule) or a low dose caffeine capsule before the runs, but without informing you which until you have done each several times. Try this procedure also after abstaining from caffeine for 4–5 days. Keep a record of your feelings and times after the runs so that you can compare differences.

Does drinking coffee, tea, or other caffeinated beverages provide any health benefits or pose any significant health risks?

The health effects of caffeine have been studied for nearly half a century. Early epidemiological research linked coffee or caffeine consumption with the development of a variety of health problems, including cancer, heart disease, osteoporosis, and birth defects. According to a 2014 Institute of Medicine workshop summary, our understanding of health outcomes associated with consumption of caffeine in dietary supplements, foods and energy drinks remains incomplete. Specific areas of future research include dose–response relationships; caffeine sensitivity in certain populations such as children and adolescents; and interactions between caffeine and other ingredients. In a review of 201 meta-analyses, Poole and others concluded various levels of increased coffee consumption were associated with *lower* all-cause mortality, cardiovascular mortality, cardiovascular disease, and various cancers, neurological, mental, metabolic, and liver conditions, but *higher* risks of low-birth weight, miscarriage, preterm births and fractures in women when compared to no/low coffee consumption.

Current research involving the health effects of caffeine has used a variety of techniques, including epidemiological prospective cohort studies, randomized clinical trials, and animal models. Investigators have looked at a variety of factors, including different sources of caffeine, such as coffee versus tea, regular versus decaffeinated coffee, and even the method of preparing coffee, such as filtered versus boiled. Investigators suggest that some of the potential health benefits of coffee and tea may be attributed to substances other than the caffeine found in these beverages. For example, the effects of green tea (*camellia sinesis*) catechins on various aspects of health have been the focus of numerous recent studies and reviews. According to Musial and others, epigallocatechin gallate (EGCG) and other catechins are capable of neutralizing reactive oxygen and nitrogen species. The anti-inflammatory and anti-cancer properties of EGCG in green tea and chlorogenic acid (CGA) in coffee have generated promising results in cell studies and animal models. However, Hayakawa and others caution that EGCG has been linked to increased esophageal and gynecological cancers while CGA has been linked to increased bladder and lung cancers in some human studies.

The following sections highlight some of the key findings relative to the effect of caffeine on various health conditions. Most research involves the consumption of caffeinated beverages, but some involves caffeine in dietary supplements.

Energy Drinks Of special interest in recent years has been the increasing popularity of caffeinated energy drinks, particularly among youngsters and college-age adults. Temple notes that recommendations for safe caffeine consumption in children and adolescents are based on dated recommendations for adults. According to a 2022 energy drink market report (www.statista.com/topics/1687/energy-drinks/#topicHeader__wrapper), Red Bull® and Monster Energy® had a combined US market share of $5.8 billion (46 percent of total market share) despite concerns by the US Food and Drug Administration over high levels of caffeine in these products. As shown in **table 13.8,** hundreds of energy drinks and shots currently on the market have varying caffeine contents per serving or product. There may be inaccuracies between the listed and actual caffeine content in such products. For example, Attipoe and others reported a 4.3 percent difference between listed (range = 5.75–16.2 mg/fluid ounces) and laboratory analysis (range = 5.1–16.7 mg/fluid ounces) of caffeine content in seven best-selling energy drinks and concluded that this difference does not comply with FDA labeling guidelines.

There is concern about potential adverse effects of energy drinks in various populations. According to de Sanctis and others, youth-targeted marketing of energy drinks that are high in sugar and caffeine can increase the risk of obesity, high blood pressure, dental caries, and symptoms of caffeine toxicity. In an American Academy of Pediatrics committee report, Schneider and others noted that diets of children and adolescents should not include the caffeine and stimulants in energy drinks. If energy drinks are mixed with alcohol, there is an increase in risk-taking behaviors, such as driving while intoxicated. As previously noted, Marczinski and others reported greater tolerance for alcohol mixed with a high-caffeine energy drink compared to alcohol alone, with no "masking" of intoxication by caffeine in the energy drink. The inability to appraise the true level of impairment has been labeled "wide-awake drunkenness." In a meta-analysis of 26 studies, Li and others reported a 19 percent increase in the risk of pregnancy loss for each 150 mg/day increase in caffeine consumption.

Scientists indicate that additional research is needed to ascertain the potential health risks of energy-drink consumption but also suggest that the government should be proactive in protecting the public. For example, current FDA regulations limit the amount of caffeine in cola-type sodas to approximately 48 milligrams per 8-ounce serving. However, the sixteen 8-fluid ounce energy drink products listed in the caffeine database (summary data in **table 13.8**) contains an average of 132 mg of caffeine. On November 17, 2010, the U.S. Food and Drug Administration announced that caffeine is an unsafe food additive to alcoholic beverages. This action prohibits the sale of caffeine/alcohol energy drinks in the United States.

Cardiovascular Disease and Associated Risk Factors

Caffeine has been studied for its effect on various risk factors associated with cardiovascular disease. O'Keefe and others comment that coffee contains hundreds of biologically active compounds in addition to caffeine which can have potent health effects. In epidemiological studies, consumption of 3–4 cups of coffee/day have consistently been associated with lower all-cause and cardiovascular disease mortality, coronary heart disease, congestive heart failure, and stroke. According to Penson and others, high diterpene levels in coffee may adversely affect lipoprotein(a) and other lipids. However, factors such as coffee

TABLE 13.8 Caffeine content in selected energy drinks ($n = 7$) and shots ($n = 5$) with descriptive information on caffeine content for 220 energy drinks and 36 energy shots*

Product name (drink)	Fluid oz	Caffeine (mg)	Caffeine (mg/fl oz)	Product name (shot)	Fluid oz	Caffeine (mg)	Caffeine (mg/fl oz)
Red Bull	8.46	80	9.5	7-Eleven Energy Shot	2	260	130
Monster Energy Drink	16	160	10	5 Hour Energy®	2	200	100
Rockstar Zero Carb Drink	16	240	15	5 Hour Energy Extra Strength®	2	230	119.2
Bang Energy	16	300	18.8	Tweaker Shot	2	275	137.5
Redline Xtreme® Energy Drink	8	316	39.5	Vital 4U® Liquid Energy	0.5	155	310
Xyience® Energy Drink	16	160	10				
Sambazon® Amazon Energy Drink	12	120	6.7				
n	220	220	220	n	36	36	36
Mean	13.1	147.7	11.4	Mean	2	193	115
Median	12	131.5	10	Median	2	193	101
SD	3.6	76.6	5.5	SD	0.8	80	67
Max	25.4	400	39.5	Max	4.3	350	310
Min	8	0	0	Min	0.25	75	3.5

*Source: www.caffeineinformer.com/the-caffeine-database.

bean type, the ratios of different diterpines, and roasting, brewing, or boiling in coffee preparation can affect coffee diterpene content as noted by Moeenfard and Alves. Miller and others observed a significant association between regular tea consumption (≥ 1 cup/day) and reduced incidence of adverse cardiovascular events in 6,500 subjects. Regular coffee intake (≥ 1 cup/day) was not associated with adverse cardiovascular events or coronary artery disease (CAD). In general, caffeine intake was marginally associated with reduced CAD progression.

High blood pressure is another risk factor that may be affected by caffeine. As discussed in chapter 9, blood pressure is the product of blood flow and resistance to blood flow. Caffeine blocks receptors for adenosine, which is a potent vasodilator. In theory, caffeine may increase resistance to blood flow and increase blood pressure, especially in individuals who are caffeine sensitive and under stress. In their review, Kallioinen and others reported that acute caffeine consumption (range = 1.6–6.0 mg/kg; 67–400 mg) increased systolic and diastolic blood pressures by 3–14 and 2–13 mmHg, respectively, in 35 studies with follow-up measurements ranging from 30 to 270 minutes. These small increases could be related to increased premature deaths from heart disease and stroke. In a prospective study of almost 44,000 participants over almost 700,000 person-years, Liu and others reported positive associations between coffee consumption and cardiovascular and all-cause mortality. Astorino and others reported that 6 mg/kg of caffeine significantly increased resting, resistance exercise, and postexercise recovery systolic blood pressures in normotensive and prehypertensive males compared to placebo. An excessive increase in blood pressure during exercise could induce a heart attack or stroke.

However, several meta-analyses have reported a dose–response inverse association between caffeine consumption and hypertension. Xie and others detected small but significant reductions in relative risks of hypertension (0.97, 0.95, 0.92, and 0.90) for daily consumption of 2, 4, 6, and 8 cups of coffee/week, respectively, compared to no coffee consumption across 10 studies of almost 250,000 participants and over 58,000 cases of hypertension. D'Elia and others had similar findings in their meta-analysis. It should be noted that a protective effect against hypertension may be explained by one or more of the myriad compounds in coffee other than caffeine. Polymorphisms of CYP1A2 and ADORA2A, genes that respectively encode cytochrome P450 involved in caffeine metabolism and the adenosine A_{2A} receptor to which caffeine competitively binds, may also affect the hemodynamic effect of caffeine consumption, as noted by Yoshihara and others.

The notion that caffeine can increase the risk of developing an irregular heart rhythm is common, but there is little research support for this association. Kim and others reported a 3 percent decrease in overall heart rhythm irregularity with each cup of coffee consumed by 390,000 participants in a multicenter study. This decrease was not affected by CYP1A2 and ADORA2A polymorphisms known to affect caffeine metabolism. In their review of factors associated with atrial fibrillation (AF), the most common irregularity in heart rhythm treated by clinicians, Sagris and others commented that higher caffeine consumption ($\geq$2 cups of coffee/day ~140 mg caffeine) lowers AF risk compared to lower daily caffeine intake. They noted a 6 percent decrease in AF for each 300 mg/day increase in caffeine.

Some at-risk individuals should be cautious about caffeine use before exercise. van Dijk and others reviewed 14 studies and concluded that recent caffeine consumption may decrease myocardial perfusion in patients with suspected CHD. Kitkungvan and others reviewed the results of over 6,000 myocardial perfusion positron emission studies of coronary artery blood flow and reported that even low serum caffeine levels may decrease blood flow during coronary artery vasodilatory stress studies. They concluded that this effect could contribute to false positive or negative interpretations. Supervised exercise is a central aspect of cardiac rehabilitation programs for those with coronary heart disease (CHD), but the temporal proximity of caffeine consumption to exercise may be an issue. Individuals known to be caffeine sensitive, particularly those with high blood pressure, may be advised to use caution when exercising.

According to Saeed and others, high coffee intake may increase cardiovascular risk factors such as cholesterol and blood pressure and may also increase homocysteine, which is synthesized from the essential amino acid methionine, and has been reported to cause vascular damage. However, in a meta-analysis of 31 studies totaling over 1.6 million individuals with almost 184,000 cases of all-cause and 34,574 cases of CVD mortality, Grosso and others reported decreased all-cause, CVD, CHD, and stroke mortalities associated with coffee consumption of $\leq$4 cups of coffee/day in nonsmokers with no additional benefit from higher consumption. Based on contemporary research, most health professional groups, such as the American Heart Association, recommend that moderate coffee consumption, about one to two cups daily, is safe and not associated with heart disease. Individuals who are hypertensive, who are under stress, or who may have other risk factors for heart disease should consult their physician regarding the use of caffeine.

Type 2 Diabetes Several longitudinal studies and meta-analyses have reported a significant association between caffeinated beverage consumption and a reduced risk of type 2 diabetes. In their meta-analysis of 30 prospective studies including almost 1.2 million participants with 53,000 cases of type 2 diabetes (T2D), Carlström and Larsson reported a six-percent decrease in T2D with each additional cup of caffeinated and decaffeinated coffee. Although the protective association was slightly higher in caffeinated coffee, the same relationship was observed with decaffeinated coffee, which suggests that ingredients other than caffeine such as polyphenols, magnesium, chromium, lignans, and chlorogenic acid may be involved in this relationship. However, Shi and others reported reduced insulin sensitivity following acute caffeine ingestion in seven small randomized clinical trials of nondiabetic subjects. Researchers recommend randomized clinical trials as a means to determine active ingredients, which may reveal a combination effect of multiple substances.

Cancer In a 2017 umbrella analysis of meta-analyses, Poole and others summarized reduced risk of incident cancer (relative risk=0.82) in high versus low coffee consumers. Specifically, high coffee consumption was associated with decreased leukemia, melanoma, prostate, endometrial, oral, colorectal, and liver cancers. There was an increase in risk of lung cancer with high compared to low coffee intake (relative risk=1.59) that diminished after statistical adjustment for smoking status. There was no effect of high coffee consumption in lung cancer among never smokers. One meta-analysis reported a lower risk of lung cancer in high consumers of decaffeinated coffee. In another umbrella review of habitual tea drinkers, Kim and others reported convincing evidence of decreased risk of oral cancer and suggestive evidence of decreased risk of biliary tract, breast, endometrial, and liver cancers. Wang and others examined data from 45 studies of over 3.3 million participants and concluded that although coffee or tea consumption was not associated with a decrease in overall breast cancer risk in postmenopausal women, 2–5 cups of coffee/day or $\geq$5 cups of tea/day may be associated with a reduced risk of estrogen receptor breast cancer. Jiang and others observed a significant decrease in breast cancer with caffeine consumption and an even stronger risk reduction in women with the *BRCA1* gene, which increases breast cancer risk. More research is needed to better understand the cancer prevention mechanisms of caffeine and polyphenolic compounds in coffee and tea.

Cognitive Functions The extensive use of caffeine as a social drug is most likely attributable to its stimulating effect on cognitive functions. Well-documented effects of caffeine include enhanced mental energy; elevated mood; and increased alertness, attention, and cognitive function—effects that are more evident in longer or more difficult tasks or in situations of low arousal. Mancini and others reviewed 21 studies and concluded that caffeine and L-theanine in green tea can synergistically decrease anxiety and improve memory and attention while each ingredient is somewhat less effective in isolation.

In their review, Socala and others discuss compounds found in coffee such as caffeine; polyphenols such as chlorogenic acid and caffeic acid; diterpines such as cafestol and kahwehol; and trigonelline, (a niacin (vitamin B_3) derivative), which may offer potential neuroprotection against Alzheimer's disease (AD), Parkinson's disease (PD), ischemic stroke, and other cognitive dysfunctions. Panza and others commented that preventing or postponing symptoms of late-life cognitive decline are paramount given the limited therapeutic value of pharmacological agents to treat AD and PD. Overall, research findings suggest habitual caffeine intake may be associated with less cognitive decline with aging.

Ren and Chen comment that high caffeine consumption may prevent degeneration of dopaminergic neurons known to cause PD and also improve motor and cognitive function in PD patients. Caffeine appears to exert this neuroprotective effect though its role as a competitive binder to the adenosine A_{2A} receptor. In 2019, the US Food and Drug Administration approved the use of istradefylline (Nourianz®), a selective adenosine A_{2A} receptor antagonist, for the treatment of PD. In their review of modifiable factor associated with multiple sclerosis (MS), Jakimovski and others noted that competitive binding of caffeine to adenosine A_{2A} receptors on macrophage and mononuclear cell membranes reduces production of the pro-inflammatory cytokines tissue necrosis factor α (TNF-α) and interleukin 6 (IL-6). High coffee consumption is associated with a lower MS risk and a lower disability score on a validated clinical scale in MS patients.

Asthma and Pulmonary Function Research suggests caffeine may improve ventilation during rest and exercise and help manage asthma. Forte and others reviewed 21 studies of the effects of diet on asthma and reported that caffeine improves forced expiratory volume in one sec ($FEV_{1.0}$), the volume (liters) expelled in the first second of a maximal forced and sustained expiration (forced vital capacity, FVC, liters). Increased $FEV_{1.0}$ suggests decreased airway obstruction. Using National Health and Nutrition Examination Survey data, Han and others reported associations of higher concentrations of the caffeine metabolites theophylline and paraxanthine with higher percent of age, gender, and height predicted values for $FEV_{1.0}$ and FVC. In their meta-analysis of eight studies, Glaister and Gissane reported higher expired minute ventilation during submaximal exercise (5–30 minutes, 60–85 percent VO_{2max}) following caffeine ingestion compared to placebo.

Osteoporosis Factors underlying the development of osteoporosis are discussed in detail in chapter 8. Calcium loss may lead to osteoporosis. Caffeine tends to accelerate the loss of calcium from bones and lead to its excretion in the urine. In a 21-year prospective study of over 61,000 Swedish women, Hallström and others observed that high coffee consumption (≥4 cups/day) was associated with a 2–4 percent decrease in bone density, but no increase in risk of fracture. In their umbrella analysis of meta-analyses, Poole and others reported that high versus low coffee consumption was associated with increased fracture risk in women and decreased fracture risk in men. According to the National Institutes of Health Institutes of Medicine, the adverse effect of caffeine on calcium stores can be mitigated by a diet high in calcium (RDA=700–1,300 mg depending on age and gender) and Vitamin D (600 to 800 IU depending on age) as described in the *2020-2025 Dietary Guidelines for Americans*. Individuals with osteoporosis should consult with their physician about the use of calcium supplements.

Pregnancy-Related Health Problems Meta analyses have consistently reported associations between high caffeine consumption and adverse pregnancy outcomes. Chen and others reported increased risk of miscarriage while Rhee and others reported increased low birth weight. In a recent umbrella analysis, Poole and others reported high versus low coffee consumption to be associated with low birth weight (Odds Ratio=1.31), loss of pregnancy (Odds Ratio=1.46), and preterm birth in the first (Odds Ratio=1.22) and second (Odds Ratio=1.12) trimesters with no adverse effects on the third trimester. In their meta-analysis of 60 studies, Greenwood and others found small but statistically significant associations between increased caffeine intake and spontaneous abortion, stillbirth, low birth weight, and small-for-gestational-age births. However, they concluded that associations between caffeine intake in the normal range and adverse birth outcomes are modest and recommend adherence to current intake limits but no further reduction in caffeine intake.

The US Food and Drug Administration and the Academy of Nutrition and Dietetics (AND) recommend that pregnant women consider abstaining from caffeine use or, if they do drink caffeinated beverages, to do so in moderation. The American College of Obstetrics and Gynecology recommends a caffeine intake of no more than 200 mg/day during pregnancy. The *2020-2025 Dietary Guidelines for Americans* recommend no more than low to moderate amounts of caffeine (≤300 mg/day; 2–3 cups of coffee/day) during pregnancy and lactation. Caffeine consumed by the nursing mother gets into breast milk. Santos and others reported no association between maternal caffeine consumption and infant nighttime wakefulness.

Weight Control Caffeine use may stimulate metabolism, increasing the resting metabolic rate about 10 percent for several hours, an effect that theoretically could facilitate weight loss. In their review, Hursel and Westerterp-Plantenga concluded that catechin- and caffeine-rich teas have been shown to reduce body mass, improve body composition, increase exercise tolerance, and decrease risk factors associated with the metabolic syndrome. However, they concede that much is unknown about thermogenic effects, absorption, bioavailability and mechanisms of action. Lopez-Garcia and others found that increased caffeine consumption was associated with decreased weight gain over the course of 12 years, but the differences were less than 1 pound. Regular consumption of coffee or caffeine would appear to make a very minor contribution to weight control, as contrasted to proper diet and exercise. In addition, Janssens and others reported no effect of 12 weeks of green tea extract supplementation compared to placebo on resting energy expenditure, respiratory quotient, fecal fat content, fecal energy content, or body composition in Caucasian men and women. Gurley and others caution that some weight-loss products may have different effects than traditional caffeine sources because of potential adverse interactions between caffeine and botanical extracts. Excessive amounts may cause adverse effects in some individuals using it for weight loss, especially when combined with ephedrine, as discussed in the following text. Proper weight-control procedures are discussed in chapters 10 and 11.

Sleeplessness Caffeine use, particularly before retiring for the night, may delay the onset of sleep because of its stimulant effects. Inadequate sleep could be detrimental to cognitive functions the following day and, as noted in chapter 10, could be a contributing factor to weight gain. Clark and Landolt noted that caffeine can decrease total sleep time, quality, and electroencephalographic slow-wave activity while increasing wakefulness. These effects may be more pronounced in older individuals and also related to genetic differences in adenosine metabolism. However, O'Callaghan and others comment that caffeine abstinence may improve the quality of their sleep but also create withdrawal leading to decreased cognitive function and performance. The benefits of caffeine consumption may be explained by reversal of caffeine withdrawal as well as previously discussed ergogenic mechanisms gain.

In contrast, preventing sleepiness may be beneficial in some situations. For example, decreased drowsiness and increased alertness may contribute to safer automobile operation under certain conditions. Mets and others reported that 80 mg of caffeine improved performance and reduced sleepiness during a monotonous driving task. These effects may help prevent vehicle accidents related to sleepiness. In a review of caffeine use by members of the military, Chaudhary and others noted that caffeine use during sleep deprivation improved cognitive and behavioral outcomes and physical performance.

Gastric Distress Some individuals experience stomach irritation due to increased secretion of gastric acids following ingestion of caffeinated beverages. In such cases, individuals should consult their physician or avoid caffeine. In a large cross-sectional study of over 8,000 Japanese subjects, Shimamoto and others found no association between caffeine consumption and several gastric diseases.

Caffeine Naiveté Abstainers or those who consume little caffeine may experience nervousness, irritability, headaches, or insomnia with moderate doses. In a case study of a caffeine-naïve military service member who was also under physiological stress, Lystrup and Leggit documented adverse psychological and neuromuscular effects following a moderate dose (200 milligrams) of caffeine. Long-term consumption of coffee generally leads to the development of tolerance and reduction of these symptoms.

Caffeine Dependence Although not classified as an addictive drug, some individuals may develop caffeine dependence, often referred to as *caffeinism*. Various health organizations differ on their classification of caffeine dependence. The World Health Organization recognizes caffeine dependence in its *Classification of Mental and Behavioral Disorders*. The American Psychiatric Association's *Diagnostic and Statistical Manual-V* lists caffeine withdrawal as a diagnosis under caffeine-related disorders. However, caffeine dependence is not considered a serious form of drug abuse.

Mortality In their meta-analysis of 40 studies including over 3.8 million individuals and over 450,000 deaths, Kim and others reported the lowest relative risks of all-cause, cardiovascular, and cancer mortality associated with 3.5, 2.5, and 2 cups of coffee/day, respectively. The lower all-cause mortality was maintained after controlling for age, body weight, alcohol consumption, and smoking status as well as the caffeine content of coffee, which may suggest that compounds in coffee other than caffeine may contribute to this effect. However, caffeine can be profoundly toxic, resulting in arrhythmia, tachycardia, vomiting, convulsions, coma, and death. Although rare, death may result from caffeine abuse, usually from overdoses of caffeine-containing diet or stimulant pills. On January 5, 2021, a personal trainer from Great Britain died following accientally mixing and consuming a solution containing 5 grams of caffeine powder (www.bbc.com/news/uk-wales-60570470). According to Musgrave and others, poor caffeine metabolizers, individuals with liver disease, and those with cardiac conditions may also be more susceptible to caffeine mortality. Over-the-counter dietary supplements may contain substantial amounts of caffeine along with other drugs such as ephedrine, which will be discussed in the next section. If taken in excess, such combinations may be fatal. The US Food and Drug Administration has issued public health warnings about the dangers of pure and highly concentrated powdered or liquid caffeine which have been linked to at least two deaths. In April 2018, the FDA banned the sale of these highly concentrated forms of caffeine in bulk quantities directly to consumers.

Summary In general, most professional health organizations note that caffeine is regarded as a safe drug. If you are healthy and are not on medications, several cups of coffee or caffeinated beverages should pose no health problems. A moderate dosage is the equivalent of about 200–300 milligrams of caffeine per day, or the amount in about two 6- to 8-ounce cups of coffee. One or more supersized 20-ounce cups from a convenience store or coffee house would easily exceed this moderate dose. Women who are pregnant may want to consider abstention, similar to the recommendations for alcohol intake during pregnancy. Moreover, keep in mind that individuals, based on their genetic variations, may respond differently to caffeine intake. Some may be more prone to its possible adverse effects as a potent stimulant.

Key Concepts

- Caffeine is a stimulant drug and can affect a variety of metabolic and psychological processes in the body that may affect exercise performance and health.
- Research suggests that caffeine may improve performance in a variety of athletic endeavors, particularly prolonged aerobic endurance exercise. An effective dose is approximately 5 milligrams per kilogram body weight.
- In general, caffeine is regarded as a safe drug, but physicians may recommend abstinence or use in moderation for some individuals. Various health professionals define moderation as the daily caffeine equivalent of one to two cups of coffee.

Check for Yourself

- Use an automatic blood pressure monitor or have a friend record your blood pressure. While resting, record your blood pressure several times over a course of 15–20 minutes. Drink a cup of coffee or two and then record your blood pressure again, about every 15 minutes over the course of an hour, again while resting. Plot the results. Does caffeine affect your blood pressure?

Ephedra (Ephedrine): Ergogenic Effects and Health Implications

What is ephedra (ephedrine)?

Ephedra sinica, commonly referred to as **ephedra** or the Chinese herb **ma huang**, is a plant containing a variety of naturally occurring alkaloids. **Ephedrine** is considered the most active alkaloid. Its synthetic version is ephedrine hydrochloride. In 2004, the US Food and Drug Administration prohibited the sale of ephedra or ephedrine-containing dietary and weight loss supplements, mainly because such products may pose some serious health threats, as noted later. Products such as Ripped Power™ and Lipodrene are still marketed online and advertised as containing "active ephedra extract" instead of ephedrine. Pure ephedrine is regulated as a drug, and the FDA allows only very small amounts in over-the-counter drugs such as cold medications. Ephedra-free products are also marketed to physically active individuals, but they contain other stimulants, such as caffeine, discussed previously, and pseudoephedrine and synephrine, discussed later.

Does ephedrine enhance exercise performance or promote weight loss?

Although ephedrine is a powerful stimulant which could be an ergogenic mechanism, early studies reported conflicting results. Jacobs and others reported similar increases in muscular endurance during the first set of traditional resistance-training exercise following ingestion of ephedrine alone compared with caffeine plus ephedrine, suggesting no additive effect of caffeine. Williams and others found no differences in muscle strength or Wingate test between placebo, caffeine, and caffeine plus ephedra treatments.

An early review by Shekelle and others concluded that ephedrine has not been shown to consistently enhance exercise performance. In a 2021 meta-analysis of 10 studies, Yoo and others reported greater weight loss (−1.97 kilograms), increased HDL-C (2.74 mg%), decreased LDL-C (−5.98 mg%), and decreased triacylglycerol (−11.25 mg%) in ephedrine compared to placebo groups. HR was also increased by almost 6 beats/min, but systolic and diastolic blood pressures were no different. In their review, Osuna-Prieto and others discuss the possible role of ephedrine and other dietary stimulants to bind with β_3-adrenergic receptors leading to activation of brown adipose tissue which was discussed in chapter 10. If confirmed, this mechanism could lead to pharmacological interventions against obesity.

Ephedrine with caffeine In their review, Magkos and Kavouras indicated that caffeine/ephedrine combinations have been reported in several instances to confer a greater ergogenic benefit than either drug by itself. Several studies by Bell and associates have shown that caffeine/ephedrine combinations may enhance exercise performance in various exercise performance tasks, many of a military nature. Using pharmaceutical-grade caffeine and ephedrine doses approximating 4–5 mg/kg and 0.8–1.0 mg/kg, respectively, they reported significant improvements in exercise tasks such as a 30-second Wingate test of anaerobic capacity, a maximal cycle ergometer performance about 12.5 minutes in duration, and a 3.2-kilometer run wearing combat gear weighing about 11 kilograms. However, they conducted another study of 10-kilometer run performance while while subjects wore 11 kilograms of combat gear and concluded that the ergogenic effect was due to ephedrine with no additive effect of caffeine. A small number of studies from over 20 years ago support an ergogenic effect of caffeine/ephedrine supplementation involving exercise tasks of a military nature that may be applicable to enhancement of sports performance.

Pseudoephedrine Herbal pseudoephedrine may be found in some dietary supplements but is most commonly found in over-the-counter cold medications. A tablet of original prescription strength Zyrtec-D® contains 120 mg of pseudoephedrine HCl.

There is limited research on the ergogenic effects of pseudoephedrine. In two separate studies, Pritchard-Peschek and others evaluated the ergogenic effect of pseudoephedrine supplementation on aerobic endurance. In a well-designed crossover study with well-trained athletes, they reported that the ingestion of 180 milligrams of pseudoephedrine 60 minutes before a cycling time trial improved performance by 5.1 percent. According to the authors, possible ergogenic mechanisms were changes in metabolism or an increase in central nervous system stimulation. However, in a later study they reported no difference between placebo, and pseudoephedrine doses of 2.3 or 2.8 mg/kg (173 or 210 milligrams) on 30-minute cycling performance by trained cyclists. In a meta-analysis of 16 studies, Gheorghiev and others reported larger effect sizes for time-trial performance in more aerobically fit and younger subjects when pseudoephedrine was administered in larger doses (>170 milligrams) less than 90 minutes before a short-duration task. However, they conclude the efficacy of pseudoephedrine was less that a legal stimulant such as caffeine. A higher dose to achieve a possible ergogenic effect may result in a positive test based on WADA prohibited urinary limit of >150 mcg/ml. Additional research is recommended.

Do dietary supplements containing ephedra pose any health risks?

Of all dietary supplements the herbal supplement ephedra may be the most hazardous. Hydroxycut products containing ephedra, which were marketed as weight-loss, fat-burning, and body-building agents, were withdrawn from the market in 2004 due to cardiovascular risks and in 2009 because of liver toxicity. These products are still marketed with substitute ingredients which are also associated with liver damage.

Use of ephedra has been associated with other health problems. Maglione and others reported adverse psychiatric effects of ephedra use, including psychosis, severe depression, mania or agitation, hallucinations, sleep disturbance, and suicidal ideation. Andraws and others comment that ephedra alkaloids are associated with increased cardiovascular risks but have little health benefit. Even ephedra-free products may increase cardiovascular risk. Haller and others implicated ephedra in risk of seizures and noted that ephedra alone may be dangerous, but ephedra combined with caffeine exaggerates the potential adverse risks. Cohen and others reported the presence of the amphetamine isomer β-methylphenylethylamine (BMPEA) in dietary supplements containing *Acacia rigidula,* a shrub found in Texas and Mexico. The researchers called for enforcement action by the U.S. Food and Drug Administration because the safety of BMPEA has not been studied in humans.

In the early 2000s, the deaths of several prominent collegiate and professional athletes made headlines when it was discovered they were using ephedra-containing supplements during training under warm environmental conditions. The risk-taking behavior associated with sports participants is well known, so athletes taking more than the recommended dose is one of the major problems. Additionally, the purity and amount of ephedra in a product are not well controlled, particularly in products marketed on the internet. Given these possibilities and its physiological effects, ephedrine could be involved in these tragic deaths.

Synephrine Ephedra-free dietary supplements have been marketed for weight loss (**figure 13.6**). These products may contain synephrine, along with other stimulants such as caffeine. Synephrine is an extract from the Seville orange, or bitter, or sour orange. Neo-synephrine is also known as phenylephrine. Synephrine is a dietary supplement in the United States but classified as a drug in Europe.

Synephrine has a chemical structure that is similar to ephedrine and has been marketed as a safe weight-loss alternative to ephedra. The purported mechanism of action is $β_3$-adrenergic receptor binding to increase thermogenesis and decrease appetite.

There are conflicting reports of the efficacy of synephrine to promote weight loss and the safety of its use. In separate reviews, Astell and Rossato and their respective colleagues concluded that synephrine is not an effective weight-loss agent. However, Stohs and others concluded that synephrine, alone or combined with caffeine or other herbals, produced modest weight loss and did not result in adverse reactions. Gurley and others commented that synephrine alone in normal doses is innocuous, but the risk of cardiovascular stress may be increased when combined with other compounds such as caffeine. Bui and others reported increases in systolic and diastolic blood pressure for 5 hours after ingesting a 900-milligram dietary supplement containing 6 percent synephrine. This could increase risks for individuals with hypertension. In separate case studies, the use of synephrine in products has been associated with ischemic stroke by Bouchard and others and with a myocardial infarction in a previously healthy 24-year-old male as reported by Thomas and others. More research is required to fully elucidate the long-term risks and efficacy of synephrine. For individuals interested in weight loss, safer approaches are discussed in chapter 11.

Permissibility in Sports The use of ephedrine, ephedra, and ma huang in competition is prohibited by the WADA and the IOC. Pseudoephedrine, which was previously on the WADA monitoring program, is now also prohibited in competition as of 2015. Urine limits for ephedrine and pseudoephedrine are 10 and 150 mcg/ml, respectively. Synephrine was added to the 2015 monitoring program list. Since ephedra, ephedrine, and pseudoephedrine are banned in competition only, athletes may still use these agents in training. Magkos and Kavouras suggested that caffeine/ephedra mixtures may become one of the most popular ergogenic aids that athletes use in training. The use of pseudoephedrine in training may also increase. As previously mentioned, Gurley and others caution that caffeine/ephedra alkaloid combinations may have unpredictable side effects.

> ### Key Concepts
> - Ephedra, or ma huang, although classified as a dietary supplement, contains a potent stimulant drug, ephedrine.
> - In general, research suggests that ephedra or ephedrine supplementation does not enhance exercise or sports performance. However, supplementation with caffeine/ephedrine compounds has been shown to enhance performance in various exercise tasks.
> - Use of ephedra or ephedra-containing supplements has been associated with serious health problems, including psychiatric disorders, increased cardiovascular risk factors, and heat stroke in athletes that could be fatal. Ephedra, ephedrine, ma huang, and pseudoephedrine are banned for in-competition use by the WADA but may still be used by athletes in training.
> - Ingestion of stimulants containing caffeine and ephedrine, ephedra, or synephrine are permitted by the WADA for use in training only but may cause an excessive increase in blood pressure or other undesirable side effects.
>
> ### Check for Yourself
> - Search online and/or visit a local health food store or search online that primarily sells dietary supplements, including sports supplements. Ask the clerk to show you products containing ephedra-related supplements for weight loss and enhanced sports performance and if there are any health risks related to their use. Record the response for class discussion.

Jill Braaten/McGraw Hill

FIGURE 13.6 Ephedra-free dietary supplements are marketed for weight loss. Many contain synephrine, a compound similar to ephedrine (see text for discussion).

Sodium Bicarbonate: Ergogenic Effects, Safety, and Legality

What is sodium bicarbonate?

Sodium bicarbonate is an alkaline salt found naturally in the human body. It is the major component of the alkaline reserve in the blood, whose major function is to help control excess acidity by buffering acids. Thus, sodium bicarbonate is also known as a buffer salt. Its action is comparable to that of medications you may take to control an upset stomach caused by gastric acidity. Sodium bicarbonate may be purchased in a supermarket as baking soda (**figure 13.7**), and it has been marketed to athletes as part of a sports supplement.

Does sodium bicarbonate, or soda loading, enhance physical performance?

During high-intensity anaerobic exercise, sodium bicarbonate helps buffer the lactic acid that is produced when the lactic acid energy system is utilized. As discussed in chapter 3, the accumulation of excess hydrogen ions from lactic acid in the muscle cell may interfere with the optimal functioning of various enzymes and thus lead to fatigue. The normal concentration of sodium bicarbonate in the blood can help delay the onset of fatigue during anaerobic exercise. It may facilitate the removal of the hydrogen ions associated with lactic acid from the muscle cell, thereby mitigating the adverse effects of the increased acidity (**figure 13.8**). However, fatigue is inevitable if the rate of lactic acid production exceeds the capacity of sodium bicarbonate to buffer it. Theoretically, an increase in the alkaline reserve could delay the onset of fatigue. Over 90 years ago, Denning and others were among the first to recognize and report on the potential of sodium bicarbonate and other alkaline salts to improve anaerobic work capacity. Using magnetic resonance spectroscopy, Raymer and others provided support for the main ergogenic theory that sodium bicarbonate supplementation could help maintain cellular homeostasis during incremental exercise by delaying the increase in intracellular acidity and the onset of fatigue. In addition, Peart and others reported that sodium bicarbonate ingestion exerts a protective intracellular response to high-intensity exercise oxidative stress as reflected by a decrease in monocyte and lymphocyte heat shock protein 72 formation.

Alkaline salt supplementation has been studied for its ergogenic potential on all three human energy systems, but mainly the lactic acid energy system. Most studies have used a double-blind placebo-controlled crossover design in which all subjects took all treatments. Sodium bicarbonate supplementation is also known as *soda loading*, from baking soda, or *buffer boosting*, for increasing the natural blood buffer content.

Supplementation Protocol and Exercise Tasks A substantial number of well-controlled experiments and reviews (e.g., Siegler and others) have provided supportive data, not only for the

Holly Hildreth/McGraw Hill

FIGURE 13.7 Baking soda is a commercial version of sodium bicarbonate.

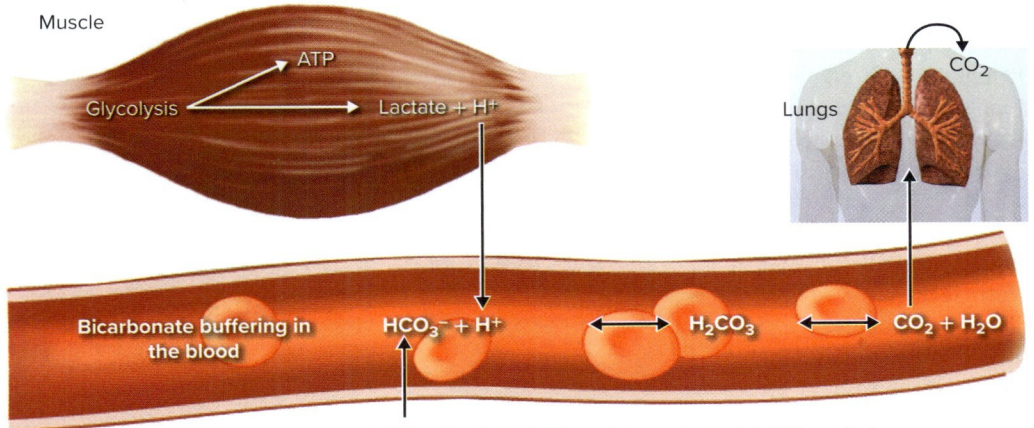

Purestock/SuperStock

FIGURE 13.8 Alkaline salts, such as sodium bicarbonate, are theorized to reduce the acidity in the muscle cell by facilitating the efflux of hydrogen ions from the cell interior, promoting a more homeostatic environment for continued muscle contraction.

underlying ergogenic mechanism, but also for the performance-enhancing effects. A relatively standard supplementation protocol has been used, but variations also have been studied. The International Society of Sports Nutrition position stand by Grgic and others concluded that 300 mg/kg body mass of sodium bicarbonate (range=200–500 mg/kg) administered 1–3 hours prior to exercise can improve performance in a variety of high-intensity single-bout and multiple-bout tasks ranging in duration from 30 seconds to 12 minutes. Higher doses do not increase the ergogenic effect and may increase side effects such as bloating, nausea, vomiting, and gastrointestinal/abdominal pain. The exercise task selected was normally one that stressed the lactic acid energy system, or about 1–3 minutes of maximal exercise. The usual experimental protocol has been to have subjects ingest the dose about 1–3 hours before the test. However, other protocols have supplemented over the course of 5 days and up to 8 weeks. A typical dosage was 150–300 mg sodium bicarbonate/kg body weight. Some studies have used sodium citrate in similar dosages, because it has also been shown to increase the alkaline reserve. More recently, **beta-alanine** (β-alanine) supplementation (~3.0–6.4 g/day for 4 weeks or longer) has also been studied due to its role in synthesis of the intracellular buffer carnosine, as noted by Hoffman and others. β-alanine was discussed in detail in chapter 6. The sections below summarize research findings on the efficacy of sodium bicarbonate and β-alanine on performance requiring major contributions from the ATP-PCr, lactic acid, and oxygen energy systems.

Effects of sodium bicarbonate and β-alanine on the ATP-PCr energy system

Most studies have reported no beneficial effects on performance in exercise bouts lasting less than 30 seconds or in resistive exercise tasks stressing strength, power, or short-term local muscle endurance. Such high-intensity tasks are of insufficient duration to significantly decrease intracellular or extracellular pH and challenge the normal alkaline reserve. McNaughton, an early researcher in this area, and Cedaro reported no ergogenic effect on maximal cycle ergometer performance in either 10-second or 30-second trials. However, Price and others reported that ingestion of sodium bicarbonate improved performance in multiple, intermittent 14-second maximal sprints (one sprint every 3 minutes) during a 30-minute cycle ergometer trial. Bishop and Claudius reported that sodium bicarbonate ingestion improved repetitive 4-second all-out cycle ergometer sprints with 100 seconds of active recovery at 35% VO_{2peak} and 20 seconds of rest over two 36-minute "halves." Although the investigators suggested that sodium bicarbonate may be a useful supplement for team-sport athletes, performance may have been somewhat dependent on the lactic acid energy system.

In their meta-analysis of studies including β-alanine and combined β-alanine/sodium bicarbonate supplementation, Saunders and others reported greater effect sizes for exercise duration between 30 seconds and 10 minutes compared to <0.5 minutes. Similarly, de Oliveira and others reported less effect of sodium bicarbonate, sodium citrate, calcium/sodium lactate, and combined buffering agents on tasks <0.5 minutes compared to longer duration tasks in their meta-analysis of 189 studies. Based on the available scientific data, β-alanine and alkaline salt supplementation do not appear to be effective ergogenic aids for exercise tasks dependent primarily upon the ATP-PCr energy system.

Effects of sodium bicarbonate and β-alanine on the lactic acid energy system

Various laboratory and field studies have reported significant improvements in tasks that primarily use the lactic acid energy system and in sports with multiple high-intensity bouts separated by brief periods of recovery. There is research support for the ability of sodium bicarbonate supplementation to increase serum pH and the concentration of bicarbonate, reduce acidosis in the muscle cell, decrease the psychological sensation of fatigue at a standardized level of exercise, and increase performance in high-intensity anaerobic exercise tasks dependent on the lactic acid energy system (**figure 13.9**).

Compared to placebo, sodium bicarbonate supplementation has been reported to be effective in single and repetitive spring running (Van Montfoort and others), judo (Artioli and others), boxing (Siegler and Hirscher), weightlifting (Carr and others), and cycling (Gough and others) tasks ≥30 seconds in duration. In contrast, other researchers have reported no effect of sodium bicarbonate on 45-second high-intensity cycling (van Someren and others) and sprint swim performance (Tan and others). Although Brisola and others reported an 8 maximal accumulated oxygen deficit (MAOD) during two supramaximal (110 percent of VO_{2max}) running bouts following sodium bicarbonate supplementation compared to placebo, there was no difference in time to exhaustion or correlation between MAOD and either 200-meter or 400-meter running time.

moodboard/SuperStock

FIGURE 13.9 Sodium bicarbonate may enhance sports performance in a variety of events dependent primarily upon the lactic acid energy system (anaerobic glycolysis), such as the 400-meter sprint in track.

Several meta-analyses support an ergogenic effect of sodium bicarbonate in tasks involving the lactic acid system. In their meta-analysis of 40 studies generating 58 effect sizes, 27 of which were in tasks ≤2 minutes, Peart and others concluded that bicarbonate loading (200–400 mg/kg) was more effective in untrained compared to trained subjects and for single-bout compared to multiple-bout activities. Greater effects were observed in performance tasks measuring total work and time to exhaustion compared to performance time or power. More recently, Calvo and others reported significant effects of sodium bicarbonate on pH (greater alkalinity), bicarbonate (greater concentration and base excess), lactate (higher concentrations), and partial pressure of carbon dioxide (greater) compared to placebo. They concluded that sodium bicarbonate supplementation significantly affects anaerobic metabolism (lactic acid system). In a 2022 meta-analysis, de Oliveira and others compared the extracellular buffers sodium bicarbonate, sodium citrate, calcium or sodium lactate, and combined buffering agents. Sodium bicarbonate was superior to sodium citrate. They reported moderate evidence of sodium bicarbonate efficacy in exercise durations of ≥0.5–<1.5, ≥1.5–<5, ≥5–<10, and ≥10 minutes. Other modulators of sodium bicarbonate efficacy included the increase in blood bicarbonate (large [≥ 4 mmol·l^{-1}] > small [<4 mmol·l^{-1}]); exercise type (capacity > performance); trained status (untrained, trained > elite); repetitive exercise bouts (3 > 2 > 1); and prior exercise (yes > no). Finally, Huerta Ojeda and others reviewed 19 studies of β-alanine supplementation on tasks requiring 60–100 percent of VO_{2max} (tasks inducing changes in ventilatory patterns). Small effect sizes were reported for time trial performance and limited-time test performance, with larger effect sizes for limited distance test performance tasks compared to placebo. They concluded that β-alanine supplementation exerted small effects on physical performance in aerobic–anaerobic transition zones. In summary, the consensus of current research indicated that sodium bicarbonate and β-alanine are effective extracellular and intracellular buffers that can enhance sustained anaerobic performance requiring the lactic acid energy system.

Effects of sodium bicarbonate and β-alanine on the oxygen energy system

Plasma volume expansion is a well-established early adaptation to endurance training. Sodium bicarbonate ingestion could theoretically result in water retention from the increased sodium as a mechanism to improve endurance performance and thermoregulation, as was discussed in chapter 9. Sodium bicarbonate supplementation has been reported to improve various cycling (Egger and others; Linossier and others; McNaughton and Cedaro; McNaughton and others; Mueller and others; Potteiger and others) and running tasks (Bird and others; Shave and others; Ööpik and others) of 4–>60 minutes requiring varying contributions of the lactic acid and oxygen energy systems to ATP synthesis. Burke noted the ergogenic potential for alkaline salt supplementation for the strenuous exercise of up to 7 minutes or even longer tasks, including intermittent high-intensity interval-type work, which is part of endurance training. Of these studies, Mueller and others reported plasma volume expansion on the first of 5 days of sodium bicarbonate supplementation.

Conversely, other studies do not support an ergogenic effect of alkaline salt supplementation on running (Potteiger and others; Vaher and others) or cycling (Schabort and others; Stephens and others) tasks performed at or above the lactate threshold. Vaher and others reported that sodium citrate ingestion (500 mg/kg) increased water retention and plasma volume but did not decrease the core temperature or improve 5-kilometer run time in a warm environment. Hollidge-Horvat and others indicated that sodium bicarbonate ingestion has been shown to increase muscle glycogenolysis during brief submaximal exercise, but Stephens and others reported no difference between sodium bicarbonate and calcium carbonate in muscle glycogen utilization or muscle lactate content during exercise. Glycogen depletion exerts an ergolytic effect on prolonged, exhaustive endurance performance. This is a finding that requires additional research.

More recent studies report equivocal findings. In a crossover study of trained male runners completing 30 minutes of treadmill running at 95 percent of lactate threshold followed by time to volitional fatigue at 110 percent of lactate threshold, Fries and others reported increased running velocity and blood lactate concentration but no difference in total time to exhaustion following sodium bicarbonate (300 mg/kg) compared to placebo. However, Durkalec-Michalski and others reported improved CrossFit workout performance, increased incremental cycle ergometer time to ventilatory threshold, and workrate at ventilatory threshold, and VCO_2 production following progressive sodium bicarbonate supplementation (37.5 mg/kg day 1 to 150 mg/kg day 10) compared to placebo. Bellinger and others compared the ergogenic effects of placebo, sodium bicarbonate, β-alanine, and buffer/placebo combinations on 4-minute cycle ergometer performance. They reported a 3.1 percent increase in average power output following the sodium bicarbonate+placebo combination but no significant additional increase following sodium bicarbonate+β-alanine (3.3 percent increase). In their meta-analysis of 17 studies, Calvo and others reported significant effects of sodium bicarbonate on pH (greater alkalinity), bicarbonate (greater concentration and base excess), lactate (higher concentrations), and partial pressure of carbon dioxide (greater) compared to placebo, but no effect on VO_2, VCO_2, or partial pressure of oxygen. They concluded that sodium bicarbonate supplementation significantly affects anaerobic metabolism (lactic acid system) but not aerobic metabolism (oxygen system). In summary, the available research supports an ergogenic effect of extracellular and intracellular buffer supplementation in tasks requiring a significant contribution to the lactic acid energy system.

Preworkout supplements

Multi-ingredient **preworkout supplements** (PWSs) are popular among active individuals and athletes. PWSs are ingested prior to exercise as part of a **nutrient timing** strategy, discussed in chapter 12 and by Aragon and Schoenfeld, which suggests that nutrients ingested before acute exercise may facilitate performance, recovery, and training adaptations. In a study of 100 PWS products, Jagim and others reported an average of over 18 ingredients in each product, with almost half (44 percent) listed as part of a "proprietary blend." The most common ingredients and average amounts per serving according to supplement facts data are beta-alanine (β-alanine, 2,000 milligrams), caffeine (254 milligrams), citrulline (4,000 milligrams), tyrosine (348 milligrams), taurine (1,300 milligrams), and creatine (2,100 milligrams). Other DWS ingredients may include nitric oxide precursors, which are discussed in chapter 6, along with β-alanine, citrulline, tyrosine, taurine, and creatine. Caffeine and β-alanine were previously discussed in this chapter.

According to Harty and others, simultaneous ingestion of the multiple ingredients in DWS may provoke different acute exercise and chronic training adaptations compared to the isolated ingestion of these ingredients. Proposed effects of PWS on acute exercise are increased muscular endurance and mood, while the regular use of PWS, combined with progressive resistance training, may increase skeletal muscle protein accretion, strength, and body composition. Effects of long-term PWS use on force production, muscular endurance, and aerobic performance are equivocal. Research comparisons of PWS products are difficult due to differences in ergogenic effects of the individual ingredients, different blends of known ingredients, and unknown composition of ingredients in the various "proprietary blends." More research is needed regarding the efficacy of PWS and the safety of long-term use. The presence of substances banned by the World Anti-doping Agency in the "proprietery blend" could lead to a positive doping test in competitive athletes.

Is sodium bicarbonate supplementation safe and legal?

The dosage of sodium bicarbonate used in most of these studies, about 300 milligrams per kilogram body weight, appears to be effective yet medically safe. Relative to possible disadvantages, several investigators have noted that some subjects developed gastrointestinal distress, including nausea and diarrhea. Excessive doses could lead to alkalosis, with symptoms of apathy, irritability, and possible muscle spasms. Some athletes experience gastrointestinal distress following an acute high dose of sodium bicarbonate. Saunders and others concluded that total work at 110 percent of maximal cycling power significantly improved only when the subjects who experienced gastrointestinal distress (4 of 21, 19 percent) were excluded from data analysis. Driller and others compared an acute dose to the same dose spread serially over 3 days. They found no difference in 4-minute power output in well-trained cyclists and concluded that serial loading is as effective as an acute dose and may be more practical and convenient for some athletes. Siegler and others compared the effects of sodium bicarbonate supplementation (300 mg/kg) administered 60, 120, and 180 minutes prior to interval sprint running exercise and reported no difference in alkalinity, but a lower risk of gastrointestinal distress with supplementation 180 minutes before exercise. Grgic and others noted that side effects are generally low and vary between and within individuals and recommend ingestion of smaller doses (<300 mg/kg), taken with a high carbohydrate meal, at an individually determined interval at least 180 minutes before exercise, and/or in capsule form to minimize adverse side effects.

Use of sodium bicarbonate currently is not prohibited by the WADA. As noted, sodium bicarbonate (baking soda) use by athletes is also known as *soda loading,* possibly to liken it to *carbohydrate loading.* As you may recall, the purpose of carbohydrate loading is to increase the storage of muscle and liver glycogen as a means to prevent fatigue in prolonged endurance events. Soda loading is viewed by some in a similar context, an attempt to increase the supply of a natural body ingredient helpful in delaying fatigue. However, because sodium bicarbonate may be regarded as a drug, it remains to be seen whether this technique will be deemed illegal. Currently there is no test to detect its use, except for urinary pH, which can also be affected by some antacids, and at present sodium bicarbonate is considered to be legal for use in sports.

> ### Key Concepts
> - Supplementation with alkaline salts such as sodium bicarbonate, sodium citrate, and beta-alanine appears to be an effective ergogenic aid in exercise tasks that depend primarily upon the lactic acid energy system (anaerobic glycolysis), such as a 400-meter dash in track.
> - Ingestion of sodium bicarbonate is generally regarded as safe but may cause acute gastrointestinal distress and diarrhea. Supplementation over a longer time frame may be effective and less likely to cause intestinal problems.

Anabolic Hormones and Dietary Supplements: Ergogenic Effects and Health Implications

The hormones testosterone, insulin, human growth hormone (hGH), and insulin-like growth factors (IGF-1, IGF-2) exert anabolic effects on body composition. As was discussed in chapter 6, specific amino acids and other dietary supplements have been utilized by athletes to increase the secretion of these anabolic hormones to stimulate muscle protein synthesis with presumed increases in strength, power, and athletic performance. Prohormones in the testosterone biosynthetic pathway and some herbal supplements have been marketed for their potential to stimulate testosterone production. In the United States, these agents are controlled under the Anabolic Steroids Control Act of 1990. In addition, anabolic steroids are listed on Schedule III of the Controlled Substances Act (CSA) of 1970. The World Antidoping Agency prohibits the use of hGH, IGF-1/2, anabolic-androgenic steroids, and anabolic prohormone dietary supplements in training and in competition.

Is human growth hormone (hGH) an effective, safe, and legal ergogenic aid?

hGH (a.k.a. somatotropin) is secreted by the anterior pituitary gland in response to the secretion of somatocrinin (a.k.a. growth hormone releasing factor) by the hypothalamus and regulated by somatostatin (a.k.a. growth hormone inhibiting hormone). hGH increases fatty acid lipolysis and decreases the use of glucose and amino acids. It also stimulates the release of IGF-1 and IGF-2 from the liver and within other tissues, such as muscle, which promotes bone and connective tissue growth. IGF-1 mediates the various metabolic effects of hGH. As discussed by Ayyar, the medical potential of hGH was realized in the 1940s. hGH was first isolated from human pituitary gland tissue in 1956. The efficacy of hGH in the treatment of hGH deficiency-related diseases in children was established in 1960 following the first hGH trial in 1958. The peptide structure of hGH was sequenced in 1972, leading to the development of recombinant hGH (rHGH) by Genentech in 1981, with FDA approval in 1985. Available data suggests that rhGH injections may help improve body composition and functional capacity in elderly men who have lower concentrations of hGH. In two meta-analyses by Rubeck and others and Widdowson and Gibney, rhGH therapy decreased body fat and increased lean body mass and VO_{2max} in hGH-deficient individuals. However, there was no improvement in strength, which may be due to reports that the increase in lean body mass following rhGH injection is due to water and not connective or contractile muscle tissue.

hGH is secreted in several isoforms in response to acute aerobic and resistance training and potentially contributes to training adaptations. However, Kraemer and others note the magnitude of increase is affected by a complex array of factors, including but not limited to training intensity, volume, and state; gender; nutritional intake; sleep; and body composition. There is scant evidence for an ergogenic effect of rhGH in individuals who are not hGH-deficient. In their meta-analysis of 11 studies, Hermansen and others concluded that rHGH increased lean body mass, decreased fat mass, and increased lipolysis in healthy young adults. However, rhGH did not increase muscle strength, VO_{2max}, or fatty acid oxidation.

According to Siebert and Rao, rhGH has been sought after by athletes since the mid-1980s as a performance-enhancing drug through smuggling; theft; illegal prescription by online physicians, pharmacies, and other online sites; and other illicit means. Despite the general lack of evidence supporting an ergogenic effect, Erotokritou-Mulligan and others noted that some athletes use rhGH in the belief they will be able to train harder and recover more rapidly. This may be unwise based on the potential health risks of rhGH and and the fact that rhGH is banned for use in training and in competition by the WADA. The potential adverse health effects of rhGH include acromegaly with increased risk of neurological and musculoskeletal problems, insulin resistance, high blood pressure, and congestive heart failure. Liu and others indicated that rhGH therapy cannot be recommended as an anti-aging therapy because of these adverse health effects. Tentori and Graziani commented that rhGH or its mediator, IGF-1, has also been associated with colon, breast, endometrial, and prostate cancers. Most researchers caution that other long-term health risks of hGH administration are unknown. A 2014 national survey of 3,705 high school students (Partnership for Drug-Free Kids, www.drugfree.org) revealed that 11 percent have used rhGH compared to only 5 percent in 2012.

As stated previously, the use of rhGH and rhIGF-1 is prohibited by the WADA in training and in competition. According to Siebert and Rao, detection of supraphysiological levels of rhGH in blood or urine is difficult due to the short half-life of rhGH (4 hours after subcutaneous injection; 22 minutes after intravenous injection) and pulsatile nature of secretion. The test must be administered within 12–24 hours after the last rhGH dose to detect doping. Criteria for a positive test involve an increase in the exogenous rhGH isoform compared to endogenous isoforms of hGH. Cox and Eichner reported that IGF-1 was detected in four of six commercial supplements marketed as all-natural and containing deer antler velvet. According to Cox and others, a finger-prick dried blood spot assays technique has shown promise in detecting IGF-1 and other biomarkers of rhGH doping. Various products are available through the internet. For example, the product Growth Hormone Support by Pure Encapsulations® contains the amino acids arginine and ornithine, which have purported but limited evidence of stimulating hGH secretion. The product Hexarelin marketed by AdvancePeps is described as "a potent hGH secretagogue that is a highly selective agonist of the growth hormone receptor." Since "human growth hormone secretagogues" are also banned by the WADA in training and in competition, such a product may lead to a positive doping test.

In the United States, rhGH is regulated as part of the 1990 Anabolic Steroids Control Act. The possession, distribution, or intent to distribute rhGH "for any use . . . other than the treatment of a disease or other recognized medical condition, where such use has been authorized by the Secretary of Health and Human Services . . . and pursuant to the order of a physician" is a felony punishable by a prison sentence of up to 5 years (www.deadiversion.usdoj.gov/drug_chem_info/hgh.pdf).

Are testosterone and anabolic-androgenic steroids (AAS) effective, safe, and legal ergogenic aids?

Testosterone, the steroid sex hormone responsible for secondary male sex characteristics, is secreted primarily by the testes (95 percent in males). The remaining 5 percent is secreted by the ovaries in females and zona reticularis cells in the adrenal cortex of both sexes. Testosterone regulates normal sexual and reproductive function, muscle mass and strength, facial and body hair, and red blood cell production. Decreasing testosterone levels in the aging male are associated with fatigue, depression, irritability, decreased self-confidence, increased risk for metabolic syndrome, type 2 diabetes, anemia, stroke, decreased sexual function, and coronary heart disease. Testosterone replacement therapy (TRT) is the focus of considerable media and advertising attention as a treatment for decreased testosterone levels, also known as "low T." Brown and Murray note that TRT is approved for the treatment of hypogonadism, sexual dysfunction, certain breast cancers, osteoporosis, and cognitive disorders. According to Bhasin, testosterone prescription sales have increased from $100 million in 2000 to $2.7 billion in 2013, with most prescriptions written for age-related decline in testosterone even though TRT is not FDA approved for this purpose. Testosterone has a short half-life and is catabolized by digestive enzymes. Although TRT can be administered orally, it is more commonly administered via intramuscular injection, transdermal patch, nasal gel, or tablet dissolved on the gum. In 31 studies of middle-aged and older men, Skinner and others reported that intramuscular injection was superior to transdermal administration in increasing fat-free mass and lower-extremity strength.

Controversy surrounds TRT regarding the possible increased risk of cardiovascular disease, prostate and breast cancers, and benign prostate hypertrophy. Meta-analyses by Isidori, Grech, Morgentaler, and their respective colleagues have reported TRT decreases body fat and total cholesterol, while increasing fat-free mass, knee extension strength, and bone mineral density with no additional risk of cardiovascular disease, prostate cancer, worsening of urinary symptoms, or sleep apnea. However, Basaria and others noted that a testosterone therapy study was terminated early because of a significantly higher rate of adverse cardiovascular events in the testosterone group than in the placebo group. In their meta-analysis, Corona and others reported decreased cardiovascular disease morbidity and mortality across 15 epidemiological studies, with no overall protective effects reported across 93 randomized clinical trials. Decreased morbidity was observed in studies including subjects with obesity, while an increased risk was observed in studies of frail subjects and/or for high TRT doses. In March 2015, the FDA required labels on testosterone replacement therapy products warning of increased risk for heart attacks and strokes.

The anabolic potential for testosterone has been understood for thousands of years through a comparison of castrated and uncastrated animals. The increased muscle mass may be attributed to muscle fiber hypertrophy in which key roles are played by the androgen receptors, as depicted in **figure 13.10**. In their position stand, Kersey and others briefly summarize the history of testosterone and **anabolic-androgenic steroids (AAS)**, oral and injectable synthetic drugs that mimic testosterone and have a longer half-life than endogenous testosterone. Testosterone was isolated in the 1930s, with the development of AAS soon thereafter. Soviet and East German weightlifters were the first to abuse AAS in international competition in the 1950s. AAS abuse continued in many countries and in many levels of multiple sports through the late 1970s. AAS testing first occurred in the 1976 Summer Olympiad in Montreal. In 1990, AAS were classified as Schedule III controlled substances according to the Anabolic Steroid Control Act. Other substances, including prohormones discussed later in the chapter, were added in 1994.

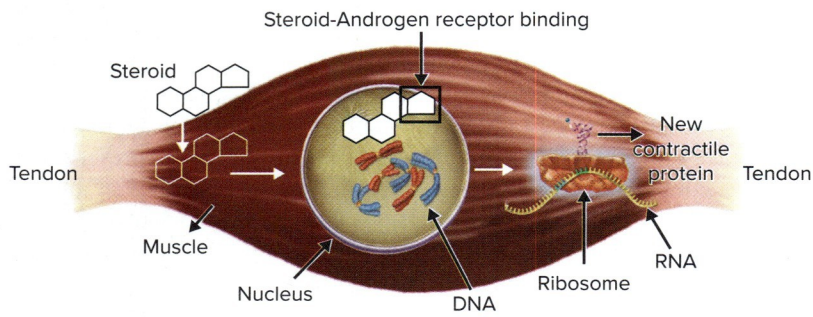

FIGURE 13.10 Anabolic steroids picked up by androgen receptors in the cell nucleus initiate the process of protein formation in cells such as muscle fibers, leading to muscle hypertrophy.

There is a dearth of well-designed, randomized, double-blind, placebo-controlled research reflecting training volumes, dosing patterns, and suprapharmacological doses employed by athletes who abuse AAS. Bioethical concerns preclude human subjects' research approval of contemporary, well-designed experimental research employing doses matching "street AAS use." In one of the earliest studies, Bhasin and others reported that weekly injections of 600 mg of testosterone ethanate increased fat-free mass, muscle size, and strength in normal men, with greater observed increases for testosterone ethanate, combined with resistance training. Rogerson and others reported that resistance-trained individuals receiving testosterone enanthate injections (3.5 mg/kg body weight weekly for 6 weeks) increased muscular strength and 10-second cycle sprint performance.

Although AAS are drugs of choice for many strength athletes and bodybuilders to improve performance and appearance, Nicholls and others concluded that factors including but not limited to gender; pursuit of muscularity; improvement of appearance; self-esteem; ego-orientation; risk-taking personality; and the influence of peers, parents, and coaches predict AAS use in adolescents. A recent update of the American College of Sports Medicine (ACSM) position stand authored by Bhasin and others notes that the administration of AAS in a dose-dependent manner significantly increases muscle strength, lean body mass, endurance, and power and that recreational AAS use appears to have surpassed athletic AAS use.

AAS abuse has also been reported by endurance athletes to facilitate recovery during high-volume training. Baume and others reported that two different anabolic steroids administered 12 times during a month of hard endurance training had no effect on standardized treadmill running performance tests or blood markers for recovery from exercise. The limited research with AAS and endurance performance in young athletes does not support an ergogenic effect.

AAS use has been associated with a number of medical and psychological problems, listed in **table 13.9,** that are also discussed in reviews by Yesalis and Bahrke, Piacentino and others, Talih and others, and in an Endocrine Society statement authored by Pope and others. Adverse cardiovascular effects include cardiomyopathy, myocardial infarction, stroke, clotting disorders, hypertension, decreased HDL-cholesterol, and polycythemia (abnormal increases in hematocrit and hemoglobin). Major mood disorders, aggressiveness, high-risk violent behavior, and psychosis may be observed with high AAS usage. AAS dependence, occurring in about 30 percent of AAS users, is related to muscle dysmorphia, which is discussed in chapter 12. Neurotoxic effects from long-term, high-dose use can cause permanent and irreversible cognitive impairments. AAS use can decrease endogenous testosterone production and lead to hypogonadism, decreased spermatogenesis, and male infertility. Intramuscular AAS injections can increase the risk of infections from unsterile needles. AAS use with high-volume weight lifting can increase the risk of rhabdomyolysis (abnormal muscle breakdown) and tendon rupture. The use of oral AAS can lead to liver damage, including elevated enzyme activity and peliosis hepatis (blood-filled cysts).

TABLE 13.9 Possible health risks associated with use of anabolic-androgenic steroids (AAS)

Cosmetic-related effects

Facial and body acne
Female-like breast enlargement in males (gynecomastia)
Premature baldness
Masculinization in females
Facial and body hair growth in females
Premature closure of bone epiphyseal growth plates in adolescents, leading to stunted growth
Deepening of the voice in females

Psychological effects

Increased aggressiveness and possible violent behavior

Reproductive effects

Reduction of testicular size
Reduction of sperm production
Decreased libido
Impotence in males
Enlargement of the prostate gland
Enlargement of the clitoris

Cardiovascular risk factors and diseases

Atherosclerotic serum lipid profile
    Decreased HDL-cholesterol
    Increased LDL-cholesterol
High blood pressure
Impaired glucose tolerance
Increased size of left ventricle
Stroke
Heart disease

Liver function

Jaundice
Peliosis hepatis (blood-filled cysts)
Liver tumors

Athletic injuries

Tendon rupture

AAS may cause premature cessation of bone growth in children and adolescents and may result in the appearance of several male secondary sex characteristics in females, some of which may be irreversible, such as deepening of the voice. Nyberg and Hallberg noted that AAS influence opioid- and dopamine-secreting reward centers in the brain. An increased sensitivity of the AAS user to opioid narcotic and stimulant drugs may lead to addiction. In a cross-sectional survey, Ip and others reported that male AAS users were significantly more likely than nonusers to have used cocaine within the last 12 months and to meet the psychiatric diagnostic criteria for substance abuse disorder.

However, many adverse health effects of AAS use appear to be reversible. Hartgens and others found that bodybuilders who cycled off AAS steroids for 3 months had similar lipoprotein profiles and liver enzymes as their non-drug-using counterparts. van Amsterdam and others indicated that severe side effects of AAS use appear only following prolonged use at high doses, and their occurrence is limited. They conclude that, based on the scores for acute and chronic adverse health effects, the prevalence of use, social harm, and criminality, AAS were ranked as a group of drugs with relatively low harm. Nevertheless, Urhausen and others reported that several years after discontinuation of anabolic steroid abuse, strength athletes (bodybuilders and powerlifters) who used AAS showed a slight concentric left ventricular hypertrophy in comparison with AAS-free strength athletes. Left ventricular hypertrophy can increase the risk of a heart attack.

In summary, position stands by the ACSM, NATA, and the NSCA, authored by Bhasin, Kersey, Hoffman, and their respective colleagues) condemn AAS use on the basis of ethics, the ideals of fair play in competition, and concerns for the athlete's health (**table 13.10**). The NSCA indicates that optimizing nutritional strategies for athletes is a key to preventing AAS use and abuse, and such strategies for strength/power athletes have been presented in chapter 12. Although a more extensive discussion of AAS is beyond the scope of this text, the ACSM, NATA, and NSCA position stands provide detailed reviews for the interested reader, as does the review by Yesalis and Bahrke.

Are anabolic prohormone dietary supplements effective, safe, and legal ergogenic aids?

Dehydroepiandrosterone (DHEA), androstenedione, and related compounds are called **prohormones** because they are precursors for testosterone production. It is theorized that prohormone supplementation might increase testosterone production, which may facilitate increased muscle mass and strength, decreased body fat, and more rapid recovery during high-volume training. Prohormones may also be derived from wild yams and other plants. The Anabolic Steroid Control Act of 2004 classified prohormones as Schedule III controlled drugs similar to AAS.

DHEA and its sulfated metabolite (DHEAS) are produced in the body by the adrenal and gonadal glands and may be converted into androstenedione with subsequent conversion to testosterone in peripheral tissues, including fat and muscle tissue. DHEA levels are high in young adulthood and gradually decrease to low levels with aging. Baker and others concluded that the efficacy of DHEA supplementation in improving strength and physical performance in older adults remains inconclusive in a meta-analysis of eight studies. Across 25 studies, Corona and others reported a small but significant reduction in fat mass in elderly men following DHEA supplementation. However, this effect disappeared after statistical adjustment for DHEA-related increases in testosterone and estradiol. Two recent meta-analyses report evidence that DHEA supplementation increases testosterone levels. Across 42 randomized placebo-controlled trials of 2,880 subjects in diverse groups, Li and others reported general increases in testosterone following DHEA

TABLE 13.10 Summary of position stands from the American College of Sports Medicine (ACSM), the National Strength and Conditioning Association (NSCA), and the National Athletic Trainers' Association (NATA) regarding use and abuse of Anabolic-Androgenic Steroids (AAS) and/or human Growth Hormone (hGH).

1	The administration of AAS in a dose-dependent manner significantly increases muscle strength, lean body mass, endurance, and power.
2	Recreational AAS use appears to have surpassed athletic AAS use.
3	AAS are classified as schedule III drugs under the 1970 Controlled Substances Act, and banned by the WADA and most sports governing bodies, and are illegal to use for athletic purposes.
4	Coaches, trainers, and medical staffs should monitor and be cognizant of visible signs of AAS use and abuse.
5	Use and abuse of AAS and hGH are associated with several notable adverse effects in men and women listed in Table 13.9 and in the text.
6	Use of AAS in prepubertal and peripubertal children may lead to early virilization, premature growth plate closure, and reduced stature.
7	Coaches, trainers, and medical staffs should be cognizant of the reasons for AAS use and abuse and deter use when possible.
8	Androgen replacement therapy is approved for the medical treatment of several clinical diseases and abnormalities.
9	hGH and other growth factors (e.g., insulin-like growth factors 1 and 2, IGF-1, IGF-2) are regulated as part of the 1990 Anabolic Steroids Control Act.
10	hGH increases lean body mass within weeks of administration, but the majority is in the water compartment, not contractile protein.
11	hGH and AAS are usually taken together.
12	hGH combined with resistance training results in only minimal gains in fat-free mass, hypertrophy, and maximal strength compared to training alone.
13	Use of hGH and AAS is appropriate for legitimate medical reasons under physician supervision (e.g., children with hypopituitarism, adults with muscle-wasting diseases).
14	Continued effort should be made to educate athletes, coaches, parents, physicians, and athletic trainers about the dangers of PED (AAS, hGH) use and emphasize optimal training and nutrition strategies and realistic goals to enhance performance.
15	The ACSM and NSCA promote efforts to document acute and long-term effects of AAS/hGH use, to discontinue the use of PEDs and to increase strategies to detect AAS/hGH use.

Sources: Bhasin, S., et al. 2021. Anabolic-androgenic steroid use in sports, health, and society. *Medicine and Science in Sports and Exercise* 53(8):1778–94.

Hoffman, J. R., et al. 2009. Position stand on androgen and human growth hormone use. *Journal of Strength and Conditioning Research* 23(5):S1–S59. doi: 10.1519/JSC.0b013e31819df2e6.

Kersey, R. D., et al. 2012. National athletic trainers' association position statement: Anabolic–androgenic steroids. *Journal of Athletic Training* 47(5):567–88.

supplementation, with greater increases in females, in response to higher DHEA doses and shorter supplementation durations, in healthy participants, and in participants with less than 60 years of age. Hu and others reported increased testosterone levels with either increased fat-free mass, decreased body mass, or decreased BMI following DHEA supplementation in nine studies (793 subjects) limited to older (≥60 yrs.) females. These findings support a possible role for DHEA to partially counter **sarcopenia**, the loss of muscle tissue associated with the aging process.

The literature is similarly equivocal in studies of DHEA supplementation to younger subjects. Acacio and others found that 6 months of supplementation with DHEA in young men elevated levels of 5α-androstane-3α-17β-diol glucuronide (ADG), a metabolite that may have a negative impact on the prostate gland. However, Liu and others reported that acute DHEA supplementation (50 milligrams) prior to a high-intensity interval training (HIIT) session increased free testosterone in middle-aged men and prevented a decline during HIIT. There was no effect on total testosterone levels. They concluded that DHEA supplementation might be beneficial to HIIT adaptations. In a field study, Taylor and others randomly assigned 48 U.S. Navy personnel to receive either DHEA or placebo supplementation before (50 mg/day for 5 days) and during (75 mg/day) 7 days of military survival training. They reported evidence of improved anabolic hormonal balance (increased DHEA/

cortisol; DHEAS/cortisol; testosterone/cortisol ratios) during a stressful field military exercise. Any role of DHEA as a supplement appropriate for older or younger, active populations, including competitive athletes, requires additional research. Regardless, the fact remains that DHEA is a Schedule III controlled substance and is banned in training and in competition by the WADA.

Androstenedione and related metabolites (4-androstenediol, 5-androstenediol, 19-norandrostenediol, 19-norandrostenedione) are immediate substrates in the production of testosterone. Proponents believe supplementation will increase testosterone production and muscle protein synthesis, with a resulting increase in strength and performance. Doses of androstenedione and related androgens in available research have ranged from 100 to 300 mg/day, with higher doses associated with reports of increased testosterone but scant evidence of muscle growth or an ergogenic effect.

In studies of males, Leder and others reported significant increases in serum testosterone in adult men following a 300-mg dose but no effect of a 100-milligram dose. Earnest and others reported small, but significant, increases in total testosterone with a 200-milligram dose of androstenedione but not androstenediol.

In a short-term study, Rasmussen and others found no effect of oral androstenedione supplementation (100 mg/day for 5 days) on serum testosterone or muscle protein anabolism in young men. In a longer study, Wallace and others reported no significant effects of androstenedione supplementation (100 mg/day for 12 weeks) on serum testosterone levels, lean body mass, or muscular strength in resistance-trained middle-aged men. Using three individual 100-milligram androstenedione doses daily (total 300 mg/day), King and others reported no significant effects on serum testosterone, body composition, muscle fiber diameter, or muscular strength in young men during 8 weeks of resistance training. In a study of young women, Brown and others reported that androstenedione intake (a single ingestion of placebo, 100 or 300 milligram at day 3 of the follicular phase) significantly increased serum testosterone concentrations. Reviews by Brown and Ziegenfuss and their respective colleagues have concluded there is no significant increase in protein synthesis, muscle growth, muscle mass, or strength following supplementation with androstenedione and related metabolites. Like DHEA, androstenedione is a Schedule III controlled substance and is banned in training and in competition by the WADA.

Androstenedione and related prohormones may be associated with increased health risks similar to AAS such as decreased HDL-cholesterol and increased LDL-cholesterol, which might increase the risk for cardiovascular disease and significant increases in estrogen hormones, which could exert feminizing effects in males, such as gynecomastia (breast enlargement). Other adverse effects on gonadal hormones may be associated with testicular shrinkage and infertility. No safety data are available regarding the long-term use of prohormones.

Prohormone use has also been banned by the WADA, the International Olympic Committee, the National Collegiate Athletic Association, and the National Football League. Although anabolic prohormones are currently banned for sale in the United States, a number of substitutes for these products are being marketed on the internet. For example, Sustanon-250® and Steel 19-NorAndro list 4-androstene-3b-ol, 17-one proprionate, and 19-NorDHEA as ingredients. Legitimate dietary supplements on sale may contain ingredients that are not declared on the label but that are prohibited by the doping regulations. Lauritzen reported that in 49 of 192 doping violations from 2003 to 2020, the athlete claimed that a dietary supplement was the source of the prohibited substance, with supporting evidence for 27 of these violations. Multiingredient preworkout supplements accounted for 20 of these violations.

Key Concepts

▶ The use of anabolic drugs or hormones to increase body weight may be effective but may also lead to a variety of health problems. The WADA prohibits the use of anabolic hormones in sports training or competition.

▶ Research has shown that prohormone dietary supplements marketed as anabolic agents, such as dehydroepiandrosterone and androstenedione, do not effectively increase muscle mass or strength. Moreover, such prohormones have been classified as controlled anabolic steroids and their use is illegal and banned by the WADA.

Check for Yourself

▶ Conduct an online search for "Anabolic Steroids," "DHEA," and "Androstenedione." Check the advertisements and information related to laws, listed ingredients, and use of such products. Share the information with your classmates.

Ginseng and Selected Herbals: Health and Ergogenic Effects

Herbs contain phytonutrients, antioxidants, vitamins, minerals, and other bioactive compounds that may be part of a healthful diet. For centuries, various herbal products have been used for purported health benefits such as treatment of depression, reduction of stress and anxiety and common cold symptoms, and weight loss. Although many herbals are regulated as drugs in some countries, most are regulated as dietary supplements in the United States. A detailed discussion of the health effects of herbals is beyond the scope of this text, but there is limited well-designed research on health outcomes associated with the use of many herbal products. For example, *Echinacea* preparations are purported to decrease the duration and severity of the common cold. A meta-analysis by Shah and others concluded that *Echinacea* decreased the incidence of developing a common cold by 58 percent and the duration of an existing cold by 1.4 days. However, in their meta-analysis of 24 double-blind studies including over 4,600 participants and 33 comparisons of various *Echinacea* preparations, Karsch-Völk and others reported some preparations showed weak and nonsignificant benefits of limited clinical relevance. They concluded that *Echinacea* preparations did not provide any benefit in treating the common cold. In their meta-analysis, David and Cunningham reported that *Echinacea* supplementation did not decrease the duration of upper-respiratory-tract infections (URTIs) compared to placebo. Although they reported a possible effect of *Echinacea* in preventing URTIs, the clinical relevance of this finding was questionable. They assessed most of the 31 studies as having at least one source of bias. In another example, Ziaei and others reported no overall effect of ginseng on favorably changing lipoprotein profiles across 28 studies of almost 1,250 participants. However, lower total cholesterol, LDL-cholesterol, and triacylglycerol levels were observed in long-term supplementation studies. In summary, some herbal supplements may potentially improve aspects of health while others are ineffective. For many herbal supplements, the research findings are equivocal.

Numerous herbal supplements are marketed for weight control. Smith and Krygsman noted that decreased appetite and weight loss reported with *Hoodia gordonii* supplementation occur at high doses and may cause severe hypertension, loss of muscle mass, and gastrointestinal discomfort. Farrington and others concluded that no evidence exists for *Garcinia cambogia, Camellia sinensis, Hoodia gordonii, Citrus aurantium,* and *Coleus forskohlii* to promote long-term weight loss in humans. Brewer and Chen noted that some herbal remedies marketed for weight reduction and other purposes may result in drug–drug and herb–drug interactions leading to a significant liver injury involving cytochrome P450.

Herbals have also been marketed as ergogenic aids for athletes. Unfortunately, with the exception of ginseng and related products, limited research has evaluated their ability to enhance exercise or sports performance.

Does ginseng or ciwujia enhance exercise or sports performance?

Ginseng and ciwujia are comparable herbs, and both have been studied for their potential effects on exercise or sports performance.

Ginseng Extracts derived from the plant family Araliaceae contain numerous chemicals that may influence human physiology, the most important being the glycosides, also known as ginsenosides. Collectively, these extracts are referred to as **ginseng**, and their physiological effects vary depending on the plant species, the part of the plant used, and the place of origin. The most common forms of ginseng include Chinese or Korean (*Panax ginseng*), American (*Panax quinquefolium*), Japanese (*Panax japonicum*), and Russian/Siberian (*Eleutherococcus senticosus*). *Eleutherococcus senticosus* is a totally different plant from Araliaceae, but it is recognized by some as a legitimate form of ginseng and its ginsenosides are also referred to as eleutherosides. The type and amount of ginsenosides present vary greatly among the different forms of ginseng. Ginseng is marketed under such labels as Active Ginseng™, Korean Panax Ginseng, Premium Authentic Panax Ginseng, Ginseng Energy Boost, and American Ginseng as a means of enhancing health and physical performance (**figure 13.11**).

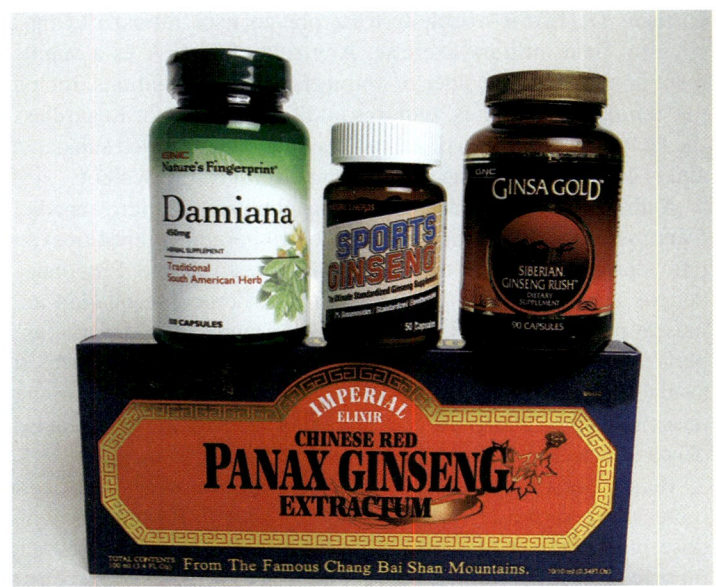

Jill Braaten/McGraw Hill

FIGURE 13.11 Various ginseng products are available as dietary supplements, some being marketed directly to athletes.

The Russians believed that ginseng helped develop resistance to the mental and physical stress associated with intense exercise training. They coined the term "adaptogens" to characterize ginseng's ability to increase resistance to the catabolic effects of stress. Purported ergogenic mechanisms of ginseng include increased cardiac function, blood flow, and oxygen transport during exercise; increased oxygen utilization and decreased lactic acid levels during exercise; enhanced muscle glycogen synthesis after exercise; and improving nitrogen and protein balance. Much early research on the ergogenic effects of ginseng supplementation included various design flaws. Later and better-designed studies generally reported no ergogenic effect of ginseng on metabolic, physiological, or psychological aspects of submaximal and maximal running (Dowling and others), cycling (Engels and Wirth), or high-intensity, interval anaerobic exercise (Engels and others). However, Ziemba and others reported that Panax ginseng improved multiple-choice reaction time before and during a cycling exercise task in soccer players. Although there was no improvement in exercise capacity, a rapid reaction time may be beneficial in some sports. In their review, Smith and others commented that certain ginsenosides may exert neuroprotective effects leading to improving cognition. This possibility merits additional research.

More recent reports suggest that ginseng may increase mitochondrial biogenesis in mice (Shin and others), attenuate eccentric exercise-induced muscle damage in men (Lin and others), modulate the gut microbiota to reduce exercise-related fatigue in rats (Zhou and others), upregulate G-coupled protein receptor signaling that is associated with increased exercise capacity (Kim and others), and increase exercise endurance (Ikeuchi and others). Ginseng is believed to influence neural and hormonal activities in the body and has been theorized to enhance the immune system. The most prevalent theory suggests that ginseng may stimulate the hypothalamus and the neuroendocrine tissue in the brain that controls the pituitary gland. The pituitary gland in turn commands other endocrine glands in the body. For example, the hypothalamus secretes corticotropin-releasing hormone (CRH). This causes the anterior pituitary to secrete adrenocorticotropic hormone (ACTH), resulting in the release of cortisol by the adrenal gland in response to stress. In essence, ginseng is theorized to enhance sports performance by allowing athletes to train more intensely and by inducing a physiological antifatiguing effect that increases stamina during competition. It should be noted that an underlying mechanism for this theorized ergogenic effect has not yet been determined and that more research is needed on the possible mechanisms identified above.

According to Sellami and others, side effects of ginseng include diarrhea, insomnia, headaches, rapid heartbeat, changes in blood pressure fluctuations, and gastrointestinal distress. Ramanathan and Penzak noted ginseng–drug interaction appears to be low despite differences in pharmacological and pharmacokinetic properties between various ginsenosides. However, they noted disparities in research findings regarding the effect of ginsenosides on cytochrome P450, a liver enzyme involved in drug metabolism. Ginseng ingested with other supplements that are banned by the WADA could lead to disqualification.

Consumers should be aware of quality control problems with ginseng supplements. Yap and others noted the need for quality assurance in product purity and ginsenoside content.

Individuals who desire to experiment with long-term ginseng supplementation should consult with their physicians, because ginseng use may exacerbate various health problems, such as high blood pressure. Given the available scientific evidence, ginseng supplements cannot be recommended.

Ciwujias Ciwujia is the common name for *Acanthopanax senticosus* (previously referred to as *Eleutherococcus senticosus*) and also known as Siberian ginseng. Ciwujia is similar to ginseng. According to Lau and others, ciwujia is used in traditional Chinese medicine as an **immunomodulatory agent** (an immune function booster). Specifically, ciwujia has been reported to increase helper/inducer T lymphocytes and natural killer cell immune components which could potentially decrease inflammation and be of benefit in cancer treatment. Ciwujia was first marketed as Endurox™ and currently as Endurox Excel®, whose advertisements suggest that it can increase fat oxidation (possibly sparing muscle glycogen), reduce lactate accumulation, raise the anaerobic threshold, lower the heart rate while maintaining the same level of workout intensity, and speed workout recovery. However, most of the claims for ciwujia are based on clinical trials with poor experimental designs.

Well-designed, peer-reviewed and published studies by Plowman, Cheuvront, and their associates reported that supplementation with Endurox™ (800 milligrams for 7–10 days) had no significant effect on heart rate, oxygen consumption, respiratory exchange ratio (a measure of fat oxidation), lactic acid accumulation, or ratings of perceived exertion during either cycle ergometer or stair-climbing exercise. The investigators concluded that their results did not verify the claims made for Endurox.

Based on the available evidence, products containing ciwujia do not appear to enhance exercise performance, and thus are not recommended for use by endurance athletes.

What herbals are effective ergogenic aids?

Caffeine, ephedra, and ginseng can be derived from various herbal sources such as guarana, kola nut, ma huang, and members of the *Araliaceae* plant family. Additionally, other herbal supplements on the market such as gamma-oryzanol, *Capsicum annuum*, *Tribulus terrestris*, *Cordyceps sinensis*, *Rhodiola rosea*, epigallocatechin-3-gallate, *Cytoseira canariensis*, and *Echinacea* are purported to have various health and performance effects.

Gamma-oryzanol (*γ-oryzanol*) is a sterol esterified with ferrulic acid that is found in various plant oils. According to Ramazani and others, *γ-oryzanol* may have antioxidant, anti-inflammatory, anticancer, and antidiabetic properties; improve lipid profile; increase testosterone; and treat symptoms of menopause. Ahn and others recently reported decreased inflammatory cytokine activity and increased aerobic capacity and strength without hypertrophy following *γ-oryzanol* in aged mice. It is marketed as a supplement to increase muscle mass and facilitate postexercise recovery despite scant evidence of these effects in human studies.

Capsicum annuum is the source of capsaicinoids which are found in chili peppers. Capsaicinoids may influence energy intake through effects on the stomach, intestinal peptides, the sympathetic nervous system, and preference for high-carbohydrate over high-fat foods. In a small meta-analysis (11 studies including 609 subjects) of the effects of *capsicum annuum* on components of the metabolic syndrome, Jang and others reported significantly lowered low-density lipoprotein levels but no effect on body mass, BMI, percent body fat, systolic or diastolic blood pressure, total cholesterol, triacylglycerol, or high-density lipoprotein cholesterol.

Tribulus terrestris is derived from the *Zygophyllaceae* plant family found in South Africa, Australia, India, and Europe. In their review, Ștefănescu and others note that *tribulus terrestris* extract contains many bioactive compounds including steroidal saponins, flavonoids, tannins, terpenoids, polyphenol carboxylic acids, and alkaloids that may increase libido, luteinizing hormone secretion, and testosterone production. *Tribulus terrestris* is also reported to possess antioxidant, antibacterial, anti-inflammatory, and antidiabetic properties. However, *tribulus terrestris* may exert a toxic effect on kidney nephrons. Moreover, alkaloid compounds in *tribulus terrestris* inhibit monoamine oxidase, an enzyme regulating brain levels of norepinephrine, serotonin and dopamine levels, and potentially create in imbalance in these neurotransmitters. Qureshi and others noted that *tribulus terrestris* might facilitate the release of nitric oxide, discussed in chapter 6, which could be physiologically meaningful, independent of any effect on testosterone. This merits further investigation.

According to Das and others, *cordyceps* is a genus of over 700 mushrooms. These *cordyceps* species include a variety of bioactive compounds that been reported to stimulate immune and antioxidant function, increase nitric oxide production to lower blood pressure, and improve lipid profile. Other reported pharmacological effects include improving glucose levels and combating depression. The literature on the ergogenic effects of *cordyceps* species is scant and equivocal. For example, Parcell and others reported that supplementation with CordyMax Cs-4®, a synthetic form of *cordyceps sinensis,* had no effect on the aerobic capacity of endurance-trained male cyclists while Hirsch and others found 3 weeks supplementation with a mushroom blend containing *cordyceps militaris* increased VO_{2max}, cycling time to exhaustion, and ventilatory threshold.

According to Pu and others, *rhodiola rosea,* a biological adaptogen similar to ginseng in the *Crassulaceae* plant family, has mild anti-inflammatory properties which contribute to improved glucose homeostasis, reduced cancer risk, cardio-protection, and neuro-protection. In their review of 11 studies, Ishaque and others noted biases and methodological flaws in the existing literature and concluded there is equivocal support for a physical performance or mental fatigue-delaying effect of *rhodiola rosea*. Lu and others reported decreased pain and muscle damage associated with training, increased antioxidant activity, and increased explosive power following *rhodiola rosea* supplementation compared to placebo in a small meta-analysis of 10 studies including 183 participants. Despite their conclusions, the authors also acknowledged that even though all studies were double-blind, many omitted details regarding treatment order and other critical design features. Sellami and others noted there is no plausible mechanism to explain an ergogenic effect for *rhodiola rosea* and that additional rigorous and better designed studies are needed.

Green tea (*camellia sinesis*) contains many antioxidant and anti-inflammatory compounds. The main antioxidant agents are catechins the most potent of which is the flavenol *epigallocatechin-3-gallate* (EGCG). According to Musial and others, EGCG and other catechins may be effective additions to standard health promotion strategies to reduce the risk of various cancers. In a recent randomized clinical trial, Roberts and others assigned overweigh males and females to EGCG, EGCG+antioxidants (quercetin and lipoic acid), or placebo groups. Following 8 weeks of standardized aerobic exercise, maximal fat oxidation and fat utilization during incremental exercise were greater and low-density lipoprotein cholesterol was lower in the EGCG+antioxidants group compared to the EGCG and placebo groups. Body composition and other markers of cardio-metabolic health were not affected by EGCG supplementation. Earlier research also reported no effect of EGCG on a 30-minute cycling time trial or on energy substrate utilization (Eichenberger and others) or on rate of fat oxidation (Randell and others). Overall, current research does not support an ergogenic effect of EGCG supplementation.

Williams and Branch noted that much of what is known about the efficacy of herbal supplements as ergogenics is based on anecdotal data and poorly controlled studies. In addition to the herbal supplements discussed here in the previous text, they concluded no significant effect of bee pollen, capsicum, ginkgo biloba, kava kava, St. John's wort, or yohimbine. Hudson and others report that 47 toxic compounds in 55 species from 46 plant families were found to potentially generate liver, cardiovascular, central nervous system, and digestive system toxicity. Athletes contemplating using herbals should consult with their health-care professional. The interested reader is referred to the review by Sellami and others.

> ### Key Concepts
> - Results from well-controlled research indicate that ginseng and related adaptogens, such as ciwujia, are not effective ergogenic aids.
> - There is limited well-controlled research regarding the effect of herbals on exercise or sports performance, and that which is available suggests that herbal sports supplements are not effective ergogenic aids.

Sports Supplements: Efficacy, Safety, and Permissibility

What sports supplements are considered to be effective, safe, and permissible?

This text has addressed the effects of a wide variety of supplements on health and efficacy, safety, and permissibility as ergogenic aids in sports or exercise performance. Carbohydrate and fat metabolites, special proteins, numerous amino acids, vitamins, minerals, caffeine, alcohol, and herbal supplements, have all been studied for their health and ergogenic potential. In the joint position statement of the ACSM, the AND and the Dietitians of Canada, Thomas and others conclude that performance and recovery from competition and training may be enhanced by research-based nutritional selections and strategies. This position statement has also classified nutritional supplements into four categories based on the supplements' legality and ability to enhance health, performance, or training when added to a healthful diet that already provides adequate kcal and essential nutrients. A similar classification system of the Australian Institute of Sport is presented in **table 13.11**. Group A supplements have research support for their efficacy and are permitted for use. Group B supplements require additional research but could be effective in certain circumstances. Group C includes supplements that are either not listed in Groups A and B or are abused Group A and B supplements. Group D supplements are banned and should not be used under any circumstances. The ergogenic effect of the supplement may also be limited to a specific type of athletic endeavor or characteristic. As examples, sodium bicarbonate may improve high-intensity, anaerobic exercise dependent primarily on anaerobic glycolysis. Females in weight-control sports may benefit from iron supplements to

TABLE 13.11 Australian Institute of Sport classification system of nutritional supplements

Group	Evidence level	Subcategory	Examples
A	Supported and permitted for use in specific situations	Sports foods	Carbohydrate/protein sports drinks/bars/gels; electrolyte replacements; whey protein
		Medical supplements	Iron, calcium, multivitamin/mineral, vitamin D
		Performance supplements	Caffeine, beta-alanine, sodium bicarbonate, beetroot juice, creatine
B	Deserving of additional research; permitted for use subject to clinical and/or research monitoring	Food polyphenols	Quercetin, montomorency, acai, gogi, curcumin
		Other	Vitamin C, Vitamin E, L-carnitine, hydroxymethylbutyrate, glutamine, eicosapentanoic acid, docosahexaenoic acid, glucosamine
C	Little proof of efficacy; not provided to athletes but may be permitted on an individual basis	Group A or B supplements not used according to approved protocols	
		Other supplements not listed in Groups A and B	Aspartate salts, bee pollen, boron, chromium, choline, condroitin, chromium, ciwujia, conjugated linoleic acids, coenzyme Q_{10}, *Cordyceps sinensis*, fat loading, gamma oryzanol, ginseng, hydroxycitrate, inosine, magnesium, medium-chain triacylglycerol, niacin, octacosanol, phosphate salts, pyruvate, *Rhodiola rosea*, ribose, selenium, tryptophan, vanadium, vitamin A, wheat germ oil, yohimbine, zinc
D	Banned by WADA or at high risk for contamination; should not be used by athletes	Stimulants	Ephedra/epidrine/ma huang, strychnine, sibutramine, methyhexanamine, others
		Prohormones	Dehydroeipandrosterone, androstenedione, 19-norandrostenedione, 19-norandrostenediol, *Tribulus terrestris*, maca root powder
		HGH releaser peptides	Glycine, ornithine, arginine, lysine
		Others	Glycerol, colostrum

Source: www.ausport.gov.au/ais/nutrition/supplements/classification.

help prevent iron deficiency and anemia, as well as from calcium supplements to help maintain bone mass. Some supplements may be considered unsafe and not recommended if that is the research-based opinion of various health organizations. With few exceptions, research does not support the efficacy of most commercial sports supplements. Moreover, some supplements may be contaminated, intentionally or unintentionally, with substances that could lead to a positive doping test. In most cases athletes can meet their nutritional needs through consumption of a well-planned, healthful diet. It should also be stressed that micronutrient supplementation is more likely to improve health and performance in subjects who are deficient in the micronutrient rather than in those who have normal biochemical levels.

APPLICATION EXERCISE

Conduct a single-subject, double-blind, placebo-controlled crossover study of caffeine supplementation. You will need three individuals: (1) an aerobically trained subject or athlete; (2) the investigator, and (3) the administrator of the placebo and caffeine treatments. The performance task should be a high-intensity 5–10-minute aerobic task such as bicycling, swimming, or running. The activity is best done indoors to control environmental conditions. Over a 5-week period, the subject will participate in five performance trials (one day each week; same day of the week; same time of day), each involving maximal performance for the selected activity. The subject should not be currently taking any supplement or stimulant and consume no caffeine at least 12 hours before or after each trial. Thirty minutes before the test, the subject should consume *either* a dose of caffeine (no more than two Vivarin® or similar over-the-counter tablets [200 mg/tablet] depending on body weight, which results in no more than 4 mg caffeine/kg of weight) *or* a placebo (same number of multivitamin tablets as caffeine tablets), each taken with water. Use the following table to determine the number of Vivarin tablets according to body weight up to a *maximum* of two tablets. The subject's eyes should be closed while taking the tablets in order to ensure supplement blinding.

	Body weight lbs (kg)					
	110 (50)	132 (60)	154 (70)	176 (80)	198 (90)	220 (100)
Caffeine dose (mg/kg)	4	4	4	4	4	4
Total caffeine dose (mg)	200	240	280	320	360	400
Maximum caffeine tablets	≤1.0	≤1.25	≤1.5	≤1.67	≤1.75	≤2.0

Record the performance times (minutes: seconds) for each week's performance. Compare the average placebo and caffeine performances for any ergogenic effect.

Week	1	2	3	4	5
Treatment	No placebo or caffeine	Placebo or caffeine	Caffeine or placebo (opposite of week 2)	Placebo or caffeine	Caffeine or placebo (opposite of week 4)
Performance time					

Review Questions—Multiple Choice

1. Of the supplements and food drugs discussed in this chapter, the WADA prohibits the use of _____ in competition *but not in training* and _____ in competition *and in training*. However, the WADA permits the use of _____ in competition and training.
   a. sodium bicarbonate/anabolic-androgenic steroids/alcohol
   b. dehydroepiandrosterone/caffeine/ephedrine
   c. ephedrine/caffeine/androstenedione
   d. human growth hormone/caffeine/ephedrine
   e. ephedrine/human growth hormone/caffeine

2. Which of the following supplements and food drugs discussed in this chapter has the greatest amount of research evidence for an ergogenic effect?
   a. alcohol
   b. human growth hormone
   c. caffeine
   d. ginseng
   e. prohormones

3. Which of the following is not a potential risk associated with use of anabolic-androgenic steroids (AAS)?
   a. breast development in males from aromatization of AAS to estrogens
   b. increased aggressiveness and risk-taking behaviors
   c. liver dysfunction
   d. increased HDL-cholesterol; decreased LDL-cholesterol
   e. tendon rupture

4. Which of the following is a physiological effect of caffeine?
   a. decreases perceived exertion to a standard rate of power output
   b. decreases metabolic rate
   c. decreases the secretion of epinephrine
   d. decreases heart rate and force of contraction
   e. decreases force of skeletal muscle contractility

5. Which of the following is *incorrect* regarding the use of sodium bicarbonate?
   a. A 70-kilogram individual should consume a dose of 35 grams immediately before exercise.
   b. Sodium bicarbonate supplementation increases the blood concentration of bicarbonate (HCO_3^-), the base in the carbonic acid/bicarbonate buffer systems, to increase alkalinity in order to buffer acidity produced with lactic acid.
   c. Sodium bicarbonate is most effective in improving prolonged high-intensity anaerobic single-bout or interval-type work of 1–2 minutes in duration.
   d. Sodium citrate is also used as a buffering agent.
   e. Side effects of sodium bicarbonate use include gastrointestinal distress.

6. If an average-size (70-kilogram, 154-pound) male adult consumes four 12-ounce beers within 2 hours, the blood alcohol concentration (BAC) would be approximately _____.
   a. ≤0.01
   b. 0.02–0.03
   c. 0.04–0.05
   d. 0.08–0.10
   e. 0.15

7. Although the deleterious effects of excessive alcohol consumption are well documented, research suggests that moderate alcohol consumption (≤1 drink/day for females; ≤2 drinks/day for males) might be beneficial to health. However, some individuals may be at increased risk for _____ even with moderate alcohol consumption.
   a. cardiovascular disease
   b. Alzheimer's disease
   c. breast cancer
   d. dementia
   e. stroke

8. Research suggests that moderate alcohol consumption, or "low-risk" drinking, may reduce the risk of CHD and all-cause mortality. Which of the following is *not* a hypothesized mechanism to reduce the risk of CHD and all-cause mortality?
   a. reduced caloric intake, induced weight loss, and prevented metabolic syndrome
   b. decreased platelet aggregability (decreased possibility of blood clots)
   c. increased blood flow to the brain
   d. a relaxation effect and reduced anxiety
   e. an increase in HDL-cholesterol
9. Which of the following is *incorrect* regarding alcohol?
   a. A 4-ounce glass of 12 percent (alcohol by volume) wine, a 12-ounce bottle of 4 percent beer, and a 1.25-ounce shot of 80-proof whiskey contain a similar content of alcohol per serving.
   b. Alcohol is an excellent energy substrate for prolonged activity.
   c. Alcohol is chemically similar to carbohydrates but is metabolized by the liver in a manner that is similar to fat.
   d. Alcohol is banned in competition by the WADA in the sport of archery.
   e. Alcohol may impair intestinal absorption of certain vitamins such as B1 (thiamine).
10. Which of the following statements is *correct*?
    a. In well-designed studies, ginseng and ciwujia have consistently been reported to improve immune function and aerobic performance.
    b. Contemporary research methodology such as magnetic resonance spectroscopy reveals that caffeine increases fatty acid availability and oxidation and "spares" the use of muscle and liver glycogen during prolonged aerobic exercise.
    c. Prohormones have consistently been reported in well-designed research to increase testosterone and increase muscle mass and strength.
    d. Alcohol consumption following hyperthermic exercise is an excellent rehydration beverage.
    e. Any alcohol consumed during pregnancy may adversely affect embryonic DNA and subsequent fetal development.

Answers to multiple choice questions: 1. e; 2. c; 3. d; 4. a; 5. b; 6. d; 7. c; 8. a; 9. b; 10. e

Critical Thinking Questions

1. Discuss both the potential beneficial and adverse health effects of consuming various amounts of alcohol.
2. Describe the efficacy, safety, and legality of caffeine supplementation as an ergogenic aid for aerobic endurance athletes.
3. Discuss the efficacy, safety, and legality of sodium bicarbonate supplementation as an ergogenic aid. In which types of sports would it appear to be most effective?
4. Compare and contrast the effects of testosterone versus its congeners DHEA and androstenedione as ergogenic aids for the development of muscle mass and strength. Discuss possible health risks associated with use of each.
5. What is ginseng, why is it purported to be an ergogenic aid, and does research support its efficacy as an ergogenic?

References

Books

American Psychiatric Association. 2013. *Diagnostic and Statistical Manual of Mental Disorders* (5th ed.). Washington, DC: American Psychiatric Association.

U.S. Department of Health and Human Services and U.S. Department of Agriculture. *2020-2025 Dietary Guidelines for Americans*.

National Academy of Sciences. Institute of Medicine. 2014. *Caffeine in Food and Dietary Supplements: Examining Safety: Workshop Summary.* Washington, DC: National Academies Press.

National Institute for Health and Welfare (THL). 2016. *Good practice principles for low-risk drinking guidelines.* Chapter 2, pp. 17–33. Montonen M., et al., eds. www.rarha.eu.

World Anti-Doping Agency Prohibited List. 2022. www.wada-ama.org.

World Cancer Research Fund/American Institute for Cancer Research. 2018. Continuous Update Expert Project Report. *Alcoholic Drinks and the Risk of Cancer.* Available at dietandcancerreport.org.

World Health Organization. 2018. *Global Status Report on Alcohol and Health.* Geneva, SUI: World Health Organization. License: CC BY-NC-SA 3.0 IGO.

World Health Organization. 1993. *ICD-10 Classification of Mental and Behavioural Disorders.* Geneva, SUI: World Health Organization.

Reviews and Specific Studies

Acacio, B., et al. 2004. Pharmacokinetics of dehydroepiandrosterone and its metabolites after long-term daily oral administration to healthy young men. *Fertility and Sterility* 81:595–604.

Aguilar-Navarro, M., et al. 2019. Urine caffeine concentration in doping control samples from 2004 to 2015. *Nutrients* 11(2):286. doi: 10.3390/nu11020286. PMID: 30699902; PMCID: PMC6412495.

Ahn, J., et al. 2021. γ-Oryzanol improves exercise endurance and muscle strength by upregulating PPARδ and ERRγ activity in aged mice. *Molecular Nutrition and Food Research* 65(14):e2000652. doi: 10.1002/mnfr.202000652.

American College of Obstetrics and Gynecology. 2010. ACOG Committee opinion no. 462 (Reaffirmed 2020): Moderate caffeine consumption during pregnancy. *Obstetrics and Gynecology* 116(2 Pt 1):467–68.

American College of Sports Medicine. 2006. American College of Sports Medicine position stand: Prevention of cold injuries during exercise. *Medicine & Science in Sports & Exercise* 38(11):2012–29.

Andraws, R., et al. 2005. Cardiovascular effects of ephedra alkaloids: A comprehensive review. *Progress in Cardiovascular Diseases* 47(4):217–25.

Anstey, K., et al. 2009. Alcohol consumption as a risk factor for dementia and cognitive decline: Meta-analysis of prospective studies. *American Journal of Geriatric Psychiatry* 17:542–55.

Aragon, A. A., and Schoenfeld, B. J. 2013. Nutrient timing revisited: Is there a post-exercise anabolic window? *Journal of the International Society of Sports Nutrition* 10(1):5. doi: 10.1186/1550-2783-10-5.

Armstrong, L., et al. 2007. Caffeine, fluid-electrolyte balance, temperature regulation, and exercise-heat tolerance. *Exercise and Sport Sciences Reviews* 35:135–40.

Artioli, G., et al. 2007. Does sodium-bicarbonate ingestion improve simulated judo performance? *International Journal of Sport Nutrition and Exercise Metabolism* 17:206–17.

Astell, K. J., et al. 2013. A review on botanical species and chemical compounds with appetite suppressing properties for body weight control. *Plant Foods for Human Nutrition* 68:213–21.

Astorino, T. A., et al. 2013. Caffeine ingestion and intense resistance training minimize postexercise hypotension in normotensive and prehypertensive men. *Research in Sports Medicine* 21(1):52–65.

Astorino, T., and Roberson, D. 2010. Efficacy of acute caffeine ingestion for short-term high-intensity exercise performance: A systematic review. *Journal of Strength and Conditioning Research* 24:257–65.

Attipoe, S., et al. 2016. Caffeine content in popular energy drinks and energy shots. *Military Medicine* 181(9):1016–20.

Attwood, A., and Munafò, M. R. 2014. Effects of acute alcohol consumption and processing of emotion in faces: Implications for understanding alcohol-related aggression. *Journal of Psychopharmacology* 28(8):719–32.

Ayyar, V. S. 2011. History of growth hormone therapy. *Indian Journal of Endocrinology and Metabolism* 15(S3):S162–S165. https://doi.org/10.4103/2230-8210.84852.

Bagnardi, V., et al. 2013. Light alcohol drinking and cancer: A meta-analysis. *Annals of Oncology* 24: 301–8.

Baker, W. L., et al. 2011. Effect of dehydroepiandrosterone on muscle strength and physical function in older adults: A systematic review. *Journal of the American Geriatrics Society* 59(6):997–1002.

Barnes, M. J., et al. 2012. The effects of acute alcohol consumption and eccentric muscle damage on neuromuscular function. *Applied Physiology, Nutrition, and Metabolism* 37:63–71.

Basaria, S., et al. 2010. Adverse events associated with testosterone administration. *New England Journal of Medicine* 363:109–22.

Baume, N., et al. 2006. Effect of multiple oral doses of androgenic anabolic steroids on endurance performance and serum indices of physical stress in healthy male subjects. *European Journal of Applied Physiology* 98:329–40.

Beck, A., and Heinz, A. 2013. Alcohol-related aggression—Social and neurobiological factors. *Deutsches Ärzteblatt International* 110(42):711–15. doi: 10.3238/arztebl.2013.0711.

Beedie, C., et al. 2006. Placebo effects of caffeine on cycling performance. *Medicine & Science in Sports & Exercise* 38:2159–64.

Bell, D., and Jacobs, I. 1999. Combined caffeine and ephedrine ingestion improves run times of Canadian Forces Warrior Test. *Aviation and Space Environmental Medicine* 70:325–29.

Bell, D., et al. 2001. Effect of caffeine and ephedrine ingestion on anaerobic exercise performance. *Medicine & Science in Sports & Exercise* 33:1399–403.

Bell, D., et al. 2002. Effect of ingesting caffeine and ephedrine on 10-km run performance. *Medicine & Science in Sports & Exercise* 34:344–49.

Bellinger, P. M., et al. 2012. Effect of combined β-alanine and sodium bicarbonate supplementation on cycling performance. *Medicine and Science in Sports and Exercise* 44(8):1545–51. doi: 10.1249/MSS.0b013e31824cc08d. PMID: 22330016.

Bellocco, R., et al. 2012. Alcohol drinking and risk of renal cell carcinoma: Results of a meta-analysis. *Annals of Oncology* 23(9):2235–44. doi: 10.1093/annonc/mds022.

Bent, S., et al. 2004. Safety and efficacy of citrus aurantium for weight loss. *American Journal of Cardiology* 94:1359–61.

Beulens, J., et al. 2006. The effect of moderate alcohol consumption on fat distribution and adipocytokines. *Obesity* 14:60–66.

Bhasin, S., et al. 1996. The effects of supraphysiologic doses of testosterone on muscle size and strength in normal men. *New England Journal of Medicine* 335:1–7.

Bhasin, S., et al. 2021. Anabolic-androgenic steroid use in sports, health, and society. *Medicine and Science in Sports and Exercise* 53(8):1778–94.

Bhasin. S. 2021. Testosterone replacement in aging men: An evidence-based patient-centric perspective. *Journal of Clinical Investigation* 131(4):e146607. doi: 10.1172/JCI146607. PMID: 33586676; PMCID: PMC7880314.

Bianco, A., et al. 2014. Alcohol consumption and hormonal alterations related to muscle hypertrophy: A review. *Nutrition and Metabolism* 11:26. doi: 10.1186/1743-7075-11-26.

Bird, S., et al. 1995. The effect of sodium bicarbonate ingestion on 1500-m racing time. *Journal of Sports Sciences* 13:399–403.

Bishop, D., and Claudius, B. 2005. Effects of induced metabolic alkalosis on prolonged intermittent-sprint performance. *Medicine & Science in Sports & Exercise* 37:759–67.

Boje, O. 1939. Doping. *Bulletin of the Health Organization of the League of Nations* 8:439–69.

Bouchard, N., et al. 2005. Ischemic stroke associated with use of an ephedra-free dietary supplement containing synephrine. *Mayo Clinic Proceedings* 80:541–45.

Brewer, C. T., and Chen, T. 2017. Hepatotoxicity of herbal supplements mediated by modulation of cytochrome P450. *International Journal of Molecular Sciences* 18(11):2353. doi: 10.3390/ijms18112353. PMID: 29117101; PMCID: PMC5713322.

Brinton, E. A. 2012. Effects of ethanol intake on lipoproteins. *Current Atherosclerosis Reports* 14(2):108–14.

Brisola, G. M., et al. 2015. Sodium bicarbonate supplementation improved MAOD but is not correlated with 200- and 400-m running performances: A double-blind, crossover, and placebo-controlled study. *Applied Physiology, Nutrition, and Metabolism* 40(9):931–7.

Brown, G., et al. 2004. Changes in serum testosterone and estradiol concentrations following acute androstenedione ingestion in young women. *Hormone and Metabolic Research* 36:62–66.

Brown, G., et al. 2006. Testosterone prohormone supplements. *Medicine & Science in Sports & Exercise* 38:1451–61.

Brown, A.S. and Murray, A. 2019. Testosterone Replacement Therapy: Controversy and Recent Trends. *US Pharmacist* 44(8):17–23.

Buckman, J. F., et al. 2013. A national study of substance use behaviors among NCAA male athletes who use banned performance enhancing substances. *Drug and Alcohol Dependence* 131(1–2):50–55.

Bui, L., et al. 2006. Blood pressure and heart rate effect following a single dose of bitter orange. *Annals of Pharmacotherapy* 40(1):53–57.

Burke, L. M., et al. 2003. Effect of alcohol intake on muscle glycogen storage after prolonged exercise. *Journal of Applied Physiology* 95(3):983–90.

Burke, L. M. 2013. Practical considerations for bicarbonate loading and sports performance. *Nestle Nutrition Institute Workshop Series* 75:15–26.

Burke, L., and Maughan, R. 2000. Alcohol in sport. In *Nutrition in Sport,* ed. R. J. Maughan. Oxford: Blackwell Scientific.

Cahill, P. A., and Redmond, E. M. 2012. Alcohol and cardiovascular disease—modulation of vascular cell function. *Nutrients* 4:297–318.

Calvo, J. L., et al. 2021. Effect of sodium bicarbonate contribution on energy metabolism during exercise: A systematic review and meta-analysis. *Journal of the International Society of Sports Nutrition* 18(1):11. doi: 10.1186/s12970-021-00410-y. PMID: 33546730; PMCID: PMC7863495.

Cargiulo, T. 2007. Understanding the health impact of alcohol dependence. *American Journal of Health-Systems Pharmacy* 64:S5–S11.

Carlström, M., and Larsson, S.C. 2018. Coffee consumption and reduced risk of developing type 2 diabetes: A systematic review with meta-analysis. *Nutrition Reviews* 76(6):395–417. doi: 10.1093/nutrit/nuy014. PMID: 29590460.

Carr, B. M., et al. 2013. Sodium bicarbonate supplementation improves hypertrophy-type resistance exercise performance. *European Journal of Applied Physiology* 113(3):743–52.

Carter, A. C., et al. 2010. The college and non-college experience: A review of the factors that influence drinking behavior in young adulthood. *Journal of Studies on Alcohol and Drugs* 71(5):742–50.

Chaudhary, N. S., et al. 2021. The effects of caffeinated products on sleep and functioning in the military population: A focused review. *Pharmacology, Biochemistry, and Behavior* 206:173206. doi: 10.1016/j.pbb.2021.173206.

Chen, L. W., et al. 2016. Maternal caffeine intake during pregnancy and risk of pregnancy loss: A categorical and dose-response meta-analysis of prospective studies. *Public Health Nutrition* 19(7):1233–44.

Cheuvront, S., et al. 1999. Effect of ENDUROX^{TM} on metabolic responses to submaximal exercise. *International Journal of Sport Nutrition* 9:434–42.

Chiva-Blanch, G., and Badimon, L. 2019. Benefits and risks of moderate alcohol consumption on cardiovascular disease: Current findings and controversies. *Nutrients* 12(1):108. doi: 10.3390/nu12010108. PMID: 31906033; PMCID: PMC7020057.

Choi, Y. J., et al. 2018. Light alcohol drinking and risk of cancer: A meta-analysis of cohort studies. *Cancer Research and Treatment* 50(2):474–87. doi: 10.4143/crt.2017.094.

Clark, I., and Landolt, H. P. 2017. Coffee, caffeine, and sleep: A systematic review of epidemiological studies and randomized controlled trials. *Sleep Medicine Reviews* 31:70–78.

Clarke, N. D., and Richardson, D. L. 2021. Habitual caffeine consumption does not affect the ergogenicity of coffee ingestion during a 5 km cycling time trial. *International Journal of Sports Nutrition and Exercise Metabolism* 31(1):13–20. doi: 10.1123/ijsnem.2020-0204.

Clements, W. T., et al. 2014. Nitrate ingestion: A review of the health and physical performance effects. *Nutrients* 6:5224–64. doi:10.3390/nu6115224.

Coathup, V., et al. 2017. Dietary patterns and alcohol consumption during pregnancy: Secondary analysis of Avon Longitudinal Study of Parents and Children. *Alcoholism: Clinical and Experimental Research* 41(6):1120–28.

Cohen, P. A., et al. 2015. An amphetamine isomer whose efficacy and safety in humans has never been studied, β-methylphenylethylamine (BMPEA), is found in multiple dietary supplements. *Drug Testing and Analysis.* doi: 10.1002/dta.1793.

Coiro, V., et al. 2007. Effects of moderate ethanol drinking on the GH and cortisol responses to physical exercise. *Neuro Endocrinology Letters* 28:145–8.

Collado-Mateo, D., et al. 2020. Effect of acute caffeine intake on the fat oxidation rate during exercise: A systematic review and meta-analysis. *Nutrients* 12(12):3603. doi: 10.3390/nu12123603. PMID: 33255240; PMCID: PMC7760526.

Conger, S., et al. 2011. Does caffeine added to carbohydrate provide additional ergogenic benefits for endurance? *International Journal of Sport Nutrition and Exercise Metabolism* 21:71–84.

Corona, G., et al. 2013. Dehydroepiandrosterone supplementation in elderly men: A meta-analysis study of placebo-controlled trials. *Journal of Clinical Endocrinology and Metabolism* 98(9):3615–26.

Corona, G., et al. 2018. Testosterone and cardiovascular risk: Meta-analysis of interventional studies. *Journal of Sexual Medicine* 15(6):820–38. doi: 10.1016/j.jsxm.2018.04.641. PMID: 29803351.

Costanzo, S., et al. 2019. Moderate alcohol consumption and lower total mortality risk: Justified doubts or established facts? *Nutrition, Metabolism and Cardiovascular Diseases* 29(10):1003–8. doi: 10.1016/j.numecd.2019.05.062.

Costill, D., et al. 1978. Effects of caffeine ingestion on metabolism and exercise performance. *Medicine & Science in Sports* 10:155–58.

Cox, H. D., et al. 2013. Quantification of insulin-like growth factor-1 in dried blood spots for detection of growth hormone abuse in sport. *Analytical and Bioanalytical Chemistry* 405(6):1949–58.

Cox, H. D., and Eichner, D. 2013. Detection of human insulin-like growth factor-1 in deer antler velvet supplements. *Rapid Communication in Mass Spectrometry* 27(19):2170–78.

D'Elia, L., et al. 2019. Coffee consumption and risk of hypertension: A dose-response meta-analysis of prospective studies. *European Journal of Nutrition* 58(1):271–80. doi: 10.1007/s00394-017-1591-z.

Das, G., et al. 2021. *Cordyceps spp.*: A review on its immune-stimulatory and other biological potentials. *Frontiers in Pharmacology* 11:602364. doi: 10.3389/fphar.2020.602364. PMID: 33628175; PMCID: PMC7898063.

David, S., and Cunningham, R. 2019. *Echinacea* for the prevention and treatment of upper respiratory tract infections: A systematic review and meta-analysis. *Complementary Therapies in Medicine* 44:18–26. doi: 10.1016/j.ctim.2019.03.011.

Davis, J., and Green, J. 2009. Caffeine and anaerobic performance: Ergogenic value and mechanisms of action. *Sports Medicine* 39:813–32.

de Hon, O., et al. 2015. Prevalence of doping use in elite sports: a review of numbers and methods. *Sports Medicine* 45(1):57–69. doi: 10.1007/s40279-014-0247-x. PMID: 25169441.

de Oliveira, L.F., et al. 2022. Extracellular Buffering Supplements to Improve Exercise Capacity and Performance: A Comprehensive Systematic Review and Meta-analysis. *Sports Medicine* 52(3):505–526. doi: 10.1007/s40279-021-01575-x.

de Salles, P.V., et al. 2021. Habitual Caffeine Consumption Does Not Interfere With the Acute Caffeine Supplementation Effects on Strength Endurance and Jumping Performance in Trained Individuals. *International Journal of Sports Nutrition and Exercise Metabolism* 31(4):321–328. doi: 10.1123/ijsnem.2020-0363.

de Sanctis, V., et al. 2017. Caffeinated energy drink consumption among adolescents and potential health consequences associated with their use: A significant public health hazard. *Acta Biomedica* 88(2):222–31.

Dean, H. 2002. Does exogenous growth hormone improve athletic performance? *Clinical Journal of Sport Medicine* 12:250–53.

Del Coso, J., et al. 2009. Caffeine during exercise in the heat: Thermoregulation and fluid-electrolyte balance. *Medicine & Science in Sports & Exercise* 41:164–73.

Delgado, C. E., et al. 1975. Acute effects of low doses of alcohol on left ventricular function by echocardiography. *Circulation* 51 (3):535–40.

Dennig, H., et al. 1931. Effect of acidosis and alkalosis upon capacity for work. *Journal of Clinical Investigation* 9(4):601–13. doi: 10.1172/JCI100324. PMID: 16693953; PMCID: PMC435718.

Doherty, M., and Smith, P. 2005. Effects of caffeine ingestion on rating of perceived exertion during and after exercise: A meta-analysis. *Scandinavian Journal of Medicine & Science in Sports* 15:69–78.

Domínguez, R., et al. 2017. Effects of beetroot juice supplementation on cardiorespiratory endurance in athletes. a systematic review. *Nutrients* 9:43. doi:10.3390/nu9010043.

Dowling, E. et al. 1996. Effect of Eleutherococcus senticosus on submaximal and maximal exercise performance. *Medicine & Science in Sport & Exercise*, 28:482–89.

Driller, M. W., et al. 2012. The effects of serial and acute NaHCO3 loading in well-trained cyclists. *Journal of Strength and Conditioning Research* 26(10):2791–97.

Durkalec-Michalski, K., et al. 2018. The effect of chronic progressive-dose sodium bicarbonate ingestion on CrossFit-like performance: A double-blind, randomized cross-over trial. *PLoS One* 13(5):e0197480. doi: 10.1371/journal.pone.0197480. PMID: 29771966; PMCID: PMC5957406.

Earnest, C., et al. 2000. In vivo 4-androstene-3, 17-dione and 4-androstene-3 beta, 17 beta-diol supplementation in young men. *European Journal of Applied Physiology* 81:229–32.

Egger, F., et al. 2014. Effects of sodium bicarbonate on high-intensity endurance performance in cyclists: A double-blind, randomized cross-over trial. *Public Library of Science One* 9 (12):e114729. doi: 10.1371/journal.pone.0114729.

Eichenberger, P., et al. 2010. No effects of three-week consumption of a green tea extract on time trial performance in endurance-trained men. *International Journal for Vitamin and Nutrition Research* 80:54–64.

El Sayed, M., et al. 2005. Interaction between alcohol and exercise: Physiological and haematological implications. *Sports Medicine* 35:257–69.

Engels, H., and Wirth, J. 1997. No ergogenic effect of ginseng (Panax ginseng C.A. Meyer) during graded maximal aerobic exercise. *Journal of the American Dietetic Association* 97:1110–15.

Engels, H., et al. 2001. Effects of ginseng supplementation on supramaximal exercise performance and short-term recovery. *Journal of Strength and Conditioning Research* 15:290–95.

Epstein, E., et al. 2007. Women, aging, and alcohol use disorders. *Journal of Women & Aging* 19(1–2):31–48.

Erotokritou-Mulligan, I., et al. 2011. Growth hormone doping: A review. *Open Access Journal of Sports Medicine* 2:99–111.

Farrington, R., et al. 2019. Evidence for the efficacy and safety of herbal weight loss preparations. *Journal of Integrative Medicine* 17(2):87–92. doi: 10.1016/j.joim.2019.01.009.

Filip, A., et al. 2020. Inconsistency in the ergogenic effect of caffeine in athletes who regularly consume caffeine: Is it due to the disparity in the criteria that defines habitual caffeine intake? *Nutrients* 12(4):1087. doi: 10.3390/nu12041087. PMID: 32326386; PMCID: PMC7230656.

Fillmore, K., et al. 2007. Moderate alcohol use and reduced mortality risk: Systematic error in prospective studies and new hypotheses. *Annals of Epidemiology* 17:S16–S23.

Foad, A., et al. 2008. Pharmacological and psychological effects of caffeine ingestion in 40-km cycling performance. *Medicine & Science in Sport & Exercise* 40:158–65.

Font, L., et al. 2013. Involvement of the endogenous opioid system in the psychopharmacological actions of ethanol: The role of acetaldehyde. *Frontiers in Behavioral Neuroscience* 31(7):93. doi: 10.3389/fnbeh.2013.00093.

Forte, G. C., et al. 2018. Diet effects in the asthma treatment: A systematic review. *Critical Reviews in Food Science and Nutrition* 58(11):1878–87. doi: 10.1080/10408398.2017.1289893.

Fox, D. J., et al. 2015. Centers for Disease Control and Prevention (CDC). Fetal alcohol syndrome among children aged 7-9 years—Arizona, Colorado, and New York, 2010. *Morbidity and Mortality Weekly Reports* 64(3):54–57.

Freis, T., et al. 2017. Effect of sodium bicarbonate on prolonged running performance: A randomized, double-blind, cross-over study. *PLoS One* 12(8):e0182158. doi: 10.1371/journal.pone.0182158. PMID: 28797049; PMCID: PMC5552294.

Galaif, E., et al. 2007. Suicidality, depression, and alcohol use among adolescents: A review of empirical findings. *International Journal of Adolescent Medicine and Health* 19:27–35.

Ganio, M. S., et al. 2011. Effect of ambient temperature on caffeine ergogenicity during endurance exercise. *European Journal of Applied Physiology* 111:1135–46.

Ganio, M., et al. 2009. Effect of caffeine on sport-specific endurance performance: A systematic review. *Journal of Strength and Conditioning Research* 23:315–24.

Gant, N., et al. 2010. The influence of caffeine and carbohydrate coingestion on simulated soccer performance. *International Journal of Sport Nutrition and Exercise Metabolism* 20:191–97.

Gardner, J. D., and Mouton, A. J. 2015. Alcohol effects on cardiac function. *Comprehensive Physiology* 5:791–802.

Garthe, I., and Maughan, R.J. 2018. Athletes and supplements: Prevalence and perspectives. *International Journal of Sport Nutrition and Exercise Metabolism* 28(2):126–38. doi: 10.1123/ijsnem.2017-0429.

George, A., and Figueredo, V. 2010. Alcohol and arrhythmias: A comprehensive review. *Journal of Cardiovascular Medicine* 11:221–28.

George, A., and Figueredo, V. M. 2011. Alcoholic cardiomyopathy: A review. *Journal of Cardiac Failure* 17:844–49.

Gheorghiev, M. D., et al. 2018. Effects of pseudoephedrine on parameters affecting exercise performance: A meta-analysis. *Sports Medicine Open* 4(1):44. doi: 10.1186/s40798-018-0159-7. PMID: 30291523; PMCID: PMC6173670.

Glaister, M., and Gissane, C. 2017. Caffeine and physiological responses to submaximal exercise: A meta-analysis. *International Journal of Sports Physiology and Performance* 5:1–23.

Gough, L. A., et al. 2017. The reproducibility of 4-km time trial (TT) performance following individualised sodium bicarbonate supplementation: A randomised controlled trial in trained cyclists. *Sports Medicine-Open* 3:34. doi:10.1186/s40798-017-0101-4.

Gonçalves, L., et al. 2017. Dispelling the myth that habitual caffeine consumption influences the performance response to acute caffeine supplementation. *Journal of Applied Physiology* 123(1): 213–20.

Graham, T. 1981. Alcohol ingestion and man's ability to adapt to exercise in a cold environment. *Canadian Journal of Applied Sport Sciences* 6(1):27–31.

Graham, T., et al. 1998. Metabolic and exercise endurance effects of coffee and caffeine ingestion. *Journal of Applied Physiology* 85:883–89.

Graham, T., et al. 2008. Does caffeine alter muscle carbohydrate and fat metabolism during exercise? *Applied Physiology, Nutrition, and Metabolism* 33:1311–18.

Grech, A., et al. 2014. Adverse effects of testosterone replacement therapy: An update on the evidence and controversy. *Therapeutic Advances in Drug Safety* 5(5):190–200.

Greenwood, D. C., et al. 2014. Caffeine intake during pregnancy and adverse birth outcomes: A systematic review and dose-response meta-analysis. *European Journal of Epidemiology* 29(10):725–34.

Grgic, J., et al. 2018. Caffeine ingestion enhances Wingate performance: A meta-analysis. *European Journal of Sport Science* 8(2):219–25. doi: 10.1080/17461391.2017.1394371.

Grgic, J., and Pickering, C. 2019. The effects of caffeine ingestion on isokinetic muscular strength: A meta-analysis. *Journal of Science and Medicine in Sport* 22(3):353–60. doi: 10.1016/j.jsams.2018.08.016.

Grgic, J. 2021. Effects of caffeine on resistance exercise: A review of recent research. *Sports Medicine* 51(11):2281–98. doi: 10.1007/s40279-021-01521-x.

Grgic, J., and Del Coso, J. 2021. Ergogenic effects of acute caffeine intake on muscular endurance and muscular strength in women: A meta-analysis. *International Journal of Environmental Research and Public Health* 18(11):5773. doi: 10.3390/ijerph18115773. PMID: 34072182; PMCID: PMC8199301.

Grgic, J., et al. 2021. International Society of Sports Nutrition position stand: Sodium bicarbonate and exercise performance. *Journal of the International Society of Sports Nutrition* 18(1):61. doi: 10.1186/s12970-021-00458-w. PMID: 34503527; PMCID: PMC8427947.

Grosso, G., et al. 2016. Coffee consumption and risk of all-cause, cardiovascular, and cancer mortality in smokers and non-smokers: A dose-response meta-analysis. *European Journal of Epidemiology* 31(12):1191–205. doi: 10.1007/s10654-016-0202-2.

Gruenewald, P., et al. 2010. A dose-response perspective on college drinking and related problems. *Addiction* 105:257–69.

Guest, N. S., et a. 2021. International society of sports nutrition position stand: Caffeine and exercise performance. *Journal of the International Society of Sports Nutrition* 18(1):1. doi: 10.1186/s12970-020-00383-4. PMID: 33388079; PMCID: PMC7777221.

Gunzerath, L. 2004. National Institute on Alcohol Abuse and Alcoholism report on moderate drinking. *Alcoholism, Clinical and Experimental Research* 28(6):829–47.

Gurley, B. J., et al. 2014. Multi-ingredient, caffeine-containing dietary supplements: History, safety, and efficacy. *Clinical Therapeutics* 37(2):275–301.

Haller, C., et al. 2004. Enhanced stimulant and metabolic effects of combined ephedrine and caffeine. *Clinical Pharmacology and Therapeutics* 75:259–73.

Hallström, H., et al. 2013. Long-term coffee consumption in relation to fracture risk and bone mineral density in women. *American Journal of Epidemiology* 178 (6):898–909.

Hamilton, D. A., et al. 2014. Effects of moderate prenatal ethanol exposure and age on social behavior, spatial response perseveration errors and motor behavior. *Behavioral Brain Research* 269:44–54. doi: 10.1016/j.bbr.2014.04.029.

Han, Y. Y. et al. 2021. Urinary caffeine and caffeine metabolites, asthma, and lung function in a nationwide study of U.S. adults. *Journal of Asthma* Oct 27:1–8. doi: 10.1080/02770903.2021.1993250.

Hartgens, F., et al. 1996. Body composition, cardiovascular risk factors and liver function in long term androgenic-anabolic steroids-using bodybuilders three months after drug withdrawal. *International Journal of Sports Medicine* 17:429–33.

Harty, P. S., et al. 2018. Multi-ingredient pre-workout supplements, safety implications, and performance outcomes: A brief review. *Journal of the International Society of Sports Nutrition* 15(1):41. doi: 10.1186/s12970-018-0247-6. PMID: 30089501; PMCID: PMC6083567.

Haugvad, A., et al. 2014. Ethanol does not delay muscle recovery but decreases testosterone/cortisol ratio. *Medicine and Science in Sport and Exercise* 46(11):2175–83.

Hayakawa, S., et al. 2020. Anti-cancer effects of green tea epigallocatchin-3-gallate and coffee chlorogenic acid. *Molecules* 25(19):4553. doi: 10.3390/molecules25194553. PMID: 33027981; PMCID: PMC7582793.

Hermansen, K., et al. 2017. Impact of GH administration on athletic performance in healthy young adults: A systematic review and meta-analysis of placebo-controlled trials. *Growth Hormone and IGF Research* 34:38–44.

Hespel, P., et al. 2006. Dietary supplements for football. *Journal of Sports Sciences* 24:749–61.

Higgins, S., et al. 2016. The effects of pre-exercise caffeinated coffee ingestion on endurance performance: An evidence-based review. *International Journal of Sports Nutrition and Exercise Metabolism* 26(3): 221–39.

Hirsch, K. R., et al. 2017. *Cordyceps militaris* improves tolerance to high-intensity exercise after acute and chronic supplementation. *Journal of Dietary Supplements* 14(1):42–53.

Hobson, R., and Maughan, R. 2010. Hydration status and the diuretic action of a small dose of alcohol. *Alcohol and Alcoholism* 45:366–73.

Hoffman, J. R., et al. 2018. Effects of β-alanine supplementation on carnosine elevation and physiological performance. *Advances in Food and Nutrition Research* 84:183–206. doi: 10.1016/bs.afnr.2017.12.003.

Hoffman, J., et al. 2009. National Strength and Conditioning Association. Position stand on androgen and human growth hormone use. *Journal of Strength and Conditioning Research* 23:S1–S59.

Hollidge-Horvat, M. G., et al. 2000. Effect of induced metabolic alkalosis on human skeletal muscle metabolism during exercise. *American Journal of Physiology* 278:E316–E329.

Houmard, J. A., et al. 1987. Effects of the acute ingestion of small amounts of alcohol upon 5-mile run times. *Journal of Sports Medicine and Physical Fitness* 27 (2):253–57.

Hoyme, H. E., et al. 2016. Updated clinical guidelines for diagnosing fetal alcohol spectrum disorders. *Pediatrics* 138(2): e20154256.

Hu, Y., et al. 2021. Impact of dehydroepiandrosterone (DHEA) supplementation on testosterone concentrations and BMI in elderly women: A meta-analysis of randomized controlled trials. *Complementary Therapies in Medicine* 56:102620. doi: 10.1016/j.ctim.2020.102620.

Hudson, A., et al. 2018. A review of the toxicity of compounds found in herbal dietary supplements. *Planta Medica* 84(9–10):613–26. doi: 10.1055/a-0605-3786.

Huerta Ojeda, Á., et al. 2020. Effects of beta-alanine supplementation on physical performance in aerobic-anaerobic transition zones: A systematic review and meta-analysis. *Nutrients* 12(9):2490. doi: 10.3390/nu12092490. PMID: 32824885; PMCID: PMC7551186.

Hursel, R., and Westerterp-Plantenga, M. S. 2013. Catechin- and caffeine-rich teas for control of body weight in humans. *American Journal of Clinical Nutrition* 98(6S):1682S–1693S.

Ikeuchi, S., et al. 2022. Exploratory systematic review and meta-analysis of Panax genus plant ingestion evaluation in exercise endurance. *Nutrients* 14(6):1185. doi: 10.3390/nu14061185. PMID: 35334841; PMCID: PMC8950061.

Ip, E. J., et al. 2011. The Anabolic 500 Survey: Characteristics of male users versus nonusers of anabolic-androgenic steroids for strength training. *Pharmacotherapy* 31(8):757–66.

Irwin, C., et al. 2011. Caffeine withdrawal and high-intensity endurance cycling performance. *Journal of Sports Sciences* 29:509–15.

Ishaque, S., et al. 2012. Rhodiola rosea for physical and mental fatigue: A systematic review. *BioMed Central Complementary and Alternative Medicine* 12:70.www.biomedcentral.com/1472-6882/12/70.

Isidori, A. M., et al. 2005. Effects of testosterone on body composition, bone metabolism and serum lipid profile in middle-aged men: A meta-analysis. *Clinical Endocrinology* 63(3):280–93.

Jacobs, L., et al. 2004. Effects of ephedrine, caffeine, and their combination on muscular endurance. *Medicine & Science in Sports & Exercise* 35:987–94.

Jagim, A. R., et al. 2019. Common ingredient profiles of multi-ingredient pre-workout supplements. *Nutrients* 11(2):254. doi: 10.3390/nu11020254. PMID: 30678328; PMCID: PMC6413194.

Jakimovski, D., et al. 2019. Lifestyle-based modifiable risk factors in multiple sclerosis: Review of experimental and clinical findings. *Neurodegenerative Disease Management* 9(3):149–72. doi: 10.2217/nmt-2018-0046.

James, J. E., and Rogers, P. J. 2005. Effects of caffeine on performance and mood: Withdrawal reversal is the most plausible explanation. *Psychopharmacology* 182:1–8.

Jang, H. H., et al. 2020. Effects of *Capsicum annuum* supplementation on the components of metabolic syndrome: A systematic review and meta-analysis. *Scientific Reports* 10(1):20912. doi: 10.1038/s41598-020-77983-2. PMID: 33262398; PMCID: PMC7708630.

Janssens, P. L., et al. 2015. Long-term green tea extract supplementation does not affect fat absorption, resting energy expenditure, and body composition in adults. *Journal of Nutrition* 145(5):864–70. doi: 10.3945/jn.114.207829.

Jones, A., et al. 2013. Habitual alcohol consumption is associated with lower cardiovascular stress responses – A novel explanation for the known cardiovascular benefits of alcohol? *Stress* 16(4):369–76.

Juliano, L., and Griffiths, R. 2004. A critical review of caffeine withdrawal: Empirical validation of symptoms and signs, incidence, severity, and associated features. *Psychopharmacology* 176:1–29.

Kalla, A., and Figueredo, V. M. 2017. Alcohol and cardiovascular disease in the geriatric population. *Clinical Cardiology* 40:444–49.

Kallioinen, N., et al. 2017. Sources of inaccuracy in the measurement of adult patients' resting blood pressure in clinical settings: A systematic review. *Journal of Hypertension* 35(3):421–41.

Kalmar, J. 2005. The influence of caffeine on voluntary muscle activation. *Medicine & Science in Sports & Exercise* 37:2113–19.

Karsch-Völk, M., et al. 2014. Echinacea for preventing and treating the common cold. *Cochrane Database of Systematic Reviews* 2(2):CD000530. doi: 10.1002/14651858.CD000530.pub3. PMID: 24554461; PMCID: PMC4068831.

Kasuda, S., et al. 2006. Inhibition of PAR4 signaling mediates ethanol-induced attenuation of platelet function *in vitro*. *Alcoholism, Clinical and Experimental Research* 30(9):1608–14.

Kawano, Y. 2010. Physio-pathological effects of alcohol on the cardiovascular system: Its role in hypertension and cardiovascular disease. *Hypertension Research* 33(3):181–91.

Kelly, Y., et al. 2010. Light drinking during pregnancy: Still no increased risk for socioemotional difficulties or cognitive deficits at 5 years of age? *Journal of Epidemiology and Community Health* (October 5).

Kendrick, Z., et al. 1993. Effect of ethanol on metabolic responses to treadmill running in well-trained men. *Journal of Clinical Pharmacology* 33:136–39.

Kennedy, O. J., et al. 2016. Systematic review with meta-analysis: Coffee consumption and the risk of cirrhosis. *Alimentary Pharmacology and Therapeutics* 43:562–74.

Kersey, R. D., et al. 2012. National Athletic Trainers' Association position statement: Anabolic-androgenic steroids. *Journal of Athletic Training* 47(5):567–88.

Kim, Y., et al. 2019. Coffee consumption and all-cause and cause-specific mortality: A meta-analysis by potential modifiers. *European Journal of Epidemiology* 34(8):731–52. doi: 10.1007/s10654-019-00524-3.

Kim, T.L., et al. 2020. Tea consumption and risk of cancer: An umbrella review and meta-analysis of observational studies. *Advances in Nutrition* 11(6):1437–52. doi: 10.1093/advances/nmaa077. PMID: 32667980; PMCID: PMC7666907.

Kim, J., et al. 2021. A network pharmacology approach to explore the potential role of Panax ginseng on exercise performance. *Physical Activity and Nutrition* 25(3):28–35. doi: 10.20463/pan.2021.0018.

Kim, E.J., et al. 2021. Coffee consumption and incident Tachyarrhythmias: Reported behavior, Mendelian randomization, and their interactions. *JAMA Internal Medicine* 181(9):1185–93. doi: 10.1001/jamainternmed.2021.3616. PMID: 34279564; PMCID: PMC8290332.

Kim, Y. S., et al. 2015. Safety analysis of *Panax ginseng* in randomized clinical trials: A systematic review. *Medicines* 2(2):106–26. doi:10.3390/medicines2020106.

King, D., et al. 1999. Effect of oral androstenedione on serum testosterone and adaptations to resistance training in young men. *Journal of the American Medical Association* 281:2020–28.

Kitkungvan, D., et al. 2019. Quantitative myocardial perfusion positron emission tomography and caffeine revisited with new insights on major adverse cardiovascular events and coronary flow capacity. *European Heart Journal Cardiovascular Imaging* 20(7):751–62. doi: 10.1093/ehjci/jez080. PMID: 31056681.

Koppes, L., et al. 2005. Moderate alcohol consumption lowers the risk of type 2 diabetes: A meta-analysis of prospective observational studies. *Diabetes Care* 28:719–25.

Kraemer, W. J., et al. 2020. Growth hormone(s), testosterone, insulin-like growth factors, and cortisol: Roles and integration for cellular development and growth with exercise. *Frontiers in Endocrinology* 11:33. doi: 10.3389/fendo.2020.00033. PMID: 32158429; PMCID: PMC7052063.

Kruisselbrink, L. D., et al. 2006. Physical and psychomotor functioning of females the morning after consuming low to moderate quantities of beer. *Journal of Studies on Alcohol* 67(3):416-20.

Kuypers, K., et al. 2020. Intoxicated aggression: Do alcohol and stimulants cause dose-related aggression? A review. *European Neuropsychopharmacology* 30:114-47. doi: 10.1016/j.euroneuro.2018.06.001.

Kwan, M., et al. 2014. Sport participation and alcohol and illicit drug use in adolescents and young adults: A systematic review of longitudinal studies. *Addictive Behaviors* 39(3):497-506.

Lankester, J., et al. 2021. Alcohol use and cardiometabolic risk in the UK Biobank: A Mendelian randomization study. PLoS One 16(8):e0255801. doi: 10.1371/journal.pone.0255801. PMID: 34379647; PMCID: PMC8357114.

Lauritzen, F. 2022. Dietary supplements as a major cause of anti-doping rule violations. *Frontiers in Sports and Active Living* 4:868228. doi: 10.3389/fspor.2022.868228.

Lawler, T. P., and Cialdella-Kam, L. 2020. Non-carbohydrate dietary factors and their influence on post-exercise glycogen storage: A review. *Current Nutrition Reports* 9(4):394-404. doi: 10.1007/s13668-020-00335-z. PMID: 33128726.

Lau, K. M., et al. 2019. A review on the immunomodulatory activity of Acanthopanax senticosus and its active components. *Chinese Medicine* 14:25. doi: 10.1186/s13020-019-0250-0. PMID: 31388349; PMCID: PMC6670126.

Lecoultre, V., and Schutz, Y. 2009. Effect of a small dose of alcohol on the endurance performance of trained cyclists. *Alcohol and Alcoholism* 44(3):278-83.

Leder, B. et al. 2000. Oral androstenedione administration and serum testosterone concentrations in young men. *Journal of the American Medical Association* 283:779-82.

Lee, N. Y., and Suk, K. T. 2020. The role of the gut microbiome in liver cirrhosis treatment. *International Journal of Molecular Sciences* 22(1):199. doi: 10.3390/ijms22010199. PMID: 33379148; PMCID: PMC7796381.

Lee, J. E., et al. 2007. Alcohol intake and renal cell cancer in a pooled analysis of 12 prospective studies. *Journal of the National Cancer Institute* 99(10):801-10. doi: 10.1093/jnci/djk181. PMID: 17505075.

Lees, B., et al. 2020. Effect of alcohol use on the adolescent brain and behavior. *Pharmacology, Biochemistry, and Behavior* 192:172906. doi: 10.1016/j.pbb.2020.172906.

Leong, D. P., et al. 2014. Patterns of alcohol consumption and myocardial infarction risk: Observations from 52 countries in the INTERHEART case-control study. *Circulation* 130 (5):390-98.

Li, J., et al. 2015. A meta-analysis of risk of pregnancy loss and caffeine and coffee consumption during pregnancy. *International Journal of Gynaecology and Obstetrics* 130(2):116-22. doi: 10.1016/j.ijgo.2015.03.033.

Li, Y., et al. 2020. A dose-response and meta-analysis of dehydroepiandrosterone (DHEA) supplementation on testosterone levels: Perinatal prediction of randomized clinical trials. *Experimental Gerontology* 141:111110. doi: 10.1016/j.exger.2020.111110.

Lieber, C. S. 2004. New concepts of the pathogenesis of alcoholic liver disease lead to novel treatments. *Current Gastroenterology Reports* 6(1):60-65.

Lieberman, H. R., et al. 2019. Daily patterns of caffeine intake and the association of intake with multiple sociodemographic and lifestyle factors in US adults based on the NHANES 2007-2012 Surveys. *Journal of the Academy of Nutrition and Dietetics* 119(1):106-114. doi: 10.1016/j.jand.2018.08.152.

Lin, C. H., et al. 2021. American ginseng attenuates eccentric exercise-induced muscle damage via the modulation of lipid peroxidation and inflammatory adaptation in males. *Nutrients* 14(1):78. doi: 10.3390/nu14010078. PMID: 35010953; PMCID: PMC8746757.

Linossier, M. et al. 1997. Effect of sodium citrate on performance and metabolism of human skeletal muscle during supramaximal cycling exercise. *European Journal of Applied Physiology* 76:48-54.

Liquori, A., et al. 1997. Absorption and subjective effects of caffeine from coffee, cola, and capsules. *Pharmacology, Biochemistry, and Behavior* 58(3):721-26.

Liu, H., et al. 2007. Systematic review: The safety and efficacy of growth hormone in the healthy elderly. *Annals of Internal Medicine* 146:104-15.

Liu, J., et al. 2013. Association of coffee consumption with all-cause and cardiovascular disease mortality. *Mayo Clinic Proceedings* 88(10):1066-74.

Liu, T. C., et al. 2013. Effect of acute DHEA administration on free testosterone in middle-aged and young men following high-intensity interval training. *European Journal of Applied Physiology* 113(7):1783-92.

LiverTox: Clinical and Research Information on Drug-Induced Liver Injury. Bethesda (MD): National Institute of Diabetes and Digestive and Kidney Diseases; 2012-. Hydroxycut. 2018 Apr 12. PMID: 31643575. www.ncbi.nlm.nih.gov/books.

Loomis, D., et al. 2016. Carcinogenicity of drinking coffee, mate, and very hot beverages. *The Lancet Oncology* 17(7):877-78.

Lopez-Garcia, E., et al. 2006. Changes in caffeine intake and long-term weight change in men and women. *American Journal of Clinical Nutrition* 83:674-80.

Lu, Y., et al. 2022. Effects of *Rhodiola rosea* supplementation on exercise and sport: A systematic review. *Frontiers in Nutrition* 9:856287. doi: 10.3389/fnut.2022.856287. PMID: 35464040; PMCID: PMC9021834.

Lucas, B. R., et al. 2014. Gross motor deficits in children prenatally exposed to alcohol: A meta-analysis. *Pediatrics* 134(1):e192-209.

Lystrup, R. M., and Leggit, J. C. 2015. Caffeine toxicity due to supplement use in caffeine–Naïve individual: A cautionary tale. *Military Medicine* 180(8):e936-40. doi: 10.7205/milmed-d-15-00045.

Magkos, F., and Kavouras, S. 2004. Caffeine and ephedrine: Physiological, metabolic and performance-enhancing effects. *Sports Medicine* 34:971-89.

Maglione, M., et al. 2005. Psychiatric effects of ephedra use: An analysis of Food and Drug Administration reports of adverse events. *American Journal of Psychiatry* 162:189-91.

Mancini, E., et al. 2017. Green tea effects on cognition, mood and human brain function: A systematic review. *Phytomedicine* 34:26-37. doi: 10.1016/j.phymed.2017.07.008.

Mandal, C., et al. 2017. Gestational alcohol exposure altered DNA methylation status in the developing fetus. *International Journal of Molecular Sciences* 18:1386. doi:10.3390/ijms18071386.

Marczinski, C. A., et al. 2018. Differential development of acute tolerance may explain heightened rates of impaired driving after consumption of alcohol mixed with energy drinks versus alcohol alone. *Experimental and Clinical Psychopharmacology* 26(2):147-55. doi: 10.1037/pha0000173.

Marxen, M., et al. 2014. Acute effects of alcohol on brain perfusion monitored with arterial spin labeling magnetic resonance imaging in young adults. *Journal of Cerebral Blood Flow and Metabolism* 34: 472-79.

McLellan, T. M., et al. 2016. A review of caffeine's effects on cognitive, physical and occupational performance. *Neuroscience and Biobehavioral Reviews* 71: 294-312.

McLellan, T., and Bell, D. 2004. The impact of prior coffee consumption on the subsequent ergogenic effect of anhydrous caffeine. *International Journal of Sport Nutrition and Exercise Metabolism* 14:698-708.

McNaughton, L., and Cedaro, R. 1992. Sodium citrate ingestion and its effects on maximal anaerobic exercise of different durations. *European Journal of Applied Physiology* 64:36-41.

McNaughton, L., and Preece, D. 1986. Alcohol and its effects on sprint and middle distance running. *British Journal of Sports Medicine* 20(2):56-59.

McNaughton, L., et al. 1999. Sodium bicarbonate can be used as an ergogenic aid in high-intensity, competitive cycle ergometry of 1 h duration. *European Journal of Applied Physiology* 80:64-69.

Meinhardt, U., et al. 2010. The effects of growth hormone on body composition and physical performance in recreational athletes: A randomized trial. *Annals of Internal Medicine* 152:568-77.

Meng, F. G., et al. 2020. Roles of peroxisome proliferator-activated receptor α in the pathogenesis of ethanol-induced liver disease. *Chemico-Biological Interactions* 327:109176. doi: 10.1016/j.cbi.2020.109176.

Merrill, J. E., and Carey, K. B. 2016. Drinking over the lifespan: Focus on college ages. *Alcohol Research: Current Reviews* 38(1):103-14.

Mets, M. A. J., et al. 2012. Effects of coffee on driving performance during prolonged simulated highway driving. *Psychopharmacology* (2012) 222:337-42.

Miller, P. E., et al. 2017. Associations of coffee, tea, and caffeine intake with coronary artery calcification and cardiovascular events. *American Journal of Medicine* 130:188-97.

Minzer, S., et al. 2020. The effect of alcohol on cardiovascular risk factors: Is there new information? *Nutrients* 12(4):912. doi: 10.3390/nu12040912. PMID: 32230720; PMCID: PMC7230699.

Moeenfard, M., and Alves, A. 2020. New trends in coffee diterpenes research from technological to health aspects. *Food Research International* 134:109207. doi: 10.1016/j.foodres.2020.109207.

Morgentaler, A., et al. 2015. Testosterone therapy and cardiovascular risk: Advances and controversies. *Mayo Clinic Proceedings* 90(2):224-51.

Morozova, T. V., et al. 2014. Genetics and genomics of alcohol sensitivity. *Molecular Genetics and Genomics* 289 (3):253-69.

Mueller, S. M., et al. 2013. Multiday acute sodium bicarbonate intake improves endurance capacity and reduces acidosis in men. *Journal of the International Society of Sports Nutrition* 10(1):16. doi: 10.1186/1550-2783-10-16.

Mukamal, K., et al. 2006. Alcohol consumption and risk for coronary heart disease in men with healthy lifestyles. *Archives of Internal Medicine* 166:2145-50.

Musgrave, I. F., et al. 2016. Caffeine toxicity in forensic practice: Possible effects and under-appreciated sources. *Forensic Science, Medicine, and Pathology* 12(3):299-303.

Musial, C., et al. 2020. Beneficial properties of green tea catechins. *International Journal of Molecular Science* 21(5):1744. doi: 10.3390/ijms21051744. PMID: 32143309; PMCID: PMC7084675.

Naimi, T., et al. 2005. Cardiovascular risk factors and confounders among nondrinking and moderate-drinking U.S. adults. *American Journal of Preventive Medicine* 28:369-73.

National Center for Statistics and Analysis. 2019. *Alcohol-impaired driving: 2018 data* (Traffic Safety Facts. Report No. DOT HS 812 864). Washington, DC: National Highway Traffic Safety Administration.

Nicholls, A. R., et al. 2017. Children's first experience of taking anabolic-androgenic steroids can occur before their 10th birthday: A systematic review identifying 9 factors that predicted doping among young people. *Frontiers in Psychology* 8:1015. doi: 10.3389/fpsyg.2017.01015.

Nyberg, F., and Hallberg, M. 2012. Interactions between opioids and anabolic androgenic steroids: Implications for the development of addictive behavior. *International Review of Neurobiology* 102:189-206.

O'Callaghan, F., et al. 2018. Effects of caffeine on sleep quality and daytime functioning. *Risk Management and Healthcare Policy* 11:263-71. doi: 10.2147/RMHP.S156404. PMID: 30573997; PMCID: PMC6292246.

O'Keefe, J. H., et al. 2018. Coffee for cardioprotection and longevity. *Progress in Cardiovascular Diseases* 61(1):38-42. doi: 10.1016/j.pcad.2018.02.002.

Ööpik, V., et al. 2003. Effects of sodium citrate ingestion before exercise on endurance performance in well-trained college runners. *British Journal of Sports Medicine* 37:485-89.

Panza F., et al. 2015. Coffee, tea, and caffeine consumption and prevention of late-life cognitive decline and dementia: A systematic review. *Journal of Nutrition, Health and Aging* 19(3):313-28.

Parcell, A., et al. 2004. Cordyceps sinensis (CordyMax Cs-4) supplementation does not improve endurance exercise performance. *International Journal of Sport Nutrition and Exercise Metabolism* 14:236-42.

Peana, A. T., et al. 2017. Mystic acetaldehyde: The never-ending story on alcoholism. *Frontiers in Behavioral Neuroscience* 11:81. doi: 10.3389/fnbeh.2017.00081.

Peart, D. J., et al. 2011. Pre-exercise alkalosis attenuates the heat shock protein 72 response to a single-bout of anaerobic exercise. *Journal of Science and Medicine in Sport* 14(5):435-40.

Peart, D. J., et al. 2012. Practical recommendations for coaches and athletes: A meta-analysis of sodium bicarbonate use for athletic performance. *Journal of Strength and Conditioning Research* 26(7):1975-83.

Penson, P., et al. 2018. Does coffee consumption alter plasma lipoprotein(a) concentrations? A systematic review. *Critical Reviews in Food Science and Nutrition* 58(10):1706-14. doi: 10.1080/10408398.2016.1272045.

Pesta, D. H., et al. 2013. The effects of caffeine, nicotine, ethanol, and tetrahydrocannabinol on exercise performance. *Nutrition and Metabolism* 10(1):71. doi: 10.1186/1743-7075-10-71.

Piacentino, D., et al. 2015. Anabolic-androgenic steroid use and psychopathology in athletes. A systematic review. *Current Neuropharmacology* 13(1):101-21.

Piazza-Gardner, A. K., and Barry, A. E. 2012. Examining physical activity levels and alcohol consumption: Are people who drink more active? *American Journal of Health Promotion* 26 (3):e95-e104.

Piazza-Gardner, A. K., et al. 2013. The impact of alcohol on Alzheimer's disease: A systematic review. *Aging and Mental Health* 17 (2):133-46.

Pirastu, N., et al. 2016. Non-additive genome-wide association scan reveals a new gene associated with habitual coffee consumption. *Scientific Reports* 6:31590. doi: 10.1038/srep31590.

Plowman, S., et al. 1999. The effects of ENDUROX on the physiological responses to stair-stepping exercise. *Research Quarterly for Exercise and Sport* 70:385-88.

Poole, R., et al. 2017. Coffee consumption and health: Umbrella review of meta-analyses of multiple health outcomes. *BMJ* 359:j5024. doi: 10.1136/bmj.j5024. Erratum in: *BMJ*. 2018 Jan 12;360:k194. PMID: 29167102; PMCID: PMC5696634.

Pope, H. G., et al. 2014. Adverse health consequences of performance enhancing drugs: An Endocrine Society scientific statement. *Endocrine Reviews* 35(3):341-74.

Potteiger, J., et al. 1996. Sodium citrate ingestion enhances 30 km cycling performance. *International Journal of Sports Medicine* 17:7-11.

Potteiger, J., et al. 1996. The effects of buffer ingestion on metabolic factors related to distance running performance. *European Journal of Applied Physiology* 72:365-71.

Prediger, R. 2010. Effects of caffeine in Parkinson's disease: From neuroprotection to the management of motor and nonmotor symptoms. *Journal of Alzheimer's Disease* 20:S205-20.

Price, M., et al. 2003. Effects of sodium bicarbonate ingestion on prolonged intermittent exercise. *Medicine & Science in Sports & Exercise* 35:1303-8.

Pritchard-Peschek, K. R., et al. 2014. The dose-response relationship between pseudoephedrine ingestion and exercise performance. *Journal of Science and Medicine in Sport* 17(5):531-34.

Pritchard-Peschek, K., et al. 2010. Pseudoephedrine ingestion and cycling time-trial performance. *International Journal of Sport Nutrition and Exercise Metabolism* 20:132-38.

Pu, W. L., et al. 2020. Anti-inflammatory effects of *Rhodiola rosea* L.: A review. *Biomedicine and Pharmacotherapy* 121:109552. doi: 10.1016/j.biopha.2019.109552.

Qureshi, A., et al. 2014. A systematic review of the herbal extract Tribulus terristris and the roots of its putative aphrodisiac and performance enhancing effect. *Journal of Dietary Supplements* 11(1):64-69.

Ramanathan, M. R., and Penzak, S. R. 2017. Pharmacokinetic drug interactions with Panax ginseng. *European Journal of Drug Metabolism and Pharmacokinetics* 42(4):545-557.

Ramazani, E., et al. 2021. Biological and pharmacological effects of gamma-oryzanol: An updated review of the molecular mechanisms. *Current Pharmaceutical Design* 27(19):2299-316. doi: 10.2174/1381612826666201102101428. PMID: 33138751.

Randell, R. K., et al. 2013. No effect of 1 or 7 d of green tea extract ingestion on fat oxidation during exercise. *Medicine and Science in Sports and Exercise* 45(5):883-91.

Rasmussen, B., et al. 2000. Androstenedione does not stimulate muscle protein anabolism in young healthy men. *Journal of Clinical Endocrinology and Metabolism* 85:55-59.

Raymer, G., et al. 2004. Metabolic effects of induced alkalosis during progressive forearm exercise to fatigue. *Journal of Applied Physiology* 96:2050-56.

Rehm, J., et al. 2008. Are lifetime abstainers the best control group in alcohol epidemiology? On the stability and validity of reported lifetime abstention. *American Journal of Epidemiology* 168:866-71.

Rehm, J., et al. 2019. Alcohol use and dementia: A systematic scoping review. *Alzheimer's Research and Therapy* 11(1):1. doi: 10.1186/s13195-018-0453-0. PMID: 30611304; PMCID: PMC6320619.

Reilly, T., and Halliday, F. 1985. Influence of alcohol ingestion on tasks related to archery. *Journal of Human Ergology* 14(2):99-104.

Reilly, T., and Scott, J. 1993. Effects of elevating blood alcohol levels on tasks related to dart throwing. *Perceptual and Motor Skills* 77(1):25-26.

Ren, X., and Chen, J.F. 2020. Caffeine and Parkinson's disease: Multiple benefits and emerging mechanisms. *Frontiers in Neuroscience* 14:602697. doi: 10.3389/fnins.2020.602697. PMID: 33390888; PMCID: PMC7773776.

Rhee, J., et al. 2015. Maternal caffeine consumption during pregnancy and risk of low birth weight: A dose-response meta-analysis of observational studies. *Public Library of Science One* 10(7):e0132334. doi:10.1371/journal.pone.0132334.

Roberts, J. D., et al. 2021. The impact of decaffeinated green tea extract on fat oxidation, body composition and cardio-metabolic health in overweight, recreationally active individuals. *Nutrients* 13(3):764. doi: 10.3390/nu13030764. PMID: 33652910; PMCID: PMC7996723.

Roerecke, M., et al. 2019. Alcohol consumption and risk of liver cirrhosis: A systematic review and meta-analysis. *American Journal of Gastroenterology* 114(10):1574-86. doi: 10.14309/ajg.0000000000000340. PMID: 31464740; PMCID: PMC6776700.

Rogerson, S., et al. 2007. The effect of short-term use of testosterone enanthate on muscular strength and power in healthy young men. *Journal of Strength and Conditioning Research* 21(2):354-61.

Ronksley, P. E., et al. 2011. Association of alcohol consumption with selected cardiovascular disease outcomes: A systematic review and meta-analysis. *British Medical Journal* 342:d671. doi: 10.1136/bmj.d671.

Rossato, L. G., et al. 2011. Synephrine: From trace concentrations to massive consumption in weight-loss. *Food and Chemical Toxicology* 49:8-16.

Rubeck, K., et al. 2009. Impact of growth hormone (GH) substitution on exercise capacity and muscle strength in GH-deficient adults: A meta-analysis of blinded, placebo-controlled trials. *Clinical Endocrinology* 71:860-66.

Sachdeva, A., et al. 2016. Alcohol-related dementia and neurocognitive impairment: A review study. *International Journal of High Risk Behaviors and Addiction* 5(3):e27976.

Saeed, M., et al. 2019. Potential nutraceutical and food additive properties and risks of coffee: A comprehensive overview. *Critical Reviews in Food Science and Nutrition* 59(20):3293-19. doi: 10.1080/10408398.2018.1489368.

Sagris, M., et al. 2021. Atrial fibrillation: Pathogenesis, predisposing factors, and genetics. *International Journal of Molecular Sciences* 23(1):6. doi: 10.3390/ijms23010006. PMID: 35008432; PMCID: PMC8744894.

Salehi, B., et al. 2018. Resveratrol: A double-edged sword in health benefits. *Biomedicines* 6(3):91. doi: 10.3390/biomedicines6030091. PMID: 30205595; PMCID: PMC6164842.

Santos, I. S., et al. 2012. Maternal caffeine consumption and infant nighttime waking: Prospective cohort study. *Pediatrics* 129(5):860-868.

Saunders, B., et al. 2014. Sodium bicarbonate and high-intensity-cycling capacity: Variability in responses. *International Journal of Sports Physiology and Performance* 9(4):627-32.

Saunders, B., et al. 2017. β-alanine supplementation to improve exercise capacity and performance: A systematic review and meta-analysis. *British Journal of Sports Medicine* 51(8):658-69. doi: 10.1136/bjsports-2016-096396.

Sawka, M. N. et al. 2007. American College of Sports Medicine position stand. Exercise and fluid replacement. *Medicine and Science in Sports and Exercise* 39(2):377-90. doi: 10.1249/mss.0b013e31802ca597. PMID: 17277604.

Schabort, E., et al. 2000. Dose-related elevations in venous pH with citrate ingestion do not alter 40-km cycling time-trial performance. *European Journal of Applied Physiology* 83:320-27.

Schneider, M. B., et al. 2011. Clinical report—Sports drinks and energy drinks for children and adolescents: Are they appropriate? *Pediatrics* 127:1182-89.

Schubert, M. M., and Astorino, T. A. 2013. Systematic review of the efficacy of ergogenic aids for improving running performance. *Journal of Strength and Conditioning Research* 27(6):1699–1707.

Scoccianti, C., et al. 2014. Female breast cancer and alcohol consumption. A review of the literature. *American Journal of Preventive Medicine* 46(W3S1):S16–S25.

Sellami, M., et al. 2018. Herbal medicine for sports: A review. *Journal of the International Society of Sports Nutrition* 15:14. doi: 10.1186/s12970-018-0218-y. PMID: 29568244; PMCID: PMC5856322.

Shah, S., et al. 2007. Evaluation of echinacea for the prevention and treatment of the common cold: A meta-analysis. *The Lancet Infectious Diseases* 7:473–80.

Shave, R., et al. 2001. The effects of sodium citrate ingestion on 3,000-meter time-trial performance. *Journal of Strength and Conditioning Research* 15:230–34.

Shekelle, P., et al. 2003. Efficacy and safety of ephedra and ephedrine for weight loss and athletic performance: A meta-analysis. *Journal of the American Medical Association* 289:1537–45.

Shi, X., et al. 2016. Acute caffeine ingestion reduces insulin sensitivity in healthy subjects: A systematic review and meta-analysis. *Nutrition Journal* 15(1):103. doi: 10.1186/s12937-016-0220-7.

Shimamoto, T., et al. 2013. No association of coffee consumption with gastric ulcer, duodenal ulcer, reflux esophagitis, and non-erosive reflux disease: A cross-sectional study of 8,013 healthy subjects in Japan. *Public Library of Science ONE* 8 (6):e65996. doi:10.1371/journal.pone.0065996.

Shin, E. J., et al. 2020. Red ginseng improves exercise endurance by promoting mitochondrial biogenesis and myoblast differentiation. *Molecules* 25(4):865. doi: 10.3390/molecules25040865. PMID: 32079067; PMCID: PMC7070955.

Shirreffs, S. M., and Maughan, R. J. 2006. The effect of alcohol on athletic performance. *Current Sports Medicine Reports* 5:192–96.

Siegler, J. C., et al. 2016. Mechanistic insights into the efficacy of sodium bicarbonate supplementation to improve athletic performance. *Sports Medicine Open* 2(1):41. https://doi.org/10.1186/s40798-016-0065-9

Siegler, J. C., et al. 2012. Sodium bicarbonate supplementation and ingestion timing: Does it matter? *Journal of Strength and Conditioning Research* 26(7):1953–58.

Siegler, J., and Hirscher, K. 2010. Sodium bicarbonate ingestion and boxing performance. *Journal of Strength and Conditioning Research* 24:103–8.

Skinner, J. W., et al. 2018. Muscular responses to testosterone replacement vary by administration route: a systematic review and meta-analysis. *Journal of Cachexia, Sarcopenia, and Muscle* 9(3):465–81. doi: 10.1002/jcsm.12291.

Smith, C. and Krygsman, A. 2014. Hoodia gordonii: To eat, or not to eat. *Journal of Ethnopharmacology* 155(2):987–91.

Smith, I., et al. 2014. Effects and mechanisms of ginseng and ginsenosides on cognition. *Nutrition Reviews* 72(5): 319–33.

Socała, K., et al. 2020. Neuroprotective effects of coffee bioactive compounds: A review. *International Journal of Molecular Sciences* 22(1):107. doi: 10.3390/ijms22010107. PMID: 33374338; PMCID: PMC7795778.

Söderpalm, B., and Ericson, M. 2013. Neurocircuitry involved in the development of alcohol addiction: The dopamine system and its access points. *Current Topics in Behavioral Neurosciences* 13:127–61.

Song, D. Y., et al. 2012. Alcohol intake and renal cell cancer risk: A meta-analysis. *British Journal of Cancer* 106(11):1881–90. doi: 10.1038/bjc.2012.136.

Southward, K., et al. 2018a. Correction to: The effect of acute caffeine ingestion on endurance performance: A systematic review and meta-analysis. *Sports Medicine* 48(10):2425–41. doi: 10.1007/s40279-018-0967-4. Erratum for: *Sports Medicine* 2018 Aug;48(8):1913–28. PMID: 30094798.

Southward, K., et al. 2018b. The role of genetics in moderating the inter-individual differences in the ergogenicity of caffeine. *Nutrients* 10(10):1352. doi: 10.3390/nu10101352. PMID: 30248915; PMCID: PMC6213712.

Spriet, L. L. 2014. Exercise and sport performance with low doses of caffeine. *Sports Medicine* 44 (S2):S175–84.

Stackelberg, O., et al. 2014. Alcohol consumption, specific alcoholic beverages, and abdominal aortic aneurysm. *Circulation* 130 (8):646–52.

Ștefănescu, R., et al. 2020. A comprehensive review of the phytochemical, pharmacological, and toxicological properties of *Tribulus terrestris* L. *Biomolecules* 10(5):752. doi: 10.3390/biom10050752. PMID: 32408715; PMCID: PMC7277861.

Steffen, M., et al. 2012.The effect of coffee consumption on blood pressure and the development of hypertension: A systematic review and meta-analysis. *Journal of Hypertension* 30(12):2245–54.

Stephens, T., et al. 2002. Effect of sodium bicarbonate on muscle metabolism during intense endurance cycling. *Medicine & Science in Sports & Exercise* 34:614–21.

Stickel, F., et al. 2017. Pathophysiology and management of alcoholic liver disease: Update 2016. *Gut and Liver* 11(2):173–88. doi: 10.5009/gnl16477.

Stohs, S. J., et al. 2012. Review of the human clinical studies involving *citrus aurantium* (bitter orange) extract and its primary protoalkaloid *p*-synephrine. *International Journal of Medical Sciences* 9(7):527–38.

Stratton, K., et al. 1996. (Eds.) *Fetal Alcohol Syndrome: Diagnosis, Epidemiology, Prevention, and Treatment.* The Institute of Medicine Report. Washington, DC: National Academy Press. doi.org/10.17226/4991.

Subramanya, S. B., et al. 2010. Chronic alcohol consumption and intestinal thiamin absorption: Effects on physiological and molecular parameters of the uptake process. *American Journal of Physiology. Gastrointestinal and Liver Physiology* 299(1):G23-31.

Talih, F., et al. 2007. Anabolic steroid abuse: Psychiatric and physical costs. *Cleveland Clinic Journal of Medicine* 74:341–44, 346, 349–52.

Tallis, J., et al. 2015. What can isolated skeletal muscle experiments tell us about the effects of caffeine on exercise performance? *British Journal of Pharmacology* 172(15):3703–13.

Tan, F., et al. 2010. Effects of induced alkalosis on simulated match performance in elite female water polo players. *International Journal of Sports Nutrition and Exercise Metabolism* 20 (3):198–205.

Tarnopolsky, M. 2008. Effect of caffeine on the neuromuscular system—Potential as an ergogenic aid. *Applied Physiology, Nutrition, and Metabolism* 33:1284–89.

Tarnopolsky, M., and Cupido, C. 2000. Caffeine potentiates low frequency skeletal muscle force in habitual and nonhabitual caffeine consumers. *Journal of Applied Physiology* 89:1719–24.

Taylor, M. K., et al. 2012. Effects of dehydroepiandrosterone supplementation during stressful military training: A randomized, controlled, double-blind field study. *Stress* 15(1):85–96.

Temple, J. L. 2019. Review: Trends, safety, and recommendations for caffeine use in children and adolescents. *Journal of the American Academy of Child and Adolescent Psychiatry* 58(1):36–45. doi: 10.1016/j.jaac.2018.06.030.

Tentori, L., and Graziani, G. 2007. Doping with growth hormone/IGF-1, anabolic steroids or erythropoietin: Is there a cancer risk? *Pharmacological Research* 55:359–69.

Thomas, D. T., et al. 2016. Position of the Academy of Nutrition and Dietetics, Dietitians of Canada, and the American College of Sports Medicine: Nutrition and Athletic Performance. *Journal of the Academy of Nutrition and Dietetics* 116(3):501–28.

Thomas, J. E., et al. 2009. STEMI in a 24-year-old man after use of a synephrine-containing dietary supplement. *Texas Heart Institute Journal* 36(6):586–90.

Tonelo, D., et al. 2013. Holiday heart syndrome revisited after 34 years. *Arquivos Brazileiros de Cardiologica* 101(2):183–89.

Traversy, G., and Chaput, J. P. 2015. Alcohol consumption and obesity: An update. *Current Obesity Reports* 4:122–30.

Ulrich, R., et al. 2018. Doping in two elite athletics competitions assessed by randomized-response surveys. *Sports Medicine* 48(1):211–19.

Urhausen, A., et al. 2004. Are the cardiac effects of anabolic steroid abuse in strength athletes reversible? *Heart* 90:496–501.

Vaher, I., et al. 2014. Impact of acute sodium citrate ingestion on endurance running performance in a warm environment. *European Journal of Applied Physiology.* doi: 10.1007/s00421-014-3068-6.

van Amsterdam, J., et al. 2010. Adverse health effects of anabolic-androgenic steroids. *Regulatory Toxicology and Pharmacology* 57:117–23.

van Dijk, R., et al. 2018. Effects of caffeine on myocardial blood flow: A systematic review. *Nutrients* 10(8):1083. doi: 10.3390/nu10081083. PMID: 30104545; PMCID: PMC6115837.

Van Montfoort, M., et al. 2004. Effects of ingestion of bicarbonate, citrate, lactate, and chloride on sprint running. *Medicine & Science in Sports & Exercise* 36:1239–43.

van Someren, K., et al. 1998. An investigation into the effects of sodium citrate ingestion on high-intensity exercise performance. *International Journal of Sport Nutrition* 8:356–63.

VanHaitsma, T. A., et al. 2010. Comparative effects of caffeine and albuterol on the bronchoconstrictor response to exercise in asthmatic athletes. *International Journal of Sports Medicine* 31(4):231–36.

Wallace, M., et al. 1999. Effects of dehydroepiandrosterone vs androstenedione supplementation in men. *Medicine & Science in Sports & Exercise* 31:1788–92.

Wang, M. Q., et al. 1995. The acute effect of moderate alcohol consumption on cardiovascular responses in women. *Journal of Studies on Alcohol* 56(1):16–20.

Wang, S., et al. 2021. Does coffee, tea and caffeine consumption reduce the risk of incident breast cancer? A systematic review and network meta-analysis. *Public Health Nutrition* 24(18):6377–89. doi: 10.1017/S1368980021000720.

Warren, G., et al. 2010. Effect of caffeine ingestion on muscular strength and endurance: A meta-analysis. *Medicine & Science in Sports & Exercise* 42:1375–87.

Welsh, E., et al. 2010. Caffeine for asthma. *Cochrane Database of Systematic Reviews* 20(1):CD001112. doi: 10.1002/14651858.CD001112.pub2.

White, A., and Hingson, R. 2013. The burden of alcohol use: Excessive alcohol consumption and related consequences among college students. Alcohol Research: *Current Reviews* 35(2):201–18.

Wicki, M., et al. 2010. Drinking at European universities? A review of students' alcohol use. *Addictive Behaviors* 35(11):913–24.

Widdowson, W., and Gibney, J. 2010. The effect of growth hormone (GH) replacement on muscle strength in patients with GH-deficiency: A meta-analysis. *Clinical Endocrinology* 72:787–92.

Wilkerson, A. H., et al. 2017. "Drunkorexia": Understanding eating and physical activity behaviors of weight conscious drinkers in a sample of college students. *Journal of American College Health* 65(7):492–501.

Williams, A. D., et al. 2008. The effect of ephedra and caffeine on maximal strength and power in resistance-trained athletes. *Journal of Strength and Conditioning Research* 22(2):464–70.

Williams, M. H. 1972. Effect of small and moderate doses of alcohol on exercise heart rate and oxygen consumption. *Research Quarterly* 43(1):94–104. PMID: 4503122.

Williams, M., and Branch, J. D. 2002. Herbals as ergogenic aids. In *Performance Enhancing Substances in Sports and Exercise,* ed. M. Bahrke and C. Yesalis. Champaign, IL: Human Kinetics.

Williams, P. 1997. Interactive effects of exercise, alcohol, and vegetarian diet on coronary artery disease risk factors in 9242 runners: The National Runners' Health Study. *American Journal of Clinical Nutrition* 66:1197–206.

Xie, C., et al. 2018. Coffee consumption and risk of hypertension: A systematic review and dose-response meta-analysis of cohort studies. *Journal of Human Hypertension* 32(2):83–93. doi: 10.1038/s41371-017-0007-0.

Yap, K. Y., et al. 2005. Overview on the analytical tools for quality control of natural product-based supplements: A case study of ginseng. *Assay and Drug Development Technologies* 3(6):683–99.

Yesalis, C., and Bahrke, M. 2005. Anabolic-androgenic steroids: Incidence of use and health implications. *President's Council on Physical Fitness and Sports Research Digest* 5(5):1–8.

Yoo, H. J., et al. 2021. Effects of ephedrine-containing products on weight loss and lipid profiles: A systematic review and meta-analysis of randomized controlled trials. *Pharmaceuticals (Basel)* 14(11):1198. doi: 10.3390/ph14111198. PMID: 34832979; PMCID: PMC8618781.

Yoshihara, T., et al. 2019. Influence of genetic polymorphisms and habitual caffeine intake on the changes in blood pressure, pulse rate, and calculation speed after caffeine intake: A prospective, double blind, randomized trial in healthy volunteers. *Journal of Pharmacological Sciences* 139(3):209–14. doi: 10.1016/j.jphs.2019.01.006.

Ziaei, R., et al. 2020. The efficacy of ginseng supplementation on plasma lipid concentration in adults: A systematic review and meta-analysis. *Complementary Therapies in Medicine* 48:102239. doi: 10.1016/j.ctim.2019.102239.

Zhang, Y., et al. 2015. Caffeine and dieresis during rest and exercise: A meta-analysis. *Journal of Science and Medicine in Sport* 18(5):569–74.

Zhou, S. S., et al. 2021. Ginseng ameliorates exercise-induced fatigue potentially by regulating the gut microbiota. *Food and Function* 12(9):3954–64. doi: 10.1039/d0fo03384g. PMID: 33977937.

Ziegenfuss, T., et al. 2002. Effects of prohormones supplementation in humans: A review. *Canadian Journal of Applied Physiology* 27:628–46.

Ziemba, A., et al. 1999. Ginseng treatment improves psychomotor performance at rest and during graded exercise in young athletes. *International Journal of Sport Nutrition* 9:371–77.

APPENDIX A

Energy Pathways of Carbohydrate, Fat, and Protein

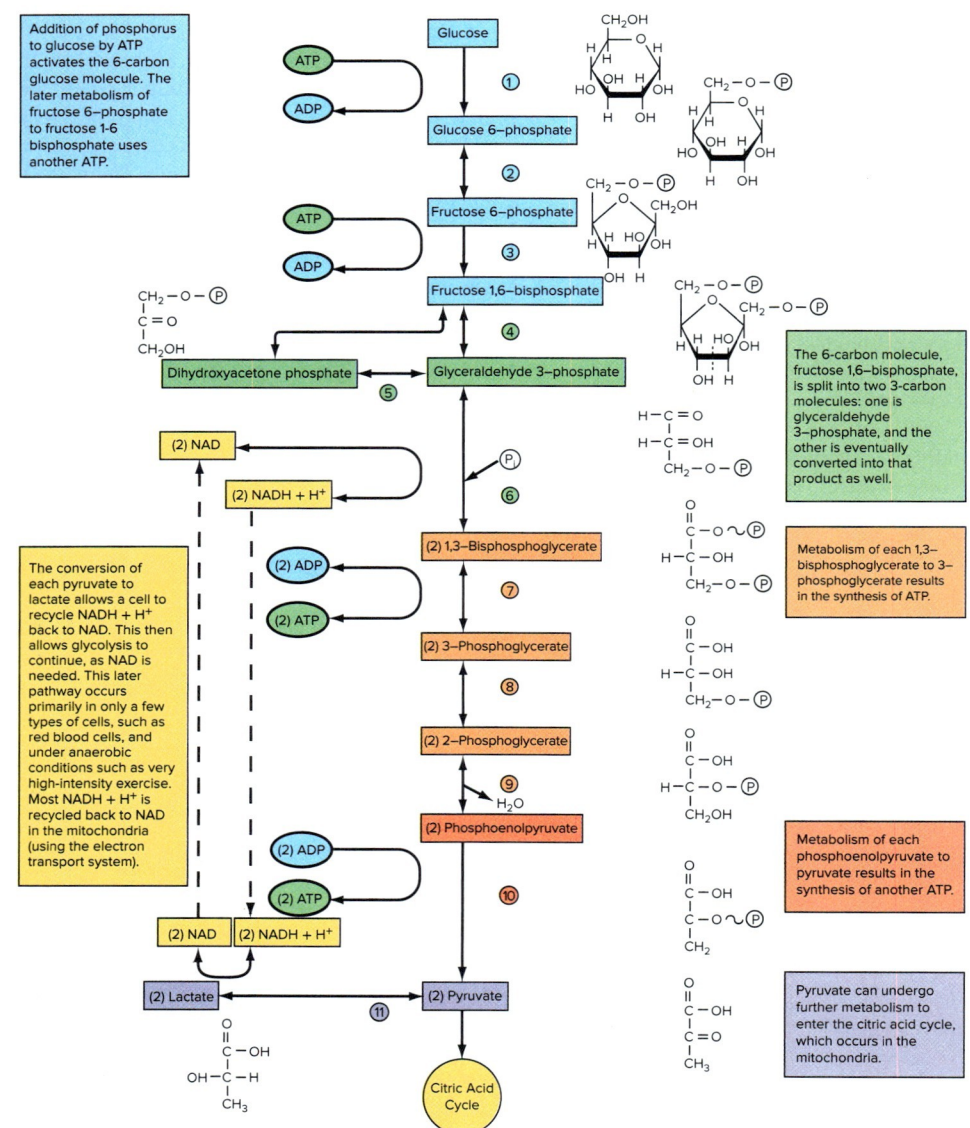

FIGURE A.1 Detailed depiction of the individual chemical reactions that constitute glycolysis—glucose to pyruvate. Glycolysis takes place in the cytosol of the cell. The enzymes in the cytosol that participate at the following steps are (1) hexokinase, (2) phosphohexose isomerase, (3) phosphofructokinase, (4) aldolase, (5) phosphotriose isomerase, (6) glyceraldehyde-3-phosphate dehydrogenase, (7) phosphoglycerate kinase, (8) phosphoglycerate mutase, (9) enolase, and (10) pyruvate kinase. Sometimes (11) lactate dehydrogenase is used to recycle NADH + H$^+$ back to NAD (anaerobic glycolysis). Ⓟ represents a phosphate group.

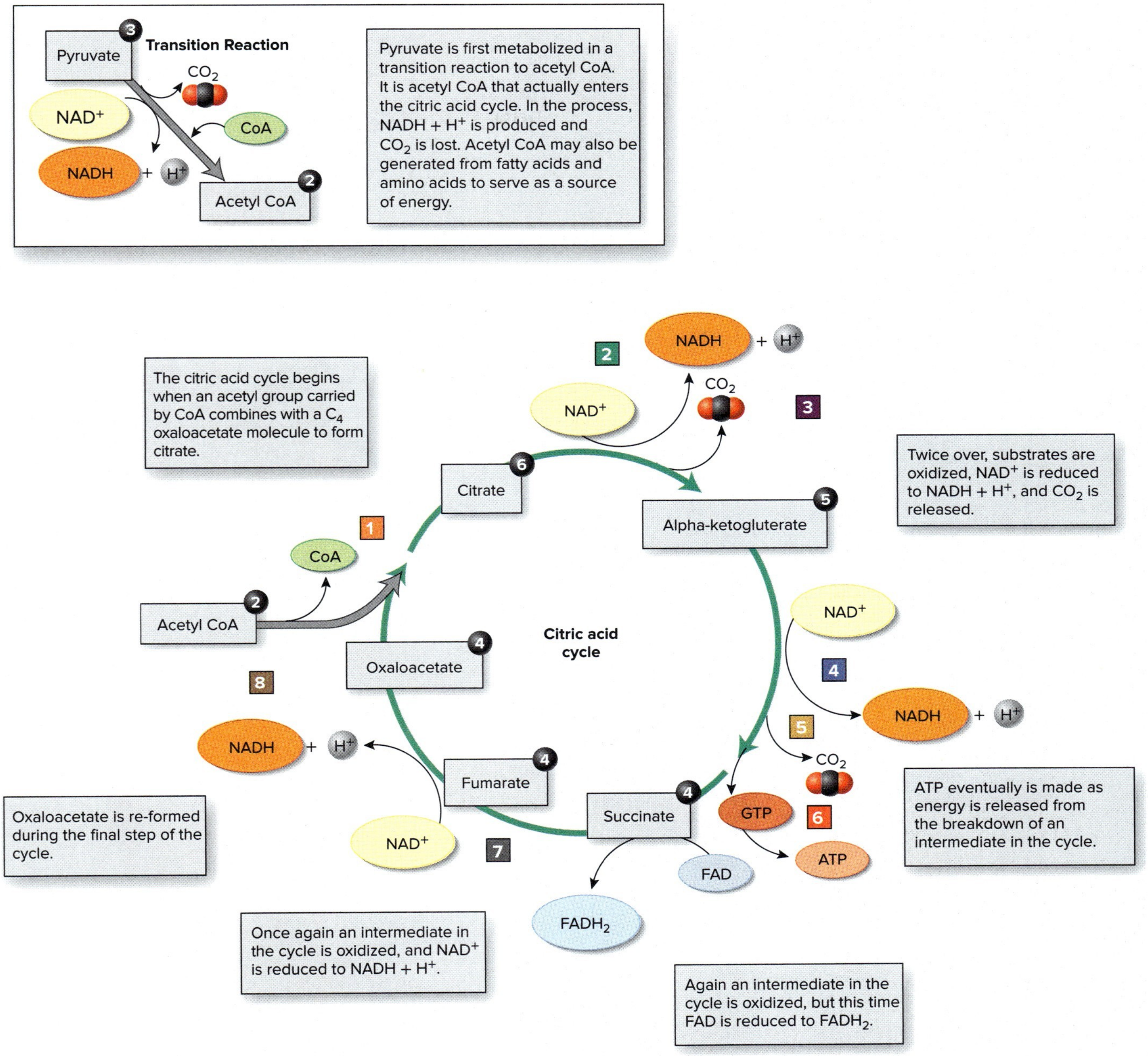

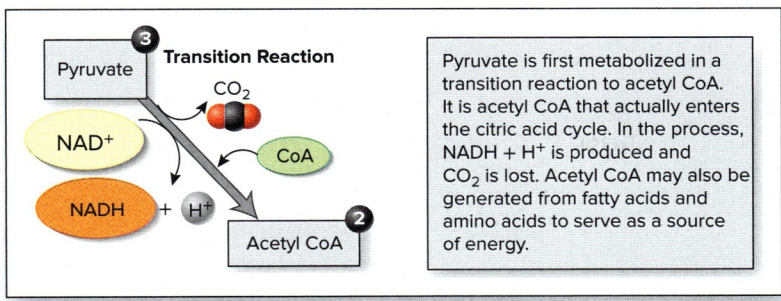

FIGURE A.2 The transition reaction and the citric acid cycle. The net result of one turn of this cycle of reactions (squares, steps 1–8) is the oxidation of an acetyl group to two molecules of CO_2 and the formation of three molecules of NADH + H$^+$ and one molecule of $FADH_2$. One GTP molecule also results, which eventually forms ATP. The citric acid cycle turns twice per glucose molecule. Note that oxygen does not participate in any of the steps in the citric acid cycle. It instead participates in the electron transport system, where the vast majority of the ATP is formed (see figure A.3). The numbers in the circles represent the number of carbon atoms.

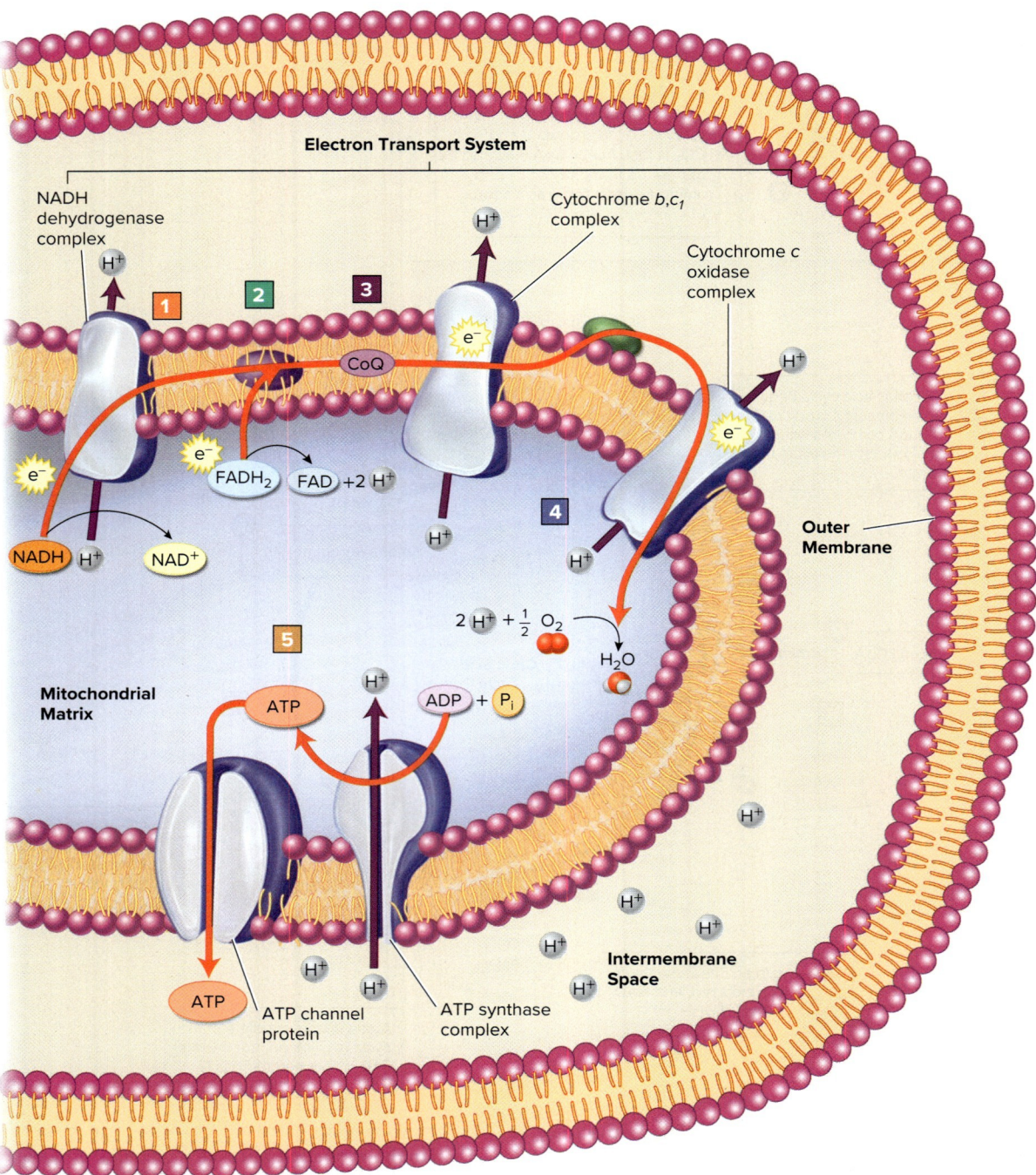

FIGURE A.3 Organization of the electron transport system. As electrons move from one molecular complex to the other, hydrogen ions (H^+) are pumped from the mitochondrial matrix into the intermembrane space (steps 1–4). (Note that each mitochondrion has an inner and outer membrane.) Hydrogen ions flow down a concentration gradient from the intermembrane space into the mitochondrial matrix; ATP is then synthesized by the enzyme ATP synthase (step 5). ATP leaves the mitochondrial matrix by way of a channel protein.

APPENDIX A Energy Pathways of Carbohydrate, Fat, and Protein

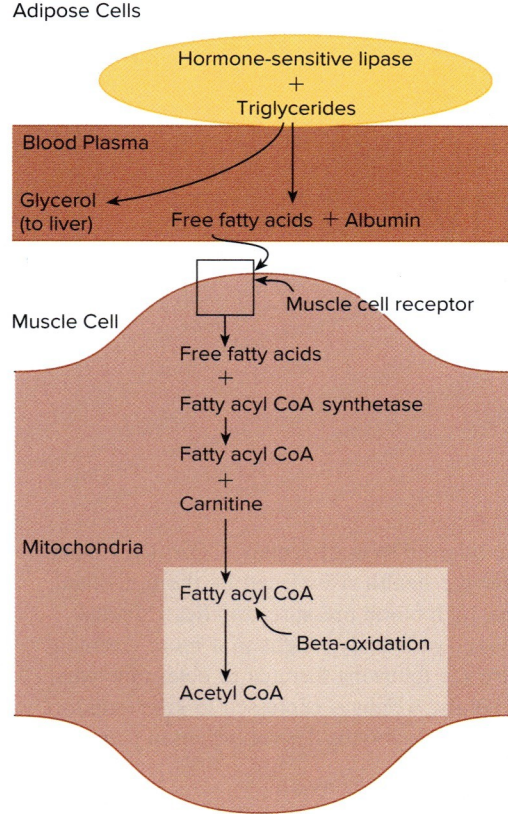

FIGURE A.4A Energy pathways for fatty acids. Triglycerides in the adipose tissue may be catabolized by hormone-sensitive lipase, with the fatty acids being released to the plasma and binding with albumin; the glycerol component is transported to the liver for metabolism. A receptor at the muscle cell transports the fatty acid into the muscle cell, where it is converted into fatty acyl CoA by an enzyme (fatty acyl CoA synthetase). The fatty acyl CoA is then transported into the mitochondria with carnitine (in an enzyme complex) as a carrier. The fatty acyl CoA, which is a combination of acetyl CoA units, then undergoes beta-oxidation, a process that splits off the acetyl CoA units for entrance into the Krebs cycle.

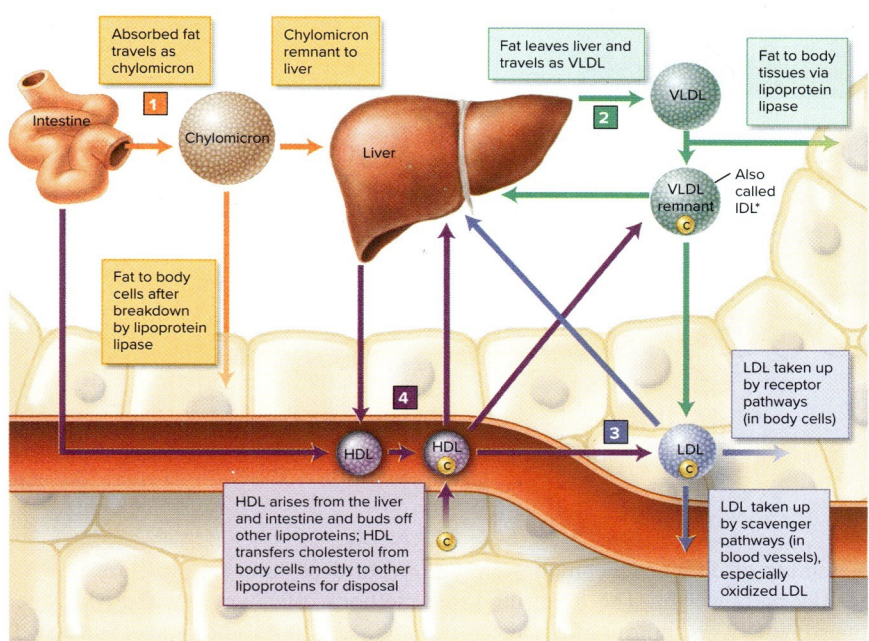

FIGURE A.4B Lipoprotein interactions. (1) Chylomicrons carry absorbed fat to body cells. (2) VLDL carries fat taken up from the bloodstream by the liver, as well as any fat made by the liver, to body cells. (3) LDL arises from VLDL and carries mostly cholesterol to cells. (4) HDL arises from body cells, mostly in the liver and intestine as well as from particles that bud off the other lipoproteins. HDL carries cholesterol from cells to other lipoproteins and to the liver for excretion.

*Intermediate-density lipoprotein.

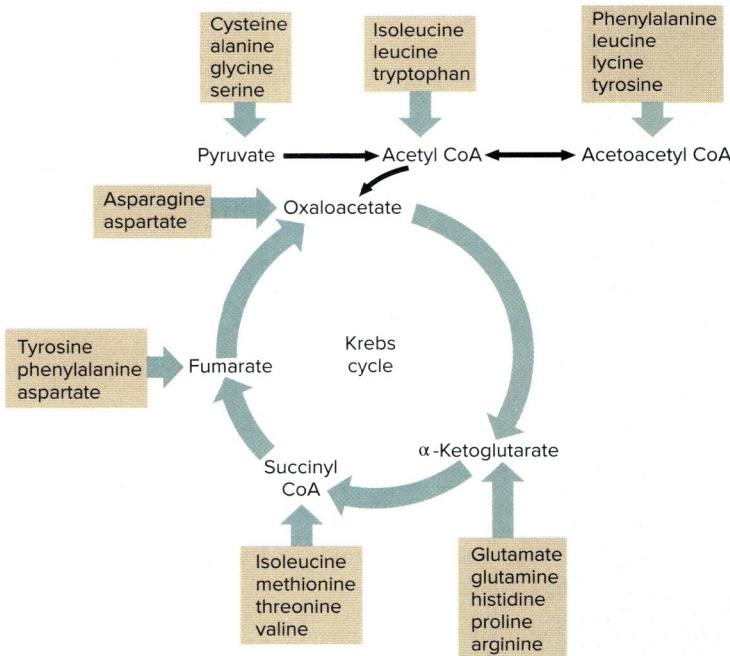

FIGURE A.5 Metabolic fates of amino acids. Various amino acids, after deamination, may enter into energy pathways at different sites.

APPENDIX B

Determination of Healthy Body Weight

A number of different techniques are utilized to determine a healthy body weight. The following three methods offer you an estimate of an appropriate body weight. Method A is based on the body mass index (BMI). Method B is based on body-fat percentage. Method C, the waist circumference, does not determine a desirable body weight but provides an assessment of desirable body-fat distribution.

Method A

The BMI uses the metric system, so you need to determine your weight in kilograms and your height in meters. The formula is

$$\mathrm{BMI} = \frac{\text{Body weight in kilograms}}{(\text{Height in meters})^2}$$

Dividing your body weight in pounds by 2.2 will give you your weight in kilograms. Multiplying your height in inches by 0.0254 will give you your height in meters.

$$\text{Your weight in kilograms} = \frac{(\text{Your weight in pounds})}{2.2} = \underline{\qquad}$$

$$\text{Your height in meters} = (\text{Your height in inches}) \times 0.0254 = \underline{\qquad}$$

$$\mathrm{BMI} = \frac{\text{Body weight in kilograms}}{(\text{Height in meters})^2} = \underline{\qquad}$$

Or you may use your weight in pounds and height in inches with the following formula:

$$\mathrm{BMI} = \frac{\text{Body weight in pounds} \times 705}{(\text{Height in inches})^2}$$

A BMI range of 18.5–24.9 is considered to be normal, but a suggested desirable range for females is 21.3–22.1 and for males is 21.9–22.4. Individuals with BMI values between 25.0 and 29.9 are classified as overweight; 30 and 34.9 as Class 1 obesity; 35 and 39.9 as Class 2 obesity; and, 40+ as Class 3 obesity, also referred to as extremely obese. The higher the BMI, the greater the health risks faced by the individual, particularly diabetes, high blood pressure, and heart disease.

If you want to lower your body weight to a more desirable BMI, such as 22, use the following formula to determine what that weight should be; the weight is expressed in kilograms, so multiplying it by 2.2 will give you the desired weight in pounds.

$$\text{Kilograms body weight} = \text{Desired BMI} \times (\text{Height in meters})^2$$

Here's a brief example for a woman who weighs 187 pounds and is 5'9" tall; her BMI calculates to be 27.7, so her weight poses a health risk. If she wants to achieve a BMI of 23, she will need to reduce her weight to 155 pounds.

$$\text{Kilograms body weight} = 23 \times (1.753)^2 = 70.6$$

$$70.6 \text{ kg} \times 2.2 = 155 \text{ pounds}$$

To calculate your desired body weight:

$$\text{Kilograms body weight} = (\text{Your desired BMI}) \times (\text{Your height in meters})^2$$

$$\text{Kilograms body weight} = \underline{\qquad} \times \underline{\qquad}$$

$$\underline{\qquad} \text{ kg} \times 2.2 = \underline{\qquad} \text{ pounds}$$

Keep in mind that the BMI does not discriminate between muscle mass and body fat, so a high BMI may reflect an increased muscle mass and body fat may actually be relatively low. Conversely, an individual with a low BMI may have a higher level of body fat if muscle mass is small. The BMI also does not account for regional fat distribution.

Method B

For this method, you will need to know your body-fat percentage as determined by the skinfold measurement procedure described in **table B.1** or another appropriate technique. You will also need to determine the body-fat percentage you desire to have. You may use **table 10.3** as a guideline.

TABLE B.1	Generalized equations for predicting body fat
Measure the appropriate skinfolds for women (triceps, thigh, and suprailium sites) or men (chest, abdomen, and thigh sites) as illustrated in **figures B.1–B.4**. You may use either the appropriate formula or the gender-specific tables below to obtain the predicted body-fat percentage.	
Women*	**Men****
BD = 1.0994921 − 0.0009929 (X_1) + 0.0000023 (X_1)^2 − 0.0001392 (X_2)	BD = 1.10938 − 0.0008267 (X_1) + 0.0000016 (X_1)^2 − 0.0002574 (X_2)
BD = Body density	BD = Body density
X_1 = Sum of triceps, thigh, and suprailium skinfolds	X_1 = Sum of chest, abdomen, and thigh skinfolds
X_2 = Age	X_2 = Age
	To calculate percent body fat, plug into Siri's equation. $$\left(\frac{4.95}{BD} - 4.5\right) \times 100$$

*From Jackson, A., Pollock, M., and Ward, A. 1980. Generalized equations for predicting body density of women. *Medicine & Science in Sports & Exercise* 12:175–82.
**Jackson, A., and Pollock, M. 1978. Generalized equations for predicting body density of men. *British Journal of Nutrition* 40:497–504.

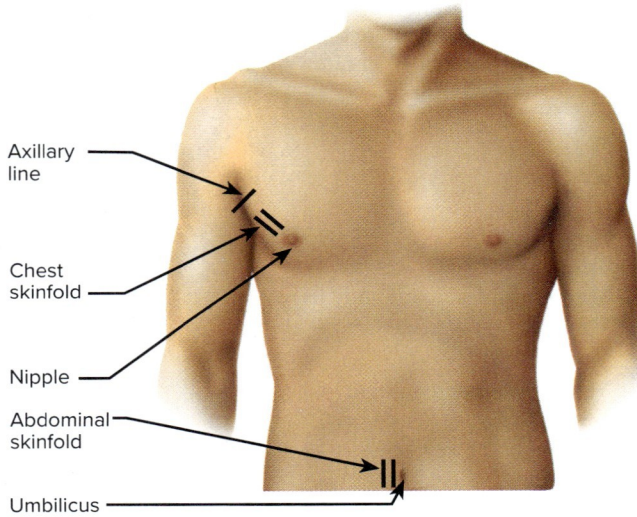

FIGURE B.1 The chest and abdomen skinfold. Chest—a diagonal fold is taken on the subject's right side between the axilla and the nipple. Use a midway point. Abdomen—a vertical fold is taken on the subject's right side about 2.5 centimeters (1 inch) to the side of the umbilicus.

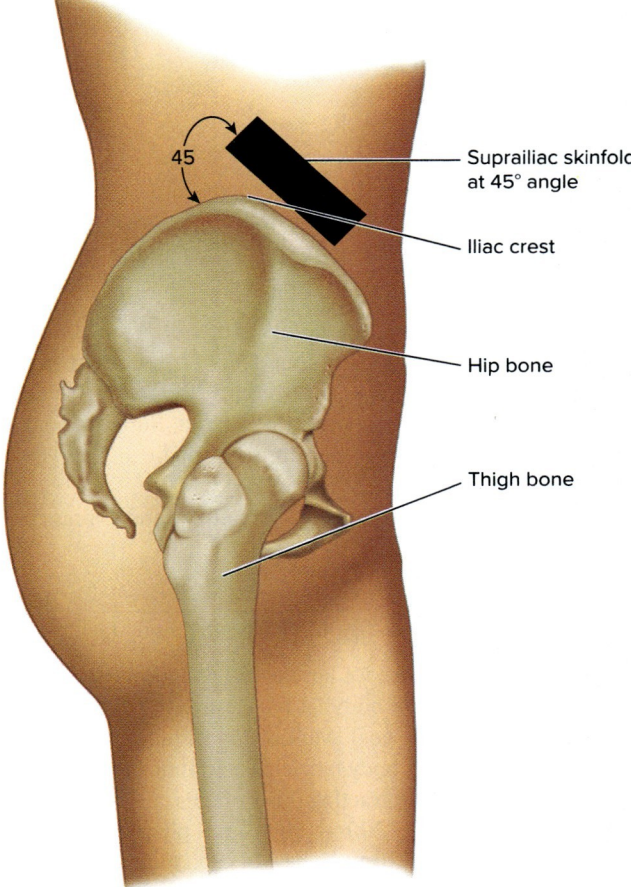

FIGURE B.2 The suprailiac skinfold. A diagonal fold is taken at the subject's right anterior axillary line at about a 45-degree angle just above the crest of the ilium.

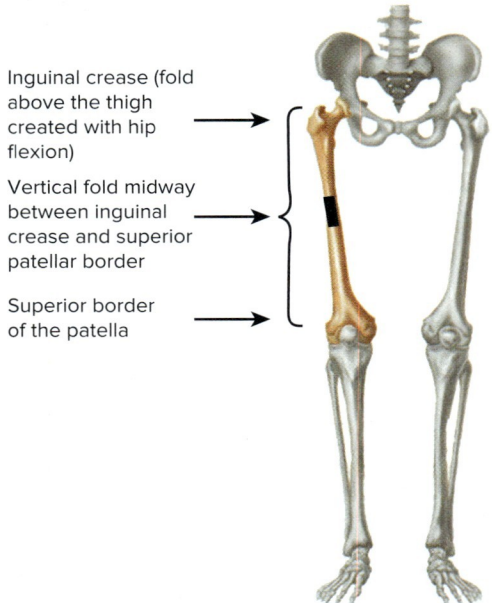

FIGURE B.3 The thigh skinfold. A vertical fold is taken on the front of the subject's right thigh midway between the inguinal crease and the superior border of the patella.

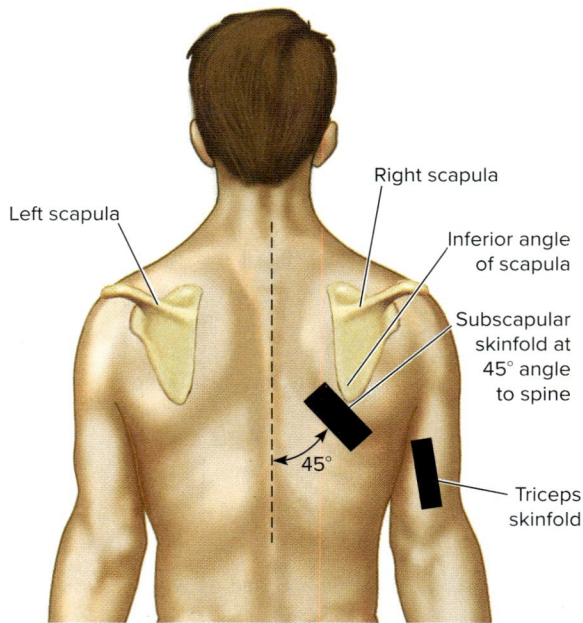

FIGURE B.4 The triceps and subscapular skinfolds. In the triceps skinfold, a vertical fold is taken over the subject's right triceps muscle one-half the distance from the acromion process to the olecranon process at the elbow. The subscapular skinfold is taken just below the lower angle of the subject's right scapula, at about a 45-degree angle to the spinal column.

You will need to do the following calculations for the formula:

1. Determine your current lean body weight (LBW). Multiply your current body weight in pounds by your current percent body fat expressed as a decimal (20 percent would be 0.20) to obtain your pounds of body fat. Subtract your pounds of body fat from your current weight to give you your lean body weight (LBW).
2. Determine your desired body-fat percentage and express it as a decimal.

$$\text{Desired body weight} = \frac{\text{LBW}}{1.00 - \text{Desired \% body fat}}$$

As an example, suppose we have a 200-pound male who is currently at 25 percent body fat but desires to get down to 20 percent as his first goal. Multiplying his current weight by his current percent body fat yields 50 pounds of body fat ($200 \times 0.25 = 50$); subtracting this from his current weight yields a LBW of 150 ($200 - 50$). If we plug his desired percent of 20 into the formula, he will need to reach a body weight of 187.5 to achieve this first goal.

$$\text{Desired body weight} = \frac{150}{1.00 - 0.20} = \frac{150}{0.8} = 187.5$$

Your current body weight _____
Your current percent body fat _____
Your pounds of body fat _____
Your LBW _____
Your desired percent body fat _____

$$\text{Desired body weight} \frac{\text{LBW}}{1.00 - ?} = \underline{\qquad} = \underline{\qquad}$$

Method C

The waist circumference is a measure of regional fat distribution. Using a flexible (preferably metal) tape, measure the narrowest section of the bare waist as seen from the front while standing. Wear tight clothing. Do not compress skin and fat with pressure from the tape. The waist measurement may be used as a simple screening technique for abdominal obesity. Females with a waist of 35 inches or over, and men with a waist of 40 inches or over may be at increased risk.

Waist girth _____

Use both your BMI and waist measurement in **table 10.1** to evaluate your risk of associated disease.

Percent fat estimate for men: sum of chest, abdomen, and thigh skinfolds

Sum of skinfolds (mm)	Age to last year								
	Under 22	23–27	28–32	33–37	38–42	43–47	48–52	53–57	Over 57
8–10	1.3	1.8	2.3	2.9	3.4	3.9	4.5	5.0	5.5
11–13	2.2	2.8	3.3	3.9	4.4	4.9	5.5	6.0	6.5
14–16	3.2	3.8	4.3	4.8	5.4	5.9	6.4	7.0	7.5
17–19	4.2	4.7	5.3	5.8	6.3	6.9	7.4	8.0	8.5
20–22	5.1	5.7	6.2	6.8	7.3	7.9	8.4	8.9	9.5
23–25	6.1	6.6	7.2	7.7	8.3	8.8	9.4	9.9	10.5
26–28	7.0	7.6	8.1	8.7	9.2	9.8	10.3	10.9	11.4
29–31	8.0	8.5	9.1	9.6	10.2	10.7	11.3	11.8	12.4
32–34	8.9	9.4	10.0	10.5	11.1	11.6	12.2	12.8	13.3
35–37	9.8	10.4	10.9	11.5	12.0	12.6	13.1	13.7	14.3
38–40	10.7	11.3	11.8	12.4	12.9	13.5	14.1	14.6	15.2
41–43	11.6	12.2	12.7	13.3	13.8	14.4	15.0	15.5	16.1
44–46	12.5	13.1	13.6	14.2	14.7	15.3	15.9	16.4	17.0
47–49	13.4	13.9	14.5	15.1	15.6	16.2	16.8	17.3	17.9
50–52	14.3	14.8	15.4	15.9	16.5	17.1	17.6	18.2	18.8
53–55	15.1	15.7	16.2	16.8	17.4	17.9	18.5	19.1	19.7
56–58	16.0	16.5	17.1	17.7	18.2	18.8	19.4	20.0	20.5
59–61	16.9	17.4	17.9	18.5	19.1	19.7	20.2	20.8	21.4
62–64	17.6	18.2	18.8	19.4	19.9	20.5	21.1	21.7	22.2
65–67	18.5	19.0	19.6	20.2	20.8	21.3	21.9	22.5	23.1
68–70	19.3	19.9	20.4	21.0	21.6	22.2	22.7	23.3	23.9
71–73	20.1	20.7	21.2	21.8	22.4	23.0	23.6	24.1	24.7
74–76	20.9	21.5	22.0	22.6	23.2	23.8	24.4	25.0	25.5
77–79	21.7	22.2	22.8	23.4	24.0	24.6	25.2	25.8	26.3
80–82	22.4	23.0	23.6	24.2	24.8	25.4	25.9	26.5	27.1
83–85	23.2	23.8	24.4	25.0	25.5	26.1	26.7	27.3	27.9
86–88	24.0	24.5	25.1	25.7	26.3	26.9	27.5	28.1	28.7
89–91	24.7	25.3	25.9	26.5	27.1	27.6	28.2	28.8	29.4
92–94	25.4	26.0	26.6	27.2	27.8	28.4	29.0	29.6	30.2
95–97	26.1	26.7	27.3	27.9	28.5	29.1	29.7	30.3	30.9
98–100	26.9	27.4	28.0	28.6	29.2	29.8	30.4	31.0	31.6
101–103	27.5	28.1	28.7	29.3	29.9	30.5	31.1	31.7	32.3
104–106	28.2	28.8	29.4	30.0	30.6	31.2	31.8	32.4	33.0
107–109	28.9	29.5	30.1	30.7	31.3	31.9	32.5	33.1	33.7
110–112	29.6	30.2	30.8	31.4	32.0	32.6	33.2	33.8	34.4
113–115	30.2	30.8	31.4	32.0	32.6	33.2	33.8	34.5	35.1
116–118	30.9	31.5	32.1	32.7	33.3	33.9	34.5	35.1	35.7
119–121	31.5	32.1	32.7	33.3	33.9	34.5	35.1	35.7	36.4
122–124	32.1	32.7	33.3	33.9	34.5	35.1	35.8	36.4	37.0
125–127	32.7	33.3	33.9	34.5	35.1	35.8	36.4	37.0	37.6

Source: A. S. Jackson and M. L. Pollock, "Practical Assessment of Body Composition," May 1985, in *Physician and Sportsmedicine*. Reprinted with permission of McGraw-Hill, Inc.

Percent fat estimate for women: sum of triceps, suprailium, and thigh skinfolds

Sum of skinfolds (mm)	Age to last year								
	Under 22	23–27	28–32	33–37	38–42	43–47	48–52	53–57	Over 57
23–25	9.7	9.9	10.2	10.4	10.7	10.9	11.2	11.4	11.7
26–28	11.0	11.2	11.5	11.7	12.0	12.3	12.5	12.7	13.0
29–31	12.3	12.5	12.8	13.0	13.3	13.5	13.8	14.0	14.3
32–34	13.6	13.8	14.0	14.3	14.5	14.8	15.0	15.3	15.5
35–37	14.8	15.0	15.3	15.5	15.8	16.0	16.3	16.5	16.8
38–40	16.0	16.3	16.5	16.7	17.0	17.2	17.5	17.7	18.0
41–43	17.2	17.4	17.7	17.9	18.2	18.4	18.7	18.9	19.2
44–46	18.3	18.6	18.8	19.1	19.3	19.6	19.8	20.1	20.3
47–49	19.5	19.7	20.0	20.2	20.5	20.7	21.0	21.2	21.5
50–52	20.6	20.8	21.1	21.3	21.6	21.8	22.1	22.3	22.6
53–55	21.7	21.9	22.1	22.4	22.6	22.9	23.1	23.4	23.6
56–58	22.7	23.0	23.2	23.4	23.7	23.9	24.2	24.4	24.7
59–61	23.7	24.0	24.2	24.5	24.7	25.0	25.2	25.5	25.7
62–64	24.7	25.0	25.2	25.5	25.7	26.0	26.2	26.4	26.7
65–67	25.7	25.9	26.2	26.4	26.7	26.9	27.2	27.4	27.7
68–70	26.6	26.9	27.1	27.4	27.6	27.9	28.1	28.4	28.6
71–73	27.5	27.8	28.0	28.3	28.5	28.8	29.0	29.3	29.5
74–76	28.4	28.7	28.9	29.2	29.4	29.7	29.9	30.2	30.4
77–79	29.3	29.5	29.8	30.0	30.3	30.5	30.8	31.0	31.3
80–82	30.1	30.4	30.6	30.9	31.1	31.4	31.6	31.9	32.1
83–85	30.9	31.2	31.4	31.7	31.9	32.2	32.4	32.7	32.9
86–88	31.7	32.0	32.2	32.5	32.7	32.9	33.2	33.4	33.7
89–91	32.5	32.7	33.0	33.2	33.5	33.7	33.9	34.2	34.4
92–94	33.2	33.4	33.7	33.9	34.2	34.4	34.7	34.9	35.2
95–97	33.9	34.1	34.4	34.6	34.9	35.1	35.4	35.6	35.9
98–100	34.6	34.8	35.1	35.3	35.5	35.8	36.0	36.3	36.5
101–103	35.3	35.4	35.7	35.9	36.2	36.4	36.7	36.9	37.2
104–106	35.8	36.1	36.3	36.6	36.8	37.1	37.3	37.5	37.8
107–109	36.4	36.7	36.9	37.1	37.4	37.6	37.9	38.1	38.4
110–112	37.0	37.2	37.5	37.7	38.0	38.2	38.5	38.7	38.9
113–115	37.5	37.8	38.0	38.2	38.5	38.7	39.0	39.2	39.5
116–118	38.0	38.3	38.5	38.8	39.0	39.3	39.5	39.7	40.0
119–121	38.5	38.7	39.0	39.2	39.5	39.7	40.0	40.2	40.5
122–124	39.0	39.2	39.4	39.7	39.9	40.2	40.4	40.7	40.9
125–127	39.4	39.6	39.9	40.1	40.4	40.6	40.9	41.1	41.4
128–130	39.8	40.0	40.3	40.5	40.8	41.0	41.3	41.5	41.8

Source: A. S. Jackson and M. L. Pollock, "Practical Assessment of Body Composition," May 1985, in *Physician and Sportsmedicine*. Reprinted with permission of McGraw-Hill, Inc.

Units of Measurement: English System— Metric System Equivalents

APPENDIX C

The Metric System and Equivalents

To measure ingredients, a standardized system known as the System Internationale (SI) has been established that is interpreted on an international basis. The SI is based on the metric system. However, in the United States, we also employ another set of measure and weight, the English system. In the field of dietetics, both systems are employed. The following tables give the quantities of the measures besides stating equivalents. With this information, it is possible to calculate in either system of measure and weight.

Household Measures (Approximations)

For easy computing purposes, the cubic centimeter (cc) is considered equivalent to 1 gram:

$$1 \text{ cc} = 1 \text{ gram} = 1 \text{ milliliter (ml)}$$

For easy computing purposes, 1 ounce equals 30 grams or 30 cubic centimeters.

1 quart	=	960 grams
1 pint	=	480 grams
1 cup	=	240 grams
½ cup	=	120 grams
1 glass (8 ounces)	=	240 grams
½ glass (4 ounces)	=	120 grams
1 orange juice glass	=	100–120 grams
1 tablespoon	=	15 grams
1 teaspoon	=	5 grams

Level Measures and Weights

1 teaspoon	=	5 cc or 5 ml 5 grams
3 teaspoons	=	1 tablespoon 15 cc 15 grams
2 tablespoons	=	30 cc 30 grams 1 ounce (fluid)
4 tablespoons	=	¼ cup 60 cc 60 grams
8 tablespoons	=	½ cup 120 cc 120 grams
16 tablespoons	=	1 cup 240 grams 240 ml (fluid) 8 ounces (fluid) ½ pound
2 cups	=	1 pint 480 grams 480 ml (fluid) 16 ounces (fluid) 1 pound
4 cups	=	2 pints 1 quart 960 cc 960 ml (fluid) 2 pounds
4 quarts	=	1 gallon

Units of Weight

		Ounce	Pound	Gram	Kilogram
1 ounce	=	1.0	0.06	28.4	0.028
1 pound	=	16.0	1.0	454	0.454
1 gram	=	0.035	.002	1.0	0.001
1 kilogram	=	35.3	2.2	1,000	1.0

Units of Volume

		Ounce	Pint	Quart	Milliliter	Liter
1 ounce	=	1.0	0.062	0.031	29.57	0.029
1 pint	=	16.0	1.0	0.5	473	0.473
1 quart	=	32.0	2.0	1.0	946	0.946
1 milliliter	=	0.034	0.002	0.001	1.0	0.001
1 liter	=	33.8	2.112	1.056	1,000	1.0

Units of Length

		Millimeter	Centimeter	Inch	Foot	Yard	Meter
1 millimeter	=	1.0	0.1	0.0394	0.0033	0.0011	0.001
1 centimeter	=	10.0	1.0	0.394	0.033	0.011	0.01
1 inch	=	25.4	2.54	1.0	0.083	0.028	0.025
1 foot	=	304.8	30.48	12.0	1.0	0.333	0.305
1 yard	=	914.4	91.44	36.0	3.0	1.0	0.914
1 meter	=	1,000	100	39.37	3.28	1.094	1.0

1 kilometer = 1,000 meters = 0.6216 mile
1 mile = 1,760 yards = 1.61 kilometers

Units of mechanical, thermal, and chemical energy (approximate equivalents)

	Foot-pounds	Kilogram-meters	Kilojoules	Watts*	Kilocalories	Oxygen**
1 foot-pound	1	0.138	0.00136	0.0226	0.00032	0.000064
1 kilogram-meter	7.23	1	0.0098	0.163	0.0023	0.00046
1 kilojoule	737	102	1	16.66	0.239	0.047
1 watt*	44.27	6.12	0.06	1	0.0143	0.0028
1 kilocalorie	3,088	427	4.18	0.00024	1	0.198
1 liter oxygen**	15,585	2,154	21.1	351.9	5.047	1

Note: Read all tables across, such as 1 watt equals 44.27 foot-pounds per minute; 1 foot-pound equals 0.0226 watt.

* Watts are units of power expressed per minute.

** Equivalents are based upon 1 liter of oxygen metabolizing carbohydrate. Energy equivalents would be slightly less on a mixed diet of carbohydrate, fat, and protein. For example, 1 liter of oxygen would equal only 4.82 kilocalories on such a mixed diet.

APPENDIX D

Approximate Energy Expenditure (Kcal/Min) by Body Weight Based on the Metabolic Equivalents (METs) for Physical Activity Intensity

The values presented in the following table are based on the metabolic equivalents (METs) of a wide variety of physical activities, including common activities of daily living as well as physical activities and sports as presented in the Compendium of Physical Activities, located at sites.google.com/site/compendiumofphysicalactivities/. Use this table as follows:

1. Consult the website for your activities and record the MET values for each. For some activities, such as walking, there are numerous choices from which to select, such as speed of walking, walking uphill, walking with a backpack, or walking with Nordic poles. Be sure to select the most similar activity to the one you performed.
2. Find the row that is closest to the MET value(s) for your chosen activity(ies).
3. Find the column with your approximate body weight.
4. The intersection of your MET row and your body weight column contains two values:
   a. **net** (gross − resting) energy expenditure kcal/min
      = (METs − 1) × 3.5 ml/kg/min × kg ÷ 1,000 ml/liter × 5 kcal/liter
   b. **gross** (net + resting) energy expenditure kcal/min
      = METs × 3.5 ml/kg/min × kg ÷ 1,000 ml/liter × 5 kcal/liter

For example, a male weighs 187 pounds (85 kg) and runs at a pace of 7 miles/hour [8.5 min/mile] (Compendium activity 12070, 11.0 METs). The estimated net and gross energy expenditures are 14.9 and 16.4 kcal/min. A 30-minute run would result in a gross and net expenditure of 492 (= 16.4 kcal/min × 30 min) and 447 (= 14.9 kcal/min × 30 min) kcal, respectively.

Fuse/Corbis/Getty Images

APPENDIX D — Approximate Energy Expenditure (Kcal/Min) by Body Weight Based on the Metabolic Equivalents (METs) for Physical Activity Intensity

METs		kg	45	48	50	53	55	58	60	63	65	68	70	73	75	78	80	83	85	88	90	93	95	98	100
		lbs	99	105	110	116	121	127	132	138	143	149	154	160	165	171	176	182	187	193	198	204	209	215	220
1.0			0.8	0.8	0.9	0.9	1.0	1.0	1.1	1.1	1.1	1.2	1.2	1.3	1.3	1.4	1.4	1.4	1.5	1.5	1.6	1.6	1.7	1.7	1.8
1.5	Net		0.4	0.4	0.4	0.5	0.5	0.5	0.5	0.5	0.6	0.6	0.6	0.6	0.7	0.7	0.7	0.7	0.7	0.8	0.8	0.8	0.8	0.9	0.9
1.5	Gross		1.2	1.2	1.3	1.4	1.4	1.5	1.6	1.6	1.7	1.8	1.8	1.9	2.0	2.0	2.1	2.2	2.2	2.3	2.4	2.4	2.5	2.6	2.6
2.0	Net		0.8	0.8	0.9	0.9	1.0	1.0	1.1	1.1	1.1	1.2	1.2	1.3	1.3	1.4	1.4	1.4	1.5	1.5	1.6	1.6	1.7	1.7	1.8
2.0	Gross		1.6	1.7	1.8	1.8	1.9	2.0	2.1	2.2	2.3	2.4	2.5	2.5	2.6	2.7	2.8	2.9	3.0	3.1	3.2	3.2	3.3	3.4	3.5
2.5	Net		1.2	1.2	1.3	1.4	1.4	1.5	1.6	1.6	1.7	1.8	1.8	1.9	2.0	2.0	2.1	2.2	2.2	2.3	2.4	2.4	2.5	2.6	2.6
2.5	Gross		2.0	2.1	2.2	2.3	2.4	2.5	2.6	2.7	2.8	3.0	3.1	3.2	3.3	3.4	3.5	3.6	3.7	3.8	3.9	4.0	4.2	4.3	4.4
3.0	Net		1.6	1.7	1.8	1.8	1.9	2.0	2.1	2.2	2.3	2.4	2.5	2.5	2.6	2.7	2.8	2.9	3.0	3.1	3.2	3.2	3.3	3.4	3.5
3.0	Gross		2.4	2.5	2.6	2.8	2.9	3.0	3.2	3.3	3.4	3.5	3.7	3.8	3.9	4.1	4.2	4.3	4.5	4.6	4.7	4.9	5.0	5.1	5.3
3.5	Net		2.0	2.1	2.2	2.3	2.4	2.5	2.6	2.7	2.8	3.0	3.1	3.2	3.3	3.4	3.5	3.6	3.7	3.8	3.9	4.0	4.2	4.3	4.4
3.5	Gross		2.8	2.9	3.1	3.2	3.4	3.5	3.7	3.8	4.0	4.1	4.3	4.4	4.6	4.7	4.9	5.1	5.2	5.4	5.5	5.7	5.8	6.0	6.1
4.0	Net		2.4	2.5	2.6	2.8	2.9	3.0	3.2	3.3	3.4	3.5	3.7	3.8	3.9	4.1	4.2	4.3	4.5	4.6	4.7	4.9	5.0	5.1	5.3
4.0	Gross		3.2	3.3	3.5	3.7	3.9	4.0	4.2	4.4	4.6	4.7	4.9	5.1	5.3	5.4	5.6	5.8	6.0	6.1	6.3	6.5	6.7	6.8	7.0
4.5	Net		2.8	2.9	3.1	3.2	3.4	3.5	3.7	3.8	4.0	4.1	4.3	4.4	4.6	4.7	4.9	5.1	5.2	5.4	5.5	5.7	5.8	6.0	6.1
4.5	Gross		3.5	3.7	3.9	4.1	4.3	4.5	4.7	4.9	5.1	5.3	5.5	5.7	5.9	6.1	6.3	6.5	6.7	6.9	7.1	7.3	7.5	7.7	7.9
5.0	Net		3.2	3.3	3.5	3.7	3.9	4.0	4.2	4.4	4.6	4.7	4.9	5.1	5.3	5.4	5.6	5.8	6.0	6.1	6.3	6.5	6.7	6.8	7.0
5.0	Gross		3.9	4.2	4.4	4.6	4.8	5.0	5.3	5.5	5.7	5.9	6.1	6.3	6.6	6.8	7.0	7.2	7.4	7.7	7.9	8.1	8.3	8.5	8.8
5.5	Net		3.5	3.7	3.9	4.1	4.3	4.5	4.7	4.9	5.1	5.3	5.5	5.7	5.9	6.1	6.3	6.5	6.7	6.9	7.1	7.3	7.5	7.7	7.9
5.5	Gross		4.3	4.6	4.8	5.1	5.3	5.5	5.8	6.0	6.3	6.5	6.7	7.0	7.2	7.5	7.7	7.9	8.2	8.4	8.7	8.9	9.1	9.4	9.6

METs			Body mass																						
		kg	45	48	50	53	55	58	60	63	65	68	70	73	75	78	80	83	85	88	90	93	95	98	100
		lbs	99	105	110	116	121	127	132	138	143	149	154	160	165	171	176	182	187	193	198	204	209	215	220
6.0	Net		3.9	4.2	4.4	4.6	4.8	5.0	5.3	5.5	5.7	5.9	6.1	6.3	6.6	6.8	7.0	7.2	7.4	7.7	7.9	8.1	8.3	8.5	8.8
	Gross		4.7	5.0	5.3	5.5	5.8	6.0	6.3	6.6	6.8	7.1	7.4	7.6	7.9	8.1	8.4	8.7	8.9	9.2	9.5	9.7	10.0	10.2	10.5
6.5	Net		4.3	4.6	4.8	5.1	5.3	5.5	5.8	6.0	6.3	6.5	6.7	7.0	7.2	7.5	7.7	7.9	8.2	8.4	8.7	8.9	9.1	9.4	9.6
	Gross		5.1	5.4	5.7	6.0	6.3	6.5	6.8	7.1	7.4	7.7	8.0	8.2	8.5	8.8	9.1	9.4	9.7	10.0	10.2	10.5	10.8	11.1	11.4
7.0	Net		4.7	5.0	5.3	5.5	5.8	6.0	6.3	6.6	6.8	7.1	7.4	7.6	7.9	8.1	8.4	8.7	8.9	9.2	9.5	9.7	10.0	10.2	10.5
	Gross		5.5	5.8	6.1	6.4	6.7	7.0	7.4	7.7	8.0	8.3	8.6	8.9	9.2	9.5	9.8	10.1	10.4	10.7	11.0	11.3	11.6	11.9	12.3
7.5	Net		5.1	5.4	5.7	6.0	6.3	6.5	6.8	7.1	7.4	7.7	8.0	8.2	8.5	8.8	9.1	9.4	9.7	10.0	10.2	10.5	10.8	11.1	11.4
	Gross		5.9	6.2	6.6	6.9	7.2	7.5	7.9	8.2	8.5	8.9	9.2	9.5	9.8	10.2	10.5	10.8	11.2	11.5	11.8	12.1	12.5	12.8	13.1
8.0	Net		5.7	6.0	6.3	6.6	7.0	7.3	7.6	7.9	8.2	8.5	8.9	9.2	9.5	9.8	10.1	10.4	10.8	11.1	11.4	11.7	12.0	12.3	12.7
	Gross		6.5	6.8	7.2	7.6	7.9	8.3	8.6	9.0	9.4	9.7	10.1	10.4	10.8	11.2	11.5	11.9	12.2	12.6	13.0	13.3	13.7	14.0	14.4
8.5	Net		5.9	6.2	6.6	6.9	7.2	7.5	7.9	8.2	8.5	8.9	9.2	9.5	9.8	10.2	10.5	10.8	11.2	11.5	11.8	12.1	12.5	12.8	13.1
	Gross		6.7	7.1	7.4	7.8	8.2	8.6	8.9	9.3	9.7	10.0	10.4	10.8	11.2	11.5	11.9	12.3	12.6	13.0	13.4	13.8	14.1	14.5	14.9
9.0	Net		6.3	6.7	7.0	7.4	7.7	8.1	8.4	8.8	9.1	9.5	9.8	10.2	10.5	10.9	11.2	11.6	11.9	12.3	12.6	13.0	13.3	13.7	14.0
	Gross		7.1	7.5	7.9	8.3	8.7	9.1	9.5	9.8	10.2	10.6	11.0	11.4	11.8	12.2	12.6	13.0	13.4	13.8	14.2	14.6	15.0	15.4	15.8
9.5	Net		6.7	7.1	7.4	7.8	8.2	8.6	8.9	9.3	9.7	10.0	10.4	10.8	11.2	11.5	11.9	12.3	12.6	13.0	13.4	13.8	14.1	14.5	14.9
	Gross		7.5	7.9	8.3	8.7	9.1	9.6	10.0	10.4	10.8	11.2	11.6	12.1	12.5	12.9	13.3	13.7	14.1	14.5	15.0	15.4	15.8	16.2	16.6
10.0	Net		7.1	7.5	7.9	8.3	8.7	9.1	9.5	9.8	10.2	10.6	11.0	11.4	11.8	12.2	12.6	13.0	13.4	13.8	14.2	14.6	15.0	15.4	15.8
	Gross		7.9	8.3	8.8	9.2	9.6	10.1	10.5	10.9	11.4	11.8	12.3	12.7	13.1	13.6	14.0	14.4	14.9	15.3	15.8	16.2	16.6	17.1	17.5
10.5	Net		7.5	7.9	8.3	8.7	9.1	9.6	10.0	10.4	10.8	11.2	11.6	12.1	12.5	12.9	13.3	13.7	14.1	14.5	15.0	15.4	15.8	16.2	16.6
	Gross		8.3	8.7	9.2	9.6	10.1	10.6	11.0	11.5	11.9	12.4	12.9	13.3	13.8	14.2	14.7	15.2	15.6	16.1	16.5	17.0	17.5	17.9	18.4
11.0	Net		7.9	8.3	8.8	9.2	9.6	10.1	10.5	10.9	11.4	11.8	12.3	12.7	13.1	13.6	14.0	14.4	14.9	15.3	15.8	16.2	16.6	17.1	17.5
	Gross		8.7	9.1	9.6	10.1	10.6	11.1	11.6	12.0	12.5	13.0	13.5	14.0	14.4	14.9	15.4	15.9	16.4	16.8	17.3	17.8	18.3	18.8	19.3
11.5	Net		8.3	8.7	9.2	9.6	10.1	10.6	11.0	11.5	11.9	12.4	12.9	13.3	13.8	14.2	14.7	15.2	15.6	16.1	16.5	17.0	17.5	17.9	18.4
	Gross		9.1	9.6	10.1	10.6	11.1	11.6	12.1	12.6	13.1	13.6	14.1	14.6	15.1	15.6	16.1	16.6	17.1	17.6	18.1	18.6	19.1	19.6	20.1
12.0	Net		8.7	9.1	9.6	10.1	10.6	11.1	11.6	12.0	12.5	13.0	13.5	14.0	14.4	14.9	15.4	15.9	16.4	16.8	17.3	17.8	18.3	18.8	19.3
	Gross		9.5	10.0	10.5	11.0	11.6	12.1	12.6	13.1	13.7	14.2	14.7	15.2	15.8	16.3	16.8	17.3	17.9	18.4	18.9	19.4	20.0	20.5	21.0
12.5	Net		9.1	9.6	10.1	10.6	11.1	11.6	12.1	12.6	13.1	13.6	14.1	14.6	15.1	15.6	16.1	16.6	17.1	17.6	18.1	18.6	19.1	19.6	20.1
	Gross		9.8	10.4	10.9	11.5	12.0	12.6	13.1	13.7	14.2	14.8	15.3	15.9	16.4	17.0	17.5	18.0	18.6	19.1	19.7	20.2	20.8	21.3	21.9
13.0	Net		9.5	10.0	10.5	11.0	11.6	12.1	12.6	13.1	13.7	14.2	14.7	15.2	15.8	16.3	16.8	17.3	17.9	18.4	18.9	19.4	20.0	20.5	21.0
	Gross		10.2	10.8	11.4	11.9	12.5	13.1	13.7	14.2	14.8	15.4	15.9	16.5	17.1	17.6	18.2	18.8	19.3	19.9	20.5	21.0	21.6	22.2	22.8
13.5	Net		9.8	10.4	10.9	11.5	12.0	12.6	13.1	13.7	14.2	14.8	15.3	15.9	16.4	17.0	17.5	18.0	18.6	19.1	19.7	20.2	20.8	21.3	21.9
	Gross		10.6	11.2	11.8	12.4	13.0	13.6	14.2	14.8	15.4	15.9	16.5	17.1	17.7	18.3	18.9	19.5	20.1	20.7	21.3	21.9	22.4	23.0	23.6

(continued)

APPENDIX D — Approximate Energy Expenditure (Kcal/Min) by Body Weight Based on the Metabolic Equivalents (METs) for Physical Activity Intensity

METs		kg	45	48	50	53	55	58	60	63	65	68	70	73	75	78	80	83	85	88	90	93	95	98	100
		lbs	99	105	110	116	121	127	132	138	143	149	154	160	165	171	176	182	187	193	198	204	209	215	220
14.0	Net		10.2	10.8	11.4	11.9	12.5	13.1	13.7	14.2	14.8	15.4	15.9	16.5	17.1	17.6	18.2	18.8	19.3	19.9	20.5	21.0	21.6	22.2	22.8
	Gross		11.0	11.6	12.3	12.9	13.5	14.1	14.7	15.3	15.9	16.5	17.2	17.8	18.4	19.0	19.6	20.2	20.8	21.4	22.1	22.7	23.3	23.9	24.5
14.5	Net		10.6	11.2	11.8	12.4	13.0	13.6	14.2	14.8	15.4	15.9	16.5	17.1	17.7	18.3	18.9	19.5	20.1	20.7	21.3	21.9	22.4	23.0	23.6
	Gross		11.4	12.1	12.7	13.3	14.0	14.6	15.2	15.9	16.5	17.1	17.8	18.4	19.0	19.7	20.3	20.9	21.6	22.2	22.8	23.5	24.1	24.7	25.4
15.0	Net		11.0	11.6	12.3	12.9	13.5	14.1	14.7	15.3	15.9	16.5	17.2	17.8	18.4	19.0	19.6	20.2	20.8	21.4	22.1	22.7	23.3	23.9	24.5
	Gross		11.8	12.5	13.1	13.8	14.4	15.1	15.8	16.4	17.1	17.7	18.4	19.0	19.7	20.3	21.0	21.7	22.3	23.0	23.6	24.3	24.9	25.6	26.3
15.5	Net		11.4	12.1	12.7	13.3	14.0	14.6	15.2	15.9	16.5	17.1	17.8	18.4	19.0	19.7	20.3	20.9	21.6	22.2	22.8	23.5	24.1	24.7	25.4
	Gross		12.2	12.9	13.6	14.2	14.9	15.6	16.3	17.0	17.6	18.3	19.0	19.7	20.3	21.0	21.7	22.4	23.1	23.7	24.4	25.1	25.8	26.4	27.1
16.0	Net		11.8	12.5	13.1	13.8	14.4	15.1	15.8	16.4	17.1	17.7	18.4	19.0	19.7	20.3	21.0	21.7	22.3	23.0	23.6	24.3	24.9	25.6	26.3
	Gross		12.6	13.3	14.0	14.7	15.4	16.1	16.8	17.5	18.2	18.9	19.6	20.3	21.0	21.7	22.4	23.1	23.8	24.5	25.2	25.9	26.6	27.3	28.0
16.5	Net		12.2	12.9	13.6	14.2	14.9	15.6	16.3	17.0	17.6	18.3	19.0	19.7	20.3	21.0	21.7	22.4	23.1	23.7	24.4	25.1	25.8	26.4	27.1
	Gross		13.0	13.7	14.4	15.2	15.9	16.6	17.3	18.0	18.8	19.5	20.2	20.9	21.7	22.4	23.1	23.8	24.5	25.3	26.0	26.7	27.4	28.2	28.9
17.0	Net		12.6	13.3	14.0	14.7	15.4	16.1	16.8	17.5	18.2	18.9	19.6	20.3	21.0	21.7	22.4	23.1	23.8	24.5	25.2	25.9	26.6	27.3	28.0
	Gross		13.4	14.1	14.9	15.6	16.4	17.1	17.9	18.6	19.3	20.1	20.8	21.6	22.3	23.1	23.8	24.5	25.3	26.0	26.8	27.5	28.3	29.0	29.8
17.5	Net		13.0	13.7	14.4	15.2	15.9	16.6	17.3	18.0	18.8	19.5	20.2	20.9	21.7	22.4	23.1	23.8	24.5	25.3	26.0	26.7	27.4	28.2	28.9
	Gross		13.8	14.5	15.3	16.1	16.8	17.6	18.4	19.1	19.9	20.7	21.4	22.2	23.0	23.7	24.5	25.3	26.0	26.8	27.6	28.3	29.1	29.9	30.6
18.0	Net		13.4	14.1	14.9	15.6	16.4	17.1	17.9	18.6	19.3	20.1	20.8	21.6	22.3	23.1	23.8	24.5	25.3	26.0	26.8	27.5	28.3	29.0	29.8
	Gross		14.2	15.0	15.8	16.5	17.3	18.1	18.9	19.7	20.5	21.3	22.1	22.8	23.6	24.4	25.2	26.0	26.8	27.6	28.4	29.1	29.9	30.7	31.5
18.5	Net		13.8	14.5	15.3	16.1	16.8	17.6	18.4	19.1	19.9	20.7	21.4	22.2	23.0	23.7	24.5	25.3	26.0	26.8	27.6	28.3	29.1	29.9	30.6
	Gross		14.6	15.4	16.2	17.0	17.8	18.6	19.4	20.2	21.0	21.9	22.7	23.5	24.3	25.1	25.9	26.7	27.5	28.3	29.1	29.9	30.8	31.6	32.4
19.0	Net		14.2	15.0	15.8	16.5	17.3	18.1	18.9	19.7	20.5	21.3	22.1	22.8	23.6	24.4	25.2	26.0	26.8	27.6	28.4	29.1	29.9	30.7	31.5
	Gross		15.0	15.8	16.6	17.5	18.3	19.1	20.0	20.8	21.6	22.4	23.3	24.1	24.9	25.8	26.6	27.4	28.3	29.1	29.9	30.8	31.6	32.4	33.3
19.5	Net		14.6	15.4	16.2	17.1	17.9	18.6	19.4	20.2	21.0	21.9	22.7	23.5	24.3	25.1	25.9	26.7	27.5	28.3	29.1	29.9	30.8	31.6	32.4
	Gross		15.4	16.2	17.1	17.9	18.8	19.6	20.5	21.3	22.2	23.0	23.9	24.7	25.6	26.4	27.3	28.2	29.0	29.9	30.7	31.6	32.4	33.3	34.1
20.0	Net		15.0	15.8	16.6	17.5	18.3	19.1	20.0	20.8	21.6	22.4	23.3	24.1	24.9	25.8	26.6	27.4	28.3	29.1	29.9	30.8	31.6	32.4	33.3
	Gross		15.8	16.6	17.5	18.4	19.3	20.1	21.0	21.9	22.8	23.6	24.5	25.4	26.3	27.1	28.0	28.9	29.8	30.6	31.5	32.4	33.3	34.1	35.0

Body mass

Source for equations to calculate gross and net estimated energy: ACSM Guidelines for Exercise Testing and Prescription. 11th Ed. 2021. Philadelphia, PA: Wolters Kluwer. [Net and Gross values calculated using Microsoft Excel then copied and pasted into Word.]

Glossary

A

Acceptable Macronutrient Distribution Range (AMDR) A range of dietary intakes for carbohydrate, fat, and protein that is associated with reduced risk of chronic disease while providing adequate nutrients.

acclimatization The ability of the body to undergo physiological adaptations so that the stress of a given environment, such as high environmental temperature, is less severe.

active transport A process requiring energy to transport substances across cell membranes.

activity-stat hypothesis A hypothesis that physical activity is regulated by the central nervous system in order to stabilize energy expenditure.

added sugars Refined sugars added to foods during commercial food processing.

adenosine triphosphate *See* ATP.

Adequate Intake (AI) Recommended dietary intake comparable to the RDA but based on less scientific evidence.

adipokines Substances released from adipose (fat) cells that function as hormones in other parts of the body.

adiposopathy (sick fat) Pathological function of fat tissue that increases risk of the metabolic syndrome.

aerobic glycolysis Oxidative processes in the cell that liberate energy in the metabolism of the carbohydrate glycogen.

aerobic lipolysis Oxidative processes in the cell that liberate energy in the metabolism of fats.

air displacement plethysmography (ADP) A procedure to measure body composition via displacement of air in a special chamber; comparable to water displacement in underwater weighing techniques to evaluate body composition.

alanine A nonessential amino acid.

alcohol use disorders (AUD) As defined by the National Institute of Alcohol Abuse and Alcoholism (NIAAA), a "medical condition characterized by an impaired ability to stop or control alcohol use despite adverse social, occupational, or health consequences."

alcoholism A rather undefined term used to describe individuals who abuse the effect of alcohol; an addiction or habituation that may result in physical and/or psychological withdrawal effects.

aldosterone The main electrolyte-regulating hormone secreted by the adrenal cortex; primarily controls sodium and potassium balance.

alpha-ketoacid Specific acids associated with different amino acids and released upon deamination or transamination; for example, the breakdown of glutamate yields alpha-ketoglutarate.

alpha-linolenic acid An omega-3 fatty acid considered to be an essential nutrient.

alpha-tocopherol The most biologically active alcohol in vitamin E.

AMDR *See* Acceptable Macronutrient Distribution Range.

amino acids The chief structural material of protein, consisting of an amino group (NH_2) and an acid group ($COOH$) plus other components.

aminostatic theory A theory suggesting that hunger is controlled by the presence or absence of amino acids in the blood acting upon a receptor in the hypothalamus.

ammonia A metabolic by-product of the oxidation of glutamine; it may be transformed into urea for excretion from the body.

anabolic-androgenic steroids (AAS) Drugs designed to mimic the actions of testosterone to build muscle tissue (anabolism) while minimizing the androgenic effects (masculinization).

anabolism Constructive metabolism, the process whereby simple body compounds are formed into more complex ones.

anaerobic glycolysis Metabolic processes in the cell that liberate energy in the metabolism of the carbohydrate glycogen without the involvement of oxidation.

android-type obesity Male-type obesity in which the body fat accumulates in the abdominal area and is a more significant risk factor for chronic disease than is gynoid-type obesity.

androstenedione An androgen produced in the body that is converted to testosterone; marketed as a dietary supplement.

angina The pain experienced under the breastbone or in other areas of the upper body when the heart is deprived of oxygen.

anorexia athletica (AA) A form of anorexia nervosa observed in athletes involved in sports in which low percentages of body fat may enhance performance, such as gymnastics and ballet.

anorexia nervosa (AN) A serious nervous condition, particularly among teenage girls and young women, marked by a loss of appetite and leading to various degrees of emaciation.

anorexiant A pharmacological agent that suppresses appetite.

antidiuretic hormone (ADH) Hormone secreted by the pituitary gland; its major action is to conserve body water by decreasing urine formation; also known as vasopressin.

antipromoters Compounds that block the actions of *promoters,* agents associated with the development of certain diseases, such as cancer.

apolipoprotein A class of proteins associated with the formation of lipoproteins. A variety of apolipoproteins have been identified and are involved in the specific functions of the different lipoproteins.

apoptosis Cell death occurring as a normal part of growth and development.

appestat Term used for the neural center in the hypothalamus that helps control appetite by stimulating either hunger or satiety.

arteriosclerosis Hardening of the arteries. *See also* atherosclerosis.

atherosclerosis A specific form of arteriosclerosis characterized by the formation of plaque on the inner layers of the arterial wall.

athletic amenorrhea The cessation of menstruation in athletes, believed to be caused by factors associated with participation in strenuous physical activity.

ATP Adenosine triphosphate, a high-energy phosphate compound found in the body; one of the major forms of energy available for immediate use in the body.

ATP-PCr system The energy system for fast, powerful muscle contractions; uses ATP as the immediate energy source, the spent ATP being quickly regenerated by breakdown of the PCr. ATP and PCr are high-energy phosphates in the muscle cell.

B

basal energy expenditure (BEE) Energy required to sustain basic metabolism over 24 hours; measured in a supine position at room temperature.

basal metabolic rate (BMR) The measurement of energy expenditure in the body under resting, postabsorptive conditions, indicative of the energy needed to maintain life under these basal conditions.

behavioral group therapy A form of psychotherapy in which a therapist works with several people at the same time. Behavioral therapy focuses on identifying and helping to modify behaviors that are potentially self-destructive and unhealthy.

beta-alanine A nonessential amino acid, also labeled as β-alanine, that is a part of the peptides carnosine and anserine; beta-alanine is purported to have ergogenic potential.

beta-carotene A precursor for vitamin A found in plants.

beta-oxidation Process in the cells whereby 2-carbon units of acetic acid are removed from long-chain fatty acids for conversion to acetyl CoA and oxidation via the Krebs cycle.

binge eating disorder (BED) Condition in which individuals demonstrate some of the same behaviors as those with bulimia nervosa, such as eating more quickly and until uncomfortably full, but do not purge.

bioavailability In relation to nutrients in food, the amount that may be absorbed into the body.

bioelectrical impedance analysis (BIA) A method to calculate percentage of body fat by measuring electrical resistance due to the water content of the body.

biotin Water-soluble vitamin that is a component of the B complex vitamins. Important as a coenzyme necessary for energy metabolism.

blood alcohol concentration (BAC) The concentration of alcohol in the blood, usually expressed as milligram percent.

BMI *See* body mass index.

body image The image or impression the individual has of his or her body. A poor body image may lead to personality problems.

body mass index (BMI) An index calculated by a ratio of height to weight, used as a measure of obesity.

brown adipose tissue Adipose tissue that is designed to produce heat; small amounts are found in humans in the area of vital organs such as the heart and lungs.

bulimia nervosa (BN) An eating disorder involving a loss of control over the impulse to binge; the binge-purge syndrome.

C

caffeine A stimulant drug found in many food products such as coffee, tea, and cola drinks; stimulates the central nervous system.

calorie (kcal) deficit Consuming fewer kcal that the body expends.

Calorie A measure of heat energy. A small calorie represents the amount of heat needed to raise 1 gram of water 1 degree Celsius. A large Calorie (kilocalorie, kcal, or C) is 1,000 small calories.

calorimetry The science of measuring heat production.

carbohydrate-electrolyte solutions (CES) Fluids containing water, various forms of carbohydrate such as glucose and fructose, and various electrolytes such as sodium, chloride, and potassium, in a solution designed to maintain optimal hydration and energy during exercise.

carbohydrate loading A dietary method used by endurance-type athletes to help increase the carbohydrate (glycogen) levels in their muscles and liver.

carbohydrates A group of compounds containing carbon, hydrogen, and oxygen. Glucose, glycogen, sugar, starches, fiber, cellulose, and the various saccharides are all carbohydrates.

cardiorespiratory fitness A component of physiological fitness. Refers to the capacity of the circulatory and respiratory systems to supply adequate oxygen to skeletal muscles during physical activity.

carnitine A chemical that facilitates the transfer of fatty acids into the mitochondria for subsequent oxidation.

carnosine A peptide found in muscle tissue theorized to possess ergogenic potential via its buffering effect on lactic acid during high-intensity exercise; carnosine contains the nonessential amino acid beta-alanine.

carotenemia Condition characterized by excessive intake of beta-carotene, leading to an accumulation of beta-carotene in fat tissues and a yellowing of the skin.

catabolism Destructive metabolism whereby complex chemical compounds in the body are degraded to simpler ones.

cellulite Lumpy fat that often appears in the thigh and hip region of women. Cellulite is simply normal fat in small compartments formed by connective tissue but may contain other compounds that bind water.

cholecalciferol (vitamin D3) The product of irradiation of 7-dehydrocholesterol found in the skin. *See also* vitamin D3.

cholesterol A fat-like, pearly substance, an alcohol, found in all animal fat and oils; a main constituent of some body tissues and body compounds.

choline A substance associated with the B complex that is widely distributed in both plant and animal tissues; involved in carbohydrate, fat, and protein metabolism.

chronic fatigue syndrome Prolonged fatigue (over 6 months) of unknown cause characterized by mental depression and physical fatigue; may be observed in endurance athletes.

chronic training effect The structural and metabolic adaptations that occur in the body in response to exercise training over time; the adaptations are specific to the type of exercise, such as aerobic endurance training or resistance training.

chylomicron A particle of emulsified fat found in the blood following the digestion and assimilation of fat.

circuit aerobics A combination of aerobic and weight-training exercises designed to elicit the specific benefits of each type of exercise.

circuit weight training A method of training in which exercises are arranged in a circuit or sequence. May be designed with weight training to help convey an aerobic training effect.

cirrhosis A degenerative disease of the liver, one cause being excessive consumption of alcohol.

cis **fatty acids** The chemical structure of unsaturated fatty acids in which the hydrogen ions are on the same side of the double bond.

ciwujia A Chinese herb theorized to be ergogenic.

coenzyme An activator of an enzyme; many vitamins are coenzymes.

compensated heat stress A condition in which heat loss balances heat production, a set body temperature is maintained, and the individual can continue to exercise in warm environmental conditions.

complementary proteins Combining plant foods such as rice and beans so that essential amino acids deficient in one of the foods are provided by the other, in order to obtain a balanced intake of essential amino acids.

complete proteins Proteins that contain all nine essential amino acids in the proper proportions.

complex carbohydrates Foods high in starch, such as bread, cereals, fruits, and vegetables, as contrasted to simple carbohydrates such as table sugar.

concentric contraction The phase of an isotonic muscle contraction where actin and myosin filaments in the sarcomere slide past each other and result in shortening of the muscle as the load is moved or lifted against gravity.

conduction In relation to body temperature, the transfer of heat from one substance to another by direct contact.

conjugated linoleic acid (CLA) Isomers of linoleic acid, an essential fatty acid. CLA is found in the meat and milk of ruminants and is theorized to have health and exercise performance benefits, such as promotion of weight loss.

convection In relation to body temperature, the transfer of heat by way of currents in either air or water.

core temperature The temperature of the deep tissues of the body, usually measured orally or rectally; *see also* shell temperature.

Cori cycle Cycle involving muscle breakdown of glucose to lactate, lactate transport via blood to the liver for reconversion to glucose, and glucose returning to the muscle.

coronary artery disease (CAD) Atherosclerosis in the coronary arteries.

coronary heart disease (CHD) A degenerative disease of the heart caused primarily by arteriosclerosis or atherosclerosis of the coronary vessels of the heart.

coronary occlusion Closure of coronary arteries that may precipitate a heart attack; occlusion may be partial or complete closure.

coronary thrombosis Occlusion (closure) of coronary arteries, usually by a blood clot.

cortisol A hormone secreted by the adrenal cortex with gluconeogenic potential, helping to convert amino acids into glucose.

creatine A nitrogen-containing compound found in the muscles, usually complexed with phosphate to form phosphocreatine.

crossover concept The concept that as exercise intensity increases, at some point carbohydrate rather than fat becomes the predominant fuel for muscle contraction.

cruciferous vegetables Vegetables in the cabbage family, such as broccoli, cauliflower, kale, and all cabbages.

cytokine Small proteins or peptides produced in cells, such as adipokines by adipose tissue cells, that possess hormone-like functions on other cells in the body; the immune system secretes a number of different cytokines.

D

Daily Value (DV) A term used in food labeling; the DV is based on a daily energy intake of 2,000 kcal and for the food labeled presents the percentage of the RDI and the DRV recommended for healthy Americans.

DASH diet The Dietary Approaches to Stop Hypertension diet plan that is designed to reduce or prevent an increase in blood pressure.

deamination Removal of an amine group, or nitrogen, from an amino acid.

dehydration A reduction of the body water to below the normal level of hydration; water output exceeds water intake.

dehydroepiandrosterone (DHEA) A natural steroid hormone produced endogenously by the adrenal gland. May be marketed as a nutritional sports ergogenic as derived from herbal precursors.

delayed onset of muscle soreness (DOMS) Soreness in muscles experienced a day or two after strenuous eccentric exercise, such as running downhill. Prolonged, excessive eccentric exercise may lead to small muscle tears, and the pain is believed to occur during the repair process when swelling activates pain receptors.

DHEA *See* dehydroepiandrosterone.

Diabesity™ Term coined to highlight the relationship between the development of diabetes following the onset of obesity.

dietary fiber Nondigestible carbohydrates and lignin that are intrinsic and intact in plants.

dietary folate equivalents (DFE) Used in estimating folate requirements, adjusting for the greater degree of absorption of folic acid (free form) compared with folate naturally found in foods. One microgram food folate equals 0.5–0.6 microgram folic acid added to foods or as a supplement.

dietary guidance systems Food guides that provide practical ways for consumers to select certain foods and food components to promote health and meet dietary recommendations.

dietary-induced thermogenesis (DIT) The increase in the basal metabolic rate following ingestion of a meal. Heat production is increased.

Dietary Reference Intake (DRI) Standard for recommended dietary intake, consisting of various values. *See also* AI, EAR, RDA, AMDR, and UL.

dietary supplement A food product, added to the total diet, that contains either vitamins, minerals, herbs, botanicals, amino acids, metabolites, constituents, extracts, or combinations of these ingredients.

differential susceptibility The potential for an individual's genetic predispositions for a health-related outcome to be modified, for better or worse, by environmental influences.

dipsogenic drive Physiological stimuli promoting fluid consumption in response to fluid loss (dehydration).

disaccharide Any one of a class of sugars that yield two monosaccharides on hydrolysis; sucrose is the most common.

disordered eating Atypical eating behaviors such as restrictive dieting, using diet pills or laxatives, bingeing, and purging. In general, disordered eating behaviors occur less frequently or are less severe than those required to meet the full criteria for the diagnosis of an eating disorder.

dispensable amino acids *See* nonessential animo acids.

doping Official term used by the WADA and the International Olympic Committee to depict the use of drugs in sports in attempts to enhance performance.

dual energy X-ray absorptiometry (DEXA, DXA) A computerized X-ray technique at two energy levels to image body fat, lean tissues, and bone mineral content.

dual-intervention point model A theoretical model of weight gain that describes competition between physiological control factors which defend an upper point of weight or fat mass and environmental factors favoring either weight gain or loss.

E

eating disorders A psychological disorder centering on the avoidance, excessive consumption, or purging of food, such as anorexia nervosa and bulimia nervosa.

eccentric contraction The phase of an isotonic muscle contraction where actin and myosin filaments in the sarcomere slide past each other and results in lengthening of the muscle as the load is moved or lowered with gravity.

eicosanoids Derivatives of fatty acid oxidation in the body, including prostaglandins, thromboxanes, and leukotrienes.

electrolyte A substance that, when in a solution, conducts an electrical current.

electron transfer system A highly structured array of chemical compounds in the cell that transport electrons and harness energy for later use.

empty kcal Foods and beverages typically high in refined sugar, fat, or alcohol that have considerable energy content but are relatively poor sources of other nutrients such as vitamins and minerals.

energy availability The amount of energy (kcal) available to the human body, including energy to support physical activity. EA = Food Energy (kcal) Intake − Exercise Energy (kcal) Expenditure.

energy The ability to do work; energy exists in various forms, notably mechanical, heat, and chemical in the human body.

enzymes Complex proteins in the body that serve as catalysts, facilitating reactions between various substances without being changed themselves.

ephedra *Ephedra sinica*, a source of ephedrine.

ephedrine A stimulant with somewhat weaker effects than amphetamine; found in some commercial dietary supplements; also known as ephedra.

epidemiological research A study of certain populations to determine the relationship of various risk factors to epidemic diseases or health problems.

epigenetics The study of various factors, including diet and exercise, that may influence gene expression without modifying the DNA, that influence gene expression. For example, exerkines produced during exercise may influence DNA activity, such as activating or deactivating genes, without changing the DNA sequence.

epigenome A structure located just outside the genome that may be influenced by various factors, such as nutrients in the foods eaten; activates or deactivates DNA and subsequent genetic and cellular activity that may have either positive or negative health effects.

epinephrine A hormone secreted by the adrenal medulla that stimulates numerous body processes to enhance energy production, particularly during intense exercise.

ergogenic aids Dietary, supplemental, pharmacological, psychological agents; training strategies; and/or advances in materials technology used by competitive athletes to improve performance.

ergolytic An agent or substance that may lead to decreases in work productivity or physical performance.

ergometer A device, such as a cycle ergometer, to measure work output in watts or other measures of work.

essential (indispensable) amino acids Amino acids that must be obtained in the diet and cannot be synthesized in the body. Also known as indispensable amino acids.

essential fat Fat in the body that is an essential part of the tissues, such as cell membrane structure, nerve coverings, and the brain; *see also* storage fat.

essential hypertension Elevated blood pressure that cannot be attributed to a specific pathological reason or cause. Also known as primary hypertension.

essential nutrients Those nutrients found to be essential to human life and optimal functioning.

ester Compound formed from the combination of an organic acid and an alcohol.

Estimated Average Requirement (EAR) Nutrient intake value estimated to meet the requirements of half the healthy individuals in a group.

Estimated Energy Requirement (EER) The daily dietary intake predicted to maintain energy balance for an individual of a defined age, gender, height, weight, and level of physical activity consistent with good health.

ethanol Alcohol; ethyl alcohol.

euhydration *See* normohydration.

evaporation The conversion of a liquid to a vapor, which consumes energy; evaporation of sweat cools the body by using body heat as the energy source.

exercise A form of structured physical activity generally designed to enhance physical fitness; usually refers to strenuous physical activity.

exercise-associated collapse Postexercise dizziness due to hypotension from venous pooling.

exercise intensity The tempo, speed, or resistance of an exercise. Intensity can be increased by working faster, doing more work in a given amount of time.

exercise metabolic rate (EMR) An increased metabolic rate due to the need for increased energy production; during exercise, the resting energy expenditure (REE) may be increased more than 20-fold.

exercise-associated muscle cramps Painful muscular cramps or tetany following prolonged exercise in the heat without water or salt replacement.

exertional heat stroke Heat stroke that is precipitated by exercise in a warm or hot environment.

experimental research Study that manipulates an independent variable (cause) to observe the outcome on a dependent variable (effect).

extracellular water Body water that is located outside the cells; often subdivided into the intravascular water and the intercellular, or interstitial, water.

F

facilitated diffusion Process whereby glucose combines with a special protein carrier molecule at the membrane surface, facilitating glucose transport into the cell; insulin promotes facilitated diffusion in some cells.

fat-free mass The remaining mass of the human body following the extraction of all fat.

fatigue A generalized or specific feeling of tiredness that may have a multitude of causes; may be mental or physical.

fat loading Practices used to maximize the use of fats as an energy source during exercise, particularly a low-carbohydrate, high-fat diet.

fat substitutes Various substances used as substitutes for fats in food products; two popular brands are Simplesse and Olestra.

fatty acids Aliphatic acids containing only carbon, oxygen, and hydrogen; they may be saturated or unsaturated.

female athlete triad The triad of disordered eating, amenorrhea, and osteoporosis sometimes seen in female athletes involved in sports where excess body weight may be detrimental to performance.

ferritin The form in which iron is stored in the tissues.

fetal alcohol effects (FAE) Symptoms noted in children born to women who consumed alcohol during pregnancy; not as severe as fetal alcohol syndrome.

fetal alcohol syndrome (FAS) The cluster of physical and mental symptoms seen in the child of a mother who consumes excessive alcohol during pregnancy.

folate (folic acid) A water-soluble vitamin that is essential in preventing certain types of anemia and, during pregnancy, is important in the prevention of neural tube defects.

free radicals Atoms or compounds in which there is an unpaired electron; thought to cause cellular damage.

fructose A monosaccharide known as levulose or fruit sugar; found in all sweet fruits.

functional fiber Isolated, nondigestible carbohydrate that has beneficial effects in humans.

functional foods Food products containing nutrients designed to provide health benefits beyond basic nutrition.

G

galactose A monosaccharide formed when lactose is hydrolyzed into glucose and galactose.

gamma-tocopherol Form of vitamin E found naturally in foods, including vegetable oil. Compared to alpha-tocopherol, has a lower biological activity.

general model A theoretical model of weight gain that combined the set and settling point theories in describing regulation of intake as functions of social/environmental versus and genetics and feedback control factors.

ghrelin Hormone released by an empty stomach to stimulate the appetite.

ginseng A general term for a variety of natural chemical plant extracts derived from the family Araliaceae; extract contains ginsenosides and other chemicals that may influence human physiology.

glucagon A hormone secreted by the pancreas; basically it exerts actions just the opposite of insulin, i.e., it responds to hypoglycemia and helps to increase blood sugar levels.

glucogenic amino acids Amino acids that may undergo deamination and be converted into glucose through the process of gluconeogenesis.

gluconeogenesis The formation of carbohydrates from molecules that are not themselves carbohydrate, such as amino acids and the glycerol from fat.

glucose A monosaccharide; a thick, sweet, syrupy liquid.

glucose-alanine cycle The cycle in which alanine is released from the muscle and is converted to glucose in the liver.

glucose polymers A combination of several glucose molecules into more complex carbohydrates.

glucostatic theory The theory that hunger and satiety are controlled by the glucose level in the blood; the receptors that respond to the blood glucose level are in the hypothalamus.

gluten intolerance A sensitivity to gluten; the immune system recognizes gluten as a foreign substance but does not induce an allergic response.

glycemic index (GI) A ranking system relative to the effect that consumption of 50 grams of a particular carbohydrate food has upon the blood glucose response over the course of 2 hours. The normal baseline measure is 50 grams of glucose, and the resultant blood glucose response is scored as 100.

glycemic load (GL) A ranking system relative to the effect that eating a carbohydrate food has on the blood glucose level, but also includes the portion size. The formula is

$$GL = \frac{(GI) \times (\text{grams of nonfiber carbohydrate in 1 serving})}{100}$$

glycerol Glycerin, a clear, syrupy liquid; an alcohol that combines with fatty acids to form triglycerides.

glycolysis The degradation of sugars into smaller compounds; the main quantitative anaerobic energy process in the muscle tissue.

gynoid-type obesity Female-type obesity; body fat is deposited primarily about the hips and thighs. *See also* android-type obesity.

H

HDL High-density lipoprotein; a protein-lipid complex in the blood that facilitates the transport of triglycerides, cholesterol, and phospholipids.

health claims FDA-approved claims found on food labels that relate consumption of a given level of a nutrient or nutrients to prevention of a certain chronic disease.

health-related fitness Those components of physical fitness whose improvement have health benefits, such as cardiovascular fitness, body composition, flexibility, and muscular strength and endurance.

heart rate maximum reserve (HR max) The maximum number of heart contractions per minute in response to maximal exercise. Usually predicted according to age-adjusted equations (e.g., 220 − age).

heat-balance equation Heat balance is dependent on the interrelationships of metabolic heat production and loss or gain of heat by radiation, convection, conduction, and evaporation.

heat exhaustion Weakness or dizziness from overexertion in a hot environment.

heat index The apparent temperature determined by combining air temperature and relative humidity.

heat stroke Elevated body temperature of 105.8°F or greater caused by exposure to excessive heat gains or production and diminished heat loss.

heat syncope Fainting caused by excessive heat exposure.

hematuria Blood or red blood cells in the urine.

heme iron The iron in the diet associated with hemoglobin in animal meats.

hemochromatosis Presence of excessive iron in the body, resulting in an enlarged liver and bronze pigmentation of the skin.

hemolysis A rupturing of red blood cells with a release of hemoglobin into the plasma.

hepcidin A hormone produced in the liver that helps regulate serum iron levels mainly by its effects on intestinal absorption. Elevated serum iron levels will stimulate hepcidin synthesis by the liver, which will inhibit iron absorption by the intestines. Conversely, decreased levels of serum iron will inhibit hepcidin synthesis and increase intestinal iron absorption.

hidden fat In foods, the fat that is not readily apparent, such as the high fat content of cheese.

high blood pressure *See* hypertension.

high-density lipoprotein *See* HDL.

high-intensity interval training (HIIT) An exercise format involving short bouts of intense, near maximal, anaerobic exercise followed by a short recovery period. Exercise protocols are designed to provide health benefits comparable to more prolonged aerobic endurance exercise, but less overall time commitment.

HMB Beta-hydroxy-beta-methylbutyrate, a metabolic by-product of the amino acid leucine, alleged to retard the breakdown of muscle protein during strenuous exercise.

homeostasis A condition of normalcy in the internal body environment.

homocysteine A metabolic by-product of amino acid metabolism; elevated blood levels are associated with increased risk of vascular diseases.

human growth hormone (HGH) A hormone released by the pituitary gland that regulates growth; also involved in fatty acid metabolism; rHGH is a genetically engineered form.

hunger center A collection of nerve cells in the hypothalamus that is involved in the control of feeding reflexes.

hydrostatic weighing An underwater weighing technique used to estimate body composition (body fat and lean body mass)

hyperglycemia Elevated blood glucose levels.

hyperhydration The practice of increasing the body-water stores by fluid consumption prior to an athletic event; a state of increased water content in the body.

hyperkalemia An increased concentration of potassium in the blood.

hyperplasia The formation of new body cells.

hypertension A condition with various causes whereby the blood pressure is higher than normal.

hyperthermia Unusually high body temperature; fever.

hypertrophy Excessive growth of a cell or organ; in pathology, an abnormal growth.

hypervitaminosis A pathological condition due to an excessive vitamin intake, particularly the fat-soluble vitamins A and D.

hypoglycemia A low blood sugar level.

hypohydration A state of decreased water content in the body caused by dehydration.

hypokalemia A decreased concentration of potassium in the blood.

hyponatremia A decreased concentration of sodium in the blood.

hypothermia Unusually low body temperature.

I

immunomodulatory agent A substance that stimulates or suppresses the immune system and may help the body fight cancer, infection, or other diseases.

incomplete proteins Protein foods that do not possess the proper amount of essential amino acids; characteristic of many plant food sources of protein.

inosine A nucleoside of the purine family that serves as a base for the formation of a variety of compounds in the body; theorized to be ergogenic.

insensible perspiration Perspiration on the skin not detectable by ordinary senses.

insulin A hormone secreted by the pancreas, involved in carbohydrate metabolism.

intercellular water Body water found between the cells; also known as interstitial water.

intermittent fasting A dietary strategy in which a person eats only at specific times. There is a period of fasting (not consuming kcals) followed by feeding (consuming kcals); there are many different protocols in terms of the timing between fasting and feeding.

intracellular water Body water that is found within the cells.

intravascular water Body water found in the vascular system, or blood vessels.

intrinsic factor Substance made by the cells of the stomach and necessary for the digestion and absorption of vitamin B12.

intuitive eating A dietary approach that supports a positive body image, self-esteem, and well-being. Key principles of intuitive eating include focusing on foods as being nourishing and enjoyable rather than "bad" or "diet" foods.

ions Particles with an electrical charge; anions are negative and cations are positive.

iron-deficiency anemia Anemia caused by an inadequate intake or absorption of iron, resulting in impaired hemoglobin formation.

iron deficiency without anemia A condition in which the hemoglobin levels are normal but several indices of iron status in the body are below normal levels.

ischemia Lack of blood supply.

isokinetic contraction A muscle contraction with a constant angular shortening velocity throughout the entire range of motion.

isoleucine An essential amino acid.

isometric contraction A muscle contraction in which myosin binds to actin in the sarcomere to create muscle tension and force production, but there is no muscle shortening or movement.

isotonic contraction The most common form of muscle contraction consisting of a shortening phase as the load is lifted against gravity (see concentric contraction), then a lengthening phase as the load is lowered with gravity (see eccentric contraction).

isotonic solutions Two solutions separated by a semipermeable membrane that have equal osmotic pressures. Normal osmolality is approximately 290 mosmol/kg.

J

joule A measure of work in the metric system; a newton of force applied through a distance of 1 meter.

K

ketoacidosis A metabolic state where uncontrolled ketone production causes metabolic acidosis. Ketoacidosis is a pathologic state that requires medical attention.

ketogenesis The formation of ketones in the body from other substances, such as fats and proteins.

ketogenic amino acids Amino acids that may be deaminated, converted into ketones, and eventually into fat.

ketones Organic compounds containing a carbonyl group; ketone acids in the body, such as acetone, are the end products of fat metabolism.

ketosis Elevated blood or urine ketone levels without metabolic acidosis. Ketosis can be a normal response to dietary changes such as low-carbohydrate diets or fasting.

key-nutrient concept The concept that if certain key nutrients are adequately supplied by the diet, the other essential nutrients will also be present in adequate amounts.

kilocalorie (kcal) A unit of energy. A large Calorie; see also Calorie.

kilojoule One thousand joules; 1 kilojoule (kJ) is approximately 0.25 kilocalorie.

Krebs cycle The main oxidative reaction sequence in the body that generates ATP; also known as the citric acid or tricarboxylic acid cycle.

L

lactic acid system The energy system that produces ATP anaerobically by the breakdown of glycogen to lactic acid; used primarily in events of maximal effort for 1–2 minutes.

lactose intolerance Gastrointestinal disturbances due to an intolerance to lactose in milk; caused by deficiency of lactase, an enzyme that digests lactose.

lactovegetarian Vegetarian who include milk products in the diet as a form of high-quality protein.

LDL Low-density lipoprotein; a protein-lipid complex in the blood that facilitates the transport of triglycerides, cholesterol, and phospholipids.

lean body mass The body weight minus the body fat, composed primarily of muscle, bone, and other nonfat tissue.

lecithin A fatty substance of a class known as phospholipids; said to have the therapeutic properties of phosphorus.

legumes The fruit or pods of vegetables, including soybeans, kidney beans, lima beans, garden peas, black-eyed peas, and lentils; high in protein.

leptin Regulatory hormone produced by fat cells; when released into the circulation, it influences the hypothalamus to control appetite.

limiting amino acid An amino acid deficient in a specific plant food, making it an incomplete protein; methionine is a limiting amino acid in legumes, whereas lysine is deficient in grain products.

linoleic acid An essential fatty acid.

lipids A class of fats or fat-like substances characterized by their insolubility in water and solubility in fat solvents; triglycerides, fatty acids, phospholipids, and cholesterol are important lipids in the body.

lipoprotein A combination of lipid and protein possessing the general properties of proteins. Practically all the lipids of the plasma are present in this form.

lipoprotein (a) Serum lipid factor very similar to the LDL, being in the upper LDL density range and containing apolipoprotein (a); high levels are associated with increased risk for CHD.

lipostatic theory The theory that hunger and satiety are controlled by the lipid level in the blood.

liquid meals Food in a liquid form designed to provide a balanced intake of essential nutrients.

long-haul concept Relative to weight control, the idea that weight loss via exercise should be gradual, and one should not expect to lose large amounts of weight in a short time.

low energy availability A state in which exercise energy expenditure (kcal) is greater than food energy (kcal) intake.

M

macronutrient Dietary nutrient needed by the body in daily amounts greater than a few grams, such as carbohydrate, fat, protein, and water.

ma huang A Chinese plant extract theorized to be ergogenic; contains ephedrine, a stimulant.

macrocycle The longest window of time in periodized training, usually the entire competitive season or a calendar year.

major minerals Those minerals essential to human health and required in amounts of 100 mg or greater per day; calcium, chloride, magnesium, phosphorus, potassium, sodium, and sulfur.

malnutrition Poor nutrition that may be due to inadequate amounts of essential nutrients. Too many Calories leading to obesity is also a form of malnutrition. *See also* subclinical malnutrition.

male athlete triad The triad of low energy availability, reproductive suppression (reduced testosterone and abnormal sperm), and poor bone health seen in male athletes.

maximal oxygen uptake *See* VO_2 max.

medium-chain triglycerides (MCTs) Triglycerides containing fatty acids with carbon chain lengths of 6–12 carbons.

menadione Synthetic form of vitamin K.

mesocycle An intermediate time block within the macrocycle, usually consisting of several weeks. Successive mesocycles in periodized training generally progress from higher to lower total volume and lower to higher intensity and a greater focus on technique.

meta-analysis A statistical technique to summarize the findings of numerous studies in an attempt to provide a quantitatively based conclusion.

metabolic aftereffects of exercise The theory that the aftereffects of exercise will cause the metabolic rate to be elevated for a time, thus expending kcal and contributing to weight loss.

metabolic syndrome The syndrome of symptoms often seen with android-type obesity, particularly hyperinsulinemia, hypertriglyceridemia, and hypertension.

metabolic water The water that is a by-product of the oxidation of carbohydrate, fat, and protein in the body.

metabolism The sum total of all physical and chemical processes occurring in the body.

metalloenzymes Enzymes that must have a mineral component, such as zinc, to function effectively.

METs A measurement unit of energy expenditure; 1 MET equals approximately 3.5 ml O_2/kg body weight/minute.

microcycle The smallest time block in the mesocycle, usually a week of training.

micronutrient Dietary nutrient needed by the body in daily amounts less than a few grams, such as vitamins and minerals.

millimole One-thousandth of a mole.

mindful eating A dietary approach that focuses on an individual being more aware of food and their experiences while consuming snacks and meals. This includes using all senses (i.e., sight, smell, taste) when selecting, preparing, and consuming foods and beverages.

mindfulness A mental state in which an individual focuses their awareness on the present moment.

mineral An inorganic element occurring in nature.

mitochondria Structures within the cells that serve as the location for the aerobic production of ATP.

monosaccharides Simple sugars (glucose, fructose, and galactose) that cannot be broken down by hydrolysis.

monounsaturated fatty acids Fatty acids that have a single double bond.

morbid obesity Severe obesity in which the incidence of life-threatening diseases is increased significantly.

motivational interviewing A counseling technique utilized as a part of weight-loss programs. Through this patient-centered counseling approach, trained health-care providers support behavioral changes that may be beneficial in the management of overweight and obesity.

muscle hyperplasia An increase in the number of muscle fibers as a mechanism for an observed increase in muscle size following high-intensity weight-training. Generally not considered as a major mechanism in human muscle hypertrophy.

muscle hypertrophy An increase in the cross-sectional area and size of existing muscle fibers as a mechanism for an observed increase in muscle size following high-intensity weight-training. Considered the major mechanism in human muscle hypertrophy.

muscle tissue accretion Observed greater rate of muscle protein synthesis (anabolism) compared to muscle protein breakdown (catabolism), leading to increased muscle growth, usually associated with optimal nutrient intake and weight training.

myocardial infarction Death of heart tissue following cessation of blood flow; may be caused by coronary occlusion.

MyPlate The graphic and associated recommendations representing the healthful food guidelines presented by the United States Department of Agriculture.

N

neuropeptide Y (NPY) Neuropeptide produced in the hypothalamus; a potent appetite stimulant.

niacin Nicotinamide; nicotinic acid; part of the B complex and an important part of several coenzymes involved in aerobic energy processes in the cells.

niacin equivalents (NE) A unit of measure of niacin activity in a food related to both the amount of niacin present and that obtainable from tryptophan; about 60 mg tryptophan can be converted to 1 mg niacin.

nitrogen balance A dietary state in which the input and output of nitrogen are balanced so that the body neither gains nor loses protein tissue.

nonessential (dispensable) amino acids Amino acids that may be formed in the body and thus need not be obtained in the diet; also known as dispensable amino acids. *See also* essential (indispensable) amino acids.

nonessential nutrient A nutrient that may be formed in the body from excess amounts of other nutrients.

nonexercise activity thermogenesis (NEAT) Thermogenesis, or heat production by the body, that accompanies physical activity other than volitional exercise.

nonheme iron Iron that is found in plant foods. *See also* heme iron.

normohydration The state of normal hydration, or normal body-water levels, as compared with hypohydration and hyperhydration.

nutraceuticals Nutrients that may function as pharmaceuticals when taken in certain quantities.

nutrient Substance found in food that provides energy, promotes growth and repair of tissues, and regulates metabolism.

nutrient content claims Claims approved by the FDA for foods or beverages containing certain amounts of nutrients. Examples include low-fat, high-fiber, or sugar-free.

nutrient density A concept related to the degree of concentration of nutrients in a given food.

nutrient timing Ingestion of carbohydrate, protein, electrolytes, and other ingredients in close temporal proximity to (shortly before and/or following) exercise in order to improve performance, shorten recovery time, and maximize training adaptations, including muscle protein synthesis.

nutrition and health misinformation Nutrition and health claims shared either intentionally or unintentionally that are not supported by well-controlled research studies.

nutrition The study of foods and nutrients and their effect on health, growth, and development of the individual.

nutritional labeling A listing of selected key nutrients and kcal on the label of commercially prepared food products.

Nutrition Facts panel Included as part of a food label to provide specific information on the nutrients provided by a food or beverage.

O

obesity An excessive accumulation of body fat; usually reserved for those individuals who are 20–30 percent or more above the average weight for their size.

omega-3 fatty acids Polyunsaturated fatty acids that have a double bond between the third and fourth carbon from the terminal, or omega, carbon. EPA and DHA found in fish oils are theorized to prevent coronary heart disease.

omega-6 fatty acids Polyunsaturated fatty acids that have a double bond between the sixth and seventh carbon from the terminal, or omega, carbon. Linoleic acid is an essential omega-6 fatty acid.

onset of blood lactic acid (OBLA) The intensity level of exercise at which the blood lactate begins to accumulate rapidly.

osmolality Osmotic concentration determined by the ionic concentration of the dissolved substance per unit of solvent.

osmotic demyelination destruction of the myelin sheath covering neurons in the brain stem.

osteomalacia A disease characterized by softening of the bones, leading to brittleness and increased deformity; caused by a deficiency of vitamin D.

osteoporosis Increased porosity or softening of the bone; primarily caused by a deficiency of calcium.

over-use injuries Muscle or joint injury, including stress fractures, that occur due to repetitive trauma resulting from training too much and/or taking on too much physical activity too quickly.

ovolactovegetarian Vegetarian who consume eggs and milk products as sources of high-quality animal protein.

ovovegetarian Vegetarian who include eggs in the diet to help obtain adequate amounts of protein.

oxygen system The energy system that produces ATP via the oxidation of various foodstuffs, primarily fats and carbohydrates.

P

pantothenic acid Water-soluble vitamin that is a component of the B complex vitamins. Important as a part of Coenzyme A (CoA), a coenzyme necessary for energy metabolism.

partially hydrogenated fats Polyunsaturated fats that are not fully saturated with hydrogen through a hydrogenation process. *See also trans* fatty acids.

peak bone mass The concept of maximizing the amount of bone mineral content during the formative years of childhood and young adulthood.

periodization A technique applied to resistance training, as well as other forms of exercise training, that modifies the amount of exercise stress placed on the individual over the course of time. Various cycles, such as the microcycle, mesocycle, and macrocycle, are designed to allow the body to adapt to exercise stress in ways beneficial to performance enhancement.

pescovegetarian Vegetarian who eat fish, but not poultry, or other animal meats.

phosphatidylserine Like phosphatidylcholine, a naturally occurring phospholipid found in cell membranes; as a dietary supplement, it is theorized to possess ergogenic potential.

phosphocreatine (PCr) A high-energy phosphate compound found in the body cells; part of the ATP-PCr energy system.

phospholipids Lipids containing phosphorus that in hydrolysis yield fatty acids, glycerol, and a nitrogenous compound. Lecithin is an example.

physical activity Any activity that involves human movement; in relation to health and physical fitness, physical activity is often classified as structured and unstructured.

physical activity energy expenditure (PAEE) Energy expended during physical activity and exercise.

Physical Activity Level (PAL) Increase in energy expenditure through physical activity based on energy expended through daily walking mileage or equivalent activities; National Academy of Sciences lists four PAL categories: Sedentary, Low Active, Active, and Very Active.

physical fitness A set of abilities individuals possess to perform specific types of physical activity. *See also* health-related fitness and sports-related fitness.

phytonutrients Substances found in plants, also known as phytochemicals, that help protect plants from threats, such as bugs. Although not essential nutrients, such as vitamins, some phytonutrients may elicit healthful effects in humans.

plant-based diet A diet that focuses predominately on plant-based foods, including fruits, vegetables, grains, legumes, and seeds.

plaque The material that forms in the inner layer of the artery and contributes to atherosclerosis. It contains cholesterol, lipids, and other debris.

polysaccharide A carbohydrate that upon hydrolysis will yield more than ten monosaccharides.

polyunsaturated fatty acid Fat that contains two or more double bonds and thus is open to hydrogenation.

population attributable risk The proportion of the incidence of a disease in a population (total new cases) associated with exposure to a particular risk factor; difference between disease risk in the total population and disease risk in population members not exposed to the risk factor of interest.

postexercise hypotension A commonly observed decrease in resting blood pressure following an exercise session which may persist for several hours.

power Work divided by time; the ability to produce work in a given period of time.

preworkout supplements A nutrient mix consisting of beta-alanine, caffeine, creatine, and various amino acids usually ingested in close temporal proximity to acute exercise to improve performance, shorten recovery time, and facilitate training adaptations such as muscle protein synthesis.

principle of exercise sequence Relative to a weight-training workout, the lifting sequence is designed so that different muscle groups are utilized sequentially so as to be fresh for each exercise.

principle of overload The major concept of physical training whereby one imposes a stress greater than that normally imposed upon a particular body system.

principle of progressive resistance exercise (PRE) A training technique, primarily with weights, whereby resistance is increased as the individual develops increased strength levels.

principle of recuperation A principle of physical conditioning whereby adequate rest periods are taken for recuperation to occur so that exercise may be continued.

principle of specificity The principle that physical training should be designed to mimic the specific athletic event in which one competes. Specific human energy systems and neuromuscular skills should be stressed.

promoters Substances or agents necessary to support or promote the development of a disease once it is initiated.

proof Relative to alcohol content, proof is twice the percentage of alcohol in a solution; 80-proof whiskey is 40 percent alcohol.

protein hydrolysate A high-protein dietary supplement containing a solution of amino acids and peptides prepared from protein by hydrolysis.

protein-sparing effect An adequate intake of energy kcal, as from carbohydrate, will decrease somewhat the rate of protein catabolism in the body and hence spare protein. This is the basis of the protein-sparing modified fast, or diet.

proteinuria The presence of proteins in the urine.

Prudent Healthy Diet A diet plan based upon healthful eating principles that is designed to help prevent or treat common chronic diseases in the United States, Canada, and Mexico, particularly cardiovascular disease and cancer.

purines The end products of nucleoprotein metabolism, which may be formed in the body; they are nonprotein nitrogen compounds that are eventually degraded to uric acid.

R

radiation Electromagnetic waves given off by an object; the body radiates heat to a cool environment.

reactive hypoglycemia A decrease in blood glucose caused by an excessive insulin response to hyperglycemia associated with a substantial intake of high-glycemic-index foods.

Recommended Dietary Allowance(s) (RDAs) The levels of intake of essential nutrients considered to be adequate to meet the known nutritional needs of practically all healthy persons.

regional fat distribution Deposition of fat in different regions of the body. *See also* android- and gynoid-type obesity.

relative energy deficiency in sport (RED-S) A syndrome of health and performance impairments resulting from an energy deficit.

repetition maximum (RM) In weight training, the amount of weight that can be lifted for a specific number of repetitions.

resting energy expenditure (REE) The energy required to drive all physiological processes while in a state of rest.

resting metabolic rate (RMR) *See* basal metabolic rate (BMR) and resting energy expenditure REE.

retinol Vitamin A.

retinol equivalents (RE) and **retinol activity equivalents (RAE)** Measures of vitamin A activity in food as measured by preformed vitamin A (retinol) or carotene (provitamin A). 1 RE or RAE equals 1 microgram of retinol or 3.3 IU.

rhabdomyolysis Abnormal catabolism of skeletal muscle with the release of large proteins into the bloodstream which can cause kidney damage.

rHGH *See* human growth hormone.

riboflavin (vitamin B2) A member of the B complex.

rickets Condition in children that is associated with a deficiency of vitamin D and abnormal bone growth.

risk factor Associated factor that increases the risk for a given disease; for example, cigarette smoking and lung cancer.

S

salsolinol A compound formed from acetaldehyde, a metabolite of alcohol metabolism, and the neurotransmitter dopamine, which may explain suppression of inhibition, increased self-confidence, reduced anxiety, and a perceived decrease in sensitivity to pain that are sometimes attributed to alcohol consumption.

sarcopenia Loss of fat-free (muscle, bone) mass occurring during the aging process.

sarcopenia Loss of muscle mass associated with the aging process.

sarcopenic obesity A form of obesity characterized by a high body fat percentage and sarcopenia, which includes low skeletal muscle mass and poor muscle function.

satiety center A group of nerve cells in the hypothalamus that responds to certain stimuli in the blood and provides a sensation of satiety.

saturated fatty acid Fat that has all chemical bonds filled.

secondary amenorrhea Cessation of menstruation after the onset of puberty; primary amenorrhea is the lack of menstruation prior to menarche.

Sedentary Death Syndrome (SeDS) Term associated with a sedentary lifestyle and related health problems that predispose to premature death.

semivegetarians Individuals who refrain from eating red meat but include white meat such as fish and chicken in a diet stressing vegetarian concepts.

set-point theory The weight-control theory that postulates that each individual has an established normal body weight. Any deviation from this set point will lead to changes in body metabolism to return the individual to the normal weight.

settling-point theory Theory that the body weight set point may be increased or decreased through interactions of genetics and the environment; an environment rich in high-fat foods may lead to a higher set point so that the body settles in at a higher weight and fat content.

shell temperature The temperature of the skin. *See also* core temperature.

sick fat *See* adiposopathy.

simple carbohydrates Usually used to refer to table sugar, or sucrose, a disaccharide; also other disaccharides and the monosaccharides.

skinfold technique A technique used to compute an individual's percentage of body fat; various skinfolds are measured and a regression formula is used to compute the body fat.

sodium bicarbonate $NaHCO_3$; a sodium salt of carbonic acid that serves as a buffer of acids in the blood, often referred to as the alkaline reserve.

sodium loading Consumption of excess amounts of sodium; endurance athletes may use sodium loading in attempts to increase plasma volume, improve blood flow, and enhance aerobic endurance.

specific heat The amount of energy or heat needed to raise the temperature of a unit of mass, such as 1 kilogram of body tissue, 1 degree Celsius.

sports anemia A temporary condition of low hemoglobin levels often observed in athletes during the early stages of training.

sports bars Commercial food products targeted to athletes and physically active individuals containing various concentrations of carbohydrate, fat, and protein; some products contain other nutrients, such as antioxidants.

sports nutrition The application of nutritional principles to sport with the intent of maximizing performance.

sports-related fitness Components of physical fitness that, when improved, have implications for enhanced sports performance, such as agility and power.

sports supplements Dietary supplements marketed to athletes and physically active individuals.

steady-state threshold The intensity level of exercise above which the production of energy appears to shift rapidly to anaerobic mechanisms, such as when a rapid rise in blood lactic acid exists. The oxygen system will still supply a major portion of the energy, but the lactic acid system begins to contribute an increasing share.

storage fat Fat that accumulates and is stored in the adipose tissue. *See also* essential fat.

strength-endurance continuum In relation to strength training, the concept that power or strength is developed by high resistance and few repetitions and that endurance is developed by low resistance and many repetitions.

structured physical activity A planned program of physical activities usually designed to enhance physical fitness; often referred to as exercise.

subcutaneous fat The body fat found immediately under the skin; evaluated by skinfold calipers.

sweat gland fatigue Progressive decreases in sweat production following high and prolonged sweat rates, leading to less effective thermoregulation.

T

thermic effect of food (TEF) The increased body heat production associated with the digestion, assimilation, and metabolism of energy nutrients in a meal just consumed.

thiamin (vitamin B1) A member of the B complex.

Tolerable Upper Intake Level (UL) The highest level of daily nutrient intake likely to pose no adverse health risks.

tonicity Tension or pressure as related to fluids; fluids with high osmolality exhibit hypertonicity, whereas fluids with low osmolality exhibit hypotonicity.

total daily energy expenditure (TEE) The total amount of energy expended during the day, including REE, TEF, and TEE.

total fiber Sum of dietary fiber and functional fiber.

trabecular bone The spongy bone structure found inside the bone, as contrasted with the more compact bone on the outside.

trace minerals Those minerals essential to human nutrition and needed in amounts of less than 100 mg per day.

***trans* fatty acids** Unsaturated fatty acids in which the hydrogen ions are on opposite sides of the double bond.

triglycerides Fats formed by the union of glycerol and fatty acids.

U

ultra-processed foods (UPFs) Food items made from substances extracted from foods, including added sugars, starches, fats, and hydrogenated fats, as well as added flavors, stabilizers, and artificial colors. Examples include frozen meals, salty snacks, and packaged cookies.

uncompensated heat stress A condition in which heat loss is insufficient to offset heat production during exercise in the heat, the body temperature continues to rise, and exhaustion eventually occurs.

unstructured physical activity Many of the normal, daily physical activities that are generally not planned as exercise, such as walking to work, climbing stairs, gardening, domestic activities, and games and other childhood pursuits.

urea The chief nitrogenous constituent of urine and the final product of the decomposition of proteins in the body.

USDA 85-15 guide Recommendations supported by the USDA Dietary Guidelines for Americans that include consuming at least 85 percent of kcal from nutrient-dense foods and beverages meeting specific food group recommendations. The remaining 15 percent of kcals can then be consumed from other food sources.

V

Valsalva phenomenon A condition in which a forceful exhalation is attempted against a closed epiglottis and no air escapes; such a straining may cause the person to faint.

vegan diet A vegetarian who eats no animal products.

vegetarian diet A dietary pattern that involves abstaining from the consumption of meat, fish, and poultry. Diet focuses on plant-based foods.

very low-Calorie diets (VLCD) Diets containing less than 800 Calories per day.

very low-density lipoprotein *See* VLDL.

visceral fat The deep fat found in the abdominal area; measurement of this fat requires special techniques, such as MRI.

vitamin A Fat-soluble vitamin that is necessary for immune health, night vision, and growth. The alcohol form of the vitamin, retinol, is the most active form in the human body.

vitamin B1 Water-soluble vitamin known as thiamin. Part of the coenzyme thiamin pyrophosphate (TPP), which is necessary for energy metabolism.

vitamin B2 Water-soluble vitamin known as riboflavin. Part of the coenzymes flavin mononucleotide (FMN) and flavin adenine dinucleotide (FAD), which are necessary for energy metabolism.

vitamin B6 Water-soluble vitamin that includes pyridoxine and other related compounds. Part of the coenzyme pyridoxal phosphate (PLP), which is necessary for energy metabolism, particularly protein metabolism.

vitamin B12 Water-soluble vitamin also known as cobalamin. A part of coenzymes found in all body cells and is essential for DNA synthesis.

vitamin C Ascorbic acid; the antiscorbutic vitamin.

vitamin D Fat-soluble vitamin essential for normal bone health. Often referred to as the "sunshine vitamin."

vitamin D3 The prohormone form of vitamin D, also known as cholecalciferol, formed in the skin by irradiation from the sun. Released into the blood and eventually converted by the kidney to the hormone form of vitamin D.

vitamin E (alpha-tocopherol) Fat-soluble vitamin that is important as an antioxidant.

vitamin K Fat-soluble vitamin necessary for the blood clotting process.

VLDL Very low-density lipoproteins; a protein-lipid complex in the blood that transports triglycerides, cholesterol, and phospholipids; has a very low density.

VO$_2$ max Maximal oxygen uptake; measured during exercise, the maximal amount of oxygen consumed reflects the body's ability to utilize oxygen as an energy source; equals the cardiac output times the arteriovenous oxygen difference.

W

waist circumference The circumference of the waist at its most narrow point as seen from the front; used as a measure of regional adiposity.

WBGT index Wet-bulb globe thermometer index; a heat-stress index based upon four factors measured by the wet-bulb globe thermometer.

weight bias Negative attitudes, judgment, and stereotyping that occur based on a person's weight. Weight bias in health care occurs when a health-care professional treats a person with overweight or obesity differently than other patients, in some cases "blaming" health issues on weight when there may be other underlying causes.

weight cycling Repetitive loss and regain of body weight; often called yo-yo dieting.

weight-loss programs Any program that is designed and marketed with specific strategies to support weight loss.

work Effort expended to accomplish something; in terms of physics, force times distance.

X

xerophthalmia Dryness of the conjunctiva and cornea of the eye, which may lead to blindness if untreated; caused by a deficiency of vitamin A.

Index

©Air Images/Shutterstock

A

AA. *See* Anorexia athletica (AA)
AAS. *See* Anabolic-androgenic steroids (AAS)
Absolute VO$_2$ max, 411
Absorption
  calcium (Ca), 283
  carbohydrates, 116, 117
  fats, 169–170
  intestinal, 339–340
  protein, 214, 215
Acacia rigidula, 514
Academy of Nutrition and Dietetics (AND), 13, 37, 51, 62, 149, 162, 255
Accelerometer, 80–81
Acceptable daily intake (ADI), 149
Acceptable Macronutrient Distribution Range (AMDR), 40, 115, 139, 162, 213, 222, 236
Accidents, 12
Acclimatization, 355, 357
ACE. *See* Angiotensin-converting enzyme (ACE) inhibitors
Acesulfame-K, 148
ACSM. *See* American College of Sports Medicine (ACSM)
ActiGraph, 81
Active couch potato, 395
Active transport, 116, 118
Activity-stat hypothesis, 391
Acupuncture, 418
Acute fatigue, 104–105
Acute high-fat diets, 177
Adaptations, exercise and, 7, 219, 220
Adaptive thermogenesis (AT), 391
Added sugars, 114–115, 437
Additives, food, 52
Adenosine triphosphate (ATP), 83, 123
Adequate Intakes (AI), 13, 40, 162, 247, 249, 250–251
ADH. *See* Antidiuretic hormone (ADH)

ADI. *See* Acceptable daily intake (ADI)
Adipokines, 7, 172, 399
Adiposopathy (sick fat), 399
ADP. *See* Air displacement plethysmography (ADP)
Adult-onset obesity, 397
Advantame, 148
Aerobic capacity, 101
Aerobic exercise
  caffeine intake on, 505
  low muscle glycogen
    brain, 128
    energy production rate, 128
    fat for energy, 128
    glycogen location, 128
    muscle fiber type, 128
  protein metabolism, 217
  training, 444
  use of fat as an energy source, 176
Aerobic exercise equipment, 97
Aerobic glycolysis, 85
Aerobic lipolysis, 87
Aerobic power, 101
Afternoon events, 67
Age
  body-fat categories by, 385
  and bone health, 286
  and physical performance, 70–71
  and tolerating exercise in heat, 356
Aggressive behavior, 497
AHA. *See* American Heart Association (AHA)
AI. *See* Adequate Intakes (AI)
Air displacement plethysmography (ADP), 381
AIS. *See* Australian Institute of Sport (AIS)
Alanine, 210
Alcohol
  alcohol dependence, 499–500
  alcoholism, 500
  blood alcohol concentration, 493
  caloric content, 493
  cancer, 498

  cardiovascular disease, 498
  cirrhosis, 496
  driving under the influence (DUI), 496–497
  ergogenic agent, 495
    energy source, 493
    exercise metabolism and performance, 494–495
    permissibility, 495
    psychological processes, 495
    social drinking and sports, 495
  on exercise metabolism and performance, 494–495
  fetal alcohol effects, 499
  fetal alcohol spectrum disorder, 499
  fetal alcohol syndrome, 499
  metabolic fate, in body, 492–493
  moderate consumption, 52, 363
  nutrient content, 492
  obesity, 499
  positive effects
    Alzheimer's disease and dementia, 500
    cardiovascular disease, 500–502
    low-risk drinking, 501
  proof, 492
  psychological problems, 496
Alcohol consumption, 497–498
Alcoholism, 500
Alpha-ketoacid, 214
Alpha-linolenic acid, 168
Alpha-tocopherol, 256. *See also* Vitamin E (alpha-tocopherol)
Alzheimer's disease, 401, 500
AMDR. *See* Acceptable Macronutrient Distribution Range (AMDR)
Amenorrhea, 417
American Academy of Pediatrics, 71, 356
American College of Sports Medicine (ACSM), 4, 255, 442, 443
  AAS use, 521, 522

  alcohol consumption, 494
  exercise and hypertension, 364
  guidelines for modifying or canceling competition, 352
  resistance-training programs, 469
American Council on Exercise (ACE), 27
American Heart Association (AHA), 4, 49, 51, 194
American Journal of Clinical Nutrition, 27
American Society for Metabolic and Bariatric Surgery (ASMBS), 451–452
Amino acids, 209
  health risks, 237–238
  supplements, 234–236
Aminostatic theory, 388
Ammonia, 214
AN. *See* Anorexia nervosa (AN)
Anabolic hormones, ergogenic effects
  anabolic/androgenic steroids, 520
  androstenedione, 521, 523
  dehydroepiandrosterone, 521–523
  human growth hormone, 518–519
  rhGH injections/therapy, 519
  testosterone, 519–521
Anabolic-androgenic steroids (AAS), 519–521
Anabolism, 88
Anaerobic capacity, 85, 101
Anaerobic exercise
  high-intensity, caffeine and, 505
  low muscle glycogen, 128–129
Anaerobic glycolysis, 85
Anaerobic power, 85, 101
AND. *See* Academy of Nutrition and Dietetics (AND)
AND Commission on Dietetic Registration, 27

Android-type obesity, 402
Androstenedione, 521, 523
Angina, 186
Angiotensin-converting enzyme (ACE) inhibitors, 359
Animal proteins, 211
 vs. plant protein, 210–211
Anorexia athletica (AA), 416
Anorexia nervosa (AN), 413–414
Anorexiant, anti-obesity drugs as, 406
Anthropometry, 383
Antidiuretic hormone (ADH), 319
Anti-obesity drugs, 406–409
Antioxidants
 and fatigue or muscle damage, 269–271
 super, 270
 and vegetarian diet, 55
 vitamins, 248, 269–271
Antipromoters, 13
Anxiety, 440
Apolipoproteins, 169
Apoptosis, 496
App-based weight-loss programs, 447–449
Appestat, 386
"Apple-shape" pattern, obesity, 402
Archimedes' principle, 377
Arginine, 235–236
Armstrong, Larry, 269
Arteriosclerosis, 186
Artificial fats intake, 197
Artificial sweeteners, 148–149
ASCVD. *See* Atherosclerotic cardiovascular disease (ASCVD)
Aspartame, 148
Asthma, 511
AT. *See* Adaptive thermogenesis (AT)
Atherosclerosis, 186, 187
 reverse, 190
 and serum lipids, 187–189
Atherosclerotic cardiovascular disease (ASCVD), 190
Athlete's Plate, 65
Athletic amenorrhea, 287
Athletic performance, weight loss and, 451–453
ATP. *See* Adenosine triphosphate (ATP)
ATP-PCr system, 84–85, 87
 energy expenditure, 92–93
 sodium bicarbonate on, 516
Australian Institute of Sport (AIS), 24
 nutritional supplements, 526

B

BAC. *See* Blood alcohol concentration (BAC)
Bad carbs, 147
Balanced diet
 foods to obtain nutrients, 42
 key-nutrient concept, 45–46
 key-nutrient sources, 45
 MyPlate food guide, 42–44, 46–47
 and nutrient density, 46
 optimal nutrition, 46–47
 overview, 42
 weight gain, 465, 467
Banned supplements, 23
Bariatric surgery, exercise before and after, 451–452
Bar-Or, Oded, 70
Basal energy expenditure (BEE), 88, 388
Basal metabolic rate (BMR), 88, 435
BAT. *See* Brown adipose tissue (BAT)
BCAAs. *See* Branched-chain amino acids (BCAAs)
BED. *See* Binge eating disorder (BED)
BEE. *See* Basal energy expenditure (BEE)
Beetroot, 231–232
Behavioral group therapy, 449
Beta-alanine, 230–231
 effects
  on ATP-PCr, 516
  on lactic acid energy system, 516–517
  on oxygen energy system, 517
 efficacy, 230–231
 safety, 231
 supplementation, 230, 516
Beta-carotene, 252, 253
Beta-hydroxy-beta-methylbutyrate (HMB), 464
Beta-oxidation, 173
BIA. *See* Bioelectrical impedance analysis (BIA)
Binge eating disorder (BED), 414–415
Bioavailability, 249, 294
Biochemical deficiency, 249
Bioelectrical impedance analysis (BIA), 382
Biofeedback, 418
Biomechanical aids, 22
Biotin
 deficiency, 266
 DRI, 266
 food sources, 266
 functions, 266
 recommendations, 266
 supplementation, 266
Blood alcohol concentration (BAC), 493, 496
Blood glucose
 exercise, 118
 and fatigue, 125–127, 130
 hormones regulating, 126
 hyperglycemia, 118
 hypoglycemia, 118
 insulin, 118, 121
 metabolism, 118–119, 121
Blood nutrient levels, energy intake and, 388
Blood pressure
 classification, 359
 high (*See* High blood pressure)
BMR. *See* Basal metabolic rate (BMR)
BN. *See* Bulimia nervosa (BN)
Board Certified Specialist in Sports Dietetics (CSSD), 27
Bod Pod, 381
Body composition
 body water, 378
 bone mineral, 378
 factors affecting, 378
 fat-free mass, 378
 overview, 376–377
 physical performance
  excess body weight, 410–412
  excessive weight loss, 412–413
 regulation of
  adult-onset obesity, 397
  contributing factors to obesity, 392–397
  energy expenditure, 388–389
  energy intake, 386–388
  energy intake and expenditure, feedback control, 389–391
  fat deposition in the body, 391–392
  prevention of childhood obesity, 398–399
  set point, 397–398
 techniques
  air displacement plethysmography, 381
  anthropometry, 383
  bioelectrical impedance analysis, 382
  dual energy X-ray absorptiometry, 382
  guidelines, 379
  hydrostatic weighing, 378, 379, 381
  multicomponent model, 383
  near-infrared interactance, 383
  skinfold technique, 381–382
  two- or three-component models, 379, 380
  total body fat, 377–378
Body mass index (BMI)
 defined, 375
 healthy body weight, 544
 range, 375
 and target body weight, 384
 values and limitations of, 375–376
 waist circumference, 546
Body temperature
 energy intake and, 388
 exercise on, 328–329
 factors influencing, 326
 heat loss, 329–330
 hyperthermia, 327–328
 normal, 325–326
 regulation by body, 326–327
Body water, 378
Body weight. *See* Weight
Body-fat percentage
 chest and abdomen skinfold, 545, 547
 guideline, 544–546
 lean body weight, 546
 and optimal weight, 384–385
 in sport, problems with rigid adherence, 383–384
 suprailiac skinfold, 545
 thigh skinfold, 546
 triceps and subscapular skinfolds, 546
Body-water compartments, 317–318
Bomb calorimeter, 79–80
Bone health
 and age, 286
 calcium supplements, 286
 and exercise, 286–287
 and protein intake, 237
 and vitamin D, 255
Bone mineral, 378
Booth, Frank, 6
Boron, 306–307
Bottled water, 316. *See also* Water
Bovine colostrum, 233
Brain damage, excessive exercise, 12
Brain health
 creatine and, 230
 exercise and diet, benefits, 16–17

Branched-chain amino acids (BCAAs), 216, 234-235
   central fatigue hypothesis, 129-130, 234-235
   supplementation
      and fTRP:BCAA ratio, 234
      and mental performance, 234
      and perceived exertion, 234
      support for hypothesis, 234-235
Brenner, Barry, Dr., 237
British Journal of Sports Medicine, 227
Brooks, George, 127
Brown adipose tissue (BAT), 389
Budget, healthful nutrition on, 44
Buffer boosting, 515
Built environment, weight gain and, 396
Bulimia nervosa (BN), 414
"Bulletproof coffee," 181
Burke, Louise, 18, 116

C

Cade, Robert, Dr., 314
Caffeine, ergogenic effects
   aerobic endurance performance, 504-505
   asthma, 511
   caffeine naivete, 512
   cancer, 510
   carbohydrate and fat oxidation, 506
   cardiovascular disease, 508-510
   cognitive functions, 510-511
   definition, 502
   delivery system, effect on, 506
   dependence, 512
   energy drinks, 508
   ephedrine with, 513
   exercise in heat, 506-507
   exercise performance, 503-504
   gastric distress, 512
   high-intensity anaerobic exercise, 505
   individuality, 507
   mortality, 512
   muscular strength and endurance, 505-506
   osteoporosis, 511
   permissibility, effect of dosage, 507
   placebo effect, 507
   pregnancy-related health problems, 511
   psychomotor responses, 504
   self-experimentation, 507-508
   sleeplessness, 511-512
   type 2 diabetes, 510
   weight control, 511
Caffeinism, 512
Calcium (Ca)
   absorption, 283
   bone health
      osteoporosis in sports, 287
      physical activity, 286-287
      supplementation, 286
   cardiovascular disease, 287-288
   deficiency
      calcium balance, 283, 284
      inadequate dietary intake, 283
      osteoporosis, 283-285
      peak bone mass, 285
      people at risk, 283
      trabecular bone, 285
   DRI, 282
   food sources, 282-283
   functions, 283
   kidney stones, 288
   obesity and weight control, 288
   plant foods rich in, 54
   RDA recommendation, 52
   recommendations, 289
   sports performance research, 288-289
   supplementation, 285-289
      and kidney stones, 288
      and sports performance, 288-289
      and weight loss, 288
Calorie, 81-82
Calorie (kcal) deficit, 435
Calorimeter, 81-82
Calorimetry
   bomb, 79-80
   definition, 79
   direct, 80
   indirect, 80
Camellia sinesis, 526
Cancer
   alcohol consumption, 498
   and caffeine, 510
   exercise and diet, benefits, 16
   and healthful dietary guidelines, 49
   and obesity, 401
   and salt intake, 361
   and vitamin D, 255-256
Capsicum annuum, 525
Carbohydrate loading
   athlete, 140-141
   daily food plan, 142
   description, 140
   detrimental effects, 144-145
   exercise performance, 143-144
   glycogen stores, muscles, 143
   methods, 141-142
   muscle glycogen concentration, 142-143
   and sodium bicarbonate, 518
Carbohydrate-electrolyte solutions (CES), 336
Carbohydrates, 467. *See also* Carbohydrate loading; Carbohydrates for exercise
   added sugars, 114-115
   caffeine, effect of, 506
   definition, 112
   diet
      adequate intake (AI), 115
      athletic training programs, 116
      Daily Value (DV), 115
      RDA, 115
      requirement, 115-116
   digestion and absorption, 116-117, 119
   disaccharides, 112, 113
   ergogenic aspects
      DHAP, 145-146
      lactate salts, 146
      multiple carbohydrate by-products, 146
      pyruvate, 145-146
      ribose, 146
   and exercise in heat, 340-341
   foods high in carbohydrate content, 114-115
   health implications
      artificial sweeteners, 148-149
      complex carbohydrates, 149-150
      fiber rich foods, 150-152
      food intolerance, 152
      refined sugars and starches, 147-148
   low muscle glycogen
      and aerobic exercise, 127-128
      and anaerobic exercise, 128-129
   monosaccharides, 112, 113
   polysaccharide, 113
   with protein, 137, 138
   simple, 112
   types, 112-114
Carbohydrates for exercise
   activities types, 124
   endurance training, 125
   as energy source, 124
   during exercise, 124-125, 135
   high-carbohydrate diet, 139-140
   hypoglycemia, 125-127
   intensity and duration of exercise, 131-133
   low endogenous carbohydrate levels, 129-130
   optimal supplementation protocol, 135-136
   physical performance
      exercise intensity and duration, 131-133
      fatigue prevention, 131
      fatigue-delaying mechanisms, 130-131
      ingested carbohydrate, 130
      initial endogenous stores, 131
      intensity and duration of exercise, 131-133
   pre-exercise
      four hours or less before exercise, 134
      immediately before, 134-135
      less than 1 hour before exercise, 134
   replenishment, prolonged exercise, 137-138
   types
      carbohydrate combinations, 136
      carbohydrate form, 136
      fructose, 136
      individuality, 137
      low-glycemic-index foods, 137
      with protein, 137
Cardiorespiratory fitness, 451
Cardiovascular disease
   alcohol consumption, 498, 500-502
   and caffeine, 508-510
   excessive protein intake, 237
   and fat intake, 186-187
   modifiable risk factors, 16
   and obesity, 400
Cardiovascular strain, 331
Carnitine, 174
   dietary sources, 183
   effects on performance, 183-184
   as ergogenic aid, 183
   and weight loss, 183-184
Carotenemia, 253
Casein, 223
Catabolism, 88
CDC. *See* Centers for Disease Control and Prevention (CDC)
Celiac disease, 152

Cellulite, 377-378
Center for Science in the Public Interest, 30
Centers for Disease Control and Prevention (CDC), 8, 444-445
  on obesity rates, 431, 432
Central fatigue, 104-105
Central fatigue hypothesis, 129-130, 234-235
Central neural fatigue, 331
Certified Nutrition Specialist (CNS), 27
CES. *See* Carbohydrate-electrolyte solutions (CES)
CFS. *See* Chronic fatigue syndrome (CFS)
CHD. *See* Coronary heart disease (CHD)
Check Your Steps Program, 52
Chemical energy, 77
Childhood obesity, 398-399. *See also* Obesity
Chloride (Cl)
  deficiency, 324
  DRI, 323
  food sources, 323
  functions, 323
Cholecalciferol. *See* Vitamin D (cholecalciferol)
Cholecalciferol (vitamin D3), 253
Cholesterol
  cardiovascular disease, 186-188
  description, 166
  fats, classification, 167-168
  foods, 166-167
  HDL, 189
  intake, reducing, 197
  LDL, 189
  ratios and tests, 189
  reverse cholesterol transport, 171, 190
  serum lipids, 190
  total, 189
Choline (vitamin-like compounds), 268-269
  deficiency, 268
  DRI, 268
  food sources, 268
  functions, 268
  recommendations, 269
  supplementation, 268
Chondroitin, 233-234
Chromium (Cr)
  deficiency, 302
  DRI, 302
  food sources, 302
  functions, 302
  recommendations, 304
  supplementation, 302-303
Chronic disease risk reduction (CDRR) sodium levels, 321
Chronic diseases
  exercise and, 6-8
  genetic predisposition, 2-3, 13
  healthful dietary guidelines and, 48-52
  preventing, 3
  and refined sugar, 147-148
  risk factor, 3
Chronic fatigue syndrome (CFS), 104
Chronic high-fat diets, 177-179
Chronic training effect, 66
Chylomicron, 169
Circuit aerobics, 484
Circuit weight training, 484
Cirrhosis, 496
Cis fatty acid, 163
  monounsaturated/polyunsaturated, 168
Citric acid cycle, 541
Citrulline, 235-236
Ciwujias, 525
CLA. *See* Conjugated linoleic acid (CLA)
Clark, Nancy, 66
Clinically manifest vitamin deficiency, 249
CNS. *See* Certified Nutrition Specialist (CNS)
Coconut oil, 181
Coenzyme, 248
Coenzyme functions, vitamins, 248
Cohort, 28
Collagen, 223, 232
Collegiate and Professional Sports Dietitians Association (CPSDA), 19, 307
Colors, food, 56
Colostrum, 227, 233
Combination devices, 81
Commercial sports foods, 68-69
Compensated heat stress, 330
Complementary proteins, 53
Complete proteins, 211
Complex carbohydrates
  description, 113
  health implications, 149-150
Concentric contraction, 480
Conduction, 326
Conjugated linoleic acid (CLA), 168, 184-185, 197
Consumer athlete
  essential nutrients and recommended nutrient intakes, 38-41
Consumer nutrition
  dietary supplements, 62-65
  food labels, 58-60
  functional foods, 61-62
  health claims, 60-61, 62
Consumer Reports, 81
Convection, 326
Cooper, Kenneth, Dr., 86
Copper (Cu)
  deficiency, 299-300
  DRI, 299
  food sources, 299
  functions, 299
  recommendations, 300
  supplementation, 300
Cordyceps sinensis, 525, 526
Core temperature, 325. *See also* Body temperature
Cori cycle, 122, 123
Coronary heart disease (CHD)
  and fat intake, 186-187
  modifiable risk factors, 16
Coronary occlusion, 186
Coronary thrombosis, 186
Cortisol, 125
COVID-19 pandemic, 374
  on dietary practices, 44-45
Coyle, Edward, 139-140
CPSDA. *See* Collegiate and Professional Sports Dietitians Association (CPSDA)
Creatine, 228-229
  brain function, 230
  efficacy, 228-229
  formulation, 229
  mechanism of action, 228
  medical applications, 230
  monohydrate, 228
  safety, 229-230
  supplementation, 228
Crossover concept, 103
CSSD. *See* Board Certified Specialist in Sports Dietetics (CSSD)
Cycling, 96-97
Cytokines, 7
Cytoseira canariensis, 525

D

Daily Value (DV), 58, 115
DASH. *See* Dietary Approaches to Stop Hypertension (DASH)
DASH (Dietary Approaches to Stop Hypertension) diet, 47, 190, 361-363
DB. *See* Dry-bulb thermometer (DB)
Deamination, 214
Death, leading causes of, 2
Dehydration
  defined, 318
  exercise in heat, 332
  involuntary, 332-333
  voluntary, 332
  and weight-loss programs, 406, 408-409
  while exercising, 333-334
Dehydroepiandrosterone (DHEA), 521-523
Delayed onset of muscle soreness (DOMS), 219
Dementia, 401
Dental caries, 147-148
Dental caries, refined sugars and, 147
Dextrose, 112
DFE. *See* Dietary folate equivalents (DFE)
DHAP. *See* Dihydroxyacetone phosphate (DHAP)
Diabesity, 400
Diabetes
  and caffeine, 510
  exercise and diet, benefits, 16
  and vitamin D, 256
Diastolic phase, 359
Diastolic pressure, 360
Dietary Approaches to Stop Hypertension (DASH), 231
Dietary fiber, 113
Dietary folate equivalents (DFE), 264
Dietary guidance systems, 42
Dietary nitrate, 231-232
Dietary Reference Intakes (DRI), 13, 39-40, 115, 147
  biotin, 266
  calcium (Ca), 282
  chloride, 323
  choline, 268
  chromium, 302
  copper, 299
  folate (folic acid), 264
  iron, 292, 294
  magnesium, 290
  niacin, 261
  pantothenic acid, 265
  phosphorus, 289
  potassium, 324
  selenium, 304
  thiamin, 259-260
  vitamin A, 252
  vitamin B2, 260
  vitamin B6, 261
  vitamin B12, 263
  vitamin C, 266
  vitamin D, 253
  vitamin E, 256
  vitamin K, 258
  zinc, 300

Dietary supplements. *See also* Sports supplements
  containing ephedra, 513
  definition, 63
  health improvement, 63–64
  issues to consider, 64
  label, 63
  omega-3 fatty acid and fish oil supplements, 182–183
  weight-gaining program, 464–465
Dietary Supplements Health and Education Act (DSHEA), 25, 64
Dietary-induced thermogenesis (DIT), 88
Dieting, and REE, 90
Dietitians of Canada, 255
Differential susceptibility, 396
Diffusion, 116, 118
Digestion
  carbohydrates, 116–117, 119
  defined, 116
  fats, 169–170
  protein, 214, 215
Dihydroxyacetone phosphate (DHAP), 145–146
Direct calorimetry, 80
Disaccharides, 112, 113
Disordered eating, 413
Dispensable amino acids. *See* Non-essential amino acids
DIT. *See* Dietary-induced thermogenesis (DIT)
DLW. *See* Doubly labeled water (DLW) technique
Docosahexaenoic (DHA) acid, 182
DOMS. *See* Delayed onset of muscle soreness (DOMS)
Doping, 23
Doubly indirect technique, 378
Doubly labeled water (DLW) technique, 80
DPHP. *See* United States Office of Disease Prevention and Health Promotion (ODPHP)
DRI. *See* Dietary Reference Intakes (DRI)
Driving under the influence (DUI), 496–497
Drugs
  weight gain and, 395–396
  weight-loss, 406–409
Dry-bulb thermometer (DB), 327
DSHEA. *See* Dietary Supplements Health and Education Act (DSHEA)
Dual energy X-ray absorptiometry (DXA/DEXA), 382–383

Dual intervention point model, 397
DUI. *See* Driving under the influence (DUI)
DV. *See* Daily Value (DV)
DXA/DEXA. *See* Dual energy X-ray absorptiometry (DXA/DEXA)

E

EAH. *See* Exercise-associated hyponatremia (EAH)
EAR. *See* Estimated Average Requirement (EAR)
Eating disorders, 413, 417–418
  alternative therapies for, 418
  anorexia nervosa (AN), 413–414
  associated with sports, 415–417
  binge eating disorder (BED), 414–415
  bulimia nervosa (BN), 414
  definition, 413
  with energy deficiency, 417
  examples of, 415
  obesity and, 440–441
  risk factors for, 417
  treatment for, 418
Eccentric contraction, 480
Eccentric exercise, 219
Echinacea, 525
EER. *See* Estimated Energy Requirement (EER)
EGCG. *See* Epigallocatechin-3-gallate (EGCG)
Eicosanoids, 172
Eicosapentaenoic (EPA), 182
85-15 guide, 436
Electrical energy, 77
Electrolytes, 280
  CDRR sodium levels, 321
  chloride (Cl), 323–324
  definition, 321
  potassium (K), 324–325
  sodium (Na), 321, 323
Electron transfer system, 86
Electron transport chain, 542
Eleutherococcus senticosus, 524
Emotional stress, obesity and, 395
Empty kcal, 43
EMR. *See* Exercise metabolic rate (EMR)
EMS. *See* Eosinophilia-myalgia syndrome (EMS)
Endogenous carbohydrates, 130
Endurance training, carbohydrates metabolism, 125, 126

Endurance-type activities, 221
Energy
  Calorie, 81–82
  chemical, 77
  definition, 77
  electrical, 77
  heat, 77
  intake, weight control and, 386–388
  International Unit System (SI), 78
  measures of, 77–82
  mechanical/kinetic, 77
  vegetarianism and, 53, 55
Energy availability, 452–453
Energy chews, 115
Energy drinks, 508
Energy expenditure
  active category, 99
  ATP-PCr energy system, 92–93
  brown adipose tissue and, 389
  cycling, 96–97
  from fat and carbohydrate, 103, 104
  group exercise, 97
  home aerobic exercise equipment, 97
  lactic acid energy system, 93
  low active category, 99
  oxygen energy system, 93
  passive and occupational, 97
  REE (*See* Resting energy expenditure (REE))
  resistance or weight training, 97
  running, 96
  sedentary category, 99
  sports activity, 97
  swimming, 96
  very active category, 99
  walking, 96
  weight control, 388–389
  white adipose tissue and, 389
Energy pathways
  electron transport chain, 542
  for fatty acids, 543
  glycolysis-glucose to pyruvate, 540
  lipoprotein interactions, 543
  transition reaction and citric acid cycle, 541
English system, 549–550
Environmental chemicals, weight gain and, 395–396
Environmental factors, obesity, 393–396
Enzymes, 248
Eosinophilia-myalgia syndrome (EMS), 237–238

Ephedra (ephedrine), ergogenic effects
  description, 513
  Ephedra sinica, 513
  exercise performance
    ephedrine with caffeine, 513
    pseudoephedrine, 513
  health risks, products with, 513–514
  ma huang, 513
  permissibility in sports, 514
  synephrine, 514
Ephedra sinica, 513
Epidemiological research (observational research), 28
Epigallocatechin-3-gallate (EGCG), 525
Epigenetics, 3
Epigenome, 3
Epinephrine, 125
EPOC. *See* Excess postexercise oxygen consumption (EPOC)
Epstein, David, 3, 17
Equine therapy, 418
Ergogenic aids
  alcohol, 495
    ciwujias, 525
    energy source, 493
    exercise metabolism and performance, 494–495
    permissibility, 495
    psychological processes, 495
    social drinking and sports, 495
  anabolic hormones (*See* Anabolic hormones, ergogenic effects)
  caffeine (*See* Caffeine, ergogenic effects)
  carbohydrates
    DHAP, 145–146
    multiple carbohydrate by-products, 146
    pyruvate, 145–146
    ribose, 146
  carnitine as, 183
  definition, 21–22
  delaying fatigue, 107
  ephedra (ephedrine) (*See* Ephedra (ephedrine), ergogenic effects)
  exercise in heat, 349
  fats
    carnitine, 183–184
    CLA, 184–185
    exercising on empty stomach, 180–181
    fat burning diets, 185

Ergogenic aids (Continued)
  glycerol portion, triglycerides, 182
  hydroxycitrate, endurance performance, 184
  ketone supplements, 185
  MCTs, 181–182
  omega-3 fatty acid and fish oil supplements, 182–183
 ginseng, 524–525
 glycerol supplementation, 350–351
 herbals, 525–526
 mechanical aids, 22
 nutritional aids, 22
 nutritional ergogenics, 21–23
  effectiveness, 22–23
  legal, 23
  popularity, 22
  safety, 23
 oxygen water enhancing exercise performance, 349
 physiological aids, 22
 pre-cooling techniques, 349
 psychological aids, 22
 sodium bicarbonate (See Sodium bicarbonate)
 sodium loading, 349–350
 sports supplements, 21, 24
 vitamin supplements
  antioxidant vitamins, 269–270
  athletes, 271–272
  multivitamin supplements, 271–272
  physically active individuals, 269
  recommendations, 271, 272
Ergolytic endurance, 494
Ergometer, 79
Essential amino acids, 210, 213
Essential fat, 377
Essential fatty acids, 194
Essential hypertension, 359. See also High blood pressure
Essential minerals, 280, 281
Essential nutrients, 13, 38–41
Essential vitamin, 248–249, 250–251
Ester, 162
Esterification, 162
Estimated Average Requirement (EAR), 40
Estimated Energy Requirement (EER), 40, 98–99, 100, 435
Estrogen, 417
Ethanol, 492
Euhydration, 318
Evaporation, 326, 329–330

Evening events, 67
Evidence-based medicine, 28
Evidence-based strategies to support weight loss, 434
Excess postexercise oxygen consumption (EPOC), 97–98
Exercise, 363
 adaptations and, 7, 219, 220
 before and after bariatric surgery, 451–452
 and alcohol, 494–495
 and bone health, 286–287
 chronic diseases and, 6–8
 defined, 4
 and diet, benefits
  brain health, 16–17
  cancer, 16
  heart disease, 16
  prediabetes, 16
 energy expenditure
  active category, 99
  ATP-PCr energy system, 92–93
  cycling, 96–97
  group exercise, 97
  home aerobic exercise equipment, 97
  lactic acid energy system, 93
  low active category, 99
  oxygen energy system, 93
  passive and occupational, 97
  resistance or weight training, 97
  running, 96
  sedentary category, 99
  sports activity, 97
  swimming, 96
  very active category, 99
  walking, 96
 energy sources used, 102–103
 excessive, health problems, 12
 human energy metabolism, 91–101
 intensity, 467
 intensity and duration, carbohydrate intake, 131–133
 as medicine, 13
 and minerals, 305–306
 moderate, 4
 muscles, energy produced, 91–92
 recovery period, 468
 on REE, 97–98
 repetition maximum, 467
 resistance training (See Resistance training)
 and serum lipid profile, 198–199

 set, 467
 vigorous, 4
Exercise addiction, 12
Exercise equipment, 79
 home aerobic, 97
Exercise Is Medicine™ (health program), 6
Exercise metabolic rate (EMR), 92
Exercise training programs
 aerobic (See Aerobic exercise)
 anaaerobic (See Anaerobic exercise)
 principles
  of individuality, 6
  of overload, 4
  of overuse, 6
  of progression, 5
  of recuperation, 5
  of reversibility, 6
  of specificity, 5
Exercise-associated collapse, 354
Exercise-associated hyponatremia (EAH), 342
Exercise-induced asthma, 12
Exercisenomics, 3
Exerkines, 7
Exertional heat stroke, 353, 354
Exogenous carbohydrates, 130
Experimental research, 28–29
Extracellular water, 317, 318

F

Facilitated diffusion, 116, 118
FAE. See Fetal alcohol effects (FAE)
FAS. See Fetal alcohol syndrome (FAS)
FASD. See Fetal alcohol spectrum disorders (FASD)
Fast-foods
 consumption, 44
 physical performance and, 69
Fast-oxidative glycolytic (FOG) fiber, 92
Fast-twitch red fiber, 92
Fast-twitch white fiber, 92
Fat burning diets, 185
"Fat burning zone," 103–104
Fat loading, 177
Fat substitutes, 165–166
Fat-free mass, 378
Fatigue
 acute, 104–106
 carbohydrate
  blood glucose levels, 125–127, 130
  endogenous carbohydrate levels, 129–130, 131

  lactic acid production, 127
  muscle glycogen, 127–129, 131
  psychological effort reduction, 130
 causes, 105–106
 central, 105–106
 chronic, 104
 definition, 104
 during exercise
  delaying onset of, 106
  energy sources, 102–103
  energy systems, 101–102
  "fat burning zone," 103–104
  nutrition related, 106–107
 peripheral, 105–106
Fats
 calories, percentage
  calculation methods, 164–165
  food label, 165
  saturated fat Calories, 166
  water content, 164
 carbohydrates from, 122
 cholesterol (See Cholesterol)
 classification
  cholesterol, 167–168
  cis monounsaturated fatty acids, 168
  cis polyunsaturated fatty acids, 168
  omega-3 fatty acids, 168
  omega-6 fatty acids, 168
  saturated fatty acids and trans fatty acids, 168
  total fat, 167
 common foods
  hamburger meat, 163
  hidden fat, 163
  meat and dairy industries, 163
  milk, 163
  plant foods, 164
  trans fatty acids, 164
 digestion and absorption, 169–170
 ergogenic aspects
  carnitine, 183–184
  CLA, 184–185
  exercising on empty stomach, 180–181
  fat burning diets, 185
  hydroxycitrate, endurance performance, 184
  ketone supplements, 185
  MCTs, 181–182
  omega-3 fatty acid and fish oil supplements, 182–183

Fats (*Continued*)
　essential, 377
　lipids, 162
　metabolism and function (*See* Metabolism)
　oxidationcaffeine, caffeine and, 506
　storage, 377–378
　subcutaneous, 377
　substitutes/replacers, 165–165
　trans fat, 58, 59
　triglycerides, 162–163
　types, 162
　vegetarian diet and, 54
Fats and exercise
　dietary effects, 175
　as energy source, 174–175
　during exercise, 176
　gender, 175–176
　ketones, 175
　limiting factors, 175
　metabolism, 176
Fat-soluble vitamins. *See also* Water-soluble vitamins
　overview, 252
　vitamin A (retinol), 252–253
　vitamin D (cholecalciferol), 253–256
　vitamin E (alpha-tocopherol), 256–258
　vitamin K (menadione), 258
Fatty acids, 162–163
　energy pathways for, 543
FDA. *See* Food and Drug Administration (FDA)
Federal Trade Commission (FTC), 25, 64
Fédération Internationale de Natation (FINA), 19
Female and male athlete triad, 416–417
　amenorrhea, 417
　disordered eating, 417
　low bone density, 417
　osteoporosis, 417
　reproductive dysfunction, 417
Female athlete triad, 287
Fenfluramine+Prentermine, 408
Fermentable Oligosaccharides, Disaccharides, Monosaccharides and Polyols (FODMAP), 152
Ferritins, 294
Fetal alcohol effects (FAE), 499
Fetal alcohol spectrum disorders (FASD), 498–499
Fetal alcohol syndrome (FAS), 499
FFA. *See* Free fatty acids (FFA)
Fiber
　content in some common foods, 115
　dietary, 113
　functional, 113
　health benefits, 150–152
　total, 113, 150
　vegetarian diet and, 55
"Fight-or-flight" response, 395
FINA. *See* Fédération Internationale de Natation (FINA)
First Law of Thermodynamics, 435
Fish intake, 195, 196
Fish oil supplements, 182–183, 195, 196
　safety issues, 183
Fitbit, 81
Flexitarians, 53
Fluid replacement, 344–347
　after competition and practice, 346–347
　before competition and practice, 344–345
　during competition and practice, 345–346
　fluid intake guidelines, 348
　in training, 347
FODMAP. *See* Fermentable Oligosaccharides, Disaccharides, Monosaccharides and Polyols (FODMAP)
FOG. *See* Fast-oxidative glycolytic (FOG) fiber
Folate (folic acid), 264–265
　deficiency, 264–265
　DRI, 264
　food sources, 264
　functions, 264
　recommendations, 265
　supplementation, 265
Follicle-stimulating hormone (FSH), 417
Food additives, 52
Food and Drug Administration (FDA), 25
　dietary supplements, 64
　ephedrine, 513
　food labeling, 58
　nutrient content claims, 59–60
　qualified health claims, 61
Food colors, 56
Food drugs
　alcohol, 492–502
　caffeine, 502–512
　ephedra (ephedrine), 512–514
　ginseng and herbals, 523–526
　sodium bicarbonate, 515–518
　sports supplements, 526–527
Food Guide Pyramid, 42
Food intolerance
　gluten intolerance, 152
　lactose intolerance, 152
Food labels
　dietary supplements, 63
　healthier diet selecting, considerations when reading, 59
　　Daily Value (DV), 59
　　functional foods, 61–62
　　health claims, 60–61, 62
　and hidden fat, 165
　legislation, 58
　nutrient content claims, 59–60
　Nutrition Facts panel, 58
　nutrition information, 58–60
Food poisoning, 52
Food safety, 52
Fox, Edward L., 84
Free fatty acids (FFA), 173, 174, 216, 261
Free radicals, 248
Fries, James, 6
Fructose, 112, 136
Fruit sugar, 112
Fruits and vegetables
　food colors of, 56
　iron and calcium rich, 54
　phytonutrient content, 55
　suggestions, 57
FTC. *See* Federal Trade Commission (FTC)
Functional fiber, 113
Functional foods, 61–62

G

Gadgets, exercise, 10–11
GAIT. *See* Glucosamine/Chondroitin Arthritis Intervention Trial (GAIT)
Galactose, 112
Gamma-oryzanol, 525
Gamma-tocopherol, 256
Gastric emptying, 338–339
Gastrointestinal distress, and caffeine, 512
Gastrointestinal losses, 218
Gastrointestinal (GI) system, 116, 117
Gatorade, 336
Gatorade®, 112, 115
Gatorade Sports Science Institute, 26
Gelatin, 232
Gender
　body-fat categories by, 385
　and physical performance, 70
　and tolerating exercise in heat, 355–356
　use of fats as an energy, 175–176
General model, 397
Genetic factors
　and chronic diseases, 2–3, 6, 13
　and nutrients, 13
　and obesity, 393
　and resting energy expenditure (REE), 89–90
　and sports performance, 3
Genome-wide association studies (GRAS), 392
Genomics, 3
Ghrelin, 389–390
GI. *See* Glycemic index (GI)
Ginseng, 524–525
GL. *See* Glycemic load (GL)
Global Status Report on Alcohol and Health (WHO), 496
Globesity. *See* Obesity
Glucagon, 125
Glucogenic amino acids, 214
Gluconeogenesis, 122, 123, 214
Glucosamine, 233–234
Glucosamine/Chondroitin Arthritis Intervention Trial (GAIT), 233
Glucose, 112
Glucose polymers, 113
Glucose-alanine cycle, 122, 217–218
Glucostatic theory, 388
Glutamine supplementation, 235
Gluten intolerance, 152
Glycemic index (GI), 117, 120, 137, 150
Glycemic load (GL), 117, 120
Glycerol, 163
Glycerol supplementation, 350–351
Glycerol-induced hyperhydration, 350–351
Glycine, 235
Glycogen loading. *See* Carbohydrate loading
Glycolysis, 85
　glucose to pyruvate, energy pathways, 540
Gonadotropin-releasing hormone (GnRH), 417
Good fats, 168. *See also* Fats
Grape sugar, 112
GRAS. *See* Genome-wide association studies (GRAS)
Green tea *(camellia sinesis)*, 526
Group exercise, 97
Gut microbiota, weight gain and, 396
Gynoid-type obesity, 402

H

Hall, Kevin, 82
Harris, Roger, 228
Hawley, John, 128, 185
HCA. *See* Hydroxycitrate (HCA)
HDL. *See* High-density lipoproteins (HDL)
HDL-cholesterol, 189
Health Canada's Natural Health Products Directorate, 63
Health claims, 60–61, 62
Healthful dietary guidelines
  alcoholic beverages, 52
  basis, 48
  dietary guidelines, risk of chronic disease, 48–52
  food additives, 52
  food safety, 52
  foods rich in calcium and iron, 52
  "healthy" fats, 50
  nutritionally adequate diet, 49
  and physical activity, 49
  plant-rich diet, 49
  protein intake, 51–52
  Prudent Healthy Diet, 49–52
  salt and sodium intake, 50–51
  sugar intake, 50
  suggestions, 52
  2020-2025 Dietary Guidelines for Americans, 48
  whole-grain products, 49–50
Healthful nutrition
  balance, variety, and moderation, 37
  on budget, 44
  essential nutrients, 38–39
  food sources from plants and animals, 45
  key-nutrient concept, 45–46
  MyPlate, 42–44
  nonessential nutrients, 39
  physical performance, recommendations for better, 65–71
  Prudent Healthy Diet, 15, 37, 49
  recommended dietary intakes, 39–41
  vegetarianism, 53–57
Health-related fitness, 4, 5. *See also* Sports-related fitness
Health-related fitness and exercise
  definition, 4
  epigenetics, 3
  epigenome, 3
  health-related fitness, 4
  physical activity, 4
  physical fitness, 4
  risk factor, 3
  training, principles, 4–6
Healthy body weight
  body mass index (BMI), 544
  body-fat percentage, 544–546
  waist circumference, 546
"Healthy" diet *(Healthy People 2030),* 15
"Healthy" fats, 50
Healthy People 2030
  dietary recommendations, 15
  objectives, 3, 14
  refined carbohydrates consumption, 147
  resistance training, 458
Heart attacks, excessive exercise, 12
Heart disease. *See also* Coronary heart disease (CHD)
  and calcium intake, 287–288
  exercise and diet, benefits, 16
Heart rate (HR)
  monitoring, 79
  and oxygen consumption, 95
Heat, exercise in
  and caffeine intake, 506–507
  carbohydrate replacement, 340–341
  dehydration and hypohydration
    involuntary dehydration, 332–333
    voluntary dehydration, 332
  ectrolyte replacement
    daily replacement, 342
    during exercise, 342
  environmental heat
    cardiovascular strain, 331
    central neural fatigue, 331
    dehydration, 332
    muscle metabolism, 331–332
  ergogenic aspects
    exercise in heat, 349
    glycerol supplementation, 350–351
    oxygen water enhancing exercise performance, 349
    pre-cooling techniques, 349
    sodium loading, 349–350
  fluid replacement
    after competition and practice, 346–347
    before competition and practice, 344–345
    during competition and practice, 345–346
    fluid intake guidelines, 348
    in training, 347
  gastric emptying, 338–339
  hyponatremia, 342–343
  individual dehydrate while exercising, 333–334
  intestinal absorption, 339–340
  maintaining water (fluid) balance
    hyperhydration, 337–338
    rehydration, 338
    skin wetting, 336–337
  salt tablets or potassium supplements, 343–344
  sweat
    composition of, 335
    determining rate, 334
    excessive, 335
  water, carbohydrate, or electrolytes, 336
Heat cramps, 353
Heat energy, 77
Heat exhaustion, 353
Heat illness, 12
  exercise in heat, 351
    acclimatization to heat, 357–358
  exercise toleration
    acclimatization, 357
    age, 356
    gender, 355–356
    obesity, 356
    poor physical fitness, 355
    previous heat illness, 356–357
  flow chart for, 352
  guidelines for modifying/canceling competition/training, 352
  hazards reduction, exercise, 357
  health hazards of excessive heat stress
    heat cramps, 353
    heat exhaustion, 353
    heat stroke, 353–354
    heat syncope, 353
  heat injuries, symptoms and treatment, 354–355
Heat index, 328
Heat injuries, 354–355
Heat shock proteins (HSPs), 7
Heat stroke, 353–354
Heat syncope, 353
Hematuria, 297
Heme iron, 294
Hemochromatosis, 299
Hemolysis, 297
Hepcidin, 295
Herbals, ergogenic aids, 525–526
HGH. *See* Human growth hormone (HGH)
Hidden fat, 163
High blood pressure, 363
  description, 359
  dietary modifications
    diet rich in fruits and vegetables, 361–363
    dietary supplements, 363
    maintaining healthy body weight, 360
    moderate alcohol consumption, 363
    reducing sodium intake, 360–361
  essential hypertension, 359
  exercise, prevent or treatment, 363–364
  measurement, 360
  postexercise hypotension, 364
  treatment, 359–360
High-carbohydrate diet, 139–140
High-density lipoproteins (HDL), 171
High-fat diets
  acute, 177
  chronic, 177–179
    ergogenic effect, 178
    no ergogenic effect, 178–179
  fat metabolism, 176
  and weight loss, 179–180
High-fat (keto) diets, 450–451
High-intensity exercise, carbohydrate supplementation
  exercise for 30 to 90 minutes, 133
  exercise for 60 to 90 minutes, 133
  resistance exercise, 132–133
  tasks greater than 90 minutes, 133
  tasks less than 30 minutes, 132
High-intensity interval training (HIIT), 96, 443
High-protein diets, 450
HIIT. *See* High-intensity interval training (HIIT)
Hippocrates, 13, 37
HMB. *See* Beta-hydroxy-beta-methylbutyrate (HMB)
HMB (beta-hydroxy-beta-methylbutyrate), 232–233
Home aerobic exercise equipment, 97

Homeostasis, 318
Hormone, 417
    energy expenditure and, 389
    functions, vitamins, 248
Hormone-sensitive lipase (HSL), 174
Household measures (approximations), 549
HR. *See* Heart rate (HR)
HSL. *See* Hormone-sensitive lipase (HSL)
HSPs. *See* Heat shock proteins (HSPs)
Human energy metabolism
    during exercise
        determining energy cost, 95–96
        energy, daily consumption, 98–101
        energy expenditure, 92–93, 96–97
        metabolic rate, 95
        muscles, energy produced, 91–92
        muscular exercise, 92
        REE, 97–98
        TEF, 98
    during rest
        energy sources, 90
        metabolic rate, 88
        metabolism, 88
        REE (*See* Resting energy expenditure (REE))
Human energy systems
    adenosine triphosphate, 83
    ATP-PCr system, 84–85
    characteristics, 84
    energy storage in body, 83–84
    and fatigue during exercise
        acute, 104–106
        causes, 105–106
        chronic, 104
        delaying onset of, 106
        energy sources, 102–103
        energy systems, 101–102
        "fat burning zone," 103–104
        nutrition related, 106–107
    lactic acid energy system, 85
    measures of energy
        commonly used measure, 81–82
        energy, 77
        physical activity and energy expenditure, 78–81
    nutrients, 87–88
    oxygen energy system, 86–87
    phosphocreatine (PCr), 83

Human Genome Project, 2–3
Human growth hormone (HGH), 236, 518–519
Hunger center, 386
Hydrodensitometry, 379, 380, 381
Hydrolysis, 87
Hydrostatic weighing, 379, 381, 382
Hydroxycitrate (HCA), 184
Hypercalcemia, 286
Hypereating, 391
Hyperglycemia, 118
Hyperhydration, 318, 336
Hyperkalemia, 325
Hyperplasia, 391
Hypertension. *See* High blood pressure
Hyperthermia, 327
Hypertrophy, 391
Hypervitaminosis, 252
Hypoglycemia
    development of fatigue, 125–127
    overview, 118
Hypokalemia, 325, 406
Hyponatremia, 342–343
    symptoms, 342
Hypothalamic-pituitary-gonadal (HPG) axis, 417
Hypothermia, 327

I

IBWA. *See* International Bottled Water Association (IBWA)
Impaired immune functions, excessive exercise, 12
Impermissible supplements, 23
IMTG. *See* Intramyocellular triacylglycerol (IMTG)
Incomplete proteins, 211
Indirect calorimetry, 80
Indispensable amino acids. *See* Essential amino acids
Ingested carbohydrate, 130
Inosine, 233
In-person weight-loss programs, 447
Insensible perspiration, 315
Insulin, 118, 121
Intensity, 467
Intercellular water, 317, 318
Intermediate-term cost, physical inactivity, 6
Intermittent fasting, 449–450
International Bottled Water Association (IBWA), 316
International Journal of Sport Nutrition and Exercise Metabolism, 19, 27, 227

International Olympic Committee, 19, 231
International Society of Sports Nutrition (ISSN), 19
International Unit System (SI), 78
Intervention studies, 28
Intestinal absorption, 338–340
Intracellular water, 317
Intramyocellular triacylglycerol (IMTG), 169, 174, 177
Intrinsic factor, 263
Intuitive eating, 438–439
Involuntary dehydration, 332–333
Ions, 280
Iron (Fe)
    bioavailability, 294
    deficiency
        deficiency in athletes, 296–297
        deficiency stages, 296
        inadequate dietary intake, 297
        iron deficiency without anemia, 297
        signs and symptoms, 296
        sports anemia, 297
        training at high altitudes, 297
    DRI, 292
    excess iron, 299
    food sources, 294
    foods rich in, 52
    functions, 294–295
    healthful dietary guidelines, 52
    plant foods rich in, 54
    recommendations, 299
    supplementation
        iron deficiency without anemia, 298
        iron-deficiency anemia, 298
        oxygen transport, 298
Iron deficiency without anemia, 296, 297, 298
Iron-deficiency anemia, 296, 298
Ischemia, 186
Isokinetic contractions, 480
Isometric contractions, 480
Isotonic contractions, 480
Isotonic solution, 319
ISSN. *See* International Society of Sports Nutrition (ISSN)

J

Jenkins, David, 198
Jeukendrup, Asker, 101, 184
Jones, Andrew, 231

Joule, 78
Journal of the Academy of Nutrition and Dietetics, 27

K

kcal intake, and weight loss, 435
Ketogenic amino acids, 214
Ketogenic diets, 175, 178, 450–451. *See also* High-fat diets
Ketones, 173, 175
    and endurance performance, 185
Key-nutrient concept, 45–46
Kidney cancer, 502
Kidney failure, excessive exercise, 12
Kidney function
    excessive protein intake, 237
    kidney stones
        and calcium, 288
        and vitamin D, 256
Kilojoule, 81
Kinetic energy, 77
Kouros, Yannis, 124
Kraemer, William, 184
Krebs cycle, 86, 261
Kris-Etherton, Penny, 186

L

Lactate salts, 146
Lactate shuttle, 127
Lactate threshold, 93
Lactic acid energy system, 85
    energy expenditure, 93
    sodium bicarbonate on, 516–517
Lactic acid production, fatigue and, 127
Lactose intolerance, 152
Lactovegetarians, 53
LBW. *See* Lean body weight (LBW)
LCFAs. *See* Long-chain fatty acids (LCFAs)
LDL. *See* Low-density lipoproteins (LDL)
LDL-cholesterol, 189
Lean body mass, 378
Lean body weight (LBW), 546
Lecithin, 167
Legumes, 211
Leisure-time activity, 3, 4
Lemon, Peter, 222
Length, units of, 550
Leptin, 390
Leucine, 217–218
Level measures and weights, 549
Levulose, 112

Limiting amino acid, 211
Linoleic acid, 168
Lipids
    absorption of, 169
    definition, 162
    energy sources, 172-173
    metabolic regulation, 172
    structure, 169, 170
Lipoprotein, 169
    HDL, 171
    interactions, 543
    LDL, 171
    lipoprotein (a), 172
    schematic of, 170
    types, 170-172
    VLDL, 171
Lipoprotein (a), 172
Lipostatic theory, 388
Liquid meals, sports, 68
Liver function
    alcohol on, 496
Long-chain fatty acids (LCFAs), 163, 174, 183
Long-haul concept, 444
Long-term cost, physical inactivity, 6
Long-term overeating, 378
Lorcaserin (Belviq®), 408
Low energy availability, 452
Low muscle glycogen
    and aerobic exercise, 127-128
        brain, 128
        energy production rate, 128
        fat for energy, 128
        glycogen location, 128
        muscle fiber type, 128
    and anaerobic exercise, 128-129
Low-density lipoproteins (LDL), 171
Low-glycemic-index foods, 137
Low-risk drinking, 501. See also Alcohol
Luteinizing hormone (LH), 417
Lysine, 236

M

Ma huang, 513. See also Ephedra (ephedrine), ergogenic effects
Macrocycle, 471
Macronutrients, 38
Magnesium (Mg)
    deficiency, 291
    DRI, 290
    food sources, 290
    functions, 290-291
    recommendations, 292
    supplementation, 291-292

Major minerals. See also Minerals
    calcium (Ca), 281-289
    description, 281
    magnesium (Mg), 290-292
    phosphorus (P), 289-290
Male athlete triad, 416
Malnourishment, 20
Malnutrition, 20-21
Maltodextrins, 113
Manore, Melinda, 184
MAOD. See Maximal accumulated oxygen deficit (MAOD)
Maresh, Carl, 269
Maughan, Ron, 65, 116, 227
Maximal accumulated oxygen deficit (MAOD), 93, 516
Maximal oxygen uptake, 93
Maximum contaminant level (MCL), 315
MCFAs. See Medium-chain fatty acids (MCFAs)
MCL. See Maximum contaminant level (MCL)
MCTs. See Medium-chain triglycerides (MCTs)
Meal planning
    on budget, 44
    MyPlate, 46-47
    RDA and, 40
Measurement, units of, 549-550
Mechanical aids, 22
Mechanical energy, 77
Medicine & Science in Sports & Exercise, 27
Meditation, 418, 438
Mediterranean diet, 49
Mediterranean Food Guide, 47
Medium-chain fatty acids (MCFAs), 169
Medium-chain triglycerides (MCTs), 169
    and endurance performance, 181-182
Menadione, 258. See also Vitamin K (menadione)
Menstrual cycles, 297
Mental health, mindfulness and, 440
Mercury, fish intake and, 196
Mesocycle, 471
Meta-analysis, 31
Metabolic aftereffects of exercise, 97-98
Metabolic equivalents (METs), 80, 93-94, 95-96, 551-554
Metabolic rate
    during exercise, 95
    muscular exercise on, 92
    during rest, 88

Metabolic syndrome, 402-403
    International criteria for components of, 403
    symptoms of, 401
Metabolic water, 315
Metabolically healthy obesity (MHO), 399
Metabolism
    alcohol, 492-493
    blood glucose, 118-119, 121
    carbohydrates
        active transport, 116, 118
        blood glucose, 118-119, 121
        digestion and absorption, 116-117, 119
        endurance training on, 125
        facilitated diffusion, 116, 118
        glycemic index (GI), 117
        glycemic load (GL), 117
        in human nutrition, 122-123
        from protein and fat, 122
        total energy, 121-122
    fats
        apolipoproteins, 169
        body lipids, 172-173
        IMTG, 169
        lipoproteins, 170-172
        liver, 169-170
        from protein and carbohydrates, 172
        sources, 169
        total energy, 173
    lipids, regulation, 172
    minerals, 280
    overview, 88
    protein
        carbohydrates and fats, 215
        functions, human nutrition, 215-216
        protein in human body, 214, 215
Metalloenzymes, 279
Metric system equivalents, 549-550
METs. See Metabolic equivalents (METs)
MHO. See Metabolically healthy obesity (MHO)
Microcycle, 471
Micronutrients, 38
Millimole, 121, 122
Mindful eating, 438, 439
    and weight loss/health outcomes, 440
Mindfulness, 438-441
    health benefits of, 438
    intuitive eating, 438-439

    and mental/physical health, 440
Minerals
    deficiencies/excesses on performance, 280-281
    defined, 279
    essential to human nutrition, 280, 281
    human energy system and, 87
    major minerals
        calcium (Ca), 281-289
        magnesium (Mg), 290-292
        phosphorus (P), 289-290
    stages of deficiency, 280
    supplements: exercise and health
        diet, 306
        exercise, 305-306
        megadoses or nonessential minerals, 306-307
        physically active individuals, 307
    and their importance to humans, 279-280
        energy metabolism, 280
        growth and development, 279
        metabolic regulation, 279-280
    trace minerals
        chromium (Cr), 302-304
        copper (Cu), 299-300
        iron (Fe), 292-299
        selenium (Se), 304-305
        zinc (Zn), 300-302
    vegetarianism and, 54
Misinformation, health. See Nutrition and health misinformation
Mitochondria, 85
Moderate exercise, 4
Monosaccharides, 112
    chemical structure, 113
Monounsaturated fatty acids, 163, 164, 193-194
Morbid hunger, 414
Morbid obesity, 392
Morning events, 67
Motion sensors, 80-81
Motivational interviewing, 447
Multicomponent model, 383
Multivitamin supplements, 271-272
Muscle glycogen
    and carbohydrate loading, 142-143
    and fatigue, 127-129, 131
    low
        and aerobic exercise, 127-128

Muscle glycogen (*Continued*)
    and anaerobic exercise, 128-129
    sparing of, 131
Muscle hyperplasia, 480
Muscle hypertrophy, 480
Muscle metabolism, 331-332
Muscles
    contractions, 480
    energy produced during exercise, 91-92
    skeletal muscle fibers, 91
    endurance training on, 125, 126
Myocardial infarction, 186
Myokines, 7
MyPlate, 15, 42-44
    definition, 42
    dietary recommendations (USDA), 15
    guidelines for healthful eating and physical activity, 44
    optimal nutrition, 46-47
    overview, 42-43
    sample recommendations, 43
    visual, 15
    website, 45
MyPyramid, 42

N

NAD. *See* Nicotinamide adenine dinucleotide (NAD)
NADP. *See* Nicotinamide adenine dinucleotide phosphate (NADP)
National Academy of Sciences, 51, 151, 162, 384
National Cancer Institute, 30
National Dairy Council, 26
National Health and Nutrition Examination Survey (NHANES), 147, 315, 403
National Institutes of Health (NIH), 8, 233, 384
National School Lunch Program, 15
National Strength and Conditioning Association (NSCA), 27
NE. *See* Niacin equivalents (NE)
Near-infrared interactance, 383
NEAT. *See* Nonexercise activity thermogenesis (NEAT)
Neotame, 148
"Net carbohydrates," 149
Neuropeptide Y (NPY), 390, 391
Newsholme, Eric, 234
NHANES. *See* National Health and Nutrition Examination Survey (NHANES)

Niacin, 260-261
    deficiency, 261
    DRI, 261
    food sources, 261
    functions, 261
    recommendations, 261
    supplementation, 261
Niacin equivalents (NE), 261
Nicotinamide adenine dinucleotide (NAD), 261
Nicotinamide adenine dinucleotide phosphate (NADP), 261
Nieman, David, 267
NIH. *See* National Institutes of Health (NIH)
Nitrate supplementation, 231-232
    efficacy, 231
    mechanism of action, 231
    safety, 231-232
Nitrogen balance, 210
Nonessential amino acids, 210
Nonessential nutrients, 39
Nonexercise activity thermogenesis (NEAT), 99, 389
    and physical inactivity, obesity, 394-395
Nonheme iron, 294
Non-nutrients, 39
Nonresearchers, 81
Nonshivering thermogenesis, 389
Normohydration, 318
NSCA. *See* National Strength and Conditioning Association (NSCA)
Nutraceuticals, 55
Nutrient, 13
    essential, 13
Nutrient content claims, 59-60
Nutrient density
    balanced diet, 46
    definition, 46
    essential to humans, 38
    foods to obtain nutrients, 42
    intake and health status, relationship, 39
    MyPlate food guide, 42-44
    vegetarianism, 54
Nutrient timing, 66, 224-225
    and performance, 225-226
Nutrient Timing (Ivy and Portman), 467
Nutrient timing, 517
Nutrigenomics, 3
Nutrition
    defined, 13
    fatigue during exercise, 106-107

    fitness and health, 2-3
    fitness and sport, 3
    and health-related fitness
        antipromoters, 13
        Dietary Reference Intakes, 13, 39-40
        DNA damage, 13
        eating right, 14-16
        exercise and diet, additional benefits, 16-17
        exercise is medicine, 13-14
        food is medicine, 13-14
        guidelines for healthy eating, 15-16
        Hippocrates on nutrition, 13, 37
        nutrition, 13-14
        promoters, 13
        Prudent Healthy Diet, 15, 37
        Recommended Dietary Allowances (RDA), 13, 39
        websites, 16
    information, food labels, 58-60
    MyPlate, 15, 42-44, 47
    during training, 65-66
    weight gain
        balanced diet, 465, 467
        calories, 460
        dietary supplements, 464-465
        protein supplementation, 461-464
    and weight loss, 434-437
Nutrition and health misinformation
    books, 26
    cautions on using the Internet, 27
    consultants, 27
    definition, 24
    government, organizations, and websites, 26
    overview, 24-25
    popular magazines, 27
    prevalence in athletics, 25
    scientific journals, 27
Nutrition Facts panel, 58
Nutrition Labeling and Education Act, 58
Nutritional aids, 22
Nutritional ergogenics, 21-23
    effectiveness, 22-23
    legal, 23
    popularity, 22
    safety, 23
Nutritional labeling, 58

O

Oatrim®, 165
Obesity. *See also* Weight gain; Weight loss
    adult-onset obesity, 397
    and alcohol consumption, 499
    android-type, 402
    and calcium supplementation, 288
    contributing factors
        built environment, 396
        drugs and environmental chemicals, 395-396
        environmental factors, 393-396
        genetic and epigenetic factors, 392-393
        gut microbiota, 396
        interaction of genetics and environment, 396-397
        personal relationships, 395
        physical inactivity and NEAT, 394-395
        sleep and emotional stress, 395
        viruses, 396
    definition, 384
    and eating disorders, 440-441
    gynoid-type, 402
    health consequences of, 432-434
    medications to treat, 406-409
    morbid obesity, 392
    physical activity and, 441
    prevalence, 374
    prevention of childhood obesity, 398-399
    social and psychological consequences, 432-433
    and tolerating exercise in heat, 356
    weight cycling and, 453
Obesity rates, 431-432
OBLA. *See* Onset of blood lactic acid (OBLA)
Observational research, 28
Occupational energy expenditure, 97
Odds ratios (OR), 28
ODPHP. *See* Office of Disease Prevention and Health Promotion (ODPHP)
Office of Disease Prevention and Health Promotion (ODPHP), 8
Olestra®, 166, 197

Omega-3 fatty acids, 163, 168, 194–197
  supplements, 182–183
Omega-6 fatty acids, 163, 168
OmniHeart diet plan, 149, 194, 404
Online/app-based weight-loss programs, 447–449
Onset of blood lactic acid (OBLA), 93
On-the-go lunches, 69
Optimal Macro-Nutrient Intake to Prevent Heart Disease (OmniHeart) diet, 361–363
OR. *See* Odds ratios (OR)
Ornithine, 236
Orthopedic problems, excessive exercise, 12
Osmolality, 319
Osmoreceptors, 319
Osmosis, 116, 118, 319
Osmotic pressure, 319
Osteomalacia, 254
Osteoporosis, 12. *See also* Bone health
  and caffeine, 511
  described, 283–284
  female and male athlete triad, 417
  low bone density and, 417
  premature, 417
  risk factors in women, 284
  "silent killer," 285
  in sports, 287
Overconsumption of food, 440
Over-exercise, 445
Overnutrition, 20
Overreaching, 104
Overtraining syndrome, 104, 219
Over-use injuries, 445
Ovolactovegetarian, 53
Ovovegetarians, 53
Oxygen energy system, 86–87
  energy expenditure, 93
  sodium bicarbonate on, 517
Oxygen system, 86
Oxygen transport, iron supplementation and, 298–299
Oxygen water, 349

P

PAEE. *See* Physical activity energy expenditure (PAEE)
Paidotribes, 491
PAL. *See* Physical activity level (PAL)
Paleolithic (Paleo) diet, 37
Panax ginseng, 524
Pantothenic acid
  deficiency, 265
  DRI, 265
  food sources, 265
  functions, 265
  recommendations, 266
  supplementation, 265–266
Partially hydrogenated fats, 163
Passive energy expenditure, 97
PCFSN. *See* President's Council on Fitness, Sports, and Nutrition (PCFSN)
PDCAAS. *See* Protein Digestibility-Corrected Amino Acid Score (PDCAAS)
Peak bone mass, 285
"Pear-shape" pattern, obesity, 402
Pedometer, 80
Percent Daily Value, 58
Performance water, 349
Performance-enhancing substances, 22. *See also* Ergogenic aids
Performance-enhancing techniques, 22. *See also* Ergogenic aids
Periodization, 471
Peripheral fatigue, 105
Permissible supplements, 23
Personal relationships, weight gain and, 395
Pescovegetarians, 53
Pharmacological aids, 22
Phillips, Stuart, 223
Phosphocreatine (PCr), 83, 85
Phospholipids
  description, 167
  food, 167
  lecithin, 167
Phosphorus (P)
  deficiency, 289
  and DRI, 289
  and food sources, 289
  functions, 289
  recommendations, 290
  supplementation, 289–290
Photosynthesis, 77, 78
Phototherapy, 418
Phylloquinone, 258
Physical activity
  ASMBS recommendations for, 452
  and energy expenditure
    calorimetry, 79–80
    classification, 98
    combination devices, 81
    DLW technique, 80
    ergometer, 79
    heart rate monitoring, 79
    METs, 80, 95–96, 97
    motion sensors, 80–81
    PAL, 80, 99
    physical activity questionnaires, 79
    Smartphone Applications (Apps), 81
  excessive, health problems, 12
  guidelines, 8–12
    across the United States, 10, 11
    key principles for developing, 8–10
    tracking, 10–11
  measurement, 78–81
  MyPlate, 44
  and obesity, 441
  program, 445–446
  Prudent Healthy Diet, 65
  and Prudent Healthy Diet, 49
  and quality of life, 6
  strategies to increase, 442–445
  structured, 4
  unstructured, 4
  and weight loss, 441–446
Physical activity energy expenditure (PAEE), 92, 99
Physical Activity Guidelines for Americans, 2018, 442, 443
Physical activity level (PAL), 80, 99
Physical activity questionnaires, 79
Physical fitness, 4
  and adverse health effects of obesity, 404–405
  and tolerating exercise in heat, 355
Physical health, mindfulness and, 440
Physical inactivity, 2
  intermediate-term cost, 6
  long-term cost, 6
  and NEAT, obesity, 394–395
  short-term cost, 6
Physical performance
  after competition, 68
  apps, 37, 69
  biotin, 266
  body composition
    excess body weight, 410–412
    excessive weight loss, 412–413
  calcium (Ca), 283
  and calcium supplementation, 288
  carbohydrate loading and, 143–144
  carnitine on, 183–184
  choline, 268
  chromium, 302–303
  chronic training effect, 66
  commercial sports foods, 68–69
  during competition, 67
  copper, 299–300
  deficiencies or excesses of vitamins, 249
  fast-food or family-style restaurants, 70
  folate (folic acid), 264–265
  gender and age, influence of, 70–71
  iron, 295–299
  liquid sports meals, 68
  magnesium (Mg), 291–292
  niacin, 261
  nutrition during training, 65–66
  pantothenic acid, 265
  and phosphorus, 289–290
  precompetition meal, 66–67
  selenium, 304–305
  sports bars, 68
  sports drinks, 68
  sports gels and candy, 68–69
  traveling for competition, 69–70
  vegetarian diet and, 56–57
  vitamin A, 252–253
  vitamin B1, 259–260
  vitamin B2, 260
  vitamin B6, 262
  vitamin B12, 263
  vitamin C, 266–267
  vitamin D, 254–255
  vitamin E, 257
  vitamin K, 258
  zinc, 300–301
Physiological aids, 22
Physiological deficiency, 249
Phytochemicals, 55
Phytoestrogens, 55
Phytonutrients (plant chemicals), vegetarianism and, 55
PINES. *See* Professionals in Nutrition for Exercise and Sport (PINES)
Plant proteins, 211
  vs. animal protein, 210–211
Plant-rich diet, 49
Plaque, 186, 188, 189
PMS. *See* Premenstrual syndrome (PMS)
Polyphenols, 501
Polysaccharide, 113
Polyunsaturated fatty acids, 163, 164
  consumption, 194–197
POMS. *See* Profile of Mood States (POMS) questionnaire
Postexercise hypotension, 364

Postexercise protein, 224–225
Potassium (K)
　deficiency and excess, 324–325
　DRI, 324
　food sources, 324
　functions, 324
　supplements, 343–344
Power, 78
PowerAde®, 112, 115
PRE. See Principle of progressive resistance exercise (PRE)
Precompetition meal, 66–67
Pre-cooling techniques, 349
Pre-diabetes, 16
Pre-event nutritional strategies, 66–67
Pregnancy, fish intake and, 196
Preliminary stage, 249
Premature osteoporosis, 417
Premenstrual syndrome (PMS), 147
President's Council on Fitness, Sports, and Nutrition (PCFSN), 8
Pre-workout supplements (PWSs), 517–518
Principle of disuse, 6
Principle of exercise sequence, 470
Principle of individuality, 6
Principle of overload, 4, 469
Principle of overuse, 6
Principle of progression, 5
Principle of progressive resistance exercise (PRE), 470
Principle of recovery, 5
Principle of recuperation, 5, 471
Principle of reversibility, 6
Principle of specificity, 5, 470
Pritikin program, 149
Professionals in Nutrition for Exercise and Sport (PINES), 19
Profile of Mood States (POMS) questionnaire, 139
Prohormones, 521, 523
Promoters, 13
Proof, 492
Protein. See also Protein and exercises
　alanine, 210
　amino acids, 209
　animal vs. plant protein, 210–211
　carbohydrates with, 137, 138
　complementary, 54
　content in some common foods, 212
　deficiency, 211, 237
　description, 209–210
　dietary guidelines, 213
　digestion and absorption, 214, 215
　excessive protein intake, 237
　glycine, 235
　intake, recommendation, 51–52
　plant and animal foods, 211
　RDA for, 66, 211
　required, 211–213
　vegetarianism and, 54
Protein and exercises
　carbohydrate
　　protein hydrolysate, 223
　　protein intake after exercise, 223
　　protein intake before exercise, 223
　daily energy intake, 222
　energy intake
　　aerobic endurance exercise, 217
　　leucine and the glucose-alanine cycle, 217–218
　　mechanisms and by-products, 217
　　protein use and importance of carbohydrate, 218
　　resistance exercise, 217
　need for dietary protein, 219–221
　　endurance-type activities, 221
　　strength-type activities, 220–221
　protein losses
　　gastrointestinal losses, 218
　　sweat losses, 218
　　urinary losses, 218
　prudent protein intakes, 226
　recovery after exercise, 218–219
　resistance-trained athletes, 221
　supplementation
　　nutrient timing and performance, 225–226
　　protein and carbohydrate intake after exercise, 225
　　before sleep, 226
　training, 219
Protein Digestibility-Corrected Amino Acid Score (PDCAAS), 210, 211
Protein hydrolysate, 223
Protein supplementation, 461–464
　amount, 461–462
　carbohydrate, 463–464
　cost, 464
　quality and quantity, 463
　timing, 462–463
Protein-related supplements
　amino acid, 234–236
　arginine, 235, 236
　beta-alanine, 230–231
　bovine colostrum, 233
　branched-chain amino acids (BCAAs), 234–235
　chondroitin, 233–234
　citrulline, 235–236
　collagen, 232
　creatine, 228–230
　dietary nitrate/beetroot, 231–232
　gelatin, 232
　glucosamine, 233–234
　glutamine, 235
　glycine, 235
　HMB (beta-hydroxy-beta-methylbutyrate), 232–233
　inosine, 233
　lysine, 236
　ornithine, 236
　taurine, 232
　tryptophan, 236
　tyrosine, 235
Protein-sparing effect, 216
Proteinuria, 218
Proteolysis, 87
Prudent Healthy Diet, 15, 37, 49–52, 64, 65, 363
Pseudoephedrine, 513
Psychological aids, 22
Psychological effort reduction, fatigue and, 130
Psychological problems alcohol consumption, 496
Psychomotor responses, caffeine, 504
Psyllium, 115
PWSs. see Pre-workout supplements (PWSs)
Pyruvate, 145–146

Q

Quality of life, obesity and, 401
Questionnaires, physical activity, 79
Quetelet's Index. See Body mass index (BMI)

R

Radiation, 326
RAE. See Retinol activity equivalents (RAE)
Randomized clinical trials (RCTs), 28
Rating of perceived exertion (RPE), 127, 130, 139, 234
RCTs. See Randomized clinical trials (RCTs)
RDA. See Recommended Dietary Allowances (RDA)
RE. See Retinol equivalents (RE)
Reactive hypoglycemia, 118
Reactive oxygen species (ROS), 248
Reactive oxygen/nitrogen species (RONS), 248
Recombinant hGH (rhGH) injections, 519
Recommended Dietary Allowances (RDA), 13
　calcium, 52, 286
　carbohydrates, 115
　diet planning, 40
　overview, 39
　for protein, 66, 211, 212
　sports nutrition, 19
　sports supplements, 23
　for vitamins, 247, 250–251
Recovery period, 468
Red meat, 51
RED-S. See Relative Energy Deficiency in Sport (RED-S)
REE. See Resting energy expenditure (REE)
Refined carbohydrates intake, 197–198
Refined sugars, 113
　chronic diseases, 147–148
　dental caries, 147–148
Regional fat distribution, 383
Rehydration, 336–337, 338
Relative Energy Deficiency in Sport (RED-S), 104, 406
Relative risks (RR), 28
Relative VO_2 max, 411
Repetition, 467
Repetition maximum (RM), 467
Reproductive dysfunction, 417
Research
　creatine supplementation, 29
　dietary recommendations, basis for, 30–31
　epidemiological research (observational research), 28
　evidence-based medicine, 28
　experimental research, 28–29
　fast-food consumption, 44
　meta-analysis, 31
　public understanding, improving, 30

579

Resistance training, 97, 443, 468, 482–483
  periodization, 471
  principle of exercise sequence, 470
  principle of overload, 469
  principle of progressive resistance exercise (PRE), 470
  principle of recuperation, 471
  principle of specificity, 470
  protein metabolism, 217
  and weight gain, 97
    basic principles of, 469–471
    combining aerobic and resistance-training, 484
    contraindications, 482–483
    gaining body weight as lean muscle mass, 471–473
    gaining weight with, 478–480
    health benefits, 483–484
    primary purposes of, 468–469
    safety concerns, 473–478
    type of, 480–481
Resting energy expenditure (REE), 388, 444
  definition, 88
  dieting and body composition, 90
  environmental factors, 90
  estimating daily, 89
  during exercise, 97–98
  genetic factors, 89–90
Resting metabolic rate (RMR), 88
Retinol, 252. See also Vitamin A (retinol)
Retinol activity equivalents (RAE), 252
Retinol equivalents (RE), 252
Reverse cholesterol transport, 171, 189
Rhabdomyolysis, 406
Rhodiola rosea, 525
Riboflavin (vitamin B2), 260
  deficiency, 260
  DRI, 260
  food sources, 260
  functions, 260
  recommendations, 260
  supplementation, 260
Ribose, 146
Rickets, 254
Rimonabant (Acomplia), 408
Risk factors, definition, 3

RMR. See Resting metabolic rate (RMR)
ROS. See Reactive oxygen species (ROS)
RPE. See Rating of perceived exertion (RPE)
RR. See Relative risks (RR)
Runners high, 12
Running, 96

S

Saccharin, 148
SAD. See Seasonal affective disorder (SAD)
Safe Drinking Water Act, 315
Safety equipment, 474
Salatrim®, 166
Salsolinol, 495
Salt intake, 50–51
Salt tablets, 343–344
Sarcopenia, 7, 237, 378
Sarcopenic obesity, 451
Satiety center, 386
Saturated fats, 437
Saturated fatty acids, 163, 164, 168
  intake, 192–193
  vegetarian diet and, 54
SCFAs. See Short-chain fatty acids (SCFAs)
Seafood Choices: Balancing Benefits and Risks, 196
Seasonal affective disorder (SAD), 147
Secondary amenorrhea, 287
Sedentary Death Syndrome (SeDS), 6
SeDS. See Sedentary Death Syndrome (SeDS)
Selenium (Se)
  deficiency, 304
  DRI, 304
  food sources, 304
  functions, 304
  recommendations, 305
  supplementation, 304–305
Selfish gene, 392
Semivegetarians, 53
Senses, energy intake and, 387–388
Serotonin, 236
Serum glucose, 118
Serum lipids. See also Cholesterol
  dietary modifications
    artificial fats intake, 197
    caloric intake adjustment, 191
    fat intake, 191–192

monounsaturated fats, 193–194
nibble food, 198
polyunsaturated fatty acids consumption, 194–197
refined carbohydrates intake, 193–194, 197–198
saturated fat intake, 192–193
trans fat intake, 193
and exercise training, 198–199
reverse atherosclerosis, 190
triglycerides, 189
Set, 467
Set-point theory, 390
Settling-point theory, 397–398
SHAPE. See Society of Health and Physical Educators (SHAPE)
Shell temperature, 325. See also Body temperature
Short, Sarah, 269
Short-chain fatty acids (SCFAs), 150, 162–163, 169
Short-term cost, physical inactivity, 6
Short-term dietary changes, 378
SHPN. See Sports and Human Performance Nutrition (SHPN)
Sick fat, 402
Silbutramine, 408
Simple carbohydrates, 112
Simplesse®, 165
Siraitia grosvenorii, 148
Skeletal muscle fibers, 91
  endurance training on, 125, 126
Skinfold technique, 381–382
Sleep disorder, and caffeine, 511–512
Sleep disorder, obesity and, 395
Slow-oxidative (SO) fiber, 91
Slow-twitch red fiber, 91
Smoothie, 68
Snow, John, 314
Social drinking, 495
Society of Health and Physical Educators (SHAPE), 4
Soda loading, 515, 518
Sodium (Na)
  deficiency and excess, 323
  food sources, 321, 323
  functions, 323
Sodium bicarbonate
  definition, 515
  ergogenic effects

ATP-PCr energy system, 516
buffer boosting, 515
carbohydrate loading, 518
lactic acid energy system, 516–517
oxygen energy system, 517
safety and usage, 518
soda loading, 515, 518
supplementation protocol, 515–516
Sodium intake, 50–51
Sodium loading, 349–350
Solar energy, 77, 78, 83
Soluble fiber, 150
Solute, 319
Specialist in Sports Dietetics (CSSD), 19
Specific heat, 329
Sport, defined, 3
Sportomics, 3
Sports
  energy expenditure, 97
Sports and Human Performance Nutrition (SHPN), 19
Sports anemia, 221, 297
Sports bars
  carbohydrate content, 115
  and physical performance, 68
"Sports confectionary," 133
Sports drinks
  carbohydrate content, 115
  and physical performance, 68
Sports gels and candy
  carbohydrate content, 115
  and physical performance, 68–69
The Sports Gene (Epstein), 3, 17
Sports Medicine, 27
Sports nutrition. See also Physical activity
  carbohydrate diet, 116
  career opportunities, 19
  certification programs, 19
  consensus statements and position stands, 19
  definition, 18
  international meetings, 19
  objectives, 18
  overview, 3
  professional associations, 19
  research productivity, 19
Sports Physiology (Fox), 84
Sports supplements, 21. See also Dietary supplements; Ergogenic aids
  banned, 23
  categories, 22

Sports supplements (Continued)
  contamination of, 23
  efficacy, 526-527
  impermissible, 23
  information on, 24
  permissibility, 526-527
  permissible, 23
  physical performance and, 69
  RDA, 23
  safety, 526-527
Sports-related fitness, 4, 17-21
  adequate nutrition for athletes, 19
  athletic performance and nutrition, 20-21
  malnourishment, 20
  malnutrition, 20-21
  optimizing sport performance, 21
  overnutrition, 20
  overview, 17-18
  undernutrition, 20
Spriet, Lawrence, 232
Starches, 113, 147-148
Steady-state threshold, 93
Stellingwerff, Trent, 179
Sterols, 166
Stevia, 148
Stomach fullness, 388
Storage fat, 377-378
Strength-endurance continuum, 468
Strength-type activities, 220-221
Stress, 440
Stress fracture, 417
Structured physical activity, 4
Subclinical malnutrition, 249
Subcutaneous fat, 377-378
Sucralose, 148
Sudden death, excessive exercise, 12
Sugar alcohols, 149
Sugar intake, 50
Super antioxidants, 270
Super foods, 21, 181
Sweat
  composition of, 335
  determining rate, 334
  excessive, 335
Sweat losses, 218
Swimming, 96
Synephrine, 514
Systolic phase, 359
Systolic pressure, 360

T

Taurine, 232
TEE. See Total daily energy expenditure (TEE)
TEF. See Thermic effect of food (TEF)
Testosterone, 519-521
Testosterone replacement therapy (TRT), 519-520
Thermic effect of food (TEF), 444
  definition, 88-89
  exercise on, 98
Thiamin (vitamin B1), 259-260
  deficiency, 259-260
  DRI, 259
  food sources, 259
  recommendations, 260
  supplementation, 260
Thrifty gene, 393
Tipton, Kevin, 183
Tolerable Upper Intake Levels (UL), 13, 40
Tonicity, 319
Total cholesterol, 189
Total daily energy expenditure (TEE), 88, 99
Total diet approach, 15
Total fat, 167. See also Fats
Total fiber, 113, 150
Trabecular bone, 285
Trace minerals. See also Minerals
  chromium (Cr), 302-304
  copper (Cu), 299-300
  iron (Fe), 292-299
  overview of, 293-294
  selenium (Se), 304-305
  zinc (Zn), 300-302
"Train low" approach, 128-129
Training
  adaptation, carbohydrate and, 128-129
  breakfast during, 66
  endurance, carbohydrates metabolism, 125, 126
  fat metabolism, 176
  fluid replacement in, 347
  at high altitudes and iron deficiency, 297
  nutrient timing, 66
  nutrition during, 65-66
  overtraining syndrome, 104, 219
  resistance training (See Resistance training)
Trans fatty acids, 58, 59, 163, 164, 168
  intake, 193
Traveling for competition, 69-70
Triacylglycerol, 162
Tribulus terrestris, 525
Triglycerides, 162-163
  and atherosclerosis, 189
  glycerol portion of, 182

TrueSport® Nutrition Guide, 65
Tryptophan, 236
2018 Physical Activity Guidelines for Americans, 8
2020-2025 Dietary Guidelines for Americans, 14, 37, 48, 50, 115, 147, 151, 162, 168, 197, 247, 281, 321, 362, 434, 436, 437, 463, 465, 496
Type 2 diabetes, caffeine and, 510
Tyrosine, 235

U

UL. See Tolerable Upper Intake Levels (UL)
Ultra-processed foods (UPFs), 436-437
Ultratrace minerals, 292
Uncompensated heat stress, 330
Undernutrition, 20
Underrecovery, 104
Underwater weighing, 379, 380, 381
Underweight, 459
  steps to gain weight, 459-460
United Nations Food and Agriculture Organization, 384
United States
  leading causes of death in, 2
United States Anti-Doping Agency, 65
United States Office of Disease Prevention and Health Promotion (ODPHP), 3
Unstructured physical activity, 4
Upper respiratory tract infection (URTI), 267
Urea, 214
Urinary losses, 218
URTI. See Upper respiratory tract infection (URTI)
U.S. Department of Agriculture (USDA), 15, 42, 45
U.S. Department of Health and Human Services (HHS), 2, 8
U.S. Pharmacopeia, 64
USDA. See U.S. Department of Agriculture (USDA)
USDA 85-15 guide

V

Valsalva phenomenon, 474
van Loon, Luc, 184
Vanadium, 307
VDRs. See Vitamin D receptors (VDRs)
Vegan, 53
Vegetarianism, 53-57
  becoming vegetarian, 56

  combining foods, protein complementarity, 54
  flexitarians, 53
  food colors, fruits and vegetables, 56
  high fiber, 55
  high vitamin and phytonutrient content, 55-56
  lactovegetarians, 53
  low energy, 55
  low fat, 54
  nutraceuticals, 55
  nutrient density, 54
  nutritional concerns
    complementary proteins, 54
    energy, 53
    minerals, 54
    protein, 54
    vitamins, 53
  ovolactovegetarian, 53
  ovovegetarians, 53
  pescovegetarians, 53
  physical performance potential, 56-57
  phytochemicals, 55
  semivegetarians, 53
  vegan, 53
Very low-density lipoproteins (VLDL), 171
Very low-kcal diets (VLCD), 409
Vigorous exercise, 4
Villi, 116, 118
Viruses, weight gain and, 396
Vitamin(s). See also Vitamin supplements; Specific vitamins
  antioxidant, 248, 249
    and fatigue or muscle damage, 269-271
    free radicals, 248
    ROS, 248
  classification, 249
  coenzyme functions, 248
  deficiencies or excesses
    biochemical deficiency, 249
    clinically manifest vitamin deficiency, 249
    hypervitaminosis, 252
    physiological deficiency, 249
    preliminary stage, 249
  description, 248
  essential, 248-249, 250-251
  fat-soluble (See Fat-soluble vitamins)
  hormone functions, 248
  human energy system and, 87
  human nutrition, 248-249

Vitamin(s) (*Continued*)
vegetarianism and, 53, 55
water-soluble (*See* Water-soluble vitamins)
Vitamin A (retinol), 252–253
beta-carotene, 252, 253
deficiency, 252–253
DRI, 252
food sources, 252
functions, 252
recommendations, 253
supplementation, 253
Vitamin B1. *See* Thiamin (vitamin B1)
Vitamin B2. *See* Riboflavin (vitamin B2)
Vitamin B6 (pyridoxine), 261–262
deficiency, 262
DRI, 261
recommendations, 262
supplementation, 262
Vitamin B12 (cobalamin), 262–264
deficiency, 263
functions, 263
recommendations, 264
supplementation, 264
Vitamin C (ascorbic acid), 266–268
deficiency, 266–267
DRI, 266
food sources, 266
functions, 266
recommendations, 268
supplementation, 267–268
Vitamin D (cholecalciferol), 253–256
deficiency, 254–255
DRI, 253
food sources, 253–254
functions, 254
recommendations, 256
supplementation, 255–256
bone health, 255
cancer, 255–256
diabetes, 256
kidney stone, 256
physical performance, 256
Vitamin D receptors (VDRs), 255
Vitamin E (alpha-tocopherol), 256–258
deficiency, 257
DRI, 256
food sources, 257
functions, 257
recommendations, 258
supplementation, 257–258

Vitamin K (menadione), 258
deficiency, 258
DRI, 258
food sources, 258
functions, 258
recommendations, 258
supplementation, 258
Vitamin supplements
ergogenic aspects
antioxidant vitamins, 269–270
athletes, 271–272
multivitamin supplements, 271–272
physically active individuals, 269
recommendations, 271, 272
VLCD. *See* Very low-kcal diets (VLCD)
VLDL. *See* Very low-density lipoproteins (VLDL)
VO_2 max, 93, 177, 267
absolute, 411
relative, 411
Volek, Jeff, 175
Volume, units of, 550
Voluntary dehydration, 332

W

WADA. *See* World Anti-Doping Agency (WADA)
Wagenmakers, Anton, 236
Waist circumference, 385–386
regional fat distribution, 546
Waist-hip ratio, 386
Walking, 96
Warm-up phase, 475
WAT. *See* White adipose tissue (WAT)
Water
adequately hydrated, 320
body, 378
body water regulation, 318–319
drinking more water or fluids, 320
extracellular, 318
functions in body, 320
intercellular, 318
intracellular, 318
needed per day, 315
overview, 314
solutes in, 315–319
storage in body, 317–318
Water balance, 316, 336–338
hyperhydration, 337–338
rehydration, 338
skin wetting, 336–337

Water intoxication. *See* Hyponatremia
Water-insoluble fibers, 150
Water-soluble vitamins. *See also* Fat-soluble vitamins
biotin, 266
choline (vitamin-like compounds), 268–269
folate (folic acid), 264–265
niacin, 260–261
overview, 259
pantothenic acid, 265–266
riboflavin (vitamin B2), 260
thiamin (vitamin B1), 259–260
vitamin B6 (pyridoxine), 261–262
vitamin B12 (cobalamin), 262–264
vitamin C (ascorbic acid), 266–268
WBGT index, 327–328
Weight
body mass index, 375
body-fat percentage, 384–385
ideal body weight, 375
regulation of
contributing factors to obesity, 392–397
fat deposition in body, 391–392
prevention of childhood obesity, 398–399
set point, 397–398
total, 435
ultra-processed foods and, 436–437
units of, 550
waist circumference, 383, 385–386
waist-hip ratio, 386
weight gain, obesity, and health (*See* Weight gain)
weight loss and health (*See* Weight loss)
Weight bias, 433
Weight control
and caffeine, 511
caloric concept of, 435
energy expenditure, 388–389
energy intake, 386–388
energy intake and expenditure, feedback control, 389–391
Weight cycling, 409–410, 453
Weight gain. *See also* Obesity
health problems
Alzheimer's disease and dementia, 401
cancer, 401

cardiovascular disease, 400
maternal health, 401
mortality and morbidity, 401
health risks in youth, 403–404
location of fat in body, 402–403
losing excess body fat, 404
nutritional considerations
balanced diet, 465, 467
calories, 460
dietary supplements, 464–465
protein supplementation, 461–464
physically fit, 404–405
resistance training
basic principles of, 469–471
combining aerobic and resistance-training, 484
contraindications, 482–483
gaining body weight as lean muscle mass, 471–473
gaining weight with, 478–480
health benefits, 483–484
primary purposes of, 468–469
safety concerns, 473–478
type of, 480–481
steps to gain weight, 459–460
underweight, 376, 459
Weight loss
and athletic performance, 451–453
and calcium supplementation, 288
and calorimeter, 82
and carnitine, 183–184
and CLA, 184–185
diet, 437
drugs, 408–409
eating disorders, 413
eating problems, sports
anorexia athletica, 416
female and male athlete triad, 416–417
ephedrine and, 513
evidence-based strategies to support, 434
and exercising on empty stomach, 180–181
female athlete triad
dehydration, 406

Weight loss (*Continued*)
　very low-Calorie diets, 409
　weight cycling, 409–410
　young athletes, 410
general dietary recommendations to support, 436
and high-fat diets, 179–180
high-fat (keto) diets and, 450–451
high-protein diets and, 450
kcal intake related to, 435
mindful eating and, 440
nutrition and, 434–437
physical activity and, 441–446
and sport-specific performance, 452–453

Weight-loss programs, 446–451
　behavioral group therapy, 449
　for children and adolescents, 449
　high-protein diets, 450
　in-person, 447
　intermittent fasting, 449–450
　online/app-based, 447–449
　selecting, recommendations when, 446, 447
　types of, 446–449
Weight training, 97
Western Ontario and McMaster Universities Arthritis Index (WOMAC) score, 232, 234

Wet-bulb globe temperature (WBGT). *See also* Body temperature
　thermometer, 327–328
White adipose tissue (WAT), 389
Whole-grain products, 49–50, 149
Williams, Mel, 233
WOMAC. *See* Western Ontario and McMaster Universities Arthritis Index (WOMAC) score
Work
　definition, 77
　measurement systems, 79
World Anti-Doping Agency (WADA), 23, 406, 491, 495, 514, 519

X

Xerophthalmia, 253

Y

Yamax Digi-Walker® SW-200, 80
Yoga, 418

Z

Zinc (Zn)
　deficiency, 300–301
　DRI, 300
　food sources, 300
　functions, 300
　recommendations, 302
　supplementation, 301–302